ANCIENT KHOTAN

ANCIENT KHOTAN

DETAILED REPORT OF ARCHAEOLOGICAL EXPLORATIONS IN CHINESE TURKESTAN

CARRIED OUT AND DESCRIBED UNDER THE ORDERS OF H.M. INDIAN GOVERNMENT BY

M. AUREL STEIN

INDIAN EDUCATIONAL SERVICE

VOL. I TEXT

WITH DESCRIPTIVE LIST OF ANTIQUES BY F. H. ANDREWS
SEVENTY-TWO ILLUSTRATIONS IN THE TEXT, AND APPENDICES
BY L. D. BARNETT, S. W. BUSHELL, E. CHAVANNES
A. H. CHURCH, A. H. FRANCKE, L. DE LOCZY
D. S. MARGOLIOUTH, E. J. RAPSON, F. W. THOMAS

GYAN

Published by

Gyan Publishing House
5, Ansari Road
Daryaganj, New Delhi-110002
Phone: 011-47034999, 9811692060
E-mail: books@gyanbooks.com

Distribution Network
gyanbooks.com
India, USA, Canada, UK, Australia, France

ISBN : 978-81-212-2424-6 (Set)
ISBN : 978-81-212-2672-1 (HB)
First Published, 1907

2nd Impression 2021

Printed at: Gyan Press, Delhi.

Ancient Khotan Vol. I
Author: M. Aurel Stein

ANCIENT KHOTAN

DETAILED REPORT OF ARCHAEOLOGICAL EXPLORATIONS IN CHINESE TURKESTAN

CARRIED OUT AND DESCRIBED UNDER THE ORDERS OF H.M. INDIAN GOVERNMENT BY

M. AUREL STEIN

INDIAN EDUCATIONAL SERVICE

VOL. I TEXT

WITH DESCRIPTIVE LIST OF ANTIQUES BY F. H. ANDREWS
SEVENTY-TWO ILLUSTRATIONS IN THE TEXT, AND APPENDICES
BY L. D. BARNETT, S. W. BUSHELL, E. CHAVANNES
A. H. CHURCH, A. H. FRANCKE, L. DE LÓCZY
D. S. MARGOLIOUTH, E. J. RAPSON, F. W. THOMAS

OXFORD

AT THE CLARENDON PRESS

1907

TO THE MEMORY

OF

COLONEL SIR HENRY YULE

R.E., C.B., K.C.S.I.

THE GREAT ELUCIDATOR OF EARLY TRAVEL

AND

A PIONEER IN THE HISTORICAL GEOGRAPHY OF CENTRAL ASIA

THIS ACCOUNT

OF THE ANTIQUARIAN RESULTS OF MY JOURNEY

ON WHICH HIS WORKS WERE MY BEST COMPANIONS

IS DEDICATED

WITH DEEP RESPECT AND ADMIRATION

FOR THE SCHOLAR, THE WRITER, AND THE MAN

INTRODUCTION

The present work is intended to describe in detail the archaeological results of the explorations which I carried out in 1900–01, under the orders of the Government of India, in the southern portion of Chinese Turkestān, and particularly in the region of Khotan. It has been my endeavour to make this account as exact and exhaustive as seemed justified by the importance of the discoveries which rewarded my journey, and by the interest of the observations then collected regarding the early geography and civilization of the country. Considering the length to which the work has grown, and also that its conclusion coincides with the commencement of a fresh and absorbing task in the same field, I may be allowed to restrict my introductory remarks to a brief record of the circumstances which led me to form my plan and enabled me to carry it into execution, and to some needful explanations concerning the conditions in which the results were finally elaborated.

Ever since 1891, when the famous birch-bark codex acquired by Colonel Bower from Kuchā became known to Indologists, my eyes had been turned towards Eastern Turkestān as a field for archaeological enterprise. But the difficulties with which then, and for long years thereafter, I had to contend, in order to assure the completion of my labours on Kalhaṇa's Chronicle of Kashmīr, while the leisure left to me by onerous administrative duties was of the scantiest, precluded any actual steps in that direction. In the spring of 1897, when the conclusion of those labours was drawing within sight, I became acquainted, through personal communications from the late Professor Bühler, with the important find of ancient birch-bark leaves, containing a Buddhist text in early Prākrit and in Kharoṣṭhī writing, of which M. Dutreuil de Rhins had acquired a portion during his stay at Khotan. These remarkable fragments, of which others had found their way to St. Petersburg, were on their publication by M. Senart at once recognized as the oldest Indian MS. then known, and they decided me to fix upon Khotan as my goal.

Other acquisitions from the same region, which had reached Calcutta, helped to assure me still further of the importance of this particular field. Dr. A. F. R. Hoernle, C. I. E., to whose long-continued efforts we owe the decipherment and elucidation of those first discoveries of ancient Sanskrit texts from Kuchā, had induced the Government of India to issue instructions to its political representatives in Kāshgar, Ladāk and Kashmīr for the acquisition of antiques from Chinese Turkestān. The most notable additions which the collection thus formed in Calcutta under Dr. Hoernle's charge received during the years 1895–7, consisted of MSS. and other remains which were said to have been obtained by native 'treasure-seekers' from sites in the Khotan oasis and adjoining parts of the Taklamakān desert. But no reliable information was ever forthcoming as to the exact origin of these finds, or the character of the sites which were supposed to have furnished them. The many questions of historical interest which from the absence of such data had to be left unanswered—and the doubts which soon arose as to the genuineness of a large portion of these acquisitions, consisting of MSS. and block-prints in a variety of 'unknown characters'—were the best proof of the urgent need for proper explorations on the spot.

I had discussed the plan of such explorations with Professor Bühler before I left Europe in 1897, and often have I thought since of the words with which that lamented scholar, during the last hour I was destined to spend in his ever stimulating company, expressed his keen interest in my plan as well as his appreciation of its practical difficulties. Not the least of these was the question how to secure the needful freedom and means. The exacting duties of the post then held by me under the Panjāb University did not allow me to take preparatory steps towards the execution of my plan until the summer of 1898, when, after having obtained in Kashmīr the necessary practical data, I asked Dr. Hoernle to help in securing for my project the support of the Indian Government. I owe it largely to Dr. Hoernle's generous interest in my proposals that the formal application which I submitted in August, 1898, met with favourable consideration. Regard for the exigencies of the post then held by me had obliged me to restrict the first scheme to a journey of only six months, to be undertaken in 1899. The total cost was estimated at Rs. 6,800, and this sum the Governments of India and the Panjāb agreed to provide by joint contributions of two-thirds and one-third, respectively.

Final sanction had scarcely reached me in the spring of 1899 when my appointment to the Indian Educational Service as Principal of the Calcutta Madrasah necessitated a postponement of the journey, and a reconsideration of the official arrangements concerning it. Fortunately the Government of Bengal, under the late Sir John Woodburn, a zealous friend of Oriental research, proved as ready to recognize the importance of my proposed explorations as had Sir Mackworth Young's administration in the Panjāb. The changed conditions of my appointment permitted an expansion of the time limit and also of the scope of my journey. The modified scheme to which the Government of India accorded their final sanction in July, 1899, provided for my deputation to Chinese Turkestān for a period of one year from June, 1900. A grant of Rs. 11,000 (£733) was placed at my disposal to meet the estimated cost of the expedition, the contribution of the Government of India being increased to Rs. 6,500, while the remaining sum, as far as it was not covered by the grant which the Panjāb Government kindly allowed to stand, was provided from provincial funds of Bengal.

It affords me gratification that, in spite of the difficulties arising from great distances and physical obstacles, and of the uncertainties inseparable from an enterprise in a new field, the whole of my programme was accomplished strictly within the sanctioned estimates of time and expense. I should probably not have been able to keep the expenses within the relatively modest limits here indicated had not the Survey of India Department offered its assistance for an important adjunct of my researches. The necessity of fixing accurately the position of ancient sites and generally elucidating the historical geography of the country brought surveys of the regions visited into close connexion with my immediate task. But I also realized from the first that only by utilizing every opportunity the journey might offer for geographical work of a more general character, could I gain that practical familiarity with the physical conditions of the country which is needed for the right interpretation of its past.

Colonel St. George Gore, R.E., C.S.I., late Surveyor General of India, readily offered for this object the aid of his Department. He kindly agreed to depute with me one of its experienced native surveyors, Bābū (now Rai Sāhib) Rām Singh, and to provide the necessary equipment of surveying instruments, together with a grant of Rs. 2,000 (£133) intended to cover the additional expenses of survey operations. With the assistance of the Surveyor, to whose excellent services I have had occasion to refer elsewhere, it became possible to carry on a continuous system of surveys by plane table, astronomical observations, and triangulation during the whole of my travels. The results of these surveys, which in the mountains I supplemented by photogrammetric survey work of my own, and to the direction and supervision of which I devoted a good deal of time and attention, will be found illustrated by the 'Map of the territory of Khotan and adjoining regions', on the scale of 8 miles to 1 inch, which accompanies this Report. Its use will, I hope, considerably facilitate a correct comprehension of the numerous questions, equally interesting for the historical student and the geographer, which I have had to discuss in relation to the ancient sites now abandoned to the desert.

For the generous consideration and aid which Lord CURZON's Government and the local administrations of Bengal and the Panjāb accorded to me, I wish to record here my deep sense of gratitude. To it I owed the long sought for chance of serving the interests of Oriental research in a new field, and with that measure of freedom which permits full concentration of scholarly efforts. But grateful, too, I must feel to Fate for having allowed me to use this chance to the full, and for having rewarded my labours on difficult ground with results which competent fellow-scholars have recognized as an ample return for the means and facilities so liberally granted to me. Of the serious difficulties and risks which had to be faced at various stages of my journey and which might easily have thwarted my aims, there is no need to speak here in detail. A perusal of my Personal Narrative[1] will suffice to show the character and extent of the obstacles which had to be reckoned with during my work in the Taklamakān.

I could not have overcome them successfully without the active co-operation of the Chinese administrators of the districts upon which I depended for whatever was needed during the winter campaign in the desert. In the Ambans Pan Dārin and Huang-Daloi, then in charge of Khotan and Keriya, I was fortunate to find reliable friends, ever ready to help me with everything in their power. I look back with all the more gratitude to their unfailing attention and kindness, as it was shown at a time when the Empire was in conflict with the European powers, and as neither honours nor material rewards were to be expected for it. The true historical sense innate in educated Chinese, and their legendary remembrance of Hsüan-tsang, the great 'monk of the T'ang period' (*T'ang sêng*) whom I claimed as my guide, proved helpful in explaining the objects of my explorations. But the sympathetic attitude which the provincial administration from the first showed towards my work, must mainly be attributed to the effective support given to me by Mr. G. MACARTNEY, C.I.E., the representative of the Indian Government at Kāshgar. Apart from the advantages which his personal influence secured to me, I derived manifold benefits from the practical help and advice with which this accomplished friend favoured me during my stay at Kāshgar. For these, and the watchful care with which he followed my explorations from afar, I must record here the expression of my lasting gratitude.

The value of the scientific results of my explorations may be left to be judged from the record presented in these volumes. Yet I feel I am justified in referring to the generous acknowledgement which those results have already received from fellow Orientalists. Their opinion, whether recorded in the formal resolution of the XIIIth International Congress of Orientalists[2], or expressed in reviews and other notices by such distinguished scholars as MM. Foucher, Sylvain Lévi, Senart, and others, have been a great encouragement to me and have helped me to bear the strain entailed by the elaboration of the results.

[1] *Sand-buried Ruins of Khotan, Personal Narrative of a journey of archaeological and geographical exploration in Chinese Turkestān.* By M. Aurel Stein. First edition, London, T. Fisher Unwin, 1903; second edition, Hurst and Blackett, 1904.

[2] See *Ruins of Khotan*, p. xiii.

The difficulties to which I refer are accounted for largely by the extent of the archaeological operations carried out, the great variety of the finds and observations made, and not less by the scope and character which I felt obliged to give to my task of recording them. The extent of the field-work is sufficiently illustrated by the fact that, taking only the territory of ancient Khotan, the old sites and remains which I surveyed in detail or excavated are scattered over an area which in a direct line stretches for more than 300 miles from west to east. The wide range of the antiquarian remains is equally striking. Dating back to periods which may be estimated to extend over at least eleven centuries, they show remarkable diversity in nature and character. Among the sites surveyed there are represented the curious wind-eroded 'Tatis', typical of the great physical changes to which the soil of the tracts bordering on the desert is subject; 'culture-strata' left behind by thickly inhabited ancient settlements and deeply buried below the loess accumulations of centuries; and in the area now overrun by the moving sands of the Taklamakān, sites with structural ruins of all kinds and in all stages of preservation. The remains discovered of temples, fortified posts, Sarais, private dwellings, &c., with the relics of ancient art-work and industrial products found in them, have been so plentiful and multifarious as to enable us to study most aspects of the early civilization which once flourished there.

The very abundance and novelty of the archaeological materials brought to light necessarily made the task of correctly recording and interpreting them one of considerable difficulty. But what greatly added to the difficulty—as it undoubtedly did also to the fascination and scientific interest of the work—was the remarkable diversity of cultural influences thus revealed in the arts of ancient Khotan. That the Buddhism which we knew from Chinese records to have been early established in the country had been imported, directly or indirectly, from India was, of course, always certain. But there was little to prepare us for such overwhelming evidence as the wealth of Kharoṣṭhī documents on wood and leather discovered at the Niya Site has furnished of the large place which Indian language and culture must have occupied in the administration and daily life of this region during the early centuries of our era[3]. That Sanskrit Buddhist literature was studied in Khotan down to the end of the eighth century A. D. has been proved beyond all doubt by the texts in Brāhmī script which I excavated in the ruined shrines and monastic dwellings of Dandān-Uiliq and Endere. At the same time, the large series of ancient stucco sculptures, frescoes, painted tablets, and relief carvings in wood, from ruins the dates of which, as determined by exact chronological evidence, range from the third to the eighth century, has conclusively demonstrated that the Graeco-Buddhist art of the extreme north-west of India found a new home in Khotan and enjoyed a long-continued local development.

The territory of Khotan had, like the rest of the Tārīm basin, been under Chinese supremacy for considerable periods both under the Han and the T'ang dynasties. The interesting finds of Chinese records on wood and paper, Chinese sgraffiti, coins, articles of industrial art, &c., at the sites excavated by me have made it abundantly clear that Chinese political control was effective, and that the influence of Chinese civilization, too, must have strongly asserted itself. The fact that Khotan occupied an important position on the ancient trade route from China to the Oxus basin no doubt helped to increase this influence. To the same connecting link we must ascribe those early relations with the distant West which the remarkable series of classical seals found impressed in clay on a number of the ancient Kharoṣṭhī documents from the Niya Site attests in so striking a fashion[4]. The imitation of Persian art, of which we find unmistakable indications five centuries later in Buddhist paintings recovered from the shrines of Dandān-Uiliq,

[3] See chap. XI, in particular pp. 363 sqq.

[4] See below, pp. 354 sqq.

forms an instructive parallel, and, like the curious find of a Judaeo-Persian document at the same site, convincingly demonstrates the continuity of these relations. Finally, the finds of Tibetan MSS. and sgraffiti in the ruined fort and temple of Endere prove that we have to reckon also with influences from the side of Tibet. From the seventh century onwards these are accounted for by historically attested invasions; but there are ethnic connexions, at present less clearly recognizable, which suggest them also for earlier periods.

But with all these varied foreign influences at work, there is ample evidence to show that Khotan, together with other territories of Eastern Turkestān, possessed during the Pre-Muhammadan epoch a well-defined civilization of its own. Neither the occasional references of the Chinese Annals nor the brief notices of Chinese Buddhist pilgrims could help us far to reconstruct, in its main aspects, that bygone phase of civilization of which all indigenous records and remains seemed to have vanished. Nor could acquisitions of chance finds, made from afar, of uncertain, if not suspected, origin and unaccompanied by precise archaeological evidence, be expected to throw true light on this cultural *milieu* and the historical changes and developments it had witnessed.

In order to secure materials that would help us to recover this interesting chapter of lost history, and to interpret it rightly, it was not enough to conduct excavations and to arrange for the deposition of what official language styles the 'archaeological proceeds' in Museums. It was at least equally important that an exact and detailed record should be kept of all observations made on the ground, and that the evidence thus secured should be published with all possible care and fullness. It is needless to emphasize for fellow-scholars the importance of this condition, which the canons of scientific archaeology impose upon any qualified worker. Yet for the benefit of others I cannot refrain from quoting the warning which an archaeological explorer of unequalled experience, Professor Flinders Petrie, has so eloquently recorded in his admirable handbook on *Methods and Aims in Archaeology*. 'To leave a site merely plundered, without any attempt to work out its history, to see the meaning of the remains found, or to publish what may serve future students of the place or the subject, is to throw away the opportunities which have been snatched from those who might have used them properly'[5].

I have often had to refer to the grave risks with which all ancient remains at the desert sites of Khotan are threatened by the destructive effect of wind-erosion, and even more at the hand of 'treasure-seekers', whose operations have in recent years been stimulated by the demands of European collectors. The thought of these risks and the knowledge that neither Khotan nor any other part of Eastern Turkestān, in spite of the conserving capacity of the climate, could ever have competed with countries like Egypt in wealth of antiquarian remains, always urged me on to unwearied exertion. The same thought and knowledge made me feel with equal keenness my responsibility in regard to the duty 'to work with the fullest care and detail in recording, to publish everything fully'[6].

To the obligation here indicated the peculiar conditions of the field in which my work lay added another closely allied one. I mean the necessity of recording in some detail my observations on the modern physical conditions and ethnography of the region. To the critical student of history, whether he deals with the written records or the antiquarian relics of the past, the powerful influence exercised by geographical conditions and surroundings must ever be present. Nor ought he to ignore the useful guidance which ethnic inheritance affords for

[5] See *Methods and Aims in Archaeology*, p. 179.

[6] Comp. Flinders Petrie, loc. cit., p. 175.

the proper comprehension of the past of a people and its civilization. Eastern Turkestān is not a part of the earth as easily accessible as Greece or Egypt or India. If in recent years much has been achieved, particularly through Dr. Hedin's labours, towards the elucidation of its topography, yet there are great tracts where the detailed knowledge of local conditions, routes, &c., must still be gained primarily on the spot. The same observation applies still more forcibly to the ethnography of the country. These considerations will explain why I often felt obliged to record modern physical and anthropological facts when discussing questions connected with earlier economic and ethnic conditions. They also justify me in looking upon the publication of my Personal Narrative, in which observations of geographical interest together with others illustrating the life and character of the population could be suitably detailed for a wider public, as a useful and even necessary preliminary to the present work.

Just as it was important for the proper investigation of antiquarian questions to keep in view the geographical and ethnographical background, so too my task presupposed constant references to whatever written records have come down to us of the early history of these regions. The great majority of these records have to be gleaned from scattered passages in the Chinese dynastic Annals and the itineraries of Chinese Buddhist pilgrims to India. The limitations of my philological knowledge, which, as I have often bitterly regretted, does not extend to Chinese, would not allow me to consult these sources in the original. Nevertheless I have thought it right to attempt the succinct sketches of the history of Eastern Turkestān, Kāshgar and Khotan which are embodied in chapters III and VII[7]. Fortunately M. Chavannes' recent publication *Documents sur les Tou-k'iue (Turcs) occidentaux* has opened to us a rich storehouse of precise and valuable information on the history of Central Asia during the period of Chinese ascendancy under the T'angs, including a great portion of those notices in critically revised and annotated translations. For the Chinese notices on Khotan I could avail myself of the full extracts which Abel Rémusat, in a small but very meritorious publication, had rendered accessible from the Encyclopaedia of the *Pien i tien*[8]. I hope those historical sketches will prove useful by the endeavour made in them to apply to the recorded historical data the light of geographical and antiquarian knowledge acquired on the spot.

The preparation of my detailed report on the lines indicated was bound to prove an exacting task; but the difficulty of securing adequate leisure made it more difficult still and has greatly delayed its completion. The period of three months' special duty which the Secretary of State had been pleased to sanction after I had, early in July, 1901, deposited my archaeological finds in the British Museum as a temporary measure, barely sufficed for the provisional arrangement of the collection and the preparation of the *Preliminary Report* published the same autumn[9]. At its expiration I was obliged to return to India, where during the following half-year my heavy official duties as Inspector of Schools in the Panjāb precluded the possibility of any scholarly occupation. Fortunately, however, the Government of India, on the proposal of the Panjāb Government, and influenced, no doubt, by the interest which H. E. Lord CURZON was kind enough to show in the results of my expedition, sanctioned, with the concurrence of H.M.'s Secretary of State, my deputation to England in May, 1902, in order that I might be enabled to elaborate those results and to make arrangements for their publication with the help of expert scholars in Europe.

[7] See below, pp. 52–72, 166–84.

[8] See *Histoire de la Ville de Khotan*, Paris, 1820.

[9] *Preliminary Report on a journey of archaeological and topographical exploration in Chinese Turkestān.* By M. A. Stein. India Office, London, 1901.

I had every reason to feel deeply grateful to the Viceroy and the Indian Government for the generous consideration thus shown to me and my work. But the period of seven months' stay in England, to which my deputation was at first limited, proved, as I had feared, wholly insufficient for the task. Two successive extensions which the Secretary of State, after reference to the Government of India, kindly sanctioned, added to the above period an aggregate of ten months. By dint of great exertions I could thus complete while in England those important portions of my task for which it was indispensable to have ready access to my archaeological finds deposited in the British Museum, as well as to the manifold expert help I needed for their detailed study and reproduction. The difficulty of these labours had been considerably increased by the uncertainty as to the length of time available for them. Yet before I left England I had managed to conclude the arrangements for the decipherment and publication of all the varied finds of MSS. and documents; to prepare and pass through the press the numerous plates of antiques; to supervise the drawing, &c., of the detailed map from original surveys which accompanies this work, and, last but not least, to bring out my *Sand-buried Ruins of Khotan*, which in many respects has formed a necessary preparation and complement to the present work.

After my return to India at the close of 1903 I devoted every moment of leisure I could spare from my duties as Inspector-General of Education and Archaeological Surveyor, North-west Frontier Province and Baluchistān, to continued work on my report. But in spite of all efforts its publication might have been delayed for years, had not the Government of India, in order to enable me to resume my explorations in Chinese Turkestān in the spring of 1906, been induced to set me free once more by the grant of six months' special duty from October of last year. It was during these busy winter months in Kashmīr, while many practical preparations for my new journey claimed attention, that the greater part of my text was written.

From the reasons which explain why this final report could not be presented earlier, I now turn with pleasure to the duty of recording my acknowledgements to those scholars and friends without whose valuable collaboration and kind help this work could not have been completed in its present form. First among them I must mention my artist friend, Mr. Fred. H. Andrews, late Principal of the Mayo School of Art, Lahore, and now Director of the Art Department, Battersea Polytechnic, whose devoted exertions have effectively furthered my labours ever since my return from Chinese Turkestān. His wide knowledge of ancient Indian art, especially of the Graeco-Buddhist style, his familiarity with the products of many old Eastern industries, and his own high artistic abilities, exceptionally qualified him to assist me in the arrangement, description, and illustration of my collection of antiques. The detailed descriptive list of these included in the report is wholly his work, though, having myself checked and revised it, I must accept full responsibility for its contents, and in particular for the data as to the origin of individual antiques. Without the painstaking accuracy and fullness of Mr. Andrews' descriptions it would probably have been impossible for me to complete my report at a time when some six thousand miles separated me from my collection. The value of Mr. Andrews' descriptive list will be all the more appreciated by students of the early arts and industries of Central Asia since, according to a decision of the Indian Government, the whole of the archaeological finds will now be distributed between the museums of Calcutta and Lahore and the British Museum. As comparison of the original objects will thus be difficult, their use for future researches will mainly depend on the details furnished by the descriptive list and on the illustrations given in the plates. For the preparation of the latter and for the investigation of all questions affecting the technical and art aspects of my finds, Mr. Andrews' expert help and advice have been of the utmost value to me. His unwearied assistance did not cease after I left England, and has greatly facilitated the passing through the press of these volumes. Besides the sketches and diagrams reproduced among the illustrations, I owe to Mr. Andrews' artistic skill the black and white drawing which has furnished the vignette for the title-page. My volumes could scarcely issue under a more felicitous emblem than this faithfully restored enlargement of the figure of Pallas Promachos as seen in several of the ancient seal impressions in clay excavated by me.

I shall always associate the remembrance of the labours given to the elucidation of my archaeological finds with the British Museum, where they found their first resting-place. I may hence fittingly record in this place my obligations to its authorities who readily accorded them shelter. Both in the Coin Department and in that of Mediaeval Antiquities, where my collection was successively accommodated, I enjoyed exceptional facilities for its arrangement, &c., Mr. C. H. Read, Keeper of the latter Department, proving in particular a most helpful adviser.

I owed these facilities in the first place to the kind help of my friend Prof. E. J. Rapson, then Assistant Keeper of the Coin Department, who during my absence in India charged himself also with the care of my collection. It was with a feeling of keen relief that I received the same friend's offer to undertake what from the first I had recognized as the most difficult of the philological tasks resulting from my explorations,—the decipherment and publication of the mass of Kharoṣṭhī documents on wood and leather recovered from the Niya Site. No scholar could have been better qualified for this arduous work. Prof. Rapson's preliminary papers show the advance he has already made in the task[10]. I owe it to Prof. Rapson's kindness that in chap. XI I have been able to indicate some results of his labours possessing special antiquarian bearings. The full publication of these documents, for which Prof. Rapson has been fortunate enough to secure the assistance of two distinguished *savants*, M. E. Senart, Membre de l'Institut, the decipherer of so many important Indian epigraphical records, and M. l'Abbé A. Boyer, had to be reserved for a separate volume which is to follow this report at a date that cannot yet be specified.

For the MSS. and documents written in Brāhmī characters, and showing texts partly in Sanskrit, partly in two non-Indian languages, Dr. A. F. R. Hoernle, C.I.E., was from the first marked out as the most competent editor. The eminent Indologist has completed his edition of these finds. It would have been difficult to provide for their publication in connexion with this report except by a separate volume; and as Dr. Hoernle has recently undertaken, under the auspices of the India Office, the editing of a series in which all previous MS. acquisitions from Eastern Turkestān entrusted to his care are to be published, the inclusion in it of those Brāhmī texts has appeared more convenient. In the meantime Dr. Hoernle has placed at my disposal a valuable analysis of the Brāhmī documents and inscriptions, which has been of great help in describing these finds in chapters IX and XII[11], and for which I wish to offer here my special thanks.

Among the Appendices contributed to my report there is none for which I feel more deeply grateful than that in which M. Éd. Chavannes, Membre de l'Institut, has with masterly clearness analysed my finds of Chinese documents and inscriptions. It is not for me to appraise the exceptional qualifications which have rendered M. Chavannes the leading authority on all Chinese sources of information concerning the history and geography of Central Asia. The great help I have derived from his annotated translations of those finds, as well as from

[10] See below, pp. 326 sqq., 364 sqq.

[11] See below, pp. 257 sqq., 295 sqq., 439 sq.

the long series of his earlier publications, will be evident to my readers from every chapter. But in addition I owe a debt of gratitude to M. Chavannes for a service of friendship which, I know, must have cost heavy sacrifices of time and labour. In order to secure in my work that uniformity in the transcription (according to Wade's system) of Chinese names and terms, the want of which in the various publications consulted for Chinese records had often been a sore difficulty to me, and with a view to assuring general correctness in Chinese references, he kindly charged himself with a detailed revision of the latter, both in my MS. and in a final proof.

For my finds of Tibetan MSS. and sgraffiti (Appendix *B*) I had the good fortune to secure as joint editors two scholars of critical thoroughness, Dr. L. D. Barnett, of the British Museum, and Rev. A. H. Francke, of the Moravian Mission, Ladāk. Dr. Barnett's preliminary notice will have shown to Orientalists how well qualified he was to elucidate the importance of those MS. finds for the history of Tibetan language and literature [12]. By his exceptional knowledge of Tibetan rock-carvings and of the colloquial language, the Rev. Mr. Francke has been able to deal successfully with the sgraffiti which would have baffled any Tibetan scholar in Europe. To Mr. F. W. Thomas, the learned librarian of the India Office, my sincere thanks are due for having kindly supplemented the Tibetan notices of Khotan otherwise accessible to me by the valuable extracts from the Kanjur reproduced in Appendix *E*. From another and widely distant field of Oriental research has valuable help come to me in the learned analysis which Professor D. S. Margoliouth has given of the curious Judaeo-Persian document from Dandān-Uiliq reproduced in Appendix *C*.

The classification of my collection of coins could not have been carried out by more competent scholars than Dr. S. W. Bushell, c.m.g., the distinguished Sinologist, and Prof. E. J. Rapson. Though I had endeavoured, as far as my means of reference would permit, to determine on the spot all coins actually found by me, yet the inventory list in Appendix *D*, and the plates illustrating it were prepared mainly from the materials which those two friends were kind enough to place at my disposal. To Dr. S. W. Bushell I am still further indebted for the liberality with which he has allowed me to draw upon his store of antiquarian and historical knowledge in questions connected with Chinese lore. To Dr. Percy Gardner, Professor of Archaeology at Oxford, I must offer my best thanks for having accorded to me the same favour in regard to objects of classical and semi-classical art contained in my collection.

The two interesting Appendices *F*, *G*, in which Professors A. H. Church, f.r.s., and L. de Lóczy have recorded the results of their respective analyses of the stucco materials, and the loess and sand specimens brought back by me from Khotan sites, deserve all the more my grateful acknowledgements, as they clear up questions in which the archaeologist can derive guidance solely from natural science. The same is the case with the important investigations which Prof. J. Wiesner, Director of the Institute for Plant-Physiology, Vienna University, has devoted to the ancient paper materials represented among my antiquarian finds [13]. Nor ought I to omit grateful mention here of the valuable assistance I have received in the anthropological field, linking as it were the domains of historical and natural science, from the instructive analysis to which my friend Mr. T. A. Joyce, of the British Museum, Secretary of the Anthropological Institute of Great Britain, has subjected the anthropometric materials I collected on my journey [14].

[12] See *J.R.A.S.*, 1903, pp. 109 sqq.; below, pp. 426 sq.

[13] See J. Wiesner, *Ein neuer Beitrag zur Geschichte des Papieres*, in *Sitzungsberichte* of the Imperial Academy, Vienna, histor.-philos. Klasse, 1904; comp. below, pp. 307 sq., 426.

[14] See *On the physical anthropology of the oases of Khotan and Keriya.* By T. A. Joyce, in *Journal of the Anthropol. Institute*, 1903, pp. 304–24; comp. below, pp. 144 sqq.

A reference to chapter vi, where I have discussed the racial origin of the Khotan population, will show how illuminating these investigations are to the historical student.

For the publication of my report, the liberal arrangements sanctioned by the Secretary of State have proved of great help, and for them I may be allowed to record my special gratitude. At my request the preparation of the collotype plates illustrating antiques was entrusted to Mr. W. Griggs, whose name has been made familiar by the labours of a lifetime to every student of Indian art. I appreciate all the more gratefully the personal care and attention which Mr. Griggs has bestowed on the work as I know that much of this, as in the case of other scholarly publications to which he has lent his skill, must have been a labour of love. The drawing of the map of Khotan and the adjoining regions was entrusted to Mr. J. W. Addison, map draughtsman of the Royal Geographical Society, and effected in a fashion which has been commended by competent judges.

I had every reason to feel gratified when the Delegates of the Clarendon Press, with a grant from the India Office, agreed to undertake the printing and publication of the work, together with the preparation of the remaining plates. My hope that the resources of that great *officina* would facilitate the early issue of the work, and in a fitting form, has been justified by experience. For the typographical care bestowed on these volumes and for much kind attention besides, I wish to express here my thanks to Mr. C. Cannan, Secretary to the Delegates, and Mr. H. Hart, the University Printer.

Owing to the great distance separating me from the Press, it was impossible for me to see more than one proof, and even for this the available time was very scanty. I could not have hoped to assure the requisite degree of correctness, had not the self-sacrificing kindness of friends come to my help. Mr. J. S. Cotton, editor of the *Imperial Gazetteer of India*, kindly revised my manuscript before it went to the press, with regard to those literary requirements of which he is so experienced a judge. Mr. Vincent Smith, the distinguished Indologist, was good enough to read a proof and favoured me with a number of useful corrections. Mr. T. W. Arnold, Professor of Arabic at University College, London, charged himself with seeing the final revises, a task for which I could not have found a friend of wider Oriental attainments nor one more willing to help. To Mr. Arnold also is due the elaborate Index, which I was unable to compile myself, but which I should not have liked to entrust to any other hands. But to none are my obligations greater than to my friend Mr. P. S. Allen, of Corpus Christi College, Oxford, who from a scholarly task of absorbing magnitude, in a field wholly Western, spared time to correct my proofs with most painstaking thoroughness. He also took entire 'charge' of the book when I had passed beyond the reach of regular postal communications. I look upon these proofs of friendly devotion as the best compensation for the difficulties I had to overcome while preparing my work.

During that happy year of travel through the mountains and the deserts which once saw the passage of Hsüan-tsang, the great pilgrim, of Marco Polo, and of him 'who sought lost Cathay and found Heaven,' the thought of the great scholar who had first with true critical intuition traced their tracks and those of many another early traveller through Central Asia, was ever with me. From his immortal volumes, which have accompanied me everywhere, I never failed to derive guidance and encouragement, whether I turned to them in camp after long hours of rough travel or in my improvised study after desk labour yet more tiring. In dedicating this work to the memory of Sir Henry Yule I claim no small privilege. But if the interest of researches on ground that was cherished by the Master, and the endeavour to carry them on in his spirit can atone for whatever there is of defects and dryness in the presentation of their results, I hope that this tribute will not be deemed inappropriate.

Fate has willed that I should write these closing lines at the foot of the mighty Kun-lun, and yet within sight as it were of the desert, on my way back to Khotan which I have so often longed to see again, and at the commencement of an enterprise which is to take me yet further afield. Grateful do I feel to Fate—and to those who dispense its decrees—for the opportunity which enables me to proceed to a fresh task at the very time when the old one has reached its end. Gladly do I now forget all the efforts that were needed to assure me this freedom, and gladly do I see in its achievement a reward for the labours recorded in these volumes.

M. AUREL STEIN.

CAMP KÖK-YĀR,
July 24, 1906.

ADDENDA ET CORRIGENDA

P. 10, l. 41. The identification of Kao Hsien-chih's route with that leading over the Barōghil and Darkōt has been confirmed in all particulars by the observations I made on my visit to these passes in May, 1906. From Sarhad to the Yasīn side of the Darkōt Pass three marches are reckoned, Chikmar-robāt and Vedinkōt forming the recognized intermediate halting-places.

P. 14, l. 9. *For* Kāshkār *read here and elsewhere* Kashkār.

P. 16, l. 29. The name reproduced in Chinese transcription by *A-shê-yü-shih-to* probably survives in that of the modern village of *Shuist*, on the upper Mastūj river, near which I traced, in May, 1906, extensive areas of former cultivation, said to have been abandoned for a long time owing to the advance of glaciers. Their re-occupation is now slowly proceeding.

P. 35, l. 1. The remains of *Kiz-kurghān* have since been surveyed by me, on May 30, 1906. They were found to consist of massive bastioned walls of stone and brick, manifestly of ancient construction, which crown the west rim of an isolated rocky spur rising circ. 600 ft. above the Tāghdumbāsh river. The top of the spur is very difficult of access, even on that side, and protected on all others by wholly unscalable precipices. The position is one of exceptional natural strength, and the ruins surveyed prove its occupation as a place of refuge at an early period.

P. 39, l. 29. Crossing the Chichiklik plateau on June 4, 1906, I traced near its centre the foundation walls of a large stone-built quadrangle, now partly occupied by decayed graves. The recent construction, by the Chinese authorities, of some shelter-huts in the immediate vicinity proves the need of a hospice in this locality. The physical aspects of this inhospitable plateau agree accurately with Hsüan-tsang's description.

P. 62, l. 33 marg. *For* Hsüan Tsung *read* Hsüan-tsung.

P. 91. l. 9. The anthropometric data collected by me in July, 1906, concerning the Pakhpo people support the ethnic affinity here indicated. I have not been able to trace the use of a separate language among them.

P. 91, l. 19. During my stay at Kōk-yār, in July, 1906, I received information about a number of caves which are venerated as Mazārs in the mountains drained by the Tiznaf river.

P. 103, l. 14. The new district established since 1903, comprising Gūma, Moji, Sanju, &c., has received the official designation of *P'i-shan*, showing that the identification here assumed is accepted by the Chinese authorities.

P. 111, last line, and p. 285, l. 3, *a. f.* *For* Pi-mo *read* P'i-mo.

P. 113, l. 12. *For* fig. 3 *read* fig. 31.
M. 001. k. *For* LI *read* IL.

P. 114. T.M. 003. j. *Add* See Plate XLII.

P. 130, note 23. On revisiting Karanghu-tāgh in August, 1906, I ascertained that the story of Abā Bakr's riches thrown into the Yurung-kāsh on his flight to this valley still lingers in local tradition. It is also popularly known in Khotan.

P. 131, l. 5. *For* Ujāt *read* Ujat.

P. 136, l. 7. *For* Cherchen *read* Charchan, *and so elsewhere.*

P. 166, n. 1. *For* 150 *read* 151.

P. 172, l. 26. *For* Si-shan *read* Hsi-shan.

P. 188, l. 31. *Read* of M. Petrowsky, late Russian Consul General, &c.

P. 190, note 2. *Read* on M. Petrowsky's large collection.

P. 208, l. 25. *For* Y. 011 *read* Y. 0011; l. 28, *for* B. 001. 1. *read* B. 001. l.

P. 213. Y. 0011. i. *For* XLIV *read* XLVII.
Y. 0011. j. *Add* See Plate XLVI.

P. 218 (col. *a*), l. 31. *For* Plate 36 *read* Plate XLVII.

P. 230, l. 2. *For* D. IV. 5 *read* D. x. 4; n. 31, *for* sec. v. *read* sec. iv.

P. 236, n. 1. *For* Plate LV *read* Plates LVI–LVII, LXXXVIII.

P. 238, l. 2. *For* Yāmen *read* Ya-mên *and so elsewhere.*

P. 246, l. 3, a.f. *For* R. iv *read* R. ii.

P. 252, l. 36. While this chapter was passing through the press Mr. F. H. Andrews informed me of his having discovered, among the objects brought back from Tibet by the late expedition, and now deposited in the British Museum, a suit of mail, evidently old, which shows the plates or scales arranged in the reversed order exactly as on the Kubera image of the Dandān-Uiliq shrine and in the Gandhāra example. The plates are not rivetted but ingeniously threaded, by a method which is illustrated in the accompanying diagram of Mr. Andrews. The disposition of the holes, one at the rounded end and three on each of the longer sides, is exactly the same as in the piece of hard leather, N. xv. 005 (see diagram, p. 411), and thus strikingly confirms the suggestion made regarding the latter (p. 374) that it belonged to a scale armour.

P. 257, l. 16. *For* king *read* kind.

P. 269, n. 15. *For* D. III. 13 *read* D. III. 12; *for* D. VI. 4, 6 *read* D. VI, 2, 6.

P. 288. D. I. 11. *Add* See Plate LV.

P. 297. D. VI. 3. *Add* See Plate LXVII.

P. 298. D. VII. 3. d. *Add* See Plate CXVI.

P. 302. D. (T.) 05. *Add* See Plate LV.

P. 335, l. 26. N x. 2. c is not shown in Plate XCVII.

P. 343, n. 4, l. 2. *For* 10+80+190 *read* 10+86+190.

P. 374, l. 26. For the discovery of a Tibetan coat of mail which conclusively proves that the piece of leather must have formed part of a scale armour, see above, *Add.* to p. 252, with illustration.

P. 376, l. 20. *For* N. xix. l *read* N. xix. 1.
l. 5, *a.f.* *For* N. xx. l *read* N. xx. 1.

P. 377, n. 6. *For* p. 335 *read* p. 334.

P. 378, l. 13. On my recent passage through Chitrāl and Mastūj I repeatedly traced the same flower ornament in the architectural wood-carving of those valleys.

P. 393. N. iv. 80. *Add* See Plate LXXI.

P. 394. N. iv. 133. *Add* See Plate XCIII.

P. 395. N. v. 6. *Add* See Plate CII.

P. 398. *For* a reproduction of N. x. 1 in fuller detail see p. xxiv.
N. x. 2. c is not shown in Plate XCVII.

P. 399. N. xii. 3. *Add* See Plate LXX.

P. 402. N. xv. 59. b. *Add* See Plate CXIII.

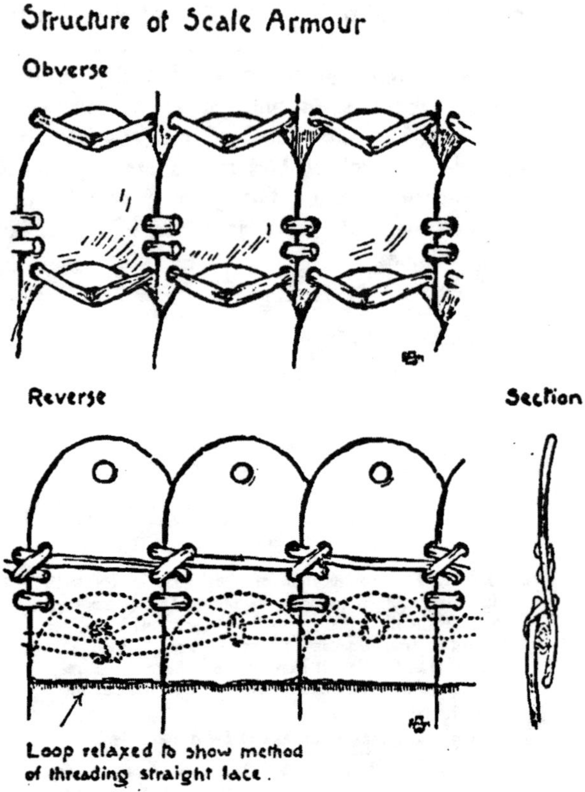

P. 404. N. xv. 122. *Add* See Plate XCVII.

P. 407. N. xv. 207. *Add* See Plate CXIII.

P. 410. N. xv. 001. a. *For* (interior) *read* (exterior), *for* (exterior) *read* (interior).

P. 411. N. xv. 001. c, f, g, h, j. *Add* See Plate LXXIV.

P. 413. N. xx. 03. *Add* See Plate LXIX.

P. 425, l. 11, a.f. *For* 31 *read* 31 a.

P. 438. E. i. 4. *Add* See Plate CIX.

P. 439. E. i. 39. *Add* See Plate CIX.

P. 441. E. i. 025 does not appear on Pl. LXXVII.

TABLE OF CONTENTS

APPENDICES

LIST OF ILLUSTRATIONS

LIST OF ABBREVIATED TITLES

Beal, *Si-yu-ki* . . . . . Si-yu-ki. Buddhist records of the Western world, translated from the Chinese of Hiuen Tsiang (A.D. 629). By Samuel Beal, in two volumes. London, Trübner & Co., 1884.

Biddulph, *Hindoo Koosh* . . . Tribes of the Hindoo Koosh, by Major J. Biddulph, B.S.C., Political Officer at Gilgit. Calcutta, Office of the Superintendent of Government Printing, 1880.

Bretschneider, *Mediaeval Researches* Mediaeval Researches from Eastern Asiatic sources. Fragments towards the knowledge of the geography and history of Central and Western Asia from the 13th to the 17th century, &c. By E. Bretschneider. 2 vols. London, Trübner, 1888.

Bushell, *Early History of Tibet* . The early history of Tibet, from Chinese sources. By S. W. Bushell, M.D., Physician to H.B.M. Legation, Peking. Reprinted from the *J.R.A.S.*, N.S., vol. xii.

Chavannes, *Les Livres chinois* . . Les livres chinois avant l'invention du papier, par Édouard Chavannes. Extrait du Journal asiatique, numéro de Janvier–Février 1905. Paris, Imprimerie Nationale. 1905.

Chavannes, *Turcs occid.* . . . . Documents sur les Tou-kiue (Turcs) occidentaux. Recueillis et commentés par Édouard Chavannes, Membre de l'Institut, Professeur au Collège de France, avec une carte. Saint-Pétersbourg, Académie Impériale des Sciences, 1903.

Chavannes, *Voyage de Song Yun* . Voyage de Song Yun dans l'Udyāna et le Gandhāra (518–522 P.C.). Traduit par M. É. Chavannes, Membre de l'Institut. Extrait du Bulletin de l'École Française d'Extrême-Orient, Juillet–Septembre, 1903. Hanoi, F. H. Schneider, 1903.

Chavannes–Lévi, *L'Itinéraire d'Ou-K'ong* Voyages des pèlerins Bouddhistes. L'itinéraire d'Ou-k'ong (751–790). Traduit et annoté par MM. Sylvain Lévi et Éd. Chavannes. Extrait du Journal asiatique (Sept.–Oct. 1895). Paris, Imprimerie Nationale, 1895.

Curzon, *The Pamirs* . . . . The Pamirs and the Source of the Oxus. By the Right Hon. George N. Curzon, M.P., with map and illustrations. London, the Royal Geographical Society, 1899.

Elias and Ross, *Tārikh-i-Rashidī* . The Tarikh-i-Rashidi of Mirza Muhammad Haidar, Dughlát. A history of the Moghuls of Central Asia. An English version edited with commentary, notes, and map, by N. Elias. The translation by E. Denison Ross. (First edition.) London, Sampson Low, Marston & Co., 1895.

Forsyth, *Yarkand Mission Report* . Report of a Mission to Yarkund in 1873, under command of Sir T. D. Forsyth, K.C.S.I., C.B., Bengal Civil Service. With historical and geographical information regarding the possessions of the Ameer of Yarkund. Calcutta, Foreign Department Press, 1875.

Foucher, *Géographie du Gandhâra* . Notes sur la géographie ancienne du Gandhâra (commentaire à un chapitre de Hiuen-Tsang), par A. Foucher, maître de conférences à l'École des Hautes-Études. Extrait du Bulletin de l'École Française d'Extrême-Orient. Hanoi, F. H. Schneider, 1902.

Foucher, *L'Art du Gandhâra* . . L'art gréco-bouddhique du Gandhâra. Étude sur les origines de l'influence classique dans l'art bouddhique de l'Inde et de l'Extrême-Orient, par A. Foucher, Docteur-ès-lettres. Tome Premier. Paris, Imprimerie Nationale. Ernest Leroux, éditeur, 1905.

Foucher, *L'Iconographie bouddhique* . Étude sur l'iconographie bouddhique de l'Inde d'après des documents nouveaux. Par A. Foucher. Vol. I, Paris, Ernest Leroux, éditeur, 1900. Vol. II, 1905.

FRANKE, *Zur Kenntniss der Türkvölker*	Beiträge aus chinesischen Quellen zur Kenntniss der Türk-Völker und Skythen Central-Asiens. Von O. Franke. Aus dem Anhang zu den Abhandlungen der königl. Preuss. Akademie der Wissenschaften von 1904. Berlin, Verlag der königl. Akademie der Wissenschaften, 1904.
GRENARD, *Mission D. de Rhins* .	Dutreuil de Rhins (J. L.). Mission scientifique dans la haute Asie, 1890–1895. Vols. I–III. Deuxième et troisième parties, par F. Grenard. Paris, Ernest Leroux, 1897–8.
GRÜNWEDEL, *Mythologie* . . .	Mythologie des Buddhismus in Tibet und der Mongolei. Führer durch die lamaistische Sammlung des Fürsten E. Uchtomskij, von Albert Grünwedel, Dr. phil., Leipzig, F. A. Brockhaus, 1900.
GRÜNWEDEL-BURGESS, *Buddhist art* .	Buddhist Art in India. Translated from the 'Handbuch' of Prof. Albert Grünwedel, by Agnes C. Gibson. Revised and enlarged by Jas. Burgess, C.I.E., LL.D. London, Bernard Quaritch, 1901.
HEDIN, *Reisen in Z.-A.* . . .	Die geographisch-wissenschaftlichen Ergebnisse meiner Reisen in Zentralasien, 1894–1897, von Dr. Sven Hedin. (Ergänzungsheft No. 131 zu 'Petermanns Mitteilungen.') Gotha, Justus Perthes, 1900.
HEDIN, *Through Asia* . . .	Through Asia. By Sven Hedin, in two volumes. London, Methuen & Co., 1898.
HOERNLE, *Report on C.-A. ant.* . .	A Report on the British collection of Antiquities from Central Asia. By A. R. Rudolf Hoernle, C.I.E., Ph.D. Part I. (Extra-Number to the Journal of the Asiatic Society of Bengal for 1899.) Calcutta, Baptist Mission Press, 1899.
	Part II. (Extra-Number 1 to the Journal of the Asiatic Society of Bengal, Vol. LXX, Part I, 1901.) Calcutta, 1902.
JULIEN, *Mémoires*	Mémoires sur les contrées occidentales, traduits du sanscrit en chinois, en l'an 648, par Hiouen-Thsang, et du chinois en français par M. Stanislas Julien. Vols. I, II. Paris, Imprimerie Impériale, 1857–58.
JULIEN, *Méthode pour déchiffrer* .	Méthode pour déchiffrer et transcrire les noms sanscrits qui se rencontrent dans les livres chinois, inventée et démontrée par M. Stanislas Julien. Paris, Imprimerie Impériale, 1861.
JULIEN, *Vie de Hiouen-Thsang* . .	Histoire de la vie de Hiouen-Thsang et de ses voyages dans l'Inde, depuis l'an 629 jusqu'en 645, par Hoei-li Yen-Thsong; suivie de documents et d'éclaircissements géographiques tirés de la relation originale de Hiouen-Thsang; traduite du chinois par Stanislas Julien. Paris, Imprimerie Impériale, 1853.
LEGGE, *Fâ-hien*	A record of Buddhistic kingdoms, being an account by the Chinese monk Fâ-hien of his travels in India and Ceylon (A.D. 399–414) in search of the Buddhist books of discipline. Translated and annotated, with a Corean recension of the Chinese text, by James Legge, M.A., LL.D. Oxford, at the Clarendon Press, 1886.
LÉVI, *Notes chinoises sur l'Inde* .	Notes chinoises sur l'Inde. Par M. Sylvain Lévi, Professeur au Collège de France. IV.—Le pays de Kharoṣṭra et l'écriture Kharoṣṭrī. Extrait du Bulletin de l'École Française d'Extrême-Orient. (Juillet–Septembre, 1904.) Hanoi, F. H. Schneider, 1904.
MARQUART, *Ērānšahr* . . .	Ērānšahr nach der Geographie des PS. Moses Xorenac'i. Mit historisch-kritischem Kommentar und historischen und topographischen Excursen von Dr. J. Marquart. Abhandlungen der königlichen Gesellschaft der Wissenschaften zu Göttingen, philologisch-historische Klasse, Neue Folge, Band III, Nro. 2. Berlin, Weidmannsche Buchhandlung, 1901.
RAPSON, *Specimens*	Specimens of the Kharoṣṭhī inscriptions discovered by Dr. Stein at Niya in Chinese Turkestan. Tentative transcriptions by E. J. Rapson, M.A. [Printed for presentation at the XIVth Congress of Orientalists, 1905.]
RÉMUSAT, *Ville de Khotan* . .	Histoire de la ville de Khotan, tirée des annales de la Chine et traduite du chinois; suivie de recherches sur la substance minérale appelée par les Chinois PIERRE DE IU, et sur le Jaspe des anciens. Par M. Abel Rémusat. Paris, imprimerie de Doublet, 1820.
RICHTHOFEN, *China*	China. Ergebnisse eigener Reisen und darauf gegründeter Studien. Von Ferdinand Freiherrn von Richthofen. Erster Band. Berlin, Verlag von Dietrich Reimer, 1877.
RITTER, *Asien*	Die Erdkunde von Asien, von Carl Ritter. Band V. Berlin, G. Reimer, 1837.

ROCKHILL, *Life of the Buddha* . .	The Life of the Buddha and the early history of his order. Derived from Tibetan works in the Bkah-hgyur and Bstan-hgyur. Followed by notices on the early history of Tibet and Khotan. Translated by W. Woodville Rockhill. London, Trübner & Co., 1884.
STEIN, *Prelim. Report* . . .	Preliminary Report on a journey of archaeological and topographical exploration in Chinese Turkestan. By M. A. Stein. Published under the authority of H.M.'s Secretary of State for India in Council. London, 1901.
STEIN, *Rājat.*	Kalhaṇa's Rājataraṅgiṇī, a Chronicle of the kings of Kaśmīr. Translated, with an introduction, commentary, and appendices, by M. A. Stein. Vols. I, II. London, A. Constable & Co., 1900.
STEIN, *Ruins of Khotan* . . .	Sand-buried ruins of Khotan. Personal Narrative of a journey of archaeological and geographical exploration in Chinese Turkestan. By M. Aurel Stein. (First edition.) London, T. Fisher Unwin, 1903.
WATTERS, *Yuan Chwang* . .	On Yuan Chwang's travels in India, by Thomas Watters, M.R.A.S. Edited, after his death, by T. W. Rhys Davids and S. W. Bushell. Two vols., with two maps and an itinerary by Vincent A. Smith. London, Royal Asiatic Society, 1905.
WIESNER, *Neuer Beitrag zur Geschichte des Papieres.*	Ein neuer Beitrag zur Geschichte des Papieres, von J. Wiesner. Sitzungsberichte der kais. Akademie der Wissenschaften in Wien, philosophisch-historische Klasse, Band CXLVIII. Wien, 1904, in Kommission bei Carl Gerold's Sohn.
YULE, *Cathay*	Cathay and the way thither; being a collection of medieval notices of China. Translated and edited by Col. Henry Yule, C.B. In two vols. London, Hakluyt Society, 1866.
YULE, *Marco Polo*	The book of Ser Marco Polo the Venetian concerning the kingdoms and marvels of the East. Translated and edited, with notes, by Colonel Sir Henry Yule, R.E., C.B., K.C.S.I., corr. inst. France. Third edition, revised throughout in the light of recent discoveries by Henri Cordier (of Paris); with a memoir of Henry Yule by his daughter Amy Francis Yule. In two volumes. London, John Murray, 1903.

REPRODUCTION IN FULLER DETAIL OF THE FRAGMENT OF CARPET, N. X. I, PAGE 398.

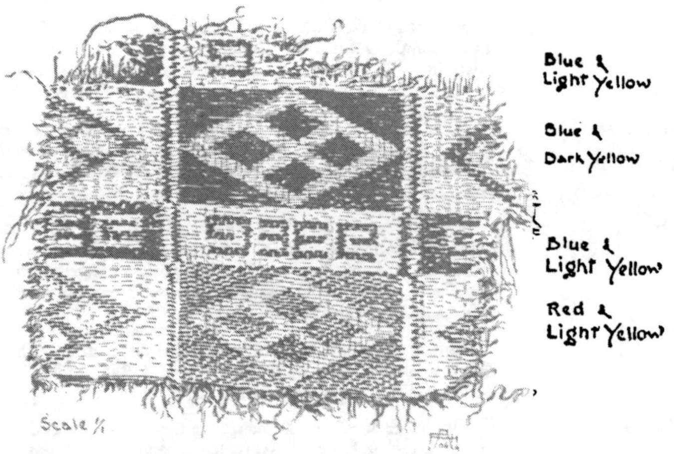

CHAPTER I

FROM KASHMĪR TO THE PĀMĪRS

Section I.—THE GILGIT ROAD AND KIṢANGAṄGĀ VALLEY

KASHMĪR, where in 1898 the first plan of my explorations had been formed, and where two years later, in May 1900, I completed the practical preparations of my journey, was by reason of historical associations and geographical position alike a most appropriate starting-point for my archaeological enterprise. I may therefore fitly commence this detailed account of my antiquarian observations from the same ground. Kashmīr, the starting-place.

The labours which I devoted during the preceding ten years to the elucidation of Kalhaṇa's *Chronicle of Kashmīr*, and the manifold historical researches necessitated by this task, had made me eagerly look out for whatever traces survived of ancient relations between Kashmīr and the Central-Asian regions lying beyond its great mountain barriers. The information to be gleaned from the sources accessible to me was scanty, even in regard to the tracts immediately adjoining Kashmīr, which were undoubtedly subject from early times to its cultural influence. But the results of the close study I bestowed on the ancient topography of Kashmīr, assured me of at least one interesting fact, which at the time of my start (May 31, 1900) appeared like an auspicious omen: there could be no doubt from the first that the route I was to follow through the mountains on leaving the Kashmīr Valley for the north had a claim to considerable antiquity and historical importance.

Practical advantages as well as geographical interests had made me decide for the route through Gilgit and Hunza, and over the Tāghdumbāsh Pāmīr, as the most suitable line of approach to Kāshgar, my immediate goal in Chinese Turkestān. The 'Gilgit Transport Road', which I was authorized to use for the first portion of my journey, from Kashmīr to Gilgit, dates in its present form only from the years 1890–92, when the placing of an Imperial garrison in Gilgit and in the adjoining valleys leading towards the Eastern Hindukush necessitated the construction of a military road fit for laden transport during at least a portion of the year. But the route which it follows, between and over the high ranges separating the Kashmīr Valley from that of the Indus, is marked out by nature as the most accessible line of communication from Kashmīr to the Dard territories northward, and there is historical evidence to prove its use at an early period. Antiquity of the 'Gilgit Road'.

In the *Memoir on the Ancient Geography of Kashmīr*, accompanying my annotated translation of Kalhaṇa's *Rājataraṅgiṇī*, I had occasion to point out the historical interest of the mountain routes leading from the north shore of the Volur Lake into the part of the Upper Kiṣaṅgaṅgā Valley which is known as Gurēz[1]. Only through the valley of the Burzil stream, here joining the Kiṣaṅgaṅgā from the north, is it possible to gain the high passes which cross the snowy watershed between the Kiṣaṅgaṅgā and the Indus and give access to Astōr and thence to the other Dard Valleys. The Kiṣaṅgaṅgā itself, though draining an extensive mountain area, flows Routes into the Kiṣaṅgaṅgā Valley.

[1] Kalhaṇa's *Rājataraṅgiṇī*, translated, with an Introduction, Commentary, and Appendices, by M. A. Stein, London, 1900, II. pp. 406 sqq.; also note on *Rājat.* vii. 1171.

throughout its upper course in an extremely confined valley. To the west, some twenty miles below the junction of the Burzil stream, this valley for a considerable distance becomes an uninhabited and almost inaccessible gorge; it cannot be crossed nor is it reached there by any route from Kashmīr. Eastwards again the great *massif* of the hoary Haramukh (Skr. *Haramukuṭa*) peaks, and the glacier-crowned range continuing it towards the head of the Sind Valley, raise an effective barrier between Kashmīr and the Kiṣangaṅgā. It is crossed only by a few difficult mountain tracks, over which even the hardy Dard hillmen cannot take their ponies, except during brief periods in particularly favourable seasons and then unladen.

Passes of Trāgᵃbal and Dudᵃkhut.

The geographical facts here briefly indicated make it quite certain that in ancient times the line of communication between Kashmīr and the Dard territories must have led through Gurēz, and over that part of the watershed which separates Gurēz from the north shore of the Volur Lake. Starting from a point due south of the main village of Gurēz, the summit ridge for about twenty miles westward shows a height of from 12,000 to 13,000 feet above the sea. The route which is followed by the present 'Gilgit Transport Road', and which has been in general use for military purposes since Sikh times, crosses the range by the Trāgᵃbal or Rāzdiangan Pass, at an elevation of close on 12,000 feet. Another route which leads over the Dudᵃkhut Pass, about eight miles further to the east, is distinctly referred to in the *Rājataraṅgiṇī* in connexion with an invasion of the Dards from the Kiṣangaṅgā Valley towards the close of the eleventh century A. D. In my comments on this, and in some other references, I have shown that the position of the hill fort of Dugdhaghāta, which the Dards had occupied and which King Harṣa of Kashmīr vainly attempted to recover, in order to close the route to their inroads, can still be traced on the summit of the Dudᵃkhut Pass to which it has left its name[2]. Though somewhat less exposed than the Trāgᵃbal route, and evidently preferred in ancient times on this account and as being more direct, the Dudᵃkhut Pass is equally liable to heavy and early snowfall. Like the corresponding portion of the present 'Gilgit Road', it can scarcely have been kept open for regular use by laden animals or large bodies of men during more than four months of the year.

Both the Trāgᵃbal and Dudᵃkhut routes ascend the southern slopes of the range from the stream of the Baṇḍᵃpōr Nullah, which flows into the Volur Lake from the north, and which, among the Brahman population of Kashmīr, is still known by its ancient name of *Madhumatī*. Close to the point in the valley where the two routes diverge lies the village of Baṇḍᵃkōṭh, once the site of a castle (*kōṭh*, Skr. *koṭṭa*) occupied by the 'Malik' or feudal chief who in Muhammadan times watched over the passes giving approach to Kashmīr from the Kiṣangaṅgā. It is very probable that in the same neighbourhood there also existed one of those fortified frontier watch-stations which in the time of the Hindu rulers closed all routes leading into the Kashmīr Valley, and which, under the designations of *dvāra*, 'gate', or *draṅga*, figure so prominently in the Sanskrit Chronicles of Kashmīr[3].

'Northern Gate' of Kashmīr.

It is true that the watch-station of this particular locality is not specifically mentioned by either Kalhaṇa or the later Sanskrit Chronicles. But there can be no doubt that it was known to the Chinese pilgrim Wu-k'ung, who visited Kashmīr from Gandhāra, the present Peshawar District, and stayed there during the years 759–763 A. D. The topographical data

[2] Compare note on *Rājat.* vii. 1171.

[3] The character and historical importance of these ancient watch-stations or 'Gates' of Kashmīr have been demonstrated by me in my *Memoir on the Ancient Geography of Kashmīr*, § 40 (*Rājat.*, II. pp. 391 sqq.); see also notes *Rājat.* i. 122; iii. 227, &c.; *J.A.S.B.*, 1895, Part I, pp. 382 sqq. The system of *rāhdārī* maintained in connexion with these watch-stations, by which persons leaving or entering the Valley were required to produce special permits or passes, survived almost to our own days.

regarding Kashmīr contained in his itinerary have been fully discussed by me elsewhere [4]. Wu-k'ung correctly described the kingdom of Kashmīr as enclosed on all sides by mountains forming its natural ramparts. Only three roads had been opened through them, and these were secured by gates. The road in the east, leading to T'u-fan or Tibet, corresponds plainly to the present route over the Zōjī-Lā to Ladāk and thence to Tibet. By another road, which started from 'the western gate' and went to Ch'ien-t'o-lo or Gandhāra, the route following the Jehlam river in its course to the west is clearly meant. The third road, in the north, is described as reaching into P'o-lü. Its identity with the route represented in its main direction by the present 'Gilgit Road' was beyond all doubt, even before Professor Chavannes' lucid analysis of the original Chinese records had conclusively proved that the term P'o-lü or Pu-lu applied to modern Gilgit as well as to Baltistān.

Of the territory into which that route first takes us, one important historical fact can still be established from Kalhaṇa's Chronicle. Several passages relating to events on this side of the Kashmīr frontier show plainly that the hill tract formed by the drainage area of the Upper Kiṣangaṅgā was occupied, in ancient times as now, by a population belonging to the Dard race [5]. The tenacity with which the ethnographic watershed has been here maintained during long centuries creates a reasonable presumption that the occupation by Dard tribes of the mountain territories extending north-westwards of Kashmīr up to Chitrāl and the Hindukush likewise dates back to a very early period. The Dards of the Kiṣangaṅgā Valley.

We must attach the more value to this evidence because the indications which Sanskrit texts and the works of classical geographers have preserved for us as to the early seats of the Dards (Skr. *Darad*, *Dārada*; the Δαράδραι of Ptolemy, Dardae of Pliny, &c.), are by no means as exact as we should wish. From the account given by Kalhaṇa of events among the Dards of the Kiṣangaṅgā Valley near his own time, it may be concluded that this hill tract then formed a small chiefship. The division of the territory inhabited by the Dards into a considerable number of more or less independent principalities has prevailed also during recent periods of their history, and is amply accounted for by geographical features. Mountain ranges of exceptional ruggedness, covered by eternal snows on their summit lines and culminating in some of the highest peaks of the world, render communication between the narrow valleys difficult even for the hardy hillmen who have succeeded during so many centuries in maintaining here their struggle with a harsh climate and a barren soil. Frequently the confined valleys themselves present almost equal difficulties, where they narrow to rock-bound gorges filled completely for considerable periods of the year by snow-fed rivers. The formidable nature of the mountain-barriers in this region is reflected in the notice which Ptolemy records of 'the Daradrai in whose country the mountains are of surpassing height' [6]. Nor is there need to search far in the accounts of modern travellers in order to find ample illustration of the difficulties and dangers so graphically described by the old Chinese pilgrims who made their way along the precipitous rock walls of the Dard Valleys [7].

I have fully discussed elsewhere the results of that remarkable seclusion which Kashmīr enjoyed during a long period of its history, owing to the natural strength of its mountain Kalhaṇa's *Daraddeśa*.

[4] See 'Notes on Ou-k'ong's account of Kaśmīr', in *Proceedings of the Imperial Academy*, Vienna, 1896, vol. cxxxv. pp. 22 sq.; also *Rājat.*, II. pp. 357 sq.

[5] Compare, regarding the Dards and their relations with Kashmīr, my notes on *Rājat.* i. 312; vii. 911; also II. pp. 431, 435.

[6] See Ptolemy, *Geogr.*, VII. i. 42.

[7] Compare Fâ-hien's *Travels*, translated by Legge, p. 26; *Voyage de Song Yun*, transl. by Chavannes, p. 28; Hsüan-tsiang's *Si-yu-ki*, transl. by Beal, i. p. 134 sq.

ramparts and the rigorous system of guarding the frontier-passes[8]. In view of this jealously maintained isolation it cannot surprise us to find that Kalhaṇa's horizon northward was practically limited by the petty hill state on the Upper Kiṣangaṅgā and its tributaries, which he designates as *Daraddeśa*, 'the Darad land.' The seat of its chiefs, referred to by the name of *Daratpurī*, 'the town of the Darads', may have stood at the present Gurēz[9]. This, the main settlement of the valley, was the residence of the Nawābs who ruled the tract until the Sikh conquest. On account of the advantages offered by its position in the widest and most central portion of the valley, Gurēz was probably a place of some importance in earlier times, though among its rude log-huts and near the rubble-built fort no remains of antiquity could be traced by me.

Once beyond this point the guidance of the Chronicler fails us completely for the historical topography of the route. It is, indeed, significant, as I have pointed out elsewhere, that 'Kalhaṇa, when describing the home of the "Mleccha" chiefs from the north who, in his own time, invaded Kashmīr together with the Darads of the Kiṣangaṅgā Valley (viii. 2762–4), can treat us only to details of the mythical geography of the Himālaya regions'[10]. Judging from Kalhaṇa's usual accuracy in matters topographical we may feel convinced that he would not have withheld from us the old Sanskrit designations of Astōr, Gilgit, and the other Dard valleys from which the northern allies were undoubtedly drawn, if those territories had been as familiar to him as they are now to the educated Kashmīrī.

Alberūnī's notice of Dard valleys.

Considering this restricted knowledge prevailing in the Kashmīr of Hindu times, and the complete inaccessibility of Kashmīr itself to all Muhammadans, it is remarkable that Alberūnī, more than a century before Kalhaṇa, should have succeeded in obtaining any information, however scanty, about that northern region. In my *Memoir on the Ancient Geography of Kashmīr*, I have shown the value attaching to the chapter of Alberūnī's *India*, which contains his description of Kashmīr derived from indigenous sources[11]. He refers there to the mountains of 'Bolor and Shamīlān' as visible for two marches on the left of the traveller who enters the Kashmīr Valley from the western 'Gate', i. e. through the gorge of Bārāmūla. At present we can trace neither the name *Shamīlān* nor the designation *Bhattavaryān* given by Alberūnī to the tribes which inhabit those mountains, and whose 'king has the title of Bhatta-Shāh'. But the position indicated, and the use of the term Bolor, which has been applied to Gilgit and Baltistān for centuries, plainly show that Alberūnī's informant meant the mountain-region of the Dards, forming the confines of Kashmīr to the north and north-west. This is confirmed by the subsequent mention of 'Gilgit, Aswīra, and Shiltās' as the chief places of those tribes; for here it is impossible to mistake a reference to the modern Gilgit, Hasōra, and Chilās[12].

SECTION II.—CHINESE HISTORICAL RELATIONS WITH GILGIT

Nowhere among the Dards do we meet with written records of any antiquity or other evidence which could throw light on the early history of the region traversed by the 'Gilgit Road'. Nor are there archaeological remains to offer us guidance; at least I did not succeed in tracing any going back to pre-Muhammadan times until Gilgit itself was reached. But

[8] See my introduction to *Rājat.*, I. pp. 30 sq.

[9] Compare my note to *Rājat.* vii. 911; regarding Gurēz and the neighbouring portions of the valley, see *Ruins of Khotan*, pp. 15 sqq.

[10] See *Rājat.*, I. p. 31, note 6.

[11] See *Rājat.*, II. pp. 359 sqq., §§ 12–14.

[12] See Alberūnī, *India*, transl. Sachau, i. p. 20; for a possible explanation of the term *Bhatta*, see my note, *Rājat.*, II. pp. 363–4.

a remarkable course of historical events, for a short period about the middle of the eighth century A. D., invested this barren mountain-land at the extreme north-west confines of India with considerable political and strategical importance to the Chinese Empire. The record preserved of those events in the official Annals of the Chinese T'ang dynasty gives us a glimpse of Gilgit history at that interesting phase, and strikingly illustrates the value which has always attached to the route connecting Gilgit with Kashmīr.

Gilgit in the Chinese Annals.

It is only quite recently, and solely through the brilliant researches of Professor Édouard Chavannes, that the Chinese records containing these curious references to Gilgit and the neighbouring mountain-territories have been rendered accessible to the Western student. They have been translated and also elucidated with rare critical acumen in M. Chavannes' *Documents sur les Tou-kiue occidentaux*. In the ninth chapter of the last and purely historical part of his work M. Chavannes has given a lucid exposition of the main events which occurred in the territories between the Oxus and the Indus during the period that witnessed the furthest extension of Chinese power westwards, from the middle of the seventh to the middle of the eighth century of our era[1]. A few facts gleaned from this account will suffice to make clear how such a remote region as Gilgit and the adjoining valleys, even as far as Kashmīr, was brought within the sphere of Chinese activity.

Gilgit between Chinese and Tibetans.

After the destruction of the empire of the Western Turks (658–659 A. D.) China secured suzerainty over a vast region, which comprised both Eastern and Western Turkestān and extended across the Oxus Valley and to territories south of the Hindukush and within the ancient confines of India[2]. The rising power of the Tibetans, who from 670 to 692 A. D. occupied Kāshgar and the Tārīm Basin, combined with dynastic troubles in China itself, reduced Chinese dominion over those distant territories in the West to a more or less nominal condition by the end of the eighth century. But the victorious advance of the Arabs, resumed with fresh vigour during 705–715, under the leadership of Qutayba ibn Muslim, obliged the small states along the Oxus and in Sogdiana to turn to the Imperial Court for protection.

The danger threatening from Muslim conquest was increased by the Tibetans, who, after having been driven back from the Tārīm Basin, were now endeavouring to join hands with the Arabs, their allies, across the Pāmīr region. Against these aggressions from west and south, China during the reign of the energetic Emperor Hsüan-tsung (713–755 A. D.) directed a long series of well-sustained efforts, both diplomatic and military. In the course of the struggle with the Tibetans Gilgit acquired strategic importance; for through it led, then as now, the nearest and most practicable route giving access from the Upper Indus, via Yasīn and the Barōghil Pass, to the central Pāmīrs and the Oxus Valley. It thus became an object of constant and special care for the Imperial Government to keep this gate of invasion firmly closed against the Tibetans, who probably from very early times were in possession of Ladāk, and had thence pushed down along the Indus into Baltistān or Skardo, the immediate neighbour of Gilgit on the south-west.

'Great P'o-lü' of T'ang Annals.

On examining the notice 'on the Great and the Little P'o-lü' as now presented by M. Chavannes from the T'ang Annals, we cannot fail to recognize, with the learned translator, that it is this territory of Baltistān which is meant by the term 'Great P'o-lü'[3]. It is correctly

[1] See Chavannes, *Turcs occid.*, pp. 287 sqq.

[2] Compare the detailed analysis of the Chinese administrative divisions established between the Oxus and Indus, *Turcs occid.*, pp. 277 sqq., 280; also S. Lévi, *Journal asiat.*, 1900, xv. pp. 305, 315.

[3] See Chavannes, *Turcs occid.*, pp. 149 sq. The identification of 'Great P'o-lü (Po-liu)' with Ladāk, suggested by Sir H. Yule, *Cathay*, p. lxx, and previously also accepted by me (*Rājat.*, II. p. 435), was based on an imperfect extract of this notice from the T'ang Annals, given by Rémusat, *Nouveaux mélanges asiat.*, i. p. 194; it cannot any longer be maintained.

described as being right to the west of the Tibetans and adjoining 'the Little P'o-lü', i.e. Gilgit, while westwards it had as its neighbour 'the territory of Wu-ch'ang (Udyāna) which belongs to Northern India'. In regard to the latter indication it must be remembered that the ancient Udyāna, as shown by the accounts of the Chinese pilgrims and other evidence, comprised not only the whole of the Swāt Valley but also the mountains on the right bank of the Indus, even beyond the river's great bend to the south.

At the time of Hsüan-tsang's visit (circa 631 A. D.) the frontiers of Udyāna in the north-east evidently extended to the Valley of Darēl (*Ta-li-lo*), almost opposite Chilās, and possibly even further up the Indus [4]. On the other hand, we find to the present day the half-Tibetan Baltīs, who have given their name to Baltistān, settled along the Indus to within a short distance above the point where the Gilgit river joins it. Thence toward Chilās and Darēl the rocky gorge of the Indus is even more confined and barren than elsewhere, and must in early times have held a very small population. Thus we can fully account for Udyāna being named in the Chinese description as the western neighbour of 'Great P'o-lü' [5].

We learn from the notice in the Annals that 'the Great P'o-lü' had become subject to the Tibetans some time before the middle of the eighth century. But both the Annals and the official records preserved in the encyclopaedia *Ts'ê fu yüan kuei* show that previously, in 717 and 721, two successive rulers of Great P'o-lü had acknowledged the suzerainty of China, and received the decree of royal investiture from the Imperial court. We have no means of ascertaining to what extent the ethnic character of the present Baltī population, Tibetan in language and largely also in race, is due to this Tibetan occupation. Possibly the close ethnic connexion of Baltistān with Ladāk is of far earlier date, and may have facilitated the extension of Tibetan power along the Indus Valley.

The ancient local term reproduced by the P'o-lü of the Chinese records still survives in the designation *Paloyo* which, as I ascertained on my passage through Gilgit, is applied by the Dards of those parts to their eastern neighbours, the Baltīs.

'Little P'o-lü' of the T'ang Annals.

The territory of 'Little P'o-lü' has been identified by M. Chavannes with the modern Gilgit [6]. A short analysis of the topographical and historical data furnished by the Annals proves that this identification is undoubtedly correct, if the term Gilgit is extended so as to include the valleys drained by the Gilgit river in its upper course, and in particular the important mountain tract to the north-west known as Yasīn. 'Little P'o-lü' is described as having 'Great P'o-lü' or Baltistān on its south-east, at a distance of three hundred li; Ku-shih-mi or Kashmīr lay five hundred li to the south of it; while at the same distance to the north there was the town

[4] See *Si-yu-ki*, transl. Beal, i. p. 134. Also Fâ-hien's *Travels*, transl. Legge, p. 24, speaks of *T'o-li* (*T'o-leih*), which is plainly shown by the context to be identical with Hsüan-tsang's *Ta-li-lo*, or Darēl, as being immediately within the borders of Northern India. This evidently indicates dependence on, or ethnic connexion with, Udyāna, to which country the traveller next proceeded and which he further on specifies as the commencement of North India.

[5] The country of *Po-lu-lo*, which Hsüan-tsang reached after leaving *Ta-li-lo*, by going to the east up the course of the Indus, for a distance of about 500 li, or circ. 100 miles, was manifestly Baltistān. The direction and character of the route ('by the help of flying bridges and footways made of wood across the chasms and precipices') distinctly point to this. Also the description of the territory as 'long from east to west and narrow from north to south,' as rich in gold and silver and possessing a climate continually cold, would apply more strictly to Baltistān than to Gilgit. But it is well to remember that the term 'Bolor' or 'Bolur,' which Hsüan-tsang's *Po-lu-lo* is undoubtedly meant to represent, had a much wider application, including at one time or other all the mountain tracts on the southern slopes of the eastern Hindu-kush, from Kāfiristān to Skardo. Compare, regarding this somewhat loose and once much-disputed term, Sir H. Yule's note, *Marco Polo*, i. pp. 178 sq., giving a lucid summary of exhaustive researches, and the judicious remarks of Elias, *Tārīkh-i-Rashīdī*, p. 385, note. The origin of the term is doubtful; and so, too, its philological relation to P'o-lü.

[6] See *Turcs occid.*, pp. 150 sq.

So-lei, belonging to the territory of *Hu-mi*, which has been recognized with certainty as the present Wakhān on the Upper Oxus[7]. A glance at the map shows that the bearings, as well as the relative distances, here indicated are in remarkable agreement with the geographical position of that central and most productive part of the main valley to which the name of Gilgit properly applies.

Gilgit in the eighth century.

The historical notices regarding 'Little P'o-lü', rendered accessible by M. Chavannes' researches, afford proof of the political importance which the Chinese attached to the Gilgit Valley as the main line of communication between the Upper Indus and the Oxus, and of the persistent efforts made by them during the reign of the Emperor Hsüan-tsung (713–755 A. D.) to close this route against Tibetan invasion. At the commencement of that reign the king of Little P'o-lü, Mo-chin-mang, came to pay homage at the court of Hsüan-tsung, who accorded him protection and constituted his country into a military territory called *Sui-yüan*. 'Owing to its proximity to the Tibetans his country suffered much from them; the Tibetans declared to him: "It is not your kingdom which we covet, but we wish to use your route in order to attack the Four Garrisons (i. e. the present Chinese Turkestān)"'[8].

In 722 A. D. Mo-chin-mang, having been deprived of nine townships by the Tibetans, applied for help. The Imperial Commissioner resident at Pei-t'ing thereupon declared: 'P'o-lü is the western gate of [the territories of] the T'ang dynasty; if P'o-lü is lost [to us] then the western countries will all become Tibetan', and he dispatched a force of four thousand chosen troops, under the prefect of Kāshgar (Su-lê), to his help. Mo-chin-mang thus succoured defeated the Tibetans and recovered the nine townships. The same historical work, the *Tzŭ chih t'ung chien*, which has preserved these details, records a fresh attack upon P'o-lü by the Tibetans fifteen years later. On that occasion, 737 A. D., Chinese action took the form of a diversion from the centre of the empire, which resulted in a great defeat of the Tibetans west of the Kuku-Nor and relieved P'o-lü[9].

After Mo-chin-mang's death 'Little P'o-lü' was ruled in succession by his sons Nan-ni and Ma-hao-lai (Ma-lai-hi). The Imperial edict concerning the latter's investiture, in 741 A. D., is still extant among the records extracted by M. Chavannes[10]. He too seems to have died early, and his successor Su-shih-li-chih was won over by the Tibetans, who induced him to marry a Tibetan princess and thus secured a footing in his territory. In consequence 'more than twenty kingdoms (i. e. little hill states) to the north-west' of Little P'o-lü are said by the Annalist to have become subject to the Tibetans. Their customary tribute no longer reached the Imperial court.

[7] See *Turcs occid.*, p. 279, also pp. 152, 154, where good reasons have been advanced for the assumption that the name which the Chinese represent by *So-lei* was really that of the main branch of the Oxus, now known as Panja. For another mention of this river made in connexion with Lien-yün, a stronghold which in all probability occupied the position of the present Sarhad at the point where the route from Yasīn over the Barōghil Pass strikes the Āb-i-Panja, compare below, p. 8. Whether the town So-lei stood in the same place seems doubtful.

[8] See *Turcs occid.*, pp. 150 sq. By the term 'Four Garrisons', the territories of Kāshgar, Khotan, Kuchā, and Kara-shahr, then occupied by Chinese forces, are meant; comp. ibid., p. 113 note.

[9] See extract given *Turcs occid.*, p. 151, note 2.

[10] See for a translation of this document *Turcs occid.*, pp. 211 sq.

SECTION III.—THE CHINESE OCCUPATION OF GILGIT AND THE ROUTE TO KASHMÎR

Kao Hsien-chih's expedition into Yasin, 747 A.D.

The situation thus created obliged the Chinese Government to make special efforts for the recovery of their lost hold upon the Gilgit Valley. Three expeditions led against 'Little P'o-lü' by the Protector of 'the Four Garrisons' proved fruitless. But a fourth, entrusted in 747 A.D. by Imperial decree to his Deputy, Kao Hsien-chih, a general of Korean origin, was crowned with complete success. The military operations connected with this expedition deserve to be briefly considered here, in view both of the historical importance of the result and of their interest for the ancient topography of these regions[1].

Kao Hsien-chih was specially appointed by the Emperor Hsüan-tsung to take charge of the campaign against the Tibetans in Little P'o-lü, under the title of 'director of camps' with a force of ten thousand cavalry and infantry[2]. Starting from An-hsi, the present Kuchā, he reached Su-lê or Kāshgar in thirty-five days, apparently via Maralbāshi. Twenty days later his force arrived at the military post of the Ts'ung-ling mountains, which occupied the position of the present Tāsh-kurghān in Sarikol[3]. The march thence, through the valley of Po-mi or the Pāmîrs into Shighnān, called by the Chinese 'the kingdom of the five *Shih-ni*', occupied forty days.

His march across the Pāmîrs.

Kao Hsien-chih then divided his troops into three columns, which were to march by different routes to the attack of the fortress of Lien-yün held by the Tibetans. The localities by which the routes for two of these columns are indicated (*Pei-ku*, 'the northern gorge', and *Ch'ih-fo-t'ang*, 'the hall of the red Buddha') cannot be traced at present; but the route of the third column, under Kao Hsien-chih himself, which is described as leading through the kingdom of Hu-mi or Wakhān, undoubtedly lay up the valley of the Oxus. Lien-yün itself, where the several columns effected their junction on the appointed day, the thirteenth of the seventh month, had the river of P'o-lei or So-lei in front of its walls. M. Chavannes has shown good reason for assuming that this name designates the main branch of the Oxus now known as the Āb-i-Panja, and that Lien-yün occupied a position corresponding to the present village of Sarhad, but on the opposite or southern bank of the river. From Sarhad starts the well-known route which leads southwards over the Barōghil Pass to the headwaters of the Mastūj river, to this day representing the easiest line of access from the Upper Oxus to Chitrāl as well as to Gilgit.

The position of this locality also readily accounts for a concentration of the attacking forces from three different directions, such as the Chinese record indicates. Besides the main column ascending the Āb-i-Panja Valley from Kala Panja and the lower part of Wakhān, another might with advantage have co-operated from the opposite direction by descending from the headwaters of the Āb-i-Panja. These could be reached without serious difficulty from Tāsh-kurghān over the Tāghdumbāsh Pāmîr and the Wakhjîr Pass, or via the Naiza-tāsh Pass

[1] The details of Kao Hsien-chih's memorable march over the Pāmîrs and across the Hindukush are contained in his biography, which has been reproduced from the T'ang Annals and annotated by M. Chavannes in his *Turcs occid.*, pp. 152 sqq. The learned translator deserves special credit for having recognized that Kao Hsien-chih's route lay over the Barōghil and Darkōt Passes. He did not, however, attempt to identify the several localities mentioned along the route across, and south of, the Hindukush. These, therefore, have required special discussion here.

[2] The biography notes particularly that 'at that period the foot-soldiers all kept horses (i.e. ponies) on their own account'. The use of these animals for the transport of food supplies must have increased the mobility of the Chinese force.

[3] Compare *Turcs occid.*, p. 125; also below, pp. 28, 36.

and the Little Pāmīr. The third column, which used 'the northern gorge', may possibly have followed the Shākh-Dara, or Ghund-Dara Valley of Shighnān, to the Oxus branch coming from the Great Pāmīr or Victoria Lake, and may, by subsequently crossing the watershed towards the Āb-i-Panja, have descended upon Sarhad from the north. It is highly probable that the division of the Chinese forces, as well as Kao Hsien-chih's move into the distant Shighnān, were intended to reduce the difficulties about supplies and transport which for so large a body of troops must have been as serious then, in the inhospitable Pāmīr region, as they would be now.

The topographical indications contained in the description of the battle by which the Chinese general rendered himself master of Lien-yün and of the route it was intended to guard, fully confirm these identifications. The stronghold itself was occupied by a thousand men, and the river in front being in flood offered a serious obstacle. Kao Hsien-chih, however, with a picked body of mounted men, succeeded in crossing unopposed and without loss. He at once led his troops, whom this success had filled with confidence, to the attack of the main force of the enemy, eight or nine thousand in number. They were posted fifteen li, or about three miles, to the south, where advantage had been taken of the mountainous ground to erect palisades. The Chinese general, having gained the heights, i.e. turned the fortified line, engaged the defenders in a fight which ended in their complete defeat with heavy loss and the precipitate flight of the survivors during the night. It is evident that this battle was fought at the entrance of the defile which leads to the Barōghil Pass[4]. *Capture of Lien-yün and the Barōghil Pass.*

Kao Hsien-chih decided to leave behind in his camp certain high officers who were opposed to a further advance, together with three thousand men worn out by the previous hardships, and ordered them to guard Lien-yün. With the rest of his troops he pushed on, and after a three days' march arrived on the summit of 'Mount T'an-chü'; 'from there downwards there were precipices for over forty li (circ. eight miles) in a straight line.' By a trick Kao Hsien-chih prevailed upon his wavering troops to effect their descent into the valley, and after three more marches was met by 'the barbarians of the town of A-nu-yüeh' offering their submission. The next day he occupied A-nu-yüeh, where the commander of his advance guard had previously entrapped the five or six chief dignitaries of the king of Little P'o-lü who were supporting the Tibetans. These he executed, and then hastened to break the bridge which led over the So-yi river at a distance of sixty li (circ. twelve miles) from A-nu-yüeh. 'Scarcely had the bridge been destroyed in the evening when the Tibetans, on foot and mounted, arrived in great numbers, but it was then too late for them to attain their object. This bridge had the length of an arrow shot; it had taken a whole year to construct it. It had been built at the time when the Tibetans, under the pretext of using its route, had by deceit possessed themselves of (Little) P'o-lü.' *Chinese invasion of Yasīn.*

It is easy to trace on the map the successive stages of the Chinese general's progress. All details recorded of it agree accurately with the route that leads over the Barōghil saddle (12,460 feet above the sea) to the sources of the Mastūj river, and then, crossing south-eastwards the far higher Darkōt Pass (15,200 feet), descends along the Yasīn river to its junction with the main river of Gilgit. Three days are by no means too large an allowance of time for a military force accompanied by baggage animals to effect the march from the Oxus *Crossing of Darkōt Pass (Mt. T'an-chü).*

[4] The account of this encounter vividly recalls the attempts made by Tibetan strategy in 1904 to bar the advance of the Tibet Mission Force near Guru and again on the Karo-Lā. In each case the main Tibetan force was content to await the attack behind stone walls erected across the open ground of the valley, leaving it free to their opponents to occupy the commanding heights on either side. Considering how scanty timber must at all times have been about Sarhad, it appears probable that by the 'palisades' mentioned in the translation walls or sangars constructed of loose stones are really meant.

to the summit of the Darkōt Pass, considering that the ascent to the latter lies partly over the moraines and ice of a great glacier. The Darkōt Pass corresponds exactly in position to the 'Mount T'an-chü' of the Annals, and possibly preserves the modern form of the name which the Chinese transcription, with its usual phonetic imperfection, has endeavoured to reproduce. The steep southern face of the pass, where the track descends close on 6,000 feet between the summit and the hamlet of Darkōt, over a distance of about five or six miles, manifestly represents 'the precipices for over forty li in a straight line' which dismayed the Chinese soldiers on looking down from the heights of Mount T'an-chü.

From the foot of the pass at Darkōt a march of about twenty-five miles brings us to the village of Yasīn[5], the political centre of the valley, in which we can safely recognize the town of A-nu-yüeh. Not only does the time indicated for Kao Hsien-chih's march to A-nu-yüeh fully agree with the three marches which are reckoned at the present day for the journey between Yasīn and the watershed, but it is also evident that we have in A-nu-yüeh a fairly accurate rendering of the name *Arniah* or *Arniya* by which Yasīn is known to the Dards of the Gilgit Valley[6].

Bridge over the Gilgit river.

The correctness of this identification is confirmed by the reference to the bridge over the river *So-yi*, which is mentioned in the above account as sixty li beyond A-nu-yüeh. This river So-yi is named in the Annals' notice of Little P'o-lü as the one on which Yeh-to, the capital of the kingdom, stood, and there can be no doubt that the main Gilgit river is meant. Now a glance at the map shows that, descending the valley from Yasīn, we reach the Gilgit river at a distance of about twelve miles, which agrees exactly with the sixty li of the Chinese record[7]. It is equally clear that as the further route towards Gilgit proper and Baltistān leads along the right or southern bank of the Gilgit river, the Tibetan reinforcements hurrying up from that direction could not arrive at Yasīn without first crossing the main river. Hence the importance to the Chinese invaders of destroying the bridge[8].

Return of Kao Hsien-chih.

Once in possession of the Yasīn Valley, the Chinese general induced the king of Little P'o-lü to surrender, and subsequently pacified the whole territory. Leaving behind a garrison to hold it, and taking along his royal prisoner he returned within two months to Lien-yün. Then he regained the Pāmīr, whence he dispatched news to the Imperial Court announcing his victory. That the fame of Kao Hsien-chih's remarkable expedition must have spread far and created a deep impression, is shown by a closing remark of the Annals, the historical significance of which M. Chavannes has duly noted: 'Then the *Fu-lin* (Syria), the *Ta-shih* (i. e. Tāzī or Arabs) and seventy-two kingdoms of divers barbarian peoples were all seized with fear and made their submission[9].'

It was, no doubt, the greatness of the natural obstacles overcome in the course of the victorious march across the Pāmīrs and the Hindukush which gave to this success of the Chinese

[5] See Biddulph, *Tribes of the Hindoo Koosh*, p. 55.

[6] See Biddulph, *Tribes of the Hindoo Koosh*, p. 62 (where 'Arinah' is a misprint for Arniah). The omission of an equivalent of *r* in the Chinese rendering of this name has its parallel in the Chinese *A-hsi-lan*, *ta-kan*, which reproduce the Turkī names *Arslān*, *tarkān*; comp. *Turcs occid.*, pp. 239, 317.

[7] Compare for this distance also *Tribes of the Hindoo Koosh*, p. 55.

[8] All communication between Yasīn and Gilgit is now carried on over the bridge which spans the Gilgit river near Gūpis at the mouth of the Yasīn Valley. A wire suspension bridge, constructed by Col. Aylmer, R.E., has, since the Chitrāl campaign of 1895, replaced the difficult rope bridge which previously existed here. Like all similar rope bridges between Kashmīr and the Hindukush, the old bridge consisted of twigs twisted into ropes, and required frequent repairs to retain even a moderate degree of safety. It is very probable that the Chinese record, by its 'pont de rotin,' intends to designate a bridge of this peculiar description. The modern fort of Gūpis marks the strategic importance of the position.

[9] See *Turcs occid.*, pp. 151, 154, 296.

arms its peculiar lustre. And it is from more than one point of view interesting to find that other serious difficulties, which from the physical character and position of the country must always beset military operations in the Gilgit region, are reflected in a contemporary notice of the Chinese records.

Chinese garrison in Little P'o-lü.

Among the important diplomatic documents embodied in the great encyclopaedia *Ts'ê fu yüan kuei*, and now rendered accessible by M. Chavannes' researches, there is found the record of a representation which Shih-li-mang-ch'ieh-lo, the ruler (*jabgu*) of T'u-ho-lo or Tokhāristān on the Oxus, addressed in 749 A.D. to the Imperial Court. Its contents throw light on the situation prevailing in the Gilgit Valley immediately after Kao Hsien-chih's expedition[10]. The prince, whose territory comprised the present Badakhshān, with some tracts on the northern bank of the Oxus[11], complains of a neighbouring hill-state called Chieh-shuai[12], which, relying on the protection of its high mountains, had allied itself with the Tibetans. Its chief 'knows that the territory of P'o-lü is limited, its population dense; that the area available for cultivation is small, and consequently when garrison troops are placed there, the supplies fail. It then becomes necessary to purchase salt and rice in Kashmīr (*Ku-shih-mi*) and it is thus that the difficulty is met. The traders' caravans must, on going and coming back, all pass by the kingdom of Chieh-shuai; its king has therefore accepted the presents offered by the Tibetans, who claimed to establish a stronghold in his territory with a view to getting possession of the important route that leads into P'o-lü. Since Kao Hsien-chih opened up P'o-lü, there are 3,000 more troops there, and P'o-lü has been crushed by it. The King of Chieh-shuai, in alliance with the Tibetans, has taken advantage of the exhausted condition of P'o-lü and decided to invade it.'

Expedition against Chieh-shuai (750 A.D.).

In order to meet this danger, the ruler of Tokhāristān then proposes a bold plan of operations, which, if supported by the Imperial authorities, would allow him to conquer 'the Great P'o-lü' and the countries east of it. This would bring him in a straight line to Khotan, Kara-shahr and beyond, where the Tibetans could then no longer maintain themselves. Besides the dispatch of Chinese troops into Little P'o-lü, he solicits that the King of Kashmīr, as a loyal ally of China, be encouraged by an Imperial edict and special honours to lend to the enterprise the aid of his great resources in troops and supplies. The Emperor is said to have responded to this appeal of the T'u-ho-lo prince, and an edict preserved by the same source records the investiture in 750 A.D. of Su-chia as king of Chieh-shuai in place of his rebel brother P'o-t'ê-mo[13].

A briefer notice of these events, extracted by M. Chavannes from another work, the *Tzŭ chih t'ung chien*, also mentions the defeat, in 750 A.D., of Chieh-shih by Kao Hsien-chih, the capture of its chief P'o-t'ê-mo, and the installation of Su-chia in his place[14].

Chinese intervention succeeded on this occasion in relieving Gilgit and Tokhāristān from the threatening invasion of the Tibetans. But it is certain that the complete defeat of Kao Hsien-chih by the Arabs in the year following (751 A.D.) and the consequent rapid decline of the Imperial power in the 'Western Countries' must have forced the Chinese to abandon

[10] See *Turcs occid.*, pp. 214 sq.; comp. also p. 296 on the historical import of this document.

[11] *T'u-ho-lo*, a great and important territory, then largely inhabited by the descendants of the Hephthalites or White Huns, has long been identified with the Tokhāristān of early Muhammadan writers; compare, e.g., Sir H. Yule, *J.R.A.S.*, N.S., vi. pp. 94 sqq.; also my note on *Rājat.* iv. 166. The valuable data concerning T'u-ho-lo, contained in the Annals and other Chinese records of the T'ang period, have been fully discussed by M. Chavannes, *Turcs occid.*, pp. 155 sqq.

[12] The name thus transcribed, 羯帥, appears in a slightly different form as Chieh-shih 羯師 in the T'ang shu; see *Turcs occid.*, pp. 158 sq., and below, p. 13.

[13] See Chavannes, *Turcs occid.*, pp. 215 sq.; also the passage of the T'ang Annals, ibid. p. 158, mentioning the defeat of the Chieh-shih people who were endeavouring to bring about an attack on Tokhāristān by the Tibetans.

[14] See *Turcs occid.*, p. 214, note 2.

their position in Gilgit, and the relations maintained through it with such distant territories as Kashmīr, Kābul, and Udyāna[15].

Supplies for Gilgit garrison.

The record just summarized is the last glimpse we obtain of Gilgit from Chinese sources. It is hence particularly gratifying to find so clear a reference to that great difficulty attending any military occupation of the Gilgit Valley, its dependence for supplies on Kashmīr. The situation there described as resulting from the presence of a Chinese garrison must, to those acquainted with the modern history of Gilgit, vividly recall the efforts and sacrifices which the maintenance of a military force from the Kashmīr side has entailed ever since the Sikhs, soon followed by the Dōgras, established their first footing in the Dard mountains across the Indus[16]. The Chinese record supplies us, in fact, with striking evidence how little the lapse of a thousand years and momentous political changes can affect the conditions of life and communication prevailing in these mountain regions, or the aspects of military activity which equally depend on unchanging geographical features.

Transport of supplies from Kashmīr.

The difficulties which the letter of the T'u-ho-lo ruler so graphically represents, and of which the Tibetans were scheming to take advantage, are exactly those with which the Kashmīr rulers, and in more recent years the military authorities of the Indian Government, have had to contend in their occupation of Gilgit. Though the Dōgra force employed in Gilgit and on the line of communication leading to it, probably never much exceeded 3,000 men, their maintenance, being wholly dependent on food supplies brought from Kashmīr, involved a severe strain on the Mahārāja's resources, and often led to serious suffering and loss of life among his Kashmīrian subjects[17]. The very restricted area available for cultivation in the Gilgit Valley renders it practically impossible to feed a garrison on local produce; nor can surplus supplies be secured from any base nearer than Kashmīr. The barrenness of the mountain region which the route from Kashmīr traverses for a distance of over 200 miles would alone suffice to make the regular transport of supplies a matter of great difficulty; for whether pack animals or men are employed, the food needed by them in transit has likewise to be provided from Kashmīr. In addition, account must be taken of the natural obstacles presented by the two high ranges which have to be crossed *en route*, and the passes of which, owing to the heavy snowfall, rarely remain open for laden traffic during more than four months of each year.

The placing of a small Imperial garrison in Gilgit in 1890 was rapidly followed by the construction of the 'Gilgit Transport Road', with all the resources of modern European engineering, and by the introduction of systematic transport arrangements under the control of the Indian Commissariat Department[18]. These changes have made it possible to dispense altogether with human labour for transport purposes, and have greatly reduced the risk with which the use of this route, whether for the annual provisioning of Gilgit, or on occasion of

[15] Compare for these diplomatic relations maintained during the first half of the eighth century, M. Chavannes' lucid exposition, *Turcs occid.*, pp. 295 sq.; as to the events connected with Kao Hsien-chih's defeat and the collapse of Chinese rule in Western Turkestān, ibid. pp. 297 sqq., also pp. 142 sqq.

[16] An excellent sketch of the operations of the Sikhs and Dōgras which brought Gilgit and the adjoining Dard tracts into dependence on Kashmīr, and thus within the sphere of Indian political interests, is given, up to the year 1875, by Drew, *Jummoo and Kashmir Territories*, pp. 437 sqq. For the events which have since led the Indian government to assume a more direct share in the occupation of Gilgit and the adjoining Dard valleys, and to extend its political control to the Hindukush watershed, publications on the Hunza (1891) and Chitrāl (1895) campaigns may be consulted.

[17] For characteristic facts concerning the hardships which the requisitions of carriage for transport to Gilgit inflicted upon the village population of Kashmīr until the construction of the 'Gilgit Road,' compare Sir Walter Lawrence, *Valley of Kashmir*, p. 413.

[18] As to the working of these arrangements and the curious manner in which they recall to the historical student certain features of the watch kept in ancient times over these mountain-routes of Kashmīr, compare my *Ruins of Khotan*, pp. 12 sq.

military emergencies, was always beset from breaks of the road, premature snowfall, and similar incidents of Alpine travel. Yet a journey along 'the Gilgit Road', or the perusal of an account of its marches as given in my Personal Narrative, will suffice to demonstrate how serious even now are the physical difficulties presented by the task of keeping a military force in Gilgit supplied from so distant a base and over such mountains.

In the facts connected with the present occupation of the Gilgit Valley we have the best illustration of the reasons which rendered it vital for the Chinese hold over 'Little P'o-lü' that the route towards Kashmīr should be kept absolutely safe and open for food convoys [19]. We find the same fact attested with equal clearness in the record which the T'ang Annals, in their notice on *Ku-shih-mi* or Kashmīr have preserved of the letter addressed to the 'Celestial Kāgān', the Chinese Emperor, by King Mu-to-pi, i.e. Muktāpīḍa, on his succession to the Kashmīr throne (733 A. D.) [20]. Muktāpīḍa, who requests his investiture by Imperial decree, as accorded before in 720 A. D. to his brother and predecessor Candrāpīḍa (Chên-t'o-lo-pi-li), particularly claims for his kingdom the merit of having, on the occasion when the troops of 'the imperial Kāgān' arrived in Little P'o-lü, come to their assistance by the dispatch of convoys of supplies, though their number amounted to two hundred thousand men [21]. It is not certain which Chinese expedition into the Gilgit region is alluded to here; but it is evident that the situation in which that service was rendered must have closely resembled the one prevailing in the Gilgit Valley immediately after Kao Hsien-chih's success. Perhaps Muktāpīḍa's reference, as suggested by M. Chavannes, is to the operations by which the Chinese in 722 A. D. relieved Mo-chin-mang, king of Little P'o-lü [22]. Assistance of Kashmīr rulers.

It remains to locate that hill-state of Chieh-shuai, which by its alliance with the Tibetans so seriously endangered the Chinese position in Gilgit. M. Chavannes has not attempted to identify it, nor does the letter of the Tokhāristān ruler which reports that alliance, furnish evidence as to the situation of 'Chieh-shuai,' except such as is conveyed in the statement that it could block the route of supplies from Kashmīr to the Gilgit Valley. But some further indications, I believe, may be gleaned from another Chinese record which M. Chavannes' researches have rendered accessible. The notice on the T'u-ho-lo country, or Tokhāristān, extracted by him from the T'ang Annals, mentions an attack planned on that territory by the Chieh-shih people in alliance with the Tibetans, and its frustration through the help of Chinese troops [23]. As the latter is said to have been secured upon the application of the Jabgu Shih-li-mang-ch'ieh-lo, the same whose letter of the year 749 A. D. we have already considered, it is certain that the Chieh-shuai of that document and Chieh-shih mentioned here are identical [24]. Position of Chieh-shuai (Chieh-shih).

Now the territory of Chieh-shih is distinctly described as bordering on Tokhāristān, and this makes it highly probable that M. Chavannes is right in recognizing it in the mountain tract which a subsequent passage of the notice of Tokhāristān mentions under an abbreviated

[19] 'Considérant que le roi de *Kou-che-mi* (Cachemire) a été fidèle et loyal envers les Chinois, qu'il a en outre beaucoup de soldats et de cavaliers, que son territoire est vaste et que la population y est dense, que les vivres y sont en abondance, j'espère tout spécialement que la bonté impériale conférera au roi de Kou-che-mi (Cachemire) un édit écrit pour l'encourager, qu'on lui donnera des vêtements et des présents, ainsi que des ornements précieux et des ceintures pour faire que (ce roi) soit touché et reconnaissant de la bonté sainte (de l'empereur) et qu'il redouble de fidélité et de loyauté.' From the letter of the Tokhāristān chief, quoted p. 11; see Chavannes, *Turcs occid.*, p. 215.

[20] See *Turcs occid.*, pp. 166 sq. The date of Muktāpīḍa's investiture is supplied by the imperial decree reproduced by M. Chavannes, ibid. p. 209, from the *Ts'ê fu yüan kuei*.

[21] This passage had been erroneously rendered in A. Rémusat's *Mélanges asiat.*, i. p. 196, where the notice of the T'ang Annals concerning Kashmīr was first translated. My reference to the Chinese data about Muktāpīḍa, in *Rājat.* iv. 126, note, should be rectified accordingly.

[22] See *Turcs occid.*, p. 167, note 2; and above, p. 7.

[23] See *Turcs occid.*, p. 158.

[24] For this very slight graphic difference between the two transcriptions, see above, p. 11, note 12.

form of the name, Chieh[25]. This tract of Chieh (*recte* Chieh-shih) lay 'in the middle of the Ts'ung-ling mountains; to the west and south it is bordered by (the territory of) Shê-mi; to the north-west are the I-ta or Hephthalites'. As the T'ang Annals explicitly state of Tokhāristān that it was the seat of the Hephthalites[26], and as the portion of Tokhāristān south of the Oxus is undoubtedly represented by the province of Badakhshān, it is plain that Chieh-shih or Chieh must be looked for to the south-east of the latter, i. e. in the direction of the valleys drained by the Chitrāl river.

Shê-mi identified with Kāfiristān.

The reasons which make me inclined to identify Chieh-shih (Chieh-shuai) with the portion of the main Chitrāl Valley, properly known as Kāshkār or Chitrāl, are briefly these. The territory of Shê-mi 賒彌, which is said to border Chieh-shih on the west and south, is also mentioned under the identical name in the account of the Buddhist pilgrim Sung Yün, and in the corresponding notices which the Wei Annals have preserved from the record of his fellow-pilgrim Hui-shêng. After a stay in the country of the Yeh-tas or Hephthalites, they passed (519 A. D.) into the small mountain tract of Po-chih, and hence into the territory of Shê-mi. There they gradually emerged from the Ts'ung-ling mountains, and proceeded to Udyāna which lay south of Shê-mi[27]. The position indicated for Po-chih (to the south-west of Wakhān) and certain features mentioned of its mountains show clearly that it comprised the headwaters of the Varduj or Kokcha river, south of Zebak and towards the Hindukush watershed. From there it is possible to reach by a number of passes the cluster of valleys to the south of the great snowy range which since mediaeval times has been known by the general name of Kāfiristān. Owing to the constant war waged against its inhabitants by the Muhammadan hill-states around, this great alpine region has remained more or less a *terra incognita* until quite recent times. But there is nothing to prove that those valleys were similarly closed to traffic in earlier periods, while it is certain that a route leading down the easternmost of them to the Kūnar or Chitrāl river and hence across Dīr into the Swāt Valley would form a shorter and in all probability an easier line of communication than a route crossing one of the high passes east of Lake Dufferin and thence descending to Kāshkār or Chitrāl proper[28].

[25] See *Turcs occid.*, p. 159, with note 3 concerning the varying Chinese representations of the local name.

[26] See *Turcs occid.*, pp. 157 sq.

[27] For an accurate translation of the passages mentioning Shê-mi in Sung Yün's account and in the Wei Annals see now Chavannes, *Voyage de Song Yun*, p. 28. Dr. Marquart, *Ērānšahr*, pp. 244 sq., had already rectified some serious mistakes in Beal's version, *Si-yu-ki*, pp. xcii sq. Compare also M. Chavannes' abstract, *Turcs occid.*, p. 159, note 4.

[28] It appears to me that a route such as I suppose Sung Yün to have followed on his journey from Badakhshān to Udyāna or Swāt would help to solve a geographical difficulty in Marco Polo's narrative which has puzzled no less an authority than Sir Henry Yule. In Chapter xxx, Marco Polo tells us 'that ten days' journey to the south of Badashan there is a province called *Pashai*, the people of which have a peculiar language, and are Idolaters of a brown complexion. They are great adepts in sorceries and the diabolic arts. They are a pestilent people and a crafty; and they live upon flesh and rice. Their country is very hot.' The traveller then proceeds to tell us 'of another country which is seven days' journey from this one towards the south-east, and the name of which is Keshimur (Kashmīr)'; see Yule, *Marco Polo*, i. p. 164.

Sir Henry Yule was undoubtedly right in assuming that Marco Polo had never personally visited these countries and that his account of them, brief as it is, was derived from hearsay information about the tracts which the Mongol partisan leader Nigūdar had traversed, about 1260 A. D., on an adventurous incursion from Badakhshān towards Kashmīr and the Punjāb. In chapter xviii, where the Venetian relates that exploit (see Yule, *Marco Polo*, i. p. 98, with note, p. 104), the name of Pashai is linked with *Dir*, the territory on the Upper Panjkōra river, which an invader wishing to make his way from Badakhshān into Kashmīr by the most direct route would necessarily have to pass through.

The name *Pashai* is still borne to this day by a Muhammadanized tribe closely akin to the Siāh-pōsh, settled in the Panjshīr Valley and in the hills on the west and south of Kāfiristān. It has been very fully discussed by Sir Henry Yule (ibid., i. p. 165), who shows ample grounds for the belief that this tribal name must have once been more widely spread over the southern slopes of the Hindukush

If the territory of Shê-mi be identified with Kāfiristān, we obtain a satisfactory explanation why, in the notice on Chieh (Chieh-shih), Shê-mi is mentioned as bordering the latter both on the west and the south. The Siāh-pōsh or 'Kāfir' tribes, which have given to Kāfiristān its mediaeval and modern designation, inhabit not only the valleys due west of Chitrāl, but supply also the bulk of the population in the Valley of the Kūnar or Chitrāl river itself, along that little-known portion of its course from Kala Darōsh southwards to Asmār. These settlements, though Muhammadanized under Chitrālī or Pathān rule, form a living proof of the fact that Kāfiristān originally included a considerable portion of the main valley due south of Kāshkār [29].

Chieh-shih identified with Chitrāl.

The open and fertile part of the main valley, containing the large villages which bear collectively the name of Chitrāl and form the political centre of the Kāshkār or Chitrāl State, answers remarkably well to the description given in the T'ang Annals of the mild climate and rich produce of Chieh (Chieh-shih). And in view of the topographical arguments already adduced for this identification, we need not hesitate to suggest also that it was the local name Kāshkār, or an earlier form of it, which the Chinese endeavoured to reproduce by Chieh-shih or Chieh-shuai. The application of the term Kāshkār to the territory of Chitrāl is well attested from Muhammadan sources, and its use is still current throughout those regions [30]. Chieh-shih, as an attempt to represent Kāshkār by Chinese sounds, would have a parallel in the name Ch'ia-sha which Hsüan-tsang gives to the present city and oasis of Kāshgar, in Chinese Turkestān [31].

as far as they are comprised in the limits of Kāfiristān. If the great commentator nevertheless records his inability to account for Marco Polo's application of 'the name Pashai to the country south-east of Badakhshan', the reason of the difficulty seems to me to lie solely in Sir Henry Yule's assumption that the route heard of by the traveller, led 'by the Doráh or the Nuksán Pass, over the watershed of Hindukúsh into Chitrál and so to Dir'.

Though such a route via Chitrāl would, no doubt, have been available in Marco Polo's time as much as now, there is no indication whatever forcing us to believe that it was the one really meant by his informants. When Nigūdar 'with a great body of horsemen, cruel unscrupulous fellows' went off from Badakhshān towards Kashmīr, he may very well have made his way over the Hindukush by the more direct line that passes to Dīr through the eastern part of Kāfiristān. In fact, the description of the Pashai people and their country, as given by Marco Polo, distinctly points to such a route; for we have in it an unmistakable reflex of characteristic features with which the idolatrous Siāh-pōsh Kāfirs have always been credited by their Muhammadan neighbours.

It is much to be regretted that the Oriental records of the period, as far as they were accessible to Sir Henry Yule, seem to have retained only faint traces of the Mongol adventurer's remarkable inroad. From the point of view of Indian history it was, no doubt, a mere passing episode. But some details regarding it would possess special interest as illustrating an instance of successful invasion by a route that so far has not received its due share of attention.

[29] Compare Biddulph, *Hindoo Koosh*, pp. 64 sq.

[30] Compare Biddulph, *Hindoo Koosh*, pp. 59 sqq.; Raverty, *Notes on Afghánistán*, 1888, pp. 152 sqq.

[31] In regard to the above proposed identification of Chieh-shih (Chieh-shuai) with Chitrāl, it is necessary to consider briefly some other Chinese geographical notices which have previously been assumed to refer to Chitrāl. Both Cunningham and V. de Saint-Martin had expressed the belief that the mountainous territory of *Shang-mi* 商彌, which Hsüan-tsang describes as situated beyond a great range to the south of Ta-mo-hsi-t'ieh-ti or Wakhān, was identical with the Chitrāl Valley (see *J.A.S.B.*, xiv. p. 433, and Saint-Martin, *Mémoire analyt.*, p. 426). Their suggestion has received the weighty support of Sir Henry Yule, who observes that the yellow arsenic or orpiment mentioned in Hsüan-tsang's account is still a characteristic product of Chitrāl (*J.R.A.S.*, N.S., vi. 114). The further fact that Shang-mi, with the alternative name of *Chü-wei*, is mentioned in a brief notice of the T'ang Annals first extracted by A. Rémusat, did not escape Sir Henry Yule's attention. But without an exact and reliable rendering of the passage, as now supplied in M. Chavannes' work (*Turcs occid.*, p. 129, note 2), it was impossible to perceive that the name had in reality a much more restricted application.

The notice of the Annals tells us: '*Chü-wei* 俱位 is also called *Shang-mi*; its capital is in the town of *A-shê-yü-shih-to*; it is situated amidst the great snowy mountains, north of the river of P'o-lü. This country is cold; it produces the five cereals, wine and pomegranates. During the winter people live in caves. The inhabitants of this kingdom have always assisted the Little P'o-lü in spying out the Middle Kingdom (China)'.

The river of P'o-lü must be the Gilgit river, and a glance at the map shows that the territory meant by *Chü-wei* or *Shang-mi* corresponds exactly to the present Mastūj and the

It was, as we have seen, the danger threatening the route between Kashmīr and Gilgit which induced the Chinese to lend their aid against the hill-state of Chich-shih. Chitrāl seems distant from that route; yet it does not need elaborate argument to prove that the co-operation of the Chitrāl chief would have made it far easier to the Tibetans to close that route of supplies for Gilgit, and thus to render the Chinese position untenable. The route followed by the 'Gilgit Road' is well protected against inroads from the east by the height and heavy snowfall of the ranges that flank it on that side, and the uninhabitable nature of the elevated valleys and plateaus between them [32]. The great glacier-crowned spurs descending from the Nanga-Parbat *massif* and the inaccessible character of the gorges that drain them, form a similar bulwark westwards as far as Astōr; but it is different beyond that place where the route descends to the Indus.

Chilās and the Kashmīr–Gilgit route.

From the adjoining territory of Chilās, lower down on the Indus, raids upon the route are possible both along the river itself and by certain tracks leading over the northernmost spurs of Nanga-Parbat. Such irruptions into the Astōr Valley were frequently made by the Dards of Chilās, who have always been renowned for their bravery and love of plunder, and were sometimes extended as far as the Kiṣangaṅgā Valley, Baltistān, and northward into Gilgit [33]. These raids continued up to the time of the Dōgra occupation of Gilgit, and were only stopped when Mahārājā Gulāb Singh's troops, about 1851, succeeded after great difficulties in invading Chilās and reducing its chief stronghold. But the Chilāsīs preserved their independence

Yārkhūn Valley. This valley, through which flows the main feeder of the Chitrāl river, lies due north of the headwaters of the Gilgit river near the Shandur Lake; and the name *Khō*, by which it is known below Mastūj, its chief place, is manifestly the modern derivative of the name intended by the Chinese transcription *Chü-wei* (compare for the name *Khō*, which is also given to the largest section of the Dard population of Kāshkār Bālā, Biddulph, *Hindoo Koosh*, pp. 59, 62). Further local research may yet help us to trace also the name of the capital *A-shê-yü-shih-to*, as recorded in the Annals.

The fertile and thickly populated Mastūj, with some adjoining parts of Kāshkār Bālā or Upper Kāshkār, has, under the still ruling Khushwakte family, enjoyed independence from Chitrāl or Kāshkār proper for a considerable period. This division, which is partly accounted for by geographical features, appears to me reflected in the distinction between Shang-mi (or Chü-wei) and Chieh-shih as indicated by the notices of the T'ang Records. The name Shang-mi is, I believe, properly applied only to the valleys drained by the Mastūj branch of the Chitrāl river. But we could well understand, from analogous cases elsewhere, how Hsüan-tsang, who did not personally visit this region, could have been led by his informants to comprise under this designation the whole Chitrāl valley, including that lower part which formed the separate territory of Kāshkār or Chieh-shih. In the first place, Mastūj or Shang-mi lay nearest to the route which the pilgrim followed through Wakhān. It must further be remembered that both Mastūj and Kāshkār may at the time of his journey have been under one ruler, as Chitrāl traditions record for more recent periods (see for such traditions retaining curious traces of earlier historical events, *Hindoo Koosh*, pp. 149 sqq.).

Dr. J. Marquart, in his exhaustive monograph on the historical geography of Tokhāristān, has likewise discussed the Chinese notices concerning Shang-mi, with his usual learning and thoroughness. But the fresh data contained in M. Chavannes' publication were not then before him, and he could thus not give adequate attention to the geographical reasons against the identity of Shang-mi and Sung Yün's Shê-mi as assumed by him. On the other hand, he rightly emphasizes the fact that the pilgrim Wu-k'ung correctly names *Chü-wei* immediately after *Hu-mi* or Wakhān. [Since the above note was written the 'Errata supplémentaires' (lithographed) which M. Chavannes has added to his *Turcs occid.* (1904), show that the distinguished Sinologist has also recognized the identity of Chü-wei or Shang-mi with 'la région de Mastoudj et Tchitrāl'. He previously located Chü-wei in Yasīn; see *Turcs occid.*, p. 129, note 2.]

It may be added that Wu-k'ung, in 752 A.D., travelled from Hu-mi or Wakhān to Chü-wei and thence, through two localities (*Ho-lan* and *Lan-so*) not yet identified, to *Yeh-ho*, in which M. Chavannes recognizes, with good reason, *Yeh-to*, the capital of Little P'o-lü; comp. *L'Itinéraire d'Ou-k'ong*, p. 12, and *Turcs occid.*, p. 129, note 2.

[32] The extent of uninhabited ground on either side of the route, bare of all resources for the support of human life, and either actual glacier area or under heavy snow for the greater part of the year, is well illustrated by the snow and race maps in Drew's *Jummoo and Kashmir*.

[33] Compare, for an interesting account of the Chilāsīs and their predatory expeditions, which are still remembered in the adjoining regions, Biddulph, *Hindoo Koosh*, pp. 15 sqq.; see also Drew, *Jummoo*, pp. 398, 404.

notwithstanding a nominal tribute, and their turbulent disposition and the danger it represented for the newly-constructed 'Gilgit Road' ultimately led, in 1892, to the occupation of their territory by an Imperial garrison.

The Chilāsīs, in race, language, and ethnic peculiarities, are closely allied to the other Dard communities which, organized in small republics, extend far down both banks of the Indus, and collectively form the tract known as Shīnkārī or the Kōhistān of the Indus Valley. Though this region is still inaccessible to the European traveller, it is certain that the several sections of the valley are not separated from each other by great natural barriers. Hence constant relations are kept up between these Dard communities which used to combine when threatened by an external foe. In fact, tradition at Chilās distinctly asserts that in pre-Muhammadan times the whole of Shīnkārī was under the rule of one Rājā[34].

We have seen already that, according to the indications furnished by the T'ang Annals and Hsüan-tsang, this tract in the Indus Valley was at one time politically dependent on the kingdom of Udyāna. The supremacy exercised from the Swāt Valley may, at the period when we hear of the alliance between the Tibetans and Chieh-shih, have been replaced by predominant influence from the side of Chitrāl. Both from Chitrāl proper and from Upper Kāshkār or Mastūj the Indus Valley can readily be reached by a number of routes leading across the headwaters of the Swāt and Panjkōra rivers, and the remarkable extension which in recent years the Khān of Dīr's power has taken in the direction of the Swāt Kōhistān and the Indus Valley presents a curious parallel. Chilās and the other Dard communities along the Indus, if left to themselves, could without great difficulty have been overawed by the Chinese garrison placed in Gilgit and Yasīn; but when controlled and supported by a neighbouring hill-state of such resources as Chitrāl, they were bound to become a serious menace to the Kashmīr-Gilgit route, which they flanked, and upon which the maintenance of that garrison depended.

Section IV.—ANCIENT REMAINS IN GILGIT AND HUNZA

The total absence of reliable records makes it impossible at present to trace more of the early history of Gilgit and the adjoining regions than is revealed for a brief period by the illuminating notices of the Chinese Annals. A patient study of local traditions and ethnology, and a systematic search for ancient remains would, no doubt, bring to light materials likely to help us in restoring some aspects of the life and culture that prevailed there during pre-Muhammadan times. But for such labours the series of rapid and often trying marches which, between June 11 and 28, carried me through the valleys of Gilgit and Hunza to the Hindukush watershed, as described in chapters II and III of my Personal Narrative, left no opportunity.

Among the few relics of Buddhist worship which are extant above ground in the main valley of Gilgit, and to which Major J. Manners Smith very kindly drew my attention, I was able to visit only the great rock-carved relief at the entrance of the Kergah Nullah, some four miles above Gilgit Fort and not far from the village of Naupūr. This relief, which appears to have been first described by Colonel Biddulph[1], occupies a conspicuous position, more than

Rock-sculpture near Gilgit.

[34] For much interesting information collected by Colonel Biddulph regarding the Indus Kōhistān, see *Hindoo Koosh*, pp. 8 sqq.

[1] See *Hindoo Koosh*, pp. 109 sqq., where a rather primitive sketch of the relief is reproduced in lithography. The attempted identification of this figure with the colossal Buddha image seen by Fa-hsien in Darēl, requires no serious consideration. A somewhat indistinct photographic reproduction of the rock-carving is given in the *Pamir Boundary Commission Report*, p. 32.

thirty feet above the ground, on a precipitous rock-face. It shows the figure of a colossal Buddha, about nine feet in height, carved in low relief within a shallow niche of trefoil shape (see Fig. 1).

Buddha is represented standing with the right hand and forearm raised across the breast, in the gesture which, in Buddhist convention, is known as the *abhayamudrā* ('the pose of assuring safety'), while the left hangs down grasping the edge of the robe. The robe is indicated only at the sides of the figure from the hip downwards, and leaves the limbs entirely bare, with the result that the statue at first sight recalls the representation of a naked Jina rather than a draped Buddha.

Yet, coarsely modelled and executed as the image is, examination of the photograph clearly shows that a fully-draped figure was intended. Broad pleats or bands appearing around the neck, and also over the right forearm, mark the folds which result from the classical draping of the robe, as invariably exhibited by Buddha statues in this posture both in Gandhāra and Khotan sculpture. The type which the Gilgit sculptor endeavoured to reproduce, with all the imperfections of his art, is unmistakably the same as appears in numerous stucco reliefs of Dandān-Uiliq, specimens of which may be seen on Plates LIV and LV. Earlier representations of it are to be found in numbers, both among Graeco-Buddhist sculptures from Gandhāra and the great stucco reliefs excavated by me at the Rawak Stūpa near Khotan[1].

Pose of Buddha statue.

It is significant that nowhere among these early instances of the type do we find the right hand brought across the breast as in the Gilgit relief. While the outward turn of the palm, which is a characteristic feature of the Abhayamudrā, could be obtained in this position only by an artificial twisting of the forearm, it is a perfectly natural gesture when the forearm is extended forward, with more or less of an upward slant, as seen in the sculptures, of which illustrations are quoted in note 2. The same strained pose of the right arm and hand is found, however, in a small relief representation of Buddha which occurs as part of the stucco wall decoration in one of the Dandān-Uiliq shrines, and a specimen of which, D. I. 11, has been reproduced in Plate LV.

In either case the change in the position of the forearm is manifestly due to the artist's inability to show in low relief the natural pose of the original model. The Dandān-Uiliq stucco figure certainly belongs to the second half of the eighth century, and the resemblance between it and the Gilgit rock-carving, in the clumsy device just discussed, must warn us against seeing in the crude modelling of the latter a mark of special antiquity. The pointed form of the trefoil arch enclosing the image seems to me, in fact, distinct evidence of a later date, judging from what a close study of that architectural element in Kashmīrian monuments has enabled me to ascertain as to the successive development of its forms. We have another sign of late workmanship in the exaggerated prominence which is given to the edges of the drapery, particularly that falling from the proper left arm. Its appearance there might easily mislead the untrained observer into a belief that the left hand, instead of supporting the robe (as clearly seen in the Dandān-Uiliq and Rawak reliefs), was holding 'a staff or some kind of weapon'[3].

[2] Compare, e.g., Grünwedel, *Buddhist Art*, pp. 169, 170, 174 (with the interesting Chinese Buddha, in wood, p. 177, which is traditionally derived as a replica from Udayana's famous statue); for corresponding relief statues at Rawak, see Figs. 68, 69, Plate XIV, &c.

M. Foucher, who has lucidly discussed the *Abhayapāṇi-mudrā* in his *Iconographie bouddhique*, pp. 68 sq., points out its frequency both in the miniature representations of particular Buddha images, as contained in Nepalese MSS., and among extant Buddhist sculptures of Northern India; for illustrations comp. Pl. II and Figs. 7, 8.

[3] See Biddulph's description, *Hindoo Koosh*, p. 110. For similar exaggerated treatment of the drapery in imitation of ancient forms, comp. Grünwedel, *Buddhist Art*, p. 171 and Fig. 125.

1

ROCK-CARVED RELIEF NEAR NAUPŪR, GILGIT.

2

CASTLE OF MĪR OF HUNZA AT BALTIT.

The deeply-cut square holes, arranged outside the trefoil niche in the form of a pentagon, undoubtedly served to support a wooden framework which once protected the image.

Colonel Biddulph has already noticed the remains of an ancient irrigation work, which can be traced close by, along the right or eastern side of the mouth of Kergah-Nullah. The square sockets which are seen there, sunk into the steep rock face at a uniform level and for a considerable distance, were manifestly designed to hold stout pieces of timber, on which a wooden trough could be fixed to conduct water from the stream of the Nullah for the irrigation of fields in the main valley. The comparatively high level at which this conduit runs above the rock-strewn bottom of the Nullah is easily accounted for by its purpose of carrying water to slopes which could not be reached by irrigation cuts taken from the stream at its actual mouth. Ancient water-conduit.

That the construction of this watercourse belongs to an early period, certainly pre-Muhammadan, is proved beyond doubt by the excellent cutting of the sockets, which shows a command of stone-craft long lost among the population of these valleys. We may well suppose that in a period of greater material culture Gilgit was more thickly inhabited than it is now, and that in consequence portions of ground on the hill sides and on alluvial plateaus more elevated than any now under cultivation were utilized for fields and required to be irrigated. Throughout the Gilgit valley it is only the water obtained from the side-streams that renders agriculture and the growing of fruit trees possible[4].

In connexion with these few notes on ancient remains in Gilgit, I may mention that the ruined mounds which Major J. Manners Smith and other officers on duty in Gilgit have noted at Hanzil and Jutiāl in all probability represent the remains of Stūpas. I was not able to visit them, but the photograph of the mound near Hanzil (a village about nine miles above Gilgit Fort) which is reproduced in the report of the Pamir Boundary Commission, distinctly suggests this origin[5]. The fact that in either place the ruin shows only rough masonry of unhewn stone would render it difficult to arrive at any conclusion as to the date of construction without systematic excavation. Ancient mounds in Gilgit.

The valley of the Hunza river, through which I passed from Gilgit to the Tāghdumbāsh Pāmīr, is, alike by the stern grandeur of its peaks and glaciers, the natural difficulty of its communications, and the strange mixture of races and languages among its population, a mountain region of exceptional interest. But the same ice-crowned ranges and almost equally formidable gorges which have rendered Hunza until our days so secure against foreign invasion, have also effectively barred the valley from ever serving as a real line of communication or otherwise acquiring historical importance. Valley of Hunza.

We have striking evidence of this isolation in the survival of Burisheski, the tongue spoken in Hunza proper, which has no relation whatever to any of the great language families (Indian, Īrānian, Turkī, Tibetan) that meet close to this easternmost point of the Hindukush watershed. Nor can it be doubted that these secluded communities, in their customs, traditions, and economic conditions, must have preserved much that would prove highly instructive to antiquarian students.

[4] See Drew, *Jummoo*, p. 407; compare also the remarks, p. 404, illustrating the reduction of arable land in Astōr which has followed long-continued decay of the artificial watercourses.

[5] See *Report of the Pamir Boundary Commission*, p. 33. The decidedly circular shape of the mound makes it very improbable that the remains can be those of a 'frontier tower of ordinary fashion', as assumed by one of the members of that Commission. Such towers are regularly built square throughout the whole of the Dard region. On the other hand, I fail to see how 'the construction of the masonry' can be adduced as an argument against 'the theory of Buddhist construction', considering that neither the masonry nor any other details of construction of pre-Muhammadan ruins in the valleys between the Hindukush and Kashmīr have as yet been examined by a trained archaeologist.

For such passing observations as my hurried marches along the difficult tracks of Hunza allowed me to gather, I must refer to the pages of my Personal Narrative[6]. But in respect of the only monument of antiquity which I was able to trace along the route, where it winds round the foot of the great Mount Rakiposhi (Fig. 3), the brief description there given requires to be supplemented by some more exact details. Close to the hamlet of Thōl, on the first of the highly cultivated little plateaus which are passed in succession on the left river bank above the historic gorge of Nilth, there stands a ruined Stūpa, shown with its west face in the accompanying photograph (Fig. 4). Considering its exposed position, the relatively good preservation of this monument of Buddhist worship is remarkable. The only serious damage it has sustained is of very recent date, being a characteristic result of the 'advance of modern civilization': the masonry of the base on its southern face has been removed to save a détour of a few feet to the hill road constructed since the occupation of Hunza territory at the close of the campaign of 1891.

Ruined Stūpa near Thōl.

The Stūpa is constructed of unhewn slabs, more or less flat, set in rough horizontal layers with a fairly hard plaster. The surface of the rough masonry originally bore a coating of similar plaster, patches of this material being found still adhering to the north face of the upper courses of the base and to the more sheltered mouldings. The lowest story of the base forms a square of 10 ft.; and has a height of 3 ft. 9 in. from the present ground level. On the top of this base, but receding by about 1 ft. from its edge, rests a second story, which is also square, and with its projecting cornice, about 14 in. high, rises to the same height of 3 ft. 9 in. The next story is octagonal and, including its cornice, 3 ft. 10 in. high. It supports a circular drum, about $1\frac{1}{2}$ ft. in height, also surmounted by a cornice. From the latter finally springs the dome of the Stūpa proper. This, in its original shape, appears to have been hemispherical, but its top has been broken, and the extant masonry reaches only a height of about $3\frac{1}{2}$ ft. from the sloping plinth of the drum. Judging from what subsequent experience showed me in the case of Stūpa ruins examined in the Turkestān plains, it may be supposed that treasure-seeking operations for the relic deposit likely to be contained in the centre of the little dome were the main cause of this injury. The total height of the structure, allowing for this loss, cannot have been less than 18 ft. and probably approached 20 ft.

Architecture of Thōl Stūpa.

This relatively great elevation, in proportion to the dimensions of the square base, strikingly distinguishes the Thōl Stūpa from the Stūpas examined by me in Chinese Turkestān. Those among the latter which were found sufficiently well-preserved to permit exact measurements, like the Stūpas of Mauri-Tim, the Niya Site, Endere and Rawak, in their original state showed a total elevation approximately equal to the side of their square base where it rested on the ground, i.e. to the greatest dimensions of the ground-plan. Striking differences are to be found also in the introduction of an octagonal above the square stories of the base, and in the boldly projecting and massive cornices by which these several stories and the circular drum beneath the dome are surmounted. For neither of these distinguishing features am I able to adduce exact parallels from Stūpa ruins with which I am acquainted in the North-West of India and the adjoining frontier regions. But there appears in them something curiously recalling the style and general character of the Chortens of Sikkim and Ladāk, and suggestive of Tibetan influence.

Mountain tracks in Hunza.

The small alluvial plateaus fringing at intervals the deep-cut bed of the Hunza river between Nilth and Baltit are the only portion of the valley where a track fit for regular use by laden animals could be maintained without recourse to modern methods of engineering. Above the village of Baltit, probably from early times the seat of Hunza chiefs[7], the route towards the

[6] See *Ruins of Khotan*, pp. 32 sqq.

[7] For a photograph of their picturesque old castle see Fig. 2.

3

MOUNT RAKIPŌSHI, SEEN FROM ALIĀBĀD, HUNZA.

4

RUINED STŪPA AT THŌL, HUNZA.

Tāghdumbāsh Pāmīr enters the long and narrow gorge in which the river has cut its way through the axial range of the Hindukush. The tracks that lead through it present, for six trying marches, all the difficulties of true Alpine climbs, including passages over large glaciers. During a considerable portion of the year they are altogether impassable except for men on foot. The transport of loads over these mountain tracks becomes possible only by the employment of the hardy hillmen, who manage to subsist on the few patches of cultivable ground to be found amidst the barren masses of rock and ice which bound the course of the main stream and the still more confined gorges of its affluents. The extreme sterility of these mountains restricts the population of 'Little Guhyāl', as this region is now called on account of its Wakhī settlers from Wakhān or Guhyāl, to a few hundred families[8]. It seriously taxes the resources of these scattered settlements to supply even porters for the occasional European traveller who is permitted to pass through the Hunza Valley. In the face of such natural obstacles, it is impossible to conceive how the latter could ever have served as a route for trade and general traffic.

After leaving behind Misgar, the northernmost hamlet of Hunza, the natural difficulties of the route decrease. The valley widens as we approach the watershed which separates the headwaters of the Hunza river from those of the Oxus on the one side and from the Tāghdumbāsh Pāmīr on the other. Lord Curzon, in his exhaustive Memoir on the Pāmīrs, has duly emphasized the important geographical fact that the water-parting in this part of the Hindukush lies considerably to the north of the axial range and is also far lower[9]. This helps to account for the relative ease with which the Kilik and Mintaka passes, giving final access to the Tāghdumbāsh Pāmīr, can be crossed, even with laden animals, during the greater part of the year.

[8] For photographs of Wakhīs and Kanjūtīs settled in 'Little Guhyāl', see Figs. 5, 6.

[9] See Curzon, *The Pamirs and the Source of the Oxus*, p. 40.

CHAPTER II

SARĪKOL AND THE ROUTE TO KĀSHGAR

SECTION I.—THE GEOGRAPHICAL POSITION AND ETHNOGRAPHY OF SARĪKOL

Meeting of ethnic areas near Kilik Pass.

THE great alpine valley of the Tāghdumbāsh Pāmīr, to which I crossed on the 29th of June, 1900, over the Kilik Pass (15,800 feet above the sea, Fig. 8), was certain to prove interesting ground from more than one point of view. At its head I was close to the point where the drainage areas of the three great river systems of the Indus, the Oxus, and the Tārīm meet, the representatives as it were of the still greater ethnic areas of India, Īrān, and Turkestān. That the view from the height of the Khushbēl Peak, my first survey station, simultaneously comprised the confines of British India, Afghānistān, Russia, and China, was the best illustration of the abiding influence which geographical facts must exercise over political developments even in this desolate region. The wide expanse of level or easily undulating grazing land which the same view disclosed at the bottom of the Tāghdumbāsh Valley, formed a striking contrast to the rocky and almost impassable gorges of Hunza, and could not fail to impress me at the outset with the advantages for communication which the Pāmīrs offer, notwithstanding their elevation and severe climate. But it was only in the light of subsequent observations, gathered on my marches along the whole length of the Tāghdumbāsh Pāmīr from the Wakhjīr Pass and the Oxus source down to Tāsh-kurghān, that I fully realized the historical interest of the route which leads through this valley.

Position of Tāghdumbāsh Pāmīr.

In order to explain the importance which may be claimed for the Tāghdumbāsh Pāmīr as an early line of communication between Chinese Turkestān and the Oxus Valley, it is not necessary to consider the historical topography of the whole Pāmīr region. Lord Curzon, in his justly famous memoir on 'The Pamirs and the source of the Oxus', has furnished a lucid and critical summary of that much-discussed subject, and to it the reader may be referred with confidence for information on all general questions[1]. An important geographical fact, aptly recognized in Lord Curzon's analysis[2], helps to limit the scope of our inquiry. The Tāghdumbāsh Pāmīr, though it shares the designation and also, in its upper portion, the main physical characteristics of the other Pāmīrs, lies in a different watershed from the rest, and is thus plainly marked as part of a different system. We find this physical division emphasized at the present day by the political boundaries, which leave the Tāghdumbāsh the only Pāmīr under Chinese jurisdiction, and we may safely attribute to it a determining influence upon the earlier historical conditions of this mountain tract.

Importance of Tāghdumbāsh route.

While all the other Pāmīrs are situated within the drainage area of the Oxus, the waters of the Tāghdumbāsh Pāmīr discharge themselves eastwards into the great Turkestān Basin.

[1] Reprinted from the *Geographical Journal*, 1896; for an analysis of early travellers' routes see in particular pp. 63 sqq.

[2] See *The Pamirs*, p. 19.

The river of which they are the main feeders, and which takes its best known name from Tāsh-kurghān, the chief place it passes, breaks through the great meridional range flanking the Pāmīrs on the east, and ultimately joins the Yarkand river or Zarafshān. The collection of valleys which the river of Tāsh-kurghān drains, together with some minor alpine tracts adjoining them towards the Upper Yarkand river, constitutes the well-defined mountain district now known as Sarīkol[3]. Topographical facts and historical evidence alike prove that the position occupied by the present fort of Tāsh-kurghān and its neighbouring villages has from very early times been the political centre of this whole territory. At Tāsh-kurghān ends the open valley of the Tāghdumbāsh Pāmīr, and the importance of the latter as a route is mainly due to the exceptionally easy access it provides to the central tract of Sarīkol, over more than a hundred miles' distance. In view of this close connexion, it will be well to begin with a review of the main data which throw light on the ancient topography of Sarīkol as a whole.

Small in extent, and devoid of natural resources, the territory of Sarīkol derives its importance solely from the advantages of its position with regard to the routes which from early times have connected the Upper Oxus Valley with the oases to the south of the Turkestān Desert, and hence with China. All the routes leading from the Oxus in that direction, whether they ascend through Rōshān, Shighnān, or Wakhān, have to cross the water-parting on the east of the Pāmīrs, and subsequently to surmount the still more elevated meridional range culminating in the Muztāgh-Ata Peaks, which forms the link between the T'ien-shan system in the north, and the extreme points of the Hindukush and Kun-lun in the south[4]. The chief valleys of Sarīkol extend between this meridional range and the watershed which fences in the true Pāmīrs on the eastern side. To this position they owe their relatively great width and also their prevailing direction, the valley of the Tāghdumbāsh descending mainly from south to north and that of Tagharma from north to south. Every route that crosses the Pāmīr watershed to the south of the Muztāgh-Ata *massif*, is compelled to debouch at one point or the other into either of these great valleys; in the opposite direction the same holds good of all the routes that lead from the drainage area of the Yarkand river westwards into the Oxus Basin. The Tāghdumbāsh and Tagharma Valleys meet at the point where the Tāsh-kurghān river takes its sudden turn to the east; and Tāsh-kurghān, only some eight miles further south, is the place towards which all the above routes, whether from east or west, naturally converge.

Importance of Sarīkol.

It is easy to show that this centre of the Sarīkol District must at all times have been

[3] The above spelling, *Sarīkol*, reproduces the pronunciation of the name as generally heard by me during my passage through the district, from both its Tājik (Īrānian) and its Kirghiz inhabitants. I am unable to decide how far this modern pronunciation conforms to the etymology of the name; for the derivation is not quite certain. If Mīrzā Ḥaidar, whose mention is the oldest I can at present trace (*Tārīkh-i-Rashīdī*, ed. Elias, pp. 297, 312), is justified in writing *Sarigh-kul*, the name would be Turkī, meaning 'Yellow-Lake.' As *kol* is a common variation of *kul* through all dialects of Eastern Turkī, and as *ī* for *igh* would easily be accounted for by the well-known phonetic processes of assimilation and subsequent 'supplementary lengthening', the modern form of the name could be readily explained on the basis of this etymology. But the form *Sirikol* (*Sirikul*) is also met with in Oriental records, and has found acceptance among numerous European geographers, perhaps on account of its supposed semi-Persian etymology (*Sir-i-kul*, 'head of the lake'). It is curious to note the same uncertainty of spelling in the case of one of the several alternative native names for Wood's Lake on the Great Pāmīr; see Curzon, *The Pamirs*, p. 43.

[4] Though accurate surveys of this meridional range have been accomplished in recent years, the admirably lucid account which Baron Richthofen, in his *China*, vol. i. pp. 195 sqq., has given of its salient features and its importance for the general orography of Central Asia, may still be consulted with advantage. The merit of having first systematically explored those portions of the range which lie between the Tāsh-kurghān river in the north and the junction of Hindukush and Kun-lun in the south, belongs to Captain H. H. P. Deasy, whose book, *Three Years in Tibet and Eastern Turkistān* (1901), furnishes also a description of the smaller Sarīkol valleys comprised in those portions; see chapters viii, xiii, xiv.

Routes through Tāsh-kurghān.

an important goal for travellers. From whichever side we may approach Sarikol, there is an inhospitable belt of high mountain land to be crossed first, practically devoid of permanent habitations, and throughout incapable of furnishing supplies and places of shelter to caravans. The elevated Pāmīr region stretching westwards can never, during historical times, have permitted of cultivation. The routes which, starting from Tagharma, connect Sarīkol with Kāshgar to the north-east and Yarkand to the east, lead by a succession of high passes over barren spurs of the great meridional range with narrow uninhabited gorges between them. It is true that the difficult and rarely frequented tracts which cross the mountains between the Tāsh-kurghān and Yarkand rivers in the direction of Karghalik and Kök-yār, pass through some of the minor Sarīkolī settlements. But the produce raised on their isolated plots of cultivable land does not suffice even for the maintenance of the small pastoral population scattered over this region. Finally, if the route be followed which leads northward past Muztāgh-Ata and then descends along the Yamān-yār river into the Kāshgar plain, as described in my Personal Narrative[5], an even greater distance has to be traversed before permanent habitations are reached.

Resources of Sarīkol.

This situation of Sarīkol, in the midst of desolate mountain tracts and yet at the junction of important routes, necessarily invests with exceptional value whatever natural resources the district possesses. Given an adequate population, and an administration capable of protecting it, these resources would, undoubtedly, be far larger than they are at present. The lower part of the Tāghdumbāsh Valley, for a distance of forty miles from below Tāsh-kurghān to the hamlet of Dafdār, shows a remarkably broad and uniform expanse of fertile ground at its bottom. Cultivation is now restricted to a small continuous stretch of the valley above and below Tāsh-kurghān, and to certain isolated settlements, such as Dafdār and Pisling, recently started at points higher up the valley where it is easy to obtain water for irrigation from side streams.

Earlier settlements in Tāsh-kurghān Valley.

But the uniform tradition of the inhabitants, as reported to me during my stay at Tāsh-kurghān, asserts that the compact cultivated area once extended much higher up the valley. In support of this belief, reference was made to the remains of extensive irrigation canals traceable along the foot of the mountains, especially on the east side of the valley, as well as to deserted village sites, such as Bāzār-dasht, found at a considerable distance beyond the limits of the present belt of village land. I believe that these statements may be accepted as based on substantial facts. On the one hand, the almost unbroken stretch of alluvial land which I passed on my route from Dafdār to Tughlān-shahr, seemed only to wait for systematic irrigation in order to yield the crops for which the climate is adapted. On the other hand, it is impossible to doubt that the periodical raids from slave-hunting Kanjūtīs and Shighnīs, to which we know Sarīkol to have been subjected before and after Yāqūb Bēg's rebellion, and which did not cease completely until the British occupation of Hunza, must have resulted in a partial depopulation of the country[6].

From the devastating effects of these raids, the upper portions of the Tāghdumbāsh Valley, lying nearest, necessarily suffered most. But they extended also to the Tagharma Valley, which, in its well-watered central flat, about twelve miles long by seven broad, offers ample ground for agricultural settlements[7].

[5] See *Ruins of Khotan*, chaps. v–vii.

[6] For the modern history of Sarīkol, compare Col. T. E. Gordon, *Roof of the World*, pp. 109 sqq., where an instructive account is given of the general conditions of the main valleys at the time of his visit (1874).

[7] The destructive results of Kanjūtī raiding in Tagharma are illustrated by the information recorded by Capt. H. Trotter in 1874 about a fort village in the south-east of the plain, the Tājik population of which had been reduced from fifty to four families within less than a generation; see *Yarkand Mission Report*, p. 269.

It is difficult to judge now, when the whole tract is slowly recovering from the results of long-continued aggression, what its population and resources may have been during periods when the support of a strong paramount power, established in Eastern Turkestān, assured to Sarīkol safety from its predatory neighbours to the west and south. The political value of this hill tract as an advanced post guarding the frontier towards Badakhshān, Gilgit, and the Pāmīrs was, as we shall see, realized early by the Chinese. But far more than the spasmodic assertion of sovereignty, which, as far as the Chinese administration was concerned, is likely to have been often purely nominal [8], the racial tenacity of the inhabitants themselves must have helped to preserve to the little alpine chiefship its continued existence and historical individuality.

Ethnic affinity of Sarīkolīs.

The population of Sarīkol, apart from the nomadic Kirghiz herdsmen who visit its grazing grounds, consists of hill Tājiks, who by physical appearance and language alike are unmistakably proved to belong to the so-called Galcha stock. The Sarīkolī tongue, first recorded by the late Mr. R. Shaw with his usual accuracy and care, is very closely allied to Wakhī, the language of Wakhān, of which it may be said to represent a mere dialect. The Sarīkolīs whom I saw showed all the racial characteristics of the Īrānian 'Tājiks' who form the bulk of the population in the Oxus region [9], and the knowledge of Persian common among all classes indicates the influence exercised by the culture of Īrān even in these distant valleys. To the prevalence of Īrānian traditional lore in the local legends I shall have occasion to refer later. The antiquity of these legends, as attested by Hsüan-tsang, the Īrānian elements in the local nomenclature, as well as the absence of any historical tradition to the contrary, render it highly probable that the population of Sarīkol has from early times borne the same ethnic character as at present.

Ethnic individuality of Sarīkol.

The geographical position of Sarīkol is such that it could never have lain on the route of one of those great waves of invasion which from time to time have swept across the more accessible parts of Central Asia, and in their progress have radically changed the ethnology of those regions. Sarīkol, with its severe climate [10]—the elevation of the main valleys sinks nowhere below 10,000 ft.—and its limited area of productive land, could not possibly attract occupation and settlement by a powerful invader. The predatory inroads to which the main valleys were exposed, during recurring periods, from the petty hill-states to the south and west, no doubt subjected the more peaceable population of Sarīkol to great vicissitudes. But considering the passing character of these raids, and the ready shelter offered by the many secluded side-valleys, they were not likely to threaten it with extermination. Experience in any case shows that the Sarīkolī settlements during recent times have recovered from severe trials of this kind, including even a forcible transportation of the greater part of the population to Kāshgar, which occurred during Yāqūb Bēg's rule [11]. It is manifest that this small Īrānian community, placed in such close proximity to a great Turkī-speaking population, and sharing its political fortunes, could not have preserved its language and racial characteristics intact without the isolating effect of its mountains and that tenacious attachment to inherited lore which is peculiar to hill people.

[8] The relations which existed during the period preceding Yāqūb Bēg's rebellion between the hereditary chief of Sarīkol and the Chinese authorities were characteristically indicated in the nominal tribute paid by the former and the valuable presents received in return. The payments in silver and gold made to the Sarīkol chief were represented 'as a subsidy for the military protection of the frontier and the road towards Badakhshān'; see Gordon, *Roof of the World*, p. 110.

[9] For a photograph see *J. Anthr. Inst.*, 1903, Pl. XXVII, where the individuals in the back row and the central figure in the front row are Wakhīs settled in Sarīkol, the rest Kirghiz. Wakhīs and Sarīkolīs proper are not distinguishable in outward appearance.

[10] The character of the climate of Sarīkol is well indicated by the saying which Col. T. E. Gordon recorded from the mouth of its governor at the time of his visit 'that there are only two seasons, summer and winter, the former lasting three months, the latter nine'; *Roof of the World*, p. 114.

[11] See Gordon, *Roof of the World*, p. 111.

Early extension of Galcha stock eastwards.

The hillmen of Sarīkol at the present day form the extreme outpost of Īrānian nationality towards the east. But if we may judge from several important indications, settlements of an Īrānian-speaking race must in ancient days have extended much further eastwards, especially in the direction of Khotan. The documents in Brāhmī script which my excavations brought to light from the ruined temples of the Dandān-Uiliq site, and which, with other records of the same character previously obtained from the same locality, have been partly deciphered by Dr. Hoernle, make it appear highly probable that the language spoken in the eighth century by the indigenous population of Khotan was of Īrānian origin[12]. On the other hand, the anthropometric data collected by me in the Khotan region, according to the careful analysis made of them by Mr. A. T. Joyce, in independence of any historical or linguistic arguments, plainly mark in the Khotanese population of to-day the prevalence of a racial element closely related to the hill-tribes generally designated as *Galchas*, who are settled in the highlands of the Oxus and Zarafshān[13].

Connexion with ancient population of Khotan.

The ethnographical importance of these observations is confirmed by the statement of the Chinese historical record to be discussed below, dating back to the period of the T'ang dynasty's rule over Eastern Turkestān (7th to 8th century A.D.), which describes the external appearance and language of the people of Sarīkol as identical with that of the Khotanese[14]. In view of this convergence of linguistic, anthropological, and historical proofs, we can scarcely avoid the conclusion that the modern Sarīkolīs represent but the remnant of a larger Galcha population which once spread as far as the territory of Khotan, but further east has since undergone considerable racial amalgamation and abandoned its language for Turkī. The great difference in geographical position, and consequently in accessibility between the valleys of Sarīkol and the oases of the plain of the northern foot of the Kun-lun range would amply account for the thoroughness with which this transformation has proceeded in the latter region.

The Pakhpo tribe.

Finally, it may be pointed out in passing that an ethnic link between the Īrānian Sarīkolīs and the present population of those oases is, perhaps, to be found in the small and little known hill-tribe of the Pakhpos, who partly as herdsmen, partly as cultivators, dwell in the narrow valleys near the headwaters of the Tiznaf and Yarkand rivers. Dr. Bellew, to whom we owe what scanty information has so far been recorded about this curious people, describes them as of 'pronounced Caucasian features' and very fair[15]. He was much struck by the difference of their physical type from that of other races he had come into contact with in Eastern Turkestān. He notes that the few Pakhpos whom he met and was able to examine, denied having any language of their own apart from Turkī; but he remarks also upon the extreme shyness of these hillmen, which 'led them to conceal all information regarding themselves'. I myself was unable to proceed near enough to their hills, south of Karghalik, to obtain an opportunity of meeting Pakhpos. But during my short stay at Karghalik, Dāud Bēg, a local official, who had for some years held charge of the tract they mainly inhabit, distinctly told me that, though all Pakhpos know Turkī, and though its use is extending owing to frequent intermarriages with people of the Karghalik oasis, another language is talked by them among themselves which

[12] Compare *Ruins of Khotan*, pp. 309 sq.; Hoernle, *Central Asian Antiquities*, ii. pp. 32 sqq.; below, chap. IX. sec. V.

[13] See below, chap. VI. sec. IV., on the racial origin of the people of Khotan, and Mr. Joyce's paper in *J. Anthr. Inst.*, 1903, pp. 322 sqq.

[14] See Chavannes, *Turcs occid.*, p. 124.

[15] Compare *Yarkand Mission Report*, pp. 61 sq. It must be remembered that Dr. Bellew, with the main portion of Sir D. Forsyth's mission, did not visit the Sarīkol valleys, and consequently could scarcely be expected to notice any similarity in physical appearance that might exist between Sarīkolīs and Pakhpos. The photographs of Pakhpos reproduced on p. 46 of the *Yarkand Mission Report* show types which to me do not appear to differ materially from those ordinarily met among Sarīkolīs and Wakhīs.

is understood by Sarīkolīs, and is apparently a dialect of the latter's tongue. If this information should prove to be correct, our presumption in favour of the Pakhpos being another remnant of a Galcha population, formerly more widely spread, would become an established fact [16].

Section II.—EARLY CHINESE ACCOUNTS OF SARĪKOL

The important position occupied by Sarīkol in respect of the routes leading across the Pāmīr region accounts for the ample information to be gathered from early Chinese records regarding its historical topography. The narrative of the pilgrim Hsüan-tsang, who on his return journey from India to China traversed Sarīkol, supplies most of the details. It was in his itineraries that the old name of this mountain chiefship, which the Chinese transcriptions variously render as Chieh-p'an-t'o, Han-p'an-t'o, &c., was first correctly identified by General A. Cunningham [1]. But for a systematic review of all available data we shall find it convenient to take as a basis the official description of the territory which the T'ang Annals furnish, and which has now become accessible in M. Chavannes' translation.

Sarīkol in T'ang Annals.

This account records for the Sarīkol tract the several names *Ho-p'an-t'o* 喝槃陀 or *Han-t'o* 漢陀, or *K'o-koan-t'an* 渴館檀 or *K'o-lo-t'o* 渴羅陀, and indicates its position with unmistakable clearness. The kingdom was reached by proceeding from *Su-lê* or Kāshgar to the south-west through the gorge of *Chien-mo*, which must correspond to the present Gez defile, and lay at a distance of 600 li, or approximately six days' march. This estimate agrees remarkably well with the route which leads from Tāshmalik, at the south-western edge of the Kāshgar oasis, along the Gez or Yamān-yār river, and past Muztāgh-Ata to the head of the Tagharma Valley, and which I myself followed in the opposite direction on my journey from Tāsh-kurghān to Kāshgar. *Ho-p'an-t'o* is correctly described as being situated directly to the west of *Chu-chü-po*, which is represented by the modern Karghalik; in the north it touched the territory of *Su-lê* or Kāshgar, in the west *Hu-mi* or Wakhān, while to the north-west there adjoined the territory of *P'an-han* in which, with M. Chavannes, we may recognize Farghāna. The administrative centre of the territory lay in the middle of the Ts'ung-ling or 'Onion mountains', which are said to encircle the whole of the kingdom [2]. This designation has been generally applied by the Chinese to the meridional range or ranges which buttress the Pāmīr region on the east [3], and divide it from the Tārīm Basin. The position of the Ho-p'an-t'o capital, as here marked with reference to them, fits exactly the present Tāsh-kurghān. The river *Si-to*, on which the capital is stated to have stood, can, in view of Hsüan-tsang's account to be noticed below, be no other than the river of Tāsh-kurghān.

The people of Sarīkol in the T'ang period.

Of the people of Ho-p'an-t'o the Annals record, as already stated, that their appearance and language were identical with those of the people of Yü-t'ien or Khotan. They are further described as strong and given to violence. Murder and brigandage were alone liable to be

[16] It deserves to be noted that Sung Yün distinctly mentions the language and customs of the people of *Chu-chü-po* as closely resembling those of the Khotan people; see Chavannes, *Voyage de Song Yun*, p. 20. Chu-chü-po corresponds to the present district of Karghalik, including the valleys towards the headwaters of the Yarkand river, in which the Pakhpo settlements are found; see Chavannes, *Turcs occid.*, pp. 123 note, 311.

[1] See his paper in *J.A.S.B.*, 1848, vol. xvii, referred to by Yule, *J.R.A.S.*, 1872, 'Notes on Hwen Thsang's Account of Tokhāristān,' p. 117.

[2] See Chavannes, *Turcs occid.*, p. 124.

[3] It will be seen below that the term *Ts'ung-ling* was used during the T'ang period as a special designation of the Sarīkol territory itself; as to the meaning and origin of the name, comp. Richthofen, *China*, i. p. 221.

punished with death; for other offences compensation by fine was allowed. Taxes were paid in clothing materials, an observation illustrated by what a native informant, in the Report of the Yarkand Mission, notes of the barter carried on in Sarīkol with cotton goods imported from the plains [4]. The military force of the territory was reckoned at a thousand men. Its ruling family is stated to have originally come from Kāshgar, and to have transmitted its power from generation to generation.

During the period corresponding to 435–439 A. D., under the later Wei dynasty, Ho-p'an-t'o is said to have first entered into relations with China. 'In the ninth year Cheng-kuan (635) it sent an envoy to do homage at the Imperial court. During the period K'ai-yüan (713–741 A. D.) China conquered and pacified this kingdom; it established there the military post of Ts'ung-ling, which is the extreme point under military occupation on the frontier of An-hsi', i. e. of the Chinese protectorate then comprising Eastern Turkestān. M. Chavannes, in his comments on this notice, points out that another passage of the T'ang Annals distinctly identifies the 'military post of Ts'ung-ling' with the ancient kingdom of *Chieh-p'an-t'o*; and it is under this appellation or simply 'Ts'ung-ling' that we find Sarīkol repeatedly mentioned in the Chinese records translated by him [5].

Fa-hsien's notice of Sarīkol.

The earliest Chinese travellers of whose visit to Sarīkol we have any record are the pilgrims Fa-hsien and Sung Yün. But in the case of the former our knowledge is not only exceedingly brief, but dependent on a conjectural identification [6]. Fa-hsien and his fellow-pilgrims, when proceeding, about 400 A. D., from Khotan towards India, reached first the kingdom of *Tzŭ-ho* 子合. A notice of the T'ang Annals translated by M. Chavannes plainly shows Tzŭ-ho to be identical with the territory known under the T'ang as Chu-chü-po, i. e. the present district of Karghalik [7]. From there the pilgrims 'went south for four days, when they found themselves among the Ts'ung-ling mountains, and reached the country of Yu-hwuy, where they halted and kept their retreat' [8]. The name *Yü-hui* 於麾, otherwise wholly unknown, presented a puzzle until M. Chavannes, by a slight emendation, restored it to *Yü-mo* 於摩, an abbreviated form of the name *Ch'üan-yü-mo* 權於摩, under which Tāsh-kurghān is mentioned in the *Pei shih*. We shall have occasion to follow elsewhere the ingenious and convincing arguments by which M. Chavannes further traces Fa-hsien's route from Tāsh-kurghān to *Chieh-ch'a* or Kāshgar, where he appears to have gone in order to rejoin some companions before attempting the passage of the Pāmīrs. But it may be noted that the four days' march south of Tzŭ-ho or Karghalik to where the Ts'ung-ling mountains were entered, could well be explained on the assumption that Fa-hsien's party for the journey to Tāsh-kurghān chose a route which first took them to Kök-yār, south of Karghalik, and from there westwards into Sarīkol through the mountains adjoining the course of the Upper Yarkand river [9].

Sung Yün's passage through Sarīkol.

The account which the next Chinese traveller, the pilgrim Sung Yün, has left us of his passage through Sarīkol (519 A. D.), is less laconic. But a want of proper sequence in the

[4] See *Yarkand Mission Report*, p. 56 ('the rate is one sheep for thirty yards of cloth. . . . . . No coin is current in Sarīgh Kúl, everything is by barter').

[5] See Chavannes, *Turcs occid.*, p. 125 note; also the passages quoted, s. v. Ts'ung-ling, p. 373, especially the one bearing on Kao Hsien-chih's expedition to Gilgit, p. 152 note.

[6] I am indebted for the first information concerning Fa-hsien's probable route from Khotan to the confines of India to M. Chavannes, who was kind enough to communicate to me, in a letter dated Sept. 4, 1903, the main points as discussed by him in a note of his forthcoming translation of Sung Yün's itinerary. [See now *Voyage de Song Yun*, p. 54, note 3.]

[7] See Chavannes, *Turcs occid.*, p. 123.

[8] Compare Fâ-hien's *Travels*, transl. Legge, p. 21.

[9] The route here suggested would partly coincide with the one discussed below in connexion with Sung Yün's account of Sarīkol; see note 13, p. 30.

details—a defect to be observed also in other portions of Sung Yün's narrative—formerly made it difficult to ascertain which of his data refer to Sarīkol itself and which to the traveller's subsequent route across the Pāmīrs. The annotated translation of Sung Yün's itinerary recently published by M. Chavannes, enables us to recognize clearly some characteristic observations bearing on Sarīkol, and to trace the probable line of route[10].

Sung Yün, like Fa-hsien, travelled westwards through Khotan and entered the limits of the kingdom of *Han-p'an-t'o* 漢盤陀 from the side of Chu-chü-po, which M. Chavannes has correctly identified with the modern district of Karghalik[11]. Going west for six days the traveller ascended the Ts'ung-ling mountains, in which we must recognize the ranges separating the valley of the Raskam Daryā from the plains about Karghalik. After three more days to the west he arrived at the 'town of Po-yü' or 'Po-mêng.' Thence he ascended, in three days, the mountains called *Pu-k'o-yi* (meaning literally 'the mountains on which it is impossible to find a rest'), where the cold was extreme and the snow was lying summer and winter. On these mountains a lake or the site of a lake was pointed out to the traveller as the former habitation of a poisonous Nāga, who for his misdeeds had been banished through an early king's magic. By difficult mountain routes he thence reached in four days the capital of the kingdom of Han-p'an-t'o. This territory was supposed to occupy the very summit of the Ts'ung-ling mountains and to be the centre of heaven and of earth.

Sung Yün notes that its inhabitants trained the watercourses to irrigate their crops, and when told that the cultivators of China depended on rain alone for their produce, they would not credit such a story. East of the city of Han-p'an-t'o was the river Mêng-chin, which flowed to the north-east towards Su-lê or Kāshgar. On the high Ts'ung-ling mountains neither herb nor tree would grow. On their west side all rivers were flowing westwards. At the time of Sung Yün's passage, in the eighth month of the Chinese year, the temperature had already turned cold; the north wind was driving away the wild geese, and snow was falling over the whole region. Sung Yün's description of *Han-p'an-t'o.*

The description of the territory which Sung Yün here gives is not as detailed as we might wish, but there can be no doubt that it fits accurately the central valleys of Sarīkol, as recognized by M. Chavannes[12]. The mention of the river flowing to the north-east, the bed of which lay east of the city of Han-p'an-t'o, clearly indicates for the latter the site of the present Tāsh-kurghān, described also by Hsüan-tsang. A reference to the brief description of the Tāgh-dumbāsh River Valley given above will suffice to show that Sung Yün's observations on the climate and character of this mountain region, and on the system of irrigation by which parts of the valley are made to yield crops, are quite accurate.

The most direct and frequented route from Karghalik to Tāsh-kurghān leads through the mountains towards the Raskam Daryā. After crossing the latter near the group of Sarīkolī villages known as Tong, it continues to the Kandahār (or Khandār) Pass, from which Tāsh-kurghān is reached in four marches. It appears very probable that this was the route which Sung Yün followed. Not having visited the route myself, and knowing only its western portion from the descriptions of Dr. Hedin and Captain Deasy, I must restrict myself to pointing out that the position of the village group of Tong would well correspond to Sung Yün's 'town of Po-yü (or Po-mêng)', and that the Kandahār Pass, being about 16,600 ft. above the sea and Identification of Sung Yün's route.

[10] See Chavannes, *Voyage de Song Yun*, pp. 20 sqq.

[11] See Chavannes, *Turcs occid.*, pp. 123 note, 311; also below chap. IV. sec. i.

[12] See Chavannes, *Voyage de Song Yun*, p. 21, note 4.

very steep, seems to represent the Pu-k'o-yi mountains as regards height, difficulty, and relative distance [13].

Hsüan-tsang's journey over the Pāmirs.

Hsüan-tsang, whose itinerary has already been referred to as our main authority concerning ancient Sarīkol, reached the district from the opposite direction, i. e. from the west. Returning after his long travels in India towards his native land about the summer of 642 A. D., Hsüan-tsang passed through Badakhshān into the kingdom of *Ta-mo-hsi-t'ie-ti*, which undoubtedly corresponds to the present Wakhān [14]. In the valley of *Pa-mi-lo*, reached by the pilgrim after seven marches to the north-east, his earliest European interpreters could not fail to recognize the Pāmīr region. The salient features of his description, as Lord Curzon duly notes, 'stand out as an unmistakable picture of the Pāmīr country . . . . . . , and leave a doubt only as to the particular valley or Pāmīr by which the traveller crossed it [15].' With Lord Curzon and the majority of Hsüan-tsang's commentators, I believe that there is preponderant evidence in favour of the route which leads through the Great Pāmīr and past Lake Victoria; for only in the latter can we find a real approach to the position and size of 'the great Dragon Lake' which the pilgrim passed 'in the middle of the Pa-mi-lo Valley' [16].

Hsüan-tsang's route into Sarīkol.

Lord Curzon has observed that this identification is distinctly supported by what we are told of Hsüan-tsang's immediately succeeding marches. 'On leaving the midst of this (Pa-mi-lo) valley and going south-east, along the route, there is no inhabited place. Ascending the mountains, traversing the side of precipices, encountering nothing but ice and snow, and thus going 500 li, we arrive at the kingdom of Chieh-p'an-t'o 朅盤陁 [17].' The south-eastern direction, here indicated, of the further journey to Sarīkol could not be accounted for if Hsüan-tsang was supposed to have travelled by the Little Pāmīr and past its lake, the Chakmaktin-Kul. A look at the map shows that a traveller proceeding from the latter towards Sarīkol would have to follow the valley of the Ak-su for some distance to the north-east before he could reach a practicable route across the watershed range to the Tāsh-kurghān river.

No such difficulty arises if we assume that Hsüan-tsang's journey lay over the Great Pāmīr. From the latter, two main routes are open to the traveller whose goal is the inhabited centre of Sarīkol. He can either make his way in a generally eastern direction to the Naiza-tāsh Pass, the descent from which, in the Shindī valley running north-east, would bring him straight to Tāsh-kurghān. Or he may direct his route first into the valley of the Ak-su river, where it leaves the Little Pāmīr, and thence reach the upper portion of the Tāghdumbāsh

[13] The route here indicated may be conveniently traced in the map accompanying Dr. Hedin's *Reisen in Zentral-Asien* (drawn by Dr. Hassenstein) and in the *Map of Portions of Western China and Tibet explored by Captain H. H. P. Deasy*, published by the Survey of India Department, 1900. For a description of the route from Tāsh-kurghān eastwards as far as Unkurluk, see Hedin, *Through Asia*, ii. pp. 702 sqq. Dr. Hedin, who took six days from Tāsh-kurghān to Langar, one of the Tong villages, describes the summit or ridge of the Kandahār Pass (16,610 feet) as being 'as sharp as a knife' and experienced near it heavy snowfall on Sept. 19–20, 1905. I am unable to refer at present to Captain Deasy's *Three years in Tibet*, which contains an account of the same mountain tract.

I have not been able to trace any information as to the existence in this region of a lake corresponding to that mentioned by Sung Yün. But this could scarcely surprise us if, as I suspect, Sung Yün's reference is really to a spot which no longer contained a lake but only the legendary site of one. The story of the Nāga Suśravas, whose earlier lake habitation is placed by Kashmīr legend near the ancient city of Cakradhara, and who is believed to have subsequently banished himself to a distant mountain lake (Suśram Nāg, near the Amarnāth Peak), seems to offer an exact parallel. The story is told by Kalhaṇa, *Rājat.* i. 203–70, and has been discussed by me in my notes, *Rājat.* i. 201–3, 267.

[14] For a critical review of all available data bearing on the early topography and designations of Wakhān, Marquart, *Ērānšahr*, pp. 223 sqq., should be consulted.

[15] See *The Pamirs*, p. 69.

[16] *Si-yu-ki*, transl. Beal, ii. p. 297.

[17] See *Si-yu-ki*, transl. Beal, ii. p. 298; transl. Julien, ii. p. 209.

Pāmīr by going southwards, the watershed range being crossed here by a series of passes of which the Payik or Bayik Pass is the most frequented and easiest[18].

The initial portion of the route here sketched may be accomplished either by traversing the so-called Nicolas Range, which divides the Great and Little Pāmīrs, through the well-known depressions of the Benderski Pass (14,705 ft.) or the Urta Bēl Pass (14,090 ft.) or by turning the eastern end of that range over the probably still lower Kizil-Rabāt Pass[19]. Whichever the passes crossed, the general direction of the second route is south-east. This exact accord with the definite statement of Hsüan-tsang may, in view of his remarkable and often proved accuracy in the matter of bearings, be considered by itself a sufficiently strong argument for the belief that the pilgrim actually followed this route. But there are other indications, too, to support it.

Season of Hsüan-tsang's Pāmīr journey.

Several allusions in Hsüan-tsang's itinerary from Badakhshān to Khotan, as well as statements contained in his *Life*, show that he effected this portion of his return journey during the spring and summer[20]. At this period of the year the Great Pāmīr route is much preferred to that over the Little Pāmīr or by the Wakhjīr Pass, inasmuch as the ascent along the Pāmīr river, coming from Victoria Lake, offers none of the difficulties to be overcome in the Āb-i-Panja Valley, when the swelling of the stream renders the tracks in the narrow gorge between Sarhad and Langar almost impracticable for laden transport[21].

Now it is noteworthy that at the same season and through the same cause, viz. the melting snows of the mountains, similar difficulties arise beyond the Naiza-tāsh Pass which offers the nearest approach to Tāsh-kurghān from the side of the Great Pāmīr. The descent from the pass to the open Tāsh-kurghān Valley leads, for a distance of upwards of twenty miles, through the narrow rocky defile of the Shindī stream. This carries a very considerable quantity of water even early in the season, and having to be crossed and re-crossed in numerous places, according to the testimony of a very competent observer, renders 'the road exceedingly difficult for laden horses'[22]. On the other hand, the Payik Pass, though its elevation (15,078 ft.) is slightly greater than that of the Naiza-tāsh (14,920 ft.), is entirely free from the risk of such

[18] Compare for these passes, Lord Curzon's *The Pamirs*, pp. 57 sq.; *Report of Pamir Boundary Commission*, pp. 41 sqq.

[19] Compare for these lines of communication, Lord Curzon's *The Pamirs*, pp. 55 sq.; *Report of Pamir Boundary Commission*, pp. 17, 22.

[20] In the *Life* (transl. Beal, p. 196) we are told that Hsüan-tsang and his fellow-travellers, after traversing other parts of the old Tukhāra country (Tukhāristān), stopped in *Po-lo-chang-na* or Badakhshān 'on account of the frost and snow, for a month and some days'. Only after this long halt, which no religious or other obligations called for, did they march on to the Kokcha Valley or Yamgān (Yin-po-chien) and Kurān or Zebak (Ch'ü-lang-na) *en route* for Wakhān. It is thus clear that they awaited the spring in the sheltered central valley of Badakhshān (probably near the present Faizābād, see Yule, *J.R.A.S.*, 1872, pp. 109 sq.) before setting out for the passage of the Pāmīrs. Subsequently, when the pilgrim, after a twenty days' stay in Chieh-p'an-t'o or Sarīkol, was crossing the mountains to the north-east, probably by the Chichiklik route, we read of an attack by robbers, in the course of which 'the elephants being driven about in the pursuit, were engulphed in the water and perished'. This incident, related in the *Life*, p. 200, clearly points to the time of the summer floods. The gorges passed by that route only then contain enough water to account for such a loss (comp. Gordon, *Roof of the World*, p. 108).

In Hsüan-tsang's own narrative we find various references to climatic conditions of the mountains around Sarīkol which distinctly suggest personal observation during the season above indicated. Thus of the valley of Pa-mi-lo we are told that 'the snow falls both in summer and spring-time', while again in describing the passage already referred to through 'the eastern chain of the Ts'ung-ling mountains', north-east of Sarīkol, the traveller points out that 'even at the time of the great heat the wind and the snow continue' (see *Si-yu-ki*, transl. Beal, ii. pp. 297, 303; Julien, ii. pp. 207, 215).

[21] Speaking of this gorge Col. Gordon observes (*Roof of the World*, p. 129): 'In summer the swelling of the stream makes this road extremely difficult, and it is then that the Great Pāmīr route is followed in preference.' Compare also *Yarkand Mission Report*, p. 270.

[22] See Captain (now Colonel) H. Trotter's remarks in *Yarkand Mission Report*, p. 270.

obstruction and, as noted by a good authority, 'must on the whole be classed as exceptionally easy [23].'

The distance of 500 li, or approximately five marches, which the *Hsi-yü-chi* records for the journey from the midst of the Pa-mi-lo Valley to the kingdom of Chieh-p'an-t'o, agrees well with the route sketched above. According to the latest map of this part of the Pāmīr region prepared by the Survey of India, the journey from the eastern end of the Great Pāmīr by the Kizil-Rabāt and Payik Passes to where the Payik Valley debouches into the Tāghdumbāsh Pāmīr would cover about eighty-four miles, while a march of only some sixteen miles further down would bring the traveller to the village of Dafdār and the commencement of the cultivated part of the main Sarīkol Valley.

Route over the Wakhjīr Pass.

Past Dafdār and the approach to the Payik Pass there leads also the third alternative route connecting Sarīkol and Wakhān, to which we had occasion to refer in the opening of this chapter. It ascends the whole length of the Tāghdumbāsh Valley, and then crosses the Wakhjīr Pass at its head to the source of the Āb-i-Panja branch of the Oxus (Fig. 7). That part of the valley which properly bears the designation of Tāghdumbāsh Pāmīr may be said to extend from the Wakhjīr Pass to a short distance above Dafdār, where the river makes its sharp bend to the north. Forming an unbroken continuation of the central valley of Sarīkol, the Tāghdumbāsh Pāmīr, with the broad grassy flat along its bottom for a distance of upwards of fifty miles, looks as if created by nature for a convenient thoroughfare from Sarīkol to Wakhān.

The Wakhjīr Pass, close on 16,200 ft. above the sea, is, it is true, higher than either the Naiza-tāsh or the Payik Pass, and is certainly deep in snow in mid-winter. Its approaches, both from the east and the west, are, however, remarkably easy. Similarly, the descent in the Āb-i-Panja Valley to Bōzai-Gumbaz, where the Little Pāmīr route joins in, and further down to Langar offers no difficulty of any kind. In accord with these observations we find it attested by Col. H. Trotter's inquiries, made in 1874, that the Wakhjīr Pass was in former times 'much used by the Bajaori merchants who used to go from Badakhshān to Yarkand by the Tāghdumbāsh and Tung Valley roads'. Considering that for centuries past the commerce of the mountain region north and south of the Hindukush has to a very large extent been in the hands of these enterprising traders from Bajaor, the evidence recorded by Col. Trotter as to their usual route over the Tāghdumbāsh Pāmīr possesses some historical interest [24].

Supply difficulties on the Pāmīrs.

At any period of regular trade intercourse between Sarīkol and Wakhān, the Tāghdumbāsh Valley route was bound to receive attention on account of a topographical fact which deserves brief notice. There is one feature of the Pāmīrs which seems to have impressed travellers of all ages with equal thoroughness: it is the total absence of permanent habitations and the want of all local resources. We can trace this feeling of utter desolation and sterility through the accounts of the Chinese pilgrims no less than in the records of European travellers since the days of Marco Polo and Goëz. From the elaborate arrangements for supplies and shelter which were deemed necessary on every occasion when large parties of western visitors such as Col. Gordon's expedition or the Pāmīr Boundary Commission had to cross, or to camp on, the Pāmīrs [25], we may judge of the difficulties which must always have attended the movements of trading caravans or military bodies across that region.

In view of this serious obstacle in the matter of supplies, it must always have been an important consideration to travel, if other physical conditions permitted, by a route on which

[23] See Sir Th. Holdich in *Report of Pamir Boundary Commission*, p. 41.

[24] See *Yarkand Mission Report*, p. 270.

[25] Compare, e.g., Gordon, *Roof of the World*, pp. 124, 163 sq.

7

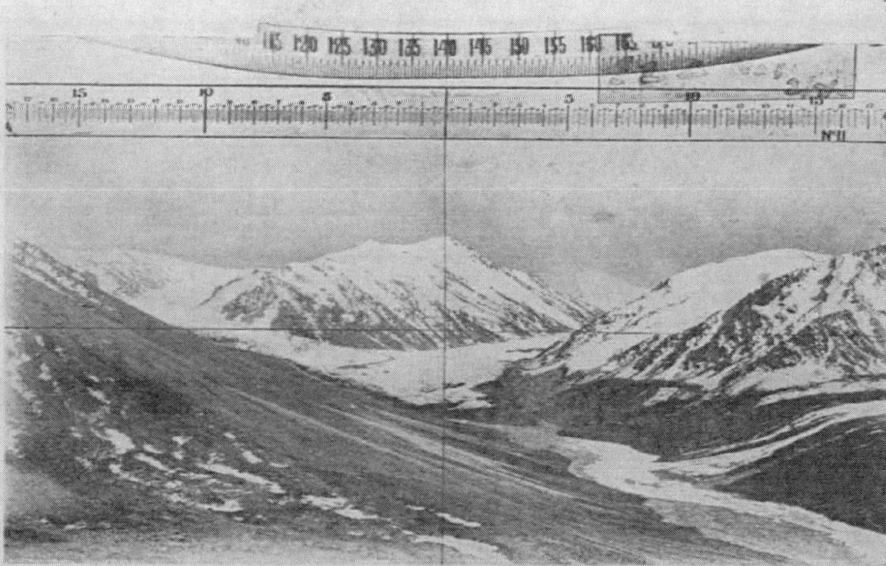

GLACIERS AT SOURCE OF OXUS, SEEN FROM APPROACH TO WAKHJĪR PASS.
(PHOTO-THEODOLITE VIEW.)

8

KILIK PASS, LOOKING SOUTH FROM KHUSH-BĒL SPUR.

the distance over wholly uninhabited ground was reduced as much as possible. Now in this respect the route over the Wakhjīr Pass offers an undoubted advantage. If starting from Langar, the highest point in the Āb-i-Panja Valley where village remains and traces of old cultivation have been observed [26], a journey of scarcely more than 106 miles would suffice to bring the traveller to Dafdār, and thus to the commencement of cultivated ground in the main Sarīkol Valley. The shortest route from the same starting-point over the Little Pāmīr and the Naiza-tāsh Pass to Tāsh-kurghān measures fully 120 miles, while on the road that leads from the latter place by the Great Pāmīr to the central part of Wakhān, some 180 miles have to be traversed before a permanently inhabited locality (Langar-Kisht) is reached [27].

SECTION III.—HISTORICAL SITES OF SARĪKOL

From this review of the Tāghdumbāsh Valley route, which was needed to complete our survey of the old lines of communication between Sarīkol and Wakhān, we may turn once more to Hsüan-tsang's itinerary, and to the account it furnishes of Sarīkol itself or 'the kingdom of Chieh-p'an-t'o'. Before we examine the details which possess a definite antiquarian bearing, it will be well to note how closely the pilgrim's general description of the territory agrees with the natural features observed to this day.

Hsüan-tsang's description of Sarikol.

In Chieh-p'an-t'o, to which is ascribed an approximate circuit of 2,000 li, or about twenty days' marches, 'the mountain chains run in continuous succession, the valleys and plains are very contracted. There is very little rice cultivated, but beans and corn grow in abundance. Trees grow thinly, there are only few fruits and flowers. The plateaux are soppy, the hills are waste, the towns are deserted.' [1] The account we receive of the inhabitants is in keeping with these stern surroundings. Their manners are described as 'without any rules of propriety', and very few of them gave themselves to study. 'They are naturally uncouth and impetuous, but yet they are bold and courageous. Their appearance is common and revolting; their clothes are made of woollen-stuffs.' Notwithstanding the coarseness which Hsüan-tsang's description indicates, and which the material conditions of life among a hill community thus placed adequately account for, the people of Chieh-p'an-t'o had received their share in the benefits of Buddhist religion and culture. 'They know how to express themselves sincerely, and they greatly reverence the law of Buddha.' There were some ten convents, in which about five hundred monks studied 'the Little Vehicle according to the school of the Sarvāstivādas'; and in regard to the letters in use we are told that they much resembled those of the *Ch'ia-sha* country or Kāshgar.

Traditions of Chieh-p'an-t'o.

Hsüan-tsang, according to the statement of his biographer Hui-li, remained in Sarīkol for about twenty days [2], and to this halt we probably owe the detailed information he gives of the traditions and legends attaching to the royal family and the capital of the country. The account recorded by him as to the origin of the former is of particular interest, for it proves the high antiquity which popular belief in Sarīkol then ascribed to the race of its rulers, while we can still trace a characteristic feature of the story in a legend actually surviving at the identical locality.

[26] See Gordon, *Roof of the World*, p. 129. It also deserves notice that firewood, another requirement of the traveller, is plentifully found up to Langar, see ibid. and *Report of Pamir Boundary Commission*, p. 16.

[27] Compare Gordon, *Roof of the World*, pp. 153, 163.

[1] See *Si-yu-ki*, transl. Beal, ii. pp. 298 sq.; Julien, ii. p. 209. The latter version has: 'Les plaines hautes et basses sont désertes, les villes et les villages sont [presque] inhabités.'

[2] See *Life of H. T.*, transl. Beal, p. 200; *Vie de Hiouen-Thsang*, transl. Julien, p. 274.

After describing the reigning king of Chieh-p'an-t'o as a man of upright character, a pious Buddhist, and a lover of learning, Hsüan-tsang proceeds to tell us that 'since the establishment of the kingdom many successive ages have passed. The king gives himself the title *Chih-na-t'i-p'o-ch'ü-tan-lo* (*Cīna-deva-gotra*), meaning descendant of China and the sun-god[3]. Formerly this country was a desert valley in the midst of the Ts'ung-ling mountains. At this time a king of *Po-la-ssŭ* (Persia)[4] took a wife from the Han country (China). She had been met by an escort on her progress so far as this, when the roads east and west were stopped by military operations. On this they placed the king's daughter on a solitary mountain peak, very high and dangerous, which could only be approached by ladders, up and down; moreover, they surrounded it with guards both day and night for protection. After three months the disturbances were quelled. Quiet being restored, they were about to resume their homeward journey, when the lady was found to be enceinte'.[5] Thereupon the king's envoy held council with his companions how to meet the consequences of this disgrace. From an attendant he learned that a spirit, coming from the sun's disk and mounted on horseback, every day at noon visited the princess. Afraid of the punishment awaiting him on return to his own country, the envoy decided to seek safety by remaining and gaining time.

Legend of foundation of dynasty.

'On this he built on the top of a rocky peak a palace with its surrounding apartments[6]; then having erected an enclosure round the palace of some 300 paces, he located the princess there as chief. She established rules of government and enacted laws.' In due time she bore a son possessed of extraordinary beauty and miraculous powers, who made his strength felt in the countries around and everywhere gained recognition for his laws. When this king died of old age he was buried in a stone chamber constructed in the caverns of a great mountain, about a hundred li to the south-east of his capital. There his body was believed to remain undecayed, though shrivelled up and thin, looking as if he were asleep, and worshipped with regular oblations.

'From that time till now his descendants have ever recollected their origin: that on their mother's side they were descended from the king of Han, and on their father's side from the race of the sun-god; therefore they style themselves "descendants of the Han and sun-god"'[7]. Hsüan-tsang adds the observation that the members of the royal family resembled in their bodily appearance the 'people of the Middle Country', i. e. China, but notes that the dress worn by them was that of barbarians. For this he evidently means to account by stating that 'in after ages these people fell under the power of barbarians'[8].

Legend of Kiz-kurghān.

We are unable even approximately to gauge what shreds of historical fact may be interwoven with the legend which Hsüan-tsang heard about the family ruling Sarīkol at the time of his visit. But we can scarcely doubt that this legend was widespread and firmly rooted in popular belief; for we find an unmistakable trace of it preserved in local tradition to the present day. To

[3] I take the translation of this sentence from Julien, whose version is evidently more accurate than that of Beal (see *Mémoires*, ii. p. 210). The simple *deva*, 'god,' in the transcribed Sanskrit title is amplified by the Chinese text into 'sun-god'.

[4] Thus Beal, *Si-yü-ki*, ii. p. 300. Julien, in translation and index (ii. pp. 210, 524, resp.), spells the name *Po-li-sse* 波利斯; but the graphic variation of the second character, *li* 利, from the character *la* 剌 in *Po-la-sse*, the usual Chinese transcription of the name *Parsa or Persia, is so slight that there can be little doubt as to the latter country being intended.

[5] Julien, *Mémoires*, ii. pp. 210 sq., translates: 'Au bout de trois mois les brigands restèrent tranquilles. L'ambassadeur voulut alors emmener la princesse de Chine dans les états de son maître; mais la jeune fille se trouvait déjà enceinte.'

[6] Julien, *Mémoires*, ii. p. 212: 'Alors il bâtit, sur le sommet de la montagne, un palais et un hôtel.'

[7] See above note concerning this title *Cīna-deva-gotra*.

[8] See *Mémoires*, transl. Julien, ii. p. 213; *Si-yu-ki*, transl. Beal, ii. p. 302.

the remains of ancient stone walls perched on precipitous cliffs, which rise above the left bank of the Tāghdumbāsh river some seven miles above Dafdār, there clings the story commonly known to Sarīkolīs and Kirghiz alike, that king Naushīrwān, an ancient Persian ruler, had once placed there his daughter for safety. On account of this story the ruins bear the popular designation *Kiz-kurghān*, meaning in Turkī 'the tower of the daughter (or princess)'.

Owing to an unfortunate chance I did not hear of the existence of these ruins until after I had reached Tāsh-kurghān (July 7, 1900), having failed to notice them two days earlier when I passed the site on my march down the valley. Want of time did not permit me to retrace my steps a distance of some fifty miles, however anxious I was to examine them personally. But I was at least able by repeated inquiries to make sure of their approximate position, and of the widely spread knowledge of the legend attaching to them.

Position of Kiz-kurghān ruins.

There can be no doubt that we have in the Kiz-kurghān legend a genuine relic of the fuller tradition current in Hsüan-tsang's days, and consequently we are justified in attributing historical significance to the place in which we find it localized. In the light of the story as recorded by Hsüan-tsang, popular tradition could not have supposed the Kiz-kurghān ridge to have been the temporary place of safety selected for the Chinese princess while the road westwards was blocked, unless at the time when the story was current a main route in that direction passed up the Tāghdumbāsh Pāmīr at the entrance of which the ridge rises. Kiz-kurghān, like the modern fortified post of Ghujak-bai (Ujad-bai of the maps), which nearly faces it on the opposite bank of the river, must, in fact, be passed by all travellers who wish to reach the Upper Oxus, whether by the Wakhjīr or the Payik Pass. Thus the legend localized at Kiz-kurghān affords direct evidence that the Tāghdumbāsh Pāmīr was used as a general line of communication in ancient times, and further helps to support the assumption explained above that Hsüan-tsang himself travelled by it on his way to Tāsh-kurghān.

Site of ancient capital at Tāsh-kurghān.

From the account of the *Hsi-yü-chi*, already quoted, we learn that the palace in which the Han princess and her miraculously conceived son, the founder of the dynasty, first established themselves and ruled Sarīkol, was built 'on the top of a rocky peak'. In the absence of any notice to the contrary, we may assume that the site of this royal palace was within the capital of Chieh-p'an-t'o which Hsüan-tsang visited, and of which he tells us that it 'rests on a great rocky crag of the mountain, and is backed by the river Śītā'[9]. The position here indicated agrees so closely with that of the present Tāsh-kurghān that the identification of the latter with the old capital of Chieh-p'an-t'o, first proposed by Sir Henry Yule, may be considered as certain[10].

The modern Chinese fort of Tāsh-kurghān and the ruined town around it occupy part of a long rocky plateau or terrace which is washed along its east foot by the Tāghdumbāsh river. That the latter is meant by Hsüan-tsang's *Hsi-to*, is clear from an earlier passage of the *Hsi-yü-chi*; there this name, a transcription of the Sanskrit Śītā, is applied to the Yarkand river, of which the Tāghdumbāsh river is one of the main feeders[11]. Hsüan-tsang's indication is borne out by the account of Sung Yün, who, as already noted, also speaks of the capital of Han-p'an-t'o or Sarīkol as having a river along its east side. Though he calls this river by a different name, Mêng-chin, yet he leaves no doubt as to its identity with the Tāghdumbāsh river; for he mentions that it flows to the north-east in the direction of Sha-lei or Kāshgar[12].

[9] See *Mémoires*, transl. Julien, ii. p. 209; *Si-yu-ki*, transl. Beal, ii. p. 298.

[10] See Yule, Introduction to Wood's *Journey to the sources of the Oxus*, p. xlviii; for a first account of my observations confirming this identification, see *Prelim. Report*, pp. 11 sq.

[11] See *Mémoires*, transl. Julien, ii. p. 208; *Si-yu-ki*, transl. Beal, ii. p. 298.

[12] See Chavannes, *Voyage de Song Yun*, p. 23.

Advantages of Tāsh-kurghān site.

In a previous section I have already explained the topographical facts which make the vicinity of Tāsh-kurghān the political centre of the Sarīkol region, and the natural meeting place for all the routes from the Tārīm Basin to the Upper Oxus. In the fertile riverine flat to which the bottom of the main Sarīkol valley widens out just before the river takes its sharp turn to the east and enters the narrow gorge of Shindī, there is no position offering greater advantages for a settlement, capable of defence and yet easily accessible, than the site of Tāsh-kurghān. It occupies a narrow but well-defined plateau of conglomerate cliffs rising immediately above the broad bed of the river and at the extreme eastern edge of a fertile plain, from three to four miles broad, which spreads from the foot of the high range westwards. The numerous irrigation canals fed by the Shingun river, which comes down from the Naiza-tāsh Pass and debouches here into the valley, render this plain the most cultivated part of Sarīkol. It is probable that the silt brought down by these canals has helped to reduce not inconsiderably the difference in level between the site of Tāsh-kurghān town and the adjoining plain. But seen from the wide expanse of meadow land to the east and south over which the river spreads itself in numerous branches, the commanding position of the plateau is fully realized; the rise of its steep banks to a height of about a hundred feet suffices to render the walls that crown them a conspicuous object from afar.

Walls of Tāsh-kurghān.

The area enclosed by these walls, as seen in the plan (Plate XIX), may be thoroughly described as an irregular quadrangle, having a circumference of about one mile. It comprises the highest portion of the plateau, being separated from its continuations both to south and north by shallow depressions, in which the drainage of the plain behind finds its way towards the river. A small portion of this area, on the east side facing the river, is occupied by the modern Chinese fort visible in Fig. 10. With the exception of the portion where its high and carefully plastered walls of sun-dried bricks hide the earlier foundations, the edge of the plateau shows everywhere the remains of massive stone walls now crumbling to ruin (see Fig. 9). Only unhewn stones of varying size seem to have been used in their construction. Large blocks are to be found, particularly in the foundations; these, however, owing to the quantity of débris encumbering the slopes, are traceable only at a few points. The walls are best preserved on the north and west faces; elsewhere they show wide gaps, attributed to a severe earthquake which is said to have occurred some thirty years before my visit.

The interior of the circumvallated area is strewn with the crumbling remains of houses, found particularly thick towards the south side. These rubble-built dwellings were tenanted as long as the risk of raids from Hunza made it impossible for the scanty colonies of cultivators to live near their fields. Since peace has come to Sarīkol and the present fort of Tāsh-kurghān was occupied by the Chinese, in 1892, new villages have sprung up near all the cultivated places, and the stronghold has become deserted. When the earthquake of 1895 shook down most of the dwellings there was no need to rebuild them.

The walls of the town had already suffered by earlier earthquakes, and from general decay, which continued unchecked during the disturbed conditions prevailing throughout Sarīkol for some time previous to Yāqūb Bēg's rebellion, and again after its collapse[13]. Rebuilt undoubtedly again and again after successive periods of neglect, and always of unhewn stone, the walls now in ruins cannot afford any distinct criterion of age. But the high mounds of débris over which the extant walls rise, in some places to a height of over twenty-five feet, plainly indicate that these fortifications mark the lines of far more ancient ones.

[13] For a brief account of the state of ruin which the town of Tāsh-kurghān presented in 1874 to members of Col. Gordon's party, see *Yarkand Mission Report*, pp. 267 sq.

9

INTERIOR OF RUINED TOWN, TĀSH-KURGHĀN, LOOKING NORTH-EAST TOWARDS SHINDĪ DEFILE.

10

CHINESE FORT WITHIN RUINED TOWN, TĀSH-KURGHĀN.

Local traditions of Tāsh-kurghān.

Local tradition in Sarīkol uniformly asserts the great antiquity of Tāsh-kurghān town, known by its old name as *Varshadeh*, and ascribes its foundation to Afrāsiyāb, the king of Tūrān, who figures so prominently in Persian epic legend[14]. The fact that I could not hear of finds of old coins nor of other antiquities at the site in no way discredits the traditional belief in its early date; for the continued occupation of the site must have resulted in any ancient remains becoming deeply buried beneath accumulations of rubbish, from which the scanty rainfall of this region is quite insufficient to bring them again to light. Nor need we feel in doubt about the identity of Tāsh-kurghān with Hsüan-tsang's capital of Sarīkol, on account of the discrepancy which the actual circumference of the walls shows from the measurement of twenty li, or about four miles, recorded by the Chinese pilgrim. In days of greater population and prosperity the area occupied by the town must have extended further, either over the lower portions of the plateau already referred to, or over part of the adjoining plain.

Structures mentioned by Hsüan-tsang.

This difference in extent renders it difficult to arrive at any definite opinion as to the position which the royal palace and the various structures mentioned by Hsüan-tsang are likely to have occupied. We have already seen that the palace, in which, according to the legend above reproduced, the Han princess with her miraculously born son was supposed to have been established, is said to have had 'an enclosure of some 300 paces'. If double paces are meant this measurement would curiously accord with the circumference (about 1,300 ft.), shown by the present fort built within the ruined town, probably on earlier foundations.

'When Aśoka Rājā was in the world, he built in the very centre of this palace a Stūpa. Afterwards, when the king changed his residence to the north-east angle of the royal precinct, he built in the position of this old palace, for the sake of Kumāralabdha (T'ung-shou), a convent remarkable for the height and largeness of its towers and pavilions. The statue of Buddha (placed in this convent) was of majestic appearance. The venerable Kumāralabdha was a native of Takṣaśilā. From his childhood he showed a rare intelligence, and in early life gave up the world, &c.'[15] Hsüan-tsang then records the spiritual excellences which made Kumāralabdha renowned as a great teacher in the north, like Aśvaghosha in the east, Deva in the south, and Nāgārjuna in the west. 'The king of this country (Chieh-p'an-t'o), therefore, having heard of the honourable one and his great qualities, raised an army, made his troops attack Takṣaśilā and carried him off by force. He then built this convent and thus manifested the admiration with which he (Kumāralabdha) inspired him.'

The account here given is of interest as showing that even little Sarīkol, in its alpine isolation, could boast of a tradition connecting one of its convents with a great luminary of the Buddhist Church in India[16]. But it does not help us to fix the position of the convent itself, or the old palace, the site of which it was believed to mark. I had no opportunity of visiting the interior of the Chinese fort, and I doubt whether in this great pile of solid clay and sun-dried brick, which the building operations of successive ages have helped to raise to a height of sixty to eighty feet above the ground level of the adjoining old town, the remains of any ancient structure, however imposing originally, could now be traced without extensive excavations.

[14] Compare *Yarkand Mission Report*, p. 269.

[15] I follow Julien's translation, *Mémoires*, ii. pp. 213 sq., which seems more exact than that of Beal, *Si-yu-ki*, ii. p. 302.

[16] The somewhat violent procedure by which the ruler of Chieh-p'an-t'o is said to have possessed himself of the person of the great doctor curiously reminds one of the stories told along the Afghān frontier of the queer methods by which Pathān clans made sure of retaining much-venerated 'Pīrs'. Compare, e.g. in Darmesteter, *Lettres de l'Inde*, the story of the saint who was murdered in order that the village which he had blessed with his presence while alive might more securely enjoy in the future the miracle-working benefit of his dead body.

Ancient mound near Tāsh-kurghān Fort.

Outside the fort, but within the ruined town walls, the accumulated masses of débris effectively hide whatever substructures of ancient buildings the ground may contain. But beyond the north-western face of the circumvallation, and at a distance of about 150 yards, there rises from the level ground of the shallow nullah that here cuts through the conglomerate plateau, a circular mound of manifestly artificial origin, which looks like a much-decayed Stūpa (see Plate XIX). The present height of the mound is above 30 feet, and its diameter over 200 feet, which shows that the original structure must have been of considerable dimensions. The mound is composed of rough stones, with layers of mortar between them, and for the sake of the latter it is being dug into by the Chinese soldiers garrisoning the fort. The manner of construction which these diggings reveal seems to agree closely with that observed in the internal structure of many an ancient Stūpa built in parts of the Punjāb and the Indian north-west frontier where stone material was readily obtainable. Yet, notwithstanding the proof of antiquity thus afforded, it appears to me doubtful, having regard to the absence of other ruined remains in the immediate vicinity, and to the situation of the mound, whether its suggested identification with the Stūpa which tradition ascribed to Aśoka would be justified[17].

Legend of Afrāsiyāb.

Thus no clear indication remains of the buildings which the tradition of Hsüan-tsang's days connected with the original residence of the founder of the dynasty. But of the legends which were told of the latter a trace survives to this day in the name Afrāsiyāb, given to the high and conspicuous mountain spur which projects into the valley east and south-east of Tāsh-kurghān. We have noted already that the present tradition of Sarīkol knows king Afrāsiyāb, the legendary king of Tūrān in the Īrānian epic, as the founder of Varshadeh or Tāsh-kurghān. On the other hand, it may be remembered that the legend heard by Hsüan-tsang located the mysterious resting-place of the first king of Chieh-p'an-t'o in the cavern of a great mountain, a hundred li, or a day's march, to the south-east of his capital. Though the legend in this form is no longer heard, it is evident that it accounts for the application of the name of Afrāsiyāb to the mountain spur rising in the position indicated by the pilgrim.

Īrānian folk-lore in Sarīkol.

In view of what has been explained above about the racial character of the Sarīkol population, it cannot surprise us that whatever I could learn in the way of local traditions shows close dependence on the legendary lore of Īrān. We have already seen how the name of Naushīrwān, the hero of the classical Persian epic, is introduced by popular tradition into the ancient legend of Kiz-kurghān. Other names famous in Persian romance are associated with the local tradition related to me about an ancient irrigation canal, which formerly carried the waters of the Tāghdumbāsh river from near Dafdār along the foot of the hills towards Tughlān-shāhr, a large collection of hamlets opposite to Tāsh-kurghān. The walls supporting this canal, which are now breached in many places, are said to be built of hewn stones. The story goes that Shīrīn, a lady who lived at Varshadeh, told her lover Farhād that she would accept his suit if he could conduct a watercourse to the fields of Tughlān-shahr large enough to sweep down a cow. Farhād built the canal now in ruins; but though its current was not sufficiently strong to fulfil the lady's condition, Farhād attained his object by placing the hide of a cow filled with straw in the water, which easily carried it down its course.

Time did not permit me to trace the remains of this old irrigation work which, no doubt, enabled much of the fertile open ground, now wholly deserted, along the right river bank above Tughlān-shahr to be cultivated. As evidence of the large population which this tract is supposed to have once supported, I was told that a piece of ground situated between Yurgāl

17 See *Prelim. Report*, p. 11.

Gumbaz and Tughlān-shahr, and known as *Bāzār-dasht*, still retains the remains of shops built in rows and of sun-dried bricks, showing that the spot was once occupied by a Bāzār.

In connexion with the reminiscences of Persian lore just mentioned, it may be pointed out that the name *Varshadeh*, used as an old designation of Tāsh-kurghān by the inhabitants speaking Sarīkolī, also clearly bears the stamp of Īrānian origin.

Apart from the capital we find two more localities specified in Hsüan-tsang's account of Sarīkol. One of these, 'a great rocky scarp' with two rock chambers, in each of which an Arhat was shown plunged in complete ecstasy, yet with his body undecayed in spite of long centuries, cannot be identified. It is described as situated some 300 li to the south-east of the city, and may, therefore, be looked for among the high ranges which are drained by the Vacha river, the nearest affluent of the Tāsh-kurghān river from the south-east. Sacred sites named by Hsüan-tsang.

The other locality was a hospice for travellers, or *Puṇyaśālā*, which the pilgrim reached after leaving the capital towards the north-east and marching for 200 li across mountains and along precipices [18]. It is described as being situated in 'a space comprising some hundred *ch'ing* (thousand Chinese acres), in the midst of the four mountains belonging to the eastern chain of the Ts'ung-ling mountains'. 'In this, both during summer and winter, there fall down piles of snow; the cold winds and icy storms rage. The ground, impregnated with salt, produces no crops, there are no trees, and nothing but scrubby underwood. Even at the time of the great heat the wind and the snow continue. Scarcely have travellers entered this region when they find themselves surrounded by vapour from the snow [19]. Merchant caravans, in coming and going, suffer severely in these difficult and dangerous spots.'

Hsüan-tsang then relates 'an old story' how once a great troop of merchants, with thousands of followers and camels, perished here by wind and snow. An Arhat of Chieh-p'an-t'o, having failed to rescue them in time, collected all the precious objects left behind by the caravan and constructed a house, in which he accumulated ample stores. He also bought land in the neighbouring territories, and with its proceeds provided houses in the bordering towns for the accommodation of travellers.

Taking into account the direction of the pilgrim's route after leaving Tāsh-kurghān and the distance indicated, it is clear that the position of the hospice must be identified with the Chichiklik plateau, a broad elevated valley, which the main route from Sarīkol towards Kāshgar and Yarkand crosses at a distance of two marches from Tāsh-kurghān. Chichiklik is the point which all travellers in the above direction must traverse by whatever passes (Yangi-Dawān, Yambulak, or Chichiklik-Dawān) they may surmount the second of the great mountain ranges which radiate southwards from Muztāgh-Ata. A reference to the detailed account of this much-frequented route given by Sir D. Forsyth's Mission [20], or to Lord Curzon's Map of the Pāmīrs, will illustrate this remark. Chichiklik plateau.

The important position of Chichiklik, as the natural halting-place between the passes leading over the first and second of those ranges, and its high elevation, explain the provision of a hospice at this point. Though I have not been able to trace any exact observation as to the height of Chichiklik, it may safely be concluded from Colonel Trotter's description and the elevations ascertained for the passes by which it is approached on either side (Kök-moinak Pass, 15,670 feet, to south-west, and the almost imperceptible watershed to the north-east, 14,480 feet) [21], that the

[18] See Julien, *Mémoires*, ii. p. 215; Beal, *Si-yu-ki*, ii. p. 303.

[19] Julien has 'au milieu des vapeurs et des nuages'.

[20] Compare *Yarkand Mission Report*, pp. 267, 433.

[21] See *Yarkand Mission Report*, pp. 267, 357 sq.

Chichiklik plain cannot be less than 14,000 feet above the sea. Whatever the exact equivalent of Hsüan-tsang's measurement of a hundred *ch'ing* may be, it is clear that the reference is to an open level space; and as such spaces are rare indeed along the route leading from Sarīkol through the mountains north-eastward, we can understand why this feature of Chichiklik, a plain about $1\frac{1}{2}$ miles in diameter, should have found special mention in the pilgrim's narrative.

Whether the remains of a large Sarai such as Hsüan-tsang's narrative seems to imply can still be traced at Chichiklik, I must leave for investigation by some future visitor to the spot. But the substantial truth of the description he gives of the forbidding nature of this high plateau, and of the hardships there suffered by travellers, is strikingly brought home to us by what we read of the experiences of a pious traveller, who followed his track across Chichiklik nearly a thousand years later.

Route of Goëz through Sarīkol. Benedict Goëz, the lay Jesuit, whom missionary zeal sent, in 1603, from the court of Akbar in search of fabled Cathay, travelled like Hsüan-tsang from India to the Upper Oxus by way of Kābul and Badakhshān. The record of the journey, compiled after Goëz' death from his notebook, is very scanty, and leaves us in doubt as to the route by which he crossed the Pāmīrs. But we can clearly trace his steps to Tāsh-kurghān, when we read how, after a twenty days' journey through desolate high mountains, he and the large 'Kāfila' of merchants to which he had attached himself, 'reached the province of Sarcil (Sarīkol), where they found a number of hamlets near together. They halted there two days to rest the horses, and then in two days more they reached the foot of the mountain called Ciecialith (Chichiklik). It was covered deep with snow, and during the ascent many were frozen to death, and our brother himself barely escaped, for they were altogether six days in the snow here. At last they reached Tanghetar (Tangitar), a place belonging to the kingdom of Cascar (Kāshgar). . . . In fifteen days more they reached the town of Yakonich (Yaka-arik). . . . After five days more our Benedict reached the capital which is called Hiarchan (Yarkand)'[22].

The several stages of the itinerary here given have long ago been correctly identified by Sir Henry Yule; and laconic as Goëz' record is, it amply shows that the dread of the Chichiklik plateau as reflected in Hsüan-tsang's story was by no means unfounded.

Section IV.—FROM SARĪKOL TO KĀSHGAR

Route followed to Kāshgar. On leaving Tāsh-kurghān (July 10, 1900), my steps, like those of Hsüan-tsang, were directed towards Kāshgar. But the route which I followed, between the eastern buttresses of the Pāmīrs and the great Muztāgh-Ata range, was chosen mainly for its geographical interest. Leading through elevated alpine valleys, and further down through the narrow and terribly barren gorge of the Yamān-yār river, regions which could never have supported permanent settlements of any size, it was not likely to offer scope for antiquarian observations. In chapters v–vii of my Personal Narrative I have given a detailed description of this route, which took me past Muztāgh-Ata to Lake Little Kara-kul, and then round the foot of the great glacier-crowned range northward into the Gez defile, finally debouching at Tāshmalik into the open plain of Kāshgar. Though scarcely more difficult than the usual route over the Chichiklik Pass and by Yangi-

[22] See Sir H. Yule's translation of Trigault's *De Christiana expeditione apud Sinas*, embodying Goëz' notes as put together by Ricci, in *Cathay*, ii. p. 562. 'Yakonich,' which Sir H. Yule could not identify, manifestly contains the misspelt name of the large village Yaka-arik, south-west of Yarkand, passed on the route from Chichiklik.

Hisār, it is certainly longer and leads for a considerably greater distance over ground which is devoid of cultivation or permanent habitations[1].

Marco Polo's route to Kāshgar.

It is the latter fact which makes me believe that Professor H. Cordier was right in tracing by this very route Marco Polo's itinerary from the central Pāmīrs to Kāshgar. The Venetian traveller, coming from Wakhān, reached after three days a great lake which may be either Lake Victoria or Lake Chakmak, at a 'height that is said to be the highest place in the world'. He then describes faithfully enough the desert plain called 'Pamier', which he makes extend for the distance of a twelve days' ride, and next tells us: 'Now if we go on with our journey towards the east-north-east, we travel a good forty days, continually passing over mountains and hills, or through valleys, and crossing many rivers and tracts of wilderness. And in all this way you find neither habitation of man, nor any green thing, but must carry with you whatever you require[2].'

This reference to continuous 'tracts of wilderness' shows clearly that, for one reason or another, Marco Polo did not pass through the cultivated valleys of Tāsh-kurghān or Tagharma, as he would necessarily have done if his route to Kāshgar, the region he next describes, had lain over the Chichiklik Pass. We must assume that, after visiting either the Great or Little Pāmīr, he travelled down the Ak-su river for some distance, and then crossing the watershed eastwards by one of the numerous passes struck the route which leads past Muztāgh-Ata and on towards the Gez Defile. In the brief supplementary notes contributed to Professor Cordier's critical analysis of this portion of Marco Polo's itinerary, I have pointed out how thoroughly the great Venetian's description of the forty days' journey to the E.N.E. of the Pāmīr Lake can be appreciated by any one who has passed through the Pāmīr region and followed the valleys stretching round the Muztāgh-Ata range on the west and north[3]. After leaving Tāsh-kurghān and Tagharma there is no local produce to be obtained until the oasis of Tāshmalik is reached. In the narrow valley of the Yamān-yār river, forming the Gez Defile, there is scarcely any grazing; its appearance down to its opening into the plain is, in fact, far more desolate than that of the elevated Pāmīr regions.

Length of Marco Polo's route.

In the absence of any data as to the manner and season in which Marco Polo's party travelled, it would serve no useful purpose to hazard explanations as to why he should assign a duration of forty days to a journey which for a properly equipped traveller need not take more than fifteen or sixteen days, even when the summer floods close the passage through the lower Gez Defile, and render it necessary to follow the circuitous track over the Tokuz-Dawān or 'Nine Passes'. But it is certainly worth mention that Benedict Goëz, too, speaks of the desert of 'Pāmech' (Pāmīr) as taking forty days to cross if the snow was extensive,

[1] The total distance from Tāsh-kurghān to Kāshgar (Old Town) by Chichiklik and Yangi-Hisār is reckoned by Col. Trotter at 168 miles. Cultivated ground is reached after a march of about 111 miles at the large village of Ighizyār (compare *Yarkand Mission Report*, pp. 432 sq.). The length of the route Tāsh-kurghān–Little Kara-kul–Gez Defile–Kāshgar is estimated by me on the basis of my notes and our survey at about 190 miles, if the direct track in the bed of the lower Yamān-yār river is followed. When the summer floods render this passage impracticable, as at the time of my journey, the trying and devious track over the Tokuz-Dawān ('Nine Passes') must be taken, adding about forty-six miles to the total length. There is practically no cultivation to be found until we arrive at Tāshmalik, about forty miles' distance from Kāshgar by the nearest route. Notwithstanding its greater length, the Gez–Kara-kul route is held to be preferable in the winter season; for the Ulūgh-Rabāt Pass, only a little over 14,000 feet in height, is then apparently far less formidable an obstacle than the Chichiklik Pass, and an easy descent in the Gez Defile is also assured at that season. But owing to the difficulties about supplies, only small parties ever frequent it.

[2] See Yule, *Marco Polo*, i. pp. 171 sq., with Prof H. Cordier's notes, ibid., i. pp. 175, 182.

[3] Compare Yule, *Marco Polo*, ii. pp. 593 sq.

an accord already noted by Sir H. Yule[4]. It is also instructive to refer once more to the personal experience of the missionary traveller on the alternative route by the Chichiklik Pass. According to the record quoted above, he appears to have spent not less than twenty-eight days in the journey from the hamlets of 'Sarcil' (Sarīkol, i.e. Tāsh-kurghān) to 'Hiarchan' (Yarkand)—a distance of some 188 miles, now reckoned at ten days' march[5].

Reference has already been made to the fact that the usual route from Sarīkol to Kāshgar leads over the Chichiklik Pass and thence by Yangi-Hisār; I believe we can show that Hsüan-tsang, too, followed this route. In our previous analysis of the pilgrim's itinerary we traced his steps as far as the Chichiklik plateau, which he correctly places at 200 li, or two daily marches, from the capital of Sarīkol. The record of the *Hsi-yü-chi* next tells us that the pilgrim 'going east from this (the locality of Chichiklik) descended from the eastern chain of the Ts'ung-ling mountains, passed dangerous defiles, traversed deep valleys, and followed paths full of precipitous places. Attacked in turn by wind and snow, he travelled 800 li or so, emerged from the Ts'ung-ling mountains and arrived in the kingdom of Wu-sha'[6].

Hsüan-tsang's notice of *Wu-sha*.

This territory is described as being about 1,000 li (or ten days' journey) in circuit, and bordering on the south on the river Hsi-to (Śītā or Yarkand river). The chief town, the position of which is left undefined, must have been a comparatively small place, as its circumference is estimated at only ten li, or about two miles. 'The soil is rich and productive; it is regularly cultivated and yields abundant harvests. The trees of the forests spread their foliage afar, and flowers and fruits abound. The country produces jade of different sorts in great quantities; white jade, black, and green. The climate is soft and agreeable; the winds and rain follow in their season; the manners of the people are not much in keeping with the principles of politeness. The men are naturally hard and uncivilized; they are greatly given to falsehood, and few of them have any feeling of shame. Their language and writing are nearly the same (Julien: *ressemblent un peu*) as those of *Ch'ia-sha* (Kāshgar). Their personal appearance is low and repulsive. Their clothes are made of skins and woollen stuffs. However, they have a firm faith in the law of Buddha and greatly honour him. There are some ten convents, with somewhat less than 1,000 priests. They study the Little Vehicle according to the school of the Sarvāstivādins. For some centuries the royal line has been extinct. They have no ruler of their own, but are in dependence on the country of Chieh-p'an-t'o (Sarīkol)[7].'

Identification of *Wu-sha*.

The topographical data which this description of Wu-sha contains are meagre, and we cannot supplement them from the Chinese historical records of the T'ang period; for, owing probably to its long-continued subjection to Chieh-p'an-t'o, or Sarīkol, the territory of Wu-sha does not appear to be separately named in them[8]. Fortunately a comparison of the map with the little

[4] Sir H. Yule's note on a letter of Goëz, *Cathay*, ii. pp. 563 sq.; Lord Curzon's *The Pamirs*, p. 73.

[5] Compare above, p. 40; *Yarkand Mission Report*, pp. 433 sqq.

[6] I have followed above Julien's translation, *Mémoires*, ii. p. 216, which appears more accurate. Beal gives the direction as north-east, makes the pilgrim descend the Ts'ung-ling mountains eastward, and states the length of the whole journey as 100 li, which is manifestly an error; see *Si-yu-ki*, ii. p. 304.

[7] Compare Beal, *Si-yu-ki*, ii. pp. 304 sq.; Julien, *Mémoires*, ii. pp. 216 sq.

[8] A note of M. Chavannes, in his lucid translation and analysis of the records of Sung Yün's travels (*Voyage de Song Yun*, p. 20, note 3) permits us, perhaps, to trace the name *Wu-sha* 烏鎩 (spelt by M. Chavannes *Wou-cha*) in a more extended application. Discussing the various Chinese designations for Sarīkol (Sung Yün's Han-p'an-t'o), he quotes there the following notice from the *Pei shih*, a text published 644 A.D.: 'Le royaume de K'iuan-yu-mo est l'ancien royaume de Wou-tch'a 烏秅. Le roi réside dans la ville de Wou-tch'a. (Ce pays) est au sud-ouest de Si-kiu-pan (Karghalik) et est à 12,970 li de Tai (près de Ta-t'ong-fou, prov. de Chan-si).'

By comparing another notice of the *Pei shih*, which describes the kingdom of K'o-p'an-t'o in the same relative position, M. Chavannes comes to the very justifiable conclu-

that Hsüan-tsang tells us suffices to fix the limits of Wu-sha with fair accuracy[9]. We saw that the river Hsi-to, i.e. the Zarafshān or Yarkand river, formed its southern border. From a subsequent notice we learn that on the north it was adjoined by the kingdom of *Ch'ia-sha* or Kāshgar, which the pilgrim reached from Wu-sha after a march of 500 li northward across stony hills and desert plains. Within the limits thus indicated, the map shows us a narrow belt of cultivable ground, extending in detached oases between the foot of the spurs which descend from Muztāgh-Ata and the westernmost portion of the great sandy desert. In the extreme south-east of this area we find the great Yarkand oasis, while the fertile tract of Yangi-Hisār, next in importance and size, marks the north-western end of this belt.

Yarkand part of Wu-sha.

The region of the present Yarkand, situated at the point where the greatest river of Eastern Turkestān debouches from the mountains, must at all times have enjoyed exceptional advantages in the matter of irrigation. It may consequently be assumed to have been in Hsüan-tsang's days, as it is now, the most populous and fertile portion of the territory above defined. Yet it is difficult to believe that Yarkand possessed in ancient days the political importance which it has claimed during recent centuries, especially during periods of Muhammadan rule. The Chinese records, which permit us to trace from the times of the Han dynasty onwards the fortunes of Kāshgar, Karghalik, and Khotan as separate and fairly powerful states, do not know Yarkand as the centre of a distinct territory. Mirzā Ḥaidar, whose history of his race, the *Tārīkh-i-Rashīdī*, is by far our best Muhammadan authority on Eastern Turkestān, distinctly tells us that in the days of his ancestors 'Yarkand was a companion city to Yangi-Hisār'. It was his uncle Mirzā Abā Bakr who first made Yarkand his capital, and turned it into what it has since remained, the largest and probably richest of the 'Six Cities' of Turkestān[10].

In view of these facts we cannot feel certain about the identification of the chief town of Wu-sha with Yarkand, tempting as it would otherwise appear. The distance to Ch'ia-sha, which Hsüan-tsang puts at 500 li, or five days' march, and which, after his usual practice, may be taken as the distance from capital to capital, would well agree with the five (rather long) marches ordinarily reckoned nowadays between Yarkand and Kāshgar[11]. Less so the direction, which is actually north-west instead of being north, as recorded by the pilgrim.

Hsüan-tsang's route to Kāshgar.

It is the direction of march here indicated, as well as the relatively short distance recorded between the Chichiklik plateau and the confines of Wu-sha, which make me believe that Hsüan-tsang followed the direct route towards Kāshgar, via Chihil-Gumbaz, Ighizyār, Yangi-

sion that K'iuan-yu-mo (Ch'üan-yü-mo) and K'o-p'an-t'o are both designations of the identical territory of Sarïkol or Tāsh-kurghān, an opinion already expressed by a Chinese commentator. The close similarity between the names Wou-tch'a (Wu-ch'a) of the *Pei shih* and Wou-cha (Wu-sha) of Hsüan-tsang (Beal: U-sha) is incidentally noted by M. Chavannes. May we not reasonably account for it by the assumption that the name variously spelt *Wu-sha* and *Wu-ch'a* properly applied to the territory described by Hsüan-tsang, but was at one period extended also to Sarïkol, with which the former was politically connected during centuries?

[9] V. de Saint-Martin showed just perception for the bearing of the topographical evidence when, in *Mémoire analytique*, p. 427, he proposed to locate Wu-sha at the present Yangi-Hisār (called by him *Inggachar*, after the nomenclature of the modern Chinese sources used by Timkowski and Klaproth; comp. Ritter, *Asien*, v. pp. 400, 417). The identification was too narrow, but as close as the cartographical materials available in 1858 would permit.

On the other hand, the identification with 'Och-Takht Soleyman', i.e. with Osh in Farghāna (!), which Julien's translation suggests, and which Beal has thought fit to reproduce (*Mémoires*, ii. p. 216; *Si-yu-ki*, ii. p. 304), shows curious disregard for plain geographical facts, and rests on nothing but a deceptive similarity of sound.

[10] See *Tārīkh-i-Rashīdī*, p. 296. Regarding Yarkand, comp. below, chap. v. sec. i.

[11] Compare *Yarkand Mission Report*, p. 430, which puts the total distance from Yarkand to the 'Old City' of Kāshgar at 120 miles.

Hisār, and perhaps did not personally visit the chief town of the territory through which this route took him. From Ighizyār, a large village where the main route from Sarīkol towards Kāshgar first touches the area of permanent settlements, Kāshgar is reached by going almost due north. The same general direction is followed also on the two preceding marches, which lead through the outer hills. As to the distance, it deserves to be noted that the 800 li, or eight marches, recorded by Hsüan-tsang for the difficult passage 'through dangerous defiles and deep valleys' (in the course of which a serious mishap befell his caravan), would agree far better with the seven ordinary stages reckoned between Chichiklik and Ighizyār, approximately aggregating seventy-eight miles, than with the equally difficult but far longer route towards Yarkand. This branches off from the Tāsh-kurghān–Kāshgar route at Chihil-Gumbaz, three marches beyond Chichiklik, and does not reach permanently inhabited ground until Yaka-arik, at a distance of about 119 miles from Chichiklik [12].

After leaving Ighizyār the traveller towards Kāshgar passes for a considerable distance over barren stony slopes and low broken hills before reaching Yangi-Hisār, while the first march beyond takes him over a good deal of sandy desert. Thus Hsüan-tsang's description of the road from Wu-sha to Ch'ia-sha would agree with the actual scenery on this part of the route.

In the general description which Hsüan-tsang gives of the products, climate, &c., of Wu-sha, the reference to the jade found there alone calls for notice. I am unable at present to adduce exact evidence as to jade-mining within the limits to which I believe the territory of Wu-sha to have extended. But it must not be supposed that the find-places of this much-prized stone are limited in Turkestān to the Khotan region, which has become chiefly famous for it. Modern Chinese accounts of 'the New Dominion', according to a communication kindly made to me by Dr. Bushell, distinctly mention jade-mining operations in the mountains near the Yarkand river, and it remains to be seen whether some of the places meant are not situated to the north of the river, within the region of the ancient Wu-sha. The only notice of this kind at present accessible to me, in the form of an extract from the *Hsi yü wen chien lu* given by Ritter, mentions such a locality, 'Mirdschai,' at a distance of 230 li from Yarkand; but I am unable to trace its position [13]. In any case, it is well to remember that, according to the testimony of a most competent geologist, 'there is no reason to doubt the existence of jade along the whole of the Kun-lun range, as far as the mica and hornblendic schists extend.' [14]

Legend of a mysterious Arhat.

A pious legend, which Hsüan-tsang relates at some length in connexion with Wu-sha, offers special interest; for if I am right in tracing the locality to which it clung, it affords fresh evidence of the tenacity with which popular tradition and worship in Turkestān, as elsewhere in the East, survive all religious and ethnical changes.

[12] According to Hui-li's 'Life', Hsüan-tsang's party was attacked by robbers on the fifth day after the start from Chieh-p'an-t'o. The merchants in his company bolted up the mountain slopes, and several elephants fell into a stream and perished; comp. *Vie de H.-T.*, pp. 274 sq. After the robbers had left, Hsüan-tsang is said to have advanced slowly along with the merchants. The mention of the loss of the elephants in a stream, as well as the date given, points to the adventure having happened in the narrow defile of the Tangitar stream, along which most of the route lies between the Chichiklik and Pas-Robāt Passes; comp. *Yarkand Mission Report*, pp. 266 sqq. For the relative distances, comp. ibid., pp. 432 sqq.

[13] See Ritter, *Asien*, v. pp. 381 sq., after Timkowski's translation from the edition of 1776. [Since the above remarks were written I have learned from my former servant Muhammadju, a Yarkandī, that jade pebbles are regularly 'fished' out of the Yarkand river bed in flood times near the village of Kōzumal, situated between Yarkand and Yaka-arik.]

[14] Compare *Yarkand Mission Report*, p. 466.

MUZTĀGH-ATA PEAK, SEEN FROM SHAMĀLDA RIDGE.

MUZTĀGH-ATA PEAK, SEEN FROM LAKE LITTLE KARA-KUL.

'Two hundred li, or so, to the west of the city [of Wu-sha],' thus the *Hsi-yü-chi* tells us[15], 'there is a great mountain. This mountain is covered with brooding vapours which hang like clouds above the rocks. The crags rise one above another, and seem as if about to fall where they are suspended. On the mountain top there rises a Stūpa of a wonderful and mysterious construction. This is the old story: many centuries ago this mountain suddenly opened; in the middle was seen a Bhikshu, with closed eyes, sitting; his body was of gigantic stature, and his form dried up; his hair descended low on his shoulders and enshrouded his face.'

Hsüan-tsang goes on to relate how a hunter once caught sight of the Arhat and told the king, who came in person to see him and pay him worship. A monk of his following explained that it was an Arhat absorbed in that complete ecstasy which produces extinction of the mind, and indicated the means by which he might be safely roused from his trance. When these had been applied, the Saint, 'looking down on them from on high for a long time,' inquired about Kāśyapa, his master, and about Śākyamuni Tathāgata. On hearing that they had both long ago attained their Nirvāṇa, he remained for a long time with his head bowed. Then he rose in the air and miraculously created a fire which consumed his body. His burned bones, which fell to the ground, were collected by the king, who raised a Stūpa over them.

Story of Muztāgh-Ata.

No one who has passed some time within sight of the great Muztāgh-Ata Peak, and has witnessed the superstitious awe with which its majestic ice-dome is regarded by the Kirghiz in all the valleys around, could fail to be reminded by Hsüan-tsang's story of the legends which cluster around that 'Father of ice-mountains'. According to the simplest form of the legend, which I myself heard from the Kirghiz with whom I came into contact during my brief surveying excursions around Muztāgh-Ata, a hoary 'Pīr' resides on the glacier-crowned and wholly inaccessible summit. Long, long ago, the eyes of adventurous hunters beheld him. Other Kirghiz legends, which Dr. Hedin heard during his prolonged stay in the same region, in 1894–5, represent the sacred mountain as 'one gigantic Mazar or burial-mound of saints', in which dwell among others the souls of Moses and Ali, &c.[16]. Stories of mysterious help rendered by these sacred dwellers of Muztāgh-Ata are interwoven with what popular tradition remembers of the struggle between the Khwājas of Kāshgar and the Chinese which found its tragic conclusion on the Pāmīrs (1759 A.D.). On the top of Muztāgh-Ata Kirghiz belief places an ancient city, whose inhabitants live on for ever in enjoyment of unblemished happiness, &c.

The great height of Muztāgh-Ata (24,321 feet according to the latest triangulation), and the dominating position it occupies, make its glittering dome visible far away in the plains about Yarkand and along the road towards Yangi-Hisār[17], whenever the dust-haze peculiar to the air of the plains clears away sufficiently. It is true that such occasions are rare, but this, perhaps, renders the vista of the distant icy peak all the more impressive to the imagination. Its direction as seen from Yarkand is almost due west—the same direction which Hsüan-tsang indicates for the great peak with its mysterious Stūpa in relation to the chief town of Wu-sha. The distance recorded by him, '200 li or so', is, indeed, beyond all proportion too small, seeing that in a direct line no less than 118 miles separate Muztāgh-Ata from Yarkand, or 65 miles from Yangi-Hisār. But it must be remembered that neither Hsüan-tsang's narrative nor his 'Life' indicates a personal visit to the chief town of Wu-sha or an actual sight of the Stūpa mountain. If the pilgrim heard the legend *en route*, while moving through the mountains relatively near to Muztāgh-Ata, so serious an underestimate of the distance would be less surprising.

[15] Compare *Si-yu-ki*, transl. Beal, ii. p. 305; *Mémoires*, ii. pp. 217 sqq.

[16] See Hedin, *Through Asia*, pp. 218 sqq.

[17] Compare *Yarkand Mission Report*, p. 286.

Muztāgh-Ata Peak suggesting a Stūpa.

Whatever opinion may be held on this point, it is certain that the remarkable shape of the huge dome of ice rising above all other mountains must have vividly suggested to Buddhist eyes the idea of a gigantic Stūpa. This striking form, which even more than its height distinguishes Muztāgh-Ata from all ice-crowned peaks of the Pāmīrs, is illustrated by the photographs I took of its west and north faces from the Shamālda ridge (Fig. 11) and Lake Little Kara-kul (Fig. 12). But it is even more notable when this mighty mountain mass is seen from a greater distance, as when I first sighted its glittering dome on my way to Tāsh-kurghān, some fifty miles away as the crow flies [18]. It is to be hoped that a telephotographic view taken by some future traveller may yet do full justice to this aspect of the 'Father of ice-mountains'.

The worship of natural objects bearing resemblance to the traditional form of Stūpas came as readily to Buddhists as that of *svayaṃbhū* or 'self-created' images to Hindus of all periods [19]. In the latter case it may be noted that the origin of the peculiarly shaped rocks, peaks, &c. thus worshipped, is invariably traced back to some miraculous event or similar extraordinary occasion. It seems probable that we have to interpret in this sense the feature of the Buddhist legend, which connected the creation of 'a Stūpa of a wonderful and mysterious character' with the miracle witnessed by an ancient king.

[18] See *Ruins of Khotan*, pp. 68 sq.

[19] Compare, regarding the worship of 'Svayaṃbhū' images, &c., in Kashmīr, my note on *Rājat.* ii. 136; for other references see the index of my *Rājat.* translation, s. v. *svayaṃbhū*.

CHAPTER III

HISTORICAL NOTICES OF KĀSHGAR

SECTION I.—THE OLD NAMES OF KĀSHGAR

Stay at Kāshgar.

KĀSHGAR, which I reached on July 29, 1900, after emerging from the difficult Gez Defile at the fertile oasis of Tāshmalik[1], detained me longer than any other single place in Chinese Turkestān. My prolonged stay, extending over fully five weeks, was mainly accounted for by the numerous practical tasks which demanded attention before I could set out for the proper goal of my explorations. In chapter VIII. of my Personal Narrative I have described these preparations in some detail, and explained the important bearing they had upon the success of my subsequent work. It will, therefore, suffice to state here that they included the careful organization of the caravan required for my travels about Khotan and in the desert, as well as the divers steps needed to familiarize the Chinese officials with the purpose of my intended explorations and to secure their goodwill. In all these preparations the experienced advice and personal help of my friend Mr. G. MACARTNEY, C.I.E., the Indian Government's political representative at Kāshgar, was of the utmost value to me.

During my stay I did not fail to examine closely whatever ancient remains of the pre-Muhammadan period survive at and about Kāshgar. Unfortunately such remains are out of all proportion scanty compared with the antiquity of the site and the historical importance of the territory of which Kāshgar has been the capital during successive ages. This importance makes it desirable to survey the accessible records bearing on the kingdom and city previous to the Muhammadan conquest before I proceed to an account of extant archaeological remains. The fact that these records are almost exclusively Chinese, and in consequence accessible to me only from secondary sources, will excuse the limitations of this survey as regards both completeness and accuracy of detail.

Chinese names of Kāshgar.

Though Kāshgar has been known to the Chinese under varying names during successive periods, yet the identity of the locality to which these names apply has never been a matter of doubt. We may see in this fact a proof of the continuity of Chinese historical knowledge

[1] This is the present, and, so far as I can judge, the correct pronunciation of the local name, such as I invariably heard it in the course of repeated inquiries both on the spot and at Kāshgar. The form *Tashbalik* or *Tashbulak* figures apparently in all modern maps, though the records of the surveys of the Jesuit Fathers (Espinha, D'Arocha, Hallerstein) whom the Emperor Ch'ien-lung sent to Chinese Turkestān after its conquest in 1759 A. D., as reproduced by De Mailla (*Histoire générale de la Chine*, 1777–81, xi. p. 575; comp. Ritter, *Asien*, v. p. 418) had furnished the correct form. *Tashbalyk* is found also in Dr. Hassenstein's excellent map accompanying Dr. Hedin's *Reisen in Zentral-Asien*, though the text (see p. 366) shows the correct *Taschmalik* or *Taschmelik*, as already recorded in *Yarkand Mission Report*, p. 38. It appears to me very probable that *Tashbalik* and *Tashbulak* are distortions due to a kind of 'popular etymology' which endeavoured to find the Turki word *balik* 'fish' or *bulak* 'spring' asa component part of the local name not readily explicable otherwise.

The name of *Kāshgar* itself is spelt variously as كاشغر, كاشگر, قاشغر, كاشكر in Muhammadan works, and modern Tūrki pronunciation and spelling similarly vary. I have adopted the form *Kāshgar* as representing a pronunciation now commonly heard throughout Eastern Turkestān and coming nearest to the quasi-official spelling *Kashgar* used in Indian and Russian publications; comp. p. 48, note 8.

concerning Kāshgar. From the time of the Former Han Dynasty, when the states of Central Asia were first opened up to the political influence of China, down to the T'ang period the region of the present Kāshgar was generally known by the name of 疏勒, read *Su-lê* or *Shu-lê* (according to Wade's transcription)[2]. The name *Sha-lê* 沙勒 is given to Kāshgar in the accounts of the journeys of the pilgrims Sung Yün, Kumārajīva (circa 400 A.D.), Fa-yung (420 A.D.), Dharmagupta (circa 593–595 A.D.) and Wu-k'ung; the latter's itinerary distinctly records the identical application of *Sha-lê* and *Su-lê*[3].

In Hsüan-tsang's *Hsi-yü-chi*, and in a passage of the T'ang Annals evidently reproduced from it, we meet with the name *Ch'ia-sha* 佉沙 (transcribed *K'ia-cha* by Julien, *K'iu-cha* by M. Chavannes, and *K'a-sha* by Dr. Franke), in which it is impossible to mistake the phonetic rendering of some form connected with the present Kāshgar[4]. The name of the town *Chia-shih* 迦師, transcribed *Kia-che* by M. Chavannes and *K'a-shih* by Dr. Franke, which the same passage of the T'ang Annals refers to as the residence of the king of Kāshgar, is evidently also closely related in origin[5]. Earlier attempts to reproduce that old local name may be recognized in the *Ch'i-sha* 奇沙 of the pilgrim Chih-mêng (circa 404 A.D.), and in Fa-hsien's *Chieh-ch'a* 竭叉, which, by M. Chavannes' brilliant and convincing analysis of the itineraries of these two pilgrims, have been shown to be identical with Kāshgar[6].

Mediaeval names of Kāshgar.

The form Kāshgar itself is attested in Muhammadan sources from the earliest period when the Arab conquest reached this part of Central Asia[7]. Spelt and pronounced in a variety of fashions (as *Kāshghar*, *Kāshgar*, *Qāshqar*, &c.) in accordance with the phonetic latitude allowed by Turkī languages in respect of certain consonants[8], this name has continued in sole use ever since Eastern Turkestān was lost to the T'ang dynasty at the close of the eighth century. Chinese records subsequent to that period invariably designate the territory and town by transcriptions of the name Kāshgar, though its identity with the earlier Su-lê was never lost sight of by the learned in China[9].

Chinese etymologies of name.

That the several names of so important a territory should have formed the subject of learned etymologies and conjectures both in the East and in the West can scarcely surprise us. Two early attempts to account for the name Su-lê (or Shu-lê) are found in Chinese Buddhist texts, and have recently been discussed at considerable length and with much critical care by

[2] The various Chinese designations of Kāshgar have been critically discussed in Dr. O. Franke's paper 'Kaschgar and die Kharoṣṭhī,' *Sitzungsber. der kön. preuss. Akad. der Wissenschaften*, Berlin, 1903, pp. 184 sqq.

The form *Su-lê* or *Shu-lê* is found also in an interesting notice of the three great routes connecting China with the West, which goes back to the information collected by P'ei Chü (circ. 605–606 A.D.) and which Baron Richthofen, *China*, i. p. 530, note, has extracted. The form *Liu-la*, given in this extract as the name of Kāshgar, is, according to the information kindly supplied to me by M. Chavannes, based upon a misreading of the usual 疏勒.

[3] Compare Chavannes, *Voyage de Song Yun*, pp. 23, 54, 57, 62, and *L'Itinéraire d'Ou-K'ong*, p. 26.

[4] Compare Julien, *Mémoires*, ii. pp. 219, 509; Chavannes, *Turcs occid.*, pp. 121, 339 (where the pronunciation *K'ia-cha* (Ch'ia-sha) is quoted from K'ang-hsi's Dictionary); Franke, *Sb.P.A.W.*, 1903, p. 186.

[5] See Chavannes, *Turcs occid.*, p. 121; Franke, *Sb.P.A.W.*, 1903, p. 186.

[6] Compare Chavannes, *Voyage de Song Yun*, pp. 54 sq. also below, chap. III. sec. iv.

[7] The capture of Kāshgar by the Arab general Qotayba, in 715–716 A.D., is recorded by Ṭabarī (tenth century), and references to Kāshgar are found in other Muhammadan authors from the tenth century onwards; compare Bretschneider, *Mediaeval Researches*, ii. pp. 45 sq.

[8] For these variations of spelling compare Shaw, 'Vocabulary of the Turkī language,' in *J.A.S.B.*, Extra No. 1878, p. 164; also above, note 1.

[9] See for such transcriptions (*K'o-shih-ha-li* in Yüan shih of 1369 A.D., *Ha-shih-ha-rh* in Ming Annals of seventeenth century), Bretschneider, *Med. Researches*, ii. pp. 148, 245; also the modern Chinese records extracted in Ritter, *Asien*, v. pp. 409 sqq.

Dr. O. Franke and Prof. R. Pischel [10]. Their interest lies mainly in the fact that the etymologies are based on alleged Sanskrit forms of the name, and thus bear evidence of local Buddhist production. The first occurs in a gloss composed within the T'ang period on the Avantaṃsaka Sūtra, as well as in a Buddhist religious compilation by a certain Hsi-lin, a native of Kāshgar. The name *Shu-lê* is here declared to be an abbreviation from *Ch'ia-lu-shu-tan-lê*, a Sanskrit term meaning '(land) possessed of bad character', and said to be applied to a mountain in that territory. 'The character of the inhabitants of that country is full of roughness and perversity, and hence the name.' [11]

*Ch'ia-lu-shu-tan-lê, and the name *Kharoṣṭra.*

Prof. Sylvain Lévi, who was the first to bring to light this curious derivation of Shu-lê, suggested that *Ch'ia-lu-shu-tan-lê* was in reality meant as a transcription of **Kharoṣṭra*, and that the latter was an ancient name of Kāshgar from which the Kharoṣṭhī script of North-Western India received its designation [12]. Ingenious as this conjecture is, I do not think it can be maintained in the face of the numerous and strong arguments which can be advanced against it, as shown at length in the above quoted papers of Dr. Franke and Prof. Pischel [13]. It appears highly improbable that a distant territory like Kāshgar, which remained outside Indian cultural influences until the advent of Buddhism about the commencement of our era, should have given its name to an Indian script already found in common use along the Indus and parts of the Punjāb in the third century B.C. The traditional derivation of the term Kharoṣṭhī, as uniformly attested in Chinese Buddhist texts (from the Skr. name *Kharoṣṭha*, meaning literally 'donkey's lip'), cannot be reconciled with a supposed form **Kharoṣṭra.* On the other hand, the actually recorded *Ch'ia-lu-shu-tan-lê* might, perhaps, ingeniously be explained as a transcription of some such Sanskrit word as **Kaluṣāntara*, **Kaluṣadhara*, or **Kaluṣottara*, which would give us the meaning 'possessed of bad character,' as pre-supposed by the above-quoted etymology of the name Shu-lê.

That this etymology is in reality nothing more than a learned concoction of some Sanskrit-knowing Buddhist scholars is the opinion held by Dr. Franke and, it seems to me, with good reason. It is a significant fact that the Annals of the Former and Later Han Dynasties, which mention Kāshgar frequently from the second century B.C. onwards, on the basis of trustworthy contemporary records, know the territory only by the name of *Su-lê* (Shu-lê). The derivation of this name as a contraction from a fuller Sanskrit form, would pre-suppose a belief that Kāshgar had, for a considerable period before the second century B.C., been occupied by a population of Indian speech and culture. Such an assumption is at variance with whatever historical information is at present accessible; and hence we are led to conclude that the alleged etymology is but a scholastic pun, originating at a later period, when Sanskrit, as the ecclesiastical language of Central-Asian Buddhism, was widely spread through Eastern Turkestān [14].

[10] In their articles 'Kaschgar und die Kharoṣṭhī,' published in *Sitzungsberichte der kön. preussischen Akademie der Wissenschaften*, Berlin, 1903, pp. 184–96 and 735–45.

[11] See Franke and Pischel, *Sb.P.A.W.*, 1903, pp. 735 sq.

[12] See *Bulletin de l'École d'Extrême Orient*, ii. pp. 246 sqq. [Since the above remarks were written Prof. S. Lévi has, in a very learned and suggestive paper, 'Le pays de Kharoṣṭra et l'écriture Kharoṣṭrī', ibid., vol. iv., 1904, taken up the question again on a broader basis, abandoning the assumption of *Ch'ia-lu-shu-tan-lê* as an old name of Kāshgar, but recognizing in that form the transcription of a geographical designation **Kharoṣṭra* of much wider application.]

[13] See particularly *Sb.P.A.W.*, 1903, pp. 185 sq., 193 sq., 736–9, 745.

[14] It is probable (as pointed out by Dr. Franke, *Sb.P.A.W.*, 1903, p. 737) that this 'popular etymology' of learned origin was meant as a gibe at the character of the people of Kāshgar, who seem to have enjoyed from early times a reputation for rough manners and deceit. See below, p. 70, for Hsüan-tsang's uncomplimentary remarks; also Chavannes, *Turcs occid.*, p. 121. Marco Polo notes regarding the people of 'Cascar': 'The natives are a wretched, niggardly set of people; they eat and drink in miserable fashion.' See Yule, *Marco Polo*, i. p. 182.

We cannot attach greater value to a second etymology of Su-lê, found in a gloss on Hsüan-tsang's account of Kāshgar, which represents that old name as a corruption of *Shih-li-chi-li-to-ti*[15]. Here, too, a Sanskrit derivation is evidently intended, but the meaning of the alleged original is not vouchsafed to us. Nor need we regret this reticence in view of the manifest artificiality of the explanation[16].

Before leaving, however, this old name Su-lê, it may be well to point out that Dr. Franke suggests *Sulek* or *Surak* as an earlier pronunciation of the Chinese name, and that such a reading possibly finds support in the form *Shulik*, by which Kāshgar is designated in early Tibetan records[17].

Western etymologies of the name.

The etymological speculations of Western scholars have naturally turned upon the name Kāshgar, by which alone the territory and its chief town have been known since the earliest Muhammadan conquest. Resemblance in sound has tempted a series of distinguished scholars, extending from De Guignes and D'Anville to Lassen and V. de Saint-Martin, to connect the name of Kāshgar with the Κάσια ὄρη of Ptolemy[18]. But Baron Richthofen, in his masterly analysis of this portion of Ptolemy's geography, has conclusively demonstrated that the Kasia mountains represent the great Kun-lun range, and that, in view of the considerable distance which separates this from Kāshgar, the assumption of a direct connexion between the two names must fall to the ground[19].

It was no less a scholar than Eugène Burnouf who first coupled the equation Κάσια ὄρη : Kāshgar, with another equally problematical conjecture deriving both names from that of the Khaśa race frequently mentioned in Sanskrit texts[20]. The weakness of this identification, from the geographical point of view, did not escape the critical judgement of Baron Richthofen,

[15] See Julien, *Mémoires*, ii. p. 219 : 'Anciennement ce royaume s'appelait *Sou-le*; c'était le nom de sa capitale. L'orthographe correcte est *Chi-li-ki-li-to-ti* (Çrī-krītati). Le mot *Sou-le* est corrompu.' As regards the authenticity of this gloss, compare Franke, *Sb.P.A.W.*, 1903, p. 738.

[16] Julien's transcription **Śrī-krītati* permits of no appropriate explanation as far as the second part is concerned; Skr. *Śrī* can safely be recognized in the first part, since the two initial characters 室 利 are regularly used by Hsüan-tsang to render that term; see Julien, *Mémoires*, ii. p. 504. Dr. Franke proposes to transcribe **Śrīkirīṭadhi* 'holding the diadem of Fortune,' and recognizes in this an attempt to provide an auspicious alternative for the somewhat derogatory etymology previously discussed; comp. *Sb.P.A.W.*, 1903, p. 738.

[17] Compare Franke, *Sb.P.A.W.*, 1903, p. 187; and concerning the Tibetan name the authorities there quoted Wassiljew, *Buddhismus*, p. 55; Rockhill, *Life of the Buddha*, p. 240 n.

[18] For references to the works of De Guignes, D'Anville, Humboldt, Lassen, compare Richthofen, *China*, i. p. 485, note 2; see also Ritter, *Asien*, v. p. 413, quoting Pater Georgi. It is from Richthofen, *China*, i. p. 494 n., that I take the reference to V. de Saint-Martin's view as to the location of the Kasia mountains near Kāshgar, set forth in his 'Étude sur la géographie grecque et latine de l'Inde' (*Mémoires présentés à l'Académie des Inscriptions*, vol. vi. 1860), in the chapter on 'la Sérique de Ptolémée' (pp. 258–286), which is at present not accessible to me.

[19] See Richthofen, *China*, i. pp. 484 sqq. In note 4, p. 485, the possibility of a derivation of *Kāshgar* from *Kāsh*, the Turkī term for jade, is learnedly discussed. The evidence in support of such a derivation appears to me very scanty, seeing that Kāshgar lies far away from the region which has been the source of the jade supply of Turkestān from ancient times, Khotan and the valleys of the Upper Yarkand river. Nor can Kāshgar be rightly assumed to have ever been the chief centre for the once flourishing jade trade. In ancient times, when a great commercial route led along the southern edge of the Tārīm Basin direct from Khotan to Lop-Nor and Sha-chou, Khotan itself was, no doubt, this centre. When that route fell into disuse during the Middle Ages, Yarkand became naturally the main mart, as clearly indicated in Benedict Goëz' narrative; see Yule, *Cathay*, ii. p. 564.

It would be far more tempting to seek, with Baron Richthofen, a similar relation between the names of the *Kāsh* stone and of the *Kasia* Mountains. Traders like the agents of the Macedonian Maës, upon whose information the account of Serica given by Marinus and reproduced by Ptolemy mainly rests, may well have recorded the local names of that distant region as they heard them from the 'Great Yüeh-chih' people in the more accessible Oxus region, who probably spoke a language of the Turkī-Mongolian family (compare for the linguistic relationship of the 'Great Yüeh-chih' my remarks in *Indian Antiquary*, xxxiv. p. 84).

[20] Burnouf's suggestion was contained in a note contributed to Humboldt, *Asie Centrale*, i. p. 115. His derivation of Kāshgar from a supposed form **Khaśagairi*

who has rightly pointed out how far removed from Kāshgar are the known seats of the Khaśas[21].

The name *Khaśa* has been used in Sanskrit literature for the designation of hill-tribes settled in widely different parts of the Himālaya regions, and is often applied very vaguely. But fortunately the territorial limits are well defined for those Khaśas of the extreme north-west of India, who in Sanskrit writings of quasi-geographical character are ordinarily associated with the Dards (*Darad*, *Dārada*), and who alone could possibly be thought of in connexion with Kāshgar. By a detailed analysis of the numerous passages in Kalhaṇa's Sanskrit Chronicle of Kashmīr, which mention these Khaśas, I have proved that they occupied the valleys encircling Kashmīr on the south and west. By their settlements in the latter direction, on the Jehlam and Kiṣangaṅgā rivers, they were thus the immediate neighbours of the Dards holding the valleys draining into the Indus[22]. Supposed connexion with *Khaśa*.

In view of what we now know of the mighty mountain ranges, and the equally great barriers of distance which separate Kāshgar from any known seats of Khaśas, a reference to Burnouf's conjecture would scarcely have been needed had it not been recently revived, though on a different ground, by so distinguished an Indologist as Professor Pischel. In his notes discussing the alleged origin of the Kharoṣṭhī script from Kāshgar, he suggests that the *Khāsya* or *Khāṣya* writing, which is mentioned in a formal list of scripts given in the Lalitavistara, must mean the writing of K'ia-sha (Ch'ia-sha) or Kāshgar[23].

It is true that the *Khāṣyalipi* figures in the list between the *Daradalipi* and *Cīnalipi*, i.e. the writings of the Dards and Chinese. But even if we credit the author of that Buddhist mythological poem with the intention or ability of following any strict geographical order in his enumeration of scripts (which comprises also numbers of purely apocryphal names), it appears far more probable that he intended a reference to the Khaśas, whose name appears widely spread along the whole Himālaya range, from the Dards in the extreme north-west to Assam, than to Kāshgar, a small and distant Central-Asian state altogether beyond the geographical horizon of ancient India[24]. In this connexion it is well to remember that we have no evidence whatever of the local name corresponding to the present Kāshgar having been applied in ancient times in a more extended sense to Eastern Turkestān generally. The term 'Kashgaria', used in this sense, seems to be of modern Russian origin, probably due to Yāqūb Beg's short-lived domination, and has no equivalent in indigenous use.

('hill of the Khaśas' with Avestic *gairi* 'hill' in the second part) was adopted by Lassen, *Indische Alterthumskunde*, i. p. 1020, and V. de Saint-Martin, in the paper quoted above (see p. 50, note 18).

[21] Compare Richthofen, *China*, i. p. 485 note.

[22] See my translation of the *Rājataraṅgiṇī*, I. pp. 47 sq., note on i. 317; II. pp. 433 sq. The modern Khakha tribe, settled in the hill tracts immediately to the west and north-west of Kashmīr, derives its name as well as its descent from the Khaśas of the Chronicle.

[23] Compare O. Franke and Pischel, 'Kaschgar und die Kharoṣṭhī,' *Sb.P.A.W.*, 1903, p. 195.

[Prof. S. Lévi, in his paper 'Le pays de Kharoṣṭra', *Bull. de l'École d'Extr. Or.*, 1904, p. 40, has also discussed this question. From the Chinese texts quoted by him it appears that Jinagupta, in the seventh century, connected the name of Kāshgar with the Khaśas. The fact that the etymology occurred at so early a date to a Buddhist scholar of Indian origin, does not, however, in my opinion, prove that there was any geographical or historical foundation for the connexion of local names so widely separated. Nor is it justifiable to ignore the absence of any aspiration in the initial consonant of the name of Kāshgar.]

[24] How limited that horizon really was, and how vague was Indian knowledge of the regions beyond the great mountain walls of the Himālaya, is well illustrated by what I have had occasion to demonstrate in the case of Kalhaṇa. Though the Kashmīrian Chronicler, in dealing with the history and antiquities of his own country, displays thorough and extensive acquaintance with its topography, and must be credited with a far keener sense for matters geographical than we can trace in any other Sanskrit author, yet his knowledge of the regions to the north of the great mountain ramparts is remarkably limited. Even the valleys of the Dards on the Upper Indus, like Astōr, Gilgit, Chilās, near as they are, are seen in Kalhaṇa's narrative through a mythical

SECTION II.—KĀSHGAR DURING THE HAN EPOCH

Northern route from China.

The earliest account of Kāshgar is found in the Annals of the Former Han Dynasty. We owe it to that great expansion of Chinese power into Central Asia which commenced under the Emperor Wu Ti (140–87 B.C.) after the famous journey of Chang Ch'ien (circa 139–127 B.C.) had extended Chinese knowledge of the 'Western Regions' to the Oxus and the confines of Persia. Appropriately enough Su-lê is first mentioned by the Annals in connexion with the northern of the two great roads which led from the Yü Gate (Yü-mên-kwan) and the region of Sha-chou to the Oxus and Yaxartes [1]. This northern road, described as passing through Turfān (the ancient Kao-ch'ang), and thence along the southern foot of the T'ien-shan range as far as Su-lê, undoubtedly corresponds to the route via Sha-chou, Hāmi, Turfān, Kara-shahr, Kuchā, and Ak-su, which still remains the most frequented line of communication between the interior of China and Kāshgar. From Su-lê 'this road passing westward across the Ts'ung-ling mountains, goes on to Ta-wan, K'ang-chü, and the Yen-ts'ai country', regions which were long ago identified with the ancient Sogdiana [2]. The well-known route from Kāshgar over the Terek Pass into Farghāna forms the direct and easiest approach to them from the whole of the Tārīm Basin.

Notice in the Han Annals.

The specific notice of the Annals concerning Su-lê informs us that the kingdom had its capital in the city of Su-lê, distant 9,350 li from Ch'ang-an, i.e. the modern Hsi-an-fu. 'The kingdom contains 1,510 families, comprising a population of 8,647, with 2,000 trained troops [3].' The mention of a variety of Imperial officials stationed there shows that Su-lê, under the Chinese protectorate, finally organized about the middle of the first century B.C., formed an administrative centre of some consequence. The particular reference made to its 'market for goods' indicates the early commercial importance of Su-lê, and this is amply accounted for by the statement that 'the road to the Ta-Yue-chih, Ta-wan and K'ang-chü lies direct west.' For Chinese trade in the direction both of Sogdiana and Bactria, of the rapid growth of which after Chang Ch'ien's mission the Han Annals give us a series of interesting glimpses, Kāshgar must certainly have formed a convenient emporium [4]. The road distances recorded in the same notice eastwards to Wu-lei, the seat of the Chinese Governor-General (2,210 li), and southwards to So-chü (560 li), afford us little help, since the position of these two localities cannot be determined with certainty [5].

According to an interpretation recently suggested by Dr. Franke [6], we should have to recognize a reference to the foundation of Su-lê in a passage of the Former Han Annals which mentions the migration of the well-known Central-Asian tribe of the 塞, whose name is variously transcribed *Sök*, *Sz'*, *Sse*, *Ssu*, *Sa*, *Se*, &c., and whose identity with the Sacae of classical records and the Indian Śakas has long ago been established. We are told there how the king of the Sök or Sz', driven forth by the westward pressure of the Great Yüeh-chih, moved south and

haze, like a kind of Ultima Thule. Compare my remarks in the introduction to my *Rājatarangiṇī* translation, I. pp. 31, 35, also note on *Rājat.* viii. 2762–64.

[1] See, for a lucid analysis of the historical notices concerning these two ancient routes, Richthofen, *China*, i. pp. 459 sqq.

[2] Compare Wylie's translation of *Notes on the Western Regions* from the Former Han Annals, in *Journal of the Anthropological Institute*, x. (1880), p. 21; as regards the position of *Ta-wan* (with its capital near the present Uratepe) and of *K'ang-chü*, see Richthofen, *China*, i. pp. 450 sq.

[3] See Wylie, ibid., x. p. 49.

[4] Compare Wylie, x. pp. 44 sqq.; Richthofen, *China*, i. pp. 455 sqq., 463 sq.

[5] Regarding Wu-lei (Wylie, *Wu-luy*) see Richthofen, *China*, i. p. 460, note 2; Wylie, ibid., p. 23.

[6] See 'Kaschgar und die Kharoṣṭhī', *Sb.P.A.W.*, 1903, pp. 739 sq.

conquered Chi-pin. 'The race of the Sök scattered far and formed a series of states. North-west of Su-lê, all tribes included in the states of *Hsiu-hsün* and *Chüan-tu* are descended from the old Sök[7].' Neither Dr. Franke's translation of the passage, nor the context of the whole notice on Chi-pin, from which the above is extracted, would seem directly to imply that Su-lê itself was founded by the Sök or Sz', though it is probable that the latter's power previous to their ejection by the Great Yüeh-chih (circa 150 B. C.) extended over the Kāshgar region.

In consequence of the internal troubles which during the reign of the usurper Wang-Mang (9–23 A. D.) preceded the downfall of the Former Han Dynasty, Chinese control over Eastern Turkestān ceased for more than half a century. When under Ming Ti, the second Emperor of the Later Han Dynasty (58–75 A. D.), Chinese power once more began to assert itself in the 'Western Regions', Su-lê was found to have passed, along with a dozen other small states, including So-chü (Yarkand), under the domination of the king of Yü-t'ien or Khotan. In 70 A. D. Pan Ch'ao, the famous Chinese general, vanquished the Khotan ruler; and by 76 A. D. the whole of the Tārīm Basin, with the exception of the territories corresponding to the modern Kara-shahr and Kuchā, acknowledged Chinese sovereignty[8]. Kāshgar under Later Hans.

The records of the Later Han Dynasty, which relate to the tributary kingdoms of the West, and which, according to Dr. Franke's statement, contain a detailed account of the history of Su-lê during that period (25–220 A. D.)[9], have not yet been rendered accessible in full and reliable translations. We are hence unable to ascertain whether the route of the justly famous Pan Ch'ao in 95 A. D. lay through Kāshgar, as in the case of Kao Hsien-chih's equally memorable exploit (747 A. D.) which we had occasion to discuss in a previous chapter[10]. Pan Ch'ao's conquests.

Pan Ch'ao's victorious campaign marks the greatest extension which Chinese power ever attained westwards. More lasting than its political results was the influence which it exercised on the development of direct trade intercourse between China and the Roman Orient, and on the spread of more definite knowledge about each other in the two great empires of the far East and the West. On the side of the West we must reckon it particularly fortunate that the opportunity then offered for recording information from enterprising traders who penetrated into China, or at least into its outlying dominions in the Tārīm Basin, was utilized by a geographer of true genius and exceptional width of vision, Marinus of Tyre. This record has been preserved for us by Ptolemy, mainly in his chapter on Serikē, as he calls the regions stretching from Scythia beyond Mount Imaus to the unknown easternmost ocean. After having been for generations the subject of much fruitless speculation, the account gathered by Marinus of the land of the silk-bringing Seres, and of the trade-route leading to it from the west, was elucidated with critical acumen by Baron Richthofen[11]. This analysis by the hand of a master renders it possible for us to touch here briefly upon a remarkable itinerary underlying that account, and to indicate its bearing upon Kāshgar.

[7] The earlier translation of Wylie, *J. Anthr. Inst.*, x. p. 34, agrees substantially with Dr. Franke's rendering. The names Hsiu-hsün and Chüan-tu are transcribed *Heu-seun* and *Keuen-tüh* by Wylie, *Hiu-siun* and *Kwan-to* by Richthofen, *China*, i. p. 448, note 2. The Chi-pin conquered by the Sacae has long been considered identical with the territory of Kābul. MM. Chavannes and S. Lévi (*L'Itinéraire d'Ou-k'ong*, pp. 36 sqq.) have expressed the view that Kashmīr is meant. Notwithstanding the authority attaching to these two scholars, and the tempting nature of some of the philological evidence advanced by them, I retain my doubts on the point, mainly on geographical grounds.

[8] For a summary of the Chinese records concerning the history of Eastern Turkestān during the first century of our era, see Richthofen, *China*, i. pp. 468 sq.

[9] See 'Kaschgar und die Kharoṣṭhī', *Sb.P.A.W.*, 1903, p. 186.

[10] See above, pp. 8 sqq. For Pan Ch'ao's expedition compare Klaproth, *Tableaux historiques*, pp. 65 sqq.

[11] See Richthofen, *China*, i. pp. 477 sqq., and concerning the itinerary of Maës, ibid., pp. 496–500.

Maēs' route to the Seres.

According to Ptolemy's express statement, Marinus gathered his geographical data about the land of the Seres from the account given by Maēs, a Macedonian merchant 'called also Titianus', who had sent agents into the country. The route of these agents, which we may safely assume to have been a caravan road in frequent use for the silk-trade from China into the Parthian Empire, led from Hierapolis on the Euphrates through Hekatompylos, Areia, and Margiana (Merv) to Baktra. From there the route passed first northward to the mountain district of the Komedi, and then along it to the south-east 'as far as the ravine that opens into the plain country'. In this ravine the traveller had to ascend for 50 schoeni northward when 'he arrives at the Stone Tower, after which the mountains recede to the east and unite with Imaus, the range that runs up to the north from Palimbothra'.[12] Another passage of Ptolemy places to the east of the Stone Tower, and in the axis of Mount Imaus itself, the station or Sarai (ὁρμητήριον) 'whence traders start on their journey to Sera'.[13]

The 'Valley of the Komedi.'

It is the joint merit of Sir H. Yule and Sir H. Rawlinson to have demonstrated beyond all doubt the identity of the mountain tract of the Komedi with the Chü-mi-t'o of Hsüan-tsang on the one hand and the 'land of the Kumēdh' of early Muhammadan writers on the other[14]. It thus became possible to locate with certainty 'the valley of the Komedi' in the mountains which divide the Wakhshāb river and the adjacent alpine tracts of Karategin from the course of the Oxus. From Karategin a direct and comparatively easy line of communication leads along the Wakhshāb up to the rich grazing grounds of the wide Alai plateau. Ascending the latter to its eastern end, it then crosses the watershed range between the Oxus and the Tārīm at its lowest point, the Taun-murun Pass[15]; and a short distance below, near the head-waters of the Kāshgar river, it joins the great route which connects Kāshgar with Farghāna over the Terek Dawān.

Position of the 'Stone Tower.'

Baron Richthofen, by a series of convincing arguments, has proved how closely the description of the road followed by Maēs' agents agrees with the route just indicated[16]. An exact location of the famous 'Stone Tower' (λίθινος πύργος) is not possible at present, and can be hoped for only from antiquarian investigations effected on the spot[17]. In regard, however, to the

[12] See Ptolemy, *Geographia*, I. cap. xii. 7 sqq. By Imaus is meant the watershed range between the Oxus and the Tārīm; compare Richthofen, *China*, i. p. 484.

[13] Ptolemy, *Geogr.* VI. cap. xiii. 1. It deserves to be noted that the traders' station is placed due east of the 'Stone Tower', Lat. 43° being indicated for both.

[14] See Yule, in *J.R.A.S.*, 1873, pp. 97 sq.; for other references compare Richthofen, *China*, i. p. 497, note. The connexion between the Komedi of Ptolemy and *Chü-mi-t'o* of Hsüan-tsang had first suggested itself to Sir A. Cunningham; but as in the case of numerous other conjectural identifications, which in the end proved equally happy, he was not able to support it by critical evidence or to locate the territory. On the other hand, the discussion of the Ptolemy passage in *Cathay*, i. p. cxlix, is still of value, as showing how Sir H. Yule, by a chain of sound critical reasoning, had been led to Karategin as the probable position of the Komedi, even before information became available as to the survival of the local name into Muhammadan times.

[15] This is the form in which I heard the name pronounced on the spot; General Kuropatkin writes 'Toongooboo-roon,' see *Kashgaria*, translated by Major W. E. Gowan (Calcutta, 1882), p. 31. In the Russian Survey Maps the spelling *Taun-murun* seems to have been adopted.

[16] See *China*, i. pp. 496 sqq.

[17] The question as to the position of the 'Stone Tower' had been discussed at length by a number of distinguished scholars, from Humboldt to Sir H. Rawlinson, before the identification of the mountains of the Komedi supplied a definite clue to the direction of Maēs' route. For a synopsis of these conjectured locations, which range from Tāshkend in the north-west to Tāsh-kurghān in the south-east, see Richthofen, *China*, i. p. 498, and McCrindle, *Ancient India as described by Ptolemy* (Bombay, 1885), p. 12.

I avail myself of this opportunity to correct the error into which I fell by accepting Sir H. Rawlinson's identification of the λίθινος πύργος with Tāsh-kurghān, in my *Preliminary Report*, p. 12, and also in *Ruins of Khotan*, p. 71. When writing the original diary notes on which both the above records of my journey are based, my scope for reference on the subject was restricted to Dr. McCrindle's above-quoted publication on Ptolemy, in the copious notes of which the well-established identification of the Komedi has unfortunately remained unnoticed.

The more recent conjectures concerning the position of the 'Stone Tower,' which Dr. Marquart quotes in his *Ērānšahr*, traders' station which Maës' account mentions to the east of the Stone Tower and on the road starting for Sera, I think that unchanging geographical conditions afford us some guidance. Baron Richthofen has justly pointed out that this station must be looked for close to the watershed crossed by the above route, since Ptolemy places it in the line of the Imaus, which undoubtedly corresponds to the range buttressing the Pāmīr region on the east, and dividing the drainage areas of the Oxus and the Tārīm. He has also rightly observed that the point where the much-frequented route coming from Farghāna over the Terek Dawān is joined by the route from the Wakhshāb valley was the most likely position for such a station.

Position of Irkeshtam.

I am glad that my return journey from Kāshgar to Farghāna in June, 1901, hurried though it had to be, allowed me to acquaint myself at first hand with the topographical features of the ancient trade route leading from Kāshgar to Osh. The observations then gathered entirely support the view put forth by that eminent geographer; for we could scarcely expect any place on that route to agree better with the conditions for a traders' station, such as Ptolemy's notice implies, than Irkeshtam, now occupied as the Russian customs station and fortified frontier post on the road to Kāshgar. It is situated at the point where the routes coming from Farghāna and the Alai unite, and only a short distance below the Taun-murun Pass, by which the latter route crosses the watershed. Irkeshtam, which now enjoys the distinction of a small fort and a Cossack garrison, offers advantages apart from its position at the junction of two great routes. Situated at an elevation of 8,500 (Russian) feet above the sea, it is not only a relatively sheltered place; but, owing apparently to a more abundant supply of moisture from across the ranges to the west and north, the valleys immediately adjacent are far less deficient in grazing and fuel than the barren gorges of rock and detritus through which most of the route further down to Kāshgar leads[18]. The inclusion of Irkeshtam within the Russian frontier-line (which elsewhere in these parts keeps to the watershed between the Kāshgar river on the one side and the Oxus and Syr Daryā on the other) has hence its adequate reason.

Invasion of Great Yüeh-chih.

The route which we have just traced forms the most direct connexion between Kāshgar and the ancient Baktria. And it is curious that the earliest notice we possess of Kāshgar after Pan Ch'ao's victorious expedition westwards relates to its invasion from that very region. From the Annals of the Later Hans we learn that in the period 107–113 A. D. An-kuo, the ruler of Su-lê, had been obliged to send his near relative, prince Ch'ên-p'an, as a hostage to the king of the Great Yüeh-chih, whose dominion at that period, as we know from other sources, comprised not only their old seats in Baktria, but also Kābul and a considerable portion of north-western India. Subsequently, after the death of An-kuo, the arms of the Yüeh-chih deposed his successor and established Ch'ên-p'an on the throne of Su-lê, during the years 114–120 A. D.[19]

p. 155, appear to be based mainly on attempts to utilize Ptolemy's artificially deduced longitudes and latitudes for the determination of ancient localities in distant parts of Asia, where, as we know, the scantiness and unreliable character of the information at his disposal excluded the possibility of even approximately correct cartographic construction. The emphatic warning uttered by Sir H. Yule (*Cathay*, i. p. cli) on the subject of the deceptive nature of these definitions deserves to be taken to heart by students of Ptolemy's Asiatic Geography.

[18] Compare my *Ruins of Khotan*, p. 496. General Kuropatkin's remarks, *Kashgaria*, trans. Gowan, p. 32, show that Irkeshtam was, even before the Russian customs station was established there, an important stage on the route. Those coming from the direction of Kāshgar regularly halted there, to make arrangements for the conduct of their caravans across the Terek Dawān when the snow lay deep on the latter. Excepting some Chinese military posts lower down the Kizil-su, Irkeshtam is the only permanently-inhabited locality between Gulcha and the Kāshgar plain, and certainly the most considerable of the halting-places.

[19] Compare Specht, 'Études sur l'Asie Centrale,' ii., in *Journal asiat.*, 1897, p. 97; Marquart, *Ērānšahr*, p. 283; Franke, *Sb.P.A.W.*, 1903, p. 740; for earlier references, Klaproth, *Tableaux histor.*, p. 166; Ritter, *Asien*, v. p. 421.

Chinese hostages at Kapiśa.

MM. Specht and Marquart have rightly pointed out how well this notice of the Annals agrees with the story which Hsüan-tsang, in his description of Chia-pi-shih (Kapiśa), records of Kaniṣka, the founder of the Yüeh-chih dominion in Gandhāra, having extended his power to the east of the Ts'ung-ling mountains, and received hostages at his court from the western dependencies of China[20]. It appears, in fact, probable that, as ingeniously suggested by Dr. Marquart, the monastery at Chia-pi-shih, which an old local tradition reproduced by Hsüan-tsang asserts to have served as the summer residence of these hostages from the confines of China, received its name *Sha-lo-chia*[21] 沙落迦 from that princely hostage of Kāshgar. For the form **Shālaka* or **Shāraka*, which the Chinese transcription may be assumed to represent, would explain itself easily as a derivative from *Sha-lê*, the alternative old name of Kāshgar already noticed, which is likely to have sounded **Shalek* according to the earlier pronunciation of the Chinese characters[22].

Introduction of Buddhism.

According to a statement of Klaproth, gathered apparently from Chinese sources, the interference of the Yüeh-chih in the affairs of Kāshgar, towards 120 A.D., resulted in the introduction of Buddhism into that territory[23]. The Chinese authority for this statement has not yet been traced; but Buddhism undoubtedly flourished in the Yüeh-chih dominions on both sides of the Hindukush, and the prolonged sojourn in them which the Kāshgar prince, subsequently elevated to the throne, had made as a hostage may well, after his elevation to the throne, have facilitated the spread of Buddhist propaganda in that part of the Tārīm Basin. This assumption would agree with the tradition recorded by Hsüan-tsang, which makes the princely hostages from the states east of the Ts'ung-ling, including Sha-lê or Kāshgar, reside in a Buddhist convent, and connects their stay with the reign of Kaniṣka, the renowned patron of Buddhism.

Hīnayāna system at Kāshgar.

To whatever period the first establishment of the Buddhist Church in Kāshgar may prove to belong, it is far more probable that it was brought from the side of Baktria than from that of Khotan. In the latter territory, which would have been the only possible alternative channel, we know for certain that the prevailing if not the sole form of doctrine and worship was the Mahāyāna or 'Great Vehicle'[24]. In Kāshgar, on the other hand, we find the predominance of the Hīnayāna School or the 'Little Vehicle' equally strongly marked since the time of Fa-hsien[25]. Now it deserves to be noticed that, wherever the evidence of Hsüan-tsang's

[20] See *Mémoires*, i. p. 42; *Si-yu-ki*, transl. Beal, i. p. 56.

[21] See *Vie de Hiouen-Thsang*, p. 71. Dr. Marquart has discussed the name at length in *Ērānšahr*, pp. 283 sq., where a possible reference to the same Buddhist convent by the Muhammadan geographer Ya'qūbī is also noticed.

[22] Compare, regarding the pronunciation *lek* for 勒 (now sounded *lê*), Franke, *Sb.P.A.W.*, 1903, p. 187.

[23] Compare Klaproth, *Tableaux histor.*, p. 166; also Dr. Franke's instructive summary of Chinese notices concerning the spread of Buddhist teaching eastwards from the Yüeh-chih empire, in *Sb.P.A.W.*, 1903, pp. 740 sqq. Dr. Franke calls attention to a Tibetan text translated by Dr. Rockhill embodying traditions of Khotan or Li-yul, which mentions that a princess of Ga-hyag, who became the wife of King Vijayasiṃha of Khotan, helped to spread Buddhism in Shu-lik. The date of this king cannot be determined; compare Rockhill, *Life of the Buddha*, p. 240; also below, Appendix *E*.

[24] Hsüan-tsang tells us that, of the 5,000 monks residing in the convents of Khotan, all (according to Julien and Beal's translations; 'most' according to Dr. Franke's interpretation, *Sb.P.A.W.*, 1903, p. 742, note) studied the 'Little Vehicle'; see *Mémoires*, ii. p. 223. *Fa-hien*, transl. Legge, p. 16, also speaks of 'several myriads of monks, most of whom are students of the Mahāyāna.'

[25] *Mémoires*, ii. p. 220; *Si-yu-ki*, ii. p. 307; *Fâ-hien*, transl. Legge, p. 23. For the identification of Fa-hsien's *Chieh-ch'a* (*K'eeh-ch'â*, Legge) with Kāshgar see below, p. 67.

The close agreement between Fa-hsien's and Hsüan-tsang's data as regards the two great schools extends also to *Tsŭ-ho: Cho-chü-chia* which, as we shall see below chap. IV. sec. iv., must be identified with Karghalik. There the prevalence and flourishing condition of the Mahāyāna is accounted for by the vicinity of, and old connexion with, Khotan; see *Fâ-hien*, transl. Legge, p. 21, and *Mémoires*, ii. p. 221. On the other hand, the Hīnayāna or 'Little Vehicle' predominated in the regions along the great route leading eastwards of Kāshgar, according to the uniform testimony of both Fa-hsien and Hsüan-tsang;

quasi-ecclesiastical survey is available, we find the Buddhist establishments of the territories along the northern foot of the Hindukush, from Balkh (*Po-ho*) to Sarīkol (*Chieh-p'an-t'o*), attached to the 'Little Vehicle'[26].

Considering the geographical position of these territories in relation to Kāshgar and the long subjection of almost all of them to Yüeh-chih rule, we may well attach historical significance to this observation. Nor would it be right, perhaps, to ignore altogether the curious fact that the 'Little Vehicle' was followed in the ancient convent of Chia-pi-shih, where the princely hostages from the western dependencies of China were believed to have resided, and where their memory was still cherished down to the time of Hsüan-tsang. In the numerous other convents of Chia-pi-shih, as elsewhere in the tracts south of the Hindukush, the Mahāyāna School seems to have enjoyed unquestioned predominance[27].

Section III.—EASTERN TURKESTĀN UNDER THE T'ANGS

Decay of Chinese power under the Later Hans.

The rapid decay of Chinese power in Central Asia, which commenced under the Emperor An Ti (107–125 A.D.), accounts for the extreme scantiness of the information to be gleaned from Chinese sources about the political conditions of Kāshgar and Eastern Turkestān generally during the next five hundred years. The revolt of the Uigur tribes in the region of Turfān and Hāmi threatened the Chinese dominion in the Tārīm Basin at its north-eastern end about the same time that Kāshgar fell under Yüeh-chih influence. A nominal protectorate appears to have continued during the second century A.D. But even this must have ceased during the disturbed reign of the last Han Emperor (Hsien Ti, 190–220 A.D.) and the period of the Three Kingdoms (221–265 A.D.), so far as Kāshgar and other territories in the distant north-west were concerned[1]. According to a notice quoted by Ritter, the power of the kings of Su-lê during the latter period extended over a number of smaller states situated to the south and west, while, on the contrary, a brief extract from the Wei Annals given by Rémusat represents Su-lê as being at the same epoch dependent on Yü-t'ien or Khotan[2].

China became united again under the Emperor Wu Ti (265–290 A.D.), who appears to have made efforts to re-establish Chinese influence in the south of the Tārīm Basin[3]. But the dynasties which followed each other in rapid succession until the advent of the house of T'ang (618 A.D.) were too weak or too much absorbed by the task of internal consolidation to resume a policy of conquest beyond Sha-chou, the westernmost district of Kan-su. During

compare *Fâ-hien*, transl. Legge, pp. 14 sq., and *Mémoires*, i. pp. 2, 4, 10.

This exact accord between what we may call the ecclesiastical surveys of the two pilgrims is of special interest in view of the considerable interval which separates their visits. If the territorial distribution of Mahāyāna and Hīnayāna in Eastern Turkestān had not changed during a period of nearly two and a half centuries, we may justly conclude that it went back far earlier, probably to the time when Buddhism first reached these territories from the south and the west respectively.

[26] See *Mémoires*, ii. pp. 30 (Balkh), 35, 37 (Bāmiān), 210 (Sarīkol). In *Huo* (Julien, *Houo*) which corresponds to the present Kunduz, both the Great and Little Vehicles were studied; see ibid., ii. p. 193.

[27] See for the convent of the hostages, *Mémoires*, i. pp. 41 sq.; *Si-yu-ki*, transl. Beal, i. p. 57; *Vie de H.-T.*, pp. 71 sq.

It is curious that in the Punjāb tract which, according to the tradition recorded by Hsüan-tsang, served as the winter residence of Kaniṣka's hostages from the confines of China, and which was believed to have received from them its name *Chih-na-po-ti* (Skr. *Cīnapati?*), we also find Hīnayāna monks in possession of the chief monastery (called Tāmasasaṃghārāma); see *Mémoires*, i. p. 199; *Si-yu-ki*, transl. Beal, i. pp. 173 sq.

[1] Compare Richthofen, *China*, i, p. 472.

[2] See Ritter, *Asien*, v. p. 421; Rémusat, *Ville de Khotan*, p. 10.

[3] See below chapter XI. sec. iv.

the reign of Wên-ch'êng Ti (452–466 A.D.) we hear of an embassy sent by the king of Su-lê to the Imperial court to present a sacred relic, the reputed dress of Buddha which proved incombustible[4]. Early in the following century Kāshgar figures among the numerous territories of Eastern Turkestān, which the Annals of the Liang Dynasty and the *Pei shih* mention as subject to the Yeh-ta or Hephthalites. By the middle of the fifth century this race, of probably Turkish origin, had founded a powerful empire in the Oxus Basin, whence they carried their conquests down to Gandhāra and beyond the Indus in the south, and as far as Khotan and Kara-shahr in the east[5].

Dominion of the Western Turks.

Between the years 563 and 567 the empire of the Hephthalites succumbed to the attack of the Western Turks (called T'u-chüeh by the Chinese) under their great Khākān Istāmi, the Dizabul or Silzibul of the Byzantine historian Menander, whose aid Khusrū Anūshirwān, the Sassanian, had called in against these dangerous foes of his dominion[6]. All the territories north of the Oxus previously subject to the Hephthalites now passed under the sway of the Western Turks. Soon their power extended also over the old Hephthalite possessions south of the river and beyond the Hindukush, which the waning strength of the Sassanians proved helpless to retain. From their encampments placed in favourite valleys of the T'ien-shan north of Kuchā and Kāshgar the Khākāns of the Western Turks exercised their sway over dominions probably exceeding in extent even those of the Hephthalites[7]. The Chinese records show that the subject states were left in charge of their hereditary local rulers, but each under the control of a Turkish *Tudun*, who watched over the collection of the tribute[8]. There can be little doubt that the political conditions of Kāshgar, and probably of most states in Eastern Turkestān, were of the type here indicated.

Establishment of the T'ang Dynasty.

The establishment of the T'ang Dynasty, 618 A.D., marks the beginning of a new and glorious epoch in the history of Chinese relations with the 'Western regions'. During the reign of its founder Kao-tsu (618–626 A.D.), and during the first years of his energetic successor T'ai-tsung (627–649 A.D.), the struggle with the Northern Turks, whose attacks threatened the very existence of the empire, prevented the prosecution of a vigorous policy westwards. While that struggle lasted the Chinese court was eager to seek the help of the Western Turks, who were then united under the rule of a powerful Kagan (Khān), called *T'ung shih-hu* in the T'ang Annals[9]. He was still reigning when Hsüan-tsang, in 630 A.D., passed through the vast region which acknowledged his sovereignty. The gracious reception which he accorded to the pilgrim

[4] See Ritter, *Asien*, v. p. 422.

[5] Compare, for the Chinese records bearing on the Hephthalites and their conquests, Chavannes, *Turcs occid.*, pp. 224 sqq.; *Voyage de Song Yun*, p. 24. Sung Yün, who in 520 A.D. visited both the Yeh-ta seats in Badakhshān and the king (Mihirakula) representing their power in Gandhāra, distinctly attests Khotan as the eastern limit of the vast dominion tributary to them; see *Voyage de Song Yun*, p. 26. The *Pei shih*, which derives its notices about the Hephthalites from Sung Yün's mission, mentions *Sha-lê* (Kāshgar) along with Sogdiana, Khotan, Bokhāra, and over thirty smaller states, as among the 'Western countries' subject to them.

[6] A rich storehouse of precise information on the history of the Western Turks has been opened up to us by M. Chavannes' translations and critical exposition of the Chinese records concerning them in his *Documents sur les Tou-kiue (Turcs) occidentaux*, 1903. A masterly analysis of the results obtained from these records and other sources is given in the *Essai sur l'histoire des Tou-kiue occidentaux*, forming Part IV of that work, pp. 217–303. Regarding the conquest of the Hephthalite dominions, see particularly pp. 226 sqq.

[7] Interesting glimpses of these royal encampments and their barbaric splendour have been preserved for us by the Byzantine embassies which were sent to the court of the early Khākāns between 568 and 577 A.D. for the sake of alliances against the Persians; see Chavannes, *Turcs occid.*, pp. 235 sqq.

[8] Compare Chavannes, *Turcs occid.*, pp. 52, 263 sqq.

[9] See Chavannes, *Turcs occid.*, pp. 263 sqq. The Chinese *Shih-hu* (Chavannes, *Che-hou*, Beal, *Yeh-hu*) is a transcription of the Turkish title *Jabgu*, which appertained to the sovereign chiefs of the Western Turks; compare Chavannes, *Turcs occid.*, p. 24, note 1.

sufficed to open for the latter all roads as far as Kapiśa[10]. In that very year T'ung Shih-hu Kagan was murdered, and with his death the confederacy of the ten tribes constituting the Western Turks broke up into two great groups, each apparently under a succession of separate Kagans, and engaged in constant hostilities.

Reassertion of Chinese influence.

The Chinese, who in the same year, 630 A.D., had succeeded in defeating and finally subjugating the Northern Turks, did not fail to profit by the internal feuds of the Western T'u-chüeh, and gradually set about to reduce the territories which once formed the western dependencies of their empire. In 640 an Imperial army crossed the great desert and occupied Kao-ch'ang or Turfān, whose prince had already, in 630, paid homage at the Chinese court, but had subsequently proved refractory. The Protectorate of An-hsi was established in that territory, strategically so important for an advance westwards, and a garrison placed there[11]. In 644 A.D. the king of Kara-shahr, who had in 632 sent an embassy to the Imperial court, but had afterwards sought independence by an alliance with the Western Turks, was vanquished and carried off as a prisoner. Already, in 641, the internal dissensions of the Turks had given the Emperor T'ai-tsung the opportunity of installing a nominee of his own, Yi-p'i shih-kuei Kagan, as chief over the western division of the tribes. Envoys sent by him with tribute arrived at the Imperial court. When the Turkish Kagan, in the year 646, asked for the hand of a Chinese princess, T'ai-tsung claimed in return the states of Kuchā, Khotan, Kāshgar, Karghalik (Chu-chü-po), and Ts'ung-ling or Sarikol as a marriage gift[12].

Reconquest of the Tārīm Basin.

The possession of the whole of the Tārīm Basin which this claim implied was not to be attained by such means. T'ai-tsung then set about to conquer the territories he had failed to secure through diplomacy, and had reduced Kuchā before he died in 649 A.D.[13]. The king of Khotan, too, impressed by this victory of the Chinese arms, hastened to make his submission and to resume allegiance to the Empire, the tradition of which had probably never ceased completely in that state[14]. Under Kao-tsung, who succeeded to the imperial throne (650–683 A.D.), the policy of expansion westwards was maintained with vigour, and soon led to unexpected triumphs. After the death of T'ai-tsung a Kagan called Ho-lu, who had previously sought refuge within the north-western border of the Empire, revolted, and within a short time established his ascendancy over the whole of the Western Turk tribes. But this revival of Turkish power was short-lived. In a series of campaigns extending from 652–658 the Imperial armies, valiantly supported by the Uigurs, vanquished in succession the Karluks and other allies of the Turks, and finally defeated the latter under Ho-lu himself in a decisive battle on the Ili. With the capture of Ho-lu, who had fled to Tāshkend, in 658 A.D., the dominion of the Western Turks came finally to an end, and China could proceed to annex officially the whole of the vast region over which the Turkish Kagans had ruled, or at least asserted a nominal sovereignty[15].

Organization of the conquered territories.

The administrative organization which the Chinese proceeded to give to their huge conquests, extending to Kābul in the south and to the confines of Persia westwards, is detailed in the T'ang Annals, forming a document of the greatest interest for the history and geography of Central

[10] Hsüan-tsang found 'Shih-hu, the Kagan of the Turks', on a hunting expedition near Tokmak, west of Lake Issik-kul. The account of this visit preserved by the pilgrim's biographer vividly records the impressions he received of the power of the Turkish chief and the magnificence of his *entourage*. See *Vie de H.-T.*, pp. 55 sqq.; and Chavannes, *Turcs occid.*, p. 194, where Hsüan-tsang's Shih-hu Kagan was first identified.

[11] See Chavannes, ibid., pp. 105 sqq., 266.

[12] See ibid., pp. 31 sq., 59, 266.

[13] See ibid., pp. 116 sq., 267.

[14] See ibid., p. 126.

[15] See ibid., pp. 59–67, 267 sq.

Asia[16]. Here we are concerned only with the territories of Eastern Turkestān. The Protectorate of An-hsi which, as we have seen, had in 640 A.D. been established in Turfān (Kao-ch'ang) for the purpose of extending Chinese control over these territories, was in 658 A.D., immediately after the final victory over the Turks, definitely transferred to Kuchā (Ch'iu-tzŭ)[17].

The 'Four Garrisons.'

The passage of the T'ang Annals tells us that this Protectorate was intended to govern *Yü-t'ien* (Khotan), *Suei-shih* (Tokmak), and *Su-lê*, the whole of these territories (including Kuchā itself) being henceforth known as the 'Four Garrisons'. There can be no doubt that this term 四鎮 included all Eastern Turkestān, not merely the territories actually enumerated as seats of the 'Four Garrisons'. The official list of the latter subsequently underwent a change, by the substitution of Kara-shahr for Tokmak, at a date variously stated as 670 or 719 A.D.; but the application of the term itself, in the sense above indicated, remained the same as long as the supremacy of the T'angs in the Tārīm Basin lasted[18].

Subjection of Kāshgar.

The fact that Kāshgar figures from the first in the official list of the 'Four Garrisons' shows the importance which the Chinese attached to this state. The Imperial decree finally ordering the administrative organization of Su-lê and Chu-chü-p'an or Karghalik, along with a number of Trans-Oxus states, was issued in the year 659[19]. But other passages in the T'ang Annals prove that the actual establishment of Chinese authority at Kāshgar was not effected immediately. The authority of the Kagans, who after the capture of Ho-lu had been placed by the Emperor in charge of the defeated tribes of the Western Turks, appears to have been very weak. Tu-man (also designated as *A-hsi-chieh ch'üeh-ssŭ-chin*), the chief of one of the Nu-shih-pi tribes belonging to the Western Turks, rose in rebellion, and carrying with him the states of Su-lê, Chu-chü-po (Karghalik), and Ho-p'an-t'o or Ts'ung-ling (Sarīkol), attacked and conquered Khotan. A Chinese force was sent against him, and succeeded in defeating and capturing him in the year 660 somewhere on the Upper Yaxartes[20].

First embassies from Kāshgar.

We have already had occasion to note that the states subject to the Western Turks, but outside the tracts actually occupied by their semi-nomadic tribes, retained their local rulers. The desert regions of the Tārīm Basin and the small oases interspersed between them were by their physical conditions effectually protected against such occupation; and accordingly the Chinese conquest found Kāshgar and the other territories of Eastern Turkestān under the rule of indigenous princes whose allegiance to their suzerain must have depended mainly upon the

[16] The geographical chapter of the T'ang Annals furnishes a detailed list of the protectorates, governments, and districts established in a portion of the territories annexed after the conquest of the Western Turks. This list, first brought to notice by A. Rémusat in his 'Remarques sur l'extension de l'Empire chinois du côté de l'Occident' (*Mémoires de l'Académie des Inscr.*, vol. viii. 1827), has been exhaustively analysed and supplemented from other sources by M. Chavannes, *Turcs occid.*, pp. 67–71, and again pp. 270 sqq. Unfortunately the original 'treatises with maps on the Western countries', which were presented to the throne in 658 and 661 A.D. by the officer entrusted by imperial order with the survey and organization of the newly-annexed territories (*Turcs occid.*, pp. 119, 156), have not been preserved.

[17] See Chavannes, ibid., p. 118.

From a record in the encyclopaedia *Ts'ê fu yüan kuei* translated by M. Chavannes, in *Notes addit. sur les Tou-kiue occid.*, p. 19, it is seen that a Protectorate of the 'Four Garrisons' was first established by T'ai-tsung at Kuchā in 648 or 649, after the reduction of that territory in the first-named year. Kao-tsung, however, on his accession in 650 A.D., decided to abandon this advanced garrison, and consequently ordered the Protectorate of An-hsi to be re-established at Kao-ch'ang or Turfān.

[18] The history of the term 'Four Garrisons' has been discussed with critical thoroughness by M. Chavannes, *Turcs occid.*, pp. 113 sq., note.

The variation in the records concerning the date when Kara-shahr took the place of Tokmak may possibly be due to the fact that popular usage had anticipated the official alteration of the list consequent upon the abandonment of Tokmak in 719 A.D. by the Chinese. Kara-shahr, by its position within the region defined by the T'ien-shan and Kun-lun, certainly fitted better than Tokmak near Lake Issik-kul into a list of names which probably soon acquired a geographical significance distinct from political conditions.

[19] Compare Chavannes, ibid., pp. 141, note, 268, note.

[20] See ibid., pp. 72 sq., 307 sq.

latter's power to enforce it. In this light it can in no way surprise us to find, from the T'ang Annals, that embassies from Kāshgar as well as from the other states of Eastern Turkestān reached the Imperial court, with what Chinese historiographers were probably justified in recording as offers of submission, long before Chinese supremacy was actually extended over the whole of those regions. The notice on Su-lê contained in the T'ang Annals mentions the first embassy of the king of Kāshgar in the year 635, and a second, bringing products of the country as tokens of submission, in 639[21]. We may safely recognize in the date of the earlier mission, and in the dates recorded for the first embassies from other states of Eastern Turkestān, a proof of the deep and widespread impression which the Chinese triumph over the Northern Turks in 630 A. D. and the simultaneous disruption of the Western tribes must have produced among the tributaries of the latter[22].

First irruptions of Tibetans, 662–692 A.D.

Chinese authority over Kāshgar and the rest of the 'Four Garrisons' was destined soon to undergo severe checks. Already in the year 662 a rebellion broke out among the Western Turks, and a Chinese army sent to repress it, while marching to the south of Kāshgar, suffered humiliation at the hands of a force of the Tibetans whom the Kung-yüeh, a northern tribe from the Ili region, had incited to invasion[23]. The Tibetans who, after their conquest of the Kuku-Nor region (663 A. D.), rapidly rose to be formidable rivals of the Chinese power in Central Asia, availed themselves of the difficulties caused to the latter by the internal dissensions of the Turkish tribes and their successive defection. Urged on by the Kung-yüeh and the ruler of Kāshgar, they attacked, in the year 665, Khotan which had to be relieved by Chinese troops[24].

After a terrible defeat inflicted on the Imperial forces in 670 A. D., north of the Kuku-Nor, the Tibetans were free to make themselves masters of the 'Four Garrisons'[25]. It is true, a Chinese record tells us of the Kung-yüeh and the king of Su-lê, frightened by the approach of an Imperial army, having offered their submission in the year 673[26]. But this success can only have been ephemeral; for the plain statements of the T'ang Annals show that the Tibetans completely subjected Kāshgar during the years 676–678, and in fact retained possession of the whole of the 'Four Garrisons' from that time until 692[27]. In that year a victorious expedition under the Chinese general Wang Hsiao-chieh re-established the Protectorate of An-hsi at Kuchā, and by maintaining there a garrison of thirty thousand men recovered once more the territories previously subject to it[28].

Chinese struggle against Arabs and Tibetans.

For more than half a century after this re-conquest the 'Four Garrisons' served as the base from which the Chinese endeavoured to assert their influence over the territories to the west and south-west. In these endeavours they had to face two powerful enemies, the Arabs in the west and the Tibetans in the south, and the success with which on the whole this double struggle was maintained during the reign of the Emperor Hsüan-tsung (713–762 A. D.) renders this period one of great historical interest. M. Chavannes, whose merit it is to have first rendered accessible and elucidated the ample Chinese accounts referring to this period, has

[21] See Chavannes, *Turcs occid.*, pp. 121 sq., where the Notice on Su-lê has been translated from chap. ccxxi. of the *T'ang shu*.

[22] According to the testimony of the T'ang Annals, the first embassy from Kara-shahr (Yen-ch'i) arrived in 632 A. D., from Khotan (Yü-t'ien) in 632, from Sarīkol (Ho-p'an-t'o) in 635, from Chu-chü-po (Karghalik) in 639; see Chavannes, ibid., pp. 111, 126, 125, 121, resp. Turfān (Kao-ch'ang) and Kuchā (Ch'iu-tzŭ), whose royal families appear to have been more closely connected with China, had already sent missions after the accession of Kao-tsu (618 A.D.), and hastened to give fresh evidence of their devotion in 630; comp. Chavannes, *Turcs occid.*, pp. 103, 115.

[23] Compare Chavannes, ibid., pp. 122, note, 280.

[24] See ibid., p. 122, note.

[25] See ibid., pp. 114, note, 280 sq.

[26] See ibid., p. 122, note. [27] Comp. ibid., pp. 119, 122.

[28] Compare ibid., pp. 119, 179 note.

critically focussed its varied historical aspects by a masterly analysis in chapters viii and ix of his *Essai sur l'histoire des Tou-kiue occidentaux*[29]. Our references to the events which affected the Tārīm Basin and Kāshgar may therefore be brief.

For more than two decades after the recovery of the 'Four Garrisons', Chinese policy towards the West appears to have been purely defensive. The internal conditions of the empire during the usurpation of the Empress Wu (684–705 A.D.), and still more the revival of the power of the Northern Turks under the great chief Kapagan Kagan, called *Mo-cho* in the Chinese Annals (691–716 A.D.), fully account for this. In the year 699 all the tribes of the Western Turks acknowledged the sovereignty of Kapagan Kagan, and it was consequently to the latter and not to the Chinese court that the princes of the Oxus region and Sogdiana turned for help against the rising tide of Arab invasion[30]. During the years 705–715 a series of brilliant expeditions under the leadership of the famous Qutayba ibn Muslim carried the Muhammadan arms victoriously across Balkh to Bukhāra, Samarkand, and Farghāna. The armed intervention of the Kagan of the Northern Turks, residing far away on the banks of the Orkhon, failed to afford protection. Thus when, in the year 714, the Chinese, evidently in pursuance of the more active foreign policy initiated by the Emperor Hsüan-tsung, recovered their supremacy over the Western Turks, they soon found themselves face to face with the Arabs[31].

Qutayba's expedition against Kāshgar, circ. 715 A.D.

The revolt of Qutayba against the Khalīfa Sulaymān, and his subsequent death in the year 715, appear to have occurred most opportunely for the ends of Chinese policy. On the Muhammadan side Ṭabarī's Chronicle informs us of a victorious expedition of Qutayba to Kāshgar, 'a city near the Chinese frontier', effected evidently at the commencement of Sulaymān's reign (715–717 A.D.)[32]. On the other hand Chinese historical records show that in the year 715 the Chinese general Chang Hsiao-sung, setting forth from the Protectorate of An-hsi or Kuchā, and supported by neighbouring tributary tribes, succeeded in forcibly reinstating the king of Pahan-na or Farghāna, whom the Arabs and Tibetans in alliance had driven from his territory. This initial success of the Chinese was promptly followed by offers of submission from a number of states in the 'Western Regions', extending from *Ta-yüan*, or Tāshkend, to *Chi-pin* or the Kābul Valley[33].

The 'Four Garrisons' under Hsüan Tsung.

The rebellion of the Turgäsh tribe of the Western Turks brought once more, in 717, the allied Arabs and Tibetans into the territory of the 'Four Garrisons', where the towns of Ush-Turfān and Ak-su, on the route between Kāshgar and Kuchā, were besieged by them[34]. The Chinese sgraffito discovered by me in the Endere temple, with its date of 719 A.D., shows that at about the same time Tibetan inroads continued into the southern portion of the Tārīm Basin[35]. But Chinese diplomacy soon succeeded in checking the danger from the side of the Turgäsh; and though effective control over the territory properly belonging to the Western Turks was not recovered by the Chinese until 738, the reign of Hsüan-tsung was a period of prolonged consolidation for their power within the 'Four Garrisons'. Without the firm position thus secured in the Tārīm Basin it would have been impossible for Chinese policy to exercise that remarkable activity, diplomatic and military, in distant territories to the west and south-west, which a series of interesting records and diplomatic documents attest for the years 719 to 751[36].

[29] See *Turcs occid.*, pp. 281, 299.

[30] See ibid., pp. 282 sq., 288 sqq.

[31] Compare ibid., pp. 283, 290.

[32] Compare Bretschneider, *Mediaeval Researches*, ii. p. 46, quoting Zotenberg's translation of Ṭabarī, iv. p. 198.

[33] See *Turcs occid.*, pp. 148 note, 291.

[34] See ibid., p. 284, note.

[35] See below chap. XII.

[36] See regarding this extension of Chinese influence, *Turcs occid.*, pp. 292–7.

Chinese activity outside 'Four Garrisons,' 719–751 A.D.

During that period the suzerainty of China was acknowledged by all the states which Muhammadan invasion threatened, from Kashmīr to the Oxus and Yaxartes, and even to distant Tabaristān on the shores of the Caspian Sea. Imperial decrees, royal titles, and occasional missions were intended to strengthen the princes of these states in their resistance by a kind of moral support. But the Chinese administration of the 'Four Garrisons' was ready for more vigorous measures, whenever it was necessary to bar the Tibetans from access to the route which might have enabled them to join hands with the Arabs, their old allies. This route led through Gilgit and Yasīn, the Little P'o-lü of the T'ang Annals, to the Upper Oxus; and we have already had occasion, in chapter I, to discuss in detail the expeditions which the Chinese undertook from the 'Four Garrisons' to prevent this strategically important territory from falling into the hands of the Tibetans[37]. In 722 A.D. it was the Sub-Delegate of Kāshgar who, at the head of four thousand troops, succoured the king of Little P'o-lü, and helped him to expel the Tibetans. Similarly Kao Hsien-chih, when setting out in the year 747 to wrest Little P'o-lü from the Tibetans, commenced his famous march over the Pāmīrs and Hindukush from Kāshgar.

Defeat of Kao Hsien-chih, 751 A.D.

Kao Hsien-chih, under whose leadership the troops of the 'Four Garrisons' in 750 a second time triumphantly penetrated south of the Hindukush[38], was destined also to bring about the event which marked the commencement of the rapid decline of Chinese power in the 'Western Regions'. Having forcibly intervened in the same year (750 A.D.) in the affairs of the tributary kingdom of Tāshkend, he treacherously put to death its ruler and excited the wrath of the people by his cupidity and oppression. The king's son stirred up the neighbouring populations against the Chinese and called in the aid of the Arabs. In the summer of 751 Kao Hsien-chih marched against the latter and their native allies; but his auxiliaries of the Karluk tribes revolted against him, and taken thus in front and rear Kao Hsien-chih, near the town of Talas (now Auliata), suffered a crushing defeat, from which Chinese authority never recovered[39].

Internal troubles of China.

A succession of disasters which about the same time overtook the imperial arms in Yün-nan, at the other end of the empire, and the subsequent rise of a dangerous pretender to the throne, made it impossible for Hsüan-tsung and his son Su-tsung, in whose favour he abdicated in 757, to direct their attention to the exposed dominions in the West[40]. On the contrary, the frontier garrisons had to be denuded of troops in order to succour the dynasty threatened in its very existence[41]. We have a significant indication of this in the fact that among the forces which finally recovered the Imperial capital of Ch'ang-an (Hsi-an-fu) in 757, contingents of soldiers from the Protectorates of An-hsi, or Kuchā, and *Pei-t'ing* (near Guchen)[42], from Farghāna, and even from the Arabs are mentioned[43]. The Tibetans, who had helped to foment these grave troubles, did not fail to turn to account the opportunity offered. From the years 758–759 onwards they gradually overran the regions of Ho and Lung, corresponding to the present province of Kan-su and the extreme west of Shan-si. From the time when they were completely established there, about 766 A.D., all direct communication between China and the Protectorates of An-hsi and Pei-t'ing was interrupted[44].

Concerning the final phase of Chinese dominion in these distant western territories, the T'ang Annals furnish a series of brief but interesting data, which M. Chavannes has lucidly

[37] See regarding these expeditions, above pp. 7 sqq.

[38] See above, p. 11; *Turcs occid.*, p. 214, note.

[39] Compare *Turcs occid.*, pp. 142 sq., 297.

[40] See ibid., pp. 298 sq.

[41] Compare Bushell, *Early History of Tibet*, p. 41.

[42] The administrative centre of Pei-t'ing occupied a position about eighteen miles to the west of the present Guchen, and corresponding to that of *Bēsh-balik* ('Five cities') of Mongol times; see Chavannes, *Turcs occid.*, pp. 11 sq. The territories governed from Pei-t'ing lay mainly north of the T'ien-shan.

[43] *Turcs occid.*, p. 299.

[44] See Bushell, *Early History of Tibet*, p. 41; Chavannes *Turcs occid.*, p. 114 note, and below Appendix *A*.

End of occupation of 'Four Garrisons.'

set forth in the concluding part of his contribution, printed below in Appendix *A*, on the Chinese records from Dandān-Uiliq[45]. They show that, notwithstanding their isolation from the rest of the empire, the Chinese governors of the 'Four Garrisons' and of Pei-t'ing succeeded in maintaining their authority over the territories confided to their charge for a comparatively long period. Kuo Hsin, the resident governor of the 'Four Garrisons,' and his colleague of Pei-t'ing, who commanded at Hāmi and Turfān, after having been completely cut off for fifteen years, succeeded in 781 in sending envoys to the Imperial court through the friendly territory of the Uigurs.

The biography of Kuo Hsin shows that the Emperor Tê-tsung liberally rewarded the faithful governors and all their officers by grants of higher rank and other honours for having maintained Chinese authority under such difficulties[46]. But effective succour such as the governors had, no doubt, eagerly solicited, was not to be obtained from the enfeebled Empire. In fact, another historical record shows that in 784 the Emperor seriously considered the recall of Kuo Hsin and his colleague Li Yüan-chung. This step was checked by the representations of his ministers, who urged the advantage of retaining at least a semblance of Chinese authority in the Tārīm Basin and in the region adjoining it on the north-east. The time for the final disappearance of this authority was close at hand, and it is a fortunate chance which allows us to verify its survival during these last few years from the testimony of a contemporary witness.

Wu-k'ung's visit to 'Four Garrisons,' A.D. 786–789.

This is furnished by the itinerary of the Chinese pilgrim Wu-k'ung, who, after a residence in India of more than thirty years, made his return journey to China during the years 786–789 through the 'Four Garrisons' and Pei-t'ing[47]. Coming from Tokhāristān and the territory of Chü-mi-chih, or the ancient Komedi, he reached Su-lê or Sha-lê some time in 786[48]. He mentions there, besides the king P'ei Lêng-lêng, the deputy governor Lu Yang, who judging by his name appears to have been a Chinese official. In Khotan he found as king Wei-ch'ih Yao, otherwise known to us from Chinese records, with the deputy-governor Chêng Chü. In Ch'iu-tzu, i.e. Kuchā, Wu-k'ung duly mentions Kuo Hsin, 'Great Protector of An-hsi', with a string of high titles, at the head of the administration. In Kara-shahr (Wu-k'ung's *Wu-ch'i*), too, a Chinese deputy-governor is noticed besides a local ruler. Pei-t'ing, the administrative centre of the Protectorate adjoining the 'Four Garrisons' to the north-east, was still under Chinese administration when Wu-k'ung was setting out from there in 789 in the suite of an Imperial delegate returning to China. The route through the Gobi was closed, no doubt by the Tibetans. Hence the devious route through the territory of the Uigurs had to be taken by the travellers, and it was only in 790 that Wu-k'ung arrived at Ch'ang-an, the Imperial capital.

[45] Compare below chap. IX. sec. vi.

[46] The Imperial decree reproduced in the biography of the *Chiu T'ang shu*, and translated by M. Chavannes (see below Appendix *A*), is lavish in terms of generous acknowledgement of the merits of the two governors, who besides other distinctions were granted the title of 'Great Protectors' of An-hsi (Kuchā) and Pei-t'ing (Guchen), respectively. In the case of their staffs the years of service spent under such exceptional conditions were to be counted seven times over for purposes of promotion. Cheap rewards, indeed, considering that the Chinese official system never knew pensions, and that few, if any, of those faithful officers could hope to regain their own land from their posts of exile.

[47] See MM. S. Lévi and Chavannes' *L'Itinéraire d'Ou-Kong*, reprinted from *Journal asiat.*, Sept.-Oct., 1895, pp. 26 sqq.

[48] This approximate date is deduced from a computation of the periods indicated for Wu-k'ung's halts at Kāshgar, Khotan (Yü-t'ien), Kuchā (*An-hsi, Ch'iu-tzü*), and Kara-shahr (*Wu-ch'i, Yen-ch'i*), and the time necessarily spent on the road, previous to his start in 789 from Pei-t'ing for the territory of the Uigurs. It is seen from the itinerary that Wu-k'ung made prolonged halts in every one of the cities counted among the 'Four Garrisons,' staying five months in Kāshgar, six in Khotan, over twelve in Kuchā, and three in Kara-shahr.

In that very year, or at the latest in 791 A. D., the Protectorate of Pei-t'ing was taken by the Tibetans, and from that time onwards, as a passage of the *Tzŭ chih t'ung chien* quoted by M. Chavannes tells us, nothing more is heard of An-hsi or the 'Four Garrisons'[49]. With this event Eastern Turkestān disappears from the horizon of the Annalists of the T'ang dynasty, and obscurity falls over its history for more than a century. Tibetan occupation of the Tārīm Basin.

At first the whole of the Tārīm Basin appears to have passed under the predominance of the Tibetans, who for a time even became dangerous neighbours to their old allies, the Arabs, in the region of the Upper Oxus. But between 860 and 873 A. D. their supremacy was broken by the Uigurs, who established a powerful kingdom comprising the region once ruled from Pei-t'ing and extending westward as far as Ak-su[50]. Khotan apparently regained independence. The rest of the territories once comprised in the 'Four Garrisons' are found subject to Turkish princes of the Karluk tribe, ordinarily residing at Balāsāghūn near Lake Issik-Kul, when early in the tenth century some scanty information about this region again becomes available from Muhammadan sources. Between the years 926 and 941 A. D. occasional missions from these princes to the Chinese court are mentioned in the Annals of the Liao[51].

Very soon after the last date must be placed the conversion to Islām of the ruler who held the territories from the Issik-Kul to Kāshgar, and who under the name of Satok Boghra Khān is celebrated in popular tradition as the pious establisher of Muhammadanism throughout Eastern Turkestān[52]. To the effective link with the West thus established and the prominent part taken by Satok Boghra Khān's successors in the disruption of the Samanide dominions in Central Asia we owe the steadily increasing flow of data which Muhammadan records henceforth supply about Kāshgar and the adjoining regions. But the period to which they refer lies beyond the scope of our present inquiry. Islām established at Kāshgar.

SECTION IV.—NOTICES OF KĀSHGAR DURING THE T'ANG PERIOD

After our brief survey of the history of Eastern Turkestān generally during the second period of Chinese ascendancy, we may now turn to the data which are furnished about Kāshgar by the records of this period.

In the first place it will be well to review the general information derived from the notice of the Annals on Su-lê[1]. '*Sou-le* est appelé aussi *K'iu-cha*[2]. Il a cinq mille li de tour; il est à plus de neuf mille li de la capitale; il s'y trouve beaucoup de déserts sablonneux et peu de terrain cultivable. Les habitants aiment la tromperie. Quand un enfant est né, eux Kāshgar described in the T'ang Annals.

[49] See below, Appendix *A*. From M. Grenard's remarks, *Journal asiat.*, 1900, Jan.-Févr., p. 24, it appears that the Turkish tribe of the Karluk helped the Tibetans to occupy the region of Turfān and to defeat the Uigurs, the allies of the Chinese.

[50] The period from the final destruction of Chinese supremacy in Eastern Turkestān to the appearance of the Muhammadanized Turkish rulers of Balāsāghūn and Kāshgar as claimants for the inheritance of the Samanide empire, has been treated with lucidity and care in M. F. Grenard's paper, 'La légende de Satok Boghra Khan et l'histoire,' *Journal asiat.*, Jan.-Févr., 1900, pp. 24 sqq. Regarding the Uigur kingdom established, with its capital Kara-khōja, near Turfān, see ibid., pp. 28 sq.

[51] See Grenard, loc. cit., pp. 36 sq.

[52] Regarding the historical data about this conversion, which appears to have been largely brought about by political motives, and certainly led to far-reaching political consequences, see Grenard, *Journal asiat.*, 1900, Jan.-Févr., pp. 38 sqq.; for the legendary account, as preserved in the popular Turkī Ta<u>dh</u>kira of Satok Boghra Khān, compare ibid., pp. 7 sqq.

[1] See Chavannes, *Turcs occid.*, p. 121.

[2] The first syllable of the name 佉沙 is to be read *Ch'ia*, the whole (pronounced *Ch'ia-sha*) being a transcription of the earliest form underlying the modern name Kāshgar; see above, p. 48.

aussi[3] lui compriment la tête pour qu'elle prenne une forme aplanie. Ces gens se tatouent le corps; ils ont l'iris des yeux verdâtre. Le roi a pour nom de famille *P'ei*; il se surnomme lui-même *A-mo-tche*; il réside dans la ville de *Kia-che*; les Tou-kiue lui ont donné en mariage une de leurs filles. Il a deux mille soldats d'élite. (Ce pays) a coutume de sacrifier au dieu céleste.'

Titles of royal family.

The description here given of the country of Kāshgar and its people is substantially the same as that found in Hsüan-tsang's *Hsi-yü-chi*, of which it is a manifest abstract. The name of the ruling family (*P'ei* 裴), which the concluding portion of the notice records, is actually found in the Imperial decree reproduced by M. Chavannes, which in 728 A.D. conferred the royal title on An-chih, chief of Su-lê[4]. It was borne also by the high Kāshgar dignitary P'ei Kuo-liang, who in 753 A.D. came to pay homage at the Imperial court[5]. The title *A-mo-chih* 阿摩支 is similarly attested by the royal decree already referred to, and was undoubtedly shared by the rulers of Khotan during the eighth century. We find it not only in a record of the encyclopaedia *Ts'ê fu yüan kuei* relating to the investiture of Wei-ch'ih Fu-chih 'qui avait les titres de *A-mo-tche* de Yu-t'ien', as king of Khotan in 728 A.D.[6], but also in an official Chinese document dating from the year 768 A.D., which was excavated at Dandān-Uiliq, and which will be discussed below[7].

The name of the capital *Kia-che* (*Chia-shih*, 迦瑟), which I am unable to trace elsewhere, is, notwithstanding its different spelling, manifestly connected with the name Kāshgar (*Ch'ia-sha*). The king to whom the Turks are said to have given a princess in marriage, and whose name is not specified, may be supposed to have been the prince actually reigning at the time of the Chinese annexation. The estimate of the number of his armed force is the same as that recorded in the Han notice on Su-lê, while in the mention of the worship of the 'god of heaven' we must, according to M. Chavannes, recognize a reference to the Zoroastrian cult[8].

Historical references to Kāshgar.

The general description of Su-lê is followed by brief references to the embassies from this kingdom in the years 635 and 639 A.D., which we have already had occasion to mention. From M. Chavannes' extracts from the encyclopaedia *Ts'ê fu yüan kuei* we learn that in 698 A.D. (after the recovery of the 'Four Garrisons' from the Tibetans) P'ei Yi-chien, king of Su-lê, sent a mission to the Imperial court to offer his tribute[9]. About the year 705 we find mention of a palisaded camp in the territory of Su-lê, to which Kuo Yüan-chên, Protector of An-hsi, retired for safety from the attacks of rebel T'u-chüeh tribes[10]. In 728 A.D. P'ei An-chih, who already had the title of *A-mo-chih* of Su-lê, received his investiture as king of Su-lê by an Imperial brevet, the text of which is preserved among the diplomatic documents of the *Ts'ê fu yüan kuei*[11]. A notice of the T'ang shu relating to the Turgäsh tribes of the T'u-chüeh shows us, in 739 A.D., the governor of the Chinese garrison of Su-lê forcibly interfering in their affairs as far as Talas (Auliata)[12]. In 753 A.D., soon after Wu-k'ung, on his way to Gandhāra, had passed through Kāshgar, we hear of certain high dignitaries from Su-lê presenting their homage at the Imperial court[13]. After this date specific references to Kāshgar in the T'ang Annals cease.

[3] The Annals previously record this custom as prevailing among the people of Kuchā.

[4] See *Turcs occid.*, p. 208.

[5] See ibid., p. 122; Chavannes, *Notes addit. sur les Turcs occid.*, p. 85.

[6] See *Turcs occid.*, p. 207.

[7] See below chap. IX. sec. v.; also App. *A*.

[8] *See Turcs occid.*, p. 121, note 4; p. 135, note 1; *Journal asiat.*, 1897, Jan-Févr., pp. 60 sq.

[9] See *Notes addit. sur les Turcs occid.*, p. 25.

[10] *Turcs occid.*, p. 189.

[11] See ibid., pp. 122 (where the name of the ruler is given as An-ting), 208; *Notes addit. sur les Turcs occid.*, p. 48.

[12] See *Turcs occid.*, p. 84.

[13] Compare Chavannes, *Notes addit.*, pp. 85 sq.; *Turcs occid.*, p. 122. Two of the dignitaries, *P'ei Kuo-liang* and *A-man-êrh-ho chü-pi-shih*, are designated as in charge of the

But we have seen that even as late as 786 A.D. Wu-k'ung, returning from India, found a Chinese deputy-governor at Su-lê by the side of a king, P'ei Lêng-lêng, who, as his family name shows, still belonged to the old local dynasty [14].

Authentic and precise as these data are, they are at the same time scanty. It is, therefore, fortunate that we are able to supplement them from the notices left to us by those Buddhist pilgrims who visited Kāshgar on their way to or from China. The information derived from the oldest of these has but recently become available to us through M. Chavannes' researches, and in particular his ingenious identification of Fa-hsien's Chieh-ch'a [15]. The biography of Kumārajīva records a visit of this Indian pilgrim to *Sha-lê* or Kāshgar about 400 A.D., and specially mentions that he placed there on his head the alms-bowl (*pātra*) of Buddha which was believed to possess the miraculous quality of changing its weight. Now the Chinese monk Chih-mêng, who proceeded to India via Lop-Nor and Khotan in the year 404 A.D., and whose biography M. Chavannes first brought to light from the Japanese Tripiṭaka, witnessed the identical miracle when handling Buddha's alms-bowl, which was shown to him in the kingdom of Ch'i-sha 奇沙. The identity of the miraculous experiences recorded by these two pilgrims, and the close approach of the form *Ch'i-sha* to the name *Ch'ia-sha* 佉沙, given to Kāshgar by Hsüan-tsang and the T'ang Annals [16], leave no doubt that Chih-mêng's notice refers to Kāshgar. **Fa-hsien's notice of Kāshgar.**

Now Chih-mêng saw at Ch'i-sha, besides Buddha's alms-bowl, also his spittoon, which he describes as being made of a stone of variegated colour; and it is the mention of this relic in Fa-hsien's account of Chieh-ch'a 竭叉 which supplies the most convincing argument for M. Chavannes' identification of the latter territory with Kāshgar. We have already, in the chapter dealing with Sarīkol, traced the route which Fa-hsien and his fellow-pilgrims followed from Yü-t'ien or Khotan to Tāsh-kurghān [17]. In the latter locality, which Fa-hsien mentions by the name *Yü-mo*, abbreviated from Ch'üan-yü-mo [18], they halted to keep the season of 'retreat', probably during the summer of 402 A.D. 'When this was over, they went on among the hills for twenty-five days, and got to Chieh-ch'a.' [19] The direct route of travellers bound for India from Tāsh-kurghān would, no doubt, have lain to the south-west, across the Pāmīrs, and not northward to Kāshgar. But the necessity of rejoining the companions who had preceded him to Chieh-ch'a direct from Khotan [20], and the desire to visit an important religious centre, suffice, as M. Chavannes justly observes, to account for Fa-hsien's détour to Kāshgar. Yet the pilgrim's narrative allows us to discern a further and, perhaps, even more cogent cause **Fa-hsien's route to Kāshgar.**

districts of *Kien* 建 and *Kin* 金, which seem to have belonged to the territory ruled from Kāshgar.

[14] See *L'Itinéraire d'Ou-K'ong*, p. 26; above p. 64.

[15] Compare, regarding the identification of Fa-hsien's *Chieh-ch'a* and Chih-mêng's *Ch'i-sha* with Kāshgar, Chavannes, *Voyage de Song Yun*, p. 54 sq.

[16] See above p. 48.

[17] See above p. 28.

[18] Compare *Voyage de Song Yun*, p. 55 note, for M. Chavannes' convincing emendation of *Yü-hui* 於麾 into *Yü-mo* 於摩; also above p. 28.

[19] Compare *Travels of Fâ-hien*, transl. Legge, p. 22 (where the name of the territory is spelt *K'eeh-ch'â*).

[20] See *Travels of Fâ-hien*, p. 18. The fact of these fellow-pilgrims having left Fa-hsien at Khotan in order to proceed in advance to Chieh-ch'a is rightly considered by M. Chavannes as an indication that Chieh-ch'a could not have been separated from Khotan by great distances and formidable obstacles; 'Ce n'est pas au moment d'entreprendre la partie la plus difficile d'un voyage qu'une caravane se divise.' Such distances and obstacles would certainly have confronted the travellers if Chieh-ch'a had been situated, as all previous interpreters supposed, to the south of the Hindukush. The only locality in that direction with which the name *Chieh-ch'a* might possibly suggest a connexion is *Chieh-shih*, identified above with Chitrāl (see pp. 14 sq.). But such an assumption is precluded by the clear statement that Fa-hsien and his companions from Chieh-ch'a 'went westwards towards North India', and only 'after being on the way for a month succeeded in getting across and through the range of the Onion mountains' (*Travels of Fâ-hien*, p. 24).

for the northward digression of the pious company, 'It happened that the king of the country was then holding the *pañcha-parishad*, that is, in Chinese, the great quinquennial assembly. When this is to be held the king requests the presence of the Śramans from all quarters of his kingdom. They come as if in clouds; and when they are all assembled, their place of session is grandly decorated, &c.'[21] The glowing description which follows of the splendour of the assembly and of the lavish offerings made to it by the king and his ministers of 'all sorts of precious things, and articles which the Śramans require', plainly shows that the attractions of this exceptional occasion were not likely to be neglected by a party of poor monks wholly dependent on charity for their progress on a distant journey[22].

Fa-hsien's description of Kāshgar.

Fa-hsien's description of Chieh-ch'a agrees well with what we otherwise know of Kāshgar old or modern. The remark that 'the country, being among the hills and cold, does not produce the other cereals, and only the wheat gets ripe'[23], is illustrated by the fact of rice, the only Turkestān cereal requiring more warmth, not being cultivated to any extent in Kāshgar, but imported from Yarkand or Ak-su[24]. Nor is it difficult for any one who has experienced the sudden transition in this region from the hot days of the late summer to a chilly and winterlike autumn, just about harvest-time, to understand the custom next related by Fa-hsien. 'After the monks have received their annual portion (of the wheat), the mornings suddenly show the hoar-frost, and on this account the king always begs the monks to make the wheat ripen before they receive their portion.'[25]

Buddha's spittoon and alms-bowl.

The pilgrim then proceeds to mention 'a spittoon which belonged to Buddha, made of stone, and in colour like his alms-bowl'. This alms-bowl was seen by Fa-hsien in Puruṣapura or Peshāwar, where it was a chief object of pious worship, and is described by him there as 'of various colours, black predominating, with the seams that show its fourfold composition distinctly marked'[26]. While we thus find Fa-hsien's account of the sacred spittoon in full accord with Chih-mêng's above-quoted description, there yet arises the question why Fa-hsien at Chieh-ch'a should pass over in silence the alms-bowl which both Chih-mêng and Kumārajīva, within a few years of his visit, had seen at Kāshgar. The answer which M. Chavannes suggests to this question appears to me in all respects adequate. Fa-hsien, too, may well have seen the alms-bowl shown at Kāshgar; but as he subsequently at Peshawar saw that sacred relic in a specimen which, from the antiquity of the legends attaching to it and the magnificence of the enshrining monastery, must have appeared to him the only authentic one, he would naturally be induced to preserve a judicious silence as to the Kāshgar counterpart[27]. On the other

[21] *Travels of Fâ-hien*, p. 22.

[22] For Fa-hsien's discriminative appreciation of the hospitality and charitable contributions accorded to himself and his companions, compare, e.g., his remarks (ibid., p. 15) on the niggardly conduct of the people of *Woo-e* (Wu-yi) and the consequent return of some of his party to *Kao-ch'ang* (vicinity of Turfān), 'to obtain there the means of continuing their journey.' Nor need we underrate his partiality for witnessing the display of great religious functions as exemplified by his remaining behind at Khotan for three months in order to see the brilliant procession of images from the chief monasteries; see ibid., pp. 18 sq.

[23] *Travels of Fâ-hien*, p. 23.

[24] See *Yarkand Mission Report*, p. 504 note.

[25] In the vicinity of Yarkand, which undoubtedly enjoys a warmer climate than Kāshgar, I found in 1900 the harvest proceeding close to the end of September. Yet already by the 24th of that month the weather became cloudy and the temperature distinctly cold; comp. *Ruins of Khotan*, pp. 171 sq.

[26] See *Travels of Fâ-hien*, p. 35; compare note 4 for references as to the Buddhist legend which accounted for the variegated colour of the sacred *Pātra* by its miraculous composition out of four distinct stone bowls presented to Buddha by the four deities of the Quarters. Compare also *Voyage de Song Yun*, p. 55, note 2, and Hsüan-tsang, *Mémoires*, i. p. 482.

[27] It would require a separate monograph to trace the worship of Buddha's alms-bowl at the numerous places where pious belief of various periods and different nations located it. A legend heard by Fa-hsien in Puruṣapura (Peshāwar) told of the attempt which a king of the Yüeh-chih had made in old times to carry off the sacred relic,

hand, he has not failed to inform us of Chieh-ch'a, like so many other places, having boasted of 'a tooth of Buddha, for which the people have reared a Stūpa, connected with which there are more than a thousand monks and their disciples, all students of the Hīnayāna'[28].

Industrial products of Kāshgar.

The only reference which Fa-hsien makes to the industrial products of Chieh-ch'a concerns materials of dress. These are said to comprise different kinds of fine woollen cloth (thus Legge; 'felt', according to Beal's translation) and of serge; apart from these 'the dress of the common people is of coarse materials, as in our country of Ch'in'. Hsüan-tsang, too, notices the felts and excellent fabrics as well as the fine and skilfully woven carpets of Kāshgar, which even at the present day continues to export considerable quantities of rough but durable cotton goods[29].

Visits of Fa-yung and Tao-yo.

Of the two Chinese pilgrims Fa-yung and Tao-yo, we know that they passed through Kāshgar on their way to India about 420 A.D. and the middle of the fifth century, respectively. But the relations they wrote of their travels are lost, and their biographies do not supply details about their visits to Sha-lê or Su-lê[30]. Also of Dharmagupta, an Indian Buddhist scholar and a native of Lāṭa or Gujarāt, who travelled to China and ended his life there, we only know that he reached Sha-lê from Kapiśa through Badakhshān, Wakhān, and Sarīkol, and that he resided there in the royal temple for two years (apparently about 580–582 A.D.)[31].

Hsüan-tsang's description of Kāshgar.

Hsüan-tsang, the next pious visitor of whom we have knowledge, has fortunately left us some accurate details concerning Ch'ia-sha or Kāshgar[32]. He tells us that its territory was about 5,000 li, or fifty marches, in circuit, that it contained plenty of sandy desert ground and but little cultivable soil. The latter, however, was very productive, and flowers and fruits abounded[33]. The climate is described as agreeable and temperate, winds and rain arriving

and how the miraculous increase of its weight had baffled his efforts; see *Travels of Fâ-hien*, p. 34. The Hindu pilgrim Fa-wei, in a notice quoted by a Chinese Buddhist author who died in 527 A.D., still speaks of Buddha's *Pātra* as worshipped in the country of the Great Yüeh-chih, in a magnificent Stūpa, which apparently was the identical structure mentioned by Fa-hsien in connexion with the relic (compare *Voyage de Song Yun*, p. 55, note 1). Hsüan-tsang, on his visit to Puruṣapura, circa 630 A.D., only saw the ruins of a Stūpa which had enclosed the *Pātra* for several centuries. After having circulated in various countries, it was then alleged to exist in Persia (*Mémoires*, i. p. 106). Sir H. Rawlinson believed that he traced this very relic in an ancient stone bowl at Kandahār, known as 'the Alms-pot', and enjoying miracle-working repute among Muhammadan devotees (*J.R.A.S.*, xi. p. 127).

On the other hand, we know from Marco Polo's detailed and interesting story how 'Sagamoni's (Śakyamuni) dish from which that personage used to eat, which is of a very beautiful green porphyry' was worshipped on Adam's Peak in Ceylon, until, in 1284, Kūblai, 'the Great Kaan', dispatched an embassy and had it brought with other sacred relics to the 'city of Cambaluc'; see Yule, *Marco Polo*, ii. pp. 319 sq. Sir H. Yule, in his detailed notes on the passage (ibid., pp. 328 sqq.), has done justice to the historical interest attaching to the legendary accounts of 'this Pātra [which] is the Holy Grail of Buddhism'.

The Chinese biographer of Chih-mêng has not failed to call attention to the discrepancies in the accounts of various pilgrims in regard to their routes and to the places where they mention the sacred alms-bowl and the skull-bone of Buddha, another famous relic. His critical conclusion is this: 'On peut voir par là qu'il n'y a pas rien qu'une route pour aller en Inde et que l'os du crâne et le bol se déplacent d'une manière surnaturelle et vont parfois dans des lieux divers' (*Voyage de Song Yun*, p. 56).

[28] The prevalence of the Little Vehicle School at Kāshgar is attested also by Hsüan-tsang, *Mémoires*, ii. p. 220; the significance of this fact has been discussed above, pp. 56 sq.

[29] Regarding the cotton fabrics of Kāshgar, compare *Yarkand Mission Report*, p. 479. Marco Polo (Yule, i. p. 181) also informs us that the inhabitants of 'Cascar grow a great deal of cotton'. Cotton is now cultivated throughout Eastern Turkestān. The production of felts at present chiefly centres at Khotan, and to some extent in the mountains south of Karghalik.

[30] Compare Chavannes, *Voyage de Song Yun*, pp. 57, 59; also p. 5, note.

[31] See *Voyage de Song Yun*, p. 62. Dharmagupta arrived at Ch'ang-an, the capital of the Sui, in 590 A.D.; his recorded halts, after Kāshgar (at Kuchā, Kara-shahr, Turfān, and Hāmi), amount to an aggregate of seven years.

[32] See *Mémoires*, ii. p. 220; *Si-yu-ki*, ii. pp. 306 sq.

[33] Marco Polo, too, specially notices 'the beautiful gardens and vineyards, and fine estates' of the people of 'Cascar'; see Yule, *Marco Polo*, i. p. 181; ii. p. 594.

The region of Kāshgar still rejoices in abundance of excellent fruit, and well deserves the praise which Mirzā with regularity. 'The disposition of the men is fierce and impetuous, and they are mostly false and deceitful. They make light of decorum and politeness, and esteem learning but little.'

The uncomplimentary character here given to the inhabitants of Kāshgar has a curious *pendant* in Marco Polo's testimony, who calls the natives of 'Cascar' 'a wretched, niggardly set of people; they eat and drink in miserable fashion'.[34] Without being able to adduce from personal observation evidence as to the relative truth of the latter statement, I believe that the judgements recorded by both those great travellers may be taken as a fair reflex of the opinion in which the 'Kāshgarliks' are held to this day by the people of other Turkestān districts, especially by the Khotanese. And in the case of Hsüan-tsang at least, it seems probable from his long stay in, and manifest attachment to, Khotan that this neighbourly criticism might have left an impression upon him.

Hsüan-tsang's account of Kāshgar people.

Hsüan-tsang describes the people of Ch'ia-sha as 'common and ignoble' in appearance, and the colour of their eyes as greenish; he also mentions their custom of painting their bodies. More important, from an anthropological point of view, is the observation (made by Hsüan-tsang also in the case of the inhabitants of Ch'ü-chih or Kuchā)[35] that it was their custom 'when a child is born to compress his head with a board of wood'. Hsüan-tsang's praise of the textile productions of Kāshgar has already been referred to[36]. The passage relating to the writing of Kāshgar does not appear to be clearly worded. So much, however, seems certain that the written characters were of an Indian type, that is, in all probability a variety of the Brāhmī script[37]. On the other hand, we are told that 'their language and pronunciation are different from that of other countries'.[38]

Buddhism at Kāshgar.

That Buddhism at Kāshgar was at the time of Hsüan-tsang's visit in a flourishing condition, at least as far as the number of its followers and their zeal were concerned, is plainly shown by the remaining portions of the pilgrim's notice. 'They have a sincere faith in the religion of Buddha, and give themselves earnestly to the practice of it. There are several hundreds of Saṃghārāmas, with some ten thousand followers; they study the Little Vehicle and belong to the Sarvāstivādin school. Without understanding the principles, they recite many religious chants; therefore, there are many who can say throughout the three Piṭakas and the *Vibhāṣā*.' Perhaps the want of scholarly application which the concluding words indicate, is the reason why Hsüan-tsang does not deign to specify a single one of those numerous religious establishments, and also why his biographer, so eloquent in the case of Khotan and other great religious centres, is satisfied with the bare mention of Ch'ia-sha[39]. Yet it deserves to be noted that Su-lê was

Ḥaidar, a good judge in such matters, like his kinsman Bābar, gives to it on this account. 'Again most of the fruits of that country (Kāshgar) are very plentiful. Among others the pears are especially good, and I never saw their equal anywhere else; they are, in fact, quite incomparable . . . . Moreover, its fruits have an advantage over the fruits of other countries, in that they are less unwholesome. . . . . During the autumn it is not the custom to sell fruit in the provinces of Kāshgar and Khotan, nor is it usual to hinder any one from plucking it. Nay more, it is planted along the roadsides, so that any one who wishes to do so may take of it.' See *Tārikh-i-Rashidī*, ed. Elias and Ross, p. 303.

[34] See Yule, *Marco Polo*, i. p. 182.

[35] *Mémoires*, i. p. 4; *Si-yu-ki*, i. p. 19.

[36] See above, p. 69.

[37] Julien, *Mémoires*, ii. p. 220, translates: 'Leur écriture est une imitation de celle de l'Inde'; Beal, *Si-yu-ki*, ii. p. 307: 'For their writing (*written characters*) they take their model from India, and although they (i.e. *the forms of the letters*) are somewhat mutilated yet they are essentially the same in form.' An alternative translation offered in a footnote of Beal is manifestly not in keeping with the context.

[38] It is curious to find Marco Polo, too, mentioning of Kāshgar: 'The people of the country have a peculiar language' (i. p. 182). From the fact of the Kudatku Bilik having been written at Kāshgar in the eleventh century, we should have been led to conclude that Turkī was in Marco Polo's days, as it is now, the language current in Kāshgar.

[39] See *Vie de H.-T.*, p. 277.

one of the two places to which Hsüan-tsang sent emissaries when endeavouring, before his departure from Khotan, to replace the sacred texts he had lost in crossing the Indus[40].

Visits of later pilgrims.

After Hsüan-tsang's time I can trace records of only two Buddhist visitors to Kāshgar. From a Chinese text quoted by M. Chavannes it appears that in 741 A. D. an Indian monk called Dharmacandra endeavoured to regain his country from China by way of Kāshgar. Stopped on his further journey through Shih-ni, or Shighnān, by a local insurrection, he was obliged to return to Kāshgar, and thence proceeded to Khotan, where he died[41]. Wu-k'ung passed through Kāshgar eleven years later on his way to Gandhāra[42]. When he travelled back towards China about 786 A. D. he stayed at Su-lê for five months, and we have already had occasion to indicate the valuable evidence which his brief notice affords for those last years of Chinese control over the territory of the 'Four Garrisons'[43].

Introduction of Christianity and Zoroastrianism into China.

It is certain that the extension, under the T'angs, of Chinese power westwards not only benefited intercourse with the Indian home of Buddhism and its old seats in Central Asia, but also facilitated the spread into China of other religions from the West. Both Christianity and the Zoroastrian cult simultaneously profited by this opening for missionary enterprise. We know that in 621 A. D. the first fire-temple was erected at Ch'ang-an, and ten years later the cult of 'the celestial god', i. e. of Ormazd, was preached in the Empire by the Magian Ho-lu[44]. The famous inscription of Hsi-an-fu, the old Ch'ang-an, attests the arrival of the first Nestorian missionary A-lo-pên in 635 A. D., with sacred books and images, and the official approval of the doctrine preached by him in 638 by an Imperial edict[45]. It also shows that at the time when it was inscribed (781 A. D.), the connexion of the Nestorian communities under the 'Bishop and Pope of Tzinisthan' (China) with the Patriarchal see of their Church, in distant Persia, was still maintained[46].

It cannot be doubted that Eastern Turkestān was the route by which these first propagandists and their clerical successors reached China; and it is a significant fact that the last nomination of a Nestorian metropolitan of China, by the Patriarch Timotheus (778–820 A. D.), coincides with the period when that region was finally lost to the T'angs[47]. It is only reasonable to assume that the territory through which Nestorian Christianity and Zoroastrianism had been transmitted to China itself possessed communities attached to those religions. But at present direct evidence is available only as regards the latter, in the form of two notices of the T'ang Annals which mention the cult of 'the celestial god' (Ormazd) at Kāshgar and Khotan, respectively[48].

Christianity at Kāshgar.

Neither historical records nor antiquarian remains have as yet come to light to illustrate the early spread of Christianity in the Tārīm Basin, previous to the conquest of this region by Islām. But by the middle of the thirteenth century we find Kāshgar (under the name of Chasemgarah) mentioned in the list of the metropolitan sees of the Nestorian Church[49]. Also

[40] See *Vie de H.-T.*, p. 285.

[41] See *Turcs occid.*, p. 163 note.

[42] See *L'Itinéraire d'Ou-k'ong*, p. 10.

[43] See ibid., p. 26; above, p. 64.

[44] See Chavannes, *Turcs occid.*, p. 301; *Journal asiat.*, 1897, Jan.-Févr., pp. 61 sq.

[45] Compare Yule, *Cathay*, i. pp. lxxxviii–ci, where the available data for the history of Nestorian Christianity in China have been lucidly discussed; see also Richthofen, *China*, i. pp. 549 sqq.

[46] It is noteworthy that Mar Idbuzid, the priest and Chorepiscopus of the capital, who erected the inscription, names as his father 'Milis of blessed memory, Priest of *Balkh*, a city of *Thokharestan*'; Cathay, i. p. xciii.

[47] Compare Richthofen, *China*, i. p. 554; Yule, *Cathay*, i. p. xci.

[48] See Chavannes, *Turcs occid.*, pp. 121, 125; also above p. 66, and below chap. VII. sec. iv. Other extracts given by M. Chavannes from the Annals show that in Samarkand (*K'ang*), also, Buddhism and Mazdeism flourished side by side (ibid., p. 135), and that the Chinese historians were fully aware that the 'cult of the celestial god' had its home and ecclesiastical centre in Persia; see ibid., p. 170.

[49] See Yule, *Cathay*, i. p. 179, ccxlv. In a later list of

Marco Polo, when passing here on his way to China (circ. 1273–1274 A. D.), notes of 'Cascar' that 'there are in the country many Nestorian Christians, who have churches of their own'. In Yarkand, too, he found Nestorian and Jacobite Christians [50].

Whether the establishment of a Nestorian archbishopric at Kāshgar was a result of that new wave of conversion, as Sir Henry Yule has aptly styled it [51], which set in after the eleventh century, with the christianization of large numbers among the Turkish and Mongolian tribes, we do not know. But if any conclusion may be drawn from the recent discovery of Nestorian cemeteries at Tokmak and Pishpek west of Lake Issik-Kul, with tombstones dating from 858 to 1339 [52], it seems probable that Kāshgar also had its Nestorian community long before its elevation to a Metropolitan see; for close political relations, as we have seen already, connected Kāshgar with the region around Tokmak both during the time of the Chinese protectorate, and again in the tenth and eleventh centuries, under the predominance of the Karluk Khāns of Balāsāghūn [53].

sees dating from about 1349 A.D., Kāshgar figures under the name of *Kashimghar*. The earlier list given by Elias, Metropolitan of Damascus (893 A.D.), does not contain Kāshgar, though duly mentioning the see of Samarkand (under the name of *Kand*, which manifestly is a form of the old name of Sogdiana, rendered by the Chinese as K'ang). This, as we know from other Syriac records, was established in the first quarter of the eighth century at the latest (Yule, *Cathay*, i. p. xc).

[50] See Yule, *Marco Polo*, i. pp. 182, 187.

[51] See Yule, *Cathay*, i. pp. xcvii sq.

[52] See the notes of M. Bonin, *Journal asiat.*, Mai-Juin, 1900, p. 587, referring to an article of Deveria in the *Journal asiat.*, 1896, which is not accessible to me at present.

[53] Compare above, pp. 60, 65.

CHAPTER IV

THE ANCIENT REMAINS OF KĀSHGAR AND THE OASES OF YARKAND AND KARGHALIK

Section I.—THE STŪPAS OF KURGHĀN-TIM AND KIZIL-DEBE

None of the early notices about Kāshgar, above reviewed, contains any definite indication as to the position of its capital. But the oldest Muhammadan description of Kāshgar I can trace—and one exceptionally trustworthy from the intimate acquaintance of its author, Mirzā Ḥaidar, with this territory—clearly shows that the position of the city of Kāshgar early in the sixteenth century was the same as now[1]. The absence of any evidence or tradition to the contrary would in itself suffice to justify this assumption; but fortunately we possess direct proof of the antiquity of the site in antiquarian remains of undoubtedly pre-Muhammadan origin. Site of ancient Kāshgar.

These consist of the large ruined mound of *Kurghān-Tim*, near the left bank of the Tümen

[1] Mirzā Ḥaidar, *Tārīkh-i-Rashīdī*, p. 295, speaks of Kāshgar as situated on the River *Tīman*, i.e. the present *Tümen*, which, as the map shows, actually flows round the north-western and north-eastern faces of the city. He places the River Kara-Tāzghun ('in the dialect of Kāshgar *Tāzghun* means a river') to the south of it, half way between Kāshgar and Yangi-Hisār. This, again, is in perfect accordance with the actual position of Kāshgar, the river in question (also known as Kara-su, by which name it is marked on my map) being still reckoned as midway between the two towns.

He further tells us that the Timan river 'flows between the ancient citadel of Kāshgar which Mirzā Abā Bakr destroyed, and the new one which he built on the banks of this river'. Mirzā Ḥaidar, who in 1514 helped to dethrone Mirzā Abā Bakr, refers elsewhere to this citadel as having been built by that tyrant shortly before his overthrow, to hold one thousand horse and foot (ibid., p. 304). Considering that this stronghold is said to have been improvised within seven days, we can scarcely feel surprised at its having disappeared without leaving any trace in remains or tradition. It may be supposed to have stood somewhere between the north-western face of the present city wall and the right bank of the river. On the other hand, the name *Kurghān*, 'fort,' and the tradition of having once been a fortified position, still clings to this day to the suburb which lines the left river bank opposite that part of the city. It is from its proximity to this suburb that the ruined Stūpa described below has received its name *Kurghān-Tim*.

The account of Kāshgar which Ritter (*Asien*, v. p. 412) quotes, through Klaproth's mediation, from the *Jahān-numā* of the Turkish geographer Ḥājī Khalfa (circ. 1640 A.D.) is manifestly a somewhat imperfect reproduction of the remarks of Mirzā Ḥaidar.

Mirzā Ḥaidar knew Kāshgar well; for much of his youth, when he served his kinsman Sultān Saʿīd Khān, Abā Bakr's successor, during the years 1514–30, must have been spent in that city. His general description of Kāshgar and the surrounding territories, written long after he had left that region to become virtual ruler of Kashmīr, is tinged with the glow of happy personal recollections, and is withal true in its particulars. He would not have been a Moghul if he had failed to extol the abundance and excellence of the fruits of Kāshgar in the passage already quoted (see p. 69, note 33). But we recognize the impress of more individual feelings in the words in which he characterizes the life of the city. Those who have lived in Kāshgar for any time can scarcely read them without being touched.

'The inhabitants of towns who go there regard Kāshgar as a wild country, while the people of the steppes consider it a refined city. It is a sort of Purgatory between the Paradise of towns and the Hell of deserts. . . . In a word, it is free from the discord of men and the trampling of hoofs, and it is a safe retreat for the contented and the rich. Great blessings accrue to the pious now, from the blessed saints who lived there in time past. From two pious persons, out of many I have seen, I have heard that when people migrate from that country to some other they cannot find the same peace of mind, and they remember Kāshgar [with regret]. This is the highest praise.' *Tārīkh-i-Rashīdī*, p. 303.

river, and opposite to the north-western face of the city, and a smaller and much more decayed mound called *Kizil-Debe*, situated on the left bank of the Kizil river, about two miles to the south of the city. The true character of these mounds as remains of ancient Stūpas appears to have been first recognized by M. N. Petrovsky, late Russian Consul-General at Kāshgar. The notice he gave of them, in an article published in the Journal of the Imperial Russian Archaeological Society, has not been accessible to me either during or after my visit to Kāshgar, and I owe my first introduction to those interesting remains to the scholarly interest shown in them by Mr. G. Macartney.

Mound of *Kurghān-Tim*.

The first-named ruin, which shares with all ancient mounds of this character the general Turkī designation of *Tim* 'mound', but is, owing to its vicinity to the suburb of Kurghān, specifically known as *Kurghān-Tim*, forms by its height and situation a very conspicuous object. It rises at a short distance above the steep loess banks which line the bed of the Tümen river to the north, and, owing to the breadth of the latter and the lower level of the southern bank, is visible from a considerable distance. The position of the mound is almost due north of Chīnī-Bāgh, Mr. Macartney's residence, outside the north-western wall of the city, and its distance from the latter is approximately one mile. The top of the mound stands seventy feet above the ground level of the neighbouring fields, but as an examination of the eroded ground at the southern foot of the mound showed, it rises in reality to fully eighty-five feet above the lowest course of masonry at present traceable.

Viewed from a distance the mound presents a roughly hemispherical appearance, while closer approach shows its present condition as that of a shapeless mass of much-decayed masonry. Its surface has in most parts suffered greatly by the disintegrating action of rain and wind. In the plan presented in Plate XX an attempt has been made to distinguish those parts of the mound where the masonry of sun-dried bricks can still be traced from those which, owing to external decay, present a surface scarcely different from that of a natural loess bank. The northern face of the mound, as seen in the photograph reproduced in Fig. 14, shows an almost vertical cleavage through the whole of the upper portion, which looks as if due to artificial cutting or else to an earthquake. The surface thus laid bare makes it easy to ascertain that the whole mound consisted originally of sun-dried bricks of large size, laid in regular courses with thin layers of mud plaster to act as binding material. The southern face has apparently undergone more gradual decay, and consequently displays on its surface less of the original masonry.

Original shape of Stūpa.

The far advanced ruin of the whole structure makes it impossible to ascertain its original dimensions and constructive features. There can be no doubt as to the remains being those of a Stūpa, built with a remarkably large dome, possibly of hemispherical shape. But as to the shape and size of the base on which this dome must be assumed to have risen, or as to the centre of the dome, the survey of the mound did not furnish any distinct indication. Judging from the evidence of all other Stūpa ruins subsequently examined by me in Eastern Turkestān, it seems probable that the base was square and arranged in three stories. Yet the plan shows a notable variation in the extent of the area over which masonry remains can now be traced above ground, its greatest length measuring about 160 feet from east to west, compared with about 130 feet from north to south.

The difficulty of tracing any indications of the original shape and dimensions of the Stūpa base is due mainly to the complete decay of the outer faces of the whole structure, and the impossibility of distinguishing, in the crumbling mass of soft bricks, between masonry retaining its original position and other that has manifestly slid down to its present low level through

13

RUINED STŪPA AND MOUND OF MAURĪ-TIM, SEEN FROM SOUTH-WEST.

14

RUINED STŪPA OF KURGHĀN-TIM, SEEN FROM NORTH.

cuttings made in the dome[2]. The undoubted rise in the level of the ground adjoining the mound also adds to the difficulty.

A careful examination of the ground shows that the lowest courses of masonry remains traceable on the western and southern faces of the mound are situated fully fifteen feet below the level of the flat ground, partly under cultivation, which adjoins the foot of the mound on the other sides, and which has been marked with zero in the plan. The difference in the levels of the lowest remains exposed on the several sides is clearly due to erosion, which on the western and southern faces has created small ravines approximately from 25 to 42 feet broad, and lined by steep banks on the sides opposite to the mound. That this erosion itself mainly proceeds from the action of water, was manifest on examining the formation of these small ravines. The greatest portion of the mound drains towards the west and south, and the rainwater collected from its slopes would have quite sufficient force to wash out the loose soil at the foot of the less soluble brickwork or to prevent its accretion. It is possible that the action of wind has to some extent aided the erosion; but as in the cultivated areas of the Tārīm Basin the force and prevalence of the winds are neither so great nor their effect on the configuration of the ground so striking as in the great desert region which those oases fringe, this co-operation of wind erosion is not easily gauged at present. Rise in level of surrounding ground.

It is possible to speak with far greater assurance on the subject of the accumulation of soil which has raised the surface of the flat ground adjoining the northern and eastern faces of the Stūpa by at least fifteen feet above the original foot of the latter. I found here the first indication of that remarkable rise in the general ground-level of the cultivated areas of Eastern Turkestān of which my subsequent observations in the region of Khotan furnished such unmistakable evidence. The close examination I made of the Yōtkan site, where the 'culture-strata' of the ancient capital of Khotan lie buried under a layer of alluvium to a depth of nine to twenty feet, together with Professor Lóczy's microscopical analysis of specimens of this soil, clearly demonstrates that this rise of level in ground kept under cultivation for many centuries is largely the result of silt deposit from the water used for irrigation. The evidence in support of this view will be found detailed in the chapter where I have discussed that important and instructive site[3]. There, too, the question of the extent to which the accretion of soil on irrigated areas is aided by sub-aerial deposit, i.e. the retention of the dust so plentifully present in the atmosphere of Eastern Turkestān for the greater part of the year, has been duly considered.

The analogy of the local conditions, as observed at Kurghān-Tim and the ancient sites within the Khotan oasis, in regard to the accumulation of soil above the original ground-level, is complete in all essential points. The whole of the fertile alluvial land of the left bank of the Tümen river is kept under intensive cultivation. The water needed for this is plentifully supplied by canals which take off from the river-bed some distance above the town. This water, by an elaborate network of channels, is conducted at frequent intervals during the spring and summer to each of the carefully terraced fields which extend along the high river bank and for a considerable distance northward. Owing to the masses of fine detritus which the river washes down in its rapid course through the disintegrated outer hill ranges, its colour, whenever I saw it, was a rich reddish brown or chocolate. Most of this silt remains suspended in the water during its passage along the 'Üstangs' and 'Ariks' of the irrigation system, and is not finally deposited until the water holding it has soaked into the soil of the field to which Irrigation of adjoining ground.

[2] Such fallen masses of masonry may, perhaps, be recognized in the foreground of the photograph showing the northern face of the mound (Fig. 14).

[3] See below chap. VIII. sec. ii.

it is conducted. The frequency and regularity of these deposits, together with the fineness of the silt itself, account for the total absence of stratification which I noticed in all the banks of earth laid bare by erosion in cultivated fields around Kurghān-Tim and elsewhere in the Kāshgar oasis. We shall see that the same observation holds good of the great layer of fertile soil which covers the débris of the ancient Khotan capital, and that its testimony to the origin of this soil is decisive.

Finds of ancient pottery.

Long, however, before the study of identical conditions elsewhere had enabled me to explain the rise of the ground-level near Kurghān-Tim, the fact itself was brought home to me by other evidence. According to a communication made to me by Mr. Macartney at the time of my first visit to the mound, fragments of ancient pottery and also some large earthenware jars had, since his arrival at Kāshgar in 1890, been found embedded in the soil near the ruin, several yards below the present surface. It appears that these finds had been made at some point of the eroded ground on the southern face of the mound. There is scarcely room to doubt that these objects, whatever their age, correspond in character and in their manner of preservation to the pottery débris, terra-cottas, &c., derived from the 'culture-strata' of Yōtkan.

It is evident that the extent of the rise of the level around or above ancient remains, being dependent on several factors necessarily differing in various localities, cannot by itself furnish precise indication of relative age. But it may be noted that a rise of fifteen feet would by no means be inconsistent with the assumption that the Kurghān-Tim Stūpa was abandoned, and its immediate vicinity converted into fields, about the time when Islām was introduced into Kāshgar; for at Yōtkan, too, where the covering layer of alluvial soil varies from nine to twenty feet in thickness, the finds of coins go down to issues of the Sung Dynasty (960–1126 A. D.). Excavations around the mound might possibly have brought to light some direct evidence of the latest date up to which the Stūpa had remained an object of worship; but a variety of practical considerations prevented me from undertaking them[4].

Interior structure of Stūpa.

Fortunately, however, the exposed portion of the ruin itself shows unmistakable proof of the antiquity of the structure at the time of its abandonment. On the north-eastern side, where the cutting into the dome is almost vertical, it is easy to distinguish an inner core of solid masonry, clearly separated from an outer mantle, for a vertical distance of about 35 feet. The photograph taken of the northern face of the mound, and reproduced in Fig. 14, shows distinctly the line of junction between the inner hemispherical dome and the outer mantle, at the point marked *A* in the ground-plan, and indicated in the photograph by a ten-foot rod. There can be no doubt that this outer mantle of masonry dates from later repairs, by which the original dome of sun-dried bricks was encased in fresh masonry and consequently enlarged. That this method of repair and extension was not infrequently applied to Stūpas, is proved by the construction of the famous Manikyāla Stūpa in the north-western Punjāb where the successive enlargements are attested by separate coin-deposits, and by other examples[5].

Form of original dome.

It is due to the protection afforded by the outer mantle that the outline of the original dome can be clearly traced at this place. Judging from the view presented by the photograph,

[4] Excavations such as could reasonably be expected to yield useful evidence would, owing to the depth of the original ground-level, have demanded considerable sacrifice of time and money. On the other hand, the utterly ruined condition of the visible remains did not offer much hope of the lower and more accessible portions of the structure having preserved enough of their original outlines, &c., to justify the heavy expense of clearing them. There was besides a quasi-diplomatic difficulty about such an undertaking. The Chinese local officials, when consulted on the point, stated that a prior right to excavation had been reserved to M. Petrovsky, the Russian Consul-General, and there were circumstances which seemed at the time to render a specific application to the latter inexpedient.

[5] Compare *Archaeol. Survey Reports*, ii. pp. 159 sq.; Foucher, *L'Art du Gandhāra*, pp. 78, 92 sqq.

the original dome appears to have been as nearly as possible hemispherical, with its centre at some point of a horizontal line which may be projected into the interior of the mound from the approximate surface-point marked with the letter *C* in the photograph. Taking measurements from the photograph, which, however, can be accepted only as roughly approximate, the radius of the original hemispherical dome may be assumed to have been about 37 feet. Adding to this the thickness of the outer mantle, which on the north-east side is seen to have been at least 13 feet, we arrive at 50 feet as an approximate estimate for the radius of the enlarged dome. Supposing this also to have been hemispherical like the original one, its height could not have been less than 50 feet.

Proportion between dome and base.

The uppermost portion of this enlarged dome has, no doubt, disappeared through dilapidation. But, even allowing for this loss in height, there remain in the extant ruin scarcely more than fifty feet of elevation for the base supporting the dome. This proportion between dome and base, if correctly estimated, seems to present a striking contrast to all other Stūpas surveyed in the course of my Turkestān explorations; for in these the height of the domed portion ordinarily represents only about one-third of the total elevation [6]. Unfortunately the Stūpa ruins still sufficiently well-preserved for accurate measurement are too few and their localities too widely scattered to permit of any safe conclusion being drawn in regard to architectural chronology. There remains then only the hemispherical shape of the original dome, which, if the parallel offered by the changes in the architectural forms of Indian Stūpas can be relied upon for guidance, may be accepted as an indication of considerable antiquity [7].

Materials of construction.

The approximate dimensions, as ascertained above, show that the Kurghān-Tim Stūpa, even in its original form, must have exceeded in size any of the Stūpas the remains of which I was able to trace between Kāshgar in the west and Endere in the east [8]. Like all these, it was constructed of sun-dried bricks laid in regular horizontal layers. The bricks of the original dome, wherever they could be measured *in situ* with approximate accuracy, showed an average size of eighteen inches square with a thickness of five inches. The bricks of the outer mantle seem to be less uniform in size and on the whole smaller. The material used is the same loess soil which prevails over the greatest portion of the plains of Eastern Turkestān, and which, whether in the form of sun-dried bricks or solid mud-walls, serves to this day everywhere for the construction of buildings. The manner of fabricating these bricks, with a liberal admixture of straw and chaff, and from pits dug as close as convenient to the structure for which they are to serve, seems to have remained practically unchanged through the course of centuries. But the size of the bricks now used is generally much smaller than that displayed by structures of the pre-Muhammadan period.

Ruined mounds of *Kizil-Debe.*

The second ancient ruin in the immediate vicinity of Kāshgar lies on the opposite side of the city, to the south, and just above the old left bank of the Kizil river, from which it probably has received its name of *Kizil-Debe* ('the Red Mound') [9]. Its distance from the

[6] Compare, regarding the Stūpas of Maurī-Tim, Tōpa-Tim, Niya River Site, Endere and Rawak, below, pp. 81 sqq., 104 sq., chaps. xi., xii., xiv.

It may, however, be noted that in the Mauri-Tim Stūpa the height of the dome, together with that of the drum immediately below it, and of the same diameter, accounts almost exactly for one half of the total elevation.

[7] Compare M. A. Foucher's luminous observations on the evolution of the Stūpa forms, *L'Art du Gandhāra*, i. pp. 65 sqq.

[8] The largest of these Stūpas, that of Rawak near Khotan shows a diameter of 42½ feet in its dome against one of about 74 feet in the original dome of Kurghān-Tim before its enlargement.

[9] The Turkī word *debe*, also pronounced *döbe*, is used throughout Eastern Turkestān for any small hillock or mound, whether natural or artificial in origin. The same word, in the forms *tepe*, *tōpe*, is common in local names of Western Turkestān, e. g. in Gök-tepe, famous in Turkoman memory.

nearest point of the south-eastern face of the city wall is about two miles, while the confluence of the Kizil and Tümen rivers is a little over one mile below. The remains consist of two mounds situated close together, as shown in the plan, Pl. XXI. The larger of the mounds closely resembles in material and construction the Kurghān-Tim ruin, but is far less imposing, as the extant mass of masonry rises to a height of only 27 feet above the level of the adjoining fields.

Size and shape of ruined Stūpa.

There can be no doubt that this mound, too, represents the remains of an ancient Stūpa. While the complete decay of the upper portion does not allow of any estimate as to the size and shape of the dome, we can, however, trace here with relative ease the outlines of the base. A look at the plan shows that this base was originally a rectangle approximately orientated. Taking the lowest course of masonry remains actually exposed on the level of the adjoining fields the maximum measurements are about 130 feet from east to west, and about 108 feet from north to south. But it must be noted that the ground to the south slopes steeply down to the edge of the river bank, which here lies fully 20 feet below the level of the fields, and at a distance of only about 25 feet from the foot of the mound. Hence it appears highly probable that the masonry originally extended further on this side, and that its outlying portion exposed to erosion has slid down the steep slopes, instead of being kept in place by the gradual rise of the adjoining soil. Taking into account this peculiar configuration of the ground, it seems likely that the original shape of the Stūpa base was a square.

The actual rise of the ground-level on the three sides where the ruins are surrounded by fields could not be correctly estimated without extensive excavations. But there can be no doubt that its cause is the same which we have traced above in the case of the Kurghān-Tim Stūpa, viz. accumulation of silt, aided probably by sub-aerial deposit. It would not be safe to accept the actual edge of the river-bank immediately below the mound as an indication of the original level, for it is possible that the low level of this portion of the river-bank (20 feet below the fields) is partly the result of erosion caused by the rainwater draining off the mound mainly in that direction.

Condition of mound.

The villagers of the hamlet close by declared the mound to have been used as a watch-station by Tungānī rebels during the last Muhammadan rising at Kāshgar (1863), a purpose for which it was well fitted by its position between the 'Old City' and the Yangi-Shahr or 'New City', that formed then, as now, the Chinese cantonment. Whether the extensive excavations visible on the top and sides of the ruin go back to that period or are still more recent, I was unable to ascertain. These cuttings have laid bare the masonry, which consists of sun-dried bricks, measuring on the average seventeen inches square, with a thickness of three inches. At various points fragments of bricks of a bright red colour crop out from the layer of débris and decomposed clay (loess) which covers the surface; but these, too, are merely sun-dried.

Kichik-Debe.

At a distance of only 50 feet to the west of the mound just described rises another much lower one, known locally as *Kichik-Debe*, 'The Little Mound.' It is of circular shape, with a diameter of about 125 feet on the present ground-level, and rises to a height of only 12 feet. No brickwork can be traced on the surface of this mound, which is covered throughout with loose earth containing numerous graves. Human skeletons had been extracted from these on the occasion of some diggings made here under Mr. Macartney's order a few years before my visit, and the account given by the villagers left no doubt as to these graves being of Muhammadan origin. It is probable that an ancient débris mound had been utilized here as a Muhammadan burial-place. The same is the case with many a piece of waste ground around Kāshgar, once occupied by buildings, and now owing to its higher level incapable of cultivation. As to the original character of the mound it is impossible to advance any certain opinion. It is possible that it marks the position of some shrine or monastic building attached to the Stūpa.

Section II.—THE RUINS NEAR KHĀN-UI

It was from the Chinese City Prefect or Hsien-kuan of Kāshgar that I first learned of the existence of ancient remains on desert ground beyond the hamlet of *Khān-ui*, some twenty miles to the north-east of the 'Old City' of Kāshgar. A tradition communicated to me by that amiable official connected the site with a Chinese settlement supposed to have existed there during the time of the Han Dynasty. But whether this tradition rested on some historical information or was merely, as I suspected, derived by a kind of 'popular etymology' from the name of the neighbouring hamlet *Khān-ui* (literally meaning 'the Khān's residence'), I was not able to ascertain. The march which brought me, on Sept. 4, 1900, to Khān-ui, through the fertile village tract of Bēsh-karīm, has been described in my Personal Narrative. There, too, I have given an account of the picturesque shrine of Bū (Bībī) Mairyam Khānum, situated some three miles to the west of Khān-ui, which marks the resting-place of the saintly daughter of Satok Boghra Khān, the royal champion of Islām, famous in the local traditions of Kāshgar[1].

At a distance of about one mile to the north-east of Khān-ui hamlet all cultivation ceases; and after another mile in the same direction, over absolutely barren 'Dasht', I found an extensive area of low denuded loess banks covered in plentiful patches with small decayed fragments of ancient pottery, glass, and slag. These remains, though far too small to show any characteristic decorative features, supplied unmistakable evidence of ancient habitation. But of the buildings from which they originated no other indication could be traced. The walls of sun-dried bricks or mud, of which these buildings must have mainly been constructed, had long ago disappeared—undoubtedly, as subsequent experience at so many other sites showed me, through the erosive action of wind and sand. The bare surface of relatively hard loess on which these fragments rested bore plain evidence, too, of the destructive forces at work here. It had, no doubt, been lowered considerably below the level of the ground at the time when the site was occupied; but the total absence of constructive remains makes it impossible to estimate the extent of reduction in level. In the shallow depressions separating these loess banks, there were to be found here and there small accumulations of sand forming rudimentary dunes. I now regret not having secured at the time specimens of this sand; for their microscopical analysis would probably have furnished clear proof as to whether this 'sand' is disintegrated loess produced by erosion on the spot, or fine detritus washed down from the hills northward. **Débris-area of *Hāsa-Tam*.**

This area, strewn with diminutive débris, extends probably over more than a square mile, and is known by the name of *Hāsa-Tam*[2]. Popular tradition, as related to me by Sōp Niāz Bōwa, the old Aksakāl of Bēsh-karīm, supposes an 'old town' (*kōne-shahr*), the capital of a 'Chinese Khākān', to have stood here until Satok Boghra Khān destroyed it. Traces of an old canal, by which the town is believed to have received its water, are said to exist south of Hāsa-Tam; but the traditional connexion of this canal with the Yamān-yār river, far away to the south, is incompatible with geographical facts. **Traditions about *Hāsa-Tam*.**

Proceeding eastward of Hāsa-Tam for a distance of three miles, over ground where stretches of wind-swept bare loess alternate with low dunes of moving sand, I reached the ruins known as *Tōpa-Tim* ('the Sandy Mound'). They proved to consist of a conspicuous mound, roughly circular in shape, built of sun-dried bricks, and the much decayed remains of a great quadrangle closely adjoining it on the west. The mound, which rises to a height of 28 feet, and on the **Ruins of *Tōpa-Tim*.**

[1] See *Ruins of Khotan*, pp. 142 sqq.

[2] *Tam* means 'wall' in Turkī. I am not aware of the significance of *Hāsa*. The word is apparently frequent as a local name, and is in at least two places, near Gūma and Moji, applied to deserted village sites; see below chap. v. sec. ii, iv.

present ground-level shows a circumference of about 350 feet, could easily be recognized as a ruined Stūpa. The ravages of time had completely effaced the outlines of its constructive features; but the masonry of sun-dried bricks, of which it was formed, was plainly revealed at several points below the covering crust of crumbling débris, and its interior appears to have remained undisturbed. The bricks, closely resembling in make those of the Kāshgar mounds, measure on the average 14½ inches square, with a thickness of 3 inches.

Ruined quadrangle.

The quadrangular structure adjoining measures about 260 feet along its two longer sides, running in the direction of north-west to south-east, and about 170 feet along its shorter sides, from north-east to south-west. Its eastern corner is separated from the foot of the Stūpa mound by a space of only 45 feet. Being unable to trace any masonry *in situ* along the low mounds which mark the sides of the quadrangle, I concluded that these represent the remains of broad walls or ramparts, which were built solid in stamped clay (loess), after the fashion still prevailing throughout Turkestān, and were once surmounted by superstructures now completely destroyed[3]. The plentiful fragments of fairly hard bricks, charred wood, and plaster which mingled with the sand covering the surface of these low mounds, seem to be distinct traces of such superstructures. The extant remains of the quadrangular enclosure rose 8 to 10 feet above the sandy level of the interior, which evidently had been an open courtyard; the breadth of the mounds measured on the top varied from about 40 feet on the north-western face to less than 20 feet on the south-eastern face, where the entrance gate may possibly have been situated. Judging from the position and size of the quadrangle, it appears highly probable that it was intended for the accommodation of the monastery once attached to the Stūpa.

Dry river-bed near Hāsa-Tam.

From Hāsa-Tam I had already sighted, through the dust-laden atmosphere, a second group of ruins known as *Maurī-Tim*, situated almost due north. I reached them on the 5th of September, after a ride of about four miles to the north-north-west of Tōpa-Tim. On the way a deep-cut river-bed was crossed, about 500 feet broad, but now almost completely dry. I was assured that only after exceptionally heavy rain in the mountains does this 'Sai', which descends from the west and is the direct continuation of the stream passing Üsten (or Little) Ārtush, carry any water so far down its course. Ordinarily its water, mainly supplied by the drainage of the Kara-Jilga Valley above that large village, is completely absorbed by irrigation channels taken off higher up. The fields of Khān-ui to the south, and those of Eski to the north of the river-bed, are now the lowest cultivated areas supplied by these canals. In ancient times canals from the same source undoubtedly reached the ancient sites now under discussion, and probably extended yet further to the south-east. Even at the present day there must be abundance of water in the area actually irrigated to account for the swamps which Colonel Trotter noticed at the point where the line of drainage of the Ārtush stream joins the Kāshgar river, some twenty-five miles below Bū Mairyam Khānum[4]. Apart from the water, which at times of exceptional floods may reach that point in the actual river-bed just described, I have no doubt much of the water absorbed in irrigation comes there to the surface again in the form of springs. We shall have occasion to note the same process of percolation and its attending result, the formation of marsh-land, in the case of all the rivers reaching the southern edge of the Taklamakān Desert from the mountains above Khotan and Keriya.

[3] The plan of construction here conjectured is well illustrated by the actually surveyed remains of the great quadrangle of *Kara-dong*, which served the purposes of a fortified Sarai or post on the ancient route leading along the Keriya river; see Plate XXXVIII.

[4] Compare *Yarkand Mission Report*, p. 261, where it is noticed that no water ever reaches Khush-toghrak, the point of junction above referred to, from the Ārtush river during the hot and irrigating season.

The ruins of Maurī-Tim occupy a conspicuous position at the extreme point of an isolated tongue of high ground which slopes down towards the edge of the plains from a low bleak ridge of conglomerate. The latter stretches eastwards of Little Ārtush, and ultimately loses itself in the sandy expanse some miles to the east of the site. Considering the exposed situation, the good preservation of the ruins—and in particular of the Stūpa which forms their most characteristic feature—was a pleasant surprise. The small gravelly plateau on which the ruins rise, extending for a length of about 500 feet from north-west to south-east, and only about 100 to 150 feet broad on the top, could never have been utilized for cultivation. While the structures were thus safe from being partially buried through a gradual rise of the adjoining ground, they lacked the protection which the accumulation of drift-sand would have afforded at true desert sites. Position of *Maurī-Tim* ruins.

It appears to me that the relative immunity enjoyed by these ruins from the destruction which has overtaken so many similar remains in these regions must be attributed to a combination of two circumstances. On the one hand, the Maurī-Tim ruins seem to have been sufficiently far removed from the dangers which the vicinity of inhabited settlements or much-frequented roads threaten to all old structures. On the other hand, their position under the shelter of a hill range has manifestly protected them from the full erosive force of the desert-winds and the sand which they move.

The plan of the site (see Plate XXII) shows that its chief remains consist of a Stūpa at the end of the small plateau already referred to, and of a large oblong mound some 200 feet behind it[5]. The Stūpa, views of which from the south-west and south-east are reproduced in Fig. 13 and Plate I respectively, rises on a square base, formed by three successively receding stories, the lowest measuring 40 feet on each side, and having an elevation of three feet above the ground. The next two stories are each $5\frac{1}{2}$ feet high, and recede from the one next beneath by three feet. The highest story is followed by a circular base, five feet high, having a diameter of 24 feet; and this again bears a drum, five feet in height, and decorated with bold projecting mouldings at top and bottom. From the drum springs finally the Stūpa dome, which at its foot shows a diameter of 17 feet, identical with that of the drum. The top of the dome is broken, but as its extant masonry rises to a height of about 14 feet, as shown by the section in the plan, it is clear that the dome or cupola cannot have been hemispherical, but must have shown a bulb shape[6]. *Maurī-Tim* Stūpa.

It will be seen from the section that the Stūpa in its present state still rises to a height of 38 feet. Its apparent height is considerably increased by the fact that it stands on the edge of a narrow plateau, which is itself raised 30 to 40 feet above the adjacent ground, and towards the west and south falls off with steep slopes. The square base of the Stūpa shows a close Dimensions and orientation of Stūpa.

[5] Nearly a year after my visit to Maurī-Tim I learned from M. N. Petrovsky that an account of its ruins as well as of the remains in the vicinity of Kāshgar had been published by him in the Proceedings of the Imperial Russian Archaeological Society. I regret that my ignorance of the Russian language did not allow me to consult this article myself, and that pressure of other tasks has prevented me from obtaining indirect information about its contents at a time when that publication would have been otherwise accessible. I must, therefore, content myself with making a bare reference to that article, published, as M. Chavannes informs me, in *op. cit.*, Oriental Section, vol. ix., 1895, pp. 147–99.

[6] I take this opportunity to correct the erroneous application of the term 'hemispherical' dome in the description given in *Prelim. Report*, p. 18, as well as some slight discrepancies in the measurements. These are due to the fact that, within the very brief time allowed for the preparation of the 'Preliminary Report', it was impossible to have finished drawings of my numerous plans prepared under my supervision. The lithographic prints rapidly produced for immediate use could not be revised by me, and have hence proved occasionally slightly inexact. Their use was discarded by me when engaged on the final publication.

approach to correct orientation. The same arrangement will be observed in the Stūpas of Endere and Rawak described below, while in the case of the only other two Stūpas sufficiently well preserved for an exact restoration of the ground-plan, those of Tōpa-Tim and the Niya Site, the sides of the square base instead of the corners were found to be orientated[7]. This difference in arrangement is of interest, but in view of the small number of examples it would scarcely be safe to draw from it any chronological or other argument.

Construction of Stūpa. The whole of the Stūpa was originally coated with a fairly hard plaster of whitish colour, which still adheres on the south-east face, as Plate I shows, to some portions of the base where projecting mouldings have offered protection. On the same side, which evidently had suffered least from 'weathering', I also found in perfect preservation some of the woodwork employed to support the bold mouldings of plaster at the top and bottom of the drum. The tamarisk sticks used for this purpose were in numerous places still firmly embedded in the brickwork, and often had retained their covering of bark. Where the outer surfaces had lost their plaster coating, but were otherwise in good preservation, they displayed solid masonry of sun-dried bricks in regular horizontal layers, each separated from the other by a setting of clay about two inches in thickness. The bricks measured 16 to 17 inches square, with a thickness of $3\frac{1}{2}$ inches. The deep cuttings into the base and dome made on different levels from the north-west and south-west faces enabled me to ascertain that the same materials and the same manner of construction had been used throughout the Stūpa.

Central shaft of Stūpa. The fact that the stability of the whole structure has scarcely as yet been impaired by the great cutting effected on the south-west side, evidently long ago, is a proof of solid construction[8]. This cutting has laid bare a small shaft, 2 feet 8 inches square, running vertically through the centre of the Stūpa dome and drum. It may have continued below the latter into the base, but the wide cutting made there has destroyed all trace of it. Above, within the dome, this shaft widens into a little chamber, four feet square, which extends to the top of the extant masonry of the dome, and thus shows a height of $5\frac{1}{2}$ feet. No indication is left of the original height of this little chamber, nor of the manner in which it may have been closed from above.

It can scarcely be doubted that this central shaft or chamber was originally built to hold a relic deposit. It was found in the Stūpas of the Niya Site, Endere, and Rawak, that is, in all of the structures of this class seen by me in Chinese Turkestān which were sufficiently well

Spoliation of Stūpas. preserved to permit of detailed examination. Invariably I found this place of deposit in Stūpas laid open by cuttings made in search of treasure. Nor can it surprise us that such conspicuous structures were thus systematically opened and 'explored', probably at an early date, in view of the wholesale spoliation of ancient remains which has gone on in Eastern Turkestān during Muhammadan times, as attested for various localities, from Kāshgar to Khotan, by curious records of Mirzā Ḥaidar[9]. At the Maurī-Tim Stūpa, as elsewhere, we can see that the hope of treasure had prompted operations even after the centre of the Stūpa had been reached, and

[7] See Plates XIX, XXIX, XXXVII, XL. It will be noticed that, in the case of the Stūpas of Mauri-Tim, Endere, Rawak, and the Niya Site, the plans of which were prepared on the spot and with the help of a properly orientated plane-table, the north to south line indicated by the bearings of the magnetic compass (a reliable instrument in the use of which proper allowance was made for the ascertained degree of magnetic variation) falls slightly to the west of the orientation line indicated by the plans of the structures themselves. I am at present unable to suggest a satisfactory explanation for this deviation, which by its regular recurrence invites attention. It cannot be due to magnetic variation, as the latter was at the time of my journey for Eastern Turkestān about 4° 30′ to the *East* of the true North.

[8] This cutting is clearly seen in the view reproduced in Fig. 13.

[9] See *Tārīkh-i-Rashīdī*, pp. 255 sqq., for the extensive excavations for treasure carried on under Mirzā Abā Bakr's orders at ancient sites of his dominions; also below chap. VIII. sec. 5.

such relics as it may have once contained had been successfully abstracted. From the foot of the north-west face and on the ground-level, a passage, apparently of more recent date, has been tunnelled deep into the solid masonry of the base.

In the central shaft of the Maurī-Tim Stūpa we have an exact counterpart of that small square chamber which I found, more injured, but still clearly recognizable, within the domes of the ruined Stūpas of Takhtaband in Bunēr, and of Bālār near ancient Taxila[10]. Judging from somewhat vague descriptions, it may be assumed to have existed also in the Manikyāla Stūpa, and in the numerous Stūpas of the Kābul Valley which were 'explored' before and during the several Afghān campaigns[11]. But the identity of this feature only helps to emphasize still more clearly the close agreement which exists in regard to general architectural arrangement between all Turkestān Stūpas examined by me, and the corresponding structures extant in the Kābul Valley and on the Indian north-west frontier. The Stūpas of Maurī-Tim, Tōpa-Tim, Niya, Endere, and Rawak all show the dome, which is the essential and unvarying feature of every Stūpa, raised on a square base, and this again arranged in three stories. The relative proportion between these stories varies considerably in the several structures, and so also do the shape of the dome and the relative height of the cylindrical portion or drum which intervenes between the top story of the base and the cupola proper. But the square shape of the base and its threefold arrangement are constant features, and the former in any case is characteristic also of the great majority of Stūpas in the border-lands of India and Afghānistān. On the other hand the round base, which belongs to an earlier stage of Stūpa construction, is represented in those territories only by a few examples, and seems completely absent in Eastern Turkestān[12].

Architectural type of Turkestān Stūpas.

An interesting notice of Hsüan-tsang, the true significance of which was first recognized by M. A. Foucher, informs us that the Turkestān Stūpas, with their square bases arranged in three stories, strictly conform to the manner of construction prescribed by a sacred tradition current in Buddhist territories beyond the Indus. In describing certain small Stūpas in the vicinity of Po-ho or Balkh, the pilgrim relates how Buddha taught his first two disciples the right way of venerating some relics of his person he had given them. First he took his Saṅghāṭi robe, and folding it into a square shape, laid it on the ground; over it he placed his Uttarāsaṅga, and next over this his Saṅkakṣikā. On the top of these garments he put his begging-bowl turned upside down, on which again he raised his mendicant's staff. 'Thus he placed them in order, making thereby [the figure of] a Stūpa. The two men taking the order, each went to his own town, and then, according to the model which the holy one had prescribed, they prepared to build a monument, and thus was the very first Stūpa of the Buddhist religion erected[13].' There can be no doubt that, as explained by M. Foucher, the three garments, folded into squares, with the largest below and the smallest on the top, represent the three

Traditional account of Stūpa construction.

[10] Compare my *Report on an Archaeological tour with the Bunēr Field Force*, p. 40; pl. viii; also *Indian Antiquary*, 1900, p. 145. In both these ruins the chamber was a cube of approximately 7 feet.

[11] See Cunningham, *Archaeol. Survey Reports*, v., pl. xxii; Foucher, *L'Art du Gandhāra*, i. p. 86.

[12] [For all points bearing on the development of Stūpa architecture in India M. Foucher's lucid and amply illustrated remarks, *L'Art du Gandhāra*, i. pp. 62–98, must now be compared. The arrangement of the Turkestān Stūpas conforms closely to the type of what M. Foucher treats as 'Les Stūpa "Transition"'; see ibid., pp. 72 sqq.]

[13] See *Si-yu-ki*, transl. Beal, i. pp. 47 sq. The correct interpretation of the several features of a Stūpa symbolized in the legend has been given in the note contributed by M. Foucher to M. Chavannes' *Voyage de Song Yun*, p. 17, n. 5. The reference in the latter text explained by Hsüan-tsang's passage is to the first Stūpa in Khotan which, according to the injunction of its legendary founder, Vairocana, is said to have been erected in the shape of a '*Pātra* renversé'; compare below, chap. VIII. sec. v. [See now also Foucher, *L'Art du Gandhāra*, pp. 63 sq.; Watters, *Yuan Chwang*, i. p. 112.]

receding square terraces of the Stūpa base; the begging-bowl turned upside down the dome; and the mendicant's staff the pinnacle, which, mostly of wood and fitted with *Chattras* or umbrellas, adorned all Stūpas.

Development of architectural features in Stūpas.

It would be an interesting task to trace the regions and periods of Buddhist worship in which the orthodox system of Stūpa construction prescribed in this legend actually prevailed; but the time and space available for the present work does not permit of such inquiries. Nor does it appear to me safe, in view of the very limited number of Turkestān Stūpas still sufficiently well preserved for accurate restoration and approximately datable, to draw any conclusion at present as to the development of their main architectural features. It may be assumed that in Turkestān, as in India, the strictly hemispherical dome, so aptly illustrated in the above legend by the begging-bowl turned upside down, represents the oldest shape adopted for relic memorials, and that bulb-shaped domes with exaggerated drums are later. But only a detailed study of all Stūpa remains and Stūpa models still extant on the Indian north-west frontier and beyond could furnish reliable indications as to the chronology of these changes, and even then the data deduced could not be applied with implicit assurance to monuments in widely separated regions.

Oblong mound behind Stūpa.

About 200 feet behind the Stūpa, to the north-west, rises a great oblong mound (see Pl. I), consisting of solid masonry of sun-dried bricks, to a height of 34 feet above the same low gravelly ridge which bears the Stūpa. The whole exterior is far too decayed to permit of an exact reconstruction of the plan and dimensions; nor can its original character and purpose be determined with certainty. Three stories, with traces of cells or large niches built into the brickwork, are roughly recognizable on the south-east face, which here, too, seems to have suffered least from exposure. The plan shows that the length of the structure from south-west to north-east was about 86 feet on the ground-level. Its original breadth seems to have been about 71 feet on the same level, but is more difficult to ascertain, even approximately, owing to the accumulation of débris in front of the south-east face. This is possibly due to the former existence here of a series of platforms or flights of stairs leading up to the higher stories.

Similarity to structures at Turfān.

It would be difficult to form even a conjectural opinion as to the original purpose of this structure, were it not for the indications furnished by some better preserved ruins of a similar type with which the investigations of Russian and German scholars in the vicinity of Turfān have recently made us acquainted. These present themselves as square structures, displaying on each face a series of vaulted niches, arranged in a succession of receding stories, which still contain remains of large stucco statues of Buddhas or Bodhisattvas. The brief description of these ruined buildings, which I take from Dr. D. Klementz's report on his preliminary survey of antiquarian remains at Turfān[14], together with the illustrations given there of two of them, situated at Astāna and Syrkheb, show plainly that their outlines resemble those of a truncated pyramid, suggested by the present appearance of the Maurī-Tim mound. Their dimensions, too, seem in several cases to approach closely those recorded for the latter[15]. Of the chamber or hall which two at least of the Turfān structures contain in their interior, I failed to trace any indication in the case of the Maurī-Tim ruin. Yet I have little doubt that it was built for identical purposes of worship. These quadrangular structures, with their rows of Buddhist

[14] See *Nachrichten über die von der kais. Akademie der Wissenschaften zu St. Petersburg im Jahre* 1898 *ausgerüstete Expedition nach Turfan*, i. pp. 25, 30 sq.; the illustrations on p. 31 and Plate i of the ruins of Astāna and Syrkheb show the niches for statues ranged in two and five stories respectively.

[15] Thus for the structure of Yar-khoto a height of 36 feet is mentioned; the one of 'Idykut-schari' is described as being 70 feet long on one of the faces of the ground-floor, with a total height of 46 feet.

images, appear to have taken the place, *mutatis mutandis*, of the chapel courts which surround the Stūpas of numerous ruined Vihāras in ancient Gandhāra, e.g. Jamālgaṛhī, Takht-i-Bahī, Rānīgaṭ, and show rows of statues, each placed in a separate niche or small chapel. The publication of more detailed information, which may now be expected on the basis of Professor Grünwedel's researches, will, I hope, permit us to form a definite opinion on this point[14].

Scattered over the area between the Stūpa and the mound just described are the remains of two small square cellas, marked *D*, *E* on the plan, and of another small building, *C*, of oblong shape. The low crumbling walls show masonry of sun-dried bricks. Those of *D* and *E* measure outside 26 and 18 feet, respectively, on each face, while the third has a width of 15 feet. The size of the bricks is approximately the same as noted in the Stūpa. The ruin of a fourth small structure, *F*, forming an oblong of 27½ by 21 feet, and containing two rooms, occupies a somewhat lower gravel terrace to the south of the Stūpa, separated by a narrow ravine from the end of the ridge. None of the little buildings preserve indications of their original use, whether as chapels or habitations for attendants; nor did their examination furnish guidance as to the date of the Maurī-Tim ruins. Ancient Chinese copper coins are said to turn up in the vicinity, but no authentic finds came to my notice. That the ruins go back to pre-Muhammadan times, is, of course, quite certain, and is fully recognized by local tradition, which sees in the 'Gumbaz', i.e. the Stūpa, the watch-tower of the fabulous 'King of Chīn and Māchīn' who resided in the 'ancient city' of Hāsa-Tam before Hārūn Boghra Khān destroyed it. Smaller ruins at Maurī-Tim.

Before closing my account of the ancient remains in the neighbourhood of Khān-ui, I must refer to a curious structure situated on the barren 'Dasht' some three miles to the south-west of Maurī-Tim (see Fig. 17). Walls built of clay cast in moulds, and showing a perceptible inward slant, form an oblong, of which the western and eastern faces measure about 23 feet, and those to the north and south about 22 feet. The walls, which still rise to a height of 16½ feet above the ground, are 3½ feet thick, and clearly show horizontal layers, about 1½ feet in height, corresponding to the wooden moulds in which, after the fashion still prevailing throughout Turkestān, the clay used in their construction was stamped. The foot of the walls, especially at the corners, has, as the photograph shows, suffered damage owing to the erosive action of the wind-driven sand. The structure does not appear to have ever been roofed, nor was it provided with a door, access to the interior being now obtained by a narrow hole cut into the east wall. Ruin of *Kaptar-Khāna.*

The name *Kaptar-Khāna*, 'the Pigeon House', by which the ruin is popularly known, is derived from the little niches, surmounted by pointed arches and looking like pigeon-holes, which line the whole of the inner sides of the walls. They are arranged in six rows, each row containing 15 holes on the western and eastern sides, and 13 holes on the shorter sides to the north and south. Each hole is about 11 inches high, with a width and depth of 10 inches, and appears to have been originally intended for the reception of human bones; for the ground inside is thickly strewn with bones, and has, according to the local information received by me, always been seen in this condition. Nothing at or near this desolate structure afforded evidence as to its date, though the relatively good preservation of the walls seems to preclude the assumption of great antiquity. Its shape and apparent purpose curiously recall a columbarium. Neither Buddhist custom as developed in India nor Muhammadanism would allow of such a disposal Receptacles for human bones.

[14] The general resemblance pointed out by Dr. Klementz (ibid., pp. 31 sq.) between the plan of those Turfān structures and the great temple of Bōdhgayā cannot be denied; but to give to the latter the character of a Stūpa is certainly erroneous.

of human remains; nor would it be in agreement with orthodox Zoroastrian practice. Unless, therefore, we may recognize in this strange ruin a relic of the times when the territory of Kāshgar contained a considerable community of Nestorian Christians, we must look to the Chinese side for some explanation of its doleful purpose.

Section III.—THE OASIS AND CITY OF YARKAND

March to Ordam-Pādshāh.

On the 11th of September, 1900, I finally set out from Kāshgar for the journey to Khotan. Avoiding the ordinary caravan route, I chose for the march to Yarkand the track which crosses the region of moving sands around the popular shrine of Ordam-Pādshāh, and joins the main road from Kāshgar and Yangi-Hisār at the oasis of Kizil. The account contained in my Personal Narrative[1] renders it unnecessary to describe here my route or the appearance of that much-frequented pilgrimage place in the desert. Before reaching the latter I received information of two old sites (*kōne-shahr*) of some kind. One of these, known as *Kizil-Debe*, and situated about five miles to the north-west of the Yakshamba or Sunday Bāzār of Khān-arik, was visited by Surveyor Rām Singh. He reported having seen there only waste ground strewn with such minute débris as is observed at Hāsa-Tam, and scanty traces of a low platform or wall built of clay. The other old site, called *Bai-khān*, and situated approximately 16 miles to the south-east of the Bāzār of the Achchik oasis, I vainly attempted to reach on my march to Ordam-Pādshāh. My failure in this attempt, explained in my Personal Narrative, need scarcely give rise to much regret; for, according to the descriptions received, this locality, too, shows only the fragmentary débris characteristic of those completely eroded sites known as 'Tatis', which we shall soon have occasion to notice in the Khotan region.

Shrine of Ordam-Pādshāh.

Ordam-Pādshāh, a desolate Ziārat surrounded on all sides by high dunes of drift-sand, owes its sanctity to the pious belief that it marks the spot where the holy Sultān Arslān Boghra Khān, a grandson of Satok Boghra Khān, with a host of faithful followers of Islām, while engaged in prayer, succumbed to a treacherous attack of the infidel army from Khotan. The legend of the royal martyr is told at length in the Tadhkira-i Satok Boghra Khān, full extracts from which have been given by Dr. Bellew[2]. No ancient remains of any kind appear to exist near this pilgrimage place; yet in view of the recurring proofs as to the continuity of local worship in these regions, the question seems justified whether Ordam-Pādshāh, like many a Mazār about Khotan, has not inherited its fame from some earlier Buddhist shrine in the same locality[3].

Ordam-Pādshāh lies undoubtedly on the most direct line connecting Kāshgar with the oasis of Yarkand; and consequently the tradition, which asserts that the main road between these two places once passed there and by the neighbouring shrine of Hazrat-Bēgim, may possibly rest on an historical basis[4]. We shall presently see that the assumption of such a shorter

[1] See *Ruins of Khotan*, pp. 149 sqq.

[2] See *Yarkand Mission Report*, pp. 127 sq.; also Grenard, in *Journal asiat.*, 1900, xv. p. 13.

[3] For a detailed description of the physical conditions of Ordam-Pādshāh, comp. Hedin, *Reisen in Z.-A.*, pp. 232 sq.

[4] See Hedin, ibid., p. 3. The belief there expressed that in old times the belt of vegetation and settlements might have extended in this region further eastwards than at present would receive archaeological support if Dr. Bellew's statement as to the existence of a débris-strewn ancient site near Hazrat-Bēgim (see *Yarkand Mission Report*, p. 38) could be verified. When encamped at Hazrat-Bēgim on September 14, 1900, I received no information as to such a site in the vicinity, and the manifestly erroneous route which the map of the *Yarkand Mission Report* marks for Dr. Bellew's excursion to Ordam-Pādshāh makes the actual position of the reported site appear doubtful.

route would help to account for the distance recorded by Hsüan-tsang for his journey from Ch'ia-sha or Kāshgar to Chê-chü-chia. The latter territory, which the pilgrim's itinerary places at a distance of 500 li to the south-east of Kāshgar, has been identified by V. de Saint-Martin, and after him by others, with Yarkand; but Hsüan-tsang's description, as we shall see, plainly shows that Chê-chü-chia must be looked for further south, in the direction of the present Karghalik.

Antiquity of Yarkand.

Though Yarkand is in all probability a place of considerable antiquity, it is difficult to trace back its name or even its existence as a town of importance previous to the Mongol conquest in the thirteenth century. In the Chinese Yüan-shih Annals relating to the period of Mongol dominion it is frequently mentioned under the names of *Yeh-li-ch'ien*, *Ya-êrh-k'an*, &c., along with the cities of Kāshgar and Khotan[5]. Marco Polo, too, on his way from Kāshgar to Khotan, duly notices the 'Province of Yarcan', where he found Nestorian and Jacobite Christians among a population generally Muhammadan, and was struck by the prevalence of goitre, still to this day a characteristic of its inhabitants[6]. Mirzā Ḥaidar supplies, perhaps, the strongest proof of the antiquity of Yarkand by the mention he makes of the excavations which Mirzā Abā Bakr carried on at the site of 'the old town,' and which were believed to have yielded him great treasures[7]. Mirzā Ḥaidar does not specify the exact position of the area thus excavated, but is careful to note that 'it is not known whether the old town was called Yarkand or whether it had another name'[8]. He adds that in the days of his ancestors 'Yārkand was a companion city to Yāngi-Hisār'.

Yarkand during the T'ang period.

I believe that we may attribute special significance to the latter remark; for it helps to explain why neither Hsüan-tsang, our safest guide in the old topography of this region, nor the records of the T'ang period furnish any definite notice which can be referred to Yarkand. We have seen above that Hsüan-tsang's territory of Wu-sha, which lay to the east of the Muztāgh-Ata range, extended on the south to the Śītā, i.e. the Yarkand river[9]. It must thus have comprised not only the greater part of the present district of Yangi-Hisār, but also the whole of the Yarkand oasis, which tradition, as well as modern administrative division, confines to the small but fertile tract on the left bank of the river after its emergence from the mountains[10].

[5] See Bretschneider, *Med. Researches*, ii. pp. 47 sq. I follow the ordinary modern pronunciation in spelling the name as *Yarkand*. The form *Yārkand* is frequent in Muhammadan texts. For an etymology see below, note 8.

[6] See Yule, *Marco Polo*, i. p. 187.

[7] Comp. *Tārīkh-i-Rashīdī*, pp. 256 n., 257, 296.

[8] See ibid., p. 296. The name Yarkand bears a Turkī look, and may contain in its first part the Turkī word *yār*, which is used throughout Eastern Turkestān for the designation of ravines deeply cut into alluvial (loess) soil, and is of common occurrence in Turkestān local names (see Hedin, *Reisen in Z.-A.*, p. 356. The meaning 'alluvial terrace' there given to *yār* does not accord with the application of the word as observed by me from Kāshgar to Khotan). The second part is clearly the Turkī word (probably of Īrānian origin) pronounced *kand* or *kant*, meaning 'cultivated land, agricultural settlement'; comp. Shaw, *Turkī Vocabulary*, p. 168.

[9] Compare above, pp. 42 sq.

[10] This inclusion of the Yarkand oasis in Wu-sha accounts for the mention of jade made in Hsüan-tsang's above-quoted description of that territory; see above, p. 44. Mirzā Ḥaidar (*Tārīkh-i-Rashīdī*, p. 298) when describing the Yarkand river's course near the town, distinctly states that jade is found in it, and this statement is confirmed by modern information. According to a notice which Mr. Macartney kindly obtained for me from Sun Ssŭ-yeh, Chinese Munshī of the Kāshgar Agency, jade used to be dug until 50 or 60 years ago near a village, the name of which is given as *I-kan-ch'i* 一桿奇, and which is described as situated to the south of Yarkand city, near or on the Yarkand river. From Muhammadju, my Yarkandī servant, I learn that jade pebbles are 'fished' out of the river-bed in flood time near the village of Kōzumal, on the left bank of the Yarkand river, and apparently some 16 miles to the SSW. of the city. Jade mines are known to exist at several points in the mountains drained by the Zarafshān or Yarkand river, and it is thus easy to realize that jade pebbles occur also in the rubble beds deposited by the river after its entry into the plain, just as they are found in the beds and near the banks of the Yurung-kāsh and Kara-kāsh rivers of Khotan. The *Hsin-chiang-chih-lio*, a treatise of the eighteenth century, notes for the river passed south of Yarkand the popular designation of *Yü-ho* 'jade river'; *J. as.*, 1846, viii. p. 245.

Taking into account that the territory of Wu-sha, according to Hsüan-tsang's definite statement, was at his time in dependence on the rulers of Chieh-p'an-t'o or Sarīkol, it is easy to understand why the city corresponding to the present Yarkand, whether its capital or only one of its chief centres, was passed over in silence by both Hsüan-tsang and the T'ang Annalists. The importance of Yarkand or its earlier representative must at that period have been still further impaired by the existence, immediately to the south, of the separate kingdom variously designated in our Chinese sources as Chu-chü-po, Chu-chü-p'an, or Chê-chü-chia, which intervened between Wu-sha and Khotan, and, as will be seen below, must be identified with the present district of Karghalik.

The *So-ch'ê* of Han Annals.

During the supremacy of the Han Dynasty the latter territory was split up among a number of small chiefships, and possibly we may connect with this political condition the fact that the Annals of that early period show us a relatively important kingdom, the Chinese designation of which, 莎車, is variously transcribed as *So-ch'ê*, *Sha-keu*, *So-khiu*, *So-kü*, *So-kiü*, &c, in a position that probably corresponds to the present Yarkand as assumed by Chinese geographers[11]. The account of the *Ch'ien Han shu*, as extracted by Wylie, places the kingdom of So-ch'ê, with its capital of the same name, at a distance of 560 li from Su-lê or Kāshgar, the direction being variously stated as south or east instead of the true one which lies between these two[12]. A still clearer topographical indication is contained in the notice that the territory of 子合 *Tzŭ-ho* (Tseu-ho or Tse-ho in M. Chavannes' transcription), which can safely be identified with Karghalik and the valleys above it, adjoined So-ch'ê from the south[13]. The population of So-ch'ê is estimated at 16,373, nearly double that of Su-lê, with 3,049 trained troops, which indicates a principality of some consequence. A mention of some dynastic changes during the reign of the Emperor Hsüan Ti (73–49 B.C.) seems to imply special relations of its rulers with the powerful Wu-sun nation, while from the list of Chinese official posts there located it may be concluded that, in the administrative system intended to maintain the protectorate of the Hans in the country, some importance was attached to So-ch'ê. At the commencement of the Later Han period So-ch'ê is said to have exercised supremacy over Khotan[14].

Commercial importance of Yarkand.

My stay at Yarkand did not bring to my notice ancient sites in its vicinity nor any antiquarian objects of pre-Muhammadan origin. On the other hand, it helped me to realize the exceptional advantages which the town enjoys as a commercial centre. Yarkand undoubtedly owes its old-established prosperity and its flourishing trade to its position at the point where the great routes from Khotan, Ladāk, and the Oxus are joined by those leading to Kāshgar and the north-eastern part of the Tārīm Basin. The abundance of local produce favours the growth of a large town population, and this, with its quasi-cosmopolitan colonies drawn from all parts of the Oxus Valley, from Ladāk, Baltistān, Afghānistān, and the border regions of China, reflects the true causes of Yarkand's importance. 'Hiarchan, the capital of the kingdom of Cascar', thus Benedict Goëz tells us, 'is a mart of much note, both for the great concourse of merchants, and for the variety of wares. At this capital the caravan of Cabul merchants reaches its terminus; and a new one is formed for the journey to Cathay[15].' Though the

[11] Compare, besides Wylie's paper quoted below, Bretschneider, *Med. Researches*, ii. p. 47; Ritter, *Asien*, v. pp. 405 sq., where earlier extracts from the Han Annals by De Guignes, Rémusat, P. Hyacinth are quoted; also Julien, *J. as.*, 1846, viii. p. 245; Franke, *Zur Kenntniss der Türkvölker*, pp. 51, 70.

[12] See Wylie, 'Notes on the Western Regions,' *J. Anthrop. Inst.*, x. pp. 47 sq.

[13] Compare Wylie, ibid., p. 31. The reference to the relative position of *So-ch'ê* and *P'u-li* is less instructive, since the latter territory, which seems to have been situated somewhere on the Upper Yarkand river, is placed in different passages either 740 li to the south-west or 540 li to the west; see ibid., pp. 47, 32.

For *Tzŭ-ho* (Tseu-ho) compare Chavannes, *Voyage de Song Yun*, p. 19, note 4; and below, p. 92.

[14] See Rémusat, *Ville de Khotan*, p. 5; below, chap. VII. sec. iii.

[15] Yule, *Cathay*, ii. p. 563.

political centre has shifted from Yarkand since the re-establishment of Chinese rule, the above description still holds good; and we may well conclude that the flourishing condition of the city which Marco Polo's account also indicates, was maintained from early times independent of political predominance.

Section IV.—KARGHALIK IN CHINESE RECORDS

Oasis of Karghalik.

We have seen already that Hsüan-tsang's itinerary, after leaving Kāshgar, took him over a distance of 500 li in a south-easterly direction, to the kingdom of Chê-chü-chia. Since the pilgrim distinctly mentions the crossing of the Śītā or Yarkand river *en route*, there can be no doubt that Chê-chü-chia must be identified with the present district of Karghalik. The latter may be roughly described as comprising the hill tracts which slope down from the snowy ranges confining the valley of the Raskam Daryā or Upper Yarkand river, and at their northern foot are bordered by the lower course of that river and the sands of the Taklamakān. Where the streams draining these hills debouch into the plain, either to carry their water into the Yarkand river, or to be absorbed in irrigation, we find oases of remarkable fertility formed on the rich alluvial soil. The largest of these, with a town of considerable size and importance, is the oasis of Karghalik, watered by the Tiznaf river and a number of smaller streams; it is joined on the north by a belt of highly cultivated ground with numerous villages, which extends unbroken for about 25 miles from Karghalik town to the right bank of the Yarkand river.

Hsüan-tsang's *Chê-chü-chia.*

It remains doubtful whether the capital of Chê-chü-chia, which Hsüan-tsang describes as having a circumference of twelve li, and defended by high and solid walls[1], is to be located at the site of the present Karghalik or perhaps further north, where, not far from the Yarkand river, we find another large and flourishing settlement at Posgām-Bāzār. But it is certain that his description of the territory, the circuit of which he estimates at a thousand li, or ten marches, accurately fits the district of Karghalik. 'La population est nombreuse; les montagnes

[1] I follow Julien's translation, *Mémoires*, ii. p. 221. This seems more accurate than that of Beal, *Si-yu-ki*, ii. p. 307, which makes the capital 'hemmed in by crags and mountain fastnesses.' In order to find a position approximately corresponding to this description, we should have to look for the old capital at least two or three marches further south, in the direction of Kök-yār. But apart from the fact that the confined nature of that valley would scarcely be suited for a capital, such a location could not be reconciled with the distance of 500 li, or five marches, indicated by Hsüan-tsang between Kāshgar (*Ch'ia-sha*) and Chê-chü-chia. We know that the pilgrim, in estimating distances from one territory to the neighbouring one, regularly records that from capital to capital. Now the estimate of five marches from Kāshgar may be accepted as approximately right, if the capital of Chê-chü-chia is to be placed at Karghalik, or rather some distance further north. But in order to reach Kök-yār in five days, a distance of over 180 miles from Kāshgar, the pilgrim would have had to make forced marches of a kind with which we cannot reasonably credit him.

In this connexion we may note a point in Hsüan-tsang's itinerary from Kāshgar to Chê-chü-chia which is in need of elucidation. The *Mémoires*, ii. p. 220, tell us: 'En sortant de ce pays, il fit environ cinq cents li au sud-est, passa la rivière *Si-to* (Śītā), franchit un grand passage de montagne couvert de sable, et arriva au royaume de *Tcho-keou-kia* (Chê-chü-chia).' If strict topographical sequence must be assumed in the description of the route, we should have to look for the great 'passage de montagne couvert de sable' to the south of the Zarafshān, and we could in that case identify it only with the 'Dasht' at the foot of the hills which the most direct route from Kāshgar to Karghalik (passing through Yaka-arik and avoiding Yarkand city) skirts just north of Karghalik. But taking into account the general topography of the route, it appears to me far more probable that Hsüan-tsang's reference is really to the curious ridge of much-decayed hills which runs into the desert south of Ordam-Pādshāh, and which, at the point where the pilgrim's track from the latter place crosses it, bears the 'Ziārat' of Ulūgh-Nishān (see *Ruins of Khotan*, p. 158; Hedin, *Reisen in Z.-A.*, p. 3). The old route from Kāshgar to Yarkand and Karghalik is, as shown above, pp. 86 sq., likely to have followed this, the shortest line; and the unexpected passage of an isolated ridge in the midst of a level expanse of sandy desert may have impressed the pilgrim, just as it surprises the modern traveller.

et les collines se touchent. De vastes plages sont couvertes de sable et de pierres. Ce royaume est voisin de deux fleuves; la culture des grains et des arbres fruitiers y est florissante. Il abonde surtout en raisins, en poires et en prunes. Le vent et le froid règnent en toute saison[2].' The latter remark applies with particular force to the higher parts of the valleys comprised in the district. But it is easy to recognize also the previous reference to the vast slopes of pebble-strewn 'Dasht' which form the glacis of the outer ranges, and to the plains of moving sand extending further northward. The two rivers alluded to are manifestly the Zarafshān and the Tiznaf; the oases irrigated from them and their tributaries still abound in rich orchards and arbours. Coming from the north, just as Hsüan-tsang did, I was struck by the variety and luxuriant growth of the fruit and other trees to be seen around Karghalik. The rural scenery thus created reminded me more of Kashmīr than any other I saw in the country (see Fig. 15)[3].

Buddhism at Chê-chü-chia.

Hsüan-tsang is less laudatory in his account of the people. 'The men are passionate and cruel; they are false and treacherous, and in open day practise robbery . . . . Their politeness is very scant, and their knowledge of literature and the arts equally so. They have an honest faith, however, in the three precious objects of worship, and love the practice of religion. There are several tens of Saṅghārāmas, but mostly in a ruinous condition; there are some hundred followers who study the Great Vehicle[4].' The latter statement plainly shows that the Buddhist establishments of the country had seen their most flourishing days by the time of the pilgrim's visit; yet he notes the interesting fact that the canonical texts of the Mahāyāna school as preserved by them were more numerous than elsewhere. 'Parmi les lieux où est parvenue la loi du Bouddha, il n'en est aucun où la doctrine du Mahāyāna soit aussi florissante. Elle embrasse dix recueils renfermant chacun cent mille Ślokas. Depuis qu'elle a été introduite dans ce pays jusqu'à nos jours, elle s'est étendue d'une manière remarquable[5].'

The writing of Chê-chü-chia.

Whatever may have been the cause of this special affluence in canonical texts, it is certain, in view of what we had occasion to remark above of the territorial division between the two great Buddhist schools in the 'Western Regions', that this predominance of the Mahāyāna doctrine in the convents of Chê-chü-chia indicates close connexion with the Buddhist Church as established at Khotan[6]. In full agreement herewith we learn that the written characters here used were the same as those of Ch'ü-sa-tan-na or Khotan. But of the spoken language Hsüan-tsang tells us that it was different. Hsüan-tsang's statement as to the writing of Chê-chü-chia is borne out and explained by identical notices of Sung Yün and the T'ang Annals to be discussed below[7]. These inform us that the written characters of that territory were the same as those of the *P'o-lo-mên* or Brahmans; and we know in fact by the manuscript finds from the ancient sites of Khotan that it was the Brāhmī script of India, with but slight local modifications, which prevailed in Khotan during the T'ang period.

Language spoken at Chê-chü-chia.

Hsüan-tsang's remark, however, regarding the spoken language is at variance with the earlier notices of Sung Yün and his fellow-traveller, Hui-shêng, to be referred to below, in which the language of the people of Chu-chü-po (i. e. Chê-chü-chia) is declared to be similar to that of Yü-t'ien or Khotan[8]. We know that these two pilgrims visited both Khotan and Karghalik about the year 519 A. D., and their observation as to the language they heard about Karghalik

[2] See *Mémoires*, ii. p. 221. Beal, ii. p. 307, translates: 'Stony districts spread in every direction,' which seems less accurate.

[3] See *Ruins of Khotan*, pp. 174, 179.

[4] Beal, *Si-yu-ki*, ii. p. 308. [5] *Mémoires*, ii. p. 222.

[6] See above, pp. 56 sq.

[7] Compare pp. 91 sq.; Chavannes, *Voyage de Song Yun*, p. 20; *Turcs occid.*, p. 124.

[8] See below, p. 92; Chavannes, *Voyage de Song Yun*, p. 19, note 4.

15a

MECCA PILGRIMS FROM KĀSHGAR.

15

MOSQUE AND TANK AT YETIMLUKUM, KARGHALIK.

16

YETIMLUKUM MAZĀR WITH CEMETERY, NEAR KARGHALIK.

receives confirmation from the close affinity noted by the T'ang Annals between the languages of Sarīkol and Khotan[9]. This affinity is fully accounted for by the Īrānian character of both, as proved for Khotan by the documents from Dandān-Uiliq in the so-called 'unknown language'[10]; and we find no difficulty whatever in assuming that the territory of Karghalik, intervening between Sarīkol and Khotan, similarly held a population speaking an Eastern Īrānian dialect.

Pakhpos, a remnant of early population.

When discussing above the evidence of the T'ang Annals as to the ancient tongue of Sarīkol and Khotan we had occasion to refer to the Pakhpo tribe in the more remote mountain tracts of the Karghalik district. Though their language still awaits investigation, there is little doubt that, on anthropological grounds, these hill-men must be classed with the Galcha or hill Tājik population found to the present day in Sarīkol and the valleys further west[11]. If we may recognize in the Pakhpos a remnant of the original population of Karghalik, in this light Hui-shêng's statement becomes intelligible. The different ethnic conditions noted by Hsüan-tsang may have been a result of the domination of the Yeh-ta or White Hans, which Hui-shêng particularly mentions, and may have been confined to the lower and more accessible parts of the district. Through these alone the pilgrim is likely to have passed on his way to Khotan.

Buddhist legend of Chê-chü-chia.

He refers, indeed, to a high mountain mass on the southern border of the district, with its vegetation stunted through intense cold[12]; but it is clear that his mention of it is solely due to a pious legend manifestly heard at a distance. Numerous 'niches and stone chambers' on the flanks of the great mountain were believed to have been miraculously frequented by Arhats from India, who obtained Nirvāṇa there. Popular belief in Hsüan-tsang's time still supposed certain deep rock caverns to be tenanted by three Arhats whose minds had become extinguished in complete ecstasy, while the hair of their withered bodies was still growing.

***Chu-chü-po* in T'ang Annals.**

M. Chavannes was the first to recognize that the notices in the T'ang Annals and Sung Yün's itinerary of a kingdom variously designated as *Chu-chü-po* or *Chu-chü-pan*, immediately to the west of Khotan, referred to the same territory which Hsüan-tsang calls Chê-chü-chia[13]. The special notice of the T'ang-shu, as extracted by M. Chavannes from chapter CCXXI[14], mentions this kingdom between those of Su-lê (Kāshgar) and Ho-p'an-t'o (Sarīkol), and tells us: 'Le *Tchou-kiu-po* (Chu-chü-po) est appelé aussi *Tchou-kiu-pan* (Chu-chü-pan); c'est le royaume de *Tse-ho* (Tzŭ-ho) de l'époque des Hans. Il s'est annexé et possède le territoire des quatre peuples appelés *Si-ye* (Hsi-yeh), *P'ou-li* (P'u-li), *I-nai* et *To-jo*[15]. Il est à mille li droit à l'ouest de Yu-t'ien (Khotan) et à trois cents li au nord des Ts'ong-ling. À l'ouest il touche au Ho-p'an-t'o (Tach-Kourgane); à neuf cents li vers le nord, il se rattache à Sou-le (Kachgar); à trois mille li au sud se trouve le royaume des femmes. Il a deux mille soldats d'élite; il honore la loi de Bouddha; l'écriture y est la même que celle des P'o-lo-men (Brahmanes-Hindous).'

Territories comprised in Chu-chü-po.

The position here indicated clearly coincides with that of the present Karghalik district and Hsüan-tsang's Chê-chü-chia, though the estimated distances to Khotan and Kāshgar are somewhat in excess of those given in the *Hsi-yü-chi*. The reference to several earlier territories

[9] Compare above, pp. 26 sq.

[10] See below, chap. IX. sec. v.

[11] Compare above, p. 25.

[12] See *Mémoires*, ii. p. 222. Here, too, Julien's translation seems more accurate than that of Beal, ii. p. 308.

[13] See *Turcs occid.*, p. 123, note 1; *Voyage de Song Yun*, p. 19, note 4. For references to the passages offering the various forms of the name *Chu-chü-po* (*Tchou-kiu-po*) 朱俱波, 朱駒波 and *Chu-chü-pan* (*Tchou-kiu-pan*) 朱俱般, 朱駒半, &c., compare *Turcs occid.*, p. 366.

[14] See *Turcs occid.*, p. 123.

[15] A passage of the *Pei shih* quoted by M. Chavannes (*Voyage de Song Yun*, p. 19, note 4), which mentions the same territory under the name of *Hsi-chü-pan* (*Si-kiu-pan*) 悉居半, similarly identifies it with the ancient kingdom of *Hsi-yeh* (*Si-ye*), which was also called *Tzŭ-ho* (*Tse-ho*).

which had become absorbed in Chu-chü-po supports the identification; for notices in chapter XCVI of the *Ch'ien Han shu* concerning Tzŭ-ho, Hsi-yeh, P'u-li, and I-nai, leave no doubt that these petty territories were situated close together to the south of the present Yarkand[16]. The passage referring to Hsi-yeh records that its ruler bore also the title of king of *Tzŭ-ho*; and in agreement with this indication of a special connexion between the two places, modern Chinese geographers identify Hsi-yeh and Tzŭ-ho with the adjacent village tracts of Yul-arik and Kök-yār[17]. Whether this identification rests on historical fact or only on learned conjecture, it does not supply us, as recognized by M. Chavannes, with any cogent reason for placing the political centre of the united territory of Chu-chü-po so far to the south as Kök-yār. This place is separated by a considerable distance from the direct route between Kāshgar and Khotan, and since Hsüan-tsang's itinerary leads us to look along this route for the capital of the kingdom as it existed in his days, the neighbourhood of the present Karghalik appears a far more likely position.

Fa-hsien's *Tzŭ-ho*. The mention of *Tzŭ-ho* 子合 as an integral part of the kingdom of Chu-chü-po is of interest, as enabling us to trace with certainty the route which Fa-hsien followed immediately after leaving Khotan. From the narrative of his travels we learn that, starting from the latter place, it took the pilgrim twenty-five days to reach Tzŭ-ho, where he found the ruler devoted to Buddhist law, and around him 'more than a thousand monks, mostly students of the Mahāyāna'.[18] Fa-hsien and his party, after a stay of fifteen days at Tzŭ-ho, went south for four days before they found themselves among the Ts'ung-ling range. From this statement it becomes probable that in his time, too, the capital of the territory lay in a relatively open position to the north of the Kun-lun range. We have already seen that Fa-hsien's immediate goal after leaving Tzŭ-ho was Sarīkol, which he probably reached by the route passing through the mountains south of the Zarafshān[19].

Sung Yün's notice of Chu-chü-po. Sung Yün also, travelling westwards from Khotan in 519 A.D., passed along the road which crosses the Karghalik district. The account of his journey refers to the latter as the kingdom of *Chu-chü-po*, and informs us that it produced cereals in plenty[20]. Its inhabitants lived in the mountains and used leavened wheat for their food. As the killing of animals was forbidden among them, the only meat they partook of was that of animals which had died a natural death. Of their customs and language it is noted that they resembled those of Yü-t'ien or Khotan, while their writing was the same as that of the P'o-lo-mên or Brahmans. The circumference of the kingdom was estimated at five days' journey, which, compared with Hsüan-tsang's thousand li, seems to indicate that the territory had not then attained its later dimensions[21].

Notices in *Pei shih*. Sung Yün's fellow pilgrim, Hui-shêng, is the source of a briefer notice which the *Pei shih* (composed about 644 A.D.) furnishes of the same territory, called here *Chu-chü* 朱居[22]. Besides

[16] For a translation of the notices concerning these petty chiefships, see Wylie, *J. Anthrop. Inst.*, x. pp. 31 sq. Various discrepancies as to the relative positions, &c., of the localities mentioned make a detailed analysis of these passages impracticable for those who are unable to consult and control the original texts. It is, however, clear that agricultural produce in all these territories was small and had to be supplemented from So-ch'ê (i.e. the oasis of Yarkand). Of Tzŭ-ho it is particularly mentioned that jade is found there.

[17] Compare *Voyage de Song Yun*, p. 19, note 4.

[18] See Legge, *Travels of Fâ-hien*, p. 21.

[19] See above, p. 28.

[20] Compare *Voyage de Song Yun*, pp. 19 sq.

[21] The same conclusion must be drawn from the statement that Sung Yün, who had entered the kingdom of Chu-chü-po on the 29th day of the seventh month, passed at the commencement of the eighth month into the territory of Ho-p'an-t'o or Sarīkol. It is evident that the latter extended in the year 519 further to the east than in Hsüan-tsang's time. For the probable route followed by Sung Yün to Sarīkol, see above, pp. 29 sq.

[22] See Chavannes, *Voyage de Song Yun*, p. 19, note 4.

confirming the essential points in Sung Yün's account, it informs us that the people all worshipped Buddha, and that they were subject to the Yeh-ta or Hephthalites.

Of the political relations of the territory the Chinese Annals furnish some further information. In another passage of the *Pei shih*, which mentions the district by the name of Hsi-chü-pan (*Si-kiu-pan*), and to which reference has already been made in a note above[23], we find it recorded that an embassy from there arrived at the Imperial court at the commencement of the *t'ai-yen* period (435–439 A. D.), and that tribute was subsequently received without interruption. The previously mentioned notices of the *Ch'ien Han shu* show that the small hill states, which later on became absorbed in Chu-chü-po, all acknowledged the authority of the Governor-General of the Western Regions during the period of the Han supremacy. The subsequent subjection of Chu-chü-p'an to the power of the Hephthalites is attested also by the Annals of the Liang dynasty, covering the period 502–556 A. D.[24]

When the Western Turks succeeded the Hephthalites as the paramount power in Central Asia, Chu-chü-po undoubtedly was among the many states subject to their sovereignty. In consequence we find it included, along with Kuchā, Khotan, Kāshgar, and Sarīkol, in the region which the Emperor T'ai-tsung demanded in 646 from Shê-kuei, Kagan of the Western Turks, in exchange for the hand of a Chinese princess[25]. But already, in the year 639, the ruler of Chu-chü-po, in accord with the ruler of Kāshgar, had deemed it advisable to dispatch an embassy with products of his territory to the Imperial court when the victorious advance of T'ai-tsung's troops towards Turfān foreshadowed the Chinese conquest of the Tārīm Basin[26]. This conquest was actually completed in 659, when the rising of the Turkish chief Tu-man, who had attached to himself the states of Kāshgar, Karghalik (Chu-chü-po), and Sarīkol, was subdued[27]. Thus we find the kingdom of Chu-chü-pan mentioned among the units of the administrative organization which the Chinese in the same year established for the region controlled by the 'Four Garrisons'[28]. No further reference to this territory is made in the Chinese records accessible to me; nor did my short stay at Karghalik acquaint me with any ancient remains which could throw light on its history.

Political Relations of Chu-chü-po.

[23] See above, p. 91, note 15.

[24] See *Turcs occid.*, p. 224.

[25] Compare *Turcs occid.*, pp. 32, 59, 266; also above, p. 59.

[26] See *Turcs occid.*, p. 121; and above, p. 61.

[27] See *Turcs occid.*, pp. 72 sq.; above, p. 60.

[28] Compare *Turcs occid.*, pp. 141, 268; above, pp. 59 sq.

CHAPTER V

THE ROUTE FROM KARGHALIK TO KHOTAN: ITS ANCIENT TOPOGRAPHY AND REMAINS

Section I.—BY THE DESERT EDGE TO KHOTAN

Road from Karghalik to Khotan.

The direct road from Karghalik to Khotan, along which I travelled between the 2nd and 12th of October, 1900, leads in an almost straight line south-eastwards along the edge of the Taklamakān. By far the greater part of the route lies over absolutely barren ground, gravel 'Dasht' or else bare loess steppe, fringed on the north by the moving sands of the desert. Yet at intervals small oases are met with, which offer convenient halting-places and render this route far more convenient for trade than it could otherwise be. We find them écheloned along the route wherever the streams which drain the snowy rampart of the Kun-lun range north of the Kara-kāsh Valley emerge from the glacis-like gravel waste at the foot of the hills and strike the edge of the desert. In order to appreciate the importance of these small oases from the point of view of historical topography, as well as the interest of their antiquarian remains, brief reference must be made to the most striking of their physical features.

Physical features of oases on route.

The water which the streams of the Kilian, Sanju, and Duwa valleys carry down from the mountains during the floods of the spring and early summer is, on its passage over the sloping 'Dasht', locally designated as 'Sai', distributed in a number of channels, and lower down is almost wholly absorbed in the irrigation of the small but flourishing settlements of Gūma, Moji, Zanguya, and Piālma. Only a small portion of this flood-water, characteristically called *ak-su*, or 'white water', from the colour of the melting snows, is finally allowed to escape into the desert beyond, where its shallow courses are soon lost amidst the dunes of moving sand. The limited but more constant supply of water, known as *kara-su*, 'black water', which springs and pools furnish at points where the subterraneous drainage from those valleys and from the 'Sai' rises to the surface, is still more carefully husbanded for purposes of cultivation.

The situation of these settlements between the barren Dasht and the dunes of the desert, which ever threaten to overrun and envelop them, is shared by most of the oases of Eastern Turkestān. Yet there is an important physical feature which distinguishes them, and which, owing to its antiquarian bearing, as illustrated by my explorations, deserves to be specially noticed. The great oases of the Tārīm Basin, which we have hitherto had occasion to discuss or to mention, Kāshgar, Yangi-Hisār, Yarkand, Karghalik, and a number of other important ones like Khotan, Keriya, Ak-su, &c., owe their water-supply, and with it their very existence, to large rivers which carry water far beyond these irrigated areas. But the oases passed on the main route between Karghalik and Khotan, and others visited by me further east along the edge of the Taklamakān, practically mark the extreme point reached by the streams which bring them fertility.

Now my explorations at ancient sites east of Khotan demonstrated that such 'terminal oases', as we may appropriately call them, possess an archaeological interest considerably in excess of their geographical importance. Through a variety of powerful causes, partly physical and partly historical, which we shall have to examine in detail hereafter, these 'terminal oases' are liable to undergo considerable changes in position and extent at different periods[1]. Owing to the relative frequency of these shiftings within historical times, and the close vicinity of the desert sand which preserves whatever it buries, ancient sites, with structures intact or other recognizable remains, can be traced far more readily in the vicinity of such oases than in the case of oases of the other type. There the variations of the boundary line between oasis and desert have ordinarily been small in proportion to the whole occupied area, and ancient sites within the latter have always been subject to the danger either of being constantly built over and thus effaced, or else of being effectively hidden under the silt deposit resulting from prolonged cultivation[2].

Archaeological interest of 'terminal oases'.

It is only if we bear in mind the general observations here indicated that we can fully recognize the interesting evidence which is afforded for the ancient topography of the Karghalik–Khotan route by the antiquarian remains I was able to trace near almost every one of the small oases along it. These remains are not in themselves of striking archaeological value; but they prove conclusively, in my opinion, that, though the area irrigated from each of the above-named streams has undergone considerable change in lateral directions, i.e. to the west or east, the line formed by the terminal oases and followed by the trade route has not shifted for the last thousand years, and probably for a yet longer period. I shall give an account of those ancient remains in the order in which I met with them on my journey towards Khotan. Before, however, proceeding it will be useful to note the historical references we possess for this route, and also to indicate briefly the conditions which seem to explain the stability of the line of oases which it follows.

Line of route through 'terminal oases'.

The need of such an explanation presents itself forcibly before us when we compare this stability with the great changes which, as the account of my subsequent explorations will show, have taken place within historical times in the position of the oases fringing the edge of the Taklamakān east of Khotan, and thus also in the direction of the once important route eastwards dependent on these oases. My explorations at Dandān-Uiliq, the Niya Site, Endere, and Uzun-Tati, establish the fact that there, during the first eight centuries of our era, a series of oases, manifestly of the 'terminal' type, extended much further north into the great desert than the present line of small cultivated areas belonging to the Keriya district. There is also good reason to believe that the once much frequented route which connects Khotan with Cherchen, Lop-Nor, and westernmost China, though perhaps not following throughout the line of terminal oases, led considerably further north than it does at the present day.

Line of oases east of Khotan.

This striking difference between the persistence of the route west of Khotan and the great variation of the line followed by the route eastwards can, I think, be adequately explained if we take into account the diversity of physical conditions. Without going into details, which would require a long digression, I may call attention to some important geographical facts. A look at the map will show that the belt of sterile 'Sai', which separates the line of small

Diversity of physical conditions E. and W. of Khotan.

[1] Compare below the remarks on the terminal oases of Yārtunguz and Endere, chap. xii., Domoko and Gulakhma, chap. xiii.; also on the ancient sites representing terminal oases, of Dandān-Uiliq, chap. ix., Niya, chap. xi., Uzun-Tati, chap. xiii.

[2] Compare below the explanations given for the silt deposits covering the remains of the ancient capital of Khotan, chap. viii. sec. ii.

oases from the foot of the mountains, is far broader to the east than to the west of Khotan. The varying width of this glacis of gravel and detritus washed down from the mountains is in direct relation to the magnitude of the ranges which it skirts. Whereas the mountains to the south of the Karghalik–Khotan route are only offshoots of an outer range of the Kun-lun system, there rises behind the belt of Dasht east of Khotan the mighty rampart of the northern main range.

Water-supply of oases.

Now this difference of the orographic background has a very marked effect upon the water-supply of the strip of fertile loess ground which intervenes between the 'Sai' and the high dunes of the desert, and which that water-supply alone renders capable of cultivation. The main range east of Khotan, with its ample and partly glacier-fed drainage, sends forth a number of relatively large rivers, like those of Chīra, Keriya, Niya, Yārtunguz, and Endere. But much of their water disappears on their passage over the broad 'Sai', the uniform surface of which facilitates the constant formation of fresh divergent channels during the summer floods. The water thus absorbed reappears near the lower edge of the 'Sai' in abundant springs and marshes which, as we shall have occasion to detail hereafter, form an important factor in the irrigation system of all existing oases between the Yurung-kāsh and Niya rivers. The information collected by me at the oases stretching from Chīra to Keriya plainly shows that the level at which this supply of spring-water becomes available for irrigation is subject to frequent and great fluctuations, and the evidence of settlements deserted or newly started within the last few generations confirm the local traditions on this point[3]. It is clear that similar conditions must have prevailed throughout the historical period. If we further consider that the large amount of water carried down by those rivers must in times of prosperity and denser population have afforded wide scope for systematic irrigation works, which were abandoned subsequently during periods of decay, the great variations in the line of oases within this region become intelligible.

The conditions found in the string of small oases along the route west of Khotan are distinctly different. On the one hand the amount of water provided by the streams of Kilian, Sanju, and Duwa, is far more limited; for none of them is fed by glacier-sources, and their drainage area is restricted to comparatively narrow valleys. On the other hand, probably owing to the narrowness of the 'Sai' belt, springs are of relatively rare occurrence, and their supply of water is not a serious factor for irrigation. We can thus understand why none of those streams could in historical times ever have carried their water far beyond the northern edge of the 'Sai', which the line of the existing oases closely follows; and also why those minor shiftings of the cultivated area from south to north, and vice versa, which east of Khotan seem a regular feature wherever irrigation is largely dependent on springs, cannot be traced here.

Irrigation dependent on flood-water.

It is, in fact, mainly upon the floods fed by the melting snows of the spring, and by occasional heavy rain in the mountains, that the oases from Gūma to Piālma depend for their irrigation[4]. It is likely that in early periods, when the population was denser and labour plentiful, means were found to distribute the water of these floods more extensively over the loess terraces, which are found at numerous points between the existing oases, but are not now reached by canals or natural watercourses. The extensive débris-strewn areas, known as 'Tatis', which are passed on the road between Gūma and Moji, support this assumption. But of ancient sites at any distance to the north of the present line of oases I could, notwithstanding careful

[3] Compare, below, the information collected regarding the oases of Domoko and Gulakhma, chap. XIII.

[4] See Hedin, *Reisen in Z.-A.*, pp. 11–18, where ample details are recorded about the facts of cultivation, &c., in these oases.

inquiries, discover no trace; and as a fairly wide belt of desert in that direction shows jungle growth and is regularly visited by hunters, we may attach some value to this negative evidence[5].

Antiquity of present route.

These general topographical considerations, combined with the evidence of the archaeological remains to be noticed, convince me that the present route between Karghalik and Khotan practically coincides with what throughout historical times must have been the direct and most frequented line of communication between Khotan and the regions westwards. The earliest historical reference to it we possess is furnished by the *Ch'ien Han shu*, which in its 'Notes on the Western Regions' places So-ch'ê (Sha-keu) or Yarkand at a total distance of 770 li from Khotan, the direction being first west for 380 li to P'i-shan (Pe-shan) and thence north-west for 390 li to So-ch'ê[6]. The total distance here indicated agrees remarkably well with the eight marches reckoned at present for the journey between Yarkand and Khotan, a distance of about 192 miles by the map[7]. The bearings equally agree, the route from Khotan to the oasis of Gūma, which plainly corresponds to *P'i-shan*, bearing generally west by north or west-north-west, while the bearing of Yarkand to Gūma is almost exactly north-west.

Early travellers on the route.

Fa-hsien and Sung Yün may both be assumed to have travelled by this route westwards; but the records of the journeys of neither furnish any details[8]. The earlier pilgrim and his party are said to have taken fully twenty-five days before they reached Tzŭ-ho (Tseu-ho), in the neighbourhood of the present Karghalik[9]—a superabundant allowance of time, about an explanation of which it would be useless to speculate. During the T'ang period this route from Khotan westwards must have seen many a traveller like Wu-k'ung, of whose passage through Khotan Chinese records have preserved some notice. But the only accounts of the route itself which I can trace—and they are brief—are to be found in an itinerary of the T'ang Annals and in Hsüan-tsang's narrative.

The route described in T'ang Annals.

The former, extracted by M. Chavannes, acquaints us with a number of localities and certain distances between Yü-t'ien (Khotan) and Kāshgar[10]; but unfortunately none of the local names appear at present to be traceable elsewhere, and consequently their identification cannot be certain. Restricting ourselves to that portion of the itinerary which must correspond to the route under discussion, we note that it takes us from Khotan first to the *Wei* passage (葦關) situated 50 li to the west. Thence after a further journey of 620 li, first westwards and then to the north-west, during which *P'o-hai* 渤海 and the river *I-kuan* 繄館 are passed, the town of *Chih-chih-man* 郅支滿城 is reached, also called the district of *Chi-nan* 磧南州.

[5] The fact that the profession of 'treasure-seeker', so common about Khotan, is not represented in the oases west of it, points significantly to the same conclusion. Whatever coins and small antiques were brought to me had avowedly been found at the old sites examined by me close to the present oases.

[6] Comp. Wylie, *J. Anthrop. Inst.*, x. pp. 30 sq., where, however, the distance from P'i-shan to So-ch'ê (Sha-keu) is given as 380 li. The *Hsin-chiang-chih-lio*, as translated by Julien, *J. as.*, 1846, viii. p. 245, in discussing the passage of the Annals, states the distance from P'i-shan to So-ch'ê (So-kiu) as 390 li, which seems more accurate. Regarding the identity of *P'i-shan* (P'i-chan) with Gūma, see below, p. 103.

[7] These marches are not of uniform length, nor are their stages fixed by necessity, account having to be taken of the provision of water, the facility of supplies, and similar considerations important in a journey mainly over barren ground. The most convenient stages at present for a traveller starting from Yarkand would generally be Jilkucha (Yakshamba Bāzār), south of Posgām, Akim-Langar, Chōlak-Langar, Gūma, Zanguya, Piālma, Zawa, city of Khotan. It must be borne in mind that the old capital of Khotan lay at Yōtkan, some six miles further to the west than the present city.

[8] See *Travels of Fâ-hien*, p. 21; Chavannes, *Voyage de Song Yun*, p. 19.

[9] See above, p. 92.

[10] Comp. *Turcs occid.*, p. 123, note 1. The same notice, in a less accurate form, had been translated by Julien from the *Hsin-chiang-chih-lio*, together with certain comments of the author of this compilation, which, however, are of little value; see *J. as.*, 1846, viii. p. 245.

M. Chavannes has already recognized that, judging by the distances indicated, Chih-chih-man corresponds to Hsüan-tsang's Chê-chü-chia, and has thus to be looked for in the vicinity of the present Karghalik. This identification is supported by the measurement of 560 li subsequently given for the total distance between Chih-chih-man and Kāshgar, and by the mention of the 'Bitter Springs' and the 'Yellow Canal' to the north-west of the former place. By the 'Bitter Springs' might be meant the extensive spring-fed marsh which is passed near the left bank of the Tiznaf river, on the road from Karghalik to Posgām, while in the 'Yellow Canal' we probably have a reference to one or other of the large canals which carry the water of the Yarkand river through the oases of Posgām and Yarkand. Nothing definite can be suggested as to the identity of *P'o-hai* and the river *I-kuan*, since their distances from Khotan are not mentioned in the itinerary. We are less in the dark as regards the 'Wei passage', which, judging from its situation, 50 li to the west of the Khotan capital, and the term *kuan*, 'frontier pass, guard-house', used for its designation, was manifestly a fortified station on the main road westwards near the edge of the Khotan oasis. We probably have its modern representative in the fort of Zawa-Kurghān, which was erected for a similar purpose during the short reign of the rebel Habībullah, near the point where the road from Karghalik enters the oasis [11].

Hsüan-tsang's route.

Turning to Hsüan-tsang's account, we learn from both the *Hsi-yü-chi* and the 'Life' that the journey from Chê-chü-chia to Ch'ü-sa-tan-na, or Khotan, took the pilgrim 800 li or eight marches eastwards [12]. The *Hsi-yü-chi* adds that the road was 'skirting along the high mountain passes and traversing valleys'. We have seen already that Hsüan-tsang's kingdom of Chê-chü-chia corresponds to the present district of Karghalik, and that its capital, from which the above measurement is in all probability taken, may be looked for in the vicinity of the town of Karghalik. The indicated length of the journey, eight days, agrees with the assumption that Hsüan-tsang travelled by the line of the present high road; for marching with heavy baggage on camels, as I did in October, 1900, I learned from practical experience that the distance of about 155 miles (by the map) between Karghalik and Khotan could not easily be covered in less than eight days. The caravan of the pious traveller, loaded as we know it to have been with plentiful collections of MSS. and sacred objects, would certainly have needed that time for the journey.

An approximate gauge as to his rate of travel is afforded by the fact that he places the spot where the sacred rats were worshipped by all wayfarers, and which, as we shall see below, is undoubtedly marked by the modern shrine of Kumrabāt-Pādshāhim, on the present Karghalik–Khotan route, at 150–160 li to the west of the Khotan capital [13]. The distance from this spot to Yōtkan, the site of the old capital, is 24 miles by the route as shown on my map, and as the road is quite easy, except for the 4 or 5 miles of drift sand to the east of Kumrabāt-Pādshāhim, this cannot possibly represent more than 30 miles of actual walking distance. A second locality on the route, the town of P'o-ch'ieh-i (*Po-kia-i*), which Hsüan-tsang places at 300 li to the west of the Khotan capital, and which he reached after crossing the western frontier of the kingdom, probably corresponds to the oasis of Piālma, approximately 56 miles by road from Yōtkan [14].

[11] Julien, *J. as.*, 1846, viii. p. 245, renders *Wei-kuan* by 'barrière des roseaux'. If this rendering is justified, could we recognize in the 'Gate of reeds' an allusion to the wide reed-covered expanse through which the Yawa-Üstang flows just after passing Zawa-Kurghān?

The term *kuan* and the position indicated recall to my mind the part which the *Dvāras* or fortified frontier watch-stations have played in ancient Kashmīr; see my translation of the *Rājatarangiṇī*, i. 122 note, and II. p. 391.

[12] See *Mémoires*, transl. Julien, ii. p. 223; transl. Beal, ii. p. 308; *Vie*, p. 278.

[13] See *Mémoires*, ii. p. 232; transl. Beal, ii. p. 315; comp. below, sec. v.

[14] See *Mémoires*, ii. p. 230; transl. Beal, ii. p. 314; comp. below, p. sec. v.

The description which the *Hsi-yü-chi* gives of the character of the route is easily accounted for if the wording is capable of the translation given by Beal, 'skirting along the high mountain passes and traversing valleys.' The mountains south of Kilian, Sanju, and Duwa, together with parts of the snowy main range on the Upper Kara-kāsh river, are fully visible from the route whenever the atmosphere is sufficiently clear. The streams from the mountains where they are crossed by the route show broad and well-marked flood-beds, in parts deeply cut into the loess terraces, as near Gūma, Chudda, Piālma. Further towards the foot of the hills, the valleys from which these streams debouch and the ravines they have cut through the conglomerate strata are also in view [15]. Thus Hsüan-tsang's reference to 'valleys' is quite justified. It is true that Julien renders the first words of the passage by 'il franchit de hauts passages de montagne'. But it must be observed that this description, if taken literally, would be equally inapplicable to any route by which Hsüan-tsang could possibly have travelled within eight days from the central part of the Karghalik District to Khotan [16]. **Hsüan-tsang's description of route.**

Section II.—THE OASIS OF GŪMA

The first trace of remains of archaeological interest was met with at the end of my first march from Karghalik. When nearing the massively built rest-house of Kosh-Langar, erected during Yāqūb Bēg's reign in the midst of a completely sterile steppe, my attention was attracted by a mound visible from afar over the level horizon, which was known to my Karghalik guide by the general designation of *Tim* [1]. It was reached after a ride of about $1\frac{1}{2}$ miles to the north-north-east of Kosh-Langar, and proved to consist of a solid mass of sun-dried bricks measuring about 75 feet in circumference at the base and rising to a height of about 30 feet. The appearance of the ruin suggests an originally conical shape for the upper portion, which would agree with the assumption that the mound marks the remains of a Stūpa. But its state of decay is too far advanced to permit of any certain conclusion. The bricks seem irregular in size, but generally smaller than those in the ruins of Kurghān-Tim and Maurī-Tim. The people frequenting the lonely station are inclined to recognize in the ruin a watch-tower of great antiquity, corresponding to the 'Potais' built by the Chinese at intervals of ten li, approximately equivalent to two miles, along the greater portion of the modern high road between Kāshgar and Khotan [2]. The fact that I could not trace any pottery débris or other remains in the vicinity of the ruin speaks against the site having once been permanently inhabited; yet the very nearness of this ruin to the present route and halting-place is an indication how little the line of the former is likely to have changed for centuries. **Ruin near Kosh-Langar.**

Gūma, which I reached on the 4th of October, after two more marches over dreary barren Dasht, is the largest of the oases on the route from Karghalik to Khotan. It occupies an extensive loess terrace, which here overlies the sterile glacis of gravel and detritus washed down from the mountains. The greatest breadth of the terrace seemed about four miles, while its length in the direction from south to north, as marked by unbroken cultivation, is at least eight miles. Only a narrow strip of gravelly 'Sai', some two miles broad, separates this terrace **The oasis of Gūma.**

[15] Compare Hedin, *Reisen in Z.-A.*, pp. 10–17, passim.

[16] It might be thought that Hsüan-tsang first travelled nearer to the hills by the caravan road leading to Sanju, and thence joined the present main road via Zanguya. But this route would have been some twenty-four miles longer than the direct one and also could not appropriately be described as anywhere surmounting 'hauts passages de montagne'.

[1] See for this term, above, p. 74.

[2] See *Ruins of Khotan*, p. 159.

on the south-east from another equally large loess area which bears the villages of the Mokuila tract administratively united with Gūma.

Irrigation of Gūma.

Both the Gūma and Mokuila oases owe their fertility to the water brought down by the river of Kilian. Where the latter emerges at the foot of the hills on to the Dasht glacis, it divides into a number of channels, partly natural, partly artificial. As these diverge a kind of delta is formed, with its base extending from Chōlak-Langar in the north-west to Kakshal-Tati in the south-east, a direct distance of about 32 miles. But the distribution and importance of these channels is very unequal. To the west of Gūma they are few, and the limited quantity of flood-water they carry for brief periods would not suffice for regular cultivation of such loess soil as may be available between the Dasht and the moving dunes of the desert. Gūma itself enjoys an abundant supply of water from a number of 'Üstangs' or canals; and the surplus, together with the occasional summer floods which pour down in broad torrent beds (*sil*) on both the west and the east of the oasis, has in recent years led to the creation of several detached settlements within the area of sandy jungle to the north-east of the oasis.

Mokuila tract.

The tract of Mokuila is less favoured by facilities for irrigation. The water obtainable from the easternmost channels fed by the Kilian river is not sufficient to fertilize more than a small portion of the loess ground which stretches eastwards of the flood-water bed marked as 'Tazgun R.' on the map[3]. Consequently we find here strips of fertile village-lands, like those of Aramelle, Chōtla, Kakshal, broken by stretches where the naturally arable loess soil either lies bare and is undergoing erosion, or else is slowly being overrun by low dunes of fine drift-'sand'. We shall see hereafter that this drifting 'sand', highly productive wherever brought under irrigation, is itself composed largely of disintegrated loess[4].

The physical features here outlined will help to give the right 'setting' to the antiquarian observations which I have had occasion to make about Gūma. These observations were at first of a quasi-negative character; but this, as subsequent experience showed, scarcely detracts from their interest. In my Personal Narrative I have indicated the reasons which induced me to make a halt at Gūma on the 5th of October for the purpose of antiquarian inquiries. Among the purchases of Central-Asian antiquities made for the Indian Government by Mr. Macartney and Captain (now Major) S. H. Godfrey, paper MSS. and 'block-prints', all in 'unknown characters', had since 1895 turned up more and more frequently and in increasing bulk. These, and similar acquisitions which had reached public collections at St. Petersburg, London, Paris, and probably elsewhere through European collectors at Kāshgar, were all supposed to have been discovered at sand-buried sites about Khotan[5].

Inquiries for Islām Ākhūn's 'find-places.'

Islām Ākhūn, the Khotan 'treasure-seeker', from whom most of these strange texts were acquired, had, in statements recorded at Kāshgar by Mr. Macartney, and reproduced in Dr. Hoernle's Report on the 'British Collection of Central-Asian Antiquities', specified a series of localities from which his finds were alleged to have been obtained. Islām Ākhūn described these places as old sites in the desert north of the caravan route between Gūma and Khotan, and furnished

[3] *Tāzgun* or *Tāzghun* is, in reality, a generic designation applied to rivers and streams which periodically carry flood-water. To distinguish the many 'Tāzguns', the names of the oases which they pass, or of the valleys from which they issue, would have to be added.

[4] With this brief sketch, based on personal observations and notes, it will be useful to compare the ample details recorded regarding the oasis of Gūma in Hedin, *Reisen in Z.-A.*, pp. 11 sqq.

[5] For a systematic synopsis of these 'finds', so far as they were acquired for the Indian Government, see Dr. Hoernle's *Report on the British Collection of Antiquities from Central Asia*', *J.A.S.B.*, Extra No. 1, 1899, pp. iii sqq.; also *J.A.S.B.*, 1897, pp. 237 sqq. For MSS. of this type purchased at St. Petersburg, comp. *Nachrichten über die Expedition nach Turfān*, 1899, p. 48, pl. 8; for similar acquisitions elsewhere, *J.A.S.B.*, 1899, Extra No. 1, pp. vii, xx, 59; Hedin, *Through Asia*, p. 760 (with illustrations in German edition).

elaborate details as to the route by which he was supposed to have reached them in the course of his various 'expeditions'[6]. Doubts had already arisen as to the genuineness of the finds sold as the result of these expeditions[7], and information that reached me at Kāshgar about Islām Ākhūn had greatly strengthened these doubts. But it was at Gūma that I first touched ground where it was possible to test the 'treasure-seeker's' statements by direct local inquiries.

The information which I readily obtained on the morning after my arrival from the assembled Bēgs and Yüz-bāshis (village headmen), was precise as to the existence of an extensive débris-covered area known to all as a 'kōne-shahr', close to the road between Mokuila and Moji, the next oasis eastwards. But nobody had ever heard of the discovery of 'old books' at this or any other site. Of a series of localities which Islām Ākhūn had named in the detailed itinerary of one of his desert journeys east of Gūma, and which figured in his accounts as the sites of some remarkable discoveries[8], only two were known to them. These, also, Kara-kul Mazār and Karatāgh-aghzi, instead of being far away among the great sands as described by that mendacious traveller, were declared to be close at hand, on the very outskirts of the Gūma oasis. So I decided to probe this part of the alleged itinerary by immediate inspection.

Kara-kul Mazār and Karatāgh-aghzi.

For a full description of the ride which took me to the places named, I may refer to my Personal Narrative[9]. Here it will suffice to state that Kara-kul Mazār ('the Mazār of the Black Lake') proved to be the resting-place of an unknown saint, marked only by the usual erection of poles bearing yaks' tails, rags, and similar ex-votos. The semicircle of sandhills, on which this humble shrine rises, lies at a distance of a little over three miles to the north-east of the Gūma Bāzār. The name is derived from a small lake of saline water close by, which is fed by flood-water of the 'Tāzgun' passing on the east. Of the vast ruined graveyard in which Islām Ākhūn had alleged that he had made a remarkable MS. find[10], no trace could be discovered anywhere. Karatāgh-aghzi[11], which also figured in Islām Ākhūn's narrative, was found to be a flourishing new colony of Gūma, only three miles further to the north-east. I reached it by following the broad bed of the 'Tāzgun', now dry except for the water of some scanty springs rising near Gūma. The people of Karatāgh-aghzi, whom I closely questioned, knew nothing of the ruined sites which Islām Ākhūn claimed to have visited in the desert eastwards, and still less of the discoveries made there[12].

Cultivation at Karatāgh-aghzi, as well as at some other small colonies further along the course of the flood-water bed, was said to have been begun only some fifteen years before my visit. In addition to the water supplied by floods, irrigation is secured from the springs which rise apparently at several points in the broad bed cut by the 'Tāzgun'. The soil, now bearing splendid crops, and partly covered with luxuriant groves of poplar, mulberry, and other trees,

[6] See *J.A.S.B.*, 1899, Extra No. 1, pp. xvi–xxii.

[7] Compare *J.A.S.B.*, 1899, Extra No. 1, pp. 57 sqq., where the question of forgery was argued with much ingenuity, but with results necessarily inadequate from want of local evidence.

[8] e.g., skulls with pillows of MSS. attached; sepulchral monuments full of MSS. &c.; comp. *J.A.S.B.*, 1899, Extra No. 1, pp. xvi, sqq.

[9] *Ruins of Khotan*, pp. 184 sqq.

[10] See *J.A.S.B.*, 1897, pp. 238, 253; 1899, Extra No. 1, p. xviii.

[11] This is the correct form of the name usually pronounced in the abbreviated form *Karatāghiz*, through a kind of 'haplography'; *aghzi* ('mouth of', 'opening of') is a very common termination of local names in Turkestān. The form and explanation of the name given in *J.A.S.B.*, 1899, Extra No. 1, p. xvi, is based on a misapprehension; so also 'Kara Targaz' in Capt. Deasy's map.

[12] Their ignorance of these localities, named Aktala-tuz, Kara-yantak, Kök-gumbaz in Islām Ākhūn's itinerary, was scarcely surprising, considering that the ingenious forger had borrowed the names for his invented sites from localities about Sampula, far away in the eastern part of the Khotan oasis. For a full account of Islām Ākhūn's manufacture of 'ancient books' and his final confession, see below, chap. xv.

seemed in no way to differ from the fine 'sand' which was seen outside the little oasis and along the edges of the torrent bed heaped into bare dunes and hillocks of varying height. It was a striking demonstration how easily irrigation could transform this seemingly sterile waste into rich agricultural land.

Drift-'sand' at Hāsa.

On my ride to Kara-kul Mazār I had been shown the little hamlet of *Hāsa*, only about 1½ miles to the south-east of Gūma Bāzār, where dunes 20 to 30 feet high, advancing from the north, are gradually overwhelming the villagers' holdings. I regret I did not think then of securing specimens of this 'sand' for microscopical examination. But every topographical consideration points to the conclusion that these destructive dunes were composed of the same 'sand' which extends round the northern edge of the oasis, and which proves so fertile wherever it can effectively be brought under irrigation. My observations of that day around Gūma have been confirmed by those subsequently made at other oases. They showed clearly that extension of cultivation on one side may proceed simultaneously with an advance of the dunes over arable land on the other, and that in this constant struggle between oasis and desert the facilities for irrigation, depending on human activity perhaps as much as on natural level and available water-supply, form the determining factor. We shall have frequent occasion to return to these points when discussing the changes during historical times in the conditions of oases further east.

The view of the newly reclaimed lands of Karatāgh-aghzi, and the fact of a belt of jungle extending along the flood-water channels for a considerable distance further northward, naturally suggested inquiries whether remains of old settlements could not be traced in that direction. But all information I could gather from cultivators and local headmen was in the negative. I am all the more inclined to accept this testimony because it agrees with the experience gained by Captain Deasy on a curious expedition, the full story of which I learned subsequently[13].

Fiction of ancient site beyond Gūma.

It appears that Captain Deasy, having been in Mr. Macartney's company about the time when Islām Ākhūn, early in 1898, furnished to the latter the above mentioned itinerary, together with some of his remarkable 'finds', was induced to propose to the 'treasure-seeker' a visit to one of the old sites described by him beyond Gūma. Islām Ākhūn, though in reality he had never been to any such sites, could not refuse to act as guide without the risk of arousing suspicions about the truth of his story, and thus spoiling the market for his 'finds' among the Europeans of Kāshgar. So much against his will the impostor had to start for the desert from Gūma in April, 1898. Recruiting two local 'guides' who knew as little of ancient sites as himself, he conducted the party for two weary marches north of Karatāgh-aghzi into the desert. Nothing, of course, was found in the belt of sandy jungle and among the dunes beyond; and when the supply of water carried began to run low, Islām Ākhūn thought it safest to abscond during the night and to return to Khotan[14].

On the basis of the above inquiries I consider it safe to assume that no ancient remains exposed to view exist in the vicinity of Gūma except those beyond Mokuila, which will be

[13] I gathered the first details of this abortive treasure-seeking adventure from Miān Jaswant Singh, who, before serving as follower of my surveyor Rām Singh, had accompanied Captain Deasy's party in a similar capacity (see *Ruins of Khotan*, p. 8). When subsequently I had the chance of making Islām Ākhūn's personal acquaintance and succeeded in extracting from this versatile rogue a full confession of his various frauds (as related below in chap. xv), a clearer light was thrown on the motives, and on various amusing incidents, of the adventure. The facts of this fruitless trip into the desert have since been briefly related in Captain Deasy's book *In Tibet and Chinese Turkestan*, pp. 161 sqq.

[14] It was, perhaps, in order to compensate himself for these unwonted hardships that Islām Ākhūn forged a note pretending to be in Captain Deasy's handwriting, with which on his return he obtained some money from the Afghan Aksakāl of Khotan! For this impudent fraud he received, however, condign punishment from Chinese justice, by having to wear the wooden collar for a considerable period, as related in Captain Deasy's pages.

described presently. This negative fact is quite compatible with the opinion that the oasis of Gūma has been under cultivation from early times. The observations to be detailed below regarding the Khotan oasis, prove beyond all doubt that an area constantly irrigated for centuries is, under the physical conditions prevailing between the foot of the Kun-lun range and the Taklamakān, bound to be covered by a steadily rising deposit formed of silt and loess dust[15]. All ancient remains are so effectively hidden by the rapid growth of this overlying deposit that, as the fate of the culture-strata of Yōtkan convincingly demonstrates, nothing but the formation of deep ravines in the soil, or else long-continued subsequent denudation through the erosive action of the wind, can ever reveal their existence.

Early occupation of Gūma.

I did not see any indication of either of these two agencies being at work now within the cultivated area of Gūma, while all round its northern outskirts the accumulation of drifting 'sand'[16] is actively proceeding. Hence I hold that the absence of any archaeological finds within or on the outskirts of the oasis is no evidence against ancient occupation.

The *P'i-shan* of Chinese Annals.

Gūma, together with Mokuila and Moji, must be assumed to have formed part of the 'kingdom' of *P'i-shan* (Pe-shan) 皮山, which the 'Notice of the Western Regions' in the Han Annals places half-way between Khotan and Yarkand (So-ch'ê), at 380 li distance from either[17]. The population was reckoned at 3,500 persons, showing that the territory was a small one. The statement made in the Notice that India adjoins it to the south, is a manifest reference to the Karakorum route, which is reached by going due south via Kilian or Sanju. The T'ang Annals speak of P'i-shan as a small territory absorbed by the Khotan kingdom[18].

The inquiries made on my behalf among the people of Gūma for antiques produced no result, notwithstanding the offer of liberal reward for any acceptable article and the manifest *empressement* shown by the local authorities to facilitate acquisitions. The small octagonal bronze seal (G. 001), apparently of Chinese make (Plate L), which was sold to me at Karghalik as having been found near Gūma, is of uncertain date and origin, and could in no case be relied upon as evidence, since it had passed through the Bāzār channel.

SECTION III.—THE TATI OF KAKSHAL

Débris area near Mokuila.

The morning of the 6th of October saw me on my way towards Moji, the next caravan stage, filled with keen expectation as to the remains of 'kōne-shahrs' which, according to my Gūma informants, were to be seen *en route*. After we had emerged from the shady lanes of the southern part of the Gūma oasis, and had crossed a stretch of stony Dasht about three miles broad, I came upon the first of these on the right bank of the wide river-bed, now entirely dry, which the day before I had followed lower down to Karatāgh-aghzi. For a considerable distance along the right bank, which, where crossed by the road rises some twenty feet above the sandy bottom of the flood-water channel, the ground was thickly strewn with small fragments of coarse red pottery. No ornamented pieces could be found, but the exceptional hardness and occasional fine grain of these potsherds showed that they belonged to a remote period. The width of the area over which these scattered fragments could be traced was not great, as it extended only for about a third of a mile from the river-bank to where the soft sandy soil, unmistakably

[15] Compare below, chap. VIII. sec. ii; *Ruins of Khotan*, pp. 263 sqq.

[16] I use the expression 'sand' here and elsewhere without any prejudice as to the real geological nature and origin of the substance composing the dunes.

[17] Compare Wylie, *J. Anthrop. Inst.*, x. pp. 30 sq.; also above, p. 97.

[18] See Chavannes, *Turcs occid.*, p. 125.

disintegrated loess, had been brought under cultivation. To the south of the road the pottery-strewn ground soon disappeared under low ridges of fine sand. But to the north, where the surface was clear of these, the débris area could be seen extending along the bank for a considerable distance—according to the statements of my local informants, fully two or three miles. It was evident also that much of the original ground westwards must have been washed away by the floods of the river-bed, which are here clearly working to the east and steadily cutting into the right bank. The extent of the area covered with broken pottery thus indicated the site of a large and thickly inhabited settlement.

Potsherds resting on loess banks.

In vain did I search the ground for any other remains but the patches of scattered potsherds. Wherever I examined these, I invariably found them resting on the bare loess with never a trace of walls or more substantial remains below. The fragments usually lay thickest on the top of small banks of relative hardness which rose here and there above the surrounding ground, the latter also composed of loess but showing a more disintegrated surface. The sides of these loess-banks, as also the eroded scarp of the river-bank, were steep in many places and hence easy of examination; but neither pottery nor other remains could be traced anywhere in the strata exposed to view. This observation rendered the abundance of pottery débris on the surface all the more curious, and could not fail from the first to puzzle me greatly. The explanation did not, however, present itself until I had observed similar conditions elsewhere. I may therefore leave their discussion for the account I shall have to give presently of a larger and still more significant site.

Immediately to the south-east of the débris area just described the road crosses, near the hamlet of Chōtla, the long belt of cultivated land forming the oasis of Mokuila. This belt is here only about a mile in breadth, but everything points to a great extension of cultivation being possible eastwards, if only additional water were made available for irrigation. The scrub-covered steppe over which the road leads for another two miles shows soft loess soil, turning into fine dust wherever broken up by traffic. Patches of this naturally fertile area had been recently taken under cultivation, especially around an isolated holding which bears the characteristic name of Yangi-arik, 'the New Canal.' It was at this point that I first sighted, at a distance of about $1\frac{1}{2}$ miles to the north-east, the ancient mound of which my Gūma informants had spoken as *Tōpa-Tim*. The designation of 'Tim' at once suggested a Stūpa, and I made haste to reach it. My first attempt proved a failure; for the deep-cut flood-bed of the Sughaz-yār, which, by the advice of my guide, I crossed a little beyond Yangi-arik and then followed northward, soon turned into a cañon-like ravine and cut us off absolutely from the mound on the opposite bank when we at last got abreast of it.

Cultivation near Yangi-arik.

Stūpa ruin of *Tōpa-Tim*.

As soon as I had crossed the ravine near the road and approached the mound along the left bank through the fields of Yangi-arik, it was easy to see that Tōpa-Tim ('the Earth Mound') preserved all the features of an ancient Stūpa. The solid masonry of sun-dried bricks had undergone great decay on the surface, manifestly through atmospheric influences, and only on the north-western side could the outlines of the several stories of the base be traced clearly. Careful measurements taken here, as well as along the more dilapidated portions of the ruin, enabled me to reconstruct the plan as shown in Plate XX. It will be seen from this that the general arrangement and dimensions of the structure resembled in several respects those of the Maurī-Tim Stūpa. The base was formed by three receding stories, of which, however, the lowest was marked only by a mass of decomposed débris sloping up to a height of 5 feet from the present ground-level. On this portion of the base, the size and shape of which can only be conjecturally restored, rests a second story, 41 feet square and 5 feet high. Next comes a circular story or drum of the same height, with a diameter of 35 feet. Above this again rises the dome, which now reaches a height of $14\frac{1}{2}$ feet. As the diameter of the dome is 29 feet, this height if original would make its shape exactly hemispherical; but since the top is much broken this must remain doubtful.

The total extant height from the present ground-level is $29\frac{1}{2}$ feet, which, in relation to the greatest dimension of the base, 47 feet square, gives a much smaller proportion between vertical and horizontal measurements than observed in the case of the Maurī-Tim Stūpa (about 38 feet to 40 feet). It must, however, be borne in mind that, not having the time or needful labour for trial excavations around the Stūpa, I was unable to ascertain definitely whether the original level of the surrounding ground was not, perhaps, considerably lower than the present level. If accretion of silt had been proceeding over the adjoining ground, as has undoubtedly happened in the case of the Kurghān-Tim Stūpa[1], we might well suppose another story of the square base to lie hidden beneath the surface.

Condition of adjoining ground.

On the whole, however, I am not inclined to favour such an assumption. For a distance of 200 to 300 yards around the Stūpa and eastwards up to the edge about 70 yards distant of the above mentioned deep 'Yār' or ravine, the ground is strewn with potsherds which seemed old, and with fragments of stones. It seems natural to connect this débris with ancient habitations of less solid construction, which existed around the Stūpa when it was still an object of worship. Judging from the observations referred to in discussing Kurghān-Tim, and to be detailed hereafter in connexion with the site of Yōtkan[2], only long-continued irrigation could have led to heavy deposit of silt and consequent rise of the ground-level. In such a case we should expect the débris to have long ago been buried out of sight under layers of fertile soil.

But the assumption of the ground-level having remained practically unchanged since the latest date we can assign to the Stūpa, i.e. the period immediately preceding the introduction of Islām at the end of the tenth century, is not without its problems. With such striking evidence as the adjacent site of the 'Kakshal Tati' furnishes of the powerful erosive action of the winds in this region, it seems difficult to understand how the ground adjoining the Stūpa, if unprotected by cultivation, could have escaped being considerably lowered in the course of nine centuries. We shall see hereafter, when discussing the remains of Endere, that since that site was abandoned early in the eighth century wind erosion has proceeded there so effectively as to lower the unprotected ground around the local Stūpa some ten feet below the original level, as marked by the base of the extant ruin[3]. Climatic conditions, no doubt, affecting the frequency, direction, and strength of the desert winds, may differ materially now, and may possibly have differed still more during past periods, at such widely distant portions of the Taklamakān. Yet it must be noted, in regard to the destructive effect of erosion upon the outer surface of structures of sun-dried bricks, that the Stūpa of Endere has suffered scarcely more than the one of Tōpa-Tim[4].

Wind-erosion and protection by drift-sand.

It will require a far more systematic study of all local conditions than was possible in the course of hurried journeys, and probably also accurate climatic observations extending over considerable periods, before such a semi-geological, semi-archaeological question can be safely answered. In the meantime I may suggest that the possibility of the ground near Tōpa-Tim having been protected against erosion by dunes, which kept it covered for centuries but

[1] See above, pp. 75 sq.
[2] See above, p. 75; also chap. VIII. sec. ii, below.
[3] See below, chap. XII; *Ruins of Khotan*, pp. 421 sq.
[4] For a view of the Endere Stūpa, see Fig. 50; I regret not to be able to illustrate 'Tōpa-Tim', since the two negatives taken for that purpose have deteriorated too much to permit of reproduction.

subsequently receded again northward, ought not to be lost sight of. Such an assumption would at least furnish a ready explanation of the fact that the surface of the Stūpa is best preserved on the north-western side. It is well known that the semi-lunar dunes in the western portion of the Taklamakān, in conformity with the prevailing winds, mostly show a line of progression from north-west to south-east. It is evident that, whenever these dunes encounter obstacles in the form of buildings, they will pile up sand highest against their north-western face and consequently assure to that better protection than to the rest[a].

Diggings of 'treasure-seekers'.

Like all other Stūpas, Tōpa-Tim must at one time or other have attracted the attention of 'treasure-seekers'. But the digging made on the top of the mound was apparently not carried deeper than about seven feet. The square well and chamber, which other Stūpa ruins examined by me show in the centre line of the dome, was not traceable here; but this might possibly be due to the much-decayed condition of the dome and the excavation just referred to. On account of the 'weathering' of the surface it was impossible to find any trace of the plaster coating which once probably covered the whole structure. Nor could I ascertain the size of the bricks used in it, as this would have required a cutting to be made into the masonry, and would in all probability have led to further destructive digging by others bent on more practical objects.

The 'Tati' of Kakshal.

On my first attempt to reach Tōpa-Tim, I had noticed on the right bank of the Sughazyār a débris-strewn area far more extensive than the one seen to the west of Mokuila. Stretching to the north of the caravan route it seemed to cover fully three square miles, perhaps more. To the south of the road and again eastwards at a distance of about two miles from the ravine the débris disappeared under low dunes of drift-sand, invaders from the desert. The lateness of the hour at which I reached this area prevented me from determining its precise limits northward. My guides called it the 'Tati' of *Kakshal*, applying to it the name of the nearest village in the Mokuila oasis.

The relics of ancient habitations—and as such they could not fail to be recognized even by the most casual wayfarer—lay here scattered in patches of varying extent and thickness. They comprised, besides pottery fragments of all sorts, small pieces of stone and burned brick, slag, broken bones, much corroded bits of metal and similar hard refuse. The potsherds were without exception of remarkable hardness, but generally coarse in texture. The overwhelming majority of them showed varying shades of red and, where lying thick, imparted a reddish glow to the whole ground. But pieces of black pottery were not infrequent. Fragments showing any ornamentation, or giving clear evidence as to the shape of the vessels to which they had belonged, were very rare, and the specimens of this type brought away and described in the list below were the most distinct among those picked up during a search of over an hour. The surface of all the potsherds was peculiarly rough, looking as if it had been subjected to 'grounding'—a manifest mark of the force of erosion which had been at work here. Fragments of bones, apparently of animals, turned up in many places, just as they would be found now in most of the rubbish-heaps of Turkestān villages. In one much-eroded piece my men recognized the knuckle-bone of a sheep, used probably, as nowadays, as a substitute for dice.

Ornamented potsherds.

The occurrence of slag was restricted to particular patches. The most striking of these was a loess bank rising some fifteen feet above the general level of the ground, not far from

[a] Compare *Yarkand Mission Report*, pp. 27 sq.; Hedin, *Reisen in Z.-A.*, p. 233; *Ruins of Khotan*, p. 155. The conditions here indicated were again and again observed by me at buildings exposed to the onset of drift-sand, from Ordam-Pādshāh to Kumrabāt-Pādshāhim.

the ravine, where a considerable quantity of slag was scattered over the surface between bits of smelted ore, half-burned bricks and stones apparently containing iron-ore. Fragments of charcoal, too, could be traced among this refuse, evidently marking the spot where a smelting-furnace had once existed.

Conditions of 'Tatis'.

More interesting than these scanty remains, and at first decidedly puzzling, were the conditions in which they presented themselves. I had, however, occasion to observe the same conditions again and again at similar old sites which I subsequently traced at numerous points beyond the limits of the present cultivated area in the Khotan region, and which are all known locally by the general designation of 'Tati'[6]. And this renewed observation of identical features, together with the lessons taught by my explorations at far better preserved sites, such as those at Dandān-Uiliq and beyond the river of Niya and Endere, gradually furnished a convincing explanation.

Débris resting on eroded loess.

The most striking feature noticed at Kakshal Tati, as well as at all other 'Tatis', was that the débris rests on nothing but natural loess, either firm or more or less disintegrated into the condition of dust or extremely fine 'sand'. It was always easy to ascertain that the soil underneath contained neither walls nor other structural remains of any kind; for the isolated small terraces or banks of loess which rise here and there above the general level of a Tati, and on the top of which the fragments usually lie thickest, invariably displayed on their bare sides the natural soil without any trace of ancient deposits or distinctive strata. At Kakshal Tati the highest of these flat-topped banks rose about fifteen feet above the dust-filled depressions around them; the average elevation of most was, however, from eight to twelve feet.

Results of wind-erosion.

In the formation of these banks, as in all other features of such sites, it was impossible not to see evidence of the powerful erosive action of the winds and sand-storms which sweep the desert and its outskirts with great frequency during the spring and summer. The remarkable force of these desert storms or 'Burāns' has been commented upon by all travellers in Eastern Turkestān. Though my stay in the desert region did not continue beyond the early season of March and April, I had ample opportunities, as my Personal Narrative shows, of gaining practical experience of their vehemence[7]. Only the materials above described could, by their hardness and weight, survive, sinking lower and lower as the ground beneath gets more and more eroded, while everything in the shape of mud walls, timber, &c., used in the construction of Turkestān houses, has long ago decayed and been swept away. Even the potsherds and other fragments which have withstood destruction, bear plain evidence of the ever recurring onset to which they have been exposed, in their small size and the peculiarly rough surface already noted.

Lowering of ground by wind-erosion.

It is evident that such a process of erosion at sites of old settlements built on friable loess soil could not have gone on during the long centuries since their abandonment without also considerably lowering the ground-level. But the erosion could never proceed uniformly over a whole area, and of this we have evidence in the banks of loess already referred to, which are seen rising like small plateaus or islands above the more disintegrated parts of a 'Tati'. They may owe their relative protection to a variety of special features, such as the greater density of the débris with which they are ordinarily covered, or to their having been once occupied by structures which, though now completely vanished, may yet by the weight of their crumbling ruins have effectively shielded the soil beneath. We shall see hereafter that the

[6] See for a list of such sites the Index, s. v. *Tati*.

[7] See, e. g., *Ruins of Khotan*, pp. 428 sqq., 451 sq.; for information collected from native sources as to Burāns ordinarily observed between Karghalik and Khotan, comp. Hedin, *Reisen in Z.-A.*, pp. 9–18.

ruins of the timber and plaster buildings of the Niya River Site were almost always found occupying such terraces, while all round them depressions were formed through erosion. To whatever specific cause or combination of causes we may have to attribute the preservation of loess banks at Tatis, it is certain that we have in them evidence of the original ground-level. As such, they are equally interesting to the antiquarian and the geologist, and it seems appropriate to apply to them the designation of 'witnesses' (or 'Zeuge') familiar to geological terminology.

Loess banks as 'witnesses'.

These 'witnesses' are, of course, of special value wherever we find on or near them archaeological remains which are approximately datable; for such remains clearly mark the highest chronological limit for the commencement of that process of erosion of which the results are seen in the depressions formed around. Such datable remains must on 'Tatis' proper be necessarily of rare occurrence. Owing to the fact that they are ordinarily within easy reach of localities which have remained inhabited, these sites at all times have been particularly exposed to destruction from human agency.

Destruction by human agency.

My observations at deserted villages of modern date near Domoko may serve as a typical illustration to show how soon after their abandonment such sites were bound to be cleared of everything of any value or that might serve as building material or fuel[8]. Timber has always played an important part in the construction of Turkestān houses. It is easy to see how its early abstraction must have converted all remains of ordinary dwellings into crumbling mud heaps within a short period after their abandonment. The regular visits of 'treasure-seekers', whose profession appears to have flourished at all times about Khotan, not only helped to complete the destruction of all perishable materials, but gradually led to the removal of any small objects of value, such as coins, metal ornaments, cut stones, &c., which might have been hidden in the ground by design or accident, and which by their harder substance were proof against decay or erosion.

Remains surviving at 'Tatis'.

Other ancient objects, such as written records, sculptures in plaster, wood-carvings, &c., which the deserted settlements probably once contained, and about which the 'treasure-seeking' visitors of 'Tatis' in former days certainly never troubled, would now, no doubt, be most valuable to us for the archaeological dating of these sites. But we have seen already that the physical conditions absolutely preclude the possibility of such relics surviving[9]. Fragments of ancient pottery, indeed, strew the ground of all 'Tatis' in abundance; but in the present state of our knowledge it is impossible to attempt a proper classification of these potsherds with a view to ascertaining the chronological relation of particular types, materials, &c. Even hereafter, when some progress may be made by the detailed examination of ancient pottery at sites where it has been protected by drift-sand or silt deposit, the proper classification of potsherds found at Tatis will present particular difficulties, owing to the small size of the fragments and the effacement which the surface of the rare decorated pieces has undergone through the 'grounding' already referred to.

Period of Kakshal Tati remains.

Thus the only relics at Tatis likely to afford guidance as to the period from which they date, are coins, cut stones, and small objects of metal work. Unfortunately I was unable to

[8] See below, chap. XIII.

[9] In the accounts given by Islām Ākhūn of his alleged discoveries, his written or block-printed 'old books' in 'unknown characters' often figured as having been found exposed on the bare surface at certain desert sites. As far as these sites had any existence at all, Tatis were meant such as the forger had occasion to see near Gūma, Ak-sipil, Hanguya, &c. Those 'books' were often volumes of fairly large sheets of paper, with fly leaves and margins intact, held together by clumsy pegs or even mere rolled-up strips of paper. An observant visit to any Tati would have sufficed to demonstrate the absurdity of the belief in such flimsy 'antiques' surviving even for a single season the fierce winds that sweep these eroded sites.

obtain any such article from the Tatis on either side of Mokuila. Possibly the fact of their being situated on a high road, and close to populous villages, has accelerated the exhaustion of the few relics the soil may once have contained. In the absence of conclusive evidence, such as coin finds might afford, I am unable to express any definite opinion as to the period when the settlements once existing here were abandoned. Judging, however, from the extent of the erosion, down to a depth of 15 feet, which the ground-level has undergone at Kakshal Tati, as shown by 'witnesses', the antiquity of this site at least becomes very probable. The only place where I met with evidence of erosion equally deep was the Niya River Site, abandoned in the second half of the third century of our era. But as already observed, the climatic conditions under which erosion proceeds may vary at sites so widely distant, and we have no means of tracing the changes which these conditions may have undergone during past centuries. The close vicinity of the Tōpa-Tim Stūpa, on the other side of the Sughaz-yār, also points to a pre-Muhammadan origin for the remains of Kakshal Tati.

Since we have no certain clue to the date when Kakshal Tati changed from a cultivated and thickly inhabited area into a waste, it would be useless to formulate any definite opinion as to the cause. The change certainly presupposes a loss of the water-supply needed for irrigation; but this loss might have resulted just as well from some political upheaval which reduced the population or otherwise interfered with the systematic upkeep of canals, as from some alteration in the natural conditions affecting the water-supply. Yet, judging from the great quantity of flood-water which the Sughaz-yār annually carries, as proved by its remarkably deep-cut bed and the villagers' statements, it seems difficult to believe that natural causes could ever within historical times have prevented water from the Kilian river being brought to Kakshal Tati. However this may be, it is important to note that this ruined area extends along the actual high road, and thus furnishes additional evidence for the antiquity of the line followed by the latter.

Cause of abandonment of site.

OBJECT FROM THE VICINITY OF GŪMA.

G. 001. Bronze octagonal seal. 1⅝″ diameter; purchased at Karghalik; said to have been found near Gūma. Design—an open lotus, with a snake on each side. Design precisely like a stencil; at back a large shank pierced. See Plate L.

OBJECTS FROM KAKSHAL TATI.

G. (Yangi-arik) 001. Fragment of coarse red pottery vessel, upper portion. Well baked and very hard. The curve from shoulder to neck and outwards again to rim, is very angular and strong. The rim is thickened, slightly chamfered and moulded. Width 3⅝″; height 2$\frac{1}{16}$″; thickness ¼″-⅜″. See Plate XLI.

G. (Yangi-arik) 002. Fragment of coarse red pottery vessel, very hard. Roughly decorated with incised patterns. A series of oblique notches form a band at junction of neck with shoulder, and below this a rough lattice pattern. Height 3¾″; width 3⅜″; thickness about ¼″.

G. (Yangi-arik) 003. Eight fragments of red pottery, more or less coarse, unglazed. The texture of most of the fragments is fretted and sponge-like on the surface. One piece, clearer than the others, exhibits fine colour and texture.

Two fragments of black pottery of very fine texture.

Fragment of dark chocolate-coloured substance, resembling lac.

Piece of 'slag'.

Y. A. (Yangi-arik) 001. Fragment of coarse red pottery in two pieces, having portion of an animal (dog?) roughly scratched in the surface while moist, with a coarse implement. Thickness varying from ⅛″ to ⅜″. See Plate XLI.

Section IV.—ANCIENT REMAINS AT MOJI

At the Kakshal Tati we left behind the easternmost limit of the delta of the Kilian river. The road then crosses strips of sandy desert, first over low dunes and then amidst small conical sandhills covered with tamarisks. After some five miles, at the village of Chudda, we struck the edge of the series of oases which are irrigated from the Sanju river. The largest of these bears the name of Moji, and consists of a number of small villages spread out from north to south, with considerable patches of unirrigated, and hence barren, loess ground between them. I was aware, from Dr. Hedin's narrative, of the existence of an old site to the north of the central village containing the Bāzār of Moji, and accordingly arranged for a halt on the 7th of October. On the morning after my arrival bags full of old Muhammadan copper coins, as well as some metal seals, bronze ornaments, &c., were brought to me by the local Bēg and village headmen as having been found at that site, which they designated as *Togujai* or *Tögujai*.

Site of Togujai.

Proceeding to this locality, situated at a distance of about three-fourths of a mile to the north of Gūma Bāzār, I found an area of bare loess ground extending over about one quarter of a square mile furrowed by the broad channels of a flood-water bed, and the banks thus formed covered with pottery débris. Where the soil had not been fissured by the action of the flood-water, there were but few potsherds to be seen on the surface. But where the banks of loess had been washed by the current, their slopes were thickly strewn with pottery fragments of all kinds, small pieces of glass and stone, and similar débris. Examining several loess banks the sides of which had been exposed more or less vertically, I found in them fragments of pottery, as well as animal bones and layers of ashes and charcoal, embedded at a depth varying from three to five feet below the present surface. It is from these same strata that the coins are washed out, of which such large collections had been shown to me. They are usually picked up in the broad stony bed of the main channel, commonly designated as 'Sai', after the floods of the spring and early summer have passed by. Rows of little burrows showed where the débris exhumed by the floods of the last season had been searched and washed by people looking for 'treasure'. There, too, about half a dozen thin copper coins of the same type as those previously brought to me were collected by the villagers whom I employed to search for ornamental pieces of pottery, &c.

Coins found at Togujai.

These coins, together with the vast majority of the pieces purchased by me from the site, belong to a hitherto unidentified Muhammadan ruler who appears to style himself on the reverse as Sulaymān Khāqān. This tentative reading was first published by Dr. Hoernle from a number of pieces contained in the 'British Collection of Central-Asian Antiquities'[1]. Nothing appears to be known of the time or genealogy of this king, and I must leave it to numismatists to determine what clue, if any, may be furnished by the type and standard of the coins. Specimens of two varieties represented among ninety pieces I obtained at Moji are shown in Pl. XC, Nos. 45, 46, 47. The abundance of these pieces at the Togujai site is certainly remarkable, especially when compared with their rarity elsewhere. Among the many collections of miscellaneous coins offered for sale at Khotan and mostly secured from Yōtkan, I do not think that I came across more than half a dozen pieces of 'Sulaymān Khāqān'—and these might well have been obtained from Moji. In addition to these coins, the collection purchased at Moji contains twenty-three Chinese copper coins (nineteen small and four large) belonging

[1] See *J.A.S.B.*, 1899, Extra No. 1, pp. 32 sq.; also Plate I, figs. 31–35. [Prof. Rapson informs me that these coins appear to bear the name of the Caliph Al-Musta'ṣim, 1242–58 A.D.]

to issues of the Ch'ien-yüan period (758–760 A.D.) of the Emperor Su-tsung, and a small number of uncertain Muhammadan copper coins. It is highly probable that these coins, too, come from Togujai. [Nine coins have since been identified by Prof. Rapson as probably belonging to Muḥammad Arslān Khān, an early successor of Satok Boghra Khān.]

Date of Togujai site.

In view of the evidence of these coin-finds, it may be considered as certain that the remains of Togujai belong to a settlement which flourished well into the period following the Muhammadan conquest. The layer from which the pottery fragments and other débris are washed out bears a close resemblance in character to the culture-strata of Yōtkan, the site of the ancient capital of Khotan, to be discussed hereafter; like them it is undoubtedly due to the gradual accumulation of rubbish. But the slight depth of the Togujai layer and the absence of earlier coins suggest that the occupation of the site, or at least that which is traceable in remains, does not date back to more than a few centuries before the advent of Islām. On the other hand, we have an indication of the lower chronological limit in the deposit of earth which covers this layer. The observations subsequently made at Yōtkan prove that this deposit must be attributed to prolonged irrigation of the ground after it had ceased to be used for dwellings[2]. Within living memory of the villagers the area had been nothing but a waste, and there seemed to be evidence that during this later period its surface had in fact undergone some erosion. Combining thus the evidence afforded by the present condition of the site with that of the deposit of silt overlying the débris layer, we are led to conclude that even the latest portion of the latter must have rested in the ground for a considerable period, probably not less than four or five centuries.

Pottery of Togujai.

It is of interest to arrive even at an approximate determination of the highest and lowest limits of time, as this helps us to classify the remarkably varied and curious specimens of pottery which the site yielded, besides other small objects in glass and metal. The glazed pottery fragments described in the list below, and partly reproduced in Plate XLII, claim particular attention owing to their relatively good technique and artistic colouring. None of the earlier sites explored by me yielded glazed ware; and hence it may be questioned whether the art of glazing was known, or to any extent practised, in Eastern Turkestān until towards the commencement of the Muhammadan period[3]. On the other hand, some of the unglazed pieces show in their decoration the survival of motives which are already met with in the wood-carving of the Niya River Site (third century A.D.), and can be traced back to Graeco-Buddhist art[4].

Absence of structural remains.

As at Yōtkan, I was much struck by the complete absence of any traces of structural remains below the ground. But this feature is easily explained by what we shall have to show hereafter as to the effect of irrigation and concomitant percolation upon sun-dried bricks, plaster, and timber—the sole building materials of Turkestān. There is another point of contact between the conditions presented by the remains of Yōtkan and those of Togujai. Just as the discovery of the site of the ancient capital of Khotan was brought about by the accidental formation of a ravine which laid bare its deeply buried 'culture-strata', so Togujai, too, would in all probability have kept its old and more interesting débris hidden but for the channels scooped out by the flood-water.

Burial-ground of *Hāsa*.

From Togujai I proceeded over waste ground, covered in parts by light drift-sand (or loess dust), to an old burial-ground known as *Hāsa*, about $1\frac{1}{2}$ miles to the north-east of Moji Bāzār. Dr. Hedin, whose special attention had been directed to this site, rightly observes that the

[2] See below, chap. VIII. sec. ii.

[3] The only other piece of glazed terra-cotta in my collection comes from Uzun-Tati, the ancient *Pi-mo*, a site which, as we shall see, was inhabited up to the end of the thirteenth century of our era; see chap. XIII.

[4] See, e.g., description of T.M. 004. a, below.

graves are Muhammadan[1]. On examining one large but low mound, from the sides of which skeletons were protruding, I found that it consisted of closely packed rows of graves. These were lined with mud-bricks and covered on the top with small planks of wood. On having one of the graves opened, I found in it the remains of a child wrapped in the rough cotton material common all over Turkestān, and known as 'Khām'. The head was turned towards the west in conformity with Muhammadan custom, and the feet tied with a bandage just as is the fashion nowadays according to the assertion of my Turkī followers. There could be no doubt that this cemetery had served for the burial of orthodox Muslims. Local tradition, in fact, maintains that it is the resting-place of 'Shahīds' who fell fighting the infidels. But I could not trace any definite legend, and my informants, who had shown no particular scruples about opening a grave for my inspection, readily expressed their disbelief when they saw the body of a small child in the supposed grave of a holy warrior.

Period of graves.

Two observations make it appear very probable that this cemetery is old, and approximately of the same period as the remains of Togujai. From one of the partially exposed graves, Niāz Ākhūn, my Tungānī interpreter, picked up a small brass ornament representing a flying bird, exactly similar to one which had been brought to me while searching the débris at Togujai (marked M. 001. d and shown in Plate LI). I found on inquiry that it is still customary among the people to deposit small articles of this kind on the graves of their relatives.

Erosion of site.

The other observation, bearing on the configuration of the ground containing these old graves, is also of interest, but did not strike me until I had gained further experience elsewhere. The graves, as far as I could observe, invariably occupied low mounds not unlike the loess-banks or 'witnesses' described above in connexion with the Tati of Kakshal, but usually larger and more rounded in shape. The fact that the contents of the graves nearest the edge of these mounds were generally exposed to view proves that the latter were subject to erosion. The ground between the mounds showed a surface of soft disintegrated loess, strewn here and there with fragments of old pottery, but not thickly. It thus resembled exactly the depressions covered with loess dust which at Tatis mark the maximum erosion of the ground. The elevation of the mounds containing graves above the rest of the ground was not great, perhaps nowhere more than 8 to 10 feet.

Kepek-gholuk Mazār.

Comparing my notes and recollections of this old burial-place, I am led to the conclusion that we have in the mounds really portions of the original ground level which the compactness of the graves has preserved from the erosion proceeding all round. This conclusion is confirmed by what I saw at Kepek-gholuk Mazār, a little shrine about half a mile to the east of the most conspicuous mound of Hāsa. It consists of a collection of staffs and flags, erected on a small well-defined loess-bank, and is believed to mark the resting-place of a saint. A photograph I retain of this 'Mazār' (Fig. 18) shows very clearly the effects of wind-erosion on the sides of the loess-bank. The latter rises about ten feet above the surrounding flat ground, and manifestly owes its preservation to tombs placed on its top. Pieces of much-decayed timber sticking out on the upper slope probably belong to tombs that have been cut away through erosion.

Possible cause of abandonment.

Drift-sand in low dunes now encircles Hāsa from the north and north-east, forming a fit setting to the dreary scene; yet there is no proof that the site of Togujai, from which this burial-ground probably received its occupants, was abandoned owing to an advance of the desert. Even at the present day there are plots of ground under cultivation north of Togujai, and the inroads regularly made there by flood-water show that irrigation would not be impossible.

[1] See *Through Asia*, pp. 738 sq.

18

KEPEK-GHOLUK MAZĀR, NEAR MOJI.

17

KAPTAR-KHĀNA RUIN NEAR KHĀN-UI, BĒSH-KARĪM.

19

RUINED MOUND OF KARAKĪR-TIM, NEAR PIĀLMA.

When I passed through Moji again, on my return in May of the following year, plenty of water capable of spreading fertility was running to waste in a number of flood-beds. The impression I received, amidst the detached areas of cultivation at Moji, was that the present population of the oasis, reckoned at 500 households, but probably considerably less, is by no means equal to making full use of the available water and arable land.

OBJECTS FROM TOGUJAI SITE BROUGHT BY MOJI VILLAGERS.

M. 001. Coins.

90 Muhammadan copper coins, among them 78 coins of 'Sulaymān Khāqān' (unidentified), in two varieties; see Plate XC, Nos. 45, 46, and *J.A.S.B.* 1899, Extra No. 1, Pl. 1, fig. 3.

19 Chinese copper coins of Ch'ien-yüan period (758–759 A.D.), small.

4 Chinese copper coins, do., large.

28 Muhammadan copper coins, mostly of 'Sulaymān Khāqān' (?); see Plate XC, No. 47.

M. 001. a. Bronze seal, 1½" × ¾", bearing badly engraved Chinese (?) characters. Shank at back, pierced. See Plate L.

M. 001. b. Seal or ornament of bronze, 1⅛". A four-petalled flower. Shank at back broken off. See Plate LI.

M. 001. c. Bronze seal, 1¼" × 1¼" (nearly). Divided into three panels, those at top and bottom being narrow, and the centre nearly a square. The device in the centre is a Svastika, and in the top and bottom portions a conventional vine, with undulating stem and well drawn leaves. Shank at back broken off. See Plate L.

M. 001. d. Bronze seal or ornament representing a flying bird, and similar to that found at Hāsa (H. M. 1.), but reversed. Shank broken off back. See Plate LI.

M. 001. e. Fragment of bronze Vajra, 1⅛" × 9/16". See Plate LI.

M. 001. f. Portion of bronze buckle or clasp; has two rivets attached for fastening. 1⅝" × 1". See Plate LI.

M. 001. g. Portion of bronze buckle or clasp; appears to have been decorated with a floral design in relief. See Plate LI.

M. 001. h. Crescent-shaped lead ornament, 7/16" × 7/16".

M. 001. i. Fragment of an arm of a figure from shoulder to above elbow. Soapstone (?). It has an amulet with a three pointed leaf-like ornament. On the shoulder drapery is indicated. 1 3/16" × ⅜"; 3/16" thick.

M. 001. j. Fragment of arm of figure carved in same stone as M. 001. i. and probably portion of the same arm. This piece is from below elbow to wrist. It has bangles at wrist. ⅝" × ½"; appr. ⅛" thick.

M. 001. k. Small intaglio. Carnelion. Aphrodite standing, drapery from waist downwards; holds sceptre. Held by Prof. P. Gardner to belong to third century A.D., Roman work. See Plate LI.

PIECES OF TERRA-COTTA, GLASS, ETC., PICKED UP AT TOGUJAI SITE.

T. M. 001. a. Fragment of coarse red pottery, well fired. Apparently portion of a handle of a vessel. On upper surface a small (⅜") stamped device resembling a rosette.

T. M. 001. b. Fragment of light red pottery, coated inside and out with a white slip, imperfectly glazed, on which is painted a design in two tints of olive green and terra-cotta colour, on one side only. See Plate XLII.

T. M. 001. c. Small fragment of red pottery, coated with slip and light yellow glaze.

T. M. 001. d. Small fragment of red pottery as T. M. 001. c. but showing sgraffito pattern scratched through slip to red paste before being glazed.

T. M. 001. e. Small fragment of thin red pottery, coated on inside with thick and very beautiful turquoise blue glaze having an egg-shell surface.

Miscellaneous fragments of glass vessels of various thickness and tints from pale cream to bottle green. Many of the thinner fragments are as thin as paper, and have acquired a beautiful iridescence. Several pieces have raised lines and portions of handles or ornaments such as are common in Roman glass. In some cases the thickened rims are hollow. See Plate LII.

Two pieces of charcoal.

Small metal fragment, perhaps portion of a coin.

T. M. 001. f. Fragment of terra-cotta pottery, bearing traces of pale greenish glaze, indifferently fluxed, and slightly sulphured at edge; 1½" × 2¼".

T. M. 001. g. Fragment of terra-cotta flat dish, made on the wheel. It seems to have been covered with a thin white slip, over which is a fine rich yellow glaze with crude pattern in umber. On the back remains portion of the foot. 2" × 2 1/16". See Plate XLII.

T. M. 001. h. Upper portion of terra-cotta vessel, with traces of slip and indifferent dark green glaze. The portion of mouth shows a diameter of about 1⅞" from which, after narrowing to a short neck, it swells out in elliptical curve to a diameter of about 3¼". Height (of fragment) 1¾", width 1⅝". See Plate XLII.

T. M. 001. i. Fragment of terra-cotta vessel, glazed with fine dark green glaze. 1¼" × 1¼".

T. M. 001. j. Similar fragment. 1¼" × ⅞".

T. M. 001. k. Fragment of centre of terra-cotta dish, covered inside with slip and yellowish glaze, on which is painted a bold pattern in yellow. 1¼" × 1¼". See Plate XLII.

T. M. 001. l. Flat terra-cotta whorl. Diameter 1⅛"; hole ⅜"; thickness $\frac{3}{16}$". See Plate XLI.

T. M. 001. m. Fragment of similar whorl. Diameter ⅞", hole ⅜".

T. M. 001. n. Burnt tarsal-bone of sheep or goat. Length 2⅛".

T. M. 001. o. Burnt fragment of bone. 1" × ⅜".

T. M. 001. p. Burnt fragments of bone (?).

T. M. 001. q. One shell-like bead, black with white lines painted in enamel. Length ⅜". See Plate LII.

Two fragments of similar beads, differing slightly in shape, and one having a few green spots as well as the white decoration. See Plate LII.

T. M. 001. r. Fragments of burnt bone. ½", $\frac{5}{16}$" × ⅛".

T. M. 001. s. Four fragments of glass in tints of green and grey. One piece showing portion of the rim of the vessel to which it belonged, and is hollow-tubular. See Plate LII.

T. M. 002. a. Fragment of red pottery, unglazed, probably part of foot of a vase, or of a lid. Decorated with incised, long, daisy-like petals radiating from centre, on each of which an incised dot and line. See Plate XLII.

T. M. 002. b, c. Two fragments of red pottery, glazed with yellowish brown glaze. They may be portions of handles of vessels. They have flattened irregular pear-shaped faces, on which is roughly moulded a kind of anthemion pattern. They were connected with narrower parts at one side and at back. See Plate XLII.

T. M. 002. d. A piece of black slag.

T. M. 003. a. Fragment of dark grey pottery, the colour being due probably to prolonged and hot firing. It shows marks on the inside of having been turned on the wheel, and on the outside has impressed a well-executed band of ornament consisting of a series of roundels between two lines. Thickness ⅜". There are traces of a thin whitish slip on the outside. See Plate XLII.

T. M. 003. b. Fragment of black pottery, very heavily fired. On the outside are various incised patterns arranged in bands. Thickness about ½".

T. M. 003. c. Fragment of coarse red pottery, apparently part of a lid. It shows portions of two concentric bands of impressed patterns, consisting of a series of anthemion-like designs with a space between each impression. The designs appear to vary slightly in the two series. An incised line divides the bands. The under side is very coarse and rough. See Plate XLII.

T. M. 003. d. Fragment of red pottery, glazed, similar to T. M. 002. b, c, but with better executed anthemion. See Plate XLII.

T. M. 003. e. Fragment similar to T. M. 003. d, but coarser, and damaged in execution.

T. M. 003. f. Fragment of pale buff, coarse pottery; the inside very rough, the outside smooth and decorated with incised lines running round, between which a cleverly executed nebular pattern. Thickness ½". See Plate XLII.

T. M. 003. g. Fragment of coarse red pottery, glazed green. It is of curious shape, but too fragmentary to determine its original complete form and use. See Plate XLII.

T. M. 003. h. Smaller fragment of object similar to T. M. 003. g, but glazed brown.

T. M. 003. i. Small fragment of coarse red pottery bearing incised pattern.

T. M. 003. j. Fragment of vessel of red pottery, glazed over a white slip, and ornamented with green grey pattern, probably under glaze.

T. M. 004. a. Three fragments of thick (⅝") pottery, of red, unglazed clay. Probably portions of a large vessel. They are decorated with a bold, well-designed and well-executed band ornament as follows:—several slightly incised lines, encircling vessel, made with blunt point; then a plain band, 1⅛" wide. Below this a deeply incised guilloche band ½" broad, then a narrow fillet, and then a broad band (about 2½") consisting of conventional four-petalled flowers closely resembling those on the wooden chair discovered at N. iii. (See Plate LXVIII). In the triangular spaces between the two upper petals is a bead with a hole in centre, and in each of the quadrangular spaces between the flowers there seems to have been a bunch of four or five beads. Where the points of petals of one flower meet those of the next, is a hole in the surface. See Plate XLI.

T. M. 004. b. Fragment of thick coarse red pottery (⅜" and 1" at rim), portion of a large vessel. The rim is flat at top, about 1½" broad, and turns over into the outer form of vessel in a sharply receding curve. The outer edge of the rim presents a half round moulding below which is a scallop decoration. Below this are a few roughly incised lines—part of decorative band. The rim forms a projecting lip also on the inner side as though to prevent liquid slopping over. See Plate XLI.

T. M. 004. c. Fragment of coarse red pottery, from ⅜" to ⅝" in thickness; probably part of large vessel. It bears a guilloche ornament rather lightly incised, which may possibly have been impressed with a stamp. A few shallow lines are observable running round vessel.

T. M. 004. d. Two fragments of coarse red pottery, ⅜″ thick, exhibiting boldly executed and well-designed band of incised ornament consisting of guilloche (similar to that in T. M. 004. a.) and below this, inverted heart-shape, within which a kind of fleur-de-lis. The spaces between heart-shapes are occupied by smaller and more simple similar patterns. See Plate XLI.

T. M. 004. e. Fragment of coarse red pottery, nearly flat. Perhaps portion of a lid of a vessel. Thickness ½″ nearly. It bears a series of patterns, each formed by impressing an L-shaped stamp four times in such a manner as to form a cross. There are two circular bands of these divided by a line, and a sort of ray pattern formed by making lines of dots radiating roughly from the centre of the object. All very crude. See Plate XLI.

T. M. 004. f. Fragment of coarse red pottery, $\frac{5}{16}$″ thick. It bears a crude kind of cable ornament in relief, running round, and on the fragment is shown a loose end depending from the band. There are traces also of roughly and lightly incised ornamentation, apparently made with a toothed tool. See Plate XLI.

T. M. 004. g. Fragment of coarse red pottery, ⅜″ thick, nearly flat. Probably part of a lid. At edge scallop pattern.

T. M. 004. h. Fragment of coarse red pottery, $\frac{5}{16}$″ thick, with roughly and lightly incised pattern suggesting a lotus-petal design between two lines. See Plate XLI.

T. M. 005. a. Fragment of coarse red pottery, ¼″ thick. Bears cable pattern in relief similar to that of T. M. 004. f.; also incised scallop and line running round.

T. M. 005. b. Fragment of pottery, coarse, red. Bears scallop in relief, and roughly incised scratches to represent leaf branches. See Plate XLII.

T. M. 005. c. Fragment of pottery, coarse, red. Diaper pattern, consisting of incised triangles and in centre of each a dot. Very crude. See Plate XLI.

T. M. 005. d. Fragment of pottery, coarse red, discoloured (or perhaps purposely darkened) on upper surface. Probably portion of lid. Crude pattern formed by impressing a simple stamp twice ☉, scattered about surface, and two lines probably running round.

T. M. 005. e. Fragment of red pottery, coarse, glazed with green glaze, which is imperfectly fluxed to the body. The edge bears a roughly incised pattern. It is probably portion of a lid.

T. M. 005. f. Fragment of coarse red pottery, about ⅜″ thick. The outer surface decorated with broadly incised patterns, the scheme of which is not quite clear from the small fragment. See Plate XLI.

T. M. 005. g. Fragment of handle and rim of vessel in coarse red pottery, rudely representing an animal's head, probably that of a cow. See Plate XLII.

T. M. 005. h. Fragment of coarse red pottery, probably portion of a lid. The edge is turned upwards and then over into scallops. The top surface is ornamented with a circle of incised rings, within which a circle of rays springs from a plain circle. See Plate XLI.

T. M. 005. i. Fragment of coarse red pottery, apparently portion of a small *lōta*. It shows the rim which is thickened, also traces of a coating of unglazed colour—a sort of thin slip—on both inside and outside. The outside bears a crudely incised scratchy pattern in the slip only, consisting of more or less parallel lines running round, and a waved line running over them.

OBJECT FOUND IN GRAVE AT HĀSA.

H. M. 1. Small Bronze ornament in form of flying bird, about $\frac{1}{16}$″ thick, having a large shank at back, pierced with hole for pin to secure the ornament or for cord; comp. M. 001. d.

Section V.—FROM MOJI TO THE KHOTAN OASIS

An easy march of some fourteen miles, mostly over stony 'Sai' covered with low scrub, brought me to Zanguya, a small but fertile oasis. According to the information supplied to me, cultivation shifts every year in turn to one of the four great plots into which the village lands are divided[1]. Inadequacy of the water-supply for the available canals is given as the reason. But two great flood-beds are passed some three and four miles west of Zanguya, where there is practically no cultivation. From a statement recorded by Dr. Hedin it appears that the loss of this water is keenly felt by the villagers, whose attempts, however, at regulating embankments prove ineffectual against any larger flood. Here, too, it is clear that a denser population under a more active administration would soon find the way to increase the productiveness of the oasis. *Oasis of Zanguya.*

[1] Compare also Hedin, *Reisen in Z.-A.*, p. 16.

Site of *Kul-langar.*

Antiquarian evidence of more extensive cultivation in former times is furnished by an old site called *Kul-langar* ('the rest-house by the tank'), which I was shown about two miles to the north-west of the central village. I found there the ground near the edge of the cultivated area strewn with pottery fragments over about half a square mile. I could discover no ornamented pieces among them, nor was there any striking evidence of erosion. Among the débris, on and below the ground, bones, layers of ashes, and charcoal were plentiful. Hence the soil is considered valuable for manuring, and is carried away from numerous pits. In one of these pits I found the strata of decayed refuse and ashes extending to a depth of six feet below the surface. At the northern end of the site, high above a watercourse now dry, are two empty tanks enclosed by circular mud *bands.* The larger one measures about 60 yards in diameter. Old copper coins are said to be picked up occasionally on the site, but I could not obtain any of them. The potsherds looked coarse, though little affected by erosion. The general impression I brought away from Kul-langar was that of a village site abandoned at a not very distant period.

Kizil-Tam.

The march from Zanguya to Piālma, the next oasis towards Khotan, passes over ground which throughout shows the character of true desert. At a distance of about two miles from the edge of the cultivated area, low dunes of coarse drift-sand begin to cover the bare loess; some distance further south-east it changes into a hard pebble 'Sai'. About six miles from the centre of Zanguya, at a spot called *Kizil-Tam* ('the red wall'), I found a small plot of ground thickly strewn with fragments of old pottery, of a fine red colour, very hard, but entirely without ornamentation. The surrounding drift-sand may cover more. Close by is passed the dry bed of a stream, which probably carries occasional flood-water from the outer hills of Duwa.

Karakir-Tim.

Already at Moji I had heard of a ruined mound called *Karakīr-Tim* ('the mound of the black ridge'), which was to be seen not far from the road to Piālma. At Sai-Langar, a lonely rest-house some 13 miles from Zanguya, where water is obtainable only from a deep well, we struck off the caravan route, and proceeding east by north over heavy dunes of coarse sand for two miles reached the mound. It proved to consist of a solid mass of sun-dried bricks badly decayed on its surface, yet furnishing evidence by its size and proportions that it represented the remains of a Stūpa. Considerable portions of the masonry have broken away, especially on the south side, while elsewhere the outlines were hidden under deep masses of crumbling débris.

A careful survey of the mound by means of the plane-table (see Plate XIX) showed that at its foot it formed a square of approximately 65 feet. But owing to the sand which from a high dune close by reaches the northern and western sides, and on account of the débris-covered slopes on the other sides, the actual outlines of the base where it touched the ground could nowhere be traced with absolute clearness. Higher up the contours were still more irregular, as will be seen from the plan. Nevertheless, the outlines visible in a photograph (Fig. 19) taken from the west still give a faint indication of what must once have been a dome surmounting a square base. The highest elevation of the mound above what appears to have been the original ground-level, as measured at the south-east corner, is about 22 feet. The top now forms an elongated small flat, showing that a considerable portion of the original superstructure must have crumbled away. In this respect, as well as in its general decay, the mound closely resembles the Stūpas of Kurghān-Tim and Kizil-Debe. Owing to the large masses of fallen masonry, I doubt whether even an extensive clearing of the débris would permit of approximately exact measurements being obtained of the several stories, &c., of the base and dome.

Condition of ruined mound.

The bricks, which were exposed in many places, showed a fairly uniform size of 16 inches square, with a thickness of 3½ to 4 inches. These dimensions agree accurately with those recorded above for the bricks of the Maurī-Tim Stūpa. It is probable that the slightly better preservation of the north-western side of the mound is due to the same cause which we have already had occasion to discuss in connexion with the Tōpa-Tim ruin; namely, accumulation of sand on the side most exposed to the prevailing winds. Whether a great débris-filled gap between the solid masonry remains on the south side (see plan) was due to excavation or to some natural cause of decay, I was unable to ascertain. As a special feature I may mention that large stones rounded by the action of water were noticed by me embedded as a regular layer in the brickwork some five feet above the ground near the south-west corner. The ruin rests on hard gravel which, however, is exposed to view only on the south and east. Elsewhere coarse drift-sand covers the ground, rising immediately to the west and north in long dunes up to a height of fifteen feet or so. This sand is absolutely sterile, forming a striking contrast to the dark line of trees of the Piālma oasis visible far away on the horizon. The ground on the east side of the mound, being clear of sand, was found strewn with small potsherds, but to no great distance; no ornamented or otherwise defined pieces were traced among them.

From Karakīr-Tim to the western edge of the Piālma oasis, a distance of close on six miles to the south-east, I could nowhere trace signs of cultivation, old or recent; considering the character of the soil, which is hard pebble 'Dasht', close up to the flood-bed of the Duwa stream west of the oasis, it is difficult to believe that cultivation was ever carried on here. The isolated position of the ruined Stūpa thus appears distinctly curious. Nor could any other remains be traced near it. Piālma itself is a small oasis, counting, according to my information, only about one hundred households, and dependent for its water supply on a stream which comes from the mountains of Duwa. As the drainage area can scarcely include any mountains with permanent snow-beds, the supply of water available for irrigation is uncertain and often scanty[2].

Position of ruined Stūpa.

Piālma is now the last settlement passed by the route outside the western limit of the Khotan district, the present administrative frontier between Karghalik and Khotan being marked by two half-decayed pillars on the road some miles from the edge of the Piālma oasis. In Hsüan-tsang's time Piālma itself must have been reckoned within the territory of Khotan; for there can be little doubt that the town of P'o-ch'ieh-i (*Po-kia-i*) 勃伽夷, which, according to the 'Life,' the pilgrim first reached after entering the frontier of the kingdom, and where he made a halt of seven days before proceeding to the capital, is to be looked for near the present Piālma[3]. From the *Hsi-yü-chi* we learn that Hsüan-tsang placed this locality at a distance of 300 li, or three marches, to the west of the Khotan capital[4], and this reckoning, according to the daily stages usual at the present day, takes us to Piālma. The distance from Piālma to Yōtkan, about 48 miles by my map, say 56 miles actual measurement, allowing for small détours and the heavy sand encountered on part of the route, agrees with this. If a resemblance in the sounds of Chinese transcriptions can be trusted, Piālma or *P'o-ch'ieh-i*, might possibly be meant also by *P'o-hai* 渤海, which the above-quoted itinerary of the T'ang Annals mentions as the second locality to the west of Yü-t'ien[5].

Oasis of Piālma.

[2] For details about the Piālma oasis, see Hedin, *Reisen in Z.-A.*, pp. 17 sq.

[3] See *Vie de Hiouen-Thsang*, pp. 279 sqq.—My previous suggestion as to the possible location of P'o-ch'ieh-i (*Pò-kia-i*) at Moji, *Ruins of Khotan*, p. 191, was based on an inadequate consideration of the distance and of the position indicated by Hsüan-tsang's 'Life'.

[4] See *Mémoires*, ii. pp. 230 sq.; *Si-yu-ki*, transl. Beal, ii. p. 314. Beal's version and that of Rémusat, *Ville de Khotan*, p. 45 seem to be more exact than that of Julien, who makes Hsüan-tsang arrive at P'o-ch'ieh-i (*Po-kia-i*) from the capital, though in reality he travelled in the opposite direction.

[5] Comp. Chavannes, *Turcs occid.*, p. 123, note 1.

Shrine and legend of *P'o-ch'ieh-i.*

P'o-ch'ieh-i contained a shrine with a miraculous statue of Buddha, the legend of which is told at length both in the *Hsi-yü-chi* and in Hsüan-tsang's 'Life', as gathered 'from ancient tradition'. Once the pupil of an Arhat, living in Kashmīr, when dangerously ill, asked for a cake of rice of a peculiar sort[6]. His saintly teacher obtained this for him from Kustana or Khotan, whereupon the Śrāmaṇēra, who much relished the dish, prayed to be reborn in that country. Having obtained his wish and become king of Khotan in his new birth, he crossed the snowy mountains and attacked Kashmīr. A battle between him and the ruler of Kashmīr was averted by the Arhat who, showing to the Khotan chief the clothes he had worn of yore as a Buddhist disciple, revealed to him his previous existence and induced him to desist from his attack. Before retiring to Khotan the king presented himself before the Buddha statue to which he had paid worship in his former birth, and took it homewards with his army. When the statue arrived at P'o-ch'ieh-i it refused to be moved further. Thereupon the king constructed a convent around the statue and placed upon its head his own diadem adorned with precious stones. This diadem, ever spreading a brilliant light, was still seen by Hsüan-tsang on the head of the statue. The latter is described as being 'seven feet high and marked with all the distinguishing signs of beauty', and its appearance as 'imposing and dignified'[7].

Traditions indicated by legend.

This legend is of interest in several respects. It proves that Kashmīr was credited with having supplied to Khotan statuary of ancient date, a fact throwing light on the channel, or one of the channels, through which Khotan art derived its unmistakable connexion with the Graeco-Buddhist art of Gandhāra. It is the only trace we possess of a Khotan tradition pointing to an invasion of Kashmīr across the great mountain barrier of the Karakorum. Legendary as this tradition may be, it deserves to be noted in view of the political connexion between the two territories which has been supposed to have existed at an early historical period[8]. Topographical interest, too, attaches to the legend; for it seems to indicate that Piālma was believed to have already in ancient times marked the frontier of Khotan for those coming from the west or south. The easiest route and, in fact, the only one frequented by caravans, from Khotan to the Karakorum Passes and Ladāk leads via Piālma and Zanguya to Sanju. From there the Upper Kara-kāsh Valley is gained over the Sanju Dawān, the direct route through the difficult gorge formed by the middle course of the Kara-kāsh river being quite impassable except in the depth of winter, and then only for men and possibly unladen animals[9].

Position of Hsüan-tsang's shrine at P'o-ch'ieh-i.

As to the position of the convent which contained the miraculous statue from Kashmīr I have no definite opinion to offer; I regret not to have made inquiries while passing through Piālma for any modern shrines in the vicinity. These, in view of the well-established tenacity of local worship proved for the Khotan region, might perhaps furnish some clue. The location of the Vihāra at the ruin of Karakīr-Tim, of course, readily suggests itself. The distance of the latter from the site of the old capital of Khotan would accurately agree with Hsüan-tsang's estimate of 300 li; the isolated position of the ruin might find a natural explanation in the traditional belief of a miracle which fixed the statue at a particular spot of

[6] Julien translates 'riz sur' in *Mémoires*, ii. p. 231; 'riz imprégné de vinaigre' in *Vie de Hiouen-Thsang*, p. 280. Beal, ii. p. 314, speaks of a 'cake of sown rice,' whatever that may mean; in *Life*, p. 204, he suggests the rendering 'sour meal'. Rémusat, *Ville de Khotan*, p. 46, has 'riz fermenté'. Some particular delicacy of Khotan must be meant.

[7] See *Mémoires*, ii. pp. 230 sqq.; *Si-yu-ki*, transl. Beal, ii. pp. 314 sq.; *Vie de Hiouen-Thsang*, pp. 279 sqq. The *Si-yu-ki* describes the statue as seated, the *Vie* as standing. But Beal, *Life*, p. 204, has 'sitting figure'.

[8] See Dr. Hoernle's hypotheses in *J.A.S.B.*, 1899, Extra No. 1, pp. 13 sq.

[9] There is a more southern route from Khotan to Sanju, followed by Grombchevski in 1895, via Ujāt, Pujiya and Duwa; but it is longer and certainly not as easy as the regular caravan route via Piālma.

the route, irrespective of any stages observed by the ordinary traveller. But in the absence of direct evidence all this must remain mere conjecture.

A long march on the 10th of October brought me from Piālma to the confines of the great oasis of Khotan. Up to Ak-Langar, the regular stage some sixteen miles from Piālma, the route lies over an absolutely barren plain, first of hard loess then of gravel. From Ak-Langar onwards, where water is obtained only from a very deep well, the route passes for a distance of some ten miles through a belt of drifting dunes. Forming regular semi-lunes of the usual shape and direction, these dunes rise to quite respectable heights, up to twenty feet and more, and extend far away to the south of the route. In the midst of this belt of drift-sand, a southern inlet as it were of the great sand ocean, the traveller reaches a remarkable shrine known as the Mazār of *Kum-rabāt-Pādshāhim*, 'My Lord of the Sands Station'. Several wooden houses and sheds serve as shelter for thousands of pigeons, which give to the shrine its popular name of *Kaptar-Mazār*, 'the Pigeons' Sanctuary'. The fluttering hosts, which are perfectly tame, are maintained by the offerings of travellers and the proceeds of pious endowments consisting of 'Waqf' lands in the Khotan oasis[10]. Mazār of *Kum-rabāt-Pādshāhim.*

According to the legend, as told to me by the son of one of the seven 'Shaikhs' who have hereditary charge of the shrine, the sacred pigeons are the offspring of a pair which miraculously appeared from the heart of Imām Shākir Pādshāh when this champion of Islām met death here in battle with the infidels, i. e. the Buddhists of Khotan. Many thousands had fallen on both sides, and it was impossible to separate the bodies of the 'Shahīds' who had died for the Faith from those of the 'Kāfirs'. Then at the prayer of one of the surviving Musalmans the bodies of those who had found martyrdom were by a miracle collected on one side, and two doves came forth to mark the remains of the fallen leader. One settled on his head, the other at his feet. From gratitude, all travellers who pass by this road offer food to the holy birds, either bringing corn for the purpose or else buying it from the store of the shrine, as I myself did in compliance with the pious custom. I was assured that birds of prey never succeed in killing a pigeon, but die in the attempt. The legend was repeated to me in the same form by Aḥmad Shāh, one of the old Shaikhs whom I subsequently met near Zawa, and is said to be recorded in a Tadhkirah or legendary, of which, however, I could not obtain a copy. Legend of the 'Pigeons' shrine'.

The absolute desolation of the surroundings made the pretty spectacle of the fluttering swarms doubly impressive; and face to face with the time-honoured practice to which they owe their maintenance, I could not fail to be reminded of what Hsüan-tsang tells us of a local cult curiously similar on the road leading to Khotan from the west. Some 150 or 160 li before reaching the capital, 'in the midst of the straight road across a great sandy desert,' the pilgrim describes 'a succession of small hills' which were supposed to have been formed by the burrowings of rats[11].

Of these rats popular legend related that they were 'as big as hedgehogs, their hair of a gold and silver colour', and that they were seen following a rat chief who daily emerged from his hole. In old days a general of the Hiung-nu, who had come to ravage the border Hsüan-tsang's legend of the sacred rats.

[10] A small spring which issues in a hollow at the foot of a loess-bank exposed amidst the dunes a little to the south of the shrine supplies the water needed by the resident custodians and their winged protégés. Does the presence of water in the midst of this arid waste of dunes account for the first origin of the local worship established here?

[11] See *Mémoires*, ii. p. 232 sqq.; *Si-yu-ki*, transl. Beal, ii. pp. 315 sq. Julien's translation does not indicate that in Hsüan-tsang's time the miraculous rats were still believed to be visible, but merely records their presence in the past. For a translation of the passage as reproduced in the Pien i tien, see Rémusat, *Ville de Khotan*, pp. 47 sqq.

with several hundred thousands of men, was believed to have encamped his host near these rat-mounds. The king of Khotan, so the story went, with a force of only some few myriads, prepared to meet the enemy, but despairing of success turned his thoughts in prayer to the miraculous rats for help. 'That night the king of *Ch'ü-sa-ta-na* (Kustana, Khotan) in a dream saw a great rat' which promised him succour and victory for the morrow. The king thereupon set out with his forces before dawn and unexpectedly fell upon the enemy. When the Hiung-nu hastily prepared to resist the attack 'they found that the leather of their armour, and their horses' gear, and their bowstrings, and all the fastenings of their clothes, had been gnawed by the rats'. The terrified Hiung-nu were defeated with great slaughter, and their chief killed. 'The king of Ch'ü-sa-ta-na (Kustana, Khotan), in gratitude to the rats, built a temple and offered sacrifices; and ever since they have continued to receive homage and reverence, and they have offered to them rare and precious things. Hence, from the highest to the lowest of the people, they pay these rats constant reverence, and seek to propitiate them by sacrifices. On passing the mounds they descend from their chariots and pay their respects as they pass on, praying for success as they worship. Others offer clothes, and bows, and arrows; others scents, and flowers, and costly meats. Most of those who practise these religious rites obtain their wishes; but if they neglect them, then misfortune is sure to occur.'

Survival of local legend. Klaproth has already noticed the curious resemblance between this legend and the story which Herodotus tells us of the destruction of Sennacherib's Assyrian host on the frontier of Egypt, effected by field rats in the same way, and also foretold in a dream[12]. But yet more interesting is the survival to the present day, and in the identical place, of the local worship which the legend told to Hsüan-tsang was intended to explain; for the locality which the pilgrim describes corresponds exactly to the position of the 'Kaptar-Mazār' relative to the ancient Khotan capital. The distance from the site of Yōtkan to the modern shine is fully 24 miles by the map, which, taking into account the increase of actual road measurement due to small détours and the heavy going over sandy ground, adequately represents the one and a half marches implied by Hsüan-tsang's estimate of 150 to 160 li. The shrine still greets the traveller 'in the midst of the straight road across a great sandy desert', exactly as the pilgrim saw it.

Sandhills near *Kaptar-Mazār*. Immediately to the east of it the road enters a tract where the dunes are plentifully interspersed with conical sandhills covered by tamarisk growth. These sandhills are typical features on those strips of the desert where the drift-sand is so plentiful as to form relatively high dunes, but where at the same time the ground water is still near enough to permit the tamarisk shrubs to send their roots down to it and thus to keep themselves alive on the top of the sand-cones which they bind together[13]. Only at this point of the whole route between Khotan and the west are these curious hillocks to be seen. I, therefore, do not hesitate to recognize in them Hsüan-tsang's 'succession of small hills formed by the burrowing of rats'. The explorer of the Taklamakān and the roving hunter or 'treasure-seeker' who visits wide areas of the desert have, indeed, occasion to become familiar in many places with these peculiar formations. But to the ordinary wayfarer who keeps to the main roads, and to the average stay-at-home inhabitant of the oasis they were bound to appear in old days, just as now,

[12] See *Mémoires relatifs à l'Asie* (1826), ii. 297 sqq., where an amusing parallel, capable of a less miraculous interpretation, is quoted from the official report of a Russian commandant about the destruction of certain military magazines at Okhotsk, in 1806.

[13] For a diagram and a lucid explanation of the formation of such tamarisk-covered cones, see Hedin, *Reisen in Z.-A.*, p. 34. For illustrations showing hillocks of this kind see below, Figs. 38, 39, 44, 52; also Pl. VI; for other references to them, compare Index, s. v. *tamarisk*.

distinctly strange and puzzling. Thus the mysterious origin which the popular legend heard by Hsüan-tsang ascribed to them becomes easily intelligible.

Pigeons substituted for sacred rats.

It is still more evident that the manner in which the pigeons kept at the shrine are propitiated with food-offerings by all modern wayfarers marks a survival of the pious practice of Buddhist times which was accounted for by that legend. As Hsüan-tsang's rats, so now the holy pigeons which have taken their place are supposed to recall a great victory. There is nothing surprising in this substitution of pigeons, visible in body yet of supposed miraculous descent, for the wonderful rats of the Buddhist story which pious superstition, even in Hsüan-tsang's days, apparently did not pretend to see any longer [14]. For to Muhammadan popular lore the pigeon is a sacred bird *par excellence*, and the custom of feeding pigeons at Mosques and Ziārats is widely spread throughout Islām [15]. Nor is there anything very remarkable in the great victory of a Khotan king over an invading host of barbarians having been converted by the Muhammadanized legend into a sanguinary contest between the infidels of Khotan and those pious emissaries of Islām whom local tradition believes to have first carried, sword in hand, the Faith into Buddhist Khotan.

Muhammadan local worship at Khotan.

We shall have occasion to recur hereafter to numerous legends, almost all localized at particular Ziārats of Khotan, which relate incidents connected with the first conversion of Eastern Turkestān, and which in many instances can be proved to have been grafted on sites already held sacred as places of worship and pilgrimage in Buddhist times. M. Grenard has collected a considerable number of these legends, and subjected them to a critical analysis which, from the point of view of the study of folklore and history of religion, is very instructive [16]. It is true that Imām Shākir Pādshāh, who is supposed to have suffered a martyr's death among the sands of Kum-rabāt-Pādshāhim, does not figure in the orthodox though apocryphal list of the twelve Imāms whom popular belief credits with having first planted the Faith in Khotan. But he shares this fate with many another alleged saintly 'Shahīd' of Khotan, to whom local tradition nevertheless clings with pious attachment. The details told of his last struggle are neither scantier than, nor very different from, those related in the Ta<u>dh</u>kirahs of the recognized warrior saints of Khotanese legend.

Painted panel representing chief of sacred rats.

How popular all over ancient Khotan the legend of the sacred rats must have been is strikingly illustrated by the fact that among the painted panels, originally deposited, no doubt, as votive offerings, which came to light in the course of my excavations among the ruined temples of the Dandān-Uiliq site, there is one that represents a rat-headed figure crowned with a diadem, and clearly marked as an object of worship by the attitude of an attendant figure. For a detailed description of this panel, D. iv. 5, reproduced in Plate LXIII, I must refer to the chapter which deals with the discoveries of that site; but I may point out here that the identity of this figure with the rat-chief mentioned in Hsüan-tsang's story can scarcely be doubted. The fact that the pictorial representations discovered at that site apparently comprise every one of the local legends recorded by Hsüan-tsang, seems to me a clear indication that the story of the holy rats, like that of the introduction of silkworms or of the minister's self-sacrifice to the river-goddess, must have been among the most cherished folk-tales of ancient Khotan [17].

[14] Julien's translation relating to the popular story says: 'Dans ce désert, il *y avait* des rats, &c.'; *Mémoires*, ii. p. 232. This seems to me preferable to Beal's version, which is in the present tense.

[15] e.g., at the popular shrine of Sheikh Junāyat in Peshāwar it is customary for all those who come to invoke the saint's help to offer corn to the swarms of pigeons which abide near his tomb.

[16] See Grenard, *Mission Dutreuil de Rhins*, iii. pp. 3–46; also ii. pp. 240 sqq.

[17] See below, chap. ix. sec. v.

Tenacity of local worship in Khotan.

It was of no small interest to me to be met thus on the very confines of Khotan by a striking instance of that tenacity of local worship which my subsequent researches showed for almost all sacred sites of Buddhist Khotan. Remembering how helpful it had been to me in Kashmīr, and elsewhere in the north-west of India, to find the position of ancient Buddhist or Hindu shrines I was in search of almost invariably marked by Muhammadan Ziārats, I felt justified in accepting this observation on my very entry into Khotan territory as an auspicious omen.

Continuity of physical conditions.

But it was also a comforting assurance to me thus to receive further convincing evidence how limited in reality are the changes which the physical conditions of the ground traversed by the western route to Khotan, and the direction of this route itself, are likely to have undergone during historical times. All along the route from Karghalik onwards we have found that the ancient remains still traceable lie close to the line of the actual road. Nowhere did I meet with any antiquarian indication to support the oft repeated assertion that the area of sandy desert has materially advanced to the south during recent periods. Here, near the very end of the route, we have direct evidence that the curious zone of drift-sand traversed immediately before my entry into the Khotan oasis bore the same natural aspect as in Hsüan-tsang's time and probably for long centuries earlier. Proofs of such continuity cannot be otherwise than encouraging to the student of the historical geography and ancient culture of this region; for they assure him that through whatever changes the population, its political conditions and civilization may have passed, the natural *milieu* and its determining physical factors cannot have altered so much during historical periods as to vitiate seriously any conclusions that may be drawn from ascertained antiquarian facts.

CHAPTER VI

THE KHOTAN OASIS: ITS GEOGRAPHY AND PEOPLE

SECTION I.—THE OASIS IN ITS GEOGRAPHICAL FEATURES

THE oasis of Khotan has from early times been the largest and most important cultivated territory in the south of the Tārīm Basin. To this fact we owe the ample information which the Chinese records furnish as to its ancient history. For the correct understanding of the antiquarian observations and finds which my explorations in this region have yielded, a preliminary review of the data available concerning the history of Khotan is essential. Before, however, we proceed to this it will be advisable to consider, if only in broadest outline, those factors to which may be ascribed a determining influence upon that history—the geography of the Khotan oasis, and the character and origin of its people.

Geographical interest of Khotan.

The territory properly known as Khotan would form an admirable subject for a geographical monograph. It presents on the one hand most of the natural features typical of the oases which fringe the great desert of Chinese Turkestān, while on the other hand its position, size, and historical past invest it with an individuality of its own. Though the number of qualified observers who have visited Khotan since the time of Johnson (1865), and recorded the results of their inquiries and surveys, is not small, the time for such a monograph does not yet seem to have come. And even if the materials available were more complete and accurate than they are, the scope of the present work would not permit of an attempt to treat the subject on such lines. I shall, therefore, content myself with briefly indicating those main geographical features which determine the physical conditions of the oasis, and which must be steadily kept in view when studying its history and antiquities.

Accounts of modern Khotan.

What illustrative details I was myself able to collect bearing on the topography of the oasis, its natural resources, cultivation, and general conditions of life, will be found recorded in my Personal Narrative[1]. Of earlier accounts by European travellers it may suffice to mention the two most notable. M. Grenard, who as companion of Dutreuil de Rhins had occasion to spend the greater part of two winter seasons (1891–3) in the town of Khotan, has given a valuable record of his own and his chief's observations, particularly detailed in all that relates to the resources of the oasis and the civilization, industries, &c., of its present inhabitants[2]. Dr. Sven Hedin, who made the town of Khotan his headquarters on two occasions in January and June of 1896, collected a mass of very useful data concerning the soil, products, and modern administration of the oasis and the great rivers which irrigate it[3].

Geographical advantages of Khotan.

The Khotan oasis owes its natural wealth and its importance entirely to the advantages of its geographical position. The mighty terrace of fertile loess which it occupies extends for

[1] See *Ruins of Khotan*, pp. 196 sqq., 247 sqq., 482 sqq.

[2] See Grenard, *Mission D. de Rhins*, i. pp. 91 sqq., 148 sqq.; ii. pp. 95 sqq. and chap. vi–x, passim; iii. pp. 325 sqq. (meteorology).

[3] Comp. Hedin, *Reisen in Z.-A.*, pp. 22 sqq., 202 sqq.

an unbroken length of forty miles along the foot of the outer hills of the Kun-lun range, and is at all times assured ample irrigation from the Yurung-kāsh and Kara-kāsh rivers which debouch into the plain immediately above it. These two rivers are the largest of those which carry the drainage of the main range of the Kun-lun northward into the Tārīm Basin. Some idea of the size of the mountain area drained by them can be formed from the fact that the source of the Yurung-kāsh, as ascertained by M. Dutrueil de Rhins and Captain Deasy on the high Ak-sai-Chīn plateau south-eastwards, is separated by a direct distance of over two hundred miles from the headwaters of the westernmost affluent of the Kara-kāsh, north of the Karakorum Pass.

Explorations in Kun-lun Mts. south of Khotan. Much of the orography of the great ranges which extend between these extreme points still awaits detailed exploration; but the expedition which I was able to undertake from Khotan southwards into the forbidding mountain region of Karanghu-tāgh and towards the headwaters of the Yurung-kāsh revealed the fact that the crest line of the magnificent snowy range, which the latter river drains first from the south and then from the north for a distance of one hundred and fifty miles, maintains an average elevation of close on twenty thousand feet, except at the point where the Yurung-kāsh has forced its passage through in a stupendous gorge behind the great Muz-tāgh Peak (23,890 feet above the sea)[4]. The map embodying the cartographical results of that expedition, and the panoramic views obtained by me with the photo-theodolite and reproduced in a separate publication of the Royal Geographical Society, will best help to realize the extent of the glaciers and slopes covered with permanent snow which feed the Yurung-kāsh and its tributaries. The Kara-kāsh river drains a great portion of the same main Kun-lun range from the south, and in addition the vast uplands known as the Ling-zi Thang and Soda plains which divide the Kun-lun from the Karakorum section of the Himālayan system. The Kun-lun range, near the point where the Kara-kāsh, after a long sweep north-westwards, breaks through it in a hitherto unexplored defile, rises, as our survey showed, to peaks well over 23,000 feet. The course of the Kara-kāsh, from its sources down to where it enters the plain near the Khotan village of Ujāt[5], is certainly longer than that of the Yurung-kāsh, and the belief of the people of Khotan in the greater volume of water carried by it during the summer is probably well founded[6].

Rivers of Khotan. The two rivers of Khotan bring down a vast volume of water during the months when the sun is powerful enough to melt the snow and ice of the high ranges. This explains why they alone, after their junction some eighty miles to the north-north-east of Khotan town, are able to penetrate through the whole breadth of the Taklamakān and to join the Tārīm, while all other rivers that enter the desert from the south get lost amongst its sand dunes. To these two great rivers the oasis of Khotan owes not only its ample irrigation but also, as geological **Loess soil of Khotan.** evidence conclusively shows, the fertility of its soil and, in fact, its very existence. Prof. Lóczy's analysis of the soil specimens brought back by me, in conjunction with my observations on the spot, proves that the loess of the oasis is of distinctly riverine type, composed of that fine sand and mud which the rivers of Khotan carry down annually in enormous quantities from the disintegrated slopes of the mountains. Most of the loess must be ascribed to subaerial deposit, the lighter constituents of this alluvium having been carried away by the winds from the immediate vicinity of the river-beds, and subsequently retained wherever the ground possessed

[4] For the survey results of this expedition compare my map, and for a description of the region explored chapters XIII, XIV of *Ruins of Khotan* (pp. 206–43).

[5] See below, chap. VIII. sec. i.

[6] Compare Hedin, *Reisen in Z.-A.*, p. 28.

sufficient moisture and vegetation to bind the dust and fine sand thus transported. This process continued during long periods has formed the mighty loess bed which throughout the Khotan oasis now overlies the coarse gravel of an earlier 'Sai'.

Seeing the dust-laden atmosphere which envelops Khotan, like other oases of the Tārīm Basin, for the greatest part of the year, and the great frequency of its sand-storms, it is easy to realize that this subaerial deposit still continues at the present day and must constantly heighten the soil[7]. But there can be no doubt that, ever since systematic irrigation has been carried on within the oasis, this process must have been greatly aided by the deposit of river silt; for of this every cultivated piece of ground receives annually a considerable quantity in the muddy water which the irrigation canal brings to it, and which is allowed to remain until it evaporates or is absorbed in the soil. When discussing later the strata of fertile soil covering the remains of the ancient Khotan capital at Yōtkan, I shall have occasion to demonstrate in detail the remarkably rapid rise of the ground-level brought about by this twofold deposit[8]. *Subaerial and silt deposits.*

It appears probable that the rivers of Khotan while thus creating as it were the soil of the oasis, have also produced that configuration of the ground which now greatly facilitates the distribution and full use of their waters.

A look at my map shows that, owing to the presence of fertile soil almost to the foot of the outer hills, cultivation begins at the very *débouchement* of the two rivers and thence spreads out fan-like over a steadily broadening area northward. Excepting the narrow strip of pebble-covered ground, probably nowhere more than three to four miles in width, which stretches along the southern edge of the oasis in the space left between the two rivers (some thirteen miles only), we do not find here the barren stony 'Sai' that elsewhere separates, like a forbidding glacis, the foot of the hills from the cultivable area. The advantages which this configuration of the surface assures must strike any observer who has paid attention to the time-honoured systems of irrigation prevailing throughout the whole region. Wherever, as in the case of all oases to the east of Khotan, a broad desert zone of stony detritus, gravel, or coarse sand, up to seventy miles in width, intervenes between the emergence of the rivers from the outer hills and the nearest large loess deposits, the utilization for irrigation purposes of the available supply of water offers considerable difficulties. *Absence of 'Sai'.*

Owing to the uniformity of level which the glacis-like 'Sai' presents, the rivers and streams passing over it necessarily show a tendency to spread themselves in numerous shallow channels. The artificial control of the rivers in their passage through this sterile zone, when in summer flood, would be a task far beyond the engineering resources of the country, and also in many cases beyond the available labour supply. Hence, as we have seen above, much of the water is lost in side channels which cannot be made to feed irrigation canals with the needful regularity[9]. Much of it, too, evaporates before the cultivable loess area is reached, and probably still more sinks into the pervious upper stratum. It is true that part of the water lost through the last cause subsequently reappears in springs;[10] but as these are subject to periodical fluctuations in level and position, and besides lack the fertilizing alluvium brought

[7] The people of Khotan seem to be well aware of this process and to ascribe to it a fertilizing effect; see Johnson's remarks in *J. R. Geogr. Soc.*, 1867, p. 6; also Geikie, *Textbook of Geology* (4th ed.), i. p. 439.

[8] Compare chap. VIII. sec. ii.

[9] See above, pp. 96, 115.

[10] Compare below, chap. XIII, the remarks about the springs from which the small oases between Chīra and Keriya derive their irrigation, also the springs feeding the river below Keriya. See also Hedin, *Reisen in Z.-A.*, pp. 39 sq., 177.

down from the mountains, it is evident that irrigation dependent on them can nowhere suffice for the creation of a large oasis.

Irrigation of Khotan oasis.

Irrigation within the Khotan oasis has to contend with none of these difficulties and losses. It begins as soon as the deep-cut valleys of the two rivers widen out sufficiently for strips of cultivation on either side.[11] The large canals, taking off from the rivers at points as close below their *débouchement* as the situation and level of the tracts to be irrigated demand, lie throughout within the cultivated area and can thus be kept in working condition without great efforts. The relative height at which the heads of these canals are established, and the marked slope of the great loess beds, make it possible for their water to be distributed not only over the whole area between the two rivers for a length of close on twenty miles, but also to be carried for considerable distances to the east and west of this 'mesopotamia'. Thus the large village of Lop on the eastern edge of the oasis lies fully sixteen miles from the nearest point of the Yurung-kāsh which supplies its irrigation. Similarly on the west the water of the Yawa-Üstang, a more or less natural channel fed from the Kara-kāsh, could easily be utilized for irrigation at a lateral distance quite as great, if only there were a sufficiency of population for such extended cultivation.

Abundance of available water-supply.

With the exception of a few enclaves left uncultivated either through want of agricultural labour or else owing to the appearance of springs, which when inadequately drained render the soil marshy[12], the whole of the area extending from Zawa in the west to Lop in the east, and varying in width from eight to twenty miles, presents the appearance of a fertile and thickly populated oasis. But it is certain that the cultivation of this tract, large even within its present limits, does not by any means absorb the whole supply of water available. Even in the early spring, when the rivers are at their lowest and water is particularly needed for the first crops, irrigation demands never quite exhaust the river-beds. Little if any of the spring water (*kara-su*) which comes to the surface in the above-mentioned marshes, and in the numerous 'Yārs' to be discussed below, is used to irrigate fields. Of the water carried in the enormous floods which descend the two rivers from June to August, only an insignificant amount can be utilized in the extant irrigation canals, while the rest is allowed to fill the broad beds of the rivers and to pass on into the desert.

Wider limits of ancient cultivation.

How much of this abundance of water might be spread by flood-canals over the sandy expanse north of the present oasis, and how much of this great area might be thus reclaimed for cultivation, must in the absence of systematic observations and surveys remain a matter for conjecture. But in view of the extensive ruined sites examined by me to the north-east of the present oasis, up to a distance of fully sixteen miles from the nearest point of the latter, and of similar traces towards the north-west, it is certain that the extent of cultivated ground in the ancient oasis of Khotan must have been very much greater[13]. Taking the conditions of the present day it may be safely asserted that it is not deficiency of water, but mainly the inadequate number and slow growth of the population, coupled with certain shortcomings of

[11] Popuna on the Kara-kāsh and Kara-yantak on the Yurung-kāsh, respectively, are the villages from which continuous cultivation begins.

[12] Marshy ground is found, e. g., at Uzun-sholok, west of Khotan town; at Halāl-bāgh (*Aiding-Kul*), south of Hanguya; and north of Zawa. Waste patches of ground overrun by light (and fertile) 'sand' are, e. g., the Palamās-kum near Tasmache and the ground about Hanguya-Langar. Traces of earlier cultivation are plentiful in both the latter localities, and re-colonization is now slowly proceeding.

[13] See below, chapter xiv, regarding the extensive and almost contiguous 'Tatis' of Ak-sipil and Hanguya spreading over square miles; also the ruins of Rawak and the débris area of Jumbe-kum. For the 'Tati' of Kara-döbe to the north-west, see chapter xv.

the administrative system, which prevent effective expansion beyond the area actually irrigated[14].

Cultivation threatened by drift-sand.

An ample supply of water for irrigation, the presence of fertile loess soil, and a sufficient population are, no doubt, the essential conditions without which it is impossible to assume a greater expansion of the oasis either in the past or in the future. It would, however, be a mistake to believe that the limits of cultivation in a territory situated like Khotan could ever in historical times have depended upon these factors alone. Wherever my search for ancient sites took me within the Taklamakān desert, I found evidence that by far the greater portion of the area over which its dunes move is formed by beds of naturally fertile loess. But it was equally clear that what cultivation had once existed there, or was still proceeding on the edges of the desert, could be carried on only in constant struggle with a formidable opponent, the drift-sand of the desert. It will probably take long years of painstaking study and observation on the spot, before all the physical facts connected with the gradual changes and movements of the desert dunes in different parts of the Taklamakān are elucidated with some approach to scientific accuracy. And as these movements are mainly influenced by the prevailing winds, which themselves, like other phenomena dependent on climatic conditions, may have undergone material changes in the course of long centuries, it appears very doubtful whether even the results of such observations could be relied upon to give in all respects a true view of the conditions prevailing at earlier periods.

Abandonment of cultivated ground.

Notwithstanding this reservation, which applies particularly to the assumed rate of progress in the general advance of the desert southwards, it seems highly probable that the edges of the oasis presented in ancient times the same abrupt transition between cultivated ground and desert which we note there at present. Wherever on my tours I moved northward beyond the limits of the Khotan oasis, I invariably found dunes, often of respectable dimensions, fringing in closest proximity the last irrigated fields. The drifting sand is thus ever at hand to overrun any ground from which for one reason or another cultivation recedes, and under special conditions it may even prove a successful aggressor in the face of human resistance. The causes for the abandonment of once cultivated ground may vary greatly, but are always easy to understand. Inadequate supply of water, due to neglect of the irrigation canals or to natural changes affecting them; shrinkage of the population and of the available labour; reduced cultivation, owing to political troubles or maladministration—are all sure to make their effect felt first on the outskirts of the oasis, where the struggle of the agriculturists must always be most severe.

Drift-sand composed of loess.

In the absence of detailed and prolonged observations, it seems less safe to express an opinion as to cases where the intrusion of the desert within the former limits of the oasis may have been the result of purely natural causes. Yet one important fact at least can be clearly established. The careful microscopical examination which Prof. L. Lóczy was kind enough to undertake of the sand specimens brought back by me from localities once included in the Khotan oasis, but buried under dunes for probably close on fifteen hundred years (Ak-sipil, Rawak), proves that the constitution of this sand differs in no essential respect from that of the alluvial loess which forms the fertile soil of the oasis. Like the latter, it consists mainly of distinctly angular quartz-grains, plentifully mixed with mica-flakes and to a less extent with fine dust, all manifestly products of the detritus which forms through disintegration in the Kun-lun Mountains, and which the rivers of Khotan wash down. The entire absence of

[14] This point has been noticed already by a number of earlier travellers, with regard to Khotan as well as other oases; comp. e.g. Grenard, *Mission D. de Rhins*, ii. p. 180; Johnson, *J. R. Geogr. Soc.*, 1867, p. 6.

fully rounded and polished grains, such as only long-continued subaerial attrition can produce, clearly distinguishes the substance of the dunes around Khotan (and I may at once add, at all the ancient sites examined by me elsewhere) from the true drifting sand of other Central-Asian deserts[15].

Natural fertility of drift-sand.

Whether the material constituting these dunes has always been gathered and shifted by the winds directly from the alluvium of the river-beds, or whether a considerable portion of it may have to be traced back to beds of alluvial loess eroded by the action of the desert winds, is a question which, however interesting from a geological point of view, we need not consider at present[16]. It is only important to note here that the danger which these dunes represent for cultivation on the outskirts of the Khotan oasis does not arise from any sterility in the 'sand' itself, but from the obstacles which its appearance in great masses necessarily offers to continued irrigation.

Zone of desert jungle.

The desert of drift-sand which so abruptly skirts the northern edge of the oasis is for a considerable distance not altogether devoid of vegetation. As the subsoil water, especially during the time of the summer floods and the periods immediately following, is relatively near the surface, tamarisks and some scrub manage to grow between the dunes. Where occasional inundations from the rivers or the canals penetrate into this zone their beds remain clothed with the hardy Kumush grass for years. The wild poplar (Toghrak), too, would no doubt manage to grow plentifully in this adjacent belt of desert did not the constant demand lead to the speedy cutting of any young trees that can serve for fuel or timber. Thus it is only several marches below the northern edge of the oasis that we meet, on the banks of the united Khotan river, with that luxuriant belt of jungle which flanks the river bed down to its junction with the Tārīm. It is probable that the same causes had already in ancient times made the desert in the vicinity of the great oasis look even more bare and desolate than it would be by nature.

Mountain region S. of Khotan.

Turning now to the south, we find the oasis bordered by a mountain region which in some respects is more barren and forbidding than the true desert itself. The account contained in my Personal Narrative of the expedition I made towards the headwaters of the Yurung-kāsh[17] renders unnecessary any detailed description of the inhospitable ranges which succeed each other from the gravel-covered slopes of the outer hills to the glacier-crowned watershed towards the Ak-sai-Chīn and the high plateaus of the Upper Kara-kāsh. Nor do the views of scenery which are included among the illustrations of the present work stand in need of much explanation. That they are truly typical of this sombre mountain world, which looks as if

[15] Prof. Lóczy notes particularly the striking difference in this respect between all my 'sand' specimens from the Khotan region and those he collected himself from the dunes of the Gobi near An-si-fan and Tung-huan-hsien in Kan-su.

[16] I regret that my want of previous geological training prevented me from collecting on the spot such observations and specimens as would suffice for a full determination of this question. As points which may possibly have a bearing on the question, I may note that in the vicinity of the Khotan oasis, and elsewhere, too, in the Taklamakān, I found the dunes usually highest along the river courses (see below my accounts of Ak-sipil and Rawak; also *Ruins of Khotan*, pp. 329, 412, 447, 451); and further that the appearance to the naked eye of the sand of the high dunes around Rawak and Ak-sipil suggested the prevalence of coarser grains and a still greater abundance of mica than in the 'sand' composing the dunes at other old sites.

In view of the geological problems to which my attention has been called since those explorations, it is a matter of special regret to me that the small collection of specimens which I was able to submit to Prof. Lóczy does not include any sand from dunes situated in parts of the Taklamakān far away both from the rivers and the ancient sites. It thus remains doubtful whether the above noted absence of true drift-sand applies to the whole of the desert region visited by me.

[17] See *Ruins of Khotan*, pp. 206–43.

ERODED RANGES TOWARDS KARA-KĀSH RIVER GORGE,
SEEN FROM ABOVE YAGAN-DAWĀN.

KUN-LUN RANGE WITH GLACIERS OF NISSA VALLEY,
SEEN FROM ABOVE BRINJAK PASS.

wholly composed of ice, crumbling rocks, and dusty detritus, will be seen on comparing them with the extensive photo-theodolite panoramas which I brought back from that trying region. Fig. 21 shows the ice-covered main range above the valleys of Nissa and Karanghu-tāgh as seen from a ridge above the Brinjak Pass (elevation about 16,000 feet)[18]. The view reproduced in Fig. 20 may help to convey some idea of the extraordinary maze of disintegrated ridges and deep-cut ravines which extends between the gorges of the Yurung-kāsh and Kara-kāsh rivers where they approach the plains[19].

The whole region drained by the rivers of Khotan presents an aspect of rugged desolation which is not easily surpassed by any mountainous territory in Asia. The same extremes of heat and cold which produce the rapid disintegration of all exposed rocks, coupled with the remarkable dryness of the climate, probably account for the exceptionally restricted plant growth observed in these barren mountains. Compared with their denuded slopes, which support only scanty tufts of hardy scrub, even where their detritus has long ago decayed into loose earth, the vegetation of the inhospitable Pāmīrs would appear luxuriant.. What little cultivation is possible in the few narrow valleys where oats will still ripen would not suffice for the maintenance even of the extremely sparse population of hardy 'Tāghliks' who permanently inhabit them. Nothing illustrates, perhaps, better the poverty of this region than the fact that the settled population, including the malefactors exiled to Karanghu-tāgh from Khotan, according to my information scarcely exceeds four hundred souls within the area of some 9,000 square miles extending to the north of the main range[20]. As this number includes the hillmen who look after most of the flocks of sheep and yaks grazing in these mountains, it is evident that the amount of available pasture must in proportion to the area be very limited at all seasons[21]. Barren desolation of mountains.

There does not appear to be anything to justify a belief that the physical conditions in the Khotan mountains have undergone a material change during historical times. The Chinese descriptions never mention them specifically, even though the jade which the rivers bring down from them would naturally have induced a reference if the region itself had possessed any economic value. That no important route could ever have passed through it may, in view of my surveys, be considered certain. Routes from the Upper Kara-kāsh to Karanghu-tāgh and Khotan, such as the one which Hayward had heard of and the other which Johnson is supposed to have followed, may exist, though I did not succeed in tracing any knowledge of them among the hillmen examined[22]. But it was clear from the configuration of the

[18] Compare *Ruins of Khotan*, p. 229.

[19] Compare *Ruins of Khotan*, pp. 233 sq.

[20] For a group of Tāghliks and exiled criminals at Karanghu-tāgh, see Fig. 23.

[21] Some of the valleys about Nissa are said to be visited during the summer by shepherds from the Khotan oasis. But it appears that the flocks driven up to these alpine grazing grounds are far less important than those owned by Khotan 'Bais' (small capitalists) in the riverine jungle tracts along the Khotan Daryā.

[22] The sketch-map by which Johnson illustrated his journey (1865) from the Upper Kara-kāsh to Khotan (see *J. R. Geogr. Soc.*, 1867, p. 1) cannot be reconciled with the true topography of the region from Karanghu-tāgh southwards. In it the hamlet of Karanghu-tāgh appears shifted some twenty miles further north than its real position, and the Yurung-kāsh river is given a wholly impossible course. By the insertion of a great bend, which in reality does not exist, the valley of the Yurung-kāsh is shown in this map again some twenty miles south of Karanghu-tāgh, i.e. approximately in its true position, but with the river flowing to the south-east, a direction exactly opposite to the true one. In this second portion of the valley a locality called *Khushlash-langar* is marked at a map-distance of some twenty-three miles to the south of Karanghu-tāgh. In reality the few huts known by that name are situated only one and a half miles to the south-east of Karanghu-tāgh, half way between the village and the left bank of the Yurung-kāsh. A strange kind of duplication seems to have occurred in Johnson's map, for which I am unable to offer a satisfactory explanation. Of the other local names recorded on Johnson's route from the Kara-kāsh to Karanghu-tāgh, some are unknown to the

mighty range southwards that only tracks passable for men or, perhaps, also for yaks, could be found across it. An historical reference to Karanghu-tāgh and its 'mountains of blinding darkness', fully confirms this conclusion. For Mirzā Ḥaidar, when describing the desperate flight of his uncle Abā Bakr from Khotan towards 'Tibet', i. e., Ladāk (A. H. 920), distinctly mentions how the dethroned tyrant, on reaching the Karanghu-tāgh valley, was obliged to kill his ponies and mules, and to abandon all the treasures carried on them, since they could not be taken further on the difficult track which alone offered escape[23].

Section II.—AGRICULTURE AND INDUSTRIES IN KHOTAN

Fertility of oasis.

From this rapid glance at the mountain background of Khotan we may now return to the oasis itself. The picture which it presents, wherever we may pass through it, is one of remarkable fertility. Almost from the heads of the first canals that take off at the *débouchement* of the rivers, down to the edge of the desert, stretches an unbroken expanse of carefully banked fields, varied only by thickly studded hamlets and villages, by their fruit-gardens, and by the fine avenues of poplars and willows which line every road and canal. Variations in the productiveness of the fields exist, of course, in the different tracts; they are influenced by the distance from the main canals, and by the time and amount of the allotted water-supply dependent on this; by the appearance of subsoil water or drift-sand, and similar features. But good crops are generally assured, whatever the produce for which the soil may be utilized, with due regard to local conditions[1].

Agricultural produce.

Wheat, rice, millet, oats, and above all Indian corn are the staple cereals; and the early harvest of spring sowings and the abundant water-supply of the summer permit of Indian corn being grown almost everywhere as a second crop. The rich fields of lucerne (*bidā*) supply plentiful fodder. Cotton is grown largely. Equally important is the cultivation of mulberry trees; for Khotan, as we shall see, has been a home of sericulture since ancient times, and is still its main centre for the whole of Eastern Turkestān. Khotan is not distinguished for its vegetables. Fruit-trees, on the other hand, flourish all over the oasis, and their produce is so plentiful that a considerable export in dried fruit takes place, especially towards Ak-su and

Tāghliks and some are applied to localities in an entirely different situation. All these discrepancies are the more puzzling since Johnson was a professional topographer who even without the use of a plane table could have secured an approximately correct record of whatever route he actually followed.

[23] The *Tārīkh-i-Rashīdī*, pp. 323 sq., 327 sq., contains a very interesting account of Abā Bakr's flight to Karanghu-tāgh. 'As the roads were difficult, it would have been hard, —nay, impossible for him to carry off all the property he had with him; he therefore collected it all together, and set it on fire.' Apart from the many loads of valuables thus destroyed, the flying ruler was believed to have thrown immense quantities of gold and silver ornaments, vases, &c., and his saddle-bags full of gold dust 'from the bridge into the River Ak-Tāsh which flows through the middle of [the valley of] Karanghu-tāgh'.

Ak-Tāsh is a synonymous designation of the Yurung-kāsh, the river of 'White Jade', jade being 'the stone' κατ' ἐξοχήν in Khotan; the bridge meant is, no doubt, the one by which the Yurung-kāsh, flowing here in a deep chasm only some seventy feet wide, is crossed before reaching Karanghu-tāgh village (see *Ruins of Khotan*, p. 213). Considering the depth of the rocky bed and the rapidity of the tossing river which fills it, the difficulty Abā Bakr's Moghul pursuers had in recovering even a small part of these riches (no doubt, greatly exaggerated) is easily appreciated. I regret that at the time of my journey I was unaware of Mirzā Ḥaidar's record of this interesting historical episode. Hence I could not ascertain whether any popular tradition of it still survives in this forlorn region.

The *Zafar-nāmah*, of Timūr's historian Sharīf-ud-dīn, refers to '*Karangutak*, a very steep and rugged mountain, to which the inhabitants of Khotan and the neighbouring places fly for refuge in time of war'; see Bretschneider, *Mediaeval Researches*, ii. p. 233, note.

[1] For useful notes on Khotan agriculture, comp. Grenard, *Mission Dutreuil de Rhins*, ii. pp. 173 sqq.

the smaller oases eastwards. My finds at the sites explored in the desert proved that the fruit-trees now most widely cultivated, such as the apricot (*ürük*), peach (*shaftālū*), olive (*jigda*), apple (*alma*) and mulberry (*üjme*), must all have been very common in ancient times. Almonds, walnuts, melons, and figs abound at present. Khotan is famous for its grapes (*tāl*), which are specially cultivated at Ujāt, a large village at the *débouchement* of the Kara-kāsh, and are widely exported in the form of raisins. Ancient specimens of them, too, turned up in the course of my excavations (at Kara-dong).

Administrative divisions of oasis.

As in the other portions of the Tārīm Basin where cultivation depends entirely on irrigation, we find everything that relates to the periodical distribution of canal-water between the various villages and holdings regulated by minute customs which probably go back to considerable antiquity. Some interesting notes on this subject, which would well deserve closer study, have been recorded by Dr. Sven Hedin[2]. Where irrigation is the mainstay of the whole economic organization, it is only natural that the administrative divisions of the territory should conform to the arrangement of the main canals. We find accordingly the Khotan oasis divided for purposes of revenue and general administration into a series of long-stretched cantons representing the areas irrigated by each of the chief canals. Just as most of the canals themselves are undoubtedly of very early date, these cantons, now known by the Turkī designation of 'Ming' ('a thousand') and each under a 'Ming-bāshī' or Bēg, are likely to have had their counterparts in ancient times. I have hence thought it useful to enumerate them, in a footnote[3], in their traditional order, irrespective of minor changes and groupings made by Chinese administration since the reconquest of the 'New Dominions'.

[2] See *Reisen in Z.-A.*, pp. 20 sqq.

[3] Beginning from the west we have first on the left bank of the Kara-kāsh the canton of *Zawa* and *Kuya*, irrigated by branches of the same main canal; next *Mākuya* (irrigated from the canal which in the map has been erroneously shown with name Kuya-Üstang); the small tract of *Kayāsh*; and the considerable canton known after its canal as *Bahrām-su* and including the town of Kara-kāsh. *Kara-sai* to the north-east is a small detached tract which derives its irrigation from the Yawa-Üstang, a quasi-natural water-course fed chiefly by the marshes below Zawa.

Between Kara-kāsh and Yurung-kāsh there follow in the same direction the cantons of *Sipā* and *Borazan*, both irrigated from the Kara-kāsh; the second is the most central portion of the Khotan oasis and contains the site of its ancient capital. Eastwards we have *Tosalla*, like the preceding two, a large and fertile canton; and the small subdivision of *Ilchi* or *Khotan-Shahr*, which is mainly composed of the present capital and its immediate vicinity.

East of the Yurung-kāsh lies the important canton which, together with its lively town, is called *Yurung-kāsh*, after the adjacent river. The well-populated tracts of *Sampula* and *Lop*, distinguished for their carpet-weaving and other flourishing industries, form the easternmost part of the oasis. They have been attached in recent years to the charge of the Chinese district magistrate residing at Keriya (no doubt for fiscal convenience, since without them the territory administered by this officer would be as unprofitable as it is extensive), but form in every respect an integral portion of Khotan. This is far less the case with *Tawakkēl*, a separate little oasis which was established in the first half of the last century by colonists from Khotan. It is situated on both sides of the Yurung-kāsh some thirty miles below Ilchi town, and is now included as a distinct Bēgship or 'Minglik' in the district of Khotan.

The extensive mountain region of Khotan never within traditional recollection appears to have ranked as a separate canton, a fact easily understood in view of its poverty and extremely scanty population. Slices of it are attached to the different Khotan cantons, an arrangement which may have its convenience for the allotment of the grazing—such as it is.

It will be convenient to show in this place also the conventional reckoning of houses for each 'Ming' or canton, since, whatever its value for statistical purposes may be, it throws light on the relative population of each tract and allows some conclusion as to the extent of the area under cultivation in each. The figures tabulated below were communicated to me in April, 1901, by Islām Bēg, my former Darōgha, to whom the Amban of Khotan had then recently entrusted the Bēgship of Kayāsh, and by M. Badruddīn, the headman of the Afghān traders in Khotan, both competent informants on matters of this kind. It must, however, be understood that these figures represent only a rough estimate accepted for the sake of convenience by those interested in the management (*recte* farming) of the revenue of these cantons, and are in no way officially recognized. The figures in the second column refer to an earlier reckoning of a similar character, vaguely ascribed to 'the time of old Chinese rule' (*kōne Khitaining waqti*), i.e. the period preceding Yāqūb Bēg's rebellion. It is generally believed that, compared with

Industrial importance of Khotan.

We should obtain a very imperfect idea of the Khotan oasis and its material resources if our estimate were based solely on the agricultural activity of its inhabitants. Khotan has in modern times been undoubtedly the chief industrial centre within Eastern Turkestān [4], and everything points to the conclusion that it occupied this position from a very early period. The mining of oriental jade might well be mentioned in the first place among the industrial occupations of Khotan; for it is the oldest of which we have distinct historical notices, and its product has more than anything else made Khotan famous throughout the East.

Jade 'fishing' and mining.

The jade of Khotan (the *yü* of the Chinese, called *kāsh* in Turkī) is obtained from the beds of its rivers either by searching and diving for pieces of the precious stone among the pebbles which the rivers wash down during the summer floods, or else by regular digging in the beds of rubble deposited along the banks of the Yurung-kāsh. Such beds are worked immediately above the *débouchement* of the river, and also lower down within the Yurung-kāsh canton along a deserted channel [5]. The kinds of jade most prized, white or of other light colours, are furnished by the bed of the Yurung-kāsh, which owes to this fact its actual name '[the river of] White Jade'. Jade of green or other dark colour is more frequently found in the Kara-kāsh river, and accounts for the latter's name '[the river of] Black Jade' [6]. High up in the valley of the Kara-kāsh, not far from the Karakorum route, green jade is quarried from a mountain side; but this rock jade is reckoned of inferior quality and does not appear to be mentioned in the early Chinese accounts [7].

Export of Khotan jade.

Khotan has, probably since the earliest times, been the chief source of supply of the *yü* stone or oriental jade, which in China even more than elsewhere in the East has been, and still remains, one of the most valued of precious stones. It is hence with good reason that

those days, the population and prosperity of the oasis have greatly increased owing to the improved system of administration. An average of eight persons is the conventional estimate for each house (*ui*).

Canton.	Modern Estimate of Houses.	Old Estimate of Houses.
Zawa-Kuya	2000	1150
Mākuya	2000	1150
Kayāsh	1500	1100
Bahrām-su (Kara-kāsh)	2000	1200
Kara-sai	1000	300
Sipā	3000	1200
Borazan	3000	1500
Tosalla	3000	1150
Ilchi (Khotan-Shahr)	1000	1000
Yurung-kāsh	3000	1100
Tawakkēl	1000	400
Sampula-Lop	5000	—
Total number of houses	27,500	12,000

My informants could not state the old traditional estimate for Sampula and Lop, which, owing to the administrative separation above mentioned, had ceased to come within the sphere of local official interests; but they were agreed in giving 12,000 houses as the conventional figure for the administrative unit comprising the whole Khotan oasis as it existed under Chinese rule in pre-rebellion days.

[4] For a still useful synopsis of Khotan industries, compiled from Surveyor Rāmchand's notes, see *Yarkand Mission Report*, pp. 446 sqq. Interesting notes on a number of the more important industries, with excellent illustrations, will be found in Grenard, *Mission D. de Rhins*, ii. pp. 184 sqq.

[5] I have described the jade-diggings situated near the *débouchement* of the Yurung-kāsh, a short distance above the old site of Chalma-kazān, in *Ruins of Khotan*, pp. 252 sqq. For an account of the jade-pits of *Kalta-kumat*, not far from Tam-öghil, see Hedin, *Reisen in Z.-A.*, p. 28.

[6] The Yurung-kāsh and Kara-kāsh rivers are apparently first referred to by the Chinese equivalents of their present names in the Annals of the Posterior Tsin Dynasty (936–947 A.D.); see Rémusat, *Ville de Khotan*, p. 81, and for later references, ib. pp. 107, 112. By the third river, that 'of green jade', which these accounts mention between the two former rivers, must be meant the broad side channel of the Kara-kāsh known as *Yangi-Daryā* (see map), which passes through the Sipā canton and carries water during the flood season; see below, chap. VIII. sec. iii. The white and green jade of Khotan is distinguished by Hsüan-tsang, *Mém.* ii. p. 223, Beal, ii. p. 309.

For *Gomaṣ̌i*, the ancient name of the Kara-kāsh, see below, chap. VIII. sec. i.

[7] For a Chinese account of this jade-quarry extracted from the *Hsi yü wen chien lu* (1778 A.D.), see Ritter, *Asien*, v. pp. 381 sqq.; a detailed description of it has been given by Dr. Stoliczka in *Yarkand Mission Report*, pp. 464 sqq. The manner in which this inferior jade is obtained was correctly related by Goëz, who speaks of the place as 'Cansanghi Cascio', i.e. *Kān sang-i-kāsh*; see Yule, *Cathay*, ii. p. 565.

A. Rémusat attached to his *Histoire de la Ville de Khotan* the learned treatise in which the historical notices relating to the stone and its extensive use were for the first time clearly traced[8]. A reference to these researches, and to the useful data which Ritter added to them, will suffice to explain the great importance which must be claimed for the mineral in the economic history of Khotan[9]. Apart from the direct profits which its export, continued during thousands of years, must have carried to the oasis, it is clear that it must always have acted as a powerful factor for drawing trade to Khotan.

The advantages which jade offered for commercial investments, owing to its portable nature, its relatively high price, and its assured market in China, were certain to be appreciated both by merchants from China trading westwards and by foreign traders proceeding to the Middle Kingdom. It is only necessary to recall the enormous distances and the natural difficulties of caravan journeys from the interior of China to either the Tārīm or Oxus, in order to realize how important it must have been for the Chinese merchants, who during the Han period and later carried their goods to Central Asia, to invest their sale profits in merchandise which, like jade, admitted of easy transport and was safe from deterioration in transit. Trade stimulated by jade exports.

That the same was true also for the Western traders proceeding to China is a fact attested by a classical witness. Benedict Goëz, who on his memorable journey to discover Cathay proved his capacity as a sensible man of business quite as much perhaps as his missionary zeal, when describing his long stay at Hiarchan or Yarkand in 1603–4, tells us: 'There is no article of traffic more valuable, or more generally adopted as an investment for this journey, than lumps of a certain transparent kind of marble which we, from poverty of language, usually call jasper. They carry these to the Emperor of Cathay, attracted by the high prices which he deems it obligatory on his dignity to give; and such pieces as the Emperor does not fancy they are free to dispose of to private individuals. The profit on these transactions is so great that it is thought to compensate for all the fatigue and expense of the journey. . . . These marbles (with which the empire is now overflowing) are called by the Chinese *Iusce*[10]. Ben. Goëz on Khotan jade.

Sericulture deserves to be mentioned next among the industries of Khotan, in view of its antiquity and continued importance. Khotan is the chief silk-producing district of the Tārīm Basin, and probably of the whole of Turkestān. The produce, which in one form or another affords occupation to the bulk of the population, is nowadays chiefly exported in the form of spun silk. But the weaving of silk fabrics also flourishes, and must, in view of the Chinese records and of my finds, be considered an ancient industry of the oasis. We shall Silk of Khotan.

[8] See 'Recherches sur la substance minérale appelée par les Chinois pierre de Iu et sur le Jaspe des anciens', in *Ville de Khotan*, pp. 119–239.

[9] See Ritter, *Asien*, vol. v. pp. 380 sqq.

[10] See Yule, *Cathay*, ii. p. 564. By *Iusce* is, of course, meant *yü shih*, 'yü stone', the Chinese name of jade. Goëz himself acted up to the commercial advice here conveyed; for he got the loan of some six hundred gold pieces which he had made to the mother of 'the Prince of Quotan' (Khotan), paid back 'in ample measure with pieces of that valuable stone' on a visit he personally paid to Khotan. Before his tragic end at Su-chou on the confines of Kan-su, the brave Jesuit had reason to appreciate the wisdom of this investment; for when 'obliged to dispose of his large piece of jade for little more than half its value', in order to maintain himself and his party during their detention, 'he got for it twelve hundred pieces of gold'; see Yule, *Cathay*, ii. p. 585.

I may refer here in passing to a modern illustration of the extent to which caravan trade with Eastern Turkestān depends on the possibility of securing easily portable articles for export from the country. Indian trade with Chinese Turkestān flourished after the Forsyth Mission as long as the Hindu merchants engaged in it were able to invest their sale profits on Indian goods in the highly valuable and easily transported *charas* drug. Since the importation of the latter into India has been charged with almost prohibitive duties, the Turkestān trade has considerably declined. In their search for profitable return consignments to replace *charas* the Indian traders are now obliged to turn to Khotan silk; but its supply is far less steady than that of the noxious drug and the profits are by no means so substantial.

have occasion to discuss below the time-honoured tradition which Hsüan-tsang heard about the first introduction of sericulture into Khotan. It proves that in his time this industry was already so old as to be credited with a legendary origin, and that it was certainly derived from China. How much of the silk which the nations of the classical West received through Persia from the distant half-mythical land of the Seres really came from Khotan we shall never know with any certainty. But the antiquity of sericulture in Khotan and the early commercial importance of the oasis seem to give some support to the conjectural location at Khotan of Ptolemy's *Issedon Serica*, as first proposed by Baron von Richthofen[11]. There appears to me still more reason to believe that the same great scholar, and before him Sir H. Yule, were right in suggesting Khotan as the territory probably meant by the country of *Serinda*, whence, according to the story related by Procopius, two monks in the time of Justinian (about 550 A.D.) first introduced the silkworm into the Byzantine empire[12].

Cotton industry.

Khotan produces more cotton than any other district of Eastern Turkestān, and the export of cotton goods, chiefly in the form of *Khām*, a rough but very durable fabric, is very considerable. Cotton weaving is carried on by a large portion of the agricultural population as a supplementary occupation. M. Grenard estimated the number of households engaged in it at about 12,000, and the value of the output at about 6,700,000 francs[13]. Though the early Chinese records do not appear to make special mention of the cotton goods of Khotan, yet fine fabrics represented among my finds at the sites of Niya and Endere clearly establish the antiquity of this industry[14].

Carpets and felts of Khotan.

Wool is plentifully obtained from the flocks grazing in the mountains and in the riverine forests, but its use is restricted almost exclusively to the fabrication of carpets and felts. For both of these manufactures Khotan is famous throughout Turkestān, and their preeminence in early days is proved by the specific reference Hsüan-tsang makes to them[15]. The carpets of Khotan, sometimes made in silk, are widely exported, and are met with throughout Central Asia and far into China. They are easily recognized by their characteristic designs and patterns, which seem to retain certain ancient motives pointing partly to Indian, partly Chinese origin[16]. We shall see hereafter that the specimens of ancient rugs found by me at the site of Niya though made in cotton, attest the high development of carpet *technique* for the third century of our era[17]. The same site has furnished the oldest samples of those felts or 'Numdahs' (*namad* in Persian, *kīgiz* in Turkī), of which Khotan exports nowadays very large quantities both to Kashmīr via Ladāk and northward. M. Grenard estimated the number of felt workshops in Khotan at approximately a thousand, the industry being conducted by households[18].

Manufacture of paper.

Another industry extensively carried on in Khotan, which deserves to be specially mentioned here, is that of paper. It is now manufactured exclusively from the bark of a mulberry-tree

[11] Comp. Richthofen, *China*, i. pp. 487 sqq. For a valuable analysis of the notices relating to early silk trade through Central-Asia, see ib. pp. 442 sqq., 474 sqq. They have been discussed briefly, but with his usual lucidity, by Sir H. Yule, *Cathay*, i. pp. xliv sqq.

[12] Compare Yule, *Cathay*, i. p. xlvi, and for a translation of the original text, pp. clix sqq.; Richthofen, *China*, i. 529 sqq., 550. Procopius' mention of 'many nations of the Indians' in connexion with *Serinda* points more than ever to Khotan, since we have learned how thoroughly and how early Indian influence established itself there. The story as related by Theophanes of Byzantium (end of sixth century) speaks only generally of the country of the Seres from which a certain Persian brought away the eggs of silkworms hidden in a walking-stick; see Yule, *Cathay*, i. p. clx. This feature of the story may be pure folk-lore; but it curiously recalls the Khotan legend told by Hsüan-tsang how the first eggs of silkworms were brought there hidden in the headdress of a princess from China; see below, chap. IX. sec. iv.

[13] See *Mission Dutreuil de Rhins*, ii. p. 191.

[14] See below, chap. XI, XII; also Plates LXXVI, LXXVII.

[15] Comp. *Mém*, ii. p. 223; Beal, ii. p. 309.

[16] For a plate showing Khotan carpets and for some notes on the industry, comp. *Mission D. de Rhins*, ii. p. 195.

[17] Comp. chap. XI. sec. ii.; also Plate LXXV.

[18] See *Mission D. de Rhins*, ii. p. 194.

which, as shown by the examination Professor J. Wiesner was kind enough to make of specimens transmitted to him, is the *Broussonetia papyrifera* or Paper mulberry-tree. This tree, which supplies the bark mainly used for paper in China, does not appear to grow to any extent in other parts of Chinese Turkestān, and its bark being particularly suited for the purpose, Khotan enjoys a practical monopoly in the local manufacture of paper. The pulp used is made up of thoroughly well-macerated fibres, and the paper itself possesses a good deal of strength and toughness.

Antiquity of Khotan paper.

The oldest datable paper MSS. which my excavations in the Khotan region brought to light must be ascribed to the period from circa 719 to 791 A.D. But the exhaustive investigations which Prof. J. Wiesner has devoted to the analysis of the ancient paper materials represented in the collection of Central-Asian MSS. under Dr. Hoernle's care, as well as those excavated by me, prove beyond doubt that the manufacture of paper in Eastern Turkestān had been carried on for centuries before that period, and in the course of time had undergone some remarkable developments. We shall have occasion hereafter to refer to various interesting results of Prof. Wiesner's researches [19]. In the present place only one point need be noted. Though the rough mechanical pounding to which the bark used for the earlier papers had been generally subjected made the determination of the fibres contained in them often very difficult, yet Prof. Wiesner has been able to identify in a number of cases fibres of the *Broussonetia papyrifera*. Some of the MSS., where such were found, belong to the site of Dandān-Uiliq, while others were obtained from Kuchā, and probably belong to an earlier period (fifth to seventh century A.D.). The paper of those from Dandān-Uiliq (eighth century A.D.) is certainly of Khotan manufacture: and in view of what has been stated as to the restricted occurrence of that tree outside Khotan, the same assumption suggests itself as regards the MSS. from Kuchā.

Industries in pottery, glass, metal.

Ceramic art, which, judging from the finds of Yōtkan, must have reached a high degree of perfection in ancient Khotan, has greatly deteriorated, and the pottery now produced is inferior even when judged from the most utilitarian point of view. The manufacture of glass, which must be assumed to have flourished in old Khotan, is now wholly unknown [20]. On the other hand, the metal-workers of Khotan, of whose skill in ancient days very few specimens have so far been recovered, produced in relatively modern times much excellent work, and enjoyed well-merited fame throughout Turkestān. I refer in particular to the Khotan workers in brass and copper, to whose hands may be traced most of the artistically wrought old metal ware, such as water-jugs, basins, trays, &c., which I have had occasion to see and admire at Yarkand, Kāshgar, and also in Western Turkestān [21].

That this art is of ancient date in Khotan is proved by a notice of the Liang Annals, which specially refers to the skill of the inhabitants of Yü-t'ien in the manufacture of copper vessels [22]. The style of the rich ornamentation employed is throughout distinctly Persian. While this once highly developed art industry has sunk to a commonplace level during the last century, the gold and silver smiths of Khotan retain to this day a good deal of skill;

[19] The results of Prof. Wiesner's investigations, as far as based on MSS. in Dr. Hoernle's collection, were published in his *Mikroskopische Untersuchungen alter ostturkestanischer Papiere, &c.*, *Denkschriften der mathem.-naturwiss. Classe* of the Imperial Academy, Vienna, vol. lxxii. 1902. Those derived from the analysis of paper MSS. excavated by me are contained in his paper 'Ein neuer Beitrag zur Geschichte des Papiers', published in vol. cxlviii (1904) of the *Sitzungsberichte der philos.-histor. Classe*, Imperial Academy, Vienna.

[20] Vases of glass are mentioned as presents from Khotan to the Chinese court in 518 A.D.; see Rémusat, *Ville de Khotan*, p. 17.

[21] A good specimen of the old brass trays, with the curious open work which seems to have been peculiar to Khotan during the last few centuries, is reproduced in M. Grenard's plate, *Mission D. de Rhins*, ii. p. 188.

[22] See Rémusat, *Ville de Khotan*, p. 16.

and their products, especially in a kind of filigree work, still find their way through the whole of Turkestān[23].

Gold production of Khotan.

I have left to the last the mention of the gold produce of Khotan; for if we except the little gold washed from the sand of the Yurung-kāsh, the precious metal with which the name of Khotan is often associated is found in its natural state only at places situated at considerable distances from the oasis. The gold mines of Surghak, Kapa, near the headwaters of the Cherchen river and on the high plateaus of the Arka-Tāgh towards Tibet, may well have been worked in ancient times. But no mention is made of them in the old Chinese notices of Khotan; and it is doubtful whether, with the exception of the first-named place (on the upper course of the Niya river), any of these localities ever fell within the political boundaries of the Khotan kingdom. That the gold extracted from them must have helped to increase the commercial importance of Khotan, as the nearest emporium for its disposal, may, on the other hand, be considered as certain.

Section III.—THE POPULATION OF KHOTAN: ITS DISTRIBUTION AND CHARACTER

Having acquainted ourselves, in main outline, with the geographical *milieu* and the material resources of the Khotan oasis, we may now complete our survey by a brief review of the people in their distribution, general character, and probable ethnical connexion. I shall not attempt to include in this sketch an account of the present conditions of life and social organization. On the one hand, the conditions at Khotan in most respects resemble so closely those found throughout the oases of Eastern Turkestān that the descriptions already available in the narratives of former visitors amply suffice for information on all general points. In this respect I may refer particularly to the excellent exposition of the subject which M. Grenard has given, based largely on observations made during a prolonged stay within the Khotan oasis[1]. On the other hand, I shall have plentiful occasion hereafter, when describing the results of my excavations at ancient sites in the desert, to notice those particular features of modern life in the extant oases which help to illustrate them, and which themselves are proved by those finds to be of early origin.

Number of population.

It is not easy to form an approximately correct estimate of the present population of Khotan and, of course, still more difficult to guess the limits to which its number may have extended during early periods of greater prosperity. In 1873, the official figure communicated to Sir D. Forsyth's Mission was 129,500 souls which, considering the undoubted diminution of the population due to the troubles of the Muhammadan rising and the exactions of Yāqūb Bēg's reign, may not have been too low an estimate[2]. General Prżewalssky, on his passage through Khotan

[23] For some specimens see *Mission D. de Rhins*, ii. p. 184, plate.

[1] See particularly chapters i–viii, x, xi, of vol. ii *Mission D. de Rhins*. For the critical student it will be well to bear in mind that much of the valuable information presented applies far more closely to Khotan than to the rest of Eastern Turkestān. In the endeavour to give a thoroughly readable and graphic picture of life and people in this wide region local distinctions have not always been brought out as clearly as needed for thorough inquiry into particular features. Reference to the *Yarkand Mission Report*, pp. 80 sqq., is still useful for a variety of data regarding daily life. Many interesting facts can be gathered also *passim* from Dr. Hedin's *Through Asia* and his *Reisen in Z.-A.* Having been obliged to spend most of my time in explorations over desert ground, and to concentrate my attention on things of the past within the oases, I could record in my 'Personal Narrative' only casual observations and rapid glimpses of everyday life.

[2] See *Yarkand Mission Report*, p. 62.

in 1885, estimated its population at 300,000, while General Pjewzow, five years later, put the number at about 160,000. That the figures communicated to Dr. Hedin in 1896, giving an aggregate of 524,000 souls for the Khotan oasis exclusive of the oases of Lop and Sampula, were greatly exaggerated, was recognized by that traveller himself[3]. I have already, in a note to the preceding section, indicated the present conventional estimate, which assumes a total of 27,500 households for the whole of the Khotan oasis[4]. Accepting the local reckoning of eight persons as an average for each household, we should arrive, on the basis of this estimate, at an approximate total of 220,000. Judging from the impressions I received of the relative density of the population in various parts of the oasis, from the extent of industrial exports and similar indications, I should not consider this estimate in any way excessive. But I am wholly unable to judge how closely it approaches to the truth[5].

Towns of oasis.

In distinction from other oases of the Tārīm Basin where urban life is concentrated in a single town, Khotan possesses at present three places to which the designation of town can be justly applied, viz. Khotan or Ilchi, Yurung-kāsh, and Kara-kāsh. In size and commercial importance these three towns do not differ greatly; but as the seat of the local administration Ilchi occupies the first place, and consequently claims the general designation of Khotan which has been applied from ancient days to both the oasis and its capital. None of the three towns equals Kāshgar or Yarkand in wealth or population; but their number when compared with the size of the oasis is a further indication of the prominent part which industrial occupations play in the economic organization of the people. In addition to these towns, the cantons of Borazan, Tosalla, and Sampula possess local commercial centres of their own, represented by the Bāzārs of Bīzin, Imām-Mūsā-Qāsim and Sampula, all of them much frequented on their weekly market days[6].

Protection against ethnic invasions.

We have previously seen how far back in history and antiquarian remains we can trace the industries and crafts which to this day distinguish Khotan. This observation alone would justify the hope that we might similarly find ancient peculiarities of character and race still surviving in the present inhabitants. But there are weighty geographical and historical considerations, too, which lend support to this expectation. It is certain that, among all the larger oases of the Tārīm Basin, Khotan was best protected by nature against any violent and thoroughgoing ethnic changes. The mighty mountain ranges on the south may not have protected it altogether from invasion by small bands of bold adventurers—we know that the Karakorum in the opposite direction did not prevent successful inroads of Turkī soldiers of fortune into the Upper Indus Valley and even Kashmīr—but Kun-lun and Karakorum combined form an impassable barrier to any large ethnic movement. The great desert northward left no possible route of invasion, except along the bed of the Khotan river; and this, owing to the extremely scanty resources offered by the narrow belt of desert jungle adjoining it for a distance of close on 300 miles, could never have served as a thoroughfare for a real migration. On the east, towards Lop-Nor, the desert belt stretching to the foot of the Kun-lun was, indeed, broken by small oases, as it is at present. But east of Niya the distances between these oases are so

[3] See *Reisen in Z.-A.*, p. 24.

[4] See above, p. 131, note 3.

[5] That the conventional estimate I heard does not err on the side of excess, seems clear, e.g. from the number of 1000 households given in it for the town of Khotan. M. Grenard, who with M. Dutreuil de Rhins had for two years made Khotan town his winter quarters, and who had spent there longer time than any other European, assumed for it a population of 26,000; see *Mission D. de Rhins*, ii. p. 96.

[6] Compare, regarding the main Bāzārs of the oasis, and the system by which the weekdays are divided between them, *Ruins of Khotan*, p. 486.

great, and the grazing and produce offered by them so extremely limited, that no great host could either in modern or ancient days have attempted the passage[7]. What obstacles the desert route from the west offered we have had occasion to note in the previous chapter.

Khotan unsuited for nomads.

But what possible inducements were there for any nomadic tribe or nation to penetrate these desert defences and settle down in Khotan? Only painstaking cultivation, based on the experience of many generations, could produce and maintain the oasis. Of ground such as the various nomadic races from the Mongolian plains and the Tibetan plateaus need for their maintenance, the region around Khotan could never in historical times offer any. Even the hardy Kirghiz, those last representatives of the Turkī race retaining its original nomadic condition, who still graze their flocks in most of the mountains encircling the Tārīm Basin, have given a wide berth to the inhospitable region that borders Khotan on the south and thence stretches eastwards into Tibet.

Tārīm Basin not on road of migrations.

It is because the conditions here briefly indicated apply with almost equal force to all the oases fringing the great desert of Eastern Turkestān that we find each great wave of Central-Asian migration, of which historical records have come down to us, moving along the northern foot of the T'ien-shan or even further north. When tracing above the general history of Eastern Turkestān, we have seen that all the tribes which followed each other on that great route of invasion westwards—Yüeh-chih, Huns, Hephthalites, Western Turks—were probably able in their time to exact tribute and a more or less nominal submission from the various small states of the Tārīm Basin while Chinese power was too weak to protect them.[8] But nothing leads us to suppose that this political dependence during limited periods affected the internal organization of these states and their ethnic composition, any more deeply than the presence of Chinese garrisons in Han times and again for over a century under the T'ang dynasty. Nor is the brief period of Tibetan predominance likely to have been accompanied by far-reaching effects in this direction. For over a century after Tibetan power had been finally broken by the Uigurs Khotan seems to have again enjoyed independence.

Effects of Muhammadan conquest.

It is only when we reach the time of the conquest of Khotan by the successors of Satok Boghra Khān that the possibility of a radical change in the ethnical character of the population seriously presents itself. The spread of Muhammadanism and of the Turkī language testify to the powerful effects which followed the establishment of the Karluk Turk family of Satok Boghra Khān in undisputed possession of the whole of Eastern Turkestān. Do these great subversions imply that the racial character of the people brought under the new rule was also fundamentally altered?[9] Our historical records for the period which witnessed the conversion to Islām, and for the centuries immediately following, are far too scanty and vague to furnish an answer to this question. Analogies drawn from other territories of Asia, where an indigenous

[7] Karīm Bēg, of Yarkand, an old retainer of Niāz Hākim Bēg, gave me a vivid account of the difficulties encountered by the small force of three hundred horsemen which the governor dispatched in 1877 via Cherchen and Lop-Nor to join the conquering Chinese army at Kara-shahr with his submission. Most elaborate arrangements for water and fodder were needed to assure their safe passage. That the conditions of travel were not much better for caravans in old days is clearly proved by the narratives of Hsüan-tsang and Marco Polo, who both followed this route.

[8] See above, p. 58.

[9] M. Grenard, who has devoted two interesting and suggestive chapters (ii, iii, of vol. ii) to the discussion of the ethnical origins of the population of Eastern Turkestān generally, has treated this question of the 'Turkization' of the earlier inhabitants with much care and ingenuity (vol. ii. pp. 50 sqq.). His observations on the levelling influence of Islām, the successive reinforcements supplied to the Turkish element by the later invasions of the Kara-Khitai, Mongols, Jungars, &c., are very judicious. But scarcely enough allowance has been made for the considerable differences likely to exist from the beginning in the ethnical constitution of a population scattered over so vast an area, and to the strongly marked individuality of each oasis explained by differences of geographical position and historical development.

population representing an earlier civilization passed under Turkish rule and amalgamated with its conquerors, are, no doubt, instructive and illuminating; but by themselves they do not supply a safe basis for judging what the ethnical results of that political change may have been in a region so remote and so peculiarly situated as Khotan. The only course by which, it seems to me, we can hope to arrive at sound conclusions as regards this important question, is to compare the physical and psychical characteristics of the present population in each part of Eastern Turkestān with whatever data we possess for earlier periods. Fortunately the materials for such a comparison are not altogether wanting in the case of Khotan.

Hsüan-tsang's account of the Khotanese.

We may take in the first place the chief features of character peculiar to the Khotanese, as it is easy to test these by the direct statements contained in early Chinese records. The description which Hsüan-tsang has left us of the people of old Khotan claims special importance, not only because it is the most detailed, but also because the pilgrim, owing to the long stay he made in the oasis, apparently over eight months, had excellent opportunities for studying the nature and ways of its inhabitants.[10] According to Hsüan-tsang, their manners and customs showed a sense of propriety and justice. The inhabitants were soft by nature and respectful; they loved to study literature, and distinguished themselves by their skill and industry. The people were easy-going, given to enjoyments, and lived contented with their lot. Music was much practised in the country, and men loved the song and the dance. Few of them wore garments of wool and fur; most dressed in taffetas and white cloth. Their appearance was full of urbanity; their customs were well regulated, and they greatly honoured the law of Buddha.

Other Chinese notices.

The features of character here ascribed to the Khotanese can almost all be traced in other Chinese notices. Fa-hsien had already described Khotan as a pleasant and prosperous kingdom, and its numerous and flourishing population as attached to the law of Buddha and very fond of music[11]. The Liang Annals mention the religious devotion of the people of Yü-t'ien, their extremely reverential habits, and their manufacturing skill[12]. They add the interesting fact that the women of Khotan were freely admitted to society, even in the presence of strangers. Sung Yün speaks of them as wearing trousers and girdles, and riding about on horseback just like the men[13]. In the Annals of the Northern Wei dynasty, under whose protection this last pilgrim travelled, we find the religious devotion of the people of Khotan equally praised; but they also refer to less commendable features, such as defective politeness and justice, and the frequency of thieves, adulterers, and other debased persons[14]. The ardour of the Khotanese in matters of cult is referred to by the T'ang Annals as well as their cleverness and insinuating ways of speech. The people are described as eager for pleasures, fond of dancing and singing, and skilled in textile arts[15]. At the same time we receive a glimpse of the debauched habits prevailing from the special mention of brothels which served as a source of revenue[16].

Marco Polo on the Khotanese.

Marco Polo's account of Khotan and the Khotanese forms an apt link between these early Chinese notices and the picture drawn from modern observation. It is brief but accurate in all details. The Venetian found the people 'subject to the Great Kaan' and 'all worshippers of Mahommet'. 'There are numerous towns and villages in the country, but Cotan, the capital, is the most noble of all and gives its name to the kingdom. Everything is to be had there in plenty, including abundance of cotton [with flax, hemp, wheat, wine, and the like]. The people

[10] See *Mémoires*, ii. pp. 223 sq.; *Vie de H.-T.*, p. 288.

[11] See *Travels of Fâ-hien*, tr. Legge, p. 16.

[12] See Rémusat, *Ville de Khotan*, p. 16.

[13] Comp. Chavannes, *Voyage de Song Yun*, p. 16.

[14] See Rémusat, *Ville de Khotan*, pp. 19 sq.

[15] See Chavannes, *Turcs occid.*, pp. 125 sq.

[16] See Chavannes, *Turcs occid.*, p. 115.

have vineyards and gardens and estates. They live by commerce and manufactures, and are no soldiers[17].' Nor did the peculiar laxity of morals, which seems always to have distinguished the people of the Khotan region, escape Marco Polo's attention. For of the 'Province of Pein', which, as we shall see, represents the oases of the adjoining modern district of Keriya, he relates the custom that 'if the husband of any woman go away upon a journey and remain away for more than twenty days, as soon as that term is past the woman may marry another man, and the husband also may then marry whom he pleases'[18].

No one who has visited Khotan or who is familiar with the modern accounts of the territory, can read the early notices above extracted without being struck at once by the fidelity with which they reflect characteristic features of the people at the present day. Nor is it necessary to emphasize the industrial pre-eminence which Khotan still enjoys in a variety of manufactures through the technical skill and inherited training of the bulk of its population.

Religious devotion of Khotanese.

Devotion to religious cult is another feature which has survived with undiminished intensity, though its objects have been transformed on the surface. We know that Buddhist Khotan resisted the introduction of Islām longer than any other part of Eastern Turkestān. Yet Khotan now proudly claims the first place in this whole region as the land of 'Shahīds'. The supposed resting-places of these holy martyrs, Ziārats and Mazārs of all sorts, stud the oasis and its vicinity more thickly than anywhere else. Pious imagination of a remarkably luxuriant growth has transplanted into the region of Khotan the tombs of the twelve Imāms of orthodox Shiite creed, together with a host of other propagators of the faith whose names are known to local legend only[19]. We have already had occasion to note that many of these Ziārats mark the position of earlier Buddhist shrines, and thus afford proof of the tenacity of local worship. Among the shrines of Khotan territory there are several, like the 'Tombs of the Four Imāms' (at Tört-Imām), the Ziārats of Imām-Mūsā-Qāsim and Imām-Ja'far-Sādiq, which annually attract crowds of pilgrims from all parts of Turkestān. But I doubt whether this widespread fame of certain pilgrimage places is more than the direct result of the pious zeal with which the Khotanese themselves worship and frequent their local shrines, large and small.

Islām in Khotan.

The people of Khotan in this respect may truly be called a *gens religiosissima*. But just as in Kashmīr, which in matters of local worship might have served as the prototype, Islām otherwise sits lightly on the popular mind. The lax observance of many tenets of religious law does not disturb the consciences of a population easy-going by nature and wholly absorbed in the things of this life. Since Sunnism now reigns supreme, and the peace of souls and clerical influence is undisturbed by any rival creed, religious fanaticism finds no scope and consequently is wholly dormant. In all these points I much doubt whether the Buddhism of old Khotan, rich in shrines and religious pomp, showed aspects essentially different.

Character of modern Khotanese.

A peculiar softness of temperament, good-natured ease in language and manners, and a disposition even more pronounced than in other parts of Eastern Turkestān to make the most of what pleasures the humblest life can offer, still distinguish the Khotanese as in the days when

[17] See Yule, *Marco Polo*, i. p. 188. The words in brackets, undoubtedly genuine and added by the traveller himself, have been introduced from Ramusio's version.

[18] Comp. ib., i. p. 191.

[19] The Khotan legendaries of these Imāms form the subject of a meritorious study by M. Grenard; see *Mission D. de Rhins*, iii. pp. 1–46. He has not failed to recognize that the worship of the Muhammadan saints had its root in an earlier cult, see ib., ii. pp. 242 sq.; but his suggestions as to the cult directly replaced having been a kind of ancestral worship are scarcely justified, seeing how clearly we can establish the connexion between numerous well-known Ziārats and Buddhist shrines which must have immediately preceded them in the same localities.

Mirzā Haidar specially notes the numerous sacred tombs of Khotan, but expresses himself very sceptically about the traditions concerning the 'martyrs' who were supposed to be buried there; see *Tārīkh-i-Rashīdī*, p. 298.

Hsüan-tsang drew his picture. It can be safely asserted that display of temper and bluffness of words and manners are far rarer among the people of Khotan than, e.g., among those of Kāshgar. The observations which I had occasion to make on this point, while watching during months the conduct of my followers from the different territories, are in entire agreement with popular verdict as current in the country itself. It charges the 'Kāshgarliks' with quick temper and obstinacy, while the Khotanese are credited with hypocritical suavity[20]. Just because the development of the population of the Tārīm Basin under conditions determined by the same geographical and historical influences must necessarily tend towards uniformity, such differences as we have just observed, though they may now appear mere *nuances*, must be of special interest to the historical student.

Fondness for amusements.

The fondness for enjoyments, which Hsüan-tsang and the T'ang Annalists noted in the Khotanese of their days, still survives as fresh as ever. Of the multifarious occasions which serve rich and poor alike for the organization of fêtes and amusements, M. Grenard's description gives a sufficient idea[21]. Though it is difficult to make comparisons in a matter of this sort without prolonged experience, I certainly carried away the impression that a Khotanese crowd or party need less of a *tamāshā* to feel thoroughly happy than people elsewhere in Turkestān, while their abandonment to the enjoyment of the hour, whether it be a religious festival, family feast, pilgrimage, &c., or merely a novel sight, seemed always distinctly keener.

Love of music and dancing.

At all feasts and entertainments music, singing, and dancing are the chief diversions; and for them the Khotanese of to-day still show the same love and insatiable appetite which Hsüan-tsang and other Chinese observers noted in their ancestors. Music and singing are dear to the people all over Eastern Turkestān; but probably nowhere are small bands of professional musicians more numerous and the demand for them greater than in Khotan[22]. It is a point of special interest that dancing, elsewhere practically confined to professional dancers, women or boys, is a pastime freely indulged in by Khotanese of both sexes and of all classes. This local peculiarity deserves the more notice since the practice must have encountered the opposition of the clergy ever since the introduction of Islām, and is certainly in strange contrast to the general gravity and decorum fostered by Muhammadan tenets.

Adventurous occupations.

This lively hankering after pleasures of all sorts must often lead to complete neglect of serious material interests. It thus probably accounts to a great extent for the strikingly large number of cases in Khotan where men of all trades and classes, after having ruined themselves, seek refuge in more adventurous occupations. Of such Khotan has always offered a pleasing variety to those low down in their luck and averse to any constant exertion. Gold-washing, jade-digging, and 'treasure-seeking' at old sites in the desert are all callings which imply the certainty of great privations and hardships with a very limited possibility of substantial profit. Yet their attractions are great, like those of a lottery, and usually suffice to prevent their votaries from ever returning permanently to settled lives. In this way the riches which the Khotan region keeps, or is believed to keep, buried below the soil have probably since early times exerted a demoralizing influence on social organization.

Want of physical courage.

Notwithstanding the relatively large number of men following adventurous pursuits, the oasis of Khotan and the adjacent tracts are remarkably clear from crimes of violence. Considering how little the existing administration could do to prevent them, we may ascribe this fact, partly at least, to that law-abiding disposition of the population which the early Chinese observers noted.

[20] Comp. Grenard, *Mission D. de Rhins*, ii. p. 91.

[21] See Grenard, ib., ii. pp. 85 sqq.

[22] Regarding the music practised, the instruments used, &c., see the full and interesting data given by M. Grenard (*Mission D. de Rhins*, ii. pp. 135 sqq.), whose information is of special value as having been chiefly gathered at Khotan.

But it is, perhaps, equally due to the conspicuous want of physical courage which characterizes the modern Khotanese, almost as much as the Kashmīrī whom he resembles in other features of character. Marco Polo's blunt statement that the people of Khotan were no soldiers probably held good also in earlier periods, though no direct evidence has come down to us. In any case, it is certain that since the Muhammadan conquest the people of Khotan never took a share in deciding their own political fate. Thus, to quote a modern instance, the successful rebellion of 1863 against Chinese rule was almost solely the work of turbulent Andijānīs and other foreigners. Yāqūb Bēg subsequently gathered its fruit without Khotan even attempting to defend its own nominal ruler Habībullah. And when Yāqūb Bēg's dominion finally collapsed, the return of the infidel Chinese was probably welcomed nowhere with more relief than among the faithful of Khotan, who still remember with dismay the attempts to exact military service from them for the benefit of Muslim independence.

Position of women. In no respect is the tenacious survival of the social characteristics of old Khotan more striking than as regards the position of women. The relative independence and freedom from seclusion which women enjoy throughout Eastern Turkestān have been often pointed out, and attract at once the attention of visitors accustomed to the conditions of other Oriental countries, especially where Muhammadan customs prevail[23]. Throughout Eastern Turkestān women are free to accompany their husbands or near male relatives on almost all occasions when the latter leave their homes, whether for business or pleasure; everywhere they freely partake in the reception of visitors, in all transactions of business, and in festive gatherings, whether public or private. But nowhere are the restrictions which are supposed to accompany these privileges, such as the use of the veil in public and the company of husbands or male relatives, so constantly ignored as in Khotan. Most of the buying and selling on market days is done by women, who at all times crowd the Bāzārs far in excess of men, and ordinarily dispense with any male assistance. Veils are rarely seen even among the women of the well-to-do middle class, and when worn usually serve the purpose of ornament rather than to screen the face.

Easy morals of Khotan. The extraordinary facilities of divorce prevailing in all parts of Eastern Turkestān and the consequent laxity of the marriage tie assure to women a remarkable degree of legal and social independence. That this independence is abused in Khotan more than anywhere else may be concluded from the proverbial reputation for immorality which its women enjoy through all neighbouring regions[24]. As none of the latter pride themselves upon any high standard of morals this reputation acquires special significance. That licentiousness of every sort has thriven in Khotan from early times is proved not only by the Wei Annals above quoted, but also by the large proportion of obscene representations among the terra-cotta figurines found at Yōtkan.

Attitude towards study. Among all the qualities and accomplishments which Hsüan-tsang attributes to the Khotanese of his days there is one only for which a modern visitor to Khotan would look in vain—a liking for literary studies. The cultivation of literature, whether theological or profane, has apparently been at a low ebb in Eastern Turkestān throughout the Muhammadan period, and has remained so to the present day. Khotan in this respect makes no exception. An analysis of the political and cultural conditions prevailing in the whole country would easily account for this neglect of literary interests; but with this we are scarcely concerned here. Nevertheless,

[23] Regarding the legal and social position of women, comp. M. Grenard's detailed description, *Mission D. de Rhins*, ii. pp. 112–127.

[24] Comp. e.g. *Yarkand Mission Report*, p. 447; Grenard, *Mission D. de Rhins*, ii. p. 122.

I was more than once surprised to find even in most unlikely places, and among people of the humblest calling, individuals who had learned to read and write Turkī[^24]. The amount of knowledge imparted by the schools attached to many mosques and Ziārats is, indeed, very limited; yet the fact of their being frequented when no material gain can be secured from instruction seems to indicate the possibility that education under an earlier civilization and under more favourable political conditions was once more widely spread and aimed at a far higher standard.

Leaving this one point apart, we are justified in asserting, as the result of the inquiry just concluded, that all the distinctive features of character and ethnic disposition which our Chinese records indicate for the people of ancient Khotan can be traced unimpaired in the present inhabitants of the oasis. Inasmuch as such a striking survival can be explained most readily as a result of direct inheritance, the inquiry so far distinctly supports the conclusion that the racial character of the Khotan people has not undergone any fundamental change since the period of the T'ang dynasty.

Section IV.—THE POPULATION OF KHOTAN: ITS PHYSICAL CHARACTERISTICS AND RACIAL ORIGIN

We must now proceed to examine whether the conclusion arrived at in the previous section is borne out by the physical characteristics of the people. In this part of our inquiry we shall have to follow a different method; for while I succeeded in bringing back a small but carefully collected amount of exact anthropometrical materials relating to the present inhabitants of Khotan and the neighbouring oasis of Keriya, there is as regards the earlier population no trustworthy evidence with which to compare these data. With the exception of a single brief remark to be mentioned below, the early Chinese records do not notice in any way the physical appearance of the people of Khotan. Archaeological evidence also fails us; for those pictorial and sculptural representations which have come to light from old sites, and which might possibly be supposed to be modelled on contemporary local types, are too few to be of use in regard to anthropological questions. Finally, my explorations have not yielded any human remains which could with certainty be assigned to any particular historical period.

Want of early anthropological data.

In view of these circumstances the only safe course is to examine in the first place the conclusions based on anthropological evidence which competent analysis has drawn from the measurements taken by me on inhabitants of Khotan and Keriya. A comparison of these conclusions with the available historical and antiquarian information may then show whether they can be reconciled with the assumption of the race having remained unchanged in its main constituents since the close of the pre-Muhammadan epoch. The necessity for concentrating all my labours on archaeological tasks, and the fact that but a very small portion of my time could be spent within the present inhabited area of the Khotan region, rendered it impossible for me to secure anthropometrical data from more than a limited number of individuals. Considering the restricted amount of material thus available, I have reason to feel grateful that

Anthropometrical data collected.

[^24]: I thus found, e.g. that a considerable proportion of the children of the shepherd families which wander with their flocks in the lonely jungles along the Keriya river had been at school for varying periods at the Ziārat of Burhānuddīn Pādshāhim or in Keriya Town. Yet the appearance and conditions of life of these people might at first sight be easily mistaken by a traveller for those of semi-savages.

its detailed examination and analysis was undertaken by such a competent scholar as Mr. T. A. Joyce, Secretary of the Anthropological Institute of Great Britain. The results of his investigations, published in the *Journal of the Anthropological Institute*[1], have so important a bearing on a number of questions connected with the historical past of the Khotan region that I regret having been unable, from considerations of space, to reproduce the paper embodying them in an Appendix to the present work.

Mr. Joyce's investigations.

Referring to this publication for all evidence, I restrict myself here to a succinct statement of Mr. Joyce's results, as far as possible in his own words. Before, however, proceeding to this I think it advisable to call attention to two facts upon which he lays stress in the introductory part of his paper[2]. 'With regard to the deduction therefrom of a theory as to the race or races which enter into the composition of the inhabitants of these oases, it must be admitted that the material is somewhat small for the attainment of any degree of certainty.' Hence 'all conclusions must be regarded as purely tentative, to be confirmed or refuted by further researches'. It was with special regard to this fact that I considered it necessary to recommend to my collaborator the advisability of keeping his investigation of the anthropometrical materials at present available entirely clear of any indications, positive or conjectural, which might be deduced from historical and similar extraneous data. For it is obvious that any anthropological conclusions drawn from such restricted materials could only have weight with critical students if the risk of any unconscious bias were excluded. A reference to the body of Mr. Joyce's paper will show how careful he has been to base his conclusions on purely anthropological evidence.

Racial affinities of modern population.

The results of his investigations on these lines are briefly the following. A scrutiny of the physical characteristics of the inhabitants of Khotan and Keriya, with a view to connecting them with one or other of the great racial stocks of Asia, proves the general absence of those traits which are recognized as distinguishing the Mongolian[3]. On the other hand, it is noted as regards the people of Khotan 'that in their complexion, character and colour of hair and eyes, and cephalic index, is shown a very close correspondence with the Galchas; in fact, they seem to bear a general resemblance rather to this than to any other type'[4]. The Galchas or Alpine Tājiks are taken by Ujfalvy, the chief authority on the subject, as fairly pure representatives of Lapouge's *Homo Alpinus*, and are certainly of 'Aryan' stock. To their race belong the mountain tribes settled in the high valleys between the Hindukush on the south and the Alai on the north, who speak closely allied Eastern Īrānian dialects[5]. Among these are the Wakhīs and Sarīkolīs, whose close approach to the Khotanese in outward appearance is clearly traceable in my photographs[6].

Certain differences, however, are at the same time noticed between the Galchas and the Khotanese with regard to stature, nasal form, and facial index, together with a tendency in the latter to darker tints in hair and eyes. In looking for an explanation of these differences Mr. Joyce arrives at the inference that in the case of the Khotanese there has been, besides a slight admixture of Turkī blood, an admixture also of Tibetan. The tendency towards the characteristics traceable to a Tibetan admixture, such as short stature, platyrrhiny, black hair

[1] See *Journal of the Anthropological Institute*, 1903, vol. xxxiii. pp. 305–324, with Plates xxvi, xxvii.

[2] See ibid., pp. 305 sq.

[3] Comp. ibid., p. 315.

[4] See ibid., p. 322.

[5] For an interesting summary of observations on the Galchas of the Zarafshān Kōhistān, who are the physically best-known representatives of the race, see Ujfalvy, *Observations sur les Tadjiks des montagnes*, Bulletins de la Société d'Anthropologie, 1887.

[6] Comp. Fig. 5 (also Pl. xxvii in *J. Anthrop. Inst.*, 1903), with Figs. 22, 23. This general resemblance of the two types struck me at the time when I first became acquainted with Khotanese.

22

TAWAKKĒL LABOURERS BROUGHT TO DANDĀN-UILIQ.

23

TĀGHLIKS AND EXILED KHOTAN CRIMINALS AT KARANGHU-TĀGH.

and dark eyes, becomes more marked in the people of Keriya, who otherwise 'seem to show, as might be expected from their neighbouring position, a very strong similarity with the Khotanese'.

Principal observations on racial character.

Mr. Joyce thus sums up the principal facts deduced with regard to the people of Khotan and Keriya. 'Both are, in the main, of so-called "Aryan" stock, the chief factor being Lapouge's *Homo Alpinus*. There is, however, in each case an admixture of Turkī blood and a further admixture of Tibetan. The latter appears to be stronger at Keriya than at Khotan, and at the same time here Mongolian influence begins to make itself felt.'

'The Pamir valleys, as far as Asia is concerned, seem to be the locality where *Homo Alpinus* appears in his greatest purity. In the Galcha he appears with a slight Turkī and Iranian admixture. In the Khotanese the Iranian is replaced by a Tibetan element, and further east, among the inhabitants of Keriya, true Mongolian traits are just beginning to appear.'

These last-named slight traces of Mongolian blood, present only in a diluted form among the inhabitants of Keriya, are attributed by Mr. Joyce to a more recent admixture and need not be specially considered here. The other racial constituents traced by the above analysis are, however, important; and the question arises whether we can account for them without assuming a radical change in the population since the pre-Muhammadan period. This, I believe, we are able to do with the help of the indications furnished by our available historical and philological materials.

Affinity between Khotanese and Galchas.

Let us take in the first place the chief racial element which enters into the composition of the people of the oasis, that of Lapouge's *Homo Alpinus*. We have seen that the typical representatives of this element in Asia are the Galcha tribes of the Pāmīr Valleys who speak Eastern Īrānian dialects, and to whom belong also the closely allied Wakhīs and Sarīkolīs. The close approach in outward appearance between the latter people and the present Khotanese has already been noted. Now it is a fact of no small significance to find the T'ang Annals distinctly asserting of the inhabitants of Ho-p'an-t'o or Sarīkol that 'their external appearance and language are the same as those of the people of Yü-t'ien (Khotan)'[7]. I had occasion in a previous chapter to discuss this important statement[8]. I also pointed out that it receives confirmation from an exactly corresponding observation made by Sung Yün and Hui-shêng concerning the population of the territory of Karghalik, which forms the natural link between Sarīkol and Khotan[9], and which still contains, in the Pakhpo inhabitants of its mountains, an ethnic element unmistakably allied to the Galchas[10].

Linguistic evidence.

To these historical attestations of an ethnic connexion between Khotan and the easternmost territories still possessing a Galcha population, we can add a weighty piece of philological evidence. As will be seen in a subsequent chapter[11], the ancient site of Dandān-Uiliq has furnished, both to me and to 'treasure-seeking' natives who previously 'explored' it, a considerable number of documents written in Brāhmī characters and in all probability belonging to the eighth century of our era. The language of these documents has been proved by Dr. Hoernle, their first decipherer, to be an Indo-Īrānian dialect having its nearest congeners in the Galcha dialects of the Pāmīr region[12]. There is every reason to believe from the character of these documents that the language in which they are written was the one actually

[7] See Chavannes, *Turcs occid.*, p. 124.
[8] Compare above, pp. 26 sqq.
[9] See above, pp. 90 sq.
[10] Compare above, p. 26.
[11] See below, chap. ix. sec. v.
[12] See Hoernle, *Report on Central-Asian antiquities*, ii. pp. 32 sqq.

spoken by the contemporary inhabitants[13]. Since this fact fully accords with the ethnic relationship which the Chinese records just quoted indicate between the Khotanese and the Galchas of the same and the immediately preceding periods, it is clear that the prevalence of *Homo Alpinus* in the anthropological composition of the present Khotan people must be attributed to direct inheritance from the pre-Muhammadan population.

Admixture of Turkī blood.

It is different with the admixture of Turkī blood. That this admixture has taken place only since Khotan was converted to Islām is clearly shown by historical and philological evidence. We know that until the close of the tenth century, when that conversion took place, Khotan under its Buddhist rulers remained independent of the Karluk Turk dominion established in the north-west of the Tārīm Basin. During the prolonged struggle which preceded the introduction of Islām, an appreciable influx of Turkī people into Khotan is highly improbable. The same observation applies to the ninth century, when Khotan with a great portion of Eastern Turkestān was under Tibetan control, while the total absence of Turkī words in the Brāhmī documents of Dandān-Uiliq, already mentioned, excludes the possibility of the population of Khotan having received any Turkī element down to the close of the eighth century.

Restriction of Turkī element.

The admixture of Turkī blood, which must thus be ascribed to the period since the conquest of Khotan by Satok Boghra Khān's family, is shown by Mr. Joyce's analysis to have been relatively small[14]. This may appear surprising, in view of the universal adoption of the Turkī language in Khotan, as throughout the oases of the Tārīm Basin, but it is in reality easily accounted for. In this region, as in other Central-Asian territories where subsistence is possible solely by the laborious cultivation of irrigated lands, or else by industries and commerce, the Turks, nomads by origin and habits, appeared primarily only as soldiers. By their superior military qualities and organization they were able even in small numbers to place their chiefs in undisputed sway over the far more civilized but peace-loving people of the ancient oases. They likewise succeeded, by a peculiar faculty for ethnic attraction, often illustrated in the case of Turkish conquests, in making subject populations rapidly adopt their own language and willingly accept their political predominance. On their own part these Turks could not escape gradual amalgamation with the people of the oases whom they ruled and protected, and in the end they became wholly absorbed in them[15]. But their numbers were far too small to affect fundamentally the racial character of the population.

The latter remark applies probably with even greater force to the Kara-Khitai, Moghuls and Kalmaks or Oirat (Eleuths), all tribes of Mongolo-Turkī race, who during subsequent periods exercised political predominance in Eastern Turkestān. Retaining their nomadic habits longer than the Turks, these nations had their main seats of power to the north of the T'ien-shan, in the country known after them as Moghulistān or Zungaria. In view of what we know of the manner in which their temporary power was exercised in the settled portions of Turkestān, they could but slightly have strengthened the Turkī element in so distant an oasis as Khotan[16].

[13] In this connexion the survival of certain Īrānian words in the Turkī now spoken in the Khotan region deserves attention. These words (e. g. *sag* 'dog') are unknown to the people in other parts of the Tārīm Basin; comp. Grenard, *Mission D. de Rhins*, iii. p. 64.

[14] See *J. Anthrop. Inst.*, xxxiii. p. 323.

[15] The political effects of the Turkī conquest of the Tārīm Basin have been judiciously analysed and described by M. Grenard, *Mission D. de Rhins*, ii. pp. 71 sqq. It must, however, be noted that the influence which the varying local conditions presented by the different territories exercised upon the progress and extent of this 'Turkization' has scarcely been sufficiently realized.

[16] Mr. N. Elias has called attention to the very small size of the 'armies' with which, according to Mirzā Ḥaidar's accounts, the various Moghul Khāns warred against each other in Eastern Turkestān or conducted their raids into neighbouring territories. The chiefs of these tiny forces were, indeed, Moghuls of various clans; but 'the tribal following which each chief could muster was a mere handful',

Taking the Tārīm Basin as a whole M. Grenard is probably right in assuming that the process of 'Turkization' was furthered far more by gradual absorption in the settled population of nomadic Kirghiz from the adjacent mountains, than by any of those great invasions from Mongolia.

This reinforcement of the Turkī element by Kirghiz transformed into cultivators is likely to have had its importance for the northern oases like Kāshgar, Ak-su, Kuchā, which are within easy reach of the Kirghiz grazing-grounds in the valleys of the Alai and T'ien-shan. But it could scarcely have affected Khotan, which is far removed from those tracts and has nothing in its own mountains to attract or support nomads. Differences of this kind may well have helped to keep the Turkī admixture in the Khotan population far lower than it appears to be in the northern oases. Such a variation in the extent of Turkī infusion would best account for the distinct though not easily defined disparity in outward appearance which struck me whenever I had occasion to compare Khotanese and people of Kāshgar. I am unable to test this impression by exact anthropological data, such as measurements on Kāshgarliks would have supplied; but I may in support of it suggest a comparison between the Khotanese seen in Figs. 22, 23 and the Mecca pilgrims from Kāshgar whose photograph (Fig. 15. a) I recently secured on their passage through Peshāwar.

Infusion of Kirghiz blood in Kāshgar.

It now remains only to examine what traces, if any, our historical and philological materials may preserve of that Tibetan element which Mr. Joyce recognizes as the second and larger admixture in the racial composition of the people of Khotan and Keriya. Before, however, proceeding to this scrutiny it will be well for us to realize clearly what is meant by this 'Tibetan' element. Mr. Joyce himself plainly tells us that 'the question as to what is the Tibetan type has not been satisfactorily answered'.[17] Not only have there been up to the present practical obstacles to a systematic study of the race in any but the outlying frontier districts, but there appears to be good reason for believing that the population of Tibet, notwithstanding the unity it shows in political organization, civilization, and language, comprises several disparate racial types which cannot easily be reduced to a common origin[18]. While 'it is certain that the majority of the present inhabitants are Mongolians', there is also evidence to show that 'the Tibetans are not all wholly Mongolian'. This is specifically true of a widespread type described by Mr. Rockhill, 'which he terms the "Drupa type," and which he regards as "comparatively pure".' As its characteristic features are indicated: stature under average; brachycephalic head; high cheek-bones; thick and broad nose; black wavy hair, little on face; brown eyes[19]. It is to a relatively large admixture of this particular type of Tibetans that Mr. Joyce attributes those unmistakable deviations from the Galcha type which the measurements taken prove for the people of Khotan and still more for those of Keriya, and which the admixture of Turkī blood alone would be insufficient to explain[20].

Tibetan admixture.

It is true that the confines of the geographical area known to us as Western Tibet extend to the main Kun-lunrange which borders Khotan and the smaller oases connected with it on the south. But it would argue a grave misconception of the true geographical facts

Natural barriers between Khotan and Tibet.

while the rest of the 'armies' (e.g. of the 4,700 men with whom Sultān Said wrested 'the Six Cities' of the Tārīm Basin from Abā Bakr), was made up apparently of Turkish and other adventurers; comp. *Tārikh-i-Rashidī*, Introd. pp. 65 sqq. From Mirzā Ḥaidar's accounts we also see how little the peaceful population of the oases was concerned in the results of these raids and counter-raids, its share in the events being confined at all times to feeding the victors and providing them with revenue.

[17] See *J. Anthrop. Inst.*, xxxiii. p. 318.

[18] On this point the remarks of M. Grenard, *Mission D. de Rhins*, ii. pp. 323 sqq., are worthy of special note.

[19] See *J. Anthrop. Inst.*, xxxiii. pp. 318 sq.

[20] Comp. ibid., p. 323.

were we to accept this apparent territorial vicinity as an adequate explanation of the racial connexion above indicated. No inhabited part of Tibet lies nearer to the oases than two hundred miles by the map, and the intervening belt of high mountains and equally barren plateaus, reaching a width of over seven hundred miles in the direction of Central Tibet, is too formidable a natural barrier to have ever permitted continuous intercourse between the populations on either side. Hence the idea of any gradual intermingling of the two peoples must be dismissed as highly improbable on physical grounds.

Tibetan influence in historical period.

It seems almost equally difficult to account for so large an admixture of Tibetan blood by any ethnic movements that could possibly have taken place within the historical period. We are sufficiently well informed by the Chinese records about the main political events which affected Eastern Turkestān and Khotan in particular, since the beginning of our era, to know that the Tibetans did not begin to play a part in them until the last third of the seventh century[21]. These early Tibetan invasions were soon checked by a reassertion of Chinese power, and though we hear of subsequent irruptions, it was only some hundred and twenty years later that the Chinese hold over the Tārīm Basin was finally brought to an end by the Tibetans[22]. As regards Khotan itself, documentary evidence supplied by finds from Dandān-Uiliq proves that Chinese influence was not upset by the Tibetans until after 790 A.D.[23] Considering that the previous invasions were scarcely more than mere raids, and these, it seems, mostly directed against the extreme east of the 'Four Garrisons' from the side of Tsaidam, they cannot reasonably be credited with having left so deep an imprint on the racial character of the people of Khotan. The period of Tibetan predominance in Eastern Turkestān, which lasted from the end of the eighth century to circ. 860 A.D.[24], appears also far too short for this. After that date the creation of the powerful Uigur kingdom in the east of the Tārīm Basin finally closed what must have always been the natural avenue of Tibetan invasion, the routes from Tsaidam.

Early ethnic tradition of Khotan.

If we weigh properly the geographical and historical facts here briefly indicated, we are necessarily led to consider the possibility of the 'Tibetan' infusion in the present population of Khotan and Keriya reaching back to a far earlier epoch. And it appears to me that there are indications which distinctly support this solution. The most important among these is the old local tradition of Khotan, to be discussed in detail in the next chapter, which ascribes the creation of the kingdom to joint occupation of its territory by two colonies originating from India and China, respectively[25]. The genuineness and relative antiquity of this Khotanese tradition is attested by its being recorded in a substantially accordant form both in Hsüan-tsang's *Hsi-yü-chi* and in certain Tibetan texts[26].

We need not consider here any of the quasi-historical details which this tradition pretends to record, nor examine to what extent it may help to explain peculiar features in the archaeology and linguistic remains of ancient Khotan. Leaving these points for subsequent consideration, we must note only the essential fact that indigenous tradition, dating back to a period long before Tibetan influence could have made itself felt politically, recognized as the earliest settlers of Khotan two entirely distinct elements, one which was believed to have come from the side of China and the other from the extreme north-west of India. That the tradition, as far as it relates to an immigration from the latter side, possessed some historical

[21] Comp. Chavannes, *Turcs occid.*, pp. 122 note, 280 sq.

[22] See above, pp. 64 sq.

[23] See below, chap. IX. sec. vi.

[24] See above, p. 65.

[25] See below, chap. VII. sec. ii.

[26] See *Mémoires*, ii. pp. 224 sqq.; Beal, ii. p. 309; Rockhill, *Life of the Buddha*, pp. 233 sqq.

foundation has been demonstrated in a striking manner by the numerous documents in an early Indian script and language which I discovered at the Niya Site. As will be seen later, the character of these documents renders it certain that the administration of the territory was during the third century of our era carried on in an Indian language, and that this language was familiar to a considerable portion of the population [27].

Similar evidence cannot be adduced at present for that early immigration from the side of China which the same tradition alleges. But there are certain facts and statements which point to some substratum of truth in that direction also. In the first place, we have the interesting and much-discussed passage in the notice of Khotan given by the Annals of the Northern Wei (386-532 A.D.), which tells us that 'the people of all territories west of Kao-ch'ang (the present Turfān) had deep-lying eyes and prominent noses, and that the inhabitants of this territory (i.e. of Khotan) were the only ones who did not present a very strange appearance but had rather a Chinese look' [28]. 'Deep-lying eyes and prominent noses' are certainly the features which would strike a Chinese observer most in physiognomies of the so-called 'Aryan' type, and we may safely conclude from their mention that this was the type then already prevailing in Eastern Turkestān. Alleged immigration from China.

The appropriateness of this description renders the exception made in regard to the people of Khotan all the more significant. But can we accept with equal confidence the statement made about the Chinese look of the latter? Would not the presence of those features in a less pronounced fashion have sufficed to suggest to Chinese eyes something more familiar in Khotanese looks, especially when the local tradition of an early immigration from the East was remembered? These are questions to which in the present state of our knowledge we must hesitate to give even conjectural answers. Yet if we assign to the observation of the Wei annalists this qualified bearing, and at the same time assume that the tradition related to an element in its population which had come from the side of China but was in reality rather Tibetan than Chinese, an explanation is furnished which would equally satisfy tradition and the evidence of anthropological facts.

There can be little doubt that even the least Mongolian type of Tibetan, Mr. Rockhill's 'Drupa type' above described, with its broad thick nose and high cheek-bones, must approach nearer to the Chinese than to any 'Aryan' type. Chinese tradition regards the Tibetans as descendants of certain tribes who were exiled from China to the Koko-Nor region in a prehistoric period [29]; in the legend of Khotan, too, the immigrants are represented as being led by a prince exiled from China. We need not attach any weight to this point of similarity. Yet it might help to explain how an immigrant element akin to the present Tibetans, which had reached Khotan through Tsaidam at a very early period, could in later tradition figure as having been of Chinese origin. That the presence of such an element would have been far more noticeable in the racial appearance of the Khotan population at the time of the Wei dynasty than at present does not need special explanation. Chinese view of Tibetans.

The same ethnic element would also account for an important fact not yet noticed here, the appearance in our oldest Khotan records, the Kharoṣṭhī documents from Niya, of numerous words, mostly titles or terms, which are certainly neither Indian nor Īrānian nor Turkī, but often suggest a Tibetan origin [30]. The very fact that some of these words, though technical Linguistic traces of Tibetan connexion.

[27] See below, chap. XI. sec. vi.

[28] Comp. Rémusat, *Ville de Khotan*, p. 20; Franke, *Zur Kenntniss der Türkvölker*, p. 19; for earlier speculations on the passage comp. e.g. Ritter, *Asien*, v. p. 362; Richthofen, *China*, i. p. 48 note.

[29] See Grenard, *Mission D. de Rhins*, ii. pp. 327 sq.

[30] According to a communication from Prof. Rapson and Dr. Barnett.

in character, must relate to objects of ordinary life, suggest that they belong to a language which had once been commonly used in this region. The character of the 'unknown' language which is found in a number of manuscript texts and fresco inscriptions from Dandān-Uiliq and Endere, and which may well prove to be the source of those non-Indian terms in the older Kharoṣṭhī documents, points to the same conclusion. For Dr. Hoernle, who has carefully analysed a number of similar texts, probably translations of Sanskrit originals into the language of the country, has from the general character of the language derived the impression 'that its identity has to be looked for in the direction of the monosyllabic Tibetan rather than of the Turkī or Mongol languages'[21]. Finally, we may note in passing that the hypothesis just indicated would furnish also the best explanation for a number of ancient local names of the Khotan region which suggest Tibetan origin, and which yet are anterior to the historical period of Tibetan invasions[22].

I need scarcely point out that most, if not all, that relates to the ethnic evolution of the Khotan people previous to the Muhammadan period, must in the present state of our knowledge remain distinctly conjectural. Yet I believe that the comparison we have just concluded between the actual anthropology of the oasis and the extant data concerning its earlier inhabitants may help us in interpreting correctly historical records and archaeological facts.

[21] See *Report on Central-Asian antiquities*, ii. p. 13.

[22] See Grenard, *Mission D. de Rhins*, ii. p. 54.

CHAPTER VII

HISTORICAL NOTICES OF KHOTAN

Section I.—THE EARLY RECORDS AND NAMES OF KHOTAN

In the introductory remarks of the preceding chapter I have already referred to the reason for which it will be necessary to review the records of the early history of Khotan before I proceed to give an account of the antiquarian observations and discoveries which have resulted from my explorations in that region. I do not propose to include among the records to be discussed here the evidence furnished by archaeological finds; for however important this evidence may prove to be, its value can only be established in connexion with a critical survey of the archaeological operations which have supplied it. Apart from these data, to be verified and interpreted hereafter, our historical notices of Khotan during the pre-Muhammadan periods are derived almost exclusively from two sources; the writings of Chinese annalists and travellers, and certain Tibetan texts which appear to be based largely on legends and traditions originally chronicled in Khotan itself.

Chinese records of Khotan.

A variety of reasons combine to make the Chinese notices of far higher value for historical research. In the first place, they represent for the greatest part contemporary records embodying either official information or the observations of thoroughly trustworthy travellers; they are in almost all cases chronologically fixed, and present the great advantage of relating mainly to matter-of-fact aspects of ancient Khotan, its secular affairs and political events. Whatever political and cultural reasons may explain the fact, it appears certain that the relations of Khotan with China for over a thousand years previous to the introduction of Islām were closer and more continuous than those of any other portion of Eastern Turkestān.

Publications from Chinese records.

Whether it was this copiousness of historical details, or the fascination of a territory which until half a century ago was to Western students one of the least known and least accessible parts of Asia, we must feel grateful to the reason which induced A. Rémusat to publish, in 1820, his *Histoire de la ville de Khotan*, containing a complete translation of all Chinese notices concerning Khotan as collected in Book LV of that vast encyclopaedia, the *Pien i tien*[1]. It is not for me to appraise the merits of this publication of the great French Sinologist; but it cannot be questioned that it has done more than any other work to direct attention to the exceptional historical interests attaching to Khotan[2]. The materials furnished in it are still the most valuable to which the non-Sinologist student of Khotan has access at present. M. Chavannes' work on the Western Turks has, of course, supplied a mass of additional and critically sifted data of considerable interest for Khotan history during the T'ang period[3]. Nevertheless, it is probable

[1] Compare Rémusat, *Ville de Khotan*, pp. iv. sq.; for a specific mention of the *Pien i tien*, curiously enough left unnamed in the Introduction, ib., p. 33 note.

[2] The stimulating influence of Rémusat's publication is illustrated by the use made of it in the exhaustive notice which Ritter, in 1838, devoted to Khotan and its history; see *Asien*, v. pp. 343–89.

[3] See Chavannes, *Turcs occid.*, pp. 125 sqq. and, for incidental notices, Index, s. v. *Khotan* and *Yu-t'ien*.

that even thus the rich storehouse of notices on ancient Khotan which the Chinese Annals and similar historical works contain is still far from being exhausted [4].

Tibetan notices of Khotan. The Tibetan notices of Khotan history bear a very different character. They are contained in four texts forming part of the great canonical collections of Tibetan Buddhism, and claiming to contain Buddha's predictions concerning events which were to take place in Li-yul or Khotan. The text described as 'The Annals of Li-yul' seems the largest and most important among these. To Mr. W. W. Rockhill, scholar and diplomatist, belongs the merit of having first called attention to these texts, and of having furnished systematically arranged abstract translations from them in his *Life of the Buddha* [5]. Mr. Rockhill's analysis makes it appear very probable that they are translations or adaptations from works written in the language of Khotan. That these works reproduced traditions, legendary or historical, then current in Khotan becomes evident from the substantial agreement which the stories related in the Tibetan texts about the foundation of the Khotan kingdom, about the origin of various Buddhist sanctuaries, &c., show with the corresponding legends recorded by Hsüan-tsang.

Sources of Tibetan notices. There is no definite indication of the date when the translations were made or the original texts composed. Yet it seems probable that the latter could not have differed much in character and contents from those Khotan chronicles to which Hsüan-tsang alludes, and from which his accounts of various ancient shrines and traditions are likely to have been borrowed directly or indirectly. The Tibetan texts may thus claim the merit of having preserved for us at least a reflex of what traditional records Khotan itself possessed during the centuries immediately preceding the Muhammadan era. But it does not require a prolonged inquiry to realize that the information supplied by them is, notwithstanding its indigenous origin, greatly inferior in historical value to that of our Chinese sources. A perusal of Mr. Rockhill's extracts shows that, apart from the legendary traditions recorded about the origin of the Khotan dynasty and people, and about that of certain prominent Buddhist shrines, the 'Annals of Li-yul' acquaint us only with a long string of royal names diversified by occasional reference to pious foundations and matters of doctrine. The same holds good also of the information given about the last twenty-eight kings who are not specified by Mr. Rockhill, but whose names will be found in the supplementary extracts which Mr. F. W. Thomas, the learned librarian of the India Office, has been kind enough to place at my disposal [6].

Tibetan list of Khotan kings. In the present state of our knowledge it is difficult to form any definite view as to the critical value to be attached to this long traditional list of kings or to any specific part of it. Leaving aside the legendary founder of the dynasty, who is represented as a son of King Aśoka, no chronological indications are furnished to us. None of the royal names, which are all Indian and formed with *Vijaya-* as their first part, has so far been found on coins or in the ancient documents discovered. If the latter portion of the list is authentic the reigns specified in it must partly at least fall within the period of the T'ang dynasty, for which the Chinese Annals give us relatively ample information about the rulers of Khotan. The repeated references made to dynastic relations with China and to Chinese ministers at Khotan seem to confirm

[4] Compare Dr. Bushell's remark in Hoernle, *Report on C.-A. antiquities*, ii. Supplem. p. 3.

[5] See Rockhill, *Life of the Buddha*, pp. 230–48.

The texts extracted are: (1) 'The Annals of Li-yul', from the *Tanjur*, vol. xciv (*u*), foll. 426–44; (2) 'The Prediction (*vyākaraṇa*) of Li-yul', ibid. foll. 420–5; (3) 'The prediction of the Arhat Saṅghavardhana', ib. foll. 412–20; (iv) 'The Gośṛṅga Vyākaraṇa' from the *Kanjur*, vol. xxx, foll. 336–54. In the case of the last-named text, which seems to form a kind of 'Māhātmya' or legendary of the sacred site of Gośṛṅga (Kohmārī) to be discussed below (chap. VIII. sec. i; comp. S. Lévi, *Notes Chinoises sur l'Inde*, iv. p. 39), the colophon distinctly records that it 'was translated into Tibetan from the language of Li-yul'; see *Life of the Buddha*, p. 231.

[6] Compare, for these extracts, Appendix *E*.

this assumption [7]. Yet owing to the Khotan rulers of this period appearing only under Chinese names in the T'ang Annals, it is impossible to attempt the identification of any of them with those bearing Indian names in the Tibetan list. All that can be suggested at present with some probability is that the family name of *Wei-ch'ih*, which is given to the period by the T'ang Annals and Wu-k'ung [8], may be a rendering of the Sanskrit *Vijaya-*, which appears invariably as the first part of the names recorded in the Tibetan list. The occasional references which we meet to the relations of 'Li-yul' with other territories besides China afford little or no clue, since the latter can scarcely ever be identified under the strange disguise of their Tibetan names [9].

The difficulties and obscurities here indicated make it impossible to treat the information gathered from the Tibetan accounts in connexion with the strictly historical data supplied by our Chinese records. Nor can I attempt to subject that information to a separate critical analysis. Whatever in the Tibetan notices I have been able to recognize as throwing light on the early legends, sacred sites, and historical relations of Khotan, will be duly noticed hereafter in connexion with other available evidence, while for a complete list of the kings and the main facts noted regarding them in the 'Annals of Li-yul' I must refer to the synopsis furnished by Mr. Thomas in Appendix *E*. Use made of Tibetan notices.

An ancient legend connects the name of *Khotan* with the very origin of the kingdom. The transformations of this name reflect, as it were, the succession of cultural and political influences which the territory has undergone from an early period. Hence our review of the historical notices concerning Khotan may fitly commence with a brief survey of the varying forms of its name. We find the most important of them conveniently enumerated in an interesting passage of the *Hsi-yü-chi* which, omitted by Julien and Beal, has been duly noticed and translated by M. S. Lévi [10]. Hsüan-tsang, at the conclusion of his account of Chê-chü-chia, tells us: 'After marching 800 li, one arrives at *Ch'ü-sa-tan-na* 瞿薩怛那. In Chinese this means "breast of the earth". This is a popular interpretation. In current speech one says *Huan-na* 渙那. The Hsiung-nu say, *Yü-tun* 于遁. The Hu say, *Huo-tan* 谿 (for 豁) 旦. The Yin-tu [Hindus] say, *Ch'ü-tan* 屈丹. Formerly one sald, *Yü-t'ien* 于闐 [11].' Ancient names of Khotan.

The name *Ch'ü-sa-tan-na*, which Hsüan-tsang mentions first, is undoubtedly intended to reproduce the Sanskrit *Kustana*, 'breast of the earth', as recognized by Chézy and Rémusat [12]. It is subsequently accounted for in the legend which Hsüan-tsang records of the first king of Khotan and his miraculously born son, and which we shall have occasion to discuss presently. The pilgrim himself calls the meaning given to the name 'a popular interpretation', and it can scarcely be doubted that the form 'Kustana' itself is only the result of a learned 'popular etymology' which endeavoured to provide an orthodox Sanskrit derivation for the old local Sanskrit name *Kustana*.

[7] See the notices concerning the kings numbered 9, 10, 13 in Mr. Thomas' supplementary list, App. *E*.

[8] See Chavannes, *Turcs occid.*, p. 126; also below, p. 173.

[9] The same difficulty is experienced about the identification of Khotan localities mentioned in the Tibetan extracts with names which seem more puzzling even than Chinese transcriptions of local names.

[10] See S. Lévi, *Notes chinoises sur l'Inde*, iv. p. 44. The passage had been previously reproduced by A. Rémusat, *Ville de Kholan*, p. 35, from a note of the Chinese editor of the Pien i tien, but without an indication of its origin.

[11] The same forms of the name are enumerated in the passage which opens the T'ang Annals' notice on Yü-t'ien, and which is manifestly an abbreviation of the above passage in the *Hsi-yü-chi*; see Chavannes, *Turcs occid.*, p. 125. Here the forms *Ch'ü-sa-tan-na*, *Huan-na* and *Ch'ü-tan* are mentioned together, without a distinction as to their use. The barbarians of the North (*Ti*) instead of the Hsiung-nu are credited with the form *Yü-tun*, and the form *Huo-tan* appears in a different graphic form as 豁怛; for the latter form comp. also *L'Itinéraire d'Ou-k'ong*, p. 26.

[12] See Rémusat, *Ville de Kholan*, p. 36, note.

name. Apart from the inherent improbability of a territory so far removed from the ethnic frontiers of India having first been named in Sanskrit, there is the significant fact that the word *ku* in the meaning of 'earth', which the above etymology assumes, is an artificial creation of Sanskrit lexicography, unknown to the genuine old language. That this learned Sanskritization of the local name is, however, far older than Hsüan-tsang's time, can now be shown conclusively from the Kharoṣṭhī documents of the third century A. D. discovered by me at the Niya Site. These, in their Prākrit queerly mixed with Sanskrit phrases, actually employ the form *Kustana* or *Kustanaka*, by the side of the far more frequent form *Khotaṃna*[13].

Early use of name *Khotan*.

The latter form, together with its variants *Khotana* and *Khodana*, is of special interest[14]; for it plainly attests the fact that the name of Khotan, whatever its origin, had already in the first centuries of our era assumed a phonetic form closely approaching to, or practically identical with, the present one. This enables us to account without any difficulty for all the varying reproductions of the local name recorded by Hsüan-tsang. Taking first the form *Ch'ü-tan*, the use of which is ascribed to the Hindus, it is evident that it transcribes a name which must have sounded very much like the present *Khotan*; for the syllable *Kho* we must, in accordance with Hsüan-tsang's transcriptional system, expect a Chinese character sounding *ch'ü*[15]. In *Huan-na*, which Hsüan-tsang mentions as the form used in current speech, we may, perhaps, recognize a local pronunciation of the name, in which the initial aspirate was reduced to a spirant and the *t* between the two vowels *o* and *a* reduced to an hiatus, as it would have been by regular phonetic change in a Prākritic language. The form *Khodana*, attested in the Kharoṣṭhī documents, would seem to indicate the first step in this process. *Huo-tan* is described as the form used by the Hu, i. e. the people of Central Asia whom the Chinese in the time of Hsüan-tsang seem to have comprehended under this general term[16]. As Chinese characters sounding *huo* are used for transcribing a variety of syllables, such as *kha, khā, ha, hu, ō*, &c.[17], it is difficult to indicate the pronunciation really meant. But it is clear that it could only have been a phonetic variation of the form *Khotan*.

The name *Yü-t'ien*.

Hsüan-tsang in his list mentions last the name *Yü-t'ien* 于闐 which, he says, was used formerly. But in reality this is the name by which Khotan is invariably designated in all the Chinese dynastic histories from the period of the Former Hans down to that of the Mings. It appears never to have quite gone out of official use even after the reconquest of Eastern Turkestān by the Manchu dynasty in the eighteenth century[18]. That Yü-t'ien was intended as a phonetic

[13] I owe the information as to the occurrence of the form *Kustanaka*, to Prof. E. J. Rapson.

[14] For specimens of passages containing the forms *Khotaṃna* and *Khodana*, compare tablets N. i. 104 + 106; N. xv. 318; N. iv. 108, tentatively transcribed and translated by Prof. Rapson in his *Specimens of Kharoṣṭhī inscriptions* presented to the Fourteenth Oriental Congress, 1905.

[15] Owing to the great rarity of Sanskrit words containing the Akṣara *kho*, Julien's list of Chinese transcriptions gives no instance of *ch'ü* (*k'iu*) or *chü* (*kiu*) as the phonetic rendering of *kho*; but examples of *ch'ü* (*k'iu*) to represent Skr. *go, gho, ku*, &c. are frequent; see *Méthode pour transcrire les mots sanscrits*, pp. 130 sq. The regular transcription of Skr. *kha* by *k'ia* presents an exact parallel; see ibid., p. 122.

[16] As to the varying application of the term *Hu* in Chinese literature, comp. Lévi, *Notes chinoises sur l'Inde*, iv. pp. 43 sqq.

[17] Compare Julien, *Méthode pour transcrire*, &c., pp. 108 sqq. (Julien's *ho* represents the sound transcribed by M. Lévi *Houo*, in Wade's system *Huo*).

[18] In all extracts from the Annals collected in the Pien i tien, as translated by Rémusat, the name Yü-t'ien is used, down to a notice relating to 1573–1619 A. D.; see *Ville de Khotan*, p. 107; also Bretschneider, *Mediaeval Researches*, pp. 47 sq. From the latter authority it is seen that during the Mongol period (1280–1368 A. D.) the name Yü-t'ien remained quite familiar, though the Chinese historians of the period 'generally try to render the name Khotan, which was in use with the Mongols'. Such renderings are the forms *O-duan, Wu-duan, Wa-duan, Hu-t'an* quoted by Bretschneider. The first three indicate a pronunciation resembling the Tibetan *U-then*, for which see below.

The *Hsi yü wen chien lu*, of 1777 A. D. gives the name as *Khotian*, but mentions Yü-t'ien also; see Klaproth, rendering of the name of Khotan, as it sounded in the time of the Former Han Dynasty or even earlier, appears highly probable. The possibility of the first character *yü* 于 being used as a transliteration of the character 玉, which means 'jade', need cause no doubt on this point, since, as Dr. Bushell has kindly pointed out to me, the selection of syllabic characters, of more or less appropriate meaning, for proper names is commonly practised by the Chinese, apparently as a kind of *memoria technica*. Khotan having always owed its fame with the Chinese mainly to its precious jade, the use of this particular character in the phonetic transcription of the indigenous name would be easily accounted for[19].

Other Chinese names.

The form *Yü-tun* 于遁, which Hsüan-tsang mentions as that used by the Hsiung-nu, and which the T'ang Annals ascribe more generally to 'the barbarians (*Ti*) of the North', manifestly represents a phonetic adaptation of the name closely approaching the official Yü-t'ien. It seems probable that this same Yü-tun is meant also by Wu-k'ung's itinerary when it mentions 千遁 *Ch'ien-tun* as an alternative name of Khotan[20]; for the *Ch'ien* of the first syllable can scarcely be anything else but a slight graphic error for 于 (*yü*). The other forms of the name which this later pilgrim records, *Yü-t'ien*, *Huo-tan* (here spelt 豁丹), and *Ch'ü-sa-tan-na*, are already known to us from Hsüan-tsang and the T'ang Annals. Wu-k'ung distinctly tells us that *Ch'ü-sa-tan-na* is the Sanskrit name 'meaning in the Chinese the kingdom of the breast of the Earth'.

Tibetan forms of name.

It is easy to recognize a phonetic reproduction of the name of Khotan also in *U-then*, which the Tibetan legend mentions as the name of the great city in the country of *Li-yul*[21]. This Tibetan U-then corresponds closely with the forms *O-duan*, *Wu-duan*, &c., quoted in a previous note as early Mongol renderings, as well as with the modern Chinese *Ho-t'ien* 和闐 and Manchu *Ho-thian*[22]. On the other hand, I am unable to connect the name *Li-yul*, literally meaning 'country of Li', by which the Tibetan texts above referred to, as well as other Tibetan works, invariably designate the Khotan region, with any indigenous term. The identity of Li-yul and Khotan, first correctly indicated by Wassilieff, has been placed beyond all doubt by Mr. Rockhill's publication; but the connexion suggested by the latter scholar between the Tibetan Li[-yul] and the name of Ilchi given to the present chief town of Khotan appears highly doubtful, considering that there is no evidence for this latter name being of any antiquity. I am unable to trace it in any records earlier than the eighteenth century[23]. Even then its

Mémoires rel. à l'Asie, ii. pp. 289 sqq. [The present official style for Khotan is 和闐 *Ho-t'ien*, while the ancient name *Yü-t'ien* is now officially applied to the new district of Keriya, according to information kindly given by Mr. G. Macartney.]

[19] I have thought it necessary specially to refer to this point as Richthofen, *China*, i. pp. 482, 486 note, has expressed the view that *Yü-t'ien* was a purely Chinese designation given to Khotan on account of its famous product, the jade. The great geographer appears to have based this opinion on the name *Yü-chih* (*Yü-tche*) 于寘, by which the report on the celebrated mission of Chang Ch'ien, as reproduced by Ssŭ-ma Ch'ien, designates the region of Khotan; comp. *China*, i. p. 450. M. Chavannes, however, to whom I referred the point, has been kind enough to explain to me that the reading *Yü-chih* is not correct. In the name as given in chap. cxxiii of Ssŭ-ma Ch'ien's history, the second character is, indeed, nowadays pronounced *chih* (*tche*); but the Chinese commentator distinctly notes that it should be pronounced as *t'ien* or *tien*. Thus, in reality, the name recorded by Ssŭ-ma Ch'ien is identical with the orthodox form 于闐 *Yü-t'ien*.

M. Chavannes considers the old Chinese name to be a purely phonetic transcription, and does not believe that 于 could be intended to represent the character *yü* 玉 meaning 'jade'.

[20] See Lévi and Chavannes, *L'Itinéraire d'Ou-K'ong*, p. 26.

[21] See Rockhill, *Life of the Buddha*, p. 233.

[22] Comp. above, p. 154, note 18; Rockhill, *Life of the Buddha*, p. 235, note 2.

[23] The earliest mention of Ilchi (*Ilitsi*) that I can find is in the report of the Chinese general who conquered Khotan with other districts of the Tārim Basin in 1759; see Ritter, *Asien*, v. p. 514.

use cannot have been general, since it is not to be found in the *Hsi yü wen chien lu*, the Chinese description of the Western Countries, published in 1777, though the various towns of the Khotan territory are there specified in some detail[24]. In Khotan itself the name Ilchi is now rarely heard. The designation 'Khotan' is generally applied κατ' ἐξοχήν to the chief town of the whole oasis, in accordance with a usage which finds its parallel in many territories of Central Asia. We may in this case safely assume it to be of very early date, since neither the Chinese Annals nor any of the old travellers, from Fa-hsien onwards, ever record any other name for the capital than Khotan[25].

Section II.—THE LEGENDARY TRADITIONS OF KHOTAN

Historical records of Khotan begin from the period of the Former Han Dynasty when, during the reign of the Emperor Wu ti (140–87 B.C.), the territory was first brought into political relations with China. The Chinese notices of that period tell us nothing of the previous history of the state. But Khotan itself preserved traditions relating to an earlier epoch and connected with the very foundation of the kingdom. Some record of these traditions has been handed down to us by Hsüan-tsang and by the Tibetan texts already described. The legendary character of these traditions becomes evident on a perusal of the two narrations. Yet their substantial agreement proves that the legends were old and of genuine local growth. A closer examination may, perhaps, still reveal to us traces of that substratum of historical fact upon which those traditions originally grew up.

Hsüan-tsang's legend of origin of Khotan.

Hsüan-tsang's *Hsi-yü-chi* tells the story of the origin of the Khotan kingdom and its dynasty, after referring to its actual ruler and his claim to descend from the god Vaiśravaṇa (*Pi-sha-mên*) or Kubera[1]. The pilgrim does not state distinctly the source from which he obtained this story. But the reference he makes in a preceding passage to the chronicles possessed by the Khotanese[2], and still more the close correspondence between his account and the Tibetan 'Annals of Li-yul' which must be based on Khotan tradition, render it highly probable that Hsüan-tsang here reproduces information received directly or indirectly from some Khotanese chronicle. His story runs thus in its main points[3]. In old times this country (of Khotan) was waste and uninhabited. The god Vaiśravaṇa (*Pi-sha-mên*) came to take up his abode there. King Aśoka's eldest son, who dwelt in Takṣaśilā (*Ta-ch'a-shih-lo*), having had his eyes put out, his royal father in anger sent one of his ministers with the order to banish the men of great family to the north of the snowy mountains and to establish them there in a desert valley. When the men thus banished arrived at the western frontier of this territory (of Khotan) they placed one of their chiefs at their head and gave him the title

[24] See Klaproth, *Mémoires rel. à l'Asie*, ii. pp. 289 sqq.

[25] It is only as a conjecture that I suggest the possibility of Ilchi being the original name of the locality to which the head quarters of the territory were transferred after the abandonment of the site of the ancient capital at Yōtkan. This transfer must be supposed to have taken place after the eleventh century; but the period cannot be fixed more accurately at present; see below, chap. VIII. sec. iii.

[1] See *Mémoires*, ii. pp. 224 sq.; *Si-yu-ki*, transl. Beal, ii. pp. 309 sqq.

[2] The reference does not appear in Julien's translation nor in that of Beal (which is based on the former); but it is found in Rémusat's rendering of that portion of the Pien i tien which faithfully reproduces the whole of Hsüan-tsang's notice of Khotan. See *Ville de Khotan*, p. 37: 'Ils ont des chroniques et leurs caractères sont, ainsi que leurs lois et leur littérature, imités de ceux des Hindous, &c.' Rémusat justly adds in a note: 'Les traditions qu'on trouvera rapportées plus bas prouveront la vérité de cette assertion: elles ne peuvent en effet être tirées que des chroniques du pays'.

[3] In the résumé given here I follow Julien's version except where otherwise indicated.

of king. At that time the son of the emperor of the eastern region (i.e. China), having been exiled, was dwelling upon the eastern frontier of the same territory (of Khotan). Urged by the people under him he also assumed the position of king.

Some time passed before the two colonies came into contact[4]. One day the two kings, while hunting, met each other in a desert. Having questioned each other as to their descent, they came to dispute about the supremacy. After angry speeches, they were about to resort to arms, when upon some one's representation they agreed each to return to his own place and to meet again in battle upon an appointed day. Having turned back, each engaged in training his warriors. Upon the appointed day the two forces were arrayed face to face. When the attack had been sounded at break of day, the western king was defeated; pursued northward, he was caught and beheaded[5].

The eastern king profiting by his victory, collected the scattered fugitives, transferred his residence to the middle of the territory and fortified it with walls. But having no (surveyed) site he feared that his plan might not succeed. When he had issued a proclamation calling for a surveyor, there appeared a heretic clothed in ashes who carried a great calabash full of water on his shoulder, and claimed to know the right method of surveying. He then walked round with the water of his calabash running out until he had completed an immense circle, whereupon he rapidly disappeared Following the traces of the water the king laid the foundations of a city and soon accomplished his plan. This became the capital of the kingdom, and it is here, says Hsüan-tsang, that 'the ruler actually resides with his court. Though the walls be not of a great height, it would be difficult to take it by assault. From ancient times to the present day nobody has ever been able to conquer it'. **Foundation of capital.**

The king built more towns, firmly established his rule and secured peace to the people, but had arrived at extreme old age without obtaining an heir. Fearing the extinction of his line he went to the temple of Vaiśravaṇa and fervently prayed for a son. Thereupon the head of the god's statue opened at the top, and there came forth a young boy. The king took him and returned with him to the palace; but while the kingdom rejoiced, the child would not drink milk and the king feared for its life. So he returned to the temple and asked the god for means to nourish the child. Then the earth in front of the statue suddenly swelled up into a shape resembling a woman's breast and the divine child drank from it eagerly. Arriving at adult age, he shed glory on his ancestors by his wisdom and courage and extended far the influence of his laws. He raised a temple in honour of Vaiśravaṇa as his ancestor. 'From that time to the present day', thus Hsüan-tsang concludes his account, 'the kings of Khotan have succeeded in regular order, have transmitted to each other the royal power, and have ruled without interruption. That is the reason why the god's temple to this day is full of rare and precious objects and is visited constantly to receive worship and offerings. From the first king having been nourished by a breast issued from the earth (*ku-stana*), the kingdom has derived its name.' ***Ku-stana*, 'the breast of the earth'.**

Before we proceed to compare the tradition here presented with the version contained in the Tibetan text, attention may be drawn to the different elements which can clearly be distinguished in the story. That part of it which relates to the miraculous birth and nourishment of Kustana, the legendary founder of the dynasty, is but an instance of that widely spread class of folklore which, based upon 'popular etymology' of a local name, provides a territory or town with its eponymous hero and with appropriately invented legends concerning **'Popular etymology' of *Ku-stana*.**

[4] Thus Rémusat, *Ville de Khotan*, p. 38.

[5] Thus Rémusat, loc. cit., p. 38, and Beal, loc. cit., p. 310.

him. All that can safely be asserted in regard to this piece of folklore is that the 'popular etymology' upon which it rests presupposes an early Sanskritization of the true old name of Khotan into *Kustana*. And, as we have seen above, this Sanskritized form is proved by documentary evidence to have been in current use in the third century of our era.

Worship of Vaiśravaṇa at Khotan.

The connexion of this mythic eponymous king with the god Vaiśravaṇa or Kubera possesses historical interest only in as far as it shows that the worship of this divinity of the Indian pantheon must have been popular in Khotan from an early period. Kubera, the Hindu god of wealth, king of the demons or Yakṣas, is a figure familiar to Buddhist mythology of all countries from Gandhāra to Japan. As he is worshipped pre-eminently among the four 'Lokapālas' or guardians of the world as the ruler of the North, his adoption by Khotan as the *genius loci* of the kingdom seems thoroughly appropriate[4]. We shall see that Vaiśravaṇa claims this position yet more conspicuously in the Tibetan version of the legend, and we have Hsüan-tsang's explicit testimony for the popularity and wealth of his shrine in the Khotan capital.

That the same tradition which credited the Khotan dynasty ruling in the seventh century of our era with a divine origin, should have assumed it to have ruled without a break and in direct descent from the time of its founder, is natural enough. We need scarcely insist on the fact that the Annals of the Later Hans record a succession of usurpers on the throne of Khotan for the years 55–75 A. D., in order to justify critical doubts as to this unbroken continuity of the reigning house. But these doubts cannot impair the significance of that salient feature of the tradition which ascribes the first occupation of Khotan to the joint settlement of two colonies drawn from the extreme north-west of India and from China.

Critical value of Khotan tradition.

Several considerations combine to invest this part of Hsüan-tsang's story with a value quite distinct from that of the rest. In the first place, it deserves attention that the tradition on this point has no direct connexion with the eponymous legend of Kustana. It is further noteworthy that, though the immigration of the Indian colony is placed in the time of Aśoka, the most renowned of the patrons of Buddhism, the introduction of Buddhist worship into Khotan is distinctly attributed, as we shall see, by both Hsüan-tsang and the Tibetan 'Annals of Li-yul' to a later period. It is thus evident that the introduction of an Indian element into the story of the genesis of Khotan could not have been a specifically Buddhistic fiction intended merely, as otherwise might be supposed, to account for the flourishing condition of the Buddhist church in Khotan. Finally, it is characteristic that, while according to the story as heard by Hsüan-tsang the union of the two colonies in Khotan was brought about by the victory of a prince exiled from China, yet there is nothing in the tradition indicating a tendency to flatter Chinese notions. Considering the close relations which had already united Khotan to China for centuries and the rapid reassertion of Chinese influence throughout Eastern Turkestān at the time of Hsüan-tsang's visit, the absence of any such tendency or of any attempt to connect some event of Chinese history with the Khotan tradition seems to encourage the belief that the latter possessed in part at least a genuine foundation.

Story of foundation of capital.

In the story told of the mysterious way in which the site for the capital of Khotan was chosen we can, of course, recognize only a piece of folklore; but this, too, is of interest,

[4] Compare, regarding Kubera: Vaiśravaṇa, also called Yakṣarāja, Dhanapati, &c. Grünwedel-Burgess, *Buddhist art in India*, pp. 45, 136 sq. For his representation in an interesting sculpture excavated in one of the Dandān-Uiliq shrines, D. II., see below chap. IX. sec. iii.

[I may point out here in passing that the distinctly non-'Aryan' features, which have caused a famous Gandhāra statue of Kubera in the Lahore Museum (see fig. 88 in Grünwedel-Burgess, loc. cit.) to be mistaken for that of an 'Indo-Scythian king', would find their best explanation in this special connexion of the god with the Scythian North].

because it bears a curious resemblance to the legend recorded in Kalhaṇa's Kashmīr Chronicle for the foundation of Pravarasena's capital, the present Śrīnagar. I have discussed this local legend of ancient Śrīnagara at length in my comments on the *Rājataraṅgiṇī* and in my 'Memoir on the ancient geography of Kashmīr' [7]. It will suffice to point out here that in it king Pravarasena II, who was anxious 'to ascertain in a supernatural way the right site and auspicious time for the foundation of his new capital', and the demon who after an adventurous encounter fulfilled his desire by laying down a miraculous measuring line, play approximately the same part as the founder of the city of Khotan and his mysterious mendicant surveyor.

Khotan legend in Hsüan-tsang's 'Life'.

Seeing the length and detail in which Hsüan-tsang has recorded for us the story of the origin of the Khotan kingdom, it must at first appear puzzling that the 'Life' of the pilgrim presents us with a summary of it differing in a very essential point. After reproducing Hsüan-tsang's account of the people of Khotan and their king in almost literal accord with the *Hsi-yü-chi*, the 'Life' tells us [8]: 'The first ancestor of the king was the eldest son of King Aśoka and resided in the kingdom of Takṣaśilā (Ta-ch'a-shih-lo). Having been exiled, he went to the north of the snowy mountains, where he led a nomad life, seeking water and pastures for his flocks. Having arrived in this country [of Khotan] he established there his residence. As he had no heir, he went one day to pray at the temple of Vaiśravaṇa, &c.' The rest of the story is then told in full agreement with the *Hsi-yü-chi*, though more briefly.

It seems difficult, in view of the greatness of the discrepancy, to assume that the different version here presented was due to a mere mistake on the part of the biographer Hui-li or of Yen-ts'ung, who completed and edited the work. Our doubt on this point must grow stronger when we find that this version coincides in the essential point with the story as told in the Tibetan 'Annals of Li-yul', which also makes an exiled son of Aśoka ultimately establish his kingdom in Khotan. Without a thorough critical analysis of all those passages of the 'Life' which differ from statements recorded in the *Hsi-yü-chi*—a task wholly beyond the competence of any one who is not a Sinologist—it seems scarcely safe to express a definite opinion as to the most likely explanation of this striking discrepancy. Is it possible that the biographer or his editor, both of them contemporaries and pupils of Hsüan-tsang [9], were aware of a different version of the Khotan tradition, communicated perhaps through the 'Master of the Law' himself? Evidence in this direction would undoubtedly add much to the critical interest of the Tibetan record to which we now turn.

Tibetan version of Khotan legend.

In the 'Annals of Li-yul,' which Mr. Rockhill's extracts have rendered accessible, the account of Khotan history opens characteristically with the Buddhist adaptation of a legend which we find also elsewhere among the popular lore of territories in and beyond the Himālaya [10]. Li-yul, originally an inhabited country, was converted into a lake by its Nāgas or Spring-deities, whom the bad treatment accorded by the people to certain Ṛṣis had angered. When Buddha visited Li-yul with a number of his disciples he enveloped the lake with rays of light; these gathered into 353 [10a] illuminated water-lilies, which marked the same number of Buddhist shrines to be built thereafter in the country. Finally the united rays encircled three times the space within which 'there will be built a great city with five towers (?) called *U-then*'. Then Buddha directed his disciple Śāriputra to pierce the lake with the butt end of his staff and Vaiśravaṇa to do the same with the end of his pike. Dwelling for seven days on Mount Gośīrṣa (called Gośṛṅga in other Tibetan texts and also by Hsüan-tsang) the Blessed One then

[7] See my notes on *Rājat.* iii. 339–349; also II. pp. 442 sqq.

[8] See *Vie de Hiouen-Thsang*, p. 279; Beal, *Life*, p. 205.

[9] Compare Julien, *Vie de Hiouen-Thsang*, pp. lxxvii sqq.

[10] See Rockhill, *Life of the Buddha*, pp. 232 sq.

[10a] [This is the figure of the text as verified by Mr. Thomas; Rockhill, loc. cit., p. 232, has 363.]

predicted that after his death the lake would dry up and become a country called *Li-yul*, with the city of 'U-then' (*Hu.then* in Mr. Thomas' transcription) as its capital.

Resemblance of Khotan and Kashmir legends.

Mr. Rockhill has compared with this tradition the legends relating to the draining of the lakes which were believed to have once occupied the present valleys of Kashmīr and Nepāl, as well as Bod-yul or Tibet. Its resemblance to the ancient legend which represents Kashmīr as having been originally the lake of Satīsaras deserves special attention, in view of that close connexion of Khotan legendary lore with Kashmīr which we shall have occasion to note hereafter. I have fully discussed the Kashmīr legend, as told by the Nīlamata Purāṇa and also by Kalhaṇa, in my 'Memoir on the ancient geography of Kashmīr'[11]. A reference to it will show how closely the rôle here ascribed to Buddha and his two companions corresponds to the part played in the Kashmīr legend by Brahman and the gods Viṣṇu and Balabhadra, who upon the former's command drained the lake by piercing the mountains. Buddha's stay on Mount Gośṛṅga seems to reflect the position taken up by Brahman and his divine host on the lofty peaks of the Naubandhana Tīrtha in the mountains south-east of Kashmīr. This close relation between the Khotan legend and the one told of Kashmīr in the Nīlamata Purāṇa is all the more noteworthy because the Kashmīr legend had assumed in Buddhistic lore a materially different form, as seen from the account of Hsüan-tsang and Mr. Rockhill's extracts from the Tibetan canon[11a]. Thus the transfer of the Kashmīr legend to Khotan cannot be attributed to any specifically Buddhist channel or agency.

Tibetan story of foundation of Khotan.

The 'Annals of Li-yul' begin the story of the foundation of Khotan with a reference to King Aśoka, who is said to have ruled over India 234 years after the death of Buddha[12]. 'At that time the lake had dried up, but Li-yul was uninhabited.' In the thirtieth year Aśoka's consort bore a son, whom the king, alarmed by the soothsayers' prediction that he would be king in his father's lifetime, caused to be abandoned. 'But when the child had been abandoned, there arose a breast on the earth from which he derived sustenance.' For this reason he was called Kustana or 'breast of the earth' (*sa.nu*)[13]. This child was miraculously carried off by Vaiśravaṇa to the king of China (*Rgya*), who had 999 sons, but wanted one more to complete a thousand, and brought up the boy. Kustana having found out his true origin 'wanted a kingdom for himself', and accordingly when twelve years old he 'got together a host of 10,000 men, and with them went to seek a home in the west, and while thus employed he came to Me-skar of Li-yul'.

About that time Yaśas[14], a minister of Aśoka, had been obliged to leave India as 'his relatives had become obnoxious to the king; so he left the country with 7,000 men, and sought a home to the west and to the east, and thus he came into the country below the river of U-then'[15]. Two followers of Kustana who had run away from Me-skar, came at *Tola*

[11] See Stein, *Rājat.*, II. pp. 388 sq.; also i. 26–27 with note.

[11a] See *Mémoires*; i. pp. 168 sqq.; Rockhill, loc. cit., pp. 167 sq.

[12] If this date refers to the year of Aśoka's accession, it approaches closely to the probable chronology of Buddha and Aśoka as ascertained by recent researches. Aśoka's accession took place about 272 B.C. and circ. 487 B.C. appears to have been in Aśoka's time the accepted date for Buddha's death; see V. Smith, *History of India*, pp. 40, 137. That the chronology of the 'Annals of Li-yul', such as it is, was derived from sources different from those which Tibetan Buddhism generally relied on, becomes evident from the fact that, according to Mr. Rockhill, none of the Northern Buddhist texts known to him place Aśoka the Pious later than a hundred years after Buddha; comp. *Life of the Buddha*, p. 233, note 4.

[13] *Life of the Buddha*, p. 234. The spelling *Kusthana*, regularly followed in Mr. Rockhill's extracts is evidently an error of transcription. The text, according to Mr. Thomas, has always *sa.nu*. The meaning indicated shows that Skr. *Kustana* is intended.

[14] Mr. Rockhill gives two forms of this name *Yaśa* and *Yāśas*; but it is evident that the Sanskrit name *Yaśas* is meant, borne by a person who figures in the traditional story of Aśoka as current in Northern Buddhism.

[15] Mr. Rockhill notes that here and elsewhere the rivers of Khotan are referred to by the expression *shel-tchab* or

upon a tract of the uninhabited land which looked inviting, and thence visited Yaśas' encampment to the south of it. When Yaśas had learned who their chief was, he sent a message to Kustana: 'Let us here unite and establish ourselves in this district of U-then; and thou shalt be king and I minister.' 'Then Kustana came with all his followers and met Yaśas in the country south (of the U-then river) which is called *Hang-gu-jo*.'

Location of first colonies.

The prince and the minister could not agree about the location of their home, and a quarrel between their hosts was imminent. But this was averted by the appearance of Vaiśravaṇa and Śrīmahādevī, to each of whom a temple was built on that very spot and who were henceforth honoured as the chief guardians of the realm. Kustana having been made king and Yaśas minister, 'the Chinese followers of Prince Kustana were established on the lower side of the U-then river and in the upper part of *Mdo me-skar* and *Skam-shed*. The Indian followers of the minister Yaśas were established on the upper bank of the river (*shel-tchu gong-ma*), and below Rgya and Kong-dzeng. Between the two (? *shel-tchu dbus*) they settled, the Indians and Chinese indiscriminately. After that they built a fortress[15a].'

In a subsequent passage 234 years are said to have elapsed from Buddha's Nirvāṇa to the time when Li-yul was founded, Kustana being then aged nineteen[16]. 'Li being a country half Chinese and half Indian, the dialect of the people (*hphral-skad*) is neither Indian nor Chinese (i.e. a mixture of the two). The letters resemble closely those of India (*Rgya*)[17]. The habits of the people are very similar to those of China. The religion and the sacred (*clerical*) language are very similar to those of India.'

Before we attempt to define what quasi-historical indications, if any, can be gleaned from this curious medley of legends, it will be well to ascertain whether any of the geographical points alluded to can be cleared up. I must frankly confess that I am as little able as Mr. Rockhill to locate *Me-skar*, *To-la*, *Skam-shed* and *Kong-dzeng*, places the names of which look queerly discordant from whatever we know of modern or ancient local nomenclature in the region of Khotan. Possibly if information could be secured as to how these Tibetan forms may have really sounded at the time when the 'Annals of Li-yul' were compiled, our task in this respect would become less hopeless. But 'the country south (of the U-then river) which is called *Hang-gu-jo*' certainly recalls the name of the present tract of *Hanguya*, situated in the extreme east of the oasis and to the north of Sampula (see map)[18].

Identification of *Hang-gu-jo*.

The location of Hang-gu-jo 'south of the U-then river' looks puzzling at first, seeing that both the rivers of Khotan, whether the Yurung-kāsh or the Kara-kāsh be meant, follow the general direction from south to north. But the Yurung-kāsh has certainly a trend towards the

shel-tchu, literally meaning 'crystal stream'. Mr. Rockhill's suggestion of a literal translation of a local term being probably intended, makes me think that possibly *shel-tchu* (*śel.chu*) may contain a rendering of *kāsh*, 'jade', found in the river names Yurung-kāsh and Kara-kāsh, or rather of some earlier equivalent of the same.

15a [For Mr. Thomas' version, which materially modifies this passage, see App. *E*.]

16 See *Life of the Buddha*, p. 237. Mr. Rockhill points out in a note that this dating does not seem to be in agreement with what a previous passage (p. 233) asserts of King Aśoka having been ruler over India 234 years after the Nirvāṇa. The Tibetan text, however, does not specify whether this date refers to the commencement or any other period of Aśoka's reign. As the latter is said to have extended over fifty-four years (*Life of the Buddha*, p. 233), an interpretation appears possible according to which the first passage was also intended to refer to the year of the legendary foundation of Khotan. This would fall within the reign of Aśoka, as assumed by the tradition of the 'Li-yul Annals', Kustana's birth being placed in the thirtieth year of Aśoka's reign and his occupation of the throne of Khotan nineteen years later. It is scarcely necessary to point out that neither the length of reign here ascribed to Aśoka nor the other dates of this tradition can claim historical value.

17 See *Life of the Buddha*, p. 236. [Mr. Thomas points out that *Rgya* by itself generally means China; see App. *E*.]

18 For Hanguya, see *Ruins of Khotan*, p. 444; below, chap. xiv.

north-east where it leaves the present limits of the oasis, and this trend must have been even more pronounced at some earlier period, if the broad 'Sai', which is passed in the desert area between the sites of Ak-sipil and Rawak and which I traced northward as far as Jumbe-Kum, marks indeed, as I believe it does, an ancient bed of the river [19]. The extensive 'Tati' around the ruin of Arka-kuduk-Tim lies just north of Hanguya, and must have formed an important site of this tract in pre-Muhammadan times [20]. A look at my map shows that, assuming the Yurung-kāsh to have followed during this period the more easterly course just referred to, the description of Hanguya as being south of the Khotan river would not have been altogether inappropriate.

Topographical allusions of legend.

The territorial division which the legend indicates between the settlements of Kustana's Chinese followers and those of the Indian exiles seems rather fantastic at first glance. Yet a reference to the real topography of Khotan helps to clear up the meaning. Though unable myself to consult the original text, I may hazard the conjecture that by 'the lower side of the U-then river' and 'the upper bank of the river' (*shel-tchu gong-ma*) which the legend distinguishes, the two branches of the Khotan Daryā, viz. the Yurung-kāsh and the Kara-kāsh, are really intended. Whatever the equivalent Khotanese terms may have been which the Tibetan text is evidently endeavouring to reproduce with the literalness usual among Tibetan translators, such an interpretation is clearly suggested by the subsequent mention of the space 'between the two', scil. rivers (*shel-tchu dbus*), in which the followers of Kustana and of Yaśas are said to have settled indiscriminately. As the former were supposed to have come from China, i.e. the east, it was natural that tradition should have assigned to them seats on the eastern branch of the Khotan river, i.e. the Yurung-kāsh, while for the Indians, whose approach to Khotan could only be assumed to have lain from the west (Karakorum route), the land west of the Kara-kāsh would appear equally appropriate. In short, I am inclined to believe that the story about the threefold division of the territory is directly based on the geographical fact of the Khotan oasis being separated by the two rivers which traverse it into three main portions, viz. the one east of the Yurung-kāsh, the one west of the Kara-kāsh, and the interfluvial tract [21].

Agreement of legends in *Hsi-yü-chi* and 'Annals of Li-yul'.

On comparing the Tibetan legend with the story as heard and recorded by Hsüan-tsang, we cannot fail to be struck by the close agreement they show in those points which for us are of essential interest. Both versions ascribe the origin of the Khotan kingdom to a joint settlement of colonies drawn from China on the one side and India on the other; both make the chief of the eastern colony assume the royal power over the amalgamated population; and in both we find the events leading to the foundation of the kingdom connected with the time of Aśoka. The only important difference between the two versions concerns just that part of the tradition in which we have already recognized an element of pure folklore, viz., the eponymous legend of Kustana. Whether the form which the Tibetan version gives to this part of the legend was known in the time of Hsüan-tsang or not, must remain uncertain. But on internal grounds it can scarcely be doubted that it represents a later development; for it is evident that the name Kustana must have been invented in the first place as a 'popular

[19] This 'Sai', with its streak of scrub, is shown on the map and was followed by me southwards to near the village of Suya. It is quite possible that the old river-bed which the jade pits of Kalta-Kumat, near Tam-ōghil village, indicate (see Hedin, *Reisen in Z.-A.*, p. 28), has its continuation further north in this 'Sai'.

[20] Regarding this large ruined area north of Hanguya, see below, chap. XIV.

[21] It deserves to be noted that the Gośṛṅga-vyākaraṇa, which gives the story of the first settlement of Khotan in a briefer form, refers to the Indian immigrants as being divided from Kustana's Chinese subjects by a stream; see Rockhill, *Life of the Buddha*, p. 234 note. [For extracts of this text see Mr. Thomas' App. *E*. The same should be consulted also for a different version of the main passage which distinguishes a 'lower' and an 'upper' river.]

etymology' for the name of Khotan, and in that case we must expect to find also the miraculous story about 'the breast of the earth' localized in that territory. In this form the story is actually recorded by the *Hsi-yü-chi*, while the point is lost sight of in the Tibetan version, which makes the miracle take place in India and brings Kustana into relation with Khotan only in a roundabout fashion.

It is not difficult to discern a motive for this transformation of the legend. Given the fact that tradition connected the foundation of Khotan with an Indian immigration about the time of Aśoka, it was quite natural that the legend under Buddhist influences should take the further step of deriving the first king from Aśoka's stock. The miraculous transfer of Aśoka's exposed son to China supplied the simplest expedient for introducing this feature into the traditional story without otherwise disturbing it in any essential point.

Buddhist colouring of Tibetan version.

The Buddhistic tendency, which is so clearly discernible in this modification of the original tradition, helps to make still more conspicuous the total absence of a similar colouring in the legend as told by the *Hsi-yü-chi*. There could have been nothing particularly flattering to the Buddhists of Khotan in the legend that they were partly descended from people whom the righteous wrath of Aśoka had banished from the extremity of his empire; nor can that part of the story which represented this Indian colony as worsted and absorbed by one from China be reasonably assumed to have been invented with a view to pleasing Buddhist or Indian notions.

I think that this complete absence of Buddhist colouring in the earlier version of the story as recorded by Hsüan-tsang possesses considerable significance. It strongly suggests the probability that the tradition was earlier than, and independent of, the introduction of Buddhism into Khotan, and that its main features had some historical background. Distant and obscure, indeed, the historical facts may seem which gave origin to the tradition, and we may, perhaps, never hope to establish them in full critical clearness. But it appears to me that the assumption of a nucleus of truth in the essential features of the tradition affords the best explanation for certain characteristic facts in the ethnic and cultural history of Khotan as we know it in the light of recent discoveries.

Historical background of legend.

The evidence to be discussed in a subsequent chapter of the Kharoṣṭhī documents from the Niya Site proves beyond all doubt that an Indian language closely allied to the old Prākrits of North-western India was in daily use for administrative purposes throughout the Khotan region about the middle of the third century A.D. Considering the character of these hundreds of documents, dealing with all the varied affairs of practical life and social organization, it is impossible to assume that their language should not have been widely, perhaps universally, known within the territory. The conclusion to be drawn from this current use of an Indian language is greatly strengthened by the Kharoṣṭhī script of the records; for we know that within India this script was peculiar to that region of which Taxila and the adjoining Gandhāra were the historical and cultural centres for centuries before and after the commencement of our era.

Early use of Indian language in Khotan.

Neither the language nor the script of these documents can be satisfactorily accounted for by the spread of Buddhism alone, which, so far as our available evidence goes, brought to Central Asia only the use of Sanskrit as its ecclesiastical language, and the writing in Brāhmī characters. But the current use in Khotan of both a Prākrit dialect and of the Kharoṣṭhī script becomes at once intelligible if we recognize a substratum of historical fact in the old local tradition heard by Hsüan-tsang, which asserted a partial occupation of Khotan by Indian immigrants from the region of ancient Taxila.

Date of supposed Indian immigration.

It is impossible in the present state of our knowledge to hazard any suggestion as to the probable date of this assumed settlement from the extreme north-west of India. But it deserves attention that both Hsüan-tsang and the Tibetan 'Annals of Li-yul' agree in making this event take place before the introduction of Buddhism into Khotan. The Tibetan text places the latter, with a semblance of accuracy which we have no means to control, in the reign of King Vijayasaṃbhava, 170 years after the establishment of the kingdom [22]. Hsüan-tsang relates substantially the same legend about the Arhat Vairocana, who was believed to have first preached the Law in Khotan, and about the oldest Buddhist convent established by him, but does not specify 'the former king' whom the saint converted [23]. On the other hand, we owe to him the explicit and significant statement that the Arhat came from Kashmīr. We shall see hereafter how much in the local traditions of Khotan and in archaeological evidence also suggests the special dependence of Khotan on Kashmīr in the matter of Buddhism and other importations from India.

Indian elements in Khotan preceding Buddhism.

Without anticipating conclusions which will have yet to be established, it is impossible here to indicate the many interesting archaeological facts which become more easily explicable if we assume that the population of Khotan had already, before the introduction of Buddhism, undergone the infusion of a strong Indian element and the cultural influences accompanying it. The history of India during the centuries which could have witnessed such a migration across the Karakorum or Hindukush is too obscure for us to expect there any trace of such an event. But if we can trust Hsüan-tsang, popular tradition must be believed to have long retained some vague recollection of it in the very region from which these Indian settlers were believed to have come.

Local legend of Taxila.

In his account of Takṣaśilā or Taxila the pilgrim relates at length the legend of Prince Kuṇāla, Aśoka's eldest son, who while governing that frontier province was believed to have been deprived of his eyesight through the intrigue of a wicked step-mother [24]. A Stūpa, still traceable near Shāh-ke-Ḍhērī, the site of ancient Takṣaśilā, was shown to Hsüan-tsang as marking the spot of this tragic event. When Aśoka had learned of his son's cruel fate, thus the story ran, he punished the guilty ministers and councillors by death or exile. Among the people who had been under Kuṇāla's charge, 'the most powerful were banished to a sandy desert, to the north-east of the snowy mountains [25].' It cannot be doubted that a reference to Khotan and its Indian colony is intended. The story of Kuṇāla's blinding can only be treated as folklore [26]. But just as the belief of a son of Aśoka having governed at Takṣaśilā can be shown to have had a foundation of historical fact [27], so is it possible that the part of the legend relating to a forced emigration from that territory to Khotan also rested on some genuine tradition.

If such a local tradition really existed at Taxila independently of that surviving at Khotan, it would, no doubt, greatly strengthen the probability of the quasi-historical character of the latter. Unfortunately we cannot hope for decisive evidence on this point. We may give credit to Hsüan-tsang for having in his account of Taxila reproduced only traditions he had actually heard there and not having treated us to information anticipated as it were from

[22] See Rockhill, *Life of the Buddha*, p. 237.

[23] Compare *Si-yu-ki*, transl. Beal, ii. p. 312, where the rendering 'a former king of the country' seems more appropriate than Julien's 'le premier roi de ce royaume', *Mémoires*, ii. p. 227.

[24] See *Mémoires*, i. pp. 154 sqq.; *Si-yu-ki*, i. pp. 139 sqq.

[25] See *Mémoires*, i. p. 161; *Si-yu-ki*, i. p. 143.

[26] See Mr. V. Smith's remarks in his *Asoka*, 'Rulers of India' Series, p. 190 note. It deserves to be noted that the legend of Kuṇāla is related in much the same form by Buddhist Sanskrit texts preserved in Nepāl, the Divyāvadāna and Aśokāvadāna, which at least shows the antiquity of this piece of folklore; see Burnouf, *Introduction à l'histoire du Buddhisme*, pp. 358, 400 sqq.

[27] Compare V. Smith, *Asoka*, p. 44.

what he subsequently gathered at Khotan. Still there remains room for suspicion whether the Taxila story, so far as it localized the banished people at Khotan, was not itself directly or indirectly a reflex of the Khotan tradition. There must have been communication between Buddhist Khotan and the Taxila region, whether through Kashmīr or Gandhāra, and we know well how far legends and traditions will travel.

A settlement of immigrants from the extreme north-west of India, such as Khotan tradition assumes, would necessarily have left its mark in the racial composition of the population. But two circumstances render it difficult for us to distinguish this mark in present anthropological facts. On the one hand, we have no certain knowledge of the racial characteristics of the population which inhabited the region of Taxila or the neighbouring territories during that early period. On the other hand, the points of anthropological affinities between the Galchas and the present Indian populations nearest to Khotan, the Dards and Kashmīrīs, are so numerous that, if we assume the supposed immigrants to have resembled the latter, their admixture could scarcely have modified in any very striking way the type of the *Homo Alpinus* which our inquiry in the preceding chapter has shown to enter so largely into the racial composition of the present Khotanese people. Nevertheless, I may note here that I was frequently struck by a certain curious resemblance in general appearance of features between the Khotanese and the Kashmīrīs, a resemblance difficult to define yet all the more noteworthy on account of the unmistakable peculiarity of type presented by the Kashmīrīs [28]. Racial traces of Indian immigration.

We have seen that the Khotan tradition, as recorded by Hsüan-tsang and the 'Annals of Li-yul', is equally explicit about an early immigration from the side of China, which was believed to have given to Khotan its first ruler and half of its original population. We have no means of applying to this part of the tradition the test of historical evidence. But I have had occasion in the preceding chapter to show that certain anthropometrical and philological observations distinctly point to an early ethnic event underlying also this part of the tradition [29]. In view of the explanations there given, a mere reference will suffice here to that early infusion of Tibetan or quasi-Mongolian blood which seems to have left so material an impress on the racial character of the Khotan population. We have further seen that the presence of such an element was sufficiently marked to make Chinese observers of the fourth or fifth century of our era ascribe to the people of Khotan a certain resemblance to their own race [30]. The fact that this element is more strongly represented in the population of the eastern part of Khotan territory (Keriya), is also significant; for it seems to confirm what the Tibetan version of the tradition relates about the territorial division between the earliest settlements of the two colonies. Tradition of Chinese element at Khotan.

The linguistic indications furnished by the non-Indian terms occurring in the Kharoṣṭhī documents, and by the 'unknown' language of certain Brāhmī MSS. probably containing Khotanese translations of Sanskrit texts, have also been noted above [31]. It only remains to call attention to the archaeological evidence which certain artistic remains of ancient Khotan supply. In some terra-cotta figurines from the Yōtkan site, in the painted tablets and frescoes of the ruined shrines of Dandān Uiliq, and in some of the relief sculptures of the Rawak Vihāra, we meet again and again with representations of quasi-Mongolian faces strangely contrasting with other features in the treatment of the human figure which are unmistakably derived from Indian models as presented by the style of Gandhāra. These peculiar traits will be duly noticed in the detailed descriptions of the relics of old Khotan art as recovered in the Traces of eastern element in Khotan art.

[28] Compare Mr. Joyce's remarks on this resemblance, *J. Anthrop. Inst.*, xxxiii. p. 317 sq.

[29] See above, pp. 148 sqq.

[30] Compare above, p. 149, for the record of the Northern Wei Annals.

[31] See above, pp. 149 sq.

course of my explorations. I need only refer here to the illustrations of the most striking specimens to be found in the Plates mentioned in the footnote[32].

Considering that the art of Buddhist Khotan can be shown to have remained to the last under the predominating influence of Indian models, the early appearance of such peculiar features as high cheekbones, oblique eyes, and short flat noses, can scarcely be accounted for otherwise than as a concession to a type actually represented among the local population. It can certainly not be explained as due to the imitation of Chinese models; for apart from the fact that China itself in matters of Buddhist art was mainly a borrower from India and Khotan, the features in question appear far too early to be attributed to any reflex movement from that side.

From the inquiry just concluded it results, I think, with considerable probability that the traditions about the origin of the population of Khotan, though legendary in their details, contain some genuine reminiscence of early ethnic movements. When these movements took place, and the exact manner in which they affected the earliest history of Khotan, we may never be able to determine. But even the dim outlines in which we can trace their effects, may help us to realize better the recorded historical facts about old Khotan, as well as the cultural *milieu* to which its antiquarian relics belong.

Section III.—KHOTAN IN CHINESE RECORDS, FROM THE HAN TO THE SUI DYNASTY

The earliest historical notices which the Chinese records furnish of Khotan begin from the reign of the Emperor Wu ti (140–87 B.C.), when Chang-ch'ien's mission had opened up the 'Western regions' to Chinese trade and political influence. These notices, and the great mass of the later ones, are contained in the dynastic Annals, from those of the Former Hans onwards. They have, as already stated in the introductory portion of this chapter[1], been translated by A. Rémusat from the chronologically arranged extracts of the *Pien i tien*, and are thus conveniently accessible for reference. Not being able myself to have recourse to the original sources, I shall reproduce these notices as briefly as possible, restricting my comments to such points as can be elucidated from my study of geographical and antiquarian facts bearing on Khotan.

Khotan during Former Han period.

The Annals of the Former Hans tell us that the first embassy from Yü-t'ien was received during the reign of Wu ti[2]. This statement is confirmed by what the T'ang shu relates of the kings of Khotan having handed down to each other the edicts and tokens of investiture received from the Middle Kingdom uninterruptedly ever since that Emperor's time[3]. The 'Notice of the Western Regions' embodied in the Han Annals states that the capital of the kingdom, known as the 'Western City,' was 9,670 li distant from the imperial capital Ch'ang-an

[32] For terra-cotta heads of the type referred to, see Plates XLIII–XLV, also Plate XI of Dr. Hoernle's *Report on C.-A. antiquities*, ii; for pictorial representations, Plates LIX, LXIV, LXXIX, also Fig. 30; for heads sculptured in stucco, Plates LV, LXXXIV, LXXXV, where the difference from heads of Buddhas and Bodhisattvas treated in the Gandhāra style is particularly noteworthy. The characteristic features of the type referred to (high cheekbones, prominent oblique eyes, flat noses) appear also in the human heads of the grotesque figures carved in wood and adorning an ancient armchair from the Niya Site; see Plate LXX.

[1] See above, p. 150.

[2] See Rémusat, *Ville de Khotan*, pp. 1 sq.

[3] Compare Chavannes, *Turcs occid.*, p. 125.

(now Hsi-an-fu)[4]. The population was reckoned at 3,300 families with 19,300 people[5], the number of soldiers being 2,400. Several high dignitaries, among them commandants of the Western and Eastern cities, are enumerated. To the south lay the territory of the Jê-Ch'iang nomads, who appear to have held the plateaus and valleys along the whole Kun-lun range from Tsaidam westwards, while to the north was that of Ku-mo (Ush-Turfau). The abundance of jade is specially noticed; the statement that all rivers west of Yü-t'ien flowed into the western ocean and those east into Lop-Nor betrays geographical confusion, of which the 'Notice of the Western Regions' shows other unmistakable signs[6].

Small territories east of Khotan.

That Yü-t'ien during the period of the Former Hans must have been a relatively weak state becomes evident from the mention which the 'Notice of the Western Regions' makes of the three small territories of *Jung-lu* 戎盧, *Yü-mi* 扜彌, and *Ch'ü-lê* 渠勒 as separate 'kingdoms' to the east of it[7]. The T'ang Annals distinctly tell us that Yü-t'ien had absorbed these territories, together with P'i-shan, since the Han period. They further show that *Yü-mi*, also called *Ning-mi* 寧彌, or *Chü-mi* 拘彌, must be identified with the oases which extend between Chira and Keriya[8]. Jung-lu and Ch'ü-lê were petty tracts at the foot of the mountains, south-east and south of Yü-mi[9]; they cannot be located with the same certainty as the latter, but correspond manifestly to the present 'Tāgh' ('hill') district, comprising the small submontane settlements which extend east and west of Polu. Considering how close these oases and hill tracts lie to Khotan, and how limited their resources are compared to those of the Khotan oasis, their political dependence upon any firm rule established at the latter would follow almost as a matter of course. The same observation applies to P'i-shan, in which we have already recognized the small oases about Gūma and Moji[10]. From the way in which Yü-mi is described in the Former Han Annals it seems safe to conclude that this little state was at that time not only independent, but also quite as powerful as Yü-t'ien.

Khotan in Later Han Annals.

The Annals of the Later Hans show clearly that it was only after the middle of the first century A.D. that Yü-t'ien rose to political importance for the Chinese. We are there informed that towards the end of the reign of the Emperor Kuang-wu ti (25–57 A.D.) King Yü-lin of Khotan had become subject to the powerful king of *So-ch'ê*, i.e. the territory of Yarkand, and had been reduced to the rank of the ruler of *Li-kuei*[11]. During the period comprising the years 58–73 A.D., however, a general of Yü-t'ien, called Hsiu-mo-pa, revolted and made himself independent as ruler of the territory. His nephew and successor Kuang-tê in turn conquered So-ch'ê and made Khotan so powerful that thirteen 'states' to the north-west, as far as Kāshgar,

[4] See for this extract *Ville de Khotan*, p. 2, and Wylie, *J. Anthrop. Inst.*, x. p. 30. It seems probable that the designation 'Western City' corresponds to that of 'city of Western mountains' (*Hsi-shan*) used for the Khotan capital in the T'ang Annals; see *Turcs occid.*, p. 125.

[5] Rémusat, loc. cit., p. 2, has '2300 maisons ou familles'.

[6] Wylie (loc. cit., p. 30 note) saw in this erroneous statement an indication that the territory of Khotan extended westwards to the Pāmīr watershed, and looked for the site of the capital of Yü-t'ien in the neighbourhood of Yarkand. This explanation is in direct contradiction to the text of the 'Notice', which specifies quite a series of independent 'kingdoms' between Khotan and Sarīkol; see above, pp. 91 sq.

[7] See Wylie, loc. cit., p. 29.

[8] Compare *Turcs occid.*, p. 125. The identification of Yü-mi with the present Keriya, which M. Chavannes, ibid., p. 128, note 1, adopts from a modern Chinese geographical work, the *Hsi yü shui tao chi*, is somewhat too limited. I shall discuss the position of Yü-mi below, chap. XIII., when dealing with Keriya.

[9] The pettiness of these 'states' is indicated by the recorded population of 1,610 and 2,170 persons, respectively. Yü-mi, on the other hand, is credited in the 'Notice of the Western Regions', with 20,040 people, and a larger number of trained troops than Yü-t'ien itself; see Wylie, loc. cit., p. 29.

[10] See above, p. 103.

[11] See Rémusat, *Ville de Khotan*, p. 3. For the probable location of *So-ch'ê* (Wylie's *Sha-keu*), at, or in the vicinity of, Yarkand, see above, p. 88. The name of *Li-kuei* cannot be traced by me elsewhere. [Mr. Thomas tentatively suggests some connexion with *Li-yul*.]

acknowledged his sovereignty. Khotan and Shan-shan (about Lop-Nor) are described as the two territories which then formed the keys to the southern route leading to China.

Khotan's submission to Pan Ch'ao, 73 A.D.

Kuang-tê was on the throne of Khotan when Pan Ch'ao, in 73 A.D., as generalissimo of the imperial forces commenced his great conquests westwards. The rulers of Shan-shan and some other states had already submitted, but Kuang-tê, encouraged by the Hsiung-nu, who had sent one of their commanders to So-ch'ê, hesitated to tender his allegiance. The execution of some emissaries whom he sent to Pan Ch'ao with an insolent demand, and the submission of Shan-shan, however, quickly removed all thought of resistance. Kuang-tê himself attacked and killed the Hsiung-nu general and made his submission to Pan Ch'ao. A Chinese garrison was established in his states[12].

Various notices in the Later Han Annals make it clear that Chinese power in the Tārīm Basin was slowly but steadily waning during the second century of our era. The efforts of Pan Yung, Pan Ch'ao's son, seem at first to have prevented a break-up. His victory over Yen-ch'i (Kara-shahr), 127 A.D., made Khotan, like the other states of Eastern Turkestān, renew its allegiance[13]. In 129 A.D. Fang-ch'ien, king of Yü-t'ien, killed the ruler of Chü-mi (Yü-mi) and gave the territory to his own son. Two years later he sent a tribute-bearing embassy to the imperial court, which had demanded the retrocession of Chü-mi[14].

Khotan troubles, 151–152 A.D.

Events recorded for the years 151–152 A.D. afford us a glimpse of the modest limits within which Chinese control was maintained at Yü-t'ien. When the governor-general Chao P'ing, representing the imperial authority, had died there, Ch'êng-kuo, the chief of Chü-mi, endeavoured to create trouble for Chien, king of Khotan, by a report that the Chinese representative had been poisoned by the royal physicians. Wang Ching, who was sent to succeed in command at Yü-t'ien, was induced by the same intriguing chief to make a treacherous attack upon Chien, in the course of which the latter was killed. In the rising which followed, the people of Khotan, under the leadership of the local governor Shu-p'o, revenged the death of their king by killing the Chinese commander, together with his guards. Shu-p'o then endeavoured to secure the crown, but was himself killed, while An-kuo, the son of Chien, succeeded to the Khotan throne. Ma Ta, the governor of Tun-huang or Sha-chou, who had been privy to Wang Ching's high-handed proceedings, then prepared to punish Yü-t'ien for its resistance, but the emperor Huan ti prohibited the attempt, and the governor of Tun-huang had to rest content with a fictitious satisfaction. The self-confidence of the people of Khotan is said to have been greatly strengthened thereby[15]. Under the last reign of the Later Han dynasty, that of Hsien ti, embassies from Khotan are mentioned in the years 202 and 220 A.D.[16]

Khotan after downfall of Han dynasty.

During the epoch of the Three Kingdoms (220-264 A.D.) Yü-t'ien appears to have been under a powerful ruler, for the Annals mention the states of Jung-lu, Yü-mi, and Su-lê as dependent on it. Nevertheless we hear of an embassy to the imperial court in 222 A.D.[17], and it is evident from the Chinese documents found at the Niya Site, which belong either to the closing years of this epoch or the very commencement of the Tsin period, that Chinese influence must have asserted itself in one form or another even after the downfall of the Hans.

During the period of the Tsin dynasties (265–419 A.D.) Chinese power can only have made itself felt in Eastern Turkestān spasmodically. It is true that we have among the Chinese

[12] See Rémusat, loc. cit., pp. 4 sq.

[13] Compare Franke, *Zur Kenntniss der Türkvölker*, pp. 70 sq.

[14] See Rémusat, loc. cit., p. 6.

[15] See Rémusat, loc. cit., pp. 6 sqq.

[16] Comp. Rémusat, loc. cit., p. 9.

[17] Rémusat, loc. cit., p. 10. This is probably the same embassy to which the *Pien i tien's* extract from the Liang Annals refers as having been sent in the reign of Wen ti (220–226 A.D.). The king of Khotan is there named *Shan-hsi*; see Rémusat, loc. cit., p. 17.

records on wood from the Niya Site one at least which belongs to the commencement of this period, bearing the date of the fifth year of the emperor Wu ti (269 A.D.). But otherwise I can trace no direct evidence of Chinese influence in the Khotan region. The compiler of the *Pien i tien* has nothing to quote about Yü-t'ien from the Tsin Annals, and only offers us a reproduction of Fa-hsien's notice. The pilgrim, who had reached Khotan about 400 A.D. after a trying journey apparently from Kuchā, found Buddhism in a very flourishing condition, and describes the glories of its monastic establishments in some detail[18].

Fa-hsien's account of Khotan.

Of the territory he tells us that it was 'a pleasant and prosperous kingdom, with a numerous and flourishing population. The inhabitants all profess our Law, and join together in its religious music for their enjoyment. The monks amount to several myriads, most of whom are students of the Mahāyāna'. Fa-hsien specially refers to the hospitable arrangements made in the Saṅghārāmas for the reception of travelling monks, and notices the custom of erecting small Stūpas in front of each family's dwelling. His description of 'the houses of the people throughout the country standing apart like (separate) stars' is a manifest allusion to the widely scattered groups of homesteads which in Khotan, as in other oases of the Tārīm Basin, make up the villages.

The *Gomatī* monastery, in which Fa-hsien with his companions lodged, contained 3,000 monks of the Mahāyāna school, and appears to have ranked at that time as the first of the great religious establishments in the kingdom. Its position cannot be identified at present, but M. Sylvain Lévi has shown that *Gomatī* was an ancient designation of the Kara-kāsh river[19]. Fa-hsien specially praises the excellent order with which the inmates conducted themselves, and then treats us to a detailed description of the great procession of sacred images which was annually celebrated in the spring, and for the sake of which he prolonged his stay for three months. On successive days, from the commencement of the fourth month, the images of the fourteen great monasteries were solemnly conducted into the city[20]. The splendidly adorned cars on which these images were moved are described as 'more than thirty cubits high' and looking 'like the great hall of a monastery'[21]. Outside the city gate each car was met by the king, who having put off his crown came barefooted to offer homage with flowers and incense, while the queen and her ladies gathered above the gate scattering flowers as the car moved through. The description of this religious festival, which lasted for fourteen days, vividly recalls the bygone splendours of Indian *rathotsavas*. Fa-hsien closes his account with a description of another great shrine, known as 'the King's New Monastery'. As its position can be determined with some probability we shall have occasion to recur to this in the next chapter.

Buddhist celebrations.

Khotan notices in Northern Wei Annals.

The notices on Khotan which the *Pien i tien* furnishes for the period of the Northern Wei Dynasty (386–534 A.D.)[22], appear to be taken partly from Sung Yün's narrative, and to a still greater extent from the *Pei shih*, a work published about 644 A.D. Those relating to the character of the country, its people, sacred shrines, &c., were, as M. Chavannes has shown, mainly borrowed by the author of the *Pei shih* from the summary account of Hui-shêng, the companion of Sung Yün on his pilgrimage (circa 519 A.D.)[23]. Leaving these to be discussed in connexion with Sung Yün's account, we may mention first some earlier historical facts

[18] See *Fâ-hien's Travels*, transl. Legge, pp. 16–20.

[19] See *Notes chinoises sur l'Inde*, iv. p. 40.

[20] The duration of fourteen days distinctly indicated for the festival shows that the number fourteen given by the Chinese copies of the text is more correct than the number four taken by Legge from the Corean version; see also Rémusat, *Ville de Khotan*, p. 14.

[21] Rémusat's version, p. 13, indicates the more modest height of three *chang* or 30 feet.

[22] See Rémusat, *Ville de Khotan*, pp. 18–28.

[23] Compare Chavannes, *Voyage de Song Yun*, p. 2.

recorded by the *Pei shih*. In 445 A.D. Khotan was invaded by Mu-li-yen, chief of the T'u-yü-hun, whom a Chinese army had driven from the Tangut country, and who took refuge westwards [24]. He is said to have killed the king of Khotan and to have effected great carnage [25]. Embassies from Yü-t'ien and gifts of presents are recorded for the years 457, 466, 467, 468 A.D. Towards the close of the reign of Hsien Wên ti (circa 470 A.D.) an envoy called Su-mu-ch'ieh arrived from Khotan to ask imperial aid against an invasion of the Juan-juan, whose cavalry was ravaging the territory to the very gates of its capital [26]. The emperor refused the requested help, on the ground of the great distance, and contented himself with sharing the hope of his ministers that Khotan would hold out behind its walls against the nomad hordes little versed in sieges.

Some time before this the king of Khotan, Ch'iu-jên-chê, had retained an ambassador of the king of Persia who was proceeding with elephants and rare presents to the Chinese court; but a remonstrance from the latter had effected his deliverance, and tribute is said to have been paid subsequently with regularity. Though a real dependence on China, then divided between the rival dynasties of North and South, cannot be supposed, embassies from Khotan are mentioned in the years 502, 507, 512, and 513 [27]. Also the Annals of the Liang Dynasty (502–556 A.D.) record such missions in the years 509, 513, 518, and 541, and it is interesting to note that among the presents brought by them were figure vases in glass and a Buddha statue of jade carved in foreign lands [28].

Khotan described in Liang Annals.

The description of Yü-t'ien which the *Pien i tien* reproduces from the Liang Annals is brief, but not without interest [29]. It mentions its numerous rivers, which lose themselves in the sands, the jade found in them, the ample produce of cereals and vines. The fruits and vegetables of the country are compared to those of China. The capital is designated as 'the city of the Western mountains'. Of the people it is stated that they are much devoted to the cult of Buddha, and very courteous in manners, to the extent of kneeling down whenever they meet. The mention made of the skill of the workers in brass and of the liberties enjoyed by the women in social intercourse has already been noted. The reference to pieces of wood used for writing has received illustration by my finds at the Niya Site, and will be discussed later. A few minor points agreeing with notices of other Annals will also be noted further on.

Sung Yün's account of Khotan (519 A.D.).

Sung Yün, whose account we find reproduced *in extenso* by the *Pien i tien*, reached Khotan in 519 A.D. from the direction of Shan-shan [30]. He relates at length the legend to be discussed in the next chapter about the first Stūpa of Khotan, erected miraculously by Vairocana. Of the territory he only tells us that it measured 3,000 li from east to west. Its king wore a gilt headdress resembling in shape a cock's crest, from which there descended behind as an ornament a band of silk two feet long and five inches broad [31]. On ceremonial occasions there were players of various musical instruments, besides bearers of swords and other arms, in his following. Of the women of Yü-t'ien, Sung Yün notes that they wore girdles, short vests and trousers, and rode on horseback like men [32]. The dead were burned, their bones subsequently collected

[24] Regarding the T'u-yü-hun and their seats in the Kuku-Nor region, compare Chavannes, *Voyage de Song Yun*, p. 11, note 4.

[25] See Rémusat, loc. cit., pp. 18, 21; *Voyage de Song Yun*, p. 16 note.

[26] See Rémusat, loc. cit., pp. 25 sqq.

[27] Ibid., p. 27.

[28] Ibid., p. 17.

[29] Ibid., p. 16.

[30] See *Voyage de Song Yun*, pp. 15 sqq.

[31] The notice of the Liang Annals describes the royal headdress as of gold and in shape resembling that of the Tartar princes; see Rémusat, loc. cit., p. 16. Rémusat's text of Sung Yün's passage mentions *two* bands of silk.

[32] Rémusat's version makes the women ride on camels as well as on horses; see loc. cit., p. 22. The point is of some interest with reference to certain pictorial representations from Dandān Uiliq. See below, chap. IX. sec. vi.

and buried under small Stūpas. The mourners cut their hair and lacerated their faces, but abandoned mourning as soon as their hair had once more grown to a length of 4 or 5 inches. The dead king's body alone was not burned, but buried in a coffin at a distant deserted spot, where it was customary to erect a funeral shrine over it and to perform sacrifices at regular periods.

Hui-shêng's notes on Khotan (519 A.D.).

The account of the *Pei shih*, which is fully reproduced by the *Pien i tien* for the Northern Wei period, and from which extracts are given again in the notices of the Northern Chou and Sui Annals, is probably to a great extent based, as already stated, upon the notes kept by Hui-shêng, Sung Yün's companion [33]. It places, correctly, Yü-t'ien at a distance of 1,500 and 1,000 li from Shan-shan (Lop-Nor) and Chu-chü-po (Karghalik) respectively; 1,400 li are reckoned northward to Kuchā; while to the south the 'Land of the Women', the Strīrājya of Indian legend, is said to be 3,000 li distant. The circuit of the kingdom is estimated at 1,000 li, and that of the capital at eight or nine li. Five large towns and some dozens of smaller places were comprised in it.

The rivers of Khotan.

The river *Shou-pa* 首拔河, from which jade was obtained, and which undoubtedly is meant for the Yurung-kāsh, flowed at a distance of 30 li to the east of the city of Yü-t'ien. A subsequent passage, manifestly drawn from some other source, gives to this river the names of *Shu-chih* 樹支水 or *Chi-shih* 言十式水, and indicates its distance as 20 li. To the west of the city, at a distance of 15 li, was another great river called *Ta-li* 達利水, which united with the former and also flowed northward. It corresponds, of course, to the Kara-kāsh, the Yangi Daryā branch of which lies actually only a little over three miles to the west of Yōtkan, the site of the ancient capital [34]. If we accept the figure of 30 li given in the first passage the distance to the eastern river is indicated with equal accuracy; for the Yurung-kāsh flows within seven miles from the eastern edge of Yōtkan. The distances here recorded are of interest as showing how little in reality the river courses within the oasis have changed during the last thirteen centuries.

The soil of the territory is described as favourable to the five kinds of cereals, as well as to mulberry and hemp. Reference is made to the wealth of the mountains in jade, and to the good breed of horses and camels. Murder alone was punished with death. The customs and products much resembled those of Kuchā. Buddhism flourished, and its shrines and monastic establishments abounded. The king was pious above all, and never failed on fast days to clean a sanctuary himself and to make his offerings. The reference to particular shrines and objects of worship which follow we shall have occasion to notice below. The curious remarks which the compiler of the *Pei shih* adds, perhaps from some other source, about the physical appearance of the people of Yü-t'ien and the defects in their character, have been discussed above at some length [35].

Khotan under sovereignty of White Huns.

Sung Yün and the *Pei shih* agree in enumerating Khotan among the numerous states of the Tārīm and Oxus basins which, at the time of the former's journey, acknowledged the sovereignty of the White Huns. From a notice of the Annals of the Liang dynasty it may be concluded that this dependence continued during the whole period of the latter (502–556 A.D.) [36].

[33] For M. Chavannes' translation, see *Voyage de Song Yun*, pp. 15 sq.; for the extracts of the *Pien i tien*, comp. Rémusat, loc. cit., pp. 18 sqq., 28 sqq.

[34] Regarding this eastern branch of the Kara-kāsh, see below, chap. VIII. sec. iii, also p. 179.

[35] See above, pp. 139, 149.

[36] Compare *Voyage de Song Yun*, pp. 26, 24, note 3; Chavannes, *Turcs occid.*, p. 224.

The Hindu pilgrim Jinagupta, who in 555 A. D. passed through Yü-t'ien on his way to China, contents himself with a bare mention of this territory[37]. The Annals of the Northern Chou (557–581 A.D.) judging from the extracts of the *Pien i tien*, have no information to offer about Khotan except a *réchauffé* of some notes from the *Pei shih* and the brief mention of an embassy in the year 574 A. D.[38] Extracts from the *Pei shih* also make up most of what the *Pien i tien* has to tell us of Khotan in the time of the Sui dynasty (581–618 A.D.)[39]. But we learn besides that the family name of the king was Wang, and his title (or according to M. S. Lévi's explanation, his name) Pei-shih-pi-lien[40]. His headdress is described as of a fabric worked with gold and lined with sable. It was forbidden to look at the king's hair, as this was popularly believed to cause a bad harvest. A mission with tribute is recorded for the year 615.

SECTION IV.—KHOTAN DURING THE T'ANG PERIOD

In chapter III a brief account has already been given of that effective reassertion of Chinese power in the Tārīm Basin which took place under the reign of the T'ai tsung, the second T'ang emperor (627–650 A. D.), and which had a determining influence on the political destinies of that region until the close of the eighth century. We have seen that from the year 648 onwards Khotan figured as one of 'the Four Garrisons' representing the main seats of Chinese authority in Eastern Turkestān. With the political importance of the territory thus recognized it is easy to understand the relative wealth of data which are furnished about Khotan by the Annals of the T'ang dynasty. We are fortunately able to consult them now in the translation of the 'Notice on Yü-t'ien' given by M. Chavannes from the original text of the T'ang shu[1].

Khotan described in T'ang Annals.

The remarks on the varying names of Khotan with which this notice opens have been discussed already, and so also those bearing on the former territories of Jung-lu, Yü-mi, Ch'ü-lê and P'i-shan, which since the Han period had been absorbed in the kingdom of Khotan[2]. The capital was still known by the designation of the 'City of the Western mountains' (*Si-shan*), as mentioned in the Han Annals[3]. The number of selected troops was estimated at four thousand. The river carrying jade is duly mentioned, and we are told the folktale that the people used to discover pieces of the precious stone in it by observing the spots where the reflexion of the moonlight was strongest. The account given of the character of the inhabitants, their industrial skill and their amusements, has been discussed already[4]. It may, however, be noted that the Annals describe not only Buddhism as flourishing but also 'the cult of the celestial god', by which the Zoroastrian religion is to be understood[5].

A subsequent passage, which appears to be identical with one previously noted from the account of the Liang Annals[6], relates to the ceremonious manners of the Khotanese which made them kneel down on meeting each other. Each time they received a private letter they

[37] See Chavannes, in *T'oung-pao*, Ser. II, vol. vi (1905), p. 341.

[38] Rémusat, loc. cit., p. 28.

[39] See Rémusat, loc. cit., pp. 30 sqq.

[40] Comp. Hoernle, *Report on C.-A. antiquities*, i. p. 7 note.

[1] See *Turcs occid.*, pp. 125–8.

[2] Compare above, pp. 153, note 11, 167.

[3] See above, p. 167.

[4] See above, p. 139.

[5] Compare Chavannes, *Turcs occid.*, p. 125, note 3, and for evidence on the point, ibid., p. 170.

[6] See above, p. 170; Rémusat, *Ville de Khotan*, p. 16.

were supposed to raise it above their head before opening it, a mark of respect, as M. Chavannes points out, reserved in China to official letters from the emperor. The reference made to pieces of wood which were used at Yü-t'ien for writing with instead of brushes, is of special interest in view of my finding such wooden pens among the ruins of the Niya Site[7]. The use of seals engraved in jade which the same passage asserts, need not surprise us, though I am not aware of any authentic find from Khotan sites to illustrate it.

Dynasty reigning at Khotan.

Of the kings of Yü-t'ien the Annals note that 'ever since the time of the Emperor Wu ti (140–86 B.C.) to our own days they have handed down, each to his successor, the edicts and insignia of investiture which had been bestowed upon them by the Middle Kingdom'. We need not accept this statement as evidence of that unbroken succession which, as we have seen, Khotan tradition in Hsüan-tsang's time assumed for the ruling house; nor are investitures, however nominal, likely to have been obtained from the Imperial court with regularity during the centuries when Chinese political influence had ceased to extend to the 'Western Regions'. Yet it seems safe to conclude from it that some recollection of the ancient political tie with China had survived in Khotan even during that period, and that popular traditions concerning it were, perhaps, more cherished there than in other territories of the Tārīm Basin.

The Annals distinctly tell us that the family name of the ruling dynasty was *Wei-ch'ih* 尉遲 and in fact the names of almost all Khotan rulers whom the T'ang Annals mention are formed with this praenomen[8]. I am unable to judge whether Wei-ch'ih could possibly represent a real Chinese name adopted by the ruling family, just as the personal names by which its individual members are referred to in the T'ang Annals are unmistakably Chinese. Otherwise a connexion might be suggested between *Wei-ch'ih* and the word *Vijaya-* which invariably forms the first part in the long string of royal names recorded by the Tibetan 'Annals of Li-yul'[9]. The question can scarcely be decided until non-Chinese documents are forthcoming to furnish us with the indigenous names of some Khotan rulers during the T'ang period.

Hsüan-tsang's stay at Khotan.

Before we proceed to examine the detailed historical data furnished by the T'ang Annals, it will be convenient to compare with the general notes just extracted the description which Hsüan-tsang has left us of Khotan and its people. The pilgrim, as we have seen, reached the oasis from the direction of Karghalik in the year 644 A.D., at a time when the predominance of the Western Turks was already broken and Chinese influence had begun to make itself felt throughout the Tārīm region. Fame was preceding the Master of the Law here as elsewhere towards the close of his great wanderings, and the king of Khotan as a devout Buddhist came to meet him at the very border of his territory[10]. Escorted by the king's son and state officers he then, after three more marches, reached the capital where another solemn reception

[7] See below, chap. xi. sec. v., and for illustrations Plate CV; also Chavannes, *Les livres chinois avant l'invention du papier* (*J. asiat.*, 1905), p. 74.

[8] See below, pp. 175 sqq., and for a list, Chavannes, *Turcs occid.*, p. 375 (Index).

[9] On phonetic grounds *Wei-ch'ih* might well be accepted as a transcription of *Vijaya*. For characters read *wei* which reproduce Skr. *vi*, compare Julien, *Méthode pour transcrire*, &c. pp. 224 sq.; for example of *ch'ih* (*tche*) representing Skr. *ji*, *je*, *dyā* (probably pronounced *jyā*), see ibid., pp. 201 sqq.

[10] It is not quite clear, from the account of the 'Life' as presented in Julien's translation (*Vie de H.-Th.*, p. 281), whether the first meeting with the king took place at P'o-ch'ieh-i (*Po-kia-i*) which, as shown above, p. 117, may be looked for near Piālma and which is described (*Vie*, p. 279) as being just within the frontier of Yü-t'ien, or at the border of the oasis proper, i.e. about Kum-rabāt-Pādshāhim. The point, of no great consequence, would be cleared up if we knew whether by 'the second day' at the end of which the pilgrim is said to have arrived within 40 li of the capital was meant the day next after the meeting or the second day of march from P'o-ch'ieh-i (*Po-kia-i*). Considering the distance from Piālma and Hsüan-tsang's preference for easy marches, the first supposition seems more likely.

awaited him. He was lodged in the convent of the Sarvāstivādin School. The 'Life' tells us that he spent altogether seven to eight months in Khotan, and that he awaited there the emperor T'ai tsung's reply to the petition he had addressed to him through a merchant caravan for permission to re-enter China.

To this long stay, no doubt, we owe the full and interesting account which the pilgrim's Memoirs furnish of the sacred sites of Khotan, its traditions and legends. We shall have occasion hereafter to analyse it in detail, as the safest guide we possess to the *topographia sacra* of Khotan and its legendary lore. In the present place we can occupy ourselves only with the pilgrim's general notice of the kingdom and its inhabitants which precedes that account. Though we have been obliged to discuss some of its characteristic passages in the preceding chapter, the interest attaching to this picture of Khotan as Hsüan-tsang saw it justifies the reproduction of the notice in full[11].

Hsüan-tsang's description of Khotan.

'The country is about 4,000 li in circuit; the greater part is nothing but sand and gravel (i.e. an arid waste), the arable portion of the land is very contracted. The latter is suitable for the cultivation of cereals, and produces abundance of fruits. The manufactures are carpets, felts of fine quality, and fine-woven light silks. Moreover, it produces white and dark jade[12]. The climate is soft and agreeable, but there are tornadoes which bring with them clouds of dust. The manners and customs show a sense of propriety and justice. The inhabitants are mild by nature and respectful, they love to study literature, and distinguish themselves by their skill and industry. The people are easy-going, given to enjoyments, and live contented with their lot. Music is much practised in the country, and men love the song and the dance. Few of them wear garments of wool and fur; most dress in light silks and white cloth[13]. Their appearance is full of urbanity, their customs are well regulated. [They have chronicles][14]. Their written characters [as well as their laws and literature] resemble the Indian model[15]; the forms have been somewhat modified, the differences, however, are slight. The spoken language differs from that of other territories. The law of Buddha is held in great esteem. There are about a hundred Saṅghārāmas containing some five thousand monks, most of whom study the doctrine of 'the Great Vehicle'[16].

Accuracy of Hsüan-tsang's description.

The preceding chapter has already shown us how accurately Hsüan-tsang's statements agree with other early records, as well as with the observations of the present day, in all that relates to the natural features of the Khotan oasis and the character, customs, and industrial occupations of its inhabitants. For the remaining remarks of the notice archaeological evidence proves equally confirmatory. The manuscript finds at Dandān-Uiliq and other sites, whether of Sanskrit or indigenous texts, have demonstrated the correctness of what the pilgrim says about the writing of Khotan and the Indian basis of its literature. The numerous remains of Buddhist

[11] I follow Julien's version, *Mémoires*, ii. pp. 223 sq. (see also *Vie*, p. 278), taking the English wording as far as possible from Beal's reproduction *Si-yu-ki*, ii. p. 309. Compare also Rémusat, *Ville de Khotan*, pp. 35 sqq.

[12] Beal translates: 'white and green jade'.

[13] It is highly probable that by 'the white cloth' ('drap blanc', Julien; 'white linen', Beal; 'laine blanche', Rémusat) is meant the white cotton cloth or 'Khām' of Khotan, which is still the regular clothing material for the great mass of the people; see above, p. 134.

[14] This important passage is found only in Rémusat's rendering, *Ville de Khotan*, p. 37.

[15] The words in brackets are taken from Rémusat's version; Julien omits them. Beal translates: 'and their mode of forming their sentences.'

[16] Julien and Beal's translations make all monks of Khotan followers of the Mahāyāna; this error has been corrected by Dr. Franke, *Sb.P.A.W.*, Berlin, 1903, p. 742. The mention of the Sarvāstivādin convent in Hsüan-tsang's 'Life' (*Vie*, p. 282), and the reference to the introduction of this school in the 'Annals of Li-yul' (Rockhill, *Life of the Buddha*, p. 239), prove that the Hīnayāna was also represented at Khotan, though the Mahāyāna system was, no doubt, in overwhelming preponderance.

shrines brought to light in the course of my explorations, and the relative ease with which the sacred sites mentioned by Hsüan-tsang can still be traced, illustrate in the same way the truth of all he has to say about Buddhist worship in Khotan.

Kings of Khotan 632–648 A.D.

Of the reigning king Hsüan-tsang tells us that he was extremely courageous and warlike, and greatly venerated the law of Buddha. Referring to the king's claim to be descended from the god Vaiśravaṇa, the pilgrim then proceeds to relate the legend of the first settlement of Khotan and of Kustana's birth, which we have already analysed in detail. Neither Hsüan-tsang nor the Annals give us the name of the king ruling in 644. It thus remains doubtful whether he was identical with one or other of the chiefs whose names are recorded for the years 632 and 648. At the earlier date the T'ang Annals mention an embassy from Wei-ch'ih Wu-mi, bringing presents to the imperial court, and note that this king was originally subject to the T'u-chüeh (Western Turks)[17]. In 635, or, according to the Chiu T'ang shu, in 639 A. D., the same ruler sent his son, who was enrolled in the imperial guard[18]. After the subjugation of Kuchā in 648 A. D., Fu-tu Hsin, the prince reigning at Khotan, felt intimidated, and sent his son with three hundred camels as an offering. The subsequent arrival at Khotan of a Chinese officer, Hsieh Wan-pei, with some light cavalry, sufficed to induce the king to proceed himself to the imperial court to present his submission. The emperor Kao tsung, who had in the meantime ascended the throne (649 A.D.), bestowed honorary appointments in the imperial guards upon him and his son Shih-hu (Jabgu) Tien, and dismissed him with rich presents[19]. Embassies from Khotan are also mentioned for the years 636, 642, 644, 645 A. D., but without the name of the king who sent them[20].

Khotan included in the 'Four Garrisons', 648/9 A. D.

In 648 or 649 A. D. Khotan was officially placed under the protectorate of An-hsi, then established at Kuchā, and thus became one of the 'Four Garrisons' (including also Kuchā, Kāshgar, and Tokmak), which were to be controlled from that centre[21]. We have seen already with reference to Kāshgar that the Chinese administrative organization extending over the whole of the Tārīm Basin did not interfere with the rule of individual territories by local dynasties. This observation is borne out also by whatever historical notices we gather about Khotan during the period while this Chinese protectorate lasted.

The latter may be said to have been finally established in 659 A. D., when the last remnants of the supremacy of the Western Turks were swept away by the imperial armies[22]. In the same year a Chinese force vanquished the Turkish chief Tu-man, who, at the head of the states of Kāshgar, Chu-chü-po (Karghalik), and Ho-p'an-t'o (Sarīkol), had revolted and attacked the kingdom of Yü-t'ien[23]. Some years later another rebellion, in which the tribe of the Kung-yueh and the ruler of Kāshgar appear to have been principally concerned, received support from the Tibetans. The latter then seem to have made their first attempts in the direction of Eastern Turkestān, and it is interesting to note that it was the king of Khotan who had to be succoured by Chinese troops from their attack in the year 665[24]. The mention made on this occasion of

[17] See *Turcs occid.*, p. 126. According to the *Ts'ê fu yüan kuei* (Chavannes, *Notes add. sur les Tou-kiue*, p. 4; *Ville de Khotan*, p. 67), the gift consisted of a jade girdle.

[18] *Turcs occid.*, p. 126, note 2.

[19] Chavannes, *Turcs occid.*, pp. 126, 178; *Notes addit.*, p. 18.

[20] Embassies in the first three years are quoted by the *Pien i tien* (*Ville de Khotan*, pp. 67 sq.) from the *Ts'ê fu yüan kuei*, but M. Chavannes' extracts from this encyclopaedia show a mission only for the year 645; comp. *Notes addit.*, p. 13.

[21] See *Turcs occid.*, pp. 118, 268; *Notes addit.*, p. 19; also above, p. 60.

[22] Compare Chavannes, *Turcs occid.*, p. 268; above, p. 60.

[23] See *Turcs occid.*, p. 72.

[24] *Turcs occid.*, p. 122, note.

a Chinese commandant of Yü-t'ien shows that the territory had already received an imperial garrison[25].

Reign of Wei-ch'ih Fu-tu Hsiung (circ. 674–690 A.D.).

About the year 674–675 A.D. Fu-tu Hsiung, king of Khotan, is recorded to have personally paid homage at the Chinese court, accompanied by his sons, younger brothers, and high dignitaries. In return for services against the Tibetans, the emperor paid him the compliment of constituting his territory into the Government of *P'i-sha* 毗沙, called after the god Vaiśravaṇa (P'i-sha-mên in Chinese transcription), and appointing him its governor. The Annals mention the division of the territory into ten districts, but unfortunately fail to give us their names[26]. On the death of Wei-ch'ih Fu-tu Hsiung, his son Ching was placed on the throne by the Empress Wu in the year 691[27]. An embassy from this ruler in the year 717, according to the *Ts'ê fu yüan kuei*, brought as presents two horses trained for the game of polo, a wild camel of remarkable fleetness, and a leopard[28].

It must have been under his reign that the garrison town of *K'an* 坎城, belonging to Khotan, and three hundred li to the east of the capital, was, circ. 705–706 A.D., raided with rich reward in plunder by the Turkish chief Ch'üeh-cho, as recorded in the biography of Kuo Yüan-chên, then 'Great Protector of An-hsi'[29]. Far more serious, however, during the latter part of his reign must have been the danger to Khotan from the Tibetans. The latter are known from the Annals to have annually harassed the Chinese borders from 714 A.D. onwards; and in the interesting Chinese sgraffito of the Endere shrine, dated in the year 719, we have definite evidence that the inroads of these formidable adversaries extended also to the confines of the protected state of Khotan[30]. In the same year an imperial decree bestowed upon Yü-t'ien, as on the other three states then counted among the four garrisons (Kāshgar, Kuchā, Kara-shahr), the right of levying duties from the merchants of the 'Western Regions' and of utilizing the proceeds for their own purposes[31].

Tibetan inroads (circ. 714–719 A.D.).

A notice of the Tzŭ chih t'ung chien acquaints us with a Khotan king whose name does not appear in the T'ang Annals. It relates that Wei-ch'ih T'iao, who in secret alliance with the T'u-chüeh and other Hu tribes was preparing a revolt, was in 725 seized and executed by the Deputy Protector of An-hsi[32]. The successor, whom the latter is said to have nominated, was in all probability Wei-ch'ih Fu-shih-chan, whose brevet of investiture as king of Khotan, dated in the first month of the sixteenth year of the K'ai-yüan period (728 A.D.), is actually among the documents translated by M. Chavannes from the *Ts'ê fu yüan kuei*[33]. The brevet mentions at the head of the king's official titles that of '*a-mo-chih* 阿摩支 of Yü-t'ien'.

A-mo-chih, title of Khotan ruler.

This title of *a-mo-chih* is of interest, as by it we find the ruler of Khotan designated in a Chinese document from Dandān-Uiliq dated 768 A.D.[34]. It appears also in a decree of the same year which bestowed the throne of Kāshgar on An-chih, *a-mo-chih* of Su-lê[35].

Khotan rulers from circ. 736 to 760 A.D.

Fu-shih-chan was succeeded by Fu-tu Ta, apparently about 736 A.D., as an imperial decree of that year records the grant of the title of princess to his consort, the dame Chih-shih[36]. Wei-ch'ih Kuei, whom the Annals name as the next ruler of Khotan, must have succeeded

[25] See *Ville de Khotan*, p. 69.

[26] See Chavannes, *Turcs occid.*, p. 127; *Notes addit.*, p. 23.

[27] For the date, see *Notes addit.*, p. 24.

[28] Compare *Turcs occid.*, p. 127; *Notes addit.*, p. 34.

[29] See *Turcs occid.*, p. 185. This garrison town of *K'an* might possibly be identical with the *Kan chou* which the notice of Kao Chü-hui's mission (938–942 A.D.) mentions as a town east of Yü-t'ien; see Rémusat, *Ville de Khotan*, p. 79. For a probable identification see below, chap. XIII. sec. ii.

[30] Compare below, chap. XII. sec. ii.; Bushell, *The early history of Tibet*, p. 26.

[31] *Turcs occid.*, p. 114.

[32] *Turcs occid.*, pp. 82 note, 311.

[33] See *Turcs occid.*, pp. 127, 207.

[34] See below, chap. IX. sec. v.; Hoernle, *Report on C.-A. antiquities*, ii., p. 24.

[35] See *Turcs occid.*, p. 208.

[36] See *Turcs occid.*, p. 127; *Notes addit.*, p. 56.

within a few years, for a similar decree in favour of his wife is recorded for the year 740[37]. Of Wei-ch'ih Kuei's son and successor Shêng we possess somewhat fuller data, a special biography being devoted to him in the T'ang Annals[38]. We learn from them that during the period T'ien-pao (742–755 A. D.) this Khotan ruler came to present offerings to the emperor and received from him an imperial princess in marriage. After returning to his territory he helped Kao Hsien-chih to attack and vanquish Sa-pi-po-hsien. In the latter M. Chavannes has recognized with good reason the chief of Little P'o-lü or Gilgit-Yasīn, against whom, as we have already seen, Kao Hsien-chih directed his famous campaign across the Pāmīrs in 747 A. D.[39]. In the latter year, as well as in 748, embassies from Khotan with presents for the imperial court are recorded[40]. Shêng testified his attachment to the imperial house still further by leaving, in 756, his territory in order to support with five thousand horse the emperor Su tsung in his desperate struggle against the pretender An Lu-shan. Shih-hu (Jabgu) Yao, his younger brother, to whom he had entrusted the charge of his state, was in 760 appointed second in command of the 'Four Garrisons', with the task of carrying on the government of the Khotan kingdom.

Wu-k'ung's visit to Khotan, circ. 786 A. D.

Shêng himself died in China, while his brother was still ruling, about 786, when Wu-k'ung passed through Khotan. The pilgrim's itinerary duly mentions him by the name of Wei-ch'ih Yao, and by his side the deputy-governor Chêng Chü, evidently a Chinese official[41]. The special interest of Wu-k'ung's notice lies in the fact that it relates to the very close of the period of T'ang dominion in Eastern Turkestān. When sketching the history of this dominion in chapter III, we have already seen how the advance of the Tibetans east of the Tārīm Basin had, during the last third of the eighth century, rendered more and more difficult the maintenance of Chinese authority by the officials and troops left behind in the 'Four Garrisons'[42].

End of Chinese supremacy, circ. 791 A. D.

Wu-k'ung's record and the Chinese documents brought to light from the ruins of Dandān-Uiliq, bearing dates from 768–790 A. D., afford valuable testimony to the fact that Chinese administrative influence made itself felt at Khotan to within a year of the date (790–791) when all connexion between the empire and the 'Four Garrisons' was finally broken by the Tibetan occupation of Pei-t'ing (Bĕshbalik). We know that the direct route between the southern oases of the Tārīm Basin and Kan-su had become closed long before through the advance of the Tibetans. A sidelight is thrown on the resulting insecurity of communication by a curious story which closes the notice of Yü-t'ien in the T'ang Annals. It relates how Chu Ju-yü, a palace official sent in 780 A. D. to Khotan to purchase jade articles for the emperor Tê tsung, was on the return journey robbed of his precious acquisitions by marauding Hui-ho (Uigurs)[43].

Section V.—LATER CHINESE RECORDS OF KHOTAN

From the year 790–791 Chinese records cease to furnish any information on the region once comprised in the 'Four Garrisons' for nearly one and a half centuries. As the notice of the Posterior Tsin Annals to be discussed presently tells us, 'the troubles which agitated China

[37] *Turcs occid.*, p. 127; *Notes addit.*, p. 61. The Annals give the name of the princess as *Ma*; the record of the *Ts'ê fu yüan kuei* calls her *Wei*, perhaps an abbreviation of the royal family name.

[38] *Turcs occid.*, p. 127, with note 4.

[39] See above, pp. 8 sqq.

[40] *Ville de Khotan*, p. 71; Chavannes, *Notes addit.*, p. 80.

[41] See Chavannes and S. Lévi, *L'Itinéraire d'Ou-K'ong*, p. 27.

[42] See above, pp. 63 sqq.

[43] See Rémusat, *Ville de Khotan*, pp. 72 sq.

precluded even a thought of the barbarians in the four quarters of the world. Those even who [before] had never ceased to be in relations with China, became barely known by name, and nothing was learned of the commencement and end of the reigns of their princes.'[1] We know that Tibetan predominance replaced Chinese control throughout the Tārīm Basin from the close of the eighth century, but there is nothing to give light as to the manner and extent in which it asserted itself generally, or how it affected Khotan.

Tibetan predominance in Tārīm Basin.

The rising power of the Uigurs appears from about 860 A.D. onwards to have gradually pressed the Tibetans southwards, and to have established itself in their place in the north-eastern part of the Tārīm Basin. But the kingdom founded by the Uigurs, with its main seat about Turfān and Urumchi, did not extend westwards beyond Kuchā, aud certainly stopped short of the limits of Khotan[2]. There is nothing to indicate that the latter had lost its local dynasty during the period of Tibetan ascendancy, and it is likely that the latter was checked and balanced to some extent by the activity of the Turkish tribes holding the valleys and plateaus of the T'ien-shan, whom the Chinese rule had controlled but never completely enfeebled. Whatever degree of independence Khotan retained during this period, it is certain that Tibetan encroachments succeeded in raising an effective barrier to any direct intercourse between its rulers and the Chinese court until the year 938.

Chinese mission to Khotan, 938–940 A.D.

It was then that the arrival of an embassy with presents from Khotan was again recorded by the Annalists of the 'Five Dynasties'. In response the emperor Kao tsu, of the Posterior Han Dynasty, granted to Li Shêng-t'ien, the ruler of Khotan, the title of king, and dispatched a mission composed of Chang K'uang-yeh, Kao Chü-hui, and some other officers, to notify to him this imperial favour. The account of this mission, as reproduced by the *Pien i tien*, is a document of considerable geographical and historical interest[3]. The detailed way in which the route followed through Kan-chou, Su-chou, Sha-chou, and then through the desert to the eastern confines of Khotan territory is there described, shows best, perhaps, how completely this great line of communication, once the high road from China to its Central Asian dominions, had fallen into oblivion during the preceding two centuries. This is not the place to examine the valuable topographical and ethnographic data which the report of Kao Chü-hui furnishes[4]. It must suffice to mention that the mission from Ling-chou, on the Huang-ho, up to the frontiers of Yü-t'ien found everywhere from place to place Tibetan tribes and encampments, and the people of Yü-t'ien engaged in a constant struggle against the Tibetans[5]. The great change which had taken place in the communications between China and Khotan since the time of the early T'angs is illustrated also by the time which the mission took to effect its object. Two years were spent on the journey to Khotan, and having started about the close of 938 the party did not return until 942 A.D.[6] The slowness of this progress, due, no doubt, to unfavourable local conditions, will be appreciated if the fact is recalled that in 644–645 it had taken the messenger dispatched by Hsüan-tsang to the imperial court at Hsi-an-fu only seven or eight months to return to Khotan with the emperor's reply[7].

Renewed diplomatic relations.

It was, no doubt, the hope of securing help against the Tibetans which had induced the Khotan ruler to revive the old allegiance of his state to China. Accordingly, Chang K'uang-yeh

[1] Compare *Ville de Khotan*, p. 74.

[2] Compare Grenard, *Mission D. de Rhins*, ii. pp. 47, 49.

[3] See Rémusat, *Ville de Khotan*, pp. 74–81.

[4] For earlier analyses, compare Ritter, *Asien*, i. pp. 212 sqq.; v. pp. 375 sqq.; Richthofen, *China*, i. p. 536.

[5] *Ville de Khotan*, p. 81.

[6] See *Ville de Khotan*, p. 75. (The seventh year there named is, no doubt, that of the regnal period *T'ien-fu* (936–944 A.D.), not the seventh year from the mission's start as assumed by Richthofen, *China*, i. p. 536.)

[7] See *Vie de H.-T.*, pp. 285, 288.

and his companions were on their arrival at Khotan at once met by Li Shêng-t'ien's demand for a treaty. Its conclusion explains the renewed presentations of tribute which are recorded for the next few decades, however limited the practical advantages which Khotan could derive from the distant empire.

The report of the mission represents king Li Shêng-t'ien as wearing clothes and headdress resembling those used in China. His palace, called *Chin-ts'ê-tien*, had all its structures facing eastwards, and among them a pavilion called that of 'the Seven Phoenixes'. We are next told of the grape wine of Yü-t'ien and of two other 'wines', violet and blue in colour, which Kao Chü-hui did not know the composition of, but found much to his taste. Rice prepared with honey, and millet cooked with cream were among the local dainties. The inhabitants wore clothes of linen and silk, and cultivated flowering trees in their gardens. They paid worship to the spirits, but most of all to Buddha. Li Shêng-t'ien in his palace was always surrounded by fifty monks wearing violet robes. The year of his reign corresponding to 940 A. D. was designated as the 29th year T'ung-ch'ing—a remark which manifestly relates to the use of some local era. Neither this nor the names *Yin chou*, *Lu chou*, *Mei chou*, noted as local designations for the south-eastern portion of his territory, can at present be elucidated[8]. Report of Mission on Khotan in 940 A. D.

The jade district was placed in the mountains 1,300 li to the south of Yü-t'ien, a location which would not be far wrong if the jade mines of the upper Kara-kāsh valley are meant. The statement that the Huang-ho rises in the Khotan mountains, is due to that piece of mythical geography still current in China, which supposes a subterraneous connexion between the Lop-Nor and the headwaters of the great river. Regarding the jade-carrying rivers of Khotan, we have a statement in which fiction and exact topography are curiously mixed. One river coming from the jade mountains is said to divide itself at Yü-t'ien into three branches. The easternmost was called the river of White Jade, the one to the west the river of Green Jade, and the westernmost the river of Black Jade, each according to the colour of the jade it was supposed to carry. It is clear that we have in the first and the last designations the exact equivalents of the present Turkī river-names Yurung-kāsh and Kara-kāsh, while the second was manifestly applied to the eastern branch of the Kara-kāsh, now known as 'Yangi-Daryā'[9].

Another and evidently fuller extract from the report of the same mission, which Rémusat has reproduced from the portion of the *Pien i tien* containing 'anecdotes on Yü-t'ien', supplements the description of the jade river and its three branches by some interesting details[10]. We read there that the jade river rises in the Kun-lun mountains, and after flowing westwards for 1,300 li enters the confines of Yü-t'ien near 'the mountain of the Ox-head'. The latter designation, as we shall see below, was borne by the Kohmārī Hill, the site of a famous sanctuary on the right bank of the Kara-kāsh where it enters the plains. The 'river of White Jade' is placed by this extract 30 li to the east of the city of Yü-t'ien, the 'Green Jade River' 20 li to the west of the latter, and the 'Black Jade River' 7 li further to the west. The distances indicated prove to be remarkably exact if referred to Yōtkan, and afford useful evidence for the location at the latter site of the ancient capital of Khotan. We are also told of the custom which precluded the people from searching for jade after the summer floods until the king in the autumn had personally visited the dry bed of the river and thus formally opened what was called 'the jade harvest'. Topographical data of report.

[8] The second part of these names may, perhaps, contain the Chinese term 州 *chou* 'city'.

[9] For a reference to this river-bed, for which, in spite of its modern name 'The New River', a respectable antiquity must be claimed, see above, p. 171.

[10] *Ville de Khotan*, p. 112.

Embassies from Khotan, 942–971 A.D.

In 942 a fresh envoy, the commander Liu Tsai-shêng, arrived from Khotan with presents, among them jade pieces to the weight of a thousand pounds. Other embassies are recorded for the years 947 and 948 [11]. Li Shêng-t'ien was still reigning when three later tribute-bearing missions arrived from Khotan in the years 961, 965 and 966 [12]. In the case of the first we read of presents offered by Mo-ni, the spiritual councillor of the king, apart from the royal tribute. It is interesting to note, with regard to the assertion of the ecclesiastical element in these missions, that the embassy of 965, which brought specially rich presents in jade, horses, camels, &c., was accompanied by two Buddhist monks from Yü-t'ien. The Chinese Buddhist pilgrim Tao-yüan also availed himself of this opportunity for regaining his native land after travels in the 'Western Countries'. In 966 it was a son of Li Shêng-t'ien, called Tê-tsung, who presented the tribute of Yü-t'ien.

In 969 the king Nan-tsung-ch'ang is named as the sender of a mission conducted by Chih-mo-shan, and accompanied by one of the Buddhist monks who had previously visited the imperial court. A magnificent piece of jade weighing not less than 237 pounds was to be offered to the emperor on condition of his sending some one to fetch it. In 971 it was again a Buddhist priest (Chi-hsiang) who brought a letter from the king of Khotan, offering to send in tribute a dancing elephant which he had captured in a war against the kingdom of Kāshgar [13].

Struggle of Khotan against Muhammadan conquest.

The mention of this war is the only indication we receive from Chinese sources of the great struggle which finally resulted in the conquest of Khotan by the Turkish rulers of Kāshgar and its conversion to Islām. But though the Sung Annals are silent on the events which closed the epoch of Buddhist Khotan and fundamentally changed the cultural history of the territory, they help us indirectly to determine their chronological limits. The next notice furnished by the *P'ien i tien's* extracts relates to an embassy from Yü-t'ien in the year 1009; and of this we read that it had been sent 'by the king or *hei-han* of that territory', and that the ambassador bringing the tribute was a *hui-hu*, i.e., probably a Muhammadan Turk, called Lo-ssŭ-wen [14]. The title Hei-han used here is a transcription of the Turkish title Khākān (Khān) and, in conjunction with the nationality of the envoy, leaves no doubt as to the change which must have taken place since 971 in the race and religion of the rulers of Khotan [15].

Muhammadan conquest of Khotan.

The conquest of Khotan for Islām plays a very prominent part in the legendary traditions still current throughout Eastern Turkestān. But unfortunately the strictly historical information to be gathered from Muhammadan sources about this important event is extremely scanty. Judging from the critical researches which M. Grenard has devoted to the elucidation of the history of the early Turkish dynasty of Kāshgar [16], the accounts of Muhammadan historians appear to furnish only one definite fact, viz. that Khotan in 1006 was held by Yūsuf Qadr Khān, a brother or cousin of Abū'l-Ḥasan Naṣr Īlik Qarā Khān, the then ruling head of the Turkish dynasty of Kāshgar and Balāsāghūn. Of the manner in which the latter extended its power to Khotan we are told nothing; but the mention of Yūsuf Qadr Khān is of value, for it proves that the legendary account to be found of this conquest in the Tadhkirah of Satok Boghra Khān, the first Muhammadan ruler of the family and the hero of popular tradition throughout Turkestān, does not altogether lack elements of historical truth.

[11] *Ville de Khotan*, p. 82.

[12] *Ville de Khotan*, pp. 83 sqq. and p. 102 which shows Li Shêng-t'ien as the sender.

[13] *Ville de Khotan*, p. 86.

[14] See *Ville de Khotan*, pp. 86 sq.

[15] The significance of the notice has been already pointed out by M. Grenard, *Mission D. de Rhins*, ii. p. 50; *J. asiat.*, 1900, xv. p. 64.

[16] See his valuable paper *La légende de Satok Boghra Khan*, in *J. asiat.*, 1900, xv. pp. 1–79.

Traditions of wars against Buddhist Khotan.

According to this text, of which M. Grenard has published an exhaustive analysis in the paper already quoted, the struggle with the infidels of Khotan commenced during the reign of Ḥasan Boghra Khān[17]. The latter corresponds to the historical Abū'l-Ḥasan Naṣr, though the legend represents him as a son of Satok Boghra Khān, while his real relation was that of great-grandson[18]. The ruler of Khotan, who in the legend usually figures under the title of 'Jagālū Khalkhālū of Māchīn', together with Chuqtah Rashīd and Nuqtah Rashīd, chiefs or ministers of the same country, attacked Kāshgar and were only repulsed after a severe siege. In the pursuit which followed, 'Alī Arslān Khān and other pious champions from Satok Boghra Khān's race were slain by the infidels. Yarkand was then converted by the strength of the sacred word and joined the cause of Islām. Subsequently, while Ḥasan Boghra Khān was warring in Western Turkestān to re-establish the Faith there, Kāshgar fell off from Islām. The Sultān then sent his brother Qadr Khān to Madā'in to implore the help of the Four Imāms, and hastened back himself to retake Kāshgar. He vanquished the infidels under Chuqtah Rashīd and pursued them to Yangi-Hisār, but was himself killed there by Nuqtah Rashīd. At this juncture Yūsuf Qadr Khān appeared at Kāshgar, with a great host of pious warriors sent by the Imāms. The infidels retreated to Khotan, where Yūsuf Qadr Khān attacked them with forty thousand men. After a siege of twenty-four years the city was taken and Jagālū Khalkhālū killed. Yūsuf Qadr Khān is then said to have reigned in peace. But the story of the Four Imāms, which is reproduced as a supplement to the legend, relates a fresh rising of the infidels at Khotan, in the course of which those four saints suffered martyrdom. In the end Qadr Khān returned with an army and accomplished the definite subjection of Khotan.

Historical data about conquest of Khotan.

Leaving aside all legendary details, we may safely assume with M. Grenard that the account given in the Tadhkirah indicates a long and difficult contest on the part of Ḥasan Boghra Khān (Abū'l-Ḥasan Naṣr) with the Buddhist kingdom of Khotan. The Kāshgar ruler had attained the throne in 993 A. D., and his struggle with Khotan was probably, as M. Grenard suggests, the cause why he delayed until 999 to assert by force his claims to the Trans-Oxus provinces of the Samanide empire which had long been awaiting disruption[19]. The final conquest of Khotan, effected under Sultān Abū'l-Ḥasan's orders by his brother or cousin Yūsuf Qadr Khān, may well have taken place immediately after the success obtained in Western Turkestān towards the year 1000, the date actually indicated by the Tadhkirah of the Four Imāms[20]. In 1007 Yūsuf Qadr Khān assisted his kinsman with forces brought from the newly conquered parts of Eastern Turkestān in the disastrous battle fought against Maḥmūd of Ghazni near Balkh. It is on this occasion that his lordship over Khotan is distinctly mentioned by the Arab historians[21].

Severity of struggle with Khotan.

In the absence of all definite indications, we are reduced to conjecture as to the particular historical causes which seem to have made the extension of Turkish rule and of Islām over Khotan a specially arduous undertaking. It is possible that a war of conquest, commenced, no doubt, in the first instance mainly from dynastic motives[22], developed into a religious struggle in which Khotan may have received aid from its Buddhist neighbours on both the south and the east. Tibet itself, though no longer a well-organized aggressive power, may in face of a

[17] See Grenard, loc. cit., pp. 12 sqq.

[18] Compare Grenard, loc. cit., p. 50.

[19] See Grenard, loc. cit., p. 67.

[20] See Grenard, loc. cit., p. 68.

[21] Compare the texts of Al-'Utbī and Ibn-al-Athīr quoted by Grenard, loc. cit., p. 69.

[22] Abū'l-Ḥasan Naṣr, in a letter addressed to Sabuktagin of Ghazni about 996 (see Grenard, p. 55), claims to seek for glory alone in fighting the infidels. But the purely dynastic aims which guided this champion of Islām and his more famous ancestor Satok Boghra Khān in their policy of aggression have been well exposed by M. Grenard; see loc. cit., pp. 41 sqq., 63.

common danger have supported its old adversary through the chiefs of apparently Turkish race established in Ladāk and other Western portions of its territory[23]. The possibility of help from the powerful Uigur kingdom, which retained its Buddhism for centuries later and was the hereditary enemy of the Khāns of Balāsāghūn, is not to be ignored altogether; and, finally, inherited wealth and the great natural resources of the Khotan region may well have enabled its rulers to organize a strenuous resistance by means of mercenary forces, however unwarlike the population actually cultivating the oasis is always likely to have been.

Part of Buddhist church in political struggle.

In the face of such obscurity as hides from our view the internal condition of the kingdom and its political relations during the two centuries preceding the introduction of Islām, we must rest content with two observations. In the first place, it is evident that the frequent missions to the Chinese court from 938 onwards, with their rich offers of tribute, may be attributed to the need felt for protection against a growing external danger. In the second place, the repeated appearance of Buddhist ecclesiastics in these missions lends support to the belief that the Buddhist church was then an important political factor in the state and, perhaps, one which largely helped to make resistance against Muhammadan aggression the more determined and effective.

For Khotan, which had undergone the influences of Indian culture and of Buddhism longer than probably any part of Central Asia, and where transplanted elements of Chinese civilization, too, seem to have found a more congenial soil than elsewhere in the 'Western Regions'[23a], the Muhammadan conquest must have signified a thorough break in the continuity of historical development. No records of any kind help us to realize its immediate effects upon the condition and social organization of the people. But there can be no doubt that the change must have meant the loss of much that had once given importance to the small state at the foot of the Kun-lun.

Resumption of trade relations with China, eleventh century.

It still retained the advantages of its position on a great route between China and the West, and efforts to utilize these for trade purposes may account to a great extent for the frequent references which the Sung Annals make to embassies from Khotan during the eleventh century. Already in the notice of the 'tribute' offered from Yü-t'ien in 1025 articles are enumerated which manifestly were introduced merely as objects of commerce[24]. The rich counter-presents accorded for such articles as were accepted by the emperor, and the manifold privileges enjoyed by the envoys during their stay in China, formed undoubtedly the main inducement for the frequent visits of such embassies. From the period 1068–1077 they are said to have become so numerous that not a year passed without their arrival, sometimes even twice in the same year; but the same notice also plainly shows that these so-called 'embassies' were often in reality little more than mercantile ventures[25]. They are represented as coming often without any credentials; they openly brought goods for sale in the markets of the empire and realized large profits. As their entertainment at public expense and the free transport of

[23] M. Grenard (loc. cit., pp. 64 sqq.) seems inclined to assume that Khotan in the tenth century had passed under the sovereignty of a Turkish dynasty established in Ladāk. But the few scattered notices of Muhammadan geographers on which this opinion is based are far too hazy to permit of sound historical conclusions. The passages of Mas'ūdī and Idrīsī which may possibly relate to Khotan only show how very vague Muhammadan knowledge of those regions was. In view of this a reference to Khotan as a part of Tibet proves little or nothing. The application of the term 'Turk' by Al-Bērūnī to the chiefs of the valleys north and north-east of Kashmīr can also not be relied on very far. For the vague use of the term and the utter inadequacy of early Muhammadan knowledge about Khotan and Tibet, compare Richthofen, *China*, i. pp. 565 sq.

[23a] The designation of Khotan as 'Chīn-u-Māchīn' (see above) has its significance.

[24] See Rémusat, *Ville de Khotan*, p. 90.

[25] Ibid., pp. 92 sqq.

their goods imposed a recurring strain on the population, an imperial decree is recorded in 1078 restricting admission to the empire to duly accredited missions.

The extremely varied list of goods which this last notice enumerates among the usual Yü-t'ien 'tribute' is of interest as showing how many articles from the Far West then found their way to China through the commercial mediation of Khotan. Whether the trade relations thus renewed continued unbroken after the close of the Sung period (1126 A. D.), and the invasion of Eastern Turkestān by the Kara-Khitai which almost coincided with it, does not appear from the Chinese records so far rendered accessible[26]. Nothing is known to us of Khotan during the period (circ. 1125–1208), when the Kara-Khitai held sway over Eastern Turkestān. Buddhists though they were, there is no evidence to show that their rule impaired the position which a century and a quarter of zealous repression must have created for Islām at Khotan. From 1218 Eastern Turkestān became part of Chingiz Khān's Mongol Empire, and Khotan may be assumed to have benefited by the facilities for trade intercourse which the vast extent of this empire created. Kara-Khitai and Mongol supremacy.

When Marco Polo visited Khotan on his way to China, between the years 1271 and 1275, the people of the oasis were flourishing, as the Venetian's previously quoted account shows[27]. His description of the territories further east, Pein, Cherchen and Lop, which he passed through before crossing 'the Great Desert' to Sha-chou, leaves no doubt that the route from Khotan into Kan-su was in his time a regular caravan road. Marco Polo found the people of Khotan 'all worshippers of Mahommet' and the territory subject to 'the Great Kaan', i.e. Kublai, whom by that time almost the whole of the Middle kingdom acknowledged as emperor. While the neighbouring Yarkand owed allegiance to Kaidu, the ruler of the Chagatai dominion, Khotan had thus once more renewed its old historical connexion with China. Marco Polo's visit to Khotan.

The ampler flow of Muhammadan records may, perhaps, render it possible to trace some details of Khotan history during the troubled times of the fourteenth century, which saw the power of the successors of Kublai in China waning and Turkestān divided between two lines of the house of Chagatai. But this period and those following lie far beyond the scope of our historical sketch. We may, therefore, close with a brief reference to the Notice of Yü-t'ien in the Ming Annals, the last of those extracted in the *Pien i tien*[28]. The record there given of some 'embassies' from Yü-t'ien, which arrived during the years 1420–1424, is of interest on account of the unreserved exposition of their true character as commercial ventures[29]. The remarks imply that after a period of interruption the trade of the West towards China was once more resuming its ancient channel through Khotan. How long this revived use of the old route lasted we do not know. But we can still discern, perhaps, the probable cause which had led trade back to it. In 1421 the imperial court had been visited by an embassy, this time not a sham one, from Shāh Rukh, the Moghul prince of Herāt. When this mission, briefly referred to also in the Ming Notice, was travelling homewards in 1422, the ambassadors, on account of troubles Notice of Khotan in Ming Annals.

[26] The *Pien i tien* furnishes no extracts on Yü-t'ien for the periods of the Southern Sungs and the Yüan or Mongol Dynasty.

[27] See above, pp. 139 sq.; Yule, *Marco Polo*, i. p. 188.

[28] See Rémusat, *Ville de Khotan*, pp. 100 sqq.; the notice is also translated by Bretschneider, *Mediaeval Researches*, ii. pp. 246 sqq.

[29] 'Those foreigners are very fond of Chinese productions, especially silk, and derive benefit from exchanging them with goods they bring from their countries. Thus the foreign merchants were in the habit of coming to China under the false pretext of carrying tribute. They brought with them camels, horses, jade, and other things. When they had entered China, the government provided them with boats and carts to travel by rivers or by land', &c.; see Bretschneider, loc. cit., p. 247. For Goëz' description of the organization of these sham embassies which proceeded to Peking from Central Asia in his own time, comp. Yule, *Cathay*, ii. pp. 564, 582.

in Mongolia, as their original relation tells us, took the unfrequented southern route which led them through the desert to Khotan[30]. In Goëz' time, however, the jade of Khotan no longer reached China by the direct route, but through Yarkand and along the foot of the T'ien-shan; and down to the present day that once frequented road via Lop-Nor and the desert to Sha-chou has remained practically unknown to the indigenous trader and the Chinese administration alike[31].

[30] See Yule, *Cathay*, i. p. ccxi.

[31] M. Grenard, *Mission D. de Rhins*, p. 170, gives an interesting account of the efforts made in 1893 by an enterprising Amban of Keriya to rediscover the ancient route leading from Lop-Nor to Sha-chou and to facilitate travel between Keriya and Lop-Nor. Traffic is still extremely scanty from Keriya to Chāklik, the small Chinese station in the Lop-Nor region. The route through the desert beyond to Sha-chou does not appear to have been followed in modern times by any traveller except J. Martin, who died without leaving an account of it beyond the few notes which M. Grenard recorded from his mouth.

CHAPTER VIII

ANCIENT SITES OF THE KHOTAN OASIS

Section I.—THE HILL OF GOŚṚṄGA

My archaeological inquiries and explorations in the Khotan region, the account of which may now be resumed, were in the first place directed towards the identification of those ancient sites within the oasis, of which some notice is to be found in the records discussed in the preceding chapter. All these sites, including that of the ancient capital, belong to the *topographia sacra* of Khotan, and consequently it is natural that Hsüan-tsang should prove our principal and most reliable guide for their location.

Topographia sacra of Khotan.

It was no small advantage that I was able to commence my survey of the sites described by the pilgrim from a point, the identity of which was from the first placed beyond all doubt by unmistakable natural features. M. Grenard had already recognized that Hsüan-tsang's Hill of Gośṛṅga, with its sacred cave and shrine, situated to the south-west of the capital, could be no other than the Kohmārī hill which rises above the Kara-kāsh river near the extreme south-west of the oasis, and in its conglomerate cliffs contains a small cave held sacred to this day as a Ziārat[1]. My surveying expedition into the Kun-lun range south of Khotan, for which I had been obliged to set out within a few days after my first arrival[2], had by the 11th of November, 1900, brought me down to Ujat, where the valley of the Kara-kāsh debouches into the fertile plain of the oasis. I could not have desired a more appropriate place from which to start my archaeological survey of the oasis; and the fatigues resulting from the trying mountain journey just completed did not keep me from visiting on the next day the neighbouring hill of Kohmārī.

Hsüan-tsang's Mt. Gośṛṅga.

Hsüan-tsang's Memoirs tell us that 'to the south-west of the capital about twenty li or so is Mount Ch'ü-shih-ling-ch'ieh (or *Gośṛṅga*, meaning in Chinese "the cow's horn")[3]. This hill has two peaks, steeply scarped and very pointed. In the valley which separates them[4], there has been built a convent; in this is placed a statue of Buddha which constantly spreads around a brilliant light. In ancient days Tathāgata came to this spot and delivered a concise digest of the Law for the benefit of the gods. He prophesied that in this country there would be founded a kingdom, and that its inhabitants would respect and honour his Law and zealously follow the doctrine of the Great Vehicle'.

[1] See Grenard, *Mission D. de Rhins*, iii. pp. 142 sqq.

[2] For an account of this expedition, rapid and difficult, but amply rewarded by geographical results, see *Ruins of Khotan*, pp. 206–43.

[3] See *Mémoires*, ii. pp. 229 sqq.; *Si-yu-ki*, transl. Beal, ii. p. 313. I follow Julien's version except where otherwise indicated.

The meaning of the Sanskrit Gośṛṅga is given in a note of the original by *niu-chio* 牛角; *niu* can mean both ox or cow. The latter translation is adopted in the name, in accordance with S. Lévi's rendering, see below, p. 186.

[4] I have reproduced above the translation given by Rémusat (*Ville de Khotan*, p. 43) which appears to come nearest to a description of the actual features of the locality meant. Julien translates 'Il est surmonté de deux pics, et, de quatre côtés, il est comme taillé à angles droits. Entre la vallée et les flancs de cette montagne, on a construit un couvent, &c.' Beal has: 'there are two peaks to this mountain, and around these peaks there are on each side a connected line of hills. In one of these valleys there has been built a *Saṅghârâma*, &c.' It would seem that the wording of the original text presents some ambiguity.

'In the scarp of the "Cow's Horn" (Gośṛṅga) hill there is a dwelling carved in the rock[5], where there is an Arhat plunged in the ecstasy which destroys thought; he awaits the coming of Maitreya Buddha. During several centuries constant homage had been paid to him. In recent times the rocks[6] have fallen in and obstructed the entrance. The king of the country sent soldiers to remove the fallen rocks, but swarms of black wasps attacked the men with poisonous stings. Hence to the present day the rock entrance has remained closed'.

Gośīrṣa mountain of Tibetan texts.

Hsüan-tsang's reference to the prophecy which Buddha was believed to have made on Mount Gośṛṅga about the future foundation of Khotan and its attachment to the Law has enabled Mr. Rockhill to recognize in this sacred site the Gośīrṣa mountain, where the legend recorded by the Tibetan 'Annals of Li-yul' makes Buddha stay for seven days on his visit to the lake of Li-yul and predict its future conversion into the land of Khotan[7]. The designation of *Gośīrṣa* (cow's head) mountain, which this Tibetan legend gives to the site, is close enough to the *Gośṛṅga* of Hsüan-tsang's text, and we find distinct mention in the 'Annals of Li-yul' of a temple and monastery at the site. In the passage relating the prediction, the Blessed One is said to have 'remained for seven days for the weal of mankind in the temple to the left-hand side of the great figure on the Goçircha mountain, where there is now a little tchaitya'[8]. We read subsequently, in the account of Vijayavīrya's reign, that 'this king built on the Oxhead mountain (*Goçircha*), the Hgen-to-shan vihāra'[9].

Identity of Mounts Gośṛṅga and Gośīrṣa.

The identity of the sacred hill thus variously designated as Mount Gośṛṅga and Gośīrṣa is now fortunately established with certainty through some interesting passages which M. Sylvain Lévi has recently brought to light from Chinese and Tibetan sources, and at the same time interpreted with critical accuracy. The Chinese translation of the Sūryagarbha-sūtra made by Narendrayaśas between the years 589 and 619 A.D., in a list of holy spots (*pīṭha*) sanctified by the presence of a Bodhisattva, mentions the residence and Caitya of the saint *Chü-mo-so-lo hsiang* 瞿摩娑羅香, or Gomasālagandha, near Mount Niu-t'ou (Oxhead: Gośīrṣa), on the steeply-scarped bank of the river in Yü-t'ien[10]. In the Tibetan version of this text the same passage is reproduced thus: 'In the Kha-śa country, at the place of "the Earth's breast" (*Sa'i-nu-ma,* i.e. Kustana), on the mountainous bank of the Gomatī, close to Mount "Cow's Horn" (*Glaṅ-ru,* i.e. Gośṛṅga), there resides Go-ma-sā-la Gandha'[11]. A comparison of the two versions leaves no doubt as to the application of the names Gośṛṅga and 'Cow's Head' to the same sacred hill, and further supplies a valuable indication by placing the saint's residence on the steep bank of the river called Gomatī.

We find another reference to this river, which clearly corresponds to the Kara-kāsh, coupled with Mount Gośīrṣa in a passage of the Yin-i, a Buddhist exegetical work composed during the eighth to ninth century[12]. In a brief notice on Yü-t'ien, to which we shall have again to

[5] Rémusat's 'Sur l'escarpement du mont' suits the actual locality better than Julien's (and Beal's) 'Dans les cavernes du mont'. The expression rendered by Julien's 'une chambre creusée dans le roc' and Beal's 'great rock-dwelling', is manifestly the same as the one which Hsüan-tsang regularly uses for the designation of natural caves, and which in Julien's version usually appears as 'grande maison en pierre'; comp. e.g. the descriptions of the sacred caves near Rājgir (Rājagṛha, in Bihār) *Mémoires*, ii. pp. 14 sq., 27, 32, &c., with my notes on these caves in *Ind. Ant.*, 1901, pp. 57 sqq.

[6] Julien: 'les bords de la montagne'; Rémusat, 'les roches', which are manifestly what is meant.

[7] See above, pp. 159 sq.; Rockhill, *Life of the Buddha*, pp. 232 sq.

[8] The form *Goçircha* used here is manifestly only an imperfect transcription of Skr. *Gośīrṣa* 'cow's head'.

[9] See Rockhill, loc. cit., p. 238. According to a communication from Mr. Thomas, the correct form of the name as given in the text is *Hgeḫu.lo.ṣan*. The original permits equally of the rendering of the name by 'Cow's head mountain'. See also App. *E*.

[10] Compare S. Lévi, *Notes chinoises sur l'Inde*, iv. pp. 31, 40.

[11] See S. Lévi, loc. cit., iv. p. 40.

[12] Compare S. Lévi, loc. cit., p. 39.

24

MAZĀR OF KOHMĀRĪ.

25

ENTRANCE TO SACRED CAVE OF KOHMĀRI HILL.

refer, this text states: 'In the territory of this kingdom there is Mount Cow's Head 牛頭 (Gośīrṣa). A divine spirit from time to time comes to this mountain and resides there; this mountain has a jade river; ordinarily the river brings down in its bed splendid jade. The king of the territory collects these stones regularly and comes to offer them from afar to the Eastern court.' This passage makes it quite clear that the 'montagne de la Tête de Bœuf', which in an extract from the report of the Chinese mission of 938 A.D., reproduced by Rémusat and quoted above, is placed at the point where the jade river reaches the confines of Yü-t'ien, is no other than Hsüan-tsang's Mount Gośṛṅga[13].

Hill of Kohmārī.

A description of the hill of Kohmārī and its Ziārat will best show how closely their position and character agree with all the indications furnished by the Chinese and Tibetan notices just reviewed. Opposite to the large village of Ujat, famous for its grapes, there rises immediately above the eastern bank of the Kara-kāsh a bluff conglomerate ridge to a height of about 250 feet above the river-bed. It forms the last offshoot of a detritus-covered spur descending towards the plain from the eroded range which separates the Kara-kāsh and Yurung-kāsh valleys where they approach the Khotan oasis. The part of the ridge known as Kohmārī falls off towards the river with an almost vertical cliff face. In order to reach its top I had first to cross the gravel-filled bed of the Kara-kāsh, here about a mile broad, but dry at the time of my visit except for a few small channels, in a south-easternly direction to just below the little village of Nussia. Here the river face of the ridge is less steep, showing on its surface mostly gravel and loose stones, and a rough road ascends its slope in the direction of Kohmārī. From the top, where the precipitous portion of the ridge is reached, the track turns eastwards for a short distance and then winds between dune-like hillocks of gravel back to the brink of the cliff.

Mazār of Kohmārī.

Close to the latter there lies, on a kind of saddle between two low hillocks of gravel, the Mazār which is worshipped as the resting-place of a saint popularly known as 'Khōja Mahēb Khōjam'[14]. His supposed tomb is marked by a stone-heap about five feet high, enclosed by a rough wooden fence and surmounted by the usual bundle of staffs with flags, pieces of cloth, yak-tails, and similar ex-votos of the pious[15]. The tomb is faced on the north by a low wooden mosque with a verandah, built, I was told, in the reign of Yāqūb Bēg. A path descending the west face of the cliff to about fifty feet below the top leads to a narrow terrace, which forms the approach to the small cave held sacred as the supposed residence of the saint. The terrace is partly artificial, resting on rough walls which have been built out with rubble taken from the conglomerate. Most of the space thus gained is occupied by a low wooden structure built against the cliff, which serves for the accommodation of the 'Shaikhs' or hereditary attendants of the shrine.

Sacred cave of Kohmārī hill.

Immediately to the south of this structure lies the entrance to the cave. It is reached by a rough wooden ladder, as seen in the photograph (Fig. 25). The cave, which may, perhaps, be due in part to artificial excavation, consists of two stories. The lower one runs from north-west to south-east, with a total length of about 39 feet. It slopes slightly upwards from the entrance and keeps an average height of from 8 to 10 feet, except in the innermost recess, which is much lower, but has a funnel-like opening on the top to a total height of 10 feet. The greatest width is about 14 feet. From this story a rough ladder leads to a small upper chamber, which communicates with the story below by means of four holes

[13] See Rémusat, *Ville de Khotan*, p. 112, and above p. 179.

[14] Muḥibb Khwāja is evidently the correct form of the name intended.

[15] See Fig. 24.

from 1 to 2 feet in diameter. This upper chamber is only about 13 feet long, with a greatest width of 8 feet. Its height is not more than 4–5 feet, except where a narrow fissure leads through the rock upwards to a height of about 15 feet. The end of this fissure, which can be scaled with some trouble, was visible by the light of a candle. The upper story of the cave is shunned by the common people, who fear to ascend into its darkness, but no special sanctity attaches to it.

The walls of the cave show everywhere the natural conglomerate, with its well-rounded pebbles closely packed and varying in size up to one foot or more in diameter. A hard crust of black soot covers the wall surfaces throughout. The legend ascribes the black colour of the walls to the smoke with which the infidels are supposed to have here killed the saint. But in reality it is manifestly a result of the fires which the pilgrims who come to sit and pray in the cave are wont to light in winter to keep themselves warm. Beneath a thin layer of dust and ashes which I had cleared out, the bottom of the cave proved to consist of hard pebbles firmly embedded.

Find of Dutreuil de Rhins MS.

I examined the interior of the cave with particular care, as the report furnished to MM. Dutreuil de Rhins and Grenard represented it as the find-place of an ancient Kharoṣṭhī MS. of the Dhammapada on birch-bark, fragments of which were acquired by them in 1892 and published by M. Senart five years later[16]. The publication of this eminent savant makes it unnecessary to point out here in detail the importance of this find, since known as the Dutreuil de Rhins MS., from both the palaeographical and philological points of view. Whether written in India or outside it, these fragments certainly constitute the oldest MS. so far known of an Indian literary product. M. Grenard's account shows that the leaves were delivered to him and his companion on two successive visits to the locality by natives who professed to have found them with some other antiques (a bowl of well-finished pottery and a small figure carved in wood and enclosed in a casket) inside the grotto. Neither to M. Dutreuil de Rhins, who first visited the place and received some fragments, nor to M. Grenard, who about a month later secured what he believed to be the rest of the find, was the exact spot of discovery shown. The men who sold the precious leaves to the French travellers appear to have hindered them from a personal inspection of the cave by alleging religious objections. They certainly took care to hide the fact that a larger portion of the fragments had been sold by them to agents of the late M. Petrowsky, Russian Consul General at Kāshgar, through whose mediation it subsequently reached St. Petersburg.

Doubts as to find-spot.

Though the visit of the French explorers was still remembered by the Shaikhs, nothing was known to them or the villagers of the alleged discovery in the cave. On the other hand I found the guardians of the sacred spot ready enough, for a consideration, to show me the cave, including its mysterious recesses. The careful examination I was thus able to make of the whole interior did not reveal any spot where the objects described by M. Grenard could have lain effectively hidden for centuries. There were no traces of any recent opening visible anywhere in the rock walls, though the coating of old soot would have easily revealed such; nor could the thin layer of dust and ashes above the hard rock at the bottom have covered and preserved such bulky articles as a bowl and casket. These considerations make it distinctly improbable that the antiquarian relics now in Paris and St. Petersburg were actually discovered within the cave. They may have been found in the vicinity and the cave indicated as the

[16] Compare Grenard, *Mission D. de Rhins*, pp. 142 sqq.; Senart, *Les Fragments Dutreuil de Rhins*, *J. asiat.*, 1898, Sept.–Oct.; for earlier notices see *Comptes rendus de l'Académie des inscriptions*, IVe série, t. xxv. pp. 251 sqq.

place of discovery merely to obviate further search by the travellers. Not having been able to ascertain either from the published accounts or by my inquiries at Khotan the names of those who supplied the relics, I had no means of following up the matter further. I hence fear that, as in the case of so many remains acquired through native agency in this region, the question as to the true origin of the Dutreuil de Rhins MS. must ever remain undecided.

Kohmārī identified with Mt. Gośṛṅga.

The uncertainty about this find cannot detract from the archaeological interest of the site; for every feature proves the identity of the latter with the Mount Gośṛṅga or Gośīrṣa of our Chinese and Tibetan records. The position occupied by the present Mazār of Kohmārī, on the top of a hill rising precipitously above the bank of the Kara-kāsh just where it reaches the Khotan oasis, agrees accurately with all topographical indications in the notices of the *Hsi-yü-chi* and other texts above quoted. The south-westerly bearing from the Khotan capital which Hsüan-tsang mentions for Mount Gośṛṅga, coupled with the fact that this capital is plainly described by successive notices of the Chinese Annals as situated between the Yurung-kāsh and the Kara-kāsh, necessarily places the sacred hill by the banks of the latter. The short distance from the capital, 20 li, named by the *Hsi-yü-chi* shows unmistakably that the hill must have been at the very edge of the oasis. Now there is no eminence on the banks of the Kara-kāsh lower down than Kohmārī, nor, for a considerable distance higher up, any which comes quite close to the river bank and rises precipitously above it. From Popuna downwards to its debouchure at Ujat the river is flanked either by cultivated flats or else by gravel slopes of a very easy gradient, except just at Kohmārī.

Local worship continued on Kohmārī.

The existence in the steep scarp of this hill, i.e. in the very position described by Hsüan-tsang, of a cave venerated to the present day as the abode of a saint completes the evidence in conclusive fashion. In the narrow fissure running into the rock from the upper story of the cave we may safely recognize the passage which in the legend related to the pilgrim is represented as having been blocked by fallen rocks and thus having hidden the Arhat[17]. The tenacity of local worship is proved once more by the tradition which has substituted for Gomasālagandha, the Buddhist saint once worshipped here, a Muhammadan martyr of the Faith. The alleged resting-place of the latter on the top of the ridge, between two small hillocks, may well occupy the place of the convent and Caitya which, according to the testimony of the *Hsi-yü-chi* and Sūryagarbha-sūtra, we must suppose to have stood on Mount Gośṛṅga.

The shrine and cave of Kohmārī still form a favourite place of pilgrimage for the faithful of Khotan, and the well-fed, contented look of its Shaikhs shows that their income derived from pious offerings is substantial. The intercession of holy 'Mahēb Khōjam' is believed to be particularly efficacious when the low state of the rivers makes the cultivators of certain tracts fear inadequate irrigation and consequent failure of crops. On this account quasi-official recognition, in the form of a liberal offering from the Amban Pan Darin, was said to have been recently accorded to the shrine. Is it possible that this belief in a connexion between worship at Kohmārī and the supply of flood-water in the rivers had its distant origin in the

[17] It is scarcely necessary to point out that it was manifestly the existence of such a natural fissure, suggesting access to some mysterious inner space impassable beyond a short distance, which originated the legend. We meet with a similar piece of folklore in Hsüan-tsang's account of the sacred cave near *Yaṣṭivana* in the vicinity of ancient Rājagṛha or Rājgir in South Bihār, which I have identified with the *Rājpinḍ* cave close to the present Jeṭhian (see my 'Notes on an archaeological tour in South Bihār', *Ind. Ant.*, 1901, p. 65). A high fissure running upwards from a corner of this cave was supposed to have once given access to the magic city of the Asuras (comp. *Mémoires de H.-Th.*, ii. pp. 14 sq.).

legend which, as related by the 'Annals of Li-yul', made Buddha symbolize at this spot the future draining of the waters covering Khotan?[18]

The name *Kohmārī.*

In conclusion, the name now borne by the hill deserves brief notice. As pronounced by the people it sounds *Komārī*, but the written form adopted for it in the short Tadhkirah of the shrine and probably elsewhere is *Kohmārī*. According to a statement of M. Grenard, *Kohmārī* is interpreted to mean 'the snake of the mountain'—evidently a 'popular etymology' which takes the Persian words *kōh* 'mountain' and *mār* 'snake' for its foundation, without troubling itself about the non-Persian form of the supposed compound. M. S. Lévi ingeniously recognized in this etymology, whether true or not, a reminiscence of the Gṛhapati (Chi-li-ho-po-ti) Nāga whom the Sūryagarbha-sūtra names as inhabiting the site[19]. I am unable to express any definite opinion as to the real derivation of the name, except that it manifestly belongs to that extensive portion of the local nomenclature of Khotan which is of pre-Turkī origin. If it were shown that the Eastern Irānian tongue once current in Khotan possessed the phonetic equivalents of Persian *kōh* and *mār*, the notice of the Sūryagarbha-sūtra would permit us to hazard the conjecture that the name originally meant 'the hill of the snake (i.e. Nāga)'.

SECTION II.—THE CULTURE-STRATA OF YŌTKAN

Apart from Mount Gośṛṅga, the position of which was indicated by unmistakable natural features, no attempt could be made to locate any of the other ancient sites of the Khotan oasis mentioned in our texts without previously determining the position of the ancient capital, for it is invariably with reference to the latter that Hsüan-tsang records those bearings and distances which alone can now guide us to the identification of minor sites. We have seen above that, according to the *Hsi-yü-chi*, Mount Gośṛṅga lay 20 li to the south-west of the capital, and that various notices of the Chinese Annals placed the latter between the Yurung-kāsh and the Kara-kāsh rivers. These data would not by themselves have sufficed for an exact location of the site had not a fortunate chance in recent times brought to light ample archaeological evidence which supplements them and removes all doubt.

Supply of antiques from Yōtkan.

To M. Grenard belongs the merit of having first clearly recognized that the little village of Yōtkan in the Borazān canton, which had furnished a constant supply of antiques, such as ancient pottery, coins, gems, &c., to Khotan traders, as well as to previous European travellers, stands on ground once occupied by the ancient capital[1]. A steady flow of antiques from Yōtkan has since M. Grenard's visit in 1891 reached public collections in Europe and India, largely through the purchases of the British and Russian representatives at Kāshgar, or else through acquisitions of European visitors to Khotan; but the information available as to the exact conditions in which those relics were found in such remarkable quantities, and the general character of the site, remained of the scantiest[2].

[18] Compare above, p. 159; Rockhill, *Life of the Buddha*, p. 233.

[19] Compare *Notes chinoises sur l'Inde*, iv. p. 40 note.

[1] See Grenard, *Mission D. de Rhins*, iii. pp. 127 sq. 128 sq.

[2] For brief notices of the locality compare Hedin, *Through Asia*, pp. 759 sq.; Hoernle, *Report on C.-A. antiquities*, i. pp. xii sq. (based on communications from Mr. Macartney and the Swedish Missionaries of Kāshgar). In all these the site is inaccurately designated as 'Borazan', in reality the name of the canton to which the village of Yōtkan belongs. Dr. Hedin's chapter on 'Borasan and its archaeological remains' (loc. cit., pp. 759 sqq.) contains, besides good illustrations of antiques coming from Yōtkan, extracts from a paper published by M. Kiseritsky on the late M. Petrowsky's large collection of similar acquisitions. The original paper, quoted as being in the Journal of the Imp. Russian Archaeological Society, is not accessible to me.

Realizing the importance of systematic examination of the locality, I had, during my first brief stay at Khotan, on October 15 paid a rapid preliminary visit to Yōtkan; at the same time I arranged for the collection of such antiques from the place as happened to be for sale among the villagers or in the hands of Khotan traders. The proceeds of this search awaited me at Khotan on my return from the mountains; and after giving to my men and animals a few days of much-needed rest I hastened to proceed to Yōtkan for a thorough survey of the site. I reached it on November 25 from the south, after a brief visit to the jade-pits at the debouchure of the Yurung-kāsh Valley, and devoted the next four days to a careful study of the antiquarian aspects of the site and its immediate vicinity.

The name Yōtkan is borne by a group of detached hamlets, of which the northern one, called Khalche, is situated, as the map of the Khotan oasis and the large scale one of Plate XXIII show, nearly five miles from the western gate of the Yangi-shahr or Chinese town of Khotan, and approximately west by north of it. Through the centre of the area formed by the village lands of Yōtkan there passes a ravine, deeply cut into the loess soil and known as the *Yōtkan-Yār*. Where the course of this ravine approaches the houses of Khalche it is adjoined on the south by the site which has furnished whatever relics have so far been recovered of the old Khotan capital. **Position of Yōtkan.**

The remarkable appearance presented by this site, and the no less curious nature of the operations to which finds of antiques are there due, were bound to attract from the first my special interest. A preliminary survey showed the site to consist of a large depressed area sunk from 20 to 30 feet below the level of the surrounding ground, and forming, as seen in the large scale map (Pl. XXIII), an irregular oblong with sides each about half a mile in length. Except where the Yōtkan-Yār enters and leaves it, this depression proved to be bordered everywhere by steep banks cut into what at first seemed natural loess, but which on closer examination showed, besides pure soil, layers full of pottery débris, ashes, decayed wood, and other decomposed matter. Along considerable portions of the enclosing banks there were plentiful signs of recent diggings and 'washings'. It was soon made clear to me by the villagers' statements that it was solely the working of these banks, or more exactly of the deeply embedded layers just mentioned, which produced the annual yield of antiques from Yōtkan; while within the depression itself, now mostly occupied by rice-fields and marshy ground, finds were wholly unknown.

The first careful inspection of the whole site sufficed to convince me that this great depression or basin was indeed, as the villagers plainly told me, the result of long-continued excavations, such as were still proceeding each summer. Neither the extent nor the shape of the depressed area could possibly be accounted for by the erosive action of flood-water carried in the Yōtkan-Yār, as had been somewhat vaguely assumed by some earlier visitors to the site when they spoke here of 'frightful ravages in the soil'. But as soon as I had realized that this strange-looking site corresponded in its present features to the huge open-air pit of some mine or quarry, and had like it been created by systematic diggings, two fresh questions at once confronted me. What was the object for which such extensive and necessarily laborious excavations had been carried on, and when and how had they originally been started? **Excavated area of Yōtkan.**

To the first question the evidence of the work which on my visit in October I found actually proceeding supplied a conclusive answer: it was for gold in the first place and next for chance finds of small objects of value, such as pieces of worked jade, gems, ornaments, &c., that the banks enclosing the site were annually being dug into and the soil from their débris-filled layers 'washed' by a varying number of diggers. To obtain a reliable answer to the second question, which touched upon the past of the site and had a direct antiquarian bearing, might have been a more difficult matter. But fortunately I was in time with my inquiries; the commencement of the excavations did not reach back further than the memory of a generation still living.

Origin of excavations.

The statements which I collected from a number of intelligent old villagers both at Yōtkan and in the vicinity, and which I took every opportunity to test during my stay, threw light on a number of interesting points and were fully accordant in essentials. From them I ascertained that no finds of any kind indicating that an ancient site was buried here below the ground had been made until the time of Niāz Ḥākim Bēg, still well remembered throughout Khotan as the first governor under Yāqūb Bēg. Two or three years after his appointment, which took place about 1866, the small canal conveying water from the Kara-kāsh river for the irrigation of the Yōtkan fields began to cut for itself a deeper bed in the soft loess, that is, to turn into a 'Yār'. This is the origin of the ravine, which begins about one-and-a-half miles to the west of Yōtkan at the village of Chalbāsh, and after passing Khalche and the Yōtkan site joins the 'Yārs' of Kāshe about a mile to the north-east.

The 'Yār' of Yōtkan.

The archaeologist has good reason to feel grateful to the Yōtkan-Yār, for without its formation the remains of the old Khotan capital might have been left buried for ages to come. It was only when the flood-water escaping in the newly-formed ravine had created a small marshy depression (*kul*) a little to the south-east of Khalche, that the villagers accidentally came across little bits of gold amidst old potsherds and other débris. The latter objects possessed, of course, no interest for them; but the gold naturally excited the cupidity of the villagers, many of whom had, like the rest of the poorer agricultural population of Khotan, at one time or another tried their luck at 'prospecting' for jade in the river-beds or else at gold-washing on the Yurung-kāsh and in the mountains. So they set to work washing the soil near the incipient Yār, and the proceeds were so rich that they came to the governor's knowledge.

Niāz Ḥākim Bēg's excavations.

Niāz Ḥākim Bēg was an administrator of considerable enterprise. He sent to Yōtkan large parties of diggers, whom he employed like the men I found working for small capitalists in the jade-pits along the Yurung-kāsh bed[3]. The owners of the fields which were gradually cut away by these 'washings' received compensation. Subsequently the excavations were continued by private enterprise, the usual arrangement being that the owners of the soil and the diggers share the proceeds equally. The earth excavated from the banks has to be washed, just like the old deposit of gold-carrying streams. The larger supply of water needed for this purpose caused the Yōtkan canal to cut its bed deeper and deeper until it formed the existing Yār, the bottom of which is from 29 to 30 feet below the level of the fields (see Fig. 26). Finally the canal had to be diverted to a higher level; but springs came to the surface at the bottom of the ravine, and these with others rising within the excavated area account for its swampy condition. In the recollection of old villagers the land of Yōtkan was everywhere a level flat; there were no springs or swampy ground—nor any knowledge or tradition of the 'old city' (*kōne-shahr*) below.

Earliest notice of excavations.

We shall see presently that this negative fact has its antiquarian interest; and it may, therefore, be useful to point out that the accidental discovery of the Yōtkan site, about the date indicated by the villagers' statement, can be established independently by a record made as long ago as 1874. The still useful notes which Pandit Rāmchand, one of the native surveyors with Sir Douglas Forsyth's Mission, collected when passing through Khotan in the spring of

[3] See *Ruins of Khotan*, pp. 253 sq.

26

NORTH-WEST CORNER OF EXCAVATED AREA AT YŌTKAN, WITH ENTRANCE TO YŌTKAN-YĀR.

27

SOUTH-WEST BANKS OF EXCAVATED AREA, YŌTKAN, NEAR ALLAMA HAMLET.

that year contain distinct reference to a 'discovery of buried ruins' said to have lately 'been made quite close to the city of Khotan (Ilchi) at a distance of about four miles to the north-west'[4]. Pandit Rāmchand evidently did not see the place himself, and what he relates about the way in which the first find was made ('a golden ornament said to have been a cow') clearly reflects the more romantic form which the story would assume in Bāzār gossip. But there can be no doubt that his notice refers to Yōtkan, thus proving that the site could not have been known for many years earlier[5]. The surveyor's record also confirms the villagers' statement about excavations for gold made under the governor's orders.

The systematic digging and washing for gold, which can thus be shown to have been carried on at Yōtkan for upwards of thirty years preceding my visit, is quite sufficient to account for the great extent of the excavated area as shown by my survey. The work was still being vigorously continued on the north side and along part of the banks to the west during the annual period when sufficient water is available for washing. Though this had passed by with September, and the diggers had departed, I found no difficulty in ascertaining the methods by which the operations are carried on and the character of their produce. Like the search for jade in the river-beds after the annual floods, like washing for gold in the flood-deposits of the Yurung-kāsh, or digging for it at the pits of Surghāk, Kapa, &c., work at Yōtkan attracts annually for a short period numbers of the poorer agriculturists of the oasis who look to these occupations as a kind of time-honoured lottery. The small quantity of gold which washing of the soil in the banks ordinarily produces is just sufficient to be 'paying', i.e. to provide the diggers, after compensation to the owners of the fields cut away, with the equivalent of the modest wage which prevails in Khotan for unskilled labour. But in recent years antiques, such as ornamented pieces of pottery, engraved stones, small objects of metal, and coins, have come to be counted as a kind of secondary product—and there always remains the chief attraction, the chance of finding valuable objects in gold, silver, or jade. Operations of diggers.

The gold usually found appears in the form of tiny flakes of leaf-gold of which I was able to secure samples, and less frequently in that of minute pieces of what looks like very thin gold plating such as might have been used in ornaments. It is easily distinguished by the villagers from the gold-dust (*kepek-āltun*) washed from the river-beds. No coins or solid ornaments of gold and silver were admitted to be found now, though the discovery of such precious articles was readily acknowledged for the early years of the working, and near the original spot where it started. Whatever the chance of making such finds in the banks of soil now under exploitation, the diggers as well as the villagers would have reason to be reticent about them. The purchase of all gold produced in the Khotan and Keriya districts forms a monopoly of the Chinese administration, and as the rates paid by the latter are very considerably below those readily offered by traders, concealment of the true quantity of produce and secret disposal of it to private individuals are common throughout the territory[6]. This practice would naturally be resorted to most readily in the case of valuable Gold washed at Yōtkan.

[4] See *Yarkand Mission Report*, p. 449.

[5] The fact that Johnson, in the account of his visit to Khotan, in 1865, does not indicate any knowledge of Yōtkan may probably be accepted as evidence that the site was then as yet undiscovered. He certainly seems to have made inquiries after ruins and ancient remains; and notwithstanding the often sketchy nature of his notes, he would not have been likely to pass over a find-place so easily accessible if it had then been known; comp. *J. R. Geogr. Soc.*, xxxvii. p. 5.

[6] The system was maintained also under Yāqūb Bēg's reign, as Pandit Rāmchand's notes show. Of the excavations ordered at Yōtkan by the governor (Niāz Ḥākim Bēg) he specially mentions 'the diggers are paid for any gold they may excavate at 110 tangas the Ser', the market value then being 138 tangas; comp. *Yarkand Mission Report*, p. 449. The information agrees accurately with what was told to me at Yōtkan of the tax levied by Niāz Ḥākim Bēg on the produce of private diggings at that period.

finds of worked objects in gold, in regard to which official control must be specially difficult. However this may be, there is material evidence of such finds in the fact that, on my later visit to Yōtkan in April, 1901, I succeeded in purchasing a tiny figure of solid gold of excellent workmanship, representing a monkey, that had been found during the previous year's washings. I have no doubt that any larger articles in gold are melted down speedily after discovery, in order to facilitate concealment and disposal.

Presence of gold accounted for.

Occasional finds of this kind cannot surprise us in the débris layers of a site which, during long centuries, was occupied by the capital of a flourishing kingdom rich through trade and mineral produce. On the other hand, it seems at first difficult to account for the prevalence of gold in that common form of leaf above described and over so large an area. Yet it is this alone which in reality constitutes the 'paying' basis of the Yōtkan diggings. The use of leaf-gold on an extensive scale in the decoration of Buddhist buildings and sacred objects offers, I think, the most likely explanation. From the detailed description which Fa-hsien has left us of the splendid Stūpa and temple he visited at what was then known as 'the King's New Monastery', a little to the west of the Khotan capital, it is certain that many parts of sacred buildings in ancient Khotan were richly overlaid with leaf-gold[7]. Apart from this distinct evidence of Fa-hsien, reproduced in the footnote, the example of Gandhāra remains, and the testimony of actual finds brought to light by my explorations lead us to conclude that the gilding of sacred images and other objects of worship must have been largely practised in Buddhist Khotan[8]. Whether applied to plaster or direct to the wood, which as we shall see, played so large a part in the construction of Khotan buildings, secular as well as religious, much of this leaf-gold must have fallen off and mingled with the dust, whenever the structures erected in such perishable materials as were known to old Khotan crumbled away or otherwise met with destruction. It is equally clear that the leaf-gold once lost in this way could not be recovered until the soil was washed by the method now followed.

Culture-strata of Yōtkan.

The examination of the stratum from which this gold is obtained proved very instructive. It varies considerably in thickness and depth at different points of the banks enclosing the excavated area, but is everywhere easily distinguished by its brownish colour and peculiar composition. It consists of decomposed rubbish and humus, in which are embedded fragments of ancient pottery, plain or ornamented, bones of animals, pieces of much decayed wood, ashes, and thin layers of charcoal—all indications that we have here the débris which accumulated on a site occupied by dwellings during long periods. Everywhere the excavations have been carried down until the natural loess soil was reached; hence the thickness of the upper stratum could be established without difficulty. Along the banks on the south and south-west, where

[7] 'Seven or eight le to the west of the city there is what is called the king's New monastery, the building of which took eighty years, and extended over three reigns. It may be 250 cubits in height, rich in elegant carving and inlaid work, covered above with gold and silver, and finished throughout with a combination of all the precious substances. Behind the tope there has been built a Hall of Buddha, of the utmost magnificence and beauty, the beams, pillars, venetianed doors, and windows being all overlaid with leaf-gold. Besides this, the apartments for the monks are imposingly and elegantly decorated, beyond the power of words to express'; *Fâ-hien's Travels*, transl. Legge, pp. 19 sq. The distance and bearing indicated make it highly probable that this great shrine must be identified with the So-mo-jê convent of Hsüan-tsang and thus be located at the present village of *Somiya*, close to the west of Yōtkan; see below, sec. v. From the great height mentioned it is clear that the structure was of wood, like the famous Vaiśravaṇa temple in the capital itself.

[8] For the numerous instances of gilding applied to Buddhist statuary as well as to Stūpas, comp. Foucher, *L'Art du Gandhâra*, pp. 82, 198 sqq. For traces of gilding in remains of Khotan sites, see below, chap. x. sec. i.; chap. xiv.

(It is curious that a passage of the *Rājataraṅgiṇī*, which mentions the gilt parasol placed over a Śiva temple in the time of Kalaśa (1063–89), ascribes instruction in 'the art of putting gold on copper' to an artist from the Turuṣka country, i.e. Turkestān; see my *Rājat.* vii. 528–531.)

the excavations touch the gardens and fields of Allāma (see Fig. 27), it shows an average of from 5 to 8 feet. In the banks on the west it seems to increase gradually as they approach the vicinity of the Yōtkan-Yār (see Fig. 26). North of the latter the 'culture-stratum' attains its greatest thickness, being fully 13 to 14 feet deep in the banks worked immediately below the houses of Khalche, where the proceeds in antiques, such as terra-cotta figurines, coins, small fragments of carved stone, &c., are the richest. It was chiefly here and along the north-western edge of the area that recent working was proceeding. The banks to the south and east, where the relatively thin culture-stratum had ceased to be 'paying', showed signs of having been abandoned for a considerable number of years, as proved by the fruit-trees planted just below them within the excavated area.

Contents of culture-strata.

It was easy to see that, notwithstanding the uniformity in the character of the culture-stratum, there were variations in the relative frequency of the objects embedded in it. In some places the pottery fragments were closely packed, forming small layers or patches easily distinguishable on the surface of the banks, whereas elsewhere only isolated pieces could be picked out. In the same way bones, ashes, &c., are far more plentiful at certain points than at others. It is clear that the varying depth of the culture-stratum must be ascribed mainly to the different length of the periods during which particular localities were occupied, and to the different character of the uses to which they had been put. Of the latter fact the variations just noted afford direct evidence.

Absence of structural remains.

In one respect all portions of the culture-strata exposed to view show a regrettable uniformity; nowhere did I come upon traces of structural remains, nor could I hear of such having been found during previous excavations. The nearest approach to such remains was represented by a layer of closely packed large pebbles, which the banks on both sides of the Yōtkan-Yār, just where it enters the excavated area from the west, showed in a thickness of about 5½ feet and about 10½ feet above the present bottom of the ravine. The stones must have been brought from some river-bed, but whether they served for a platform, foundation of a wall, or some similar purpose could not be determined. Above this layer I noticed the occurrence of ancient pottery fragments, while none were exposed below it*.

I believe that this absence of structural remains is to be accounted for by the fact that, owing to the total want of suitable stone, sun-dried bricks and clay supplemented by timber must have been in old days, as now, the only materials available for the construction of houses in the Khotan region. All of the mud walls of buildings that had not crumbled into dust was bound to decay completely in the course of the centuries during which, as we shall presently have occasion to demonstrate, the site was taken up for cultivation and the soil kept constantly moist by the percolation of irrigation water. The same fate overtook, of course, whatever of the wood once contained in the more substantial buildings had not been extracted and utilized by successive occupiers of the site. It might have been different if the old town, of which the 'culture-strata' of Yōtkan now retain the sole relics, had been overwhelmed by some sudden catastrophe and its site once for all deserted. Then we might expect to find the original materials and even some structural outlines of buildings preserved in a recognizable form under the soil. The assumption of such a catastrophe, a great flood which was supposed to have destroyed the old town, has indeed been hazarded by some of the earlier European visitors to Yōtkan, who attributed to it the thick cover of earth now

* Tokhta, an old villager of Allāma, talked of a square well, lined with wood, having been struck during the early excavations; but it seemed very doubtful from his description whether this belonged to the time of the 'culture-strata'.

hiding the relics[10]. But it only needed a careful examination of this very feature to dispel the notion of the conjectural catastrophe.

Layers of pure soil above débris.

The strata containing the old remains are everywhere covered by a considerable layer of pure soil, which from the first interested me greatly on both antiquarian and geological grounds. It shows a fairly uniform thickness from 9 to 11 feet in the banks near Allāma, while in those exposed south of Khalche it varies from a minimum of 11 feet to a maximum of about 20 feet at different points. This layer is easily distinguished by its light colour from the 'culture-strata' below, and is absolutely free from remains which might indicate subsequent occupation of the site. The latter fact is realized by the villagers, who everywhere would readily show the limit up to which the old stratum extends, and above which there begins what they call *bēkār yer* ('unprofitable ground'), i.e. the natural soil, yielding neither gold nor antiques. Wherever I examined the earth overlying the 'culture-strata' it proved exactly the same in substance and colour as that which is to-day turned up by the plough of the Yōtkan cultivator. Another important fact, which the careful examination of this layer established with equal clearness, is the total absence in it of any traces of stratification.

These observations conclusively prove that the present position of the culture-strata of Yōtkan cannot possibly be the result of a great flood or series of floods which have overwhelmed the site; for such a catastrophe, if it had ever occurred, would have been bound to leave its mark in stratified deposits of varied character. But the facts thus observed on the spot did more than dispose of a vaguely suggested hypothesis; they also led on to what I still hold to be in the main the true explanation of the deep cover under which the remains of the old town have rested.

Origin of covering earth layers.

In my 'Preliminary Report' I did not hesitate to express my belief that this great layer of pure soil, above what we can show to be the débris accumulations of the ancient capital, was 'due mainly to silt-deposit, the necessary result of intensive and long continued irrigation'[11]. The reasons for this conclusion were derived solely from what local observation had shown me within the Khotan oasis. Having since my return fully realized the geological interest involved in the question and its close relation to the much-discussed problems of loess, I feel specially gratified that the opinion then formed on the spot has stood the test of the analysis to which an exceptionally qualified geologist, Prof. L. Lóczy, has been kind enough to subject the specimens of soil brought back by me from Yōtkan.

Silt-deposits from irrigation.

Without anticipating the information which Professor Lóczy's instructive communications in 1903 enabled me to gather with regard to a very important subsidiary factor, I may be allowed to state in the first place the case as it presented itself to me solely on the basis of my local impressions, and as I have already recorded it in my Personal Narrative[12]. Cultivation in the Khotan oasis, as everywhere else in Eastern Turkestān, demands constant and ample irrigation. The Kara-kāsh river, from which the water for the Yōtkan fields is drawn in the spring and summer, carries down during this season enormous quantities of disintegrated soil from the eroded ranges it drains. Its water, full of suspended sediment, and hence of a dark yellow or chocolate colour, is allowed to stand on the carefully terraced fields until it is completely absorbed or evaporated. As all the fine detritus or mud borne down by the river-water thus finds its way to the fields and is ultimately deposited there, the accumulation of silt over the latter must be constant and relatively rapid. Since the conditions are exactly the same in the

[10] See Hoernle, *Report on C.-A. antiquities*, i. p. xiii; *Mission D. de Rhins*, iii. p. 126.

[11] See *Preliminary Report*, p. 29.

[12] *Ruins of Khotan*, p. 263.

case of the tracts irrigated from the Yurung-kāsh, the level of the cultivated portions of the oasis is everywhere bound to rise steadily. There may be local differences in the rate of this rise, in accordance with the varying quantity of water supplied by the several canals, the distance of the irrigated lands from the canal heads, &c. But considering how near the Yōtkan fields are to the debouchure of the Kara-kāsh and thus to the region where the river collects most of this silt on its passage through the outer ranges, the supply of silt cannot fail to be particularly ample there. It seems safe to assume that land so favourably situated, in a most fertile and easily irrigated part of the oasis, cannot have been allowed to remain long uncultivated after it had once ceased to be the site of a town. Seeing that by the chronological evidence of the latest of the Yōtkan remains close on eight hundred years must have passed since the final abandonment of the site, the thickness of the deposit now observed above the strata of débris presents nothing surprising.

Raised level of irrigated ground.

Observations which I had occasion to make again and again after my first visit to Yōtkan fully support this explanation. Everywhere within the oasis I noticed that the main roads were sunk considerably below the surrounding level where they pass through land which has been long under cultivation; while elsewhere on waste land, near newly tilled fields or within the villages, they kept flush with the adjoining ground. This low position of the roads was so uniformly observable and so marked within the old portions of the Khotan oasis that I soon learned to distinguish them by this test alone from any areas which had been rendered arable in more recent times by the extension of irrigation on the edges of the oasis or by the reclamation of waste ground within it. Thus, e.g., in the Tawakkēl oasis, lower down on the Yurung-kāsh, which otherwise exactly reproduces the general conditions of cultivation, &c., of the Khotan oasis, but had been colonized only some sixty years before my visit, the level of the fields was nowhere more than about a foot or so above that of the roads. It is impossible not to seek for a natural cause of the far more deeply sunk position occupied by all roads within areas of old cultivation. None I could think of seemed more probable than that the level of the fields is constantly rising by irrigation, while that of the roads cannot undergo any marked variation.

Low level of roads and cemeteries.

That erosion of the soft soil by traffic could not be held to account for this striking and constant difference in level, seemed clear from two observations. On the one hand, little frequented paths leading to small holdings, fields, &c., often showed quite as low a level as constantly used high roads. On the other hand, I was soon struck by a still more characteristic fact—the low position of all the old cemeteries that are surrounded by fields. Cemeteries of any age are easily recognized by their extending around some Mazār or shrine, and in their case I invariably found that the ground-level lay considerably, up to 10 feet and more, below that of any adjoining fields[13]. Erosion by traffic or winds cannot be thought of here; for the closely packed and often massive tombs protect such cemeteries effectively against either[14]. This curious position becomes, however, at once intelligible if we recall the fact that the fields are constantly receiving a deposit of silt from irrigation, while the cemeteries are naturally kept clear of water and consequently of this accretion.

Professor Lóczy's analysis of soil specimens.

I have already noted that the soil of the layer which covers the 'culture-strata' of Yōtkan, as far as I could judge by the naked eye and without geological training, presented exactly the same appearance as the loess that forms the arable ground throughout the oasis. Being

[13] For a striking instance, comp. below, sec. v, the description of the old graveyard around the Ziārat of Bōwa-Kambar.

[14] For the view of a typical cemetery of this kind, comp. Fig. 16.

aware that a widely accepted theory, first propounded by Richthofen, ascribed the formation of loess (or 'adobe') to the long-continued drifting and deposit of fine dust by wind, i.e. to purely 'aeolian' action[15], I was anxious that the view I had been led to form of the character and origin of the covering-layer at Yōtkan should be tested by geological evidence. I accordingly secured a considerable number of specimens of soil taken from various portions of this layer as well as from the 'culture-strata' below. In 1903 I had the satisfaction of submitting these specimens, together with a number of others brought away from ancient sites in the desert, to Professor L. Lóczy, of the University of Budapest, who, owing to his extensive travels and researches in the loess regions of Western China, and his subsequent investigations into the loess formations of the Danube Basin, was specially competent to deal with them. The result at which the distinguished geologist and geographer has arrived, by the careful microscopical examination of these specimens and by their comparison with other similar materials available to him, will, I hope, be soon made accessible [see now Appendix *G*]. In the meantime I am enabled by Professor Lóczy's kindness to indicate here the main facts as summed up in his letter dated December 4, 1903.

Riverine loess at Yōtkan. The specimens submitted from Yōtkan show all the characteristics of riverine loess, i.e. a deposit formed of river-carried fine silt. It consists chiefly of distinctly angular grains of quartz, showing scarcely any trace of that attrition which is undergone by particles long subject to movement by wind, such as those contained in true drift-sand. Mica flakes, evidently originating from the granite and gneiss of the mountains drained by the Khotan rivers, are found plentifully in it. On the other hand, the deposit contains very little fine dust and practically no clay; it is but slightly calcareous, and lime concretions are rare. The constituents revealed by microscopical analysis are thus all such as might be expected in the sediment of the Khotan rivers; but there is reason to believe that the deposition of this sediment in loess layers such as seen at Yōtkan has not throughout taken place direct from the water, but to a very considerable extent also by subaerial ('aeolian') transport from the river-bed. The fine silt carried up from the latter by the wind must necessarily reach first the tracts immediately adjoining the river course, and its retention there must be all the more certain and constant the moister the soil is kept by irrigation. Thus cultivated ground in the oasis is assured steady accretion in level by a simultaneous deposit both from the water brought to it and from the atmosphere, the material accumulating by either process being always the same silt of the rivers.

Loess in culture-strata. The explanation thus furnished makes it easy to understand why Professor Lóczy's examination has revealed no difference whatsoever in mineral constituents between the layer of pure riverine loess and the culture-strata of Yōtkan. Whatever of the latter does not consist of still recognizable hard débris or decomposed vegetable and animal matter, can only be loess soil once used for sun-dried bricks and mud walls or else river-silt in the form of dust brought to the site while it was still inhabited. In either we must expect the identical mineral composition which the more recently formed loess of the oasis presents. The difference in colour is caused solely by the abundance of decomposed organic materials in the ancient culture-strata. It is evident that the definite explanation thus obtained of the deep layer of earth that covers the remains of the old Khotan capital applies equally to other ancient sites within the oasis which have left no trace above the present ground-level and which thus still await discovery and excavation[16].

[15] Comp. Geikie, *Text-book of Geology*, i. pp. 439 sq.

[16] We shall see hereafter that there is at least one other site of Khotan—I refer to Tam-ōghil—which has actually yielded antiques from strata buried like those of Yōtkan under a deep deposit of loess; comp. below, chap. xiv.

But quite as important from the antiquarian point of view is another geological fact which Professor Lóczy's examination of the specimens submitted to him has clearly established. He has been able to show that not only is there almost complete uniformity in composition between the recently formed loess of Yōtkan and the soil of the ancient sites in the desert such as Dandān-Uiliq, Niya River Site, Rawak, but that even the dunes of moving 'sand' now surrounding and partly covering these sites contain nothing but naturally fertile river-silt brought either direct from the river-beds or derived from eroded loess beds. *Identity of loess at Yōtkan and at ancient desert sites.*

We shall have occasion, when discussing the conditions prevailing at the sites just named, to recur more than once to the antiquarian significance of these observations. In the meantime it may suffice to point out that, if archaeological research in the Khotan region has every reason to be grateful for the guidance obtained from geological observations, it can also in its own field offer evidence which ought to prove useful to the geologist. The scrutiny of the antiques yielded by the Yōtkan culture-strata supplies, as the next section will show, a sufficiently exact *terminus a quo* for the formation of the loess bed overlying them. It thus enables us to gauge the rate at which the deposit of fresh soil is proceeding. We shall later on see that equally valuable testimony is afforded in the opposite direction by the remains at those ancient sites in the desert where the level of the ground has been greatly lowered through wind erosion. Here again it is archaeological investigation alone that can lead to a correct estimate of the time needed for this geologically interesting process. *Geological interest of observations.*

Section III.—THE SITE OF THE ANCIENT CAPITAL

Having acquainted ourselves with the physical aspects of the Yōtkan site, we may now proceed to examine its archaeological character and the remains which it yields. There can be no doubt that the 'culture-strata' brought to light beneath the fields of Yōtkan owe their origin to the débris accumulated during centuries on the site of an ancient town. The Yār which passes through Yōtkan from west to east, and the excavations to the south of it resulting from gold-washing operations, enable us to form some idea of the extent and position of this old town.

The banks of the Yār cease to yield any gold or antiques about 200 yards to the south-east of the houses of Khalche. Accordingly digging has stopped there. To the south the diggings, after extending for nearly half a mile from Khalche, have been abandoned just below the houses and gardens of the Allāma portion of the village, as the coveted gold could not be obtained in paying quantities from the banks reached there. It is along the banks on the west and north-west of the excavated area that the work of washing still continues vigorously each season; and it is under the fields lying in that direction, and particularly northward, that the remaining parts of the old town are likely to have been situated. It is important to note that the Yārs of Kāshe, which, as the detailed map shows, intersect the ground to the south and east at distances varying from a quarter to three-quarters of a mile from the edge of the excavated area, nowhere cut through layers containing any old remains. Seeing that these Yārs are at least as deep as the Yōtkan-Yār, and that any discoveries along their banks would certainly have been followed up by similar diggings, the negative evidence thus furnished is sufficient to exclude the idea of the town having ever extended in those directions. *Extent of ancient site.*

M. Grenard was undoubtedly right when he recognized in the town site disclosed by the Yōtkan remains a part of the area once occupied by the old capital of Khotan. It is, however, *Alleged tradition of 'old town.'*

impossible to accept as evidence of this identification either the alleged local tradition of Yōtkan or that attaching to the mound of Naghāra-khāna, 2½ miles to the south-east, which M. Grenard —on insufficient grounds, I believe—has assumed to have belonged to the ancient capital[1]. The careful inquiries made by me among the old villagers of Yōtkan conclusively prove that until the accidental discovery of the first finds in Niāz Ḥākim Bēg's time nothing whatever was known of the 'old town' beneath the ground, nor of any tradition existing about it. Even now the villagers' opinion about the origin of the remains is of the vaguest. In fact, the only statement made to me about an ancient city having once stood here was clearly traceable to the view expressed by a former European visitor. A tradition of a somewhat more genuine kind certainly locates an old town, the residence of the heathen ruler of Khotan, in the immediate vicinity of Naghāra-khāna. But it does not in any way connect the latter site with Yōtkan, and its own derivation is, as we shall see, subject to doubts[2].

Evidence for location of ancient capital.

In reality the proof for the location at Yōtkan of the old Khotan capital lies in the exact agreement of the site with the topographical indications furnished by the Chinese Annals, in the character and abundance of its antiquarian relics and—last but not least—in the ease with which we are able to identify from this starting point the positions assigned by Hsüan-tsang to the prominent Buddhist shrines he visited in the vicinity of the capital. The evidence derived from the latter sites must be left for discussion in the last section of this chapter.

Topographical record of 938 A. D.

Among the notices which the Annals contain relative to the position of the capital of Yü-t'ien, those derived from Kao Chü-hui's report on the Mission of 938 A. D. are certainly the clearest and most accurate. We have already seen that they place the White Jade river or the Yurung-kāsh at a distance of 30 li to the east of the city, the Green Jade river, corresponding to the present Yangi-Daryā branch of the Kara-kāsh, 20 li to the west of it; and the Black Jade river or Kara-kāsh another 7 li further to the west[3]. A look at the map of the Khotan oasis will show how closely the distances here indicated agree with those actually measured between Yōtkan and the nearest points of the rivers. Taking Khalche as our starting point, we have a little over seven miles to the left bank of the Yurung-kāsh near the present Ilchi. The 30 li of the report would, according to the average equation of 5 li to the mile, take us within a mile of the actual river-bed; and it is certainly noteworthy that this is exactly the width of the low-lying waste ground, once undoubtedly part of the Yurung-kāsh flood-bed, which separates the eastern edge of Khotan town from the river. The remains of an old embankment marked in Dutreuil de Rhins' plan of Khotan show clearly that the floods of the Yurung-kāsh must even in recent times have approached the town quite closely[4]. Turning to the western river-beds the agreement is, if possible, still more striking. From Khalche to the nearest point on the Yangi-Daryā measures four miles, which corresponds exactly to the 20 li of the report; while the distance of 1½ miles separating the Yangi-Daryā from the right bank of the Kara-kāsh is as close an approach to the 7 li of the notice as we could hope for.

The close accord between the actual distance separating the Yurung-kāsh and Kara-kāsh rivers and the aggregate of the measurements recorded in the report of the mission of 938 A. D. attests the accuracy of the latter, irrespective of the position assumed for the old capital. The firm evidence thus secured makes it of less consequence that earlier records of the Annals

[1] See *Mission D. de Rhins*, iii. pp. 127, 138.

[2] Compare for Naghāra-khāna and the lake of Aiding-Kul, section v.

[3] See above, p. 179.

[4] See *Mission D. de Rhins*, i. p. 96. The river is now endeavouring to shift its bed again further to the west, and embankments have to be maintained in order to keep it out of the stretch of ground which evidently once formed part of its flood-bed; see Hedin, *Reisen in Z.-A.* p. 26.

concerning the position of the capital show discrepancies which greatly reduce their critical value. The notice of the *Pei shih* on Yü-t'ien, dating about 644 A. D., puts in one place the river Shou-pa which carries jade 30 li to the east of the capital[5]. This clearly refers to the Yurung-kāsh and is quite correct. But a subsequent passage, which mentions the same river by the name of Shu-chih or Chi-shih, gives 20 li as the distance to it east from the capital. To the west of the latter another great river, called Ta-li, which unites further on with the Shu-chih, is mentioned here at the distance of 15 li. One or other branch of the Kara-kāsh must be meant, but evidently not much reliance can be placed on the distance, since the text of the same notice, as reproduced in the Northern Wei Annals and extracted by Rémusat from the *Pien i tien*, gives it as 50 li. So much, however, seems clear that, according to the notice of the *Pei shih*, we must look for the site of the capital at some point situated between the two rivers and probably a little nearer to the Kara-kāsh than to the Yurung-kāsh. The topographical indication thus supplied strongly points to the conclusion that the position meant was the same as the one more accurately defined by the report of the mission of 938 A. D. We know that the *Pei shih's* notice of Yü-t'ien was directly based on the information brought back by Hui-shêng, Sung Yün's companion on his journey to India (518–522 A. D.)[6]. Thus it appears safe to conclude that the position which Kao Chü-hui's report indicates, and which clearly corresponds to the Yōtkan site, was occupied by the Khotan capital four hundred years earlier.

Earlier Chinese notices of capital.

But the notice furnished by the author of the *Pei shih*, or rather by Hui-shêng, is of value in another direction. It tells us that the capital had a circuit of 8 or 9 li[7],—a piece of information about which all our other accessible records are silent and which yet possesses considerable archaeological interest. As the circuit indicated represents only $1\frac{1}{2}$ to 2 miles, it is certain that the extent of ground covered by the old capital was relatively very limited. No doubt the measurement given need not be taken to represent an approximately exact survey; but that it indicates the capital to have been very restricted in area, becomes quite evident when we compare this estimate with the circuit of 20 li which Hsüan-tsang mentions for the chief place of even so poor a territory as Chieh-p'an-t'o or Sarīkol[8], or the dimensions (12–13 li length and 4–5 li width) which the same traveller records for the capital of Kashmīr, ancient Śrīnagara[9].

Circuit of ancient capital.

It would be possible to suggest a number of conjectural explanations for the fact that the official capital of a rich and important territory was confined to such narrow limits. But whatever the true cause may have been, it is certain that the measurement indicated by the *Pei shih* agrees singularly well with what we have found above to be the actual dimensions of the Yōtkan site. So far as excavated it shows a circumference of roughly two miles. Only to the north or north-west can the existence of further débris-strata hidden below ground be assumed with any probability; and even if the area awaiting excavation in that direction should prove quite as large as the area already exploited to the south, the circuit of the site would thereby be extended only to about three miles[10].

Limited area of site.

[5] See Chavannes, *Voyage de Song Yun*, p. 15, note 9. The same notice as found in the Northern Wei Annals and the *Pien i tien*, has been translated by Rémusat, *Ville de Khotan*, pp. 18 sqq. Part of it is reproduced also in the Annals of the Northern Chou; see Rémusat, loc. cit., p. 30.

[6] Comp. Chavannes, *Voyage de Song Yun*, p. 2.

[7] See Chavannes, *Voyage de Song Yun*, p. 15, note 9.; Rémusat, loc. cit., pp. 19, 28.

[8] See *Mémoires*, ii. p. 209; Beal, ii. p. 299, above, p. 37.

[9] See *Mémoires*, i. p. 167; Beal, i. p. 148.

[10] If the legend related by Hsüan-tsang of the miraculous way in which the limits of the capital were originally laid down (see above, p. 157) can be interpreted as having some direct connexion with the shape of the area occupied by the town, the probability of any considerable extension of the débris-layers northward would be small, for that legend clearly represented the area circumscribed by the mysterious water-carrying mendicant as a circle (comp. *Mémoires*, ii.

Abundance of antiques. It does not require much reflection to realize that it is just the limited extent of the Yōtkan site which makes the abundance of antiques yielded up by it all the more striking. And this very abundance, as well as the character of the remains, supplies strong evidence in support of the assumption that the site is indeed that of the ancient Khotan capital. Only at the site of a town that possessed a wealthy population and rich public buildings and shrines is it possible to account for débris-layers such as those of Yōtkan, which even the systematic digging operations of more than thirty years have failed to clear of their gold and other valuable contents. The culture-strata of Yōtkan are shown by unmistakable evidence to contain the débris accumulations of a thousand years, if not more. Where else but at the capital are conditions of affluence likely to have continued unchanged for so long a period?

Buildings of ancient capital. The available Chinese records tell us little about the capital of ancient Yü-t'ien, but it is characteristic that almost all of these notices contain some reference to buildings of considerable magnificence. The most interesting among them is a passage of the Yin-i, composed by Hui-lin in the eighth or ninth century, which M. S. Lévi has brought to light, and to which we have had already occasion to refer in connexion with Mount Gośṛṅga [11]. It shows that the famous temple of Vaiśravaṇa, which the legends related by Hsüan-tsang and the Annals of Li-yul represent as the oldest shrine of the kingdom [12], stood in the city of Yü-t'ien, and that it was 'a tower built in wood seven stories high'. The god was supposed to reside on the summit. Hsüan-tsang particularly extols the richness of the temple as it existed in his time. It is evident that we must think here of a structure imposing in its architecture notwithstanding its perishable materials, like the great temples and Stūpas built of wood which the Chinese pilgrims admired at Puruṣapura and elsewhere in the north-western borderland of India [13].

As regards the royal palace we learn from the account of the Chinese mission sent in 938 that it comprised a number of buildings all facing to the east, and among them a pavilion called that of the seven Phoenixes [14]. An earlier notice, found in the Liang Annals, specially mentions the frescoes adorning the king's palace [15]. That the city was enclosed by walls we know from Hsüan-tsang. He adds that they were of no great height, though the capital passed for never having been taken by force [16]. The ancient fortifications still existing at the sites of Endere, Kara-dong, Ak-sipil show that we may safely assume those walls to have consisted mainly of ramparts of stamped loess. At the gate of the city there used to take place the solemn reception by the king and the ladies of his court of the procession of great image cars of which Fa-hsien has left us so vivid a description [17].

Character of ancient remains. The culture-strata of Yōtkan do not help us to reconstruct a picture of the city which once occupied the site. But in view of the observations detailed above we have no difficulty

p. 226). This shape would be approached more closely by the irregular but approximately equilateral oblong of the actually excavated area than by one having, say, double its length without an increase in width.

It is, of course, doubtful whether the legend can bear so strict an interpretation. Yet it deserves to be noted that the legend related in the 'Annals of Li-yul,' discussed above, p. 159, also makes the mysterious light sent forth by Buddha *encircle* three times the spot where the 'great city with five towers called U-then' was to be built in a future age; see Rockhill, *Life of the Buddha*, p. 233. That legends of this kind may sometimes reflect exact details of local topography is a fact illustrated, e.g., by the old Kashmirian legend about the foundation of Pravarapura, in which the reference to the demon's knee has been shown by me to be based on a topographical fact still clearly visible in the configuration of the present Śrīnagar; see my *Rājat.*, I. pp. 100 sq.

[11] See above, p. 186; S. Lévi, *Notes chinoises sur l'Inde*, p. 39.

[12] See *Mémoires*, ii. p. 227; Rockhill, *Life of the Buddha*, p. 236.

[13] See Foucher, *L'Art du Gandhāra*, pp. 83 sqq.

[14] Rémusat, *Ville de Khotan*, p. 80, where the name of the palace is given as *Chin ts'ê tien* 金册殿.

[15] Rémusat, loc. cit., p. 16.

[16] See *Mémoires*, ii. p. 226.

[17] *Travels of Fâ-hien*, transl. Legge, p. 19.

in understanding the condition to which its débris has been reduced, and the peculiar character of the objects which alone can still be recovered from it. In a soil kept constantly moist through percolation from the irrigated top layer none but objects of exceptionally hard substance, such as terra-cotta, stone, metal, or bone, could escape decomposition in the course of centuries, and to this material the antiquarian relics of the site are in fact entirely confined. That these objects are mostly of very small size or fragmentary is also readily explained. They belong to the rubbish gradually accumulated at a site which probably for a period of more than a thousand years preceding the Muhammadan conquest had never ceased to be occupied by houses, and which even when the end came is not likely to have at once been abandoned completely. Objects of larger size and any practical utility were thus sure to be removed from the débris and to be utilized elsewhere.

It would have been undoubtedly a matter of considerable antiquarian interest to trace the exact distribution of the still recognizable remains through the layers of different depth. But a systematic search for such relics would have cost time, even if the convenient and effective method of excavation customary with the local diggers could have been followed. Their *modus operandi* consists of cutting away vertically the banks exposed at the edge of the excavated area with the help of little watercourses conducted over them. The soil carried off is subsequently washed in rough sieves, when, besides the tiny flakes of gold which attract the diggers, coins, gems, fragments of decorated pottery, and similar small objects can be readily recognized and extracted. The season of abundant water supply from the river, when alone such washing operations can be carried on, lasts only from July to September, and had long passed at the time of my visit. Hence my efforts at the site had to be restricted to the acquisition of as much of the last season's output of antiques as had not yet left the hands of the villagers.

Antiques purchased at Khotan.

In addition, I purchased through Munshī Badruddīn, the Ak-sakāl or headman of the Afghān traders at Khotan, and some other agents, whatever antiques were traceable in the Bāzārs of Khotan. It is certain that the latter receive by far the greater part of their supply in antiques such as ancient coins, cut stones, ornamented pottery, directly from the diggings of Yōtkan, and relatively little from those professional 'treasure-seekers' who annually in the winter months visit the old sites in the desert around the oasis, or else from chance finds elsewhere. For this reason it has appeared to me advisable to include in the list of antiques given at the end of the next Section those purchased by me at Khotan. But full certainty can never be obtained about the find-place of any objects which have once found their way into a Turkestān Bāzār, and consequently it would not be safe to base far-reaching arguments on any individual article procured through such channels. The distinctive marking of all batches of antiques purchased by me either at Yōtkan or in Khotan town will render this exercise of critical caution easy. It is scarcely necessary to point out that, even in the case of antiques bought at Yōtkan or avowedly brought to me from that site (marked with *Y.* in the list), the evidence as to their origin cannot have the same value as that of finds yielded by systematic exploration under my own eyes.

Coin finds at Yōtkan.

In respect of no other class of antiques would such full authentication be so important as in that of coin finds, for in the present state of our knowledge we must depend entirely on them for exact chronological evidence as regards the period during which the site of Yōtkan was occupied by the Khotan capital and its débris-layers deposited. Unable to secure this evidence by actual excavation, I took care to make my purchases of old coins from the site as far as possible at Yōtkan itself. The quantities of such coins annually washed out of the ancient strata seem to be considerable, and I had no difficulty in obtaining on my several visits to Yōtkan an aggregate of close on 200 specimens. All are of copper; I did not succeed in tracing any gold or silver coins, nor even information about such.

I much regret that neither during my journey nor subsequently after my collections had been deposited at the British Museum did I obtain sufficient leisure for an accurate numismatic analysis of these purchases; nor could I claim full competence for it, seeing that the vast majority of the coins secured from Yōtkan belong to purely Chinese issues. A considerable number are too much worn or corroded for exact determination. Of the rest, however, 150 have been classified with the kind help of Dr. Bushell and Prof. Rapson (see App. *D*); and the result of this preliminary analysis is sufficiently definite to be discussed here in its chronological bearing. Numismatic details in support of it will, I hope, be made accessible hereafter.

Sino-Kharoṣṭhī coins.

In the first place it is important to note that the collection includes over forty pieces of the Sino-Kharoṣṭhī type, bearing Chinese legends on the obverse and Indian Prākrit ones in Kharoṣṭhī characters on the reverse. A far larger number of these interesting coins has been examined by Dr. Hoernle, who, in Part i. of his 'Report on the British Collection of Central-Asian antiquities', has set forth the results of his painstaking researches with great lucidity and thoroughness[18]. Dr. Hoernle recognizes in these coins a Chinese currency of Khotan issued by its native rulers at a period later than 73 A.D., when, as we have seen above, Khotan appears to have first passed under effective Chinese supremacy. 'It is distinctly a Chinese currency, because the standard of the coins is Chinese, inscribed in Chinese language and characters. . . . On the other hand, the reverse of the coins bears the symbols and names of the native kings in native (Indian) characters—a fact which indicates both that native kings still continue to reign, and that the language and characters used by the native administration were Indian.' The use of the types of the bare horse and more rarely of the Bactrian camel (both turned to the right) on these coins leads Dr. Hoernle to suggest some special connexion between the Khotan rulers and the 'Scythian' kings of north-western India, Maues, Azes, and their successors, who are assumed on numismatic grounds to have ruled in the Punjāb and adjoining tracts from about 50 B.C. to 80 A.D.; on their coins the type of the bare horse occurs, with a great variety of others. In view of the obscurity which still hides from our view the origin of those 'Scythian' rulers of India and the true extent of their dominion, this resemblance in the coin types scarcely warrants at present far-reaching historical conjectures. But the justness of the main conclusions above indicated as to the character of the Sino-Kharoṣṭhī coins of Khotan and their historical significance cannot be subject to any reasonable doubt.

Find-place of Sino-Kharoṣṭhī coins.

My collection includes specimens of both the large and the small Sino-Kharoṣṭhī coins described by Dr. Hoernle (intended to represent according to their Chinese legends a weight in copper of one Liang and four Chu and of six Chu, respectively). Several varieties of the latter class distinguished by the arrangement of the Kharoṣṭhī legends, can also be recognized. But the specimens, as will be seen from those reproduced in Plates XLIX, LXXXIX, 2–4, seem too much worn to give hope of their proving useful for the determination of the legends and of the royal names they contain[19]. On the other hand, it is an observation possessing numismatic interest that whatever specimens of the Sino-Kharoṣṭhī currency I could secure or see were with some doubtful exceptions confined to the Yōtkan site. Considering the large number of ancient

[18] See *Report on C.-A. antiquities*, i. pp. 1–16, with Plates I, III.

[19] Dr. Hoernle distinguishes five varieties of the royal name: *Gugramada*, *Gugradama*, *Gugramaya*, *Gugramoda*, *Gugrañda*. None of these names can be traced otherwise. The recurrence of the identical *Gugra-* in the first part of these names finds, however, its parallel (as noted by Dr. Hoernle) in the Tibetan list of royal names all beginning with *Vijaya*; see *Report on C.-A. antiquities*, i. pp. 8, 16.

coins which were obtained by me from other sites, either in the course of my own explorations or through 'treasure-seekers', this negative fact acquires significance. We may, perhaps, conclude from it that this local currency of Khotan could have had but a limited circulation outside the capital, and further that its issues could not have extended over a very long period. Is it possible to assume that its reduced convenience for transport as compared with the square-holed purely Chinese pieces had something to do with this?

Period of Sino-Kharoṣṭhī coinage.

The form of the Kharoṣṭhī script as presented by these coins agrees closely with that which prevailed in India under Kuṣana or Yüeh-chih rule, and this circumstance has led Dr. Hoernle to assume that the Sino-Kharoṣṭhī coins could not be placed later than the end of the second century A.D. The Kharoṣṭhī documents on wood and leather discovered by me at the Niya Site have, however, since proved that the use of the Kharoṣṭhī script in this particular form continued in the Khotan region up to the second half of the third century of our era. Similarly, the finds of Chinese records at the same site, including one dated in 269 A.D., clearly demonstrate that Chinese influence in Khotan did not cease with the close of the Later Han Dynasty (220 A.D.). Hence the question as to the terminal date of the period to which the Sino-Kharoṣṭhī coin issues belong must be kept open for the present.

Indo-Scythian coins.

Of coins representing types which in their original issues were probably earlier than this local currency of Khotan my Yōtkan purchases contain an interesting variety. The oldest approximately datable one is a copper piece of Kujula-Kara-Kadphises (reproduced in Plate LXXXIX, 1), one of the earliest Kuṣana rulers, whose reign probably falls close to the beginning of our era[20]. Kaniṣka, the greatest of the Kuṣana kings, is represented by five pieces in copper, all much worn, but one still showing the figure and legend of the moon-god MAO. It is noteworthy that Dr. Hoernle's collection also contains a relatively large number of Kaniṣka coins obtained from Khotan[21].

Early Chinese coins.

Among the Chinese coins from Yōtkan are three well-preserved square-holed pieces without any legend, and, perhaps, more of the same kind among those classified for the present as 'illegible'. Of Chinese round coins, those bearing no legend belong to the earliest type. But as the type appears to have been current under both the Former and the Later Han dynasties our specimens afford no safe chronological evidence[22]. The one coin of the usurper Wang Mang bearing the legend *Hou Ch'üan* (see Plate LXXXIX, 11) belongs to an issue dating from 14–19 A.D. But here, too, it must be remembered that the type appears to have been current during the period of the Later Hans as well. The same remark applies to the *wu-chu* currency represented by three specimens.

The well-preserved Chinese coin, bearing two characters on the obverse and apparently of iron, shown in Plate LXXXIX, 5, presents special interest, as it seems to be unique, and has struck Dr. Bushell as the most ancient of the whole series of Chinese coins secured by me. The distinguished Sinologist has favoured me with the following note on it: 'From style, material, and script I would attribute it to the Former Han dynasty. The first character is certainly 于 *yü*, which I take to stand for Yü-t'ien (Khotan). The second appears to be an archaic form of 方 *fang*, meaning "territory, quarter", &c. This with some reserve—at first I thought it might be a form of 先 *hsien*, which used to be written 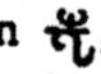, but the middle transverse stroke is wanting. The coin is unknown to Chinese numismatists, and must be of local mintage, like the interesting *wu chu* piece figured in *Mission Dutreuil de Rhins*, iii, p. 132.'

[20] Prof. Rapson refers for this coin to Cunningham, *Num. Chron.*, 1892, Pl. IV. 9. Regarding Kujula-Kara-Kadphises, see Rapson, *Indian Coins*, p. 17.

[21] See *Report on C.-A. antiquities*, i. pp. 27 sq.

[22] See Hoernle, loc. cit., i. p. 18.

Coins of T'ang dynasty.

My Yōtkan acquisitions comprise no Chinese coins later than the issues just mentioned, until we come to the T'ang period. Curiously enough there is only one piece with the legend *Kai-yüan t'ung pao* (see Plate LXXXIX, 24), first introduced by Kao tsu (618-626 A.D.), among the Yōtkan coins, though specimens of 'cash' bearing this nien-hao were frequently found by me at other sites, having been minted, as Dr. Bushell informs me, all through the dynasty. On the other hand, there are not less than fifty-eight specimens (see Plate XC, 28–30) of the various issues showing the name of the Ch'ien-yüan period (758–759 A.D.) of the emperor Su tsung. The next reign, that of Tai tsung, is represented by nine 'cash', all belonging to the Ta-li epoch (766–779 A.D.) (see Plate XC, 32), a relatively large number considering that coins of this epoch, as Dr. Bushell informs me, are rare in China itself.

Coins of Sung period.

The events discussed above, which put an end to Chinese control over Eastern Turkestān under the reign of Tai tsung's successor, Tê tsung (780–804 A.D.), fully account for the absence of any later T'ang issues among the Yōtkan coins. It is the more interesting to find among them six coins belonging to the Sung dynasty. Two of these (see Plate XC, 36, 38) are 'cash' of the periods Yüan-yu (1086–93 A.D.) and Shao-shêng (1094–97), respectively, of the emperor Chê tsung, while a third (Plate XC, 39) belongs to the Ts'ung-ning epoch (1102–06 A.D.) of his successor Hui tsung. The rest seem issues of the Yüan-fêng period (1078–85 A.D.). The evidence of these coins obliges us to conclude that the site of the ancient Khotan capital continued to be inhabited down to the beginning of the twelfth century. We need the less hesitate about accepting that evidence as the extracts reproduced by Rémusat from the Sung Annals, and briefly referred to in the preceding chapter, plainly indicate a resumption of close relations, at least commercially, between Khotan and China for the period comprised in the last reigns of that dynasty (1063–1124 A.D.)[23].

Rarity of Muhammadan coins.

Considering that the site could thus not have been abandoned for at least a century after the Muhammadan conquest, it is curious that only eight Muhammadan coins passed into my hands at Yōtkan. One of them probably belongs to Muḥammad Arslān[24], the rest are uncertain. Also among the coins said to have come from this site, but purchased at Khotan, there are only six doubtful specimens, which may belong to some Muhammadan mintage. It thus seems safe to conclude that Muhammadan coins must be rare at the site; and further that the Chinese 'cash', which seems to have been practically the only currency of Buddhist Khotan, after the Han times, must have continued in use for some time after the Muhammadan conquest.

However this may be, we cannot expect numismatic evidence to enlighten us as to the cause which led to the abandonment of the old capital; and as reliable records of the local history of Khotan fail us completely during the centuries which followed its conversion to Islām, there seems little hope of a definite solution of the problem.

Section IV.—ANTIQUES ACQUIRED FROM YŌTKAN AND IN KHOTAN

I have, in the preceding Section, explained the reasons for treating in one place whatever antiques were acquired by me at the site of the ancient capital or in Khotan town. For the same reasons no distinction has been made between such acquisitions in arranging the plates intended to illustrate the classes of objects which are represented among them. Two considerations justify me in restricting myself to brief explanatory remarks on these plates. In the

[23] Compare *Ville de Khotan*, pp. 91–100; above, pp. 182 sq.

[24] See above, p. 111.

first place the detailed list of the objects, which has been prepared by Mr. Fred. H. Andrews, and which will be found reproduced below with few additions and alterations, contains not only an accurate and full description of every article, but also indications of such special points of interest as may deserve further study by experts. In the second place, the acquisitions made by me do not differ in general character from the large collection of Khotan antiques which have accumulated through successive purchases on behalf of the Indian Government, and which at the hands of Dr. Hoernle have undergone investigation as thorough as it is learned. For all questions which bear upon the relationship between the art of these antiques and that of India and Greece, and for many others raised by their technique, probable use, &c., I may therefore refer to Dr. Hoernle's analysis in the second part of his 'Report on the British collection of Central-Asian antiquities'[1]. Even if sufficient leisure had been at my disposal for the purpose, I could not have hoped to supplement his observations on essential points without reference to collections and publications now far beyond my reach.

Ancient terra-cotta vessels.

Fragments of ancient pottery, plain or decorated, form the commonest of the antiques found at Yōtkan. But complete vessels of a size meant for actual use are rare, and hence the almost intact specimens shown in Plate XLIII are of special interest. The small jug (Y. 0024), perfect and of simple but remarkably graceful design, resembles a Greek Oenochoê. Another small vessel (Y. 0028), complete but for the lip, is curious for the clever decoration of its handle showing a monkey playing the Sitār—a motive amply represented among the terra-cotta figurines of Plate XLVI. The fine jar, with almost spherical body (Y. 0027. a), is interesting for the spiral fluting which decorates the body. The small bifrons vessel (Y. 0030), showing on one side a male, on the other a female head, is remarkable for the close approach of its design to that of Greek Bifrons vases; at the same time the peculiar features of the heads, the treatment of the hair, &c., unmistakably reproduce a local type which is found represented also in numerous terra-cotta miniatures. The portion of a coarse terra-cotta vessel (Y. 0022) shows in its simple incised ornaments and mouldings motives common to many of the small pottery fragments found in a more or less eroded condition on the Tatis near Gūma and around the Khotan oasis. The garland-holding female figures seen on the neck of the large jar (Y. 0023) were favourite decorative subjects with the Khotan artist. Specimens in terra-cotta recur on Plate XLV (B. 001. d; Kh. 003. b) and were frequently found among the stucco wall-decorations excavated at Dandān-Uiliq (Plate LVI). They are evidently meant for Gandharvīs, and their treatment imitates similar figures frequent in the Graeco-Buddhist reliefs of Gandhāra.

***Appliqué* pottery ornaments.**

The moulded ornaments of the jars were invariably *appliqué* work, i. e., made separately and attached before firing. Plates XLIV and XLV show numerous pieces of this kind, most of them broken off. Grotesque heads, human or animal, are frequent among them. Particularly noteworthy is the satyr-like mask (Kh. 003. k) decorating a handle, for its clever naturalistic treatment and distinctly classic look. There is a suggestion of Neptune-like features also in the grotesque mask, Y. 0017 (Plate XLIII). Full figures or small groups in relief were similarly used for decoration, as seen in Y. 0019, B. 001. c, Kh. 003. c, Kh. 003. m, and Mac. 001. The latter piece, the richly ornamented neck of a jar, shows an unmistakably Indian motive in the two halo-crowned figures seated on lotus-pedestals with an elephant to support them as a *vāhana*, while the other *appliqué* decoration, with its two parrots kissing above a bunch of grapes, recalls the use of these elements of decorative art in Gandhāra sculptures, the Ajanta frescoes and, in similar combination, in Coptic embroideries.

[1] See *Report on C.-A. antiquities*, ii. pp. 42–55.

Terra-cotta heads and masks.

Of the heads worked in the round (B. 001. f, g, h, i; Y. 0012), it may also be supposed that they served the purpose of vase decoration, though it is not easy to determine the exact position they occupied. Of special interest is the large female head (Y. 0031) which clearly suggests modelling after a well-defined local type, recognizable also in some small worshipping figures of the Dandān-Uiliq frescoes (see Plates III, IV). The very oblique eyes are in curious contrast to the thoroughly 'Aryan' look of the other well-shaped features. The arrangement of the hair and the elaborate ornaments manifestly reproduce local fashion of the period[2]. The fragmentary mask (B. 001. a), shows an entirely different type of head. With its large and strongly aquiline nose, its eyes near together and but slightly oblique, it recalls some marked features in the fine Gandhāra relief in the Lahore Museum known as an 'Indo-Scythian king'. Though this sculpture in reality is intended to represent Kubera, the guardian of the northern region, the 'Scythian' type of its portrait-like head is unmistakable[3]. Of the grotesque animal figures represented by Y. 009. a, b, Dr. Hoernle has already shown that they served as decorative handles of vases[4].

Terra-cotta figurines.

Plates XLVI and XLVII illustrate a class of Yōtkan antiques largely represented in former collections, that of terra-cotta figurines. The subjects are chiefly figures of animals shown in the round, among which monkeys largely predominate. The variety of treatment shown by these miniature representations is very remarkable; but whether the monkey figures are treated in careful naturalistic fashion or but roughly modelled, interest is imparted to them by the clever way in which human expression is given to their faces, or human gestures and poses humorously indicated. We see them most frequently playing on musical instruments, as if to mark the weakness prevailing among the Khotan people, ancient and modern, for such entertainment. Among the instruments we find the *sitār* (Y. 0011. b, d, i), a kind of *rabāb* not unlike the one of which a portion was excavated by me at the Niya Site (Y. 0011. k), more often the syrinx or Pan's pipe (Y. 011. c, e, f, j; Y. 0013. e), and various kinds of drum (Y. 0012. n, Kh. 003. m). To the special interest possessed by the occurrence of the syrinx, which seems quite foreign to Indian art, attention has already been called by Dr. Hoernle[5]. We see monkeys sitting in meditation, praying, laughing, or feasting (Y. 0014. b, B. 001. 1, Y. 0011. g, Y. 0014. a, B. 002). But far more common are ithyphallic representations (Y. 009. i, q, Y. 0012. a. iii, Y. 0013. c, Y. 0025) or else groups in amorous embrace (Y. 0012. a. i, a. ii, z, y, Y. 009. r). The relatively large proportion of such poses and scenes curiously reflects the reputation for sensuality enjoyed by the Khotan population since ancient times[6].

Terra-cotta grotesque monkeys.

The skill with which the character of the monkey is expressed in most of these figurines, down to such details as the direction of hair growth (Y. 0012. d) or the shape of the skull (Y. 0013. b, B. 002), is all the more curious since the monkey can never in historical times have been found in a free state or otherwise been common in Khotan. In the figures of camels or horses (Y. 0012. p, t, Y. 0013. d, Y. 009. l, B. 001. j, Y. 009, c, l) we find, on the other hand, a far more conventional style of work. We have a human subject in the infant in swaddling clothes (Y. 0012. c), and various miniatures of household vessels in Y. 002. b. i, Y. 0012, w, x.

[2] The male head reproduced in Plate XI. 1, of Dr. Hoernle's *Report*, part ii, which may have formed the neck of a large vessel, is a curious *pendant*; for the similar treatment of the hair in both heads compare the bifrons vessel, Y. 0030, Plate XLIII.

[3] See for this interesting sculpture Grünwedel-Burgess, *Buddhist art*, p. 137, fig. 88; also above, p. 158.

[4] See *Report on C.-A. antiquities*, ii. p. 40; also frontispiece.

[5] See *Report on C.-A. antiquities*, ii. p. 49.

[6] See above, pp. 139, 142.

Of the pottery and terra-cotta illustrated by Plates XLIII–XLVII, it cannot reasonably be doubted that they are all typical products of the local art of ancient Khotan. It is very different in regard to the small objects in stone, metal, and bone which are shown in Plates XLVIII–LI. On some of the miniature stone carvings in the first-named Plate the style of work as well as the material makes it highly probable that they were made in distant Gandhāra and brought to Khotan through the link of common Buddhist worship. Among the intaglios of Plate XLIX there are at least two (Kh. 002, Kh. 001. a) which are unmistakably the work of engravers in the classical West, while others point to Īrān or India as their places of origin. Considering how easily small objects of value make their way by trade or otherwise over vast areas, it is impossible to feel sure that the rest of these cut stones, or the metal seals and miniatures in the other Plates, were all produced at Khotan. But this doubt in no way detracts from their antiquarian interest; for it is evident that, if originally brought from elsewhere, these small antiques found at Yōtkan must be tangible evidence of the varied art influences which were at work in ancient Khotan. Objects of foreign origin.

This is particularly true of the small but remarkably well-carved reliefs (Kh. 003. g, Kh. 005, B. D. 001. a) in a slatey stone which Plate XLVIII shows. They are characteristic specimens of that Graeco-Buddhist art of Gandhāra which, as we shall see, contributed more than any other influence to shape the development of Buddhist art in Khotan. The detailed description contained in the list renders it unnecessary to account here for the peculiar shape of these reliefs. They undoubtedly formed part of the decoration of miniature shrines or Stūpas, such as we must suppose to have been in use for purposes of private worship or else as votive offerings[7]. The most interesting of these pieces is Kh. 003. g, which shows on its obverse a scene famous in Buddhist legend, but not hitherto found among the reliefs from the ruined sites of Gandhāra. The Bodhisattva, having forsaken his princely abode to take up the life of a mendicant, is cutting with the sword in his right hand his long lock (*cūḍā*), the end of which is grasped by his left. M. Foucher, who at once recognized the scene on being shown its reproduction in the Plate, has since discussed the legend of the texts and its extremely rare representation in Buddhist iconography with his usual lucidity and acumen[8]. Miniature reliefs.

The other two small reliefs represent scenes familiar in Gandhāra sculpture, the birth of Buddha (Kh. 005) and Buddha's temptation by Māra (B. D. 001. a). In the former the clever adaptation of the relief on the obverse to the contour furnished by the larger scale carving of the reverse is deserving of notice. For the remaining small sculptures shown in Plate XLVIII, four of them seated Buddhas or Bodhisattvas, Indian origin is less clearly indicated; but the influence of Gandhāra style is recognizable in all of them, and the materials (soapstone and slate) are such as we commonly meet with in the sculptures of the Peshāwar valley. The small ivory group in the round (Kh. 008), on the other hand, might well be Khotan work; the treatment of its subject, a male figure embracing a female, seems closely allied to that observed in certain terra-cottas, and the fashion of hair and dress similarly shows points of contact[9].

Among the intaglios from Yōtkan reproduced in Plate XLIX the standing Eros (Kh. 002), and the quadriga with charioteer (Kh. 001. a) are manifestly Roman work of the early centuries A. D. Intaglios.

[7] Miniature Stūpas and fragments of panels which once adorned miniature chapels abound among the collections of Graeco-Buddhist sculpture at Lahore and elsewhere, though I am unable at present to refer to pieces on an equally small scale.

[8] See *L'Art du Gandhâra*, pp. 363 sqq. I owe it to Mr. F. H. Andrews to mention that his description of the little relief, with the identification of the scene represented in it, was written long before I had an opportunity of submitting the Plate for M. Foucher's friendly scrutiny.

[9] See, e.g., Hoernle, *Report on C.-A. antiquities*, ii. Pl. XI, fig. 23.

Impressions from intaglios very closely related to the former in style and character (Eros, Pallas, Heracles) appear in the clay seals on some of the Kharoṣṭhī documents discovered by me at the Niya Site, which date from about the middle of the third century[10]. Of the rest of these cut stones Professor Percy Gardner, to whom I am indebted for a close examination of them, holds that they belong to the second and third centuries A. D., and that most of them are rather oriental than Roman. The large intaglio (Y. 008. b) is interesting on account of its careful delineation of a warrior of Indo-Scythic type arrayed in an elaborate assortment of arms. Y. 008. a shows an inscription of 'unknown' characters above its device, two lions fighting over a prostrate bull. In this connexion it may be mentioned that the intaglio (I. 001) also shows a legend hitherto undeciphered. Its characters closely resemble the corrupt Greek letters found on the so-called 'Scytho-Sassanian' coins, and the features of the king's head surrounded by the legend point equally to origin in the Indo-Īrānian border lands.

Of the seals from Yōtkan or Khotan shown in Plate L, all with one exception are in bronze, and either square or oblong. The best designed among them is, perhaps, Y. 009. k, showing a cow with figure milking. Seals of exactly similar make are numerous in the collection reported on by Dr. Hoernle[11], and it is thus probable that we have in them work of the local metal-workers whose skill the Chinese Annals specially mention. B. D. 001. b, a bell-shaped seal in bone or ivory, shows a Tibetan inscription; since the piece was purchased from a Khotan trader, its antiquity cannot be vouched for independently of such evidence as the design or writing may afford. Among the small miscellaneous objects in metal or stone which Yōtkan and Khotan have contributed to Plate LI, few call for special mention. The tiny figure of a monkey (Y. 004) in solid gold is interesting on account of its excellent workmanship, and as the only specimen of work in precious metal which can with certainty be traced to that site. The octagonal weight in bronze (Y. 003), likewise obtained at Yōtkan, may prove of value if we ever obtain definite indication as to the system of weights current in ancient Khotan. Its weight (425 grains) approaches, as Professor Percy Gardner points out, that of three staters. The miniature figures in bronze representing Buddha or a Bodhisattva (Y. 002. a, Kh. 003. j) are too much corroded to permit of any certain conclusion as to their age or origin.

ANTIQUES FROM YŌTKAN PURCHASED AT SITE.

Y. 002. a. Bronze seated Buddha figure showing distinct traces of having been gilt. Detail of figure much lost by wear. Behind the head a very elongated aureole. The general appearance of the miniature is Chinese. $\frac{7}{8}'' \times \frac{5}{8}''$. See Plate LI.

Y. 002. b. Three small charms.

- i. **Miniature vase in terra-cotta.** See Plate XLV.
- ii. **Duck carved in a greenish stone.** See Plate LI.
- iii. **Tortoise or frog of glass.** See Plate LI.

Y. 002. c. Bronze seal, $\frac{7}{8}''$ square, with Svastika device; shank with hole at back. See Plate L.

Y. 003. Octagonal piece of bronze, probably a weight, with edges of upper and lower octagonal faces chamfered. The upper and lower faces are stamped, each four times, with a label containing four roundels or disks. Side facets also stamped with the same pattern placed so as to form a zigzag line round the object. Weight 425 grains (3 staters?). $\frac{5}{8}'' \times \frac{1}{2}''$. See Plate LI.

Y. 004. Miniature monkey in solid gold. It is squatting on its haunches, its hands together and forearms resting on knees. It is well proportioned, has small ears, thin waist, no tail; hairiness suggested by tiny dots indented all over it. $\frac{7}{16}''$ high, $\frac{3}{16}''$ broad. See Plate LI.

Y. 005. Small round intaglio, carnelion; head of wolf or bear in profile. Diameter $\frac{5}{16}''$. See Plate XLIX.

Y. 006. Elliptical intaglio, garnet, roughly cut, goat to R. proper, feeding from a tree; on its back a bird (crow?). $\frac{9}{16}'' \times \frac{7}{16}'' \times \frac{3}{16}''$. See Plate XLIX.

[10] See Plate LXXI.

[11] See *Report on C.-A. ant.*, i. Pl. III.

Y. 007. a. Bronze seal, 1" square, representing lion (?) in profile seated to R., with open mouth and with conventional bushy tail curling away from back. Perforated shank at back broken; comp. Hoernle, *Report on C.-A. ant.*, i. Pl. III., No. 61. See Plate L.

Y. 007. b. Fragment of stone relief, representing torso, R. p. arm and upper portion of L. arm of seated figure. It is nude, save for a heavy necklet and pendant, an armlet with decorative circular jewelled bosses on each arm, near the shoulder, and a bangle at wrist. Judging from the very carelessly carved hand remaining, the pose was the *dhyāna-mudrā*. Height 1½", width 2". See Plate XLVIII.

Y. 007. c. Portion of bronze clasp, through which remains of two iron rivets. The shape is elliptical, one end terminating in a circular ring. Design chased on elliptical part is arabesque; back plain. Length 1", width ½".

Y. 008. a. Intaglio, garnet, below an Indian cow couchant to L.; on either side and meeting above cow, a lion rampant. Over, inscription in unknown characters partially erased. Shape of seal elliptical. 1⅝" wide, ⅞" high, Front convex, back flat. See Plate XLIX.

Y. 008. b. Intaglio, carbuncle. Warrior of Indo-Scythic type, front, head to L.; fully armoured. On the head is a kind of cassis, with a curtain of plate mail hanging from the back half, fitting over the ears, passing under the chin and forming a gorget. He appears to wear a tunic or cuirass of either scale armour or quilted material, and a skirt reaching below the knees of a similar nature, but not quite the same. The shoulders and upper part of the arms are guarded by jointed armour, the forearm and just above elbow being either bare or covered with close-fitting sleeve. Legs and feet in close-fitting top boots. Carried in left hand and resting on shoulder is a round-headed mace. In right hand a spear. Across front of thighs, with hilt to right, is slung a long straight sword. At back of thighs, and parallel to the sword, is slung what may be a quiver. Behind that and showing one corner to left of figure, bow in case. The seal is well cut and polished. Height ⅞", width ⅝", thickness $\frac{5}{16}$". Front highly convex, back flat. See Plate XLIX.

Y. 009. a. Terra-cotta ornament showing grotesque animal, and forming probably handle of vase (see Hoernle, *Report on C.-A. ant.*, ii. p. 44); fragment. The head and ears are evidently suggested by the camel. Rising above the forehead is a single, straight, horn-like projection (broken). The neck is arched and maned as that of a horse. On the left shoulder is a raised circular boss, rudely ornamented, from which springs a wing directed backwards and upwards. Seven quills are countable, divided by incised lines, each feather having series of dots dug into it. The ornament was cast in two halves in moulds; then the insides of the halves were roughened and stuck together, after which they were fired. Nearly the whole of one side of this specimen is broken away. It is also broken across just below the wing and across the upper part of what appears to be a fore leg. Length 3", height 2¼". See Plate XLV.

Y. 009. b. Terra-cotta ornament; grotesque animal closely resembling Y. 009. a. The head of this is more bird-like. The horn-like projection on top of head is intact on one side, and is in three tiers. The decorative boss on shoulder is of different pattern. No wing is visible, the fragment being broken closer to the boss; and also higher up the leg. The end of beak is broken, and nearly the whole of one side is broken away. 3" x 2". See Plate XLV.

Y. 009. c. Terra-cotta miniature horse, legless. It carries a saddle of the usual oriental type—high in front and at back, which is most elaborately secured. Two straps or bands run from each side at the back, and disappear beneath the bushy tail. One at each side in front passes to front of chest. One at each side in front again passes to a strong headstall, on which seems to be a bit. Over the saddle is a saddle-cloth with ornamental border. The horse is fairly well modelled, but the front of the head is very flat and the eyes simply circles with a dot in the centre of each. The frontal band of the headstall rises to a point in the centre above the eyes and ears. The mane is bushy and upright and clipped. Although the legs are missing, the object does not appear to have been broken, as the under surfaces are smooth. Length 2⅝", height 1", width ⅝". See Plate XLVI.

Y. 009. d. Terra-cotta spout broken from some vessel. It is crudely decorated with eyes, nose, and ears (one ear broken away) and is brought up to a kind of knob to represent a forehead (cf. T. M. 005. g.). The hole running through is rectangular at the end towards the vessel, and circular at the mouth. Length 1⅞", width (at mouth) ⅝". See Plate XLV.

Y. 009. e. Fragment of terra-cotta monkey, sitting on his heels. Arms and L. leg missing (broken). Height 1¾".

Y. 009. f. Fragment of terra-cotta monkey, upper half. On head a cap with rolled brim; left hand held over mouth and nose. Height 1½". See Plate XLVII.

Y. 009. g. Terra-cotta monkey's head, broken off at neck. Round neck appears portion of a rope or cord. Head apparently modelled as a separate piece. Modelling very rough. Width $1\frac{5}{16}$", height 1½". See Plate XLVII.

Y. 009. h. Terra-cotta grotesque, two monkey-like figures riding astride a horse. The head of the rear figure is missing and the left hind leg of horse is broken off. The horse bears a general resemblance to Y. 009. c, but is much less carefully modelled. There is no harness whatever. Chest and shoulders of animal are evidently meant to be very hairy, being marked with little digs of a modelling stick. The mane is very thick and arched, and is erect, a feature having quite a Roman appearance.

The foremost of the two riders appears to be without clothing. The R. arm and hand are raised as though holding or caressing the horse's mane. The left arm hangs down and in front, and the hand appears to grasp a narrow bag-like object which is thrown across the horse in front of the rider. This may be intended for the front of a saddle. The rider's head is grotesque and set low on the square shoulders.

The rear figure, also astride, appears to have an object in front of it, which may be intended to represent the back of the saddle. A fur cloak or perhaps just an animal's skin hangs from the neck and shoulders, in five long plaits or sections, one of them resting on what appears to be the horse's tail twisted up into a ball. The two arms grasp the foremost figure. Length 2", height 2", breadth $\frac{13}{16}$". See Plate XLVI.

Y. 009. i. Terra-cotta fragment of squatting monkey, legs missing. The workmanship of this example, although rapid, is very clever. The character of the monkey is perfectly preserved, while human interest is given by the expression of the face, and the handling of the *sitār* it is playing. The hair on its head is expressed by a few strokes of the sharp point or blade of a modelling tool; yet it is unmistakably the growth of the hair of the monkey. The eyes are apparently put in with a stamp. The mouth is cut with a thin instrument, with a hole dug at each corner giving a humorous expression. Height $1\frac{1}{4}$", width 1". See Plate XLVI.

Y. 009. j. Terra-cotta grotesque ornament, from a jar. Half human and half lion head in relief. The heavy lion's upper lip is marked with holes to represent the large bristles and softens off on each side into sweeping human moustache curling upwards. The nostrils of the broad nose are directed forwards; the eyes are deep sunk under heavy brows and are modelled as round beads with a hole in the centre of each. They are cleverly placed and the surrounding part well modelled. The brows are heavily contracted, and above is human hair parted in the centre and brushed right and left. The single ear unbroken stands out sharply from the temple and is not well placed. The whole is evidently cast in a mould. The hole in the L. proper nostril goes through to the back of the cast, as does also one at the right proper angle of the mouth. Height $1\frac{7}{8}$", width 2". See Plate XLVII.

Y. 009. k. Square bronze seal, representing a cow to left being milked by a seated figure. A calf tries to suck and has its head directed upwards and to the R. towards udder. The cow's head is turned full face as though to watch calf. Well proportioned and of good design. There is a small hole through thin part; perhaps a fault in casting. Shank at back, broken. $1\frac{1}{16}$" square. See Plate L.

Y. 009. l. Terra-cotta Bactrian camel, badly modelled. Made in two halves, in moulds, and joined before firing. The tail, which is twisted laterally, was put on separately. Portions of a broad girth remain. Long hair on front of neck indicated by coarse toolings. There is other harness indicated on the legs. The four legs are represented by two thickened pieces, one fore and the other hind. On the top of the head is a conical protuberance similar to those on Y. 009. a. Length $2\frac{5}{8}$", height $2\frac{1}{2}$", thickness $\frac{7}{8}$". See Plate XLVII.

Y. 009. m. Fragment of terra-cotta grotesque monkey. Portion of torso and head only. The face is very owl-like and the eyes have been put in with a stamp. The head is turned to R. proper. The arms appear to have been outstretched. Height $\frac{5}{8}$", width $\frac{9}{16}$".

Y. 009. n. Terra-cotta bird, made to string on a thread as a bead. Judging from its general form and the large erect crest it is probably intended to represent the *hoopoe*. The beak is missing. The bird is well-shaped, with wings (which are broken at end) folded over body. There are markings indicating feathers, and transverse lines incised to suggest the different rows of wing feathers, and probably their colouring. No legs. Length $\frac{15}{16}$". See Plate XLVI.

Y. 009. o. Terra-cotta grotesque monkey squatting on his heels. With both hands he holds something to his mouth, probably a syrinx. In his lap is an object probably meant to represent a sort of short kilt. Height $1\frac{3}{8}$", width $\frac{3}{8}$". See Plate XLVII.

Y. 009. p. Terra-cotta grotesque head, probably ornament from a jar. It is a purely animal head something like that of a hyaena, but all the teeth (shown in the upper jaw, the mouth being open) are short and regular. The end of the nose is broken. Heavy, curled moustaches furnish the upper lip; the eyebrows are treated similarly and have the appearance of buffalo horns, two holes being placed near their bases which meet above the nose. The eyes seem to have been applied after the head came from the mould, as were also the several holes. There is a hole in the centre of the curled ends of the horn-like eyebrows. Two large, long and thick ears issue from beneath the eyebrows. The receding forehead is in two large ridges, and closely curled hair surrounds the piece, the centre of each curl having a hole. There is no chin, and the lower lip is modelled. The back is hollowed out, and the impressions of the maker's thumbs and fingers are clearly registered. Height $2\frac{1}{8}$", width $2\frac{1}{8}$", depth $1\frac{3}{8}$". See Plate XLIV.

Y. 009. q. Terra-cotta grotesque monkey seated astride a phallus. Naturalistic type and cleverly executed. The figure is a female, having well-developed human breasts. Portion of the tail remains. The legs are gathered up so that the heels touch the buttocks, and the phallus upon which the figure is astride runs backwards, forms a ring, and brings its head straight up between knees of figure and as high as the waist. From below the legs of the figure issue downwards a second pair of legs, and encircling the upper curve of phallus bring their feet together in front. The arms of the figure hang down, the hands resting on the inner angles of the thighs. Above them the umbilicus is strongly marked. (For treatment of monkey comp. Y. 0012. d, Y. 0012. v, B. 002). Height $1\frac{1}{4}$", width $\frac{1}{2}$", depth $\frac{5}{8}$". See Plate XLVII.

Y. 009. q. i. Terra-cotta grotesque monkey seated astride a phallus. Similar to Y. 009. q but less carefully made. There is a single pair of legs only. The head of phallus on which the hands rest reaches to face of monkey, so that the sex of monkey is indistinguishable. Height $\frac{7}{8}$", width $\frac{5}{16}$", depth $\frac{1}{2}$". See Plate XLVII.

Y. 009. r. Terra-cotta grotesque. Two monkeys, naturalistic type, embracing amorously. The head of one is missing, and the lower portions of legs of both. One has a long tail of hair hanging from the top of the head to the root of the tail proper. The tail is very broad at the root, spreading over the upper surface of buttocks. The tail of the other figure is similar. Their legs are entwined; and the arms of the one pass under the arms of the other, the hands of each supporting the other behind the shoulders. Hair is indicated only on head and tails. They both seem to wear bangles on their wrists. Cleverly modelled. Height $1\frac{7}{16}$", width $\frac{5}{8}$", depth $\frac{5}{8}$". See Plate XLVII.

Y. 0010. Terra-cotta human head of very low type. The face is rather broad, with large eyes obliquely placed, nose short, flat, and broad. Full and pointed beard. Bushy eyebrows. Very low forehead and bull-necked. Mouth coarse, ears high. The whole very clumsily executed. Height $1\frac{1}{4}$", width $1\frac{1}{8}$", depth 1". See Plate XLVI.

Y. 0011. a. Terra-cotta grotesque head of crocodile. Lower jaw missing. The end of the snout, which is moderately broad, bears the protuberance such as the *Gavialis gangeticus* has. The ears are shown as external organs. The armour of the crocodile is expressed by a series of transverse ribs. The eyes are too wide apart, and on the sides instead of the top of head. A kind of eyebrow is indicated by the usual dashes representing hair; and a collar encircles the neck and runs behind the ears. On the head between the eyes is a mark made with a circular stamp. Length $1\frac{7}{8}$", width $\frac{3}{4}$". See Plate XLVI.

Y. 0011. b. Terra-cotta grotesque monkey, hairless and nude, seated upon a long object (a log?) with legs dangling, playing the *sitār*. The L. proper arm is broken off. The head is of owl-like type. Height $1\frac{1}{4}$", width $1\frac{3}{8}$". See Plate XLVI.

Y. 0011. c. Terra-cotta grotesque monkey, hairless, and with narrow band thrown over lap; seated on something, the legs dangling and seeming to be only to knees. The two hands are raised, grasping a syrinx upon which it plays. Face of owl-like type. A hole has been dug between the legs. There is another on the back which is probably due to some impurity in the clay. Height $1\frac{1}{4}$". width $\frac{11}{16}$". See Plate XLVII.

Y. 0011. d. Terra-cotta grotesque monkey, having hair indicated on the body and head. It is a fragment, to the middle only; face naturalistic. The figure is playing a kind of *sitār* with four strings. The left hand shows fingers and thumb grasping the neck of the instrument, while the fingers of the R. hand are roughly suggested by the usual digs of the modelling tool. Height 2", width $1\frac{1}{2}$", depth $\frac{7}{8}$". See Plate XLVI.

Y. 0011. e. Terra-cotta grotesque monkey; naturalistic type. It grasps in both hands a very large syrinx, very clearly shown. The six pipes are held together by a single band round the centre. The monkey is squatting but the legs are not formed, the body ending in a roughly quatrefoil-shaped base enabling it to stand. Hair is indicated all over the body. The mouth very cleverly suggests the peculiar formation of the upper lip assumed by a person blowing downwards into a pipe or flute. Height $2\frac{3}{8}$", width $1\frac{1}{4}$", depth $\frac{7}{8}$". See Plate XLVI.

Y. 0011. f. Terra-cotta grotesque monkey, fragment to middle, in pose similar to Y. 0011. e, but from different mould. The syrinx is smaller and shows two cinctures. Six pipes are shown. The arms and hands are more carefully modelled than in Y. 0011. e, as is also the head, although it has not such a successful blowing expression. This example has a very satyr-like appearance. Clay deep red and fine texture, similar to Y. 001. j. Height $1\frac{5}{8}$", width $1\frac{1}{8}$", depth $\frac{3}{4}$". See Plate XLVII.

Y. 0011. g. Terra-cotta grotesque monkey. Part below hips broken away. Naturalistic type, but with abruptly protruding mouth and snout. Ears very low, narrow, and forward. Body hairy. The hands are placed together, the finger-tips touching, in an attitude of supplication or humility. Round the wrists are bands. This specimen seems to be untouched after leaving the mould. Height $1\frac{3}{8}$", width $\frac{7}{8}$". See Plate XLVII.

Y. 0011. h. Fragment of terra-cotta grotesque monkey. Head and part of neck. Type very similar to Y. 0011. g, but mouth not so prominent, and eyes smaller. Height $\frac{13}{16}$", width $\frac{5}{8}$", depth $\frac{11}{16}$". See Plate XLVI.

Y. 0011. i. Terra-cotta grotesque monkey. Owl-face type. Appears to be squatting, but R. proper leg is missing. Both hands are raised to mouth. Height $1\frac{1}{8}$". See Plate XLIV.

Y. 0011. j. Terra-cotta grotesque monkey, in sitting posture, the legs dangling from the knees. In the lap across middle is some object which may be clothing: otherwise nude and hairless. Owl-face type. Hands raised holding syrinx to lips. Skin markings from the potter's fingers are visible on this. Clay dark red and of fine texture. Height 1".

Y. 0011. k. Terra-cotta grotesque monkey, naturalistic, seated cross-legged, playing four-stringed *rabāb*. The whole piece is well modelled, and shows interesting detail in the *rabāb*. Height $1\frac{1}{2}$", width $1\frac{3}{8}$". See Plate XLVI.

Y. 0012. a. Terra-cotta grotesque monkey, naturalistic type. R. arm broken off. Figure seated, the legs being shortened to mere stumps above knees. A well-defined plaited kilt in front of hips. The back covered with hair. The left arm hangs at side, the hand resting on the thigh in front. Height $1\frac{9}{16}$", width $\frac{3}{4}$". See Plate XLVI.

Y. 0012. a. i. Terra-cotta grotesque, similar to Y. 0012. y, but both heads are present, snout touching snout. Neither has a pigtail or tail proper. Height $1\frac{3}{8}$". See Plate XLVII.

Y. 0012. a. ii. Terra-cotta grotesque, similar to Y. 0012. y, but both heads are present as in Y. 0012. a. i., and one figure has pigtail. Height 1". See Plate XLVII.

Y. 0012. a. iii. Terra-cotta grotesque monkey; naturalistic type. Seated, the legs mere stumps. R. hand laid flat on breast, with outspread fingers, L. hand *veretrum tenens.* The face is very pointed, nose and mouth projecting considerably. Hair of head brought to a conical point. Body hairy, and girdle shown at back only of waist. Cf. Y. 0013. c; A. 001. a. Height $1\frac{3}{8}$". See Plate XLVII.

Y. 0012. b. Terra-cotta grotesque monkey. Similar to Y. 0012. a, but L. hand rests on side of thigh and a small cylindrical object is held under the arm against the side. The R. arm is flexed across the body and the R. hand rests upon the near end of the cylindrical object. Latter marked by incised lines running parallel to its length. Height $1\frac{1}{4}$", width $\frac{3}{4}$". See Plate XLVI.

Y. 0012. c. Terra-cotta grotesque human infant, swathed from neck to ankles in a binding which encircles it five times. The feet are also covered. There is a slightly owl-like treatment of the eyes and brow (which is very low), but distinctly different from the monkey type. The figure is evidently intended to lie flat, as the back is unmodelled. Length $1\frac{1}{4}$", width $\frac{3}{8}$". See Plate XLVI.

Y. 0012. d. Terra-cotta grotesque monkey. Naturalistic type. Very minutely modelled, particularly the head and face. It is seated cross-legged, the left foot resting on right thigh. The R. leg is broken off at the knee. Both arms are missing. Care is shown to represent the various directions of the hair-growth—not always correctly. The figure is nude as to drapery. Height $\frac{15}{16}$". See Plate XLVI.

Y. 0012. e. Terra-cotta grotesque monkey. Owl-like type, but face very deformed. It is seated on its heels, but with the toes pointed directly downwards. It plays upon a kind of *sitār*, the L. arm being stretched at full length holding the neck of the instrument. The head is turned slightly, as though to allow the L. ear to catch the sound. Height $\frac{7}{8}$". See Plate XLVI.

Y. 0012. f. Terra-cotta grotesque monkey, squatting, owl-like type. Arms broken off. Object in lap serving as kilt or apron. Very roughly modelled. Height 1".

Y. 0012. g. Terra-cotta grotesque monkey, owl-like type; long body; squatting on heels. Apron-like kilt. Arms and R. leg broken off. Height $1\frac{1}{8}$", width $\frac{1}{2}$".

Y. 0012. h. Terra-cotta bird. Resembling Y. 009. n, but not pierced for thread, and more coarsely modelled. Length $\frac{15}{16}$". See Plate XLVI.

Y. 0012. i. Terra-cotta bird; variation of Y. 0012. h. Length 1". See Plate XLV.

Y. 0012. j. Terra-cotta grotesque bird. Wings broken off short. The body is composed of a flattened piece, curved into a hook shape, the under surface of the upward curved tail part being marked with digs of the modelling tool, and resembling slightly the re-curved body of a scorpion. The other end is rounded, and is fashioned into a head having a well-formed beak but a flattened and spreading face rising into a transverse crest. It is marked with incisions. Beneath the chest is a sort of pedestal, flattened at the bottom so that it will stand. Length $1\frac{5}{16}$", height $1\frac{5}{16}$". See Plate XLVI.

Y. 0012. k. Terra-cotta grotesque monkey, naturalistic type. Head and torso only. Cf. Y. 0011. g. Height $1\frac{3}{8}$", width $\frac{3}{4}$". See Plate XLVII.

Y. 0012. l. Terra-cotta grotesque monkey, owl-like type. Legs broken; head to R. proper. In front of chest is held a rather shapeless object, flat on upper and lower sides; one hand rests on the top, and the other below supports it. Figure hairless. Skin marks from potter's fingers visible. Height $1\frac{3}{16}$", width $\frac{5}{8}$". See Plate XLVI.

Y. 0012. m. Terra-cotta grotesque monkey, naturalistic type. Arms missing; legs short stumps; sitting posture; body badly formed; hair indicated on head and front of body. Kilt-like apron. Height 2". See Plate XLVI.

Y. 0012. n. Terra-cotta grotesque monkey; naturalistic type. The figure is lying on its back on a kind of couch formed by the lower parts of two legs placed close together. A pillow looking like an Indian drum rests across the knees of the legs forming the couch, and supports the head of the reclining figure. The monkey is nude and hairy, and holds between its two hands, and resting on its abdomen, a small drum. The legs forming couch appear to be clothed in knee breeches and top boots. Length $1\frac{1}{2}$", width 1", height 1". See Plate XLVI.

Y. 0012. o. Terra-cotta head with quaint head-dress, features crudely expressed. Coronet from upper part of which jug-handle-shaped piece curves over to back of head; in its loop a small pillow-shaped pad or billet. Hair drawn back and down towards nape of neck; tied, and tail turned sharply up covering lower part of 'jug-handle.' A band passes up from below ears to upper part of tail. Coronet or hair curves round above ears to cheeks. Length $1\frac{1}{2}$". See Plate XLVI.

Y. 0012. p. Terra-cotta grotesque dog (Chinese pug). L. fore leg missing. Roughly modelled. The tail is curled over the back. The R. ear is directed forward (the L. ear missing). A series of dots made with the modelling tool decorate the back, from the neck to the waist. Length 1", height $\frac{3}{4}$". See Plate XLVI.

Y. 0012. q. Terra-cotta bull's head, much worn. L. horn missing. The form of the head is very natural, but the ears are placed above and before the horns. The horn that remains is very thick and directed backwards and upward. Length from tip of nose to level of points of horns $1\frac{3}{4}$". See Plate XLVII.

Y. 0012. r. Grotesque seated figure in terra-cotta; seems to be human; head missing. The hands rest on knees. Abdomen rather distended. The feet are mere stumps and on the shins are leaf-like markings. Height $1\frac{7}{8}$", width $1\frac{1}{4}$", depth $1\frac{1}{8}$". See Plate XLIV.

Y. 0012. s. Terra-cotta mask, human face, cast from a mould. The eyes are large and long. Eyebrows, indicated by dots, well arched. Cheeks full and round. Nose small. Small drooping moustache. Lower lip full. Chin small and round. Forehead receding. Top of head bald. Between eyebrows two small parallel curved marks. Hair indicated on the R. proper side of head. The L. side is broken away. The whole face is very weak, but of natural proportions excepting the eyes, which are too large. Height $1\frac{3}{8}$", width $1\frac{1}{8}$". See Plate XLVI.

Y. 0012. t. Terra-cotta grotesque Bactrian camel. Harnessed as, and generally similar to, Y. 009. l. No tail and no conical protuberance on head. Height $1\frac{3}{4}$", length $1\frac{3}{4}$". See Plate XLVII.

Y. 0012. u. Terra-cotta grotesque horse's head and neck, used as the spout of a vessel. The mouth is open showing a row of small top teeth, and from between the lips issues a round tube through which the liquid was poured. The upper lip is recurved and the nostrils placed on the front of the nose. Hair is indicated on the edge of each lower jaw-bone. Mane erect, neck well arched. On that part of the fragment which joined the body of the vessel are several deeply indented holes, evidently to give tooth for the connecting clay. The colour is a good red and the material fine. Length $2\frac{1}{4}$", width 1". See Plate XLVI.

Y. 0012. v. Small terra-cotta grotesque monkey. L. leg broken away; also arms. Naturalistic type. Hairy; obviously male; seated. Height $\frac{7}{8}$". See Plate XLVI.

Y. 0012. w. Miniature terra-cotta vessel, amphora-shaped, but without the handles. It has on one side a projecting ring by which it was suspended. It has a band of simple zigzag ornament round the shoulder. The body is conical, point downwards, and the shoulder rounded. The neck is narrow and trumpet-shaped. Height $\frac{7}{8}$", diameter $\frac{7}{16}$". See Plate XLVI.

Y. 0012. x. Miniature terra-cotta vessel. Handle broken away. The body, in vertical section, is elliptical, the foot short and conical; the neck long and conical, broad end uppermost. Height $\frac{3}{4}$", diameter $\frac{9}{16}$". See Plate XLVI.

Y. 0012. y. Terra-cotta grotesque. Two monkeys, owl-like type, embracing amorously. One head is missing, the other has pigtail. Pose similar to Y. 009. r. Modelling very crude. Height $1\frac{7}{8}$". See Plate XLVII.

Y. 0012. z. Terra cotta grotesque, similar to Y. 0012. y, and similarly broken. Height $1\frac{3}{8}$". See Plate XLVII.

Y. 0013. a. Terra-cotta grotesque seated monkey. Naturalistic type; no arms, and legs mere stumps. Hair on head represented as very thick. Hair also indicated on front of body, and over middle an apron. On breast are ends of what appear to have been ribbons of some kind. Through stump of one leg a hole pierced right through to back of figure; and at this part a roughened depression (exhibiting potter's finger marks) as though the figure had been attached to something. Height 2". See Plate XLVI.

Y. 0013. b. Terra-cotta grotesque monkey, naturalistic type; seated on his heels, he supports with both hands in his lap a shallow bowl. Head tilted slightly to L. proper. Hair only on head. Height $2\frac{3}{8}$", width $1\frac{3}{8}$", depth $1\frac{3}{8}$". See Plate XLVI.

Y. 0013. c. Terra-cotta grotesque monkey; naturalistic type. Hairy; squatting with R. hand raised to breast, one finger on chin, the other missing, but probably pose was similar to Y. 0012. a. iii. Height $1\frac{5}{8}$", width 1". See Plate XLVI.

Y. 0013. d. Terra-cotta grotesque Bactrian camel. Similar to Y. 009. l, but more roughly modelled. A band goes across between the humps, and supports a load on each side. No other trappings are shown. Height $1\frac{5}{8}$", length $1\frac{3}{8}$". See Plate XLVI.

Y. 0013. e. Terra-cotta grotesque miniature monkey, Naturalistic type. Wearing apron, and playing syrinx. Height $\frac{5}{8}$". See Plate XLVII.

Y. 0014. a. Terra-cotta grotesque monkey. Naturalistic type; hairless, top of head conical. It is seated and has R. hand resting on R. knee, the L. hand to mouth apparently to hide a laugh which is cleverly indicated. Height 1", width $\frac{11}{16}$". See Plate XLVI.

Y. 0014. b. Terra-cotta grotesque monkey, seated, resting with elbows on knees, its head between its hands. Naturalistic type; hairy; head conical. Height $\frac{7}{8}$". See Plate XLVI.

Y. 0015. Fragment of terra-cotta vessel, having moulded *appliqué* enrichments, consisting of grotesque lion's head, with large round ears, projecting eyes and heavy moustache-covered upper lip. Surrounding the face is conventional closely curled hair. Below this runs a fillet, with vertical ribbing. Between this and a similar fillet runs a plain band about 1" broad, slightly sunk, on which are *appliqué* elliptical jewel-like bosses, consisting of a convex elliptical centre surrounded by a thin fillet and outside by a border of circular beads. Height 4⅜", width 3⅛", thickness circ. ½". See Plate XLIV.

Y. 0016. Fragment of terra-cotta vessel, made on wheel and bearing *appliqué* enrichments. Between vertical incised lines in pairs, a grotesque human face, of sottish and bibulous aspect, with distorted ears; round, fleshy nose; heavy, frowning brows over prominent bead-like eyes; a large, coarse, grinning mouth showing upper teeth and inflated cheeks. The face is surrounded by a narrow fringe of hair. Immediately above the head fragment of a bunch of grapes. Height 2¼", width 3$\frac{9}{16}$", thickness circ. ½". See Plate XLIV.

Y. 0017. Circular terra-cotta appliqué ornament, detached. It represents a grotesque Neptune-like head, having deep sunk eyes, arched eyebrows with a kind of *tilaka* mark between; broad, fleshy nose, inflated cheeks, grinning mouth, with long, flowing moustaches and flowing beard, spreading fanlike as it descends. Prominent, pointed ears, and a flat disk-like object on the head having a ring at each side. Resting vertically on edge, upon this disk, is a kind of 'Cakra' with centre hole in which is a bead, and surrounding this centre, between faintly incised radiating lines, nine holes. The background is composed of incised lines radiating from the centre and running from behind the face, off the edge. Diameter 2". See Plate XLIII.

Y. 0018. Terra-cotta fragment of human face, nearly life size. Only a portion of R. proper eyebrow and prominent circular eye, with a little of face to L. of it, remain. Comp. B. 001. a, C. 004. Height 1½", width 2". See Plate XLIV.

Y. 0019. Fragment of terra-cotta vessel, showing wheel marks. In the centre is an *appliqué* figure fully dressed in boots, trousers, long-sleeved tunic reaching below the knees, and a kind of cape in points, one of which falls over each shoulder and the third over breast. This cape, the hem of the tunic, and the bottoms of the trousers are decorated with an incised line, above which a row of little dots. At each lower side of the tunic hangs a globular object—perhaps a bell. In the L. hand is an object like a scroll, upon the end of which the R. hand rests. The chin is visible, and portions of large, massive, round earrings.

To the R. of this figure is a small, squatting figure, cross-legged, playing a syrinx. To the left a standing figure to same scale, the weight of the body on the R. leg, the left being slightly bent. This figure appears to be nude, with the exception of a cap on the head. It is playing the *sitâr*. On all these figures the potter's finger marks appear. Three incised parallel lines run round the object. Clay of fine texture. Height 3½", width 3⅞", thickness ⅜" to 1". See Plate XLIV.

Y. 0020. a. Terra-cotta whorl, decorated on the upper surface with a circle of small incised rings. The form of the whorl is circular, with sides and under surface elliptically convex, the upper surface slightly concave. Diameter 1⅜", thickness ⅜", diameter of hole ¼".

Y. 0020. Terra-cotta whorl, similar to Y. 0020. a, in dark grey clay. No decoration. Diameter 1½", thickness ⅞", diameter of hole ¼".

Y. 0021. Four fragments of plain pottery, showing marks of the potter's wheel. The clay is fine and baked to a good red. Length of pieces from 1½" to 3".

Y. 0022. Fragment of coarse terra-cotta vessel, consisting of mouth, neck, and part of shoulder of large vase. On one side of the neck is the broken stump of a handle, and opposite a hole and surrounding scar where probably another handle was attached. The rim is simply moulded and turns onwards in a gradual curve which begins about half-way up the neck. Length of neck 4". At the junction of neck and shoulder is a fillet moulding, $\frac{7}{16}$" broad, channelled on its face. Above this roughly incised, double nebulée, about 1" broad, then a pair of incised lines, then a 1" band of roughly incised lattice lines, and finally two more parallel incised lines. All these details encircle the neck, but break off where the handles have been. Round the shoulder, immediately below fillet is a repetition of the nebular pattern, below which a pair of incised lines, then a band of lattice, two more incised lines, and the nebulée again. On the shoulder, and nearly 3" below the hole in the neck (centre to centre), is a similar hole, obviously the other point of insertion of this handle. There is a considerable piece broken out of the rim and neck.

Height of fragment 6⅝", external diameter of mouth 5⅜", width of narrowest part of neck 3½", greatest width of shoulders 7⅞". Thickness of material about $\frac{5}{16}$". See Plate XLIII.

Y. 0023. Neck and mouth of large terra-cotta vessel. The fragment is circular and trumpet-shaped, but curving outwards also at the lower end to meet the shoulders. The rim is turned over outwards and worked into a flat fillet on its outer aspect, slightly overhanging. Just within the mouth is an incised line. The neck has round its middle five parallel incised lines. At equal intervals are the broken stumps of three broad handles (about 1½" broad). In each of the spaces between these stumps is an *appliqué* moulded upper half of a female figure carrying in uplifted hands a heavy garland of pearls (?) and probably representing a Gandharvī (comp. for similar figures in stucco D. I. 02, D. XII. 5, &c., in Plate LVI). The eyes are large, the other features small and refined. The hair is dressed loosely on the top of the head in a manner resembling the fashion of to-day in Europe and of ancient Greece. A heavy necklace surrounds the neck and another hangs on the breast. On the wrists are bangles. All these figures have been cast in the same mould. Diameter of mouth 6⅛", height 4", narrowest part of neck about 3⅞". See Plate XLIII.

Y. 0024. Terra-cotta jug with single handle, resembling the Greek Oenochoē. It is perfect, except for a small superficial flake chipped from the body. The form of the body is composed of elliptical curves, narrowing gradually from the shoulder to the bottom, which is flat. The shoulder is full and turns over quickly to the slender neck. The neck is trumpet-shaped, pinched in on one side to form a spout. The simple and graceful handle falls from just beneath the mouth on the side opposite the spout, to the shoulder. The neck joins the shoulder in a soft concave curve, slightly depressed below the surface of the shoulder. The decoration is extremely simple. It consists of two incised lines running round the outer surface of the mouth; four lines round the narrowest part of neck; and two, lightly incised, where the curve of the shoulder meets that of the side. The surface of the shoulder is divided into four unequal segments by pairs of incised lines radiating from the base of neck to the lines at outer curve of shoulder. On each side of these double lines is a row of dots. One segment (the smallest) contains the junction of handle with shoulder. In each of the other three, which are practically equal, is drawn, by a pair of incised lines, a rather flat arc, extending from junction to junction of the lines encircling shoulder with those radiating from the neck; in the segment so formed is a lattice pattern, lightly incised, and in each division a dot.

Height $5\frac{5}{8}''$, width $4''$, narrow part of neck $1''$, spout to back of mouth $2\frac{3}{8}''$. Height of body to spring of neck $3\frac{1}{2}''$. See Plate XLIII.

Y. 0024. a. Two grotesque monkeys embracing amorously. Owl type. One head missing. Similar to Y. 0012. y. Height $1''$. See Plate XLVII.

Y. 0025. Grotesque monkey in grey terra-cotta, squatting with L. leg drawn up so as to support L. elbow, L. hand supporting head. R. hand *veretrum tenens*. R. leg partly outstretched and hanging down. Tail swings round almost to R. knee. Nipples and navel carefully and distinctly marked. The whole figure is carefully modelled and finished in detail. The back is hairy, also the front of the legs. Height $1\frac{7}{16}''$, width $1\frac{1}{4}''$. See Plate XLVII.

Y. 0026. Soapstone miniature figure of angel in attitude of adoration. The head is slightly advanced and the face turned slightly upward. The hands are together, as are the knees and feet. The figure kneels, and is sitting upon its heels. A pair of wings, folded together, spring from shoulders and extend downwards to soles of feet. The figure appears to be nude. The face is of same type as the infant in Y. 0012. c. Height $1\frac{1}{8}''$, depth $\frac{9}{16}''$.

Y. 0027. a. Body of fine terra-cotta vessel, with portion of neck and stem. The body is nearly spherical, slightly elongated vertically. Round upper part, at junction of tapering neck, is a flat band $\frac{3}{8}''$ broad, having a raised narrow fillet at each edge. The band between the fillets is decorated with *appliqué* hemispheres almost touching each other. Neck is plain, also stem, of which very little remains. Surface of body is peculiarly ribbed. The potter, having shaped and smoothed the surface, appears to have drawn a slightly convex cutting tool over it from top to bottom *spirally* twenty-nine times, making twenty-nine very shallow, slightly spiral flutes, lying close together, so that a sharp rib or ridge divides one flute from the next, as in the flutings of a Doric column. At one side, just above the middle, is an *appliqué* bull's head which evidently supported the handle. On the opposite side a hole is broken. The stem was hollow and, like the neck, was evidently made separately. The whole piece shows wheel-marks inside. Height nearly $7''$, width of body $5\frac{1}{2}''$, narrowest part of neck $1\frac{1}{4}''$, stem $1\frac{1}{4}''$, thickness of material about $\frac{3}{16}''$. See Plate XLIII.

Y. 0027. b. Small steatite carvings (charms?), roughly made. See Plate LI.

i. Monkey seated on haunches. Height $\frac{3}{4}''\times\frac{5}{16}''$.

ii. Grotesque animal, perhaps meant for frog (fish?), with triangular hole in centre, $1\frac{3}{16}''\times\frac{7}{16}''$.

iii. Fragment of human figure (female?), legs apart, apparently nude but for girdle. Head and feet missing, $\frac{9}{16}''\times\frac{1}{4}''$.

Y. 0028. Terra-cotta vessel, lemon-shape, broken at lip, probably used as an Ampulla. It has one small round handle at its upper half, and as a decoration a moulded *appliqué* monkey (naturalistic) is seated on top, leaning its back against the body of vessel and playing the *sitār*. The body of vessel is divided into six segments by deeply incised lines extending from the neck to the base. Between these the surface is corrugated by equally deeply incised transverse lines, ridges and depressions being of about equal thickness. The bottom is small and flattened but not accurately enough to enable the vessel to stand. Height $4''$, width $2\frac{7}{8}''$, width of neck about $1\frac{1}{4}''$, diameter at base $1''$, thickness of material $\frac{3}{16}''$. See Plate XLIII.

Y. 0029. Buddha figure carved in soapstone. Seated, face destroyed. The figure wears a single robe covering both shoulders, arms and legs. Ears elongated, hair smooth. Pose *dhyāna-mudrā*, on a lotus with three whorls of petals. The back is rough and flat, and seems to have been drilled in one place, from below, probably for a means of attaching the piece to something. Work of Gandhāra type, but rough. Height $2\frac{7}{8}''$, width $2''$, thickness $\frac{5}{8}''$. See Plate XLVIII.

Y. 0030. Terra-cotta bifrons vessel, with slender neck, probably used as an Ampulla. Both faces are broad, with long and slightly oblique eyes, small noses and mouths. One has a conventional moustache, quite small and twisted cable-like. This is obviously a male face, and the other is probably a female. The hair of both is treated similarly to that in B. 001. f., but on the female face the hair hangs down to the level of the mouth and is slightly turned forward at the ends. Ears are roughly modelled, and above ears are what appear to be the lower ends of handles. Neck of vessel rises from top of head, and has two cinctures surrounding its junction with head. The narrow part is about ⅞" in diameter and widens out, trumpet-shape, towards the mouth, where it is much broken. Foot broken off, leaving a hole in bottom of vessel. Vessel moulded in two halves stuck together. Height 3½", width 2¾", depth (face to face) 2½". See Plate XLIII.

Y. 0031. Terra-cotta female head. Very oblique eyes, small mouth, slightly protruding chin. The hair is brought low on forehead, is parted in centre, brushed gently R. and L. and clipped quite even, level with eyebrows, with slight break at the parting. Long hair hangs straight down at back and behind ears. There seems to be a small turban on the top of the head, the embroidered and shaped ends of which hang down on front fringe. Moderately long earrings adorn the well-formed ears, and a large and well-modelled rosette, with two dependent streamers, is placed at the back of the head. There is a round hole immediately above this rosette about ⅛" in diameter, and there is another about the same size at the neck. Inside of head hollow. It was evidently joined to something at the neck. Height 3¼", width 1⅞". See Plate XLV.

ANTIQUES ACQUIRED AT KHOTAN TOWN.

B. 001. a. Terra-cotta fragment of mask of human face. Top of head, ears, and chin missing. Nose strongly aquiline and large. Bushy eyebrows, well arched, spring from above the nose. Eyes large, near together, slightly oblique, but too full to suggest a Mongolian type. Between curves of eyebrows a deeply indented *tilaka* mark, and the forehead around and above it is wrinkled. Nostrils parallel and narrow. On upper lip, which is extremely short, the bushy moustaches are brushed upwards, right and left. Lower lip, narrow. Cheeks full and rather high. General form of face long and narrow. On the back are potter's finger-marks. The nose appears to have been worked upon after it left the mould. Length 2⅝", width 1½". See Plate XLIV.

B. 001. b. Terra-cotta grotesque monkey, squatting upright; hairy; naturalistic type; legs merely short stumps. Arms broken away. Navel strongly marked. Fragment of R. proper hand remaining, apparently *veretrum tenens.* Height 2⅜", width 1⅛". See Plate 36.

B. 001. c. Fragment of terra-cotta vessel, bearing *appliqué* moulded half-length figure. The face, which is a good deal worn, is oval, and the eyes horizontal. On the head is a curious bag-like cap the end of which hangs down to L. proper shoulder and has a tassel or bell attached. In the hands, in front of breast, are held a pair of cymbals connected by a cord. Sleeves enveloping the arms heavily quilted. Round neck are necklaces. Height 1¾", width 2⅛", thickness about 3/16". See Plate XLV.

B. 001. d. Terra-cotta appliqué figure, Gandharvī, broken away at middle. It closely resembles those on Y. 0023, the principal difference being that this example wears a rich tiara, and folds of drapery hang down on either side of face. The type of face recalls that of Rawak stucco heads. Height about 2", width 1⅝". See Plate XLV.

B. 001. e. Terra-cotta grotesque, probably handle of a jar, representing nondescript animal's head and neck. It vaguely resembles a goat's head, and the neck seems to be covered with close woolly hair. It has been moulded in two halves and stuck together before burning. Height 1¼", width ¾", depth 1¼". See Plate XLVII.

B. 001. f. Terra-cotta grotesque head, in archaic style, probably part of a vase decoration. Face very broad and short. Eyes very long and prominent; nose broad; lips thin; chin round and prominent. Eyebrows (emphasized by a series of small indentations suggestive of hair) very arched. Hair quite conventional. It is sharply defined round the face like a lawyer's wig, pointed between the eyebrows, and following the curve of the brows sweeps round outer angles of eyes and forward to just below malar bone where it is cut off square to the ear lobes. A centre parting is indicated by an incised line, and the growth of the hair R. and L. is indicated by incised lines perfectly hard and straight. That part which covers cheek is shown as brushed forward. At back of head the hair hangs perfectly straight to nape of neck, from each side of which it is cut in a sloping line to about middle of ears. On the nape is the broken stump of a handle (?), the other end of which is probably shown by a broken crown-like fragment on the top of head. The neck portion is quite a rough tenon-like piece, which evidently fitted into a socket. Ears moderate in size, but projecting and roughly formed. The example is cast in two halves. Cf. Y. 0012. o, Y. 0031, B. 001. g. Height 2 1/16", width 1½", depth 1½". See Plate XLV.

B. 001. g. Terra-cotta head very similar to B. 001. f but not so carefully finished. The back half is missing, and the reverse of fragment shows scorings for the purpose of giving a tooth for connecting clay. Compared with B. 001. f, the features are more pronounced and the eyes more oblique. Height 1⅝", width 1⅜". See Plate XLV.

B. 001. h. Terra-cotta grotesque head somewhat resembling B. 001. f, but not so archaic, and coarser. The eyes are nearly horizontal, nose *retroussé*, chin receding, eyebrows very pronounced. The wig-like hair is open slightly in the centre of forehead, and round the head is a garland of roundels which may represent curly hair or flowers. Above this is a plaited band running round the head. From lower part of back of head issues a handle-like tail, which curves upwards to just above the level of plaited band. Near the upper end it is tied round with a band. Between end of tail and back of head rises a large drooping feather, very closely resembling in form an ostrich feather. This hangs free over top of head, and falling forward, curls over and joins plaited band in front. The ears appear to be tied in tightly across their middle, so that they are divided each into two projecting portions. Height $2\frac{3}{8}$", width $1\frac{5}{8}$", depth $1\frac{3}{8}$". See Plate XLV.

B. 001. i. Terra-cotta head, closely resembling B. 001. h, but all above plaited head-band broken away. The tenon in place of neck is scored as though to be fastened into a socket. Height $2\frac{1}{4}$", width $1\frac{1}{2}$". See Plate XLV.

B. 001. j. Terra-cotta grotesque Bactrian camel, closely resembling Y. 009. l. There was probably a figure or some other load between the humps originally. Very roughly made. Height $2\frac{1}{2}$", length $2\frac{3}{8}$". See Plate XLVII.

B. 001. k. Terra-cotta grotesque monkey, owl-like type. Legs broken off. A flat object is held horizontally at chest, R. hand above it and L. hand below supporting it. Across middle a kilt-like object, which however does not seem to serve any purpose of decency. Height $1\frac{3}{8}$", width 1". See Plate XLVII.

B. 001. l. Terra-cotta grotesque monkey. Naturalistic type, rather carefully made. The figure is sitting on its heels, the legs being a natural length, which is unusual with these grotesques. The two hands (broken off) were resting together in front of breast. Body entirely nude and hairy; probably that of a female. Height 1". See Plate XLVI.

B. 002. Terra-cotta grotesque monkey, upper half; holding in both hands a bowl as though drinking. Naturalistic type; hairy. Eyes distinctly looking downwards into bowl. Height 1", width $\frac{3}{4}$", depth $\frac{5}{8}$". See Plate XLVI.

B. D. 001. a. Fragment of sculptured slate, forming part of the angle of some larger piece, and sculptured on two surfaces at right angles to each other. The shape of the piece is curious. The more perfect side presents a shape vertical on L. proper side, and vertical on opposite side half-way up, when it turns in a very shallow curve and meets upper end of L. side. The lower side is horizontal and dentilated, and forms a dividing member between the upper and lower (broken away) panels. The contiguous surface appears to have been of the same shape. A groove or rabbet is on outer side to R., where probably another portion of design joined. The subject, carved in high relief in the Gandhāra style, is Buddha's temptation by Māra and his daughters. In a hollow, or cavea, is seated the Bodhisattva, right shoulder and breast bare, and wearing a single robe. On his head appears to be an ornament. His R. hand resting across his leg is directed towards the ground (*bhūmisparśa-mudrā*); in L. he holds the corner of his robe. Three female figures —two to L. and one to R.—exhibit their charms, the pose of one closely resembling that of the Nāginī in the fresco of Cella D. II. at Dandān-Uiliq. In the foreground to R. proper is the crouching figure of Māra carrying *vajra* in R. hand, his head turned towards Buddha. To L. another crouching male figure, bearing a club or sword in R. hand, the face being turned towards Buddha. Of lower panel, two small heads only are visible. In upper cavea of contiguous side are visible a tree, and the upraised R. arm and the shoulder, undraped, of a central figure (Buddha?). To R. proper in foreground is a half-kneeling figure in an attitude of adoration. The carving is fine, but not very finished. Height $1\frac{7}{8}$", width 1", width of broken face $\frac{5}{8}$". See Plate XLVIII.

B. D. 001. b. Bone or ivory seal, bell-shaped. The device is a grotesque lion with upraised tail, in outline. Above this an inscription in Tibetan characters, read by Dr. Barnett རྒྱ་ཁྲིལ་བཟང་ *rgya k'ril bzang* (literally 'abundant heap-good.') Height $\frac{5}{8}$", diameter $\frac{9}{16}$". See Plate L.

B. D. 001. c. Signet ring of chalcedony or pale carnelion (according to Prof. Miers' examination). On flattened upper surface cut in intaglio male bust to R.; hair rolled, wearing earring. $1\frac{1}{8}$" × $\frac{7}{8}$". See Plate XLIX.

B. D. 001. d. Elliptical seal of onyx, bearing, in intaglio, head of Persian type, to R., wearing moustache and hair rolled; drapery over shoulders. $\frac{5}{8}$" × $\frac{1}{2}$" × $\frac{1}{8}$". See Plate XLIX.

B. D. 001. e. Elliptical seal of onyx, bearing, in intaglio, a galloping Indian bull to R., with tail uplifted. $\frac{5}{8}$" × nearly $\frac{1}{2}$". See Plate XLIX.

B. D. 001. f. Circular seal in carnelion, bearing, in intaglio stag with branching horns couchant to R. Very roughly cut. Diam. $\frac{7}{16}$". See Plate XLIX.

B. D. 001. g. Circular seal in onyx, bearing, in intaglio, male (?) head to R. Hair smoothed close to head, with a cincture of loose hair passing completely round forehead, behind ears and below occiput, giving a cap-like appearance. Diam. $\frac{9}{16}$". See Plate XLIX.

B. D. 001. h. Elliptical seal in agate, bearing, in intaglio, bust to R., with moustache, earrings, and mane-like head-dress extending to nape of neck. Bust draped. $\frac{1}{2}$" × $\frac{7}{16}$" See Plate XLIX.

B. D. 001. i. Elliptical seal in carnelion, representing, in intaglio, a very roughly cut straight-horned sheep or goat to R. $\frac{5}{8}$" × $\frac{1}{2}$" (nearly). See Plate XLIX.

B. D. 001. j. Elliptical seal in onyx, bearing, in intaglio, a female head to R.; on top of head is a mask of a man's face, tilted upwards and backwards in the manner of a visor, the peak of a helmet apparently coming from beneath and covering brow of female. The mask wears a helmet, the frontal portion of which is visible, and a mane-like plume, which issues from it, falls in position of female's hair. The type of heads is Persian. In front of female are four Kharoṣṭhī (?) characters written in a vertical column. Engraved surface $\frac{5}{8}$" × $\frac{1}{2}$". See Plate XLIX.

B. D. 001. k. Elliptical convex seal in carbuncle, flat on the under side. Device, in intaglio, a wild boar, crined, couchant to R. 9/16" × ¼". See Plate XLIX.

B. D. 001. l. Grotesque figure carved in steatite, monkey-like, seated with outstretched legs, L. fore-arm lying on L. leg; R. hand raised to mouth. Tail lies straight beneath legs. Height 1", width ⅜", depth (front to back) ⅞". See Plate LI.

B. D. 001. m. Square bronze seal, shank at back broken. A bird standing to L. with wings uplifted, perhaps meant for a goose. Much corroded. ⅞" square. See Plate L.

B. D. 001. n. Bronze seal, broken at corners and shank. It appears to have been circular, with four decorative excrescences equidistant from each other giving a square contour to the whole. Two of the projections are missing. The device seems to be a primrose-like flower in intaglio. Diameter about ⅝".

I. 001. Intaglio, carbuncle; elliptical, plano-convex; broken at lower end. On convex surface, in intaglio, finely cut portrait head of Indo-Scythian type three quarters to L. The face is long and narrow; prominent eyes, well-marked eyebrows, delicate nose and mouth, slightly prominent chin. A circular jewel is on front of head, from behind which fall strings of jewels, or plaited hair. The ears are jewelled. A jewelled and massive necklace is visible on proper R., and what appears to be ornaments on R. shoulder. A flowing end of drapery floats out over it. R. and L. is an inscription running round edge of surface, resembling the debased Greek characters of 'Scytho-Sassanian' coins. ⅝" × ½". See Plate XLIX.

Kh. 001. a. Elliptical seal of sard bearing, in intaglio, quadriga of horses to R., with driver standing on chariot (Helios?) behind. 7/16" × ⅜". See Plate XLIX.

Kh. 001. b. Elliptical seal of chalcedony, bearing, in intaglio, boar's head to R. ⅜" × 5/16". See Plate XLIX.

Kh. 002. Elliptical seal of nicolo, bearing, in intaglio, winged Eros, standing to R., with wreath in R. hand and arrow in L. Carefully engraved and well-proportioned. 9/16" × 7/16". See Plate XLIX.

Kh. 003. a. Moulded terra-cotta spout in the form of a grotesque face. The eyebrows are bushy and meet in a knot; eyes are projecting roundels; nose prominent, swollen, and *retroussé*; cheeks puffed; lips thick, grinning and slightly parted, the L. proper half of the mouth being pierced through to form spout. There is no chin, and curling hair surrounds face. 1¾" × 1⅜". See Plate XLIV.

Kh. 003. b. Fragment of fine terra-cotta vessel bearing an *appliqué* figure in bas-relief. It represents a female, Gandharvī, rising from a lotus, all the body below the middle being concealed by the flower. The hands are raised on either side to the level of the ears, and each grasps an end of a garland which hangs in a festoon in front of the body below the breasts. Wrists adorned with bangles, and just below the flattened table-like centre of lotus is a row of beads. The face was much damaged before firing, the nose being flattened; but the expression of the mouth is smiling; the eyes are rather prominent and full. On the head is an inverted lotus serving as a cap. The pose and general design closely resemble the stucco Gandharvī figures of D. I. and D. XII; comp. Plate LVI. Height 2⅜", width 1⅞". See Plate XLV.

Kh. 003. c. Fragment of terra-cotta vessel, bearing a flying or dancing figure. Hands raised and joined together above head. The head is inclined slightly to R. and downwards. Legs (R. broken off) crossed, left in front of right, suggesting rather dancing than flying. Over shoulders hangs a very full and loose cloak. Size 1½" × 2". See Plate XLV.

Kh. 003. d. Fragment of terra-cotta vessel, ornamented with raised moulding dented at short intervals; below this an *appliqué* moulded ornament composed of two circular concentric rows of beads with a fillet between, round a centre larger bead. Size 1⅜" × 1¾".

Kh. 003. e. Terra-cotta appliqué ornament; grotesque animal's head resembling that of a Chinese pug dog. Size 1⅜" × 1⅜". See Plate XLIV.

Kh. 003. f. Terra-cotta appliqué ornament, showing animal's head similar to Kh. 003. e, but with large ears, and much less carefully modelled. Dark grey clay due to over-firing. This fragment has been subjected to great heat after being broken. Size 1⅜" × 1½". See Plate XLVII.

Kh. 003. g. Relief fragment on dark slate-stone richly carved on both sides, and apparently part of a miniature shrine. This appears to be a portion of a L. proper wing of the object, and shows that it was divided architecturally into panels or caveae, each stage being supported on cantilevers as in many of the Gandhāra sculptures. On obverse is represented the scene of the Bodhisattva (Buddha) cutting his hair (comp. Foucher, *L'Art du Gandhâra*, pp. 363 sqq., where this relief is mentioned). The figure of B. is shown nearly full face, turned slightly to R. proper, seated cross-legged upon a draped cushion. The robe covers L. shoulder and then passes under R. armpit, leaving R. arm, breast and shoulder bare. It appears to cover the legs completely. From top of head, which is inclined slightly downwards, depends the hair (twisted into rope shape) which is grasped by L. hand. The R. hand upraised cuts the hair with a sword. To R. a kneeling figure with garland, and to L. a standing figure with turban. Above and within, right and left, two adoring figures. In the upper cavea appears to have been Buddha's alms-bowl; in the lower another figure subject of which one head only (to R.) is left. On the reverse the fragment shows relief representation of a draped figure in profile, carrying a nude child; and bearing on the back a basket, the straps passing over the shoulder and under the arm. Contour of fragment cleverly adapted to subjects on both sides. The figures on reverse are on a considerably larger scale than those on the obverse. Head of large figure is missing, and also portion below hips. Height 1½", width 1¼". See Plate XLVIII.

Kh. 003. h. Sculptured fragment of slate-like stone, in Gandhāra style. The obverse represents a seated Bodhisattva, the two hands (*dharmacakra-mudrā*) in front of breast. Face, though roughly carved, of Gandhāra type; hair arranged in four rows of small curls, above which *uṣṇīṣa*. The ears are greatly elongated and slit. The robe is the usual simple one, and leaves R. arm and breast bare. On brow *ūrṇā*. The reverse is much broken, and shows only a portion of a long quilted tunic and *langūṭī* covering L. leg of a standing figure, which would probably be about 3–4 inches high. The fragment probably belongs to a miniature shrine. Height 2", width 1½". See Plate XLVIII.

Kh. 003. i. Fragment of sculptured slate representing a figure completely clothed in a single robe, seated upon a lotus, the hands resting one within the other in the lap (*dhyāna-mudrā*). Ears elongated and slit. Face broad and eyes slightly oblique. The head appears to be enveloped in a smooth round cap. Behind head is indicated a nimbus. The whole, although roughly carved, is reminiscent of the Gandhāra style. The piece is evidently a fragment of a larger composition. There is no *uṣṇīṣa* or visible *ūrṇā*, although there seems to be a trace of the latter having been painted on. Height 2", width 1⅛". See Plate XLVIII.

Kh. 003. j. Bronze miniature of standing Bodhisattva figure; R. hand raised, palm forward, fingers up (*abhaya-mudrā*). L. arm pendant, probably holding corner of robe or *amṛta* flask. The style of figure resembles the stucco reliefs of Rawak Vihāra; comp. Plate LXXXVI. Beneath the base on which the figure stands is a chule or tang, for insertion into a socket. Height 2", width ⅝". See Plate LI.

Kh. 003. k. Portion of handle of terra-cotta vessel. The fragment is curved, and on its convex side is a bract-like outgrowing piece having a concave curve, and broadening laterally. In concavity of this portion a moulded *appliqué* satyr-like mask. Very naturalistic and classic. Surrounding the grinning face is an encircling row of rings suggesting curls, and outside that another row of circles incised in bract portion of handle. From lower lip depends a conventionally treated wedge-shaped beard, broad end downwards. The whole piece is extremely good and quite unoriental. Height 2½", width 2¼", depth 2¼". See Plate XLIV.

Kh. 003. l. Terra-cotta fragment, probably portion of handle of a vessel. Ornamented with *appliqué* grotesque crocodile head (*Gavialis gangeticus*) similar to Y. 0011. a, but in this instance eyes more naturally placed near top of head; over them heavy eyebrows. The neck tapers into a recurved point. A few incised lines and dots on the handle show that that was also decorated. Length 2¼", width ⅞". See Plate XLV.

Kh. 003. m. Terra-cotta fragment, apparently portion of a handle of some vessel. It is flat and rather thin. On outer surface an ornament consisting of a row of circles with a dot in centre of each, bordered on each side by two lines, all incised. At each side of handle, standing on portion of body of vessel remaining, and with backs resting against narrow edges of handle, are two *appliqué* figures of musicians with drums. They are fully clothed in long *kurta* and boots. One holds a drum with its flat sides vertical, and the other with the sides horizontal, one hand resting on the upper surface, and the other on the lower. The body of the vessel was decorated with incised lines. Both figures are headless. Height 1⅞", width 1¼", thickness of handle $\frac{5}{16}$", width of handle $\frac{11}{16}$". See Plate XLV.

Kh. 003. n. Terra-cotta grotesque monkey, female. Naturalistic type. Lower limbs broken off. Well-developed breasts, R. proper being held by R. hand. L. hand pressed to L. parietal part of head (cf. Y. 0025). Head coarsely modelled. Hair indicated on head and arms only. Height 1½", width 1⅛", depth $\frac{9}{16}$". See Plate XLVII.

Kh. 003. o. Terra-cotta grotesque bear, embracing, as though trying to climb, a tree stump. The head resembles rather the owl-like monkey type, but the short thick tail and characteristic lower limbs clearly indicate a bear. Height 1⅞", width ⅝", depth ⅞". See Plate XLVII.

Kh. 003. p. Fragment of relief figure carved in steatite. Portion from waist upwards and feet missing. The figure is draped from waist to ankles in loose, clinging robe. L. proper hand rests on L. hip, and a bangle is on wrist. Pose, that of Māyā at the nativity of Buddha. The back is plain and flat, and has a hole drilled half-way through. Height 1⅞", width ⅞", thickness $\frac{7}{16}$". See Plate XLVIII.

Kh. 003. q. Sculptured fragment in steatite. Human head. The face, considerably worn, is hairless and of Gandhāra type; eyes very slightly oblique. The hair, which is wavy, grows to a point in the centre of forehead. Shape of face a full oval; mouth delicate and smiling. A kind of high coronet surrounds the hair, consisting of a flat band which inclines slightly forward at its upper edge and is ornamented with three bosses, centre one oval and those at the sides circular. Each boss is divided into four sectors by two crossed incised lines. In each sector is an incised triangle, whose sides are equidistant from each boundary of sector. Between the bosses is a kind of double bud, or *vajra*-like form—common in Greek and Roman ornament, and in Graeco-Buddhist sculpture. From the two side bosses to the back the band is plain, but in the centre of the back is a knot and two depending embroidered ends of drapery, so that the bosses appear to be stiff ornaments attached to a band of soft material, a kind of *taenia*. Height 1⅜", width 1", depth 1". See Plate XLVIII.

Kh. 005. Fragment of relief in slate. Obverse: Gandhāra style. Nativity of Buddha. In an upright conical cavea, to which is joined on R. proper a shorter cavea, stands the figure of Māyā holding by R. hand the branch of a tree bending to right above her head. She is supported on L. by an attendant. On R. a figure receives the infant which is issuing from R. side. Below in foreground a standing figure of the infant taking his first steps. Below was a second cavea, now broken away.

Reverse: Head and shoulder of a figure in profile to L. proper, with long hair, and draped. Behind figure rises a conical object (Stūpa ?) decorated with horizontal bands of ornament and supported by the R. hand of another figure, broken away. The contour of the whole fragment is the contour of these reverse details, and the form of the obverse is adapted to this. A hole is drilled in the top. Height 1", width ⅞". See Plate XLVIII.

Kh. 006. Fragment of standing figure sculptured in soapstone. The figure is nude to the hips, below which is a falling robe. The upper part of figure leans gracefully to L. proper. On neck and breast necklaces. Feet, arms, and head missing. A few rough lines suggestive of modelling on the back. Height 2¼", width ¾", thickness ⅜". See Plate XLVIII.

Kh. 007. Bronze panel, representing Gaṇeśa seated on an open lotus. R. hand raised holding an object supported on R. leg (perhaps an *ankus*), and L. resting in lap and holding a round object. Ears very large and wing-like. Height 1⅜" × ⅞".

Kh. 008. Fragment of ivory carving in the round. A male figure, beardless, with moderately short and wavy hair, and wearing earrings and necklace, dressed in a single robe caught in at the waist, and having long, tight sleeves, has his left arm round shoulders of a small female figure and his right hand, palm upwards, on her breast. His head is turned slightly to L. proper and downwards—that is towards the female. The female wears a similar long robe, but in addition a short cloak at the back which is either fur-lined or embroidered round edges. The R. hand is placed behind L. shoulder of the man, and L. hand, depending at the side, carries gracefully a small wreath. The lower extremities of man broken away, also most of R. leg of female, and all below calf of L. leg. Head and L. arm and shoulder of female missing. The proportions of the figures are good, but the workmanship is not of a high order. Colour brown. Height 3", width 1⅜", thickness ½". See Plate XLVIII.

Mac. 001. Neck of terra-cotta vase (presented by Mr. G. Macartney). Fragment, circular in section, broadening at lower end to join the shoulder of the vessel and at the upper to form the lip, which is broken away. Encircling centre part are five depressed annulets $\frac{3}{16}$" broad, the dividing fillets ⅛" to $\frac{3}{16}$" broad. Three moulded enrichments in low relief are applied at equal distances upon the neck and lie across the annulets which serve to connect them. They represent (*a*) a dancing or floating Gandharvī figure, to front, hands upraised and clasped above head, the flowing ends of a scarf depending over shoulders; necklet indicated; legs crossed as in running. General proportions Cupid-like. (*b*) Standing upon a bunch of grapes, two well-modelled, long-tailed parrots, facing each other, with beaks crossed. A somewhat similar application of parrots is found occasionally in the Gandhāra sculptures. (*c*) An elephant to right, with uplifted bifurcate trunk, each branch ending in a lotus from which issues the upper half of a figure, with hair dressed in a top-knot, the two figures being back to back and having a single nimbus extending behind both heads. In their hands, brought to chin level, each holds a garland. Although the height of the elephant is only $\frac{9}{16}$" the character of the frontal bones, invariably faithfully expressed by Indian sculptors, is even here clearly observed. The texture of the hide is indicated by a few incised dots. In the Ajantā Cave paintings dots are used for the same purpose. Above the intervals between these enrichments are pairs of half vine leaves similar to those in the plaster bas-reliefs of Dandān-Uiliq (cf. Plate LVII). Below, and marking the junction of neck and shoulder of the vessel is an encircling band of small incised rings, each ring having a centre dot. There are three interruptions in this band coinciding with the three spaces between the three applied reliefs, and it seems probable that there were originally three handles to the vessel. The aperture of the neck has been smoothed and cleared with a cutting instrument before burning. The quality of clay is very fine. Height 2⅝". Breadth at lower part 2⅜", at centre 1⅝", at upper part 1⅝". Thickness about ¼". See Plate XLIV.

Ya. 001. Bronze seal, oblong. The device is divided into two parts. The lower part a square, in which is cut a conventional flower. The upper part, oblong, in which is a figure, very rudely drawn. Figure is full face, R. proper leg bent, thigh directed outwards and lower part of leg inwards. L. proper leg stretched out to left. R. arm bent, perhaps to suggest the hand resting on R. thigh. L. arm hanging down, and in the hand an upright object which may be a spear or club. Over the right shoulder three dots of various shapes; over head a dot and below feet another. Figure probably intended to be seated on a cushion, with one leg gathered up and the other dangling. On back of seal a shank with a hole through it. 1⅜" × ¾". See Plate L.

Ya. 001. b. Bronze right hand, probably from a small Buddha figure. Length 1¼".

Section V.—BUDDHIST SITES DESCRIBED BY HSÜAN-TSANG

We must feel grateful to Hsüan-tsang for having left us unusually full descriptions of various places of Buddhist worship outside the capital, not only on account of their intrinsic interest for the *topographia sacra* and the folklore of Khotan, but also because they furnish evidence which enables us to test and, as we shall see, to confirm the location of the capital at Yōtkan. The positions of these sacred places is invariably indicated by distances and bearings from the capital. Hence I was naturally led to make my search for them in close connexion with my investigation of the latter site, and it is fitting that I should record the results in this place.

Legend of *So-mo-jê* convent. The nearest among these sanctuaries was the convent of *So-mo-jê* 娑摩若, with a Stūpa a hundred feet high in its centre, which the pilgrim visited at a distance of 5 or 6 li (a little over a mile) to the west of the royal city [1]. This is the distance indicated by Rémusat's and Beal's versions, whereas Julien's translation, either by mistake or owing to a variant, gives it as 50 or 60 li. The legend as told by Hsüan-tsang distinctly favours the nearer location. It relates that at one time an Arhat coming from a distant foreign land had taken up his abode there in the middle of a wood. The miraculous light spread around by his spiritual power was noticed by the king as he stopped at night in a double-storied pavilion of his palace. Having been informed of its cause the king proceeded to the holy man and respectfully invited him into the palace. On the Śramaṇa refusing to leave the wood, the king full of reverence built a convent for him and a Stūpa. When afterwards the king had procured a quantity of sacred relics and regretted not to have been able to insert them under the Stūpa, the Arhat directed him to have the precious objects enclosed successively in receptacles of gold, silver, copper, and stone. When this had been done and the relics had been transported by the king and his chief officers on an ornamented car to the convent,[2] the Arhat raised the Stūpa on the palm of his hand and held it while the king's workmen dug a place for the sacred deposit. Then on the work being accomplished the Arhat once more lowered the Stūpa to its original position without any damage [3].

Tibetan form of legend. That the Stūpa which had been thus uplifted and replaced in so miraculous a fashion was an object of special veneration in Khotan is evident also from the reference made to it in the 'Annals of Li-yul'. These tell us of king Vijayavīrya, the eighth successor of Vijayasaṃbhava under whom Buddhism was believed to have been first introduced: 'One day while looking out of Srog-mkhar [3a] he perceived a light brilliant as gold and silver at the spot where now stands the Hgum-stir Caitya. When the king learned that the Buddha had foretold that at that spot a Vihāra would be built, he called to his presence the Buddhist Buddhadūta, and having made him his spiritual adviser, ordered him to direct the building of the Hgum-stir Vihāra [4].' I think that, notwithstanding the somewhat different form in which the legend as to the origin of the shrine is here told, it can scarcely be doubted that the same locality is intended and that Buddhadūta corresponds to Hsüan-tsang's Arhat.

[1] See *Mémoires*, ii. pp. 235 sqq.; *Si-yu-ki*, ii. pp. 316 sq.; Rémusat, *Ville de Khotan*, pp. 50 sqq. [Also Watters, ii. p. 297, has 'five or six li'.]

[2] For sculptural representations of such solemn relic depositions, comp. Foucher, *L'Art du Gandhâra*, i. pp. 592 sqq.

[3] This legend and a somewhat similar one told by Hsüan-tsang of a Stūpa near Kapiśa (*Mémoires*, i. pp. 45 sq.; *Si-yu-ki*, i. p. 60) have been discussed by M. Foucher, *L'Art du Gandhâra*, i. p. 52, with reference to the light they throw on the purpose of Stūpa construction.

[3a] [*Srog-mkhar*, as Mr. Thomas informs me, means literally 'Life-fort'.]

[4] See Rockhill, *Life of the Buddha*, p. 238.

Fa-hsien's mention of convent.

We may in all probability recognize an earlier mention of this sanctuary in the passage of Fa-hsien which describes the gilt splendour of 'what is called the king's New monastery' and which I have already had occasion to quote in full[5]. This monastery was situated 7 to 8 li to the west of the city, a bearing and distance closely agreeing with those recorded by Hsüan-tsang for the So-mo-jê convent. Its building was supposed to have taken eighty years and extended over three reigns. The Stūpa, 250 cubits in height[6], was 'covered above with gold and silver, and finished throughout with a combination of all the precious substances'. Of the Hall of Buddha, i.e. the Vihāra built behind the Stūpa, Fa-hsien extols 'the magnificence and beauty; the beams, pillars, venetianed doors, and windows being all overlaid with gold-leaf.' Besides this, 'the apartments for the monks are imposingly and elegantly decorated, beyond the power of words to express.' Fa-hsien's reference to the rich offerings which the monastery received from 'the kings in the six countries on the east of the Ts'ung range of mountains'[7], makes it clear that this shrine must have enjoyed exceptional fame in his time. Even though he does not tell us the story of its origin and speaks of it under a different designation, it seems difficult to assume that he could have meant any other than the So-mo-jê convent still renowned in Hsüan-tsang's days for its splendour and miraculous manifestations.

Village of *Somiya.*

Judging from what previous experience had taught me of the fate which has befallen all ancient structures within the oasis, I did not expect to trace remains of what, notwithstanding all its glittering splendour, could only have been a pile of sun-dried bricks or wood doomed to rapid decay. All the more delighted was I when on the morning of Nov. 28th, in the course of inquiries preliminary to a survey of the villages west of Yōtkan, I first heard the name of *Somiya* mentioned. The village was said to be close to Yōtkan and to the west of it. Its name at once suggested a direct phonetic derivative of the ancient local name which is intended by the Chinese transcription of So-mo-jê, and to philological presumption topographical evidence soon added its weight.

Leaving the excavated area of the ancient city at its north-west corner by the route marked in the map of Plate XXIII, I reached first the hamlet of Eskente, half a mile to the west. There I was told of a 'Döbe' or mound existing near the cemetery of Somiya. The latter place I found to be situated only three-fourths of a mile further west, and to consist of some thirty scattered dwellings. I proceeded at once to the local Mazār, the reputed resting-place of three saints, and found it surrounded by an old cemetery extending over a considerable area at a level far lower than the adjoining fields. On asking from the first old villager I met for the reported 'Döbe' I was promptly taken to a field close to the north-eastern corner of the cemetery. There a little low mound, rising scarcely five feet above the surrounding ground and some 33 feet in diameter, is respected by the villagers with a kind of superstitious fear, though it shares in no orthodox way the sacred character of the neighbouring Mazār and cemetery. I soon had the oldest men of the village summoned to the spot, and in what they agreed in telling me of the mound may, I think, be traced the last lingering recollection of the ancient shrine that has left its name to Somiya.

Tradition of Somiya mound.

Shāmi Sōpe, a withered old man of about ninety, had heard from his father and grandfather, who had both died at a great age, that the little mound had ever been respected by the folk of Somiya as a hallowed spot not to be touched by the ploughshare. Some unknown saint

[5] See above, p. 194, note 7.

[6] Rémusat, loc. cit., p. 14, translating the passage from the *Pien i tien*, gives the height as 'vingt-cinq tchang.'

[7] In the 'six countries' east of Ts'ung-ling, i. e. the Pāmīrs, we possibly have an early reference to the traditional *Alti-shahr* or hexapolis of Eastern Turkestān; comp. for the latter *Tārīkh-i-Rashīdī*, Introd. p. 51.

(Ulūgh zāt), who had come from afar, is supposed to have sat at the spot, and evil would befall those who should touch the ground. The name of the saint is forgotten, and the villagers could not assert whether he rests under the mound or not. But the people of Somiya never pass without saying a prayer, and according to the testimony of Shāmi Sōpe and his forbears they have clung to this custom for the last two centuries. I take it as a sign of the antiquity of the tradition that no name is assigned to the saint whose memory lingers about the 'Döbe', whereas the names of his three companions, who are supposed to have disappeared beneath the ground at the spot now sanctified by the Mazār, are still currently known to young and old as Mullah Yahyamutti, Mullah Allāmutti, Mullah Serekmutti [8]. Nobody seemed to know of any other 'Döbe' similarly surrounded by superstitious awe in the neighbourhood.

Considering the concordant evidence of the name and position of Somiya, I think it probable that the worship of this nameless mound is the last trace left of the So-mo-jê Stūpa of Buddhist days. We may not be able to throw light on the character or origin of the name So-mo-jê, but there is evidence of phonetic analogies to support the assumption that Somiya is the direct derivative of the ancient name intended by the *Hsi-yü-chi's* transcription. The character 娑 *so* is attested elsewhere as a rendering of the syllable *so* [9]. 摩 *mo* as a rendering of Indian *ma* or *mâ* is also well known [10]. Finally 若 *jê*, which in Julien's list appears as the representative of the Indian syllables *jña*, *jñā*, *ṇa*, *nyā*, *ṇya*, *ña*, may very well have been used here to express a syllable of which the modern phonetic derivative would be *ya*, seeing that the *ya* of the local name Niya in the *Hsi-yü-chi's* transcription is rendered by the character 壤 *jang*, which is otherwise used to express the Indian sounds *jña* and *ña* [11]. The change from a form **Somaña*, such as may be conjecturally restored from the transcription *So-mo-jê*, into the present *Somiya* is slight, and can be fully accounted for by a well-known phonetic process [12]. The name *Somiya*.

The same day's search enabled me to identify in all probability another sacred site mentioned by Hsüan-tsang. In a convent known as the *Ti-chia-p'o-fu-na* Saṅghārāma, and situated a little over 10 li to the south-west of the capital, the pilgrim was shown the statue of a standing Buddha which was supposed to have miraculously come to this spot from Kuchā [13]. A Khotan minister exiled to the latter state had secured his repatriation by assiduous worship of this image, and as he continued to honour it after his return it came one night by itself. The minister then built for it this convent. The name in this case can no longer be traced [14], but exactly in the direction and at the distance indicated (about two miles) there lies the popular Ziārat of 'Bōwa-Kambar', visited by people from all parts of the Khotan district. I found it to consist of a large square cemetery enclosing the high mud-built tomb of the saint, who is believed to have acquired holiness by his devoted services as the groom of ''Alī Pādshāh'. The level of the cemetery near its centre lies fully twelve feet below the surrounding fields—a certain indication of its antiquity according to my previously detailed observations. A grove Convent of *Ti-chia-p'o-fu-na*.

[8] I do not attempt to restore the orthodox Arabic or Turkī forms of these names. The closing *-mutti*, also occasionally pronounced *-mupti*, probably represents the Arabic *mufti*.

[9] See Julien, *Méthode pour transcrire*, p. 186.

[10] Comp. Julien, loc. cit., p. 154.

[11] See Julien, loc. cit., p. 115; for *Ni-jang*: *Niya*, see chap. x. sec. iii.

[12] Julien, *Mémoires*, ii. p. 235, has assumed the Sanskrit form **Samajña* as the original of *So-mo-jê*; but quite apart from the question whether a Khotan locality in the seventh century is likely to have been known by a Sanskrit name, it must be pointed out that, though *samajña* is a compound possible in grammar, it presents no meaning acceptable for a name.

[13] See *Mémoires*, ii. p. 230; *Si-yu-ki*, ii. p. 313; *Ville de Khotan*, p. 45.

[14] Julien has tentatively restored Ti-chia-p'o-fu-na into *Dirghabhāvana* (recte °*bhavana*), but though *p'o-fu-na* as a transcription of Skr. *bhavana*, 'residence, sacred habitation' (common in the names of Buddhist shrines in Kashmīr, see *Rājat.*, II. pp. 339, 369) looks probable enough, I cannot find evidence for the rendering of *Dirgha* by *Ti-chia*.

of fine old trees faces the eastern entrance, and a row of booths testifies to the popularity of the fairs which take place here at the time of pilgrimages.

Aiding-Kul marsh.

The 29th of November, the day of my return to Khotan, was utilized by me for a visit to localities south-east of Yōtkan, which offered special interest in connexion with the *topographia sacra* of the oasis. I first followed the Yōtkan Yār down from Khalche to where it unites with the Yār of Kāshe. After crossing by a bridge the fairly deep stream thus formed, I reached on the other bank of the ravine the lands of Halāl-bāgh, a collection of large hamlets which I was anxious to examine more closely as a local tradition connects the site with the pre-Muhammadan rulers of the country. Close to the south of the central hamlet there stretches a marsh, known as *Aiding-Kul*, covering an area of about three-quarters of a square mile. It is overgrown with reeds, and traversed in parts by low sandy ridges; between these appear copious springs, which form numerous channels, and subsequently collect into a little stream at the northern end, draining towards the Yurung-kāsh.

Tradition about Halāl-bāgh.

On my visit to Halāl-bāgh I enjoyed the advantage of an intelligent guide in the person of Ibrahīm, a local Mullah and proprietor well known for his learning and piety. Though eighty-six years old at the time of my visit he was still quite active. Ibrahīm Mullah asserted with pride that it was at Halāl-bāgh that there once stood the city of the 'Khalkhāl-i Chīn-u-Māchīn', the legendary heathen ruler of Khotan, and in evidence showed me chapter and verse in the Turkī Tadhkirah of the Four Imāms, of which, as well as of the legendaries of other well-known Mazārs he possessed a copy[15]. According to the popular tradition as embodied in this text, apparently at no very remote period, the Four Imāms whose martyred bodies are supposed to rest under a famous Mazār near Polu vanquished the obstinate opponent of Islām after a long siege, and his city became a waste. The shrine of Kum-i-Shahīdān, about half a mile to the north-west of the marsh, is supposed to mark the spot where 360 faithful of the Imāms found martyrdom in the final struggle.

Story of Abā Bakr's excavations.

But Ibrahīm Mullah had information to offer beyond the brief and somewhat vague statement of the Tadhkirah. According to him, the site where the heathen city had stood was excavated for its hidden treasures under the orders of Abā Bakr, the ruler of Kāshgar and Khotan in the early part of the sixteenth century. He brought water from the Yurung-kāsh to the place to enable his labourers to wash the soil—just as is now done at Yōtkan—and in the hollow left by his diggings there formed the marsh of Halāl-bāgh. No old remains of any kind are now found, nor could I trace any genuine tradition about such discoveries ever having been made at Halāl-bāgh. On the other hand, it is a fact well attested by a detailed account in Mirzā Ḥaidar's *Tārīkh-i-Rashīdī* that Abā Bakr, the uncle of the author, carried on treasure-seeking operations on a large scale at various old sites in his dominions[16]. The methods employed must have been curiously similar to those still practised at Yōtkan, except that Abā Bakr turned prisoners' labour to use on such work. We are told by the Chronicler that 'he ordered the old cities . . . to be excavated by these [prisoners] and the earth dug from them to be washed. If there were anything big, they would come upon it in digging, while anything small [such as gems] they would find when they washed [the earth]. In this way innumerable treasures in precious stones, gold, and silver were discovered'. After relating a very fantastic story about a wonderful find of treasure thus made 'in the citadel of Khotan', or according to the version of the Turkī translator 'in the old city of Yarkand', Mirzā Ḥaidar mentions that 'several other treasures were brought to light in the old cities of Kāshgar, Yarkand, and

[15] For an abstract of the Tadhkirah in which, however, Halāl-bāgh is not specified as the site, see Grenard, *Mission D. de Rhins*, iii. pp. 38 sqq.

[16] See *Tārīkh-i-Rashīdī*, pp. 255 sqq.

Khotan'. But though describing at length the discipline, &c., under which the prisoners worked, he gives no indication as to the position of these 'old cities'.

Alleged origin of Aiding-Kul.

A vague recollection of Abā Bakr's treasures and the operations by which he secured them still lingers among the people in various parts of Eastern Turkestān[17]. But there does not appear to exist anywhere a genuine tradition about specific localities; and in the absence of all traces of ancient débris or of earlier evidence of such a tradition we may well doubt whether Halāl-bāgh was really among the places exploited by Abā Bakr, and whether it was not his reputation alone which induced local literati like Ibrahīm Mullah to connect his name with the supposed origin of the Aiding-Kul. The configuration of the marsh itself did not appear to me to lend any support to the assumption that it had an artificial origin. With its springs and low sandy hillocks it resembles closely the marshes observed by me adjoining the cultivated areas of the Keriya and Niya oases, and also north of Lop and of Tasmache in the Khotan oasis. Immediately to the east of the Aiding-Kul, a good deal of marshy ground is crossed by the road to Khotan town, while some three miles to the south the watercourse feeding it from the Yurung-kāsh is also flanked by water-logged areas about the village of Kācha (see map of Plate XXIII).

Mound of *Naghāra-khānah.*

In reality there seems good reason for believing that the marsh of Aiding-Kul existed long before Abā Bakr's time, and that we have a reference to it in a legend told by Hsüan-tsang. Close to the south-eastern shore of the marsh there rises a mound known as *Naghara-khāna*, which local tradition, as represented by Ibrahīm Mullah, assumes to have formed part of the wall enclosing the old city. M. Grenard was the first to recognize in this name, correctly pronounced as *Naghāra-khānah* and literally meaning 'the house of the kettle-drum', the trace of a local legend which Hsüan-tsang has recorded of a site in the vicinity of the ancient capital[18]. Though the conclusion he drew from this as to the position of the latter cannot be maintained in view of what has been proved above, I believe the identification of the site itself to be justified.

Legend of the Nāginī and minister.

According to the story told at great length in the *Hsi-yü-chi*[19], there was once a great stream of which the people of Khotan took advantage to irrigate their lands; Julien's and Beal's versions make its course lie to the south-east of the capital and directed north-westwards; the distance is given as 100 li in the former and as 200 li in the latter translation[20]. On the stream ceasing to flow, the king inquired as to the cause of this calamity, and he was told by an Arhat to propitiate the 'dragon', i. e. the Nāga, dwelling in the stream by sacrifices. When this had been done, a woman emerged from the stream and, explaining the arrest of the waters by the recent death of her husband, asked for a minister of noble birth to be given to her in marriage. On the Nāginī's desire being announced, a great officer of state offered to sacrifice himself for the restored flow of the water on condition that the king should found a convent. After a solemn leavetaking the minister, dressed in white and mounted on a white steed, entered the stream. When he reached the middle of the current he struck the water with his whip and immediately disappeared. A short time after his steed came to the surface again, carrying on its back a great drum of sandalwood. This was found to contain a letter to the king in which the minister asked him to suspend the drum to the south-east of the city[21]; on the approach of an enemy it would sound in advance. The waters of the stream have

[17] See *Tārikh-i-Rashīdī*, p. 256.

[18] See *Mission D. de Rhins*, iii. p. 138.

[19] See *Mémoires*, ii. pp. 239 sqq.; *Si-yu-ki*, ii. pp. 319 sqq.

[20] Rémusat's extract from the *Pien i tien* makes the stream flow south-west of the capital and at a distance of 100 li; *Ville de Khotan*, p. 57. [Watters agrees with Julien.]

[21] Thus Beal and Rémusat; Julien has 'Au sud-est du royaume,' a manifest oversight.

continued to flow ever since. Hsüan-tsang tells us that the drum of the dragon (Nāga) had long ago disappeared. But there existed at the place where it had been suspended a lake known as 'the drum-lake', and there were still seen the ruins of the convent which had been built by its side but was then without monks and deserted[22].

Location of Hsüan-tsang's 'drum-lake' convent.

Seeing that the Aiding-Kul lies exactly south-east of Yōtkan, and that apart from it there is no sheet of water in the vicinity of the site of the ancient capital which could be called a lake or even a natural pond, it is difficult to resist the conclusion that the local name of Naghāra-khānah, 'the drum house,' is connected with the old legend about the miraculous drum which, as we know from Hsüan-tsang, gave its name to the lake and in all probability also to the ruined convent he mentions by its side. The examination of the mound now known as the 'Naghāra-khānah' leads me to believe that we have in it the remains of the structure which was shown to Hsüan-tsang as a ruined convent far more probably than a fragment of an otherwise wholly untraceable city wall. The mound measures about 175 feet from east to west, with a maximum width of 72 feet: its approximately level top rises about 27 feet above the adjoining fields. I failed to trace on its slopes any marks of brickwork or stratification, the whole surface showing what looked to me like stamped loess soil. The general appearance of the 'Naghāra-khānah' reminded me of the large oblong mound once bearing a Vihāra or convent which rises behind the Maurī-Tim Stūpa and which has been fully described above[23]. It can scarcely be doubted that the interior of the Maurī-Tim structure consists of nothing but stamped loess; and that when once the facing of sun-dried bricks, which still shows traces of niches or small cellas arranged in receding terraces, has undergone further decay in the course of centuries, its surface will present the same shapeless look as the Naghāra-khānah mound does now. The latter, if our identification is right, must already in the seventh century have been a ruin of considerable antiquity, and its wholly amorphous condition at the present day could thus in no way surprise us.

Connexion of Naghāra-khānah with Hsüan-tsang's legend.

It would be of little use to offer conjectures as to the way in which the folklore story of the minister's Curtius-like sacrifice and the miraculous drum came to be attached to this ruined mound. But it may be worth noting that, just as all old mounds in this region are now popularly supposed to have been constructed as watch-stations by ancient kings, a ruined mound of such conspicuous height as the Naghāra-khānah must have shown before the ground-level around it rose and erosion reduced its elevation, was likely to exercise popular imagination in old times. The distance from the Naghāra-khānah to the excavated area of Yōtkan is only two miles, and thus the belief in its having served for the location of a great drum intended to warn the capital of an approaching enemy could readily be accounted for.

Continuance of local worship.

I am unable to make any definite suggestion as to the stream which the legend looked upon as the habitation of the Nāginī and her self-sacrificing human husband, seeing the great variations in distance and bearing which the several translations show. Possibly some point higher up on the Yurung-kāsh might have been thought of as the scene of the minister's sacrifice. That the legend itself was popular can be safely concluded from Hsüan-tsang's detailed mention. I believe that we have visible proof of this in a fresco excavated by me in one of the shrines of Dandān-Uiliq, of which I hope to show in the next chapter that it represents the characteristic figures of the legend[24]. At a sacred spot connected with such a popular legend we might expect to see local worship continued to the present day; and in fact,

[22] This purport of the passage, which seems clear enough, is brought out better in Rémusat's version (loc. cit., p. 60) than in that of Julien which Beal slavishly follows.

[23] See above, pp. 81 sqq.

[24] See below, chap. IX. sec. iii.

we do find the immediate vicinity of the Naghāra-khānah occupied by a well-known Mazār, that of the Three Ghāzīs. According to the story as related to me, the three holy men had, after the martyrdom of the Four Imāms, sought for their resting-place without finding it. They then stopped at this site when the earth closed over them [25]. The saints' tomb is marked by a relatively well-built 'Gumbaz', and forms an object of popular pilgrimage from all parts of the oasis.

Hsüan-tsang's *Lu-shê* convent.

It is also to the south-east of the capital, but at a distance nearer than the Aiding-Kul marsh, that we have to look for the site of another Buddhist convent which Hsüan-tsang mentions in connexion with an interesting local legend [26]. To the south-east of the royal city, five or six li [27], so the *Hsi-yü-chi* tells us, there was a convent known as the *Lu-shê* Saṅghārāma, founded by the queen of a former ruler, to whom tradition ascribed the introduction of sericulture into Khotan. In old times the country knew nothing of either mulberry trees or silkworms. Hearing that China possessed them, the king of Khotan sent an envoy to procure them; but at that time the ruler of China was determined not to let others share their possession, and had strictly prohibited seeds of the mulberry tree or silkworms' eggs being carried outside his frontiers. The king of Khotan then with due submission prayed for the hand of a Chinese princess. When this request had been acceded to, he dispatched an envoy to escort the princess from China, taking care to let the future queen know through him that, in order to assure to herself fine silk robes when in Khotan, she had better bring some mulberry seeds and silkworms with her.

Legend of the introduction of silkworms.

The princess thus advised secretly procured mulberry seeds and silkworms' eggs, and by concealing them in the lining of her headdress, which the chief of the frontier guards did not dare to examine, managed to remove them safely to Khotan. On her first arrival and before her solemn entry into the royal palace, she stopped at the site where subsequently the Lu-shê convent was built [28], and there left the silkworms and the mulberry seeds. From the latter grew up the first mulberry trees, with the leaves of which the silkworms were fed when their time had come. Then the queen issued an edict engraved on stone, prohibiting the working up of the cocoons until the moths of the silkworms had escaped. 'Then she founded this Saṅghārāma on the spot where the first silkworms were bred [29]; and there are about here many old mulberry tree trunks which they say are the remains of the trees first planted. From old time till now this kingdom has possessed silkworms, which nobody is allowed to kill, with a view to take away the silk stealthily. Those who do so are not allowed to rear the worms for a succession of years [30].'

[25] The names were mentioned to me as Khwāja Paklan Ghāzī, Khwāja Shaikh Atā-ulwalī Ghāzī, Khān Amīn Nukrubil Ghāzī. M. Grenard probably means the same shrine when speaking of the tomb of 'Keytous Maghrébî,' a follower of the Four Imāms, as adjoining the Naghāra-khānah; see *Mission D. de Rhins*, iii. p. 40, note 1. Much in the nomenclature of modern Muhammadan legend about Khotan seems to be subject to similar variations.

[26] See *Mémoires*, ii. pp. 237 sqq.; *Si-yu-ki*, ii. pp. 318 sq.; *Ville de Khotan*, pp. 55 sq.

[27] Thus Beal and Rémusat; regarding Julien's 'cinquante ou soixante li au sud de la capitale,' which may be merely a mistake of translation, see below.

[28] For the correct interpretation of this passage, Beal's and Rémusat's versions must be consulted.

[29] Beal's version reproduced above is substantially also that of Rémusat, loc. cit., p. 56. Julien translates: 'Aussitôt après, elle fit construire ce couvent en l'honneur de la déesse des vers à soie.'

[30] Julien's version of the concluding passage slightly differs; but there can be no doubt that the prohibition of the killing of the silkworms within the cocoon is referred to. Such a practice would necessarily be opposed to Buddhist notions about the sanctity of animal life, whatever its advantages from the point of view of the silk producer. The purport of the queen's alleged edict is curiously illustrated by a paragraph of *The Pioneer*, June 21, 1905, recording 'an unexpected and unforeseen difficulty' which has cropped up in connexion with the spread of silk production in Ceylon. The demand is now said to be almost entirely for unpierced

Panel illustrating legend.

That the legend here related about the origin of one of Khotan's most important industries enjoyed widespread popularity is proved by the painted panel (D. IV. 5) discovered by me in one of the Dandān-Uiliq shrines, which presents us, as my detailed analysis will show [31], with a spirited picture of the Chinese princess in the act of offering protection to a basketfull of unpierced cocoons. An attendant pointing to the princess's headdress recalls her beneficent smuggling by which Khotan was supposed to have obtained its first silkworms, while another attendant engaged at a loom or silk-weaving implement symbolizes the industry which the princess's initiative had founded. A divine figure seated in the background may represent the genius presiding over the silkworms [32].

Tibetan version of legend.

But the fame of the legend is also attested by the version of it which we meet with in the 'Annals of Li-yul' [33]. There we are told that king Vijayajaya, the third successor of Vijayavīrya mentioned above in connexion with the legend of the So-mo-jê convent, married Princess Pu-nye-shar, the daughter of the ruler of China. Wishing to introduce silkworms into Li-yul she commenced raising some at Ma-dza, a locality not otherwise specified. Misled by a mischievous story told him by the ministers of China, the king had the house burned down where the silkworms were being raised. The queen, however, managed to save and rear some silkworms, and when she was able to show to the king the garments she had made of the silk thus procured he greatly regretted his previous deed. 'He called from India the Bhikṣu Saṅghaghoṣa and made him his spiritual adviser, and to atone for his wickedness he built the Po-ta-rya and Ma-dza tchaityas and a great Vihāra (or, the Caitya and the great Vihāra of Ma-dza).'

Location of *Lu-shê* convent.

Different as the form of the legend here is, there can be little doubt that by the Caitya and Vihāra of Ma-dza the same sacred locality is meant where Hsüan-tsang found the Lu-shê convent. For the location of the latter the *Hsi-yü-chi's* notice is our only guide; and as the popular shrine of Kum-i-Shahīdān before mentioned occupies a position relative to Yōtkan, exactly corresponding in direction and distance to the 5–6 li to the south-east of the pilgrim's description, I believe we are justified in recognizing here another instance of the survival of old local worship, notwithstanding characteristic change in its objects. It is true that Julien's translation places the Lu-shê convent 50 or 60 li south of the capital. But even if this rendering is supported by a variant of the Chinese text and is not merely the result of inadvertence (by no means unfrequent in the *Mémoires* where figures and bearings are concerned), it could not claim much weight; for the distance of 50–60 li or 10–11 miles to the south would take us far beyond the limits of cultivable soil to the stony waste of the 'Sai' at the foot of the hills, where it is impossible to suppose that mulberries or any other tree-growth could ever have flourished within historical times. It must further be noted that the legend heard by Hsüan-tsang represented the Lu-shê convent as built on the site which had served as a halting-place for the princess coming from China immediately before her solemn entry into the royal palace. It is clear that a locality 10 miles or so to the south of the capital would not answer this description.

Legend of Vairocana's convent.

Among the sacred sites of Khotan described by Hsüan-tsang there still remains one awaiting our notice; and as it was believed to be the oldest Buddhist sanctuary of the territory, and manifestly enjoyed great fame, I should not have referred to it last but for the uncertainty of its location. This uncertainty is the more curious since the shrine is mentioned not only by

cocoons, fetching ten times the price of pierced ones. Yet Buddhist sentiment in the island is strongly averse to the killing of the worm inside, which the working of unpierced cocoons presupposes.

[31] See below, chap. IX. sec. V.

[32] This figure is male; else it might be adduced in support of Julien's translation of a passage above quoted which represents the Lu-shê convent as constructed in honour 'de la déesse des vers à soie.'

[33] See Rockhill, *Life of the Buddha*, pp. 238 sq.

Hsüan-tsang but also by Sung Yün, the *Pei shih*, and other historical texts, as well as by the 'Annals of Li-yul'.

According to the account uniformly told in the *Hsi-yü-chi* as well as in the 'Life' there stood to the south of the royal city at a distance of 10 li a large convent built by an ancient king in honour of the Arhat Vairocana (P'i-lu-chê-na), to whom was attributed the introduction of Buddhism into Khotan[34]. In old days, when the law of Buddha had not yet reached the country, the Arhat came to this spot from Kashmīr, and in a wood practised meditation. When the king, informed of his strange appearance and dress, came to see him the Arhat discoursed to him of the boons of Tathāgata's doctrine, and exhorted him to follow the Law. The king offered to believe in Buddha if he could see his appearance; and Vairocana promised fulfilment of his desire after the building of a convent. The convent was accordingly built, and a great assembly of monks had gathered, but as yet there was no *ghaṇṭā* to call together the congregation. When the king then inquired for the promised appearance there suddenly descended from heaven an image of Buddha presenting a *ghaṇṭā*. Thereupon the king became confirmed in the faith and eager for its propagation. Hsüan-tsang's biographer, on concluding the story, specially notes that this convent was thus the first built in the kingdom.

Sung Yün's version of legend.

Sung Yün has left us a record of the same shrine, which in several respects is of special interest[35]. The king of Khotan, so we are told, was not a believer in the law of Buddha when a merchant brought a Bhikṣu called Vairocana (P'i-lu-chan) to a spot under an apricot tree south of the city. When he announced to the king that he had brought a Śramaṇa from abroad the prince became very angry, and proceeded at once to see Vairocana. The latter thereupon announced that he had been sent by Tathāgata to exhort the king to build a Stūpa of the turned-up *pātra* [shape], and thus to gain religious blessings. The king promised to obey if he were allowed to see Buddha. Thereupon Vairocana sounded a bell, and Buddha sent Rāhula, who appeared to the king in the form of Buddha himself, The king prostrated himself at the sight, and built at the foot of the apricot tree a structure to serve as a temple, with a representation of Rāhula, but this structure suddenly disappeared of itself. Thereupon the king built a new Vihāra to shelter the image, in such a way that the image, though placed under the turned-up *pātra*, appeared constantly outside—a miracle which greatly benefited the spiritual state of all beholders. Near this place the shoes of a Pratyekabuddha (Pi-chih-fo) were seen, miraculously preserved without any alteration, of neither leather nor silk, but of an unknown substance.

Stūpa of the 'turned-up pātra.'

The learned comments of M. Chavannes have removed the obscurities which beset Sung Yün's account of this shrine in earlier versions. At the same time M. Foucher, by a convincing reference, already quoted in a preceding chapter, to a passage of the *Hsi-yü-chi*[36], has made it clear that by the 'turned-up *pātra*' is meant a Stūpa dome of the orthodox shape, originally hemispherical. The miracle, which appears to have made the sacred image enclosed in the Stūpa dome visible to the eyes of pious visitors on frequent occasions, is also capable of explanation in view of certain structural features which have been noted elsewhere in extant Stūpa ruins or in descriptions by the Chinese pilgrims[37]. Thus Sung Yün's notice, besides confirming Hsüan-tsang's account of the origin of Vairocana's shrine, supplements it by some interesting

[34] See *Mémoires*, ii. 227 sqq.; *Vie*, pp. 282 sqq.; *Si-yu-ki*, ii. pp. 312 sq.; *Ville de Khotan*, pp. 40 sqq.
Julien translates, 'premier roi,' but the interpretation given above from Beal and Rémusat's versions is evidently more correct; see, below, the Tibetan version.

[35] See Chavannes, *Voyage de Song Yun*, pp. 17 sq.

[36] See above, p. 83, *Si-yu-ki*, i. p. 47; also *L'Art du Gandhâra*, p. 63.

[37] Foucher, *L'Art du Gandhâra*, pp. 86 sq.

The *Tsan-mo* temple.

details as to its sacred objects. The notice which the *Pei shih* presents, probably from the records of Hui-shêng, Sung Yün's companion, and which appears to have been reproduced in substantially the same form by the Annals of the Northern Chou [38], is far briefer; but, on the other hand, it furnishes the name of the site occupied by the shrine and a more definite indication of its position. We read there that 'fifty li to the south of the city is the *Tsan-mo* temple 贊摩寺: this is the place where once the Arhat and Bhikṣu Lu-chan (Vairocana) constructed for the king of this country the Stūpa of "the turned-up pot"; on a rock there is the place where a Pi-chih-fo (Pratyekabuddha) walked with bare feet—the two impressions are still to be seen there'.

Legend of *Tsar-ma* Vihāra.

Before proceeding to consider the topographical indications furnished by this passage, it will be useful to acquaint ourselves with the form in which the legend of the oldest Buddhist sanctuary of Khotan is reflected in the Tibetan 'Annals of Li-yul' [39]. 'One hundred and sixty-five years after the establishment of the kingdom of Li-yul Vijayasaṃbhava, son of Yeula, ascended the throne, and in the fifth year of his reign the Dharma was first introduced into Li-yul. This king was an incarnation of Maitreya and Mañjuśrī. Having assumed the form of a Bhikṣu, the Ārya Vairocana, he came and dwelt in the Tsu-la grove, in the country of Tsar-ma. There he became the spiritual guide of the inhabitants of Li-yul, and taught the ignorant cattle-herders in the Li language, and invented the characters of Li. After this the Dharma appeared. Then King Vijayasaṃbhava built the great Vihāra of Tsar-ma.' When the Vihāra was finished Vairocana asked the king to sound the *ghaṇṭā*. As the latter refused unless the Tathāgata appeared and gave him a *ghaṇṭā*, 'immediately Vairocana assumed the appearance of the Tathāgata, and after having taught like the Tathāgata sixty great Śrāvakas at Tsar-ma, he gave king Vijayasaṃbhava a *ghaṇṭā*, and the king sounded it without ceasing for seven days. After that Vairocana invited the Nāga king Hu-lor to bring from Kashmīr a Caitya which contained corporal relics of the seven Tathāgatas. It came through the air, and is at present at Tsar-ma [40]. This Caitya is in the Gandhakūṭa, and is surrounded by a halo [41].'

The interest of this version, which in the main is but an amplification of the legend as told by Hsüan-tsang, lies in the indication it furnishes as to the traditional date for the introduction of Buddhism into Khotan (404 years after Buddha's Nirvāṇa), and in the name *Tsar-ma* given to the locality where Vairocana's Vihāra was built. The identity of the name here intended with the *Tsan-mo* of the *Pei shih* passage has been already pointed out by Mr. Rockhill. Nor does it seem doubtful that this Tibetan form came closer to the original local name than the Chinese transcription *Tsan-mo*, in which, as so often, the closing *r* of the first syllable has been dropped [42].

Position of Vairocana's convent.

But while the recorded forms of the name of the locality closely agree, the statements as to its position show a serious discrepancy as far as the distance from the capital is concerned. The *Hsi-yü-chi* and the 'Life' place it 10 li to the south of the latter, but the notices of the *Pei shih* and Chou shu, though agreeing in the bearing, give the distance as 50 li. My inquiries in the neighbourhood of Yōtkan failed to reveal any site to the south at the approximate distance

[38] See Chavannes, *Voyage de Song Yun*, p. 15, note 5, p. 18, note 9; for the corresponding extracts in the *Pien i tien* comp. *Ville de Khotan*, pp. 20, 29.

[39] See Rockhill, *Life of the Buddha*, pp. 237 sq. I have given the correct spelling of Sanskrit words often imperfectly transcribed in these extracts; thus also above.

[40] The translation shows the name here as a *T'sar-ma*.

[41] In a subsequent extract from the 'Prediction of Saṅghavardhana', relating a fantastic story of the emigration of all Buddhist priests from Khotan to Tibet, we find the *Ts'ar-ma* (*sic*) Vihāra mentioned as the locality where the Bhikṣus assembled before setting out on their journey; see *Life of the Buddha*, p. 241.

[42] See the remarks above, p. 10, note 6.

of two miles indicated by Hsüan-tsang which, on the ground of local tradition or worship, could be connected with the ancient Buddhist shrine. There is, indeed, a much-frequented Muhammadan shrine, the Mazār of Imām-Mūsā-Qāsim, at Kōsa due south of Yōtkan, but the distance is close on six miles, and thus far beyond Hsüan-tsang's estimate. If we adopt the *Pei shih's* distance of 50 li, or approximately 10 miles, the difficulty seems at first sight equally great, for beyond Imām-Mūsā-Qāsim's Mazār and the village of Kōsa there stretches towards the hills the barren pebble Sai, where a grove, such as both Hsüan-tsang and the Annals of Li-yul mention in connexion with Vairocana's convent, could never have existed.

The difficulties thus encountered will explain why I feel induced to discuss here the possibility of an identification which, though purely conjectural for the present, deserves some attention. Immediately before proceeding to Yōtkan I had, on November 24, visited the ancient sites of Jamada and Chalma-kazān to the south of the oasis, and not far from the left bank of the Yurung-kāsh. The former is a typical *Tati* closely adjoining the irrigated area of Jamada village (see map), where the eroded loess soil for a distance of about one mile from east to west and for about half a mile across is strewn with ancient potsherds. Old Chinese coins are said to be picked up here; but Wang-Daloi, an intelligent Chinaman trading in jade who acted as my guide here as well as at the jade-pits further south, had unfortunately mixed up his specimens with those he had collected from Yōtkan and elsewhere. Hence no certain conclusion could be drawn from his collection, which seemed to consist chiefly of 'cash' of the T'ang dynasty. On the west and south rise loess banks, which the people of the neighbourhood take for the *sipil* (wall) of the 'old town'; but probably they are only 'witnesses' of the original ground-level which have escaped the erosion that lowered the unprotected loess soil around[43]. Site of *Jamada.*

Far more extensive and, judging from the mass of débris, more important is another ruined area known as *Chalma-kazān* about five miles further south, which occupies the plain, about $1\frac{1}{2}$ miles broad, between the west bank of the Yurung-kāsh and the foot of a low gravel ridge flanking the *débouchement* of the river. It is strewn with a vast quantity of coarse potsherds, mixed here and there with small bits of broken glass and slag, and to a larger extent with stones from the river which no doubt had served for rough walls. The soil is loess, manifestly of the riverine type; and with irrigation the plain, now a barren waste, would present a very different aspect. The eroded loess banks seen at various points of the area showed no great height, from which it would appear either that the erosive power of the winds was less in the Yurung-kāsh Valley than in the open plain northward, or that the period since cultivation has ceased does not lie so far back as in the case of some of the ancient sites to be described hereafter. Site of *Chalma-kazān.*

In the middle of this area a low mound, covered with large rounded stones from the river-bed, at once attracted my attention. It had been mentioned to me before by the villagers of Jamada and Bīzil, who apply the name of *Chalma-kazān* in particular to it. Its well-defined round shape suggested a Stūpa, and a closer examination proved this to be the case. Unfortunately, others before me had guessed the nature of the structure; and a large trench cut from the east into the centre of the mound and down to the present level of the surrounding ground showed that 'treasure-seekers' had been at work. The mound in its present state has a diameter of about 98 feet and rises about 15 feet above the ground. From the excavation made it could be seen that the lower portion, which evidently served as a base, and of which a height of about 8 feet lay exposed, was constructed of closely packed layers of rough stones. The cavity which the burrowing 'treasure-seekers' had made in the centre of the mound lay above this, and disclosed Mound of *Chalma-kazān.*

[43] For such 'witnesses' see above, p. 107.

first a bed of fine loess about 3 ft. 10 in. high and above a solid layer of stones. Of this a height of close on 3 feet could be measured in the centre, but further out, towards the periphery, it seemed to curve down towards the base like a circular vaulted wall. From the kind of well or chamber thus formed in the centre, and now filled with loess soil, some deposit had apparently been extracted; but how long ago nobody seemed to know.

Coins from *Chalma-kazān.*

I could not ascertain any traditions or legends about Chalma-kazān, since the site is ordinarily visited only by the jade-diggers, whose pits commence to line the river-bank a short distance from the southern edge of the débris area which extends here to about half a mile from the mound. At the time of my visit working at the jade pits, for a description of which I must refer to my Personal Narrative[44], had mostly ceased for the winter. Neither coins nor other objects that could furnish approximate chronological indications were picked up during my visit. But among the Chinese copper coins presented to me by Wang-Daloi as coming from Chalma-kazān is a small but representative series of T'ang pieces, comprising issues of the period K'ai-yüan (713–741 A. D.), Ch'ien-yüan (758–759), Ta-li (766–779), Chien-chung (780–783), as well as a single coin of the Sung dynasty bearing the 'nien hao' of Ts'ung-ning (1102–1106 A. D.)[45]. It thus appears that the site of Chalma-kazān continued to be occupied approximately as long as that of the old capital. The only pieces of ornamented pottery I could obtain from Chalma-kazān, also presented by Wang-Daloi, and described with a few other articles in the list at the end of this section, show the same type of work as noticed at Yōtkan, and thus agree with the evidence of the coins.

Identification of site.

The distance from Chalma-kazān to Yōtkan is approximately 13 miles, and the bearing south-south-east. Comparing with this position the topographical indications above detailed, it has occurred to me whether we might not reasonably look at Chalma-kazān for the site of the Stūpa of the turned-up *pātra* and Vairocana's convent. Neither distance nor bearing would differ very much from the 50 li to the south which the notice of the *Pei shih* and Chou shu records; the latter estimate is in all probability meant only to indicate half a day's march, such as the easy journey from Khotan town or Yōtkan to Chalma-kazān would be reckoned nowadays. The mention made in that notice of the footprints of a Pratyekabuddha which were shown at that site on a rock seems to point distinctly to the valley of one or the other great rivers; for only in the beds of large rubble deposited there and not in the unbroken loess area of the oasis could be found a stone block large enough to show markings, natural or artificial, in which the eyes of the pious would recognize sacred footprints. We know from India, where the worship of such *pādukās* has always been popular among all sects, that these supposed impressions of the feet of holy personages were ordinarily of a considerable size[46]. Since the Kara-kāsh Valley immediately above its *débouchement* takes a distinctly westerly direction, only the Yurung-kāsh Valley remains in which to look for a site that could show a 'rock' of the requisite size and that at the same time might be described as to the south of the old capital.

The name *Chalma-kazān.*

Finally, attention may be called to the curious resemblance which the name *Chalma-kazān* shows to the designation of the locality of Vairocana's shrine and to that of its Stūpa. The second part of the modern local name is manifestly the Turkī word *kazān*, meaning 'pot', while *Chalma* might be regarded as the barely changed equivalent of the name **Charma*, which the

[44] See *Ruins of Khotan*, pp. 253 sq.

[45] See reproductions in Plate LXXXIX, 26; Plate XC, 33, 34, 40.

[46] Compare, e.g., the reproduction of the stone on the Upper Swāt river showing Buddha's *pādukās* as described by Hsüan-tsang, *Mémoires*, i. p. 135, published by Bühler in the *Proceed. of the Imp. Academy of Vienna*, 1898, histor.-philol. Classe, from an impression which Col. H. A. Deane had secured for me.

transcriptions *Tsar-ma* in Tibetan and *Tsan-mo* in Chinese clearly pre-suppose as their original. Can the mound of Chalma-kazān have derived its name, literally meaning the 'pot of Chalma', from the designation once borne by the chief object of worship at **Charma*, the 'Stūpa of the inverted *pātra*'? The word *kazān*, however common in its ordinary use, is not known to me as forming the second part in any local name of Turkestān, a circumstance which, in view of the great uniformity observed in Turkī local nomenclature, makes its use here as a component part all the more curious.

I do not know how, if the old site of Chalma-kazān really marks the position of Vairocana's shrine, we should have to account for the distance of 10 li noted by Hsüan-tsang. But it may be useful to point out that the pilgrim's estimate of distance for Mount Gośṛṅga supplies an unquestioned instance of similar error on the same ground. Hsüan-tsang places Mount Gośṛṅga 20 li or about 4 miles to the south-west of the Khotan capital. Here, too, we have a correct bearing; but the distance is very considerably underestimated, since, as a look at the map shows, the hill of Kohmārī, whose identity with Mount Gośṛṅga has been definitely established, lies fully 11 miles from Yōtkan [47].

OBJECTS FROM CHALMA-KAZĀN.

C. 001. Square stone button of serpentine. Carved on the obverse, a 4-petalled flower within a simple fillet border. Reverse, formed into a very flat truncated 4-sided pyramid, near apex of which two holes are drilled from opposite sides in a slightly downward direction meeting on the line of the axis, and giving a means by which the button may be sewn to anything. ¾" square. Thickness ¼". See Plate LI.

C. 002. a. Fragment of fine terra-cotta vessel; rim. Extremely rich in colour, surface fretted. Upon upper curve of rim an ornamental border consisting of roundels close together, an incised line running round beneath them. 1⅛" × ⅞". See Plate LII.

C. 002. b. Four fragments of glass, moulded, of greenish and pinkish tints. See Plate LII.

C. 002. c. Two fragments of flint. Perhaps from the glass-maker's workshop where such flint was probably used. See Plate LII.

C. 004. Fragment of grotesque face, showing about ¾ of nose, portion of mouth and strongly recurving moustache (cf. Y. 0018, B. 001. a). The nose is much exaggerated in length, but otherwise well modelled. Face evidently made in a mould, and seems to be surfaced with a finer clay than the inner substance. Height 1¼" × 1". See Plate XLIV.

[47] This discrepancy has been duly noted by M. Grenard, *Mission D. de Rhins*, iii. p. 143.

CHAPTER IX

THE RUINS OF DANDĀN-UILIQ

Section I.—THE DESERT MARCH TO THE SITE

'Treasure-seeking' at Khotan.

Already on my first arrival in Khotan I had commenced local inquiries for ancient sites in the desert particularly deserving exploration. It had soon become clear that the only hope of securing sure guidance lay in the collection of specimens distinctly traceable to specific old sites among those which 'treasure-seekers' were in the habit of visiting. 'Treasure-seeking', i.e. the search for chance finds of precious metal within the areas of ancient settlements now abandoned to the desert, has been a time-honoured occupation in the Khotan oasis, offering to certain more adventurous elements of the population the combined fascinations of a kind of lottery and a roving life. In recent years, owing to the continued demand from European collectors, the small fraternity of professional treasure-seekers have learned on their periodical visits to ancient sites to pay attention also to antiquities as a sort of secondary proceeds. Nevertheless, all the information that could be elicited at Khotan, even from persons who seemed reliable, was far too vague to permit a plan of systematic explorations to be based upon it. I accordingly gladly availed myself of the offer of Badruddīn Khān, the Ak-sakāl of Afghān traders, who from previous services rendered to Mr. Macartney was well acquainted with the fraternity, to organize and send out small 'prospecting' parties which might secure me the needed specimens.

During the short stay I made at Khotan in the middle of November, after my return from the surveying expedition into the Karanghu-tāgh mountains, these small parties turned up with their spoil. The objects brought to me from sites close to the oasis, such as Hanguya Tati, Ak-sipil, Jumbe-kum, did not seem to justify a belief that the remains there surviving were essentially different from those that might be found on ancient 'Tatis'. It was otherwise with the party which, under the guidance of Turdi, an old and, as experience showed, thoroughly reliable 'treasure-seeker' from a village of the Yurung-kāsh canton, had gone out to visit the most distant of the locally known sites, called by them *Dandān-Uiliq* ('the houses with ivory'). Among the specimens brought back by them were two small pieces of fresco inscribed with Brāhmī characters, numerous fragments of stucco reliefs which had unmistakably served to decorate Buddhist shrines, and also a small but undoubtedly genuine piece of a paper document in cursive Central-Asian Brāhmī[1].

Objects brought from *Dandān-Uiliq*.

It appeared probable, on further examination of the 'treasure-seekers', that the ruins from which they had unearthed these remains and which they described as situated nine to ten marches north-eastwards through the desert, were identical with the site which Dr. Hedin had seen on his memorable march to the Keriya Daryā, and which his narrative speaks of as 'the ancient city Taklamakān.' He had reached it by another route from Tawakkēl beyond the

[1] For a description of these finds see the List of Objects below, under D. T. 01-017; for the fresco pieces see Plate LVIII, for the stucco fragments Plate LV.

northern edge of the oasis. Any possible doubt as to the identity of Dandān-Uiliq was set at rest when a few days later I was able to examine, in the presence of Turdi, the two hunters Ahmad Merghen and Kāsim Ākhūn, who had guided Dr. Hedin on his journey and whom at my request Pan Dārin (Ta-jên), the kindly Amban of Khotan, had promptly sent for. I was thus able to arrange definitely the initial programme of my tour, by fixing upon this site as the best place for commencing systematic explorations in the desert.

Amban Pan Dārin.

The specimens brought back by Turdi had the further advantage of convincing the Amban, who previously, notwithstanding his thoroughly friendly attitude, had been somewhat sceptical as to the existence of ancient remains in the desert, that my enterprise was fully justified. Pan Dārin's scholarly type of mind and his keen historical sense made it easy for me to explain to him the direct bearing which the exploration of such ruins would have upon the elucidation of the accounts of old Khotan left to us by the great *T'ang sêng* ('The monk of the T'ang period', i.e. Hsüan-tsang) and other Chinese travellers. His effective assistance for all the needful preparations was thus more readily assured, and to it I owed largely, as results show, the possibility of successful work at the site.

Preparations at Khotan.

The week following my return from Yōtkan (Nov. 29) was spent busily at Khotan in writing up accounts of my geographical and antiquarian observations for dispatch to Europe, and in final preparations for the desert journey. In order to keep the camels, on which we should have to depend for the difficult marches through the sands, as lightly laden as possible, it was necessary to restrict the baggage to absolute essentials and to leave behind at Khotan a depôt of all stores and materials not immediately required. The elimination was no easy matter. On the one hand, no estimate could be formed of the length of time during which my explorations would keep me away from Khotan, while, on the other, it was certain that in the inhospitable regions where we were to pass the winter any deficiency in the necessary supplies and equipment might seriously affect our health and hamper my movements. It was largely through the care bestowed on transport and supplies that I was subsequently able to extend my operations so much further from the Khotan base than was originally anticipated.

March to Tawakkēl.

On the morning of December 7, a misty and bitterly cold day, I set out for the winter campaign in the desert. In order to reach Dandān-Uiliq I had decided on the route via Tawakkēl, which, though longer than the track leading straight into the desert north-east of Yurung-kāsh, which Turdi, my 'treasure-seeking' guide, was accustomed to follow, somewhat reduced the extent of actual desert-marching with its inseparable privations for man and beast. The first day brought me, partly over marshy ground where the Lasku-Üstang and other canals finally lose themselves, to Yangi-arik, a village on the edge of the oasis north of Khotan. Its fields, as the name shows, had only been brought under cultivation in recent times. After two more dreary marches along the barren left bank of the river, where nothing was to be seen but high dunes of drifting sand to the west and reed-covered strips near its winding course, we crossed the Yurung-kāsh and soon arrived at the southern end of the Tawakkēl oasis.

Oasis of Tawakkēl.

This oasis owes its existence to the construction, some sixty years before my visit, of an irrigation canal which takes off the river waters a few miles further south near Kalama-Langar. Originally settled by forcibly transported colonists from different parts of the Khotan oasis, Tawakkēl has since developed into a prosperous small canton of some thousand households. The names of the various hamlets, called Tosalla, Borazān, Yurung-kāsh, &c., still recall the tracts which supplied the original parties of settlers. The proximity of the forest belt, which accompanies the Yurung-kāsh on its course through the desert, provides a plentiful supply of wood to this outlying colony. The fact was brought home to me at the very time of my arrival by the big bonfires which the local Bēg, instructed in advance from Pan Dārin's Yāmen, had caused to be prepared to light up my way, a luxury which could not have been indulged in within the main oasis. I was struck also by the prevalence of wood in the construction of houses. Whereas in Khotan the free use of this building material is restricted to the dwellings of the well-to-do and to other substantial structures, at Tawakkēl the houses even of simple cultivators show the usual wooden framework and wattle walls covered with mud plaster. We shall see from the account of the Dandān-Uiliq ruins that this observation is not without antiquarian interest.

Conditions of cultivation. After halting at Hong I moved my camp on the following day to the Bēg's house at Āt-bāshi, some six miles further north, in order to complete there the arrangements for the party of labourers I wished to take along. In view of the observations detailed above as to the rise of the ground-level in the old cultivated area of Khotan, I was interested to note on the march that in this comparatively recent oasis the roads and waste spaces lay nowhere more than about one foot below the level of the neighbouring fields. The period of irrigation had manifestly been too short here to permit of any appreciable rise in the level of the fields. It was equally in keeping with previous observations to learn that the area of the colony might be greatly extended towards the desert by the construction of additional canals. The abundant supply of water which the river carries down during the spring and summer months would suffice to bring under cultivation large tracts along either bank of the river now covered by low dunes. The fertility of the 'sand' which forms the latter is well-known to the cultivators; and since its appearance differs in no way from that of disintegrated loess soil within the oasis, its identical origin from the fertile deposits of the river courses can scarcely be doubted. But here as elsewhere along the southern edge of the Taklamakān desert no surplus of population is available for such extended cultivation, nor is the administration capable of undertaking fresh irrigation works on a large scale.

Engagement of excavation labourers. The 11th of December was spent at Āt-bāshi in collecting a party of thirty labourers for my intended excavations, together with four weeks' food supply. Notwithstanding the good pay offered, $1\frac{1}{2}$ Miskals per diem, more than twice the average wages for unskilled labour, I should not have secured the requisite number of men had not stringent instructions to that purpose been issued by Pan Dārin; for both on account of superstitious fears and in view of the expected rigours of the winter, the cultivators were very reluctant to venture so far into the desert. Fortunately the two Tawakkēl hunters, Ahmad Merghen and Kāsim Ākhūn (father and son), were with me to inspire confidence. I had engaged them as guides; and inured to all hardships, and by the experience of their roving life intelligent far beyond the villagers' horizon, they soon proved most useful in looking after the labourers both on the march and at excavation work. I took care to select the physically fittest from among the men brought before me, and to assure their being supplied with all needful warm clothing and an adequate store of food. Each man had to bring his 'Ketman', the hoe in common use throughout Turkestān, which proved an excellent implement for work in the sand. For the carriage of their food-supplies and other impedimenta the few camels I could spare from my train of eight were not sufficient. So they were supplemented by a dozen donkeys, which offered the advantage of needing a minimum of fodder. For the camels only a quantity of oil made of rape seed could be taken along. Half a pound daily of this for each animal proved remarkably effective in keeping up their stamina during the trying desert marches, when they had to go without grazing or fodder of any kind and sometimes for days without water. Our ponies, for which the desert to be crossed offered neither sufficient water nor fodder, were sent back to Khotan, however unwelcome to my Turkī followers was the prospect of having to trudge through the sands on foot.

On the forenoon of the 12th of December, when the troop of labourers had duly collected and their loads had been arranged, the start was made for the desert, half the population of Tawakkēl being assembled to witness our departure. The first march was a short one, to a point of the scrub-covered river-bank beyond the northernmost edge of the oasis, where the animals were to be allowed a plentiful drink in the evening before entering the sands eastwards. On the following day we struck to the east, and soon picked up the track marked by the footprints of a small advance party which I had sent ahead two days previously under the guidance of Kāsim Ākhūn. He had orders to dig wells at all places suitable for camps, and after reaching Dandān-Uiliq to push on to the Keriya Daryā. From there Surveyor Rām Singh, whom three weeks earlier I had sent from Khotan on a supplementary surveying expedition into the mountains south-east of Khotan, was expected to join me. Start into desert.

I need not attempt here any detailed description of the desert zone through which the next five days' marches took me. In this desolate waste, which showed no trace of human occupation until the vicinity of Dandān-Uiliq was reached, there was no scope for antiquarian observations, while those physical features which are of general geographical interest have already been recorded by a most competent observer, Dr. Hedin, who on his journey from Tawakkēl to the Keriya river in January, 1896, had crossed the same ground[2]. Yet a brief account of this desert journey and its experiences, as already related in my Personal Narrative, may well find a place here. It will serve to illustrate the strangely forbidding surroundings which formed the scene of the greatest, and certainly the most fascinating part of my archaeological labours, and to give some idea of the physical conditions under which the work of the winter was done.

Though the dunes in the area crossed during the first two days were low, rising only to 6–10 feet, marching in the drift sand was slow work. The feet of men and animals sank deep at every step into the fine sand, resembling loess dust, and the progress of the camels was reduced to about $1\frac{1}{4}$ miles per hour. In view of the want of sufficient fodder and water they had to be saved all over-exertion; hence I soon found that the direct distance covered by a day's march could rarely exceed nine to ten miles. The tamarisk and 'Kumush' scrub, which was plentiful at first, grew rare in the course of our second march, while the wild poplars or 'Toghraks' disappeared altogether as living trees. Luckily amidst the bare dunes there rose at intervals small conical hillocks thickly covered with tamarisk growth, such as I have already had occasion to describe in the vicinity of the ancient desert site of the sacred rats, and the decayed roots of this scrub supplied excellent fuel[3]. Close to these hillocks there were usually to be found hollows scooped out of the loess soil, undoubtedly by the erosive action of wind. These hollows, which reach down to at least 10–15 feet below the level of the little valleys separating the neighbouring dunes, offer, of course, the nearest approach to the subsoil water. It was invariably in them that Kāsim's advance party had dug their wells, and in them also we chose our camping-places. The water, which was reached after digging to an average depth of 5 to 7 feet, was very bitter at the first two camps, and scarcely fit for human consumption. But as we moved further away from the river the brackishness of the wells decreased, an observation well known to my guides as generally applicable to these parts of the desert, and First marches through desert.

[2] See *Reisen in Z.-A.*, pp. 33 sqq.

[3] See above, p. 120. For a graphic diagram and an account of the formation of these tamarisk hillocks, see Hedin, *Reisen in Z.-A.*, p. 34.

already noticed by Dr. Hedin. The supply of water furnished by these wells was decidedly scanty for so large a party, and since the freezing of the damp soil overnight stopped it altogether spare water had to be collected in the evening in two of my iron tanks and stored as ice for use on the next day.

Climatic conditions. The winter of the desert had now set in with full rigour. In the daytime while on the march there was little to complain of, for though the temperature in the shade never rose above freezing-point, yet there was no wind, and the delightfully pure air of the desert, and its repose (which nothing living disturbs), could be enjoyed without discomfort. But at night, when the thermometer would go down to minimum temperatures from 0° to −10° Fahr., my small Kābul tent proved a terribly cold abode, and all my fur clothing was needed. Fortunately the ample fuel allowed big fires to be kept up by the men, who crouching around them in closely packed circles passed the nights safely.

Guidance of Turdi. On the evening of the fourth day after entering the desert, as we were pitching camp amidst desolate dunes covering dead tamarisk roots, two of the men sent on ahead returned with the report that Kāsim's party had failed to trace the ruined site we were in search of. It was now the turn of old Turdi, my 'treasure-seeking' guide, to prove his knowledge of this dreary region. He had only once in his life approached Dandān-Uiliq from this side, but he had on the march several times told me that he thought the track followed under the guidance of the Tawakkēl hunters was leading too far north. On the plain avowal of their inability to discover our goal I could see a gleam of satisfaction pass over his wrinkled face. A short conversation with the returned men sufficed for him to locate the point reached by Kāsim's party, and early next morning they were sent back with full instructions to guide Kāsim into the right direction. The guidance of our own party was now taken over by Turdi, who, with an instinct bred by the roamings of some thirty years and perhaps also inherited—his father before him had followed the fortunes of a treasure-seeker's life—found his bearings even where the dead uniformity of the sand dunes would to ordinary eyes seem to offer no possible landmark.

Approach to ancient site. On December 17 we skirted the foot of several higher ridges of sand or Dawāns, reaching approximately to 25–30 feet above the level of the depressions between them and stretching usually in the general direction from north to south, until in the evening we reached a belt of ground where dead trees were seen emerging from heavy sand. Shrivelled and bleached as they appeared, Turdi and the men could recognize among them trunks of the Toghrak or wild poplar, and some other jungle trees, unmistakable proofs that we were in an area once reached by water. Turning now to the south-east we came, about one and a half miles further, upon stretches of bare loess with an extensive line of hollows, curiously resembling a dry river course, yet undoubtedly only a result of wind erosion. In one of these steep-banked hollows we succeeded in digging a well, and thus saved ourselves a search in the dark for the spot which alone, according to Turdi's knowledge, offered water in the immediate vicinity of the ruins. On the following morning, December 18, after turning a great Dawān, this one with a main direction from north-east to south-west, Turdi guided us to the isolated remains of a much-decayed small structure to be described hereafter, and a little over two miles further south I found myself within view of the ruins which mark the site of Dandān-Uiliq.

First view of ruins. Scattered in small groups over an area which my subsequent survey (see the plan in Plate XXIV) showed to extend for about one and a half miles from north to south, with a width of close on a mile, there rose from among the low dunes the remains of buildings modest in size but of manifest antiquity. Where the sand had blown away, the walls, constructed throughout of a timber framework with thick layers of plaster over a kind of wattle, were exposed to view, broken down to within a few feet from the ground. Elsewhere in places, covered by low dunes not exceeding 6–10 feet in height, the walls could still be made out by the rows of wooden posts rising in splintered stumps above the sand. Pottery débris was seen strewing many small patches of ground where the dunes had left the loess soil bare in the vicinity of the ruins. All structural remains left exposed showed signs of having been explored by 'treasure-seekers', and the marks of the damage caused by their operations was only too evident. Yet even thus the ruins, on a first hurried inspection, furnished unmistakable evidence of their character and approximate date. The remains of frescoes visible on the much-injured walls of one of the exposed structures indicated that Buddhist places of worship were amidst the ruins. The style of these frescoes, clearly derived from Indian models, seemed to mark the last centuries preceding the introduction of Islām as the probable date when the settlement had been deserted. And already on the first day this conclusion received support from two Chinese copper coins of the K'ai-yüan period (713–741 A.D.), which were picked up under my eyes from the débris-strewn ground near the ruined buildings.

Old Turdi felt quite at home among these desolate surroundings, which he had visited so often since his boyhood. It was the fascinating vision of hidden treasure which had drawn him and his kinsfolk there again and again, however scanty the tangible reward had been of their trying wanderings. Yet the familiarity thus acquired with the silent relics of the past had developed in him an instinctive interest in all traces of the life that once reigned here. His shrewd observations on some of its more practical aspects, and his excellent memory for matter-of-fact details noticed in the course of his diggings made him distinctly useful at the site. It was evident from his communications that the conditions of the dunes were changing very slowly at this site. Consequently Turdi had no difficulty in recognizing the places where he and his companions had been at work during previous visits. Luckily their scanty resources had never allowed them to overcome the difficulty presented by the transport to this distant site of supplies sufficient for a prolonged stay, or to bring there working parties of more than a few men. Observations of Turdi.

Hence the structures more deeply buried in the sand had a good chance of escaping unopened. It was important to select these in the first place for my excavations, and I felt grateful for Turdi's good memory and topographical instinct which enabled him readily to indicate their positions. Guided by this first rapid survey I chose for my camp a place from which the main ruins to be explored were all within easy reach. Practical considerations compelled me to make my choice carefully. For in order to keep the men at work as long as possible every day they had to be spared tiring tramps through the drift-sand. Still more important was it to keep to a spot where fuel could be readily obtained in quantities sufficient for our prospective long stay. Tamarisk-covered hillocks were rare, and subsequent observation showed that we had almost reached the northernmost limit of their occurrence in this region. Hence the dead trees of ancient orchards could alone be expected to furnish an adequate supply of fuel, and they were found to be very unequally distributed over the ruined area. As soon as the spot which seemed to answer these conditions best had been decided upon, and the baggage unloaded, I hastened to dispatch the camels on their three days' journey to the Keriya Daryā, there to find the fodder they so badly needed. The donkeys, too, which had carried the men's food supplies, with the meagrest rations for themselves, were now sent back to Tawakkēl. Selection of camp.

Section II.—FIRST EXCAVATION OF BUDDHIST SHRINES

Before proceeding to a detailed account of the excavations carried on by me at the site of Dandān-Uiliq, and the archaeological finds yielded by them, it will be useful to describe in broad outlines the characteristic features of the area which contains the ruins. With the exception of occasional narrow patches already mentioned, where the hard loess soil lies exposed, the whole of it is covered by low dunes of the usual semi-lunar shape ranged in close succession. In accordance with the prevailing winds, their crest was found generally facing with its concave side to the south or south-east. The drift-'sand' of the dunes differs in no way in appearance from the fine sand that filled the structures excavated. Of the specimens brought back from the latter, Prof. Lóczy's microscopical examination has shown that their constitution is identical with that of the riverine loess found in the various layers of Yōtkan. Hence the conclusion seems justified that the dunes of Dandān-Uiliq, like those at Rawak, Ak-sipil, and elsewhere near the edges of the Khotan oasis, owe their origin to accumulation of finer river deposits carried away by the winds or else to identical materials collected by the winds from eroded loess beds.

Condition of drift-sand.

Within the limits of the area where pottery débris occurs on any bare patches of loess (marked by a dotted line in Plate XXIV) the height of individual dunes above the original level did not seem anywhere to exceed 10 feet. But beyond those limits to the north the average elevation of the dunes increased up to about 20 feet, and the small depressions between them invariably showed deep sand at their bottom. Two miles to the north lay the great ridge of drift-sand I had passed on approaching the site, with its broad swelling crest reaching an apparent elevation of about 50 feet above the ground to the south. Two similar great Dawāns were seen rising to the east and north-west of the ruined area, high enough to hide completely any ancient buildings that might have existed there. But their slopes did not approach the area nearer than about half a mile, and as the intervening ground showed no traces of ruins there is nothing to suggest that those heavy masses of sand cover ancient remains.

Wind-eroded depression.

Within the area anciently occupied only one prominent physical feature attracted attention, a clearly marked depression stretching in the direction south-east to north-west to near the spot where our well was dug in a basin-like hollow. The width of this depression varied from about 50 to 80 yards and more, and its depth from 10 to 20 feet. Its banks steeply cut into the loess soil, together with the curving course, curiously suggested a river bed. Yet a closer examination of the banks, especially those towards the west, showed at once that the force which had excavated this depression could only be wind erosion. That this potent agency was still actually at work could be seen where the great trench passes close to the group of ancient dwellings marked D. v and D. ix. Here progressive erosion had undermined the timber foundations of structures adjoining the bank, and large fragments of wood which had evidently become detached quite recently, together with other débris, were strewing the steep slope. I had subsequently frequent occasion to observe exactly the same process in its progressive stages of destruction near the ruined buildings of the Niya Site, where the surrounding ground had been gradually cut away to 15 feet and more below the original level as indicated by the ruins[1].

[1] I am inclined to believe that the depression above briefly described, and also marked in the plan, is closely related in character and origin to the large wind-cut trenches, often developed into regular parallel valleys, which Prof. E. de Cholnoky has observed in certain sandy areas of the Hungarian plains. See his paper 'On the laws of drift-sand movement' in the *Journal of the Hungarian Geological Institute*, 1902 (transl. in *Földtani közlöny*, xxxii, pp. 138 sqq.), where the great geological importance of this phenomenon for the explanation of surface formations appears to have been recognized for the first time.

Apart from the ancient structures thus attacked in their very foundations, there was ample evidence at the site of the havoc worked by erosion both at the present and during earlier periods. Wherever the tops of posts or other parts of the timber framework of houses still rose above the protecting cover of sand, they showed by their curious splintered and abraded appearance the force of the wind and driven sand to which they were exposed during the Burān periods. The same was the case with all walls which the operations of treasure-seekers or the shifting of dunes had within recent years deprived of their protective layer of sand. Finally, we have to recognize manifest results of erosion, probably from an early period, in the numerous places between the ruins where the ground was thickly strewn with fragments of coarse pottery, small corroded pieces of metal and similar débris, while no trace of structural remains survived. In view of our previous observations on 'Tatis' it seems safe to assume that these remains mark the positions occupied by less pretentious buildings, which, like the houses of Khotan cultivators at the present day, were built wholly of sun-dried bricks or stamped clay. These were likely to crumble away far more quickly than buildings with a solid timber framework covered by hard plaster, and erosion could complete their destruction before the drift-sand arrived in sufficient masses to cover them up and afford protection. This observation helps to explain, at least partly, why at a site like Dandān-Uiliq, which for various reasons may be supposed to have been occupied by a comparatively large settlement, the structural remains traceable at present are limited to small groups of timber-built structures widely scattered over the ruined area. *Effects of erosion.*

I propose to describe the excavations carried on among these structures group by group, though the limited time at my disposal did not allow me systematically to follow the same order in the course of the work. I may, however, make an exception in the case of the small structure with the clearing of which I commenced my explorations on the morning of December 19. On the previous evening I had noticed much-decayed remains of a small square building, D. I, immediately to the south-east of my camp. Turdi knew it as a 'Būt-khāna', or 'temple of idols', and well remembered having searched it in his own fashion on previous visits. The last of these must have been his 'prospecting' expedition on my behalf, for the several fragments of stucco reliefs brought back by him, and now described in the list of Dandān-Uiliq antiques (see below) as D.T. 001–014, agree exactly with those excavated by me at D. I, and in one instance supply a portion missing in one of the latter pieces. There can be no doubt either that among those who had searched the ruin before was also Dr. Hedin, for the small stucco reliefs obtained by him from a much-decayed building, where 'the sand was quite shallow', comprising seated and standing Buddhas (the latter described as draped women), as well as decorative fragments, must, as the illustrations show, have been cast from the same moulds as many of the pieces described in my list[1]. But the sand, though only 2 to 3 feet high, had not been removed, and by laying bare the foundations and floor I might expect to gain useful preliminary knowledge as to the general construction and arrangement of such shrines. *Ruined shrine D. I.*

My hope in this respect was not disappointed. A careful examination of the remains of walls brought to light on the north and west sides showed that there had been an inner oblong or square cella enclosed by equidistant outer walls, forming a kind of corridor or passage on each side about 5 feet wide. The walls to the south and east had completely disappeared to the very *Structural arrangement.*

[1] See *Through Asia*, p. 797; illustrations, p. 799. From earlier visits of Turdi date probably the fragments of stucco reliefs showing a garland-holding Gandharvī and a seated Bodhisattva, reproduced by Dr. Hoernle, *Report on C.-A. ant.*, ii. Pl. XII, Nos. 2, 3.

foundations, and of the others the original dimensions of only the western outer wall, 20 feet long, could be measured with certainty. Both inner and outer walls consisted of plaster laid on a framework of wood and reed matting, which itself was held in position by square posts fixed at regular intervals. The walls of structures subsequently excavated were far better preserved, and details as to this manner of construction may more conveniently be given in connexion with them. But it deserves to be noted that, however badly decayed the remains of walls were in D. I, traces of fresco decorations showing rows of miniature representations of Buddhas or Bodhisattvas, all uniform except in the colouring of dress and background, and evidently stencilled, could just be made out on them.

Decorative remains of D. I.

Before commencing the clearing, I had carefully examined the timber and plaster pieces which strewed the surface of the sand in the interior of the small shrine and immediately around it. Among this débris, evidently representing what former searchers had extracted but thrown away again, I was gratified to light upon fragments of coloured stucco reliefs which evidently had belonged to wall decorations [3], as well as upon the remains of a rectangular upright post, 26 inches high (D. I. 04), showing on one side the elaborately-painted figure of a Buddha or Bodhisattva standing on a blue lotus (see Plate LXV). From the shape of the post and the decorations painted on part of one other side, it is clear that it must have served as a pilaster or jamb projecting from a plaster-covered wall. Considering the exposed position in which the piece had lain the preservation of the colours was certainly remarkable, and furnished reassuring proof of the preserving power of the dry desert air. Of the rich decoration which the upper portions of the inner cella walls, long ago decayed, had once borne I could not remain in doubt when fragments of small stucco reliefs (which must have originally belonged to plaques of regular patterns) turned up in dozens from the sand covering the interior.

Stucco reliefs from walls.

The numerous replicas by which most of the decorative pieces and the small reliefs of Buddhas and attendant figures are represented indicate that the whole of this stucco work must have been done in moulds. Owing to the smallness of the fragments, it would have been difficult to form an approximate idea of the manner in which the reliefs were arranged, had we not the examples of similar decorations brought to light from the shrine D. II to be described presently. On comparing the relief fragments D. I. 11, 42, 87, 99, reproduced in Plates LIV, LV, LVII, with corresponding pieces from D. II shown in the same Plates, it becomes very probable that, like the latter, they were mostly intended to fill decorative aureoles or vesicas round some larger images now completely perished [4]. Of the general effect aimed at we may judge by comparing the elaborate aureoles or vesicas in relief which surround some colossal Buddha figures of the Rawak Vihāra (see Figs. 63, 64, Plate XVIII), though the work in the latter is separated by centuries from that of the Dandān-Uiliq shrines. But among the stucco ornaments from D. I there are also some representing leaves and other decorative motives (see Plate LVII) which might have served to adorn relief friezes or cornices.

Analysis of stucco.

The technique of these small decorative relics proved to be identical in all of the Dandān-Uiliq temples. They were treated in *appliqué* fashion, the figures or ornaments being separately moulded and subsequently attached, by a grooving or some other device intended to give tooth, to a plaster background [5]. The latter again is easily distinguished from the far coarser plaster of the wall. Prof. Church's chemical analysis, reproduced in Appendix *F*, shows that the hard stucco uniformly used in the decorative reliefs of Dandān-Uiliq shrines is essentially plaster of

[3] See D. I. 03, 9–14, in list.

[4] Compare, e. g., Mr. Andrews' remarks on D. I. 42, 94.

[5] Compare notes on D. I. 70, in descriptive list.

Paris. This accounts for the hardness of almost all the pieces from D. I and the relatively good preservation of their surface, including the colouring. The stucco of the walls, on the other hand, being soft in comparison and friable, proves to consist of loess containing a considerable proportion of true clay. A special distemper was probably applied to prepare its surface for frescoes. The fact that the stucco of the large reliefs and images at Dandān-Uiliq, as well as at the other sites, was invariably made up of loess like that of the walls, explains why, of the large image or images which the shrine D. I must be supposed to have once contained, nothing turned up in the débris but a life-size finger in red clay (D. I. 14). It is probable, however, that the colossal hand grasping a billet (D. T. 012), made of the same friable stucco, which Turdi had brought from his previous visit, was also obtained in this ruin.

Turning to individual pieces recovered from the wall decorations of D. I. we may specially notice the small relief figure D. I. 11 (Plate LV), represented in a series of replicas, of a Buddha standing with the R. hand raised in the Abhaya-mudrā attitude, but brought across the centre of the breast. D. I. 99 (see Plate LV), one of several replicas, shows a small Bodhisattva seated on a lotus and surrounded by an elaborate vesica, and D. T. 05 on the same plate a more carefully modelled head of a Buddha or Bodhisattva. The graceful figure of a garland-holding woman rising from a lotus (D. T. 02, D. I. 110, D. I. 014) which Plate LVI illustrates, is probably meant for a Gandharvī, and must have been a favourite decorative theme with the local artists; for her figure reappears in a variety of poses and sizes among the stucco reliefs from the walls of other shrines (see Plate LVI). The female head reproduced in Plate LIV from D. I. 90, one of several fragmentary specimens (comp. D. I. 18, 43, 89) must also, in view of the curious hairdress, be attributed to a Gandharvī, though of a somewhat different type. Description of stucco reliefs.

Among the fragments of relief ornaments, D. I. 42 (Plate LIV) serves to illustrate the frequently recurring half-round border of a vesica, showing rows of overlapping lotus-petals originally in alternate colours (comp. D. II. 34 in same Plate). D. I. 87 (Plate LVII) is a specimen of a curious ornament representing tongues of flames in varied colours, which seems to have regularly adjoined the border just noticed (comp. D. II. 55, 89, Plate LV). Peculiar to D. I, and distinctly graceful in design, are the floral ornaments (see Plate LVII) showing vine scrolls and grapes (D. I. 114; also D. I. 53 in Plate LIV), lotus leaves springing from a moulding of beads (D. I. 60, 74), lotus petals (D. T. 011) or conventional chrysanthemum leaves (D. I. 012). Jewel-shaped ornaments are illustrated by D. I. 10 (Plate LVII), D. I. 44 (Plate LV).

The clearing of this single small ruin not only yielded a large number of sculptural fragments showing unmistakable affinity to the Buddhist art of Gandhāra, but supplied me with the indications needed in order to start the systematic excavation of structures more deeply buried in the sand. So when on the next day I proceeded to the group of ruins situated half a mile to the south of my camp, where on my first rapid survey I had noticed a number of small buildings well covered by the sand and thus giving promise of better preservation, I was able with some assurance to gauge their construction and character, though only the broken and bleached ends of posts were visible above the sand. The arrangement of these, in the case of the easternmost structure D. II, is shown by the accompanying photograph, Fig. 28, as it appeared before excavation. Together with a bit of fresco-covered wall laid bare by some recent burrowing of Turdi's party, it suggested two small temple cellas. The excavations commenced here soon proved the surmise to be true, and in addition revealed unexpectedly rich relics of the sculptures and frescoes which had once adorned the cellas. Their constructive features and adornment proved typical of those observed in other shrines subsequently excavated at this site, and may therefore be described in some detail. Southern group of ruins.

Large cella, D. II. Construction of walls.

The large cella, as seen in Plate XXV, formed a square of 9 feet 6 inches inside, with a door opening in the centre of the north wall. The cella was enclosed by a quadrangular passage about $4\frac{1}{2}$ feet wide on the east and west sides, but broadening to 6 feet 5 inches on the north and south. This passage, which undoubtedly served for the purposes of circumambulation or *pradakṣiṇā*, had its entrance from the north, but a little to the east of the centre. The photograph reproduced in Fig. 29, which shows the cella as seen from the north after excavation, will help to illustrate the construction of the walls enclosing the cella and passage. They consisted, as in all structures subsequently explored at this site, of a wooden framework having a kind of wattle in its interstices, and coated on either side with layers of plaster to a total thickness of $6\frac{1}{2}$ inches. The principal parts of the timber framework were massive posts, about 4 inches square, set approximately 3 feet apart. These posts rested on massive beams embedded in the ground and forming a foundation; higher up they were held in position by horizontal transverse beams. Remains of the latter are clearly seen in the photograph reproduced in Plate III, showing the cella from outside its south-east corner. This also helps to indicate how beams and posts were jointed by means of dowels. The beams fixed at vertical distances varying apparently from 3 to 4 feet (the decayed condition of the upper wall makes it impossible to trace anywhere the positions of more than two lines of beams), also served to keep in position a double row of stout sticks, usually three or four to the interval between each pair of posts. Between the double row of sticks were inserted horizontal layers of reeds, clearly seen in Fig. 29, the whole forming a very stout wattle for the support of the plaster coating. The latter consisted of clayey loess mixed with coarse straw and grass, and was surfaced with a finer layer of the same materials containing an admixture of vegetable fibres to increase its consistency. Provided with a coating of distemper the wall surfaces were quite smooth and firm enough to receive fresco paintings, such as were found to cover the walls inside the cella and passage in this as well as in other shrines. The walls of the cella had suffered far less damage than those enclosing the passage; yet even they did not rise anywhere to more than 5 feet above the floor. The latter was found to be covered with plaster similar in make to that of the walls.

Remains of stucco images.

The interior of the cella was once occupied by a colossal statue made of stucco and painted, which most probably represented a Buddha. But of this only the feet remained, about 13 inches long, raised on an elaborately moulded oblong base, 3 ft. 3 in. long, 2 ft. 9 in. wide, and about 3 ft. high. As seen in Fig. 29, the top of this base was formed into a lotus pedestal on which the feet of the image rested. This pedestal was found to be covered by only about a foot of sand. Owing to this want of protection the other parts of the statue had crumbled away long ago. The fragments unearthed close to the surface comprised parts of the legs and of the lower drapery, the latter still bearing traces of its dark red colouring. They broke at the slightest touch and could not be removed. These fragments were worked in the same reddish clay, largely mixed with fine vegetable fibres, which is shown by some small stucco reliefs representing seated Bodhisattvas, D. II. 3, 6, 12 (see Plates LIV, LV); these were found within this cella, and probably belonged to some long-perished wall decorations. Of the wooden framework or core, which once supported the heavy image, only the lowest part was still intact, fixed within the left foot (as seen in Fig. 29). Each of the four corners of the cella was occupied by a draped stucco figure, probably somewhat under life-size, standing on a lotus pedestal. But of these statues only the one in the north-west corner was found intact to below the waist. The pose of the legs and the wavy drapery as seen in a photograph recall the small attendant figure R. iv. among the colossal images of the Rawak Vihāra (see Fig. 61); but the drapery is treated in a far more conventional fashion, and the execution is less careful. Of the other corner statues nothing

28

RUINS OF BUDDHIST SHRINE D. II, DANDĀN-UILIQ, SEEN FROM SOUTH BEFORE EXCAVATION.

29

RUINS OF BUDDHIST SHRINE D. II, DANDĀN-UILIQ, SEEN FROM NORTH-EAST AFTER EXCAVATION.

but the feet and portions of the lower edge of the drapery remained, and I am hence unable to make any suggestion as to their original character. The arrangement of these four corner figures closely agrees with that observed in the Endere temple (see plan in Plate XXXVI).

Painted panels.

That the cella had once contained other sculptured objects, but of a far more modest size, was shown by the finding of two small wooden boards lying detached on the top of the base and near the feet of the principal image. They were perfectly plain, but each bore the much decayed and almost shapeless remains of a pair of feet and legs in soft stucco reaching to a height of about 3 inches. These undoubtedly belonged to some small statue similar to the one subsequently excavated in the adjoining cella (see D. ii. 09 in Plate LIII). The exposed position of these little sculptures, built up on wooden boards for the sake of mobility, had caused even the wood of the latter to perish. The same cause had greatly damaged three small wooden panels (D. ii. 2, 4, 03), bearing paintings, which I found leaning against the lotus pedestal of the central image. There could be no doubt that these panels had been placed there by worshippers and represented objects or scenes of religious interest. But the decay of the wood and the damp which had reached the painted surfaces has made the outlines difficult to recognize even to the practised artist eye of Mr. Andrews. No reproduction of these painted panels could be attempted, and I must refer to Mr. Andrews' description in the list at the end of this chapter for any details which can still be made out. D. ii. 2 and D. ii. 4 were painted on both sides; the latter panel, measuring 12 by $4\frac{1}{8}$ inches, shows apparently three seated figures on either side. D. ii. 03, about $8\frac{1}{2}$ inches wide and $5\frac{5}{8}$ inches high, painted only on one side, displays two seated figures with halos, each holding what looks like a swathed infant. Curiously enough both D. ii. 2 and D. ii. 03 show remains of written leaves, which had been pasted across the surface and thus had helped to efface the painting. The paper of the leaves has decayed so badly that only a few characters of the Brāhmī script in each line can now be made out. But these suffice to show that the Brāhmī writing was of the upright Gupta type of the seventh or eighth century, as it appears in several MS. finds from the neighbouring ruin D. iii. (see Plate CVIII), and that each leaf or piece of leaf originally belonged to some MS. of the usual oblong shape, one apparently showing three lines and the other five lines per page. Insignificant in themselves these almost perished scraps of writing were of value to me at the time as raising hopes of real MS. finds.

Mural paintings.

The walls of the cella, which, judging from the size of the principal image, must have been of considerable height, were found to have been decorated inside with frescoes. The colours looked much faded and worn, especially in the extant upper portions, as if the surface had been exposed for a considerable time to climatic influences, including rain trickling down from the roof, for some time before the protecting sand invaded the building. Over a considerable portion of the extant walls the painted surface had peeled off. Nevertheless it was possible to make out that the walls west, south, and east had each been occupied in the centre by a sacred figure standing enveloped in a large vesica. As these central figures were over life-size, only the feet with a broad painted frieze below them showing lotuses and small figures of worshippers could be made out in parts. There were narrow vesicas painted around each of the four corner images, a leaf border being the most distinguishable feature in them. The western wall, the best preserved on the whole, showed between this corner decoration and the central painting remains of very graceful tracery work curiously suggestive of late Gothic, and inserted between it miniature representations of what looked like a small shrine (or prayer-flag?) containing the figures of two seated deities. Most of the outlines seemed drawn in a kind of terra-cotta colour, while the colours between had often faded into complete indistinctness.

Fresco decoration of outside walls.

The decoration of the outside of the cella walls consisted mainly of narrow fresco bands about five inches high, containing a succession of miniature representations of Bodhisattvas or Arhats seated in the 'Dhyāna-mudrā' attitude, and coarsely painted on backgrounds of different colours. In the same manner the colours of halos and robes alternated between red, yellow, and white. The whole of this decoration, which Plate III shows on the east and south sides, was in all probability produced by means of stencils. The same kind of decoration was also found on the inside of the low extant portion of the north passage wall. Amidst these conventional designs there appeared on the outside of the cella wall facing south a fresco which, though much effaced, attracted my interest as evidently representing some sacred legend, perhaps of a local character. It is just visible in Plate III to the left in the shadow. It showed three rows of youths riding Bactrian camels or else dappled horses, four or five in each row, each holding a cup in his outstretched right hand, while above one of the riders a bird is swooping down on this offering. The popularity of the subject is attested by my subsequent discovery in other shrines of the site of the well-preserved painted tablet D. VII. 5 (Plate LIX), and of the panel D. x. 5 (see Plate LXII), on both of which a similar scene is figured[6].

Frescoes on passage wall of D. II.

The walls enclosing the passage had suffered badly on all sides, and only of the southern wall did enough remain to show portions of the frescoes with which its inner face had been decorated. There were the feet and lower drapery of two life-size figures, each standing upon an open lotus, and surrounded by a vesica; below them ran a decorative frieze in which lotuses floating in the water and small human figures, perhaps meant for Nāgas, in a worshipping attitude at the feet of sacred personages above, could still be distinguished. Plate III shows the feet of the life-size standing figure to the right with a portion of the frieze below, and on the left the figure of a small seated Buddha filling the triangular space left above the frieze between the lower portions of the two large vesicas. I succeeded in removing safely the piece of painted plaster, D. II. 08, containing the latter part of the fresco, which has accordingly been reproduced with its original colours in Plate LVIII. It shows a Buddha or Bodhisattva in a dark red robe seated in the attitude of argumentative discourse (*nyāya-mudrā*) with crossed bare feet.

Inscriptions below frescoes.

The short inscription painted beneath contains six characters in the peculiarly cursive Brāhmī which appears in other inscriptions found at Dandān-Uiliq, and also in certain paper documents from the same site to be discussed hereafter[7]. They have been read by Dr. Hoernle as *dvī pī sā dāṁ śo dā*, and are presumably in the same early Eastern Īrānian dialect which Dr. Hoernle's researches have demonstrated to be the language of those documents. The other fresco piece which could be removed from the same ruined wall, D. II. 13, has also, notwithstanding its very friable condition, survived the long journey without serious damage. It shows the feet of the life-size figure to the left standing upon an open lotus and between them an inscription in seventeen cursive Brāhmī characters, some of them rendered indistinct by the peeling off of the painted surface[8]. Here, too, the language seems to be the previously mentioned Eastern Īrānian dialect, but no interpretation can be attempted. A third short inscription, in cursive Brāhmī characters, rendered almost wholly illegible by the rubbing off of the paint, was found below two small seated figures which occupied a position close to the ground near the south-east corner of the passage. Finally it may be mentioned that the small piece of painted plaster, D. T. 016 (see Plate LVIII), showing traces of Brāhmī characters which was among the objects

[6] See below, sec. vi.

[7] See below, sec. v.

[8] The reading with which Dr. Hoernle has kindly furnished me is: *bya ja ga ri ja sho bha yi ? re mā pyā da ? s* (?) ? ?. Doubtful characters are marked by ?.

first brought to me by Turdi, was declared by him to have been obtained from the south wall of this cella. Marks of recent burrowing were visible there on my arrival, but I could not ascertain the exact position from which this piece might have been removed. It seems probable that these short inscriptions had a dedicatory or invocatory character; but this is all that in our present extremely scanty knowledge of their language can be suggested as to their contents.

Section III.—ART RELICS OF SHRINE D. ii.

Small cella D. ii.

The excavations, when extended on December 21 to the remains immediately adjoining the west wall of the shrine just described, brought to light another temple cella which, notwithstanding its small dimensions, proved remarkably rich in art relics. This little chapel, which is seen in Fig. 31, measured only 12 feet 8 inches from north to south, with a width of 8 feet 8 inches (including walls), and had no enclosing passage. Its walls, built of a timber framework and plaster as previously described, had a thickness of only 4 inches, and had in consequence crumbled away to within a foot or two from the ground, except on the east side, where the closely adjoining outer wall of the larger cella gave support, and on the south, where a platform, 1 foot high and 2 feet 3 inches broad, surmounted by a massive base for the principal image, had been built along the whole length of the wall. Of the stucco image which once occupied this base only the scantiest fragments could be found; for with a pedestal raised more than 3 feet above the ground it must have long remained without the protecting cover of drift-sand. This was testified by the extremely friable condition of the few recovered fragments of soft stucco that belonged to this image. The thin coloured surface peeled off at the slightest touch; yet in one piece the fragment of a head worked in the round could be recognized, and as this measured about 7 inches from the inner corner of the right eye to the edge of the ear, the statue must have been over life-size.

Base of principal image.

The stucco base, which was 2 feet 10 inches long, 2 feet high, and 2 feet 6 inches broad, was flanked on either side of the front by the half-projecting figure of a lion, as seen in Fig. 31, and was thus manifestly intended to represent a 'Siṃhāsana'. The heads of the lions had decayed long ago, but the conventionally-treated curls indicating the manes falling over the fore part of the bodies were still clearly recognizable. The front of the base bore traces of having been painted and was decorated with a broad, low-hanging wreath. The projecting moulding running along the top of the platform had broken just below the base and showed the reed layers over which the plaster had been fixed. On the top of the base there once rose a circular drum-shaped pedestal in stucco, decorated with lotus-leaves and apparently over 3 inches in height when intact.

Reliefs from wall decoration.

In the top layer of sand which covered this base and the south-eastern corner of the cella, were found numerous replicas of small reliefs, in hard plaster of Paris, representing Buddhas or Bodhisattvas, some of them still attached to portions of a similar hard stucco background. This was decorated in relief with elaborate and gaily coloured borders closely resembling in design those found in D. i. It is evident, from an examination of these fragments, described in the list below[1], that all belonged originally to plaques which probably in the form of aureoles had adorned the uppermost part of the south wall. Plates LIV and LV contain

[1] All pieces of 'Chunam' (plaster of Paris) marked in the list as from D. ii were found in the smaller cella in the position described. Those bearing numbers beginning with zero (D. ii. 01, &c.) are pieces of which the original numbers given on the spot had become illegible owing to the effacement of the pencil marks.

illustrations of the best preserved fragments. D. II. 34 and 74 show how the small figures of Buddhas standing in the 'Abhaya-mudrā' attitude, $6\frac{5}{8}$ inches high from below the lotus pedestal to the top of the nimbus, had been arranged in ascending rows corresponding to the curve of the border, as well as their colouring. The modelling, in regard to both pose and drapery, is better than that of the corresponding relief figures from D. I. The dark grey background was decorated besides with rosettes and painted devices resembling lotuses. The border of the aureole is formed by rows of overlapping lotus-petals, each row being coloured alternately red and green (or blue). From D. II. 55 (Plate LV) it is seen how rows of flame-tongues coloured in a similar way adjoined the outside edges of these borders; for the arrangement of the flame-tongues D. II. 89, in the same plate, and the description of D. II. 24 may be referred to[2]. The small relief of a seated Buddha or Bodhisattva, D. II. 10 (Plate LIV) nearly 5 inches high, shows careful modelling, the drapery, like that of the standing figures, unmistakably recalling its distant Grecian origin through Gandhāra models[3].

Stucco images.

The fact of these small stucco reliefs being found in the south-east corner of the cella in loose sand, and several feet above the platform previously referred to, is a clear indication that the interior of the little shrine had been invaded by drift-sand while its walls were still intact to a considerable height. Comparatively well protected as this south-eastern portion of the cella was, the image of friable stucco (clay) which once occupied the corner had decayed, just as in the case of the other three corners, down to the feet. These, however, as well as the elegantly moulded lotus-pedestal of circular shape, can still be made out in Fig. 30. Luckily this corner had afforded better protection for some other adornments of the shrine. On clearing the platform between the corner pedestal and the base of the principal statue I found a small detached statue in stucco (D. II. 09), originally about 1 ft. 4 in. high, and well preserved but for the head and arms. Fig. 31 shows it placed subsequently on the main base, and in Plate LIII it has been reproduced after its transport to the British Museum. The figure which it represents, seated on a rectangular seat with its feet resting on a flat footstool, must, by evidence of the carefully indicated dress, have been intended for a Buddha. The colours of the robe, or Uttarāsaṅga, in red, of the under-garment, or Saṃghāṭi, in a brownish purple, and of a second garment—probably the Saṃkakṣikā—in green, together with those of the respective linings or borders, have survived well. For a description of the seat and the background, perhaps part of an aureole, I must refer to the list below. The small wooden board on which the image had been set up, evidently for the purpose of convenient transport, was still intact; and as the stucco, too, though merely clayey loess mixed with straw, grass, and fine fibre, had kept comparatively hard, I was able to risk its removal. Carefully packed away in one of my mule-trunks the little statue accomplished its long journey to London far better than I had expected, the cracks now visible in the coloured reproduction being mainly the result of subsequent atmospheric influences.

Painted panels.

At the foot of the main base, and leaning against it, were found four painted panels of wood, all oblong, but of varying sizes. Owing to their position near the ground the wood of these panels has much decayed, evidently through damp, and the thin layer of water-colour with which they are painted has also suffered considerably. For the same reason the removal

[2] For flames surrounding the nimbus and aureole of a Buddha, comp. the modern Japanese representation, in Grünwedel-Burgess, *Buddhist art*, p. 172.

[3] For the tenacity with which this treatment of the drapery has survived in Buddhist sculpture of China and Japan, comp. Grünwedel-Burgess, *Buddhist art*, pp. 170 sqq.; for specimens of drapery in Gandhāra sculptures of Buddha, comp., e.g., ibid., Figs. 110, 112, 117 (Abhaya-mudrā), 118, 119.

of the crust of sand and siliceous dust which adheres to the surface proved a very delicate task. The largest of these panels (D. II. 010), reproduced in Plate LXVII, but without the much effaced colours, measures 19½ by 5⅝ in., and shows five seated figures, of which the third, bearded and holding in its four arms apparently a vajra (?), lotus (?), baton and hatchet, manifestly represents some Buddhist divinity. The two figures on either side shown as playing on various musical instruments, among them a *rabāb* or mandoline, and perhaps castanets, are meant probably for divine attendants, since each is shown with a nimbus[3a]. A full description of this and the other panels will be found in the list. Another panel (D. II. 79) reproduced in Plate LXVI, 5¼ in. high and 4½ in. wide, exhibits the figure of a dancing woman drawn in full movement and with much freedom. From the head, which is thrown back, there flows downwards a quantity of black hair, while the left hand holds the loop of a sash poised in graceful curve over the head. The clever execution of the visible details make the deleted state of the greater portion of the figure all the more regrettable. A seated figure on the reverse has suffered even more, the outlines being barely traceable. D. II. 16, a narrow and much deleted panel, shows Gaṇeśa or Vināyaka, whose popularity in the Khotan region is attested by a far better preserved painted tablet from the Endere ruins[4]. In accordance with the Indian practice the figure of the elephant-headed god is shown nude except for a loin-cloth. On a fourth badly faded panel (D. II. 21) two divine figures seated on lotuses, with a vesica and nimbus behind each, are just recognizable.

These painted tablets, like all the others subsequently discovered at the bases of sacred images in ruined shrines of Dandān-Uiliq, were undoubtedly still in the same position in which they had originally been deposited as votive offerings by pious worshippers. The last days of worship at this small shrine were vividly recalled by far humbler yet equally touching relics. On the floor near the principal base, and near the corners, I discovered several ancient brooms, which had manifestly been used by the last attendants to keep the sacred objects clear of the invading dust and sand. The sand against which these humble implements were once used to wage war had been the means of preserving them. But they had become extremely brittle, and only one could be brought away safely. This specimen (D. II. 011, seen in Plate LXXIII) is about 16 in. long and constructed in a very ingenious way from stalks of some hardy grass. At their bottom these stalks were plaited into a continuous strip, subsequently rolled up tightly and bound round with twisted grass, their feathery ends being thus gathered into a convenient birch-like broom. Another curious relic, found in the south-eastern corner close to the small seated image, was a small bag of cotton cloth (D. II. 013) filled with human teeth and small fragments of bones. Were they ex-votos deposited with some superstitious object, or had they been brought here by some visitors as reputed relics? **Indications of last worship.**

The east wall of the small cella, owing to the backing afforded by the adjoining passage wall of the larger shrine, still stood to a height of about four feet; but of its fresco decoration only the lowest portions survived. In the south-east corner above the platform, only remains of the aureoles painted in a colour that now looked dark-brown, around the corner image and the one to be described presently, could be made out. The ground seemed to have been coloured a bright red brown. As the work of clearing proceeded along this wall it revealed an interesting relief statue in stucco immediately in front of the platform. As seen in Fig. 30, and in the enlargement of Plate II, it is a male figure, complete but for the head and the left **Relief image on east wall.**

[3a] [Mr. Andrews has since recognized in this panel another representation of the legend illustrated by D. x. 4; see below, pp. 259 sq.]

[4] See below, chap. XII. sec. iii.

arm, standing over the body of a prostrate foe. The figure, which measured a little over three feet from the heel to below the arm-pit, is clad in a coat of mail reaching below the knees and elaborately decorated. The small plates or scales which form the mail are shown above the waist overlapping each other in vertical rows, while below it their arrangement is horizontal and their shape different. Here the double rivets (or lacing) joining one plate above the other are accurately indicated. The gay colours of the successive rows of plates, alternately red-blue and red-green, were remarkably well preserved, and no less so all details of the jewelled ornaments which are shown along the front and the lower edge of the coat and on the girdle around the waist. From the lower edge of the coat of mail rich kilting depends. The right arm seems to have been dressed in a close-fitting sleeve, but most of the surface-coating of the friable stucco had peeled off, and the colouring could no longer be distinguished. The hand resting on the waist holds a small object which has also become indistinct through abrasion but which, as seen in the photograph, seems to suggest a money-bag.

Cognizance under feet of image.

The feet seem to be clad in wide top-boots of soft leather, just like the 'Chāruks' still worn throughout Eastern Turkestān, and originally coloured a light reddish brown. They are placed over the contorted body manifestly of a demon shown in low relief. The features of the latter's head, which alone is raised somewhat from the ground, with eyes wide open and the upper teeth displayed, express terror. The body appears to have been painted dark blue, but owing to the low position of this relief the stucco retained little of the original colouring. On the demon's right arm an armlet and bracelet can still be distinguished. The representation of the thick hair by elaborately worked spiral tufts closely recalls the distinctly Indian treatment of the hair which is frequently met with in Buddha figures of later Gandhāra style, and which has remained the orthodox one in the later Buddhist sculpture of Magadha, Nepāl, and elsewhere [5].

Representation of scale armour.

There is another detail in this little group of reliefs which takes us back to Gandhāra models. I mean the curious representation of the coat of mail in the standing figure. We find exactly the same distinctive arrangement of the plates or scales in the armour worn by two soldiers in the well-known Gandhāra relief in the Lahore Museum, showing Māra's army [6]. These figures, equipped mainly like Greeks, wear below the waist a coat of mail in which the plates are ranged in horizontal rows, while another mail garment above, resembling a *lorica*, shows scales in vertical order, and strangely enough also, as in our relief, with the rounded (or broadly pointed) ends upwards. In calling attention to this curious ranging of the scales, the reverse of that shown by the classical scale armour, Professor Grünwedel and Dr. Burgess have expressed a belief that the stone-cutter who produced this relief had shown the scales wrongly, the coat of mail being unintelligible to him [7]. Seeing how exactly the same arrangement is represented in the Dandān-Uiliq relief, the question may well arise whether these sculptures, so widely separated by distance in time and space, do not both reproduce a genuine detail of Asiatic armour. My finds in the rubbish-heaps of the Niya Site included scales of hard leather which undoubtedly belonged to armour [8]. The use of scale armour of horn among Central-Asian nations is attested from early classical times, and it is possible that for some reason or other the arrangement adopted by them for the fixing of the scales differed altogether from the western one [9].

However this may be, it seems to me highly probable that the artist has reproduced in

[5] Compare Grünwedel-Burgess, *Buddhist art*, pp. 167 sq.; for illustrations, ibid., Figs. 82, 115, 126, 127, 128.

[6] Reproduced by Grünwedel-Burgess, Fig. 48.

[7] See loc. cit., p. 97, note 1.

[8] See below, chap. xi. sec. vi.

[9] See Mollett, *Illustrated Dictionary of Art and Archaeology*, p. 197. [The surmise expressed above has now been strikingly confirmed by Mr. Andrews' discovery of an old Tibetan coat of mail showing scales arranged exactly as in our figure; see *Addenda*.]

30

FRESCO AND RELIEF SCULPTURE IN SMALL CELLA OF SHRINE D. II, DANDĀN-UILIQ.

31

SMALL CELLA OF BUDDHIST SHRINE D. II, DANDĀN-UILIQ, SEEN FROM WEST AFTER EXCAVATION.

this figure details of dress and ornament with which he was familiar from his own times. The point is of the more interest because we may with great probability recognize in the figure Kubera or Vaiśravaṇa, the most prominent of the four Yakṣa kings or Lokapālas who, as we have seen, was an object of special worship at Khotan. Vaiśravaṇa is represented in one of the pillar sculptures from Barhut (Barāhat) as standing 'on a pointed-eared thick-set demon', and has retained this Yakṣa as his cognizance or *vāhana* even in Lamaism and Japanese Buddhism[10]. His character as the god of wealth (Dhanapati) might well be indicated in our relief by the bag of gold, if that be intended by the object held in the demon-king's right hand[11]. Seeing the position occupied by this figure in one corner of the cella below the platform, it seems probable that the pedestals in the other three corners (compare Plate LXVI) once bore images of the remaining three Lokapālas.

Image identified as Vaiśravaṇa.

The cella wall immediately adjoining the relief group revealed at its base a series of small fresco paintings, which by their unconventional subjects and their spirited drawing at once attracted my attention. The one nearest on the left, as seen in the photograph (Plate II), shows a woman standing in an oblong tank of water, enclosed by a tesselated pavement and filled with lotuses. The figure, 18 inches high as far as seen above the water, is nude except for a large red headdress resembling an Indian *pagrī* and profuse ornaments round the neck, arms and wrists, and is drawn with remarkable verve, in simple but graceful outlines of true flesh-colour. The right hand, with its shapely fingers, rests against the breasts, while the left arm is curved down towards the middle of the waist. Fourfold strings of small bells (or beads?) are shown hanging in elegant curves around the hips, just as in representations of dancing-girls and other female attendants in early Indian sculpture, while, curiously enough, an elaborate vine-leaf appears where post-classical convention would place its fig-leaf. The woman's face is turned to her proper right, down towards a small figure of a male, also nude, who is shown as if trying to rise from the water by holding to her side. Further to the left appear the head and shoulders of another small male figure, just rising above the water as if in the act of swimming. In the foreground, and in front of the tank, the foot of the fresco showed in faint but unmistakable outlines a small riderless horse, closely resembling, in its dappled colour, trappings, and pose, the horse represented on the painted tablet D. VII. 5 (Plate LIX). The empty saddle of characteristic deep shape can still be made out in the original photograph.

Fresco painting on east wall.

The delineation of the lotus-flowers rising from the tank in a variety of forms, closed or half open, as well as their colours, ranging from dark blue to deep purple, seemed quite true to nature, and distinctly suggested that these sacred flowers were familiar to the painter from personal observation. In the Tao-tai's garden at Kāshgar I had seen a tank full of splendid lotuses which had been grown from seed imported from China, and in view of this pictorial representation I think it safe to assume that ancient Khotan had known the graceful plant dear to the gods of India. Considering the close historical connexion between Kashmīr and Khotan, as attested by some of the local traditions with which Hsüan-tsang's record has already acquainted us, it needed no special effort of imagination on my part to believe that the lotuses of ancient Khotan were originally derived from the lakes of the great Himalayan Valley where I had so often admired them.

Representation of lotuses.

The subject of this small fresco, as well as individual features of its figures, presents points of considerable interest. It was the appearance of the riderless horse in front of the tank

Identification of fresco subject.

[10] See Grünwedel-Burgess, loc. cit., pp. 40 note, 41; Grünwedel, *Mythologie des Buddhismus*, pp. 6, 181, where the blue colour of the demon in Tibetan representations is specially noted.

[11] Comp. Grünwedel-Burgess, pp. 136 sq.

which first suggested to me, on closer study of the photograph shown enlarged in Plate II, that the strange subject of the fresco might possibly be explained by the legend which Hsüan-tsang relates of the Nāga's widow residing in a stream east of the Khotan capital and her wooing by the self-sacrificing minister. We have already had occasion to consider the legend in detail in connexion with the 'Drum-lake Convent'[12]. Seeing that without exception the figures or scenes appearing in the decoration of the shrines excavated by me at this and other sites, whether sculptured or painted, bore a religious or mythological character, the presumption seems *a priori* justified that some legendary scene was intended here also. As far as my knowledge of Buddhist iconography goes, there is no legend affording as suitable an interpretation for our fresco as the one of the Nāga's widow. An observation of local archaeology comes to support the conjecture. The popularity of two other Khotan legends connected with particular sites, those of the sacred rats and of the silk-bringing princess, is attested beyond all doubt by pictorial representations found at Dandān-Uiliq. Hence it is natural to look in the first place among the stories of sacred local lore as recorded by Hsüan-tsang for a clue to this otherwise unexplained scene.

Figures from legend of the Nāga's widow.

Starting then from the figure of the riderless horse, it is clear that it would be most appropriately accounted for by the legend which represented the minister's horse as the bringer of his message and miraculous gift after his own disappearance. The minister himself might be recognized both in the swimming figure with head and shoulders just visible above the water, and in the small nude male which tries to rise by holding to the side of the woman. Such repetition for the purpose of indicating successive stages of a legend is a device as well known to old Indian as to mediaeval art of the West. The disproportion in size between the female and the male figures is another feature easily explained on the basis of the suggested interpretation. The divine Nāginī would, in accordance with a convention which Gandhāra art borrowed from the declining antique, be necessarily shown far larger than the mortal wooing her[13]. Finally, we may reasonably expect to see the watery home of the Nāginī indicated in the painting by a regularly enclosed tank, seeing that the same manner of representation is ordinarily resorted to by the Gandhāra sculptors when their reliefs have to show the dwelling places of Nāgas[14].

In the face of these features supporting the identification, the difficulties presented by two negative ones should not be ignored. In view of Hsüan-tsang's story, it is strange to miss in a painted representation of it the miraculous drum with which the minister's horse was supposed to have returned, and neither in the original fresco nor in its photograph could I trace any object resembling this. It is further noteworthy that the figure of the woman, however richly adorned, bears no indication of her character as a Nāginī or semi-divine being. In Gandhāra sculpture a snake-hood above the human head would have been ordinarily resorted to to mark the Nāga's consort[15], while even if this characteristic emblem was unknown to the Buddhist art of Khotan, there was at least the aureole conveniently at the artist's hand to mark superhuman origin. It must, however, be borne in mind that Kashmīr tradition knew Nāgas in a purely human shape[16]. It may also be doubted whether the legend, as heard by Hsüan-tsang, reckoned the Nāginī among divine beings proper, seeing that it related her first husband's death.

Suggestions of western models.

However this may be, it is clear that the suggested identification of the scene must be

[12] See above, p. 227.

[13] See Grünwedel-Burgess, *Buddhist art*, p. 138.

[14] Comp. Foucher, *L'Art du Gandhâra*, i. p. 388; Figs. 194–196; Grünwedel-Burgess, loc. cit., Figs. 57, 59.

[15] Comp. Foucher, *L'Art du Gandhâra*, i. p. 384; Grünwedel-Burgess, pp. 106 sq.

[16] See my note, *Rājat.* i. 30; also i. 220.

treated at present as conjectural. Two curious points, wholly independent of it, still remain to be noted. There is nothing, as far as I know, in Indian art to account for the appearance of the vine-leaf which is displayed so conspicuously in front of the lady's body. Manifestly an imitation of the post-classical fig-leaf, it presents us with an unmistakable proof of western influence on Khotan art of the period. Have we possibly to recognize another but more distant trace of this in the general pose of the principal figure? I am unable to adduce for this pose any parallel from old Indian art; but western eyes cannot fail to be struck by the curious resemblance it presents to that of the traditional Venus of late Greek sculpture. The position of the right hand is practically identical, while the left is raised only to such an extent as the use of the vine-leaf would justify. Is it mere coincidence which presents us here, in a fresco of a small Buddhist shrine, dating as we shall see from the last quarter of the eighth century, with a pendant of a famous type of classical antiquity; or is it possible to suppose links which would account for the re-appearance of that type, however disguised in form and motive, in so distant a quarter of Central Asia, and at so late a period? Only further research can give the assurance needed for a definite answer.

Remaining frescoes.

The adjoining small frescoes of the east wall offer no difficulty in their interpretation. On the right of the scene just discussed we see a seated Buddha surrounded by a medallion-shaped aureole, the whole measuring only about 4 inches. The dress is shown a deep red-brown against a green background. On the left there appears above the tank, which extends frieze-like below the remaining frescoes, a remarkably well-drawn though much effaced male figure of youthful appearance. Seated in cross-legged fashion, and dressed in a dark blue cloak that leaves the right shoulder bare and shows excellent drapery, it is manifestly that of a Buddhist scholar. His right hand holds the oblong leaves of a 'Pōthī,' on which his eyes are fixed in intent study. Beyond this figure, and likewise turned to the proper right, an old man is depicted in the act of teaching. His robe, worn in the same fashion, seems to be made up of patches of varying shades of brown, thus curiously suggesting the orthodox *cīravastra* of mendicant monks of all Indian sects. The well-shaped right hand is raised in the act of teaching, with the second and third fingers stretched out, while the palm of the left supports a closed Pōthī. The two boards of thin wood between which the leaves are placed, after a fashion still commonly observed in the case of Indian manuscripts, are distinctly marked, and can be made out in the photograph. The cleverly-drawn features of the old man's face bear an expression of complacent assurance in his teaching and full abstraction in its subject. The tank below this figure showed open lotuses floating in the water, and also two birds looking like wild geese, with necks marked dark blue and green and turned towards the teacher.

To remove any portion of these interesting frescoes proved quite impracticable, owing to the friable condition of the plaster. The faded state of the colours made it difficult even to secure a photographic record. Nevertheless, I hope the enlarged reproduction of the photograph (Plate II) will suffice to illustrate some thoroughly Indian features which these mural paintings exhibit in style of composition and the drawing of figures. There can be little doubt that, just as in the case of its sculpture, the original models of the pictorial art of old Khotan were derived from Gandhāra and the immediately adjoining region. Of the paintings which once adorned the walls of Buddhist shrines and monasteries in the Valleys of Kābūl, Peshāwar or Swāt, no direct remains have survived, and this loss invests such relics as the frescoes and the painted tablets of the Dandān-Uiliq shrines, however distant the reflex they represent, with additional interest.

Section IV.—FIRST FINDS OF ANCIENT MANUSCRIPTS

It had not needed the discovery of the pictorial representation of Pōthīs to make me look out eagerly for finds of ancient manuscripts. None turned up in the shrines excavated during the first three days, but the fragments of leaves which I had found pasted across the two painted panels previously described helped to revive my hopes. The few characters still legible on them showed a bold literary hand, and this, together with the arrangement of the lines, clearly suggested that the leaves had once belonged to some manuscript written in the form of a Pōthī.

Excavation of dwelling, D. III. The little temples so far excavated had revealed something of the cult and art which this sand-buried settlement had possessed before its abandonment. But it was manifest that indications of the conditions of every-day life and other documentary evidence would have to be looked for elsewhere. A ruined structure close by seemed by its ground-plan, as deducible from the arrangement of the wooden posts that were seen sticking out from the sand, to suggest an ancient dwelling-place, and to the excavation of this I accordingly proceeded on December 22. The structure, D. III, lay about 20 yards to the north-west of the temple-cellas last described, and owing to the height of the dune rising immediately to its south (subsequently ascertained to be fully 16 feet above the original ground-level) there was hope of finding its interior undisturbed by recent diggings.

The excavations, started from the west side, soon brought to light the top part of walls built with timber and plaster after the method previously described, but of greater thickness, viz. 10 in. Notwithstanding the increased strength of construction the south wall was found to have completely decayed, evidently before the ruin had been protected by its present cover of sand, and the drift-sand ever slipping in from the slope of the dune which rose on that side greatly added to the difficulty of clearing. The walls proved to have belonged to an oblong apartment, measuring 23 feet from east to west and close on 20 feet wide, which had evidently formed the lowest story of a dwelling-house. Judging from some of the posts still intact in the framework the original height of the apartment must have been about 10 feet. Fig. 32 shows the north-west corner in the course of clearing, together with fragments of the heavy plaster mouldings which seem to have decorated the upper portion of the walls but were now found almost wholly detached.

First MS. find. At a depth of only 2 feet from the surface a small scrap of paper showing a few Brāhmī characters turned up in the loose sand filling the north-west corner. I greeted this small fragment (placed now with D. III. 3 in the inventory list) with no small satisfaction as a promise of richer finds; and in order to stimulate the efforts of my labourers, who with the sand constantly falling in from the south had no easy task in effecting a clearance, offered a small reward for the first discovery of a real manuscript. Barely an hour later a cheerful shout from a young labourer working at the bottom of the small area so far excavated announced the finding of a 'Khat'.

Carefully extracted with my own hand and cleared from the adhering sand, it proved a perfectly preserved oblong leaf of yellowish paper, measuring about 14 by $2\frac{7}{8}$ in., which had clearly formed part of a large manuscript arranged in the form of a Pōthī. The reproduction in Plate CX shows the six lines of remarkably clear writing in Brāhmī characters of the upright Gupta type covering each side. The circular hole intended for the string, which was passed through the separate leaves of the Pōthī and served to keep them in order, is seen on the

RUINED DWELLING D. III, DANDĀN-UILIQ, IN COURSE OF EXCAVATION.

INTERIOR OF RUINED DWELLING D. III, DANDĀN-UILIQ, AFTER EXCAVATION.

left side of the leaf, as in most of the ancient manuscripts recovered from Eastern Turkestān. Dr. Hoernle's examination (see list) shows that the leaf, numbered 8, contains a text written in the non-Sanskritic (Proto-Tibetan ?) language, which he was the first to recognize in other fragmentary manuscripts previously obtained from Eastern Turkestān, in part at least from Khotan, and that the manuscript to which it belonged was probably that of a Buddhist canonical text. The writing is attributed by Dr. Hoernle, on palaeographical grounds, to the seventh or eighth century[1].

MS. of *Prajñā Pāramitā.*

The find just described was made at a depth of about 5 feet from the surface and close to the rough wooden post fixed in the floor as seen in Fig. 33. It was quickly followed by a series of other manuscript finds, all consisting of fragments, but varying in extent from relatively large portions of single leaves to batches of numerous minute pieces. Among these fragments (numbered D. III. 2–11) the large majority could at the time be easily recognized, by their conformity in paper, size, and handwriting as having originally formed part of a Sanskrit manuscript treating of Buddhist canonical matter. Dr. Hoernle's careful analysis proves that these fragments belong to the eighteenth and nineteenth chapters of a Mahāyāna text, apparently some king of *Prajñā Pāramitā*, written in Gupta characters of the seventh or eighth century on leaves of very large size, about 18 by 7 inches[2]. A leaf of this work, as restored from two pieces (D. III. 7, 8) is reproduced in Plate CVII. The obverse contains the conclusion of the eighteenth chapter (*Bhūmiparivarta*), marked also by a diagram of homocentric rings, and bears on its left hand margin the pagination number 132. Dr. Hoernle's notes show that remnants of two other Buddhist Sanskrit texts, one of them apparently containing the story of the Yakṣa chief Maṇibhadra (see Plate CIX), are represented among these fragments[3].

Position of MS. remains.

The position in which all these manuscript pieces were found, embedded in loose sand at levels varying from 4 to 3 feet above the original flooring, proved beyond all doubt that they could have got there only by accident. Their distribution in varying depths and places makes it probable that they had fallen in from an upper story, while the basement was gradually filling up with drift-sand. This assumption was borne out by small pieces of animal bones, felt, leather, and oilcakes (*kunjara*), which turned up in the same layers. The earlier any manuscript remains had reached the safe resting-place offered by the sand-covered basement, the more extensive they might reasonably be expected to be. So I watched with growing eagerness the progress made by my men on December 23 in clearing the sand nearer down to the original floor. The first find, made a little to the east of the post already mentioned and about $1\frac{1}{2}$ feet above the floor, was the portion of a document (D. III. 12) written on a sheet of coarse thin paper, about 10 by 8 inches, in cursive Brāhmī characters of the eighth century (see Plate CX). In the text Dr. Hoernle has recognized the same Eastern Īrānian language in which a number of other documents from Dandān-Uiliq are recorded. As in the majority of these, the writing is confined to one side of the sheet only, since the thin unsized paper, like the modern Khotan paper, would readily absorb the ink and let it pass through.

Carved posts.

As the work proceeded towards the centre of the room, a massive beam of poplar wood ('Terek', *Populus alba*), 11 inches thick and in almost perfect preservation, was laid bare. Its length, close on 19 feet, and its position showed that it had once stretched right across the room, undoubtedly supporting its ceiling. Two well-carved octagonal posts, each 4 feet 5 inches high and exactly alike, had turned up before (see Figs. 32, 33). Their bell-shaped capitals, 4 inches high, were surmounted by single circular bands accurately reproducing the Amalaka ornament of

[1] Comp. Hoernle, *Report on C.-A. ant.*, ii. pp. 12 sq., 18 sqq.

[2] See Dr. Hoernle's Note i. in list of finds.

[3] See Notes ii. and iii. in list below.

Indian architecture, while below were bases 8 inches square. I am unable to indicate the exact position and arrangement of these posts, but it seems probable that they had served to support the central beam.

A little beyond the latter, towards the east, the men clearing the sand just above the floor came upon a closely-packed layer of manuscript leaves (D. III. 13. a), evidently still retaining the order they had occupied in the original Pōthī. Subsequently two more small packets of leaves lying close by, and belonging to the identical MS. (D. III. 13. b, D. III. 13. c), were brought to light, practically intact, though the action of moisture to which these leaves must have once been subjected, owing to their position not far above the ground, had stuck them close together, and had made them so very brittle that their successful separation could only be accomplished in London through the expert help of the MS. Department of the British Museum. The ends of the leaves had been bent over near the usual string-hole, and had often become detached through this folding of centuries; but they could be fitted again without difficulty to their proper places.

Recovery of *Vajracchedikā* MS.

The leaves in their complete state, as shown by the specimens reproduced in Plate CVIII, measure about $14\frac{1}{4}$ by 3 inches, and show on each side six lines of Brāhmī characters of the upright Gupta type, ascribed by Dr. Hoernle to the seventh or eighth century. In the text, which is Sanskrit, the same distinguished Indologist has recognized the *Vajracchedikā*, a famous treatise of the Mahāyāna school, first edited by the late Professor Max Müller in the *Anecdota Oxoniensia*. The exact arrangement of the leaves was greatly facilitated by this discovery, and has shown to Dr. Hoernle that out of the twenty folia which this manuscript originally contained not less than fifteen have been preserved, in whole or part[4]. It is of some interest to record that, sticking closely to the packet D. III. 13. a of Vajracchedikā leaves, there turned up also some much twisted fragments of the Sanskrit manuscript in large Gupta characters described above as a sort of Prajñā Pāramitā. These fragments, which showed a far more brittle condition than the rest of the leaves described in Dr. Hoernle's Note i., had evidently become attached to the Vajracchedikā Pōthī in its earlier resting-place, and sharing its fate had found their way into the sand of the basement-story far earlier than the other pieces.

The fact that all manuscript remains found in this ruined structure (with the exception of the small document in the 'unknown' Eastern Īrānian language) proved to have religious contents, makes it appear highly probable that they are derived from the library of a monastic establishment that once occupied the structure, and, no doubt, supplied the attendant priests for the adjoining small shrines. That the basement room excavated had offered only accidental shelter to these fragmentary relics of Buddhist literature, and had originally served the more prosaic purposes of a cook-room for the little Vihāra, became abundantly clear as the work of clearing was carried on to the east wall. Built against the latter there was found a big fireplace, constructed of fairly hard plaster with an elaborately-moulded top looking like a chimney but showing no arrangement for the egress of smoke, the whole reaching to a height of over 6 feet from the floor.

Indications of ancient cook-room.

By the side of it a broad bench, also seen in Fig. 33, filled a kind of recess. Judging from a similar arrangement still observed in Turkestān houses, and from bits of coarse broken pottery found below it, this bench probably served for the handy storing of cooking utensils. In front of it, and not far from the fireplace, there stood a rough wooden tripod (visible in the photograph), such as is still used throughout the country to support large water-jars required for kitchen purposes. In the same way the short rough post with branching head, which I found fixed in the ground close to where the first manuscript leaf was discovered, certainly served to hang

[4] See his Note v. in list below.

a kettle from. Remains of ancient bones, oilcakes, and small layers of charcoal found scattered over the floor in several places corroborate the above conclusion.

Excavation of shrine D. x.

Of the remaining structures traceable at the southern group of ruins two could easily be recognized, by the characteristic arrangement of their posts emerging above the sand, as small shrines of the type with which we have already become familiar. The first cleared among them (D. x.), situated about 80 feet to the west-north-west of the little monastic dwelling last described, proved to consist of an oblong cella measuring 10 ft. 6 in. outside from north to south, with a width of 9 ft. 6 in. The entrance, as in the case of the other small temples of this group, lay to the north. A passage, 4 feet wide, surrounded the cella, but the walls of this had almost completely decayed on the north and west, and elsewhere rose only 1 to 2 feet above the original ground. The walls of the cella, too, stood nowhere higher than 4 feet. Their construction was of the usual timber framework and plaster, with a thickness of about 6 inches. The mural decorations were here of a very simple kind, consisting both inside and outside the cella almost solely of rows of small seated Buddha figures. Each row, including the vesica around the figure and the gaily coloured background, was 6 in. high. The dresses of the Buddhas were alternately coloured dark brown, red, and white, the colours of the background varying in a similar fashion. The whole of these little frescoes showed inferior work, evidently done with the use of stencils.

The sculptural decoration of the shrine must have been equally modest, and restricted to a single image occupying the octagonal base found near the south wall of the cella. Of the statue itself there remained nothing but the wooden post which had served as a core, and shapeless pieces of very friable stucco. But the base, still fairly intact on the north to a height of about 2 ft. 6 in., showed some peculiar features. It consisted below of eight facets, 1 ft. 5 in. broad at the foot, sloping inwards, and thus gradually narrowing to a width of about 1 foot. Then followed a circular moulding, $1\frac{1}{2}$ in. high, and above this eight facets of exactly similar dimensions, but arranged in the inverse direction and thus sloping outwards[4a]. These upper facets had suffered greatly, and when the sand had been cleared from behind the base it was found to have been dug into almost to the centre, undoubtedly by 'treasure-seekers' who may have suspected here some valuable deposit.

In view of the damage thus caused it was the more gratifying that my clearing revealed in front of the base, and leaning against it, a series of painted panels which had remained undisturbed just as the last attendants at the shrine must have placed them. The relatively good preservation of some of these panels is probably due to the fact that they had rested not on the actual floor of the cella, but on a rim of plaster about 3 in. high and 5 in. broad which ran round the foot of the pedestal, and was evidently intended for the placing of offerings. The most interesting among the paintings, and fortunately also one of the best preserved, is D. x. 4 (see Plate LXIII), painted on a wooden tablet 18 in. long and $4\frac{5}{8}$ in. high, which was found standing upright against the east corner of the front facet of the pedestal. In its subject I think I can recognize with certainty a spirited representation of the main features of the legend which Hsüan-tsang relates of the introduction of sericulture into Khotan and of the princess instrumental in it[5]. The second female figure from the R. proper, bearing over her rich curling locks an elaborate golden diadem, is undoubtedly meant for the princess. Our attention is forcibly drawn to the diadem by the significant gesture with which the outstretched arm and forefinger of the first figure from the R. proper, evidently a female attendant, point towards it.

Illustration of legend of silk-bringing princess.

[4a] [Mr. Andrews points out that this form of base is seen in many Chinese bronzes.]

[5] See above, pp. 229 sq.

From the silkworms' eggs hidden in the princess's headdress had been derived the cocoons which we see heaped up in a basket in front of the princess. Her eyes turned down upon them, and the right arm stretched out towards them seem clearly to indicate the protection which, according to the legend, the princess was anxious to assure, by an edict, to the silkworms within the cocoons.

Until I had correctly realized the character of the second figure and significance of the first's gesture as well as of the basket placed between them, the attitude of the fourth figure, also female, and of the object in front of her seemed very puzzling. With the subject of the panel once identified it is easy now to recognize, notwithstanding the faded colours, that the long stretched object shown in brown is meant for the wooden framework of a loom or silk-weaving gear such as is used to this day in Indian villages for preparing a warp of spun cotton threads. The instrument which the fourth figure is holding in her right hand is probably identical with the wooden implement, curiously resembling a currycomb, found in one of the ruined houses of the Niya Site, N. xx., and shown in Plate LXXIII. It use is illustrated by the exactly similar wooden instrument which is still used in India by cotton-spinning villagers for separating and combing the threads while attached to the rude frames usually found fixed by the roadside. The four-armed third figure seen seated to the left proper of the princess undoubtedly represents a male divinity, perhaps the god presiding over the silkworms. He wears high top-boots like another divinity which appears on the interesting panel (D. vii. 6) reproduced in Plate LXI, but the emblems held in his hands are scarcely distinctive enough to guide us to a certain identification[6].

Indications of Persian influence.

As regards all details in the drawing and colouring of the figures, I must refer to the description given in the inventory list and to the illustration in Plate LXIII, where, however, a full reproduction of all colours and delicately-shaded tints of the original could not be attempted. Special mention may be made here of the cleverly-arranged composition of the whole scene and the free and spirited drawing of the main figure. It is of interest to note the unmistakably Persian type of face given to the princess in spite of her Chinese origin, as rightly maintained by the legend. We have here a clear indication of the influence exercised at Khotan by Īrānian pictorial art, which is traceable also in other panels of Dandān-Uiliq. The curious miniature representation of what looks like a tower with four pointed turrets or finials, faintly visible on the left proper edge of the panel, may possibly be meant to indicate a kind of reel round which the silk thread is wound when preparing it for the warp. The petal-shaped daubs of dark pink which appear on the background bear manifestly the same character as the auspicious *sindūra* or sandal-ointment marks with which pious Hindus have ever been fond of bedaubing objects of worship or votive offerings.

Painted panel, with Bodhisattvas.

Another well-preserved tablet is D. x. 3, painted on both sides (see Plate LXIV). Its interest lies in the representation on the obverse of three seated Bodhisattvas, whose figures and attributes seem to agree closely with the typical forms given to this highly popular class of divinities by Tibetan and Mongolian Buddhism. I must leave it to experts in Northern Buddhist iconography to determine exact identifications. The two-armed figure first on the right proper holding the Vajra in the right hand is probably meant for Vajrapāṇi. The three-headed and four-armed third figure holding in one of its left hands a bow may in view of this attribute be supposed to represent some form of Mañjuśrī[7]. As Vajrapāṇi and Mañjuśrī are found very

[6] Julien's translation of Hsüan-tsang's legend, *Mémoires*, ii. p. 239, refers to the *Lu-shê* convent having been built 'en l'honneur de la déesse des vers à soie'. But other renderings give a different meaning to the passage; comp. above, p. 229.

[7] Comp. Grünwedel, *Mythologie des Buddhismus*, p. 138.

frequently combined into a triad along with Maitreya, I am inclined to recognize the latter Bodhisattva in the middle figure, especially as two of Maitreya's common attributes, the Cakra and the holy water flask, seem to be represented in his hands. The scarf-like garment wrapped around the arms and the deep red colour of the undercoat also point to Maitreya[8]. The thin Chinese-looking moustaches of the two side figures are a curious un-Indian feature. Of the four seated figures shown originally on the reverse only the two middle ones are still fairly distinguishable. The third from the right proper is shown in the 'Dhyānamudrā' attitude. It deserves to be noticed that both in this figure and in the one adjoining it the outlines marking the eyes are continued by a straight black line extending from the outer angle of the eye to the top of the ear, and giving precisely the effect of spectacles. Is it possible that an indication of the latter was intended?

Repetition of subjects in painted panels.

It is interesting to find in three of the panels subjects treated which are known to us likewise from paintings discovered in another shrine of Dandān-Uiliq. The comparison enables us to realize to what extent details of pictorial representation had become fixed even where of no apparent mythological importance. On the reverse of D. x. 5 (see Plate LXII) we meet with the scene of the horseman and bird to which reference has already been made in connexion with a fresco of D. II. A comparison of this panel with the far better preserved one D. VII. 5 (Plate LIX) shows that, notwithstanding the considerable difference in artistic execution and care, both must be directly or indirectly derived from the same prototype. The pose of rider and horse is identical in both; the uniformity of treatment extends to the dress and accoutrements and even the saddlery. The high sugar-loaf shaped cap with its 'vandyked' corners is borrowed from the camel-riding figure of D. VII. 5. Leaving other details for mention in connexion with the latter panel, attention may be called to the distinctly Persian look of both rider and horse. The elaborately painted figure of the three-headed divinity, perhaps a Tantric form of Avalokiteśvara, which we see in D. VII. 6 (Plate LX), and which we shall have occasion to discuss below, is represented by two replicas on the obverse of D. x. 5 and on D. x. 8. The latter painting (Plate LXII) is of interest notwithstanding its very poor preservation, because it shows the god combined with his Śakti, an arrangement common in Tibetan representations of the special tutelary divinities or *Yi-dam*[9]. The Persian treatment of the female head shows that Īrānian influence had affected even those sacred figures and groups which the Buddhism of Khotan, like that of Tibet, must be supposed to have received ready-made as it were and fixed in all essential details from the Mahāyāna imagery of India.

MS. fragments as pious offerings.

The manuscript finds in this shrine were slight, but the place and distribution in which they turned up make them instructive. Numerous fragments of leaves written in Brāhmī characters were found in front and east of the image base lying on the previously mentioned rim in little packets distinguished in the list as D. x. 9, D. x. 10. a, 10. b, 10. c. α, 10. c. β, 10. d. From the way in which they were placed between various painted panels it was clear that they had been deposited as votive offerings by some of the last worshippers at the shrine. The marks of red paint which I found sprinkled over the leaves fully agreed with this conclusion; for they also appeared, as we have seen, on several of the panels. Dr. Hoernle's analysis[10] shows that the fragments distributed in packets D. x. 9, D. x. 10. c. β, D. x. 10. d, all belonged to one leaf, and that similarly D. x. 10. a, D. x. 10. b were made up of portions of a large leaf, but from some other manuscript. The fragments D. x. 10. c. α, which were lying

[8] Comp. for characteristic points in Maitreya representations of Northern Buddhism, Grünwedel, *Mythologie*, pp. 126 sq.

[9] Grünwedel, *Mythologie*, p. 94.

[10] See Notes xix., xx., xxi., in the inventory list below.

as a separate little packet close above that marked with β, had formed part of a third manuscript.

These observations prove that before use as offerings detached folia of different manuscripts had here been purposely broken up so as to make several small heaps convenient for deposition. Under exactly similar conditions portions of one relatively large Tibetan Pōthī, containing the Śālistamba-Sūtra, and of a Sanskrit text were subsequently discovered by me in the Endere temple cut up into little packets, or even separate bits, which the offering worshipper had thought fit to distribute before a number of different sacred images[11]. The facts here observed provide a satisfactory explanation for the small detached fragments of manuscripts which turned up at several other shrines of Dandān-Uiliq, though no longer in their original place of deposition. Of the leaves recovered from D. x two had belonged to manuscripts written in upright Gupta characters and in the non-Sanskritic language which Dr. Hoernle tentatively designates as Proto-Tibetan, while the third manuscript, remains of which are contained in D. x. 10. c. α, was in Sanskrit and similar to the large-leaved manuscript of a *Prajñā Pāramitā* text found in D. III. Dr. Hoernle ascribes its upright Gupta writing to the seventh or eighth century.

Inscription on wooden beam. The manuscript fragments just referred to were not the only written remains of this small temple. In the sand filling the western passage outside the cella there was found at a height of about 2 feet above the original floor a fairly well preserved wooden beam, 5 ft. 11 in. long, 4 in. wide, and 5 in. thick, which must have originally been fixed in the passage wall, probably for the purpose of joining the posts as described in the case of D. II. Over a portion of the carefully-smoothed face of this beam, which I was able to remove safely (D. x. 6), there are written at intervals of ½ to ¾ in. thirty-seven characters of the same very cursive Brāhmī which is shown in the inscriptions painted below the frescoes. Though the wood is in perfectly sound condition inside the surface has become very dark, and consequently no mechanical reproduction of the inscription could be obtained. Dr. Hoernle, who has studied it in an eye-copy prepared by Dr. Barnett, and whose transcription is given below in the descriptive list, has been able to determine that it is in the same 'unknown' language of Eastern-Īrānian type which appears in the Brāhmī documents and fresco inscriptions of Dandān-Uiliq. The special interest of this inscription lies in the fact that it contains, as pointed out by Dr. Hoernle, 'a few recognizable Sanskrit words, "done" after the fashion of the "unknown language"', viz. *kuśala-mūla* and *Bodhisattva* (spelt *bo-dhyau satvyau*).

Clearing of structure D. XI. A ruined structure situated about 15 yards to the north-east of D. x. was cleared without any finds being made. The apartment excavated measured 29 by 19 feet, and probably formed the lower story of a dwelling-house. As the floor lay covered by only about 4 feet of sand the ruin is likely to have been searched previously by 'treasure-seekers'. This had certainly been the case with a smaller ruined dwelling close to D. III. on the north, which Turdi remembered having opened.

Excavation of shrine D. XII. Another structure, D. XII, situated about 70 feet to the south of D. x, proved on excavation to consist of a small cella of the construction usual at this site, measuring outside 13 by 10 feet. It had its entrance on the north, and was originally surrounded by a passage 4 feet wide. Though overlain now by 7 feet of sand the ruin showed far advanced decay, the result, no doubt, of long exposure. In consequence, the cella walls, as seen in Plate V, stood nowhere higher than 3 ft. 6 in., while those enclosing the passage had crumbled away almost to the ground. Within the cella, towards the south wall, were found the remains of an octagonal pedestal in stucco, 4 feet in diameter, resembling in shape that of D. x, but of better design

[11] See below, chap. XII. sec. ii.

and with more elaborate mouldings. The photograph shows that it rested on a low circular base decorated with lotus-petals. The whole front portion of the pedestal and base was found to have been destroyed, no doubt in the course of some 'treasure-seeking' operations, and of the statue which once had stood on it not even fragments survived. It was probably due to the same reason that no painted panels were found here. Near the north wall, however, some small reliefs in plaster of Paris were found, showing that the upper portions of the walls had once borne decorations similar to those of D. I and D. II. One of these reliefs (D. XII. 1), seen in Plate LVI, and originally coloured, represents a flying female figure, probably a Gandharvī, holding in her outstretched hands a festoon cloth. The other pieces are replicas of the Gandharvī figure D. XII. 5, illustrated in the same plate, which is shown here as rising from a large-petalled lotus. We have met with this motive already among the stucco decorations from D. I.

Mural decoration of shrine D. XII.

Of the frescoes which had covered the cella walls naturally very little remained. Each wall-face, except the one to the north, seems to have borne three life-size paintings of Buddhas or Bodhisattvas surrounded by aureoles and standing on open lotuses. The photograph reproduced in Plate V shows what remains of this mural decoration at the western end of the south wall. The triangular space left between the lowest portions of two aureoles is filled by a well-draped female figure with her head turned upwards and holding in her left a lyre. She is shown standing on a lotus, and may thus be meant for a divine attendant. The small figures, male and female, seen at the bottom of the wall, are typical representations of worshippers. Very little remained of the colouring of these frescoes apart from the outlines of the figures, usually drawn in a kind of terra-cotta colour over the greyish-white distemper.

MS. remains in shrine D. XII.

The clearing of the south-east corner revealed, close to the floor, manuscript remains of considerable extent, but unfortunately decayed into an almost unrecognizable state. The first indication of such remains was furnished by the fragment of a wooden vessel, D. XII. 6, about $7\frac{3}{4}$ by 3 in., to which were adhering, both inside and outside, layers of completely rotten paper. Writing of upright Gupta characters, arranged apparently in four lines, can just be distinguished on the blackened surface. A few inches below this I came upon a consolidated mass of 'sand,' *recte* loess dust, extending over a space of about 11 by 6 in. and firmly adhering to the floor. In it were embedded relatively thick layers of almost completely perished manuscript leaves. The latter, together with the overlying brittle crust of 'sand,' broke at the slightest touch. It proved equally impossible to disengage the leaves or to remove the whole layer. I succeeded in slicing the latter off the floor but only in fragments, and the separation of the paper-flakes proved an impossible task even in the British Museum. The few characters distinguishable on the surface show bold Gupta writing similar to that seen in Plate CVII. Embedded in the layer were two thin wooden boards, about $9\frac{1}{2}$ in. long and 4 in. wide (see Plate CVI), which had undoubtedly served as covers for one of the Pōthīs that had supplied these perished leaves. Flakes of paper still adhere to them in layers. A third board, completely decayed, which lay at the bottom touching the floor, had evidently belonged to a second Pōthī.

Moisture of soil.

The entire decay of these manuscripts can be explained only by the damp to which they must have been exposed for some time while lying on the floor of the shrine. The observation is of interest as showing that even after the abandonment of the settlement there followed a period when water must have reached it, at least periodically, in quantities sufficient to spread moisture to the basement of buildings. If the manuscripts had come to lie above a layer of the drift-sand which invaded the building, or if all supply of water had ceased immediately after their deposition on the floor, their state of preservation would have probably differed little from that of the manuscripts found in D. III.

Section V.—DISCOVERY OF DATED DOCUMENTS

Ruins D. iv, D. v.

The scattered group of ruined structures situated about half a mile east-north-east of my camp was the next to which I turned attention. The first to be explored were the closely adjoining ruins, marked D. iv and D. v in the plan, in which I could without difficulty recognize the remains of an oblong temple-cella and of a large dwelling-place. These ruins had suffered badly from erosion which, in the unprotected soil immediately to the north and east of them, had produced broad depressions to the depth of about 20 feet below the original ground-level. The ruins, owing to this lowering of the adjoining ground, as marked in the detail plan (Plate XXVI), seemed now to occupy a raised tongue of land quite clear of dunes, and nowhere retained more than two or three feet of covering sand. Above this there rose the splintered short stumps of posts which once had held the framework of wattle and plaster walls, as seen in the photograph of D. v (Fig. 35). In places the rows of these posts still clearly marked the position of dividing walls, though the walls themselves had practically disappeared down to the ground-level. Elsewhere there remained under the thin layer of sand only the massive beams, usually 6 to 8 inches in thickness, which had formed the foundation of the walls, while near the edges of the ground actually undergoing wind-erosion even these remains had disappeared, or were strewing the slopes in shapeless fragments of bleached brittle timber. The plaster work had survived only in a few portions of walls, which in the detail plan can readily be distinguished by being marked in black. Nowhere did these remnants of walls exceed a height of 2 feet.

Remains of temple-cella D. iv.

The exposed condition of these ruins had, of course, attracted the visits of 'treasure-seeking' parties, including some that Turdi had personally conducted in former years; and their burrowings had left visible marks in the débris of plaster, timber, potsherds, &c., which lay scattered about on the surface of the sand, as seen in the photographs. Notwithstanding the damage thus caused, my careful clearing of the ruins was rewarded by some interesting relics. Within the cella D. iv (Fig. 34), which measured 13 ft. 8 in. inside from east to west with a width of 11 ft., only some much decayed fragments in friable stucco were found, which seemed to belong to the drapery of a colossal figure. This had no doubt occupied the base, 3 feet square, which faced the entrance on the east and still rose to a height of 8 inches. At the foot of the west wall, behind the base, there remained traces of a painted frieze, probably resembling that illustrated from D. xii. Of the walls enclosing a passage around the cella, 5 feet wide, very little remained but the foundation beams and stumps of posts. Yet even this scanty shelter had sufficed to preserve on the west side the two painted panels, D. iv. 4 and 5, but their colours have suffered much.

Illustration of legend of sacred rats.

The obverse of the panel D. iv. 5, measuring $17\frac{1}{2}$ in. in length with a width of $4\frac{1}{4}$ in. and shown in Plate LXIII, is of special interest: for in the half-length human figure with the head of a rat and wearing a diadem, which occupies the central semi-ellipse, we have manifestly a representation of the chief of those sacred rats which were worshipped as having saved Khotan from an attack of the Hsiung-nu, and whose legend, as heard by Hsüan-tsang at the site corresponding to the present 'Kaptar-Mazār' at the western edge of the oasis, we have already discussed in detail[1]. Though the characteristic features of the rat's head are unmistakably indicated, the painter has cleverly imparted to it a human expression. The colour visible on the face and on the part of the neck left bare by the robe is distinctly yellow, thus agreeing

[1] See above, pp. 119 sq.

34

REMAINS OF RUINED SHRINE D. IV, DANDĀN-UILIQ.

35

REMAINS OF EASTERN PORTION OF RUINED DWELLING, D. V, DANDĀN-UILIQ.

with Hsüan-tsang's story which represents those sacred rats as having 'hair of a gold and silver colour'[2]. The worshipping attitude of the nude male figure on the right proper, carrying a leaf-shaped fan and looking towards the rat-headed king, clearly marks the sacred character of the latter. A third smaller figure on the left proper is almost completely effaced. On the reverse, too, five seated figures can be made out in faint outlines. It was only after the panel had been cleaned in the British Museum from the thin layer of adhering sand, and carefully examined by the trained eye of Mr. F. H. Andrews, that I realized the peculiar features of the rat-headed figure and its significance.

The other panel, D. iv. 4, measuring $10\frac{1}{8}$ by $5\frac{1}{8}$ in. (see Plate LXV), shows on the obverse the red-robed figure of a standing Buddha, apparently in the attitude of teaching, and surrounded by vesica and nimbus. The standing figure on the reverse is too deleted for any certain interpretation, but attention may be called to the peculiar circular markings which appear above the knees of this figure on legs apparently bare.

MS. fragments from D. iv.

Of manuscript finds the cella yielded from its north-east corner the fragments D. iv. 1–3, written in Brāhmī characters of the upright Gupta type, recognized by Dr. Hoernle as belonging to single leaves of two separate Pōthīs. The language is Sanskrit, and the text apparently from some Buddhist canonical work. D. iv. 6 was subsequently discovered in the north-west corner of the cella in the form of a small ball of crumpled-up paper. This proved to consist of two pieces of paper, each measuring about 11 by 5 inches, found attached to one another, and according to Dr. Hoernle's examination evidently making up one document. It is written in cursive Brāhmī characters placed widely apart like the documents to be discussed presently, but on both sides of the paper, while ordinarily, owing to the thin unsized paper used for all documents whether in cursive Brāhmī or Chinese, the writing is confined to one side only. As in all cursive Brāhmī documents from Dandān-Uiliq, the language is the non-Sanskritic one of Eastern Īrānian type. Dr. Hoernle, without being able to indicate the purport of the document, has noted in it the mention of a definite date, the 17th day of the month Mūñaṃji[3].

Condition of ruined house, D. v.

The remains of the ruined dwelling-house D. v, situated to the north of the cella and at a distance of only some 50 feet, had been exposed too long to retain any objects which 'treasure-seeking' parties might consider of value. Turdi distinctly remembered having searched them years before, and having obtained there a number of *khats* ('writings'), though he, naturally enough, could not give any information as to their character. Nevertheless, the systematic clearing of this ruin was attended by some important results. By tracing the dividing walls some seven apartments could still be made out, besides two enclosures built with walls of rushes and plaster in which we may recognize outhouses or stables. Other apartments must have disappeared through erosion, judging from the plentiful timber débris strewing the slope adjoining. Nothing but fragments of coarse plain pottery, red and grey, and pieces of wall plaster were found on the slopes and in the rooms northward. But luckily the two southernmost rooms were still under a cover of sand from 2 to 3 feet high, which the wall portions extant there had helped to retain, and under this cover some interesting relics had escaped previous searchers' attention.

In particular, the large room forming the south-eastern end of the house, and measuring 17 ft. 6 in. by 15 ft. inside, yielded an unexpected variety of written remains. The two fragmentary leaves (D. v. 1, 2) in upright Gupta writing of the seventh or eighth century, which turned up first near the north-east corner, have proved to belong to Pōthīs, probably containing

[2] See *Mémoires*, ii. p. 232.

[3] The month name *Mūñaja* has been recorded by Dr. Hoernle from the cursive Brāhmī documents previously examined by him; see *Report on C.-A. ant.*, ii. p. 35.

Buddhist canonical texts in the non-Sanskritic (Proto-Tibetan) language[4]. In the south-west corner, where the sand lay highest, were found numerous fragments (D. v. 7) belonging to a leaf from a manuscript in the same language, which evidently was written by the copyist of the Bhūmiparivarta text discovered in D. III[5]. But more interesting than these fragments of Pōthīs are the wooden tablets with cursive Brāhmī writing (D. v. 3, 4), the wooden stick with Chinese characters (D. v. 5), the Chinese document on thin water-lined paper (D. v. 6), and the large document in cursive Brāhmī (D. v. 8), signed with a Chinese monogram, which all successively emerged from the same shelter.

Chinese petition, D. v. 6.

By far the most important among these records is the Chinese document D. v. 6 (see Plate CXV), measuring about 10 by 5 inches, and recovered almost complete from the narrow roll into which it was found folded. The full annotated translation which M. Chavannes has given of this paper in Appendix *A* shows that it is a petition by one Ssŭ-lüeh, an inhabitant of *Li-hsieh*, 'in the territory of the Six Cities,' for the recovery of a donkey which had been sold 'for six thousand pieces of money' to two individuals who, after a delay of ten months, had failed to pay the money or return the animal. The persons against whom the petition is preferred are designated as 'scribes in the barbarian language' subordinate to the *a-mo-chih Shih-tzŭ*, one of them being named *A-shih-nai* (?), while the name of the second is missing. The petition is dated in the second month of the sixteenth year of the Ta-li period, which corresponds to 781 A. D. It is not merely on account of the exact date thus supplied that this document may claim special value, but also because, in conjunction with three Chinese documents which Mr. Macartney obtained in 1898 through M. Badruddīn, the Afghān Ak-sakāl at Khotan, and which were first published by Dr. Hoernle[6], it makes it possible to fix with practical certainty the name of the settlement represented by the ruins of Dandān-Uiliq, as well as the official designation of the Chinese administrative division to which it belonged.

Chinese documents from *Li-hsieh*.

The full translation which M. Chavannes has given in Appendix *A* of the three documents just referred to, and the learned notes with which he has there elucidated their contents and historical significance, render it easy to acquaint ourselves with the character and bearing of these records. Hence I may restrict myself here to a summary of the essential points. The first document (*A*) is a letter, presumably a draft, dated on the 23rd day of the third month of the third year Ta-li (768 A.D.), and is addressed by Ch'êng Hsien, the Chinese military officer commanding at Li-hsieh, to 'Wei-ch'ih, chih-lo prefect of the Six Cities and a-mo-chih.' M. Chavannes has lucidly demonstrated that Wei-ch'ih Yao, king of Khotan, whom we know to have reigned from 764 up to Wu-k'ung's visit (788–789 A. D.), is undoubtedly meant here. The grant of the title *a-mo-chih* to the king of Khotan is attested by an imperial brevet of the year 728, mentioned in a previous chapter[6a]. M. Chavannes has made it equally clear that 'the Six Cities', of which Wei-ch'ih [Yao] is described as the prefect, can only be the official designation of the territory of Khotan, comprising the capital Yü-t'ien and the five towns or districts dependent on it which the T'ang Annals mention. The letter of the commandant refers to the petition of certain people originally of Li-hsieh, who owing to repeated depredations of bandits had left the place, and who now pray for exemption from certain requisitions of grain and forced labour still exacted from them. The commandant refuses the claim to exemption, but recommends that the petitioners be enabled by special written authorization of the king to return to Li-hsieh in order to fulfil their obligations with the grain and the men

[4] See Dr. Hoernle's notes xi., xii., in inventory list.

[5] Comp. Dr. Hoernle's notes i., xiii.

[6] See *Report on C.-A. ant.*, ii. pp. 22 sqq., Plates III, IV, for facsimiles of two of the documents and abstracts of contents.

[6a] See above, p. 176.

they had left behind there. The second document (*B*), dated only by month and day, contains a military requisition from the Li-hsieh garrison, addressed by Yang Chin-ch'ing, commandant of the place, to the civil magistrate, asking for skins to re-cover drums and for quail's feathers to refit arrows. The third, dated in the seventh month of the seventh year Chien-chung (786 A.D.), records the issue of a loan of 15,000 pieces of money on certain conditions to a person who appears to have been of Li-hsieh; for a note subsequently added and dated in the tenth month of the same year records that 10,000 pieces of money have been repaid on account of that sum by an inhabitant of Li-hsieh.

With the records just described the Chinese documents excavated by me in D. v and in three other ruins of Dandān-Uiliq (D. VII, VIII, IX), show the closest agreement, not only in general appearance, paper, and style of writing, but also in character of contents and in dates. A comparison of these documents, as translated by M. Chavannes in Appendix *A*, conclusively shows that they are all formal records of public or private transactions. Records of loans, exactly resembling in their terms and provisions the third of the above documents (*C*), are the most frequent. On the other hand, we see that their dates, ranging from A.D. 781 to 790, fall within, or quite close to, the period covered by the two dated ones among Dr. Hoernle's previously discussed documents. Character of Chinese documents excavated.

In view of this complete agreement of contents, dates, and palaeographic evidence, it seems practically certain that the latter represent some of the finds of 'Khitai khats', i.e. papers with Chinese writing which Turdi well remembered to have made on a visit to the site some years previously, and which with other 'old things' he had sold to Badruddīn Khān, his usual employer at Khotan. It is impossible now to ascertain whether these particular documents, which had all three been transmitted to Mr. Macartney in one batch from Khotan[7], came from one of those rooms in the ruined house (D. v) which I found to have been roughly searched, or from some of the other ruins that had similarly been 'explored' before by Turdi's parties. But even without such specific information the comparison of D. v. 6, the first Chinese document I unearthed at the site, makes it clear beyond all reasonable doubt that *Li-hsieh*, the locality all these documents name, must be identified with the settlement or small tract to which the ruined shrines and dwellings of Dandān-Uiliq belonged.

M. Chavannes, in his first note on the document *A*, points out that the reading of the name 傑謝 *Li-hsieh* is rendered uncertain by the first character not being found in Chinese dictionaries[8]. The doubt as to its true sound could be removed only by the discovery of the local name in a printed Chinese text, or else, perhaps, by its identification in one of the Brāhmī documents from Dandān-Uiliq. Fortunately no obscurity can attach to the interpretation of the term 六城 *Liu-ch'êng* 'Six Cities', given in D. v. 6 as well as in the document *A*, as the designation of the administrative division to which *Li-hsieh* belonged. Already, on my return journey through Kāshgar, when Mr. Macartney and his Chinese literatus Sun Ssŭ-yeh kindly furnished me with transcripts and preliminary translations of the Chinese documents brought back by me, I had ascertained that the term 'Six Cities' is still well known by Chinese officials in the 'New Dominions' as an old designation of the Khotan territory. In a note *Li-hsieh*, name of ancient settlement.

[7] See Dr. Hoernle's note, *Report on C.-A. ant.*, i. p. vi., on batch M. 9, received by him in Oct., 1898. It is significant that the eight sheets of paper with cursive Brāhmī writing comprised in the same batch are all documents in the 'unknown' Eastern Īrānian language, corresponding exactly in writing and character to the Brāhmī documents discovered by me at Dandān-Uiliq along with Chinese records; see *Report on C.-A. ant.*, ii. pp. 36 sqq., and below, pp. 269 sqq.

[8] The difficulty had already been noted in Dr. Hoernle's *Report*, ii. p. 22.

'Six Cities' of Khotan. attached to the translation which Mr. Macartney had made in 1898 of the document *A* and which he then kindly communicated to me, Sun Ssŭ-yeh had identified five of the 'Six Cities' as corresponding to the present Ilchi or Khotan, Yurung-kāsh, Kara-kāsh, Chīra, Keriya, including *Lo-lasung*, a place of doubtful identity, as the sixth. Mr. Macartney himself referred to a remark in Sir Douglas Forsyth's Mission Report which mentions 'Alty Shahri Khutan' as a designation of the territory of Khotan, and supposes it to be derived 'from the six towns composing it', viz. the five above mentioned and Niya[9]. The passage quoted by M. Chavannes from the *Hsi yü t'u chih*, a modern Chinese account, fully bears out these statements but gives *Thakkaga* as the name of the sixth place, possibly meaning thereby Tawakkēl, since it is located near the confluence of the Kara-kāsh and Yurung-kāsh. Finally, M. Chavannes has clearly demonstrated that the use of such a term as the 'Six Cities' of our documents is reflected by the references made in the T'ang Annals to the five districts or towns dependent on Yü-t'ien[10].

Character of ruin, D. v. M. Chavannes' translation of D. v. 6 clearly shows that the document was a petition addressed to a civil officer of Li-hsieh, and taken down by a scribe from the complainant's oral deposition. It seems, hence, only reasonable to assume that the large dwelling (D. v) in which it was found was the residence, permanent or temporary, of that official. The conclusion would, of course, become greatly strengthened if it were possible to prove that the documents *A*, *B*, *C*, which Badruddīn had received from Turdi and transmitted to Mr. Macartney in 1898, had actually been found in the same ruin. In the absence of definite evidence on this point we may note that the parallel presented by the Chinese documents from D. VII distinctly supports this presumption. In the considerable series of papers brought to light there, all that are still more or less complete bear direct proof in their contents of having been either addressed to, or received by, monks of a Buddhist religious establishment. At the same time unmistakable archaeological indications show that the ruined structure in which I found them had been inhabited by monks.

Chinese record on wood. Leaving the historical conclusions to be drawn from the Chinese documents of Dandān-Uiliq for discussion below, I turn now to the other finds yielded by D. v. First among them may be mentioned another Chinese record, D. v. 5 (Plate CVI), not written on paper but on a stick of tamarisk wood, close on $13\frac{1}{2}$ inches long, with a width of a little over 1 inch. One side of the stick is flat, having evidently been prepared by splitting and subsequent smoothing. On this there remain very faint traces of about a dozen Chinese characters written in ink, but badly faded. Dr. Bushell, who kindly examined the object, found none of them any longer legible. The reverse, which has been partly left in its original round form and partly roughly flattened by chipping, shows on the latter portion a single column of large Chinese characters seemingly in two handwritings. Dr. Bushell interprets them as 'referring to the weight of a quantity of *Ch'ing*, amounting to one *tan* and one *tou* (the *tou* being the tenth of a *tan*)', and notes: 'The total weight would be roughly a man's burden, which makes me think that the three characters in the middle 阿閇娑 might be a man's name, if not the foreign (non-Chinese) name of the *Ch'ing*. Ch'ing is "green" or "blue"; but whether indigo, lapis lazuli, or some other commoner colouring material is meant I am not prepared to guess'[11].

[9] See *Yarkand Mission Report*, p. 33.

[10] Comp. his note 5 on Document *A* in Appendix *A*.

[11] ['Sur la face bombée de ce morceau de bois, on lit les mots: 青一石一斗五 (?) 升, qui me paraissent désigner une quantité de *grain* vert exprimée en mesures de *capacité* par les mots: "un *che*, un *teou* et cinq (?) *cheng*";—le *che* vaut dix *teou*, et le *teou* vaut dix *cheng*.—Au-dessous de ces mots, on lit, en écriture plus foncée, les mots 阿閇 (=閉) 娑 "A-pi-so" qui paraissent être un nom propre, puis les mots 青一石一斗? "un *che*, un *teou* et ? de (grain) vert."—Sur la face plane du morceau de bois, on

Brāhmī records on wood, D. v. 3, 4.

Seeing the roughness of the material, and the apparent purport of the legend which curiously recalls what might be inscribed on a mere tally, this find alone would scarcely have suggested that wood was used as a regular writing material at the time when this site was deserted. Yet the fact of such use was clearly established by the discovery in the same place of two small wooden tablets bearing writing in cursive Brāhmī characters. Both tablets (D. v. 3, 4) are oblong and rounded off at the left end. D. v. 3, as seen in Plate CVI, was provided with a small string-hole at the same end, and shows writing only on one side, arranged in four lines; it measures 9¾ by 2½ inches. D. v. 4, measuring 7¾ by 2 inches, is inscribed on both sides with four and three lines, respectively, parallel to the length of the tablet. From the transcription and notes with which Dr. Hoernle has favoured me for the writings, it results beyond all doubt that their language is the same early Eastern Īrānian dialect which Dr. Hoernle has recognized in the paper documents with cursive Brāhmī writing previously secured from Dandān-Uiliq. A number of words which occur in Dr. Hoernle's extracts from the latter reappear in these tablets, and so also the peculiar dating by day and month [12]. Signatures, too, are found attached in the form of curious marks or monograms. Seeing that the script presents no apparent differences, it may be safely assumed that these written tablets cannot be far separated in time from the corresponding paper documents of the same site.

Wooden *takhtī*.

A third and somewhat larger tablet (D. v. 10), 13¾ inches long and 3 inches broad, which turned up close by, at once attracted my interest by showing in its form and lozenge-shaped handle the closest resemblance to the *takhtī*, that traditional wooden board which in all native schools of Northern India takes the place of the slate [13]. This tablet was found blank, but there are plentiful marks of scraping to show that it had once been used for writing on. I shall have occasion to refer again to the antiquarian interest of these finds; for my subsequent discoveries at the Niya Site amply proved that we must look upon these few tablets, like the Indian 'Takhti' itself, as quasi-archaic survivals from the time when paper was unknown, and wood was the general writing material of Eastern Turkestān. But at the time of their discovery I little suspected what much more extensive finds of records on wood were awaiting me elsewhere.

Paper documents in cursive Brāhmī.

The document in cursive Brāhmī (D. v. 8), written on a sheet of thin paper, 10¾ by 7½ in., turned up almost complete in the form of a crumpled small roll; when opened out by me, with fingers half-benumbed through cold, it broke into two pieces. It was the first find of this class made in the immediate proximity of a record of ascertained date, and I may hence appropriately discuss here what observations I have to offer as regards the probable character of these documents and their antiquarian bearing. The total number of manuscript pieces in cursive Brāhmī recovered by me at Dandān-Uiliq was six, to which must be added the small fragment brought to me by Turdi previous to my start for the site [14]. Out of the six pieces three were found in ruined dwelling-places, viz. D. III, D. V, D. VIII, while the rest turned up in shrines D. IV and D. VI, which were closely adjoined by residences [15]. All fragments were found either crumpled like bits of waste paper or folded into narrow rolls just like the Chinese

discerne la fin d'une évaluation en mesures de capacité: 壹升伍合 "un *cheng* et cinq *ko*";—le *ko* est la dixième partie du *cheng*.—Immédiatement après ces mots, on lit: 六年九月 "sixième année, neuvième mois"; l'indication du jour est illisible; d'ailleurs, comme la période d'années n'est pas exprimée ici, la date ne peut être connue.'—E. CHAVANNES.]

[12] The words particularly noted by Dr. Hoernle with references to their occurrence in previously published documents are: *spâlā*, comp. *Report on C.-A. ant.*, ii. p. 37; *siḍaki*, comp. loc. cit., p. 36; *hvaṃde*, comp. p. 36; 14 *mye haḍai* 'on the 14th day', comp. p. 35.

[13] For 'Takhtīs' of exactly similar shape found at the Niya Site, comp. Plate CI.

[14] See above, p. 236.

[15] The pieces are numbered D. III. 13; D. IV. 6; D. V. 8; D. VI. 4, 6; D. VIII. 2.

documents. The paper is in all cases of a thin flimsy kind, closely resembling in outward appearance that of the Chinese documents found at Dandān-Uiliq. With a single exception (D. IV. 6), only one side of the paper is inscribed, probably because the latter was not sized, and hence would allow the ink to pass through. Of D. IV. 6 it may be supposed that its paper, like that of one at least of the Chinese documents (D. VII. 3), which Prof. Wiesner has subjected to microscopical analysis, had received a loading with starch[16].

Probable character of Brāhmī documents.

Even such cursory examination as was possible on the spot convinced me that these more or less fragmentary sheets could not have belonged to Pōthīs, but had evidently served for detached records of some kind. The cursive Brāhmī writing which they invariably showed was not to be found on any leaf which by its shape or material could be recognized as part of a manuscript book. The impression gained from the outward appearance of these papers has been confirmed by Dr. Hoernle's examination. This shows that they are documents closely agreeing in language, writing, and probable contents with the far more extensive collection of Brāhmī documents described and analysed by him in the second part of his *Report*[17]. Referring to that publication for all details of Dr. Hoernle's painstaking researches, I shall restrict myself here to those antiquarian points which my own observations on the spot may help to illustrate. The manuscript materials in cursive Brāhmī upon which Dr. Hoernle worked, and which comprised no less than thirteen sheets complete or nearly so, besides a large number of fragments, had been derived from purchases made during the years 1895–8 by Mr. Macartney and Captain (now Major) S. H. Godfrey from Badruddīn Khān, the previously mentioned Ak-sakāl of the Afghān merchants of Khotan[18]. Internal evidence of a conclusive kind, which Dr. Hoernle has duly indicated, proves that the whole of the Brāhmī documents comprised in those purchases 'came from the same locality, and even belonged to the same community'. On the other hand, the results of my inquiries at Khotan and the comparison of these documents with my own finds make it appear practically certain that this locality was Dandān-Uiliq, and that the documents described by Dr. Hoernle represent chance finds made by Turdi during his earlier visits to that site. Badruddīn Khān acknowledged from the first that, apart from the ample supply of forged manuscripts and block-printed 'old books' with which Islām Ākhūn had furnished him, all genuine acquisitions of ancient manuscripts had reached him in the form of fragments of leaves or crumpled lumps of paper brought by Turdi and his people. My old 'treasure-seeking' guide himself asserted with equal emphasis and consistency that nowhere else but at Dandān-Uiliq had he and his fraternity ever found ancient 'Khats', and that all of such that he had sold to Badruddīn Khān came from that site. In appearance, material, and, so far as he could judge, in their writing his manuscript finds had always resembled those which rewarded my search at Dandān-Uiliq.

Brāhmī documents previously acquired from site.

These statements made by the two persons through whose agency the documents comprised in Dr. Hoernle's collection had reached Kāshgar and Leh, is strongly supported by the evidence of the documents themselves. In paper, writing, and even the crumpled condition in which many of them reached Dr. Hoernle, they show the closest resemblance to my own finds. That their language, which Dr. Hoernle's philological acumen has successfully deciphered, is found

[16] See J. Wiesner, *Neuer Beitrag zur Geschichte des Papiers*, pp. 11 sq.

[17] See *Report on C.-A. ant.*, ii. pp. 30 sqq.

[18] Captain Godfrey's contribution (G. 1) which comprised a number of the documents here under discussion, is said to have been made up of finds made near Kuchā. But Dr. Hoernle has pointed out that the recurrence in them of the names of the same persons as are mentioned in documents avowedly purchased from Khotan by Mr. Macartney proves a different origin for at least part of that collection. See *Report*, ii. p. 31. Badruddīn regularly supplied antiques to Capt. Godfrey at Leh.

also in my Brāhmī documents can be definitely asserted on the basis of Dr. Hoernle's preliminary notes [19]. Without awaiting the results of his detailed analysis it is impossible to say whether the contents, too, furnish definite proof of identical origin. But in this respect we can fortunately rely on the parallel evidence of the Chinese documents. We have seen already that two of these contained in Dr. Hoernle's collection must have been written at Li-hsieh, which the testimony of my own find (D. v. 6) enables us to fix at Dandān-Uiliq, and from the same batch (M. 9) which furnished those two Li-hsieh records, came eight of Dr. Hoernle's best preserved specimens in Brāhmī script with an Eastern Īrānian language.

Language and contents of Brāhmī documents.

The minute analysis of the Brāhmī documents previously at his disposal enabled Dr. Hoernle to establish several philological facts which are of very great interest. By determining a number of words, mainly numerals and terms used in the dating, he succeeded in proving the Eastern Īrānian type of this 'unknown language' and its special connexion with the Galcha dialects of the Pāmīr region [20]. He clearly ascertained the important fact that the majority of the complete documents are fully dated, though the key to the reckoning of years has yet to be discovered [21]. A number of ingenious observations, such as the discovery of lists of names at the end of documents, accompanied by what are manifestly the marks of witnesses, the frequent occurrence of Chinese signatures or office-stamps in the same place, &c., permitted Dr. Hoernle to arrive at the undoubtedly correct conclusion that we have in them records of official or private transactions similar in character to the deeds of loan, requisition orders, &c., represented by the Chinese documents already discussed. Yet in order to invest these observations with their full historical value, it was essential that the place of origin and the period of these records should be fixed beyond all doubt. The special importance of the Brāhmī documents brought to light by me at Dandān-Uiliq lies in the fact that, few and fragmentary as they are, the certainty which exists as to all circumstances of their discovery supplies just what the student of these records needed.

Date of Brāhmī documents.

The close association of Brāhmī documents with Chinese ones, not only in D. v, but also in other ruins (D. vii, viii) to be discussed hereafter, permits us to fix the time of the former with approximate certainty. We shall see that the dates of the Chinese records discovered by me all range between the years 781 and 790, and that very distinct archaeological evidence points to the buildings containing them having been finally abandoned soon after. Insignificant in size and material as the Brāhmī documents are, it appears improbable that they should date back to a period appreciably earlier than that of the Chinese papers with which they were found. Hence the conclusion seems justified that the Brāhmī documents, too, must belong to the last quarter of the eighth century. If, in conjunction with this chronological fixing, we consider the character of those Brāhmī documents which Dr. Hoernle has fully analysed, it becomes evident that their language must have been that actually spoken by the inhabitants of the ruined settlement during the period immediately preceding its abandonment. This then was the 'barbarian language', to the use of which, by the people of Li-hsieh, the Chinese documents D. v. 6 and *A* both distinctly allude [22].

Eastern Īrānian language of documents.

I have had occasion, in a previous chapter, to point out that the fact of this language having proved to be of Eastern Īrānian origin is in full accord with what indications we other-

[19] See Dr. Hoernle's notes v., x., xiv., xvi., xvii., xviii., in inventory list.

[20] See *Report on C.-A. ant.*, ii. p. 32 sq.

[21] See loc. cit., p. 35.

[22] In D. v. 6 we read of the petitioner's complaint being laid against two 'scribes in the barbarian writing'. In *A* the Chinese commandant mentions a petition in 'barbarian writing' which he had received from the people of Li-hsieh and which he then proceeds to reproduce in Chinese.

wise possess about an important factor in the ethnic composition of ancient Khotan. In connexion herewith it is interesting to note that, by the side of the numerous Indian names of persons which Dr. Hoernle has recognized in his Brāhmī documents, a few others appear which look distinctly Persian [23]. One more observation of historical bearing deserves to be mentioned. From the fact that these Brāhmī records were found again and again in close proximity to Chinese ones, it may be inferred that the use of the local language for purposes of administration had not ceased during the period of T'ang supremacy, in spite of Chinese garrisons and an extensive infiltration of Chinese culture. Seeing that, in addition to these two languages, Sanskrit also and the non-Sanskritic language, to which Dr. Hoernle for the present gives the provisional designation of 'Proto-Tibetan', must, judging from manuscript finds, have counted students, perhaps even speakers, among those who frequented the little shrine (D. IV) and lived in the closely adjoining house (D. V), we may admire the polyglot faculties of at least the clerical and monastic section of the Khotan population in the eighth century.

Antiquarian finds in D. V.

The last dwellers in the ruin D. V had left behind some other curious relics of their daily life, besides the remnants of manuscripts and records on paper or wood that must have passed through their hands. In the same south-west corner room was found the fragmentary lower portion of a turned wooden bowl, D. V. 9, richly lacquered and painted. The elaborate floral decoration of the inside, as shown in Plate LXV, bears a Chinese look. The wood, too, seems different from any now available in Eastern Turkestān, and points to Chinese origin [24]. An article of more humble use, but undoubtedly of local manufacture, is the fairly well-preserved shoe (D. V. 11) made of plaited hemp-string which turned up in the room eastwards next to the corner room. Though the shoe shows signs of much hard wear, Mr. Andrews has been able to ascertain all details of its make. Finally, I may mention a torn piece of fine cord silk, D. V. 12 (Plate LXXVI), found in the same place, as a sample of contemporary textile work.

Original character of ruin D. V.

The variety of the finds made in D. V does not help us, I fear, to form a definite opinion as to the original character of the building. Its vicinity to the temple cella (D. IV), and the discovery in it of several fragments of Pōthīs in non-Sanskritic (Proto-Tibetan?), one apparently a Buddhist canonical text, would naturally suggest that the ruined building had served as a convent. But the contents of the Chinese document D. V. 6, and the relatively large size of the ruin compared with that of other monastic dwellings at the site, might make us think rather of the residence of a local official or other person of position to which the cella was attached as a sort of private chapel. The numerous remains of ancient fruit-trees just visible above the sand some fifty yards further to the south-west may have belonged to an orchard connected with the ruined dwelling. Some 150 yards from the latter in the same direction I picked up a well-preserved Chinese 'cash' with the legend of the K'ai-yüan period (713–741 A.D.).

Ruined structures D. XIV., D. XV.

About 250 yards to the north-east of D. V. I noticed under a scanty cover of sand, nowhere more than 2 feet, the remains of three small structures (D. XIV). Owing to their exposed condition they had evidently been thoroughly exploited long before. Turdi remembered to have found here a number of 'Khats' in the form of crumpled scraps of paper or folded rolls, probably some of the documents which went to Dr. Hoernle's collection. Of the southernmost structure, of which nothing remained but the timber forming the foundation, it was just possible to make out that it had been a cella enclosed by the usual passage. The other two structures still showed scanty remains of walls; but no finds of any kind rewarded their clearing. The same was the case with the small cella (D. XV) situated about 150 yards to the south of D. V. Here the walls,

[23] See *Report on C.-A. ant.*, ii. p. 34.

[24] For a piece of more ancient lacquer work, also a bowl and probably of Chinese make, see N. XV. 001. a., Plates LXV, LXX.

forming a square of 8 feet inside, were found standing to a height of 2 to 3 feet; but the excavation of the interior brought to light no remains of any sort, except traces of coarsely-done mural paintings made up of the usual rows of small seated Buddha figures. An isolated small structure immediately to the south was also cleared without results.

Ruin D. IX.

About 260 yards to the south-east of the last-named ruins, just across the previously-mentioned depression, lay the remains of a large but very coarsely built structure. Though much decayed through erosion and the burrowings of 'treasure-seekers', two wings situated at right angles could still be distinguished; the one facing north measured about 80 feet in length, and the one adjoining westwards about 110 feet. The walls, where still traceable, consisted merely of rushes fixed vertically and overlaid with a thin coating of plaster. Apart from fragments of coarse pottery vessels and a large wooden trough hollowed out of a poplar trunk, which the men thought to have served for storing grain, no antiquarian find of any sort resulted from the clearing of the western wing, except the small fragment of a Chinese document (D. IX. 1). This was found on the floor of the central apartment, which, covered by sand to a depth of about 5 feet, seemed alone to have remained undisturbed. Luckily the legible portion of the fragment, evidently the heading of a document, has proved to mention a date, 'the sixth year of [the period] Chêng-yüan' corresponding to 790 A. D.[23], the latest date found in the Chinese records of the site.

Section VI.—RECORDS FROM THE HU-KUO CONVENT

If the structures around D. V failed to fulfil the hopes raised by the finds in the latter ruin, the loss was compensated by the unexpectedly rich harvest of interesting relics yielded by the small isolated group consisting of the cella D. VI and the dwelling D. VII. Situated about half a mile to the south-east of my camp and within 40 feet of each other, the two little buildings were well protected under the slope of a small dune stretching in the approximate direction of north to south. The cella D. VI (see Plan XXVI), which was correctly orientated, measured inside 10 ft. 8 in. from north to south, with a width of 9 ft. 4 in. Outside its four sides ran a passage 4 feet wide, its enclosing walls, like those of the cella, built of the usual timber framework and plaster with a thickness of 6 inches. The entrance to both passage and cella lay from the north. Inside the cella, and towards its south wall, rose an oblong base of plaster measuring 3 ft. 3 in. by 2 ft. 8 in., and 9 inches high. Of the statue which once occupied it no remains could be found.

Cella D. VI.

Fresco decoration of D. VI.

Notwithstanding the deep cover of sand, about 6 feet above the original flooring, the cella walls were found standing only to a height of 3½ to 3 feet, a fact clearly indicating that they had been exposed previously to long-continued erosion. Yet what remained of the walls retained its original fresco decoration. Inside the cella each wall-face, with the exception of the one containing the entrance, was occupied by three large painted figures, of which, however, the extant wall portion showed only the feet standing on lotus cushions and the lower part of the drapery. Plate IV reproduces the greater part of the south wall, where the triangular spaces left between the aureoles around the principal figures were filled by smaller representations of Buddhas in dark red robes seated in the 'Dhyānamudrā' attitude. Below the lotus cushion bearing the central figure a boldly-moulded pedestal is shown, resembling in its structural arrangement that actually found within the larger cella of D. II. A tank with floating lotuses

[23] See M. Chavannes' translation, App. *A.*, Plate CXVI.

and swimming geese is painted around the pedestal, while at the foot of the latter appear the small figures of two worshippers. On the R. proper is seen a male figure in the act of offering lotus-flowers; his flowing garments seem to have a Chinese cut, while his hair is dressed in a fashion distinctly resembling that shown by certain of the Yōtkan figurines. On the L. proper a female figure appears kneeling and with hands folded. Her gown is shown brown with a brocade pattern, while a wide mantle of dark green colour hangs from the shoulders. The design of the painted plinth which runs round the foot of the walls, and in front of which these figures, as well as a small group of male worshippers to the right, are shown, can still be made out in the photograph.

Mural paintings of cella passage.

The walls of the passage were also covered with frescoes, but all these had suffered badly. Plate IV shows the decoration of the inner wall-face in the south passage with rows of small seated Buddha figures, each 6 inches high. Here, too, as in other mural paintings of this kind, only the colours of the dress, vesica, and nimbus varied in regular succession. On the outer wall of the west passage traces of the lotus pedestal of a large painted figure could be made out near the centre, flanked on either side by rows of small seated Buddhas. Below two of these in the lowest row on the left an inscription of eight cursive Brāhmī characters, about half an inch high and painted in black, survived, but owing to its position close to the floor I could take only an eye copy of it. Judging from this the legend would appear to have run as follows: *va di ra rmi* (?) *ga*(*gu* ?) *na ja ? ja*. There had been a long legend also below a similar row of small figures on the right, but this had been almost completely effaced by a crack in the plaster, and only the first two characters remained legible, *dvī pī*, agreeing with those at the commencement of the inscription below the fresco image, D. II. 08 (see Plate LVIII).

Painted panels from D. VI.

In front of the cella pedestal were found two painted panels, evidently as originally deposited, but the colours had faded or peeled off in numerous places. The larger panel (D. VI. 3), reproduced in Plate LXVII without its colours, 27 inches long, with a width of nearly 5 inches, shows ten figures, perhaps Bodhisattvas, seated on a lotus or Padmāsana. The arrangement and colouring of the figures, with their aureoles, &c., resemble those of the mural decoration previously described, but each figure is shown here wearing a yellow hat of Tibetan shape. The other panel (D. VI. 4), measuring $13\frac{1}{4}$ by 8 inches, and painted on both sides, shows in each six Buddha figures seated in the 'Dhyānamudrā' attitude and arranged in three vertical rows. For a detailed description I must refer to the list. The fragmentary relief (D. VI. 5; see Plate LVI) in hard gypsum (plaster of Paris), and originally coloured, was found near the floor of the cella near the north-east corner, and must be supposed to have fallen there from some stucco decoration on the upper portion of the wall. It represents the head of a female figure, evidently a Gandharvī, judging from its resemblance to the corresponding figures from the relief decoration of shrines D. I and D. XII, as seen in Plate LVI.

MS. finds of D. VI.

The first among the manuscript finds consisted of the left-hand fragment of a leaf (D. VI. 1) written in upright Gupta characters of the seventh or eighth century, which turned up in the south-west corner of the cella. It is numbered 70, and is shown by Dr. Hoernle's note iv. to have belonged to a long, narrow Pōthī containing a Buddhist canonical work in Sanskrit. In the same place was recovered the small fragment (D. VI. 2) of a document in cursive Brāhmī bearing traces of a Chinese monogram signature. The third was a narrow strip of paper (D. VI. 6) containing part of a document written in cursive Brāhmī and the Eastern Irānian language. This was found in the form of a small roll lying on the floor of the north passage.

Finds in ruined

At a distance of only 40 feet to the north of the latter a small dwelling-house (D. VII) could be made out under the sand by its posts rising well above the slope of the dune. In

dwelling D. vii.

the eastern room (see Plate XXVI), which was cleared first, nothing was found but a small fireplace built of plaster against the southern wall. But the contents of the larger room adjoining it westwards, and measuring 18 by $12\frac{1}{2}$ feet, proved all the more interesting. In the sand which filled it to a height of about 5 feet, there appeared first in the middle of the east wall the remains of a well designed fireplace, 3 ft. 8 in. broad, surmounted by bold plaster mouldings. By its side, towards the south-east corner, there was first found, lying in the loose sand about 1 foot above the floor, the small painted panel D. vii. 1, and immediately below it the Chinese document D. vii. 2, neatly folded up into a roll about 11 inches long, which opened up easily into a completely preserved sheet (see Plate CXV). Lower down, and sticking to the mud flooring, there lay two small packets, D. vii. 3 and D. vii. 4, of Chinese papers, each consisting of several folded-up rolls, mostly fragmentary. Nearer to the south wall of the corner and several inches above the floor the two well-painted panels D. vii. 5 and D. vii. 6 were discovered in excellent preservation. Finally, about 4 feet to the west of the fireplace and near to the floor, there was found the almost complete document D. vii. 7 (see Plate CXVI), also folded into a roll, 11 inches long, the specially interesting contents of which we shall have occasion to discuss presently. Owing to the complete decay of the greater part of the walls facing west and south it was impossible to determine where the entrance to this room lay. But the height of the posts still rising above the better protected east wall plainly showed that there had once been another story above it. A small structure built of rush-walls against the north wall of the house (see plan) may have served as a store-room.

Condition of Chinese documents discovered.

The annotated translations which M. Chavannes has given in Appendix *A* of all the Chinese documents found in D. vii make it easy to realize that, petty as are the affairs with which they deal, their value for the determination of antiquarian questions directly concerning the site is very considerable. Owing to the damp that must once have reached them through the mud floor, all the small rolls of folded paper comprised in the packets marked D. vii. 3 and D. vii. 4 have suffered considerably. Some documents, of which only small fragments could be recovered, may have been partially broken up either before they reached the place where they were found, or before a sufficient cover of sand had accumulated to protect them. The folded condition in which they were discovered, with the damage caused by the partial peeling off of the outer folds, is illustrated by Plate CXV, which shows the document D. vii. 4. a both before and after the unfolding. In view of these circumstances it is particularly fortunate that the two documents D. vii. 2 and D. vii. 7, having been entirely embedded in sand and well above the floor, have survived almost wholly intact. Apart from their intrinsic interest they help to settle all doubts as to the character and origin of the rest. The position in which these two documents were discovered suggests that they might have fallen into the sand, which had already commenced to fill the room, from some higher receptacle on the walls or possibly through some fissure in the flooring of the upper story.

Dated Chinese bonds.

D. vii. 2, written on a sheet of thin yellowish paper measuring about $15\frac{1}{2}$ by $11\frac{1}{4}$ inches (see Plate CXV), is a formal bond dated in the 3rd year Chien-chung (A.D. 782) for a sum of one thousand pieces of money, i.e. 'cash', which Ma Ling-chih, a soldier, has borrowed from Ch'ien-ying, a monk of the Hu-kuo temple, at a rate of interest no less than ten per cent. per mensem, pledging for this loan all his movable property. We find the identical monk named as the creditor in another bond, D. vii. 4. a (Plate CXV), dated in the same year[1], which relates

[1] In D. vii. 2 the year 782 A.D. is correctly indicated as the 3rd year Chien-chung; in D. vii. 4. a, however, as the 17th year Ta-li, though the latter period comprised only the years 766–779. Similar datings in periods already elapsed are frequent in these documents; see M. Chavannes' notes on D. v. 6, D. vii. 3. d, MS. *C*. Perhaps we may take this

to the loan of a certain quantity of corn for the fixed period of nine months to Ho Hsin-yüeh, an official. No rate of interest is specified in this case, and we are thus left in doubt as to the particular manner in which the pious money-lender—for as such we may safely recognize Ch'ien-ying—had assured his profit on the transaction. The formal pledge of the borrower's property we find here repeated. At the end of both bonds are mentioned, besides the names and ages of the borrowers, those of two of their near female relatives (mother and sister, wife and daughter), who are evidently sureties. Near the end we find, not the 'finger impressions' of the contracting parties which the text names, but ink marks which, perhaps, served the same purpose.

Chinese records of loans.

Among the fragmentary documents found in the packets D. vii. 3 and D. vii. 4 there are no less than six which, by the recurrence of the identical formulas or by other indications of the contents, have been recognized by M. Chavannes as bonds relating to loans. The dates, so far as preserved in three of them, belong either to the periods immediately following or the one preceding the Chien-chung period (780–783 A.D.), and thus fall quite close to the year 782 A.D. of the previously-quoted documents [2]. Seeing that some of these fragmentary bonds (D. vii. 4. b, 4. e) were found in the identical packet with the record D. vii. 4. a, distinctly naming Ch'ien-ying as the lender, while the rest (D. vii. 3. a, b, c, d) were comprised in one packet lying within less than a foot of it, it appears probable that these papers, too, relate to money-lending transactions of the same monk or possibly some equally businesslike brethren from the Hu-kuo convent.

Order from the *Hu-kuo* convent.

This conclusion seems the more justified as every one of the remaining four Chinese documents from D. vii contains distinct evidence of having issued from, or having been addressed to, members of a Buddhist monastic establishment. By far the most interesting piece is D. vii. 7 (see Plate CXVI), complete but for a small lacuna, containing an order from the three chief dignitaries of the *Hu-kuo* temple to certain of their dependants, among them apparently the monk Ta-yen, 'in supervising charge of the outlying [property].' The order is dated on the 27th day of the eighth month, without indication of the year, and directs in somewhat peremptory language that on its receipt at *Yang-ling* the servants of the temple should all be employed for three days in cutting the grass, while one man only should be left to look after the irrigation of the fields. The order is signed in different hands by the three monks holding the offices of chief Karmadāna, Sthavira, and Vihārasvāmin of the temple or Vihāra. It is far more likely that an order of this kind would be retained at the place to which it was addressed than at the one from which it was sent. It is further difficult to suppose that the little shrine (D. vi) could have boasted of such an array of monastic dignitaries as signed the order, and thus everything points to the conclusion that D. vii, the find-place of the document, was in reality the modest residence of the monkish caretakers who looked after the outlying landed property of the Hu-kuo temple in this neighbourhood. That a small shrine like D. vi should be attached to their residence scarcely needs explanation. If this assumption is right *Yang-ling* would have to be taken as the name of the particular village or hamlet where the lands were situated, while the designation of Li-hsieh previously discussed applied to the whole tract.

Documents on monastic affairs.

The name of Hu-kuo does not occur in the remaining three documents, which are mere detached fragments, but in each of them we meet with the mention of Buddhist monks.

as an indication of the time needed for changes in the official chronological reckoning to get known and applied in so distant a territory as Yü-t'ien.

[2] D. vii. 3. d (see Plate CXVI) is dated in the 8th year Chien-chung (*recte* 3rd year Chêng-yüan), 787 A.D.; D. vii. 3. c dates from the Chêng-yüan period, 785–804 (indication of year missing): D. vii. 3. a (see Plate CXVI) from the Ta-li period, 766–779 (year lost).

D. VII. 4. d, dated in the year 789, mentions quite a number of them (Ying-ch'ing, Hsüan-ying, I-fa, Shan-i, Fa-yu). Another list of monks, with the names P'u-?, Pao-ming, Fa-chin, Tao-ch'ao, still preserved, was contained in D. VII. 4. f, while the third fragment, D. VII. 4. c, of uncertain purport, furnishes the name of the monk *Ta-p'i*, who figures in another fragmentary Chinese record (D. VIII. 1) from a small shrine still to be mentioned, with the title of *Wei-na* or Karmadāna.

Locality of *Hu-kuo* convent.

We have no means whatever for locating the *Hu-kuo* Vihāra, whose relation to the last dwellers of this modest structure the documents just discussed have so curiously revealed to us. The Chinese designation (*Hu-kuo*, literally 'country-protecting') and the Chinese names of the monastic dignitaries, as well as of the monks specified, leave little doubt as to the nationality of that religious establishment. But that the population among which Ch'ien-ying and his brethren lived was not Chinese seems plainly indicated by the inscriptions in cursive Brāhmī which we have already referred to in describing D. VI, as well as by the finds of cursive Brāhmī documents in the same place.

Chronological value of Chinese papers from D. VII.

The very pettiness of the affairs recorded in these Chinese papers increases their value from a chronological point of view. Unimportant in character and insignificant in appearance, it is highly improbable that these records of the private transactions of a few monks and of the casual orders, &c., sent to them by their convent should date back to a period preceding by any great length of time the final abandonment of the building. We have seen that all the papers from this ruin which can be dated with accuracy belong to the years 782–789 A.D. Taking into account that the first Chinese document found in the ruined house D. V under exactly similar conditions, bears the date of 781, and that the fragment discovered in D. IX, the only other dated Chinese paper from the site, was written in 790 A. D., we are almost forced to the conclusion that the settlement to which these dwelling-houses and shrines belonged was deserted between the last-named date and the close of the eighth century of our era. In each case the papers were discovered on the original floor or quite close to it, which proves that drift-sand must have entered the rooms very soon after these petty records had been scattered about there. For light and flimsy as they are, the little paper rolls could not have resisted very long the force of the storms which pass over the country each successive spring and summer.

Painted panels from D. VII.

It is a fortunate circumstance that such unmistakable chronological evidence has been obtained in the very same structure which furnished us with the best preserved specimens of contemporary painting from this site and, perhaps, also the most interesting. I have already described the position in which the three painted panels of wood I refer to were found in the loose sand near the south-east corner of the main room of D. VII. Direct evidence in the case of two panels allows us to infer that these pictures had once been fixed high up on the wall, from which they dropped when the little dwelling was gradually being filled with sand. The panel D. VII. 5 still retains on its plain reverse the remains of dowels in several places showing how it had been affixed, while the 'Takhtī'-shaped tablet D. VII. 1 has a hole drilled through its handle by means of which it could be nailed to the wall. Perhaps it was due to this manner of fixing that this particular panel lay considerably higher above the floor than the rest. The third panel, D. VII. 6, which turned up quite close to D. VII. 5, and which being painted on both sides could not have been attached in the same manner, may have been made to stand upright on a small shelf by means of a socket or otherwise. These observations account for the excellent preservation of the wood of all these panels, and for the remarkably good condition of the colours in the case of two of them. The fact of these panels having been found in a dwelling-house and not a shrine is, of course, explained by the fact of the dwellers having been monks.

Panel of horseman and camel-rider, D. vii. 5.

The panel D. vii. 5, of which Plate LIX shows a successful reproduction in colours reduced to a two-thirds scale, may be mentioned here first, as it illustrates a legend which representations found in the shrines D. ii and D. x have rendered familiar[3]. The panel, which has a rectangular shape, with pointed arched top, is 15 inches high and nearly 7 inches broad. It shows two figures, both mounted, one above the other. The upper figure is seen riding on a high-stepping horse and holding in his right hand a patera towards which a bird, more distinctly shown in D. x. 5 as a wild duck, is swooping down in full flight. The rider, whose handsome youthful face shows an interesting combination of Indian and Chinese features, wears his long black hair tied in a loose knot at the crown, while a yellow band passes round the head holding in front a large elliptical jewel. Besides the long pink tunic, the rider wears a narrow light scarf which, descending apparently from the back of the head, is curled round both upper arms, while its ends float behind, indicating rapid movement[4]. The feet are cased in high black boots with felt soles, very much like those still worn by men of means in Chinese Turkestān, and are placed in stirrups. From the girdle hangs a long sword, nearly straight, of a pattern that appears early in Persia and other Muhammadan countries of the East. The horse, which is well drawn even to its legs and hoofs, by its colour—white with large spots of black—curiously recalls the appearance of the piebald 'Yarkandi' horse which until recent times was much fancied by natives of Northern India. Over a large 'Numdah' or felt cloth it carries a deep and narrow saddle and shows elaborate trappings, for a description of which I may refer to Mr. Andrews' detailed notes in the list of objects. We could scarcely have wished for a more accurate picture of that 'horse millinery' which in the eighth century evidently flourished throughout Eastern Turkestān quite as much as it does nowadays.

Camel-riding figure in D. vii. 5.

No less interesting is the representation of the second figure below, riding a two-humped camel, shown in full movement and with striking fidelity to nature. The rider wears over his short curly hair (partly deleted) a curious sugar-loaf hat, with its broad brim turned up into 'Vandyke' points, like that of the riding figure seen in D. x. 5 (comp. Plate LXII). Marks on the hat indicate some spotted fur as its material. The long and loose-fitting green garment worn by the rider is gathered below the knee into the wide tops of red boots or mocassins without soles, closely resembling the 'Chāruks' now used throughout the whole of Eastern Turkestān, particularly in the winter months. While the left hand is guiding the camel by a nose-cord, the right, in the same pose as that of the rider above, raises a shell-shaped cup. The elaborate fittings of the saddle, and the stirrups, show that the animal bestridden by this personage is meant for a riding camel, rarely used now in this region, but distinctly referred to by a version of Sung Yün's notice of Khotan[5]. Some freely-drawn outlines visible behind the camel's legs are meant to indicate hilly ground or else high ridges of sand, where, indeed, the camel would be far more in its place. The nimbus painted round the head of each of the riders plainly shows their holy character, and the identity of their attitude leaves no doubt as to their connexion with what was evidently a popular local legend. But to the nature of this legend I have not been able to trace any clue.

Painted panel D. vii. 6.

The panel D. vii. 6, painted on both sides and measuring 12¾ by 8 inches, is seen in

[3] See above, pp. 248, 261.

[4] The fixing of this scarf, corresponding to the *uttarīya* of Indian Buddhist iconography, at the back of the head is more clearly indicated in the figures of panels D. vii. 1, 6 (see Plates LX, LXI, LXVI).

[5] Comp. Rémusat, *Ville de Khotan*, p. 22; also above, p. 170, note 32.

Plates LX and LXI as it now appears after having been carefully cleaned at the British Museum by the artistic hands of Mr. R. F. Fry from a thin but closely adhering crust of fine sand. It presents several points of exceptional interest. The excellent preservation of the colours enables us to appreciate the pictorial merit of the work; the sacred figures represented on obverse and reverse possess particular interest for the student of Northern Buddhist iconography, as relatively early and approximately datable specimens of pictorial types which hitherto have been known only from textual descriptions or else later representations, mainly of Tibetan origin. Finally, the blending of Indian and Western influences in the art of old Khotan is illustrated by the different styles which a comparison of the obverse and reverse reveals even on cursory inspection. In the former we see a three-faced and four-armed divinity seated cross-legged on a cushion which is supported by two couchant bulls. The flesh of the divinity is shown dark blue throughout, excepting in the two side heads, of which the one on the R. proper, coloured white, bears an effeminate look, while the one opposite is dark yellow with the expression of a demon. The rich diadem of the main head, with its side ornament resembling a half-moon, the third eye on the forehead, the tiger-skin forming the *dhōtī* or loincloth, and finally the bulls shown as Vāhanas are all so many emblems recalling the Brahmanic Śiva.

Identification of Bodhisattva figured on obverse.

It is these which seem to justify our identification of the figure with one of the many forms of Avalokiteśvara. The close connexion which exists in the iconography of Indian Buddhism between this Bodhisattva and the foremost god of the Brahmanic trinity has long ago been pointed out[6]. The attributes held in the four hands, among which the Vajra, shell (*śaṅkha*), and wheel (*cakra*) can clearly be recognized, and, perhaps, also the drum (*ḍamaru*), may yet help to an exact determination of the particular form of the Bodhisattva intended. All of them, as well as the blue colour, the rich ornaments of the body, and other peculiarities already mentioned, can be traced in the detailed descriptions of various Avalokiteśvara forms which M. Foucher has recently made accessible from certain *Sādhana* texts[7]. But neither among the forms for which the essential descriptive details are excerpted there, nor among those represented in the series of old miniatures which form the subject of M. Foucher's masterly study in the first part of his *Iconographie bouddhique*, is there any combination of all the characteristic features of our picture. Nor can this be surprising in view of the almost infinite multiplication of forms which mythological fancy within Indian Buddhism from an early period created for this and other chief Bodhisattvas.

Bodhisattva on reverse of D. vii. 6.

There is nothing in the picture of the obverse that does not directly presuppose the reproduction of Indian models. The contrast offered by the figure painted on the reverse is hence all the more striking. Were it not for nimbus and vesica, and in particular the four arms with the emblems in some of them, it would be difficult to believe that we have here before us the representation of a Buddhist divinity. So completely does a quasi-secular treatment in essentially Persian style prevail in the figure and its accessories. We see the Bodhisattva—for as such we must evidently accept this sumptuously dressed sacred personage—seated cross-legged on an elaborately embroidered cushion. The head, with distinctly Persian features, shows long black hair falling down to the shoulders and a thick short beard. It is surmounted by a curious yellow head-dress, which resembles a Persian tiara with a double point falling over in spirals. Over a close-fitting yellow undergarment decorated with large flowers the figure wears a tight coat of equally rich material in dark green. From a narrow belt

[6] See Foucher, *Iconographie bouddhique de l'Inde*, i. pp. 172 sqq.; ii. p. 39; Grünwedel, *Mythologie des Buddh.*, p. 132.

[7] For an Avalokiteśvara of blue colour, see Foucher, *Iconographie bouddhique*, ii. p. 38; for the tiger-skin, the half-moon, drum, Vajra, ibid., pp. 30, 32, 34, 38.

round the waist is slung a short, broad sword in an ornamented sheath. The most curious part, however, of the attire are the high black boots exactly resembling Hessians, which reach to the knee, and are picked out in ornamental yellow lines suggesting inlaid leather-work or stitching in gold-thread. The scarf-like Uttarīya, which descends from the back of the head and is curled round both arms, is the only part of the dress that the figure shares with other representations of sacred personages in these panels. The left proper front hand holds the Vajra, while the rear one raises a spearhead; on the right proper the front hand rests on the thigh, the one behind holding a deleted object, possibly a flower.

Identification of Bodhisattva on reverse.

For the identification of this strangely attired divinity, the Vajra or mystic thunderbolt seems to afford the only clue. It is the characteristic emblem of no less than three divinities, the Dhyāni-Bodhisattvas, Vajrapāṇi, and Vajrasattva, as well as of the primordial Buddha Vajradhara[8]. With none of the representations that I can trace of these at present does our figure fully agree; but seeing that it shows neither the green colour prescribed for Vajrapāṇi, nor the blue one of Vajradhara, the conjecture may be hazarded that it is perhaps meant for a form of Vajrasattva. This Bodhisattva is ordinarily represented white, with the Vajra in one hand, and the other resting on the hip and holding the *ghaṇṭā* or bell[9]. The latter is certainly absent in our picture, yet the general pose bears some resemblance to that of the Vajrasattva seen in the miniatures which M. Foucher has published.

Persian style of picture.

The real interest of the painting lies, however, not in the eventual identification of the figure, but in the application of a design of essentially Persian style to a subject of Buddhist mythology. The immediate juxtaposition of this design to the wholly Indian one shown by the obverse of the same panel necessarily increases this interest. There is nothing to suggest that the two sides of the panel were painted by different hands, however different in origin were the models which the artist in each case followed. The representation of a Buddhist divinity in forms derived from a non-Indian art has in itself nothing surprising, since we see the identical process illustrated to the largest possible extent as far as classical art is concerned in the Graeco-Buddhist sculptures of Gandhāra. But of the influence of Persian art under which Buddhist iconography may have come in those extensive parts of Eastern Īrān where Buddhism found a home and flourished, we know as yet practically nothing. Is it possible that this curious little painting reflects features of a 'Perso-Buddhist' art such as might have developed in Balkh, that city of many splendid Vihāras, and elsewhere within ancient Ariana?

Painted panel D. vii. 1.

I have left to the last the mention of the panel D. vii. 1, the first discovered, because though remarkable, as Mr. Andrews observes, for its good drawing, free execution, and good composition, it has suffered far more than the other panels of this ruin in its colours, and hence cannot be properly judged from the reproduction (Plate LXVI)[10]. The figure is undoubtedly that of a Bodhisattva, seated on a low throne, and holding in the left proper hand a bud-shaped lotus over a long, gracefully curving stem, while the right raises a patera in front of the breast. The upper part of the body appears to be dressed in a black garment, while from behind the head descends the scarf-like Uttarīya curling round the upper arms in the fashion previously noted. The flesh is coloured pink. The shape of the lotus seems to suggest the *Utpala* or blue lotus; and as the latter is the characteristic emblem of Mañjuśrī, who usually appears to hold it in his left hand and in the pose here represented, I am inclined to believe that this Bodhisattva

[8] Compare Foucher, *Iconographie bouddhique*, i. pp. 121 sqq.; Grünwedel, *Mythologie*, pp. 98, 141.

[9] See Foucher, *Iconographie bouddhique*, i. p. 123, Plate vi. 5.

[10] In the plate the number of colours reproduced had to be limited to those absolutely needed for the indication of the main features.

is intended[11]. But it must be pointed out that, among the attributes mentioned in the descriptions of Mañjuśrī forms accessible to me, the cup here seen in the right hand does not figure. On the other hand, we can well reconcile the natural pink colour of the flesh with the saffron colour which certain of the Sādhana texts indicate for Mañjuśrī[12]. I have already had occasion to state that this panel was found fully one foot above the floor, and the good preservation of the wood also shows that the damage suffered by the colours cannot be due to moisture. This makes me think that possibly the partial disappearance of the coloured surface may be the result of long exposure to which the panel was perhaps subjected, while the nail passed through the hole of its handle kept it fixed to the wall of the deserted dwelling, as above suggested.

Section VII.—OTHER RUINS OF DANDĀN-UILIQ AND GENERAL OBSERVATIONS ON SITE

Central group of ruins.

The remaining ruins explored by me at Dandān-Uiliq can be dealt with more briefly. They consist of what I may call the central group of remains immediately to the south of my camp, and of certain isolated structures elsewhere. To the former belonged D. I, the much-decayed small cella, the débris heaps of which were examined by me immediately after my arrival, and which has already been described[1]. D. VIII, a small structure situated about 30 yards to the north-east of it, proved to contain a single room measuring 17 by 12 ft. inside. Though covered by some 6 ft. of sand, the room on excavation yielded nothing but the fragment of a paper document in cursive Brāhmī (D. VIII. 2), measuring about 6 by 4 inches, which turned up close to the floor in the north corner. Some débris of plaster and completely decayed timber which lay half-covered by sand close to the north-east wall of D. VIII, probably originated from earlier burrowings of 'treasure-seekers'. On clearing it, a small and brittle piece of crumpled paper was found containing the fragmentary Chinese document D. VIII. 1 (see Plate CXVI). M. Chavannes' translation shows that it probably formed part of a letter written by, or addressed to, the resident monk of a Buddhist shrine[2]. It derives interest from the mention made of the monk *Ta-p'i*, whom we have already met in the document D. VII. 4. c, and who is here designated by the title of *Wei-na* or Karmadāna of a monastic establishment. The reference to the contract about a female servant given as security suggests that this communication, too, may have been connected with business arising from some loan or contract.

Dwelling D. XIII.

At a distance of about 85 yards to the north-east of D. VIII were the much-decayed remains of a large dwelling (D. XIII on plan), measuring about 60 feet from east to west. According to Turdi's statement a find of silver equivalent to about Rs. 200 had been made here in the time of Niāz Ḥākim Bēg, and he, as well as others of his profession, had since tried their luck at this ruin by burrowing into the walls or floor of exposed rooms in the hope of further finds. This information was corroborated by the completely exposed and eroded state of what seemed to have once formed the north flight of rooms in the building. The rooms to the south, being apparently built on a lower level, had preserved their walls better, and were filled with sand from 5 to 7 feet in height. I had the central one, measuring about 22 by 18 feet inside, cleared,

[11] Comp. Foucher, *Iconographie bouddhique*, i. pp. 115, 119, Plate VI. 3, 4; ii. pp. 40 sqq. The *Utpala* belongs also to Vajrapāṇi, but in his case the essential Vajra would be expected.

[12] See Foucher, *Iconographie bouddhique*, ii. pp. 40 sq., 45, 47.

[1] See above, pp. 243 sqq.

[2] See Appendix *A*.

but convinced myself by the absence of any movable objects except a plain wooden cornice, as well as by other indications, that it had been ransacked long ago. On this account I did not consider it useful to extend the labour to a large apartment eastwards traceable below the sand. I had subsequently reason to regret this omission, for it was probably from among the rubbish which treasure-seekers may have thrown outside on occasion of previous diggings that the curious relics were obtained which some of my labourers brought to me after I had proceeded to Rawak[3]. The room cleared was provided with a large fireplace, but had no door, from which I concluded that it was entered from an upper floor and served as a basement, perhaps for use during winter.

Ancient orchard and courtyard.

The numerous remains of ancient trees, bleached and splintered, in the neighbourhood of this ruin may have belonged to the orchard or garden once surrounding the dwelling. Among the trunks still rising above the sand my labourers were able to distinguish apple and apricot, as well as the 'Jigda' (*Eleagnus angustifolia*), with the wood of which all Khotan cultivators are familiar from their own homesteads. A little to the west of the ruin remains of a fence made of rushes indicated an ancient enclosure such as could be traced also near the building D. v, and such as I subsequently noticed often near the ruined dwellings of the Niya Site. By scraping the sand-covered bank of a small depression formed through wind-erosion near the edge of the area which this fence appeared to have enclosed, Turdi laid bare a closely-compressed mass of straw, evidently once deposited in the corner of a fenced courtyard. The straw, though darkened and, of course, completely dried by the long centuries that had passed since its burial under the drift-sand, was in a remarkably well-preserved condition, so much so that Turdi looked upon it at first as a providential means for saving the pony which he had been ill-advised enough to bring to the site without my knowledge. The poor famished animal at the beginning swallowed ravenously this most 'desiccated' fodder stuff, without, of course, being saved thereby from the end related in my Personal Narrative[4].

Detached ruins.

The arrival of Rām Singh, who on December 24 had safely joined me (as previously arranged) from the side of the Keriya river, after completing the survey of the high snowy range between Karanghu-tāgh and Polu, enabled me to leave, for short periods, the actual excavation work under his supervision, and to reconnoitre, under Turdi's guidance, the outlying portions of the ruined site. It was in the course of these excursions that I acquired reasonable assurance that, apart from the detached ruins to be noticed, there were no remains of any size recognizable below the drift-sand. The small ruin D. xvi, nearly one mile to the south of my camp, and at the southern edge of the area over which pottery débris could be traced, proved to be the remains of a little cella nine feet square, destroyed to within a foot of the ground. Nothing but a shapeless mass of plaster in the centre, probably the last trace of a statue base, was found by clearing it. Proceeding to the north-east of my camp, all traces of old occupation, such as pottery débris and remains of trees, ceased within half a mile, the dunes rising here to 15 feet and more. Crossing the western offshoot of the 'Dawān' marked on the plan, I came upon what looked like the remains of a circular mud wall cropping out in five or six places between the closely packed dunes. No exact survey could be made, owing to the height of the latter; but from the measurements taken by me it appeared that the circular space enclosed by the walls had an approximate diameter of 260 feet. The walls, which could be traced only in the form of low mounds with horizontal layers of rushes between the stamped loess, seemed to have had a width of about 35 feet at their base. Owing to the conditions of the sand, excavation

Traces of ruined fort.

[3] See below, pp. 306 sqq.

[4] See *Ruins of Khotan*, pp. 304 sqq.

was quite beyond the limits of available labour and time; and I might have remained in doubt as to the real character of the scanty remains, had I not subsequently become acquainted with the circumvallations of the ancient forts at Endere and Ak-sipil exactly resembling this in shape and construction. Beyond the north-east segment of wall a small bit of ground left clear of sand showed fragments of much-decayed timber, with some broken pottery and little heaps of burned earth of a red colour. Elsewhere near this ruin the strips of exposed ground, which became more extensive and frequent towards the west in the direction of the previously mentioned well, showed no pottery débris. I conclude from this that the small circular fort—to such I now think it safe to attribute the remains I could trace—was situated beyond the closely inhabited area of the settlement.

Also when I proceeded subsequently to the small ruin about 2½ miles due north of my camp which I had noticed on my first arrival, the intervening ground, where bare of sand, showed no pottery débris beyond the limits approximately marked on the plan. The clearing effected by my labourers first laid bare here the remains of a small cella (D. xvii), 10 feet square. The walls had completely decayed on the south and east, while on the other sides, too, only about a foot and a half of them had been left standing. Débris of frescoed stucco lying within showed that treasure-seekers had been at work here, probably not very long before my visit. Turdi declared that he had noticed the little ruin for the first time on his recent prospecting trip, and that he had then secured here the small piece of inscribed fresco, D. T. 017 (see Plate LVIII). I could not ascertain its exact position, but as it shows what looks like the front part of the L. foot of a figure standing over a floral design, it may well have come from the low portion of wall still standing. Dr. Hoernle's reading of the inscription, in cursive Brāhmī and in the Eastern Irānian language, will be found in the list below. About 80 yards to the north-east the foundation beams of a single-roomed small structure could be made out in the ground, the walls above having completely disappeared, and a little to the south-west of this a heap of rough posts and brushwood with layers of dung seemed to indicate the position of a sheep-pen. Ruin D. xvii.

Having now completed the description of the ruins explored by me, and of the antiquarian finds they yielded, I may briefly indicate the conclusions which the observations above recorded seem to justify as to the history of the site. Whether our interest in the latter is purely archaeological, or of a more general kind connected with the geography and cultural past of the whole region, the chronological question as to the date of the ruins and the abandonment of the site must claim our attention first. Fortunately the finds of the dated Chinese documents discussed above, together with the collateral evidence of the coins, enable us to answer this question with full assurance. We have seen that those documents dating from the years 781–790 A. D., by their very character and the conditions in which they were found, prove the closing years of the eighth century as the period when the ancient settlement was abandoned by its last inhabitants. In entire accord is the evidence of the Chinese copper coins, of which altogether seventeen were picked up at the site either by myself or my men. Leaving aside those which are too effaced for identification, we have among them six coins clearly bearing the legend of the K'ai-yüan period (713–742 A. D.), two of the Ch'ien-yüan period (758–760 A. D.), while six are without legend and of the type which was current under both the Han dynasties, and probably for some centuries later[1]. Date of abandonment. Evidence of coins.

In the present state of our knowledge it is not to be expected that the manuscript finds in Brāhmī script and the manifold art remains should yield chronological indications of equal Other chronological indications.

[1] For reproductions of specimens of these coins, see Plate LXXXIX, 23, 25, 27 and Appendix *D*.

clearness. But here, too, it may be safely asserted that there is no object that would suggest for its origin a date later than the eighth century. Nor is there, with the possible exception of some fragmentary Pōthīs, any among them which is likely to prove of an appreciably earlier period than the last decades preceding the abandonment. In the case of the wall paintings, the stucco images, and other objects of worship found in the small shrines, the perishable nature of the materials precludes the idea that they were already of considerable age previous to their burial under the drift-sand, while for the documents in Brāhmī script everything, as we have seen, suggests an origin contemporary with the Chinese records.

Last relics of T'ang dominion.

The value of the latter, however, is not limited to supplying us with a firm chronological basis. They also establish the interesting fact that Chinese administrative control and a Chinese garrison within the Khotan region was a reality down to the end of the eighth century. M. Chavannes, in his concluding remarks on the Chinese records from Dandān-Uiliq, has called attention to the remarkable agreement which the dates of these documents present with what the historians of the T'ang period tell us of the end of Chinese sovereignty over the 'Four Garrisons'. The historical synopsis given in a previous chapter has already acquainted us with the fact that it was the year 791 which saw the final abandonment of the whole of Eastern Turkestān to Tibetan invasion, after a long period of isolation and struggle[6]. Thus our Dandān-Uiliq records, closing with the year 790, represent, as M. Chavannes has justly pointed out, the last trace of Chinese political influence under the T'angs in that great region[7].

Historical cause of abandonment.

But may we not, apart from this general historical interest, attribute to these relics of Chinese occupation a further specific significance for the history of the site? The letter which the Chinese commandant of *Li-hsieh* addressed to the king of Khotan in 768 A. D. (Document *A*) has shown us in most authentic form that already at that time the settlement to which Dandān-Uiliq belonged had lost a part of its population, which had retired to the main oasis owing to depredations of bandits[8]. Seeing how seriously local insecurity had affected the settlement in the immediately preceding period, it seems difficult not to connect its final abandonment after 790 A. D. with the great political upheaval of the years next following. The collapse of Chinese authority and the successful Tibetan invasion must have meant for Khotan a period of exceptional trouble; for Tibetan power is, from all that we know, likely to have asserted itself mainly in unchecked raids of large marauding bands, such as had already devastated extensive parts of Western China. In a previous chapter I have endeavoured to show that the constant struggle with the drift-sand of the desert, which the outlying parts of every oasis on the edge of the Taklamakān have to carry on, cannot be maintained successfully except with an effectively supervised system of irrigation and an adequate population. Both conditions are likely to fail during prolonged periods of political trouble, and in no part of the cultivated area are the effects of such failure bound to appear so surely and rapidly as in isolated colonies like Dandān-Uiliq.

Effect of political troubles on irrigation.

For those who are familiar with tracts like the greatest part of the Western Punjab, where cultivation is wholly dependent upon an elaborate system of canal irrigation, it needs no great effort of imagination to realize the effects of prolonged political troubles and insecurity on the cultivated area. As soon as the want of a firm central authority or a diminution in the available supply of labour stopped the effective working of the canals the reduced water-supply would force the cultivators to retire from all outlying lands however fertile by nature. Then

[6] See above, pp. 63 sqq.

[7] See Appendix *A*.

[8] See above, p. 266.

the colonies, which in our own days we have seen springing up with such astonishing rapidity within the areas reached by great irrigation works like the Chināb and Jhelum canals, would lapse within an equally short time into the arid uninhabited waste from which they had emerged.

But apart from the effect which political conditions must thus have upon cultivation in the Khotan region, there are, no doubt, physical changes, too, which gravely threaten it and represent a far more permanent danger. The question of the character and extent of these changes concerns the geographer rather than the archaeologist. Yet it is plain that we must touch on it here, if only in order to explain why the site, which was once capable of supporting a flourishing settlement, was never occupied again after its first abandonment. That the main danger to the maintenance of cultivation does not lie in the advance of the drift-sand itself, but in the reduction or failure of the water-supply needed to cope with it, must be clear from what has already been said. Therefore before we attempt to form any opinion about the physical changes which caused this ancient settlement to be definitely absorbed in the desert, it is essential that we should realize what the source and conditions of its water-supply were. Effect of physical changes.

The preliminary question thus raised presents at the first glance far greater difficulties at Dandān-Uiliq than at any other of the ancient sites I examined. Whether they were 'Tatis' near existing oases, like those between Gūma and Piālma and on the outskirts of the Khotan oasis, or marked ancient terminal oases since completely abandoned, like the Niya River Site and probably Endere, no doubt ever arises as to the source from which they received irrigation. In each case we find the river or stream, which alone could have supplied it, unmistakably indicated by existing hydrographic conditions. In the case of Dandān-Uiliq we are confronted by a series of possibilities, and only a far more exact survey of the whole desert region around than could be attempted by me, including a series of levellings, will enable us to decide the question conclusively. If we merely look at the map the position of Dandān-Uiliq about half-way between the Yurung-kāsh and Keriya rivers suggests the possibility of water having been brought to the site by canals taking off from either. Again, the map shows us that Dandān-Uiliq lies exactly in the natural line of drainage for the rivers of Chīra, Gulakhma, and Domoko, which now lose themselves in the desert due north of these closely adjoining oases. Source of water-supply of Dandān-Uiliq.

Archaeological evidence, so far as I could gather it at the site, does not help us here. Though there were little inequalities of the ground at several places left bare of sand, which looked like possible remains of small embankments for irrigation cuts, it was only in one place, to the north of D. VII (see Plate XXIV), that the trace of such a channel, about 2 feet broad, could be definitely followed for a short distance between the dunes. Its apparent direction was from SSE. to NNW., but seeing the smallness of the channel and the shortness of the distance (less than 200 yards) over which its traces could be picked up, it is manifestly impossible to base upon this any safe conclusion as to the direction and origin of the main canal from which it was fed. Ancient irrigation cuts.

As already indicated in my Preliminary Report, a series of subsequent observations about the gradual receding of cultivation near Chīra, Gulakhma, and Domoko incline me to the opinion that Dandān-Uiliq had received its water by an extension of the canals which down to a much later date irrigated the area now abandoned to the desert north of the above-named oases. The débris-covered site of Uzun-Tati which I surveyed there, and which I think can safely be identified with the *Pi-mo* of Hsüan-tsang and Marco Polo's *Pein*, is proved by unquestionable evidence to have been occupied for at least five centuries longer than Dandān-Uiliq. It lies fully twelve miles, if not more, beyond the northern edge of the Chīra oasis; Irrigation from side of Chīra and Gulakhma oases.

yet I found wild poplars still vigorously growing about the site, and the aspect of the desert beyond suggested that during the summer floods the water of the Chīra river must reach, either in small open channels or as subsoil water percolating close to the surface, much further northward. Dandān-Uiliq lies about 36 miles beyond Uzun-Tati, but I do not think this distance alone would have prevented effective irrigation, seeing that both the Chīra river and those now watering Gulakhma and Domoko come down from a snowy range rising to over 21,000 feet. Being all glacier-fed they must carry a volume of water which, if properly caught and united, would, perhaps, even now suffice to penetrate the desert to the latitude of Dandān-Uiliq. That south of the latter and at no very great distance stretches of ground still receive surface moisture at times was plainly indicated by Turdi's statement, who, when endeavouring to procure fodder for his starving pony, had come within half a day's march, say 6–8 miles, upon depressions with Kumush grass. I much regretted at the time that regard for the heavily-laden camels, and for the men who had suffered from a month of privations in the wintry desert, did not allow me to attempt the long march due southwards on leaving the site; for I have little doubt that the observations to be gathered in that direction would materially aid in deciding the question just discussed.

Other solutions of problem.

In the absence of detailed surveys and levellings even an expert in irrigation engineering might well hesitate to express an opinion as to the possibility of bringing a canal to Dandān-Uiliq from either the Yurung-kāsh or the Keriya rivers as they exist at present. Still less could we hazard an opinion in regard to earlier periods, when the conditions of the rivers, the desert sand, the climate, &c., may have differed materially. Yet so much may be safely asserted that, given river courses approximately the same as they are now, the distance over which any canal taking off from them would have had to be carried could not have been less than the distance between Uzun-Tati and Dandān-Uiliq.

Suggested old course of Keriya river.

The settlement of the question at issue would be greatly simplified if it were possible to accept the assumption put forth by Dr. Sven Hedin according to which the Keriya Daryā flowed in old times in the immediate vicinity of the ruined site[9]. This assumption of the distinguished explorer was, so far as the record of the scientific results of his journey enables me to judge, based solely on the 'traces of old riverine terraces (Uferterrassen)' which he thought to have recognized quite close to the ruins, and on the observation of a present tendency in the Keriya Daryā to shift its course eastwards. The careful survey I was able to make of the whole site renders it certain that by the 'traces of old riverine terraces' can only be meant the shallow depression previously referred to, extending along the eastern groups of ruins in the general direction from SSE. to NNW. In regard to this depression I have already had occasion to adduce evidence showing that it is manifestly due to wind erosion[10]. Numerous similar depressions, bordered by sharply-marked loess terraces and also showing the same general direction, were crossed on the route between the Yurung-kāsh and Keriya Daryā in positions where their frequent recurrence excludes the possibility of a riverine origin[11].

Difference of levels between site

I do not think it necessary to examine in detail whether the tendency of an eastward shifting of the Keriya Daryā can be shown to be of old date, since there is a plain topographical fact

[9] See *Reisen in Z.-A.*, p. 37.

[10] See above, p. 242.

[11] Dr. Hedin himself, with his unfailing eye for all topographical detail, has duly noticed these regularly recurring terraces on the marches west of Dandān-Uiliq; see *Reisen in Z.-A.*, p. 36. He has there, however, also correctly stated the reasons which make a riverine origin of these terraces distinctly improbable. The true explanation of these depressions and bordering terraces has since been furnished by Prof. de Cholnoky's observations on wind-cut trenches as recurring topographical features; see above, p. 242, note 1.

which appears to exclude the possibility of the Keriya river having ever in historical times flowed in the immediate vicinity of the site: I mean the relatively high level of the latter. The hypsometrical heights ascertained by me show that the ancient ground-level of the Dandān-Uiliq site in the immediate vicinity of the depression referred to lies 200 feet higher than the bank of the Keriya Daryā at Kochkar-Öghil, where the river is nearest. The approximate accuracy of this estimated difference of level may the more be relied upon since it agrees exactly with the difference shown in Dr. Hedin's map[12]. Now it seems to me exceedingly difficult to believe that the Keriya river should until eleven hundred years ago—geologically speaking a mere yesterday—have at this point followed a course fully 200 feet higher than the one which it follows now, and which lies in the direct line of its drainage. With nothing but soft loess or drift-sand to resist its course after the debouchure from the mountains the river must have found its bed within the depressed area where it now flows long before any historical period. Within this depression it may shift its bed to the west or east as long as its waters keep to the lowest level, but only engineering works on a large scale could ever force its whole volume to the high ground of Dandān-Uiliq[13]. and Keriya river.

From the levels recorded by Dr. Hedin and myself, it is clear that the ruins of Dandān-Uiliq with the lands once surrounding them are situated on a band of high ground which forms the watershed between the Yurung-kāsh and Keriya rivers, and this fact has undoubtedly its antiquarian interest. Most unlikely as the line of a river-course, this is just the ground which the makers of an irrigation canal would for preference follow, as any properly contoured canal map of the Punjab or similarly irrigated area would show at a glance. By keeping to the line of highest level, a canal not only commands the maximum area of land capable of irrigation, but is also more easily protected against breaches by natural drainage channels and similar disturbing influences. Watershed near Dandān-Uiliq.

In view of the facts just set forth, it is impossible to accept the theory of a great shift of the Keriya Daryā as an explanation why the site of Dandān-Uiliq was deserted and has remained unoccupied ever since. The immediate cause of the abandonment at the close of the eighth century is indicated with sufficient clearness by the antiquarian evidence already specified; but there remains still the question as to the physical causes which have helped to render that abandonment permanent. The question seems to me closely connected with a far larger one, that of the desiccation which is assumed to have gradually been proceeding over large portions of Central Asia within historical periods. The consideration of this problem with special regard to the Khotan region is a task for which properly established archaeological data may be most helpful, but which itself can be dealt with only on a geographical basis, and it must therefore remain outside the scope of this work. The case of Dandān-Uiliq would prove of particular interest in connexion with this problem if conclusive topographical evidence were to confirm the view which I have set forth above, and which appears to me by far the most probable, viz. that a canal fed by one or several of the streams now irrigating the oases of Chīra, Gulakhma, Causes of permanent abandonment.

[12] While my hypsometrical heights for Dandān-Uiliq and Kochkar-Öghil are 4,290 and 4,090 feet, respectively, Dr. Hedin's map shows the elevations of 1,319 and 1,258 metres for his camps in practically the same positions. The corresponding figures in feet are (approx.) 4,327 and 4,127, respectively, giving the identical difference of 200 feet.

[13] We shall see below, in discussing the site of Kara-dong, that so far as antiquarian evidence goes, these shifts of the Keriya river bed seem within historical times to have been relatively very moderate. The ruins of Kara-dong, though probably considerably older than Dandān-Uiliq, are yet only a little over five miles removed from a bed which still receives water at flood times. The area between the ruins and this bed shows plentiful remains of dead forest, while these strips of *kötek* (dead forest), always marking old river-beds, are completely absent between Dandān-Uiliq and the Keriya Daryā, a distance of about twenty-eight miles.

and Domoko, once brought water to the lands of the old settlement. The deserted site of Uzun-Tati, which certainly was occupied down to the thirteenth century, and perhaps even later, furnishes unmistakable proof that the cultivated area dependent on these streams has receded considerably during the last six centuries.

Receding of cultivated area.

That the process, whatever its cause, has been continuing in this area down to quite recent times, is shown by the observations which I had occasion to make near the oases of Gulakhma and Domoko, and which will be found fully detailed in a subsequent chapter[11]. There the homesteads and cultivation of certain villages have, owing to the difficulty of carrying the irrigation water sufficiently far, shifted, within the memory of living men, as much as 6 to 8 miles further to the south. The immediate cause which led to the abandonment of Dandān-Uiliq cannot, for historical and topographical reasons, be supposed to have been identical. Yet it is certain that the crumbling ruins of homesteads which I saw at these deserted village sites of Gulakhma and Domoko, stripped of all materials that could be of use, and the miles of once cultivated ground which the drift-sand of the desert is now slowly overrunning, but on which the lines of empty canals, the little terraces embanked for irrigation, the poplar-lined roads, &c., can still be made out, are the best illustration of the process by which the habitations and lands of old Li-hsieh became finally merged in the desert.

Section VIII.—LIST OF OBJECTS EXCAVATED OR FOUND AT DANDĀN-UILIQ

OBJECTS FROM RUINED SHRINE D. I.

D. I. **D. I. 2. Stucco relief fragment.** White chunam, coloured. Replica D. I. 94. Portion of vesica and figure to level of L. p. elbow. Hair black, robe and outer ray blue, inner portion of vesica red. 3" high, 1¾" wide.

D. I. 7. Stucco relief fragment; ornament. White chunam, coloured pink, green, and grey (?). Portion of lotus-petal border of aureole. Cf. D. II. 24. Length 7$\frac{5}{16}$", width 1½". Hard, scored at back, fragment of grass adhering.

D. I. 8. Stucco relief fragment; ornament; Replica D. I. 7; coloured pink, blue, and black (?). 7½" x 1½".

D. I. 10. Stucco relief fragment. Grey chunam, coloured. Traces of pink and blue. Border of aureole (vesica?) similar to that of D. I. 42. Within this border, on field, portion of an inner border consisting of jewels composed of a centre, ovoid cartouche, round which two fillets, and a row of beads. Size of complete jewel about 2½". See Plate LVII.

D. I. 11. Stucco relief fragment; standing Buddha. Feet missing; fitted with head, D. T. 06. Grey chunam, no colour. Long red robe, green undergarment, open low at neck. R. p. hand raised to centre of breast, palm outwards. L. p. arm depending straight at side holding lightly loose end of robe. Cast bears evidence of having been worked upon with modelling tool. Height 5½", width 2⅜".

D. I. 14. Stucco relief fragment; two top joints of finger, life-size. Red clay, with traces of whitewash. Filbert shape nail, trimmed short below tip of finger. Cf. E. I. 4. Friable.

D. I. 18. Stucco relief fragment; head, female (Gandharvī). Pink chunam, traces of black or dark colour. Hair divided over temples and flowing backwards and outwards behind ears, also gathered in high loose knob on top of head and tied with narrow band. Ears elongated. Nimbus. L. p. hand raised to level of hair knob, knuckles touching nimbus, grasping end of rosary-like garland, of which two beads remain. Replica D. I. 89. Height 1⅞", width 2⅝".

D. I. 24. Stucco relief fragment; standing Buddha. Lower half; feet missing. Grey chunam, traces of colour. Replica D. I. 11. Height 3½".

D. I. 25. Stucco relief fragment; ornament. Chunam, traces of colour. Roots of two rows of flames. Contour of cross-section almost semicircular. Deeply reeded into six divisions, also two detached tongues of flame, the upper portion of each waved almost to a right angle with

[11] See below, chap. XIII. sec. ii.

lower part. Cf. D. II. 24. Large piece $5\frac{1}{2}'' \times 3\frac{1}{4}''$. Detached flames, respectively $2\frac{1}{8}''$ and $2\frac{3}{4}''$.

D. I. 26. Stucco relief fragment; head of Buddha. Chunam, traces of colour. Replica D. I. 55.

D. I. 29. Stucco relief fragment; ornament. Pink chunam, traces of colour. Portion of lotus-petal, finely modelled; front surface convex in direction of mid-rib of petal; under surface concave, showing impression of fine canvas backing and remains of canvas itself. Width of petal 3".

D. I. 30. Stucco relief fragment; ornament. Pink chunam, portion of leaf and background. Leaf bright green, background bright red. Replica D. I. 41.

D. I. 35. Stucco relief fragment; standing Buddha. Grey chunam, without colour. Upper portion; head missing. Replica D. I. 11. Height 3".

D. I. 36. Stucco relief fragment in two pieces; ornament. Red chunam; portion of lotus-petal. $2\frac{7}{8}''$ by $2\frac{5}{8}''$.

D. I. 40. Stucco relief fragment; leaf, in two pieces; chunam. Replica D. I. 41.

D. I. 41. Stucco relief fragment; leaf. Grey chunam, coloured dark; from large aureole. Replica D. I. 012. Length $3\frac{1}{2}''$.

D. I. 41. a. Stucco relief fragment; portion of Buddha torso; grey chunam. R. p. hand and arm; traces of red on robe. Replica D. I. 11. Height $2\frac{1}{2}''$.

D. I. 42. Stucco relief fragment; ornament. Grey chunam, traces of colour. Border and portion of field of aureole (prob. vesica); border consists of half-round moulding modelled to represent overlapping lotus-petals ranged horizontally, and bound with cincture at intervals of about $1\frac{1}{2}''$; alternate sections coloured pink and green (?). Field, depressed below border, has conventional chrysanthemum leaves moulded upon it, variously coloured. Cf. D. I. 10. See Plate LIV.

D. I. 43. Stucco relief fragment; female head (Gandharvi). Head and nimbus. Has been coloured pink all over; over pink, blue on nimbus; black hair. Replica D. I. 18.

D. I. 44. Stucco relief fragment; ornament. Pink chunam, traces of pink colour. Jewel, ellipsoid centre surrounded by two fillets and row of beads. At one end ribbon or cord tied in bow. Cf. D. I. 10; E. I. 1. $2\frac{7}{8}''$ by $1\frac{3}{4}''$. See Plate LV.

D. I. 52. Stucco relief fragment; ornament. Pink chunam, coloured. Leaf and portion of background. Fillet moulding to R. p. of leaf, and portion of inner background. Leaf blue; background pink; inner background brown pink. Replica D. I. 012.

D. I. 53. Stucco relief fragment; portions of grape vine scroll. Replica D. I. 114. One piece green, other pink. See Plate LIV.

D. I. 55. Stucco relief fragment; head of Buddha. Hair (modelled) black; nimbus red; face pink; robe dark colour; ears elongated. Replica D. I. 71. Height $1\frac{7}{8}''$, width $2\frac{1}{2}''$. D. I.

D. I. 60. Stucco relief fragment; ornament. Traces of pink. From moulding of beads springs in graceful curves a single row of lotus-leaves. Probably portion of border of aureole. Replica D. I. 74. Length $3\frac{1}{2}''$, width $1\frac{1}{2}''$.

D. I. 65. Stucco relief fragment; head of Buddha. Pink chunam, traces of pink colour. R. p. side of nimbus broken away. Replica D. I. 71.

D. I. 69. Stucco relief fragment; seated Bodhisattva. White chunam, coloured. Replica of D. I. 94, but colours: hair black, robe brown, seat pink, inner part of vesica pink, outer ray emerald green. The whole appears to have been on background of indigo. Point of vesica broken. Extreme height $3\frac{3}{8}''$, width 3".

D. I. 70. Stucco relief fragment; seated Bodhisattva. White chunam, traces of colour. Replica D. I. 94. Half of head cut away, and much of figure. Hair black, robe blue or dark green, seat probably white, also inner vesica; outer ray purple brown. Background red. This fragment shows three superimposed layers of plaster: (1) Moulded vesica and figure scored at back to give tooth for fixing; (2) a layer less than $\frac{1}{8}''$ thick, of slightly more coarsely ground plaster which forms background; (3) a coarse layer nearly $\frac{1}{4}''$ thick. All are quite hard. On back of fragment, i. e. on back of third layer, fresco, representing on pink ground portion of upper arm, forearm (raised) and R. p. side of figure. These portions nude, save for scarf twisted round arm just above elbow. To R. p. leaf in white. $4\frac{1}{2}''$ high, $3\frac{7}{8}''$ greatest width.

D. I. 71 + D. I. 010. Stucco relief fragments; standing Buddha. Replica D. I. 11. Feet missing. Pink chunam, traces of red on nimbus; robe dark colour. Saṅghāṭi red.

D. I. 72. Stucco relief fragment; ornament (two pieces). Portion of outer moulding of aureole, half-round section. Overlapping lotus-petals ranged horizontally; tied at intervals of about $1\frac{1}{2}''$; alternate sections coloured red and blue. Cf. D. I. 42. Length 6", diam. $1\frac{1}{2}''$. Surface friable.

D. I. 74. Stucco relief fragment; ornament. White chunam, traces of colour. Similar to D. I. 60, but within bead moulding a fillet moulding about $\frac{3}{8}''$ wide. Length $4\frac{5}{8}''$, width 2". See Plate LVII.

D. I. 80. Stucco relief fragment; ornament. Chunam, traces of colour (pale blue). Fragment of crescent-shaped ornament, with broken portion standing between the horns of crescent. Width 3", depth 2". See Plate LVII.

D. I. 84. Stucco relief fragment; Buddha. Upper portion only. Grey chunam. Traces of colour. Head missing. Replica D. I. 11. Height $2\frac{1}{8}''$, width $2\frac{1}{2}''$.

D. I. **D. I. 87. Stucco relief fragment**; ornament. Chunam, coloured. Tongues of flame, broken at second wave, coloured respectively emerald green, bright pink, and bright blue. Reeded into five divisions. Cf. D. II. 24. 4″ by 1⅞″. See Plate LVII.

D. I. 89. Stucco relief fragment; head, female (Gandharvī). Pink chunam, coloured. Nimbus dark grey-blue, hair black, flesh pink. Replica D. I. 18. Height 1⅝″, width 1¼″.

D. I. 90. Stucco relief fragment; female head (Gandharvī). Pink chunam, traces of colour. Hair black, nimbus dark grey or green. Three beads of garland. Replica D. I. 89. Height 1¾″, width 2¼″. See Plate LIV.

D. I. 94. Stucco relief fragment; seated Bodhisattva, on half-open lotus. Flowing robe concealing feet. Hands resting in lap. Face and hair indistinct. Top-knob. At back, extending upwards from knees a vesica (broken above head). Centre portion, consisting of radiating lotus-petals, was coloured pink. Outer ray and robe of figure show traces of a dark colour, probably blue. Lotus seat white. The whole, an ornament from aureole round large figure; moulded separately and affixed to background by chunam. Replica D. I. 69. 3⅜″ from top-knob to below seat; 2⅞″ broadest part of vesica.

D. I. 95. Stucco relief fragment; ornament. Replica D. T. 011. Pink chunam. Portion of two rows of lotus-petals overlapping, the centre of outer petal lying under point of second petal. Convex in direction of mid-rib. Concave below (at back); impression of fine canvas backing. Cf. D. I. 29. Width of fragment 4″, depth 3″.

D. I. 96. Stucco relief fragment; ornament; leaf. Pink chunam, traces of dark colour. Replica D. I. 40.

D. I. 97. Stucco relief fragment; head of Buddha. Replica D. I. 55. Black hair, pink nimbus.

D. I. 99. Stucco relief fragment; seated Boddhisattva. Replica D. I. 94, but with piece of background. Hair black, flesh white or pink, robe purple brown, lotus seat white, inner part vesica bright pink, outer ray emerald green. Traces of red on background. Raised fillet moulding on background to R. p. of figure; on side furthest from figure trace of blue. 4″ across length of vesica, 6⅝″ greatest width. Surface soft. See Plate LV.

D. I. 105. Stucco relief fragment; head of Buddha. Replica D. I. 71. Much worn.

D. I. 107. Stucco relief fragment; portion of lower part of standing Buddha. Robe bright pink, Saṅghāṭi grey. L. p. hand hanging straight at side lightly grasping loose end of robe. Hand pink. Feet missing. Replica D. I. 11. Height 3¼″.

D. I. 110. Stucco relief fragment; Gandharvī figure rising from centre of lotus. Draped in close-fitting vest, open low at neck, fastened by jewel (?). Loose girdle of beads at hips. Lotus is circular and completely modelled except where covered by figure. Garland of beads supported in upraised hands. Lotus coloured pink. Head and hands missing. Replica D. I. 014. Height 3½″, width 3″. Broken.

D. I. 114. Stucco relief fragment; ornament. Portion of curved border consisting of an inner part 1½″ broad, a raised fillet ⅜″, and an outer part in same plane as inner part, about 1″ wide. On inner part, in low relief a long ᔕ scroll, at each end a half vine leaf. This arrangement appears to be repeated on each side of scroll but reversed. Between half leaves in each case a bunch of grapes. Colouring in alternate sections of blue and pink. Fillet quite plain. The outer part appears to have fragment of chrysanthemum leaf. Cf. D. I. 53. 3⅝″ × 3⅝″. See Plate LVII.

D. I. 118. Stucco relief fragment; head of Buddha. Replica D. I. 71. Much defaced.

D. I. 119. Stucco relief fragment; standing Buddha (three pieces). From neck to middle. Chunam; traces of dark colour (black ?) on robe. Replica D. I. 84. Height 2½″, width 2⅜″.

D. I. 127. Stucco relief fragment; head of Buddha. Replica D. I. 71. Hair black; nimbus with traces of red. Height 1⅝″.

D. I. 128. Stucco relief fragment; feet of standing figure in relief. Pink chunam, traces of pink paint. Feet rest upon base formed of half-open lotus. Replica D. I. 03. Height of fragment 2½″, width 1¾″.

D. I. 03. Stucco relief fragment; feet of standing figure. Replica D. I. 128. Pink chunam, traces of pink paint.

D. I. 04. Wooden post; rectangular, upright, painted. Back recently sawn. Front: standing Buddha on blue lotus with yellow centre, seed-cells indicated by rings of red. Surrounding centre of flower, ring of white dots. Entire figure three-quarter to R. p. Pose, R. p. hand to breast palm out, L. p. hand pendant holding robe (?). Feet close together inclined to R. p. Head inclined slightly down, eyelids drooped. Tilaka red. Contour lines of flesh red, flesh light. Robe deep purple bordered white. Contour lines black. Under-garment white or pink unusually high above ankles, ornamented (?) with broad vertical stripes which may represent vertical folds. Background dark pink. Hair deleted and all around head and shoulders. Below lotus, separated by white line, a band of deep purple, on which three rosettes in white. R. p. edge of post, deep purple or black, with similar rosettes at intervals. Colours remarkably preserved where visible. Wood perishing from damp at edges, but section shows it perfectly sound inside. Appears to be part of an upright post, pilaster or jamb, as three sides were evidently visible. Length 26″, width 4½″, thickness 1½″ (originally 4¼″ before removal of plain portion). See Plate LXV.

D. I. 08. Stucco relief fragment; standing Buddha. Head and feet missing. Replica D. I. 11, but robe red, Saṅghāṭi green, flesh pink. Height $5\frac{1}{4}''$. Hard.

D. I. 09. Stucco relief fragment; standing Buddha. Replica D. I. 11. Lower half; feet missing. Height $3\frac{1}{4}''$.

D. I. 011. Stucco relief fragment; standing Buddha. Replica D. I. 11. Torso.

D. I. 012. Stucco relief fragment; leaf. Replica D. I. 41. Portion of field of vesica. Conventional chrysanthemum leaf (portion), coloured red; ground green. $3\frac{1}{4}''$ D. I.
long, $1\frac{7}{8}''$ wide. See Plate LVII.

D. I. 013. Stucco relief fragment; ornament. Lotus petal springing from row of beads. General shape of surface flat; under side concave. Depth from beads to point $2\frac{3}{8}''$, width $2\frac{3}{8}''$.

D. I. 014. Stucco relief fragment; Gandharvī figure. Replica D. I. 110. Scored lightly on back with tool. Much defaced and friable.

OBJECTS FROM RUINED SHRINE D. II.

D. II. 2. Painted wooden panel; rectangular; painted probably on both sides, but one side only recognizable. OBVERSE: Figure kneeling on L. p. knee. R. p. hand resting on R. p. knee, holding what appears to be a band. Above the hand appears a sheaf of some objects (arrows?). Head quite indistinct, but seems to be in profile. Nimbus. Sleeves of garment loose. Folds and all contour lines black. Very free and suggestive in treatment. REVERSE: Subject quite indistinguishable. Adhering to surface fragments of paper manuscript in upright Gupta characters; five to six lines of thick black writing, but only a few detached Akṣaras in each legible. Wood perished; shows ravages of insects. $4\frac{7}{8}''$ wide, $7\frac{1}{4}''$ high, $\frac{7}{8}''$ thick.

D. II. 3. Stucco relief fragment, coloured. Red clay, no fibre. Replica of D. II. 12, colouring similar, but lines of drapery in black round neck showing loose end thrown over L. p. shoulder distinct. Friable. $3\frac{5}{8}''$ high, $2\frac{1}{2}''$ wide.

D. II. 4. Painted wooden panel; rectangular. OBVERSE: Two seated figures, apparently nude to middle, one on either side of a central figure (?). Arms of side figures akimbo; hands probably resting on thighs. Flesh pink, outlined black. Figure to L. p. head, three-quarter face, inclined to R. p. and slightly down. Necklet. Nimbus. Very much damaged. Centre unrecognizable.

REVERSE: Apparently three figures. To L. p. seated figure inclined to R. p. L. p. hand raised to breast. Towards centre appears representation of basketwork (?), perhaps portion of seat. Much deleted. Lower corner broken. Wood perished; general condition points to damage by damp. $12''$ long, $4\frac{1}{8}''$ high, $\frac{1}{4}''$ thick.

D. II. 5. Stucco relief fragment, in several pieces. Replica D. II. 101. White chunam, perished. No colour. Brittle. Height $6\frac{5}{8}''$.

D. II. 6. Stucco relief fragment; seated Bodhisattva on pedestal. Red clay with fine fibre (perished), coloured. Full robe covering feet. Hand resting in lap. Ears elongated. Top-knob. Ovate vesica (top broken) from knees. Whitewash over the whole piece, red contour lines on L. p. ear. Hair black. Vesica flat, divided into concentric bands by black lines. Portion behind head tinted pale green, shaded to white at outer edge; next a band of pink similarly shaded. Then a narrow band white; outside this a broad band divided by radiating lines D. II.
in black: outside this again a narrow plain band. Friable. $3\frac{7}{8}''$ high, $3\frac{3}{8}''$ wide. See Plate LV.

D. II. 8. Stucco relief fragment. Replica D. II. 101. White chunam.

D. II. 10. Stucco relief fragment; seated Bodhisattva. Chunam, coloured. Seated figure on lotus, feet concealed by robe. R. p. hand raised, palm outwards; L. p. hand hangs over thigh, upon which forearm rests. Vesica rises from knees, behind figure. Nimbus. Robe and outer edge of vesica pink. Flesh light pink. Field of vesica dark (blue?) Nimbus black (?). Height $4\frac{7}{8}''$, width $4''$. See Plate LIV.

D. II. 11. Stucco relief fragment; standing Buddha. Replica D. II. 101. White chunam, coloured. Head, feet, and R. p. arm missing. Colouring: robe, pink; Saṅghāṭi, grey; flesh, pale pink; background, emerald green. Back of background scored, portion of twig and portions of grass adhering in scorings. Height $4''$, width $2\frac{3}{8}''$.

D. II. 12. Stucco relief fragment. Coloured. Replica D. II. 3. Red clay mixed with fine vegetable fibre. Seated Bodhisattva. Robe covering feet. Hands in lap, L. p. resting in palm of R. Ears elongated. Top-knob. Pointed oval vesica rising from knees. Nimbus at head. Hair and nimbus black, eyes outlined in black. Flesh white. Vesica emerald green. Soft. Height $3''$, width $2\frac{1}{4}''$. See Plate LIV.

D. II. 13. Fresco fragment on stucco with inscription. Removed from lower part of frescoed wall. Coarse mixture grey clay, straw and grass; broken in two places. Feet, life-size, of standing figure (Buddha?) resting upon open lotus. Between feet, and extending to R. p. outside foot, single line inscription in cursive Brāhmī characters, written in black with brush. Inscription between feet read by Dr. Hoernle, *bya ja ga ri ja sho bha yi ? re mā pyā dā ? s(?) ? ?*. In centre of lotus, seed-cells indicated by red circles. Below lotus sketchy lines in black. Lotus-petals pink, shaded by washes of darker pink and outlined Indian red. Feet pink, outlined bright red. Very friable. $20'' \times 7\frac{1}{2}''$. See Plate LIX.

D. II. D. II. 16. **Painted wooden panel**; rectangular; traces of painting on one side. Figure of Gaṇeśa seated on dice-pattern floor or mat in two colours. Dhōtī round hips, upper part nude, excepting for thin band of drapery on L. p. arm, possibly ornaments. L. p. hand raised to breast. Three-quarter face directed to R. p. Trunk curled to L. p. Jewelled diadem on head. Nimbus. Painting much deleted. Wood perished; very soft. 8½" high, 3¼" wide.

D. II. 17. **Stucco relief fragment.** Replica D. II. 101; standing Buddha; upper portion only. Chunam, coloured. Robe pink, flesh light pink, nimbus green. Back of background scored; fragments of grass adhering. Height 3", width 3¼".

D. II. 19. **Stucco relief fragment.** Replica D. II. 101; standing Buddha; torso. Chunam, traces of pink. Height 2½".

D. II. 20+69. **Stucco relief fragment.** Replica D. II. 101. Head of Buddha, fitted with body D. II. 69. Chunam; traces of green on nimbus, pink on robe, black on background. Height 1½", width 1¾".

D. II. 21. **Painted wooden panel**; rectangular, rounded at one end. Painted one side. Two seated figures on lotuses (Bodhisattvas?), heads inclined towards each other, three-quarter face. Elongated ears slit. Hands resting in lap. Robe made with semblance of sleeved garment, but also with end thrown over L. p. shoulder. Behind each a vesica, and at head a nimbus. Hair and eyes black, all other lines red. Vesica, L. p., green, the other doubtful. Much faded and damaged in upper part. Wood perished and soft. 5½" high, 10½" long, ⅜" to ½" thick.

D. II. 22. **Stucco relief fragment.** Replica D. II. 101. White chunam, coloured. Head missing. Robe pink; Saṅghāṭi grey; lotus and flesh white or pink. Remains of background, dark grey. Material of background, coarse hard chunam, much scored at back, with fragments of long grass adhering in scorings. Height about 5", width over all 3⅝".

D. II. 24. **Stucco relief fragment**; ornament in several pieces. White chunam, coloured. Portion of lotus-petal border of aureole, with one row of lower parts of flame-tongues attached to outer edge. Whole piece curved, forming segment of Caitya-shaped aureole. Overlapping petals divided at every third row by double cincture, and sections so formed coloured alternately red and green (or blue). Flames also alternately coloured in the same way, and grooved or 'reeded.' The several strata in this fragment do not adhere well. Hard. Length 7⅛", width 2½".

D. II. 26. **Stucco relief fragment.** Replica D. II. 101. White chunam. Portions of two figures of Buddha on common background; details of each figure as D. II. 101. traces of red on robes. Background dark grey. Soles of feet of R. p. figure level with elbows of L. p. figure. Space between ankle of one and robe at elbow of other 1⅟₁₆". R. p. figure, upper half missing. L. p. figure, feet gone. Deep scorings at back. Hard. 6½" high, 4½" wide.

D. II. 31. **Stucco relief fragment.** Replica D. II. 101. Chunam, traces of red on robe. Back scored. Height 6¼".

D. II. 34. **Stucco relief fragment**; portion of aureole in three pieces. White chunam, coloured. Border of lotus-petals; cf. D. II. 24. On grey background two Bodhisattvas, head of lower being level with knees of upper, and about 1½" away (cf. D. II. 26). Colouring as D. II. 17. To L. p. of upper figure near feet a small open lotus. Height 10½", width 8⅜". See Plate LIV.

D. II. 36. **Stucco relief fragment.** Replica D. II. 101; standing Buddha. Chunam, traces of colour. Figure against background on which to L. p., at level of shoulder, an open lotus 1⅛" diam. Height of figure, including lotus under feet, 6½"; width of fragment 4⅞".

D. II. 41. **Stucco relief fragment.** Replica D. II. 68. Feet and lotus missing. White chunam, coloured. Colouring as D. II. 68, but hair, traces of black; nimbus pale green. Background scored at back. Height over all 6", width 3⅞".

D. II. 53. **Stucco relief fragment.** Replica D. II. 101. Head of Buddha. Chunam, coloured. Nimbus light green. Height 1⅝", width of nimbus 1⅝".

D. II. 55. **Stucco relief fragment**; ornament. White chunam, traces of colour, blue, black, and red. Portion of lotus-petal border of aureole, with roots of three tongues of flame attached to its outer edge. In the flames the feet and portion of Saṅghāṭi of figure (Buddha?). Cf. D. I. 87, D. II. 89, D. II. 24. Height 4", width 4". See Plate LV.

D. II. 57. **Stucco relief fragment.** Replica D. II. 101. Chunam, coloured. Height 6⅜", width 3¼".

D. II. 60. **Stucco relief fragment.** Replica D. II. 101. White chunam, traces of pink on robe. Back of background scored; fragments of grass adhering. Height 6⅝", width at widest 3¾".

D. II. 61. **Stucco relief fragment.** Replica D. II. 101. White chunam, coloured. Portion of torso, Buddha, with R. p. forearm raised, hand uplifted palm outward, L. p. arm (portion) hanging. Robe red, hand pink. Width across arms 2¼".

D. II. 62. **Stucco relief fragments.** White chunam. **a.** Buddha, head missing. Replica D. II. 101. **b.** Head of Buddha, no colour. Replica D. II. 101.

D. II. 68. **Stucco relief fragment.** Replica D. II. 101. White chunam, coloured. Head missing. Flesh pink or white; Uttarāsaṅga bright pink; Saṅghāṭi grey (?). Background dark grey. Lotus white. Height about 5".

D. II. 71. Stucco relief fragment; Buddha standing on half-open lotus. White chunam. No colour. Replica D. I. 101, but superior modelling. Neck short. Nimbus. Knees slightly bent; L. p. foot advanced. Surface much worn. Extreme height 7½", width 2¼".

D. II. 74. Stucco relief fragment. Replica D. II. 101; standing Buddha (in three pieces). White chunam, coloured. On background to R. p. of figure a device in red and white resembling a lotus. See Plate LIV.

D. II. 76. Stucco relief fragment. Replica D. II. 101. Chunam. Figure of Buddha; feet missing. Robe pink; flesh light pink; nimbus light green; traces of red contour lines on face; black eyebrows, eyelashes, and pupils. Eyes wide open. Figure adhering to large slab of background coloured dark. Roughly scored at back. Height 7", width 5¾".

D. II. 79. Painted wooden panel; oblong, rounded at ends. OBVERSE: Dancing figure (Gandharvi?). Much deleted. L. p. breast, shoulder, and arm; neck, part of face, hair and sash visible. Traces of L. p. leg. Head thrown back, quantity of hair flowing backwards (probably to assist in expression of vivacious movement). L. p. hand raised grasping loop of sash, which is poised over head and probably is held by other end in R. p. hand; a loose end falls from L. p. hand appearing to swing by dancing movement. Eye large. Two parallel curved black lines forming double point in forehead, probably an ornament. Ear pierced and ringed about half-way up. Necklet appears to consist of three large jewels on hoop, and pendant. Ornament on upper arm, consisting of band and large semicircular ornament. Bracelet near hand. Figure so far as visible appears to be nude (excepting ornaments) and has traces of pink, shaded by darker washes. Sash ochre. All outlines black. Background light. Much freedom in execution.

REVERSE: Much deleted. Seated figure with head ornament; black hair; nimbus; scarf round L. p. upper arm. Three-quarter face to R. p. Wood perished by damp and very soft. 8½" high, 4½" wide, ½" to ⅛" thick. See Plate LXVI.

D. II. 81. Stucco relief fragment. Replica D. II. 101. Standing Buddha. Chunam coloured. Feet missing. Painted on background to lower R. p. a flower (lotus?), red with white tipped petals, stem and two leaves white. Red robe, grey Saṅghāṭi; pink flesh; hair, eyebrows, lashes, and pupils, black; nimbus green; background black (?). Height 6¼".

D. II. 89. Stucco relief fragment; ornament. White chunam, no colour. Triple row of tongues of flame (rising from border of large nimbus?); each tongue is cast separately and affixed singly to background, which is in three planes, each plane slightly in advance of the one above it, so that the tongues overlap about one-third of their length. Each tongue is about 3¼" long and ⅞" wide at root. It has three waves in its length and two deeply scored lines running from root to point. Lowest row missing excepting small portion of one tongue. Cf. D. II. 24, 55. Width about 5½", depth 4¾". See Plate LV.

D. II. 101. Stucco relief fragment, in two pieces. White chunam, traces of red colour. Buddha (?) standing on half-open lotus, L. p. foot advanced, R. p. drawn back. R. p. knee slightly bent. R. p. hand raised to level of chest, palm outward. L. p. hand slightly raised supporting loose end of long flowing upper robe. Ears elongated. Eyes wide open. Top-knob. Nimbus. Figure with lotus and nimbus moulded in one piece, and affixed to background (part of which remains attached). Cf. D. II. 26. Height from below lotus to top of nimbus 6⅝", width at shoulders 2$\frac{1}{16}$".

D. II. 01. Stucco relief fragment. Replica D. II. 101. Head. Chunam, traces of colour. Height 2½".

D. II. 02. Stucco relief fragment. Replica D. II. 101. Head. Chunam, traces of colour. Height 1½".

D. II. 03. Painted wooden panel. Rough panel, uneven thickness, painted one side. Two seated figures, each holding an infant which appears to be swathed, and has a headband. Drapery contours and flesh contours in red. Eyes of all figures, hair of seated figures, headbands of infants black. Seated figures have haloes round heads. Painting badly defaced. Between the two figures, running transversely to top, paper on which three lines of thick upright Gupta characters can be traced. On top edge beginning of five lines in same characters. Paper with writing a subsequent addition, as it covers lower part of left figure. Perished and very soft. 8½" wide, 5⅞" high; thickness varies from 1⅛" to ⅛".

D. II. 04. Stucco relief fragment. Replica D. II. 101. Standing Buddha. White chunam, coloured.

D. II. 05. Stucco relief fragment. Replica D. II. 101. Standing Buddha. White chunam, coloured. Upper half of figure only; L. p. arm missing.

D. II. 06. Stucco relief fragment. Replica D. II. 101. Standing Buddha. White chunam, coloured. Head and feet missing.

D. II. 07. Stucco relief fragment. Replica D. II. 101. Standing Buddha. White chunam, traces of colour. Lower half of figure only. Height 3⅜".

D. II. 08. Fresco fragment on stucco. Grey clay mixed with coarse straw and grass; stiffened with wood. Surfaced with same clay mixed with finer grass and fibre; distempered. Subject of painting: Seated Buddha, dark red robe, buff Saṅghāṭi, pink flesh, red contour lines, black hair. Head inclined slightly to R. p. Hands upraised in front of breast in Nyāyamudrā attitude. Feet crossed and bare. Cushion or mat as seat. Oval vesica, white shaded to buff, then black line, border of white, and second black line. Vesica-shaped nimbus, pink. To R. and L. of figure, portions of adjoining large aureoles. In front and below feet an inscription in black, consisting of seven characters in cursive Brāhmī. Below this, conventional tufts of grass and foreground outlined black and red. Ground white, to pink behind vesica. Stucco friable. Width at lower

D. II. part of fragment 18"; at upper part $6\frac{1}{4}$" (approx.); height 15", thickness $1\frac{7}{8}$". See Plate LVIII.

D. II. 09. Relief in stucco; seated Buddha or Bodhisattva. Grey clay mixed with straw, grass, and fine fibre. Male figure seated on rectangular seat, feet resting on flat footstool, the whole on a piece of board. R. p. arm hangs straight to elbow, where it is broken off. L. p. arm has forearm bent at nearly right angle to upper arm and directed forward. Wooden cores to both arms. Hand missing. Knees very wide apart; feet nearer together. Head missing. R. p. shoulder and arm bare, pink.; robe, which passes under R. p. arm and over L. p. shoulder, coloured red, with narrow border of light green at lower edge, and blue at breast. This latter may be lining. There seems to be a second garment (or lining?) across breast coloured green. Below robe, at ankles, is the Saṅghāṭi coloured red-brown with white border. Feet bare, pink. Footstool green (?). Seat light green in front, pale pink on sides and top; top divided into squares by fine, dark lines. On background, visible under R. p. arm, concentric bands and lines of various colours appear, perhaps part of aureole. Whole relief practically in the round. Very friable. Extreme height from top of board to neck 14", width to outside of knees $7\frac{1}{2}$", waist $2\frac{1}{8}$", width at shoulders $5\frac{1}{4}$", front to back of seat $5\frac{1}{4}$"; wooden board $\frac{3}{4}$" thick. See Plate LIII.

D. II. 010. Painted wooden panel; oblong. Five figures. *First* from R. p. three-quarter face to L. p., female seated or kneeling at a horizontal loom, evidently engaged in exhibiting the various processes of weaving. In upraised L. hand a brush or weaving comb (red). R. hand seems to be lowered to front end of loom, and may be manipulating a shuttle, very indistinctly shown in extreme corner of panel. Some of the outlines (red) of loom, with warp stretched across, clearly distinguishable; spindles and winding frame may be made out between this and second figure. Garment yellow tunic, short-sleeved. Red Uttarīya coming from head encircles R. arm. Hair black and long; upper eyelids outlined black, elongated at outer corners. Head nimbate. Between this figure and second a vessel containing cocoons (?).

Second figure resembles first but seems to have between legs an object like a kettle-drum; hands appar. inside it. *Third.* Seated front, cross-legged; head three-quarters to R. p.; wide, short-sleeved tunic (yellow); on legs, top boots (mocassins) apparently of leopard skin. Coat long-sleeved, grey; Uttarīya (red) over arms and R. p. leg; girdle. Flesh light; small, thin moustache and imperial, black. Eyes cast downwards, brows, eyelashes, and pupils black. Hair long, wavy, on shoulders, black. Diadem. Nimbus yellow. Four arms, R. p. front, at breast holding Vajra or cup; back, upraised holding baton; L. p. front at knee holding an object resembling a large mulberry; back, upraised holding trowel-shaped implement. This figure resembles in almost every detail the four-armed deity in D. x. 4, excepting slight differences in colour, and presence of red Uttarīya in this representation. Character of head identical with that in D. x. 4. *Fourth.* Kneeling three-quarters to R. p. on both knees, playing mandoline (*rabāb*), which has four large string keys at head. Sounding-board pierced by two Triśūla-shaped openings. The instrument is held precisely as is the modern mandoline, and with R. hand player uses a large curved plectrum. Close fitting tunic, white, low at neck, short at knees. Yellow undergarments from neck to ankles, and to wrists. Uttarīya (red) over head (?), round arms to ground. Flesh rather dark. Small decoration on forehead. Nimbus. Usual outlines and contours. *Fifth.* Three-quarters to R. p. kneeling on both knees. Short-sleeved tunic (yellow) fitting close to body. Uttarīya red, from head round arms to ground. Both arms raised to breast, holding in each a small white object connected (?) by a cord, perhaps castanets. Nimbus yellow. Wood perished; warped. Length $19\frac{1}{2}$", breadth $5\frac{3}{8}$", thickness $\frac{1}{2}$". See Plate LXVII.

D. II. 011. Grass broom. Constructed as follows: Head of suitable grass, having stalks about 16" long, are laid side by side with heads level. About 4" of the lower end of each stalk is then bent at right angles to the other portion and plaited round the next two right angle pieces in succession, the remaining end being then bent upwards to prevent slipping. When a continuous strip of about 25" is thus formed (the upper parts being, of course, unjoined) it is rolled up tightly, and bound round with twisted grass to prevent unwinding. Thus, the feathery ends being brought into a bunch, form a convenient birch-like brush or broom. Very brittle. See Plate LXXIII.

D. II. 012. Stucco relief fragment. Replica D. II. 10.

D. II. 013. a. Small cotton bag containing calcined fragments of skull-bones, jaw-bones, and molar and incisor teeth; tied round at the mouth with strip of cotton cloth.

b. Terra-cotta fragments, plain brown (3).

c. Small fragments of copper vessel (2).

MANUSCRIPT FINDS IN RUIN D. III.

D. III. **D. III. 1. MS. on paper**; Brāhmī; No. 7. One complete leaf. See Note iii. Plate CX.

D. III. 2. MS. on paper; Brāhmī; No. 1. One large fragment; left side of leaf, but number lost. See Note i.

D. III. 3. MS. on paper; Brāhmī; No. 2. Two small fragments; from left and right sides of leaf, the former numbered 96. See Note ii.

D. III. 4. MS. on paper; Brāhmī; No. 3. Four small fragments; from middle of a leaf. See Notes i., ii.

D. III. 5. MS. on paper; Brāhmī; No. 4. Numerous small fragments. See Note i.

D. III. 6. MS. on paper; Brāhmī; No. 5. Numerous small fragments. See Note i.

D. III. 7. MS. on paper; Brāhmī; No. 6. One large fragment; from left side of leaf, with number 132. See Note i. Plate CVII.

D. III. 8. MS. on paper; Brāhmī; No. 8. One large fragment; from right side of leaf. Also one small fragment. See Note i. Plate CVII.

D. III. 9. MS. on paper; Brāhmī; No. 9. Two small fragments; from middle of leaf. See Note iv. Plate CIX.

D. III. 10. MS. on paper; Brāhmī; No. 10. Numerous small fragments of (apparently) one leaf. See Note i.

D. III. 11. a. MS. on paper; Brāhmī; No. 11. Forty-four very small fragments of (appar.) one leaf. See Note i.

D. III. 11. b. MS. on paper; Brāhmī; No. 12. Forty-eight very small fragments of one leaf. See Note i.

D. III. 12. MS. on paper; Brāhmī; No. 13. One fragment of a document. See Note v. Plate CX.

D. III. 13. a. MS. on paper; Brāhmī; No. 14. Three large fragments; left sides of leaves, numbered 6, 16, 17. See Note vi. Plate CVIII.

D. III. 13. a. MS. on paper; Brāhmī; No. 15. One large fragment, from middle of leaf. See Note vi. D. III.

D. III. 13. a. MS. on paper; Brāhmī; No. 16. Numerous fragments of (apparently) one leaf. See Note i.

D. III. 13. b. MS. on paper; Brāhmī; No. 17. Six small fragments, from left side of leaves, numbered 2, 9, 10, 13, 14, 19. Also six large fragments, right sides of leaves. Also three very small fragments, from middle of a leaf. See Note vi.

D. III. 13. c. MS. on paper; Brāhmī; No. 18. Two small fragments from left sides of leaves, numbered 17 and 18. Also five large fragments, right sides of five leaves. Also two small fragments, from right sides of two leaves. See Note vi. Plate CVIII.

D. III. 13. d. MS. on paper; Brāhmī; No. 19. Numerous small fragments of (apparently) one leaf. See Note i.

D. III. 14. MS. on paper; Brāhmī; No. 20. Three fragments from middle of a leaf. See Note vii.

Notes on above by Dr. Hoernle.

Note i.—No. 1, also three pieces of No. 3, also Nos. 4, 5, 6, 8, 10, 11, 12, 16, 19. All these belong to the same MS., consisting of leaves of very large size, about 18″ × 7″, paper and script in all being the same. Language: *Sanskrit.* Characters: upright Gupta, of seventh or eighth century. Contents: a canonical work of the Mahāyāna school; apparently some kind of *Prajñā Pāramitā.* The fragments are from the 18th and 19th chapters, the former being named *Bhūmi-parivarta.* The fragments belong to (at least) three leaves, one of which bears the number 132.

Note ii.—No. 2, also one piece of No. 3, make up one almost complete leaf, numbered 96. Size: about 12″ × 3″. Language: *Sanskrit.* Characters: upright Gupta, of seventh or eighth century. Contents: apparently a Buddhist canonical work.

Note iii.—No. 7, one complete leaf, numbered 8. Size: about 14″ × 3″. Language: *non-Sanskritic* (Proto-Tibetan ?). Characters: upright Gupta, of seventh or eighth century. Contents: apparently a Buddhist canonical work. In very good preservation.

Note iv.—No. 9. Both fragments belong apparently to the same leaf. Language: *Sanskrit.* Characters: a peculiar square cursive script, resembling modern Nāgarī. Contents: apparently the story of the *Yakṣa-senāpati Maṇibhadra.*

Note v.—No. 13. Portion of a document. Language: *non-Sanskritic* (Eastern Irānian). Characters: cursive Brāhmī, of eighth century, rather illegible. Contents unknown.

Note vi.—Nos. 14, 15, 17, 18. All these belong to the same MS., size 14½″ × 3″. Language: *Sanskrit.* Characters: upright Gupta, of seventh or eighth century. Contents: copy of the *Vajrachhedikā.* As this work is published by the late Prof. Max Müller in the *Anecdota Oxoniensia,* there was no difficulty in piecing the fragments together; the result being that there are preserved the following leaves, in part or whole—Nos. 2, 6, 7, 8, 9, 10, 11, 13, 14, 15, 16, 17, 18, 19, 20. The following leaves are entirely missing: 1, 3, 4, 5, 12. Leaf 20 contains the conclusion of the MS.

Note vii.—No. 20. Apparently pieces of the same leaf. Language: *non-Sanskritic* (Proto-Tibetan ?). Characters: upright Gupta of very large size, of the seventh or eighth century. Contents unknown.

OBJECTS FOUND IN SHRINE D. IV.

D. IV. 1, 2. MS. on paper; Brāhmī; No. 21. Two small fragments of one leaf.—Note viii. No. 21. Belong to the same leaf; size about 12″(?) × 2½″. Language: *Sanskrit.* Characters: upright Gupta, of seventh or eighth century. Apparently some Buddhist canonical work. [Dr. Hoernle.]

D. IV. 3. MS. on paper; Brāhmī; No. 22. One small fragment from lower right corner of leaf.—Note ix. No. 22. Belongs to a leaf of another MS. than the preceding; for paper and script differ. Language: *Sanskrit.* Characters: upright Gupta, of seventh or eighth century. Contents unknown. [Dr. Hoernle.] D. IV.

D. IV. **D. IV. 4. Painted wooden panel**; rectangular. Lower R. p. corner broken. Obverse: In vesica a standing figure (Buddha?). R. p. hand raised to centre of breast, L. p. hanging down grasping end of robe. Feet bare. Nimbus. Robe dark red, Saṅghāṭi dark green (?). Flesh pink. Contour lines of drapery and hair black. Contour lines flesh, red. Vesica, colour perished. At bottom of panel, running behind vesica, band of dark brown; general background green (?); band of colour at top. Nimbus pink. Reverse: Standing figure, same pose as obverse, no loose drapery. Appears to wear tight fitting vest or merely ornaments. On each forearm a Vajra and bracelets, and other ornaments (indistinct) also on parts of body. Just above knees are two circular objects consisting each of three concentric circles. Too deleted to distinguish meaning. Ornamental details are visible lower down on legs. Between circular objects and lower edge of vest traces of red visible, and from a slight break in line of thigh (R. and L.) perhaps Dhōtī is indicated. Feet deleted. R. p. hand at breast, L. p. before L. thigh, palm inwards. Face almost circular. Double nimbus. Outer ring pink; inner (?). Hair (with top-knob), black. Vesica, field, blue; inner border, pink; outer, green. Ground of panel, pink, with 2" blue band at top passing behind vesica. Wood very soft. Height $10\frac{1}{8}$", width $5\frac{1}{8}$", thickness $\frac{1}{8}$". See Plate LXV.

D. IV. 5. Painted wooden panel; rectangular. Obverse: In two semi-ellipses, two half-length figures. *First*: R. p., a nude male carrying in L. p. hand a long-stemmed, leaf-shaped fan (?) resting on L. p. shoulder. Hair black, in loose top-knob, which is inclined to R. p. of head, and adorned with row of small pearls round roots of hair, and cincture of pearls binding top-knob. Head three-quarter face to L. p. L. p. hand (holding fan) raised to breast. R. p. half pendant at R. side. Large thick rings in ears. Background white (?). Flesh dark. Figure is perhaps attendant of next personage. *Second*: Rat-headed figure, head, profile to R. p. wearing diadem. Robe red with small 'powdered' pattern in white dots. Border at neck, blue. Band round arms blue. Flesh dark yellow. R. p. hand at breast, L. p. half pendant. Background pink (?). The surrounding elliptical arches to both figures are a broad Indian red band, and outside it a thin line. The background between the arches, pale pink, with diaper of simple crosses (like + signs) in Indian red. Two end portions of panel much deleted, but there appears to have been a figure to L. p. with deep red robe, black hair, light flesh. Arch over this figure cuts into that of rat-headed figure, while there is a space of more than 1" between the next two. There appears to be marks of purple dye (Sindūra?) in places. See Plate LXIII. Reverse: Traces of five seated figures in niches, hands in lap. Robe red (?); flesh light; contours red; black eyebrows, eyelashes, and pupils. Nimbus. 4th: from R. p. three-quarter face to R. p.; niche pale green; nimbus dark, outlined white. 5th: Same pose, top-knob (?); niche pink; nimbus pale green, outlined white. Spandrils between niches dark red (?). Marks of Sindūra. Wood fairly preserved. Painting very much damaged, and on reverse almost entirely deleted. $17\frac{1}{2}$" long, $4\frac{1}{2}$" wide, $\frac{3}{8}$" thick.

D. IV. 6. MS. on paper; Brāhmī; No. 23. Two documents, attached to one another, top to top.—Note x. No. 23. These two pieces appear to make up only one document, inscribed on both sides. Language: *non-Sanskritic* (Eastern Īrānian). Characters: cursive Brāhmī of eighth century. Mentions the 17th day of the month *Mūñamji*: but purport unknown. [Dr. Hoernle.]

OBJECTS FOUND IN DWELLING, D. v.

D. v. **D. v. 1. MS. on paper**; Brāhmī; No. 24. One small fragment; from middle of leaf.—Note xi. No. 24. Belongs to a leaf of a MS. in *non-Sanskritic* (Proto-Tibetan?), and very large upright Gupta characters of the seventh or eighth century. Compare No. 20 from D. III, but probably a different MS. [Dr. Hoernle.]

D. v. 2. MS. on paper; Brāhmī; No. 25. One very small fragment, from middle of leaf.—Note xii. No. 25. Belongs to a leaf of a MS. in *non-Sanskritic* (Proto-Tibetan?) and upright Gupta characters of the seventh or eighth century. Paper stiff, and coloured a yellowish tint; the fragment suggests itself as belonging to the MS. described in my *Report*, ii. p. 20, No. 3. Contents: apparently a Buddhist canonical work. [Dr. Hoernle.]

D. v. 3. Inscribed wooden tablet, $9\frac{3}{4}$" long, $2\frac{1}{2}$" high, $\frac{1}{4}$" thick, rounded off on left with small hole for string. Four lines of cursive Brāhmī characters on one side only, commencing on the round side of tablet. Language: *non-Sanskritic* (Eastern Īrānian). Contents unknown. See Plate CVI. [Dr. Hoernle.]

D. v. 4. Inscribed wooden tablet, $7\frac{3}{4}$" long, 2" high, $\frac{1}{4}$" thick. On obverse four, on reverse three lines of cursive Brāhmī characters. R. end of tablet rounded off. Language: *non-Sanskritic* (Eastern Īrānian). Dated on 14th day of month *jsimji* (?). Contents unknown. [Dr. Hoernle.]

D. v. 5. Inscribed wooden stick of tamarisk, flat on one side, upon which are faintly traceable remnants of about a dozen Chinese characters. Reverse retains in part its original round form, flattened by cutting from a little above middle to lower end. On this flattened part Chinese writing by two different hands. There appear to be five Chinese symbols in the upper section written with coarse brush and rather clumsy hand; and seven in the lower much more delicately written. Wood perishing. Length about $13\frac{1}{2}$", width about $1\frac{1}{8}$". See Plate CVI.

D. v. 6. Chinese document on paper, dated 781 A. D. (translated in Appendix *A*). See Plate CXV.

D. v. 7. MS. on paper; Brāhmī; No. 26. Six small fragments, from right side of (apparently) one leaf.—Note xiii. No. 26. All pieces belong apparently to one leaf of a MS., in *non-Sanskritic* (Proto-Tibetan?) and upright Gupta characters of the seventh or eighth century. Paper and script appear to be exactly the same as those of the MS., Nos. 1–19, from D. III., but the MS. of course, is different, because the language is different. [Dr. Hoernle.]

D. v. 8. MS. on paper; Brāhmī; No. 27. Two large fragments of one document, with Chinese signatory monogram.—Note xiv. No. 27. Both belong to one document, size $10\frac{3}{4}'' \times 7\frac{1}{2}''$. Language: *non-Sanskritic* (Eastern Īrānian). Characters: cursive Brāhmī of eighth century. Inscribed on one side only, and signed with a Chinese monogram. Contents unknown. [Dr. Hoernle.]

D. v. 9. Wooden bowl. Portion of lower part of turned wood bowl or dish, lacquered (?) and painted. Probably Chinese. INSIDE: Portion of wreath of flowers and leaves, round ornamental inner border. Flowers deep crimson and geranium scarlet. Leaves green. All outlines and small diaper background in yellow. Scarlet flowers shaded by crimson lines, leaves by black. Ground green. Ornamental border, yellow tracery. Arrangement of flowers and leaves alternate. Colours still bright, but covered by D. v.
sand and dirt which it is difficult to remove without disturbing painting. OUTSIDE: Olive green at bottom and partly up side, on which traces of pattern in black. Above this, deep terra-cotta red. The marks of turning are distinctly visible, and the grain of wood where fractured shows that the bowl is turned out of the solid. Condition of wood good, but much of the virtue of lacquer perished. $7\frac{1}{2}''$ by $1\frac{1}{2}''$. See Plate LXV.

D. v. 10. Wooden panel; 'Takhtī' shape. Traces of scraping (from erased writing). Wood soft at edges. $13\frac{3}{4}''$ long, $3''$ high, $\frac{1}{2}''$ thick.

D. v. 11. Shoe; made of plaited hemp string. Sole, now very compact and almost felted by hard wear, seems to consist of an insole of coarse string, coiled flat, and an under sole similarly laid, and perhaps plaited into the other. The two are bound together by a close warp of thinner and more finely made string. Long, free ends of warp turned up at right angles to form side of shoe, which thus consists of an arrangement of parallel cords only, lying closely together. Upper ends tightly plaited round two horizontal stout cords to form upper edge or opening of the shoe. These two cords cross in a loop in front. Much broken. Front edge of sole appears to have been cut off. Adult size.

D. v. 12. Woven fabric. Fine cord silk. Dirty buff colour. $6''$ by $4\frac{1}{2}''$. See Plate LXXVI.

OBJECTS FOUND IN D. VI.

D. VI. 1. MS. on paper; Brāhmī; No. 28.—Note xv. Left-hand fragment, $4\frac{1}{2}'' \times 3''$, with string-hole, of a leaf numbered 70. Full size, probably $18'' \times 3''$. Language: *Sanskrit*. Characters: upright Gupta of seventh or eighth century. Contents: apparently some Buddhist canonical work; on obverse, end of section 43, the beginning of the 44th section being *Smarāmy-ahaṃ kalpāṅs-afila* . . ., on reverse, end of section 51. [Dr. Hoernle.]

D. VI. 2. MS. on paper; Brāhmī; No. 29. One very small fragment of a document, with traces of Chinese signatory monogram.—Note xvi. Nos. 29 and 30. Fragments of two documents, in *non-Sanskritic* (Eastern Īrānian) and cursive Brāhmī. [Dr. Hoernle.]

D. VI. 3. Painted wooden panel; oblong. Ten seated figures (Buddhas?) on continuous Padmāsana. Heads three-quarter to centre. Robe red. Hands crossed in lap. Behind each figure a vesica (green) and nimbus (alternate?) red and white. Padmāsana yellow. Each figure wears a Tibetan hat (yellow). Painting fragmentary. $27''$ long, $4\frac{7}{8}''$ high, $\frac{1}{8}''$–$\frac{1}{4}''$ thick.

D. VI. 4. Painted panel. OBVERSE: Six Buddha figures seated in meditation, figures arranged vertically in three rows, two in each row. The rows seem to have been divided by narrow lotus-petal bands running across from D. VI.
side to side. The pairs of figures incline heads (three-quarter) slightly towards each other. One dressed in dark saffron garment edged with white, other in pink. Flesh in both, light; eyes white. Flesh contours on both, and draperies on pink figure outlined in Indian red. Hair, eyebrows, eyelashes on both and dark garment, in black. Vesica behind dark figure, red, and outside vesica Indian red. Vesica behind pink figure, some light tint now unrecognizable; outside vesica probably pink. In one case (pink figure) the Buddhic top-knob is clearly shown, probably it was present in all. Tilaka also present. The dark and light figure alternate vertically. Very badly defaced. REVERSE: Similar to obverse, but top pair missing. Background outside vesica of dark figure, light. That of pink figure, Indian red. Work very rough. The dark saffron figure is the similarly robed 'spectacled' figure of D. x. 01, but lacking the care in execution. Size $13\frac{1}{2}'' \times 8'' \times \frac{3}{8}''$.

D. VI. 5. Stucco relief fragment (two pieces). Female head (Gandharvī). Chunam; traces of black on hair and eyes. Nimbus. Height $2\frac{1}{8}''$, width $3\frac{3}{8}''$. See Plate LVI.

D. VI. 6. MS. on paper; Brāhmī; No. 30. One fragment, a long, narrow strip, of a document. (See Note xvi. above).

OBJECTS FOUND IN D. VII.

D. VII. **D. VII. 1. Painted wooden tablet,** in shape of Takhtī, with hole through handle. Painting, with head in direction of handle. Seated on low throne (picked out in black lines forming squares, alternate squares ornamented in red) from which appears to flow, in front, ample green drapery, a cross-legged male figure; head three-quarters inclined to R. p. R. p. arm flexed across body bringing hand to centre of breast. In R. p. hand is held lightly a patera having a knob-shaped foot. In L. p. hand, which rests upon L. p. thigh, is held by long, gracefully curving stem, a lotus bud, the bud being above level of head. Arms appear to be in sleeves of black garment, yellow lined, cuffs turned back. The colour has perished on body. Round upper R. p. arm appears to be armlet tinted yellow. Loosely round neck heavy necklet having three large jewels divided by pairs of smaller ones. Tinted yellow. Tightly round throat double band, yellow. L. p. ear appears to sustain ornament. Traces of black hair (or drapery). Nimbus tinted green, shaded darker towards head. From head appears to fall narrow sash R. p. on side of shoulder, round upper arm, then outwards and upwards with bold sweeping double curve, the end appearing to flutter open on level of head. On L. p. the arrangement is repeated. Traces of rich red on lower garment. All outlines and contour lines of flesh, drapery, ornaments, &c., black, excepting on sash, where are some lines in deep red. Flesh pink; sash pink and yellow; patera yellow. Lotus, traces of pink. Suggestion of head ornament outlined deep red. Wood well preserved. The painting is remarkable for good drawing, gracefulness of pose, and free execution. 6" wide, 6½" high. Handle 1¾" × 1", thickness ⅜". See Plate LXVI.

D. VII. 2. Chinese document on paper; dated 782 A.D.; translated in App. *A*. See Plate CXV.

D. VII. 3. Chinese documents on paper; see App. *A*.

a. Document dated in Ta-li period (766–779 A.D.); translated in App. *A*. See Plate CXVI.

b. Fragment of a document.

c. Dated in Chêng-yüan period (785–804 A.D.).

d. Document dated 787 A.D.; translated in App. *A*. See Plate CXVI.

D. VII. 4. Chinese documents on paper; see App. *A*.

a. Document dated A.D. 782; translated in App. *A*. See Plate CXV.

b. Several fragments of a document.

c. Several fragments of a document.

d. Fragments of a document dated A.D. 789.

e, f. Each a fragment of a document.

D. VII. 5. Painted wooden panel. Rectangular, with pointed, arched top. Ground white. Two figures, one in upper portion of panel and one in lower. UPPER: A personage of high rank seated on piebald (or dappled) horse. Figure of man has long black hair tied at crown in loose knob with yellow band. Band round head holding in front elliptical jewel. Long ear. Green nimbus. Face three-quarters to R. p. showing somewhat Chinese features; small drooping moustache. Single garment, tunic pink, with yellow neck-band. Scarf descends from back of head (apparently) and is curled round both upper arms, the two ends flying freely behind, indicating movement. Black high boots with white felt soles; stirrups. Single rein held in L. hand. R. hand raised, holding patera, towards which a bird is flying down. Suspended from girdle a sword, nearly straight, of Persian pattern. The horse appears to be ambling, and is high-stepping. Colour white with large leopard spots of black. Tail tied in loop. Bridle single, and apparently no bit in mouth. A large ornamental plate covers forehead and nose, with projecting horns—one at forehead carrying a crescent or Triśūla; another from nose carrying a red silk (?) mango-shaped knob (pompon). Many of these knobs appear on head, surcingle and crupper (cf. Ajantā cave frescoes). Deep saddle (black) with large 'Numdah.' The drawing of horse's legs and hoofs is very free and clever. LOWER: A personage of high rank seated on Bactrian (?) camel and passing through hilly country or sand dunes. Has short curly hair, wears curious sugar-loaf hat, with broad turned up brim having erect 'Vandyke' points. There are marks on hat suggesting fur with spots. Red nimbus. Face deleted. Single garment, a loose fitting tunic, green, gathered below knee into red mocassins. Small feet in shoes or boots (yellow). Scarf from back of head (?) round upper arms, both ends flying free behind to suggest forward progress of camel, which is ambling. In L. hand nose cord of camel; in R., upraised shell-shaped cup, suspended to girdle, a Persian pattern sword. 'Zīnpōsh' red with green border. Camel light brown. Long hair on under part of neck, also on front of forehead and a lock falling back over neck. General character of animal suggests the two-humped variety. Tail raised to suggest speed. Large canine tooth displayed over lower lip. All contours in black, and very free. Flesh pink, and shaded. Reverse plain, but has remains of dowels in five places, showing that the tablet was affixed to something. 15" high, 6⅞" broad. See Plate LIX.

D. VII. 6. Painted wooden panel; rectangular. OBVERSE: Seated three-faced, four-armed figure, cross-legged, on cushion decorated with dice pattern in yellow, red, and green. Flesh blue. On arms Uttarīya (yellow), armlets and bracelets (yellow); round neck, massive jewelled collar; large rings in ears; fine diadem (yellow and white) with pink flowers; hair, eyebrows, &c., black. Hair curling and long; Langōṭī tiger-skin in red and yellow. Long, loose collar, outlined only in black, hangs round neck as low as waist. Below cushion, as Vāhanas, two bulls sitting in profile, facing inwards; humps very small. Face to R. and L. of centre. R. p., three-quarter to R., effeminate, white, black hair, simple jewelled diadem. L. p., grotesque head, ferocious, dark flesh, eyeballs white, eyebrows thick and black, large mouth open. On head, close-fitting, pink Pagrī. R. p. front hand at breast holding white object (drum, *ḍamaru?*); R. p. back hand raised to level of head, *cakra*; L. p. lower hand on thigh, grasping *vajra*; L. p. upraised hand *Śaṅkha* (?). Nimbus behind all three heads, rich green; vesica, fine vermilion field, yellow border, divided by white and Indian red lines; background of upper part of panel appears to have been tinted with white three-pointed leaves. Background of lower part white. All contour lines Indian red. Reverse: Seated four-armed figure, cross-legged on red cushion, decorated with large four-petalled flowers, highly finished; petals, alternate pink and green, and red centres, and in some cases a black spot in each petal near centre. Costume: yellow undergarment ornamented with large flowers described, and covering legs. Over this a coat, green, with large flowers in black. Round waist a narrow belt, to which is slung, by two double slings, a short, broad sword in ornamented sheath; handle set at angle with blade. On feet, and rising to just below knee, high boots (resembling Hessians), black, embroidered in yellow. Ornaments in yellow encircle ankles. Uttarīya yellow, hangs over arms falling to thighs. Head three-quarter face to R. p., purely Persian. Black, almost straight hair; black, bushy eyebrows; thick, short beard; thin moustache, trained to sharp points. Lips full and sensuous, mouth smiling, slightly open, showing upper teeth. Eyeballs white, pupils black. Large, heavy rings in short ears. Ornaments in yellow on upper arms. Headdress in yellow, resembling a Persian tiara, retaining indications of diagonal pattern on front. Round lower edge of headdress is tied the Uttarīya in a loose knot at back of head, the ends falling over arms in usual way. R. p. hand front, clenched, rests on thigh; R. rear, upraised, holds flower (?). L. p. at breast, Vajra; L. p. rear, upraised spear-head. Flesh deep pink. Nimbus green, outlined white. Vesica, field red, border pale green, divided by white line. Sleeves on rear arms purple with white diaper; cuffs pink with large green flowers. In top spandrils conventional four-petalled flower and leaves, pink on white ground. Whole character of this side of panel essentially Persian. Wood in good condition. 12¼" high, 8" wide, ¼–½" thick. See Plates LX, LXI. D. vii.

D. vii. 7. Chinese document on paper, from Hu-kuo Convent, translated in App. *A*. See Plate CXVI.

OBJECTS FOUND IN D. viii.

D. viii. 1. Fragment of Chinese document on paper, in two pieces; translated in App. *A*. See Plate CXVI.

D. viii. 2. MS. on paper; Brāhmī; No. 31.—Note xvii. Two fragments of (apparently) one document, forming its upper right-hand quarter, in *non-Sanskritic* (Eastern Irānian) and cursive Brāhmī; measuring 6" × 4"; hence full size probably 12" × 8". Inscribed on one side only. Contents unknown. [Dr. Hoernle.] D. viii.

OBJECT FOUND IN D. ix.

D. ix. 1. Fragment of a Chinese document, dated 790 A.D.; translated App. *A*. See Plate CXVI. D. ix.

OBJECTS FOUND IN D. x.

D. x. 1. Painted wooden panel; rectangular. Painted one side. Very indistinct. Seated figure wearing Uttarīya (deep red); flesh white. Vesica light. R. p. hand raised to breast. 4" by 7½", ⅜" thick.

D. x. 3. Painted wooden panel. Oblong. Obverse: Three seated Bodhisattvas. *First* (R. p.): Head to R. p. three-quarter face. Six-pointed diadem. Top-knob. Hair long, flowing on shoulders. R. p. forearm bare, bracelet. R. p. hand at breast holding Vajra. L. p. hand on thigh. Upper garment close fitting; scarf round arms and behind back. Girdle yellow. Long loincloth. Nimbus yellow. Vesica blue. Hair and all contours, black; costume, flesh, and diadem, white (or pink). Vajra yellow. *Second*: Head three-quarters to R. p. Wide-sleeved, long, yellow robe, blue bands, edged white at arm-holes; deep red, long-sleeved under-coat; hair long, curling on shoulders, black; diadem white; long-flowing blue (?) drapery from beneath figure falls over front of seat (cushion?). Four arms: 1st R. p. at breast, holding Vajra; 2nd R. p. with Cakra; 1st L. p. rests on thigh; 2nd L. p. holds round object (lotus or fruit, or holy water vessel). Contour lines all black. Vesica yellow. No nimbus. *Third*: Three heads, on each yellow peaked diadem; hair black, long, waving on shoulders; long, narrow eyes; thin Chinese moustache to front face; front face three-quarters to R. p.; R. p. face to R.; L. p. face to L. Four arms: 1st R. p. at breast, holding cup; 2nd R. p. double object, looking like small tongs, with Vajra-shaped ends; 1st L. p., hand on thigh; 2nd L. p., bow. Short-sleeved garment (white or yellow), with wide arm-holes, having yellow band to each; girdle crossed at back and ends brought to front (pink); under-coat with long D. x.

D. x. sleeves, white or yellow. Cushion red and yellow (brocade). Vesica, blue field, yellow border. No nimbus. Contour lines black. Background of panel white or pink. Reverse: Four (?) figures seated in row, of which outer two are almost entirely deleted. R. p. first seated figure, half nude; R. p. hand raised (?), L. p. hand indistinguishable. Lower garment rich red with delicate flower diaper in dots of white. Head quite deleted. Second figure in full green robe, folds and contour lines cleverly drawn in black; hands in lap (Dhyānamudrā attitude); head to L. p., eyes outlined in black, the outer angle continued in a single line almost to top of ear, giving precisely the effect of spectacles. Hair of both this and third figure appears to be partially shaved. Nimbus. Third figure similar to second, but head three-quarters to R. p.; same curious elongation of outer angle of eyes. Nimbus light. Robe rich red. Fourth figure too fragmentary to make out. 10″ × 4⅜″, very soft, both colours and wood. See Pl .te LXIV.

D. x. 4. **Painted wooden panel**; rectangular. Four figures are represented, three being to the same scale, only half length. The fourth a full figure, seated, and to smaller scale. Beginning R. p., *first* figure, female, three-quarter face to L. p., dressed in simple, short-sleeved yellow vest, with pink diaper pattern, over under-coat of green, with long sleeves tight at wrist. On head a diadem, yellow, composed of two transverse fillets, holding a number of short upright fillets. Nimbus. L. p. arm outstretched to full extent, slightly raised, finger pointing to diadem of second figure. Flesh pink. Hair long and curly, black. Eyes black. All contour lines red, excepting those of folds on green sleeve, black. *Second* figure, female, three-quarter face to R. p., wearing simple yellow or pale-green vest, similar to first, spotted with simple flower pattern. Arms bare. Hair long, curly, drooping in many curls on to shoulders. Elongated ear. Face very Persian. On hair, which is dressed high, an elaborate jewel, elliptical, a centre gem surrounded by border divided by radiating lines, bound to head by fillet, and what appear to be several loose upward curling ends. Nimbus. Eyebrows and upper lashes, black. Contour lines of vest black; of face, arms, and nimbus, red. Between first and second figures a basket containing cocoons. *Third* figure, seated cross-legged on cushion (cf. D. II. 010). Head three-quarters to R. p.; pink, short-sleeved vest; green, long sleeves to under-coat. Lower part of legs and feet appear to be covered by boots. Four arms. R. p. front hand at breast, holding small object; L. p. front hand on thigh; R. p. hand back, upraised, holding spear-head (?); L. p. hand back, upraised, holding object indistinguishable. Dark coloured nimbus. Between third and fourth figures representation of a weaving frame (yellow), with horizontal warp. *Fourth* figure, standing, three-quarter face to R. p., holding in R. p. hand an instrument resembling a weaving comb or brush. Vest pink, diapered with large flowers, white, outlined red. Head similar to first figure, and has similar headdress. Eyebrows, eyes, hair, black; flesh pink; all contour lines, red. Nimbus yellow. To L. p. of figure outlines in red, possibly meant to represent the framework drum used in spinning silk thread. Whole of background of panel appears to have been painted light pink; on this petal-shaped daubs of dark pink. Wood well preserved, fairly hard. 18″ (appr.) long, 4⅝″ high, ½″ thick. See Plate LXIII.

D. x. 5. **Painted wooden panel.** Obverse: Trimūrti figure, seated cross-legged upon a low seat covered with diced material. Figure, dark blue (slate colour), and has four arms, two upraised, holding in L. p. flower with long, yellow stalk; in R. a wheel-like object. R. p. lower rests with hand on R. thigh, holding Vajra (?); L. at level of chest holds indistinguishable object. Feet a good deal damaged, but the R. one appears to have been painted twice, as though a correction of careless painting. Round loins a tiger-skin (yellow with black markings). Bracelets (simple bangles), yellow, on all four wrists. Armlets at biceps are of usual half-lotus pattern, yellow. Narrow scarf goes round arms, as usual, and falls from them down to thighs; appears to have been green. There is also a thread-like chaplet (in black) tied into bows, and appears to hang round neck as in prototype of this picture, D. VII. 6. Centre head three-quarters to R. p. Third eye present. Two wrinkles on neck. Thin, black wiry moustache; smiling mouth. Black, waving, long hair. Head crowned with high crown (yellow) having Cakra-like ornament in front. Rings in both ears. L. p. head yellow, large-mouthed, coarse and fearsome; matted hair, green. Eyebrows thick and frowning. R. head deleted entirely. All flesh outlines and features in Indian red. Eyebrows, eyelashes, drapery outlines, and Vajra in black. Aureole outlined in red. Ground of panel Indian red. Reverse: Horseman (very Persian) on dappled grey horse to R. p. Tight-fitting, long-sleeved under-garment. Short-sleeved upper-garment reaching to knees. Top-boots. Belt, and long, straight sword slung by two slings to belt. Saddle rising in front and back. Numdah, from under which passes rope-like girth. Simple bridle and surcingle. The pose of the horse is that of prototype D. VII. 5, and similar to usual Persian tiles of princes hawking. L. p. hand holds lightly single rein; R. hand upraised holds a patera. Cap high and of sugar-loaf shape, with edge turned up and 'vandyked.' Face, three-quarters to right, white and plump. Features delicate. Nose straight. Hair black, slightly bushy. Nimbus. Ear small, with small black earring. Two wrinkles on neck. Flying slightly downwards and towards figure, an unmistakable wild duck (black), with beak slightly opened. Indications of hilly country or sandhills. Flesh white. Under-garment yellow; coat greyish pink. Cap, patera, saddle, white. Numdah, nimbus, yellow. Sword, hair, and all outlines, black. 11″ × 5¼″, ½″ thick. See Plate LXII.

D. x. 6. **Inscribed wooden beam.** Oblong, with one line of writing in cursive Brāhmī characters. Original thickness sawn off to facilitate transport. Left end (commencement) has suffered by decay; right complete. Face well smoothed. Wood in perfectly good condition, except left end. Length 42", width 4", thickness about 1⅛" (originally 4").

Inscription in 'unknown' (Eastern Īrānian) language; interspersed corrupt Sanskrit (*kuśala-mūla, Bodhisattva*). Transcription from Dr. Barnett's eye-copy:—

1	2	3	4	5	6	7
?	*ñe*	*?*	*bb* (?)	[*r*]*e* (?)	*z* (?)	*jsa*

8	9	10	11	12	13
haṃ	*?*	*śrī* (?)	[*s*]*e* (?)	*l'aṃ* (?)	*ñe*

14	15	16	17	18	19	20
jsa (?)	*tte* (?)	*pū* (?)	*ña*	*ku-*	*śa-*	*l'a*

21	22	23	24	25	26	27
me	*-mū-*	*la*	*bo-*	*dhyau*	*sa-*	*tvyau*

28	29	30	31	32	33
jsa	*haṃ*	*śrī*	*ho*	*ba* (?)	[*hs*]*ū* (?)

34	35	36	37
śva (?)	*pa* (?)	*ri*	*hāṃ* [Dr. Hoernle].

D. x. 7. Painted wooden panel; rectangular. Obverse: Almost indistinguishable, rough sketch in black, outline of seated Buddha (?) with almsbowl in lap. Reverse: Sketch (very spirited and good), standing Bodhisattva. Pose free. Slightly inclined to R. p., head inclined (three-quarter face) R. p. Upper part of body slightly bent forward, throwing weight on Triśūla which is grasped by L. p. hand on level with head. Lower end of Triśūla rests upon disc upon which figure stands. L. p. foot advanced, weight of body being thrown on to R. p. leg. R. p. hand seems to rest on upper part of thigh, in front. Figure has moustache and very Chinese cast of face. Diadem. Necklet with jewel, armlets, girdle, the long ends of which float freely at each side. From below girdle, a garment which bears resemblance to old Chinese leather plate armour; on legs and feet top-boots, or boots and greaves. Upper part of body probably enveloped in close-fitting vest. Hair long, to shoulders. Nimbus and vesica. Wood well-preserved. 5" × 3", thickness ⅜".

D. x. 8. Painted wooden panel, badly damaged, with very little of colour remaining. Four-armed Trimūrti figure, seated cross-legged. Figure clothed in tight-fitting, long-sleeved white vest. Tiger-skin round middle with point rising in front as in D. vii. 6. Simple lines encircling wrists seem to indicate stitching or cuffs. Massive yellow armlet is visible on L. p. arm; L. p. lower hand rests on thigh and holds Vajra. R. lower arm is thrown round neck of Śakti who kneels on his R. thigh. The remaining two arms are upraised, with hands grasping objects (unrecognizable). Long neck cord, tied into bows at intervals. Centre face almost deleted, turned three-quarters to R.; direction of glance appears to be downwards at Śakti. There are traces of the third eye. Hair apparently long and wavy. Crown of usual Īrānian type. Legs and feet bare. Fearful face to L. p. has usual heavy mouth, eyebrows very arched and dipping low over nose. A third eye clearly visible. Face to R. effaced. The Śakti kneels on R. thigh of deity, body in profile, apparently enveloped in long, simple stola, with tight sleeves to wrist. The head is thrown slightly back and is three-quarter face. With R. hand she raises a cup to the deity. Type of Persian Houri. Well drawn eyebrows, highly arched; eyes long; mouth sensual; chin and neck full. Complexion white, pink cheeks, down the side of which ripples a lovelock. This face perfectly preserved, while remainder of painting is almost gone. Colour of Śakti's garment green (?). Outlines of flesh throughout panel Indian red; those of draperies black. Background of vesica Indian red, with border of green. Outside this Indian red. Size 10¼" × 7½"; thickness ¼". See Plate LXII.

D. x. 9. MS. on paper; Brāhmī; No. 33. Numerous very small fragments of a leaf.—Note xix. Nos. 33, 37, 38. All belong to one leaf, numbered 51, being its left portion with string-hole, 3⅜" × 5½"; probably full size 3⅜" × 15", with five lines on page. Language: *non-Sanskritic* (Proto-Tibetan ?). Characters: upright Gupta of eighth century. [Dr. Hoernle.]

D. x. 10. a. MS. on paper; Brāhmī; No. 34. One large fragment, left side with string-hole, but number lost. —Note xx. Nos. 34 and 35. All belong to one leaf, number of which, however, is missing. The existing left portion, with string-hole, measures 3½" × 6"; probably full size 3½" × 16", with five lines on page. Language: *non-Sanskritic* (Proto-Tibetan ?). Characters: upright Gupta of eighth century. N.B. The foregoing two MSS., though similar in some respects, are different in paper and script. [Dr. Hoernle.]

D. x. 10. b. MS. on paper; Brāhmī; No. 35. Numerous very small fragments of a leaf. See Note xx., above.

D. x. 10. c. α. MS. on paper; Brāhmī; No. 36. Six fragments: two larger, four very small.—Note xxi. All belong to one (or several?) large leaf of a MS. in Sanskrit, and upright Gupta characters of the seventh or eighth century; similar to the large-leaved MS. of the *Bhūmi-parivarta* (see Note i.) found in D. iii., but a different MS. to judge from paper and handwriting. [Dr. Hoernle.]

D. x. 10. c. β. MS. on paper; Brāhmī; No. 37. One large fragment, left side with string-hole, numbered 51. See Note xix., above.

D. x. 10. d. MS. on paper; Brāhmī; No. 38. Numerous very small fragments of a leaf. See Note xix., above.

D. x. 11. Painted wooden panel; rectangular, much effaced. Seated figure, Buddha (?), full face; in vesica, with nimbus. Robe appears to be blue. Vesica some light colour; nimbus dark red; hair (with top-knob) black. Background green (?). Small hole pierced in centre near top edge. Wood fairly hard. 2¼" × 4⅛", ⅝" thick.

OBJECTS FOUND IN D. XII.

D. xii. **D. xii. 1. Stucco relief fragment.** Gandharvī. Chunam, traces of colour. Flying figure, head, upper portion of body and arms as D. xii. 5. From waist downwards figure curves towards R. p., the legs being extended horizontally. Supported in outstretched hands a festoon of cloth, doubled end blowing freely out from R. p. hand. Red on background, traces of pink on flesh, and red contour lines. 6½″ long, 3¼″ wide. See Plate LVI.

D. xii. 2. Stucco relief fragment. Gandharvī. Replica D. xii. 5. L. p. arm missing below amulet. Lotus missing.

D. xii. 3. Stucco relief fragment. Gandharvī. Replica D. xii. 5. Head, R. p. hand missing. Lotus complete; coloured pink.

D. xii. 5. Stucco relief fragment. Gandharvī. Chunam, coloured. Female figure, rising from large-petalled lotus. Body rather gross and masculine. Hair tied in high, loose knob, divided over forehead and carried behind ears to shoulders; large rings in ears; necklet of flat beads; large bead pendant; bead amulets on upper arms; bracelets at wrists; girdle of beads attached apparently to a band which falls from armpit. Both hands clenched and raised to level of shoulders. Nimbus. Hair, eyebrows, eyelashes and pupils, black; flesh, pink; contour lines and halo, red. Height, from girdle to halo, 3¼″. Width, hand to hand, 3¼″. See Plate LVI.

D. xii. 6. Fragment of wooden vessel, to which are adhering, both inside and outside, portions of paper with upright Gupta characters, very black. The paper, in some cases several layers thick, is wrapped over broken edges. Length 7¼″, width 3″, thickness ⅜″.

OBJECTS BROUGHT TO RAWAK, PROBABLY FROM D. XIII.

D. xiii. **D. 001. Judaeo-Persian document on paper,** brought to Rawak on Jan. 5, 1901 (see below, pp. 306 sqq.). Fragment of a sheet 16″ high; sides badly torn. See Plate CXIX.

D. 002. Stucco fragment, with Chinese inscription in three columns, enclosed in rectangle. Painted in black on white or tint wash ground. Much deleted. Grey clay, very brittle and friable. Brought to Rawak on Jan. 5, 1901 (see below, p. 309). 6½″ × 5″ × 1⅛″.

OBJECTS BROUGHT BY TURDI.

D. (T.) **D. (T.) 01. Stucco relief fragment.** Gandharvī. Chunam, traces of pink colour. Replica D. i. 014.

D. (T.) 02. Stucco relief fragment. Gandharvī. Chunam, traces of colour. For detail of head, cf. D. i. 43; for detail of body cf. D. i. 014. Height 5 9/16″, width at elbows 3 9/16″. See Plate LVI.

D. (T.) 03. Stucco relief fragment. Gandharvī. Chunam, traces of colour. Lower half, L. p. wrist and top of nimbus missing. Replica D. i. 02. Height 3⅜″, width 3⅜″.

D. (T.) 04. Stucco relief fragment. Head, Buddha (?). Chunam, traces of colour. Replica D. i. 55. Height 1 9/16″.

D. (T.) 05. Stucco relief fragment. Head of Buddha (?). Chunam, traces of colour. Replica D. i. 55. Height ⅞″.

D. (T.) 07. Stucco relief fragment. Ornament. Chunam, traces of colour. Replica D. i. 44.

D. (T.) 08. Stucco relief fragment. Figure of Buddha. Chunam, coloured. Replica D. i. 08. Waist upwards missing, feet missing. Height 4″.

D. (T.) 09. Stucco relief fragment. Standing Buddha; upper half; head missing. Chunam, no colour. Sharp tool marks. Replica D. i. 11.

D. (T.) 010. Stucco relief fragment. Standing Buddha; torso. Chunam, no colour. Replica D. i. 11.

D. (T.) 011. Stucco relief fragment. Ornament. Chunam, no colour. Replica D. i. 95. See Plate LVII.

D. (T.) 012. Stucco relief fragment (two pieces). Hand grasping object; colossal. Red clay, traces of colour. L. p. hand grasping billet, palm upward (?); all fingers and thumb broken off; fractures show thin wooden cores. Traces of pink colour. Billet measures 5¾″ length, 1¾″ width, 1″ thickness, and is rectangular. An incised line runs round upper surface about ¼″ from edges (suggesting movable panel). 1¼″ from upper end four incised lines are drawn round three sides (fourth left rough), suggesting a ligature of cord, under which midway between long edges of top surface lies a raised circular disc 9/16″ diam. (suggesting a seal). 1″ from lower end three incised lines are drawn round billet. These two ligatures are connected on upper surface of billet by two incised lines running diagonally from edge to edge. (The purpose of these ligatures would appear to be to keep the movable panel secure, and the seal to prevent tampering with fastening). Billet broken near centre, exposing remains of ½″ wooden core. Billet coloured white, with ¼″ black edge to all surfaces (except under, which is unfinished). Very friable. See Plate LXXXVIII.

D. (T.) 013. Stucco relief fragments. Fingers (four pieces); colossal. Red clay, traces of whitewash. Portions of four fingers. One appears to belong to D. (T.) 012; others doubtful. Cf. E. 1. 04. Very friable.

D. (T.) 014. Stucco relief fragment; ornament. Chunam, traces of colour. Blue. Replica D. 1. 53. Height 1½", width 2½".

D. (T.) 015. Stucco fragment from wall, composed of grey clay mixed with coarse grass and straw. Faint traces of colour, and of inscription. 4⅛" × 3⅜" appr. Very friable.

D. (T.) 016. Fresco fragment on stucco, with inscription. Coarse mixture of grey clay, grass and straw. Traces of colour—white, black, red, and buff. On buff band edged by two curved black lines, six inscribed characters in cursive Brāhmī; mostly illegible; read by Dr. Hoernle *? dā ? ca ? sa.* Also traces of black lines over red. 8" long, 3⅞" at broadest, 1¼" thick. Very soft. See Plate LVIII.

D. (T.) 017. Fresco fragment on stucco, with inscription. Coarse mixture of grey clay, grass and straw; broken in several places. Fres·o: L. p. portion of rectangular panel, within which portion of aureole (?). Within this on purple brown ground, conventional floral design in white and pink, outlined in black. The treatment of leaves is very free. To L. p. toes of a large foot. In spandril, on red ground, roughly sketched sprigs of flowers with stem and leaves. To L. p. of rectangle a band of grey colour, on which sprigs in white. Below on band of white (vertical) a two-line inscription in cursive Brāhmī written in black with brush, and extending over grey band above; read by Dr. Hoernle (1st line) *ga ña brā lla ro ji gī ? na haṃ* (?) *brī mye* (or *hye*) *le ro jo haṃ brī hye le* 3 ? 2nd line: [*d*]*o ? ?.* Very friable. 16" × 6" (appr.). See Plate LVIII.

D. (T.) 018. MS. on paper; Brāhmī; No. 32. One large fragment, from middle of a document.—Note xviii. Fragment belongs to the middle of a document in *non-Sanskritic* (Eastern Īrānian). Inscribed on one side only, in cursive Brāhmī. Size 4" × 3". [Dr. Hoernle.] D. (T.)

D. (T.) 001. a. Bronze spoon, imperfect and in many fragments. Length, inclusive of handle, about 7". Width of bowl about 1½". Angle of bowl to handle, very wide; and lateral curvature of bowl itself, very flat. See Plate LI.

D. (T.) 001. b. Seven bronze fragments, apparently belonging to another spoon.

D. (T.) 001. c. Bronze brooch. 1⅜" × 1", elliptical. Outer edge shows bead ornament, with narrow fillet; within this a flat band of pebble or glass, divided into eight equal segments. In centre, elliptical crystal, or pebble, or glass, raised and surrounded by containing fillet, outside which a row of beads. Back plain; remains of broken shank. See Plate LI.

D. (T.) 001. d. Beads and glass fragments; pink pebble bead, pierced; porcelain (?) bead, unglazed, coloured pale turquoise; fragment of light turquoise glass; fragment of fine blue glass; water-worn fragment of slate.

D. (T.) 001. e. Two bronze rings, broken and imperfect. Very thin and light. One contains a yellowish pebble; other has setting for three stones.

D. (T.) 001. f. Half of bronze tweezers. See Plate LI.

D. (T.) 001. g. Bronze style; probably used to apply colouring matter to eyelashes. See Plate LI.

D. (T.) 001. h. Small plate, elongated, elliptical, 1⅛" × ½", with hole at each end; might be from jazarant armour. See Plate LI.

D. (T.) 001. i. Odd fragments of metal.

CHAPTER X

FROM DANDĀN-UILIQ TO THE NIYA RIVER

SECTION I.—THE RAWAK SITE

Visit to Rawak ruins.

ON the 3rd of January, 1901, the explorations at Dandān-Uiliq were completed. The camels had, in accordance with previous instructions, duly arrived from the Keriya river; and my men, who had suffered from the long exposure to the cold of the wintry desert and the brackish water, were all eager for the start to less desolate surroundings. Before, however, finally leaving for the Keriya river, I decided to visit some ruins of which Turdi had spoken as situated to the north and known to treasure-seekers by the name of *Rawak* ('High Mansion').

On the morning of the 4th of January I paid off and dismissed to Tawakkēl a portion of my little force of labourers. With the rest I set out to the north, and after marching across gradually rising broad ridges of sand for a distance representing about seven miles in a straight line, reached once more easier ground, where the appearance of potsherds on the loess between the dunes indicated the former existence of habitations. Camp was pitched at a spot between loess terraces deeply eroded by the wind, where exceedingly brackish water was reached after sinking a well to a depth of about 7 feet. On the next day Turdi guided me to where, behind a long-stretching ridge of sand, some 60 ft. high, the ruins were situated at a distance of about one mile due north of my camp. They proved to consist of two low and much-decayed mounds, which looked like the last remains of small Stūpas. But repeated diggings, together with the effects of erosion, had rendered the shape of the superstructure quite unrecognizable, and even the foundations very difficult to distinguish. So much, however, was clear on examination of the débris that the structures must have been composed of sun-dried bricks, probably about 3 inches in thickness. The smaller mound to the east showed what might have been a Stūpa base about 10 feet square, with layers of red burned loess embedded among the débris. The second mound, situated about 50 feet to the west, seemed to show the foundations of some circular structure, measuring about 32 feet in diameter.

Mounds of Rawak.

From among the débris of ancient pottery, broken glass, &c., which strew the ground near the mounds, I picked up a small fragment of remarkably hard greyish stucco (D. R. 003), on which the practised eye of Turdi at once discovered traces of a thin gold-layer. Judging from its shape, this piece is likely to have belonged to a statue that had once been gilded. The material has been proved by Professor Church's analysis (see App. *F*) to be unusually crystalline plaster of Paris. At the back there appear what look like markings left by a canvas backing. Specimens of the glass and terra-cotta fragments found here are shown in the descriptive list under D. R. 001, 4, while D. R. 002 comprises beads of glass and stone picked up on different parts of the site.

Condition of ruins.

The remains of a small structure built in timber and plaster, about 40 yards to the south of the mounds, were found completely destroyed by recent burrowings; and this has undoubtedly been the fate of whatever ruin the movement of the dunes may have exposed here in recent times. As the dunes at this site rise to heights over 25 feet, and are proportionately large, they are likely to cover more buildings; but only in one place, about a quarter of a mile north-east of my camp, did we succeed in tracing below the sand the remains of a house built with timber. Its walls, including the usual posts, had decayed by previous erosion to within a few feet of the ground. They were found now buried under the slope of a high dune, which, rising immediately above the ruin, rendered it impossible to clear more than a single room. This had formed apparently the south-east corner of the building, and measured 15 by 12 feet.

Only a single find rewarded the search, but this one had special interest. It consisted of an inscribed wooden tablet (D. R. 01), $5\frac{1}{2}$ in. long and $1\frac{1}{2}$ in. high, found broken in two pieces, and of a second tablet (D. R. 02), which originally must have fitted the former exactly, but was only recovered as a fragment. The obverse of the first tablet shows five lines of cursive Brāhmī writing, which, though the ink has faded with the decaying surface of the wood, could yet be completely deciphered by Dr. Hoernle; its reverse is left blank. The preserved portion of the second tablet has on its obverse part of a raised square socket manifestly intended for a seal, while the reverse displays three legible lines of cursive Brāhmī. Three holes drilled through both tablets in identical positions (one in the centre of the seal cavity, and two about an inch from the opposite edge) undoubtedly served for a fastening string. Plate CVI shows the obverse of both tablets, and will help to explain the arrangement. The latter did not suggest itself at once, and I doubt whether it could have been established with full certainty had not the fortunate discovery at the Niya River Site of those many Kharoṣṭhī documents on wood, to be described in the next chapter, readily furnished the clue. Inscribed wooden tablet.

On comparing the 'double rectangular tablets' in Kharoṣṭhī script, of which specimens are shown in Plates XCIV, XCV, XCVII, it will be easily recognized that the arrangement followed in writing these documents and in securing their contents against unauthorized inspection was the model from which the fashion illustrated by these Brāhmī tablets had descended. They too form, in fact, a 'double rectangular tablet', to use the terminology explained in the next chapter, the first forming the complete 'under-tablet', while in the fragmentary second tablet we have a portion of the 'covering-tablet' which took the place of an envelope. But the long period which had passed between the writing of those Kharoṣṭhī documents in the third century of our era and the time of the unique Rawak tablet, probably early in the eighth century, had witnessed minor modifications in the use of this ancient wooden stationery. The side rims of the under-tablet have disappeared in the Rawak document, and this omission has necessitated a change in the ingenious method of fastening, though the seal-socket on the covering-tablet is retained. Instead of being passed transversely over both tablets and secured in grooves below the clay seal, as in the case of the Kharoṣṭhī documents of this class, the string was here threaded in some manner through the three holes above mentioned and sealed down in the socket. The device was equally effectual for the purpose of uniting the pair of tablets and preventing unauthorized inspection of the matter recorded on the inner sides, but it is less neat and strong than the older method, and the only possible advantage of the change could have lain in the simplification of the make of the tablets and a consequent cheapening of their production. Dr. Hoernle's transcript shows that the language of this curious document on wood was the same Eastern Īrānian dialect in which the paper documents of Dandān-Uiliq are written; of its purport nothing can be asserted at present. Independently, however, of its contents, the record may claim interest as another striking proof that the quasi-archaic use of wood as a writing material by the side of paper survived down to the eighth century. Fastening of double tablet.

Date of Rawak remains.

The scanty remains of Rawak did not yield records which, like the Chinese documents of Dandān-Uiliq, would enable us to gauge with approximate accuracy the time when the site was deserted. Yet the coins picked up near the ruins afford useful chronological guidance. Of the six coins found five were small copper pieces without legends, of the type current under both Han dynasties and probably for some time later[1], while the sixth bears the mark of the K'ai-yüan period (713–741 A.D.). From the absence of any later issues, such as turned up at Dandān-Uiliq, and from the predominance of pieces without legends, I am inclined to infer that the settlement represented by Rawak was probably deserted somewhat earlier than that of Dandān-Uiliq. For this we might account without difficulty by its outlying and hence more exposed position.

OBJECTS FOUND AT RAWAK NEAR DANDĀN-UILIQ.

D. R.

D. R. 01. Inscribed wooden tablet (in two pieces); forming under-tablet of D. R. 02. Cursive Brāhmī writing, five lines, on obverse. Language, Eastern Irānian; 19–20 Akṣaras to the line. Three string-holes. Wood decaying through salt impregnation. Size $5\frac{3}{4}'' \times 3\frac{1}{2}''$. See Plate CVI.

D. R. 02. Inscribed wooden tablet, frag. of covering-tablet of D. R. 01; on reverse three lines of cursive Brāhmī characters, 19–20 per line. On obverse raised socket for seal. $\frac{1}{4}''$ from R. edge a small string-hole. Impregnated with salt. $5\frac{1}{2}'' \times 1\frac{1}{2}''$. See Plate CVI.

D. R. 001. a. Fragment of glass vessel, foot. $1\frac{1}{8}'' \times \frac{11}{16}''$. **b. Fragment of glass,** rich blue. $\frac{5}{16}'' \times \frac{5}{16}''$.

D. R. 002. Glass and stone beads, &c. a. Opaque green glass round bead. Diam. $\frac{1}{2}''$. **b.** Small pink pebble round bead. Diam. $\frac{5}{16}''$. **c.** Do., do. elongated bead. Diam. $\frac{3}{8}''$; length $\frac{7}{16}''$. **d.** Broken small blue glass bead. **e.** Small bronze rectangular plate, with rib along centre. $\frac{5}{8}'' \times \frac{7}{8}''$.

D. R. 003. Stucco fragment, probably from some sculpture. Very hard plaster of Paris, with greyish marble-like surface, and traces of gilding. On back appears a perfect matrix of the coarse canvas used as a backing. The fragment has been subjected to fire. Size $3'' \times 3''$.

D. R. 004. a. Two terra-cotta fragments with traces of dark green glaze. $1\frac{5}{8}'' \times \frac{1}{4}''$; $1\frac{7}{16}'' \times 1\frac{1}{8}''$.

D. R. 004. Miscellaneous fragments. b. Fragment of talc. $1\frac{1}{4}'' \times \frac{7}{8}''$. **c.** Eight small fragments of glass, showing moulded work. One piece has a tubular edge. **d.** Fragment of stone (crystal). **e.** Small gutta-shaped piece of glass, with hole at thin end. $\frac{5}{8}'' \times \frac{3}{16}''$.

SECTION II.—A JUDAEO-PERSIAN DOCUMENT

Objects brought from Dandān-Uiliq.

The day I spent at Rawak did not end without further finds, though it was not this site which furnished them. On my return from the find-place of the Brāhmī document on wood I was surprised to see my camp joined again by a small party from among the men I had dismissed at Dandān-Uiliq. More enterprising than the rest, they had put off the start for Tawakkēl in order to indulge freely for a day or two in the traditional pastime of 'treasure-seeking'. While thus engaged, they related, they had scraped away the sand outside the broken walls of the apartment left unexcavated at the south-east corner of the ruin D. XIII[1], and there, apparently among miscellaneous débris of stucco and timber dating from earlier diggings, they had come upon the two antiques which they now brought to me in the hope of a small reward. One was a piece of coarse friable stucco, of the kind used as wall plaster in all the Dandān-Uiliq structures, showing some much-effaced Chinese characters painted in three lines. The other find was a lump of thin brownish paper so closely crumpled that I found it quite impossible at the time to attempt its unfolding. On one edge, however, I could make

[1] For a specimen, see Plate LXXXIX, 9.

[1] See above, p. 282.

out some characters resembling cursive Hebrew. It was not until this tightly-compressed piece of ancient waste-paper had undergone careful treatment at the expert hands of Mr. Hunt, of the MS. Department of the British Museum, that it resolved itself into the relatively large but unfortunately much-mutilated fragment of a Judaeo-Persian document, seen in Plate CXIX. The extant portion of the document, with its closely-written thirty-seven lines, covers one side of a piece of paper nearly 16 inches in height and apparently preserving its original dimensions in that direction. The original width of the paper cannot be ascertained, as the fragment is badly torn on either side and presents ragged edges; the actual width varies from 4 to 8 inches.

Question of origin of finds.

In view of the forgeries practised by Islām Ākhūn, of which I had acquired convincing evidence before leaving Khotan, and the story of which I was subsequently able to expose[1], I had exercised the closest watch while the excavations were proceeding in order to be able by personal observation to authenticate any antiquarian finds. The want of equally precise testimony was hence keenly felt by me in the case of the two curious finds now brought to me after my departure. I cross-examined the several men who alleged that they had been present at the discovery, and found their separate statements to agree well. A post-factum inspection of the find-place held out little hope of additional assurance, since I knew that the loose sand and débris from which the small objects had avowedly been scraped out could not possibly retain any distinct trace of their position. A return to Dandān-Uiliq would have meant a delay of at least one or two days, and this I was obliged to avoid out of consideration for the practical difficulties likely to arise if the heavily-laden camels, which had already subsisted for five days on the scantiest of rations and practically without water, were to be kept in this condition beyond the carefully calculated programme.

Proofs of antiquity of objects.

In the introductory note prefixed to Professor Margoliouth's paper, where the Judaeo-Persian document was published for the first time[2], I have already explained the above circumstances as well as the possibilities which *a priori* presented themselves as to the real origin of these finds. I have shown there that, unless these objects were of modern origin and had been purposely taken along from Tawakkēl or Khotan to Dandān-Uiliq on the chance of an opportunity offering to sell them to me as antiques, they could only have either been found under the conditions alleged by the men, or else abstracted in the course of my excavations at some other structure and secreted for a time with a view to subsequently securing some special reward. Fortunately the first-named supposition, rendered improbable at the outset by several weighty considerations, need not be examined any further, since the expert analysis of the document itself, as recorded in detail in the above publication, has furnished conclusive proof of its antiquity in respect of both script and paper.

For the palaeographic evidence as regards the Hebrew writing of the document, it will suffice to refer to Mr. Cowley's remarks embodied in Prof. Margoliouth's paper which is reproduced in Appendix *C*. These show clearly that 'the writing is throughout more archaic than that of the Persian deed of 1021', the oldest Judaeo-Persian document previously known, and stands midway between that deed and the remains of the third and fourth centuries. The result of the microscopical examination which Prof. J. Wiesner has been kind enough to effect of a specimen taken from the margin of the document, is equally decisive. The opinion of this distinguished expert, whose detailed researches into the material of the paper MSS. and documents contained in Dr. Hoernle's and my own collections, have for the first time elucidated the early history

Microscopical analysis of paper.

[1] See below, chap. xv. sec. i.

[2] See *An early Judaeo-Persian Document from Khotan, in the Stein Collection, with other early Persian Documents.* By D. S. Margoliouth; with an Introductory Note by M. A. Stein and communications from W. Bacher, A. E. Cowley, and J. Wiesner, in *J.R.A.S.*, 1903, pp. 735 sqq.

of paper manufacture in Central Asia, will be found reproduced in the same Appendix. It proves that the paper of the Judaeo-Persian document could not be distinguished in make from the paper of other ancient documents (Chinese, Brāhmī) excavated by me at Dandān-Uiliq, while on the other hand it was found to differ entirely in substance and structure from the modern Khotan paper which has exclusively served as material for the forged MSS. and 'block prints' from Islām Ākhūn's factory.

Thus there remains only the question as to whether the document was picked up in the alleged place and manner, or obtained in the course of my excavations at some other ruin of the site. Personally I am strongly inclined to credit the men's story; for the strict supervision exercised over the labourers would have rendered the abstraction of objects, especially such a relatively bulky one as the Chinese stucco inscription, distinctly difficult. But this question fortunately does not affect in any way the conclusion to be drawn as to the age of the document, or the historical and philological interest of its contents.

Historical allusion in Judaeo-Persian letter.

From Prof. Margoliouth's translation and notes, reproduced in Appendix *C*, it will be seen that the document represents the much-mutilated fragment of a letter written by a Persian-speaking Jew, and mainly relating to certain business affairs. Owing to the extent of the lost portions, not a single line being complete, it has been impossible for Prof. Margoliouth to attempt to make out a continuous sense; but more important from an antiquarian standpoint than the general contents is the reference made in line 23 to the 'Ispahbud'. As pointed out by Prof. Margoliouth, only one of the successive rulers of Tabaristān can be meant here, whose distinctive title *Ispahbud* is well known from Muhammadan historians. As a reference to Tabaristān is thus quite certain, Prof. Margoliouth proposes to identify the *Yazīd* named in the same passage with the Muhammadan general of that name, son of Al-Muhallab, who is known from Muhammadan chronicles to have conquered parts of Tabaristān in 717 A. D. On the ground of other historical and philological observations, Prof. Margoliouth concludes that, if this identification is correct, the letter containing this reference to the relations between the Ispahbud and Yazīd is likely to have been written about 718 A. D.

Date of document.

The dating thus proposed can well be reconciled with the available archaeological evidence. The fact of the Dandān-Uiliq site having been abandoned at the close of the eighth century furnishes a definite chronological *terminus ad quem* for the Judaeo-Persian document; in the other direction no fixed chronological limit can be deduced from the antiquarian observations and finds I made. Considering the remarkable dryness of the atmosphere in the Khotan region, there is nothing to preclude the possibility of a paper document, complete or fragmentary, remaining in fair preservation for a considerable number of years, even when not protected by the sand nor specially cared for. Hence the interval of fifty years which separates the date of our earliest Chinese document, 768 A. D., from Prof. Margoliouth's conjectural dating of the Judaeo-Persian document can in no way be considered a valid argument against the latter.

Relations between Tabaristān and China.

More difficult it may seem at the first glance to explain how a document written by a Persian-speaking Jew of distant Tabaristān should have found its way into an outlying settlement of Khotan. Yet Chinese historical records furnish evidence to meet any doubts that might arise on this score. M. Chavannes' extracts from the T'ang Annals give us interesting glimpses of the diplomatic relations which the rulers of Tabaristān maintained with China during the eighth century[4]. Pressed hard by Muhammadan conquest, they naturally turned for help to the great though distant power which asserted its hold over Sogdiana and even

[4] See *Turcs occid.*, pp. 173 sq.; *Notes additionnelles* ... (*T'oung pao*, 1904, pp. 70 and note, 76, 77 and note).

on the Upper Oxus well up to the middle of the eighth century. Embassies from Tabaristān to the imperial court are recorded as late as the years 744, 746, and 754 A.D., about which time the Chinese lost Western Turkestān to the Arabs. The exact description which the Annals give of the geographical position of Tabaristān, their distinct reference to the title Ispahbud (rendered quite correctly by 'hereditary commander-in-chief of the east for Persia'), and their mention of Tabaristān's struggle against the Arabs, are all clear indications of the closeness of these political relations. It is evident that by the latter a connexion of commerce must also have been encouraged, for which the Oxus valley and Khotan was the natural and most direct route. There is little hope of our ever obtaining light as to the incidents and conditions which brought the writer or the recipient of this Judaeo-Persian epistle to Khotan. But even without such information this fragmentary paper may claim to be considered one of the most interesting relics from Dandān-Uiliq; for it is a direct witness to those relations with distant Īrān which so much in the Buddhist art of the site presupposes, and at the same time, strangely enough, it is the earliest document in modern Persian which has come to light as yet.

The second relic from Dandān-Uiliq, which reached me at Rawak, the small piece of stucco inscribed with Chinese characters and apparently cut out from a plaster-covered wall, does not call for so full a notice. Owing to the extremely friable condition of the stucco it broke into several pieces before it reached London, but the few characters that were legible when it was first brought to me have escaped further injury. According to Dr. S. W. Bushell, who was kind enough to examine the inscription, 'it has every appearance from the handwriting of being a genuine relic of the T'ang dynasty'. The writing was arranged in three columns, each about 5 inches high and originally holding about eight characters, but owing to the painted surface having peeled off over the greater part Dr. Bushell 'could only decipher the first three characters and the upper two of the second column; these are *Fo ti-tzŭ* 佛弟子 "The Disciple of Buddha", and *Kuan-shih* 觀世, sufficient to indicate a memorial of the dedication of an image of Avalokiteśvara (Kuan-shih-yin). The occasion would have been given lower down, as *wei* 爲 "on account of", occurs there where all else is gone.' Chinese inscription on stucco.

Section III.—KERIYA, NIYA, AND IMĀM JA'FAR SĀDIQ

The examination of the scanty remains at Rawak had completed the task for which I had set out just a month previously from Khotan. So on the morning of January 6 I dismissed Ahmad Merghen with the last batch of the Tawakkēl labourers, and set out with a much reduced caravan for the Keriya river. About two miles to the south-east of my Rawak camp we passed a strip of ground about half a mile broad, where broken pottery, fragments of glass, and the usual 'Tati' débris cropped up on the bare patches of loess between the dunes. Turdi called the place *Tört-Uiliq* 'the four Houses', but had never come across any structural remains on it. Beyond, all traces of ancient habitation ceased, and soon I passed also the last of the shrivelled dead trees, here a clump of wild poplars (Toghraks), with the sight of which I had become so familiar during these weeks. I had originally intended to steer due east, in order to strike the nearest point of the river, but the rising height of the dunes and the impossibility of getting at water obliged me after the first day to seek the route south-eastwards which the camels had previously followed. Even thus the two remaining marches led over truly forbidding ground. The individual sand-dunes were all between 30 and 50 ft. high, while the successive great ridges of sand or 'Dawāns', of which some seven had to be crossed, March to Keriya river.

assumed formidable proportions, rising to heights between 150 and 200 feet. By the evening of January 8 the hard-frozen river was at last safely reached near the grazing-ground of Kochkar-Öghil.

Mazār of Burhānuddīn.

I had decided first to visit Keriya, the head quarters of the District east of Khotan, before commencing other explorations, in order to secure personally the assistance of the local Amban as the indispensable condition for successful work. Kara-dong, the old site higher up the Keriya river, seemed temptingly near, but in the lonely jungle tracts along the river, uninhabited except by nomadic shepherds, it would have been impossible to raise either labourers or the badly-needed supplies. For a description of the four long marches which brought us to Keriya through the belt of Toghrak jungle and scrub accompanying the river's course in the desert I may refer to my Personal Narrative[1]. The only place of quasi-antiquarian interest was the Mazār of Sayyid Burhānuddīn Pādshāhim, reached after the first march, where I was surprised to find a small but flourishing settlement of Shaikhs (see Fig. 36), attesting the widespread popularity which the saint's tomb enjoys as a pilgrimage place for the people of the oases southwards. Nothing seemed to be known of the holy man's story except that he was connected in some way with the still holier Imām Ja'far Sādiq worshipped at a famous desert shrine beyond Niya, and that like him he had fallen as a 'Shahīd'. Considering the distance, fully 50 miles, at which the Burhānuddīn Mazār is situated from the oasis of Keriya, and the wholly deserted condition of the intervening ground, I wondered whether the worship of this far-outlying shrine may not in some way represent a survival of the attraction which the Buddhist establishments of Li-hsieh or Dandān-Uiliq had once possessed for the pious. Some cultivation is carried on near the shrine, and this could be greatly extended, as there is plenty of fertile level ground as well as abundance of water.

Stay at Keriya.

At Keriya, which I reached on January 12, I found Huang Daloi, the Amban, duly advised of my visit and its object by Pan Dārin, my Khotan friend, and full of eagerness to facilitate my explorations in every possible manner. The necessity of giving my men and camels a much-needed short rest, as well as of awaiting the arrival of the ponies which were to rejoin us from Khotan, obliged me to make a stay of five days at Keriya. Apart from numerous tasks which had accumulated during my stay in the desert, and which could only now be attended to, I used it for careful inquiries about ancient remains in the district. Of antiques none were forthcoming; for Keriya town is not itself an old place—it has acquired size and importance only since the creation of the new District of which it forms the head quarters[2]—and the 'treasure-seeking' profession does not flourish as at Khotan. But besides learning of a number of sites manifestly of the Tati type, I received information about a *kōne-shahr* or 'old town' in the desert north of Niya, which decided me to extend my explorations first in that direction. Abdullah, a respectable old cultivator of Keriya, told me of having seen about ten years earlier houses, evidently of a type similar to the structures of Dandān-Uiliq, half-buried in the sand some marches beyond the famous Mazār of Imām Ja'far Sādiq. Others, too, had heard stories of this 'old town', and the existence of the Muhammadan shrine itself seemed, in the light of former experience, to point to some earlier site of interest being found in the vicinity. So I did not hesitate to set out for it on the 18th of January.

During the four days' march which brought me to Niya there was little to be seen but

[1] See pp. 329 sqq.

[2] The district which was formed after the Chinese re-conquest of 1877 out of the eastern portions of the old Khotan Amban-ship now extends from east of Yurung-kāsh town to beyond Charchan, and thus comprises a territory, mostly desert, stretching over no less than five degrees of longitude. It is curious to note that the Chinese administration officially styles it by the name of *Yü-t'ien*, thus transferring to it the ancient Chinese designation of the Khotan oasis.

the mighty snow-covered wall of the outer Kun-lun range to the south and the high dunes of the desert northward. The barren uniformity of stony 'Sai' or low sand-dunes over which the route leads is broken only by the small oasis of Ui-toghrak, east of Keriya, and the tiny halfway settlement of Yes-yulghun[3]. At Niya I was glad to find myself once more on Hsüan-tsang's track, for there can be no doubt that the present small oasis corresponds to the pilgrim's town of *Ni-jang*. He reached it after entering the desert east of P'i-mo, which, as already noted, must be located near Uzun-Tati, north of the present oases of Chīra and Gulakhma. The distance of 200 li between the two places is certainly under-estimated, even allowing for the reduced length of the more direct route north of the present one, which Hsüan-tsang is likely to have followed. But apart from this measurement, his account of Ni-jang, 泥壤, points quite clearly to the present Niya[4]. 'This town is 3 or 4 li in circumference; it is situated in a great marsh; the soil of the marsh is warm and soft, so that it is difficult to march on it. It is covered with reeds and tangled herbage, and one sees neither roads nor tracks. There is only the route leading to the town which is practicable to some extent. On this account those coming and going must pass by this town. The king of Ch'ü-sa-ta-na [Khotan] has placed there the guard of his eastern frontier.'

Niya, Hsüan-tsang's *Ni-jang*.

The extensive marshes, which, as the map shows, are found on both sides of the Niya river just north of the present cultivated area and the main village of Niya, correspond exactly to the pilgrim's description. The route from the east winds through them for a distance of close on four miles, and westwards also the traveller would have to pass through them if he followed, as Hsüan-tsang in all probability did, a track a little more northerly than the present Niya–Keriya route. Swamps of less width, but equally thickly covered with reeds and 'tangled herbage', accompany the river-course for about twelve miles further north, and between them numerous small lakes are passed. Neither the stony 'Sai' commencing south of Niya nor the forest belt further north offers similar features, and it hence appears to me safe to conclude that Hsüan-tsang's town of *Ni-jang* must have lain within the zone just indicated, though its location may not have been exactly the same as that of the present Niya villages. The latter, counting altogether about 500 houses[5], formed the easternmost of the smaller oases which were included in the Khotan District until the recent constitution of Keriya as a separate administrative unit extending to Charchan. Thus Niya may well be said to have retained to our own days the character of a frontier station, as noted by Hsüan-tsang. *Ni-jang*, by which the *Hsi-yü-chi* and the 'Life' render the old name of Niya, has been supposed by Julien to reproduce a form approximately sounding **Ninya*, because the character 壤 in other transcriptions is used to represent the Indian *ṇya, ña, jña*[6]. The supposition that the second syllable of the old name began with a nasal sound would be greatly strengthened if my suspicion should prove right that by *Nina*, a locality repeatedly mentioned in one of the Kharoṣṭhī documents from the Niya Site which Professor Rapson has published, our Niya is meant[7]. The phonetic transition from a form **Niña* into Niya could be accounted for by a parallel phonetic process in the case of Prākrit languages[8].

Marshes around Niya.

[3] For a description see *Ruins of Khotan*, p. 340.

[4] See *Mémoires*, ii. p. 246; *Si-yu-ki*, transl. Beal, ii. p. 324.

[5] For a brief description comp. Hedin, *Reisen in Z.-A.*, p. 215; also *Ruins of Khotan*, pp. 341 sq.

[6] See *Mémoires*, ii. p. 517; *Méthode pour transcrire*, p. 115.

[7] See Kharoṣṭhī tablet N. i. 16 in Prof. Rapson's *Specimens of Kharoṣṭhī Inscriptions*, &c., p. 14. This tablet, forming together with N. i. 104 a double 'wedge' document, refers to the dispatch of a certain Ṣameka as a messenger from *Calmadana* via *Saca* and *Nina* to *Khotana*; comp. below, p. 326. I surmise that by *Calmadana* is meant the same locality which Hsüan-tsang calls *Chih-mo-t'o-na*, and which, being placed ten marches to the east of Niya, manifestly corresponds to the present Charchan.

[8] If Hsüan-tsang's *So-mo-jê*, discussed above, pp. 223 sqq.,

First Kharoṣṭhī documents on wood.

The halt which I had to make at Niya on January 22, in order to arrange for the labourers and supplies to be taken to the desert site I was about to visit, proved unexpectedly profitable from the archaeological point of view. A large jar of ancient pottery (see Pl. IX), remarkable for its good preservation and the hardness of its material, which had been found some five years earlier at that site by a Niya village headman and brought away to his home, was at first the only antiquity that Niya could show me. It measured 33 inches in diameter and height, with a mouth 10 in. wide. But in the afternoon I received proof of a kind wholly unexpected as to the great age of the desert site I was bound for. One of my two camelmen, a young fellow of an unusually inquisitive turn of mind, had in the Bāzār come across a villager possessing two inscribed tablets brought away from that site. As soon as they were produced before me I discovered to my joyful surprise that they contained Kharoṣṭhī writing, of a type which closely agreed with that prevailing during the period of Kuṣana rule in the first centuries of our era [9].

Origin of find.

The man who brought me the tablets had picked them up near Mūsa-Bēgim Öghil, the first stage on the pilgrims' road to Imām Ja'far Sādiq's Mazār. But the original finder was soon ascertained in the person of Ibrāhīm, an enterprising young villager, who had dug them out from a 'house of the old town' beyond. He had gone there a year before in search of treasure, but had found only these, to him, useless tablets. He brought away half a dozen or so, only to throw some away on the road and to give the rest to his children to play with. Of the latter tablets only one could be recovered next morning [10], though Ibrāhīm, seeing how well I rewarded the more sensible second-hand finder, had eagerly searched his house for them. He declared that he had left plenty more at the find-place. Fearing the possibility of being forestalled, I tried to hide my delight, but took care to secure Ibrāhīm as a guide and to assure him of a good reward if he could show me undisturbed the ruined structure where he made his find. The ancient bilingual coins of Khotan, and the fragments of the Dutreuil de Rhins codex, had so far been the only evidence of the use of Kharoṣṭhī writing in Central Asia. Hence the two tablets which a fortunate chance had placed in my hands were scanned by me with no small keenness in the evening. The writing on *b.* (a rectangular 'under-tablet', to use the terminology subsequently established) seemed legible enough, in spite of the faded appearance of the ink; but the very cursive form of the characters and philological difficulties since better appreciated prevented any attempt at immediate decipherment. Nevertheless even this cursory study, resumed after each day's march, sufficed to convince me that the tablets were documents with an early Indian text, and to assure me of the antiquity of the ruins. But full of expectation as I was, I little anticipated at the time what a rich harvest was awaiting me.

Marchalong Niya river.

On January 23 I set out from Niya for Imām Ja'far Sādiq with twenty labourers and a small convoy of additional hired camels to help in the transport of a month's supplies. A three days' march brought me to the shrine, the starting-point for my fresh expedition into the desert. Famous as a pilgrimage-place throughout Turkestān, it had never previously been visited by Europeans except M. Dutreuil de Rhins and M. Grenard. The route lay all along the Niya river, and through the belt of thick forest which accompanies its course from the point where the marshland immediately north of Niya is left behind to where the river dies

is intended to transcribe a name **Somaña*, or some similar form, we should have an instance of that process in Khotan itself; for the present name of the locality meant is *Somiya*, as demonstrated above, p. 225.

[9] Now marked, with the date of their acquisition, N. 22. i. 1901, a, b.

[10] Marked N. 23. i. 1901.

away in the sand. For a detailed description of the scenery passed through and observations on geographical features I may refer to my Personal Narrative[11]. Among the latter I may note here that the Niya stream, just like the Keriya Daryā, gathers volume from the springs and marshes below the oasis. These are, of course, fed by the water which has been absorbed higher up by irrigation and comes to the surface again lower down. In view of the ample ground available for cultivation from below Mūsa-Bēgim, where the jungle belt widens out to eight miles and more, I was particularly interested in the new canal which had been begun here two years before by the Keriya Amban's orders, and which could be followed along the well-marked pilgrims' road for a distance of over eight miles. The ground is everywhere fertile loess, and so level that the creation of a large colony would be easy if the effort were persisted in and an adequate population assured. Further to the north the route lay in parts along old beds of the river, all lined with luxuriant Toghrak jungle, while elsewhere patches of dead forest indicated ground which the shifting of the stream eastwards had long ago deprived of its subsoil water.

The actual river-course, where met again some six miles south of the Mazār, had dwindled down to a narrow band of ice scarcely 20 feet in width, and further on was rapidly diminishing through branches sent off on either side. Yet the trees and shrubs around seemed to increase in size and luxuriance. It was evident that here, near the river's end, the fertilizing power of its water, freely spread out during the summer floods, was strongly reasserting itself—an apt illustration, it seems to me, of the advantages which the position of a terminal oasis must always offer for cultivation.

Mazār of Imām Ja'far Sādiq.

After the days spent in this lonely woodland the collection at the Mazār of Mosques, Madrasahs, and houses for Shaikhs and pilgrims, humble as the structures are, looked impressive. It attests the popularity of the saint, whose supposed tomb occupies the top of a prominent ridge to the west of the settlement. A group of small lakes divides the latter from the hill, which rises to a height of about 170 feet. On ascending it I was much struck by various signs of the pilgrims' devotion, and of the care taken to foster it. The fine old trees at the foot of the ridge, and the large number of rough wooden arches passed by the path winding to the top, are hung with the largest and most motley collection of votive offerings I ever saw in India or Turkestān. The variety of the materials represented among the rags originating from widely distant parts of Asia and Europe which make up the mass of these ex-votos, would make this exhibition of textile fabrics a most instructive archaeological find if it were safely buried beneath the desert sand and excavated after long centuries. Little heaps of earth arranged like graves, and covering the slopes of the hill in thousands, symbolize the resting-places of the Shahīds who are supposed to have fallen here with Imām Ja'far Sādiq, their holy leader, in his last fight against the infidels of *Chīn-u-Māchīn*.

Legend of saint. Origin of local worship.

In the long Tadhkirah of the warrior saint, from which M. Grenard has translated full extracts, it is not possible to trace any special reference to the site where his worship is now localized, except that the legendary makes him succumb in the desert. This makes it appear still more probable than it would be *a priori* that the sanctity of the hill goes back to pre-Muhammadan times. I was hence particularly interested to find that a striking natural peculiarity allows us here to account for this sanctity. While the forest belt of the river is everywhere else bordered by high ridges of sand, such as I have noticed in connexion with the marches from Dandān-Uiliq to the Keriya river, I found to my surprise that the soil on the slopes

[11] See *Ruins of Khotan*, pp. 345 sqq.

of the sacred hill consists solely of stony detritus overlying reefs of salt. The presence of a real hill so far away from the foot of the mountains, and in a position surrounded by drifting dunes, would, like all striking natural features, suffice to attract local worship, as shown by the numberless 'Svayaṃbhū Tīrthas' of India, ancient and modern[12]. But the occurrence of rock salt, so rare elsewhere in this region, must have increased pious regard in this case.

Arrangements for supply of ice.

On January 26 we left this curious desert shrine, after having taken along as additional labourers a dozen or so of able-bodied men from the secluded little settlements of Shaikhs, shepherds, and mendicants dependent on the Mazār. The small watercourses into which the river splits up before it is finally absorbed in the sand reach only some three miles beyond Imām Ja'far, and as water cannot be got at the site by digging the four iron water-tanks brought by me had to be filled before the start. It would have been very difficult, if not impossible, to keep my camp, counting in the end from forty to fifty people, supplied with the indispensable amount of water so far out in the desert had not the intense cold still prevailing (on January 26 I registered a minimum of 44° Fahr. below freezing-point) permitted of its convenient transport in the form of ice, sacks and nets being improvised for the purpose. At a little farm kept by shepherds of the Mazār, near the terminal marsh of Tülküch-Köl, our ponies were left behind, and only the camels taken onwards. Some eight miles beyond the Mazār the forest, before invaded by heavy drift-sand, gave way to a wide expanse of sand-cones thickly overgrown with tamarisk scrub. From a high sandhill near the extreme limit of living trees where my camp had been pitched, I could make out a broad scrub-covered belt stretching away to the NNW. between the great sandy ridges of the true desert. It clearly marked the direction of the old extension of the river course.

Approach to ancient site.

The march of the next day, January 27, confirmed my surmise that the ancient site would be reached by following this direction. For the first five miles or so thick patches of dead forest were encountered between the tamarisk-covered hillocks. The time when its trees, mostly Toghrak, had flourished could not be very remote, for many of the lifeless trunks still retained their branches. Winding along the eastern edge of this dead forest, a dry channel, about 4 ft. deep and about 10 feet broad, could be traced for some distance. All remains of old forest disappeared when, lower down, we entered a zone of steep conical sand-hillocks 15 to 30 ft. high, rising close to each other, and all covered with tamarisk scrub on their tops. In the midst of this belt, about three miles broad from north to south, I came upon a small open area showing broken pottery and remains of an enclosure made of thickly-packed rushes. Inside it the men recognized a few much-withered trunks of planted poplars or *Terek*. It was manifestly the last trace of some ancient farm, or, perhaps, of some village site completely covered up elsewhere by the closely packed sand-cones[13]. Beyond, the ground became more open, with low bare dunes resembling those about Dandān-Uiliq. Above them rose here and there isolated sand-cones, bearing the only living tamarisk bushes that were to be seen. After marching for about five miles patches of ground between the dunes strewn with potsherds, fragments of stone, slag, &c., showed that we had reached the southern edge of the ruined area. The total distance from Imām Ja'far had been about 24 miles, less than the three marches for which my guides Abdullah and Ibrāhīm had prepared me.

[12] It is noteworthy that the small rocky hills, which as the last isolated remains of a completely decayed mountain range crop out of the desert near the courses of the Khotan and Yarkand rivers (see Hedin, *Reisen in Z.-A.*, pp. 220 sq., 242, &c.) are everywhere known by the name of *Mazār-tāgh* 'the hill of the sacred shrine.'

[13] The débris of the *Aktaz* site near Domoko was found in exactly similar surroundings; see below, chap. XIII. sec. ii.

Shrivelled trunks of ancient fruit-trees appeared rising from the low sand. Moving on northward for less than two miles I soon sighted the first two 'old houses', standing on what looked at first sight like small elevated plateaus, but which closer observation proved to be merely portions of the original loess soil that had escaped the erosion proceeding all round. These were the ruins to be described below as N. III. and N. IV.[14] A rapid inspection showed me that the mode of construction in these buildings was substantially the same as that in the dwellings of Dandān-Uiliq, but their dimensions were larger and the timber framework far more elaborate and solid. The conclusion that I instinctively drew as to the far greater antiquity of these remains was confirmed almost immediately when, in one of the outer rooms of the ruin N. III., where the sand lay only about half a foot deep, I came upon some finely-carved pieces of wood lying practically on the surface, which showed ornamentation unmistakably of the Gandhāra style. I subsequently identified them as parts of the ancient chair seen in Plate LXVIII. Marching about two miles further north, across broad swelling dunes, I arrived at the ruined structure of sun-dried bricks of which Abdullah had already spoken at Keriya as a 'Potai'. It proved, as I had expected, to be the remains of a small Stūpa, buried for the most part under the slope of a high conical sandhill. First ruins sighted.

Here, in a position conveniently central for the exploration of the scattered ruins (see Plate XXVII), I pitched my camp. The ground in the immediate vicinity showed all the characteristic traces of excessive erosion. Broad patches of bare loess displayed in profusion pieces of broken pottery, bleached and twisted trunks of fallen poplars and other garden trees, as well as much decayed remains of ancient timber that splintered and broke almost as soon as lifted. Of the buildings to which this timber débris had belonged it was impossible to trace even the roughest outlines. A few small loess-banks, regular 'witnesses', with almost vertical wind-eroded sites, which rose a short distance to the west of the Stūpa, 12–15 feet above the level of the débris-covered area, showed plainly how far the present surface of the latter had been reduced below the ancient ground-level. Numerous fragments of stone, evidently the remains of larger pieces that must have once been brought from far away for use in the houses, bore even more impressive evidence to the destructive force of the desert winds and of the extremes of climate. But as I retired to my first night's rest among these silent witnesses of ancient habitations my main thought was how many of the precious documents on wood, which Ibrāhīm declared he had left behind at the ruin 'explored' by him a year before, were still waiting to be recovered. Effects of erosion near Stūpa camp.

[14] See Figs. 40, 41.

CHAPTER XI

THE ANCIENT SITE BEYOND THE NIYA RIVER

SECTION I.—THE RUIN N. I. AND THE FIRST FINDS OF INSCRIBED TABLETS

Ibrāhīm's find-place of Kharoṣṭhī tablets.

AT sunrise of January 28, with the temperature still well below zero Fahr., I hastened to the ruined building where Ibrāhīm had a year previously picked up his ancient tablets inscribed with Kharoṣṭhī characters. According to his statement, he had left plenty more of them *in situ.* It had been impossible to hide from him the value which I attached to these records; and as he subsequently seemed to regret not having himself made a haul of them, I had him watched *en route* by Ibrāhīm Ākhūn, the excellent Darōgha whose services the Amban of Keriya had placed at my disposal, and also after our arrival, to prevent his escape or any possible interference with the spot. The mingled feelings of expectation and distrust with which I now approached it, soon changed to joyful assurance. About one mile to the east of the camp I sighted the ruin (marked N. I. on the plan) to which Ibrāhīm was guiding us, on what looked like a small terrace or plateau rising 12–15 feet above the eroded ground near by[1]. On ascending the west slope, seen in the foreground of Fig. 37, I picked up at once three tablets inscribed with Kharoṣṭhī lying amidst the débris of massive timber which marked wholly eroded parts of the ruined structure. On reaching the top I found to my delight many more scattered about in the sand within the nearest of the rooms still clearly traceable by remains of their walls. The layer of drift-sand that had spread over the tablets since Ibrāhīm had thrown them down here a year before, was so thin as scarcely to protect the topmost ones from the snow that lay about one inch deep over the more shaded portions of the ground. It dated, no doubt, from the snowfall which I had encountered on my way from Keriya to Niya eight days earlier.

Original place of discovery in ruin N. I.

Ibrāhīm seemed scarcely less elated than myself at seeing his statement confirmed, and the good reward I had promised him thus assured. He at once pointed out to me that the find-place of the relics was not in this room, marked *v. a.* in my detail plan Pl. XXVIII, where he had thrown them away in utter ignorance of their value, but in the south corner of the room *i.* immediately adjoining eastwards. There, in a little recess (*a*) about 4 feet wide, formed between the fireplace, well recognizable above the sand, and the wall dividing rooms *i.* and *v.a.* (seen on the right in Plate VI), he had come upon a heap of tablets while scooping out the sand with his hands in search of 'treasure'. The ancient documents which he appears to have found lying in horizontal rows, possibly with some sort of arrangement, on the low mud platform extending along that side of the room, impeded his burrowing, and were hence promptly thrown across the decayed wall into the next room. It was a fortunate chance which had brought me to the site so soon after his discovery. For, fully exposed to the sun and wind, these thin wooden boards could not have long retained their writing in such wonderful freshness as they had during their safe

[1] The photograph, Fig. 37, taken after excavation from the height of a large tamarisk-covered sandhill to the north-east, well illustrates the position of the ruin. The rooms of the north wing are seen in front, and behind them to the left the remains of the east wing. For the patches of snow lying on the north slopes of dunes, see below.

36

MAZĀR OF BURHĀNUDDĪN-PĀDSHĀHIM ON KERIYA RIVER.

37

FIRST FIND-PLACE OF INSCRIBED TABLETS.

RUINED HOUSE N. I, NIYA SITE, SEEN FROM NORTH-WEST AFTER EXCAVATION.

interment of many centuries beneath the sand. As it was, the sun of one year and its rain and snow, however slight in quantity, had sufficed to bleach and partly efface the fully-exposed writing of the topmost tablets.

Clearing of room N. i. After placing a guard over the room where Ibrāhīm's scattered finds lay, so as to prevent further injury or abstraction, I set the men to work to clear the room N. i. (see Plate VI), where he had first come upon them. This was an easy task, as the room measured only 16 by 14 ft., and the sand filling it was not deep. On the north-west side, near the edge of the eroded slope, it lay only to a depth of about 2 ft., which increased to about 4 ft. further in, where the better preserved south-east wall retained the drift-sand. While the clearing proceeded cautiously, I had time to examine the system of construction followed in the ruined building. As the same was subsequently found in almost all other structures of the site, it will be convenient to describe it in this place. In its general features it showed close resemblance to that noticed in the Dandān-Uiliq structures, being based on the use of timber; but there were characteristic minor differences. Massive squared beams of White Poplar or *Terek* wood, usually extending below several rooms, and in some instances exceeding 40 ft. in length, formed a kind of foundation; their thickness, which varied from 6 to 10 ins. according to the size and importance of the walls they supported, and their perfect finish and fitting always caused my workmen to wonder at the skill of those ancient carpenters. On this foundation were set wooden posts from 4 to 6 ins. square, which supported the roof and at the same time served as a frame for the walls. These, and smaller but equally well-finished intermediary posts, fixed at regular intervals usually of about one foot, were joined by heavy crossbeams on the top and light ones between. In the ruin N. I. the upper portions of the walls had decayed too far to show the crossbeams in position; but they can be seen clearly, or else the dowels that once held them, in the photograph (Fig. 41) of the better preserved large dwelling N. III. The arrangement of the upright posts is, however, fully visible in Plate VI, showing the room N. i. after excavation. To this framework, and usually on the outside of the small intermediary posts, was fixed a strong kind of matting of thin tamarisk branches woven diagonally. This again was covered on each side with layers of hard, white plaster, giving a total wall-thickness varying from 6 to 8 ins. in different structures. In the photograph (Plate VI), showing the north-east and south-east walls of room *i.* in N. I. as seen from outside before excavation, the diagonal matting, which had become denuded of its plaster covering owing to exposure above the sand, is distinctly visible. This diagonal tamarisk matting undoubtedly supplied a stronger core to the walls than the horizontal layers of reeds used for the same purpose at Dandān-Uiliq. But that the latter system was known also at the earlier period to which the ruins of the Niya Site belong is proved by its employment in several instances in the ruins N. III., N. IV., and N. V., to be noticed hereafter.

Construction of houses.

Walls of rushes and plaster. A third, far rougher, method of wall-construction, which closely resembles that still in vogue in the Khotan region, was applied apparently only to cattle-sheds, stables, and similar outhouses. It consisted of vertically-placed and closely-packed rushes covered with layers of mud plaster. Walls of this kind, in which rough wooden corner-posts supported the roof, while saplings or roughly cut branches of trees were inserted at intervals to strengthen the rushes, were found near N. I. and most of the ruined buildings of the site, and have been distinctively shown in the plans. In N. I. also the more solidly built walls had completely decayed, where not actually covered by sand, but many of the stronger posts originally holding them still rose high above the surface, in some instances 10 feet and more. Their splintered and shrivelled appearance showed the long periods for which they had been exposed to the destructive forces of the desert.

Inscribed tablets from room N. i.

As the room N. i. was gradually cleared the inscribed tablets or portions of tablets now marked N. i. 100–114 turned up at different places on the low platform of plaster along the south-west wall and towards the centre of the room. Most of them lay close to the mud floor or else above a thin stratum of sand. It was clear that they could have reached that position only by accident. Two more tablets, marked N. i. 120, 122, together with a carved piece of wood from a rail or baluster, N. i. 121 (see Pl. LXX), were found lying on the platform that lined the completely decayed north-west wall, while another inscribed tablet (N. i. 180) was the solitary find near the north-east wall. There was nothing directly to indicate that these tablets had been removed at some earlier date from the main deposit of wooden documents which Ibrāhīm had lighted upon in the southern corner of the room; yet their widely scattered positions seemed to suggest this, and the supposition has since been confirmed by Professor Rapson, who has recognized in tablet N. i. 16 the cover fitting to, and completing the text of, tablet N. i. 104[1]. As soon as I had convinced myself that the room held no other remains, I proceeded to collect carefully the tablets which had been left behind from Ibrāhīm's find. Those lying on the surface of the sand in the adjoining room were marked N. i. 1–52, in the order as they were recovered from west to east, while a search in the sand immediately below them, and in the same order, revealed those numbered N. i. 53–85.

Clearing of apartments N. v. a. and N. ii.

The subsequent clearing of this room (N. v. a.) brought to light only a low plaster platform running round three of its sides. In the same narrow room (N. ii.), however, which communicated by still traceable doorways with N. i. and N. iii., and probably served as a mere passage, three more documents were discovered, among them a remarkably well-preserved one, the 'double wedge' tablet N. ii. 2 with its clay seal and fastening still intact.

Arrangement of wedge-shaped tablets.

The hundred and odd inscribed tablets with which I returned to camp from my first day's work amid these ruins represented a harvest far more abundant than I could reasonably have hoped for. The remarkable state of preservation in which a considerable portion of them was found made it easy for me, even during the first rapid examination on the spot, to recognize certain main features in their outward arrangement; and the few hours of study which I was subsequently able to devote to them in my tent during the bitterly cold evenings soon familiarized me with some aspects of their use as an ancient writing material. Leaving aside the three tablets found in room N. ii. and some ten pieces which were oblong or otherwise of peculiar form, all tablets from N. i. were wedge-shaped and showed unmistakable evidence of having been originally arranged in pairs. Eight of such pairs of accurately fitted tablets (N. i. 9, 12, 64, 67, 78, 83, 105, 122) were still held together by a string, which passed round the square ends of the wedges and also through a hole drilled into both tablets where their left end tapered to a point. One of the complete pair of wedge-shaped tablets showed invariably on its outside surface a square socket sunk into the wood and communicating with grooves which held the string in carefully arranged cross-folds. This socket, of what I may distinguish as the 'covering-tablet', still retained the clay sealing for which it was intended, either complete or in part, not only in most of the complete pairs (N. i. 9, 12, 64, 67, 122), but also in several of the numerous 'covering' tablets found detached (N. i. 14, 17, 35, 51, &c.). Of such detached tablets altogether some 38 are clearly recognizable by their seal-socket as covering-tablets, many of them also retaining portions of their ancient string fastening.

Covering- and under-tablets.

The string-hole near the pointed end and the peculiar arrangement of the writing to be referred to presently made it easy to recognize in the rest of the detached wedge-shaped tablets

[1] See *Specimens of Khar. Inscr.*, p. 14; also below, p. 326.

from N. i. (about 41, including several fragments) corresponding under-tablets belonging to original pairs. A number of these detached tablets have since been re-united by Mr. Andrews and Professor Rapson into complete 'double-wedge' documents, e.g. N. i. 16+104 already referred to. This possibility had, indeed, from the first suggested itself, in view of the close approach in number of the two classes of wedge-shaped tablets recovered. It is probable that the covering- and under-tablets of many of these documents still lay close together, even though their original fastening may have been missing when Ibrāhīm lighted upon them. But in view of the thoroughness with which he had disturbed the original position of this collection of documents no certain opinion can be expressed on the point. The fact of the tablets subsequently excavated by me in other places of N. i. having proved to be mainly detached pieces suggests that the deposit of documents may either have been disturbed by some earlier searchers, or originally thrown down in confusion just as seems to have been the case in N. iv.

Description of wedge-shaped tablets from N. i.

The exact details of the ingenious method of fastening adopted for these remarkable documents on wood were fully ascertained by me only on the subsequent discovery of practically perfect specimens among the rich finds yielded by the ancient rubbish-heap N. xv. It will, therefore, be more convenient to leave their discussion, as well as that of other technicalities of archaeological interest connected with the use of this ancient 'stationery' on wood, for a later section (iv.). Exact details as to the size, type, and condition of each tablet found in N. i. are given in the inventory list at the end of this chapter. It will hence suffice to describe here the general appearance of the wedge-shaped tablets recovered from this particular place, and to call attention to individual pieces. The length of the tablets, which in each pair fitted each other exactly in size and shape, varies from 7¼ in. to 15½ in. (N. i. 9. a), and their width at the square end proportionately from 1¼ in. to 2¾ in. The thickness of the wood is from a quarter to three-quarters of an inch. The wood is generally of uniform thickness throughout each tablet, except near the right hand or square end of the covering-tablets, where an extra thickness has ordinarily been spared for the square seal-socket. The raised edges of the latter are always neatly bevelled down sideways towards the square and pointed ends of the tablets, while the upper and lower edges are left high for the formation of the string grooves (three on each side). The seal-socket is rectangular, in most cases square, the sides varying in proportion to the width of the tablet from 1½ to 1 inch. Plates XCVIII and C show the complete double-wedge documents N. i. 122 and N. i. 9 as seen from the Reverse of the under-tablet, with their original fastening of hemp string. In Plate IC the Obverse of N. i. 103 is reproduced; here the string crossfolded through the grooves and seal-socket is visible, owing to the clay sealing which had once been inserted in the latter having perished, as in the great majority of tablets.

Kharoṣṭhī text of wedge-shaped tablets.

The text, invariably written in Kharoṣṭhī characters and running from right to left parallel to the longer side, occupies the inner surfaces of the tablets, i.e. the Rev. of the covering- and the Obv. of the under-tablet, which in the arrangement of pairs as originally fastened were turned towards each other. The text always commences on the top of the Obv. of the under-tablet, and only its conclusion is written on the Rev. of the covering-tablet, which explains why the latter often shows only a single line of characters or is left altogether blank, as in N. i. 23, 24, 67, 43, 55. The reproduction given in Plates XCVIII, IC of the inside faces of complete wedge-shaped documents from N. xv. will serve to illustrate the corresponding arrangement in all wedge-shaped tablets of N. i[3].

[3] The inside faces of two documents (N. i. 104+16 and N. i. 105) will be found reproduced among Prof. Rapson's *Specimens*.

Entries on outside surface of 'wedges'.

On the Obv. of the covering-tablet, which bears the clay seal, and which soon proved to have served the purposes of a kind of envelope, there appear invariably brief entries in Kharoṣṭhī script, consisting of a short line to the right of the seal-socket and of a few characters to the right of the string-hole near where the wedge runs out in a point. These entries, having become effaced through exposure, do not appear clearly in the reproduction of N. i. 103 (Plate IC), though the original shows them; but in the reproductions of the wedge-shaped tablets N. xv. 24 (in the same plate), N. xv. 137 (Plate XCVIII), N. xv. 71 (Plate C) they are plainly visible. Their very position and form at once suggested that they were intended either as memoranda of the contents, or else to convey the name of the sender or addressee. The last supposition proved the right one, when subsequent examination showed that the three characters near the string-hole were to be read *datavo* 'to be given to', and that the words on the right of the seal-socket always contained a name with the appropriate case-ending. The Rev. of the under-tablet, otherwise ordinarily left blank, shows in all well-preserved tablets a single word written near the square end, which, by Professor Rapson's examination, has proved to contain the name of the person entrusted with the document. This last-named entry is seen on N. i. 9 (Plate C) and N. xv. 137 (Plate XCVIII). N. i. 122 (Plate XCVIII) is interesting as the only instance where the Rev. of an under-tablet displays this entry not written in ink but engraved into the wood, and in a transverse direction, instead of the usual one parallel to the longer side. The inscribed tablets not showing the wedge-shape which turned up in N. i., only some eleven in number, were too few and too varied to yield at the time any clue as to their character or arrangement. Hence it must suffice to mention among them N. i. 70 (see Plate CIII), which, though shaped like a wedge and provided with a string-hole was not a document of the regular type. The deep groove passing round the broad end was certainly meant for a fastening. The few characters (*datavo*) near the string-hole show that the tablet contained an address, and suggest its having been used as a label, perhaps attached to a bag. For another tablet of peculiar shape, see N. i. 56 in inventory.

Preservation of inscribed tablets.

Apart from the large number of inscribed tablets yielded by N. i., this first find-place of the site was remarkable also for the relatively good preservation of most of its epigraphic relics. Where double tablets had remained together and thus protected each other, the black ink of the Kharoṣṭhī lines written on the inner surfaces looked as fresh as if penned quite recently. On those tablets which, when thrown down by Ibrāhīm, had come to lie with their written surface fully exposed, the characters had often become bleached, evidently by the effect of sun and sand; while the ink had 'run' or become somewhat blurred through the moisture which the recent snowfall had brought. But the writing, even on these pieces, has scarcely ever become illegible. Curiously enough, some of the tablets which Ibrāhīm had left undisturbed below the sand had suffered a good deal more by warping, cracking, and similar damage to the wood. The cause in these cases was undoubtedly the position close to the floor which the tablets occupied, and which had made them liable to absorb what subsoil moisture the ground received after the abandonment of the site. That some had suffered even in ancient times becomes probable from the insect borings found in several (e.g. N. i. 12, 49, 60), and from the broken condition of others which Ibrāhīm had not disturbed (e.g. N. i. 100+109, 111, 101, 106).

Character of Kharoṣṭhī writing.

The good preservation of the great majority of that first day's finds made it easy to recognize that the tablets, though written by many different hands, showed throughout the characteristic peculiarities of that type of Kharoṣṭhī writing which is exhibited by inscriptions of the extreme North-West of India during the Kuṣana or Indo-Scythian rule. In spite of the chronological problems presented by the era used in those inscriptions, it is certain that the period of this rule falls within the first three centuries of our era. Hence, while still engaged in gathering the remarkable documents that were coming to light that day in such surprising numbers, I felt absolutely assured as to their exceptional value. At first the apprehension obsessed me, that these strange records, strikingly similar in their outward form, and almost all showing when complete an identical short formula at their commencement, might prove to be mere replicas of the same text; perhaps a prayer or an extract from some Buddhist sacred text. But such hurried comparison and study of the tablets as I was able to make during the evening following their discovery, quickly showed that their text varied greatly in both extent and matter; and as I steadily continued it during the scanty hours of rest of the next few days, evidence gradually accumulated in support of a far more encouraging supposition.

Prākrit language of documents.

The care taken about the sealing of most of the tablets from the first suggested that their contents were of a more practical nature—letters, perhaps, or documents of official character. I could not fail to realize how much the historical interest of the finds would be increased thereby; but only decipherment could establish the fact, and for the exceptional difficulties of this the experience furnished by the Dutreuil de Rhins fragments, and more than one Kharoṣṭhī inscription from the Indian North-West Frontier, had prepared me. I could not hope to attempt on the spot what, owing to the very cursive character of the writing and all the uncertainties of language and contents, has since proved so serious a task even for expert epigraphists working in their study. Hence, I had all the more reason to feel gratified when a series of philological observations bearing on the phonetic character of the characters, single or compound, and on the recurrence of particular inflexional endings, together with the tentative identification of a few detached words, proved that the language was an early Prākrit, probably of a type akin to the dialect found in the legends of the Sino-Kharoṣṭhī coins of Khotan and in the Dutreuil de Rhins codex.

Official documents.

As regards the character of the contents in the largest portion of the finds, I could remain no longer in doubt when, a day or two later, I succeeded in definitely deciphering the brief initial formula which appeared invariably at the head of the text on all wedge-shaped under-tablets, and which, curiously enough, had at first been a cause of misgivings[4]. Its wording *mahanuava maharaya lihati*, 'His Excellency the Mahārāja writes,' at once established that the documents thus prefaced conveyed official orders. But even without this explicit assurance as to the contents, there seemed enough in the first day's discoveries to justify the conclusion that, with the Kharoṣṭhī script transplanted from the extreme north-west of India, an early form of Indian speech had also been brought into use within the territories of ancient Khotan, probably from the same region. Such a fact could be accounted for only by historical events of far-reaching importance, or else by ethnic movements little suspected hitherto. The hope of recovering evidence that might help to elucidate the problem thus raised made me look with intense interest for additional epigraphical finds at the site.

Clearing of room N. iii.

My expectation of such finds soon proved well founded, when I began next morning (January 29) the clearing of the southern wing of the ruined building. Owing to the scanty cover of sand, and progressive erosion, the walls in this part of the building had suffered even more than in the row of small rooms excavated on the previous day. As shown by the plan (Plate XXVIII), it adjoined at right angles the eastern end of this row, and communicated with it by a door leading at first into a small room (N. iii.), only 10 ft. broad, which might

[4] See for this formula, always separated from the body of the text by a considerable interval, the under-tablets of N. xv. 137, xv. 24, N. ix. 1, N. xv. 71, reproduced in Plates XCVIII–C.

have served as a kind of ante-chamber. A platform, $4\frac{1}{2}$ ft. broad, built of plaster to a height of some 3 ft. above the floor, looked as if intended to accommodate attendants, an exactly similar arrangement being often observed in modern Turkestān houses belonging to people of means. What purpose the wall served, the remains of which were found dividing the small space between the western end of this platform and the wall towards room N. ii., I was unable to ascertain. It was here, in the recess marked *b*, that I found the fairly well-preserved tablet N. iii. 1, the sole object yielded by this room, and the first record written on a wooden board closely resembling in shape an Indian 'Takhtī'[5]. The piece also attracted my special interest by exhibiting on either side four narrow vertical columns of Kharoṣṭhī, which suggested either a metrical text arranged after the fashion of the Dhammapada in the Dutreuil de Rhins fragments, or else lists.

Excavation of room N. iv.

As soon as the clearing of the large apartment (N. iv.) adjoining immediately on the south had been started, inscribed tablets of all shapes and sizes began to crop up in rapid succession from the shallow sand covering it. It was a room 27 feet square, with a plaster platform 13 inches high and 4 feet broad running round three of its sides (see Plate XXVIII). The remains of eight posts arranged in a rectangle, 10 by $8\frac{1}{2}$ ft., indicated a central area, which in all probability had a raised roof, with a clerestory opening below to admit light and air, after the fashion still observed in the halls or *Aiwāns* of all Turkestān houses of any pretension[6]. As the protective layer of sand rose only about 2 feet high above the floor, little more was left of the walls than rows of much-decayed posts, which the plaster of the platforms had helped to keep upright (see Fig. 37). In consequence of this inadequate covering the first inscribed tablets which turned up in the sand above the south platform and close to the surface (N. iv. 1–6) had suffered far more than any excavated in N. i. The perished and cracked surface of their wood and their often warped and split appearance showed plainly the effects of the climatic influences, and in particular of the terrible summer heat to which they must have been exposed since the winds had carried away most of the sand that originally protected them. The fantastically-twisted fragment, N. iv. 11, reproduced in Plate CIV, which still retains a small portion of its text arranged in five lines near the square end, illustrates the destructive effects of such exposure. Another characteristic specimen of this class of badly damaged records is N. ii. 1, found in exactly similar conditions in the passage of the north row and reproduced in the same plate. Here the under-tablet of a rectangular document has been contorted almost into the shape of a half-open roll, part of the inside surface still retaining some legible lines of bleached writing. The badly-split and warped covering-tablet of a similar document (N. iv. 28) is seen in Plate CV, still recognizable by its seal-socket and string-grooves, but retaining only faint traces of writing.

Effects of exposure on tablets.

Inscribed tablets found on platform of N. iv.

After the discouraging appearance of the first finds, I had reason to feel all the more gratified when I found that even the light remaining cover of sand had sufficed to preserve in a more or less legible condition the majority of the numerous inscribed tablets that were found scattered over the platform along the southern side of the room. Those marked with numbers N. iv. 4–36 turned up either separately or in small batches lying close above the plaster flooring, and the damp once rising through the latter, no doubt, accounted for the

[5] For a similar board found at Dandān-Uiliq, see above, p. 269.

[6] M. Grenard's sketch of such an *Aiwān* in a well-to-do citizen's house (*Mission D. de Rhins*, ii. p. 97) illustrates the raised roof and the open space left below its edge all round, which, like a skylight, serves for the admission of light and air (*tungluk*). The plan given (ibid., p. 99) of a modern Khotan mansion will help to explain other corresponding features in the ancient houses of the Niya Site, the distribution of the rooms and passages, &c.

stained and discoloured surface of a number of otherwise well-preserved tablets. The compact little heap comprising the tablets numbered N. iv. 46–60 had fared better; for lying close to the centre of the south wall (in the position marked *c* in the plan) they had received some protection from what remained of the foot of the wall, and those lying nearest to the flooring had helped to keep off the damp from the rest. The documents contained in this accumulation, and some minor batches near by, had probably remained undisturbed just as they were thrown down by the last occupants of the dwelling. With regard to the few tablets (N. iv. 80–84) which were found scattered on the east platform and on the floor immediately below it, this seemed less likely, and it was manifestly not the case with the very numerous pieces that rewarded the subsequent clearing of the central area.

Inscribed tablets from central area.

Outside its edges and well above the floor there turned up the tablets N. iv. 99–106 to the south, and N. iv. 107–114 to the east. As the passages between the platforms and the posts marking the central area were being cleared, it was found that the latter was covered by a square piece of heavy tamarisk matting lying about one foot above the floor; as the size of this matting corresponded to that of the area itself, and as some light rafters were found lying below it, it had evidently once belonged to the roof over this area. The roof must have fallen in at this point when the sand had already accumulated one foot, and the twenty odd tablets found above the matting near the posts marking the southern end of the area (N. iv. 115–141) could only have got there subsequently. It is probable that they were thrown there when the abandoned dwelling was visited by some one searching its ruins after the fashion of the modern 'treasure-seekers'. It was not likely that the ancient wooden records left behind by the last occupiers as so much 'waste-paper' (to use an anachronism) would have been treated by him with more respect than Ibrāhīm had shown for the collection of tablets he had unearthed in N. i. Below the matting I found only four tablets (N. iv. 142–145), but these are in a very good state of preservation, as might be expected from the early and effective protection afforded them. One of them (N. iv. 144), a rectangular under-tablet, shows thirteen long lines of perfectly clear writing, and thus offers a text of respectable extent. A small oval-shaped platform of plaster, rising about 6 in. above the floor, which was after this disclosed within the central area, must, judging from the rim enclosing it, have served as an open fireplace.

Variety of shapes in tablets.

The epigraphical finds of N. iv. were no less remarkable for their variety in shape and size than for their number. The wedge-shaped tablets familiar from N. i. reappeared here again, two complete double-wedges (N. iv. 108, 120) being recovered, besides four detached under-tablets and seven covering-tablets of the same type, all from within the central area[7]. But in numbers they were far surpassed by inscribed boards of wood to which, notwithstanding great variations in shape and proportions, the general designation of 'oblong' seems applicable. Three among them (N. iv. 33, 46, also the fragment N. iv. 17. c) showed the familiar shape of the 'Takhtī', the handle being either pentagonal or rounded, as seen in N. iv. 33 (Plate CIII). The latter specimen, badly faded in its writing, still retains the string for which the hole always found in the handle was intended. The arrangement of the lines here, as in almost all Takhtīs where the writing runs parallel to the longer side, shows that the handle was meant to be grasped in the left hand whichever side was uppermost[8].

No classification into particular types can be attempted for the great mass of the oblong

[7] Under-tablets N. iv. 105, 107, 111, 117; covering-tablets N. iv. 80, 109, 119, 121, 134, 135, 137. For the well-preserved clay seal shown by N. iv. 80, comp. Plate LXXI.

[8] Comp. Mr. Andrews' note on N. iii. 1 in list.

Characteristics of 'oblong' tablets. tablets, owing to the striking divergence between individual pieces as regards shape, proportions, and finish; but it is easy to indicate certain generally prevailing characteristics. Most striking among these is the comparative narrowness of the pieces; for though many among them attain considerable dimensions in length, up to close on 30 in., yet none exceed $3\frac{1}{2}$ in. in width, while some of the longest are only from $1\frac{1}{2}$ to 2 in. wide, and thus approach the shape of mere sticks, e.g. N. iv. 103 (see Plate CII)[9]. Out of a total of 44 'oblong' tablets recovered in N. iv. not less than 26 show a string-hole. Useful as this hole must have been for handling and storing them, it is clear that it could not have been intended for a string, either to fasten pairs of them into a closed document, as in the case of the 'double wedges' previously described, or to unite a series of them into a sort of 'Pōthī,' or file, as might be done with palm-leaves, the shape of which these long narrow tablets curiously recall. For not only are such tablets of entirely different sizes, but the great divergence observed in the position of the hole, placed indiscriminately in the corner of a square end, near a pointed end, the centre of a long side, &c., seems to preclude any thought of regular 'filing'[10].

Probable contents of oblong tablets. The irregularity so noticeable in the outward appearance of these oblong tablets fully agrees with, and is in fact explained by, what even a cursory examination of the writing, without any attempt at decipherment, sufficed to indicate as to their probable contents. The majority of the pieces show plainly—by the irregular arrangement of their writing, in small columns which, whether parallel or running in different directions, usually conclude with numerical figures; by the appearance of various handwritings or different inks on the same tablet; tabular entries, erasures, bracketings, ink-cancellings, and similar indications—that they did not contain texts or connected communications such as letters or official orders, but probably memoranda, tabular statements, lists, accounts, and other miscellaneous records of a less formal character[11]. N. iv. 29. a (see Plate CIII) may serve as a specimen of the columnar arrangement of lines all ending with numerals. The same is clearly visible also in the tablets N. iv. 35. a, 124 (written transversely), 103 (over 25 in. long) which Plate CII shows, though their much reduced reproduction fails to bring out the writing. The text arranged in three columns (with some lines cut out by crossing) can be seen more distinctly in N. iv. 8. a (see Plate CIII). The latter oblong tablet, curving to a point at one end, closely approaches in form certain large wedge-shaped tablets found in N. iv., which from the evidence of their writing had not belonged to regular 'double-wedges', but were used in the same way as oblongs[12].

Rectangular tablets. I have left to the last the mention of two series of tablets largely represented among the finds of N. iv., which showed far greater regularity and care in writing as well as in technical

[9] See also N. iv. 22, 24, 24. a, b, 56, 118, 123.

[10] The only oblong tablets showing traces of a device, distinct from a string-hole, which was probably intended for fastening, are N. iv. 110, 133 (see Plate XCIII). The notches observed here on the Reverse manifestly mark these pieces as under-tablets; but the corresponding covering-tablets are missing, and the original arrangement can therefore not be ascertained.

[11] Columnar arrangement of the writing, usually parallel to the longer side of the tablets, but sometimes transversely, is observable in N. iv. 1, 2, 4, 8, 8. a, 9, 17, 24, 25, 29. a, 35. a, 35. b, 47, 53. a, 54 (with head-lines), 103, 113, 123 (with 19 columns), 124 (two different hands), 125, 129, 132, 142. For erasures, bracketings, and lines cancelled see N. iv. 8, 8. a, 9, 113. For different handwritings, N. i. 124.

[12] See N. iv. 6, 20, 81, 136; for a reproduction and translation of the last comp. Professor Rapson's *Specimens*. This reproduction shows very distinctly the numerous scorings, as if made with the point of a knife, which appear on the Reverse of this tablet. The same curious feature recurs in other tablets from N. iv., e.g., N. iv. 17, 55, 124 (Plate CII), 129. It is quite distinct from the scraping by which certain tablets, after having been previously written upon, were rendered available for fresh use (like palimpsests). The scorings look exactly like those which would be left by a sharp knife cutting leather. Do they indicate that tablets which had become 'waste-paper' were sometimes used in ancient 'Daftars' as convenient cutting-boards for those pieces of leather, which, as we shall see, served as subsidiary 'stationery'?

finish, but were none the less puzzling at the time of their discovery. One series consisted of tablets of rectangular shape varying in length from 6 to 16 inches, which at the narrower sides of their single inscribed surface invariably showed a raised rim about $\frac{1}{2}$ to 1 inch broad, exactly resembling a margin, as seen in the reproduction of N. iv. 139 (Plate XCVI)[13]. The writing extending between these rims in neatly written lines, always parallel to the longer side, often showed at the commencement a Kharoṣṭhī numerical figure preceded by a word which I very soon made out to be the Sanskrit or Prākrit *saṃvatsare* 'in the year'. Immediately after came the words *mahanuava maharaya* 'His Excellency the Mahārāja', as found already in the initial formula of the wedge-shaped tablets, but followed here by what I concluded to be the name of the ruling king, always showing the genitive ending *sa*. Next there appeared with equal regularity numerical figures preceded by the words *mase* and *divase*, 'in the ... month' and 'on the day'. There could be no doubt that these particular tablets contained documents fully dated. But there was nothing in this to explain the peculiar form of the tablets or their special use.

Rectangular covering-tablets.

Busily occupied as I was in directing the excavation, and in removing and numbering each of the rapidly succeeding finds, I failed to realize at the time the close connexion between the tablets just described and another class, of which N. iv. furnished even more numerous specimens. They consisted of neatly finished rectangular pieces of wood, varying in length from $3\frac{3}{4}$ to $8\frac{3}{4}$ in. and in width from $2\frac{1}{8}$ to $4\frac{1}{2}$ in., with the relatively great thickness of about $\frac{1}{2}$ to $\frac{3}{4}$ in. The flat reverse rarely contained any writing, while the obverse in its raised centre invariably showed a square or oblong seal-socket provided with string-grooves, and along its side edges one to three transversely written lines of Kharoṣṭhī[14]. It was only after the remarkable rubbish-heap of N. xv., described below, had yielded up its treasures that an explanation, as definite as it was simple, revealed itself of these curious seal-bearing tablets and of the rims of the rectangular documents to which they had once been fitted as envelopes[15].

Character of dwelling N. i.

The remains of two small rooms adjoining N. iv. on the south, to which I turned after the latter had been completely cleared, proved to be eroded almost to the floor and furnished no finds. Their condition forcibly demonstrated what would have become of N. iv. if the plaster platforms had not offered some resistance to erosion, and of its relics if my exploration had not been effected in time. The timber-débris covering the eroded slopes north and eastwards plainly indicated that all the extant ruins I had cleared were but a portion of a large dwelling-house; but apart from the writings found in it there was nothing to throw light on its last occupants. So much was certain, that all these multifarious records on wood could only be the remnants of a collection that had gradually accumulated and been left behind by chance when the place was abandoned. The inference I had from the first been inclined to draw, from the abundance of the documents and peculiarities already noticed in their appearance, as to the ruin having served as the residence of some local official, could in the absence of documentary evidence be little more than a guess. But fortunately the difficult task of decipher-

[13] Rectangular 'under-tablets,' to designate them by their subsequently determined character, are N. iv. 13, 16, 18, 30, 55, 58, 82, 84, 106, 115, 127, 139, 143, 144. The largest in size is N. iv. 58, measuring $16\frac{13}{16} \times 2\frac{1}{8}$ in. The number of lines varies from five (in N. iv. 82) to thirteen (in N. iv. 144).

[14] Rectangular covering-tablets or fragments of such are N. iv. 7, 13, 14, 17. a, 17. b, 21, 26, 28, 32, 34, 48, 83, 114, 116, 122, 128, 130, 131, 138, 140, 141, 145. It is thus seen that the number of 'envelopes' of rectangular documents found in N. iv. considerably exceeds that of 'under-tablets' of the corresponding type. N. iv. 13 probably retains also a portion of its original under-tablet, but the wood has so completely perished that the character of the two pieces could not be recognized at the time of discovery.

[15] See below, pp. 351 sqq. Plates XCIV, XCV, XCVII, CIV illustrate the arrangement of covering- and under-tablets in rectangular documents. For a curiously decayed specimen of the former, N. iv. 28, see Plate CV, and above, p. 322.

ment and interpretation with which Professor Rapson has charged himself, has proceeded sufficiently far to permit him to include tentative transcriptions and translations of four documents from N. I among the 'Specimens' presented by him to the Fourteenth Orientalist Congress, and a comparison of these distinctly supports the above conclusion.

Decipherment of four documents.

Referring to that publication for all details, I must restrict myself to the following brief notes regarding their contents. N. i. 104+16, N. iv. 108, and N. i. 105 are all three brief official orders written on double wedge-shaped tablets. The first-named is addressed to Cojhbo Bhima and Ṣoṭhaṃga Lipeya about a messenger called Ṣameka, who proceeds from Calmadana via Saca and Nina to Khotana, and who is to be provided with an escort[16]. N. iv. 108 is addressed to Ṣoṭhaṃga Lipeya alone, and directs that Cuvayalina Phummaseva going as messenger to Khotana should be furnished with certain transport and supplies. In N. i. 105 Śili and Piteya, designated by the frequently recurring title Cojhbo, are ordered to inquire into the claim raised by a certain Opgeya, apparently to a share in some property, and to send the matter up for judicial inquiry in case of dispute. Finally, in N. iv. 136, written on a large single wedge, Ṣoṭhaṃga Lipeya himself sends a long and politely worded communication to his 'beloved brothers Cojhbo Tsmaya, the scribe Aṅgaca, and the secret agent Sucama', principally about some sacrifice he is anxious to have performed at the advice of 'his reverence Kuṅgeya'. The character of the orders conveyed in the two 'double wedges', N. i. 104+16 and N. iv. 108, seems to make it clear that Ṣoṭhaṃga Lipeya, their recipient, must have been a local official of some kind who would be expected to arrange for the safe progress of official messengers. That he actually resided at or near the house represented by the ruins of N. I becomes all the more probable from N. iv. 136, which, as it bears neither a seal nor any signature or other mark of authentication, I take to be the draft of a letter sent by this official. The absence of any arrangement for fastening this tablet, and its needlessly large and inconvenient size, preclude the idea of its being a letter actually dispatched. Finally, the 'Cojhbos' Śili and Piteya, to whom N. i. 105 is addressed, may have held office by the side of Ṣoṭhaṃga Lipeya, or may have been officials who preceded or followed him in his charge.

Office rooms in N. I.

Seeing what these documents prove as to the official character of the last occupants of the house, a suggestion may be hazarded as to the particular use of the two rooms which between them yielded almost the whole of this great collection of records. That they were used as offices does not need further demonstration; but a curious difference as to the types of documents found in them deserves notice. We have seen that, while N. i. contained (with the exception of less than a dozen miscellaneous tablets) only wedge-shaped documents, the latter were in N. iv. a very insignificant minority as compared with the imposing array of oblong tablets of all kinds and rectangular tablets containing formal letters. Now I shall have occasion to show below that the wedge-shaped tablets, with their limited writing space, were chiefly, if not solely used for brief official records serving to corroborate orders entrusted to messengers and the like, while rectangular tablets and oblongs were the regular 'stationery' intended for communications and records of a more permanent character. In view of this distinction, and looking at the relative sizes of the two rooms, as well as the significant position of N. i., the much smaller one, near the approach to N. iv., I am inclined to believe that the latter, with its ample space and good lighting, served as the proper office room where 'papers' of consequence were kept and disposed of, while N. i. might have been utilized for the accommodation of the subordinate clerk who took charge of miscellaneous petty matters. That messengers' warrants corresponding to the 'Parwānas' of modern India, or orders about a preliminary police

[16] For the localities here mentioned see above, p. 311.

inquiry such as the wedge-shaped tablets above specified, belong to this category must be clear to any one familiar with the routine of an Eastern 'Daftar'.

The fact that, among all this wealth of records on wood, not a single scrap of paper had been found seemed from the first a strong confirmation of the early date which palaeographic considerations indicated for my discoveries. It was evident that the use of paper, so much more convenient than wood, however old it may have been in China, had not yet spread to Eastern Turkestān at the period when this ancient settlement was abandoned. The possibility, however, of its having known another writing material besides wood suggested itself when, among scanty débris of old pottery, ropes of twined rushes, and timber found under the sand in room N. ii. and outside it, I had come upon small cut pieces of thin sheepskin which looked as if prepared for writing. Yet it was left for another rich mine of ancient records (N. xv.) to verify the surmise which I noted down at the time. **Other writing materials.**

The undisturbed condition in which I had found the contents of N. iv., though lying so close to the surface, was reassuring proof that the ruins of this site could not have suffered so much as those of Dandān-Uiliq from burrowings of 'treasure-seekers'. But, on the other hand, with the view of the ruined building before me as seen in Fig. 37, it was impossible to ignore the fact that the havoc wrought here by wind-erosion had been distinctly greater. The extent to which the destructive power of the desert winds had asserted itself in the course of long centuries could be measured by the 12–15 feet difference of level separating the small plateau occupied by the ruin from the immediately surrounding ground. There could be no doubt that the former, protected by the walls and débris of the structure, had retained the original level, while the open surface near by had been steadily lowered by erosion. As the drift-sand carried at present over this portion of the ancient site is insufficient to fill the depressions scooped out or to cover the ruins, the raised ground bearing the latter is being steadily cut into and undermined, just as if it were exposed to the action of running water. The result finally produced by this slow but steady process of destruction is illustrated by the photograph just referred to; for the heavy timber débris which is seen there, strewing the slope of the foreground or in places overhanging it, represents the last remains of the foundations and superstructure of a part of the original building, which has completely fallen owing to the soil beneath having been eroded. **Wind-erosion near ruin N. i.**

Exactly similar conditions were observed by me around all other extant ruins of this site, the strips of ground occupied by them rising island-like above the level of the adjoining area, which in some places was found to have been eroded to a depth of 25 and even 30 feet[17]. A reference to the plans reproduced in Plates XXIX–XXXIII, XXXV, as well as to the photographs (Figs. 39, 40, 41, 44), will illustrate this. The only apparent exception was that of the northernmost ruin (N. viii) and some structures closely adjoining it, which were found deeply embedded in dunes, and where consequently the effects of former erosion were hidden from view. But even in this vicinity the loess soil, wherever bare, showed proofs of having been greatly lowered. The photograph, Fig. 46, taken of a completely-eroded ruin near N. xi will help to illustrate the final stage of destruction to which all ancient buildings are doomed in the zone lying further south and less protected by drift-sand. With such striking archaeological evidence before my eyes, I could not for a moment remain in doubt as to the true origin and significance of the broad ravines, about 15 to 30 feet deep, stretching across many places where the excavating force of the winds could freely assert itself in the bare loess. **Erosion near other ruins of site.**

[17] See the plans of N. iii (Plate XXX), N. v (Plate XXXII).

Section II.—EXCAVATION OF ANCIENT RESIDENCES, N. ii, N. iii, N. iv.

Group of ruined structures, N. ii.

The short reconnaissances on which I had sent out the Surveyor and some of my own men immediately after our arrival had shown the existence of remains of ancient dwellings at various parts of the site. The large group of ruined structures, marked N. ii on the plan (Plate XXVII), had been sighted by myself from some high dunes between N. i and the camp, and as it lay only about half a mile to the north-east of the latter I proceeded to examine it on January 30. The condition of the remains which I found here scattered over an area roughly measuring 430 ft. from north to south and 470 ft. from east to west (see sketch-plan in Plate XXIX), was an apt demonstration of the danger to ruins involved in erosion. As the panoramic view (Fig. 39) taken from the south shows, splintered posts marking the position of ancient structures were seen rising at many points between and above the light dunes covering the area. But in most of the ruins the sand lay only to a height of 1–2 feet or even less, and wherever it was removed it soon became evident that the interior as well as the walls had, long before this cover of sand came, been eroded to the very foundations. A look at the cleared structures in the foreground and on the left side of the photograph will demonstrate how little chance there was of any antiques surviving within the lines of posts that had long ceased to be walls and to afford protection. Near the edges of the area, and also within it, erosion had cut away and scooped out much of the ground, leaving the ruins bordered by sharply-marked banks, as clearly seen on the right of the photograph. In the débris scattered here over the slopes it was easy to recognize the fallen posts and foundations of structures that had otherwise completely disappeared, and to realize the final stage of destruction awaiting the rest of the ruins.

Evidence of excessive erosion.

Some low dunes, rising to about 6–8 feet above the original ground-level, stretched across the ruined area in the direction from NNE. to SSW.; but the excavation of the remains near the eastern edge, which they covered better than any others, proved that even here erosion had done its work thoroughly. In a large dwelling (see sketch-plan) the timber and plaster walls of two rooms still stood to a height of 2–3 feet, owing to the support afforded by a mud platform and a fireplace, but no objects of any kind were found in these or the other rooms. It was hence the more gratifying when, on clearing a small detached structure (N. v.) situated about 50 feet to the south-west, a considerable number of inscribed wooden tablets came to light from below the thin cover of sand, only about ½ to 1 foot deep. Three large posts, visible in the photograph, seemed to mark a central area provided originally with a raised roof; but as no trace of walls or platforms could be discovered, the interior arrangement of the room remains doubtful. Owing to the poor protection afforded against atmospheric influences the majority of the fifty odd tablets found here had withered and bleached until all trace of writing was lost. The wood of many had perished so completely that they broke at the slightest touch and could not be removed. Among the numerous tablets which turned up along the eastern part of the central area only those marked N. v. 1–6 retained part of their original writing or were capable of removal. Even in their case the surface of the wood was invariably much bleached, rotten, or otherwise damaged. The few pieces found further west, among them N. v. 7–10, had suffered equally. On the northern edge of the central area, where the sand lay slightly deeper, the tablets brought to light (N. v. 12–20) did not show quite so much the effects of exposure, but their wood, even where relatively well preserved, appears badly discoloured.

Finds of inscribed tablets in N. v.

38

RUINED STŪPA, NIYA SITE, SEEN FROM SOUTH-EAST.

39

GROUP OF RUINED DWELLINGS N. II, SEEN FROM SOUTH-EAST AFTER EXCAVATION.

Large oblong tablets from N. v.

The great majority of the tablets belonged to the class of oblongs, counting sixteen specimens among the twenty that were capable of transport. The various modifications of shape as described with regard to the finds of N. iv. are represented here also, as a reference to the inventory list will show. Writing in columns, or else in short detached items suggesting lists of names or accounts, is frequently seen in the still legible pieces. The large proportion of 'oblongs' of considerable length was here particularly striking. Among the above specimens there are seven measuring over 14 in. in length, N. v. 1 and N. v. 6 attaining the respectable dimensions of 30½ and 31½ in. respectively. N. v. 6, reproduced in Plate CII, is curious also for showing portions of its columnar text both on Obv. and Rev., written originally with a better ink than the rest, and hence presumably later additions. Owing to the dark colour assumed by the partially-perished wood the writing, though clear enough in reality, is barely traceable in the reproduction. The fact of some bark still remaining attached to the narrow sides of this tablet shows the rough methods with which this wooden 'stationery' was prepared on occasion. Among pieces that retained no trace of writing, and could not be carried away owing to their complete decay, I counted eight having a length of over 16 in., one of them showing the imposing but inconvenient dimensions of 4½ feet in length, with a width of 4½ in. Such dimensions, and the prevalence of oblong tablets with columnar writing and detached notes, point to records kept in some office. The very small number of wedge-shaped and rectangular documents[1] found might be taken to indicate that the office once housed in N. v. had little to do with correspondence. Fragments of a large pottery jar and small pieces of felt and coarse sacking (N. v. 17) were also found here.

Discovery of ancient ice-pit.

In view of the conditions already described it could not surprise me that the clearing of half a dozen more ruined dwellings yielded no further results. Everywhere erosion had done its work thoroughly, sparing neither walls nor any objects that might have been left behind within them. By the rows of posts it was just possible to distinguish the division of apartments, and the sketch-plan of Plate XXIX shows that the latter were numerous in several of these ancient houses. But the decay was far too advanced to permit the character, communication, &c., of the rooms to be ascertained. The only discovery of interest in the course of a second day's excavations was an ancient ice-pit in the outhouse of a modest dwelling-place on the western edge of the area. Here, in a small room measuring about 12 by 9 ft., the labourers came upon two unhewn trunks of Toghrak half-embedded in the floor and lying parallel close together. Abdullah, my guide from Keriya, at once suggested that we had found a *muz-khāna* or ice store-room, trunks of trees being ordinarily used now in exactly the same way for keeping the ice from touching the ground. Abdullah's conjecture was soon confirmed by the discovery of thick layers of poplar leaves filling the space of about 2 feet between the trunks, heaps of such leaves being still the usual covering for the ice which well-to-do villagers are accustomed to store for use in the summer.

Conclusion of work at N. ii.

The sand covering the dwelling just referred to was fully 4–5 ft. deep, yet owing to ancient erosion nothing was left of the walls but the bleached and withered posts that had guided me to it. The absence of walls, of course, greatly increased the difficulty of excavation, since the drift-sand was ever flowing back into the space cleared. Seeing how slow progress would be under these conditions, and how little prospect there was of adequate return for further labours, I decided, after the conclusion of the second day's work, to abstain from clearing the remaining structures of N. ii traceable on the northern side of the area, and to turn to more

[1] Rectangular tablets are N. v. 2, 15, 18; wedge-shaped, N. v. 7, 8.

promising ruins. The change already noticeable in the climatic conditions supplied a forcible reason to hurry on work. The minimum temperature of January 31 was still −3° Fahr., but at noon the thermometer rose to 45° Fahr. in the shade, and the warmth of the days was steadily increasing. It was clear that I could not depend for many weeks longer on the rigour of the desert winter to facilitate our water-supply in the form of ice, while even for other sites, where the difficulties raised by this question of water might be less serious, the available season was distinctly limited by the approach of spring with its sand-storms.

Position of ruins N. III, N. IV.

For my next excavations I selected the remains of the two large dwelling-houses which I had passed on the evening of my first arrival, about two miles due south of the Stūpa (see plan in Plate XXVII). Both ruins were prominent from their position on isolated high banks of loess, due to the erosion of the surrounding ground, and from the rows of large fallen poplars marking the ancient gardens and avenues which once surrounded them (see Figs. 40, 41). The number and size of the apartments, together with the careful construction of the timber framework, showed that they were the remains of substantial residences. In the case of the ruin to the east (N. III), the pieces of an elegantly carved wooden chair which I had previously noticed lying on the surface were a specially promising indication. As this building was also buried far deeper in sand than the other, and thus likely to prove better preserved, its excavation was commenced first. The work of clearing it occupied my band of labourers fully four days, though by this time they had been reinforced by the last able-bodied men from Imām Ja'far Sādiq, summoned in haste as soon as I realized the great extent and importance of the site.

Depth of sand in parts of N. III.

The difficulty of the work was mainly due to the depth of the sand filling the apartments that lay on the south side of the ruin and were thus, as the plan (Plate XXX) shows, well protected from the erosion attacking the edges. The large central room which, judging from its great size, 38 by 26 feet, probably served as a kind of reception hall, held no less than 9 feet of sand. The four massive beams of poplar wood, which stretched across the whole length of the hall and once supported the roof, were found in their original position resting on the sand that had accumulated beneath them, as seen in the centre of Fig. 43, which shows this part of the building before excavation. The smaller rooms adjoining the central hall westwards were also filled with deep sand, and owed to it the relatively good preservation of the timber framework of their walls, as seen in Fig. 42. The apartments lying to the north had lost most of their cover of sand by erosion, and their walls consequently were found standing only a few feet above the foundations.

Protection from fallen garden trees.

That the building had once extended further in this direction was proved by the plentiful débris of timber strewing the north slopes. That erosion had not succeeded in attacking the ruin from its southern side also was due mainly to the protective belt formed here by the large trunks of ancient poplars, which were found lying in groups and rows as they had fallen long centuries ago. A number of these trees, even in their shrivelled and splintered condition, still measured well over 50 ft. in length. The strength left in their wood was attested by the positions some of them were found occupying, raised high into the air where they lay heaped up over other trees, or else jutting out far beyond the eroded slope, as seen in Fig. 41. The weight of these rows of fallen trees had effectively protected the ground beneath, which accounts for the fact that the plateau-like raised area occupied by N. III was found to comprise not only the actual site of the ruined building, but also a considerable portion of the orchard and fenced courtyard once surrounding it (see Plate XXX).

Refuse layers over loess-banks.

Where the edges of this plateau marking the ancient ground-level had been cut away sharply and steep banks formed in the loess as towards the north-west (seen to the right in

RUINED HOUSE N. IV, WITH ARBOUR, NIYA SITE, SEEN FROM NORTH-EAST.

41

RUINS OF HOUSE N. III, WITH FALLEN GARDEN TREES, NIYA SITE, SEEN FROM WEST.

Fig. 42), a covering layer, 1–1½ feet high, formed of consolidated vegetable matter and refuse, could clearly be made out above the soil proper. I observed exactly the same around the small area covered with large fallen trees and originally fenced-in which lay close to the north of the ruin N. iv and almost facing N. iii (seen to the right in Fig. 41). It was undoubtedly a detached arbour; and here, too, the weight of the fallen masses of wood was retarding the erosion of the steeply cut loess-banks.

In order to protect the walls of the central hall during excavation from the pressure of the deep sand surrounding it on the east and south, it was necessary to clear simultaneously a broad trench outside the walls,—a task which greatly added to the actual labour. While it was proceeding, I had the smaller rooms of the north side, which held only from 1 to 5 ft. of sand, successively excavated and searched under my immediate supervision. In the room marked N. vi. the first finds were two small inscribed tablets (N. vi. 1, 2) which turned up near the eroded north end almost on the surface. While examining the layer of clay mixed with straw and dung which formed the flooring and which lay exposed on that side, I found imbedded in it three more Kharoṣṭhī documents on wood (N. vi. 3, 4, 5), at what appeared to have been the north-east corner of the room. The three tablets lay close together, about one foot below the plastered top of the flooring, and had manifestly got there before the latter had been laid down or repaired. Scattered over the surface of the actual floor were found the Kharoṣṭhī tablets N. vi. 6–11, while from below a piece of roof-matting which had fallen close to the west wall, there turned up close together N. vi. 12–15, these last all in very fair preservation. The majority of the pieces from N. vi. are covering-tablets of wedge-shaped or rectangular documents, but under-tablets of the same and records on single 'oblongs' (see N. vi. 12 in Plate CII) are also represented. Those which lay imbedded in the flooring had received a strong impregnation with salt, accounted for by the presence of ammoniac, and are hence very susceptible to damp. Seeing that only a few stray tablets were found in the other apartments, I am inclined to believe that N. vi. had served as the office room of this ancient residence.

Finds in room N. vi.

To the east of it, but separated by a narrow passage, lay a room (N. ix.), which had evidently been a kitchen. Along part of its west wall ran a platform of plaster, about 3 ft. broad and 6 in. high, provided all round with a raised rim of an additional 6 in. in height. The men, who seemed acquainted with a similar arrangement from their own houses, at once recognized in it an open kitchen grate, and pointed in confirmation to the semi-circular dip at the south end, clearly intended for the removal of ashes. Remains of several large jars of plain red pottery, and of a wooden trough found close by, supported their conclusion. In the south-east corner the well-preserved under-tablet of a wedge-shaped document, N. ix. 1 (see Plate C) was found; and two more inscribed pieces (oblongs), one of them (N. ix. 2) containing what looked like a list arranged in columns with numerals at the end, were discovered close to the fireplace or grate. Perhaps these wooden documents had found their way into the kitchen, like so much 'waste-paper,' as convenient material for kindling a fire.

Excavation of kitchen, N. ix.

A small room communicating with the kitchen by a door in its south wall proved quite empty, but the narrow closet-like apartment (N. viii.), lying between this and N. vi., yielded a series of very curious relics. In the shallow sand there was found broken in two pieces a carefully-turned shaft of poplar-wood, about 6 ft. 4 in. long and measuring about 1½ in. in diameter. At one end this shaft has a piece of strong camel-skin, about 6 in. long, tightly sewn round it. The shaft looks as if it had been made for a spear or similar weapon, but I am unable to account for the use of the piece of leather. That the closet, measuring 17 by 6 ft., must have been

An ancient store-room, N. viii.

used as a storage-room, became evident when in succession there emerged from it a bow of roughly-carved tamarisk wood, about 3 ft. long; a curved piece of willow-wood, 3 ft. long and about 6 in. broad in the middle, which looks as if it had once formed a section of an oblong shield; a donkey's saddle-tree; and, besides other broken shafts, a stout walking-stick of apple-wood. The photograph reproduced in Plate IX shows these objects with other ancient wooden implements from ruins subsequently excavated at this site.

Arms and implements from N. viii. The bow was still crisp and capable of use. The wood of what I take to be the section of an oblong shield was remarkably light and tough, about half an inch thick for about 2 feet of its length, and then thickening to 1¾ in. in the top part, where it curves inwards so as to offer better protection against downward strokes. Along the inner side edge it bore holes, no doubt meant for fastenings that joined it to another section. The donkey's saddle-tree, made of mulberry-wood, is in excellent preservation, and of practically the same pattern as that still in use throughout Eastern Turkestān. Of the two inscribed tablets found here one is the covering-tablet or envelope of a rectangular document, while the other (N. viii. 1) resembles a label and shows three columns, each with five short lines of Kharoṣṭhī which end with numerical figures and evidently contain some list. All the rooms so far described formed a small block by themselves, separated from the rest of the extant portions of the house by a passage about 4 feet broad. An arrangement similar to this is seen in the ground-plans of the dwellings N. IV and N. VII (see Plates XXXI, XXXII), and may be found reproduced with some modifications in modern residences of the country[1].

Clearing of room N. vii. Next to this block, but south of the passage, lay the room marked N. vii., which formed a kind of ante-chamber to the central hall. It measured 20 ft. square, and showed a raised platform of plaster 3 ft. 8 in. broad and 13 in. high along its north and west sides. By a door 2 ft. 9 in. broad and 6 ft. 4 in. high, which is seen in the photograph (Plate VII), it communicated with the hall. Another door, slightly wider, but only 5 ft. 4 in. high, formed the entrance from the passage westwards. The single wooden leaf which once closed it was found in good preservation, and still on its hinges, leaning against the south wall of N. vii., just as when it had been last opened. Its top is just visible in the photograph (Plate VII). A little to the east of the centre of the room two round posts were found, 10 ft. high, which had probably supported a raised portion of the roof serving as a skylight; for just between the two posts the floor showed a small oblong area sunk 6 in. below the general surface, which had undoubtedly been used as a fireplace. The smoke rising from the latter would thus find above a convenient exit. Within this fireplace lay several small torn pieces of the same fine coloured rug, which I shall have to describe presently in connexion with the central hall. On the west platform, and close to the door, were found the under-tablet of a rectangular document, broken into two pieces but otherwise well preserved, and a carefully-turned stick, probably of sandal-wood, about 11½ in. long, bearing an ivory ferule at one end and fitted also at the other for a ferule or knob of some kind. The small hole bored through the stick below the ferule does not help to indicate its original purpose.

Central hall of N. III. The central hall, the excavation of which had cost so much time and labour, was found to have been cleared completely by the last dwellers or visitors of any movable objects of practical value it might have once contained. Yet the architectural and decorative details revealed in it formed some compensation for the trouble taken. I have referred already to the four massive beams which stretched over the longer side of the hall and once supported

[1] Comp. the plan of the house of a well-to-do Khotanese which served MM. Dutreuil de Rhins and Grenard as winter quarters; *Mission D. de Rhins*, ii. p. 99.

42

WESTERN PORTION OF RUINED RESIDENCE N. III, NIYA SITE, BEFORE EXCAVATION.

43

RUINS OF RESIDENCE N. III, NIYA SITE, SEEN FROM SOUTH-EAST BEFORE EXCAVATION.

the roof. Those crossing the central part of the hall were 38 ft. long and 9 in. square, made, like the rest of the principal timber pieces in these structures, of the wood of the *Terek* (*Populus alba*). The corbel, which was fixed beneath these two central beams, was also a fine piece of wood, 7 ft. 9 in. long, 10 in. high and 8 in. thick. The post which once carried it had fallen long ago, and could not be identified with certainty, but its large circular base with a socket was found in the original position a little to the north of a raised oblong fireplace occupying the centre of the floor. It is possible that the massive and boldly-moulded pillar-head, 2 ft. 4 in. high, seen on the extreme left of the photograph (Plate VIII), as well as the square piece with carved floral design seen on the right, had formed part of this post. They were both found some 4 ft. above the floor near the east wall.

Walls of central hall in N. iii.

The walls of the hall were found standing to a height varying from 6 to 8 feet, except near the south-east corner, where a small portion of the south wall was entirely broken. Their relatively massive construction made it all the more difficult to prop them up and to prevent their falling in during excavation. The arrangement of the timber-framework and its plaster covering corresponded closely to that described in N. i, and is illustrated by the photographs (Plate VII), showing the greater portions of the north and south walls. The longitudinal walls were divided by three posts, 6 in. broad and 4 in. thick; in each of the intervals six round stakes served as supports to the wattle of matting and its plaster covering. Curiously enough, the diagonal tamarisk matting peculiar to most buildings of this site reached here only to a height of 6 ft. from the floor, while above this a horizontal reed-matting was found, just as at Dandān-Uiliq. Two horizontal beams inserted between each pair of posts helped to give strength to wattle and plaster.

Fresco ornamentation of walls.

As the work of clearing proceeded, the stuccoed walls revealed interesting remains of ornamentation in fresco. This consisted in the first place of a broad horizontal scroll, painted in dark red and black on a plaster ground of creamy white. It began about 3 ft. 8 in. below the original top of the walls, and showed alternately five-petalled flowers and plain double rings, both painted red and of 8 in. diameter. They were joined by a scroll ornament, consisting of two semi-lunes put back to back and having a double cincture in the middle. Both the flowers and the last-named ornament have their exact counterparts in motives of floral decoration common in Gandhāra sculpture[3]. Above and below the broad band thus formed, traces of which can still be seen in the photographs (Plate VII), there ran bands of black, only 2 in. broad, showing a simple floral pattern in white resembling a fern leaf. From the lower black band hung perpendicular streamers 20 in. long and 2 in. broad, at intervals of 10 in. Between each pair a wedge-shaped garland was shown in black outline, finishing below in a pendant formed by a closed lotus in dark-red colour corresponding to the colours of the flowers above. Small irregularities proved that this wall-decoration had been done by hand and without the use of a stencil.

Ancient coloured rug.

The open fireplace in the centre of the hall already mentioned was formed by a rim of plaster raised 6 in. above the floor and 6 in. broad. A dip on its north side served for the removal of ashes. The remains of embers found within were a memento of life long departed, and so also, but more curious, the large torn piece, 19 by $15\frac{1}{2}$ in., of an ancient rug which had been thrown into this receptacle. Smaller pieces of this rug had turned up within the fireplace of the adjoining room. The careful coloured reproduction given in Plate LXXV, and the technical

[3] For the five-petalled flower comp. Foucher, *L'Art du Gandhâra*, i. p. 218, figs. 96, 213, &c.

A scroll ornament exactly similar is inserted between four-petalled flowers on the plinth of the famous sculpture in the Lahore Museum showing Māra's army; see Grünwedel-Burgess, *Buddhist art*, Fig. 48.

details recorded in the inventory entry (N. vii. 3) below make it unnecessary to describe here at length this very interesting specimen of ancient textile industry. Its arrangement of transverse stripes and bands, and its decoration with simple geometrical patterns, show a remarkably close approach to the type of the Indian 'Darrie' as it exists at present. The harmoniously blended colours are also such as are familiar to those acquainted with the cotton floor coverings in ordinary use throughout India. The Svastika-like pattern which appears in all the broad stripes, and the Stūpa-like ornament, of which a portion (in green) is visible at the bottom of the red stripe, are unmistakably Indian in design, and in combination with the general make of the rug justify the conclusion that not only the sculptural and pictorial arts of Khotan, but also the more decorative branches of its textile industry had from an early date received their models from India. The brightness of the colouring has remained in all pieces of the rug, in spite of the long use proved by their torn condition and the numerous traces of mending and patching.

Ancient carved chair from N. III.

But the most thorough proof of the dominating influence which Indian art exercised on the industries represented in this ancient settlement was furnished by the ornamental wood-carving of the chair (N. vii. 4) reproduced in Plate LXVIII. Its pieces, though disjointed, lay close together on the floor of the half-eroded outer room north of N. vii. where they had been noticed by me on my first arrival. It was fortunate that the pieces (reproduced in the photograph, Plate VIII, before I had realized how they fitted together) had fallen with their carved faces towards the ground; else their elaborate carvings could not have remained in such good preservation after progressive erosion had removed the sand which once protected them. One of the four broad carved panels which were tenoned into the legs of the chair and held them together was missing. But even without this the whole can still be made to stand upright by itself as seen in Plate LXVIII and in the show case of the British Museum. The height of the legs is 23 in. and the width of the chair (approx.) 26 in. The shorter end panels are still firmly retained between their respective pair of legs by wooden pegs passing through the dowels and tenons, while the long front panel has got detached and, owing to warping and fissures, no longer fits its dowels exactly. The great interest of this ancient piece of furniture lies in its ornamental carving, which shows the closest resemblance to decorative motives familiar from Gandhāra reliefs. Referring for all details of the carving to the description given in the list, I may point out that the four-petalled flower of a shape closely approaching the large purple clematis, which forms its most frequent feature, is well known in Gandhāra sculptures. We find it there either complete and enclosed in squares forming ornamental bands, or else halved and placed within triangular spaces just as seen on the panels and legs of the chair[4]. The zigzag arrangement of these halved flowers has also its exact parallel there. The eight-petalled lotus is unmistakably Indian, while the central ornament of the front panel, with its conventionalized fruit (pomegranate ?) and leaves, recalls decorative elements on certain Indo-Corinthian capitals.

Character of ruined residence, N. III.

This close agreement with decorative motives found in Gandhāra sculptures of the first centuries of our era was welcomed by me from the first as valuable confirmation of the chronological evidence deducible from the Kharoṣṭhī writing of the tablets. In another direction, too, this piece of ancient art furniture serves as useful testimony. Though not of intrinsic value in its material, and on that account no doubt left behind when the dwelling was abandoned, it yet shows by its workmanship that those who once lived here were people in affluent circum-

[4] Comp., e.g. the sculptures from Loriyān-Tangai, reproduced by Foucher, *L'Art du Gandhâra*, i. Fig. 313; Grünwedel-Burgess, *Buddhist art*, Fig. 94. To examples familiar to me among the Gandhāra sculptures of the Lahore Museum I am unable to refer at present.

stances. The relatively large size of the house similarly helps to suggest that this was the residence of a man of means, and possibly in authority—an impression that led my labourers to christen it the 'Yamên.' In spite of the expectations thus raised or, perhaps, just because such hopes had attracted searchers to the abandoned dwelling long before the sand had completely invaded it, no objects of any kind were found in the other rooms cleared. In view of this it appeared improbable that the large area to the east of the hall, which looked like an open courtyard, should retain antiquarian relics; and as its excavation, owing to the depth of the sand, would have cost even more time than that of the hall, I decided to leave it unopened. Of the small detached structures to the west (see plan, Plate XXX), built with walls of rushes and plaster, it was easy to ascertain from the accumulations of animal refuse that they had served merely as cattle-sheds or stables.

Excavation of ruin N. iv.

The clearing of the ruined dwelling-house (N. iv) situated to the west of N. iii at a distance of a little over a furlong, proved far easier, since its broken walls retained less of drift-sand, while at the same time it was attended by results equally curious. The plan of this house and its immediate surroundings, reproduced on Plate XXXI, is a good illustration of the typical arrangement of these ancient residences. The disposition of the rooms, of the adjoining orchard and arbour, &c., recalled what I had observed in modern Khotan dwelling-places of some pretensions. The effects of erosion of the ground had not caused so much damage here as, e.g., at N. i and N. iii; on the other hand, the thinness of the walls, the plaster of which was put against horizontal layers of reeds, had caused much of the superstructure to decay and the protecting cover of sand to be reduced in consequence. Among the rooms ranged on both sides of a central passage the one marked N. x., though covered only by about one foot of sand, proved particularly rich in interesting objects. There could be no doubt that we were clearing the miscellaneous waste left behind in an ancient scriptorium or 'Daftar,' to use the comprehensive term of the modern East, when there emerged from close to the floor, besides a number of Kharoṣṭhī documents on rectangular, oblong, and 'Takhtī'-shaped tablets (see N. x. 2. c in Plate XCVII) a variety of writing implements and unused wooden 'stationery.'

Writing implements from room N. x.

The four roughly-cut wooden styles consisting of tamarisk twigs sharpened at one end, of which two specimens are reproduced in Plate CV, put me at once in mind of the remark of the T'ang Annalist that the people of Khotan 'used pieces of wood instead of brushes [for writing]'[5]. The subsequent discovery of the far more finished wooden pen from N. xv. (see same plate) disposed of any possible doubt on the point. Two oblong tablets (N. x. 9. a, b), which together with the styles were found on the low plaster platform running round two sides of the room, are blank; their curious shape, with a handle at each end, suggests their having been intended for labels. Another blank tablet (N. x. 9. c, see Plate CV), 10 in. long and only half an inch wide, presents points of special interest. Its very narrow shape and general resemblance to the slip-like pieces of wood on which the subsequently-discovered Chinese documents of N. xv. are written[6] make it appear distinctly probable that it, too, was meant to receive a single row of Chinese characters. Now N. x. 9. c differs from the inscribed pieces of N. xv. by showing a hole at each end, and these holes (one of them pierced through a small handle) could manifestly have served only for stringing purposes. We shall see later that the Chinese documents on wood found in N. xv., plentiful as they were, do not solve the question how records extending over more than one row of characters were conveniently ranged and kept in order. Is it possible that in this piece (N. x. 9. c) we have the solitary example of that kind of wooden stationery

[5] See Rémusat, *Ville de Khotan*, p. 33; now Chavannes, *Turcs occid.*, p. 125.

[6] See below, pp. 358 sqq.; for illustrations, Plates CXII–CXIV.

which would allow of Chinese documents being recorded in the traditional fashion of one row of characters per slip of wood (or bamboo), and yet spread over more than one slip without risk of confusion or partial loss? We shall have occasion to recur to this point hereafter.

Kharoṣṭhī tablets found in N. x.

Among the tablets inscribed in Kharoṣṭhī, of which altogether nine were found in this room, two may be specially mentioned here. N. x. 3 is a 'Takhtī' closely covered with fairly clear writing, which, however, owing to the darkened surface of the wood, fails to appear in the reproduction (Plate CI). The interest of this piece lies in the divergent directions of the writing, as explained in the inventory entry, which prove that, different from the usual fashion, the Takhtī handle was held occasionally in the right hand of the reader. The oblong label-shaped tablet N. x. 4 (Plate CIII), containing on one side two narrow columns of Kharoṣṭhī, is curious on account of the small twig which was found fastened into its hole, and by which it had been in all probability pinned to the wall. We thus see how closely the practical use of such wooden labels must often have resembled that of our memo. slips on paper.

Miscellaneous finds in N. x.

The three eating sticks (N. x. 06) of varying lengths, and two spindles (N. x. 01, 2) would suffice to show that the 'Daftar' had been put at times to homely uses, even if there had not been found in it a large mat made of closely-woven tamarisk rushes, about 6 ft. 3 in. long, which by the evidence of the two parallel staves placed below it must have served for a sleeping cot or hammock. An ancient broom, N. x. 07 (see Plate LXXIII), made of grass in a fashion closely resembling that described above in connexion with the find D. II. 011, shows that whoever had last taken up his quarters here was not altogether averse to tidiness. What use he could have made of the wing of a bird about the size of a small hawk (N. x. 08) is uncertain.

Ancient guitar and armchair.

Among the rooms comprised in the series towards the south-west, and mostly provided with low sitting platforms, only those marked N. xi., N. xiii., N. xiv., yielded any objects; the finds consisted in each case of one or two Kharoṣṭhī tablets, none of them showing any novel features. But in the passage (N. xii.) separating these rooms from those facing north-east some curious remains came to light. The well-preserved upper part of a guitar (N. xii. 2) about 16 in. long, reproduced in Plate LXXIII, closely resembles the 'Rabāb' still in popular use throughout Turkestān[7]. It was fitted with four strings, of which remains of three still adhere to the existing keys. The substitution of a roughly-carved stick for one of these, and the gut fastened round the fissured neck of the instrument plainly show that it had long been in a bad state of repair. Even more interesting were the remains of two elaborately decorated wooden chairs which were also recovered here. The legs of one, three in number (see Plate LXX), are 19 in. high, and represent monsters with the head of a lion, the leg of a horse, and what looks like the indication of a winged body, the whole suggesting the imitation of a 'Siṃhāsana'. But little remains of the original colouring, chiefly in pink and black, the latter colour being used over a pink ground to mark the mane. Two legs of another chair, $13\frac{1}{2}$ in. long, also shown in Plate LXX, are formed by a pair of well-carved monsters, male and female. The heads and apparently the busts are human, the parts from the waist downwards are bird-like, while the legs are those of a horse with strongly marked hoofs. The ground-colour of bright red is well preserved in parts, and over it appear traces of dark blue and black paint marking the plumage and hoofs. Composite figures of a closely similar type are well known from an early period in Indian iconography, being used at Sānchī for the representation of certain classes of demigods. Both Gandharvas and Kiṃnaras are there shown with human busts set on bird-like lower parts of the body[8]. I am unable at present to trace the above features combined with the legs of

[7] Comp. the representation of a Rabāb with five strings in *Mission D. de Rhins*, ii. p. 136.

[8] See Grünwedel-Burgess, *Buddhist art*, pp. 46 sq.

a horse. But the frequent representations of Tritons or, to use the more explicit technical term, of Ichthyo-centaurs, in Gandhāra sculpture, with human busts, wings, horses' legs, and serpentine fish tails[9], prove that even more complicated monstrosities directly borrowed from the classical West were well known to that form of Buddhist art in India which supplied the models for the art work of old Khotan.

In the same passage (N. xii.) were found several small pieces of a woollen carpet, with a carefully worked fret pattern in blue and light yellow as described in the list below, as well as a long stick-like tablet (N. xii. 1) showing columns of Kharoṣṭhī writing. Were all these miscellaneous objects and fragments thrown into, or dropped in, the passage when the final clearing of the dwelling took place before its desertion?

A curious feature of this ruined residence was the clearness with which the arrangement of the adjoining garden and of two arbours could be traced. The smaller of the latter, situated about 80 ft. to the north-east of N. x., has already been referred to, and is seen on the right of Fig. 40. Its enclosing fence of rushes, forming a square of about 45 ft., had for the greater part survived, while the bleached trunks of the poplars once growing inside, splendid specimens most of them, from 40 to 50 ft. long, were lying in heaps as they had fallen. In the other arbour, just outside the ancient rush fence to the north-west of the building, as shown in the plan of Plate XXXI, the gaunt trunks of dead poplars still rose 8 to 10 ft. from the original surface, being thus clearly visible above the drift-sand which lay here to an average depth of 4–5 ft. The trees were planted at regular intervals, which my plan accurately indicates, and grouped in a rectangular avenue and small square exactly after the fashion in which I found every well-kept 'Bostān' laid out from Kāshgar to Keriya. Within the squares little tanks are usually formed, having a platform in the centre, which serves as a convenient place for festive parties or repose in summer-time. This may have been the arrangement here also, but there was no time for clearing the ground. The difference between these rows of trunks still rising in their original positions and the heaps of large fallen trees seen near N. iii and other ruins was very striking. It suggested that there had been from the beginning considerable variations in the quantity of sand that covered different portions of the ground, the dead trees still standing having been kept erect by the early accumulation of high sand around them. **Ancient arbours.**

The area immediately to the west and south of the ruined house was found enclosed by a rush fence, mostly intact, though completely covered in some parts by the sand. The shrivelled trunks of fruit trees rising above the latter in small irregular groups showed that this had been an orchard and garden. Here, as well as among the remains of other ancient orchards which could be traced near ruined structures of this site, my diggers had no difficulty in distinguishing various fruit-trees, such as the peach, apricot, mulberry, Jigda (oleaster), with the wood of which they were familiar from their own homes. The small specimens of different woods which Plate CIV reproduces, were collected from the trees found around this ruin. To the south of the orchard two parallel fences of rushes could be made out for a short distance, marking a little country lane. At one or two places near the eastern entrance of this lane the wind had swept away the drift-sand, and here the ground displayed pottery fragments, bits of charcoal, and well preserved small twigs—the very soil on which the last inhabitants must have trodden. Searching in the sand at the bottom of the fences close by, my antique walking-stick (from N. viii.) disturbed rustling dead leaves of Terek and fruit-trees which had found a safe shelter there at a time when the bleached trunks of the orchard were still living. It was with a strange feeling, **Ancient fenced orchard and lane.**

[9] See Foucher, *L'Art du Gandhâra*, i. pp. 241 sq.

almost obliterating the sense of time, that I found myself picking up frail leaves that had fallen when the Caesars still ruled in Rome and the knowledge of Greek writing had scarcely yet vanished on the Indus.

Ruin with human remains.

In the evening of February 5, when the clearing of N. IV had been completed, I visited the spot about 600 yards to the south where on my first arrival I had noticed the slope of a small loess-bank strewn with bones, apparently human. On clearing away the sand, which lay only 1 to 2 ft. deep on the top of this bank, low foundations of mud walls were laid bare. Those of a room measuring about 17 ft. from east to west and 15 ft. across could still be distinguished; of the walls of a second room very little was left, owing to erosion of the ground. The walls showed a thickness of about 2 ft. and nowhere stood higher than $1\frac{1}{2}$ ft.; of brickwork there was no trace. Inside the area occupied by the second room, and chiefly in that part of it which had undergone some erosion, there lay in confusion human bones belonging apparently to not more than six skeletons. Most of them were broken or splintered, and of the skulls only one was fairly intact. The measurements I was able to ascertain were, maximum antero-posterior length, 185 mm.; maximum transverse breadth, 142 mm.; minimum frontal breadth, 118 mm. No timber-débris of any kind was found near this small ruin, about the true character of which I was unable to form any definite opinion. I noted, however, that the level of the ground occupied by the foundations was at least 15 ft. lower than that of N. III and N. IV, and I wondered whether this might be an indication of later—or earlier—origin.

SECTION III.—DISCOVERIES IN AN ANCIENT RUBBISH-HEAP, N. XV.

Reconnaissance to northern groups of ruins.

The excavation of the ruins so far described had furnished conclusive evidence of the fact that the extant structures of this site were mainly private dwellings which had been cleared by their last inhabitants, or else soon after their departure, of any objects of intrinsic value or practical utility. It was clear that my hopes for further archaeological finds of importance mainly depended upon any rubbish remains which might have been left behind. These hopes were greatly encouraged by the results of a rapid reconnaissance which, on February 3, I had been able to make of ruins reported north of my camp. On that occasion I visited or sighted more than half a dozen small groups of ruined structures scattered over an area measuring about $3\frac{1}{2}$ miles from south to north, and about two miles broad.

First finds at N. xv.

In a much-decayed ruin (N. v) situated about half-way to the northern limit of the site, which neither by its size nor its state of preservation would then have attracted special attention, I had come upon about a dozen of once inscribed tablets lying exposed, and in consequence entirely bleached and splintered. A search in the sand of the eroded slope to the north of the room and a little digging at the exposed edge of it had, within half an hour, put me in possession of over thirty inscribed pieces. Among the important finds which attended this rapid 'prospecting', there was a completely preserved double-wedge, N. xv. 24 (see Plate IC), in the clay seal impression of which I at once recognized a standing Pallas with shield and aegis, and two novelties which, though small in size, could not fail to excite my utmost interest. One consisted of some narrow and thin pieces of wood (now marked N. xv. 02, 08, 09, 010), all fragmentary, showing Chinese characters arranged in single columns; the other was a fragmentary document on leather torn into two pieces (N. xv. 29), one of them bearing a short entry in Kharoṣṭhī, which I could read without difficulty as the date record *mase* 4 *divase* 10 'on the tenth day in the fourth month'. Finds so varied and occurring in such rapid succession betokened a rich deposit; the ruin holding it lay fully two miles to the north of the Stūpa, and in order to secure more time for its careful excavation I decided to move my camp to it.

Survey of Stūpa ruin.

While the men were occupied in effecting the shift on the morning of February 6 I found time to make a close examination of the little Stūpa below which my first camp had stood. Its west side had been dug into and destroyed to a considerable extent; also the south face as far as exposed showed damage. From the north the vicinity of the high sand cone rising above the Stūpa, as seen in Fig. 38, made excavation difficult. On the east, too, much drift-sand had accumulated against the Stūpa, but the ground immediately southward sloped down, and thus a rapid clearing of that face became possible. This proved that the small cylindrical dome, about 7 ft. high and 6 ft. 6 in. in diameter, did not rest, as had appeared at first, on a single square base, but that this Stūpa, too, possessed its orthodox triple base, as shown by all sufficiently well-preserved Stūpa ruins in the region explored by me (see ground-plan and elevation in Plate XXIX). The topmost base was formed by a small platform 8 ft. 6 in. square and only 1 ft. high, which everywhere, except towards the north-east, had become indistinguishable. Next below rose the middle base, 13 ft. 6 in. square and 6 ft. 6 in. high, which is visible with its much damaged western and southern faces in Fig. 38. The lowest base, measuring 19 ft. 6 in. square and 6 ft. in height, was brought to light only when the excavation on the east side had, under the Surveyor's supervision, been carried down to the original ground-level. As the top of the cylindrical dome was broken, and must originally have risen several feet higher, it is clear that the total height of the structure, now 20 ft. 6 in., when intact exceeded considerably the greatest dimension of the base. The same may have been the case with the Maurī-Tim Stūpa[1].

Construction of Stūpa.

The masonry throughout consisted of sun-dried bricks, those used in the base portions measuring 22 by 17 in., with a thickness of 4 in. In the dome bricks of the same thickness but of smaller width appear to have been used, as was indeed required by the shape and reduced dimensions of this part of the structure. Owing to erosion of the surface and the plaster filling the frequent interstices their exact size could not be ascertained. On the south face the middle base had fallen off bodily, and thus displayed the inner construction. It was seen here that the innermost part of the base, 8 ft. 6 in. square, corresponding to the top platform, had been built separately, with a finished outside surface, and the middle base constructed outside it by adding walls 2 ft. 6 in. thick to each face. This construction may, perhaps, have been used throughout, the three base portions being built one outside the other as concentrical squares, with the cylindrical portion corresponding to the Stūpa dome as a circular core. Otherwise the fact mentioned might possibly be explained as an indication of subsequent reconstruction and enlargement.

Cuttings into Stūpa.

In the centre of the Stūpa dome there was a shaft, just as observed in the Stūpa of Maurī-Tim and in the Stūpas of Endere and Rawak to be described later. It only measured 1 ft. square, and had been laid open by a cutting, which treasure-seekers had made from the west into the brickwork both of the dome and the two upper bases. From a large hole dug to the centre of the dome and below it a relic deposit may possibly have been abstracted at some earlier date. Ibrāhīm on his previous visit had continued this excavation towards the centre of the middle portion of the base, but evidently without securing thereby the hoped-for 'treasure'. Another large hole had been excavated sideways into the base. In none of these burrowings could I trace any indication of the extant masonry covering some earlier Stūpa.

[1] See above, p. 81.

Changed aspect of ground.

As soon as the main features of the Stūpa had been clearly recognized I hastened to proceed to the ruin N. v, the object of my next excavations. On the way I noticed a gradual change in the appearance of the ground, the broad dunes with frequent cones covered by tamarisk scrub giving way to extensive bare loess-banks between small dunes which were rarely interrupted by sand-cones. The tamarisk growth, to which the latter owed their formation, was scanty and mostly dead. Near the line where the change became marked there was the half-eroded ruin of a small structure of rush-walls, apparently a cattle-shed.

Appearance of ruin N. v.

Promising as the finds were which my previous 'prospecting' at N. v had yielded, there was nothing in the survey of the surface remains that could lead me to anticipate how rich a deposit of ancient records I had struck in this ruin. The remains of the main structure, as seen in the photograph (Fig. 44), taken from the south and before excavation, consisted only of much decayed posts marking the walls of a few rooms and of a rush-wall strengthened by timber which had enclosed a courtyard. There was timber-débris scattered over the slopes adjoining the ruin on the north and north-east, and leading down to eroded ground from 20 to 25 ft. below the original level; but neither its quantity nor its appearance suggested that more than a modest dwelling-house had once stood here. The only feature attracting attention was the large orchard, marked by dead fruit-trees and a fence traceable in deep sand, which extended to the west and south of the structure as seen in the plan (Plate XXXII).

Condition of room N. xv.

The small dimensions of the extant remains made it easy to let the men under Rām Singh's supervision proceed with the clearing of the rooms southward and the adjoining portions of the courtyard, while I myself, with a few of the more handy labourers, among them Ibrāhīm, concentrated my attention upon the corner room, N. xv., to the north or more correctly NNW., on the edge of which towards the eroded slope my previous search had brought to light the already mentioned finds. On carefully clearing the drift-sand from the eroded slope in the direction towards what (for brevity's sake) I may call the northern side of this room, I ascertained that owing to erosion the wall of the room on this side had disappeared almost entirely. But part of its foundation beam was found still *in situ*, and thus the original dimensions of the room, 23 by 18 ft. (including walls) could be measured. Owing to the erosion of this wall a portion of the layer of rubbish it had retained within the immediately adjoining part of the room (marked *N* on plan) had slid down the slope, and from this most of the previous finds proved to have been recovered. A careful search of the remainder furnished the inscribed tablets now marked N. xv. 30–36.

Rubbish deposits within N. xv.

As soon as systematic clearing had reached the area inside the north wall it revealed layers upon layers of wooden tablets embedded in a mass of what looked like old rubbish deposits mixed up with fine dust and remains of a straw-covered roof. On the top of this fairly compact mass, and clearly distinguishable from it, there lay a thin cover of drift-sand only about 1 ft. deep. It was not from this sand, but from the consolidated mass of refuse forming a bank close on 4 ft. thick above the original mud floor, that I extracted tablet after tablet. Thus the truth soon dawned upon me. I had struck an ancient rubbish-heap, formed by the accumulations of many years, and containing also what, with an anachronism, may fitly be called 'the waste-paper' deposits of that period. Fig. 45 shows a portion of the refuse stratum filling the room as seen when the clearing had proceeded to its centre, and will help to illustrate the conditions in which the epigraphic riches of this ruin were recovered.

Composition of ancient rubbish-heap.

Throughout the room the documents on wood and leather, of which it yielded in the end over 250, were found, either separately or in relative proximity, scattered among layers containing broken pottery, pieces of matting and wood, straw, rags of felt and a variety of woven materials,

44

RUINED DWELLING N. V, NIYA SITE, SEEN FROM SOUTH BEFORE EXCAVATION.

45

ANCIENT RUBBISH HEAP N. xv, NIYA SITE, SEEN FROM WEST IN COURSE OF EXCAVATION.

pieces of leather, and other promiscuous rubbish. The earth which enveloped all this litter looked exactly like the ordinary loess, and manifestly owed its origin to the consolidation of the dust deposits which the rubbish-heap must have received in the course of its gradual formation. The dust which had been bound together here, was, of course, of the same character as that which has helped to form the loess soil. It was evident that the consistency which this accretion of varied refuse intermingled with dust had acquired in the course of long centuries, had more than anything else helped to protect it against the erosive action of the winds. It was, in fact, the compactness of these deposits, filling the whole interior of N. xv. to a height of about 4 ft., which had saved the walls of the room (seen in Fig. 45) from being reduced to mere rows of posts like almost all the other walls in this ruin. The combined conditions of cohesion and dryness which the weight of this ancient dustbin and the absorbent character of its sub-aerial dust deposits assured, account for the remarkable state of preservation shown by the great mass of the precious records embedded in it. At the same time we have a clear indication of the more unsavoury associations of a dustbin in the salt impregnation of many of the wooden tablets from N. xv., mainly due to the presence of ammoniac. It makes them now particularly sensitive to atmospheric moisture, and liable to develop stains even under the protection of the well-fitting glass panes of the British Museum.

Clearing of refuse-heap.

I had ample opportunities to familiarize myself with all the manifold contents of this remarkable refuse-heap during the three laborious days which its clearing cost me. As soon as I had realized the peculiar origin of its deposits, it became a matter of importance to keep as accurate a record as the conditions of the work and the available time would permit of the relative position in which each written document or other object of interest turned up. This might assist in tracing chronological order and in some cases possibly also internal connexion among the various documents. Accordingly, every inscribed piece had to be removed by myself and briefly tabulated after a preliminary cleaning. This was no easy task with fingers half-numbed by cold in the fresh north-east wind which was blowing during the greatest part of the time, and in the dust it raised from the dug-up refuse-heap. Its odours were still distinctly pungent and trying after so many centuries; but there was all the fascination of a great antiquarian haul to make me forget such discomforts.

Local distribution of documents in N. xv.

The four wooden posts which rose above the rubbish in the centre of the room, and which in all probability had supported an elevated portion of the roof, after the fashion of a modern *Aiwān*, afforded a convenient means for demarcating the several portions of the room successively cleared. The space to the north of the central area (marked *N* in the plan) which was first excavated and searched on February 6, in the direction from east to west, yielded the inscribed pieces N. xv. 1–79, including therein those found on my prospecting visit amidst the layer of rubbish that had slid down near the eroded north wall. The finds made in the space which extended between the east wall and the central area of the room (shown as *E* in the plan), and the rubbish contents of which proved relatively richest in documents, were marked with the numbers N. xv. 80–149. Here, as well as in the remaining two sections, the order of working was from north to south. The section between the west wall and the central area (shown as *W*), which was cleared on February 7 and 8, furnished the pieces N. xv. 150–207, while from the space between the four posts and the one immediately to the south of it (*C*) came the series numbered N. xv. 300–363[2]. The mud platform running

[2] I may mention here in passing that the few breaks which the enumeration of these distinct series in the descriptive list shows, are due either to mistakes originally made in the sequence of numbering, or else to the figures marked on the wooden tablets having become effaced or illegible subsequently. The light sand blown by the wind over the

along the greater portion of the south wall, together with the circular plaster trough, 5 ft. in diameter and 3 ft. high, which was brought to light in clearing the section marked *C*, necessarily reduced the space available for the accumulation of rubbish.

Plaster trough and platform.

Those of the men who had been to Keriya or as far as Khotan, and had seen there houses of well-to-do people, thought that this curious trough, with its top hollowed out to a depth of 10 in., was intended, like a similar contrivance still used by the women there, for keeping flowers fresh in water or under moist straw. I am unable to say whether this explanation is correct; but the fact that the hollow of the trough was found filled with sand only, clearly points to the contrivance, whatever its purpose, having continued in use even while the accumulation of rubbish within the room was proceeding. Another peculiar feature of the room deserves to be noticed here. The door, $2\frac{1}{2}$ ft. broad, opening in the south wall and still clearly distinguishable by remains of its jambs, led on to the mud platform here, instead of, as usual in these ruined dwellings, to the level floor of the room.

Accumulation of documents near walls.

It appears to me probable that the position of this entrance, as shown in the plan, Plate XXXII, accounts for the relatively greater frequency with which inscribed wooden tablets turned up near the north and east walls. When once the room had come to be used as a dustbin, it was natural that any one who emptied refuse into it should throw it from the entrance and as far as possible. The inscribed pieces of wood, which plainly represent the sweepings of an ancient office accommodated within some other portion of the house, were by their compactness and relative weight likely to be thrown further than the rest of the refuse, and often to strike against the walls opposite to the entrance. Keeping this in view we need not, perhaps, consider it altogether accidental that of the 24 documents on leather which were picked out from the litter, only 9 were found in the sections *N* and *E*, while the rest lay scattered within *C* and *W*, i. e. much nearer to the entrance. Thin pieces of leather, being light, would, when thrown (say, out of a basket), not reach so far as handy pieces of wood similarly treated.

Defective roofing of room used as dustbin.

The layers of straw and the fragments of tamarisk matting, which were found embedded deep down in the rubbish, had in all probability belonged to the roof once covering N. xv. Their position below 1 to 3 feet of miscellaneous refuse, including waste tablets and leather documents, plainly showed that while the room was used as a dustbin it must have partly at least been exposed to the sky. It is easy to understand that the roof of a room that had once been appropriated for such use would not be kept in repair, or else that an apartment which had become uninhabitable from defective roofing would be permanently utilized for rubbish. The condition of one of the rooms in the large house of Tokhta Ākhūn Bēg at Khotan, where I put up on two occasions[3], has since recurred to my mind as a very apt illustration. There a large apartment had become disused, owing to a conspicuous hole in the mud roofing which the wealthy owner never thought it worth his while to repair. Being situated conveniently near the kitchen all refuse from the latter had habitually been thrown into it, with results that obliged me to locate my cook elsewhere when I took up temporary quarters there. That the straw and fragments of matting in N. xv. actually came from the original roofing, was proved by a large piece of matting with a mass of well-preserved wheat-straw, which was found

tablets, while they lay exposed until packing became possible after the day's work, had sufficient abrading force to efface pencil marks which, for one reason or other (softness of surface, want of space, &c.), could not be made sufficiently bold and lasting. Such pieces had to be given the supplementary numbers N. xv. 01–010. This explanation holds good also of occasional breaks in the enumeration of finds from other ruins of this site.

[3] See *Ruins of Khotan*, pp. 198 sq., 268.

under several feet of sand near the end of the passage adjoining N. xv. from the south (marked in plan, Plate XXXII).

Documents damaged before deposit on rubbish-heap.

The conditions peculiar to dustbin deposits are curiously reflected by various observations concerning the documents recovered from N. xv. In the first place I may note the relative frequency of broken but otherwise well-preserved pieces. I do not mean the instances, so numerous also in other ruins, where covering- and under-tablets originally forming one record have become separated, but cases where fragments of the same document were found scattered in different parts of the room, or where individual pieces of wood and leather, though otherwise in good condition, show marks of having been broken or torn[4]. It is safe to assume that the damage in almost all these cases was caused before the pieces and fragments had been swept out of living rooms of the house and thrown on to this rubbish-heap. The thin and narrow slips of wood which served for Chinese documents were, of course, far easier to 'tear up', i.e. to break when their contents had been disposed of, than the more substantial Kharoṣṭhī 'wedges', 'oblongs', rectangular tablets, &c. This at once explains why the great majority of the Chinese 'tablets' as seen in Plates CXII–CXIV were found broken, often into very small pieces[5]. In almost all these Chinese pieces the wood was found in good preservation, which makes it still more probable that the breakage had been caused purposely before the dustbin received them. Just like our 'waste-paper', these ancient records on wood, when no longer needed by their owners, must have often come handy for lighting fires with. We have evidence of this use in certain tablets from N. xv. which bear marks of having been partly burnt[6]. Damage by knife or chisel-cuts is seen on tablet N. xv. 334 (Plate XCV); and in one instance (N. xv. 199) we have marks of the scraping by which a wooden tablet already inscribed could be rendered fit again for use, but no fresh record has been added in this case. A more minute examination of the tablets from N. xv. may bring to light more than one piece which had actually served for a palimpsest.

Protection afforded to documents.

The consolidated mass of rubbish and dust must be credited with having afforded excellent protection to the documents which had once become embedded in it. The good state of preservation shown by the great majority of the documents, whether on wood or leather, bears ample evidence to this. A brief analysis of the cases where tablets or leather documents have undergone deterioration during the long centuries since they were first thrown on to this rubbish-heap, proves that the risks to which they were there exposed were distinctly smaller than if they had been left behind on the floor of some office-room, as in N. i., N. iv., N. v., protected only by the easily shifted cover of drift-sand. The number of tablets, the wood of which has perished or become soft in the way characteristic of the effects of exposure to atmospheric influences or damp, as illustrated by so many pieces from the above find-places, is relatively very small[7]. It is also

[4] Kharoṣṭhī documents recovered in two or more fragments are N. xv. 10+80+190, 14+27, 23+150, 29. a.+b, 55+81, 76+181, 87+308, 154. a + 157, 301+302. It is probable that on further examination more of the fragmentary pieces will prove to fit each other. Broken tablets of recognizable original shape are 108, 116, 153, 175, 206, 318, 322, 340; fragments of tablets, the type of which can no longer be determined, are 80, 80. a, 142+147+148, 193. 364, 01.

[5] Out of forty-nine inscribed Chinese 'tablets' (to use this term for uniformity's sake) only five, viz., N. xv. 93, 109, 197, 197. a, 314, were recovered complete, showing the full average length of 9 to 9½ in. Among the other broken pieces small fragments, down to 1⅝ in. in length, are very numerous; see N. xv. 34, 59. b, 69, 72, 78, 82, &c. An inspection of the Chinese records on wood as reproduced in Plates CXII–CXIV will best illustrate this observation.

[6] See N. xv. 20, 34, 90, 174. The Chinese tablets being thin and narrow would have made excellent 'spills' for lighting a pipe. N. xv. 34, slightly burnt at one end (see Plate CXII), may possibly have served such a purpose.

[7] Tablets with soft or perished surface are N. xv. 6. a, 8, 19, 22, 23, 46, 57, 65, 82, 82. a, 132, 154+157, 205, 206, 335, 343, 03; similar also is the leather document, N. xv. 336.

significant that the majority of such tablets came from the northern section of the rubbish deposits, where, as we have seen, erosion had been at work for some time past and had caused exposure. It is practically only in this section that the few pieces with bleached surface or faded writing were found [8].

Documents incrusted with refuse.

On the other hand, we have the associations of the dustbin and their deleterious effects brought before our eyes in the documents, fortunately not very numerous, which have become partially effaced or illegible through incrustation with dust, straw, and promiscuous refuse [9]. There can be little doubt that the moisture which had caused this coating of dirt to adhere so firmly must have been in most cases poured on to the refuse-heap while it was still accumulating. This accounts for the fact that only particular tablets, probably lying exposed at the time or near the surface, have suffered in this way. It is very probable that the stained or discoloured surface shown by a few tablets and documents on leather has resulted from the same cause [10]. A similar explanation suggests itself for the buckled appearance of some other leather pieces [11]. I also noticed that wherever large pieces of pottery, such as seen in the foreground of Fig. 45, were found embedded in the rubbish, any documents that had come to lie immediately below them were almost invariably found in particularly good condition. Many of the records, though well-preserved and fresh in appearance have, as already noted, since their recovery shown considerable sensitiveness to atmospheric moisture. The salt impregnation thereby indicated points to the plentiful presence of ammoniac, and leads us to suspect that the room filled with refuse was often resorted to by the dwellers for even less savoury purposes. Perhaps it is only because the rubbish accumulation usually offered other and more nourishing food that the number of insect-eaten or mouse-nibbled documents is not larger [12].

Salt impregnation of documents.

Section IV.—ANCIENT DOCUMENTS ON WOOD AND LEATHER

Technicalities of ancient stationery.

The great collection of documents which came to light from the ancient rubbish-heap just described was as remarkable for the diversity in form and material of individual pieces as for the good preservation of the great majority. This makes it convenient to combine its description with a comprehensive analysis of what the epigraphic finds of the whole site have revealed, as to the outward arrangement of the main classes of documents, and as to technicalities in the ancient stationery of wood and leather used for them.

Kharoṣṭhī documents on leather.

Kharoṣṭhī documents on leather were the first striking novelty which the clearing of N. xv. yielded, and this review may hence fitly commence with them. As among the total of twenty-three pieces recovered not less than eleven are complete and three others almost so [1], the general appearance and arrangement of these documents could from the first be recognized with certainty. Judging from the complete specimens, from which those surviving in a fragmentary condition show no notable deviation, such documents consisted always of oblong sheets of carefully prepared

[8] See N. xv. 20. a, 28, 31, 32, 45, 54, 77, 99, 07.

[9] Such incrusted pieces are N. xv. 135, 138, 146, 158, 159, 160, 162, 165, 177, 320, 323, 347, 352, 361, 06.

[10] Such are N. xv. 51, 79, 114, 149, 182, 340, 350, 355.

[11] See N. xv. 43, 201.

[12] For insect-eaten pieces see N. xv. 66, 88 (leather), 197 (leather). The Chinese tablet, N. xv. 85, shows traces of having been attacked by some rodent.

[1] Complete documents on leather are N. xv. 42, 88, 112, 197, 201, 304, 305, 310, 333, 346, 350; nearly complete are N. xv. 164, 182, 319. Fragments, some of them evidently representing fairly large portions of the originals, are N. xv. 29, 43, 101, 110, 114, 149, 168, 329, 336; for specimens of such fragments (N. xv. 161, 168) see Plates XCI, XCII.

smooth sheepskin, with the Kharoṣṭhī text written on one face only and in lines parallel to the longer side, as seen in the specimens reproduced in Plates XCI–XCIII. The length of the extant specimens varies from $11\frac{1}{4}$ in. in N. xv. 319, to $6\frac{3}{8}$ in. in N. xv. 333 (see Plate XCII); N. xv. 305 (see Plate XCIII) shows the greatest width, $5\frac{3}{4}$ ins., while N. xv. 42, with a width of 2 in., is the narrowest. N. xv. 88 (see Pl. XCII) and N. xv. 350 show the longest text, each with eleven lines. Separated from the text and near the bottom edge of the obverse to the R. proper, there appears in the majority of the complete documents a brief date entry, *mase . . . divase . . .*, with numerical figures indicating the month and day[2], as seen in the complete specimens which Plates XCI–XCIII reproduce. In no case did I note a record of the year. Almost all the documents which have preserved the topmost portion of the text show there the same initial formula, *mahanuava maharaya lihati*, indicating official origin, which we have mentioned before as regularly introducing the text of wedge-shaped under-tablets[3]. Here, too, it is invariably separated from the commencement of the text proper by a considerable interval, as seen in Plates XCI, XCII. For the few documents like N. xv. 333 (see Plate XCII) which do not open with this formula, we may assume a non-official character.

Folding of leather documents.

The great majority of the documents on leather were found folded up into little rolls, and in the case of the rest previous folding was indicated by the leather being worn or otherwise marked where the folds had been. The well-preserved documents still found in their folded condition could in most cases be opened out by me without serious difficulty; but it required expert treatment at the British Museum to make these pieces of leather, bent by the folding of so many centuries, assume the flat appearance now shown in their reproductions. In Plates XCI–XCIII the lines of the folding can still clearly be seen, even where they have not, as in the case of N. xv. 88, led to the leather on the edges of the folds getting worn and broken. The reproduction in Plate XCI of the document N. xv. 310, both before and after its opening, best illustrates the method of folding usually adopted. The piece having been folded longitudinally into half its size, was next folded a second time in the same direction, in such a way that the lowest quarter of the piece, containing ordinarily only the date entry, would form one of the outside faces of the narrow roll, but with its written surface turned inwards. Thus the whole of the writing of the obverse was protected. Finally the roll thus formed was doubled over vertically, as seen in the folded state of N. xv. 310. In the latter document the bottom edge towards the left proper is seen cut off into a kind of half-attached strip, the extreme end to the left proper being turned down so as to expose just the initial formula of the topmost line. Remains of a similar cut-off strip are noticeable also, though less clearly, in some other leather pieces (e.g. N. xv. 319). This has led me to conjecture that this strip might in some way have been utilized in connexion with the fastening of such documents. What the exact manner of fastening was in the case of leather documents I have not been able to ascertain fully, but there are some observations bearing on this point which I think it useful to record here.

Method of closing leather documents.

In view of the elaborate and ingenious arrangement which, as I shall presently have occasion to explain, was adopted in order to secure the contents of communications written on 'double-wedge' or double rectangular tablets against unauthorized inspection, it appears *a priori* probable that some device must have been available for similarly protecting the contents of leather documents, especially when the latter, as the presence of the initial formula in the case of most of the extant specimens proves, conveyed official orders. An envelope after our fashion would,

[2] See N. xv. 88 (Plate XCII), 112, 164, 201, 304, 305, 310 (see Plate XCI), 319, 333.

[3] For a leather document without this initial formula see N. xv. 333 (Plate XCII).

of course, have effected that purpose; but nothing of the kind was found in the rubbish-heap of N. xv., and seeing how rich it proved otherwise in stationery sweepings of all kinds, this negative fact may claim some weight. Now, it is interesting to note that in several of the larger leather documents, but nowhere more clearly than in N. xv. 88 and 305 (see Plates XCII and XCIII) we have distinct indications of a flat oval-shaped object having once been inserted between the third and fourth fold of the document and towards the proper left of it. These indications consist of concavities in the surface of the obverse retaining the impression of the object once placed there, and of a corresponding discoloration of the surface portions thus affected[4]. It is obvious that the object inserted must have remained in that position for some time in order to leave so lasting an impression, and the conclusion that it was a seal thus becomes still more probable. The fact of the impression being found on the same side of the document as the half-detached strip of leather previously noticed suggests a connexion between the two.

Traces of former sealing.

It is possible to imagine more than one device by which the loose end of the narrow leather strip might be passed round the edges of the doubled-up roll, and ultimately passed between the two folds where the impression just mentioned appears, to be fastened down there by a seal. But in the absence of any direct evidence on the point it would serve no useful purpose to detail such devices. In the case of any such fastening as suggested, it would have been necessary either to cut the strip or to destroy the seal in order to open the document, just as we shall find it to have been the case with the sealed Kharoṣṭhī tablets. But what I am unable to make out clearly is how the reinsertion of a spurious fresh seal could be prevented, the application of a proper seal to the sealing material being made difficult by the very limited space available between the folds. But clerical dexterity, such as we have ample reason to credit the scribes of these ancient 'Daftars' with, might have overcome this difficulty, especially if the sealing material was mainly clay, as in the case of all seals which turned up at this site whether on tablets or separately. In no leather document did I find any actual remains of the seal, and hence only a chemical examination of the surface retaining the mark of the now vanished seal could, perhaps, enlighten us as to the substance used. That it could not have been one likely to affect the writing, as our sealing-wax undoubtedly would, scarcely needs special explanation.

Addressing of leather documents.

By the described arrangement of folding, the whole of the obverse surface of the document containing the text was protected. The reverse was always left blank, except for a brief line which appears in all well-preserved specimens on the fold that was exposed. Being placed on the back of the left proper side of the obverse, it cannot be seen in the reproduction of N. xv. 310 (Plate XCI). The fact that this short line invariably ends with the same word *dadavo*, 'to be given (to)', which appears on the obverse of all wedge-shaped covering tablets near the string hole, and that the two or three words preceding always close with the genitive ending in *-sa*, left no doubt from the first as to its containing the address. The recurrence of identical names, among which that of the Cojhbo Soṃjaka is most frequent[5], immediately after the initial formula of all official orders, enabled me on the spot to assure myself of the correctness of this conclusion. Owing to its exposed position on the outside surface, the writing of the address has often become faint or been partly rubbed off. But the ink on the obverse has in most cases retained remarkably

[4] In N. xv. 310 the impression appears also on the folds above named, but towards the proper right edge of the document, and the discoloration is less marked.

[5] See N. xv 88 (Pl. XCII), 305 (XCIII). For the same address on wedge-shaped tablets see N. xv. 24 (Plate IC), 137 (Plate XCVIII), 318 (in Professor Rapson's *Specimens*, p. 15).

well its original black colour, as shown by Plates XCI–XCIII, and makes the writing clearly legible even in those cases where the leather itself has become discoloured or stained[6].

Ink used in documents.

I regret not to have found an opportunity of arranging for a chemical examination of this ancient ink. But, judging from its appearance, it seems probable that it was Chinese (or Indian) ink, such as that of which a small stick was actually found by me among the rubbish layer inside the Endere Fort (see Plate CV). The ink used on the tablets, both Kharoṣṭhī and Chinese, varies considerably in quality and thickness (comp. e.g. tablets N. xv. 24, Plate IC, and N. xv. 71, Plate C), but I did not observe any indication pointing to a difference in the composition of the ink.

Leather as writing material.

The documents just described, quite apart from their contents, have a special interest as the first specimens yet discovered of leather used for early records in an Indian language. What literary evidence exists as to the use of leather for writing purposes in ancient India is extremely scanty and vague[7]. Religious objections, based on the ritual impurity of animal substance, might easily be supposed to have militated against it. Yet here we have incontrovertible evidence that, whatever those objections may have been in theory, they had no more weight in practice with the Buddhists of the Khotan region than with the orthodox Brahmans of Kashmīr, who probably from a very early period and down to our own time have been accustomed to use leather bindings for their Sanskrit codices[8]. The finish given to the leather of those ancient documents indicates extensive practice in the preparation of the material. Small pieces of blank leather of the same kind (see N. xv. 319. a), unmistakably shreds left after the cutting of full-sized sheets, and subsequently swept out of the office room, turned up among the rubbish. They show that the official residing in the ruined dwelling not only received communications on leather but also issued such. Nevertheless the relatively small number of such documents, as compared with the abundance of wooden tablets of all kinds, proves beyond all doubt that wood was by far the more prevalent writing material. Leather offered the manifest advantages of being light and easy to dispose of, besides permitting the writer to cut his sheets according to his requirements at the time. But wood was probably far cheaper, and the stationery made of it lent itself more readily to effective fastening. Whatever the reasons for this unquestionable preference may have been, it is certain that wooden tablets could not have been manufactured *in loco*, for no chippings or other remains of wooden stationery in the rough turned up in the rubbish-heap.

Wedge-shaped Kharoṣṭhī tablets.

Nothing proves better the great diversity as well as the general good preservation of the Kharoṣṭhī documents on wood yielded up by N. xv. than the relative ease with which they enabled me definitely to ascertain all technicalities connected with the use of their various classes. The wedge-shaped tablets, of which I had already recovered so many from N. i., were here even more numerous, six complete pairs in practically perfect preservation being found, besides forty-five covering- and thirty-four under-tablets. Their average dimensions agree closely

[6] The freshness of the ink was strikingly brought home to me when the leather documents were being slowly unfolded and straightened at the British Museum by the expert hands of Mr. Hunt, of the MS. Department. Warm and somewhat moist air being used for the process, special care had to be taken in order to prevent the writing impressing itself on any leather surface with which it might come in contact. On blotting paper placed over the writing it was then easy to obtain fairly clear impressions.

[7] See Bühler, *Indische Palaeographie*, p. 89.

[8] All old birch-bark MSS. of any size written in Kashmīr, some of them certainly going back to the fifteenth century, if not earlier, are invariably bound in leather or show evidence of having once been contained in such bindings. From the style displayed by the leather covers which are in my possession or have come under my observation, I think it safe to conclude that the ornaments used in this leather-work are of pre-Muhammadan origin. This implies that bookbinding in leather goes back to the Hindu period of Kashmīr.

with those ascertained in the case of the first find, the length of the wedges varying from $7\frac{1}{8}$ in. to 15 in., while their width at the square end proportionately varies from $1\frac{1}{8}$ in. to $2\frac{1}{2}$ in.

Arrangement of 'double wedges.'

The complete 'double wedges' (N. xv. 12, 24, 66, 71, 137, 140) still retaining the string with which they had been originally fastened by their senders, afford the clearest possible illustration of the manner in which documents of this type were written and dispatched. They consisted invariably of pairs of pieces, fitted exactly to match each other in size, as seen in the complete specimens N. xv. 24, 71, 137, reproduced in Plates XCVIII–C. The thickness of the two pieces, except where the raised seal-socket is placed on the covering-tablet, is also fairly uniform in each double wedge. Otherwise it varies, in some relation, perhaps, to the size of the tablets and the quality of the wood, from $\frac{7}{16}$ in. to $\frac{13}{16}$ in. It is very probable that the carpenter engaged in preparing this kind of stationery first shaped the 'double wedge' as a whole, and then

A

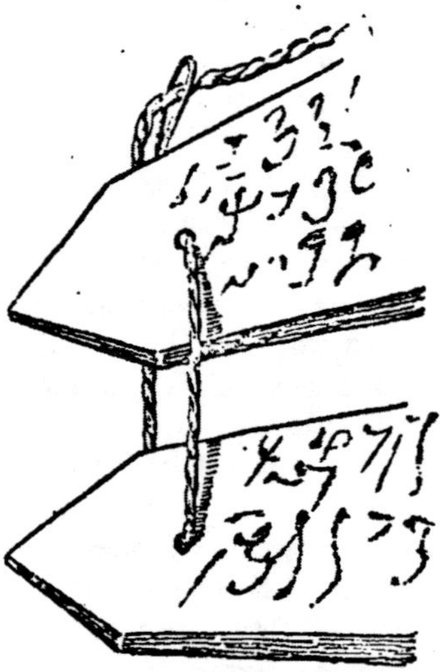

INNER SURFACES OF DOUBLE WEDGE, SHOWING FASTENING AT STRING-HOLE END.

B.

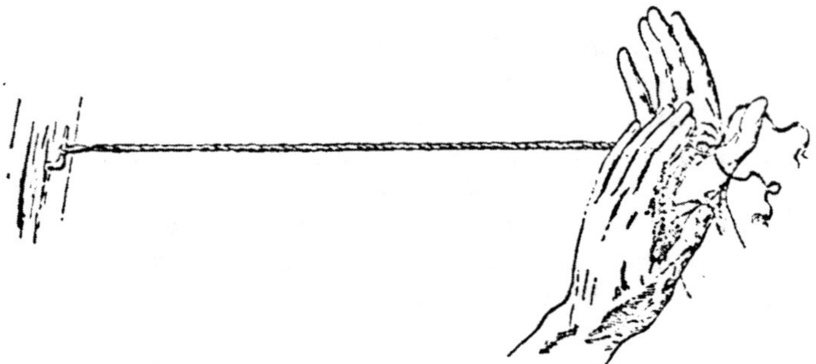

STRING TWISTED INTO LOOP FOR FASTENING DOUBLE WEDGE.

separated its two pieces by sawing or splitting. This was certainly the method adopted by the maker of N. xv. 24 (Plate IC), the inner surfaces of both covering- and under-tablet showing identical faults in the grain. One end of the double tablet thus formed was cut off square; the other runs out into a point, near which a string-hole is drilled through both pieces. The text written in lines parallel to the upper long side of the wedge, and nowhere exceeding four in number, occupies the smoothed obverse of the under-tablet, and is protected by the upper- or covering-tablet, which rests on it and serves as a kind of envelope. Whenever the length of the communication required it, the writing was continued on the reverse of the covering-tablet in such a way that, with the latter turned back, the whole of the text could be seen simultaneously and with the writing placed the right way in both tablets. The position of the covering-tablet of N. xv. 24, as reproduced in Plate IC, will explain this.

Covering-tablet as wooden envelope.

Fastening and sealing of 'double wedges.'

The wood of the covering-tablet shows greater thickness towards the square end; and in this raised portion of the outside surface, ordinarily at a distance of $1\frac{1}{4}$–$2\frac{1}{2}$ in. from the square end, a rectangular socket was roughly cut, for the reception of a clay seal. This socket or seal-cavity is seen empty in N. xv. 71 (Plate C) and N. i. 103 (IC), and filled with the clay seal for which it was made in N. xv. 137 (Plate XCVIII) and N. xv. 24 (Plate IC). Through either of the raised rims left between the seal-cavity and the edges of the covering-tablet on the long sides three grooves were cut for the purpose of securing the string that was to unite the

C.

REVERSE OF UNDER-TABLET OF DOUBLE WEDGE, WITH STRING-HOLE (a) AND FOLDS OF STRING HELD BY GROOVES (b).

D.

OBVERSE OF COVERING-TABLET OF DOUBLE WEDGE, WITH STRING-HOLE (a) AND STRING FASTENED IN SEAL SOCKET (b).

pair of tablets. The method used for fastening this string was distinctly ingenious, and will best be explained by reference to the accompanying diagrams, Fig. 45. a., drawn by Mr. F. H. Andrews, to whose technical acumen I am indebted for the elucidation of its more minute details. The string of hemp, of which more or less considerable portions are still retained by the complete wedge-shaped tablets and by many of the detached pieces[a], was first passed in a cleverly-devised fashion, which diagram *A* illustrates, through the string-hole of both covering- and under-tablet. The end of the string here inserted was previously provided with a loop by the expedient

[a] See the tablets N. i. 122, xv. 137 in Plate XCVIII; N. i. 103, xv. 24 in Plate IC; N. i. 9, xv. 71 in Plate C.

illustrated in diagram *B*, the string being doubled and converted by twisting from a single into a double ply. The covering-tablet having then been put into its proper place, its reverse lying flush with the obverse of the under-tablet, the running loop was drawn tight over the pointed ends of both tablets, and the string passed along the reverse of the under-tablet towards the position of the seal-socket, as seen in diagram *C*. Here the string was laid in regular cross-folds, invariably arranged after the fashion shown by diagram *D*, and visible also in N. i. 103 (Plate IC), over the seal-socket, the grooves communicating with the latter being used to hold the folds in their proper position. Finally, the loose end of the string, after having been passed through the groove of the upper edge nearest to the point of the tablets, was slung over the longitudinal portion of the fastening on the reverse of the under-tablet and, having been drawn tight there, tied into a knot as diagram *C* shows. The socket of the seal was then filled with clay, covering the folds of the string. When once the seal of the sender had been impressed into the clay, it became impossible to separate the pair of tablets without either breaking the seal impression or cutting the string.

Opening of sealed 'double wedges'.

The ingenious arrangement here described rendered the communication written on the inner sides of the tablets absolutely safe against unauthorized inspection and any sort of tampering. If the recipient wished to preserve the sealing and also to retain a convenient fastening for the tablets after having acquainted himself with the contents—an obvious advantage whenever such letters had to be kept for subsequent reference—he had only to cut the string near the string-hole. The under-tablet could then easily be slid out from the folds of string running beneath the seal, and after being read be passed back again into its original position just as we can do this now, after so many centuries, in the case of the double tablets (N. xv. 24, 137) reproduced in Plates XCVIII, IC.

Seal at string-hole of 'double wedge'.

There might have been cases in which it was important for the recipient of a 'double wedge' communication, after having opened and read it, to have the connexion of the two tablets attested in an unmistakable manner. An interesting instance of this kind may probably be recognized in the double tablet N. xv. 71 (Plate C), which, besides retaining all the arrangements needed for the regular fastening, including a portion of the string, shows a pendent seal of fairly hard reddish clay attached to a separate short string that passed through the string-hole of both tablets and holds them together. The string used for this purpose seems different in quality from the one which served for the regular fastening; and I conclude from this, as well as from the narrowness of the string-hole, which could scarcely have been intended for two strings, that the present fastening, with its pendent seal, was applied to the document *after* the latter had been opened. Probably this opening was effected by breaking the seal originally filling the socket, which was found empty. Else this seal in its protected position would have been more likely to survive than the wholly exposed one near the string-hole. It is scarcely necessary to point out that the latter seal was manifestly used in the same way as a seal that would be attached to the string or tape uniting several sheets of a modern legal document.

Addressing of wedge-shaped tablets.

Leaving the seals to be described further on, in connexion with those presented on rectangular tablets, I may complete this account of the wedge-shaped documents by a brief reference to the entries invariably appearing on the outside surfaces of well-preserved specimens. The obverse of the covering-tablet always shows, written from the edge of the square end towards the seal cavity, and in some instances, where this space did not suffice, extending beyond it, the name and title of the person or persons to whom the letter was addressed. Thus on the covering-tablet of N. xv. 24, 137 (Plate XCVIII, IC) we read the address *Cojhbo Soṃjakasa*, 'to the Cojhbo Soṃjaka'. This is the title and name of the official to whom, as I had already realized during the excavation of N. xv., a very considerable number (if not the majority) of the Kharoṣṭhī documents found there, whether on leather or wood, are addressed [10]. Where a document of this class was addressed to two or more persons, the line giving their names and titles would be continued beyond the seal cavity towards the pointed end, as in the case of N. i. 104+16, where the obverse names 'Cojhbo Bhima and Ṣoṭhaṃga Lipeya' as the recipients [11]. To the left proper of the string-hole, and usually quite close to it, is written the word *dadavo*, 'to be given to', which completes and defines the preceding address entry at the opposite end of the cover [12].

Entry on reverse of under-tablet.

The reverse of the under-tablet regularly shows a brief entry written close to its square end, as in N. xv. 122, 137 (Plate XCVIII) and N. i. 9 (Plate C). In the complete wedge-shaped documents translated by Professor Rapson, this entry always contains the name of the messenger or other person referred to in the text as the carrier of the letter [13]. It appears probable that the corresponding entry seen on the reverse of other under-tablets served the same purpose; our practice of showing on the cover of a letter of introduction the name of the person who is to deliver it, offers an exact parallel to this. On the reverse of N. i. 122 (Plate XCVIII) three large characters, incised in a transverse direction, take the place of this last-named entry. The text of the tablet is not accessible to me; but the characters, read by me as *Praścaya*, permit of the assumption that they, too, contain the name of the person entrusted with the document. On the reverse of N. xv. 137 (Pl. XCVIII) there appear, besides a short entry near the square end, evidently containing the name of the person who was to carry the missive, four closely written lines which, though the writing in parts is much faded, can clearly be recognized as by a different hand. M. Boyer, who has recently published a translation, with valuable comments, of this interesting document—apparently an order for the arrest and production of certain fugitives [14]—has shown that this partly deciphered record on the reverse must deal with the same object. Is it possible that we have here a disposal order or 'docket' notes by the official, the Cojhbo Soṃjaka, to whom this 'double wedge' is addressed? Of a more humble character, but also curious, are the roughly incised marks which the reverse of some other under-tablets display. In N. xv. 17. a it is a Svastika; in N. xv. 04 and 05 a small circle crossed by a stroke. There can be little doubt that they are meant for distinguishing marks, probably scratched in after the fashion of the 'Nishāns,' which Indian 'Chaprassis' are in the habit of using on our envelopes, to facilitate correct delivery by a messenger not acquainted with the writing. In the case of N. xv. 121 we find thus two small crosses scratched in on the obverse of a covering-tablet.

Incised characters.

Incised distinguishing marks.

Kharoṣṭhī documents on rectangular double tablets.

As already stated, it was only in the light of the practically perfect specimens which N. xv. yielded that I was able definitely to ascertain the manner of using and fastening the rectangular double tablets. Here, too, as in N. iv., this class of Kharoṣṭhī documents on wood was mainly represented by detached pieces wanting the corresponding covering- or under-tablets with which they must once have been fitted. Of such incomplete documents, twenty-eight covering- and nineteen under-tablets turned up here. The predominance of covering-tablets, equally marked also in the case of wedge-shaped documents, is a characteristic result of the process by which these 'waste-papers' found their way into the rubbish-heap. Opened 'envelopes' would necessarily

[10] For N. xv. 137 comp. the complete translation given by M. Boyer, *J. asiat.*, Mai–Juin, 1905, pp. 463 sqq. For other addresses comp. Professor Rapson's translations of the wedge-shaped tablets, N. xv. 12, N. i. 104+16, 105, N. iv. 108, in *Specimens*, pp. 11 sqq.

[11] See Professor Rapson's *Specimens*, p. 14.

[12] See the obverses of N. xv. 137 (Plate XCVIII), 24 (Plate IC), 71 (Plate C).

[13] See the text and translation of N. i. 104+106, 105, 108; N. xv. 12 in Professor Rapson's *Specimens*, pp. 11 sqq.

[14] See *J. asiat.*, 1905, Mai–Juin, pp. 463 sqq.

go there first, while the corresponding under-tablets might be retained for reference or otherwise disposed of. In their detached state these pieces presented a puzzling appearance; but as soon as I had found the first of the four intact rectangular double-tablets (N. xv. 151, 155, 166, 196), all difficulty disappeared[14]. From the reproductions of two of them N. xv. 155, 166 (Plates XCIV, XCV), it will clearly be seen that the under-tablet was in this case provided with a raised rim, from $\frac{1}{2}$ to $1\frac{1}{4}$ in. broad, on either of the shorter sides. Between these rims fitted exactly a covering-tablet, the obverse of which, in its raised centre, had a square or oblong socket for the reception of a clay seal.

Manufacture of rectangular tablets.

As the specimens in the Plates already referred to show, as well as the detached covering- and under-tablets reproduced in Plates XCVI and XCVII, the sizes in which these rectangular tablets were used varied greatly. The largest under-tablet found here (N. xv. 10 + 86), measured $9\frac{1}{2}$ by 4 in., while that of the complete and perfectly-preserved document N. xv. 196 (Plate XCIV) is only $4\frac{11}{16}$ in. long, with a width of $1\frac{5}{8}$ in. Among covering-tablets N. xv. 160 (Plate XCVI) is the largest, with a size of $8\frac{3}{4}$ by $4\frac{1}{4}$ in., another better-preserved one, N. xv. 154 (Plate XCVII), $7\frac{1}{2}$ by $4\frac{1}{8}$ in., approaching it closely. It is *a priori* probable that in order to assure exact fitting the pair of tablets was cut out of one piece, just as was done in the case of 'double wedges.' Of this direct proof is afforded by the identity of the grain of wood shown by both covering- and under-tablets in the complete specimens, as is distinctly visible in the reproduction of the obverse of N. xv. 155 (Plate XCIV).

Fastening and sealing of rectangular tablets.

The method of fastening provided for these rectangular double tablets was scarcely less ingenious than that of the wedge-shaped documents. Just as in the case of the latter, there were three grooves, communicating with the seal-cavity, cut through the rims which edged the latter towards the longer sides of the covering-tablet. These grooves are seen equally clearly in the covering-tablets N. xv. 155, 166 (Plates XCIV, XCV) which still retain their clay seals in the sockets, and in those which, like N. xv. 154 (Plate XCVII) and N. xvii. 3 (Plate CIV), display empty seal-cavities. A double-stranded string was passed over both tablets, drawn tight through one of the end grooves by means of a loop formed in the same way as previously described, and subsequently laid in double folds through this and the other two grooves. The seal-cavity of N. xvii. 3 (Plate CIV), which has lost its clay seal but retains the folds of the string, except the second diagonal fold parallel to the first one, will help to illustrate the arrangement. After having been folded twice through the third groove, the string was secured by a knot at the edge of the under-tablet, and its end allowed to pass loosely under the folds at the back of the latter, as seen in N. xv. 196 (Plate XCIV)—a neat little double tablet found in perfect condition and still unopened. When once the folds of string laid through the socket had been secured under a clay seal inserted there, it became impossible to separate the covering- and under-tablets without either cutting the exposed folds or completely breaking the seal. Thus any unauthorized opening and reading of the document written on the inner sides of the two tablets was effectively prevented. The method of fastening could not have varied materially in the covering-tablet N. xv. 334 (Plate XCV) which shows only two string grooves, or in N. xv. 159, which has four.

Arrangement of text in rectangular tablet.

The arrangement of the text within these rectangular tablets is fully demonstrated by the two complete documents reproduced in Plates XCIV and XCV, in both their closed and open condition. Of these N. xv. 155, showing two seal-impressions on its covering-tablet, was discovered with its fastening entirely intact. Opened in the British Museum, it shows the writing of the inner sides in remarkable freshness. N. xv. 166 (Plate XCV) had its string-folds

[14] Another complete document is made up of the pieces N. xv. 10 + 86 + 190 found in different places.

cut with a knife, having been opened before it was thrown on the rubbish-heap; but the covering-tablet closely fitted the under-tablet and thus protected its writing. The text of these rectangular documents always begins on the obverse of the under-tablet, being ordinarily arranged in lines parallel to the longer side; where necessary its continuation appears on the reverse of the covering-tablet, being written there in such a way that it can be read at the same time as the main portion when the covering-tablet is turned back upwards. Very frequently I noticed that the text of the under-tablet opens with a full date, the year (*saṃvatsare*) specified by a numeral being followed by the name and titles (*maharaya, devaputra*) of the reigning king in the genitive, as read also at the commencement of N. xv. 155, 166, xvii. 2 (Plate CIV). In keeping with this elaborate dating, which manifestly indicates formal documents intended for permanent record, is also the writing, which in most cases seem distinctly less cursive than that of the wedge-shaped tablets. The reverse of the under-tablet is ordinarily left blank; but in N. xv. 111. a it has been utilized for columnar writing, which appears here also on the obverse[16]. The writing on the covering-tablets, which may be assumed to include the address or else some brief indication of the contents, always runs transversely, i.e. parallel to the narrower sides, and commences close to their edges. It is far more extensive than on the wedge covering-tablets, and when found along both side edges usually shows reversed directions, as seen in N. xv. 154 (Plate XCVII)[17]. The number of lines varies from one to four on each side of the seal.

Addressing of covering-tablets.

Character and purpose of rectangular tablets.

Of none of the rectangular tablets have translations yet been published by the scholars who have charged themselves with the fascinating but exceptionally difficult task of deciphering and elucidating the ancient Kharoṣṭhī documents discovered by me. Without the guidance which such translations of even a few rectangular tablets would afford, and having been prevented myself, as explained below, from taking a share in the labour of their systematic decipherment, I cannot do more than call attention to certain features in the outward appearance of these documents which may hereafter, perhaps, prove to possess significance. In the first place, it is evident that tablets of this type, when made of a moderately large size, like those reproduced in Plates XCIV–XCVII, afford ampler and more conveniently arranged space for formal communications and records, whether official or private, than the wedge-shaped tablets which, when lengthened too much, were liable to become unwieldy. The greater strength of their fastening, not dependent as in the case of the 'double wedges' on the lasting of a single string or rather of half of it[18], must at the same time have made them more suitable both for transmission to a distance, and for permanent safe-keeping.

Rectangular documents found unopened.

That such safe-keeping must have been intended in the case of some at least of these rectangular documents, appears probable from the fact of not less than three of the complete rectangular double tablets from N. xv. having been found by me unopened just as when the seals were inserted. We should hardly expect to find on a rubbish-heap documents which had never been opened, unless they had appeared of no interest whatever to the recipient, or unless they had been purposely kept unopened for a long time until in the end they lost their value

[16] The piece is fragmentary, and it seems possible that in this case the wood of a rectangular letter has been re-used for memorandum purposes. Columnar writing is found also on the reverse of the covering-tablet, N. xv. 344.

[17] Here the elaborately-worded address to a certain Cojhbo appears to the L. pr. of the seal-cavity when the tablet is held in the way required for the reading of the text. This seems to have been the usual position.

[18] The explanations given above will have made it clear that, if the single ply of the string forming the loop near the string-hole broke or got worn, the wedge under-tablet could be slid out with ease from below its cover. In the case of rectangular documents six folds of double-ply string had to be cut or else the whole seal removed in order to open them.

or use. About the first possibility it is, of course, impossible to form any opinion at present; but the careful, neat way in which these tablets are made up and sealed, seems to speak against it. In support of the second alternative the double seals found on two out of these three tablets (N. xv. 151, 155, see Plate XCIV) may be mentioned; for these would suggest an agreement or contract which two parties attested and which might well be kept unopened until the time for adducing its evidence came, or until the period of its use had passed. In the case of N. xv. 196, which bears only a single seal, the assumption of its being a bond or similar record seems also possible.

Authentication of documents on wood.

I am the more inclined to favour such a conjecture because, as far as our available materials show, seals in clay were the only means used to authenticate any of these ancient Kharoṣṭhī documents. In none of those which are translated, or which I have been otherwise able to examine, do we meet with anything that could be taken for a signature. In the wedge-shaped tablets we do not even find the senders' names mentioned, their seal-impressions being evidently held sufficient to indicate the origin of the missives and attest their contents. If this assumption as to the use of seals as the sole means of authentication is true, it would seem to follow that documents which were intended eventually to serve as legal evidence had, when written on wood, to be kept sealed and unopened until produced in court; for only so long as the fastening remained intact could the seal-impressions on the covering-tablets be appealed to as proofs of the genuineness of the contents within. Inconvenient as such an arrangement might have been at times, it would at least help us to understand why documents which had ceased to possess value found their way to the rubbish-heap without ever being opened.

Clay seal-impressions.

Of the clay seal-impressions which originally must have been borne by the obverses of all covering-tablets, whether of wedge or rectangular shape, not less than twelve were recovered more or less intact from this antique rubbish-heap, besides two detached clay seals. In Plates LXXI and LXXII reproductions, in some cases slightly enlarged, are presented of them, as well as of the few equally curious specimens which are preserved on tablets found in the ruined buildings N. I and IX. Apart from their use as means of authentication, exceptional interest may be claimed for these seals; for they have furnished direct evidence of the influence of classical art in distant Khotan. It was a delightful surprise when the first intact seal-impression that turned up here, N. xv. 24 (see Plate LXXI), presented me with the unmistakable figure of Pallas Athene carrying aegis and thunderbolt, deeply impressed from a well-engraved intaglio. The identical seal appears to have been used also on the wedge-tablets N. xv. 307 (see Plate LXXI) and N. i. 17, 67, though the impressions there are not quite so well-preserved. It shows the goddess in profile standing to the left proper, helmeted, with the right arm drawn back in the act of raising the thunderbolt. The outstretched left supports an object which must be meant for the aegis. Passing over both arms and drawn behind the shoulders there appears a scarf-like chlamys, with its loose ends hanging down from the elbows. The pose of the figure, rigid but yet alert, suggests imitation of an archaic type of Athene Promachos, according to information kindly supplied by Mr. Cecil Smith. Considering the smallness of the figure, measuring only about $\frac{7}{16}$ in., and the relative coarseness of the material in which the impression was taken, the fact of these details being still distinguishable shows that the original seal must have been finely cut. A restored but faithfully conceived enlargement of it, by the artist hand of Mr. F. H. Andrews, is presented in the vignette of the title-page.

Seal with Pallas Athene.

Legend on seal-setting.

The seal, with its bold elliptical mouldings, must have been mounted on a square setting. In N. xv. 24, where the impression of this setting appears more clearly (see Plate LXXI), it is still possible to make out the traces of a legend running round the outermost moulding on the left proper. The relief characters composing it are, however, so faint that Prof. Percy Gardner, who was good enough to examine them in different lights and under the microscope, could assert of them only that they were *not* Greek. Prof. Rapson came to the conclusion that the inscription was Kharoṣṭhī, but he could only distinguish traces of a few detached characters, among them *ma* at the bottom of the impression. Traces of characters appear also on the flat part of the seal-impression on N. ii. 2 (see Plate LXXI), while on the seals of N. i. 9, iv. 80, xv. 137, the only other ones showing square settings, the clay surface corresponding to the latter is too poorly preserved to indicate whether there was a legend or not. It is thus possible to assume that the square setting, wherever provided for these seals, might have been specially intended for engraving the name of the owner. Its impression would have sufficed for the distinct indication of the sender where the latter was not named in the document.

Names shown on seal-settings.

In this connexion it is curious to note that the impressions of seals provided with settings appear without exception on wedge-shaped tablets (see N. i. 9, 17, 67; ii. 2; iv. 80; N. xv. 24, 137, 307, in Plate LXXI), i.e. on a class of documents where, judging from the specimens already translated or examined, the senders' names would not be stated in the text. On the other hand, none of the nine rectangular documents which have preserved their seal-impressions[19] show any settings. Is it possible to assume that the seals with settings and legends served as a kind of office stamp, specially designed for use on documents where the issuing official's name would not otherwise appear? Such an explanation would render it easy to understand the combination of seals of classical design and make with Kharoṣṭhī legends. While the former would be imported from the West, the settings engraved with the latter had to be prepared locally to show the names of their users. That these were officials can be safely asserted of all seals provided with settings; for the wedge-shaped tablets on which their impressions occur were used solely for the record of brief official orders.

Other classical seals.

A Pallas Athene of the same type as above described, but from a different and less carefully made intaglio, is seen in the seal-impression of the wedge-tablet N. xv. 137 (see Plate LXXI). N. ii. 2 (ibid.) shows a well-modelled male figure standing to the right proper, which from its muscular development appears to be meant for a Heracles or possibly an Eros. The latter god may be intended also in the seal of N. i. 9, if the male figure standing to the right proper is to be taken as winged. But the impression of this seal is too poorly preserved to permit of a certain opinion. Also of the seal on N. iv. 80 (see Plate LXXI) impressed in very coarse clay it is impossible to say more than that it represents a male figure of classical design.

Turning to the series of seals preserved on rectangular tablets we find two which plainly display western, i.e. Roman, workmanship. N. xv. 166 (Plate LXXII) shows a Pallas standing to the left proper, but with a figure far less carefully modelled and engraved than in the seal found impressed on N. xv. 24, 307, &c. More curious and also far better preserved is the seal of the small covering-tablet N. xv. 330, which Plate LXXI reproduces in its original size, together with an enlargement of the seal only. The figure shown by the impression within an elliptical ribbed border of mouldings is that of a youth apparently nude, seated to the left proper on a stool. The appearance of what look like folded wings at the back of the figure suggests a representation of Eros, and the strongly-curved bow seen in front of the figure and above its knees supports this identification. The good preservation of this seal is probably due to the superior quality of the clay used for it. Of seals which are rather Oriental than Roman, yet show too much of classical influence to be taken for the work of Khotan artists,

[19] N. xv. 38, 155, 163, 166, 167, 196, 330, 344; xxiii. 1 (see Plate LXXII).

we have three, impressed on the tablets N. xv. 155 and 163 (see top of Plate LXXII). The two seals borne by the latter both show male heads which may be portraits. Of the two seals which were impressed on N. xv. 155 (see also Plate XCIV) only one has survived. It shows the head and bust of a female turned to the left proper and holding in her raised right hand what may be a flower or an elliptical mirror. An elaborate headdress with flowing bands is indicated. The bust rests on a kind of base which cuts across the lowest portion of the elliptical border formed by a toothed moulding.

Oriental seals in clay.

For the remainder of the clay seals reproduced in Plate LXXII it is possible to assume local origin, the designs being distinctly Oriental. On tablet N. xv. 38, shown full size, we see a rudely-modelled male head with moustache, mane-like hair, and prominent cheek-bones. Of the two seals impressed on N. xxiii. 1, the one to the right proper, about $\frac{3}{4}$ in. high, is a male head turned to right showing a distant resemblance to the head of the Śaka king Maues, as represented on his coins. The other seal is a bird rising with wings extended. Tablet N. xv. 167 shows in characteristic juxtaposition two well-preserved square seals, of which the one to the left proper consists of four Chinese lapidary characters divided by cross-lines. I have not been able to obtain an interpretation of them. In the other we see the head of a female turned to the left and holding a flower. The elongated ear and the general treatment suggest Khotan work. The coarsely-modelled male head seen on the pendent clay seal of N. xv. 71, which has already been referred to, resembles that of N. xv. 38. The unopened small rectangular tablet, N. xv. 196[20], shown here with its whole obverse, bears a seal of geometrical design not unlike some of the metal seals from various Khotan sites which Plate L reproduces. In the small square seal of tablet N. xv. 334 (shown full size in Plate XCV) we have what are manifestly Kharoṣṭhī characters arranged in a circle. The traces, however, are not sufficiently distinct to admit of any satisfactory reading.

Detached clay seals.

The two detached clay seals, N. xv. 133. a, b, which Plate LXXII reproduces, present technical interest apart from their devices. The seal of N. xv. 133. a, showing an eagle with outstretched wings and talons, is impressed on red clay contained in a wooden case, which undoubtedly, as shown by the string grooves and the remaining portions of string, was intended to be attached to a document or some consignment. The exactly corresponding seal-case, N. xv. 74 (Plate CV), which was found empty, shows that the string was passed through a hole at the bottom. Another empty seal-case, but of octagonal prism shape, N. xv. 002, is described in the list. A good illustration of the practical use to which such seal-cases might have been put was afforded to me by the fact that the bags in which Mr. Macartney's Kāshgar office used to forward my mails, had their string fastening always sealed down in a wooden case closely resembling these ancient specimens[21]. The second detached seal, N. xv. 133. b, shows a relatively well-designed gryphon in low relief within a circle. This seal was not provided with a wooden case for protection, and its good preservation is probably a result of the very fine and dense clay used, which, perhaps, was mixed with a wax or mucilage. Fragments of woollen string are still embedded in it, indicating that it, too, must have been used somewhat after the fashion of the far rougher clay seal, N. xv. 003 (seen in Plate CV), which shows no seal mark except a circular depression but preserves a portion of its original string fastening. Here the clay seems to have been strengthened by an admixture of fibre or hair, as noticed in the case of N. xv. 166.

[20] For a view of the reverse see Plate XCIV.

[21] As far as I know the Kāshgar contrivance was Mr. Macartney's own invention, due to the necessity of providing a safe receptacle for seals which would otherwise be sure to be obliterated in the course of a long rough transit. I have since come across somewhat similar seal-cases on Indian mail routes.

In this connexion may be mentioned also the thin rectangular piece of wood, N. xv. 60 (Plate CV), which, by the evidence of its notched edges, intended to hold a string, and of the circular depression cut on one side, may be assumed to have served as a seal-cover.

Seals indicating classical influence.

The archaeological and historical importance of the seal-impressions does not require elaborate demonstration. In those made from seals of purely classical design we have the most tangible proof as yet furnished how far towards China the use of Western art products had penetrated in the early centuries of our era. Whereas in the case of engraved stones of similar make, found in the débris layers of Yōtkan, or otherwise recovered from Khotan sites, the period and the circumstances of their use must necessarily remain doubtful, we can here determine these within narrow limits by means of the very documents to which the seals were applied. The decipherment of these documents, it is true, has not yet proceeded sufficiently far to show us the exact functions or occupations, the place of residence, &c., of those who once attached their seals to them. But the date of the records can, as we shall see, be fixed with considerable accuracy, and there can be no doubt that they originated in the vicinity of the ancient site where they were discovered, or at least within the borders of the Khotan kingdom.

Diversity of seal-impressions.

The fact that these documents are so closely associated in date, language, and contents, and for the greatest part have been preserved for us in the same rubbish-heap, must make us appreciate still more clearly the remarkable diversity in origin and types which their seal-impressions exhibit. The juxtaposition of classical seals with a Chinese one on records of the same office seems the best illustration of that strange mixture of influences from the Far West and the Far East which the culture of ancient Khotan witnessed. The classical seals appear all to be Roman work, dating from the third or at the earliest from the second century of our era. The documents which have preserved their impressions must, in view of the evidence discussed in the next chapter, be assigned to the middle of the third century A. D. Hence it seems safe to conclude that the trade with the distant West, which accounts for such imports, still flourished at the period preceding the abandonment of the Niya Site.

Oblong tablets.

Of Kharoṣṭhī documents on wood there remains now to be noticed only the miscellaneous class to which, when describing above the finds of N. iv., I have applied the comprehensive term of 'oblongs'. At N. xv. this class is represented by a relatively small number of tablets (37), and none among them seem to offer novel features in their outward appearance. As a general observation, I may note that pieces of imposing size, such as N. iv. and N. v. had furnished in considerable numbers, were here completely absent; while, on the other hand, narrow tablets, lath-like or resembling mere labels, were very frequent[22]. The peculiar way in which the rubbish contents of N. xv. had accumulated may possibly account for this difference, unless we chose to seek in it some indication of the clerical work here having differed in character from that carried on in the 'Daftars' of N. iv. and N. v. Of the few Takhtī-shaped pieces, with handles, which turned up, N. xv. 185 (reproduced in Plate CI) is the largest[23]. N. xv. 76+181 (see Plate CII) is curious, as showing how readily a rough bough, probably of tamarisk, could be converted into material for recording brief memos. or items of account. That such are the contents of many of these oblongs, whatever the specific modification of their shape, is rendered very probable by the frequent arrangement of their text in small columns

[22] See in the list N. xv. 4, 20. a, 23, 28, 33, 55+81, 58, 68, 99, 120, 122, 122.a, 130, 141, 146, 172, 174, 198, 206, 335, 342, 07.

[23] N. xv. 92. a, 172 are the only other Takhtīs found here.

or in detached items usually ending with numerals[14]. That tablets of this class would often, after having been written upon, be utilized again for fresh notes or drafts, readily suggests itself; and in N. xv. 199, where part of the text on the reverse has been deleted by scraping, we have evidently an instance where this process had been begun.

Section V.—CHINESE DOCUMENTS FROM N. xv. AND THE WRITING ON WOOD

Interest of Chinese records.

The Chinese records on wood, which were among the first striking finds revealed by N. xv., form the only class of epigraphical relics from this site still awaiting description. In numbers, size, and variety of outward appearance, they cannot compare with the rich harvest of Kharoṣṭhī documents which the same ancient rubbish-heap yielded. Yet with the light which has been thrown upon them, mainly through M. Chavannes' investigations, it is easy, even for one who is no Sinologist, to realize their special antiquarian interest and their historical value.

Form of Chinese records on wood.

In regard to their material and form these Chinese records display a remarkable uniformity. With the single exception of the rectangular covering-tablet (N. xv. 345) referred to below, Chinese characters are found only on narrow and thin pieces of wood, for which the expression of 'slip' seems more appropriate than that of 'tablet'. These bear their writing invariably in a single column and on one side only. The wood used appears to be that of the Terek or *Populus alba*, as also in the great majority of the Kharoṣṭhī tablets. As seen in Plates CXII–CXIV, where the whole of the Chinese records have been reproduced in approximately full size, the width of the slips is only from $\frac{7}{16}$ to $\frac{5}{8}$ of an inch. The length of the five complete slips, found either intact (as N. xv. 314, 353, 362) or broken in two closely-fitting pieces (as N. xv. 75+328, 93. a. b), varies only from 9 to $9\frac{1}{4}$ in.; and an examination of the forty-two fragmentary documents[1], as well as of the two broken pieces found blank (N. xv. 59, 111), proves that none of the slips used for these Chinese records of N. xv. is likely to have exceeded this length. I have already referred to the probable cause which has made broken Chinese documents, varying from almost complete pieces like N. xv. 203, measuring $8\frac{5}{8}$ in., to N. xv. 72, only $1\frac{3}{8}$ in. long, so very numerous in proportion to complete ones. Owing to their narrowness and the slight thickness of the wood, such slips were far more liable to be broken by chance or 'torn up' on purpose than the substantial Kharoṣṭhī tablets.

Writing on bamboo slips in China.

The close resemblance in shape between these Chinese documents on wood and the pieces of bamboo which, according to a widely-known tradition, were used in China as writing material before the invention of paper, had become evident to me already at Khotan. The Amban Pan Dārin, my kind and learned friend, on being shown there some of the Kharoṣṭhī tablets, quite correctly indicated his own conclusion as to their approximate date by a reference to the bamboo slips which in China served for writing purposes up to the Han period. There being no bamboo in Hsin-chiang, that scholarly Mandarin thought wood had naturally been substituted. The explanations which Mr. Macartney and Sun Ssŭ-yeh, the Chinese Munshī of the Agency, kindly furnished to me at Kāshgar as to the Chinese records on wood discovered by me, left no doubt on the point. But it was only after the publication of M. Chavannes' exhaustive

[14] Columnar arrangement is observed in N. xv. 33, 68, 92. a, 130, 184; for items with numerals see, e.g., N. xv. 76+181, 122, 129, 185.

[1] See N. xv. 34, 37, 53, 59. b, 61+62, 69, 72, 73, 78, 82, 82. a, 85, 100, 101. a, 109, 116, 117, 123, 125+127, 139, 145, 152, 169, 175, 176, 188, 189, 191, 192, 203, 207, 324, 326, 337, 339, 348, 349, 351, 02, 08, 09, 010.

and lucid treatise, *Les livres chinois avant l'invention du papier*[1], that it became possible for me to realize how closely the documents recovered from N. xv. conform in all external features to the descriptions preserved in Chinese literature of that earliest writing material, slips of bamboo. I cannot attempt, nor is it necessary, to reproduce here in detail the convincing and curious evidence collected by M. Chavannes, and the critical conclusions he draws from it as regards the shape, use, &c., of these bamboo slips during the different periods of Chinese antiquity. A brief reference in each case to the facts established in his publication will suffice to prove what mainly interests us here, viz. that our Chinese documents on wood have derived all peculiarities of their outward appearance from the bamboo slips used in China itself for the corresponding class of records during the period immediately preceding the invention of paper.

The shape and size of bamboo slips.

The feature which proves this most significantly for our wooden 'slips' is their narrowness, and the consequent restriction of the text written on them to a single row of characters. M. Chavannes has demonstrated from an abundance of testimonies that the *chien* 冊, or slips in bamboo, were during all periods limited to a very slight width, never exceeding two centimetres, and often much less, and that, with the exception of some very rare cases where they bore two parallel lines of characters, writing on them was restricted to a single row and to one side only[3]. This arrangement was natural and practically inevitable in the case of slips cut from the round stem of a bamboo, hollow inside and having ordinarily only a small diameter[4]. It has no justification in records written on wood, a material which readily lends itself to cutting into shapes far more convenient for the purpose of writing. Its maintenance in our wooden slips can be explained only as a designed imitation of the bamboo slips, indicating an endeavour to preserve the traditional form established by them. We are led to the same conclusion by comparing the length of our wooden slips 9–9¼ in., as above indicated, with the standard sizes for bamboo slips as they prevailed during the Han period. M. Chavannes has shown that, with the exception of certain well-defined classes of texts, such as classical works, ritual rules, &c., for which immutable tradition had fixed special sizes, the slips used for writing measured always one Chinese foot (of ten inches) in the case of ordinary individuals, while the Emperor, to mark his superiority, had his edicts written on slips one foot and one inch long, or else employed slips of two feet and of one foot alternately[5]. I do not know whether the measure exactly equivalent to the foot of the Han epoch can be ascertained; but it is evident that the actual length of the complete Chinese records from N. xv., most of them official, but not Imperial edicts, must closely approach the normal of one foot or ten inches.

Ancient pen of wood.

M. Chavannes, in the concluding remarks of his treatise, has already emphasized the importance possessed by our Chinese documents on wood as a striking archaeological confirmation of the data which the literary sources critically examined by him furnish as to the ancient writing material of China, and the manner of its use previous to the invention of paper in 105 A.D.[6] He has recognized equally interesting evidence in the ancient wooden pen, N. xv. 21 (Plate CV), which was found in the identical rubbish-heap and has been reproduced in my *Preliminary Report*. It consists of a small twig of tamarisk wood, about 4¾ in. long, split at its pointed end and fitted at the other with a bone knob which manifestly served the purpose of a burnisher. Rougher styles made of the same wood and also intended for writing, N. x. 04, 05 (Plate CV), had been

[2] See *Journal asiat.*, 1905, Janv.–Février, pp. 5–75; quoted below from reprint.

[3] See Chavannes, *Les livres chinois*, &c., pp. 34–42.

[4] See Chavannes, ibid., p. 14.

[5] See *Les livres chinois*, p. 34, with the evidence detailed, pp. 18 sqq.

[6] See *Les livres chinois*, pp. 73 sqq.

found by me before in what was undoubtedly the office-room of the ancient dwelling N. IV[7]; these specimens are not split at their points.

Early Chinese writing with wooden styles.

Whether these wooden pens had actually served for the writing of Chinese or of Kharoṣṭhī documents, it is impossible to decide. But as M. Chavannes has observed, the peculiar form assumed by the Chinese characters on our wooden slips distinctly points to their having been written with such styles of wood. Hence there can be little doubt that the latter in their form and material correspond closely to the *pi* 筆, which, as demonstrated by M. Chavannes, consisted of a small stem of wood, usually bamboo, used for writing in ancient China, before the general adoption of the brush, invented about the third century B. C., caused the term to be appropriated for the designation of the latter writing implement[8]. The use in Khotan of wooden pens instead of the brush for writing purposes is specifically recorded by the Chinese Annals as late as the Liang and T'ang periods[9]. The quasi-archaic survival here attested is of special interest, as it furnishes an exact parallel to the long-continued use of wood as a writing material in the same region. As we shall presently see, paper had been known in China for fully one and a half centuries previous to the time when the documents of the Niya Site were still recorded on wood, and the wooden documents from Dandān-Uiliq and Rawak prove that the use of the latter had not completely ceased even as late as the eighth century[10].

Arrangement of texts on Bamboo slips.

Seeing how closely the Chinese documents of the Niya River Site conform in their appearance and material to what we know from literary sources of the form of written records in ancient China, it seems justifiable to inquire whether the antiquarian materials brought to light there may not throw light on other points connected with early writing arrangements in China. From a great number of passages discussed in M. Chavannes' treatise, it results that, as a single bamboo slip (*chien*) could afford space only for from twenty to thirty characters, a series of these slips were used for all writings of any length[11]. The number of slips being necessarily great in the case of books, they used to be arranged into *pien* or packets, ordinarily corresponding to chapter divisions. In order to keep together the number of bamboo slips forming such a packet, it was manifestly necessary to give them a fastening of some kind. Several texts quoted by M. Chavannes mention the silk cords or thin leather thongs which were used in antiquity for this purpose[12]. But they do not furnish any indication as to the particular method of fastening by which the retention of the slips in their correct order was secured. Yet it is evident that without some such method the greatest inconvenience was bound to arise, since, owing to their narrowness and probably curved surface, the slips could not by themselves lie flat one above the other and thus retain their position, as e. g. a bundle of loose leaves of paper would.

Wooden slip with string-holes.

In Indian palm-leaf MSS., where a similar need was felt, a string passed through either one or two string-holes in each leaf secured the desired order[13]. The observation of two unmistakable string-holes in a blank slip of wood (N. x. 9. c) found in the office-room of N. IV, hence suggested to me whether it might not be a specimen of the kind of slip used for Chinese records where fastening in a fixed order was intended. The fact of this slip, reproduced in Plate CV, being only half an inch broad, seems a distinct indication of its having been meant to receive Chinese writing. Its length between the holes placed near the two ends is $9\frac{1}{8}$ in., thus agreeing exactly with the average size of the Chinese slips from N. xv. The holes are in each case

[7] See above, p. 335.

[8] See Chavannes, *Les livres chinois*, pp. 65 sqq., 70, 72 sqq.

[9] Comp. Rémusat, *Ville de Khotan*, p. 16; Chavannes, *Turcs occid.*, p. 125; above, pp. 170, 173.

[10] See above, pp. 268 sq., 305.

[11] See *Les livres chinois*, pp. 40, 43 sqq.

[12] Comp. *Les livres chinois*, p. 43.

[13] See Bühler, *Indische Palaeographie*, p. 89.

about $\frac{7}{16}$ in. distant from the actual ends, one of these being shaped into a small rounded handle, perhaps intended to mark the top of the slip. As the slip was found blank and manifestly unused, the suggestion here made must, of course, remain conjectural.

Wooden tablets used in ancient China.

There is another question connected with early Chinese writing, for the elucidation of which a close study of the ancient documents on wood brought to light by me may perhaps prove useful. From the texts translated and discussed by M. Chavannes it is clear that Chinese antiquity, apart from slips of bamboo, knew also small wooden boards called *fang* 方 for writing purposes: these, however, were meant to receive only short documents not exceeding a hundred characters, and appear to have been reserved mainly for official use [14]. They were always rectangular in shape; but the fact that they are said to have been square or approximately so, as well as their having been used manifestly without a covering, preclude the idea of any special connexion between them and the form of our rectangular Kharoṣṭhī tablets. Whether the latter or any other class of tablets familiar to us from the Niya Site could possibly be connected in their shape with the wooden tablets called *tu* 牘, to which M. Chavannes refers in a note [15], I am not in a position to ascertain. But it is noteworthy that the tablets thus designated were primarily destined for letters, whether from the Emperor or private individuals, while the description given of them as being smaller than a *fang* but larger than the slip (*chien*) might well agree with a form not unlike that of one or other class of our sealed Kharoṣṭhī tablets, which undoubtedly served for correspondence.

Rectangular tablet with Chinese inscription.

However this may be, we actually possess at least one document on wood bearing a Chinese inscription, which, but for one slight peculiarity, shows in its form complete agreement with the rectangular covering-tablets of Kharoṣṭhī documents. I mean the tablet N. xv. 345, reproduced in Plate CXIV (less clearly also in Plate CV). Its obverse displays the usual seal-cavity, now empty, with the three string-grooves on each side, and above it a row of faded Chinese characters, which have been read by M. Chavannes 鄯 善 王 'the king of *Shan-shan*', with a fourth character to the right which might be read 詔 'edict' [16]. The reverse is blank, but curiously enough has its centre portion slightly raised with a narrow lower margin all round. This raised portion has the appearance of having been cut to fit a corresponding opening in an under-tablet. But the latter has not been found, and there thus remains the possibility of this interesting tablet having served as a lid to a small box [17]. In any case there can be no doubt that the arrangement of sealing and fastening was here identical with that observed in all rectangular Kharoṣṭhī documents.

Possible Chinese origin of 'wooden stationery' for Kharoṣṭhī.

In the absence of other evidence, this use of an identical method of fastening is of considerable interest. Seeing how conservative Chinese fashion has always been in matters concerning the forms of written record [18], it seems far more probable that the ingenious methods of fastening and authentication which are so amply illustrated by our Kharoṣṭhī documents on wood, wedge-shaped and rectangular, were originally derived from China, than that the Chinese *chancellerie* of the ruler of Shan-shan, a state south of Lop-Nor, and thus relatively close to the border

[14] See *Les livres chinois*, pp. 13–17.

[15] See *Les livres chinois*, p. 26, note 1.

[16] See Appendix *A*, part ii.

[17] In case the tablet really served as a lid, we might think of a box that contained slips of wood with the text of the edict apparently referred to in the legend of the cover. The total width of the cover being $1\frac{1}{2}$ in. and that of the raised central portion about 1 in., a set of slips of the usual width could just have been inserted vertically into the receptacle. The use of special cases and boxes for the guarding of written slips of bamboo is amply attested for the period of the Former Han dynasty; see Chavannes, *Les livres chinois*, p. 63, note.

[18] Compare e.g. Chavannes, *Les livres chinois*, pp. 16, 25 sqq., 26, note, 29, with note, 30 sqq., &c.

of China proper, should have adopted for official missives a form borrowed from Khotan. This conclusion, based upon the single tablet N. xv. 345, seems at present the only direct argument in favour of the assumption that the arrangements and fashions prevailing in the 'wooden stationery' of our Kharoṣṭhī documents had developed under Chinese influence. But there is indirect and quasi-negative evidence to lend support to such an opinion.

Ancient writing-materials of India.

Whatever ancient remains have come to light as yet in the Khotan region of the period to which the ruins of the Niya Site belong, or in fact, of the whole Buddhist epoch, show that the civilization then flourishing was indebted mainly to India and China for its constituent elements. Now for Chinese antiquity, the extensive and varied use of wood as a writing-material is, as we have seen, amply attested; but the same is by no means the case for ancient India, especially in the period during which Khotan received its importations of Indian religion, language, and art. There can be no doubt that in the extreme North-West of India, to which Khotan was directly indebted for everything that is of Indian origin in its culture, the easily obtainable and very convenient birch-bark (*bhūrja*) always formed the writing material most widely used for all purposes [19]. By its side the use of palm-leaves, which is attested for the whole of India from a very early period, must also be assumed to have been common [20].

Use of wood for writing in India.

On the other hand, what scanty references we possess to wood as a writing-material in early India seem to indicate that its use was restricted to specific purposes widely distinct from those of the great majority of our Kharoṣṭhī documents. Of such purposes that of the *phalaka*, or writing-board used in the schools is the best attested [21]; as already mentioned, this wooden board, which took the place of our slate, and is now known throughout Northern India as *Takhtī*, has survived to the present day. Its form may have served as a model for the oblong Kharoṣṭhī tablets with handles described above; but the reproduction of so simple an implement does not justify any conclusion as to the origin of the other far more ingenious forms and arrangements illustrated by our Kharoṣṭhī documents on wood. Nor are the few references to wooden boards, used apparently like our blackboards for public announcements, relevant in this connexion [22]. Still more specific are the uses of wood alluded to in a few isolated passages of Buddhist works [23]. To conclude from them as to the general use of wood for records and written communications in Buddhist India would be as reasonable as if the writing of amulets on birch-bark, which is still customary throughout Northern India [24], were adduced as evidence for the general use of this material in Indian records, &c., of the nineteenth century.

Wood as writing-material in Khotan.

It must be reserved to future researches—or to future finds—to decide how much of the arrangements and forms observed in our Kharoṣṭhī documents on wood was borrowed from other civilizations and how much was local development. But it is clear in any case that the economic conditions of the Khotan region must have favoured the use of wood as the chief writing-material previous to the introduction of paper. The Bhūrja tree (*Baetula Bhōjpattr*), which supplied and in part still supplies Northern India with a bark convenient for writing as well as for wrapping, is not, as far as I know, to be found to the north of the Himālaya, and certainly not on the terribly barren slopes of the Kun-lun. The difficulties of transport must have made the importation of

[19] See Bühler, *Indische Palaeographie*, p. 88. The general designation *lekhana* 'writing-material', given in Sanskrit works to birch-bark, as well as the use of the term *bhūrja* for 'document', is significant.

[20] See Bühler, ibid., p. 89.

[21] See Bühler, ibid., pp. 5, 88.

[22] See Bühler, ibid., p. 88.

[23] Bamboo cuts (*śalākā*) with the name of the owner were given to Buddhist monks for purposes of legitimation; Burnouf, *Introduction*, p. 259, note, quoted by Bühler. The wooden tablets with incised directions as to religious suicide, the preparation of which is prohibited by a passage of the Vinayapiṭaka (Bühler, ibid., p. 5), may have owed their form to some peculiar superstition.

[24] See Bühler, *Kashmīr Report*, p. 29, note.

Bhūrja from Kashmīr, the nearest area which could furnish it, very troublesome and expensive. That Bhūrja was nevertheless known in Khotan is proved by the Dutreuil de Rhins MS., and by the tiny fragment I discovered on the plaster wall of the Endere temple cella [25]. Palm-leaves must have been still more difficult to obtain, though it is probable that MSS. written on this material were at times imported from India into Buddhist establishments [26]. Leather was used for writing purposes, as my discoveries in N. xv. have proved; but its preparation was certainly more troublesome than that of wood, and its cost, too, in all probability greater. Wood was thus indicated by nature as the common writing-material in the Khotan region, and probably throughout Eastern Turkestān, just as the bamboo was in China, until the introduction of the far more convenient paper rendered its use obsolete.

Paper not used in Khotan about middle of third cent. A.D.

Why paper should not have come into use in Khotan territory even one and a half centuries after its invention in China is a question which cannot be definitely answered from our available materials. The fact itself must be considered as certain, for rich though the ruins explored by me at the Niya River Site were in rubbish remains of all kinds, not one of them yielded the smallest scrap of paper. This total absence of records on paper is all the more curious in view of the political connexion with China which did not cease, as our Chinese documents plainly prove, even after the close of the later Han dynasty. Nor can it be attributed to the possible want of the paper-mulberry tree (*Broussonetia papyrifera*), from the bark of which the modern paper of Khotan is exclusively manufactured, seeing that the alternative use of rags, hemp, and other substitutes was known in China from the very time of the first invention of paper (105 A.D.) [27]. Whatever the cause may have been, the continued use of wood during the latter half of the third century in a distant Khotan settlement cannot be a matter of surprise when it is considered that, as M. Chavannes has shown from incontestable evidence, writing on bamboo slips was still currently practised in China itself about 200 A.D. [28]

SECTION VI.—DECIPHERMENT OF ANCIENT DOCUMENTS, KHAROṢṬHĪ AND CHINESE

Study of Kharoṣṭhī documents.

In the preceding sections I have endeavoured to record an accurate description of the localities and conditions in which my main discoveries of ancient documents at this site were made, and to set forth and justify those archaeological conclusions which the observations then made and the subsequent scrutiny of the documents in their outward appearance have led me to draw as regards their form, arrangement, and probable purpose. The historical importance of the records contained in them was in the course of my excavations ever vividly before my eyes; and in the case of the Kharoṣṭhī documents I had reason to welcome each fresh find as additional help towards the elucidation of both script and language. But from the first I recognized that the decipherment and elucidation of this wealth of materials would require much time and patient labour. The nature of my philological qualifications obliged me, at the time of discovery and during the rare moments of rest on my subsequent travels, to limit my attention to the Kharoṣṭhī records. When, after their safe transfer to the British Museum, Prof. Rapson,

[25] See below, chap. XII, sec. ii.

[26] Evidence in this direction is furnished by the arrangement of the Brāhmī MSS. on paper from both Dandān-Uiliq and Kuchā, which clearly shows imitation of palm-leaves, in their Pōthī shape, string-holes, &c. Bhūrja MSS. were either rolled, as shown by the Dutreuil de Rhins MS., or else arranged in 'forms' and bound after the fashion of our books, as is invariably the case with Kashmīr Sanskrit MSS.

[27] See Chavannes, *Les livres chinois*, p. 6.

[28] See ibid., pp. 74 sq., also p. 58.

aided by M. Senart and M. l'abbé Boyer, kindly charged himself with the difficult undertaking of editing them and otherwise preparing the way for their elucidation, I was forced by the urgency and extent of the manifold tasks to which I had to devote myself, reluctantly to forego any attempt at systematic co-operation in the arduous but fascinating labours of those learned friends.

Results of decipherment so far available.

The painstaking researches carried on by them are not likely to be concluded for a long time, and no definite date can as yet be indicated for the full publication of the texts and the results achieved in their interpretation. But preliminary transcripts and renderings have been recently published of seven documents by Prof. Rapson and of an eighth by M. Boyer; and the interesting information thus rendered available, combined with the notes which Prof. Rapson kindly placed at my disposal early in 1903 for use in my 'Personal Narrative', justifies the attempt to indicate here certain main results which have an antiquarian and historical bearing, and also some of the more curious details. Knowing the exceptional difficulties with which the work of decipherment and interpretation is beset, on account of the very cursive character of the Kharoṣṭhī script and the puzzling phonetic uncertainties of the Prākrit dialect employed, I recognize that much in the above materials must necessarily remain tentative for the present. At the same time the scholarly acumen and critical thoroughness possessed by Prof. Rapson and his coadjutors make me feel confident that the statements made below will not require modification in essential points in the light of further researches.

Language and character of Kharoṣṭhī documents.

Taking then the main facts, it is a source of gratification to me that the conclusions I first arrived at regarding the language and general character of the Kharoṣṭhī documents[1] have been fully confirmed by Prof. Rapson's labours. His exact analysis of the greater part of the collection has made it absolutely certain that the language throughout is an early Prākrit, closely akin to that of the Dhammapada version contained in the Dutreuil de Rhins MS., and showing a considerable admixture of Sanskrit terms often much garbled. The occurrence of the latter is particularly frequent in the introductory and other formal parts of letters and records, i. e. exactly where the epistolary custom of modern Indian vernaculars has large recourse to phrases of the classical language. As regards a great portion of the documents there can be no doubt that they contain, as surmised by me from the first, official correspondence and records of various kinds, such as reports and orders to local officials on matters of administration and police, complaints, summonses, directions for the supply of transport, &c., to persons travelling on public business.

Kharoṣṭhī documents with official contents.

Of the wedge-shaped double tablets which form so large a proportion of the whole, it appears highly probable that they were generally, if not always, intended for the conveyance of brief orders which concerned the bearer, or in the execution of which the bearer was to co-operate. Of three documents of this class found in N. i the translations excerpted above show that they related to the escort or the transport to be supplied to official messengers who are named as carrying the order, and in the third case to a local inquiry to be made into a complaint preferred by the bearer[2]. Among three more wedge-shaped tablets translated by Prof. Rapson and M. Boyer, one (N. xv. 318) directs the provision of camels and an escort to a 'bearer of royal matters'; another (N. xv. 12) apparently relates to a messenger's claim

[1] See the remarks written immediately after the conclusion of the excavations, in my note '*Archaeological discoveries in the neighbourhood of the Niya River*', published in the July number of the *JRAS.*, 1901, pp. 569–72; compare also *Preliminary Report*, pp. 50 sq.

[2] See above, p. 326, for a summary of the contents of N. i. 104+16 (supply of escort to Ṣameka travelling as messenger from Calmadana to Khotaṃna); iv. 108 (provision of two camels for Cuvayalina Phummaseva, messenger to Khotaṃna, and of a third for his companion); i. 105 (inquiry into Opgeya's claim to some property).

for compensation arising from a journey on duty, while N. xv. 137 (the well-preserved document shown in Plate XCVIII), urges the immediate arrest and transmission to the issuing officer of certain fugitives wanted for judicial proceedings[3]. The document is of special interest as it mentions two 'letters of injunction written at length'[4] (*livistarena anadilekha*) which had vainly been issued previously for the same purpose, the document itself being designated as a *kilamudra.* To this term, used in several other documents of this class, I shall have occasion to recur presently. By the 'letters of injunction' (*anadilekha*) probably full orders are meant, like our Kharoṣṭhī documents on leather, in which this expression repeatedly occurs[5].

Private letters and drafts.

Private communications or records, too, are represented among the documents, some of the rectangular double tablets undoubtedly bearing this character[6]. The likelihood of some of the latter being agreements, bonds, or formal records of a similar kind, has been indicated already[7]. Drafts of letters seem to be frequent among the contents of the large single tablets of varying shape which have been spoken of above under the general designation of 'oblong'. Of translated documents of this kind we have, besides the previously-mentioned tablet N. iv. 136, a very interesting specimen in the well-preserved 'Takhtī' N. xvi. 2 (reproduced in Plate CI). As proved beyond all doubt by Prof. Rapson's analysis, the text consists here of three entirely distinct parts, written in varying directions, yet by the same hand[8]. As each of these parts contains a private letter from a different person (or persons) to a different address, it is quite clear that the tablet could not have been intended for transmission. As the subjects of the communications have no connexion whatever, while the elaborately polite wording of the introductory phrases, greetings, &c., show an unmistakable similarity, the conclusion seems justified that we have here the drafts of communications penned by one and the same scribe for different persons. This conclusion is supported by the appearance of the ink varying in the several communications, and thus showing that they were written down on separate occasions. Plentiful marks of scraping and traces of erased writing prove that the writing-board had served before, probably for similar purposes.

Records of accounts, lists, &c.

Of the mass of miscellaneous 'papers' written on single tablets of irregular shape, such as long wedges, lath or label-like pieces, &c., and usually in columns ending with numerical signs,

[3] For the text and tentative translations of N. xv. 12, 138, comp. Professor Rapson's *Specimens*, pp. 12 sqq., 15. N. xv. 137 has been translated and annotated by M. Boyer in *Journal asiat.*, Mai–Juin, 1905, pp. 463 sqq. The order is addressed to the 'Cojhbo Soṃjaka', the official residing, as we have seen, in N. v. It refers to the case as pending for the third year, and its wording plainly indicates dissatisfaction at the delay.

[4] For the term *livistarena anadilekha*, i.e. Skr. *lipivistareṇa ājñaptilekha*, see below.

[5] See, e.g., N. xv. 88 (Plate XCII), line 5, xv. 33 (ibid.), line 3.

[6] Compare, e.g., the address on the covering-tablet, N. xv. 154, Plate XCVII.

[7] See above, p. 354.

[8] See *Specimens*, pp. 9 sqq. In piece *A* Ogu Cinaphara and Cojhbo Cinyaśasa send information to their 'beloved brother Cojhbo Ṣaṃmasena' about a certain unpleasant affair, not precisely specified, in which he is concerned, and which, if no agreement is arrived at, 'must be taken in hand and transferred to the King's Gate (*rayadvara*) here,' i.e. taken into court. *B* is a letter conveying an eagerly expressed request from Kāla Kuṣanasena to 'the Great Cojhbo Śitaka', for the early transmission of *uṭa*, for which Professor Rapson justly suggests the rendering 'news', and which may, I think, represent Skr. *vārtā*. The quaint way in which the identical request is repeated over and over again with slight variations recalls to my mind the fashion of reiteration in which Indian correspondents of the present day indulge when they wish to emphasize the importance of any topic. The forms for Sanskrit correspondence contained in that curious Kashmīrian manual, the *Lokaprakāśa*, with other evidence that cannot be detailed here, show that this fashion is of old date in India.

Perhaps the most curious of these letters is *C*, in which the Śramaṇas Baṅgusena and Pocgayaesa 'inquire after the spiritual and bodily health again and again, many times and unceasingly' of their 'dear friend, Cojhbo Nandasena and Cataroyae', and then expatiate with pious unction upon the news of the death of a certain person. There is good reason to suppose that the ruined structure N. xvi., adjoining N. v, was a Buddhist shrine (see below, p. 375). Is it possible that one or other of these monks was the attendant of the shrine, as well as the scribe who penned the drafts?

no detailed analysis is yet available. But there are indications that accounts, lists of labourers or materials, &c., are frequent among their contents. Considering the administrative duties of the officials whose residences were the main find-places of such documents, it is very likely that some at least of these lists and accounts may have reference to taxation.

Designation of ruler in Kharoṣṭhī documents.

The ruler, in whose name the official orders, &c., are issued, is given in the initial formula of the wedge-shaped tablets only the brief title *mahanuava maharaya*, corresponding to Skr. *mahānubhāva mahārāja* 'His Excellency the Great King.' But in the introductory dating of many of the rectangular tablets, where the year is indicated with reference to the reign of a particular ruler, we find his name besides the above title coupled with the designation of *devaputra*[9], which recalls the official style of the Kuṣana or Indo-Scythian kings as observed in their inscriptions in their Indian territories. The royal names, of which two at least can be distinguished in the documents, are to be found neither on the Sino-Kharoṣṭhī coins of Khotan nor in the dynastic lists of the Tibetan texts. Yet the constant references in the tablets to Khotan (called *Khotaṃna, Khodana, Kustanaka*) show that the district containing this ancient settlement must have formed part of the kingdom of Khotan, and it seems difficult to believe that by the above titles any other ruler but that of Khotan could be meant.

Personal names in Kharoṣṭhī documents.

Many of the persons to or by whom documents are dispatched bear names which are either purely Indian, such as Bhima, Baṅgusena, Nandasena, Ṣamasena, Śitaka, Upajiva, &c.[10], or else look like Indian adaptations, e.g. Aṅgaca, Cuvayalina Phummaseva, Piteya, Śili, Saṃghila, Saṃjaka, Soṃjaka, Sucama, Sughiya[11]. But others are distinctly un-Indian, e.g. Lipeya, Opgeya, Limira, Maṃṅgaya, Tsmaya[12]. A few, like Paśaspa and Cinaphara, suggest Irānian influence in their origin or formation[13]. It is interesting to see that among the correspondents in N. xvi. 2 there appears a *Kuṣanasena*, as if to emphasize some connexion with Indo-Scythian dominion far away to the south-west.

Official titles.

In strange contrast to the names, some of the most frequently-occurring titles borne by these officials are wholly non-Indian, such as *Cojhbo, Ṣoṭhaṃgha, Kala*[14]. But the official designations familiar from ancient Indian usage are also met with, e.g. *divira* 'clerk', *cara* (or *caraka*) 'secret agent', *rayadvara-purasthita* 'president of the royal court'[15]. Letter-carriers (*lekhahāraka*) are often referred to by their Sanskrit designation, while the duty of the official messengers, for whose requirements *en route* many of the wedge-shaped tablets were intended to provide, is always spoken of by the term *dutiya*, derived from their correct Indian appellation of *dūta*[16]. The often-recurring introductory formulas, with their stereotyped greetings, elaborately constructed honorific addresses, and polite inquiries after the health and spiritual welfare of the addressees—the whole corresponding to the verbiage which Anglo-Indian custom in the reproduction of vernacular documents curtly disposes of as 'after compliments'—possess a distinct flavour of that quaint phraseology which Sanskrit epistolary style has always affected, and to which the correspondence of my Kashmīrian Paṇḍit friends has accustomed me[17]. That official custom, however, knew also

[9] See, e.g., N. xv. 155 (Plate XCIV); xv. 166 (Plate XCV).

[10] See for these names N. i. 104, 105; xv. 12; xvi. 2, in Prof. Rapson's *Specimens*, from which also the subsequent references to particular names, &c., are taken.

[11] Comp. N. i. 105; iv. 108, 136; xv. 12, 137, 318.

[12] See N. i. 105; iv. 136; xvi. 2.

[13] Comp. N. xvi. 2.

[14] For *Cojhbo* see N. i. 104, 105; iv. 108, 136; xv. 137, 318; xvi. 2; for *Ṣoṭhaṃgha* (also spelt *Ṣoṭhaṃga*), N. iv. 104, 108, 136; *Kala*, N. xvi. 2.

[15] Comp. N. iv. 136; xv. 137; for the frequently occurring term *rayadvara* (Skr. *rājadvāra*) 'royal court of justice', see, e.g., N. i. 105; xv. 12; xvi. 2.

[16] See for *dutiya*, N. i. 104; iv. 108; xv. 12.

[17] The complimentary introductions of the letters in the previously-discussed tablet, N. xvi. 2 (see Prof. Rapson's *Specimens*, pp. 9 sq.) furnish typical illustrations of this phraseology. *Priyadarśana* 'whose sight is dear', *priyadevamanuṣya* 'who is dear to gods and men', *devamanuṣyasampūjita* 'honoured by gods and men', *sunāmaparikīrtita* 'whose

a style far less ornate is amply shown by the business-like and peremptory tone adopted in some of the wedge-shaped tablets, ordering the submission of affidavits (*śavatha*) according to a specified list; the production of certain witnesses; the arrest of individuals, &c.

Antiquarian interest of petty records. The antiquarian interest possessed by many of the details which the elucidation of petty records may reveal is illustrated by a rectangular tablet dated in the ninth year of King Jitroghavarṣman. It relates to a transaction by a certain Buddhagoṣa, apparently the slave of the Śramaṇa Anadasena, concerning some household goods, pawned, perhaps, or taken in pledge. The articles are enumerated in detail, and their value indicated in a currency that may yet be determined in the light of other documents. It is of interest to find that this list, besides sheep, vessels, wool-weaving (?) appliances and some other implements, enumerates also *namadis*. We may recognize here an early mention of the felt rugs or 'Numdahs' (Persian-Turkī *namad*) which to this day form a special product of Khotan home industry, large consignments being annually exported to Ladāk and Kashmīr. Small pieces of well-worked felt were plentifully mixed up with the other rubbish contents of N. xv.[18] Numerous tablets seem to have reference to disputes about water; and though the elucidation of details must here necessarily be attended with special difficulty, we may reasonably hope for interesting sidelights to be thrown by them upon the ancient system of irrigation.

Mention of ancient local names. For the old topography, too, of this and the adjoining regions the Kharoṣṭhī documents are certain to furnish valuable materials. I have already had occasion to point out how important it is to find the antiquity of the name Khotan, practically in its present form, attested by the *Khotaṃna* (probably pronounced *Khotana*) and *Khodana* of these records[19]. By their side we find the Sanskritized form *Kustana* or *Kustanaka*, a distinct indication of the antiquity of the learned adaptation of the local name and the legend about 'the Breast of the Earth' as related by Hsüan-tsang and our Tibetan sources[20]. The explanations given above as to *Nina* and *Calmadana*[21] show the possibility of arriving hereafter at the correct identification of other old localities mentioned in the documents. Unfortunately the specific name of the ancient site itself has not been traced as yet. As if to remind us of the position which the ruined settlement must have occupied on the border of the Khotan kingdom, we meet with frequent mention of 'frontier watch-stations', designated by the Sanskrit term *draṅga*, the true significance of which I was first able to establish in Kashmīr[22]. That Buddhism was widely spread, if not actually the prevailing religion in the territory, is amply proved by the frequent references to Śramaṇas, and by passages like the one in the above-quoted tablet N. xvi. 2, enumerating various sacred categories of the Buddhist heaven[23].

good name is far-famed,' belong to the regular stock of honorific addresses. Phrases like *pracachadevata* (Skr. *pratyakṣadevatā*) 'a divinity incarnate', *atrptapriyadarśana* 'of whose dear sight there is never enough,' &c., are attempts at special flights of epistolary politeness. For complimentary inquiries or wishes about the addressee's health comp. N. iv. 136 (*divyaśariraarogi sampreṣeti bahu aneka*); N. xvi. 2, B (*arogya preṣeti bahu aprameyaṃ*); ibid., C (*divyaśariraarogi puripruchaṃti.*) The endeavour to use Sanskrit in the introductory parts of these epistles, and the thoroughgoing disregard for its grammar and spelling, are familiar features to any one acquainted with the correspondence of Hindus who have received a traditional education, but are able to express themselves only through the medium of their vernacular.

[18] For specimens, coloured and plain, N. xv. 014, 015, see Plate LXXVI.

[19] *Khotaṃna* and *Khodana* may alternate in the same document; see N. i. 104 + 16 in Prof. Rapson's *Specimens*, p. 14. For the fashion prevailing in this script of announcing a coming nasal by an Anusvāra added to the preceding syllable, see Rapson, 'On the alphabet of the Kharoṣṭhi documents,' in *Actes du XIV*e *Congrès des Orientalistes*, i. p. 12.

[20] See above, pp. 153 sq.

[21] See above, p. 311.

[22] *Rājat.*, II. pp. 291 sq. I have since found the word surviving in various localities outside Kashmīr where once a customs station or frontier post existed, e.g. at the well-known salt quarries of *Drang* in the Kohāt District of the N. W. Frontier.

[23] See above, p. 365; *Specimens*, p. 10.

Technical terms for designation of Kharoṣṭhī documents.

Not the least curious among the results so far secured by the work of decipherment is the proof that the use of the various types of ancient stationery represented was restricted to particular classes of documents for which we can still ascertain the official terminology. With unchanging regularity the double wedge-tablets are designated in their context by the term *kilamudra.* I have previously suggested its interpretation as a compound of *kila*, corresponding to Skr. *kīlaka* 'wedge', and *mudrā* 'seal'. That the application of a term meaning 'wedge [-tablet] and seal' would be appropriate in this case is clear; but I admit that, in regard to the explanation of the second portion of the term, the occasional variant *°mumtra*, noticed by Prof. Rapson, deserves special attention [24]. We have seen already that the term *anadilekha* (Skr. *ājñaptilekha*), 'letter of injunction' or 'rescript', is specially used for official orders recorded on leather. The rectangular double tablets are always referred to as *lihitaka* or 'letter'. Another technical term of this kind is *stovana*, but whether it was applied to an open communication written on one or the other type of oblong tablets, as its use in the draft *B* of N. xvi. 2 seems to suggest [25], remains at present doubtful. It is evident that the clerks of those ancient offices were quite as particular about bureaucratic distinctions of this kind as the 'Babu' of modern India, who would never make a mistake about supplying himself with 'octavo note' for his D. O.'s, foolscap for his 'fair dockets', or slips for his 'office memos.'

Historical import of use of Prākrit language.

The very nature of the contents, and the complete absence of similar records of ancient date in India itself, must render the full elucidation of the Kharoṣṭhī documents a slow task beset with exceptional difficulties. Not until the whole of these ample materials has been rendered accessible to Indologists by the textual publication to which Prof. Rapson, assisted by MM. Senart and Boyer, is now devoting his critical labours, will it be possible to form an approximately correct estimate of the multifarious aspects of ancient life and culture which they may bring back to light. But whatever revelations of detail may be in store for us, it is not likely that any of them will approach in historical interest one important fact which can now be clearly recognized. I mean the use of an Indian language throughout the mass of non-Chinese records, whether official or private. When we take into account the distinctly secular character of most of them, this evidence of the language must be considered a striking confirmation of the old local tradition, recorded by Hsüan-tsang as well as in the Tibetan 'Annals of Li-yul', and already discussed in chapter VII, according to which the territory of Khotan received a large portion of its early population by immigration from the region of Takṣaśilā or Taxila. It must certainly lead us to believe in some historical fact underlying this tradition, if in an outlying settlement of ancient Khotan we find a mass of multifarious records, accumulated by chance, and all closely connected with indigenous administration and ordinary life, written in a language which has its nearest extant congener in that of the inscriptions and coins from the extreme North-West of India during the centuries immediately before and after the commencement of the Christian era.

Kharoṣṭhī script in Taxila region.

It is an equally significant fact that the Kharoṣṭhī script which our documents present in a form very closely allied to that found in the inscriptions of the Kuṣana period [26], was

[24] See his remarks in *Specimens*, p. 13; also M. Boyer's note, *J. asiat.*, 1905, Mai–Juin, p. 466. The term, if interpreted as I proposed, would appear a somewhat barbarous compound; but some grammatical latitude might well be allowed to the old Khotan clerks, who are likely to have coined it, perhaps, as a rendering of some foreign (Chinese?) term.

[25] See *Specimens*, p. 11. The bearer of the missive is spoken of as 'letter-reader' (*lekhavacitu*).

[26] For the Kharoṣṭhī writing of the documents and its relations to the Indian forms of the script see Prof. Rapson's illuminating notes, 'On the alphabet of the Kharoṣṭhi documents,' in *Actes du XIV*[e] *Congrès des Orientalistes*, i. pp. 210 sqq.

within India peculiar to the region that had in Taxila its oldest historical centre. It does not appear possible to account satisfactorily for either the language or the script of our documents by the spread of Buddhism alone, seeing that Buddhism, so far as our available evidence goes, brought to Central Asia only the use of Sanskrit as the ecclesiastical language, and the writing in Brāhmī characters. Nor would the assumption of a temporary extension to Eastern Turkestān of Śaka or Kuṣana power from the north-western borderlands of India be sufficient to explain the transplanting of an Indian language and its adoption for ordinary use among the people; for on the one hand, such a political connexion, if it ever really existed, must, in view of the Chinese historical records, have been very transitory, while on the other hand the forces that might have effected it were themselves Central-Asian rather than Indian.

Palaeographic evidence of date of Kharoṣṭhī records.

It seems strange that ruins far away in the barbarian North, overrun by what Hindu legend vaguely knew as the mythical 'Ocean of sand'[27], should have preserved for us records of everyday life older than any written documents (as distinguished from inscriptions) that have as yet come to light in India itself. But from the first there was ample evidence pointing to this chronological conclusion. The close agreement in palaeographical features between the documents discovered and the Kharoṣṭhī inscriptions of the Kuṣana kings led me at once to the conclusion that those records on wood and leather must belong approximately to the period covered by Kuṣana rule in the extreme North-West of India. We know, in spite of all uncertainties as to the era or eras used in the inscriptions, that the domination of the Kuṣana dynasty in the Indus Valley and Punjāb must mainly fall within the first three centuries of our era. It is certain that the Kharoṣṭhī script ceased to be used soon afterwards, in those territories which had formed its Indian home; and it seems very improbable that it could have remained in current use in Khotan for a long period without undergoing perceptible changes. Valuable collateral evidence was provided by an interesting find made in the ruined building, N. VIII, which will be described below. A narrow, Takhtī-shaped tablet (N. xx. 1) excavated there shows on one side a single line of Kharoṣṭhī, while on the other side I discovered to my surprise three lines of Brāhmī characters, the only specimen of this writing from the site. Unfortunately the ink on both sides has become very faint, and of the Brāhmī text in particular it has become impossible to make out more than a few detached characters here and there. In these, however, I thought that I could recognize characteristic features of the Brāhmī writing of the Kuṣana period. There is nothing in the appearance of the tablet to suggest that the two texts were written at appreciably different times. Hence the close agreement of the palaeographical indications furnished by the Brāhmī and Kharoṣṭhī writings supplied further evidence in favour of the conclusion already indicated.

Unique document in Brāhmī characters.

Numismatic and other indications of early date.

I could not fail to recognize another proof of considerable antiquity in the use of wood as the only writing material apart from leather; for the total absence of paper among the remains of the site clearly marks a date earlier than the fourth century of our era, from which onwards the use of paper in Eastern Turkestān is palaeographically attested by MS. remains from Kuchā[28]. Finally, numismatic evidence helps to confirm the conclusion, since the coins which were picked up at different points of the site during my stay, seventeen altogether, were, with the exception of one of doubtful origin, all Chinese copper pieces current under the Later Han dynasty[29].

[27] Comp. *Rājat.* iii. 279 sqq.

[28] See Hoernle, *Report on Central-Asian antiquities*, ii. pp. 13 sq.

[29] As will be seen from App. *D*, 8 coins from the Niya Site belong to issues showing the symbols *Wu-chu* (see for specimens Nos. 12–14 in Plate LXXXIX); seven are small pieces without any legend, such as are usually attributed to the reign of Hsien ti (190–240 A. D.); specimens of these are seen in Nos. 7, 8, 10, Plate LXXXIX. No. 7 with another piece was picked up on eroded ground close to N. xv. One piece is in too poor a state to be determined. Another coin, the characters on which could not be made out by me at the

Discovery of precise date, 269 A.D., in Chinese document.

But with all these indications to guide me, there was reason to feel particularly gratified when the approximate dating thus arrived at received confirmation and precision by incontrovertible chronological evidence. It came, as I had always hoped, from the Chinese records on wood found in our ancient rubbish-heap. Their first cursory examination at Kāshgar, effected with Mr. Macartney's help through Sun Ssŭ-yeh, showed that their contents were mainly of an official nature, and that their characters bore distinct resemblance to the type of writing which Chinese palaeographers associate with the Han epoch. Their subsequent scrutiny in the British Museum by Sir R. Douglas enabled me to form a more exact idea of the nature of the brief police orders, requisitions, &c., recorded on some of these slips of wood. But it was only when in December, 1902, Dr. S. W. Bushell had occasion to examine the documents that one of them, N. xv. 326 (see Plate CXII) was discovered to be fully and precisely dated. The initial characters, as since verified by M. Chavannes, who subsequently was kind enough to undertake the systematic decipherment and translation of all Chinese records brought to light by me, unmistakably indicate the fifth year of the *T'ai-shih* period of the Emperor Wu ti, the founder of the Western Chin dynasty, corresponding to 269 A.D. The accuracy of this reading is established against any possible doubt, as M. Chavannes has proved, by the specification of a chronological detail in the rest of the dating: 'the twentieth day *ting-ch'ou* of the tenth month, of which the first day is the day *wu-wu*[30].' Thus a precise date has been found which fixes the period when this remarkable collection of documents accumulated. N. xv. 326 itself was discovered in a layer of refuse about one foot above the original floor.

Historical data in Chinese documents.

The highly instructive notes with which M. Chavannes has accompanied his full transcripts and translations of the Chinese documents on wood, reproduced in Appendix *A*, make it possible for me to attempt here a brief delineation of the general character of these records, as well as of the main historical data they furnish[31]. Among the latter, to which we may well turn first, the precise date of N. xv. 326 is undoubtedly of exceptional value, in connexion with a number of antiquarian questions. But wider historical importance may be claimed for what the whole collection proves as to the re-assertion of Chinese authority in Eastern Turkestān under the Emperor Wu ti (265–289 A.D.). The Chin Annals, as noted by Chavannes in his remarks on N. xv. 93, record of Wu ti that he caused his supremacy to be recognized in the 'Western Countries'; they also record embassies with tribute in 270 A.D., from Farghāna and Kara-shahr,

site, has since been identified by Dr. Bushell as bearing the legend *Ch'ien-yüan ch'ung pao* 'Copper money of the Ch'ien-yüan period' (758–763 A.D.). This coin would have to be accepted as a proof that the deserted site was visited during the T'ang period, if its evidence could quite be relied upon. My notes show that this particular coin was brought to me at the site by Turdi Khwāja, when he joined me from Khotan on February 7, 1901 (see *Ruins of Khotan*, pp. 406 sq.). He said he had picked it up on his march to my camp, then near N. xv., and as I understood him, since he had left Imām Ja'far Sādiq. Whether this was really the case, and at what particular part of the area he found it, I could not ascertain at the time, and the question must thus remain open. My treasure-seeking guide, always on the look-out for antiques, was honest enough on most matters, but not exactly burdened with an 'archaeological conscience'.

[30] See App. *A*, part ii, note on N. xv. 326.

[31] I regret that in the Plates CXII–CXIV, which reproduce the whole of the Chinese records on wood, they are shown in an order different from that in which Prof. Chavannes has arranged his translations in App. *A*. While the latter arrangement is based on a systematic classification of the slips according to their contents, the Plates show them as grouped according to the various types of handwriting recognized by Mr. Rionin Kohitsu, a Japanese expert in Chinese palaeography, who early in 1903 kindly offered his help for the reproduction of the documents. I regret that I could not find time previous to my departure from England to arrange for the preparation of fresh negatives when M. Chavannes' translations had become available, and was subsequently obliged by practical considerations to have the collotype plates printed from the available negatives. As the context does not suffice to establish connexion between any two slips, the criterion of handwriting may possibly yet prove of some use. On the other hand, the reproduction of the first grouping has resulted in N. xv. 328 and 75, which are the broken halves of one slip and exactly fit each other, being shown erroneously in different plates.

and in 287 A.D. from Sogdiana. But by the side of these references, and of those made to princes of the ruling families of Shan-shan, Kuchā, and Kara-shahr, who in the years 283 and 285 A.D., took up appointments at the Imperial court, the proof now furnished by our documents as to the actual exercise of Chinese authority in this region is specially welcome.

References to Chinese administration.

They clearly show that Chinese posts, under officers invested with some measure of administrative authority, must have been established in different parts of Khotan territory and probably elsewhere also. The existence of a Chinese administrative organization extending over the whole of the Tārīm Basin is clearly indicated by the mention in N. xv. 93 of 'a prefect [governing by delegation of the dynasty] *Chin* . . . great marquis invested by the *Chin*, [governor of] *Shan-shan*, *Yen-ch'i* [Kara-shahr], *Ch'iu-tzŭ* [Kuchā], *Su-lê* [Kāshgar] . . . allied to the *Chin*.' But the similarity specially noticed by M. Chavannes between the last of these titles and the formula with which investiture was accorded in 229 A.D. to Po-t'iao, 'king of the Great Yüeh-chih allied to the Wei', seems to make it doubtful whether those high dignities were not borne by some indigenous ruler acting under Chinese control. The title of the 'Chang-shih of the Western Countries', compliance with whose orders is recorded in N. xv. 328+75, occurs also in the Chin Annals towards 324 A.D. That the officer holding it at the time to which our documents belong must have exercised direct authority over the Chinese garrisons established in Khotan territory seems indicated also by the reference made in N. xv. 85 to an edict of his. Both in N. xv. 328+75 and in N. xv. 348 we find a certain *Lo-pu-yen* mentioned as the officer who has given effect to an edict; but in the latter slip, which is incomplete, the authority issuing it is not named.

Edicts of kings of *Yü-t'ien* and *Shan-shan*.

That the Chinese supremacy re-established under Wu ti contented itself with the political control of the indigenous principalities, but did not efface them, might have been concluded from the system maintained during the Han occupation and again under the T'angs. Yet it is of historical interest to find edicts both from the king of Yü-t'ien and of Shan-shan (south of Lop-Nor) distinctly referred to in our records. The covering-tablet N. xv. 345, on which the latter ruler is mentioned, has on account of its peculiar form been discussed above[32]. N. xv. 73 is incomplete, and merely contains an acknowledgement of the receipt of the king of Yü-t'ien's edict. That the latter, whatever its contents, must have been written in Chinese may be considered as certain. Close communication with the districts of westernmost Kan-su, through which then as now the great route between China and Eastern Turkestān led, is attested by several of our documents. The slip N. xv. 326, which has proved so valuable chronologically, mentions the prefect of *Tun-huang*, the ancient frontier station situated in the vicinity of Sha-chou. A title, of which the incomplete slip has preserved but a part, seems to suggest that the prefect was also in charge of military administration in that region. That 'the prefect of Tun-huang' exercised at times direct authority in Turkestān affairs, is shown by what the Later Han Annals record about his share in the Yü-t'ien troubles of 152 A.D.[33] Tun-huang is mentioned again in N. xv. 188, which gives the names of six out of the eight 'commanderies' recorded in the Chin Annals as dependent on Liang-chou[34]. As M. Chavannes' note shows, the six commanderies named on the slip correspond to modern districts in the extreme north-west of Kan-su, extending within the Great Wall from Lan-chou to Hsi-ning and Su-chou, and including also the sub-prefecture of Tun-huang between the latter and Sha-chou. In N. xv. 203 reference is again made to the prefect of Tun-huang, who is to have certain articles transmitted to him; but here, too, the text is evidently a mere fragment.

The prefect of *Tun-huang*.

[32] See above, p. 361.

[33] See Rémusat, *Ville de Khotan*, pp. 7 sqq.; above, p. 168.

[34] Two of them occur also in the fragmentary slip, N. xv. 116.

Miscellaneous official orders, &c., on Chinese slips.

Interesting information can be gleaned also from those Chinese records which do not mention otherwise known localities, officers, &c., in spite of the fact that most of the slips are in a fragmentary condition, and that few if any among them, even when intact, are likely to have formed by themselves a complete text. Their translation in M. Chavannes' Appendix clearly proves that, notwithstanding all uncertainties of detail, we can safely recognize in the great majority of them portions of official orders and reports. That judicial and police functions were exercised by the officers from whom these miscellaneous 'papers' originated is shown by the numerous pieces which refer to the arrest of, or accusations against, particular individuals[35]. The order 'to all prefects and sub-prefects to furnish escorts to a person on official duty', which N. xv. 101. a reproduces, touches an item of administrative routine familiar to us from the Kharoṣṭhī wedge-tablets. Orders of various officials (governor, prefect, deputy, *ts'ung-shih*) are alluded to in several short fragments, without their subjects being indicated by the extant text[36]. Actions arising from pecuniary obligations seem to be reported upon in some other slips[37].

Chinese 'passes' to individuals.

But far more numerous are the pieces which describe specific individuals as to age, personal appearance, dress, &c., and manifestly represent permits or passes of some kind[38]. Of special interest among these is N. xv. 53, describing 'the [*Yüeh*]-*chih*, i. e. Indo-Scythian, barbarian *Chu*, [a native] of the *Yüeh-chih* [i. e. Indo-Scythian] kingdom, aged 49 years, of middle stature, dark complexion'; and N. xv. 191, which also refers to 'a barbarian, of the Yüeh-chih kingdom', possibly the same individual. From which part of the wide Indo-Scythian dominion 'the barbarian *Chu*' came, it is impossible to guess. Kashmīr, which formed part of it, undoubtedly lay nearest. In any case we have a trace here of that foreign intercourse with China for which the territory of Khotan long formed a favourite line of transit. Some of the individuals to whom these slips may have been issued as means of legitimation are likely to have passed the post, evidently existing at this point of the settlement, more than once; for we find '*I*, also called *Nu*, 56 years of age, of middling height, with hair and beard turning grey', described in almost identical terms in two different slips[39]. Individuals are repeatedly described as wearing trousers and coats of linen, while hemp shoes, such as I found in one of the ruined dwellings of Dandān-Uiliq, are mentioned in one instance[40]. Evidently with a view to authorize the passage of certain property, N. xv. 61+62 specifies in the case of a certain individual two bullock-carts and two bullocks. The mention in N. xv. 324 of 'a white horse, spotted, with a saddle and bridle, old and common', may have served a similar purpose. The specification in N. xv. 78 of 'ginger, betel-nuts of the south' as merchandise, may possibly have some connexion with customs duties.

Dress and appurtenances of individuals named.

Character of Chinese administration as indicated by documents.

The very pettiness of the affairs treated in the majority of the Chinese records seems an indication that Chinese control at that period cannot have been restricted to a purely military occupation of the territory. What division of functions existed between the Chinese officers established at places like the Niya Site and the indigenous administration we are not likely ever to ascertain in detail. But whatever we know of Chinese methods of control in the 'Western Countries' must *a priori* lead us to suppose that the indigenous ruler and his

[35] See N. xv. 314 (which relates to the examination of a fugitive by the prefect and his deputy); 189, 362 (an order for the arrest of eight individuals); 315 and 37 (both relating to an accusation by a military officer); 176; 123; 125+127.

[36] See N. xv. 145, 010, 59, 117.

[37] Comp. N. xv. 109, 353, 123, 207.

[38] See N. xv. 53, 191, 337, 152, 08, 192, 09, 02, 339, 169, 175.

[39] See N. xv. 08, 192.

[40] See N. xv. 337, 09, 02; for a hemp shoe see above, p. 272.

administrative agents were left a preponderating share in the management of affairs directly concerning the local population. The contents of our ancient rubbish-heap seem strikingly to illustrate this view. Since Kharoṣṭhī and Chinese documents intermingled throughout its layers, there can be no doubt as to the simultaneous existence of the two administrations working side by side. Yet at the same time the overwhelming proportion of Kharoṣṭhī records proves that the range of affairs left to native management must have been far the wider.

Chronological conclusions as to rubbish accumulations.

The occurrence of the Chinese slips of wood throughout the rubbish accumulations and at greatly varying depths is of special chronological interest. We know that after the reign of the Emperor Wu ti (265–289 A. D.) Chinese political relations with the 'Western Countries' completely ceased[41]. Taking then into account that the document N. xv. 326, accurately dated in 269 A. D., was found only about one foot above the floor, and that a considerable number of Chinese records with contents of similarly official character were found fully 3 to 3½ ft. from the floor, and thus close to the surface of the extant mass of refuse[42], it is evident that we are justified in recognizing the period covered by Wu ti's reign as the time during which the whole of the records, &c., found in N. xv. must have accumulated. The chronological limits thus ascertained make it less of a loss to us that the other Chinese documents are either undated or only by month and day[43], and that we cannot at present fix the chronology of the reigns to which the fully dated Kharoṣṭhī tablets refer.

Time of abandonment of site.

Considering the unsubstantial construction of the building, it appears very improbable that N. v. could have continued to be inhabited for many years after Wu ti's time; and as its fate may safely be assumed to have been typical of that which the other structures within this portion at least of the site underwent, we are led to conclude that the end of the third century of our era must have seen the site abandoned to the drifting sand of the desert. Whether this abandonment was indirectly connected with the political and economic changes which undoubtedly accompanied the withdrawal of Chinese authority, or was due solely to natural causes affecting irrigation and therefore cultivation, is a question regarding which the available evidence would scarcely justify the expression of a definite opinion at present.

Imported industrial relics in rubbish-heap N. xv.

It was interesting to trace the close connexion with China also among the relics other than documents which this ancient rubbish-heap had preserved, and which may now be briefly reviewed in conclusion. The fragments of a well-finished circular lacquered bowl (N. xv. 001), decorated outside with black and red bands (see Plate LXV), and showing a highly-polished red surface inside (see Plate LXX), unmistakably indicate Chinese workmanship; and the cane material too, clearly points to that origin. The pieces of delicate plain silk (N. xv. 011), of which Plate LXXVI reproduces portions, may also have come from the Far-eastern portions of the empire. The fragments of plain and cut glass, N. xv. 001. b–h (see Plate LXXIV), mostly greenish in colour, showing great transparency, and very different from the coarse material found at other sites, may be the remains of Western imports. Glass was until the middle of the fifth century of our era known in China only as an import from Ta-ch'in, the Far West[44]; and it appears very improbable that the introduction of glass manufacture could have been delayed so long, if the making of glass had been an art practised in Eastern Turkestān when the latter was under Chinese control during earlier centuries. It is significant that, according to the *Pei shih*, the first makers of glass in China were traders 'from the country of the Great

[41] See M. Chavannes' remarks on N. xv. 93 in App. *A*, part ii.

[42] See, e. g., N. xv. 337, 339, 349, 351, 353, 02, 08, 09, 010.

[43] Dates by month and day only are found on N. xv. 348, 100, 207, 351.

[44] See Hirth, *China and the Roman Orient*, pp. 228 sqq.

Yüeh-chih', i. e. from the old Indo-Scythian dominion[45], for whom the land route through Khotan was a more likely line of communication than the sea route.

Articles of local manufacture.

Far more numerous, of course, were articles for which local manufacture may safely be assumed. Among the rags mixed up with the refuse were the pieces of cotton and woollen materials, showing considerable variation in texture and colour, from which representative specimens (N. xv. 012, 13, 16, 17, 18) have been reproduced in Plate LXXVI in their original colouring. The coarse cotton check (N. xv. 012), in dull yellow with blue stripes, curiously recalls the 'Jharans' in favourite use through modern India. The pieces of felt, N. xv. 014, 015 (see Plate LXXVI), the one very compact and strong, the other loose but richly coloured, show that this still flourishing branch of Khotan textile industry had already attained full development. The comparison of all these fabrics with those found in the Endere temple, from which specimens are shown in Plate LXXVII, might, in view of the four or five centuries separating the two sites, possibly prove instructive to an expert student of ancient textiles. The twisted cord (N. xv. 001. k), with its hard, bead-like masses entwined at regular intervals, has the appearance of a kind of rosary. In the horn spoon (N. xv. 001. i), we have a utensil of domestic use. A large number of sheep's knuckle-bones, often painted pink like the specimen N. xv. 001. j, showed that gambling with this simple form of dice must have had its votaries in the household. Besides these there was found also an ivory die, N. xv. 004 (see Plate LXXIV), showing four equal rectangular sides and two square ends. By its shape and size it distinctly recalls an ancient brass die from Eastern Turkestān described by Dr. Hoernle; but as it is not perforated, it could not have been used for the system of divining explained in Dr. Hoernle's remarks[46]. It may be noted that the arrangement of the dots distributed over the four equal sides differs from that seen on the cube-shaped ivory die found at Endere (see E. 001 in Plate LII). A curious small relic is the oblong piece of thick and very hard leather, N. xv. 005 (see the diagram, p. 411), rounded at one end and square at the other, pierced by three holes on each of its longer sides. Its shape and the manifest arrangement for sewing suggest the possibility of its having belonged to a piece of scale armour, where the scales of leather were fixed in the manner displayed on the Kubera statue of the Dandān-Uiliq shrine, D. II, discussed above[47]. Finally the dried date (N. xv. 001. l), may find mention here as a curious proof that even such delicacies from the distant South were obtainable by those who once lived in this ruined dwelling.

Section VII.—EXPLORATION OF RUINS N. vi–xii AND GENERAL OBSERVATIONS ON SITE

Ruined structure N. xvi.

What the remaining ruins at this site yielded in written documents must appear small in comparison with the rich finds described in the preceding sections. But their excavation was attended by some interesting results. First among these ruins was a small detached structure, N. xvi. (see plan in Plate XXXII), situated about 70 ft. to the south of the nearest extant portion of N. v., in what evidently was once an orchard attached to that residence. The walls of timber and plaster forming a square of 16 ft. had, owing to the insufficient protection of sand, which lay only to a height of about 4 ft., decayed badly. The interior of the room, which by its dimensions and isolated position at once reminded me of the small temple cellas of Dandān-Uiliq, was found to be occupied by a platform of stamped earth, 8 ft. square and

[45] See Hirth, *China and the Roman Orient*, pp. 230 sq.

[46] See *Report on C.-A. antiquities*, i. pp. 39, 44.

[47] [For a striking confirmation of this view, see *Addenda*.]

2 ft. 9 in. high. The sides of the platform were formed by a facing of solid planks rising in two steps and held in position by massive round posts. Towards the middle of the south-eastern side of the platform, and at a distance of about 2 ft. from it, were the much-decayed remains of a small semicircular plaster base, about 6 in. high, having embedded in its centre the stump of a wooden post, which looked as if it might have served as the support of a stucco image constructed after the fashion illustrated by the Dandān-Uiliq sculptures.

Find of images carved in wood.

Owing to the top of this base lying almost exposed on the surface, no remains of any kind were discovered on it. But just in front of it there turned up a small relief image carved in wood, only 6 in. long, so completely rotted by exposure as to be barely recognizable. It broke into small splinters when I attempted to move it. On the first step formed by the wooden revetment of the platform towards the north-west was found another and much better preserved wooden image, N. xvi. 1 (see Plate LXX), measuring, with its tenon-like base, 12 in. in height. It is rudely carved and shows a nude standing figure, with a top-knob on head and with hands folded at the breast in the attitude of worship (*kṛtāñjali*). The features are marked in black, now but faintly visible, while traces of red colour at the base may probably be due to daubing with Sindūra. Along the same edge of the platform were found the well-preserved Takhtī-shaped tablet (N. xvi. 2), the interesting Kharoṣṭhī text of which has been discussed in the preceding section[1], as well as a much-perished rectangular covering-tablet (N. xvi. 3). The discovery of the two figures in wood, which manifestly had connexion with worship, together with the arrangement of the platform, make it highly probable that this isolated small structure had served as a shrine, similar in character to the small cellas of Dandān-Uiliq. The fact that one portion of the text in N. xvi. 2 contains a letter from two Śramaṇas, probably a draft as explained above, tends to confirm this supposition. It may also be noted that the arrangement of the platform with its wooden facing exactly recalls that observed at the small Stūpa base which came to light in the course of my excavations at Rawak[2].

Excavation of dwelling N. vi.

About 340 yards to the south of N. v lay the remains of a small dwelling N. vi (see plan in Plate XXXIII) covered only by a slight layer of sand, and hence badly eroded. Its southernmost room (N. xvii.), which must have once been provided with a raised roof over its central area, was the only one yielding any finds. Among the five documents discovered here the pair of rectangular tablets N. xvii. 2 + 3 (see Plate CIV) were remarkably well-preserved, considering that the protecting sand lay here only about 1 ft. deep. An avenue to the east of the dwelling was clearly marked by a row of large fallen poplars.

Remains of house N. vii.

The excavation of the last-named small ruin having been effected under the Surveyor's supervision while I was still engaged in the clearing of N. xv. close by, I was able to turn, on February 9, to the remains of the relatively large dwelling, N. vii (for plan see Plate XXXIII), which occupied a conspicuous position on a loess-bank rising island-like above much-eroded ground nearly one mile to the north of N. v. A good deal of timber-débris strewing the eroded slopes to the west and north showed that the building must once have been more extensive. The same conclusion was indicated by the dimensions of the hall, seen in the plan at the south-eastern end of the extant ruin. Here the sand lay 6 to 7 ft. deep, and its clearing cost a good deal of labour. The row of smaller rooms on the north-west side, separated from the rest of the building by a passage 4 ft. broad, occupied a level fully 5 ft. higher, a circumstance which may possibly indicate construction at a different time. Here the sand-layer was much slighter, and along the outer faces of the rooms the ground was undergoing erosion. On the

[1] See above, p. 365; Rapson, *Specimens*, pp. 9 sqq.

[2] See below, chap. xiv.

floor of what remained of the small room (N. xviii.) eight Kharoṣṭhī tablets were found, which the proximity of the south-east wall had helped to protect, most of them short label-like pieces. Completely bleached and withered fragments, which lay on the eroded slope adjoining, showed that more records had once been left behind in this room. Apart from some small rags of coarse cotton and woollen fabrics (N. xviii. 01. b–f) found in this and the adjoining rooms, there turned up here a contrivance formed by two wooden discs (N. xviii. 01. a, see Plate LXX), which looks as if it had been used as a 'dead eye'. A tree-trunk, which set up upright might have served as a rough seat or table, and a large jar of coarse pottery embedded in the floor, and approaching in dimensions that seen at Niya[3], both marked in the plan, were the only other finds on this side of the ruin.

Finds of agricultural implements and produce.

The rooms to the north-east may have been used as godowns and sheds. In one of them (N. xix.), a good deal of straw turned up in excellent preservation[4], mixed with what the examination effected at the Royal Gardens, Kew, has shown to be husks of *Panicum miliaceum*. Here was found the ancient wooden pitch-fork (see Plate IX), over 5 ft. long, made solely of wooden pieces fastened with leather thongs. Here, too, lay a flat wooden bowl (see Plate IX), measuring 8 in. in diameter, containing grains of millet (*Setaria italica*). The wooden implement, reproduced in Plate LXXIII, was recognized by the men from Niya as a mouse-trap, similar to those still in use, but I had no opportunity of ascertaining its arrangement. A small and completely preserved pot of bright red clay, about 7 in. in diameter, is seen in the photograph (Plate IX). A large wedge under-tablet, N. xix. I (see Plate C), had likewise found its way among the contents of this agricultural store-room, its thick encrustation with straw and other vegetable matter still bearing evidence of this association. The charred condition of the square end suggests that it had been used as a torch or taper. The walls throughout the house showed little strength, being only 4 to 5 in. thick, and constructed, apart from the usual timber framework, of plaster laid over horizontal layers of reeds similar to those used in the Dandān-Uiliq structures. As to the status of the last owner, only the Kharoṣṭhī tablets might possibly allow us to form a conjecture.

Northernmost group of ruins.

While the clearing of this ruin was still proceeding, I made an examination of the northernmost group of remains, situated nearly 2 miles to the north-east of N. v., which I had sighted on my first reconnaissance in this direction, and subsequently, on February 10, I shifted my camp to it. The ruins here consisted of a number of small houses and cattle-sheds scattered over an area roughly 400 yards square, as seen in the plan of Plate XXXIV. Most of these structures were badly decayed and only just traceable, owing to the extensive erosion which this portion of the ground had undergone. But here and there some larger dunes, rising to 15–20 ft. in height, had accumulated, and close by two better-preserved structures had survived.

Finds in ruined house N. III.

The larger of these (N. VIII), which I first proceeded to excavate, consisted, as the detail plan in Plate XXXIV shows, of two rows of three apartments each, separated by an inner line of smaller rooms 11 ft. broad. The most interesting of the finds made here came from the northernmost room (N. xx.) which was covered by only 2 to 3 ft. of sand, and had consequently very little of its outer walls left. On the plaster platform running along the south-eastern wall I discovered the Takhtī-shaped tablet (N. xx. I) 11½ in. long, showing on one side much-faded Brāhmī writing of the Kuṣana type, and on the other side two short lines of Kharoṣṭhī, the palaeographic and chronological interest of which has been explained in the preceding section[5].

[3] See above, p. 312.

[4] Specimens of this, as well as of other cereals found at the site, are in my collection.

[5] See above, p. 369.

The well-preserved covering-tablet of a wedge-shaped Kharoṣṭhī document (N. xx. 2) was also found here. Close by there turned up the curious carved wood panel, N. xx. 01 (see Plate LXVIII), representing the upper portion of a long-tailed ibex-like animal, of spirited design though rudely executed. Judging from its size this panel, which must have adjoined a lower piece, may have formed part of a cupboard, such as will be described presently.

On the floor of the same room lay the well-carved large double bracket of wood measuring 41 by 9 in. with a thickness of 3 in., which the photograph reproduced in Plate IX shows in the centre. In constructive detail it closely resembled the smaller brackets which were found in one of the inner rooms, and which the same photograph shows on either side of it; but owing to exposure the carved surface had suffered more damage. The carving of the oblong modillions below the architrave of the bracket shows decorative motives exactly similar to those found on the ancient wooden chair from N. iii described above[a]. In one modillion we have the four-petalled and four-sepalled flower, so frequent there and well-known to us from Gandhāra sculpture, repeated four times within fillet-borders. In the other modillion halves of the same flower reappear within four triangles, which are formed by diagonally crossing fillets. The bracket had, as the socket in the centre showed, rested on the round head of a post and, no doubt, served to support a true architrave beneath the roof. It is more difficult to say what architectural use there can have been for the carved wooden finial about 6 in. long seen in the foreground of the photograph (Plate IX). Its five circular tori or corrugations, widening in succession, recall the *chattras* or umbrellas in wood or stone which were mounted in a similar fashion above the dome of Indian Stūpas. Two implements of home industry, simple in character, but none the less interesting, also turned up in the same room. The wooden instrument N. xx. 05 (see Plate LXXIII), resembling a currycomb, has already been referred to in the description of the interesting painted panel from Dandān-Uiliq (D. x. 4), illustrating the legend about the introduction of sericulture, which it helped to explain. It served, like the similar implement used at the present day, for beating together the picks of the weft in the course of weaving. The wooden boot-last, 10 in. long (seen in Plate LXXIII), differs in no way from the same contrivance used by Khotan bootmakers at the present day.

Carved wooden bracket.

The hopes raised by these interesting finds of the first day were imperfectly fulfilled by the results of the subsequent excavation of other rooms in this dwelling. It was attended with considerable difficulty, owing to the height of the sand increasing towards the centre of the dune. Only the centre room on the west face, which was filled with sand to a height of fully 10 ft., yielded any finds. From the construction of the timber framework it was clear that the extant walls belonged to a lower story, and that this had once been surmounted by a second one, of which, however, only portions of the main posts survived. In view of this evidence I think it safe to conclude that the small recess, measuring 8 by 5 ft., built within the northern corner of the room was intended for a staircase or ladder communicating with the upper floor. The fact of its being accessible by doors, from both this room and the one adjoining, supports this supposition. In the centre of the room was found a wooden cupboard of plain make, 2 ft. 5 in. long, 1 ft. 10 in. broad, and standing with its legs 2 ft. 8 in. from the ground, the plain legs themselves being 1 ft. 10 in. high. In make it otherwise resembled exactly the cupboard subsequently brought to light from N. xxii. and seen in the photograph (Plate IX); the latter specimen, however, shows in the curved shape and mouldings of its legs some endeavour at ornamentation. In both cupboards a small opening, about 8 in. square, was found at one

Upper story of house N. viii.

Ancient cupboard.

[a] See p. 335 and Plate LXVIII.

end of the longer sides, a somewhat similar arrangement being commonly observed in the cupboards used nowadays in the Khotan region for the storing of bread and other articles.

Specimens of architectural wood-carving.

Of far greater artistic interest are the two well-carved double brackets now marked N. xx. 02, 03, which turned up in the sand a few feet above the floor. Their almost identical dimensions (about 2 ft. long, with a width of close on 6 in.), and closely similar ornamentation leave no doubt that they were intended to form a pair. The reproductions shown in Plate LXIX of the under-surfaces of both these brackets, as well as of the side elevation, which is identical in both, together with Mr. Andrews' detailed and expert description of the pieces as given in the list below, will suffice to explain the design and ornamentation of these fine specimens of ancient architectural wood-carving. The four-petalled flower, noted already on the ancient chair from N. iii as an ornament unmistakably borrowed from the style of Gandhāra sculpture, supplies in several variations the chief decorative motive. Its long-continued popularity is attested by its occurrence on modern boat carvings both in Kashmīr and on the Indus and lower Jehlam. Mr. Andrews has noted the striking similarity between these pieces and the architectural carving still practised in Bhēra and the neighbouring parts of the Punjāb, as regards the manner in which the chisel has been manipulated—an observation of special interest when viewed in connexion with the remarkable continuity of the patterns. These finds at an outlying settlement of ancient Khotan thus help to prove how far back in history the style and *technique* peculiar to the modern architectural wood-carving of the Western Punjāb reach, though this region itself has not preserved for us any known pre-Muhammadan specimens of the art.

Other ruins of northern group.

Trunks of dead fruit-trees emerging from the sand to the south and east of the ruined dwelling indicated the position of what had once been a large orchard. Parallel lines of fencing traceable in the low sand for a distance of close on two hundred feet marked the direction of an ancient lane, about 12 ft. broad, running from south-west to north-east. Much-eroded remains of two houses lay at short distances to the south-east of N. viii, and another to the north-west in positions marked on the plan (Plate XXXIV). But as the protecting cover of sand was here very slight, it was difficult to trace the disposition of the rooms, and the absence of any finds was scarcely surprising. A dune rising to a height of close on 20 ft. in its central part had afforded better protection to two small ruins situated close together nearly 300 yards to the north-east. The posts of their walls were just jutting out above the slope, as seen in the background of the photograph reproduced in Fig. 47. The structures thus indicated proved to contain one and two rooms respectively, the clearing of which was effected with difficulty owing to the proximity of the high dune. No finds of any kind were made; but the construction of the timber framework, and the presence in one of the rooms of a staircase recess as described above, proved that here too the extant remains had once been surmounted by a second story. Traces of ancient fences and rush-built cattle-sheds with scattered groups of fruit-trees were noticed at several points of the area shown in the plan of the northern group of ruins (see Plate XXXIV).

Ancient tank.

But more curious than these was the clearness with which the position and arrangement of an ancient tank (see Fig. 47) could be made out about 200 yards to the north-east of the ruin N. viii. The ground there being quite clear of dunes, the embankment of the tank forming a square of about 48 ft. was distinctly traceable. On it were lying the shrivelled trunks of the large poplars that had once given shade to the water, while one tree still upright raised its gaunt, bleached trunk, as seen in the photograph, to a height of close on 12 ft. In the centre of the tank, which in spite of the drift-sand there accumulated still showed a depth of about 6 ft., a small mound of earth, about 5 ft. in diameter, and rising

46

REMAINS OF DWELLING COMPLETELY ERODED, NEAR N. XI, NIYA SITE.

47

REMAINS OF ANCIENT TREES AND TANK, NEAR DWELLING N. VIII, NIYA SITE.

$2\frac{1}{2}$ ft. above the sand, attracted my attention. My labourers at once recognized in it a feature still regularly provided in modern village tanks. When these are being excavated a small bank of earth, known as *dömbel*[1], is always left standing in the centre. As explained to me, it is used to mark the level down to which the water of the tank has first to be let out before fresh water is introduced from the feeding canal at the periodical renewals provided for. The latter are obviously necessary, since the tanks are the chief supply of drinking-water during the season when irrigation ceases.

Having previously ascertained, by personal inspection of the ground around Camp 95, and by dispatching Surveyor Rām Singh to a high ridge of dunes some three miles north whence a good view was to be obtained with field glasses, that there were no structural remains in that direction left exposed among the lower dunes, I next proceeded to the excavation of those ruins southwards which, among those traced on my previous reconnaissances, still remained unexplored. The first of them was N. x, situated about $1\frac{1}{2}$ miles to the south-south-east of N. viii, and a little over half a mile to the west of N. v. Here, owing to far-advanced erosion, only two small rooms, measuring each $9\frac{1}{2}$ by 6 ft., were still traceable, out of what must have once been an extensive building, judging from the débris of timber that strewed the eroded slopes to the east and west. The plaster walls of the two extant apartments were broken down to within a foot or so of the floor, and the sand covering the latter lay only 6–8 in. deep. Yet even under this slight protection eight fairly well-preserved Kharoṣṭhī tablets had survived close to the east wall of one of the rooms (N. xxi.). Most of them were pieces of rectangular documents. I was interested to note at the time that the years shown in the dating of the under-tablets (N. xxi. 3–5) extended from 7 to 44, thus covering a period of 37 years, on the assumption that years of the same reign are recorded. This discovery was another striking instance of the little guidance that the appearance of extant remains at this site could furnish as to the possibility of interesting finds of records. **Finds in ruined dwelling N. x.**

About half a mile to the south lay the relatively better preserved ruins of a dwelling (N. xi), covered by 2–4 ft. of sand. Among the five rooms of which the walls could still be traced, only the southernmost, measuring 25 by 17 ft. and provided on three of its sides with a sitting platform of plaster, 3 ft. broad, yielded a find. It was the wooden cupboard shown by the photograph reproduced in Plate IX and already referred to in connexion with an exactly similar specimen from N. viii. The cupboard found here was 2 ft. 10 in. long and $1\frac{1}{2}$ ft. broad, with a total height of 2 ft. 8 in. About 200 yards to the west of N. xi there rose above the low dunes a small eroded bank of loess strewn with splintered timber-débris. This and the small row of posts and rushes crowning its crest seen in the photograph (Fig. 46) were all that remained of the ancient dwelling—a typical illustration of the ultimate fate of all structures attacked by erosion. A completely bleached and abraded wedge covering-tablet (N. xxii. 1), barely recognizable by its seal-cavity, was picked up on the slope. **Excavation of ruin N. xi.**

My work on February 11 concluded with the clearing of a small dwelling-house (N. ix), situated about three-quarters of a mile further to the south, to which an interesting find had attracted my attention on the day after my arrival at the site. Ḥasan Ākhūn, the inquisitive young camel-man to whom I owed my acquisition of the first Kharoṣṭhī tablets at Niya[2], had, while looking about for 'treasure', come upon this ruin and brought away from it a rectangular covering-tablet (N. xxiii. 1) with two intact seal-impressions (see Plate LXXII), which as the first find **Find of sealed 'envelope' in N. ix.**

[1] The word may have some connexion with *döbe* (*debe*) applied to small natural mounds; see above, p. 77.

[2] See above, p. 312.

of its kind had at the time greatly puzzled me. I visited the ruin on January 30, and ascertained that the place where Ḥasan Ākhūn said he had picked up the tablet was a small triangular shelf, marked *a* in the detail plan of the ruin (Plate XXXV), in the corner of the central room. The sand filling it to a height of about 4 ft. just reached this shelf. The effect of the exposure thus undergone by the tablet is marked by its warped and slightly cracked appearance, which makes the fair preservation of the two clay seals all the more striking. One of them represents a barbarian head recalling that of Maues on certain Śaka coins from North-western India, the other a bird rising with wings extended. The tablet must have lain on the shelf with the obverse downwards and would, no doubt, have been completely destroyed by a few seasons more of exposure.

Excepting a small store of wood comprising branches of poplar, Jigda, and tamarisk stacked in a passage (see plan in Plate XXXV), no other movable objects were found on clearing this modest residence. But it was of interest to observe that the room built on to the north of the one just referred to had three of its walls constructed of stamped clay. An outer shed adjoining eastwards appears to have been constructed with rush walls, and contained the remains of what seemed to have been a circular baking-oven built in clay, about 2 ft. in diameter. To the east of the dwelling numerous trunks of dead fruit trees marked an orchard, and beyond this the line of an ancient canal about 8 ft. broad could be traced for a little over a hundred yards in the general direction S. to N. It was once bordered by a double row of fine poplars, the dead trunks of which, massive and imposing even in their splintered and withered condition, lay stretched out in the sand more or less exposed.

Ancient canal and avenue.

Ruined dwellings N. XII.

February 12, the day preceding my departure, was spent in the examination of a small group of ruins (N. XII) lying about half a mile to the west of N. XI, and close to a belt of tamarisk-covered sand-cones which here fringes the high ridge of dunes marking the western edge of the ancient cultivated area (see site plan in Plate XXVII). Being hidden in a dip behind a series of broad swelling dunes, rising to about 15 ft. above the original ground-level, these remains did not attract attention until the clearing of N. XI had been completed. They consisted of two relatively large but much-decayed dwellings situated within about 300 yards of each other, and of three small structures lying in a row a short distance to the east of them. The northern and larger of the two houses formed a compact block of rooms measuring about 62 ft. from north to south, with a breadth of about 44 ft. The covering sand lay only to a height of 2 to 5 ft., and all the woodwork of the walls bore marks of long-continued exposure. In spite of the fissured and shrunk surface traces of ornamental wood-carving, similar in style to that observed in N. VIII, survived of a doorpost and some fragments of roof-beams. A pattern closely recalling the *kingri* work so common in the architectural carving of the Western Punjāb could clearly be made out[9]. The rooms, with the exception of one at the south-west corner, were cleared completely, but no finds of any kind rewarded the labour. The excavations made in the smaller house situated to the south-east yielded no better result. The only discovery made here consisted of a very large jar let into the floor of the north-east corner room. In shape and material it closely resembled the ancient jar I had seen at Niya. The diameter was here fully 2 ft. 11 in., while the mouth measured nearly a foot across. The inside was filled with loose sand. The remaining three ruins seemed to be the remains of small dwellings or cattle-sheds roughly built of timber and rush walls. Being situated close under the lee side of relatively high dunes, their clearing would have cost considerable labour and delay with but

[9] Compare for this pattern the side elevation of the bracket, N. XX. 02, Plate LXIX, with description in list.

scant prospect of interesting results. A change in the arrangements already made for the morrow's start eastward would have implied fresh hardship to the men, exhausted by close on three weeks' continuous labour, and to the camels, equally hard-worked in the transport of our supplies of ice and water. Hence I decided to leave those structures unopened.

Miscellaneous objects picked up at site.

On their tramps between camp and the various ruins the men, always on the look-out for 'treasure', had occasion to pick up from the sand or eroded loess ground a variety of small objects in metal, stone, or other hard substance, which under my standing orders were shown to me, and acquired for suitable compensation when of any antiquarian value. Many of the miscellaneous relics thus obtained (all marked with N. 00 and entered at the end of the descriptive list) have been reproduced in Plate LXXIV, some also in Plates IL and L; and it will be convenient briefly to note here those of special interest. Among metal objects the fragments of circular mirrors cast in bronze, apparently of Chinese make (N. 009, 12. f, g, h, see Plate LXXIV) are conspicuous. The obverse bore in each case ornamentation in bold relief. The only object in gold is the small ornament N. 0017 (Plate LXXIV), which may have formed part of a pendant or earring. It is made of thin gold-plate filled with lac or wax, and has small jewels of pebble or glass set at one end. Of bronze rings we have several bearing devices cut into the metal and still recognizable, such as N. 0014. f (Plate LXXIV), and N. 0014. g, 15, 16 (Plate IL). With the exception of the well-preserved signet-ring (N. 0015), which shows a lion with a bird clinging to his breast, the devices are very crude, consisting of a few strokes only. The square bronze seal N. 0011. b (Plate L) is decorated with a geometrical design closely resembling those on some seals acquired from the Ak-sipil site. The two bronze arrowheads, N. 005, 14. c (Plate LXXIV), are of superior workmanship, and still show the arrangement by which they were secured to the shaft. Owing to oxidization few of the small iron objects found still retained recognizable shape, one of these being the buckle N. 008 (Plate LXXIV). Beads in pebble, glass, shell, and pottery were picked up plentifully, see N. 0014. d, 20. a, 21. b in Plate LXXIV. Among them a number of small gilt glass beads (N. 0020. a) were found at one spot near N. IX, and had evidently once formed a string. Glass was used also for seals, as shown by N. 006 (Plate L). The fragment of good white glass, N. 0019 (Plate LXXIV) was apparently the foot of a vessel. Whereas the glass objects may be supposed to have come from the West, it is more difficult to determine the origin of the pyramidal seal with scroll design N. 0018 (Plate L), and the Amalaka-shaped bead N. 005. b (Plate LXXIV), both made in a hard ceramic substance, green or blue, which resembles porcelain, but is correctly described as a kind of frit or celadon. According to information kindly supplied by Mr. C. Read, of the British Museum, its manufacture was known from a very early period not only in China, but also in some parts of Western Asia. Beads of exactly similar shape are found also among Roman work.

Chronological limits of miscellaneous 'finds' in sand.

None of the small objects thus picked up can be ascribed to a date appreciably later than that which, as ascertained above, saw the final abandonment of the site. But it is well to bear in mind that there is no chronological limit in the opposite direction determined by equally clear evidence. Objects which may have got lost and hidden in the soft loess ground during widely different periods would be found lying side by side after the strata containing them had been successively eroded. The importance of this observation becomes evident when we remember that ground unprotected by buildings can be proved to have undergone erosion to a depth of 15 to 25 ft. below the ground-level as it existed about the last quarter of the third century A. D. The worked flint flake (N. 003) probably of Neolithic make, picked up on the eroded ground immediately to the south of my first camp (C. 93) supplies tangible proof of the need of such caution; for it is evident that the time when it was made and used must lie many centuries further back than the period to which the structural remains of the site belong.

Limitation of archaeological evidence to structural remains.

It is thus obvious that the peculiar conditions created by erosion at this and also at other ancient sites of this region must impose limitations upon the conclusions of the archaeologist. Only where structural remains have survived can we hope to determine the period of simultaneous occupation and the approximate date when this occupation ceased. Where the ancient remains are of the 'Tati' type, i.e. consist only of pottery fragments and other small débris, finds of coins, seals, and other approximately datable objects may help to show the latest period down to which the ground was in all probability occupied or cultivated. But not until we have learned to distinguish with certainty between the potsherds of different periods, will it be possible to form any definite conclusions as to how much of the débris really dates from that latest epoch, and how much represents remains of earlier habitations. Seeing that the latter need not have been built always in the same positions which were occupied by the dwellings of the period immediately preceding the abandonment, it is evident that the mere extent of débris-strewn patches of ground by itself furnishes no safe argument regarding the character and size of the settlement which had last occupied the site. It will, therefore, be well to restrict ourselves to the positive evidence which the examination of the structural remains has supplied.

Extent of ancient site marked by structural remains.

There can be no doubt that all the ruins with which the excavations described in the preceding sections have made us acquainted were inhabited within the same period and also abandoned about the same time. The area over which they are scattered covers fully seven miles, from N. VIII in the north to where the first dead fruit-trees were met with at the approach from the south, while its greatest width, as marked by N. I on the east and N. XII on the west side measures over two miles. It is possible that in those portions of the area where the dunes are higher, ruins of substantial buildings may lie completely hidden from view; but even allowing for this, and for the complete disappearance of others through erosion, we must be struck by the relatively small number of extant ruins of any size and by their wide dispersion. We have already had occasion to note the same fact at Dandān-Uiliq, and must hence attach to it the more significance.

Small number and scattered position of substantial dwellings.

Two observations appear to me most likely to explain it. In the first place, it is evident that relatively substantial houses would necessarily offer longer resistance to the destructive forces of erosion than the mud-built dwellings of small cultivators. Hence we may recognize that the extant ruins represent only the few well-constructed residences of the ancient settlement. In the second place, the examination of any modern oasis too small to possess a town centre, such as Gūma, Moji, Chīra near Khotan, or else Tāshmalik, Khān-arik, &c., near Kāshgar[10], would show us exactly the same scattering of the population in numerous small hamlets widely separated from each other by stretches of cultivation and gardens or even by strips of desert.

Scattered hamlets and country residences in modern oases.

In none of those oases is there any agglomeration of houses that with any approach to accuracy could be called a town. The 'Bāzār' which forms their commercial centre is nothing but a street of booths practically without inhabitants on any but market days, while the well-to-do 'Bēgs' who, as Mingbāshis or in other official posts, carry on the whole of the local administration, generally reside far away in the hamlets or in isolated farms nearest to their landed property[11].

[10] Comp. *Ruins of Khotan*, p. 116, regarding Tāshmalik, where I first had occasion to learn practically how widely separated the habitations even in a populous oasis may be, and how difficult in consequence it is to find their administrative centres. See also *Yarkand Mission Report*, p. 32, for some judicious observations.

[11] Compare for descriptions of such residences belonging to Bēgs either in office or *in spe*, where I had occasion to take up temporary quarters, *Ruins of Khotan*, pp. 152, 153, 199, 257, 271.

It is to the residences of these Bēgs, often remarkably well-built, judged by Turkestān standards, and comfortable in spite of their wholly rural surroundings, that I think we must look for the modern counterparts of the more substantial ruins of the Niya Site. Thus guided by what observation at the present day teaches, we can readily understand the relatively high standard of comfort which these ruins indicate, even though the description of their surroundings as a 'town' or 'city' must appear inappropriate.

Site that of a terminal oasis.

As to the population which this ancient settlement once contained we could only form vague guesses, but a look at the map shows that in size the area for which simultaneous cultivation can be proved by definite archaeological evidence approximately corresponded to the present oasis of Gūma, Chīra, or Tawakkēl. It is still more certain that the ancient settlement, like the first two of the modern localities just named, must have been a typical terminal oasis[13]. In contrast to Dandān-Uiliq, no question whatever can arise here as to the source of its water-supply. It can only have come from the Niya river, the course of which still bears straight in the direction of the ruined site, but now finally ceases, even during the summer floods, some seven to eight miles from the southernmost point of the latter. Taking the direct distance between the northernmost structural remains explored by me and the last point down to which I could trace the flood-bed cut in recent times by an extreme off-shoot of the river, we find that the belt fertilized by the latter has receded some fifteen miles since the abandonment of the ancient site. This distance is reduced to about eleven miles if we take our measurement from the northernmost ruins to the limit of living trees as marked on the map.

Shrinkage of Niya river course.

That the abandonment of the site must be connected with this shrinkage of the river course, whatever its explanation, is an assumption which naturally suggests itself first. But the historical student, and I think also the geographer who is prepared to scrutinize his evidence critically, may well pause before committing himself to the conclusion that this undoubted recession of the river's terminal area must necessarily have been the immediate or sole cause of the abandonment. In the case of all rivers and streams that flow from the Kun-lun range into the Tārīm Basin and lose themselves in the sands of the Taklamakān, we find the final portion of their course liable to great and frequent shiftings laterally, i.e. to the east or west. The phenomenon has long ago been noted by all competent observers and finds its natural explanation in the masses of silt, turning into drift-sand, which these rivers deposit in their terminal course, and which in turn force them to divide their waters delta-like or periodically to seek new channels.

Lateral shiftings of terminal river courses.

For a typical illustration we need not look further than at what my map shows for the terminations of the Yārtunguz and Endere rivers nearest to the Niya river on the east. As I shall have occasion to mention in the next chapter, the relatively recent colony of Jigdalik-Bulung and Kala-sulaghi, forming the terminal oasis of the Yārtunguz river, had for several years previous to my visit been in danger of losing its water-supply, not from any want of water in the river but owing to the head of the river having shifted considerably to the west and thus failing to fill the canal from which the fields of Yārtunguz-Tārīm are irrigated. The people of the small colony being far too few to cope with this danger by strengthening the head of their canal or otherwise confining the vagaries of the river, have resorted to the expedient of starting fresh cultivation in the new terminal area known as Yilba-Ṣarīgh (see map). That the same river had at different periods followed courses also to the east of the present Yārtunguz-Tārīm was demonstrated when I passed the broad depressions shown by

[13] Compare for general remarks on 'terminal oases' and their characteristics, above, pp. 94 sq.

the map eastwards, which represent old main beds of the river and still at times may reached by its overflow. Almost identical conditions prevail at the end of the Endere river, where the small terminal colony of Endere-Tārīm has recently been abandoned, mainly because the river has been shifting its course from the so-called Yangi-Daryā or 'New Channel' back again westwards into its 'Old Channel'.

Possible causes of abandonment of ancient site.

With such modern changes before us, small in scale as far as cultivation is concerned, but typical in character for the difficulties which at all times must have surrounded terminal oases in this part of the Khotan region, it seems to me well worth considering whether the abandonment of the ancient site which once formed the terminal oasis of the Niya river may not partly at least have been due to some great shift in the river's end course, which the settlement for some reason or other was unable to avert or counteract. In view of this possibility, I must greatly regret that, in the midst of the incessant excavation labours which filled my stay at the site, and which rendered my presence obligatory, I could not spare time for topographical explorations to the east and west. I must, however, mention that Surveyor Rām Singh, whom I sent out to accompany Darōgha Ibrāhīm on January 28 on a part of the latter's reconnaissance westwards[13], reported that he found plenty of dead Toghraks in depressions he passed through about 8 miles west of Camp 93, and even that he sighted living trees some two miles beyond. Both observations may be accepted at present as evidence of the river having at some period followed divergent beds in that direction.

No direct archaeological evidence as to cause of abandonment.

The archaeological or historical evidence at present available is quite insufficient to allow us to form any definite opinion as to the immediate cause or causes which led to the abandonment of the site. The Kharoṣṭhī documents, it is true, seem to refer in places to difficulties about irrigation. But these are more or less constant for individual parts of the smaller oases, whether from inadequate snowfall in the mountains, destructive floods or similar causes; and of course, the localities from which those reports date have as yet to be determined. Also the indirect effect, which the final withdrawal of Chinese central authority from Eastern Turkestān after Wu ti's reign may have had on local conditions generally, must remain a matter of mere conjecture. But it will certainly strike the historical student as probably more than mere coincidence that both at Dandān-Uiliq and at the Niya Site, and also, as we shall see, at the Endere ruins, the abandonment of these ancient colonies is proved by incontrovertible evidence to have followed closely upon great political changes affecting the Khotan region. That changes of this kind may, under the peculiar physical conditions prevailing in this region, exercise an immediate and very important effect upon the extent of cultivation, and the general economic conditions of the oases, is strikingly proved by the great reduction in the population, cultivation, and industrial productiveness which undoubtedly resulted in Khotan from the troubles of the last Muhammadan rising and from Yāqūb Bēg's reign, relatively short though they were[14]. But it would be useless to speculate at present upon the extent to which troubles of a similar kind at earlier periods may have rendered the struggle for the maintenance of irrigation more precarious, or the effect of any contemporary great shifting in the river-course more important and difficult to deal with.

Physical evidence at present available.

Turning to the purely physical side of the question, it appears to me that we have at present only one absolutely clear fact to fall back upon, the shrinkage by a distance of at least

[13] See *Ruins of Khotan*, pp. 354, 370.

[14] Compare, e.g., *Yarkand Mission Report*, p. 33. Dr. Bellew, in 1874, was informed 'by a trustworthy resident of the city who had witnessed the whole revolution from the overthrow of the Chinese to the establishment of the Amīr,' 'that the whole country [of Khotan] was now completely impoverished, and that it had lost nearly half of its male population.'

fifteen miles of the river's final course and of the belt it fertilizes. It seems probable that we must recognize in this shrinkage a result of that general desiccation which large parts of Central Asia are, according to a much-discussed geological theory, believed to be slowly undergoing. But while this physical fact fully explains why the ancient site was never re-occupied, it does not prove that the original abandonment was directly and solely due to it. Were this shrinkage the only possible factor to be reckoned with, we should still be puzzled to find an explanation why the cultivation, when it could not be maintained any longer at the old site, was not shifted southwards and continued—possibly within reduced limits—near the present end of the river-course. We have seen that the fertile belt both above and below Imām Ja'far Sādiq, now occupied by luxuriant forest-growth, would permit of the creation of a terminal oasis, and that a new irrigation canal, taking off some 26 miles above the Mazār, had actually been started a few years before my visit[15]. Yet nowhere between Niya and the present termination of the river did I trace any ancient remains or hear of such having ever come to light.

Impossibility of final conclusions.

Taking these various facts into account, it appears to me that no explanation which it is at present possible to advance for the abandonment of the ancient site can claim definite acceptance as a safe basis for historical or geographical conclusions. As in most historical changes, the causes may have been far more complex than the modern inquirer is apt to assume. As regards the physical aspects of the question, it will be well to await reliable data such as can be furnished only by a detailed topographical survey, combined with prolonged and exact observations as to the volume of water carried by the river at different seasons, the rate of evaporation, &c. Nor can we hope to secure any clear knowledge of the irrigating capacity which the river may still retain at the present day, until its conditions and those of other rivers in the Khotan region have been systematically studied by a competent irrigation engineer, preferably of Indian experience.

Section VIII.—LIST OF ANTIQUES EXCAVATED AT NIYA SITE

DOCUMENTS ON WOOD ACQUIRED AT NIYA AND IMĀM JA'FAR SĀDIQ.

Documents from Niya.

Niya. 22. i. 1901. a. Rectang. under-tablet, split at one end, and pierced by insects. *Obv.* 11 ll. Kharoṣṭhī, very fine writing, closely resembling N. iv. 143; quite legible. *Rev.* blank. $8\frac{1}{2}'' \times 3''$. Well preserved.

Niya. 22. i. 1901. b. Wedge cov.-tablet; hole near point. *Obv.* Seal cav. $1\frac{3}{8}''$ from sq. end; surface bleached. *Rev.* 3 ll., very coarse and cursive Khar., $9\frac{1}{4}'' \times 1\frac{7}{8}'' \times \frac{3}{8}''$. Wood soft.

Niya. 22. i. 1901. c. Rectang. under-tablet; fragment. *Obv.* Faint traces of 5 ll. cursive Khar. *Rev.* blank. $9\frac{1}{2}'' \times 1\frac{3}{8}''$. Wood well-preserved, dark and broken.

Imām J. S. 25. i. 1901. a. Wedge under-tablet; hole near point. *Obv.* 3 ll. Khar. cursive, but very black and fine. Initial formula. *Rev.* 1 l. from sq. end. $9'' \times 1\frac{9}{16}''$. Well preserved.

Imām J. S. 25. i. 1901. b. Wedge under-tablet; hole near point. *Obv.* 4 ll. Khar., very coarse, cursive, and blotted; much faded. *Rev.* 1 l. from sq. end. $9\frac{7}{8}'' \times 2''$. Wood well-preserved.

OBJECTS FOUND IN N. I.

N. I.

N. I. 001. a. Fragment of stone, nearly black, smooth on two sides and one edge. Otherwise rough and broken. $3\frac{1}{4}'' \times 1'' \times \frac{3}{8}''$.

N. I. 001. b. Fragment of coarse pottery, with roughly incised decorative lines. $2\frac{5}{8}'' \times 1\frac{1}{4}''$.

[15] See above, p. 313.

N. 1. **N. 1. 001. c. Fragment of dark grey pottery,** portion of neck and rim of vessel; prism-shaped rim. Width $3\frac{1}{4}''$, height $1\frac{3}{8}''$.

N. 1. 002. Stone bead (jade?) pierced with small hole. Diam. $\frac{3}{8}''$.

N. i. **N. i. 1. Wedge under-tablet** (belongs to N. i. 8). *Obv.* 3 ll. Khar.; initial formula. *Rev.* blank. Well-preserved. $7\frac{3}{8}'' \times 1\frac{3}{16}''$.

N. i. 2. Takhtī-shaped tablet, portion of hole at edge $3\frac{1}{4}''$ from end of handle. *Obv.* Very faint traces of characters. *Rev.* 3 ll. Khar.; indistinct, in parts deleted. $11\frac{1}{2}'' \times 1\frac{1}{4}''$. Broken and warped.

N. i. 3. Wedge covering-tablet. *Obv.* Depression for seal $1\frac{3}{8}''$ from sq. end. Seal missing. 3 string grooves, in which fragments of string. At sq. end a few Khar. char., longitudinally; and on opposite side of seal-place a few more, rather indistinct. *Rev.* Single line of Khar. Well-preserved, $8\frac{7}{8}'' \times 2\frac{1}{4}'' \times \frac{5}{8}''$ at thickest part where seal socket.

N. i. 4. Wedge cov.-tablet (belongs to N. i. 21). *Obv.* Seal cavity $1\frac{7}{8}''$ from sq. end. Remains of clay seal. Cavity $1\frac{1}{2}'' \times \frac{7}{8}''$. 3 string grooves, in which fragments of string. At sq. end a line of Khar., legible. At point end traces of characters. *Rev.* Faint traces (?) of characters at L. pr. edge. Wood well-preserved, slightly warped; splitting at sq. end. $11\frac{3}{4}'' \times 2'' \times \frac{3}{4}''$ at seal cav.

N. i. 5. Wedge under-tablet (belongs to N. i. 11). *Obv.* 3 ll. Khar., very faint; initial formula. *Rev.* 1 short line Khar. at sq. end. Well-preserved. Portions of fibre of bark on each edge. $10'' \times 1\frac{1}{2}'' \times \frac{1}{4}''$.

N. i. 6. Wedge cov.-tablet (belongs to N. i. 62). *Obv.* Seal cav. in which portions of clay and strings placed diagonally. At sq. end single line Khar. distinct. Opposite side of seal cav. a few characters faintly traceable. *Rev.* At upper edge, single line of Khar., very distinct. $8\frac{1}{4}'' \times 1\frac{5}{8}'' \times \frac{5}{8}''$. Warped slightly transversely. Wood well-preserved.

N. i. 7. Wedge under-tablet. Short string cut with knife (?) on obverse side. *Obv.* 4 ll. Khar. written with broad pen. Initial formula; fourth line short. Generally quite legible. *Rev.* blank. Mark of red paint $\frac{1}{2}''$ from sq. end. $13\frac{1}{2}'' \times 2\frac{1}{4}'' \times \frac{3}{8}''$. Wood well-preserved.

N. i. 8. Wedge cov.-tablet (belongs to N. i. 1). Hole at point containing string cut by knife at reverse side. *Obv.* $1\frac{1}{2}''$ from sq. end, seal cav. String grooves clumsily cut; string adhering. One l. Khar. between sq. end and seal cav.; 3 char. at hole; all clear. *Rev.* Few characters at upper edge, sq. end. $7\frac{7}{8}'' \times 1\frac{1}{2}'' \times \frac{1}{2}''$. Well-preserved.

N. i. 9. Wedge tablet complete, cov.- and under-tablets, with string and seal partly broken. *Obv.* of cov.-tablet. Seal cav. with seal, $1\frac{1}{4}''$ from sq. end. Seal broken at top R. p., but device remains (see Pl. LXXI). Standing figure either winged (Eros?) or wearing chlamys, $\frac{3}{4}$ to R. p. head, profile to R. p., in cameo, deeply impressed in clay, surrounded by 3 mouldings (intaglio), shape finely proportioned ellipse. Strings remain intact under seal. Between seal and sq. end 1 line Khar. and on opposite side of seal 1 short line Khar. Near hole at point few characters. *Rev.* of under-tablet, a short line Khar. near sq. end. $9'' \times 1\frac{7}{8}'' \times 1\frac{3}{8}''$ (combined thickness of cov.- and under-tablets). See Pl. C.

N. i. 9. a. Oblong tablet, broken at R. p. end. Writing on one side. Near top edge 1 line Khar., well-written, and at R. p. end, below this line 3 very short lines in columns. $15\frac{1}{4}'' \times 2\frac{3}{4}''$. Well-preserved.

N. i. 10. Wedge cov.-tab. *Obv.* Seal cav. $1\frac{1}{8}''$ from sq. end; cav. $1\frac{1}{2}'' \times 1\frac{1}{16}''$. Clay adhering but device of seal quite gone. String is like coarse cotton fibre. Between cav. and sq. end 1 line Khar. Near hole few characters. *Rev.* 1 line Khar. $9'' \times 1\frac{7}{8}'' \times \frac{5}{8}''$.

N. i. 11. Wedge cov.-tablet (belongs to N. i. 5). *Obv.* Seal cav. with string (seal missing) $1\frac{1}{2}''$ from sq. end. 1 l. Khar. between cav. and end, faint. Few characters near hole. *Rev.* 2 ll. Khar., much faded, especially towards point. $10'' \times 1\frac{1}{2}'' \times \frac{5}{8}''$. Wood well-preserved. Fibre of bark adhering to parts of obverse.

N. i. 12. Wedge tablet complete, broken as to both tablets at point. Seal deleted. String intact round thick part. Cov.-tab.: *Obv.* 1 l. between cav. and sq. end; cav. rather roughly cut and not 'square' with length. *Rev.* blank. Scored by insect borings. — Under-tablet: *Obv.* 3 ll. Khar. *Rev.* 1 short line Khar. near sq. end. Wood fairly hard, but broken and much eaten by insects. $7\frac{3}{8}'' \times 1\frac{1}{2}'' \times 1\frac{1}{8}''$ (over both tablets).

N. i. 13. Wedge cov.-tab. (belongs to N. i. 54). *Obv.* $1\frac{3}{8}''$ from sq. end, seal cav. $1\frac{1}{2}'' \times \frac{7}{8}''$. 1 l. Khar. between seal and sq. end. Seal deleted, string remains. Few characters Khar. on other side of seal cav. A few also within about $1\frac{1}{2}''$ of hole. *Rev.* 1 straggling line Khar., very cursive, along upper edge, and a second short line from sq. end. $9'' \times 1\frac{1}{2}'' \times \frac{5}{8}''$. Wood well-preserved. Writing much faded.

N. i. 14. Wedge cov.-tablet. *Obv.* $1\frac{3}{8}''$ from sq. end, seal cav. with seal (poor impression, but appears to represent standing figure) in cameo; deeply impressed. String under seal is triple instead of double as usual. Quality of string poor, which may account for this variation. In space between seal and sq. end, 1 line Khar., very clear. Near hole a few characters. *Rev.* 3 ll. Khar., very distinct excepting at top R. p. corner; top. l. follows curve of edge of wood, 2nd l. follows first, 3rd l. short. $7\frac{1}{4}'' \times 1\frac{1}{2}'' \times \frac{7}{8}''$. Wood well-preserved.

N. i. 15. Wedge cov.-tablet (belongs to N. i. 107). *Obv.* $1\frac{1}{2}''$ from sq. end, seal cav. with remains of seal, quite deleted. Cav., which is cut carelessly, $1\frac{1}{2}'' \times 1\frac{1}{16}''$. 1 l. Khar. between cav. and sq. end. Characters near hole. *Rev.* 1 l. (short) at upper edge, very distinct. $8\frac{1}{2}'' \times 1\frac{7}{8}'' \times \frac{5}{8}''$. Wood hard, but broken towards middle.

N. i. 16. Wedge cov.-tablet (belongs to N. i. 104). *Obv.* $1\frac{1}{4}''$ from sq. end, seal cav. (seal missing, portion of string remains); between cav. and sq. end indistinct line of Khar.; also on other side of cav. Usual char. near hole. *Rev.* 3 ll. Khar., indistinct towards point. All writing very coarse and cursive. $9'' \times 2'' \times \frac{5}{8}''$. Very well preserved.

N. i. 17. Wedge cov.-tablet. *Obv.* $1\frac{3}{4}''$ from sq. end, seal; standing figure in cameo (prob. Pallas). 1 l. Khar. between seal and sq. end, very clear. Usual char. near hole. *Rev.* 2 ll. Khar., very cursive and scratchy but quite distinct, except towards point where lower line deleted. $9\frac{1}{4}'' \times 2\frac{1}{4}'' \times \frac{5}{8}''$. Wood in perfect preservation.

N. i. 18. Wedge under-tablet (?), much broken. *Obv.* 4 ll. Khar., very cursive, but clear. Initial formula. *Rev.* 1 short line. $8'' \times 2''$. Wood hard.

N. i. 19. Wedge cov.-tablet. *Obv.* Seal cav. $1\frac{1}{4}''$ from sq. end. Seal gone. String crossed as usual, intact. Cav. $1\frac{1}{4}''$ sq. 1 l. Khar. between sq. end and seal. Usual char. near hole. *Rev.* 3 ll. Khar., very cursive, 1st, 2nd long, deleted at point end, 3rd short. $9\frac{1}{4}'' \times 1\frac{7}{8}'' \times \frac{5}{8}''$. Well-preserved.

N. i. 20. Wedge under-tablet. *Obv.* 4 ll. Khar. with initial formula; 4th short. Very distinct, but faint in colour. *Rev.* Few char. at lower edge near sq. end. $9\frac{1}{2}'' \times 1\frac{1}{2}'' \times \frac{1}{4}''$. Well-preserved. Roughly finished on outside.

N. i. 21. Wedge under-tablet (belongs to N. i. 4). *Obv.* 3 ll. Khar. with initial formula. Faint, part deleted. *Rev.* 1 l. Khar. from sq. end, very faint. $11\frac{1}{4}'' \times 2''$. Warped; well-preserved.

N. i. 22. Wedge cov.-tablet. *Obv.* Seal cav. $1\frac{5}{8}''$ from sq. end. 1 l. Khar. on both sides of cav.; usual char. at hole. Traces of Sindūra (?) round cav. *Rev.* Traces of char. near lower edge from sq. end. $13\frac{1}{2}'' \times 2\frac{1}{4}'' \times \frac{7}{8}''$. Soft at edges; appears to have been exposed to weather.

N. i. 23. Wedge cov.-tablet. Unusual shape, the sides being parallel until near point when they curve to sharp point. *Obv.* $1\frac{3}{8}''$ from sq. end, seal cav. Seal missing. 1 l. Khar. between sq. end and cav. Usual char. near hole. *Rev.* blank. $9\frac{1}{8}'' \times 1\frac{1}{2}'' \times \frac{3}{4}''$. Wood perfectly preserved.

N. i. 24. Wedge cov.-tablet (belongs to N. i. 180). *Obv.* $2\frac{1}{2}''$ from sq. end, seal cav. Remains of string; cav. $1\frac{5}{8}'' \times 1\frac{1}{4}''$. No writing visible. *Rev.* blank. $13\frac{7}{8}'' \times 2\frac{1}{8}'' \times \frac{1}{4}''$. Much perished by damp.

N. i. 25. Wedge under-tablet. *Obv.* 3 (4?) ll. Khar. with initial formula. Indistinct and part deleted at sq. end. Remainder very clear; rather coarse and cursive. *Rev.* Few char. at sq. end. $14\frac{5}{8}'' \times 2\frac{1}{4}''$. Indifferently made. Warped and slightly cracked.

N. i. 26. Wedge under-tablet. Part of upper edge broken away. *Obv.* 3 ll. Khar.; portion with initial formula broken off and part of 2nd line. Coarse, very cursive. *Rev.* 1 short line near sq. end. $9\frac{1}{8}'' \times 1\frac{3}{8}''$. Hard and well-preserved, except for break.

N. i. 27. Wedge under-tablet. *Obv.* 4 ll. Khar. with initial form.; writing thin, cursive, and very clear. Perhaps a few char. missing through break at point. *Rev.* 1 l. from sq. end. Very faint. $9'' \times 1\frac{1}{4}''$. Well-preserved.

N. i. 28. Wedge cov.-tablet (belongs to N. i. 48). *Obv.* $1\frac{1}{2}''$ from sq. end, seal cav. in which traces of string. 1 l. Khar. between sq. end and cav. Usual char. at hole. *Rev.* 3 ll. Khar., cursive. Part deleted at sq. end (from bleaching); otherwise distinct. $10\frac{1}{4}'' \times 1\frac{1}{8}'' \times \frac{9}{16}''$. Well-preserved; slightly warped.

N. i. 29 + 38 + 40. Rectangular under-tablet, in 3 fragments. *Obv.* Portions of 3 ll. Khar., distinct; cursive. *Rev.* blank. $7\frac{1}{4}'' \times 1\frac{1}{2}''$ to $\frac{1}{2}''$, $8\frac{1}{4}'' \times 1\frac{7}{8}''$, $7'' \times 1\frac{1}{8}''$. Hard. N. i. 38 broken at top and L. p. end; N. i. 40 broken at L. p. end.

N. i. 30. Wedge under-tablet (belongs to N. i. 43). *Obv.* 4 ll. Khar., cursive, but well written and very clear. Initial formula. *Rev.* 1 short l. at sq. end. $9\frac{5}{8}'' \times 1\frac{1}{2}''$. Well-preserved.

N. i. 31. Wedge under-tablet (belongs to N. i. 65). *Obv.* 3 ll. Khar. with initial form. Cursive, fairly distinct. *Rev.* 1 l. from sq. end. $9'' \times 1\frac{1}{4}''$. Well-preserved.

N. i. 32. Wedge under-tablet. *Obv.* 3 ll. Khar., very cursive but distinct; initial form. *Rev.* Faint trace of 1 l. from sq. end. $8\frac{1}{8}'' \times 1\frac{1}{4}''$. Well-preserved.

N. i. 33. Wedge cov.-tablet (belongs to N. i. 41). *Obv.* $2\frac{1}{2}''$ from sq. end, seal cav. containing string which still encircles tab. 1 l. Khar. at sq. end. Usual char. at hole. *Rev.* 3 ll. Khar., much faded. $8\frac{1}{2}'' \times 1\frac{15}{16}'' \times \frac{5}{8}''$. Well-preserved, but bleached. Portion of bark at seal cav.

N. i. 34. Wedge under-tablet. *Obv.* 3 ll. Khar. with initial form.; very cursive; clear. *Rev.* Traces of l. from sq. end. $8\frac{1}{4}'' \times 1\frac{1}{4}''$. Well-preserved, but splitting.

N. i. 35. Wedge cov.-tablet. *Obv.* $1\frac{1}{4}''$ fr. sq. end, seal; device deleted. Ends of string. 1 l. Khar. between sq. end and seal. 1 char. at opposite side of seal. Usual char. at hole. *Rev.* 3 ll. Khar., very cursive. Ink has 'run' and blotted. $8\frac{5}{8}'' \times 1\frac{5}{8}'' \times \frac{3}{4}''$. Well-preserved; slightly warped.

N. i. 37. Wedge under-tablet. *Obv.* 4 ll. Khar., with initial form. Writing thin, cursive, clear in places, but elsewhere faded or destroyed by insects aud by splitting of wood. *Rev.* Few char. at sq. end. $9\frac{1}{4}'' \times 1\frac{5}{8}''$. Hard, slightly warped; split from sq. end. Portions of bark.

N. i. 39. Wedge cov.-tablet. String in hole near point. *Obv.* $2''$ from sq. end, seal cav. with fragments of seal, and string. 1 l. Khar. between sq. end and cav. Usual char. at hole. *Rev.* 4 ll. Khar., very cursive, but black and clear. Slightly damaged at top edge near sq. end. $10\frac{7}{8}'' \times 2\frac{1}{4}'' \times \frac{5}{8}''$. Well-preserved.

N. i. 41. Wedge under-tablet (belongs to N. i. 33). *Obv.* 4 ll. Khar., much faded, with initial formula; 3rd and 4th ll. almost invisible. *Rev.* Few char. at sq. end. $8\frac{1}{4}'' \times 2''$. Wood well-preserved.

N. i. **N. i. 42. Wedge cov.-tablet** (belongs to N. i. 57). *Obv.* $1\frac{5}{8}''$ from sq. end, seal cav.; frags. of seal and string. Between sq. end and cav. 1 l. Khar. At hole usual char. *Rev.* 3 ll. Khar., very cursive, black and clear throughout. $9\frac{5}{8}'' \times 1\frac{3}{4}'' \times \frac{5}{8}''$. Well-preserved.

N. i. 43. Wedge cov.-tablet (belongs to N. i. 30). *Obv.* Seal cavity $1\frac{1}{8}''$ from sq. end. Remains of clay and string. 1 l. Khar. to R. of cav. Usual char. near hole. *Rev.* blank. $9\frac{1}{2}'' \times 1\frac{5}{8}''$. Wood well-preserved.

N. i. 44. Rectang. under-tablet, with sunk centre, &c. *Obv.* 6 ll. Khar., well written, but faded in parts. *Rev.* blank. $8'' \times 2''$. Well-preserved.

N. i. 45. Wedge under-tablet (belongs to N. ii. 3). String in hole. *Obv.* 4 ll. Khar., with initial form. All distinct, but very badly and coarsely written. *Rev.* Char. near sq. end. $14\frac{1}{4}'' \times 2''$. Well-preserved; warped.

N. i. 46. Wedge under-tablet (belongs to N. i. 55). *Obv.* 4 ll. Khar. with initial form., coarsely but distinctly written. Slightly deleted in places. *Rev.* 1 l. at sq. end. $15\frac{1}{8}'' \times 2\frac{3}{8}''$. Well-preserved, but knotty; cracked at sq. end.

N. i. 47. Wedge under-tablet. *Obv.* 3 ll. Khar., faint but mostly readable, with initial form. *Rev.* Some char. faintly traceable at sq. end. $11\frac{3}{4}'' \times 1\frac{3}{4}''$. Wood well-preserved; knotty.

N. i. 48. Wedge under-tablet (belongs to N. i. 28). *Obv.* 4 ll. Khar., cursive and rather blotted, but quite readable; initial form. *Rev.* 1 l. at sq. end. $10\frac{1}{4}'' \times 1\frac{3}{4}''$. Well-preserved.

N. i. 49. Wedge cov.-tablet; tied round, over seal cav., with piece of cloth. *Obv.* $1\frac{1}{2}''$ from sq. end, seal cav., 1 l. Khar. on both sides of cav. Usual char. at hole. *Rev.* 3 ll. Khar., cursive and rather blotted. $9'' \times 1\frac{3}{4}''$. Well-preserved; one or two insect holes.

N. i. 50. Wedge cov.-tablet (belongs to N. i. 58). *Obv.* $2\frac{3}{16}''$ from sq. end, seal cav. Remains of string. 1 l. Khar. to R. of cav. much faded. *Rev.* 1 l. Khar. just discernible. $9\frac{7}{8}'' \times 1\frac{7}{8}''$. Well-preserved; bleached.

N. i. 51. Wedge cov.-tablet, with seal and string. *Obv.* $1\frac{7}{8}''$ from sq. end, cav. containing seal (not clear, but torso of standing figure). 1 l. Khar. on both sides of seal. Usual char. at hole. *Rev.* 2 ll. Khar., faint. $11\frac{1}{2}'' \times 2'' \times \frac{1}{2}''$. Well-preserved.

N. i. 52 + 72. Wedge under-tablet, found in two pieces. *Obv.* 3 ll. Khar. with initial form. *Rev.* blank. $5\frac{3}{4}'' \times 1\frac{1}{2}''$, $4\frac{1}{2}'' \times 1\frac{5}{8}''$, resp. Well-preserved.

N. i. 53. Wedge cov.-tablet. *Obv.* $1\frac{5}{8}''$ from sq. end, seal cav. (broken at top edge) in which string. To R. of cav. 1 l. Khar., very cursive. Near hole usual char. *Rev.* 2 ll. Khar. cursive; very distinct. $8\frac{5}{8}'' \times 1\frac{7}{8}'' \times \frac{9}{16}''$. Well-preserved.

N. i. 54. Wedge under-tablet (belongs to N. i. 13). *Obv.* 3 ll. Khar. with initial form. Cursive, partially deleted. *Rev.* 1 l. Khar. near sq. end. $9\frac{1}{8}'' \times 1\frac{5}{8}''$. Well-preserved; fibre of bark adhering.

N. i. 55. Wedge cov.-tablet (belongs to N. i. 46). *Obv.* $2\frac{1}{4}''$ from sq. end, seal cav. in which frags. of seal and string. Seal-clay mixed with grass. To R. of cav. faint traces of Khar. line. *Rev.* blank. $15\frac{1}{4}'' \times 2\frac{1}{4}'' \times \frac{5}{8}''$. Well-preserved; bleached.

N. i. 56. Spear-head-shaped tablet. Bluntly pointed at R. p. end, where hole; broadest part of tab. about $\frac{1}{3}$ of its length from L. p. end. *Obv.* 3 ll. Khar., cursive, clear. *Rev.* 2 ll. Khar., 1st faint, 2nd very black. $14'' \times 1\frac{7}{8}''$. Well-preserved, but discoloured.

N. i. 57. Wedge under-tablet (belongs to N. i. 42), with string. *Obv.* 3 ll. Khar., very cursive, very clear, with initial form. *Rev.* 1 l. Khar., $9\frac{1}{2}'' \times 2\frac{1}{8}''$. Well-preserved.

N. i. 58. Wedge under-tablet (belongs to N. i. 50). *Obv.* 3 ll. Khar., very faint, cursive; initial form. *Rev.* 1 l., faint, at sq. end. $10'' \times 2''$. Hard, discoloured.

N. i. 59. Wedge cov.-tablet (belongs to N. i. 120). *Obv.* $1\frac{3}{8}''$ from sq. end, seal cav. To R. 1 l. Khar. (faint), also to L. Usual char. at hole. *Rev.* 3 ll. Khar., faint but legible. $10\frac{3}{4}'' \times 2'' \times \frac{5}{8}''$. Hard, split at sq. end.

N. i. 60. Wedge under-tablet. *Obv.* 3 ll. Khar., cursive, with initial form. Part of 3rd l. deleted. Surface much discoloured. *Rev.* Traces of l. from sq. end. $9'' \times 1\frac{7}{8}''$. Hard, but slightly warped and gnawed by insects at one edge.

N. i. 61. Wedge under-tablet. *Obv.* 3 ll. Khar., very cursive, black and clear, with initial form. *Rev.* 1 l. Khar. from sq. end. $7\frac{3}{4}'' \times 1\frac{3}{4}''$. Well-preserved.

N. i. 62. Wedge under-tablet (belongs to N. i. 6), with string. *Obv.* 4 ll. Khar., cursive, fine, legible; initial form. *Rev.* Faint traces of char. near sq. end. $9\frac{1}{4}'' \times 1\frac{5}{8}''$. Wood well-preserved.

N. i. 63. Wedge cov.-tablet. *Obv.* $1\frac{1}{4}''$ from sq. end, seal cav., showing crossing strings and frags. of seal. To R. and L. of cav. 1 l. Khar., faint. Near hole usual char. *Rev.* Short l. at top edge, coarsely written. $9\frac{7}{8}'' \times 2\frac{1}{16}''$. Well-preserved.

N. i. 64. Double-wedge tablet, complete document, *Obv.* of cov.-tab. $1\frac{1}{2}''$ from sq. end, seal cav. with string and remains of clay seal. To R. and L. of seal 1 l. Khar., very thin and distinct. Usual char. near hole. *Rev.* of under-tablet 1 l. Khar. from sq. end; faint. Document has not yet been opened, but from edges having slightly warped writing inside is seen neat and black, and probably very slightly deleted. Some of orig. string fastening still holds two tabs. together firmly. $8\frac{3}{4}'' \times 2'' \times 1''$. Well-preserved.

N. i. 65. Wedge cov.-tablet (belongs to N. i. 31). *Obv.* $1\frac{1}{4}''$ from sq. end, seal cav. with crossed strings. To R. of cav. 1 l. Khar., faint but clear. Usual char. near hole. *Rev.* 4 ll. Khar., clear cursive. $9'' \times 1\frac{1}{2}'' \times \frac{7}{8}''$. Well-preserved, stained dark on Rev.

N. i. 66. Wedge under-tablet. *Obv.* 3 ll. Khar., cursive, fine, generally clear; initial form. *Rev.* 2 short ll. from sq. end. $9\frac{1}{4}'' \times 1\frac{1}{2}''$. Well-preserved.

N. i. 67. Double-wedge tablet, complete document. Cov.-tablet. *Obv.* $1\frac{1}{4}''$ from sq. end, seal cav. Seal intact with strings binding two tabs. together. Device of seal, standing figure (Pallas) to L. p.; profile; helmet, aegis. R. p. leg slightly bent at knee; cameo deeply impressed, surrounded by mouldings in intaglio. Figure is very graceful, and seems to have been finely cut. Between sq. end and seal 1 l. Khar. Usual char. near hole. *Rev.* blank.

Under-tablet. *Obv.* 3 ll. Khar. with initial form., very clear. *Rev.* 1 l. from sq. end, faded. $9\frac{3}{8}'' \times 1\frac{3}{4}''$. Well-preserved.

N. i. 68. Wedge under-tablet. *Obv.* 4 ll. Khar. with initial form., cursive but clear. *Rev.* 1 short line Khar. from sq. end. $8\frac{3}{4}'' \times 1\frac{7}{8}''$. Well-preserved.

N. i. 70. Wedge tablet of unusual shape; has no seal cav., but $\frac{3}{16}''$ from sq. end is a very deep groove in which probably string was placed. *Obv.* Usual characters at hole near point, but no other writing visible. *Rev.* blank. $5\frac{7}{8}'' \times 2'' \times \frac{7}{8}''$. Hard and well-preserved. Very roughly cut. See Pl. CIII.

N. i. 71. Wedge cov.-tablet (belongs to N. i. 87). *Obv.* 1″ from sq. end, seal cav. with frags. of string. To R. and L. of seal 1 l. of Khar. *Rev.* 1 l. Khar. from sq. end. $8\frac{3}{4}'' \times 1\frac{7}{8}'' \times \frac{5}{8}''$. Well-preserved.

N. i. 73. Wedge cov.-tablet, small frag. of sq. end only. *Obv.* 1 l. Khar. between sq. end and seal cav. Seal cav. $1\frac{1}{2}''$ from sq. end. *Rev.* 1 l. Khar., cursive, fine, faded. $3'' \times 2'' \times 1\frac{1}{2}''$. Frag. is broken at side of cav. nearest to point. It has been cut with knife after being broken. A hole drilled in cav.; sq. ends have been slightly rounded at corners. Portions of bark.

N. i. 74. Wedge cov.-tablet, point with hole broken away. *Obv.* $1\frac{3}{4}''$ from sq. end, seal cav., in which frags. of string. 1 l. Khar. between sq. end and cav. Usual char. near hole. *Rev.* 1 l. Khar., very clear and black. $8\frac{3}{4}'' \times 1\frac{3}{4}'' \times \frac{9}{16}''$. Well-preserved.

N. i. 75. Wedge cov.-tablet (belongs to N. i. 80), with string. *Obv.* 1″ from sq. end, seal cav. with remains of seal and string. 1 l. Khar. between sq. end and cav. indistinct. Usual char. at hole. *Rev.* 1 l. Khar., very clear. $8'' \times 1\frac{1}{2}'' \times \frac{1}{2}''$. Well-preserved.

N. i. 76. Wedge under-tablet, with string. *Obv.* 4 ll. Khar., rather blotted in places, but generally distinct; very cursive, with initial form. *Rev.* 1 short l. $8\frac{3}{4}'' \times 2''$. Well-preserved. Rather roughly cut.

N. i. 77. Wedge under-tablet (belongs to N. i. 82). *Obv.* 3 ll. Khar., very cursive, thin and clear; initial form. *Rev.* 1 l. Khar., faint. $10'' \times 1\frac{7}{8}''$. Well-preserved. N. i.

N. i. 78. Double-wedge tablet, complete document. *Obv.* of cov.-tab. $1\frac{3}{4}''$ from sq. end, seal cav. in which string. 1 l. Khar. to R. of cav. Usual char. at hole. *Rev.* 1 l. Khar., very cursive but fairly clear. *Obv.* of under-tablet 3 ll. Khar., partly deleted by stains; initial form. *Rev.* 1 l. $9'' \times 2'' \times \frac{1}{2}''$. Well-preserved, but much stained, and warped. Remains of string still holding pieces together.

N. i. 79. Wedge under-tablet. *Obv.* 4 ll. Khar., coarse, much faded, very cursive; initial form. *Rev.* 1 l., faded. $8\frac{1}{4}'' \times 1\frac{7}{8}''$. Wood well-preserved, but stained and warped.

N. i. 80. Wedge under-tablet (belongs to N. i. 75). *Obv.* 3 ll. Khar. with initial form., well written, but deleted and stained in parts. *Rev.* 1 l. Khar. $8'' \times 1\frac{3}{4}''$. Wood well-preserved, but stained.

N. i. 81. Wedge under-tablet. *Obv.* 3 ll. Khar., faded. Unreadable in parts; in 1st line initial form. *Rev.* 1 l. Khar., very faded. $8\frac{3}{4}'' \times 1\frac{5}{8}''$. Wood well-preserved, but much stained.

N. i. 82. Wedge cov.-tablet (belongs to N. i. 77). *Obv.* 2″ from sq. end, seal cav. in which string. Between cav. and sq. end 1 l. Khar., faded. On opposite side of cav. a long line, faded, but part decipherable. *Rev.* blank. $10'' \times 2'' \times \frac{1}{2}''$. Wood well-preserved. Sq. end of obv. cut in rather unusual shape.

N. i. 83. Double-wedge tablet, complete document, with string in hole. Cov.-tablet. *Obv.* $\frac{7}{8}''$ from sq. end, seal cav. To R. and L. of it 1 l. Khar. (very faint). Usual char. at hole.

Under-tablet. *Rev.* 1 l. Khar. $8\frac{1}{8}'' \times 2'' \times \frac{1}{2}''$. Well-preserved. String almost intact binding two parts together. Inscription inside very black, fine and clear.

N. i. 84. Wedge under-tablet. *Obv.* 4 ll. Khar., much faded, with initial form. *Rev.* blank. $8\frac{3}{8}'' \times 1\frac{3}{4}''$. Badly preserved; warped and soft on surface.

N. i. 85. Wedge cov.-tablet. *Obv.* $1\frac{1}{2}''$ from sq. end, seal cav. in which crossed strings. 1 l. Khar. between sq. end and cav. Usual char. at hole. *Rev.* 3 ll. Khar., very cursive. $9\frac{1}{4}'' \times 1\frac{3}{4}'' \times \frac{5}{8}''$. Well-preserved; slightly chipped and cracked.

N. i. 87. Wedge under-tablet (belongs to N. i. 71). *Obv.* 4 ll. Khar., cursive, well-written, rather faded, with initial form. *Rev.* Few char. from sq. end. $8\frac{3}{4}'' \times 2''$. Well-preserved.

N. i. 100 + 109 + 111. Rectang. cov.-tablet, found in 3 frags. *Obv.* 1 l. Khar. on L. p., transversely; indistinct. *Rev.* 4 ll. Khar., very clear and black. $3\frac{1}{4}'' \times 2\frac{1}{8}''$, $2\frac{1}{4}'' \times 1\frac{1}{8}''$, $2\frac{1}{4}'' \times 1\frac{1}{2}''$ resp. Writing well-preserved.

N. i. 101. Oblong tablet, fragment; hole near thick end. *Obv.* blank. A burn mark at edge. *Rev.* L. p. near hole, a short l. Khar. To R. of this a vertical line (bracket?).

N. i. To R. again 2 ll. and portions of further char. Fairly clear. 5″ × 2¼″ × ⅜″. Wood fairly hard. Broken and stained.

N. i. 103. Wedge cov.-tablet. *Obv.* 1½″ from sq. end, seal cav. in which string crossed. 1 l. Khar. to R. of cav. Usual char. at hole. *Rev.* 1 l. Khar., very cursive, faded towards sq. end. 9¼″ × 1¾″ × ⅝″. Wood well-preserved, warped. See Pl. IC.

N. i. 104. Wedge under-tablet (belongs to N. i. 16). *Obv.* 3 ll. Khar., cursive, coarse; initial form. *Rev.* 1 l. 9″ × 2″. Wood well-preserved.

N. i. 105. Double wedge tablet, complete document. Cov.-tablet. *Obv.* 1½″ from sq. end, seal cav. in which string remains. To R. and L. of seal cav. 1 l. Khar., faint. Usual char. at hole.

Under-tablet. *Rev.* 1 l. from sq. end. String remains bound round. 8⅛″ × 2⅛″ × ⅞″. Well-preserved.

N. i. 106. a, b. Wedge under-tablet, in 3 fragments. *Obv.* 4 ll. Khar., cursive but very clear and black; initial form. Portion missing at broken part. *Rev.* 2 ll. from sq. end. 11″ × 2″ (approx.). Well-preserved. Break not due to decay.

N. i. 107. Wedge under-tablet (belongs to N. i. 15). *Obv.* 4 ll. Khar., thin, black, very clear, but cursive; initial form. *Rev.* 1 l. from sq. end. 8⅜″ × 1⅝″. Well-preserved but much stained.

N. i. 108. Wedge cov.-tablet. *Obv.* 1½″ from sq. end, seal cav. with remains of seal. Design of seal oval, but no detail distinguishable. Between sq. end and seal, 1 l. Khar. very faint. Usual char. at hole. *Rev.* 1 l. fairly distinct. 9″ × 2″ × ⅞″. Wood well-preserved.

N. i. 110. Wedge under-tablet. *Obv.* 4 ll. Khar.; initial form. Writing coarse, blotted, and a good deal faded. *Rev.* 1 l. from sq. end. 14¾″ × 2¼″. Discoloured and very susceptible to damp.

N. i. 112. Oblong tablet. Hole at R. p. end. *Obv.* 2 ll. Khar., very faded and partially obliterated; initial form. *Rev.* Nothing visible. 11½″ × 1⅜″ × ½″. Very soft and stained.

N. i. 113. Oblong fragment. *Obv.* To L. p. 3 ll. Khar., rather coarse and very cursive. To R. p. 4 ll. Khar., fine and well written. *Rev.* blank. 6¼″ × 1⅞″. Much perished; very soft and surface seamed with open cracks.

N. i. 114. Rectangular fragment; appears to be portion of under-tab. with thick end roughly bevelled to general surface, and hole bored subsequently. *Obv.* 5 ll. Khar., fine, black; all fragmentary with repetition of wording. Hole at top R. p. *Rev.* 2 ll. Khar. (fragmentary). 3¼″ × 2¼″. Very thin. Wood hard.

N. i. 120. Wedge under-tablet (belongs to N. i. 59). *Obv.* 3 ll. Khar., faint and cursive; initial form. *Rev.* Few characters from sq. end. 10⅞″ × 2⅛″. Wood well-preserved, but susceptible to damp.

N. i. 121. Carved wood. Probably a rail or baluster. At one end a tenon, flush with back surface. Front elevation: centre portion, concave at each side, narrowest part being ⅞″ and broadest (above and below) 1⅝″; length 3⅞″; grooved by parallel curved lines on front, as ornament. Above centre 2 inverted truncated pyramids one above the other. Below, similar arrangement, reversed. Length 8⅜″, thickness ⅞″. See Pl. LXX.

N. i. 122. Double-wedge tablet, complete document. String in hole near point. Cov.-tablet. *Obv.* Seal cav. 1½″ from sq. end; portion of clay adhering to string in cav. 1 l. Khar. to R. and L. of cav. No char. visible near hole. *Rev.* 3 ll. Khar., very clear, but scratchily written and very cursive.

Under-tablet. *Obv.* 4 ll. Khar., with initial form., generally clear, but here and there blotted and deleted. *Rev.* Transversely at sq. end three characters cut into the wood. 10¼″ × 2″ × ½″. Excellently preserved. Wood very hard. See Pl. XCVIII.

N. i. 180. Wedge under-tablet (belongs to N. i. 24), part from hole to point broken off. *Obv.* 4 ll. Khar., bold and fairly well written; initial form. *Rev.* bleached; cracked and peeled by damp at point end. 13¼″ × 2⅛″. Wood soft.

N. ii. 1. Rectang. under-tablet. *Obv.* 4 ll. Khar., well written; much deleted in parts by exposure to damp and sun. *Rev.* 9¼″ × 2¼″. Curled and peeled by exposure. See Pl. CIV.

N. ii. 2. Double-wedge tablet, complete. Cov.-tablet. *Obv.* 2¼″ from sq. end, seal cav. 1 l. Khar. on either side of cav. Usual char. near hole. Seal intact (see Pl. LXXI). In upright elliptical cavity a standing nude male figure to R. p.; appears from its muscular development to be intended for Herakles, but may be an Eros. On flat surface surrounding seal, probably setting, are traces of few inscribed characters (Khar.?). This tablet is superficially broken near point. *Rev.* 1 l. Khar.

Under-tablet. *Obv.* 3 ll. Khar., with usual form. Few char. near sq. end. 9⅜″ × 2¼″, thickness of cov.-tablet 1⅜″. String intact excepting at hole, so that tablets can be slid apart.

N. ii. 3. Wedge cov.-tablet (belongs to N. i. 45). *Obv.* 1¼″ from sq. end, seal cav. 1 l. Khar. to R. and L. of cav. Usual char. near hole. String in grooves. Writing coarse. *Rev.* 4 ll. Khar., coarse, legible generally; very cursive. 14¼″ × 2″ × ⅝″. Perfectly preserved, but stained dark on rev.

N. iii. 1. Takhtī-shaped tablet. Pear-shaped handle, pierced, broken. *Obv.* (handle in L. hand) 4 columns Khar., very cursive. Ink very black. *Rev.* (handle in L. hand) 4 columns Khar., much deleted. 12″ × 2¼″. Much damaged by exposure, and encrusted by sand.

N. iv. 1. Oblong tablet; fragment. Hole (?) at one end, broken. *Obv.* Well-written Khar. char. in columns. Surface much cracked and discoloured. *Rev.* Surface quite perished. $13'' \times 3''$. In bad preservation.

N. iv. 2. Oblong tablet; fragment. *Obv.* Traces of Khar. char., prob. in columns. *Rev.* quite perished. $10\frac{1}{4}'' \times 2\frac{1}{4}''$. Prob. undecipherable.

N. iv. 3. Oblong tablet. Hole at one corner. *Obv.* Traces of 3 ll. Khar., very faint. *Rev.* Traces of 5 ll. Khar., very faint. $12\frac{1}{2}'' \times 3\frac{1}{4}''$. Perished, split and curling.

N. iv. 4. Oblong tablet. *Obv.* Traces of Khar. char. in column arrangement to read transversely. Surface split and perished. *Rev.* Traces of Khar. char., rather small, tabular, to read transversely. $8\frac{5}{8}'' \times 2\frac{5}{8}''$. Badly perished and split.

N. iv. 6. Wedge-shaped tablet. Hole near point, broken. *Obv.* Long l. Khar. at top. Below this at sq. end, column of 7 ll., faint but fairly legible. *Rev.* blank. $16'' \times 3\frac{1}{2}'' \times \frac{1}{4}''$. Much perished, especially at edges, and discoloured.

N. iv. 7. Rectang. cov.-tablet; fragment in several pieces. *Obv.* 2 or more ll. transversely below seal cav., almost deleted. *Rev.* 5 ll. Khar., well-written but faded; fragmentary. $5\frac{1}{2}'' \times 2\frac{1}{2}''$ (approx.). Completely perished by damp.

N. iv. 8. Oblong tablet. Hole at one corner. *Obv.* Tabular. Column of 6 ll. Khar. near hole. Irregular ll. in centre. Column of 8 or 9 ll. at opposite end. Some ll. erased by stroke. Khar. very cursive, faded. *Rev.* blank. $9\frac{1}{4}'' \times 3\frac{7}{8}''$. Perished, but uninjured otherwise.

N. iv. 8. a. Oblong tablet, curling at one end to point. Hole about $\frac{1}{3}$ from point. *Obv.* Tabular Khar. writing in 3 columns; very cursive. 1st column (from sq. end), 3 ll., very faded. 2nd column, 4 ll., 2nd and 4th erased. 3rd column, 7 ll., 1st, 3rd, and 4th erased. Last 4 ll. bracketed. *Rev.* Column of 6 ll., transversely from sq. end, very faded. $9'' \times 2\frac{1}{4}'' \times \frac{1}{2}''$ to $\frac{1}{16}''$. Dark in colour; fairly preserved. See Pl. CIII.

N. iv. 9. Parabolic tablet. Large hole in round end. *Obv.* 6 columns of 7 to 8 ll. Khar. each. 2 ll. in upper corner, bracketed. Writing, very cursive, part faded. *Rev.* 7 columns of 6 to 10 ll. each; very cursive, but clear. $7'' \times 2\frac{1}{4}''$. Soft; well-preserved.

N. iv. 11. Tablet, fragment, twisted fantastically. *Obv.* 5 ll. Khar. at sq. end. Well-written and containing numerals. Further characters are traceable, but owing to broken and twisted condition of wood prob. undecipherable. *Rev.* Traces of Khar. char., but very fragmentary. $6\frac{3}{4}'' \times 2''$ approx. Quite perished. See Pl. CIV.

N. iv. 13. Rectang. tablet in several pieces, prob. complete. Cov.-tablet. *Obv.* Seal cav. $\frac{7}{8}''$ sq. 3 ll. Khar., black, cursive, transversely at L. p. edge. *Rev.* 7 ll. Khar., well-written, black, perfectly clear except where wood is perished.

Under-tablet. *Obv.* 6 ll. Khar., well-written, clear, and black; last l. broken away. $9'' \times 2''$ (approx.). Quite perished; wood dark and twisted like a piece of leather that has been wet. N. iv.

N. iv. 14. Rectang. cov.-tablet, in 2 pieces. Seal cav. $1\frac{1}{8}'' \times 1\frac{1}{2}''$. *Obv.* 3 ll. Khar., transversely at top. Very indistinct. *Rev.* 6 ll. Khar., very cursive, clear, and black. $4'' \times 2\frac{1}{2}''$. Wood perished; much discoloured.

N. iv. 15. Oblong tablet. Hole at one corner. *Obv.* 4 ll. Khar., well written, black, cursive; all readable. *Rev.* 7 ll. Khar., finer than obv. but much deleted and faded. $6\frac{5}{8}'' \times 3\frac{1}{8}'' \times \frac{3}{8}''$. Wood much perished and stained.

N. iv. 16. Rectang. under-tablet, in many pieces. *Obv.* Traces of Khar. characters. Very fragmentary. *Rev.* blank. $8\frac{1}{4}'' \times 2\frac{1}{4}''$. Perished, stained, and encrusted with sand.

N. iv. 17. Oblong tablet. *Obv.* 4 columns of 8 and 9 ll. Khar. each. 1 l. above columns, and a short l. (numerals) beside last column. Writing faded, but nearly all legible. Scorings as though from knife point in cutting leather (?). *Rev.* Scorings from knife point. $15\frac{1}{8}'' \times 3\frac{5}{16}''$. Wood well-preserved. Much encrusted with sand.

N. iv. 17. a. Rectang. cov.-tablet, in 2 pieces (frag.). *Obv.* 2 ll. Khar., cursive, black, incomplete; transversely at L. p. edge.. *Rev.* 5 ll. Khar. blotted, cursive. $2\frac{1}{2}'' \times 2\frac{1}{4}''$. Perished and discoloured.

N. iv. 17. b. Rectang. cov.-tablet; frag. in 2 pieces. *Obv.* 3 ll. Khar. transversely at L. p. edge; incomplete. *Rev.* 2 ll. Khar. cursive, blotted, incomplete. $4'' \times \frac{7}{8}'' \times \frac{1}{4}''$. Perished and discoloured.

N. iv. 17. c. Takhtī-shaped tablet (?), fragment of; having portion of hole. Both sides blank.

N. iv. 18. Rectang. under-tablet, fragment. *Obv.* Portions of 8 ll. Khar., very black. *Rev.* Faint traces of char. near edge. $3\frac{1}{4}'' \times 2\frac{1}{2}''$. Perished, twisted, very dark, and cracked.

N. iv. 20. Wedge-shaped tablet (point gone). *Obv.* 4 ll. Khar., cursive, generally distinct. *Rev.* blank. $10\frac{3}{8}'' \times 2\frac{1}{4}'' \times 1\frac{1}{8}''$. Perished and dark.

N. iv. 21. Rectang. cov.-tablet, in 2 pieces. Seal cav. $1''$ sq. *Obv.* 2 ll. Khar., transversely at top, very indistinct. *Rev.* Traces of Khar. char., fragmentary and indistinct. $5\frac{7}{8}'' \times 3\frac{5}{8}'' \times \frac{1}{2}''$. Hard, very dark, warped and cracked.

N. iv. 22. Stick-like tablet. *Obv.* At broad end 3 ll. Khar. transversely; then 2 ll., 1 l., and 1 l., with intervals transversely; writing cursive, faint. *Rev.* blank. $11'' \times 1''$. Wood well-preserved.

N. iv. 23. Stick-like tablet, warped. *Obv.* Traces of Khar. characters—some numerals, apparently arranged in columns. *Rev.* Do., but rather more legible; about 4 ll., very cursive. $8\frac{1}{4}'' \times 1\frac{1}{2}''$. Wood fairly hard. Discoloured and sand-encrusted.

N. iv. **N. iv. 24. a and b. Stick-like tablet** in 2 pieces. Traces of characters on both sides; illegible. $9'' \times 1\frac{3}{8}''$. Wood hard, but perished. Sand-encrusted.

N. iv. 25. Oblong tablet, in 2 pieces. Hole at one end. *Obv.* Traces of Khar. characters. Scorings by knife. *Rev.* Tabular, reading transversely, dividing strokes at irreg. intervals. Very indistinct, but part legible. Scored. $24\frac{3}{4}'' \times 3\frac{1}{8}'' \times \frac{5}{16}''$. Wood rather soft on surface.

N. iv. 25. a. Oblong tablet. Very dark and stained. Traces of characters on one side. Quite undecipherable. $9'' \times 1\frac{3}{4}''$. Perished.

N. iv. 26. Rectang. cov.-tablet. Seal cav. $1\frac{1}{8}'' \times \frac{15}{16}''$. *Obv.* blank. In cav. fragments of charcoal. *Rev.* blank. $4\frac{1}{8}'' \times 2\frac{1}{8}'' \times \frac{7}{16}''$. Perished and discoloured.

N. iv. 28. Rectang. cov.-tablet, much fissured, split, and chipped. *Obv.* Seal cav. $1'' \times 1\frac{3}{8}''$. Extremely faint traces of several lines Khar., transv. at one end. *Rev.* Faint traces of 2 ll. or more Khar. All writing very cursive and too faded to decipher. The whole tablet much abraded, sand-encrusted and dark in colour. Wood very soft and brittle. $8\frac{3}{8}'' \times 3\frac{1}{2}''$. See Pl. CV.

N. iv. 29. Oblong tablet, in 2 pieces. Blank. $24'' \times 2\frac{1}{4}''$. Rotten, much encrusted with sand.

N. iv. 29. a. Oblong tablet. Hole at one end. *Obv.* 6 items Khar. arranged in 2 columns; first two char. are repeated 5 times, and the last three or four with slight variations. Khar., cursive; very distinct. *Rev.* 3 ll. very distinct. First two characters in each, same as those on Obv. $7\frac{1}{2}'' \times 1\frac{7}{16}''$. Well-preserved. See Pl. CIII.

N. iv. 30. Rectang. under-tablet, fragment. *Obv.* 2 ll. Khar., with numerals, cursive, but black. *Rev.* blank. $6\frac{1}{8}'' \times \frac{3}{4}''$. Wood well-preserved.

N. iv. 31. Oblong tablet. Hole in centre, lower edge. *Obv.* 3 ll. Khar., very cursive, indistinct and fragmentary. *Rev.* blank. $7\frac{1}{2}'' \times 1\frac{1}{4}'' \times 1''$. Wood fairly preserved.

N. iv. 32. Rectang. cov.-tablet. Seal cav. $3\frac{7}{16}'' \times 1\frac{5}{16}''$. *Obv.* 2 ll. Khar., cursive, well written, transversely, at L. p. edge. Stain of Sindūra (?). *Rev.* blank. $7\frac{1}{2}'' \times 4\frac{1}{2}'' \times \frac{11}{16}''$. Wood moderately hard; thin edges much split.

N. iv. 33. Takhtī-shaped tablet. Hole in handle, with original string. *Obv.* 4 ll. Khar., very cursive, faded. *Rev.* 6 ll. Khar., very cursive, faded. $7\frac{1}{4}'' \times 2\frac{1}{2}''$. Wood well-preserved. See Pl. CIII.

N. iv. 34. Rectang. cov.-tablet, fragment. Seal cav. $\frac{7}{8}'' \times 1''$. *Obv.* 2 ll. Khar., very cursive. *Rev.* 3 ll. Khar., very coarse and cursive. $4\frac{1}{8}'' \times 1\frac{7}{8}''$. Wood very soft. Sand-encrusted.

N. iv. 34. a. Inscribed tablet, fragment. 2 ll. Khar., cursive; numerals. $2\frac{5}{8}'' \times \frac{3}{4}''$. Very soft.

N. iv. 35. a. Oblong tablet, square at one end. *Obv.* 2 ll. Khar., cursive, part faded. *Rev.* 6 columns, 1st and 2nd seven ll.; 3rd, 4th, 5th, 6th one l. each; 6th four ll.; then a single l. $15'' \times 2\frac{3}{8}''$. Wood soft and rotten at one end. See Pl. CII.

N. iv. 35. b. Oblong tablet. Hole at one end; one corner broken. *Obv.* 4 columns. 1st imperfect; 2nd and 3rd six ll.; 4th five ll. Numerals. *Rev.* 1 column, to read transversely. $13\frac{3}{4}'' \times 2\frac{1}{8}'' \times \frac{3}{8}''$. Wood rather soft; cracked.

N. iv. 36. Oblong tablet. Hole near pointed end. *Obv.* 3 items Khar., very cursive. *Rev.* 2 ll., divided into items; very cursive and black. $10\frac{1}{2}'' \times 1\frac{1}{4}''$. Wood well-preserved. Five notches on one of the edges near square end.

N. iv. 41. Oblong cov.-tablet. *Obv.* Seal cav. $1\frac{3}{8}'' \times 1\frac{1}{16}''$, empty. 1 l. Khar. transversely at L. p. end. Short l. Khar. near cav. on same side. Traces (?) of char. to R. p. *Rev.* 6 ll. Khar., cursive, mostly clear and black. $5\frac{1}{4}'' \times 3\frac{1}{4}'' \times \frac{11}{16}''$. Now paired (?) with N. iv. 127.

N. iv. 46. Takhtī-shaped tablet in many frags. Pentagonal handle with hole. *Obv.* 3 columns with 5 and 6 ll. resp., Khar., cursive, fine, faded. *Rev.* blank. $7\frac{1}{4}'' \times 1\frac{11}{16}''$. Wood perished but clean.

N. iv. 47. Oblong tablet. Hole at top R. p. corner. *Obv.* 2 columns cursive Khar. 1st, 5 ll., 2nd, 3 ll., and a line separated from others. Clear. *Rev.* blank. $6\frac{1}{4}'' \times 2\frac{1}{4}''$. Wood very soft.

N. iv. 48. Rectang. cov.-tablet. Seal cav. $1\frac{1}{4}'' \times 1\frac{1}{4}''$. *Obv.* 2 ll. transverse at L. p. edge. *Rev.* 3 ll. Khar., cursive, indistinct. $5\frac{1}{8}'' \times 2\frac{1}{4}'' \times \frac{3}{4}''$. Wood well-preserved. Much discoloured and sand-encrusted.

N. iv. 50. Rectang. cov.-tablet, fragment. *Obv.* 1 l. Khar. at upper edge, transversely; very fine, rather indistinct. *Rev.* 3 ll. Khar., very fine; fragmentary. $7\frac{7}{8}'' \times 2\frac{3}{8}''$. Well-preserved. Sap-wood soft, heart-wood hard.

N. iv. 51. Oblong tablet, in 2 pieces. *Obv.* Faint traces of closely written columns (?), Khar., prob. illegible. *Rev.* blank. $19\frac{1}{4}'' \times 2''$. Very soft and much encrusted by sand.

N. iv. 52. Oblong tablet. Hole at top R. p. corner. *Obv.* 4 ll. Khar., carefully written, black, clear, containing numerals. *Rev.* 2 ll. Khar., cursive. $11'' \times 2''$. Well-preserved.

N. iv. 53. Inscribed tablet, frag. in 2 pieces. Traces of Khar. characters. $4\frac{3}{8}'' \times 1\frac{1}{2}''$. Perished and dark in colour.

N. iv. 53. a. Oblong tablet. Corners cut off at one end. Hole at other. *Obv.* Traces of columns of about 5 ll. each; very cursive and fragmentary. *Rev.* blank. $21\frac{1}{2}'' \times 2\frac{1}{4}'' \times \frac{3}{8}''$. Surface soft and cracked at edges. Wood much stained and sand-encrusted; very knotty.

N. iv. 54. Oblong tablet; round at one end at which hole. Small projection at opposite end. *Obv.* Column of 4 ll.; next, 2 ll. near upper edge, and 1 l. near lower; column of 5 ll.; 1 l. near upper edge, 1 l. towards lower. Khar. cursive; numerals. *Rev.* 2 columns of 3 ll. each; single l. next; thin, cursive. $13'' \times 1\frac{1}{2}''$. Well-preserved.

N. iv. 55. Rectang. under-tablet. *Obv.* 7 ll. very cursive Khar.; fairly legible. *Rev.* blank. Marks of knife scorings. $6\frac{1}{4}'' \times 2\frac{1}{2}''$. Soft; much discoloured.

N. iv. 55. b. Oblong tablet; obtuse pointed. Hole near point. Blank. $5\frac{1}{4}'' \times 1\frac{1}{2}''$. Well-preserved.

N. iv. 56. Stick-like tablet. Pointed with hole at one end. *Obv.* 3 ll. Khar., very cursive; partly deleted. *Rev.* 2 ll. do. $9\frac{5}{8}'' \times 1\frac{1}{4}'' \times \frac{1}{2}''$. Hard, Well-preserved.

N. iv. 57. Oblong tablet. Sharp pointed by concave cuts; hole at point. *Obv.* 4 ll. Khar., very cursive, faint and fragmentary. *Rev.* blank. $14\frac{1}{2}'' \times 2\frac{1}{2}''$. Well-preserved.

N. iv. 58. Rectang. under-tablet (belongs to N. iv. 141). *Obv.* 6 ll. Khar., very cursive; clear. *Rev.* blank. $6\frac{1}{8}'' \times 2\frac{1}{4}''$. Very discoloured.

N. iv. 59. Oblong tablet. Round-pointed end; hole (broken). *Obv.* 4 ll. Khar., very cursive, faint, in parts deleted. *Rev.* blank. $8\frac{1}{2}'' \times 2\frac{1}{4}''$. Very soft; discoloured.

N. iv. 60. Oblong tablet. Hole in centre of one long edge. *Obv.* 4 ll. Khar., cursive, faded, but mostly decipherable. *Rev.* blank. $6\frac{3}{4}'' \times 1\frac{1}{4}''$. Well-preserved.

N. iv. 80. Double-wedge tablet, complete. Cov.-tablet. *Obv.* Seal cav. $1\frac{1}{8}''$ from sq. end. Seal $1'' \times 1\frac{1}{4}''$. The device appears to be, in usual oval or vesica shape, surrounded by deep moulding, a slim, nude figure to R. p., probably an Eros, but much deleted. Outside vesica moulding, four spandril-shape corners, slightly impressed, and evidently forming part of the setting of the matrix seal. A few Khar. characters at sq. end, and a few on opposite side of seal cav. Usual formula at hole. All tolerably clear. *Rev.* 3 ll. Khar., very cursive; clear in places.

Under-tablet. *Obv.* 4 ll. Khar., very cursive. Very clear in places. *Rev.* 1 l. Khar. Wood well-preserved, but very dark inside, and generally sand-encrusted. $8\frac{1}{4}'' \times 2'' \times \frac{11}{16}''$ (cov.-tab.), $\frac{7}{16}''$ (under-tab.).

N. iv. 81. Wedge-shaped tablet. Hole near point. *Obv.* 3 ll. Khar., cursive and clear. *Rev.* 2 ll. Khar., cursive, faint, but decipherable. $9\frac{1}{4}'' \times 2''$. Well-preserved; very sensitive to damp.

N. iv. 82. Rectang. under-tablet. *Obv.* 5 ll. Khar., very cursive, faint, and part deleted. Stains of Sindūra (?). *Rev.* blank; stained. $9\frac{7}{8}'' \times 3\frac{1}{2}''$. Warped and split by damp and exposure.

N. iv. 83. Rectang. cov.-tablet. Seal cav. $1\frac{1}{4}''$ sq. *Obv.* Traces of 3 ll. Khar. transversely at L. p. edge. *Rev.* 1 l. Khar., clear but cursive. $6\frac{3}{8}'' \times 3\frac{3}{8}''$. Wood fairly hard. Slightly warped, split, and broken at ends; sand-encrusted.

N. iv. 84. Rectang. under-tablet, broken, part missing. *Obv.* 5 ll. Khar., faded, very cursive. *Rev.* blank. $11\frac{1}{4}'' \times 4''$. Surface much injured by exposure.

N. iv. 99. Stick-like tablet. Roughly pointed at one end; broken. *Obv.* (flat). Khar. (?) characters; irregular (cf. N. xviii. 3). *Rev.* (round). Encrusted with sand. Traces of characters. $27\frac{1}{2}'' \times 1\frac{1}{4}''$. Much perished and discoloured.

N. iv. 100 + 101. Rectang. double tablet. Cov.-tablet. *Obv.* Seal cav. $1\frac{1}{2}''$ sq.; frags. of string remaining and pieces of seal clay. 1 l. Khar. at one end. *Rev.* 3 ll. Khar. Sand adhering. $6\frac{3}{4}'' \times 2\frac{7}{8}'' \times \frac{3}{4}''$.

Under-tablet. *Obv.* 6 ll. Khar., cursive, very indistinct in places. *Rev.* blank. $8\frac{1}{2}'' \times 2\frac{3}{4}'' \times \frac{1}{2}''$. Broken at lower L. corner. Split at R. end and along long edges. Wood very dry and soft.

N. iv. 103. Stick-like tablet, cut to obtuse point at one end. *Obv.* 7 columns, 2 ll. each, Khar., cursive, clear. *Rev.* 7 columns, 1st, 4 ll., 2nd, 5th, and 6th, 3 ll. each, 3rd, 4th, and 7th, 2 ll. each. Khar., cursive. $25\frac{3}{8}'' \times 2''$, dim. to $1\frac{1}{4}''$. Fairly preserved, rather soft, warped and slightly encrusted with sand. See Pl. CII.

N. iv. 104. Oblong tablet. Hole at small end near edge. *Obv.* 5 ll. Khar., cursive, boldly written and clear. 1st l. interrupted by 2 intervals. *Rev.* 2 ll. Khar.; similar writing to Obv.; 1st l. in two portions. $11\frac{1}{4}'' \times 5\frac{3}{8}''$, dim. to $1\frac{1}{2}''$. Well-preserved.

N. iv. 105. a, b. Wedge under-tablet, in 2 pieces. Hole near point. *Obv.* 4 ll. Khar., very cursive, clear, and complete, with initial form. *Rev.* 1 l. from sq. end. $12\frac{3}{4}'' \times 2\frac{1}{4}''$. Well-preserved; slightly bored by insects.

N. iv. 106. Rectang. under-tablet. *Obv.* 7 ll. Khar., very coarse and cursive. *Rev.* blank. $9\frac{3}{8}'' \times 3\frac{1}{4}''$. Wood fairly hard. Much stained and encrusted with sand.

N. iv. 107. Wedge under-tablet. Hole near point. *Obv.* 3 ll. Khar., finely written but part deleted. *Rev.* blank. $8\frac{3}{4}'' \times 1\frac{7}{16}''$. Wood well-preserved, but stained and sand-encrusted.

N. iv. 108. a, b. Double-wedge tablet, complete. Hole near point. Cov.-tablet. *Obv.* Seal cav., $1\frac{7}{8}''$ from sq. end, contains remains of clay and string. Between sq. end and cav. 1 l. Khar., very faint. *Rev.* 1 l. boldly written Khar., very black in parts.

Under-tablet. *Obv.* 5 ll. very cursive Khar., clear and black, with initial form. *Rev.* 1 l. from sq. end. $9\frac{1}{8}'' \times 2\frac{1}{4}'' \times \frac{7}{8}''$. Well-preserved.

N. iv. 109. Wedge cov.-tablet (belongs to N. iv. 111). Hole near point. *Obv.* Seal cav. $1\frac{7}{8}''$ from sq. end. On both sides of cav., 1 l. Khar., very fine and cursive. Usual char. at hole. *Rev.* blank. Stained with Sindūra. $14\frac{1}{2}'' \times 2\frac{3}{16}'' \times \frac{9}{16}''$. Well-preserved, but stained and sand-encrusted.

N. iv. 110. Rectang. under-tablet (belongs to N. iv. 140); 3 notches in centre of each long side, for string. *Obv.* 6 ll. Khar., very coarse and cursive. *Rev.* blank. $6\frac{1}{4}'' \times 2\frac{1}{2}'' \times \frac{1}{2}''$. Well-preserved.

N. iv. 111. Wedge under-tablet (belongs to N. iv. 109). Hole near point. *Obv.* 2 ll. Khar., finely and carefully

N. iv. written, with initial form. *Rev.* 1 short line Khar. from sq. end. $14\frac{1}{2}'' \times 2\frac{3}{16}''$. Well-preserved. Slightly warped and encrusted with sand.

N. iv. 113. Oblong tablet. Hole at one end. *Obv.* 4 columns with numerals (?). Each column has 4 ll. (top line in one cut off); cursive. *Rev.* 2 columns, 1st, 5 ll., 2nd, 3 ll., with numerals (?); cursive and faint. $5\frac{3}{8}'' \times 1\frac{7}{8}''$. Well-preserved.

N. iv. 114. Rectang. cov.-tablet, fragment. Seal cav. 1" sq. *Obv.* 1 l. Khar. coarse and distinct to L. p. of cav. Indications of an extra string-groove. *Rev.* blank. $3\frac{3}{4}'' \times 2\frac{1}{2}'' \times \frac{1}{2}''$. Wood soft.

N. iv. 115. Rectang. under-tablet. *Obv.* 6 ll. coarsely written Khar., much faded. *Rev.* blank. Stained with Sindūra (?). $8\frac{1}{4}'' \times 3\frac{1}{4}''$. Much stained and bleached by exposure.

N. iv. 116. Rectang. cov.-tablet (narrowing slightly to one end). Seal cav. $1\frac{1}{4}'' \times 1\frac{3}{16}''$. Frags. of string. *Obv.* Transversely above cav., 4 ll. Khar., below 7 ll.; faded, coarsely written. *Rev.* 7 ll. Khar., black, cursive. $7\frac{1}{8}'' \times 2\frac{1}{2}'' \times \frac{1}{2}''$. Well-preserved. Centre part of Rev. (heart-wood) very dark.

N. iv. 117. Wedge under-tablet. Hole near point. *Obv.* 3 ll. Khar., generally fairly distinct, but faded and cursive; initial form. *Rev.* 1 l. Khar. from sq. end. $8\frac{3}{4}'' \times 1\frac{7}{8}''$. Wood hard, but appears to be abraded by sand.

N. iv. 118. Stick-like tablet. Hole near one end; opposite end knotted and rough. *Obv.* 3 ll. Khar., very cursive; some numerals. *Rev.* blank. One edge notched (cf. N. iv. 36). $15\frac{1}{2}'' \times 1\frac{1}{2}''$. Well-preserved.

N. iv. 119. Wedge cov.-tablet. Hole near point. *Obv.* Seal cav. $\frac{3}{4}''$ from sq. end. 1 l. Khar. on both sides of seal cav. Usual char. at hole. Coarse writing. *Rev.* 2 ll. Khar., upright, black. $9\frac{3}{8}'' \times 1\frac{3}{4}''$. Wood well-preserved.

N. iv. 120. Double-wedge tablet, complete. Hole near point. Cov.-tablet. *Obv.* 2" from sq. end, seal cav. containing remains of clay and string. Sides of cav. are decorated by a notch cut in each interval between grooves. *Rev.* blank.

Under-tablet. *Obv.* 3 ll. Khar., very cursive, stained and faded. *Rev.* Traces of char. near sq. end. $9\frac{3}{4}'' \times 2\frac{3}{8}'' \times 1\frac{1}{8}''$. Well-preserved, but stained and sand-encrusted.

N. iv. 121. Wedge cov.-tablet. Hole near point, in which string. *Obv.* Seal cav. $1\frac{1}{4}''$ from sq. end. Blank. *Rev.* 4 ll. Khar., cursive, irregular and faded in parts. $8\frac{3}{4}'' \times 1\frac{1}{2}'' \times \frac{13}{16}''$. Well-preserved.

N. iv. 122. Rectang. cov.-tablet. Seal cav. (for two seals ?) $3\frac{3}{16}'' \times 1\frac{1}{4}''$. *Obv.* 2 ll. Khar., transversely; indistinct. *Rev.* blank. $6\frac{7}{8}'' \times 4\frac{3}{8}'' \times \frac{3}{8}''$. Well-preserved.

N. iv. 123. Stick-like tablet. Hole at one end. *Obv.* in columns consisting generally of 4 ll. Khar., arranged in groups of 3 and 2 columns divided by strokes. There appear to be nineteen columns in all. Numerals occur frequently. *Rev.* blank. $29\frac{1}{4}'' \times 1\frac{13}{16}'' \times \frac{1}{2}''$. Well-preserved.

N. iv. 124. Oblong tablet. *Obv.* 2 long columns (to read transversely) evidently written by two hands. 1 l. (longitudinally) from smaller end at top edge. Well written, but cursive and rather faded. *Rev.* blank. Much scored by a knife, as though used as a board on which to cut leather. $12\frac{1}{2}'' \times 3\frac{1}{4}''$. Wood well-preserved. See Pl. CII.

N. iv. 125. Oblong tablet, fragment. *Obv.* Traces of two columns, Khar. much deleted. *Rev.* Traces of 1 l. and a column. $3\frac{7}{8}'' \times 2''$. Soft, stained and encrusted with sand.

N. iv. 126. Tablet nearly square. Hole at lower L. p. corner. *Obv.* 3 ll. Khar., well written and quite clear, containing numerals. *Rev.* blank. $5\frac{3}{16}'' \times 4\frac{1}{8}''$. Well-preserved.

N. iv. 127. Rectang. under-tablet (cf. N. iv. 41). *Obv.* 8 ll. Khar., very cursive; nearly all legible. *Rev.* blank. $8\frac{1}{2}'' \times 3\frac{1}{8}''$. Wood hard, much stained.

N. iv. 128. Rectang. cov.-tablet. Seal cav. $1\frac{1}{4}''$ sq. *Obv.* 2 ll. Khar., transverse, at L. p. edge; coarse and cursive. *Rev.* Traces of Khar. deliberately effaced by chisel cuts. $6\frac{1}{2}'' \times 3\frac{1}{2}''$. Fairly preserved.

N. iv. 129. Oblong tablet. Hole in centre near edge. *Obv.* 2 columns 5 ll. each, Khar., very cursive, generally clear. *Rev.* 5 ll. Khar., cursive, black. Scores made by knife point as though from cutting leather (?). $9\frac{1}{4}'' \times 2\frac{1}{4}'' \times \frac{3}{16}''$. Well-preserved.

N. iv. 130. Rectang. cov.-tablet. Seal cav. $1\frac{7}{8}'' \times 1\frac{1}{4}''$. *Obv.* 3 ll. Khar., coarse, cursive, and black. *Rev.* 5 ll. do. $7'' \times 3\frac{1}{4}'' \times \frac{1}{8}''$. Fairly preserved, but cracked; sensitive to damp.

N. iv. 131. a, b. Rectang. cov.-tablet; 2 pieces. Seal cav. $1\frac{1}{2}'' \times 1\frac{7}{16}''$. *Obv.* 3 ll. Khar. transversely at L. p. edge. Coarse, black. *Rev.* 2 ll. Khar., cursive, well-written. $6\frac{1}{8}'' \times 2\frac{1}{8}'' \times \frac{1}{8}''$. Wood hard; much chipped at one end.

N. iv. 132. Oblong tablet, broken. Hole in centre of one end. *Obv.* Columns. At top 2 ll. (partly missing); below, 1st column of 4 ll.; 2nd column (or columns) of 3 ll. Khar. cursive, fairly clear. *Rev.* 4 ll. $8\frac{7}{8}'' \times 2\frac{3}{8}''$. Fairly preserved. Slightly sand-encrusted.

N. iv. 133. Oblong tablet, slightly tapering to one end. Notches on long edges for string, and therefore prob. under-tab., adapted from wedge tablet. *Obv.* 6 ll. Khar., cursive, black, generally very clear. *Rev.* blank. $7\frac{1}{2}'' \times 2\frac{1}{2}''$. Well-preserved.

N. iv. 134. Wedge cov.-tablet. Hole near point. *Obv.* 2" from sq. end, seal cav. 1 l. Khar. on both sides of cav. Usual char. at hole. *Rev.* 1 l. from sq. end containing numerals; very cursive. $14\frac{3}{4}'' \times 2\frac{3}{4}'' \times \frac{3}{4}''$. Well-preserved.

N. iv. 135. Wedge cov.-tablet. Hole near point. *Obv.* Seal cav. 1" from sq. end. Traces of Khar. on both sides of cav. Usual char. at hole. *Rev.* Faint traces of char., undecipherable. $8\frac{7}{8}'' \times 1\frac{7}{8}'' \times \frac{5}{8}''$. Wood fairly hard, but stained and abraded.

N. iv. 136. Wedge-shaped tablet. *Obv.* 7 ll. Khar., cursive, faint, but quite clear. *Rev.* 2 ll. similar, clear. Surface scored as though by the point of a knife. $16\frac{1}{4}'' \times 1\frac{11}{16}''$. Well-preserved.

N. iv. 137. Wedge cov.-tablet. Hole near point. *Obv.* Seal cav. 2" from sq. end. 1 l. Khar. on both sides of cav. Char. at hole. *Rev.* 1 l. very sq. Khar., rather coarse. $10\frac{5}{8}'' \times 1\frac{11}{16}'' \times \frac{7}{8}''$. Wood fairly preserved, but much stained.

N. iv. 138. Rectang. cov.-tablet. Seal cav. $2'' \times 1\frac{1}{4}''$. *Obv.* 3 ll. Khar., squarely written, rather coarse, to read transversely at L. p. edge. 4 ll. similar writing, but done with a fine pen, transversely, at R. p. edge. *Rev.* 8 ll. Khar., similar in character to Obv., coarsely written; very clear. $5\frac{1}{4}'' \times 3\frac{1}{4}''$. Wood well-preserved.

N. iv. 139. Rectang. under-tablet. *Obv.* 10 ll. Khar., cursive and rather indistinct in places. *Rev.* Blank. $9\frac{1}{4}'' \times 4''$. Well-preserved. See Pl. XCVI.

N. iv. 140. Rectang. cov.-tablet (belongs to N. iv. 110). Seal cav. 1" sq. *Obv.* 2 ll. Khar., very black, coarse, and square, transversely at L. p. edge. *Rev.* 6 ll. similar to Obv., to read longitudinally. $6\frac{1}{2}'' \times 2\frac{1}{4}'' \times \frac{3}{8}''$. Wood remarkably preserved.

N. iv. 141. Rectang. cov.-tablet (belongs to N. iv. 58). N. iv.
Seal cav. $1\frac{1}{4}'' \times 1\frac{1}{2}''$. *Obv.* 3 ll. Khar., transversely at L. p. edge. *Rev.* 4 ll. Khar., very cursive, quite clear. $5\frac{1}{4}'' \times 2\frac{1}{4}''$. Well-preserved.

N. iv. 142. Oblong tablet. Hole at one end, broken at other. *Obv.* 1 l. Khar., much faded. *Rev.* Column of 5 ll. Khar. near hole; traces of other columns; very faint. $13'' \times 1\frac{7}{8}''$. Wood fairly hard, much stained, very sensitive to damp.

N. iv. 143. Rectang. under-tablet, split at one end. *Obv.* 12 ll. very fine Khar., perfectly clear. *Rev.* blank. $9\frac{1}{4}'' \times 3\frac{3}{8}''$. Well-preserved.

N. iv. 144. Rectang. under-tablet, split at both ends. *Obv.* 13 ll. Khar., cursive, rather blotted; otherwise perfectly clear.. *Rev.* blank. $9\frac{5}{8}'' \times 4\frac{1}{2}''$. Well-preserved.

N. iv. 145. Rectang. cov.-tablet. Seal cav. $1\frac{1}{4}'' \times 1\frac{3}{16}''$. *Obv.* Faint traces of char. reading transversely. *Rev.* Traces of about 6 ll. Khar., very faded and cursive. $5\frac{1}{4}'' \times 2\frac{3}{4}''$. Wood well-preserved.

N. iv. 01. a. Piece of coarse red pottery, from a large vessel, quite plain; found in centre area. $5'' \times 4\frac{3}{4}''$.

b. Two pieces of coarse red pottery, quite plain. $2\frac{1}{2}'' \times 1\frac{7}{8}''$ and $3'' \times 2\frac{1}{4}''$.

DOCUMENTS, ETC., EXCAVATED AT N. v.

N. v. 1. Tablet, fragment in 3 pieces, about $30\frac{1}{2}''$ long, $2\frac{7}{8}''$ broad at widest, and about $\frac{3}{16}''$ thick. Broken at both ends. Rotten and cracked by exposure; warped. No writing visible upon either side.

N. v. 2. Rectang. cov.-tablet. *Obv.* Seal cav. $1\frac{3}{16}''$ sq. Frags. of string. Blank. *Rev.* blank. $10\frac{5}{8}'' \times 3\frac{1}{2}'' \times \frac{11}{16}''$. Soft, split, and bleached.

N. v. 3. Stick-like tablet, one side rough. *Obv.* 2 ll. Khar., separate items with figures; much faded. *Rev.* blank. $16\frac{3}{4}'' \times 2'' \times \frac{3}{8}''$. Much weather-worn and soft.

N. v. 4. Stick-like tablet. *Obv.* 3 ll. Khar., disconnected items with figures (?); very black. *Rev.* 1 l. Khar., part very black. $5\frac{7}{8}'' \times 1\frac{1}{4}''$. Much perished at one edge, and bleached at point.

N. v. 5. a, b. Tablet, gladius-shaped, in 2 pieces. Hole in centre of one end; groove down centre one side. No inscription. $9\frac{1}{4}'' \times 1\frac{1}{2}''$. Soft; coated with sand.

N. v. 6. Oblong tablet. Much perished in one half. *Obv.* (perished end to R. p.) six columns Khar. 1st and 3rd (from sq. end) 3 ll.; 2nd, 2 ll.; 4th, 4 ll.; 5th, 6 ll.; 6th, 3 ll., the bottom line only perfectly clear and very black; bottom line of 1st column also very black and distinct. (These two black ll. and one on the Rev. evidently written with better ink than the rest, and perhaps when the wood was in a better condition for preserving ink). Some of the ll. fragmentary. There appears to be indication of a 7th column, but the wood is so much N. v.
perished at that end of tab. that it is uncertain. *Rev.* from sq. end, 1 l. at top, extending over 2 columns, of 2 ll. each, lower l. of 1st being very black and sharp. To R. p. of these, 1 column of 3 ll., and another of 4 (?) ll., much deleted. Beyond this, surface of wood perished and peeling. Writing generally much faded. $31\frac{1}{2}'' \times 3\frac{1}{2}'' \times \frac{7}{16}''$. Portions of bark remaining on sides. Wood quite hard and well-preserved at sq. end; the other fairly hard, but surface perished, and broken away.

N. v. 6. a, b. Stick-like tablet, in 2 pieces. Cut into kind of handle at one end. *Obv.* Traces of Khar. characters. *Rev.* blank. $16\frac{1}{4}'' \times 1\frac{1}{2}''$. Perished and much discoloured.

N. v. 7. Wedge cov.-tablet. Point broken. *Obv.* Seal cav. 1" from sq. end. No inscription. Surface rotten and peeling. *Rev.* 3 ll. Khar., very cursive, blotted, very indistinct. $7\frac{1}{2}'' \times 1\frac{3}{4}'' \times \frac{5}{8}''$. Much perished.

N. v. 8. Wedge under-tablet. Broken at end and edges. *Obv.* Traces of 3 ll. Khar., deleted. *Rev.* Traces of Khar. char. at sq. end. $8\frac{1}{4}'' \times 1\frac{3}{8}''$. Perished, bleached, and split.

N. v. 9. Wedge-shaped tablet. No hole; broken at sq. end. *Obv.* 6 ll. Khar. in columns; deleted at sq. end. *Rev.* blank. $5\frac{1}{2}'' \times 1\frac{3}{8}''$. Much damaged by exposure; soft.

N. v. **N. v. 10. Oblong tablet,** obtuse point at one end; prob. hole near lower edge at point. *Obv.* 7 ll. Khar. in columns, very black, well written. 1st l. long and partly deleted. Much discoloured. *Rev.* blank. $14\frac{1}{2}'' \times 3\frac{7}{8}''$. Much perished by damp.

N. v. 12. Wedge-shaped tablet, roughly shaped. *Obv.* 5 ll. Khar. in columns; faded towards point. *Rev.* 3 ll. in columns, much faded. $7\frac{1}{8}'' \times 1\frac{3}{8}''$. Well-preserved but discoloured.

N. v. 13. Oblong tablet, much discoloured. *Obv.* Hole at R. p. top corner. Several ll. Khar. lost in dark stains, but appear to be in columns. *Rev.* 6 ll. of well written characters visible towards R. p. end; much faded and fragmentary. $15\frac{1}{2}'' \times 4'' \times \frac{3}{8}''$. Wood well-preserved.

N. v. 14. Club-shaped tablet, flat; hole at narrow end. *Obv.* 3 ll. Khar., in columns; indistinct. *Rev.* blank. $10\frac{3}{4}'' \times 2'' \times \frac{1}{4}''$. Much perished and discoloured.

N. v. 15. Rectang. cov.-tablet. Seal cav. $1'' \times 1\frac{1}{4}''$. *Obv.* At L. p. edge, transversely, frags. of ll. *Rev.* 5 ll. Khar., black, cursive; partially deleted. $5\frac{1}{2}'' \times 2'' \times \frac{1}{2}''$. Very soft; perished; discoloured.

N. v. 16. Oblong tablet: fragment. *Obv.* 2 ll. Khar., transversely. *Rev.* blank. $1\frac{7}{8}'' \times 1\frac{1}{4}''$. Much perished; split and soft.

N. v. 17. Oblong tablet, in two pieces. *Obv.* 2 ll. Khar., cursive, much deleted. *Rev.* blank; split and discoloured. $9\frac{3}{4}'' \times 3\frac{3}{8}'' \times \frac{5}{8}''$. Rotted and soft.

N. v. 17. a. Terra-cotta frag., well baked, coarse. $3\frac{1}{2}'' \times 2\frac{1}{2}''$.

b. Piece of coarse sacking. Attached, a small frag. of felt.

N. v. 18. Rectang. under-tablet, in several pieces. $\frac{3}{8}''$ at thickest. *Obv.* Apparently 10 ll. Khar., cursive, fine, but much deleted. *Rev.* About 9 ll.; very indistinct. $12'' \times 3\frac{1}{2}''$. Perished, soft, cracked, and discoloured.

N. v. 19. Stick-like tablet. *Obv.* 3 ll. Khar., fragmentary; very indistinct. *Rev.* Traces of 2 ll. Khar. $15\frac{1}{2}'' \times 1\frac{1}{4}'' \times \frac{1}{2}''$. Perished, cracked, and discoloured.

N. v. 20. Tablet, elongated oval. Hole at narrow end. *Obv.* 7 ll. Khar., in columns. Faded but legible. *Rev.* 6 ll. Khar., in columns; faded but legible. $10\frac{3}{4}'' \times 2\frac{3}{8}''$. Well-preserved; dark in colour.

OBJECTS EXCAVATED IN N. vi.

N. vi. **N. vi. 1. Wedge under-tablet.** Hole near point. *Obv.* 4 ll. Khar., very cursive and clear, with initial form. *Rev.* 1 line from sq. end. $9\frac{3}{8}'' \times 1\frac{7}{8}''$. Well-preserved.

N. vi. 2. Rectang. cov.-tablet (?). *Obv.* bleached. In centre of each side a notch as though to receive string. At lower L. p. corner a small hole, in which is frag. of string. *Rev.* 1 l. Khar., clear, fairly well written. $4\frac{5}{8}'' \times 1\frac{3}{4}'' \times \frac{3}{8}''$. Well-preserved, but bleached.

N. vi. 3. Rectang. cov.-tablet. *Obv.* Seal cav. $1\frac{1}{4}'' \times 1\frac{1}{4}''$. 2 ll. Khar., very faded, transversely above seal cav. *Rev.* 3 ll. Khar., cursive, faded. $4\frac{7}{8}'' \times 3'' \times \frac{5}{8}''$. Hard, well-preserved.

N. vi. 4. Wedge-shaped tablet. Apparently hole near point (broken). *Obv.* 3 ll. Khar., very cursive; consisting apparently of a number of disconnected items, separated by vert. down strokes. *Rev.* 3 ll., much faded, coarse, and very cursive. $7\frac{1}{2}'' \times 1\frac{3}{8}''$. Rather perished; very susceptible to damp.

N. vi. 5. Wedge cov.-tablet. Hole near point. *Obv.* Seal cav. $1\frac{3}{8}''$ from sq. end. Much encrusted with sand. No inscription visible. *Rev.* 2 (?) ll. Khar., illegible. $9'' \times 1\frac{1}{2}'' \times \frac{5}{8}''$. Injured by damp and sand.

N. vi. 6. Rectang. under-tablet, raised ends, $\frac{1}{2}''$ and $\frac{7}{8}''$. *Obv.* 4 ll. Khar., well written, cursive. *Rev.* blank. $6\frac{1}{8}'' \times 2\frac{3}{8}''$. Well-preserved, but writing rather faded.

N. vi. 7. Rectang. under-tablet, raised ends, $\frac{5}{8}''$. *Obv.* 9 ll. fine Khar., faded in parts. *Rev.* blank. $7\frac{1}{2}'' \times 3''$. Soft, but form well-preserved.

N. vi. 8. Wedge cov.-tablet, with point end cut off; appears to have been adapted for use in manner of rectang. cov.-tabs. *Obv.* Seal cav. $1\frac{1}{2}''$ from sq. end. Clay remains; no device. *Rev.* 3 ll. Khar., much faded, very cursive. $6'' \times 2\frac{1}{8}'' \times \frac{5}{8}''$. Well-preserved.

N. vi. 9. Rectang. cov.-tablet, fragment. Both sides blank. Broken at two edges. $4'' \times 4''$. Very soft.

N. vi. 10. Wedge under-tablet, hole near point. *Obv.* 4 ll. Khar. cursive, 'condensed,' with initial form. *Rev.* 1 l. from sq. end. $10\frac{5}{8}'' \times 1\frac{7}{8}''$. Well-preserved. Hard.

N. vi. 11. Rectang. under-tablet, fragment. Raised end, $1\frac{1}{4}''$. *Obv.* 2 ll. Khar., very cursive and coarse. *Rev.* blank. $7\frac{7}{8}'' \times 2\frac{1}{2}''$. Wood well-preserved, but broken.

N. vi. 12. Oblong tablet. *Obv.* 3 ll. Khar., columnar; very black and clear. *Rev.* blank. $5\frac{1}{4}'' \times \frac{7}{8}''$. Well-preserved. Traces of Sindūra (?) at one end. See Pl. CII.

N. vi. 13. Wedge cov.-tablet; hole near point; roughly made. *Obv.* Seal cav. $2\frac{5}{8}''$ from sq. end. 1 l. Khar. on both sides of seal cav. Very cursive and coarsely written. Usual char. at hole. *Rev.* 1 l. Khar., very cursive. $13\frac{3}{4}'' \times 1\frac{5}{8}'' \times \frac{5}{8}''$. Well-preserved.

N. vi. 14. Wedge cov.-tablet; hole near point. *Obv.* $2\frac{1}{4}''$ from sq. end. Seal cav. (small). Between sq. end and seal cav. 1 l. Khar. Few characters on opposite of cav. *Rev.* 1 l. Khar. in which figures seem to occur. Well and clearly written. $14\frac{5}{8}'' \times 2\frac{1}{4}'' \times \frac{1}{2}''$. Well-preserved. Pink stain at point end.

N. vi. 15. Wedge cov.-tablet, hole near point. *Obv.* Seal cav. 2¼" from sq. end. 1 l. Khar. near sq. end; usual char. near point. *Rev.* blank. 14"×2⅜"×½". Well-preserved. Pink stain at sq. end.

N. vii. 1. a, b. Rectang. under-tablet, in 2 pieces. Thick ends (approx.) 1" wide. *Obv.* 5 ll. Khar., cursive, but well written and all legible. *Rev.* blank. 7¼"×3". Well-preserved.

N. vii. 2. Stick, round, ⅜" diameter one end and tapering to rather less than 5/16" the other. At thicker end an ivory ferule 1" long, tapering slightly. Just below ferule a small hole bored completely through stick, in which remains a fragment of string. At thinner end there has evidently been a knob or ferule of some kind, but now only a broken chule (of the wood) remains. Wood dark brown and heavy, and probably sandal. Very hard. Length 11⅝".

N. vii. 3. Portion of rug, of the regular Indian Darrie type, woven in transverse stripes and bands, and decorated with simple geometrical patterns. Warp buff in colour, and made of a more closely-twisted yarn than that used for the weft, which generally spreads enough to conceal the warp. The cross stripes are buff, blue, green, and red, and are of varying width. On broad blue band small Svastika-like pattern in red and buff. On buff band same pattern in red and blue, and red and green. On red band the pattern is in buff and blue, and buff and green. Where the patterns occur, the weft threads of the ground are arrested, and the colour required locally introduced as in tapestry. As these parts were more laborious to make, double threads are used right across the fabric on the lines where the patterns occur, thus saving time. The coarser texture of the resultant surface is clearly visible in the reproduction. 19"×15½". See Pl. LXXV. Other pieces of same rug, found in N. vii. and hall of N. III. measure 20"×12", 16"×8", 21"×6½", 9"×9", 26"×9", 3"×5"; some small fragments aggregating to 19"×10".

N. vii. 4. Wooden chair. Carved frame of lower part of chair, four legged. The legs are held together by four broad carved panels, which are tenoned into them, and dowelled. In addition to these there is a rough rail at each end between the end pairs of legs, also dowelled. At top of each leg stands a tenon (chule) upon which probably knobs or raised sides were fixed. Carving rough, nothing centred or measured, and there is much repetition in the elements composing designs. The most frequent feature is a four-petalled flower, the form of which closely resembles the large purple clematis. Generally sepals are shown between petals. Frequently, superposed upon first petals, a second whorl of smaller ones, and sometimes a third. The legs are 23" high; width at front 4", side 2½".

Inner edges of legs are straight, and they are straight in their upper part at all edges down to about 12" from foot. This lower portion slightly hollow at front and side, curving outwards again at foot. Ornament on upper part 1st leg front: between two fillets a bead band commencing along top edge L. p. to R., then at right angles downwards; then R. p. to L., down, and L. p. to R., thus forming two contiguous squares minus one side in each case. In each of these squares, a double-petalled and sepalled flower. Below these an eight-petalled lotus in circle. From below lotus running centrally down to foot and taking curve of leg, an overlapping double-leaf pattern, formed by a series of inverted V-shaped cuts. Side: fillet and bead ornament, but in zigzag line. In each triangle so formed half of double-petalled flower. 2nd leg same; 3rd leg side, same as that of other two. Front, double zigzag forming diamond and half-diamond shapes in which whole and half flowers. Below, double-leaf pattern as first leg. 4th leg front, same as front of 3rd; side, same as front of 1st and 2nd, but in place of lotus another square is formed similar to upper two, and containing similar flower. N. vi.

End panels: 7¼"×14" (approx.). Two converging bead-bands starting at lower corners and meeting at top edge at 90°. In half-square so formed, a half-clematis having triple petals, and a clumsy extra one to the two half-petals. At upper corners quarter flowers double-petalled in one case, triple-petalled in other. Front panel: divided into 3 oblong divisions by 2 approximately vertical bead-bands. The two side-divisions, narrower than high, contain double-petalled flower. The centre a conventionalized pomegranate (?) with leaves, buds, and bead ornaments. Back panel missing. Width of chair 26" (approx.), depth 16", height 23". Well-preserved; feet splitting. See Pl. LXVIII.

N. viii. 1. Oblong tablet, fragment. *Obv.* 5 ll. Khar., appar. in 3 vertical columns (?). Each item terminates with a numeral. Much faded but probably decipherable. *Rev.* blank. 4½"×1⅞". Wood well-preserved, but all edges broken.

N. viii. 2. Rectang. cov.-tablet. *Obv.* 3 ll. Khar., very cursive, seal cav. 1½"×1¼". *Rev.* 3 ll. Khar., cursive. 5⅛"×2". Wood very hard; well-preserved.

N. viii. 3. Donkey's saddle-tree of mulberry wood. Of a somewhat horseshoe shape, pierced with 6 rectangular holes probably for binding material of seat. Wood hard and well-preserved. Size across points of bow 12". Height 7". Width at centre 2"; at ends ¾". Thickness about 1". See Pl. IX.

N. viii. 4. Bow of tamarisk wood. Made from small branch or sapling, halved to near centre, where the round section is left to form a hand-piece. Ends tapered and shaped to take string. Broken at one end. Resembles the pellet-bow used by bird-scarers in India to-day. Length 3' 6". See Pl. IX.

N. viii. 5. Shaft of spear (?) broken in 2 pieces. At upper end a broad leather ferule is laced on by means of a leather thong. Each end shows a broken section. Length 4' 3½", thickness 1⅜". See Pl. IX.

N. viii. **N. viii. 6. Walking-stick of apple-wood (?),** split at one end. Hard and well-preserved. Length 2′ 9″, thickness 1″.

N. ix. 1. Wedge under-tablet. Hole near point. *Obv.* 3 ll. Khar., well-written, black, and clear; initial form. *Rev.* Few char. at sq. end. $14\frac{3}{8}$″ × $2\frac{3}{8}$″. Well-preserved. See Pl. C.

N. ix. 2. Oblong tablet, clumsy hole at one end. *Obv.* 5 ll. Khar. with numerals (?) in column at L. p. end; 2 ll. at middle. *Rev.* blank. $8\frac{3}{8}$″ × $2\frac{1}{4}$″. Wood well-preserved.

N. ix. 3. Oblong tablet. *Obv.* 7 ll. Khar., very cursive; quite clear. *Rev.* blank. 6″ × $2\frac{1}{4}$″. Well-preserved.

OBJECTS FOUND IN N. iv.

N. x. **N. x. 1. Fragments of woollen carpet,** 4 or 5 small pieces as shown in sketch.

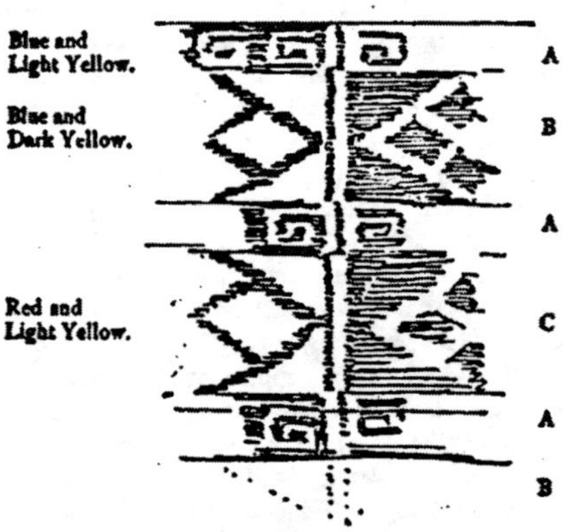

A. Fret pattern in narrow bands at intervals dividing the larger bands from each other. These narrow bands are always blue on light yellow, and light yellow on blue, the ground and pattern counterchanging as shown in sketch. *B* and *C* occur alternately. The warp of the fabric is of undyed wool, and varies in colour from dark brown to dirty white.

N. x. 2. Rectang. under-tablet, fragment; broken diagonally. Upper part missing. *Obv.* 5 ll. Khar., very cursive, part deleted. *Rev.* blank. $5\frac{1}{4}$″ × $1\frac{7}{8}$″. Wood well-preserved.

N. x. 2. b. Tablet, fragment, warped and twisted. *Obv.* 3 ll. Khar., fragmentary. *Rev.* blank. $5\frac{1}{2}$″ × 1″ (approx.). Very spongy.

N. x. 2. c. Takhtī-shaped tablet; trapezium handle, with hole. *Obv.* 3 ll. Khar., well written, part faded. *Rev.* 1 l., very clear. $6\frac{1}{2}$″ × $1\frac{1}{2}$″. Well-preserved. See Pl. XCVII.

N. x. 3. Takhtī-shaped tablet; trapezium handle, with hole. *Obv.* (handle in L. hand). 1 long l. at top edge, well-written but rather coarse. Short l. at sq. end. Short l. in R. p. half. 1 long l. at lower edge, but reverse way to that at top, so that to read, hold handle in R. hand. 2 columns of five ll. each, with interval between columns; read transversely, from sq. end downwards. All fairly clear. Characters round hole. *Rev.* (handle in R. hand). 5 columns (beginning at handle) having 6 or 7 ll. in each, and reading longitudinally. 1 l. from about centre towards sq. end. 1 long column extending from sq. end to about centre, reading transversely. $13\frac{1}{4}$″ × $2\frac{1}{2}$″. Well-preserved. See Pl. CI.

N. x. 4. Oblong tablet, rounded at small end, and pierced, with small twig sharpened at one end, serving as pin. *Obv.* 2 columns Khar. and 1 short l. *Rev.* blank. 6″ × $2\frac{3}{8}$″. Pin $4\frac{7}{16}$″ long. Well-preserved. See Pl. CIII.

N. x. 5+6. Rectang. double-tablet, complete. Cov.-tablet. *Obv.* Seal cav. $1\frac{1}{2}$″ × $1\frac{3}{8}$″. 2 ll. Khar. transversely at top. *Rev.* 7 ll. Khar., very cursive; much encrusted with sand. $4\frac{3}{8}$″ × $2\frac{5}{8}$″.

Under-tablet. *Obv.* 7 ll. Khar., black, clear, cursive. *Rev.* blank. $6\frac{1}{4}$″ × $2\frac{3}{8}$″. Well-preserved.

N. x. 7. Rectang. cov.-tablet, much broken. *Obv.* 1 l. Khar. transversely at L. p. edge. 2 ll. transv. at R. p. *Rev.* 5 ll., cursive, well-written. Many numerals (?). $6\frac{1}{8}$″ × $2\frac{3}{8}$″. Much perished and damaged.

N. x. 8. Rectang. cov.-tablet. Seal cav. $1\frac{3}{8}$″ × $1\frac{3}{16}$″. *Obv.* 2 ll. Khar., fine, black, cursive. Remains of seal clay and string. *Rev.* blank. $6\frac{5}{8}$″ × $2\frac{7}{8}$″. Wood well-preserved.

N. x. 9. a, b. Oblong tablets. Handle at each end. Blank. $6\frac{1}{2}$″ × 1″.

N. x. 9. c. Stick-like tablet; at one end small handle in which hole; hole also at opposite end. Blank. 10″ × $\frac{1}{2}$″. Well-preserved; unused (cf. above, p. 360). See Pl. CV.

N. x. 01. Spindle, made of a light wood; broken at thicker end. Length $8\frac{3}{4}$″. At thin end $\frac{3}{16}$″ thick, the point being round. It thickens to $\frac{5}{8}$″ at 3″ from thin end, and thins again to rather less than $\frac{1}{2}$″ where it is broken. There is an × mark on thick part. Hard and well-preserved.

N. x. 02. Spindle, similar in all respects to N. x. 01, but not broken at end. A hole is pierced through end farthest from thickened portion.

N. x. 03, 04. Pens. Two roughly-cut wooden styles, consisting of twigs sharpened at one end and at the other rough. Length $5\frac{3}{8}$″ and $5\frac{5}{8}$″, respectively. Hard. See Pl. CV.

N. x. 05. Pens. Two twigs sharpened at one end. The wood is very hard, and colour of bark a dark rich brown. Length about 9″. See Pl. CV.

N. x. 06. Eating sticks. Three wooden sticks, in length respectively $7\frac{3}{8}''$, $8\frac{7}{16}''$, and $10\frac{1}{4}''$, round and slightly tapering towards one end. Thicker end about $\frac{1}{2}''$. Well-preserved.

N. x. 07. Broom, made of grass, similar to D. ii. 011 (see description above, p. 294), but not so well preserved. Very brittle. See Pl. LXXIII.

N. x. 08. Wing of bird. Portion of ulna and radius, and digital part of wing, with 7 principal quills, 8 secondary (over), and a few under quills of bird about size of small hawk. Length $8\frac{1}{2}''$, width $1\frac{1}{8}''$. Dry, but fairly perfect.

N. xi. 1. Stick-like tablet. *Obv.* 3 ll. Khar., black, very cursive, coarse. *Rev.* 2 ll. do. $10\frac{1}{4}'' \times 1\frac{5}{16}''$. Well-preserved; warped.

N. xii. 1. Stick-like tablet. Hole at one end. *Obv.* 7 columns, 3 ll. each, except last, which has 2 ll. Khar., very cursive; numerals occur. *Rev.* Faint Khar. writing, arranged irregularly in columns of 3 ll. $14\frac{1}{2}'' \times 1\frac{1}{2}''$. Wood well-preserved.

N. xii. 2. Guitar (Rabāb) upper portion only. Consists of a small part of the shoulder where broken from the body, the whole of the slender and tapering neck, and the head. The section of neck is similar to that of a modern violin, and the head, hollow, is pierced laterally for four pegs, three of which still remain, one being a temporary one. Round these are still wound the remains of strings, one at least of which appears to be gut. The frets conducting strings to neck are deeply grooved. The neck is split and has an ancient binding of cord round it. A small piece of veneer remains near the shoulder, and a tiny peg or dowel shows where and how another piece was affixed. (Compare this with the instrument in painted panel D. ii. 010, above, p. 294). $17\frac{1}{4}'' \times 1\frac{3}{8}''$. See Pl. LXXIII.

N. xii. 3. Legs of wooden chairs. Three similar to each other are very Persepolitan in design. The upper part is a fine conventional lion (or tiger) head, supported on a very arched neck, the chest thrown forward and touched by protruding tongue. The lower portion is fashioned into a lion's hind leg, with solid hoof-like paw. Small wings are shown at the shoulders. The crinis forms a flat fillet-like band reaching from the wings to the forehead. Above the head is an upright projection cut into a sort of hook shape at top and back, and in its lower part immediately above the head is the mortise with dowel holes, showing how the horizontal rails of the seat were held. The hook-like terminal was evidently for the purpose of securing a cushion by means of the well-known *palang-doris* or bed ties. The fact of these hooks being painted all round, including the ends, shows them to be complete in their present state. V-shaped markings in black paint up the neck (pink) represent skin or feather markings. The leg, from the wing downwards was dark brown or black. $17'' \times 3\frac{3}{4}'' \times 1\frac{1}{8}''$. N. x.

Two legs, evidently a pair, show the leg and hoof of a horse, and above, the breast, neck, and head of a human being; in one case a female, as indicated by the contours and decorative line work on the breasts; in the other, a male, as shown by the moustache. The transition from leg to figure is effected by the shaft of the leg spreading into a half-open lotus, out of which the figures rise. There are small wings indicated. Each head has a Pagrī surmounted by a hooked projection as above described. Colours lac red, dark red, and black. $13\frac{1}{2}'' \times 2\frac{1}{2}'' \times 1\frac{3}{4}''$.

N. xiii. 1. Rectang. cov.-tablet. *Obv.* Seal cav. $1'' \times 1\frac{1}{4}''$. Transversely 1 l. Khar., very indistinct; interval in middle. *Rev.* 9 ll. Khar., well-written; indistinct in places. $6'' \times 3\frac{3}{8}''$. Very soft; broken and decayed at transverse edges.

N. xiv. 1. Wedge under-tablet. Hole near point. *Obv.* 3 ll. Khar., with initial form.; black, cursive, much covered by sand. *Rev.* 1 short l. from sq. end. Rudely incised cross near centre. $8\frac{1}{2}'' \times 1\frac{7}{8}''$. Wood well-preserved, but much encrusted with sand, and very susceptible to damp.

N. xiv. 01. Oblong tablet, fragment; obtuse point at one end, with hole; broken at other. Blank. $4\frac{1}{2}'' \times \frac{13}{16}''$. Well-preserved.

N. xiv. 02. Oblong tablet, fragment; handle at one end, with hole; broken at other. Blank. $4\frac{1}{2}'' \times \frac{13}{16}''$. Well-preserved.

OBJECTS EXCAVATED IN N. v.

N. xv. 2. Wedge cov.-tablet; hole at point; seal cav. $1\frac{1}{2}''$ from sq. end. *Obv.* 1 l. Khar., between sq. end and cav., very cursive. Usual char. at hole. *Rev.* 3 ll. Khar., very cursive and blotted. $9\frac{3}{8}'' \times 1\frac{3}{4}'' \times \frac{3}{4}''$. Wood well-preserved.

N. xv. 3. Wedge under-tablet; hole at point. *Obv.* 5 ll. Khar., fairly well written, generally clear; stained. *Rev.* Traces of characters, very faint, part undecipherable. $9\frac{1}{2}'' \times 1\frac{3}{4}''$. Well-preserved.

N. xv. 4. Oblong tablet; hole at one corner. *Obv.* 3 ll. Khar., very cursive, clear. *Rev.* 1 l. Khar., cursive, clear. $5\frac{1}{2}'' \times 1\frac{1}{2}''$. Well-preserved.

N. xv. 5. Rectang. cov.-tablet; seal cav. $2\frac{1}{8}'' \times 1\frac{3}{8}''$, in N. xv.
which remains of string. *Obv.* 4 ll. Khar., cursive, well written, transversely at L. p. edge. Below seal 1 l. cursive, clear, and a few disconnected char. *Rev.* 6 ll. Khar., cursive, black. $6\frac{1}{8}'' \times 2\frac{3}{4}'' \times \frac{1}{2}''$. Wood soft, dark, and chipped at ends.

N. xv. 6. Wedge cov.-tablet; hole at point. *Obv.* Seal cav. $1\frac{1}{4}''$ from sq. end. No inscription visible. *Rev.* Traces of several ll. Khar., much faded. $10\frac{1}{4}'' \times 2\frac{5}{16}'' \times \frac{3}{4}''$. Much discoloured and encrusted with sand. Wood soft.

N. xv. 6. a. Wedge tablet, fragment of, in 2 pieces;

N. xv. hole at point. *Obv.* 3 (?) ll. Khar., well written; faded. *Rev.* blank. $3\frac{1}{4}'' \times 1\frac{3}{8}''$. Perished.

N. xv. 7. Wedge cov.-tablet; hole at point. *Obv.* Seal cav. $2\frac{5}{16}''$ from sq. end. 1 l. Khar. between sq. end and cav. Usual char. at point. *Rev.* 1 l. Khar., cursive. $8'' \times 1\frac{3}{8}''$. Well-preserved.

N. xv. 8. Wedge cov.-tablet, frags. of string in cav. *Obv.* Seal cav. 2″ from sq. end. 1 l. Khar. between sq. end and cav. *Rev.* blank. $12\frac{5}{8}'' \times 2\frac{7}{8}''$. Well-preserved; much encrusted with sand.

N. xv. 10 + 86 + 190. Rectang. tablet, complete, found in 3 pieces. Cov.-tablet (N. xv. 190). *Obv.* Seal cavity $1\frac{1}{8}'' \times 2\frac{1}{2}'' \times \frac{3}{8}''$. At L. p. edge 2 ll. Khar. transversely, finely written. Traces of 2 ll. on R. p. edge. *Rev.* 5 ll. Khar., well-written. $7\frac{1}{2}'' \times 4'' \times \frac{3}{8}''$. Soft, discoloured, and sand-encrusted.

Under-tablet broken into 2 pieces (N. xv. 10 + 86). *Obv.* 13 ll. Khar., with several numerals, cursive, black, neatly written. *Rev.* Transversely, 8 ll., and after interval 8 more ll., Khar., very cursive and badly written by different hands. $9\frac{1}{2}'' \times 4''$. Hard and well-preserved.

N. xv. 11. Wedge tablet; hole near point. *Obv.* 3 ll. Khar., cursive, partly deleted. *Rev.* blank. $7\frac{3}{16}'' \times 1\frac{1}{2}''$. Much perished.

N. xv. 12. a, b. Wedge tablet, complete; hole at point. Cov.-tablet. *Obv.* Seal cav. $2\frac{3}{8}''$ from sq. end; remains of seal clay and string. 1 l. Khar. between sq. end and cavity, very cursive. Usual char. about 1″ from hole. *Rev.* 3 ll. Khar., very cursive.

Under-tablet. *Obv.* 4 ll. Khar., very cursive, distinct, with initial formula. *Rev.* blank. $9\frac{1}{2}'' \times 1\frac{5}{8}'' \times 1\frac{1}{8}''$ (both together). Well-preserved.

N. xv. 13. Wedge cov.-tablet, broken at sq. end; hole at point. *Obv.* Seal 2″ from sq. end; seal clay and string remain. Khar. char. at square end; fragmentary; char. traceable on opposite side of cavity. *Rev.* 2 ll. Khar., cursive, faint, 2nd l. partly deleted. $9\frac{1}{8}'' \times 2\frac{1}{16}'' \times \frac{3}{4}''$. Well-preserved.

N. xv. 14 + 27. Oblong tablet, tapering to one end, in two fragments. *Obv.* Traces of about 5 ll. very cursive Khar., mostly illegible. *Rev.* Traces of several ll. Khar., cursive, fragmentary; one l. partly legible. $9\frac{1}{4}'' \times 3\frac{3}{4}'' \times \frac{1}{4}''$. Surface perished by damp in N. xv. 14; one side hard in N. xv. 27.

N. xv. 15. Wedge under-tablet; hole at point. *Obv.* 4 ll. Khar., cursive, part faded; with initial formula. *Rev.* Few char. at sq. end. $9\frac{1}{8}'' \times 1\frac{13}{16}''$. Well-preserved.

N. xv. 16. Wedge under-tablet (belongs to N. xv. 26); hole near point. *Obv.* 3 ll. Khar., very cursive, faded; with initial formula. *Rev.* Few char. from square end. $9\frac{1}{8}'' \times 1\frac{7}{8}''$. Well-preserved.

N. xv. 17. Wedge cov.-tablet; hole near point; square end chipped. *Obv.* Seal cav. $1\frac{7}{8}''$ from sq. end. Few char. Khar. at sq. end. *Rev.* 1 l. Khar., cursive, fine, faded, towards sq. end. $10'' \times 2\frac{1}{16}'' \times \frac{1}{2}''$. Well-preserved, but appears to be sand-abraded.

N. xv. 17. a. Wedge under-tablet; hole near point. *Obv.* 4 ll. Khar., very indistinct and in parts quite faded; probably initial formula. *Rev.* Traces of 2 or more ll. from sq. end. Near centre incised Svastika, about $1'' \times \frac{3}{4}''$. $8\frac{1}{2}'' \times 2\frac{1}{4}''$. Bleached and abraded.

N. xv. 18. Wedge cov.-tablet; broken at point. *Obv.* $1\frac{1}{2}''$ from sq. end, seal cav., 1 l. Khar. at sq. end, cursive, fine, faded. *Rev.* 1 l. Khar. from sq. end, similar. $7\frac{3}{8}'' \times 1\frac{3}{4}'' \times \frac{1}{2}''$. Perished and very soft.

N. xv. 19. Wedge cov.-tablet; fragment. *Obv.* Seal cav. $1\frac{3}{4}''$ from sq. end, contains clay and string. 1 l. Khar. at sq. end, cursive, fine, faded. *Rev.* blank. $5\frac{11}{16}'' \times 2\frac{1}{2}'' \times \frac{1}{2}''$. Soft.

N. xv. 20. Rectang. cov.-tablet; fragment. *Obv.* Seal cav. $1'' \times 1\frac{3}{8}''$. Traces of Khar. char. transversely at L. p. edge. *Rev.* 5 ll. Khar. (1st has upper portion broken away), fine, cursive, faint. Lower edge and corner burnt. $4'' \times 1\frac{7}{8}'' \times \frac{1}{2}''$. Rather soft.

N. xv. 20. a. Oblong tablet; hole at one end. *Obv.* Traces of Khar. characters, faded. Most of surface bleached. *Rev.* Traces of 3 ll. Khar. at one end; cursive, faint. $7\frac{3}{8}'' \times 1\frac{5}{16}''$. Hard, but bleached and abraded.

N. xv. 22. Wedge cov.-tablet, fragment; broken at point. *Obv.* Seal cav. $1\frac{5}{16}''$ from sq. end. A few traces of characters. *Rev.* Traces of 1 l. Khar. (?); surface split and perished. $5\frac{1}{4}'' \times 1\frac{3}{4}'' \times \frac{7}{16}''$. Perished, soft, and cracked.

N. xv. 23 + 150. Oblong tablet, broken in two pieces. *Obv.* 3 ll. Khar., well written, faint in N. xv. 23, and almost entirely faded in N. xv. 150. *Rev.* blank. $5\frac{9}{16}'' \times 2\frac{1}{2}''$. Soft.

N. xv. 24. Wedge tablet, complete. Cov.-tablet remains intact, clay seal and strings in cavity $1\frac{5}{8}''$ from sq. end. Seal (see Pl. LXXI), $1\frac{1}{4}'' \times 1\frac{1}{8}''$, shows, deeply impressed from intaglio, profile figure of Athene Promachos standing to L. p. On sq. setting remains of Khar. legend in relief. 1 l. Khar., black at sq. end. *Rev.* 3 ll. Khar., cursive, nearly all very clear and fresh. Surface badly smoothed and knotty.

Under-tablet. *Obv.* 3 ll. Khar., cursive, but clear, with initial formula. *Rev.* A few char. near sq. end. Wood hard and well-preserved. Rather sand-encrusted outside. $9\frac{1}{8}'' \times 2\frac{1}{8}'' \times \frac{5}{8}''$ (covering-tab.), $\frac{1}{4}''$ (under-tab.). See Pl. IC.

N. xv. 26. Wedge cov.-tablet (belongs to N. xv. 16); hole at point. *Obv.* Seal cav. $1\frac{1}{2}''$ from sq. end. Clay and string remain in cav. 1 l. Khar. at sq. end, very cursive. Traces of char. at opposite side of cav. Usual char. about 1″ from point. *Rev.* 1 l. Khar. from sq. end very cursive and blotted. $10'' \times 1\frac{7}{8}'' \times \frac{3}{4}''$. Perfectly preserved.

N. xv. 28. Oblong tablet, obtusely pointed at one end; hole near pointed end. *Obv.* 1 l. + 2 ll. + 1 l. Khar., very faint; part faded. *Rev.* 2 ll. in several separate items; faint. $5\frac{1}{4}'' \times 1\frac{5}{8}''$. Bleached and split at one end.

N. xv. 29. Document on leather, in 2 pieces; fragmentary. *Obv.* blank. *Rev.* Date in 1 short l. Khar., with numerals, cursive, clear. $6\frac{1}{4}'' \times 2\frac{1}{2}''$.

N. xv. 30. Wedge under- (?) tablet; hole near point. *Obv.* Bleached and surface perished; no writing visible. *Rev.* Traces of char. at sq. end. $13\frac{5}{8}'' \times 2\frac{1}{4}''$. Hard, but surface perished in part; bleached.

N. xv. 31. Wedge cov.-tablet; broken at point. *Obv.* Seal cav. $1\frac{5}{8}''$ from sq. end; string in cavity. Bleached, perished, and cracked all over. *Rev.* 2 ll. Khar., cursive, fine, clear. Part towards point bleached and deleted. $7\frac{1}{4}'' \times 1\frac{7}{8}'' \times \frac{9}{16}''$. Portion of Rev. seems to have been screened from damp, sun, and sand, which have warped, bleached, and abraded the Obv. and part of Rev.

N. xv. 32. Wedge cov.-tablet; hole at point. *Obv.* Seal cav. $1\frac{11}{16}''$ from sq. end. All surface bleached and peeling. *Rev.* 1 l. Khar., very cursive. Surface part bleached and peeled. $8\frac{3}{4}'' \times 1\frac{7}{8}'' \times \frac{1}{2}''$. Hard.

N. xv. 33. Oblong tablet, tapering to both ends. *Obv.* 2 columns of 3 ll. each, and 1 l. in addition; very cursive Khar., distinct. *Rev.* 2 columns of 3 ll. each + 1 l. + 1 l. Khar., very cursive. $7\frac{3}{8}'' \times 1\frac{1}{2}''$. Well-preserved. Broken at one end.

N. xv. 34. Slip of wood, Chinese char. on Obv.; burnt at one end. $3\frac{3}{8}'' \times \frac{1}{2}''$. Well-preserved. See Pl. CXII.

N. xv. 35. Rectang. cov.-tablet. *Obv.* Seal cav. $\frac{3}{4}'' \times \frac{5}{8}''$, with one string-groove only. Blank. *Rev.* blank. $3\frac{3}{8}'' \times 1\frac{5}{16}'' \times \frac{7}{32}''$. Perfectly preserved. See Plate CV.

N. xv. 36 + 170. Wedge under-tablet broken in 2 pieces (belongs to N. xv. 119); hole near point. *Obv.* 4 ll. Khar., very cursive and indistinct. *Rev.* 1 l. Khar., faded. $8\frac{3}{4}'' \times 2\frac{1}{4}''$. Surface soft.

N. xv. 37. Slip of wood, Chinese char. on Obv.; fragment. $5\frac{5}{8}'' \times \frac{7}{16}''$. Well-preserved. See Plate CXIII.

N. xv. 38. Rectang. cov.-tablet; cracked at one end. *Obv.* Seal cav. with seal intact (see Pl. LXXII), representing rude head in profile to L. p. Mane-like hair, high cheekbones, moustache. Above seal 2 ll. Khar., black, cursive. Below seal and reverse way up, 1 l. Khar. *Rev.* 4 ll. Khar., very black and clear. $4'' \times 2'' \times \frac{5}{8}''$. Well-preserved.

N. xv. 39. Rectang. under-tablet, fragment. *Obv.* 4 ll. Khar., cursive, clear, lowest line partly broken away. *Rev.* blank. $6\frac{1}{2}'' \times 1\frac{5}{8}''$. Wood well-preserved.

N. xv. 40. Wedge cov.-tablet; broken at point. *Obv.* $2\frac{1}{8}''$ from sq. end, seal cav., 1 l. Khar. at sq. end; well written and very clear. Few char. at opposite side of cav. Usual char. towards pointed end. Stained with pink. *Rev.* blank. $10'' \times 1\frac{7}{8}'' \times \frac{1}{8}''$. Wood perfectly preserved. N. xv.

N. xv. 41. Rectang. cov.-tablet. *Obv.* Seal cav. $1'' \times 1\frac{1}{16}''$. Transversely at top, 2 ll. Khar., very cursive and clear. *Rev.* 7 ll. Khar., very cursive, fairly clear. $4\frac{1}{4}'' \times 3\frac{3}{16}'' \times \frac{3}{8}''$. Perfectly preserved.

N. xv. 42. Document on leather; complete. *Obv.* 5 ll. Khar., very cursive, very black. *Rev.* 3 items in Khar. $10'' \times 2''$ (average). Much discoloured and rotted.

N. xv. 43. Document on leather, fragment. *Obv.* 8 ll. Khar., cursive, black, part deleted. *Rev.* blank. $2\frac{1}{4}'' \times 3\frac{5}{8}''$. Rather buckled and perished.

N. xv. 45. Rectang. under-tablet; unusually thick. *Obv.* 5 ll. well-written Khar., clear in places, but in some quite deleted. Surface bleached and cracked at one corner. *Rev.* bleached, cracked, and partially peeled. $5\frac{3}{4}'' \times 1\frac{7}{8}'' \times \frac{3}{4}''$. Perished at Rev. from exposure.

N. xv. 46. Wedge cov.-tablet; fragment. *Obv.* Seal cav. $1\frac{1}{16}''$ from sq. end. String in original arrangement remains. Traces of Khar. char. at sq. end. *Rev.* Traces of numerals along lower edge. $6\frac{1}{2}'' \times 2\frac{1}{4}'' \times \frac{3}{8}''$. Bleached, perished, and cracked. Portions of bark round cav. show method of crossing string.

N. xv. 47. Wedge under-tablet (?); fragment. *Obv.* 3 ll. Khar., cursive, fairly clear, with initial formula. *Rev.* 1 l. from sq. end. $6'' \times 1\frac{1}{2}''$. Fairly preserved.

N. xv. 48. Wedge under-tablet (?); fragment. Blank both sides. $9'' \times 1\frac{7}{8}''$. Sand-encrusted, soft and much broken.

N. xv. 50 + 200. Rectang. under-tablet; fragment in two pieces. *Obv.* Seal cav. 1" sq. Upper portion missing. Lower transversely 6 ll. Khar., cursive, fine, black. *Rev.* 4 ll. Khar. (imperfect), very cursive, black and clear. $3\frac{1}{4}'' \times 3\frac{1}{4}'' \times \frac{7}{16}''$. Well-preserved.

N. xv. 51. Wedge cov.-tablet; fragment. *Obv.* 1 l. Khar. between sq. end and seal cav.; fine and black. *Rev.* 2 ll. Khar., cursive, fairly distinct. $2\frac{7}{8}'' \times 1\frac{5}{8}'' \times \frac{11}{16}''$. Discoloured and susceptible to damp.

N. xv. 52. Wedge under-tablet; fragment. *Obv.* 4 ll. Khar. (incomplete), cursive, black and clear; with initial formula. *Rev.* 2 ll. from sq. end; probably written by different hands. $4'' \times 2\frac{1}{2}''$. Wood perfectly preserved.

N. xv. 53. Slip of wood, Chinese char. on Obv. $4\frac{7}{8}'' \times \frac{5}{8}''$. Well-preserved. See Pl. CXIII.

N. xv. 54. Oblong tablet; hole at one end. *Obv.* 3 ll. Khar., cursive, faded but fairly clear. *Rev.* 2 ll. Khar., cursive, faded and part deleted. $4\frac{1}{2}'' \times 1\frac{3}{8}''$. Wood perfectly preserved.

N. xv. 55 + 81. Oblong tablet, fragment in 2 pieces. *Obv.* 5 ll. Khar., cursive, part deleted. *Rev.* blank. N. xv. 55. $9'' \times 1\frac{1}{2}''$; wood perfectly preserved. N. xv. 81. $6\frac{1}{4}'' \times \frac{1}{2}''$. Warped and bleached.

N. xv. **N. xv. 56. Rectang. cov.-tablet**; fragment. *Obv.* Traces of 3 ll. Khar. below seal cav.; very cursive. *Rev.* 1 l. Khar., incomplete. 3⅜″ × 1$\frac{3}{16}$″ × ⅜″. Well-preserved; wood dark.

N. xv. 57. Oblong tablet; broken at one end, hole at other. *Obv.* 4 ll. Khar., much faded. *Rev.* Traces of 2 or more ll. undecipherable. 5¼″ × 1¼″. Soft.

N. xv. 58. Oblong tablet; hole at each end. Faint traces of characters on one side only. 6⅞″ × 1″. Soft; sand-abraded.

N. xv. 59. Slip of wood, blank. 7″ × ½″.

N. xv. 59. b. Slip of wood, Chinese char. on Obv. 4¼″ × $\frac{7}{16}$″. Well-preserved.

N. xv. 60. Rectang. tablet, probably seal cover. *Obv.* Notched with 3 notches on each of two opposite edges as though for string. *Rev.* Roughly cut circular depression, about $\frac{3}{32}$″ deep. 2$\frac{3}{16}$″ × 1⅞″ × ½″. Rather soft. See Pl. CV.

N. xv. 61 + 62. Slip of wood, Chinese char. on Obv.; broken in 2 pieces. 5½″ × $\frac{9}{16}$″. Well-preserved. See Pl. CXIII.

N. xv. 63. Rectang. under-tablet; fragment. *Obv.* 5 ll. Khar., fine, cursive, and fairly clear. *Rev.* blank. 8⅞″ × 1½″. Wood well-preserved.

N. xv. 64. Wedge cov.-tablet; fragment. *Obv.* Seal cav. 1⅝″ from sq. end. 1 l. Khar. between sq. end and cav. Portions of bark remaining. *Rev.* 2 ll. Khar., cursive, well-written, black. 7⅜″ × 2½″ × ⅝″. Well-preserved. Surface perished on Obv.

N. xv. 65. Rectang. cov.-tablet; much broken. *Obv.* Seal cav. 1⅛″ square. At L. p. edge 1 l. Khar., very cursive and faint, transversely; at R. p. traces of Khar. char. *Rev.* Traces of 6 ll. Khar., very cursive and faint. 5½″ × 2¾″ × $\frac{13}{16}$″. Rather soft; much split, and surface perished.

N. xv. 66. Double-wedge tablet, complete; hole at point. Cov.-tablet. *Obv.* Seal cav. 1¼″ from sq. end; remains of clay. 1 l. Khar. between sq. end and cav., cursive; usual char. at hole. *Rev.* Traces of char. in centre; sand-encrusted.

Under-tablet. *Obv.* 3 ll. Khar., cursive, well-written, generally very black, with initial form. Sand-encrusted. *Rev.* Few characters at sq. end. 9½″ × 1⅞″ × ⅝″. Hard, but chipped and insect-eaten, especially Rev. of under-tablet.

N. xv. 67. Wedge cov.-tablet; broken at point. *Obv.* Seal cav. ⅞″ from sq. end. Khar. char. on each side of cav. Usual char. near hole. *Rev.* Traces of 2 ll. Khar., very cursive and faint. 8½″ × 1⅛″ × ⅜″. Fairly preserved. Found 1½ ft. above floor.

N. xv. 68. Wedge-shaped tablet; broken and curled at point. *Obv.* Column of 7 ll. and beyond that a single l. Khar., cursive, clearly written. *Rev.* blank; bleached and peeling. 14″ × 2¾″. Fairly hard except on surface. Found 1½ ft. above floor.

N. xv. 69. Slip of wood, Chinese char. on Obv.; fragment. 3″ × ½″. Well-preserved. Found 1½ ft. above floor. See Pl. CXIII.

N. xv. 71. Double-wedge tablet, complete; connected by string passing through holes at pointed ends with attached pendent clay seal. Seal (see Pl. LXXII) is in hard red clay, and contains the ends of the short string loop. The device is, in a circle, a somewhat coarsely cut male head to L. p. with bushy hair extending to nape of neck, moustache, heavy chin, straight and rather prominent nose, full eye (cf. seals B. D. 001. d, g, and h, Pl. IL). *Obv.* of covering-tab. Usual seal cavity 2″ from sq. end measures 1¾″ × 1¼″, and contains some of original string. A few Khar. characters at square end, and others on opposite side of cavity. Usual formula at hole. *Rev.* 1 l. Khar., very cursive but well-preserved.

Under-tablet. *Obv.* 4 ll. Khar. very coarsely written; a smudge at square end. Well-preserved. *Rev.* A few Khar. characters at sq. end. × scratched in wood near centre. 13¾″ × 2⅝″ × ½″ (covering-tab.), ⅜″ (under-tab.). See Pl. C. Found 3 ft. above floor.

N. xv. 72. Slip of wood. Chinese char. on Obv.; fragment. 1⅜″ × ½″. Well-preserved. See Pl. CXIV. Found 1 ft. above floor.

N. xv. 73. Slip of wood. Chinese char. on *Obv.*; fragment. 5″ × ½″. Well-preserved, writing faded towards broken end. See Pl. CXII. Found 1 ft. above floor.

N. xv. 74. Wooden seal case. Has deep string-grooves at sides of cav. There is a hole through bottom of cav. which appears to have been made by burning. No inscription. 1¾″ × 1¼″ × 1″. See Pl. CV.

N. xv. 76 + 181. Tablet formed of small bough, flattened on two sides by cutting; broken in 2 pieces. *Obv.* 2 ll. Khar., arranged in separate items, mostly terminating in numerals. *Rev.* 4 items Khar., in 2 pairs; clear, cursive. 18″ × 1″ (approximately). Perfectly preserved. See Pl. CII.

N. xv. 77. Wedge cov.-tablet; fragment; point end missing. *Obv.* Seal cav. 1⅜″ from sq. end. Few Khar. char. from sq. end. *Rev.* 1 l. Khar. (incompl.), cursive, faint. 4½″ × 1⅝″ × ⅝″. Perished and warped. Surface bleached and soft.

N. xv. 78. Slip of wood, Chinese char. on Obv.; fragment. 3½″ × ½″. Well-preserved. See Pl. CXII.

N. xv. 79. Wedge under-tablet; hole near point. *Obv.* 4 ll. Khar., very cursive, black. *Rev.* 1 l. Khar., from sq. end, very cursive and incomplete. 9$\frac{5}{16}$″ × 1$\frac{13}{16}$″. Hard, but stained and much encrusted with sand. Found 1 ft. above floor.

N. xv. 80. Oblong tablet; fragment. *Obv.* Traces of coarsely written Khar. char., very faint. *Rev.* Traces of 3 ll. Khar., very cursive, fine, and faded. 3¾″ × 1″. Soft; much encrusted with sand. Found 1 ft. above floor.

N. xv. 80. a. Tablet; fragment. *Obv.* 1 l. Khar. (incomplete), very cursive. *Rev.* blank. $1\frac{3}{4}'' \times \frac{7}{16}''$. Hard.

N. xv. 82. Slip of wood, Chinese char. on Obv.; fragment. $2\frac{1}{2}'' \times \frac{9}{16}''$. Soft. See Pl. CXII.

N. xv. 82. a. Slip of wood, Chinese char. on Obv.; fragment. $3\frac{1}{8}'' \times \frac{1}{2}''$. Soft, sand-encrusted. See Pl. CXIV.

N. xv. 83. Tablet; fragment. *Obv.* Traces of 2 ll. Khar., cursive, black, part obscured by sand. *Rev.* blank. $2\frac{1}{8}'' \times \frac{1}{2}''$. Hard, split, and abraded.

N. xv. 84. Wedge cov.-tablet; hole near point. *Obv.* Seal cav. $1\frac{1}{8}''$ from sq. end. 1 l. very finely written Khar. between sq. end and cav. Usual char. at hole. *Rev.* blank. $8\frac{9}{16}'' \times 2'' \times \frac{3}{4}''$. Well-preserved. Wood extremely hard.

N. xv. 85. Slip of wood, Chinese char. on Obv.; fragment. $3\frac{1}{4}'' \times \frac{5}{8}''$. Nibbled by rodent at one end. See Pl. CXIV.

N. xv. 87+308. Rectang. under-tablet; broken in 2 pieces. *Obv.* 7 ll. Khar., very cursive, part deleted by sand. *Rev.* blank. $8\frac{3}{8}'' \times 3''$. Hard, stained, and sand-encrusted.

N. xv. 88. Document on leather; complete but slightly damaged. *Obv.* 11 ll. Khar., very clear and black; one short l. at bottom edge. *Rev.* Traces of char. $9\frac{1}{4}'' \times 4\frac{5}{8}''$. Well-preserved, but insect-eaten in places. See Pl. XCII.

N. xv. 89. Wedge cov.-tablet; hole near point. *Obv.* Seal cav. $1\frac{7}{8}''$ from sq. end. 1 l. Khar., very cursive from sq. end; usual char. near hole. *Rev.* 1 l. Khar., very cursive, black, and badly written. $8\frac{1}{4}'' \times 1\frac{1}{2}'' \times \frac{7}{16}''$. Well-preserved; wood very hard.

N. xv. 90. Wedge under- (?) tablet; fragment. *Obv.* Traces of very cursive Khar. char. *Rev.* blank. $2\frac{7}{8}'' \times \frac{7}{8}''$. Charred at one end.

N. xv. 91. Wedge under-tablet; hole near point. *Obv.* 4 ll. Khar., very cursive, black, nearly all quite legible; initial form. *Rev.* 1 short l. Khar. from sq. end. $9\frac{3}{16}'' \times 2\frac{5}{16}''$. Hard, well-preserved.

N. xv. 92. Wedge cov.-tablet; hole near point. *Obv.* Seal cav. $1\frac{7}{8}''$ from sq. end. 1 l. Khar. from sq. end, very cursive; usual char. near hole. *Rev.* blank. $9\frac{1}{8}'' \times 1\frac{7}{8}'' \times \frac{7}{8}''$. Wood hard and well-preserved.

N. xv. 92. a. Takhtī-shaped tablet; hole at handle. *Obv.* Apparently 9 items in two columns Khar., very cursive. *Rev.* 5 items similar to Obverse. $9'' \times 2\frac{1}{4}''$. Very hard.

N. xv. 93. a, b. Slip of wood, Chinese char., complete but broken in 2 pieces. $9\frac{1}{4}'' \times \frac{5}{8}''$. Perfectly preserved. See Pl. CXIII.

N. xv. 95. Rectang. cov.-tablet. *Obv.* Seal cav. $\frac{7}{8}''$ sq.; 2 ll. Khar., transversely at L. p. edge, very faint. *Rev.* 7 ll. Khar., very cursive and faint. $4\frac{3}{8}'' \times 2\frac{7}{8}'' \times \frac{5}{8}''$. Wood hard, but susceptible to damp.

N. xv. 96+325. Wedge under-tablet; broken in 2 pieces. *Obv.* 5 ll. (incomplete) Khar., cursive, neatly written; initial form. *Rev.* 1 l., cursive, coarse, from L. p. end. $9\frac{7}{8}'' \times 1\frac{1}{2}''$. Wood very hard. N. xv.

N. xv. 97. Wedge cov.-tablet; hole near point. *Obv.* Seal cav. $1\frac{13}{16}''$ from sq. end. 1 l. Khar. from sq. end, very cursive, black. Usual char. at hole. *Rev.* blank. $10\frac{1}{8}'' \times 1\frac{7}{8}'' \times \frac{1}{2}''$. Wood very hard and well-preserved. Found $3\frac{1}{2}$ ft. above floor.

N. xv. 98. Wedge cov.-tablet; hole (broken) near point. *Obv.* Seal cav. in which remains of seal (probably Pallas Athene) and string. Between cav. and sq. end 1 l. Khar. Very cursive and indistinct. Char., usually at hole, are in this case about $1\frac{1}{2}''$ away. *Rev.* blank. $10\frac{3}{8}'' \times 1\frac{7}{8}'' \times \frac{5}{8}''$. Wood hard and well-preserved. Stained and sand-encrusted. Found $3\frac{1}{2}$ ft. above floor.

N. xv. 99. Lath-like tablet; pointed roughly one end, broken at other. *Obv.* 2 ll. Khar., very cursive, much faded and fragmentary. *Rev.* blank. $7\frac{5}{8}'' \times 1\frac{1}{8}''$. Wood hard. Encrusted with sand. Found $3\frac{1}{2}$ ft. above floor.

N. xv. 100. Slip of wood, Chinese char.; fragment. $3\frac{1}{2}'' \times \frac{9}{16}''$. Split. See Pl. CXII. Found $3\frac{1}{2}$ ft. above floor.

N. xv. 101. Document on leather; fragment. *Obv.* 7 ll. Khar., cursive, black, incomplete. *Rev.* blank. $6'' \times 2\frac{3}{8}''$. Leather fairly preserved, but torn. Found $3\frac{1}{2}$ ft. above floor. See Pl. XCI.

N. xv. 101. a. Slip of wood, Chinese char.; fragment. $3\frac{3}{4}'' \times \frac{9}{16}''$. Well-preserved. Found $3\frac{1}{2}$ ft. above floor. See Pl. CXIV.

N. xv. 102. Wedge cov.-tablet; hole near point. *Obv.* Seal cav. $1\frac{5}{8}''$ from sq. end. 1 l. Khar., from sq. end, very cursive. Usual char. at hole. *Rev.* blank. $10'' \times 2'' \times \frac{5}{8}''$. Wood hard, well-preserved, sand-encrusted.

N. xv. 103. Rectang. cov.-tablet; fragment. *Obv.* At L. p. edge 2 ll. Khar., on R. p. edge 3 ll., all transversely, very cursive, faded. *Rev.* 3 ll. Khar., coarse, cursive, and faded. $4\frac{3}{8}'' \times 1\frac{1}{8}'' \times \frac{3}{8}''$. Well-preserved. Wood seems lighter both in colour and weight than the majority of tablets.

N. xv. 105. Wedge under-tablet; hole near point. *Obv.* 3 ll. Khar., cursive, mostly clear; initial form. Stained by wet in places. Heart wood dark. *Rev.* Few char. at sq. end. $8\frac{1}{4}'' \times 1\frac{3}{8}''$. Very hard, well-preserved.

N. xv. 106. Wooden pen or style of tamarisk, split at point. Horn knob, $1\frac{3}{8}''$ long, at other end, crenellated, fitted with polished conical top which is intended to serve as burnisher. $6\frac{5}{8}'' \times \frac{3}{4}''$. See Pl. CV.

N. xv. 107. Wedge cov.-tablet; hole near point; fragment. *Obv.* 3 char. Khar., $2\frac{1}{2}''$ from point. Usual char. at hole. *Rev.* End of 1 l. Khar., very cursive, quite clear. $4\frac{3}{16}'' \times 1\frac{11}{16}''$. Wood hard and well-preserved.

N. xv. **N. xv. 108 + 113. Double-wedge tablet,** cut short and sq.-ended; written in usual manner of rectang. tablet. Cov.-tablet. *Obv.* Seal cav. $1\frac{1}{4}''$ from sq. end. 3 ll. Khar. in space between sq. end and cav. transversely, cursive. *Rev.* 5 ll. Khar., very cursive, fine, but the ink has run in some places.

Under-tablet. *Obv.* 5 ll. Khar., cursive, closely written, blurred in places. *Rev.* Few char. at broad end. $5\frac{1}{2}'' \times 2\frac{1}{4}''$. Hard and well-preserved.

N. xv. 109. Slip of wood, Chinese char.; fragment. $6'' \times \frac{1}{2}''$. Well-preserved. See Pl. CXIII.

N. xv. 110. Document on leather; fragment. *Obv.* 4 ll. (3 incomplete) Khar., cursive, black. There seem to be traces of writing erased by insects towards lower part. *Rev.* 3 words Khar. $4\frac{7}{8}'' \times 4\frac{1}{2}''$. Well-preserved. Found 1 ft. above floor.

N. xv. 111. Slip of wood, intended for Chinese char., fragment; blank. $3\frac{1}{2}'' \times \frac{1}{2}''$. Well-preserved. Found 1 ft. above floor.

N. xv. 111. a. Rectang. under-tablet; fragment. *Obv.* In centre depression 2 col. of 4 ll. each Khar., very cursive, black and clear. On R. p. thick end, transversely, 1 l. Khar. *Rev.* 4 col. 1st and 4th four ll. each, 2nd five ll., 3rd two ll. One l. transversely through 4th col. Cursive. $7\frac{1}{8}'' \times 1\frac{3}{4}''$. Wood hard and well-preserved.

N. xv. 112. Document on leather. *Obv.* 5 ll. Khar., very clear and black. A short l. at lower edge. *Rev.* 2 fragmentary ll. $9\frac{1}{4}'' \times 5''$. In good condition.

N. xv. 114. Document on leather, fragment. *Obv.* 6 ll. Khar., cursive, black, well-written. *Rev.* blank. $6\frac{1}{8}'' \times 2\frac{3}{8}''$. Stained and partly rotted.

N. xv. 114. a. Wedge cov.-tablet; 2 holes near point. *Obv.* Seal cav. $1\frac{3}{8}''$ from sq. end. 1 l. Khar. from sq. end, very cursive. *Rev.* blank. $8'' \times 2'' \times \frac{3}{4}''$. Wood hard, encrusted with sand.

N. xv. 115. Wedge cov.-tablet; hole near point. *Obv.* Seal cav. $1\frac{7}{8}''$ from sq. end. 1 short l. Khar. at sq. end; usual char. at point. *Rev.* blank. $10\frac{1}{2}'' \times 1\frac{3}{4}'' \times \frac{7}{8}''$. Wood hard and well-preserved; sand-encrusted.

N. xv. 116. Slip of wood, Chinese char.; fragment. $2\frac{1}{4}'' \times \frac{7}{16}''$. Well-preserved. See Pl. CXIII.

N. xv. 117. Slip of wood, Chinese char.; fragment. $5\frac{1}{4}'' \times \frac{7}{16}''$. Fairly preserved. See Pl. CXIII.

N. xv. 118. Wedge under-tablet; hole near point; broken and imperfect. *Obv.* 4 ll. Khar., small and cursive; faint in places. Beginning of first 2 ll. broken away. *Rev.* blank. $8\frac{1}{2}'' \times 1\frac{5}{8}''$. Hard and well-preserved.

N. xv. 119. Wedge cov.-tablet (belongs to N. xv. 36 + 170); hole near point. *Obv.* Seal cav. $1\frac{1}{4}''$ from sq. end. 1 l. Khar. between sq. end and cav., cursive, fine, usual char. at hole. *Rev.* 1 l. Khar. in centre, well-written and black. $8\frac{15}{16}'' \times 1\frac{15}{16}'' \times \frac{5}{8}''$. Wood extremely hard; centre shows dark heart wood.

N. xv. 120. Lath-like tablet; fragment; obtuse point at one end. *Obv.* 2 ll. Khar. ending in numerals; cursive. *Rev.* blank. $1\frac{3}{4}'' \times \frac{3}{4}''$. Hard, well-preserved.

N. xv. 121. Wedge cov.-tablet; hole at point. *Obv.* Seal cav. $1\frac{5}{8}''$ from sq. end. 1 l. Khar. at sq. end. Usual char. at hole. Between cav. and point two crosses scratched with sharp instrument. *Rev.* 4 ll. Khar., very cursive, and much faded. $9\frac{1}{4}'' \times 1\frac{7}{8}'' \times \frac{5}{8}''$. Wood hard and well-preserved. Found $1\frac{1}{2}$ ft. above floor.

N. xv. 122. Lath-like tablet; obtuse point at one end. *Obv.* 2 ll. Khar., cursive, black and clear. *Rev.* 3 ll. Khar., cursive, black and clear, with numerals. $7\frac{1}{8}'' \times \frac{15}{16}''$. Heart-wood dark; well-preserved. Found $1\frac{1}{2}$ ft. above floor.

N. xv. 122. a. Label-shaped tablet; hole at pointed end. *Obv.* 2 ll. Khar., cursive, very faint; sand-encrusted. *Rev.* 3 ll. Khar., very indistinct. $4'' \times \frac{3}{4}''$. Hard and well-preserved.

N. xv. 123. Slip of wood, Chinese char.; fragment. $3\frac{1}{2}'' \times \frac{5}{8}''$. Well-preserved. See Pl. CXII.

N. xv. 124. Wedge tablet; fragment; probably hole at point, broken. *Obv.* Few Khar. char. *Rev.* blank. $2\frac{3}{4}'' \times \frac{1}{2}''$. Sand-encrusted and abraded.

N. xv. 125 + 127. Slip of wood, Chinese char.; fragment in 2 pieces. $5'' \times \frac{7}{16}''$. Fairly hard. See Pl. CXIV.

N. xv. 126. Oblong tablet; fragment. *Obv.* 2 ll. Khar., very indistinct. *Rev.* blank. $2\frac{3}{8}'' \times \frac{3}{4}''$. Hard.

N. xv. 128. Rectang. under-tablet; fragment. *Obv.* Traces of Khar. char. *Rev.* blank. $4\frac{1}{8}'' \times \frac{1}{2}''$.

N. xv. 129. Oblong tablet; fragment. *Obv.* 2 ll. Khar. containing numerals. *Rev.* blank. $3\frac{1}{2}'' \times \frac{7}{8}''$.

N. xv. 130. Oblong tablet; fragment. *Obv.* 2 ll. Khar., very cursive. 1st l. divided into three by two intervals. *Rev.* blank. $8\frac{1}{2}'' \times \frac{7}{8}''$. Hard, well-preserved.

N. xv. 131. Rectang. under-tablet. *Obv.* 7 ll. Khar., very coarse and cursive, but clear. *Rev.* blank. $8\frac{3}{8}'' \times 3\frac{7}{8}''$. Hard and well-preserved.

N. xv. 132. Wedge cov.-tablet; hole at point. Piece of cotton rug attached to clay in seal cav. *Obv.* Seal cav. $1\frac{3}{8}'' \times 1\frac{1}{4}''$. Faint traces of Khar. char. at sq. end and at hole. *Rev.* Traces of 2 ll. Khar., very cursive and covered with clay and sand. $8'' \times 1\frac{3}{4}'' \times \frac{1}{2}''$. Rather soft on surface.

N. xv. 133. Wood carving, miniature grotesque lion or Chinese dog. Statant, regardant, tail reflexed over back and very soft. Mouth slightly open, showing tongue. Each pair of legs (fore and hind) made into one solid piece. Carving very rough, and inartistic. Legs broken. Length $1\frac{3}{8}''$, height $1\frac{3}{4}''$. See Pl. LXX.

N. xv. 133. a. Red clay seal contained in wooden receptacle. Device, within a circle of beads or cable, an eagle flying with outstretched wings and talons; tail broad and square at end. Workmanship inferior. The case is a roughly squared piece of wood, $2''$ long, $1\frac{5}{8}''$ wide, $1''$ thick;

sides and angles rounded, in which a cavity about $1''$ sq. has been made, with 3 coarse saw-cuts passing across it for string grooves. Portions of string remain beneath the seal. Probably attached to some document or other consignment; cf. N. xv. 74. See Pl. LXXII.

N. xv. 133. b. Clay seal, grey, roughly circular, having fragments of woollen string embedded in it. Device, in low relief, within a circle, a gryphon, rampant, regardant, with bifurcate reflexed tail. In general proportion it more resembles a hippogryph, but tail and feet seem to be those of the gryphon. Neck very long and slender, and long sweep of line from back of head to end of belly bold and good. The one visible wing is well placed, springing accurately from shoulder. Head broken away, only end of beak being visible. Clay extremely fine and dense; perhaps mixed with a wax or mucilage to bind it. Diameter $1\frac{3}{8}''$. (Cf. N. 006.) See Pl. LXXII.

N. xv. 134. Rectang. under-tablet; two long cracks. *Obv.* 7 ll. Khar., very cursive, clear. *Rev.* blank. $8\frac{1}{4}'' \times 3''$. Hard.

N. xv. 135. Wedge under-tablet; hole at point. *Obv.* 4 ll. Khar., generally illegible because of incrustations of sand, clay, and straw; initial form. *Rev.* 1 l. Khar. from sq. end. Condition similar to Obv. $9\frac{1}{4}'' \times 2\frac{1}{4}''$. Hard.

N. xv. 136. Wedge under-tablet; hole at point. *Obv.* 3 ll. Khar., very finely written; initial form. *Rev.* 1 l. from sq. end, very cursive. $11'' \times 1\frac{7}{8}''$. Hard; well-preserved.

N. xv. 137. Double-wedge tablet, complete. Seal (see Pl. LXXI), in good condition, shows Pallas Athene, deep sunk, with elliptical mouldings surrounding. Cov.-tablet. *Obv.* $2\frac{3}{16}''$ from sq. end, seal cav. 1 short l. Khar. at sq. end. Usual char. at hole. *Rev.* blank.

Under-tablet. *Obv.* 5 ll. Khar., very clear and black. *Rev.* 1 l. Khar. from sq. end. 4 ll. in different hand, very scratchy. $11'' \times 2\frac{1}{8}''$. Perfectly preserved. See Pl. XCVIII.

N. xv. 138. Rectang. cov.-tablet; chipped at corners. *Obv.* Seal cav. $2\frac{1}{2}'' \times 1\frac{1}{8}''$. 1 l. Khar. at L. p. edge, transversely, incomplete. *Rev.* 3 ll. Khar., very cursive, black, faded in parts, incomplete. $6\frac{3}{4}'' \times 3\frac{3}{8}'' \times \frac{11}{16}''$. Hard, discoloured, encrusted with sand and refuse.

N. xv. 139. Slip of wood, Chinese char.; fragment. $3\frac{5}{8}'' \times \frac{7}{16}''$. Well-preserved. See Pl. CXIII.

N. xv. 140. Double-wedge tablet, complete; hole at point. Cov.-tablet. *Obv.* Seal cav. $1\frac{1}{2}''$ from sq. end. 1 l. Khar. at sq. end. Usual char. at hole. *Rev.* 2 ll. Khar., very cursive, black and clear.

Under-tablet. *Obv.* 3 ll. Khar., very cursive, clear, except near sq. end and centre. Initial form. *Rev.* 1 l. Khar. from sq. end. $9'' \times 1\frac{3}{4}'' \times \frac{7}{8}''$. Hard, well-preserved, wood dark.

N. xv. 141. Oblong tablet, round-pointed at one end, obtuse-pointed at other. *Obv.* 3 ll. Khar., very cursive and faded. *Rev.* blank. $5\frac{1}{2}'' \times \frac{7}{8}''$. Well-preserved.

N. xv. 142+147+148. Tablet; fragment in 3 pieces N. xv.
Obv. 2 ll. Khar., cursive, black. *Rev.* blank. $5\frac{3}{8}'' \times 1\frac{1}{16}''$ Wood well-preserved.

N. xv. 143. Wedge cov.-tablet; hole at point. *Obv.* Seal cav. $1\frac{1}{2}''$ from sq. end. 1 l. Khar., very cursive and coarse, on both sides of cav. Usual char. at hole. *Rev.* 1 l. Khar., very cursive and black. $9'' \times 2'' \times \frac{9}{16}''$. Well-preserved. Most of the string remains, showing method of fastening.

N. xv. 145. Slip of wood, Chinese char.; fragment. $6\frac{7}{8}'' \times \frac{1}{2}''$. Perfectly preserved. See Pl. CXIV.

N. xv. 146. Oblong tablet. *Obv.* 3 ll. Khar., cursive, faded. *Rev.* blank. $4\frac{9}{16}'' \times 1\frac{1}{8}''$. Hard, well-preserved, but sand-encrusted.

N. xv. 149. Document on leather; fragment. *Obv.* 8 ll. Khar., cursive, black, incomplete. *Rev.* 1 l. Khar., faint. $3\frac{7}{8}'' \times 3\frac{1}{8}''$. Much stained and rotted at edge. Found 3 ft. above floor.

N. xv. 151. Rectang. double tablet, complete, retaining original fastening and unopened. *Obv.* Seals (2) much damaged. That to R. p. appears to contain more than one figure. To L. p. apparently a head to L. p., but doubtful. Seal cav. $2\frac{3}{8}'' \times 1\frac{1}{16}''$. To L. p. 3 ll. Khar., very cursive, transversely. To R. p. 2 ll. Khar., transversely. *Rev.* blank. String complete, broken in one place only. Tablets damaged at one end, and under-tablet much cracked. $7\frac{1}{8}'' \times 3\frac{1}{16}''$. Susceptible to damp.

N. xv. 152. Slip of wood, Chinese char. on Obv.; broken at upper end. $6\frac{1}{2}'' \times \frac{1}{2}''$. Perfectly preserved. See Pl. CXIII.

N. xv. 153. Rectang. cov.-tablet; broken at one corner and along one edge. *Obv.* Seal cav. $1\frac{1}{4}'' \times 1\frac{1}{8}''$. At L. p. end, transversely, 1 l. Khar., very cursive; also end of a 2nd l. (?). *Rev.* blank. $6\frac{13}{16}'' \times 2\frac{9}{16}'' \times \frac{7}{16}''$. Hard.

N. xv. 154. Rectang. cov.-tablet; seal cav. $1\frac{1}{8}''$ sq. *Obv.* 4 ll. Khar. transv. at L. p. end; 3 ll. Khar. transv. reverse way at R. p. end, very cursive. *Rev.* 9 ll. Khar., very cursive; blotted but clear. $7\frac{1}{2}'' \times 4\frac{1}{8}''$. Well-preserved. See Pl. XCVII.

N. xv. 154. a+157. Rectang. under-tablet; fragment in 3 pieces. *Obv.* 6 ll. Khar., very imperfect. *Rev.* blank. $5\frac{3}{16}'' \times 2\frac{13}{16}''$. Perished, much abraded. Found 1 ft. above floor.

N. xv. 155. Rectang. double tablet, complete, found intact with fastening and seals; opened at British Museum.

Cov.-tablet. *Obv.* Seal cav. $2\frac{1}{4}'' \times 1\frac{1}{4}''$, with two seals (see Pl. LXXII), upper, circular, broken, showing hair of male head tied in topknot. Lower, elliptical, showing female bust, turned to L. p., holding in raised R. p. hand flower or mirror (?), wearing tiara, and surrounded by elliptical mouldings. On L. p. edge 2 ll. Khar. transversely, R. pr. blank. *Rev.* 1 short l. Khar., black.

N. xv. Under-tablet. *Obv.* 7 ll. Khar. clear; black ink. *Rev.* blank. Wood well-preserved. $6\frac{1}{4}'' \times 3''$ (nearly) $\times \frac{3}{8}''$. Found 3 ft. above floor. See Pl. XCIV.

N. xv. 156. Rectang. cov.-tablet. *Obv.* Seal cav. $1\frac{3}{8}'' \times 1\frac{1}{4}''$. R. p. edge notched tooth-like. 1 l. Khar. and 2 disconnected characters, transversely, very cursive, and faint. *Rev.* 1 l. Khar., very cursive, faint, part covered by encrusted sand. $4\frac{1}{4}'' \times 1\frac{3}{4}'' \times \frac{1}{2}''$. Hard, well-preserved. Found 3 ft. above floor.

N. xv. 158. Rectang. under-tablet. *Obv.* 9 ll. Khar., very coarse and cursive. *Rev.* blank. $7\frac{3}{4}'' \times 3\frac{1}{2}''$. Soft, sand-encrusted and stained. Found 3 ft. above floor.

N. xv. 159. Rectang. cov.-tablet. *Obv.* Seal cav. $2'' \times 1\frac{1}{4}''$. Clay remains, probably of a double seal. 4 string-grooves. At top traces of 2 ll. Khar. Sand-encrusted. *Rev.* Traces of 5 ll. Khar. Surface much abraded, and sand-encrusted. $5\frac{1}{2}'' \times 3\frac{1}{2}'' \times \frac{7}{8}''$. Soft. Found 3 ft. above floor.

N. xv. 160. Rectang. cov.-tablet. *Obv.* Seal cav. $1\frac{3}{8}'' \times 3\frac{7}{16}''$. On R. p. edge 3 ll. Khar. transverse, very faint and cursive. L. p. 1 l. Khar. transverse, very faint and cursive. *Rev.* 5 ll. Khar., very cursive, distinct. $8\frac{3}{4}'' \times 4\frac{1}{4}'' \times \frac{11}{16}''$. Rather soft. Slightly sand-encrusted. See Pl. XCVI. Found 3 ft. above floor.

N. xv. 162. Rectang. under-tablet; chipped at lower edge. *Obv.* 7 ll. Khar., cursive, decipherable only in parts. Much sand-encrusted. *Rev.* blank. $5\frac{3}{4}'' \times 2\frac{1}{2}''$. Soft. Found 3 ft. above floor.

N. xv. 163. Rectang. cov.-tablet, raised in middle to $\frac{1}{2}''$ thickness. *Obv.* Two seals (see Pl. LXXII) in middle transversely, impressed in grey clay. L. seal, male head turned to L., $\frac{5}{8}''$ high, raised from concave background with plain rims; R. seal, male (?) head, turned to R., $\frac{1}{2}''$ high, raised on flat background, with higher ribbed rim $\frac{3}{4}''$. Traces of 1 line Khar. on R. p. edge transverse; also 1 l. tranverse on L. p. *Rev.* 3 ll. Khar., bold hand. $9'' \times 3\frac{1}{4}''$. Well-preserved. Found 3 ft. above floor.

N. xv. 164. Document on leather, almost complete. *Obv.* 6 ll. Khar., cursive, black; 4th, 5th, and 6th ll. incomplete. 1 short l. near lower edge. *Rev.* 3 items in Khar., rather indistinct. $10'' \times 3\frac{5}{8}''$. Well-preserved. Found 3 ft. above floor.

N. xv. 165. Rectang. cov.-tablet; broken at one corner. *Obv.* Seal cav. $2'' \times 1\frac{1}{16}''$. At L. p. edge, transversely, 3 ll. Khar., cursive, indistinct; on R. p. 2 ll. Khar., cursive, part indistinct. *Rev.* 2 ll. Khar., fragmentary. $5\frac{7}{16}'' \times 3'' \times \frac{3}{8}''$. Soft, abraded, stained, and sand-encrusted.

N. xv. 166. Rectang. double tablet, complete. Cov.-tablet. *Obv.* Seal $1\frac{1}{8}''$ sq. (see Pl. LXXI); device Pallas Athene to L. p., in cameo, surrounded by elliptical borders. Deep impression. Clay mixed with hair. Transversely at L. p. edge 2 ll. Khar., cursive, black; on R. p. 1 l. similar. Strings (which have been cut with knife) hang from R. p. side. *Rev.* 2 ll. Khar., cursive, black.

Under-tablet. *Obv.* 7 ll. Khar., very cursive, clear. *Rev.* blank. $6\frac{1}{2}'' \times 2\frac{11}{16}'' \times 1\frac{1}{4}''$ (both tablets). See Pl. XCV.

N. xv. 167. Rectang. cov.-tablet. Two raised seals (see Pl. LXXII) in grey clay, transversely, over broken strings. Seal R. p. 4 Chinese characters, lapidary, divided by cross ll., two lower appar. identical. Seal L. p., female head turned to L. p. holding flower in hand; head wearing cap (?) with parallel folds; hair-knob (?) or large button on top; R. ear elongated; lips parted, drapery below neck. On L. p. edge 2 ll. Khar. transversely; on R. p. 2 ll. Khar. transversely. *Rev.* blank. $5'' \times 3''$. Well-preserved.

N. xv. 168. Document on leather; fragment. *Obv.* 6 ll. Khar., incomplete, very cursive; black. *Rev.* blank. $4\frac{3}{16}'' \times 2\frac{3}{4}''$. Well-preserved. See Pl. XCII.

N. xv. 169. Slip of wood, Chinese char.; fragment. $1\frac{3}{4}'' \times \frac{1}{2}''$. Fairly preserved. See Pl. CXIII.

N. xv. 172. Takhti-shaped tablet; hole in handle, broken at sq. end. *Obv.* 2 ll. Khar., cursive, faint. Burn mark on handle. *Rev.* 3 ll. Khar., very cursive, faint. $6\frac{5}{8}'' \times 1\frac{1}{4}''$. Soft, sand-encrusted.

N. xv. 173 + 06. Rectang. double tablet, complete. Cov.-tablet. *Obv.* Seal cav. $1\frac{1}{4}''$ sq. At L. p. edge 2 ll. Khar. transversely, cursive, faint, sand-encrusted. Clay in cav. *Rev.* blank.

Under-tablet in 2 pieces. *Obv.* 5 ll. Khar., faint, and in some parts quite faded. Writing small and neat. Sand-encrusted. *Rev.* blank. $8\frac{1}{2}'' \times 3\frac{3}{4}''$. Soft and warped.

N. xv. 174. Oblong tablet, rounded at one end, square at other. *Obv.* 3 ll. Khar., faint and cursive. *Rev.* 4 ll. Khar., faint and cursive; sand-encrusted. $4\frac{1}{2}'' \times 1\frac{3}{8}''$. Soft and perished.

N. xv. 175. Slip of wood, Chinese char.; fragment. $1\frac{5}{8}'' \times \frac{1}{2}''$. Well-preserved. See Pl. CXIV.

N. xv. 175. a. Rectang. cov.-tablet; one corner broken. *Obv.* Seal cav. $2\frac{1}{4}'' \times 1\frac{1}{4}''$. At L. p. edge 2 ll. Khar., very cursive, part indistinct. *Rev.* blank. $6\frac{3}{16}'' \times 4'' \times \frac{7}{16}''$. Soft.

N. xv. 176. Slip of wood, Chinese char.; fragment. $2\frac{7}{8}'' \times \frac{9}{16}''$. Fairly preserved. See Pl. CXIII.

N. xv. 177. Oblong tablet, hole near centre of lower edge. *Obv.* 6 items in Khar., cursive. Sand-encrusted. *Rev.* blank. $7'' \times 2\frac{3}{16}''$. Hard.

N. xv. 178 + 183 + 186. Rectang. cov.-tablet, in 3 fragments. *Obv.* Seal cav. $1\frac{1}{2}'' \times 1\frac{1}{8}''$. 2 ll. Khar. transversely at L. p. edge, very cursive. *Rev.* 6 ll. Khar., cursive, fairly clear, deleted by hole in centre. $7\frac{3}{8}'' \times 2\frac{7}{8}'' \times \frac{5}{8}''$. Wood hard. Found 3 ft. above floor.

N. xv. 179. Wedge under-tablet; hole near point. *Obv.* 4 ll. Khar., very cursive, partly deleted, with initial form. Part broken. *Rev.* Some char. at sq. end, very indistinct. $7\frac{1}{8}'' \times 1\frac{11}{16}''$. Hard. Found 2 ft. above floor.

N. xv. 180. Wedge under-tablet; hole at point; cracked at sq. end. *Obv.* 4 ll. Khar., cursive, part deleted, with initial form. *Rev.* Few char. at sq. end. $14\frac{1}{8}'' \times 2\frac{3}{8}''$. Hard, sand-encrusted. Found $2\frac{1}{2}$ ft. above floor.

N. xv. 182. Document on leather. *Obv.* 4 ll. Khar., cursive; rather scratchy. Part deleted where leather is destroyed. *Rev.* blank. $10'' \times 2\frac{1}{4}''$. Much destroyed and discoloured. Found 6 in. above floor.

N. xv. 184. Oblong tablet; hole at one end. *Obv.* 2 columns, 6 ll. each + 2 ll. in Khar., cursive; part very black, part deleted. *Rev.* blank. $5\frac{1}{2}'' \times 2''$. Found $2\frac{1}{2}$ ft. above floor.

N. xv. 185. Takhtī-shaped tablet; hole in handle. *Obv.* 3 ll. Khar., very cursive; many numerals. *Rev.* blank. $9'' \times 3\frac{1}{2}''$. Hard, well-preserved. Found 2 ft. above floor. See Pl. CI.

N. xv. 187. Wedge cov.-tablet; hole near point (which is cut off square). *Obv.* Seal cav. $2\frac{1}{16}''$ from sq. end. 1 l. Khar. on either side of cav. Usual char. near hole. *Rev.* 2 ll. Khar., very cursive. $15'' \times 2\frac{1}{4}'' \times \frac{9}{16}''$. Hard, well-preserved. Found $2\frac{1}{2}$ ft. above floor.

N. xv. 188. Slip of wood, Chinese char.; fragment. $3\frac{3}{4}'' \times \frac{9}{16}''$. Well-preserved. Found $2\frac{1}{2}$ ft. above floor. See Pl. CXII.

N. xv. 189. Slip of wood, Chinese char.; fragment. $2\frac{3}{8}'' \times \frac{7}{16}''$. Well-preserved. Found 2 ft. above floor. See Pl. CXIII.

N. xv. 191. Slip of wood, Chinese char.; fragment. $1\frac{7}{16}'' \times \frac{9}{16}''$. Soft. Found $2\frac{1}{2}$ ft. above floor, close to N. xv. 192. See Pl. CXII.

N. xv. 192. Slip of wood, Chinese char.; fragment in 2 pieces. $4\frac{5}{8}'' \times \frac{1}{2}''$. Soft. Found $2\frac{1}{2}$ ft. above floor. See Pl. CXIV.

N. xv. 193. Rectang. tablet; fragment, cut at both ends. *Obv.* 2 ll. Khar., very cursive; incomplete; rather faded. *Rev.* blank. $2\frac{1}{16}'' \times 1\frac{3}{8}''$. Well-preserved. Found 3 ft. above floor.

N. xv. 194. Wedge under-tablet; hole at point. *Obv.* 3 ll. Khar., very cursive, black, with initial form. *Rev.* 1 l. from sq. end. $8'' \times 1\frac{1}{2}''$. Hard, well-preserved. Found $1\frac{1}{2}$ ft. above floor.

N. xv. 195. Wedge cov.-tablet; broken at point. *Obv.* Seal cav. $1\frac{3}{4}''$ from sq. end. 1 l. Khar. from sq. end; very cursive. Usual char. near hole. *Rev.* 4 ll. Khar., very cursive. Part deleted by sand. $8'' \times 1\frac{7}{8}'' \times \frac{1}{2}''$. Hard. Found 4 ft. above floor.

N. xv. 196. Rectang. double tablet, complete, unopened. Cov.-tablet. *Obv.* (see Pl. LXXII). Seal $1\frac{1}{8}''$ sq. Device, in centre, a square surrounding dot. From each outer corner of square a line diagonally, terminating at outer end in crescent, in centre of which a dot. Between these diagonals a bell shape. At L. p. edge transversely 2 ll. Khar., very cursive. At R. p. edge transversely, 2 ll. Khar., very cursive. **N. xv.**

Under-tablet. *Rev.* (see Pl. XCIV) blank. String intact showing method of tying. $4\frac{11}{16}'' \times 1\frac{5}{8}'' \times \frac{3}{4}''$. Perfectly preserved. Found $2\frac{1}{2}$ ft. above floor.

N. xv. 197. Document on leather; complete but damaged. *Obv.* 8 ll. Khar., very cursive; blotted and deleted in parts. *Rev.* 4 items, very indistinct. $8\frac{3}{4}'' \times 4''$. Much eaten and deleted. Found 1 ft. above floor.

N. xv. 198. Oblong tablet, pointed at one end, in 2 pieces. Hole near point; part missing. *Obv.* 1 l. Khar., cursive, very faint. *Rev.* blank. $8'' \times 1\frac{3}{8}''$. Soft. Found 3 ft. above floor.

N. xv. 199. Oblong tablet, round ends; hole at each end. *Obv.* 1 l. Khar., cursive, faint but legible. *Rev.* 2 ll. Khar., cursive, black, part effaced by scraping. $10\frac{1}{4}'' \times 2'' \times \frac{5}{16}''$. Hard, well-preserved. Found 2 ft. above floor.

N. xv. 201. Document on leather, complete. *Obv.* 5 ll. Khar., rather faded; cursive. 1 l. near lower edge. *Rev.* 3 items Khar. $8\frac{5}{8}'' \times 4\frac{3}{4}''$. Leather buckled, causing distortion to writing. Found 6 in. above floor.

N. xv. 202. Wedge cov.-tablet; hole at point. *Obv.* Seal cav. 2″ from sq. end. 1 l. Khar. between sq. end and cav., very cursive. Usual char. about 1″ from hole. *Rev.* 1 l. Khar., very cursive, clear. $9\frac{3}{8}'' \times 2'' \times \frac{5}{8}''$. Well-preserved. Found 6 in. above floor.

N. xv. 203. Slip of wood, Chinese char.; nearly complete. $8\frac{5}{8}'' \times \frac{9}{16}''$. Well-preserved. Found 8 in. above floor. See Pl. CXII.

N. xv. 204. Oblong tablet; notch in centre on each edge. *Obv.* 1 l. Khar., cursive, part deleted. *Rev.* blank. $5\frac{5}{8}'' \times 2\frac{3}{8}''$. Hard, well-preserved.

N. xv. 205. Rectang. cov.-tablet; fragment. *Obv.* blank. *Rev.* A few Khar. char.; fragmentary. $3'' \times 1\frac{3}{4}'' \times \frac{7}{16}''$. Soft and perished.

N. xv. 206. Oblong tablet, notch in each short edge, also probably in upper edge; broken at upper edge. *Obv.* 3 ll. Khar., cursive, fine and black, except where perished by damp. *Rev.* blank. $4\frac{5}{16}'' \times 1\frac{5}{8}''$. Rather soft.

N. xv. 207. Slip of wood, Chinese char.; fragment. $1\frac{7}{8}'' \times \frac{9}{16}''$.

N. xv. 300. Wedge under-tablet; hole at point. *Obv.* 2 ll. Khar., closely and finely written; 2nd l. black. Initial form. *Rev.* Few char. at sq. end. $9\frac{11}{16}'' \times 1\frac{1}{2}''$. Hard and well-preserved. Found 2 ft. above floor.

N. xv. 301 + 321. Wedge under-tablet; broken in two pieces. *Obv.* 4 ll. Khar., very cursive, clear. Initial form. *Rev.* Few char. from. sq. end. $8\frac{1}{2}'' \times 1\frac{1}{8}''$ and $6'' \times 1\frac{3}{8}''$. Well-preserved. N. xv. 301 found 2 ft., N. xv. 321 found $2\frac{1}{2}$ ft. above floor.

N. xv. **N. xv. 302 + 313. Rectang. under-tablet**; fragmentary, in 2 pieces. *Obv.* 4 ll. Khar., very cursive, black. *Rev.* blank. $7\frac{3}{8}'' \times 1\frac{13}{16}''$, and $3'' \times \frac{11}{16}''$, resp. Hard and well-preserved. N. xv. 302 found 2 ft., N. xv. 313 1 ft. above floor.

N. xv. 303. Wedge under-tablet; hole at point. *Obv.* 4 ll. Khar., very cursive, generally black and clear. *Rev.* Few char. at sq. end. $9\frac{1}{2}'' \times 1\frac{3}{4}''$. Hard and well-preserved. Found 2 ft. above floor.

N. xv. 304. Document on leather; complete. *Obv.* 6 ll. Khar., cursive, black and clear. 1 l. at lower edge. *Rev.* 1 short l. $8\frac{1}{2}'' \times 4\frac{1}{2}''$. One corner cut away, without loss to document. Well-preserved. Found 2 ft. above floor.

N. xv. 305. Document on leather; complete. *Obv.* 10 ll. Khar., cursive, uneven. Hole in leather; 1 short l. at bottom. *Rev.* 3 items Khar., $10'' \times 5\frac{3}{4}''$. Well-preserved. Found 2 ft. above floor. See Pl. XCIII.

N. xv. 306. Wedge cov.-tablet; hole near point. *Obv.* Seal cav. $1\frac{1}{2}''$ from sq. end. 1 l. Khar. at sq. end. Char. at hole faint. *Rev.* 4 ll. Khar., very cursive, thin and generally clear. $7\frac{3}{4}'' \times 1\frac{1}{2}'' \times \frac{1}{2}''$. Hard and well-preserved. Found 2 ft. above floor.

N. xv. 307. Wedge cov.-tablet; hole near point. *Obv.* Seal cav. 2" from sq. end. Well-preserved clay seal (see Pl. LXXI). Pallas Athene to L. p., deep sunk, in elliptical border. Between sq. end and seal 1 l. Khar. On opposite side of seal 1 l. Khar., cursive, clear. Char. at hole differ from those usually found there. *Rev.* Few characters at sq. end. $10\frac{3}{8}'' \times 2\frac{1}{16}'' \times \frac{3}{4}''$. Well-preserved. Found 2 ft. above floor.

N. xv. 309. Rectang. under-tablet; fragment. *Obv.* 4 ll. Khar., cursive, well-written, clear. *Rev.* blank. $6\frac{1}{4}'' \times 1\frac{5}{16}''$. Hard.

N. xv. 310. Document on leather; complete. *Obv.* 7 ll. Khar., very cursive, black, clear, excepting last line. Short line at lower edge. *Rev.* 3 items Khar. $8\frac{3}{4}'' \times 5\frac{1}{4}''$. Well-preserved. Found $1\frac{1}{2}$ ft. above floor. See Pl. XCI.

N. xv. 311. Wedge cov.-tablet.; hole at point. *Obv.* Seal cav. $1\frac{1}{8}''$ from sq. end. 1 l. Khar. between sq. end and cav. Usual char. at hole. *Rev.* blank. $9'' \times 2'' \times \frac{13}{16}''$. Well-preserved. Found $1\frac{1}{2}$ ft. above floor.

N. xv. 312. Wedge cov.-tablet; hole near point. *Obv.* Seal cav. $1\frac{7}{16}''$ from sq. end. 1 l. Khar. between sq. end and seal cav. Usual char. at hole. *Rev.* 4 ll. Khar., very cursive and clear. $9\frac{5}{8}'' \times 1\frac{3}{4}'' \times \frac{1}{2}''$. Well-preserved. Found $1\frac{1}{2}$ ft. above floor.

N. xv. 314. Slip of wood, Chinese char.; complete. $9\frac{1}{8}'' \times \frac{3}{8}''$. Perfectly preserved. Found $1\frac{1}{2}$ ft. above floor. See Pl. CXIII.

N. xv. 315. Slip of wood, Chinese char.; fragment. $5'' \times \frac{9}{16}''$. Well-preserved. Found 1 ft. above floor. See Pl. CXIII.

N. xv. 316. Rectang. cov.-tablet; fragment. *Obv.* 2 Khar. characters at L. p. edge transversely. *Rev.* 1 l. Khar., very cursive, faded, incomplete. $2\frac{3}{4}'' \times 1\frac{3}{8}'' \times \frac{1}{2}''$. Well-preserved; light wood. Found $2\frac{1}{2}$ ft. above floor.

N. xv. 317. Wedge under-tablet; hole near point. *Obv.* 4 ll. Khar., very cursive, black, and clear. *Rev.* 1 l. Khar., very cursive. $9'' \times 1\frac{3}{4}''$. Hard, well-preserved. Found 2 ft. above floor.

N. xv. 318. Wedge under-tablet; hole near point; broken at top edge. *Obv.* 4 ll. Khar., very cursive; quite clear; with initial form. *Rev.* 1 l. Khar., very cursive. $11'' \times 1\frac{5}{8}''$. Well-preserved. Found $1\frac{1}{2}$ ft. above floor.

N. xv. 319. Document on leather; fragment. *Obv.* 5 ll. Khar., very cursive, clear, and black. 1 short line at lower edge. *Rev.* 3 items Khar. $11\frac{1}{4}'' \times 4\frac{1}{4}''$. Well-preserved, but part missing at upper R. p. corner. Found 1 ft. above floor.

N. xv. 319. a. Cutting from prepared sheep-skin; irregular shape. Blank.

N. xv. 320. Rectang. under-tablet. *Obv.* 2 ll. Khar., cursive, black, and clear. *Rev.* blank. $5\frac{3}{4}'' \times 1\frac{3}{8}''$. Hard, much sand-encrusted. Found 8 in. above floor.

N. xv. 322. Rectang. cov.-tablet. Seal cav. $1\frac{1}{8}'' \times 1''$. *Obv.* 3 ll. Khar., very cursive, badly writen; transversely at L. p. *Rev.* 6 ll. Khar., very cursive and badly written. $5\frac{1}{8}'' \times 2\frac{1}{2}''$. Wood well-preserved; slightly chipped at ends. Found 2 ft. above floor.

N. xv. 323. Wedge cov.-tablet; hole near point. *Obv.* Seal cav. $1\frac{3}{4}''$ from sq. end. Traces of char. on both sides of cav. *Rev.* 4 ll. Khar., very cursive and faint in parts. $8\frac{1}{2}'' \times 1\frac{13}{16}'' \times \frac{11}{16}''$. Hard and sand-encrusted. Found $2\frac{1}{2}$ ft. above floor.

N. xv. 324. Slip of wood, Chinese char.; fragment. $2\frac{3}{4}'' \times \frac{3}{8}''$. Perfectly preserved. Found 2 ft. above floor. See Pl. CXII.

N. xv. 326. Slip of wood, Chinese char.; broken at lower end. $6\frac{13}{16}'' \times \frac{9}{16}''$. Perfectly preserved. Found 1 ft. above floor. See Pl. CXII.

N. xv. 328 + 75. Slip of wood, Chinese char.; complete in 2 pieces. $9\frac{1}{8}'' \times \frac{5}{8}''$. Well-preserved. N. xv. 328 found $1\frac{1}{2}$ ft. above floor. See Pl. CXII and CXIII.

N. xv. 329. Document on leather; fragment. *Obv.* 6 ll. cursive Khar., fragmentary. *Rev.* Suggestions of Khar. char. $2\frac{7}{8}'' \times 2\frac{1}{2}''$. Found 1 ft. above floor.

N. xv. 330. Rectang. cov.-tablet. *Obv.* Seal (see Pl. LXXI, enlarged) $1\frac{1}{8}''$ sq. in grey clay. Device, a figure seated on a stool, perhaps an Eros, surrounded by elliptical border of mouldings. At top 2 ll. Khar., transversely, very cursive. *Rev.* 6 ll. Khar., very cursive. Partly obscured in places by sand. $4\frac{1}{8}'' \times 2\frac{1}{8}'' \times \frac{7}{16}''$. Well-preserved. Found 1 ft. above floor. See Pl. LXXI.

N. xv. 331. Wedge under-tablet; hole near point. *Obv.* 3 ll. Khar., very cursive, thin, black, and clear; initial form. *Rev.* 1 l. from sq. end. $10'' \times 2''$. Hard and well-preserved. Found $1\frac{1}{2}$ ft. above floor.

N. xv. 332. Wedge under-tablet; hole near point. *Obv.* 4 ll. Khar., very cursive. *Rev.* 1 l. Khar. $10\frac{3}{4}'' \times 2\frac{1}{16}''$. Hard and well-preserved. Found $1\frac{1}{2}$ ft. above floor.

N. xv. 333. Document on leather; complete. *Obv.* 8 ll. Khar., very cursive, generally clear. 1 l. at lower edge, very deleted, appears to read reverse way of other text. *Rev.* blank. $6\frac{3}{8}'' \times 3''$. Well-preserved. Found 8 in. above floor. See Pl. XCII.

N. xv. 334. Rectang. cov.-tablet. *Obv.* Seal (see Pl. LXXII) $\frac{7}{8}'' \times \frac{13}{16}''$. Device appar. Khar. characters. 2 ll. Khar. transversely at L. p. edge; cursive, part deleted. Only two string-grooves. *Rev.* 7 ll. Khar., cursive, black, deleted in places. $2'' \times 4\frac{3}{4}'' \times \frac{7}{8}''$. Well-preserved, but has been purposely damaged by knife or chisel cuts. Found $2\frac{1}{2}$ ft. above floor. See Pl. XCV.

N. xv. 335. Oblong tablet, fragment; 2 notches on one long edge. *Obv.* 2 ll. Khar., very cursive. *Rev.* 2 ll. Khar., very cursive. $4\frac{13}{16}'' \times 1''$. Soft. Found 2 ft. above floor.

N. xv. 335. Document on leather; fragment. *Obv.* 5 ll. Khar., perfect; black and well-written. *Rev.* blank. $3\frac{7}{8}'' \times 1\frac{5}{8}''$. Much stained and perished. Found 3 ft. above floor.

N. xv. 337. Slip of wood, Chinese char.; fragment. $4\frac{1}{8}'' \times \frac{9}{16}''$. Well-preserved. Found 3 ft. above floor. See Pl. CXII.

N. xv. 338. Rectang. cov.-tablet; seal cav. $1\frac{3}{8}'' \times 1\frac{1}{4}''$. *Obv.* On L. p. transversely 2 ll. Khar., cursive, very faint. On R. p. transversely 1 l. Khar., cursive, very faint. *Rev.* blank. $6\frac{11}{16}'' \times 4\frac{5}{8}'' \times 1\frac{1}{8}''$. Wood hard and heavy. Found 3 ft. above floor.

N. xv. 339. Slip of wood, Chinese char.; fragment. $1\frac{7}{8}'' \times \frac{1}{2}''$. Perfectly preserved. Found 3 ft. above floor. See Pl. CXIII.

N. xv. 340. Wedge under-tablet, broken at point and upper edge. *Obv.* 2 ll. Khar., cursive, clear, fragmentary; initial form. *Rev.* blank. $8\frac{1}{2}'' \times 1\frac{5}{8}''$. Hard and well-preserved. Found 3 ft. above floor.

N. xv. 342. Oblong tablet; rounded at ends. *Obv.* 2 ll. Khar., cursive, clear. *Rev.* 2 ll. Khar., cursive, clear. $7'' \times 1\frac{3}{8}''$. Well-preserved. Found 3 ft. above floor.

N. xv. 343. Rectang. cov.-tablet. *Obv.* Seal cav. $1\frac{1}{4}''$ sq., 2 ll. Khar., transversely at L. p., well-written, clear; 1 l. Khar. transversely at R. p. *Rev.* 10 ll. Khar., small, well-written, but portions deleted through perishing of wood. $5\frac{3}{16}'' \times 3\frac{7}{16} \times \frac{1}{2}''$. Wood rather soft, encrusted with sand in places. Found 2 ft. above floor.

N. xv. 344. Rectang. cov.-tablet. *Obv.* Seal cav. $1'' \times 1\frac{1}{4}''$. At L. p. edge transversely 2 ll. Khar., very cursive. At R. p. transversely 2 columns of 4 ll. each. *Rev.* 1 l. at top, and below, 3 columns of 6, 5, and 4 ll., respectively. All very cursive. $4\frac{5}{8}'' \times 2\frac{7}{8}'' \times \frac{1}{2}''$. Well-preserved. Found 2 ft. above floor.

N. xv.

N. xv. 345. Rectang. cov.-tablet. *Obv.* Seal cav. $\frac{15}{16}'' \times 1''$. To L. p. Chinese characters; faded. *Rev.* blank. Centre portion is raised, leaving a lower margin all round of $\frac{3}{16}''$ to $\frac{5}{16}''$. This raised portion has the appearance of having been cut to fit a corresponding opening in under-tablet (in a box of which this may have been the lid). $3\frac{3}{8}'' \times 1\frac{1}{2}'' \times \frac{7}{16}''$. Wood hard and well-preserved. Found 3 ft. above floor. See Pl. CV, CXIV.

N. xv. 346. Document on leather, prob. complete. *Obv.* 6 ll. Khar., very cursive, but black and fairly clear. *Rev.* 1 l. Khar., faded. $8\frac{1}{4}'' \times 2\frac{3}{8}''$. Discoloured, broken in places and part perished. Found $1\frac{1}{2}$ ft. above floor.

N. xv. 347. Wedge under-tablet; broken at point and part of upper edge. *Obv.* 4 ll. Khar., very cursive, with initial form.; incomplete. *Rev.* blank. $8\frac{3}{4}'' \times 1\frac{9}{16}''$. Hard, encrusted with sand. Found 2 ft. above floor.

N. xv. 348. Slip of wood, Chinese char.; fragment. $6\frac{1}{2}'' \times \frac{5}{8}''$. Perfectly preserved. Found $1\frac{1}{2}$ ft. above floor. See Pl. CXII.

N. xv. 349. Slip of wood, Chinese char.; fragment. $3\frac{7}{8}'' \times \frac{9}{16}''$. Fairly preserved. Ink faded in places, but where this has happened, the char. remain in relief, the ink having served to preserve the wood it covered. Found $3\frac{1}{2}$ ft. above floor. See Pl. CXII.

N. xv. 350. Document on leather; almost perfect. *Obv.* 11 ll. Khar., cursive, well-written, black. End of ll. 3 and 4 deleted, also end of 9th l. *Rev.* blank. $6\frac{1}{4}'' \times 4\frac{13}{16}''$. Fairly well-preserved, but discoloured. Found $2\frac{1}{2}$ ft. above floor.

N. xv. 351. Slip of wood, Chinese char.; fragment. $3\frac{5}{8}'' \times \frac{7}{16}''$. Soft. Found 3 ft. above floor. See Pl. CXIV.

N. xv. 352. Wedge under-tablet, fragment; hole at point. *Obv.* 3 ll. Khar., cursive and only partly legible. *Rev.* blank. $4\frac{11}{16}'' \times 1\frac{1}{2}''$. Wood perished and encrusted with sand. Found 3 ft. above floor.

N. xv. 353. Slip of wood, Chinese char.; complete. $9\frac{1}{4}'' \times \frac{1}{2}''$. Hard and well-preserved. Found 3 ft. above floor. See Pl. CXII.

N. xv. 354. Rectang. tablet, fragment; hole at one end. *Obv.* 2 ll. Khar., very cursive. *Rev.* blank. $2\frac{7}{16}'' \times 1\frac{1}{4}''$. Cut from longer tablet. Wood soft. Found $2\frac{1}{2}$ ft. above floor.

N. xv. 355. Rectang. under-tablet. *Obv.* 8 ll. Khar., well-written but faint, and in some parts obliterated by dried clay and stains. *Rev.* blank. $6\frac{1}{2}'' \times 2\frac{3}{4}''$. Hard. Found 2 ft. above floor.

N. xv. 356. Wedge cov.-tablet; hole at point. *Obv.* Seal cav. $1\frac{3}{8}'' \times 1''$. 1 l. Khar. between sq. end and cav., black. Char. $1\frac{1}{2}''$ from hole. *Rev.* blank. $10\frac{1}{4}'' \times 1\frac{7}{8}''$.

N. xv. Hard and well-preserved. Portions of bark at lower side. Found 3 ft. above floor.

N. xv. 357. Rectang. under-tablet. Much split. *Obv.* 7 ll. Khar., very cursive and coarse. *Rev.* blank. $8'' \times 3\frac{1}{2}''$. Fairly hard. Found 4 ft. above floor, against south wall.

N. xv. 359. Wedge under-tablet; very small hole at point. *Obv.* 4 ll. Khar., very cursive and fine, with initial form. *Rev.* 1 l. from sq. end. $19\frac{1}{8}'' \times 2''$. Hard and well-preserved. Found 2 ft. above floor.

N. xv. 360. Oblong tablet; fragment. Traces of Khar. char. $3\frac{7}{8}'' \times \frac{3}{4}''$. Wood well-preserved. Ink black, but part encrusted with sand. Found $1\frac{1}{2}$ ft. above floor.

N. xv. 361. Wedge under-tablet; hole at point. *Obv.* 4 ll. Khar., well-written, black. *Rev.* 1 l. Khar., very cursive. $8\frac{3}{4}'' \times 1\frac{11}{16}''$. Hard; sand-encrusted. Found 2 ft. above floor.

N. xv. 362. Slip of wood, Chinese char.; complete. $9'' \times \frac{1}{2}''$. Perfectly preserved. Found 6 in. above platform. See Pl. CXII.

N. xv. 363. Wedge cov.-tablet; hole at point. *Obv.* Seal cav. $1\frac{1}{8}'' \times 1\frac{1}{2}''$. 1 l. Khar., well-written, between sq. end and cav. Usual char. at hole. *Rev.* blank. $9\frac{3}{8}'' \times 2\frac{1}{4}'' \times \frac{11}{16}''$. Perfectly preserved; roughly made. Found 1 ft. above platform.

N. xv. 01. a. Fragment of wedge-tablet. Hole at point. *Obv.* Traces of Khar. char. *Rev.* blank. $8\frac{1}{4}'' \times 1''$.

N. xv. 01. b. Fragment of tablet, with Khar. char. $5\frac{7}{8}'' \times \frac{3}{8}''$.

N. xv. 01. c. Fragment of slip of wood, showing very faint Chinese char. $5'' \times \frac{1}{2}''$.

N. xv. 01. d. Three small fragments of tablets showing Khar. char. Much weather-worn and dark.

N. xv. 02. Slip of wood, Chinese char.; fragment. $2\frac{7}{8}'' \times \frac{1}{2}''$. Soft. Found on 3. ii. 1901. See Pl. CXII.

N. xv. 03. Rectang. cov.-tablet; fragment in 3 pieces. *Obv.* Seal cav. $1\frac{1}{8}'' \times 2\frac{1}{8}''$, blank. *Rev.* Traces of Khar., writing indistinct. $3\frac{3}{8}'' \times 2'' \times \frac{9}{16}''$. Quite perished; dark in colour; sand-abraded.

N. xv. 04. Wedge under-tablet; hole near point. *Obv.* 4 ll. Khar., very cursive and blotted, with initial form. Dark streak of heart wood down centre. *Rev.* 1 l. Khar. from sq. end. Incised circle, about $\frac{5}{8}''$ in diameter, roughly cut, near centre; probably a distinguishing mark. $8\frac{1}{2}'' \times 1\frac{3}{4}''$. Well-preserved and hard.

N. xv. 05. Wedge under-tablet; hole near point (broken). Split in its length almost to point. *Obv.* 5 ll. Khar., fragmentary, all completely deleted at point end. Initial form. *Rev.* Traces of 1 l. Khar. from sq. end. Towards point end, a circle with straight line through centre, incised. $9\frac{1}{4}'' \times 2''$ (approx.). Hard, but bleached at point end, and abraded.

N. xv. 07. Lath-like tablet; pointed at both ends; broken at one end, hole at other. *Obv.* 3 ll. Khar. in column, very cursive, black and clear. *Rev.* blank. $6\frac{3}{4}'' \times 1''$. Bleached and abraded at one end which has probably been exposed. The other end is completely preserved except that it is chipped.

N. xv. 08. Slip of wood, Chinese char.; broken at both ends. $6\frac{1}{4}'' \times \frac{9}{16}''$. Writing very black, except at one end where wood is bleached and writing faded. Found 3. ii. 1901. See Pl. CXII.

N. xv. 09. Slip of wood, Chinese char.; broken at both ends. $5\frac{11}{16}'' \times \frac{1}{2}''$. Writing faded; wood bleached. Found 3. ii. 1901. See Pl. CXIV.

N. xv. 010. Slip of wood, Chinese char.; broken at both ends. $3\frac{3}{8}'' \times \frac{1}{2}''$. Found 3. ii. 1901. See Pl. CXIV.

N. xv. 011. Fine plain silk, resembling modern Lahore silk. A piece of dull red (would be called a 'pastel' shade at the present day) sewn to a small fragment of dull yellow. Red $4\frac{1}{2}'' \times 3\frac{1}{2}''$; yellow $2\frac{1}{4}'' \times 1''$. See Pl. LXXVI.

N. xv. 012. Coarse cotton check, in dull yellow with blue stripes. Coarsely oversewn on loosely twisted blue cord along one edge. $14'' \times 3''$. See Pl. LXXVI.

N. xv. 013. Fine cotton cloth; plain. Dull 'pastel' red. $3\frac{1}{2}'' \times 2\frac{1}{4}''$. See Pl. LXXVI.

N. xv. 014. Piece of felt, loose and very uneven; rich red; may have been used as padding between cloth. $3\frac{1}{4}'' \times 6''$. See Pl. LXXVI.

N. xv. 015. Fine hard woollen felt; very compact and strong. Dull yellow. Probably part of saddle Numdah or floor covering. $1\frac{3}{8}'' \times 1\frac{1}{8}''$. See Pl. LXXVI.

N. xv. 016. Coarse woollen cloth. Dark pink. $7'' \times 1\frac{1}{4}''$. See Pl. LXXVI.

N. xv. 017. Cotton fabric, resembling coarse harsh muslin; very loose and open. Dull violet-pink. $5\frac{1}{2}'' \times 5''$. See Pl. LXXVI.

N. xv. 018. Coarse cotton fabric, resembling Indian *kharwa*. Dull red. $5\frac{1}{4}'' \times 3''$. See Pl. LXXVI.

N. xv. 001. a. Eight fragments of lacquered bowl, circular, of Chinese workmanship. Interior, sealing wax red. Exterior, bands of black, divided by thin lines of red. Between the lacquered black bands are rough-surfaced bands, from which probably the original colour has perished, being less durable. Bowl appears to be made of thin sheet of cane (probably bamboo). Original diameter of bowl appar. about 8″. See Pl. LXX (interior), LXV (exterior).

N. xv. 001. b. Piece of crystal, irregularly oblong. $1'' \times \frac{7}{8}''$.

N. xv. 001. c. Fragment of glass, very clear, pale greenish, showing on outside a cut pattern. $1\frac{3}{8}'' \times 1''$.

N. xv. 001. d. Fragment of glass, similar, plain. $1\frac{3}{8}'' \times \frac{7}{8}''$.

N. xv. 001. e. Fragment of glass vessel, rim and side. $1\frac{1}{4}'' \times 1\frac{1}{4}''$.

N. xv. 001. f. Fragment of glass, olive-green, with raised chevreul. $1\frac{1}{4}'' \times \frac{7}{8}''$.

N. xv. 001. g. Fragment of glass vessel, handle. $\frac{7}{8}'' \times \frac{1}{2}''$.

N. xv. 001. h. Glass bead, moulded, green. $\frac{5}{8}'' \times \frac{7}{16}''$.

N. xv. 001. i. Bowl of horn spoon. Shape roughly oval; cut with a kind of raised midrib on under-surface, which runs into handle. The handle which is at an angle of about 120° with the bowl, is broken, and the stump remaining is cut across in a transverse splay, a hole being drilled through as though for a dowel to connect this portion with stem of handle. On each side of stump scraped marks as of a file or rasp. Size of bowl $1\frac{3}{8}'' \times 2''$.

N. xv. 001. j. Astragalus bone, well-preserved, and of a pink colour; has almost weight and texture of ivory. A round hole drilled through one part. $1\frac{1}{4}'' \times \frac{3}{4}''$.

N. xv. 001. k. Tangled cords of wool of various qualities and textures. The longest piece consists of the usual two strands twisted loosely together, and at intervals of about $\frac{1}{8}''$ or less, tied round very tightly with cord in which some pigment appears to have been mixed to cause it to be a hard, bead-like mass. The total result is a kind of rosary.

N. xv. 001. l. A dried date. $2'' \times \frac{7}{8}''$.

N. xv. 002. Wooden seal case. Half of irregular octagonal prism, $2''$ long, $1\frac{1}{2}''$ diam. Ends of facets chamfered. Cav. about $1''$ sq. Hole bored through centre of cav. String grooves unequal, the centre one on each side being larger than the end ones. One side of cav. broken away. Wood hard.

N. xv. 003. Detached clay seal, rough, with fragments of original string attached. Vegetable fibre and perhaps hair mixed with the clay. Seal mark simply a circular depression $1''$ diameter, without visible device. $1\frac{3}{4}'' \times 1\frac{5}{8}'' \times 1\frac{1}{8}''$. See Pl. CV. N. xv.

N. xv. 004. Ivory die, oblong, square ends. On rectangular sides are the spots, marked each by a circle and dot in centre, as though engraved with a two-pointed instrument, the dot being the mark of one point which acts as a pivot round which the other point revolves to describe the circle. Order of dots is 3 opposite 1 and 4 opposite 2. Length $\frac{7}{8}''$, width $\frac{3}{8}''$ nearly. See Pl. LXXIV.

N. xv. 005. Oblong piece of hard 'green' leather (probably camel-skin), rounded at one end, square at other, with three holes bored through either of longer

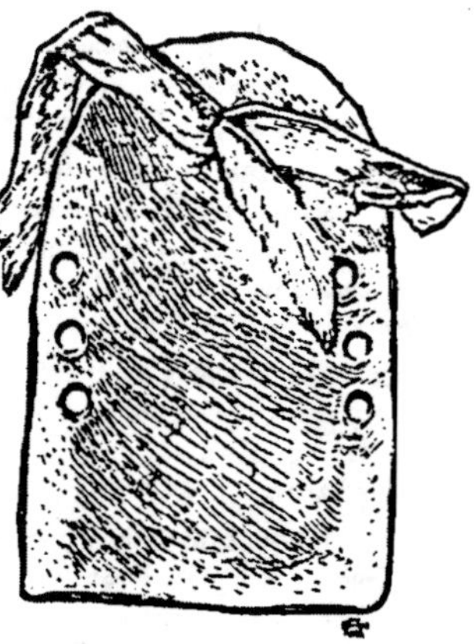

edges, and a hole at rounded end. Fixed in latter two tufts of soft, tanned leather bearing traces of red colouring. Probably belonged to scale armour. $2\frac{3}{8}''$ long, $1\frac{5}{8}''$ broad, thickness $\frac{1}{16}''$. See diagram; also *Addenda*.

OBJECTS EXCAVATED IN N. xvi.

N. xvi. 1. Figure rudely carved in wood, nude, standing on tenon-like base. Hands together at breast; top knot. Traces of red on base, and lines of features in black. Height of figure $8\frac{3}{4}''$, base $3\frac{1}{4}''$, width of shoulders $2\frac{1}{8}''$, width of base $1\frac{1}{2}''$, thickness about $\frac{1}{2}''$. Well-preserved. See Pl. LXX.

N. xvi. 2. Takhti-shaped tablet, with handle $2\frac{1}{2}''$ long, rounded at end. *Obv.* 5 ll. Khar., clear bold writing, but partly faded by damp. *Rev.* 3 ll. Khar. running towards handle, clear, very black. On lower edge in reverse direction 2 ll. Khar. by same hand, but in fainter ink. Both sides of tablet show marks of plentiful scraping. $16\frac{3}{4}'' \times 2\frac{3}{8}''$. Well-preserved. See Pl. CI. N. xvi.

N. xvi. 3. Rectang. cov.-tablet, in 3 pieces. *Obv.* No writing visible. Seal cav. about $1'' \times \frac{7}{8}''$. *Rev.* blank. $5\frac{1}{2}'' \times 3''$. Much perished and all writing (if any) deleted.

DOCUMENTS EXCAVATED IN N. vi.

N. xvii. 1. Wedge under-tablet; hole near point (broken). *Obv.* 3 ll. well-written cursive Khar., nearly all very clear and black; initial form. *Rev.* 1 l. Khar. from sq. end; faint. $13'' \times 1\frac{1}{2}''$. Well-preserved, but soft.

N. xvii. 3 + 2. Rectang. double-tablet, complete. Cov.-tablet. *Obv.* Seal cav. $1\frac{1}{4}'' \times 1\frac{1}{4}''$. Frags. of string; no inscription. *Rev.* 1 short line Khar. at top edge; very clear, well-written and black; cursive. Well-preserved. Sap wood has become very soft; heart wood is very hard. N. xvii.

N. xvii. Under-tablet. *Obv.* 6 ll. Khar., cursive, well-written; very black and clear. Slightly insect-eaten. *Rev.* blank. Perfectly preserved. $7\frac{1}{4}'' \times 2\frac{3}{16}''$. See Pl. CIV.

N. xvii. 4. Wedge cov.-tablet, much broken; point missing. *Obv.* Seal cav. $2\frac{1}{4}''$ from sq. end. Traces of writing on both sides of cav. Remains of seal. *Rev.* 1 l. Khar., fragmentary, black, cursive. $11\frac{5}{8}'' \times 2\frac{1}{4}'' \times \frac{5}{8}''$. Very rotten, discoloured, and sand-encrusted.

N. xvii. 5. Stick-like tablet, fragment. *Obv.* (flat) 1 l. Khar., cursive, coarsely written, much faded. Surface cracked and perished. *Rev.* (round) blank; bleached. $6\frac{1}{4}'' \times \frac{7}{8}''$.

OBJECTS EXCAVATED IN N. VII.

N. xviii. N. xviii. 1. Oblong tablet, obtusely pointed at one end. Hole near point, not central. Traces of Khar. char. on both sides (cf. N. xviii. 3). $3\frac{1}{4}'' \times 1\frac{1}{4}''$. Perished, curled, and bleached.

N. xviii. 1. b. Stick-like tablet; hole near pointed end. *Obv.* (flat) 3 ll. Khar., broken by intervals (cf. N. xviii. 3). *Rev.* (round) blank. $5'' \times \frac{7}{8}''$. Well-preserved.

N. xviii. 2. Stick-like tablet; hole at one end. *Obv.* (flat) 4 ll. Khar., very finely written. Broken by intervals, forming vertical columns (cf. N. xviii. 3). *Rev.* (round) faint traces of characters 1" from hole. $6\frac{3}{8}'' \times \frac{7}{8}''$. Soft, rather perished at one end.

N. xviii. 3. Stick-like tablet, hole at pointed end. *Obv.* 3 ll. Khar., fine, rather indistinct. Lines broken by intervals, items ranging in vertical columns. *Rev.* blank. $5'' \times \frac{7}{8}''$. Well-preserved.

N. xviii. 4. Wedge cov.-tablet, hole near point. *Obv.* Seal cav. $1\frac{3}{8}''$ from sq. end. 1 l. Khar. (faint) between sq. end and cav. Usual char. near hole. *Rev.* 2 ll. Khar., very cursive; fairly clear in middle. $7\frac{1}{4}'' \times 1\frac{3}{8}''$. Wood hard.

N. xviii. 5. Oblong tablet; hole at top to R. p. *Obv.* 3 ll. Khar., columnar, faint. *Rev.* 1 l. faint. (Cf. N. xviii. 3.) $4\frac{1}{4}'' \times 1\frac{3}{16}''$. Soft, but apparently complete.

N. xviii. 6. Oblong tablet; hole below to R. p. *Obv.* (?) Faint traces of Khar. characters. *Rev.* Transversely (opposite end to hole) 3 ll. Khar., very cursive. $5\frac{1}{8}'' \times 1\frac{1}{4}'' \times \frac{3}{8}''$.

N. xviii. 7. Oblong tablet, obtusely pointed at one end, sq. at other, near which hole. Khar. char. very faint, and apparently very irregular, written on both sides. $8'' \times 1\frac{3}{8}''$. Rather perished.

N. xviii. 8. Wedge under-tablet; hole near point. *Obv.* 2 ll. Khar., cursive, with initial form. *Rev.* Traces of char. at sq. end. $7\frac{1}{2}'' \times 1\frac{3}{8}''$. Well-preserved.

N. xviii. 01. a. Two wooden disks, irregularly circular, diam. 2", about $\frac{1}{8}''$ thick in centre, thinner at edges; two holes in each about $\frac{5}{8}''$ apart. Joined by cord passing through holes, ends of cord cleverly 'put away'; looks as if it might have been used as a dead eye. See Pl. LXX.

N. xviii. 01. b. Piece of knotted hemp (?) cord, similar to that used in (a).

N. xviii. 01. c. Piece of coarse woollen fabric, dark-grey and dirty-white. Probably portion of saddle-bag.

N. xviii. 01. d. Portion of cotton (?) cloth, with simple diaper pattern. Some edges neatly hemmed. Joined in one place by oversewing.

N. xviii. 01. e. Portion of plain cotton (?) cloth.

N. xviii. 01. f. Portion of plain cotton (?) cloth dyed reddish-brown (Indian *majīt* colour). All fabrics much perished and insect-eaten.

N. xix. 1. Wedge under-tablet; hole near point. *Obv.* Encrusted with sand and straw. Khar. char. visible where wood is exposed. *Rev.* Encrusted as Obv. No writing visible. $14\frac{1}{4}'' \times 2''$. Wood hard; large piece split off from sq. end; this end charred. See Pl. C.

N. xix. 2. Wooden mousetrap. Consists of a piece of wood $16'' \times 4''$ in its broadest part $\times \frac{1}{4}''$, shaped somewhat like a tailor's sleeve-board, but smaller and more tapering. Through the broad end is pierced a round hole $1\frac{9}{16}''$ in diam., and round about it are drilled 4 pairs of smaller holes. Near the thinner end is another small hole. These small holes were evidently intended to hold pegs, probably for adjusting the necessary tackle, bait, &c., and the remains of two pegs are still present. A shallow groove runs along one face of the instrument from the wider end almost to the thinner, crossing the large hole centrally. Well-preserved. See Pl. LXXIII.

OBJECTS FOUND IN N. VIII.

N. xx. N. xx. 1. Takhtī-shaped tablet; trapezoid-shaped handle, pierced with large hole. *Obv.* (handle to R. p.) 3 ll. carefully written but badly faded Brāhmī. Encrusted with sand. *Rev.* (handle to L. p.) 2 short ll. Khar., cursive, from handle. $11\frac{1}{2}'' \times 1\frac{7}{8}''$. Wood soft.

N. xx. 2. Wedge cov.-tablet; hole near point. *Obv.* $2\frac{7}{8}''$ from sq. end. Seal cav. (small); on both sides 1 l. Khar., very cursive and faint. Usual char. at hole. *Rev.* blank. $13\frac{1}{2}'' \times 2\frac{1}{4}''$. Extremely well-preserved.

N. xx. 01. Carved wood panel, portion of, representing a long-tailed ibex. Only upper portion remains, lower part apparently having been on a separate piece of wood placed below this. The pose is good, although the carving is very rude. The single, thick-based horn has double row of V-shaped cuts to represent rugosity. Mouth slightly open, showing two rows of teeth, and the tongue protrudes. Long tail curled over back in manner of heraldic lion rampant, and very thick at tip. The neck shows 3 V-shaped parallel creases. Carved part of panel slightly sunk, the two side edges, respectively 1¾" and 1¼" wide, being higher. At each of two upper corners a square of about 1¾" is cut out. Width of piece 17", height 5". Soft at edges; hard in centre; surface of back split and weather-worn.

N. xx. 02. Carved double bracket in wood, with socket in centre for round head of post. Portion of socket broken away. Both sides are equally well finished, also ends; evidence that it was exposed to view all round, and of course, underneath. Side elevation presents following appearance; architrave (i. e. the upper member of the piece) 22" long, 1⅞" wide, divided horizontally into (1) a flat bead ⅜"; (2) a carved band—*kingri* pattern, perhaps to represent overlapping leaves—1" wide, divided vertically in centre by ¼" fillet, curved vertically as though encircling wreath of leaves; (3) rectangular billet moulding slightly recessed, ¼" wide, each billet about 2¼" long, divided by ⅜" interval, containing prism shape. Below this, brought forward to level of (1), an abacus, flat on front to a depth of 1⅝", then chamfered at angle of about 30°, to a depth of about 1¼". In profile the abacus widens by another chamfer springing ⅝" above lower chamfer. Width at widest part of abacus 7¼"; above lower chamfer 6¼"; below lower chamfer 4".

On either side of abacus, and below architrave of bracket an oblong modillion, which seen in side elevation presents:—next to abacus, a cavetto (about 60°), upper end being ¼" below architrave; from this a vertical flat turning into half-round bead (depth 1"); then narrow vertical flat, and long horizontal chamfer to end. The end elevation presents return of architrave (same design as side); foreshortened modillion, and abacus. Width of upper members and abacus 5½", at modillion 5".

Plan of under surface: Centre portion, abacus in which socket 2¾" diam. (evidently not a door socket as there is no sign of wear). Side portions, modillions (about 5¼" × 6½"). On one, a 4-petalled 4-sepalled flower with round centre, petals arranged diagonally in square (narrow fillet border). On the other the square with fillet border is divided diagonally by two narrow fillets, and in each triangle thus formed is a half of flower similar to that in other. N. xx.

These patterns occur very frequently in Gandhāra sculptures, and on the boat carvings on Indus and Jehlam. The carving is all shallow, and evidently rapidly done, with bold, long cuts of the chisel. The style of manipulation is precisely that of the architectural carving as practised at Bhēra, in the Punjāb to-day, and there is a general resemblance in designs. Wood extremely well-preserved and seems quite sound. See Pl. LXIX.

N. xx. 03. Carved double bracket in wood, exactly similar to N. xx. 02 with exception of patterns on modillions. In this, one has within square fillet border, a parallel and concentric oblong with fillet border, and measuring (outside) 2⅞" × 3¼". Within this the 4-petalled flower, and in the four corners rudely cut ¼ flowers. The other has, within a fillet border, a rudely cut and very conventional representation of a bird (?) with outspread wings. At two inner corners ¼ flowers. Socket is complete, but split by shrinkage. Wood very soft all over surface, and in places broken away. Upper surface perished. Length 24", width 5¾", height 5½".

N. xx. 04. Wooden boot-last (?). Roughly made. Broken slightly at ankle. Much scored by knife-cuts on sole, as though frequently used as a block for cutting leather. On the upper surface are obvious traces of pink paint. 10" × 3⅜" (at toes), 2" (near heel). Thickness 2¼" to 1¼". See Pl. LXXIII.

N. xx. 05. Wooden implement, somewhat resembling a curry comb, used in weaving to drive together the weft picks. It is in the shape of a wedge, the short blunt teeth being at the thin edge, and the thick edge being rounded into a form to fit comfortably into the palm of the hand, and furnished with a projecting knob to give firmness to the stroke in using. An implement of similar form (but in this case possibly a brush for cleansing or sizing the threads) is seen in the hand of a figure in the painted panel D. x. 4 (Pl. LXIII). 7¼" × 4" × 1¾". See Pl. LXXIII.

DOCUMENTS FOUND IN N. x.

N. xxi. 1. Rectang. cov.-tablet, broken at top corners and perished at lower edges. *Obv.* Seal cav. ½" sq. 2 ll. Khar. transversely at L. p. edge. *Rev.* 8 ll. Khar., very cursive, black; parts missing where broken, parts deleted. 5" × 3¼". Fairly well-preserved.

N. xxi. 2 + 3. Double rectang. tablet, complete. Cov.-tablet. Seal cav. 1⅛" × 1⅜". Part of seal and string remaining. *Obv.* 3 ll. Khar. transversely at L. p. edge. *Rev.* 5 ll. Khar., cursive, rather coarse. Partially deleted by encrusted sand. N. xxi.

Under-tablet. Thickness of raised part ¼". *Obv.* 7 ll. Khar., cursive, rather coarse. Partially deleted by sand. *Rev.* blank. 8¾" × 3½". Well-preserved.

N. xxi. **N. xxi. 5. Rectang. under-tablet**, in 2 pieces (joined). *Obv.* 7 ll. Khar., faded, cursive; spotted by water-stains. *Rev.* 1 l. transversely, very cursive. $8\frac{1}{4}'' \times 3''$. Fairly preserved. Spotted with water-stains and sand.

N. xxi. 6. Elongated oval tablet; hole at small end. *Obv.* 2 ll. Khar., very cursive, complete. *Rev.* blank. $5'' \times 1\frac{1}{4}''$. Well-preserved.

N. xxi. 7+4. Rectang. tablet, complete. Cov.-tablet. *Obv.* Seal cav. $1\frac{5}{8}'' \times 3''$, evidently intended for two seals. Empty. 3 ll. Khar. to L. p., very cursive and indistinct, but probably a good deal legible. To R. p. 2 ll. Khar. *Rev.* 7 ll. Khar., cursive, partly deleted. $7\frac{3}{16}'' \times 3\frac{7}{16}''$.

Under-tablet. *Obv.* 8 ll. Khar., cursive, faint, and partly deleted. *Rev.* A few char. near R. p. end. $10'' \times 3\frac{1}{4}''$. Perfectly preserved.

DOCUMENT FOUND NEAR N. xi.

N. xxii. **N. xxii. 01. Wedge cov.-tablet.** Seal cav. about $1\frac{3}{4}''$ from sq. end. Cavity $1\frac{1}{4}'' \times 1''$; hole near point, which is split. Completely bleached and much abraded by sand. No trace of writing left. $9\frac{1}{4}'' \times 1\frac{1}{2}'' \times \frac{1}{4}''$.

DOCUMENT FOUND IN N. ix.

N. xxiii. **N. xxiii. 1. Rectang. cov.-tablet**, warped; found by Ḥasan Ākhūn on 28. i. 1901. *Obv.* Two seals transversely, in cav. $1\frac{1}{4}''$ high, $1\frac{3}{4}''$ broad, impressed in grey clay (cracked). L. seal, male $\frac{3}{4}''$ high, turned to R. in low relief. R. seal, bird rising with wings extended, from seal slightly longer than L. seal. On L. p. edge 2 ll. Khar., slightly damaged; near R. p. edge 1 l. Khar. *Rev.* blank. Slight crack. $7'' \times 3\frac{3}{4}'' \times \frac{3}{8}''$. See Pl. LXXII.

MISCELLANEOUS OBJECTS FOUND IN SAND AT NIYA SITE.

Miscell. finds. **N. 001. Fragment of bronze ornamental object.** Perhaps portion of large buckle or fastening. $2\frac{1}{4}''$ diameter; $\frac{7}{16}''$ widest part. Found near Camp 95. See Pl. LXXIV.

N. 003. Flake of flint. Straight on one side, round on opposite side, and round at one end. Opposite end was probably pointed, but is broken off. Along round side is a sharp edge. $1\frac{1}{4}'' \times \frac{1}{2}''$. Found close to Camp 93.

N. 004. Pebble whorl, conical dome shaped. Diameter $1''$; diam. of hole $\frac{1}{4}''$; thickness $\frac{5}{8}''$. See Pl. LXXIV.

N. 005. a. Bronze arrow-head. Shape of point an equilateral triangular pyramid. In centre of each triangular face an incised triangle, into which has been put some substance darker than the bronze. There is a short hexagonal neck, and a hole pierced as though for a shaft. Length $1\frac{1}{16}''$, width of triangular face $\frac{3}{8}''$. See Pl. LXXIV.

N. 005. b. Portion of Amalaka-shaped bead in porcelanous frit with turquoise-blue glaze. Exactly similar to Roman beads. Height $\frac{3}{8}''$, diam. $\frac{3}{4}''$. See Pl. LXXIV.

N. 006. Square seal of green glass, pierced for cord. On obverse sq. face in upper half, lion (or dragon) rampant, regardant to R.; in lower half, similar creature reversed. Between the two are two char., one of which *may* be intended for a snake. Cf. gryphon on N. xv. 133. b. On reverse 3 incised char. On rectangular (oblong) sides: (*a*) on either side of the hole an incised dot; (*b*) opposite *a*, two parallel incised lines; (*c*) a deep groove; (*d*) opposite *c*, two short grooves. $\frac{7}{8}''$ sq., $\frac{1}{2}''$ thick. Found south of Camp 95. See Pl. L.

N. 007. Fragment of bronze object, unidentified. $1\frac{3}{4}'' \times \frac{3}{4}'' \times \frac{1}{4}''$. Found north of Camp 94. See Pl. LXXIV.

N. 008. Iron buckle, much oxidized. The form is a kind of heart-shape, and the movable point (now fixed by oxidation) strong and tapering. $1\frac{1}{2}'' \times 1\frac{1}{2}''$. Found north of Camp 93. See Pl. LXXIV.

N. 009. Fragment of Chinese mirror, cast bronze; circular, having projecting knob in centre through which a hole is pierced. The edge is thickened; reverse plain. Obverse ornamented with narrow border of radiating flutes within the outer plain rim. Within fluted border are three concentric rings; remaining portion occupied by design in bold relief, part shown on fragment containing a conventional dragon-like creature with a long, thin, and curling tail. Cf. N. 0012. f. g. h. Diam. about $2\frac{3}{4}''$. Found near Camp 95. See Pl. LXXIV.

N. 0010. Metal fragments; beads. Two fragments of plain bronze finger ring; piece of lead wire; two fragments of small bronze hook or clasp, with hole drilled in one piece for rivet. Two small blue glass beads; one piece of broken glass bead. Pebble or piece of water-worn jade. Irregular piece of metal, probably silver. There may be a device on one side, but it is very indistinct.

N. 0011. a. Bronze seal, shank at back broken; irregular trapezium form, bearing countersunk conventional floral design. $1'' \times 1''$. Found near Camp 94. See Pl. LXXIV.

N. 0011. b. Bronze seal, much corroded. Lower portion a square of about $\frac{3}{4}''$ sides with geometrical device. Attached by four corner pieces at top or reverse, a dome-shaped piece. Hollow inside. See Pl. L.

N. 0011. c. Moulded glass beads, two fragments.

[**N. 0012. Objects found near Camp 95.**]

N. 0012. a. Small bronze bell, with attached ring; evidently cast in a mould. See Pl. LXXIV.

N. 0012. b. Portion of hinge, slightly tapering towards end opposite working end. Two holes are drilled in plate—one at each end. At working or hinge end, a tongue of metal, rather more than ⅓ of whole width, and about 1″ long, has been doubled under to form loop through which hinge pin passed. Space between tongue and back of plate to take thickness of metal, leather, or other material to which hinge was rivetted. Plate slightly convex in its length. Length 2¼″, width at wide end ¾″, at narrow end ½″. Thickness of metal about $\frac{1}{16}$″. See Pl. LXXIV.

N. 0012. c. Elliptical bronze split ring, made of wire about ⅞″ thick. In centre of one long side on its internal surface is a depression to extent of nearly half the thickness of the wire, and split in ring occurs in the middle of this thin portion. The ring was evidently intended to be attached to some object which it would hold in this depression, as the ring of a watch holds the knob to which it is attached. $\frac{11}{16}$″ × $\frac{9}{16}$″. See Pl. LXXIV.

N. 0012. d. Bronze split ring, made of round wire, beaten flat at side opposite split. In this flattened part are remains of small rivet, or piece of wire passing through a hole. $\frac{13}{16}$″ × $\frac{12}{16}$″.

N. 0012. e. Portion of bronze ring made of a flat band of metal, ¾″ wide. Diameter about ⅝″.

N. 0012. f. g. h. Fragments (3) of small bronze Chinese mirror. On each fragment are traces, on one side, of decoration in relief. Frag. h. has one edge thickened and of a circular curve, and is therefore evidently part of the edge of object. Cf. N. 009. See Pl. LXXIV. Fragment f. shows a flat circular band, raised from inner edge of which proceed short rays to another band parallel to the first but quite narrow. Beyond outer edge of broad band is part of an ornament, which seems to be designed to fill a spandril shape. f. 1″ in diam.; g. ½″; h. ⅞″.

N. 0013. Objects in metal and glass, found in sand west of Camp 93. 4 small blue glass beads (one broken). 1 fragment of a straw-coloured glass bead. 1 pink pebble or jade bead. Small portion of a thin bronze plate, containing bronze rivet. Small portion of cast bronze plate covered on both sides with closely placed small blunt points, forming a surface something like shagreen. See Pl. LXXIV.

Hemispherical head of a stud. Bronze. See Pl. LXXIV.

Bronze seal, having device on each side, countersunk, too corroded to make out. Pierced longitudinally. $\frac{9}{16}$″ × $\frac{7}{16}$″ × $\frac{3}{16}$″.

N. 0014. a. Fragment of clay slab, hard-burnt, coarse, pierced with holes. Probably used in building to fill window openings, &c., as similar slabs are now used in India.

N. 0014. b. Fragment of pottery, on which a horn-like excrescence, glazed with rich and varied green glaze. Coarse clay. **Miscell. finds.**

N. 0014. c. Bronze arrow-head, cast, having three fine blades set round the centre hollow ferule. The blades are finely curved for cutting and piercing, and extend slightly back from the open end of ferule. Ferule itself pierced by hole in each division between blades, probably to enable a pin to be inserted to hold shaft securely. See Pl. LXXIV.

N. 0014. d. String of beads of various shapes and sizes; of glass, pebble, shell, and pottery. See Pl. LXXIV.

N. 0014. e. Small bronze bell, of kind used on dog-collars to-day. Metal bead is still inside and free. See Pl. LXXIV.

N. 0014. f. Bronze ring bearing seal cut in the metal, representing a deer. See Pl. LXXIV.

N. 0014. g. Two bronze rings (one imperfect) with roughly engraved characters. See Pl. IL.

N. 0014. h. Fragment of bronze, miniature spoon or spatula. See Pl. LXXIV.

N. 0014. i. Bronze ornament, cast, in form of conventional bull's head, horns forming a crescent, with two depressions perhaps for jewels. A hole in forehead seems to have been for fastening. See Pl. LXXIV.

N. 0014. j. Convex ornamental bronze disk with scalloped edge. See Pl. LXXIV.

N. 0014. k. Bronze pellet, pierced.

N. 0014. l. Bone object, oblong, rectangular. Flat on one side, where it is partially corroded or eaten by insects. On Rev. a channel is cut, 1″ broad in widest part. At one end two holes are drilled obliquely, so that they run from the end surface to the concave surface. At opposite end one hole is drilled from obv. to rev., a few irregular notches are cut in one edge. Length 4¼″, width 1½″, thickness ¼″.

N. 0015. Bronze seal ring, countersunk, broken at thin side. Device perfectly preserved, on a well-shaped elliptical cartouche, held at ends between two broadened knob-like ends of ring. A lion statant to L. with tail recurved over back. A bird clings to his chest and appears to be looking into his open mouth. Extreme R. and L. are two S-shaped objects, probably snakes. Elliptical surface $\frac{15}{16}$″ × $\frac{10}{16}$″. See Pl. IL.

N. 0016. Bronze signet ring, having crude design engraved. Perhaps intended to be representation of a face. See Pl. IL.

N. 0017. Gold ornament; made of thin gold plate and filled, or 'loaded' with lac or wax, perhaps portion of a pendant or earring. From thick ring-like body (half missing) projects an elongated knob, joining body by a slightly narrowed necking, round which are two slightly raised fillets. From this the form is ellipsoidal, and

Miscell. finds.

towards the upper end are 4 ovoid jewels of pebble or glass in a gold setting. On one shoulder of body a similar gem. The ornament appears to have been made in two halves, joined together and filled with the lac. Height 1", width ⅜" nearly. Found south of N. VIII. See Pl. LXXIV.

N. 0018. **Pyramidal seal in green** of porcelanous frit. Face has simple (probably symbolical) scroll design, countersunk. The back near apex has a hole pierced for cord. Apex is cut off square. Face 1" square. Height ¾". Found south of N. VIII. See Pl. L.

N. 0019. **Fragment of white glass,** apparently foot of a vessel. Thickness ⅛"; diam. of circular foot 1⅞", height ¾" nearly. Found near N. X. See Pl. LXXIV.

[N. 0020. **Objects found in sand west of N. IX.**]

N. 0020. a. **A number of gilt glass beads,** resembling in shape rice grains, but smaller. A piece of sapphire-blue glass, prism-shaped, drilled as a bead. Four white glass (or shell) beads, very small. A piece of amber-coloured glass, drilled as a bead. ⅝" × ½". See Pl. LXXIV.

N. 0020. b. **Small flat slab** of slate, drilled at one end. ⅞" × 9/16" × ⅛".

N. 0020. c. **Small bell** of usual spherical kind, with loose bead inside.

N. 0020. d. **Bronze ring** in two pieces.

N. 0020. e. **Small piece of bronze,** hook-shaped.

N. 0021. a. **Piece of bronze wire** bent into hook-shape; 2 fragments of bronze.

N. 0021. b. **Metal disk** pierced near edge, with small hole for cord. Probably from a necklace or similar ornament. See Pl. LXXIV.

N. 0022. **Beads:** gourd-shaped, black clay bead (½" long); a black clay bead (7/16" long); a roughly-carved 6-petalled flower, in shell, pierced with two fine holes (9/16" diam.); thin pink pebble or jade bead (9/16" long); fragment of green glass bead, cut cubical with the angles cut away. Evidently all portions of a necklace. See Pl. LXXIV.

N. 0023. **Bronze disk,** broken at one part; concavo-convex. Marked on concave side with small square punch mark. Diam. 1 1/16". Found near N. IX.

CHAPTER XII

THE ENDERE RUINS

Section I.—THE MARCH TO THE ENDERE RIVER

I had already heard at Niya of ancient remains in the desert near the Endere river about half way towards Charchan, and subsequent information as to the existence there of a 'Potai' and other structural ruins decided me to select them for my next exploration. Their distance had supplied a special reason for completing my explorations at the Niya Site as rapidly as possible. During the greater part of my stay there had been a succession of deliciously clear days with bitterly cold nights and mornings, the minimum thermometer usually showing temperatures from 6° to 9° Fahr. below zero. It was striking evidence of the remarkable clearness of the atmosphere that early on the morning of February 11 the Surveyor's sharp eyes distinctly sighted the snowy mountains south of Niya, some 120 miles away. But I knew that such favourable conditions for desert work could not be expected to serve us much longer. I thought of the number of sites that still remained to be explored, and the great distances to be covered between them, before the season of sand-storms would put an end to my explorations, and consequently realized the necessity of setting out for those fresh fields of work as early as possible. Start from Niya Site.

The timely arrival of the camels enabled me on the 13th of February once more to start my caravan back to Imām Ja'far Sādiq. As I passed one ruin after the other familiar to me from the incessant work of the last weeks, my elation over all the discoveries they had yielded mingled with regret at having to leave this fascinating site so soon. But there was the hope of fresh discoveries awaiting me elsewhere and the possibility of return in years hence, by which time a shifting of the dunes might have helped to reveal yet other ruined structures still hidden under their protecting cover of sand.

It would have been difficult to take all my former labourers along to the new site, as the distance was great and the men were well-nigh exhausted by the hard work and exposure of the last three weeks. The fresh set of men needed could only be secured from Niya, and it was hence no small relief when, on arrival at that day's camping-place, information reached me that all arrangements had been made by the local Bēg for the timely dispatch of the fresh contingent. The next day's march brought me to Imām Ja'far Sādiq, the living forest passed *en route* proving even in its winter sleep a great change after the silent sands and ruins amidst which I had dwelt. At the Mazār I was kept busy long into the night by the dispatch of my mails to Europe and India, with the first notice of my recent discoveries, and by the settlement of all accounts with the labourers and the Shaikhs of the pious settlement. Return to Imām Ja'far Sādiq.

I had wished to reach the Endere ruins by striking straight across the desert to the east of Imām Ja'far Sādiq, instead of returning to Niya and thence marching along the Charchan Road. At first all knowledge of such a route was stoutly denied, but in the end one of the shepherds from the Mazār acknowledged that he had more than once visited flocks grazing on the Yārtunguz river, and that further guidance might be secured there. So led by him and a half-crazy devotee of the shrine who claimed to have paid a visit to those ruins, we set out eastwards on February 15. Two miles beyond the Mazār all vegetation was left behind. Then we crossed two steep Dawāns rising to about 150 feet in height, and toiled on through high sandhills for about six miles in a south-easterly direction, until large patches of gravel soil were struck where camels and ponies could march with ease. A supply of ice brought along from the Mazār enabled us to camp that evening at a spot where low tamarisk scrub and some Kumush supplied fuel. On the following day the march was resumed with a general bearing to the east-south-east. After crossing a Dawān about 100 feet high, and next a low gravel-strewn swelling of the ground traversed by occasional dunes, we passed at a distance of about six miles a belt of living Toghraks, which probably receives underground moisture from the marshes subsequently seen thirty miles southwards near the Bileklik Lake. Some five miles further a broad depression was reached with plentiful scrub and Kumush, which our guides called *Suzüje-Daryā*. It was said to receive water through occasional summer floods, manifestly coming from the stream and springs which feed the Sizütke Lake, passed on the Charchan route. After crossing another high Dawān we struck a broad tamarisk-covered belt, which was clearly marked as an old bed of the Yārtunguz river, but now completely dry. It was at once succeeded by a high sand ridge such as usually accompanies these desert rivers, and for the sake of our tired animals I was glad when, from its height, the glittering ice of the stream came at last in sight.

March to Yārtunguz river.

March down Yārtunguz river.

On February 17 we followed the Yārtunguz river, which higher up near the mountains is known as Tolanghuja, down to the area where it is absorbed by the sands. Its breadth, some thirty yards near Hēlyabēg, the grazing-ground where we had camped, gradually diminished in the many short bends which form the river's course northward. On the other hand, the strip of vegetation accompanying the banks, narrow at first, gradually widened, until at a distance of about six miles below our camping-place close Toghrak jungle was entered. A short distance further down, at a point appropriately called *Arik-aghzi* 'mouth of the canal' (misprinted Irik-aghzi in map), and marked by many fine trees, lies the modest *band* which helps to guide water into the small canal irrigating the fields of Yārtunguz-Tārīm some thirteen miles further north. The stream, now only some ten yards broad, turns from here to the north-north-west, evidently cutting its left bank further and further into the sandy ridge westwards. Two miles lower down its flow during the winter months seemed to cease. The wide expanse of forest, through which we continued our march due north, and which maintained a breadth of at least three miles, is still reached by the summer floods. Through it lies the old main bed of the river, which we found empty except for occasional shallow depressions still retaining frozen sheets of water. At Kala-sulaghi were passed the first clearings for cultivation, now again deserted and overgrown with young shrubs. About two miles further north we entered the present oasis know as Yārtunguz-Tārīm, with its holdings scattered amongst extensive fields, chiefly of wheat, Indian corn, and cotton.

Terminal oasis of Yārtunguz river.

From Abdul Karīm, an intelligent and manifestly well-to-do 'Dehkān', who welcomed me at the substantial farm of Jigdalik-bulung, I learned that the small colony had been formed only during Niāz Ḥākim Bēg's *régime*. His father had come from Faizābād in Badakshān on a pilgrimage to Imām Ja'far Sādiq, and on settling down received a grant of land here. He, with the help of some other families of settlers, made the canal, the head of which we had passed at Arik-aghzi, and their descendants still own whatever land has been brought under cultivation. They are sufficiently well-off to employ twenty to thirty labourers, and to the number of hands thus available the extent of the clearings has been restricted, though there is fertile ground easily capable of irrigation close at hand amply sufficient for several villages.

Vagaries of terminal river course.

For the last five or six years preceding my visit the terminal course of the Yārtunguz stream has shown a decided tendency to shift its main channel westwards, where it was said to lose itself at a distance of 'two or three Tāsh' from Jigdalik-bulung. In consequence of this change the existing *band* at the canal head has failed at times to divert into the latter as much water at the proper season as the area under cultivation would need. Unable from want of adequate labour to cope with the difficulty by a more elaborate 'band' or the construction of a new canal, the small settlement has found it expedient to start fresh cultivation at Yilba-sarīgh, a grazing-ground near the end of the new main channel, as an alternative in years when irrigation from the old canal is likely to prove insufficient. Yet even in the same years sudden floods may extend along the old bed and prove an equal source of trouble to the cultivators. I regret that the necessity of pushing on to Endere prevented me from giving time to a detailed inspection of the little colony and the surrounding ground; for, as I had occasion to explain in my general observations at the close of the preceding chapter, it presents, on a small scale but in a typical form, the characteristic features of a terminal oasis such as we must suppose the Niya Site to have been. So much, however, seemed clear from what personal observation and local information showed me, that given an adequate supply of labour for the systematic maintenance and extension of embankments and canals, the area of cultivation could be greatly increased. If fields were to replace the expanse of luxuriant jungle which at present covers an area at least eight miles long and from three to five miles broad, the terminal oasis of the Yārtunguz river might well present conditions approximating to those we must assume to have once existed round the ancient site below Imām Ja'far Sādiq. What portion of this area could possibly be reclaimed at the present day with the water-supply actually obtainable from the Yārtunguz river, is a question to which a safe answer would be possible only on the basis of an exact survey by a competent irrigation engineer. So much, however, may be considered as probable, that the vagaries of the river would prove a far more constant danger to a relatively large settlement than any advance of the desert dunes. According to the explicit assertion of Abdul-Karīm, who knew the ground well since his youth, the dark ridges of high sands which fringe the fertile area on the east and north-east, and form a striking contrast in colour to the yellow fields and the dry Kumush beds, had practically retained their position. The river's shift westwards was remembered to have commenced long years before the difficulty about feeding the canal was seriously felt.

Desert march to Endere river.

The inquiries which the situation of the colony naturally suggested as to any ancient remains possibly extant in the direction of the desert proved fruitless, but our energetic Darōgha succeeded at least in hunting up reliable guides who were acquainted with the Endere river and the ruins in its vicinity. It was equally useful that our supplies of foodstuffs and fodder could be replenished from the ample stock of the little settlement, and some labourers hastily recruited overnight for excavation work. On the morning of February 18 we set out across the desert to the forest belt of the Endere stream. Immediately after leaving the open plain of Yārtunguz-Tārīm a formidable 'Dawān' of sand, reaching a height of about 180 feet, had to be surmounted. Then we crossed a depression about two miles broad, covered in great patches with soda efflorescence. It marks, no doubt, an old flood-bed of the Yārtunguz river, a small saline stream still making its way along the western edge. Two more ridges of dunes, less high than the first, and strips of Kumush-covered low ground between them were crossed before we reached that evening's camping-ground in a dreary expanse covered here and there with low sand-cones. The dead tamarisk scrub which strewed their slopes sufficed for our camp fires, while the ice brought along from the Yārtunguz-Tārīm saved us the trouble of digging a well. A fairly strong wind had been blowing for the greater part of the day from the north or north-east, and the dust-haze it raised showed an ominous persistence.

Our march so far had been almost due east. But on leaving Yantak-chaval on February 19 our guides struck to the north-east. The reason for the change of direction became clear when, after crossing a great 'Dawān' some 120 ft. high, we found ourselves traversing a broad stretch of low undulating dunes where progress was relatively easy. Plenty of living tamarisk and occasional little groups of Toghraks showed that subsoil water was near, and the general impression I gained was that we were crossing here what may once have been the terminal area of a river. The latter could only have been an earlier course of the present Endere river. Towards the east this area was bordered by a long-stretched chain of high dunes, in which it was easy to recognize from afar the ridge flanking the left bank of the actual Endere river-bed. From its darker colour it was known to our guides by the appropriate name of *Kizillik*, the 'Red (sands)'[1]. For fully two miles our march lay over this great accumulation of dunes, and then we emerged on the western edge of the belt watered by the Endere stream.

Deserted terminal oasis of Endere river.

The channel which we now ascended towards the south-east was called by our guides 'the old Daryā' of Endere. Here, too, the river has in recent years shown a tendency to shift westwards, so that we found quite a respectable sheet of ice, from 10 to 20 yards in width, covering what previously had been a deserted dry bed. On the other hand, this return of the summer floods to the earlier channel was causing the 'New Daryā', said to have been formed further westwards some twenty years before, now to receive annually less and less water. It was with this change of course, and not with any diminution in the quantity of the water brought down by the river, that my guides connected the abandonment of the little colony of 'Endere-Tārīm', which had been formed on the 'New Daryā' at a point apparently four to five miles to the north-north-west of where we first struck the river. It was impossible to spare time for a visit to this modern representative of the terminal oasis of Endere, or for a survey of the river-courses down to their actual termination. Judging from the guides' statement, the summer grazing-grounds of the Endere shepherds may extend northward for a day's march beyond the now deserted 'Tārīm'.

March along Endere river.

On February 20 we marched along the 'Kōne-Daryā' up to the point where, near the grazing-ground of Kokul-toghrak, the 'New River' was branching off from it, and thence followed the main river bed upwards. Not far from a rustic Ziārat we crossed to the right bank. The river was here about 20 yards broad, holding under its ice-sheet about 2 feet of water, but the well-defined, steeply-cut banks rising 6 to 7 feet above the level of the ice indicated a considerable volume of water at other seasons. At a deserted hut of rushes marking the shepherd station (*kīchik*) of Kara-öchke-öltürgan ('where the black goat sat') we left the ponies behind, and struck into the desert south-eastwards. The belt of vegetation, the width of which, owing to the persistent haze, it was difficult correctly to estimate as long as we kept by the river bank, proved here very narrow. Hence, where we camped for the night, though scarcely more than two miles in direct distance from the river, there was only bare eroded ground with here and there low sand-cones covered by scanty dead tamarisk scrub.

[1] Compare the high *Kizil-Kum* range of dunes flanking the upper course of the Keriya Daryā (see map).

Section II.—EXCAVATION OF THE ENDERE TEMPLE

On the morning of February 21 we continued south-eastwards, and after a march of about three miles sighted in the distance what the guides called the 'Potai' of the 'Kōne-shahr'. Even from afar I could recognize through my glasses that it was a Stūpa, as I had, of course, assumed when this feature of the site was first vaguely mentioned to me at the Mazār. It was interesting to note that on approaching the Stūpa we passed through a belt of sand-cones thickly covered with living tamarisk and the scrub known as *Ak-tiken*. It was here, at a distance of about three-quarters of a mile from the Stūpa, that we found an old well, which, when cleared to a depth of 9 ft., yielded sufficient but rather brackish water. About half a mile west of the Stūpa we reached eroded ground plentifully strewn with pottery fragments, and this continued up to the ruin itself, which such typical 'Tati' débris surrounded on all sides. A rapid inspection of the Stūpa proved that the ground near it had been eroded to a depth of from 10 to 15 ft. in different places, as clearly seen in Fig. 50, which shows the ruin from the south-west side, together with the bare loess and low sands behind it[1]. Nor was I surprised to find that the ruin had been dug into in two places, no doubt in search of treasure.

Approach to Endere Site.

The contingent of labourers ordered up from Niya had arrived just when I was nearing the Stūpa. Considering the great distance, some 120 miles, from which the men had been brought, and the difficulty of communicating with them over wholly uninhabited ground, I felt not a little pleased at this well-managed concentration, which enabled me to start excavation work at once. Leaving, therefore, the accurate survey of the Stūpa for later, I pushed on with increased eagerness south-eastwards to where the remains of 'old houses' were said to exist, which held out better antiquarian promise than either Stūpa or the Tati around it. For about a quarter of a mile the pottery-strewn ground continued, the eroded bare loess soil being exposed in wide patches amidst dunes not exceeding 5 to 6 feet. Among the pottery fragments, mostly small and showing long-continued abrasion, I noticed a good deal of coarse black terra-cotta and occasional pieces retaining a green glaze. Representative specimens of these materials will be found described in the list of antiques (E. 006). Then followed dunes somewhat higher, and among them occasional sand-cones with scanty tamarisk growth. Shrivelled trunks of dead Toghraks emerged here and there from the drift-sand in the depressions between the dunes, but no trace of structural remains appeared until I had arrived quite close to the dunes surrounding the ruins which the guides from Imām Ja'far Sādiq and Yārtunguz had spoken of as 'old houses'. The rows of wooden posts rising above the sand were a familiar sight. But the high brick walls of some large building and the remnants of a massive clay rampart encircling the ruins presented a novel and very striking feature.

'Tati' around Stūpa of Endere.

Going over the ground I soon realized that the extant portions of the rampart, which was largest and best preserved on the south, had belonged to a circumvallation approximately circular and enclosing an area which the subsequent careful survey (see plan in Plate XXXVI) proved to be nearly 420 ft. in diameter (including the thickness of the walls). The large building towards the eastern segment of the circumvallation (E. iii) was found for the greater part almost clear of sand, and clear, too, of objects that could claim archaeological interest. But west of it a broad dune stretched across the interior of the fortified area, and near the centre of the latter I noticed rows of wooden posts just rising above the sand. Their arrangement in con-

First inspection of ruined fort.

[1] The line where the lowest course of masonry and the original ground-level meet is marked by the foot of the 10 ft. measuring staff; see also plan in Pl. XXXVII.

centric squares at once recalled the temple cellas with enclosing passages I had excavated at Dandān-Uiliq. A little experimental digging at the south-eastern corner of the inner square soon brought to light small fragments of soft and much-decayed stucco, which had belonged to a large-sized image. So the whole of my little force, counting over twenty 'Madigārs' from Niya, soon supplemented by shepherd-guides and every able-bodied man whom the latter could impress at the shepherd stations higher up the river, was at once set to work here.

Discovery of temple cella.

Within an hour I had conclusive proof that my surmise was right. From the sand which covered the interior to a height of about 5 feet on the east, but increased to fully 7 ft. towards the west, there emerged the timber and plaster walls of a square cella enclosed within a passage, no doubt intended for circumambulation. The outer walls of the passage had decayed almost to the ground, and a good deal of excavation was needed there before the clearing of the interior of the cella could be safely proceeded with. Nevertheless, before work ceased with nightfall more stucco fragments had furnished some indication of the general decoration of the shrine, while three finds of fragmentary paper leaves with Brāhmī writing, among them three halves of folia clearly recognizable as belonging to a Buddhist canonical text in Sanskrit (E. i. 2), had helped me to form an approximately correct idea as to the age of the ruined shrine. All these turned up on the east side of the cella, lying in loose sand from 1 to 2 feet above the original floor. On the following day the excavation proceeded sufficiently to show me all the main structural features of the small temple. But it was not until the 23rd of February that the interior was completely cleared and the rich haul of interesting MSS. and other varied finds carefully gathered.

Construction of temple cella.

The cella formed a square of 18 ft. 4 in. inside, enclosed by walls of timber and plaster having a thickness of 10 in. As seen in the detailed plan (Plate XXXVI), the walls were only roughly orientated. Their construction differed from that observed at Dandān-Uiliq and the Niya Site by having no matting within the plaster, but the timber framework was massive, as seen in Fig. 49, which shows the cella after excavation, and specially strengthened by diagonally-placed rafters. The main posts of both cella and passage stood at a uniform height of 9 ft. But it may well be doubted whether they indicate the original height of the roof, seeing that the central group of relief statues must have reached close to that height. The inner faces of the walls seemed to have received a rough coating of stucco, but showed no trace of fresco decoration. The entrance lay to the east. The enclosing passage, the walls of which were of similar construction but far more decayed, was 5 ft. wide; its plastered floor lay 3 ft. higher than that of the cella, a circumstance which had helped, no doubt, to protect the lower portions of the cella walls. On the east side the outer wall of the passage had for the greater part disappeared entirely, together with the portion of the floor facing the entrance, possibly as a result of the early burrowing effected here by treasure-seekers, to which I shall have occasion to refer below.

Sculptures in corners of cella.

The four corners of the cella were occupied by plaster images, almost wholly detached and standing each on a base meant to represent an open lotus with the petals pointing downwards. Those in the north-west, north-east, and south-west corners alone survived, and they, too, only in their lower portions, as seen in the photographs reproduced in Plates XI, XII. Of the south-east corner statue nothing remained but the small undefinable fragments which had turned up at the first trial excavation. By the side of the statues in the north-east and south-east corners there must have stood in each case a small figure (*b*, *e*, in plan), but of these only the lotus-shaped pedestals remained, the sculptures having crumbled away owing to the decay of the east wall behind them. It is impossible to say what divinities all these figures

48

RUINED BUILDING E. III, WITHIN ENDERE FORT, SEEN FROM SOUTH-WEST.

CHINESE SGRAFFITO.

49

INTERIOR OF RUINED TEMPLE CELLA E. I, ENDERE FORT, SEEN FROM EAST AFTER EXCAVATION.

represented, though the position of those in the corners suggests their having been intended as Lokapālas or 'Guardians of the regions'. The drapery resembled that of the Dandān-Uiliq figures of divinities, both relief and painted, and unmistakably showed derivation from Gandhāra models.

Technique of stucco sculptures.

The images, which appear to have been approximately life-size, were modelled in a very coarse plaster, consisting of clay mixed with straw and other vegetable matter. To support this very friable material a wooden frame was used, built crosswise and supplemented by a packing of reed, as partly seen in the photograph of the south-west statue (Plate XII). The surface of the sculptures was coated with a very friable sort of stucco, red in colour, and consisting, as the analysis with which Professor A. H. Church has kindly supplied me (see Appendix *F*) shows, of loess which contains a considerable proportion of true clay. With this stucco was generally found mixed a very fine vegetable fibre, the exact nature of which has not yet been determined. The colour once laid over this coating had peeled off to a great extent, except where protected, as within the folds of drapery. White appears to have been the colour of the robes in the north-east and north-west corner statues, while the drapery of the south-west figure showed alternate stripes of reddish-brown and dark blue. Of the upper portions of the statues only very small fragments were found, which points to the friable material having completely crumbled away before the layers of drift-sand had reached a height sufficient to protect them. Many of the fragments, especially those which contained the coarse greyish clay of the interior plaster, were so soft as to break up at the slightest touch.

Specimens of stucco ornaments.

The pieces which were safely brought away will be found described in the list (E. i. 01–015). It is significant that the great majority of them are small ornaments modelled entirely of the fine red clay, mentioned above as a coating. These had undoubtedly been used in *appliqué* fashion, and having fallen off early after the abandonment of the shrine had found safe protection under the accumulating cover of sand. Most of them turned up on or near the base of the north-west and north-east images. Such are the jewelled ornaments (E. i. 01, 02, 014, 015) which, like the strings of Maṇis (E. i. 06, 09, 010), may be supposed to have formed part of necklaces, or else to have belonged to arm-bands or tiaras. Specimens are reproduced in Plate LXXVIII. A similar ornament appears still attached to the girdle of the south-west corner statue as seen in Plate XII. The Vajra-shaped ornaments, E. i. 07, 08 (see Plate LXXVIII), may possibly have belonged to the relief decoration of a nimbus similar to those seen behind some of the Rawak Vihāra statues (see Fig. 61, Plates LXXXIII, LXXXVII). E. i. 013. a shows a corner of the hanging robe of the south-west statue, which must have broken off early, and thus has preserved its colouring well; see Plate LXXVIII. The same Plate shows the elongated ear-lobe (E. i. 013. b), which had evidently belonged to the north-west figure; a poorly preserved portion of the head and right ear of the north-east statue (E. i. 05), as well as some life-size fingers (E. i. 04, 011). In the latter, just as in the ornaments, a core of wood or cord was used to give strength to the plaster.

Central base for images.

The centre of the cella was occupied by a massive octagonal base or platform, 9½ ft. long and 7½ ft. broad, which Fig. 49 shows in the foreground. It was built of sun-dried bricks with a thick coating of plaster. Each of its facets measured 4 ft. in width, six being slightly curving, while those facing east and west were straight, as seen in the detailed plan. The base proper rose to a height of 2 ft. 8 in. above the floor. On it four life-size relief statues in stucco must once have stood, the lotus-shaped pedestals of three being still more or less intact, while that of the fourth facing eastwards had disappeared owing to the damage done to the base by an excavation from that side. In the centre, and serving as a backing to the statues, there rose a mass of sun-dried brick and plaster, the curious shape of which the detailed plan will explain. The statues once occupying each of its four main faces had completely crumbled away, except on the west and north, where the feet with parts of the draped robe below the knees survived. As the feet measured in each case 1 ft. the statues must be supposed to have been about life-size. The arrangement of the drapery remains and the shape of the stucco backing suggested that the statues were seated. There was nothing to indicate what particular Buddhas or Bodhisattvas they represented. The work was in relief much lower than that of the corner images. Remains of oval halos reaching down to the level of the feet could be distinguished on the plaster background. It appears probable that, for purposes of lighting, the cella was provided with a raised roof in the centre. With the decay of this the central group of stucco sculptures would necessarily have been much exposed to atmospheric influences, which may, together with their raised position, account for the scantiness of the remains.

Fresco decoration of central base. The east and south-east facets of the base were found to have been completely destroyed by an old excavation, undoubtedly the work of treasure-seekers, which had been carried towards the centre, but could not be wholly cleared of sand from risk to the extant structure. Of the way in which this burrowing affected the MS. finds made in the shrine I shall have to speak presently. It also caused the almost complete loss of the fresco paintings with which the portion of the base facing the entrance appears to have been decorated. On the north-east facet the fresco remains showed two rows of seated male figures, seven in each, apparently Buddhas or Bodhisattvas. The colours had badly faded, and only the outlines could be made out; even these were effaced in many places. In spite of this poor preservation the work seemed distinctly superior in drawing and execution to the corresponding type of wall-decoration in the Dandān-Uiliq shrines. Of the fresco decoration of the eastern facet I fortunately recovered an interesting fragment in the small piece of painted plaster (E. i. 012), which was found lying in the loose sand filling the old excavation made in front of the base. Its careful reproduction in Plate LXXIX shows the delicate and harmonious colouring which the little fresco piece retains in spite of its faded condition. In it the head and shoulders of a figure surrounded by a green nimbus and probably seated can be clearly made out, with portions of the robes of other figures. Mr. Andrews' description in the list below indicates interesting technical details as to the preparation of the stucco ground.

Finds of Brāhmī MSS. leaves. The hopes of MS. remains which the first afternoon's experimental digging had raised were amply realized by the numerous finds made in the course of the subsequent excavation. These finds derive particular interest, not only from the variety of languages and texts represented among them, but also from the curious conditions in which the numerous MS. portions were recovered. That the three leaves (E. i. 2) of a Buddhist text in Sanskrit, discovered on the first day about 1 ft. above the floor of the east side of the central base, had belonged to a larger Pōthī was at once clear to me, from the pagination numbers on their margins and the string holes which showed them to be left halves. I was encouraged in my hope for the recovery of more of this text when on the following morning there turned up towards the north-east, and also not much above the floor, two more packets of broken folia (E. i. 4, 5), containing between them 12 left halves and 4 right halves of the identical MS. Finally, when after excavating the rest of the cella the portion between the east side of the central base and the entrance could be completely cleared of sand, there were discovered in the cutting about 2 ft. deep which treasure-seekers had made into the floor of that portion two more packets of the same text (E. i. 39, 40), containing 8 left and 19 right halves of folia, besides 2 detached half-leaves (E. i. 41, 43). From a comparison of the total number of left and right half folia, and from the indications furnished by the pagination figures, I concluded at the time that we possessed in these leaves the greatest portion of the MS. Dr. Hoernle's exact analysis[2] has fully confirmed this conclusion, and has made it probable that all the 46 leaves of which the MS. when entire appears to have consisted are represented among the recovered pieces. Three of the folia are complete, their right and left halves having been pieced together with certainty (for a specimen see Pl. CIX). The full size of the leaves must have been about 14 by 3 in., the text being written in five lines on each side in clear upright Gupta characters, which Dr. Hoernle assigns to the seventh or eighth century. The text is a Buddhist canonical work of the Dhāraṇī class. From the uniformly faded and perished state of the lowest leaves in E. i. 5 (left halves) and E. i. 40 (right halves) it appears very probable that the Pōthī in its entire state must have lain for a long time before the left and right portions got separated. Possibly the break occurred in course of the excavation made into the central base; but there can be no doubt that this burrowing had caused the dispersion.

The experience gained at the Dandān-Uiliq ruins suggested from the first that the MS. leaves found had originally served as votive offerings. Proof for this surmise was soon forthcoming in plenty. Close to the north foot of the central base there turned up two closely-packed rolls of paper, which might have fallen from the pedestal of the image above. One contained the fragmentary leaf E. i. 6, about 9 by 3 in., written in slanting Central-Asian Brāhmī and a non-Sanskritic language, which Pl. CIX reproduces. The other roll (E. i. 7), which was secured with a paper strip still closely wound round the centre, and which could be opened only in the British Museum, resolved itself into four large folia measuring 18 by $3\frac{3}{4}$ in., written in bold upright Gupta characters (see Plate CXI). As Dr. Hoernle's examination has shown[3], they must have been taken from an extensive work composed in the non-Sanskritic language which he tentatively distinguishes as 'proto-Tibetan'[4]. The frequent occurrence of the Sanskrit term *bhaiṣajya* suggests that its subject was medical or magical. Among the remaining finds in Brāhmī there were three small pieces of leaves (E. i. 9, 33), which lay on different parts of the projecting moulding at the foot of the central base facing to the north, west, and south-west, respectively. As two of these pieces must have belonged to the same leaf, their position made it at once clear that it was torn up purposely in order to furnish fragments for deposition before different images.

Detached MS. leaves as votive offerings.

This conclusion forced itself upon me in a still more striking way when all round the foot of the central base, as well as on the floor near the pedestals of the images in the north-west, south-west, and south-east corners, there turned up besides a large and well-preserved sheet of Tibetan writing (E. i. 11; see Pl. CXVIII) and a number of miscellaneous Tibetan fragments (E. i. 15, 19, 20, 25, 31; see Pl. CXVIII), which, by the uniformity of the clear, well-formed writing and of the paper, I could easily recognize on the spot as portions of an identical Pōthī. The widely different positions in which these fragmentary pieces were recovered will be found noted in the list. Some picked up underneath the walls or on the floor might have been blown away from their original place of deposition; but most of them had evidently been retained by the accumulation of drift-sand in the place where the last owner of the MS. had intended to propitiate with them the various divinities. The fragments recovered, 27 in all, have been proved by Dr. Barnett to have originally formed part of a large Pōthī, about $17\frac{1}{2}$ in. long and $2\frac{5}{8}$ in. broad, containing the canonical text of the *Śālistamba-sūtra* embodied in the collection of the

Finds of Tibetan MSS.

Tibetan text of Śālistamba-sūtra.

[2] See his note xxiii. below, p. 439.

[3] See Dr. Hoernle's note xxvi. below, p. 440.

[4] Comp. regarding this language, represented also among the Dandān-Uiliq finds, above, pp. 271 sq.

Kanjur[1]. By a painstaking comparison with the text which the latter collection contains Dr. Barnett has succeeded in determining the correct order of all the fragments, even the smallest, and has thus been able to prepare the edition presented in Part i. of Appendix *B*. There, too, will be found Dr. Barnett's annotated transcriptions of the miscellaneous fragments of religious works (Part ii.), and a full translation, with exegetical notes, by the Rev. Mr. Francke, of the two religious poems which the completely preserved sheet (E. i. 11) has proved to contain (Part iii.).

Analysis of paper in Tibetan MSS.

Referring for details to Dr. Barnett's introductory remarks on the Śālistamba-sūtra MS. and to the *Preliminary Notice* of my Tibetan MS. finds previously published by him[6], I shall briefly indicate here the main facts as to the material, &c., of this MS., and the philological and historical interest attaching to its text. Several of the pieces consist of practically full half-leaves like E. i. 24 + 21, 10 + 13, which Plate CXVII reproduces. In the arrangement of the lines (five per page) and the string-hole the MS. does not differ materially from the early Brāhmī Pōthīs previously described. But here the writing is confined to one side of the leaf only, a peculiarity which from the first attracted my attention. It has found its explanation in the interesting results of the detailed microscopical and chemical analysis to which Professor J. Wiesner kindly subjected the paper used in these leaves[7]. His investigations have proved that the paper consists wholly of the well-macerated raw fibres of a Thymelaeacea, in all probability a *Daphne* plant, such as *Daphne papyracea*, which is still used in the preparation of modern Nepāl paper. As this plant and kindred species of Daphne are not to be found in Eastern Turkestān, it becomes highly probable that the MS. was not written in the country, but imported from Tibet. An equally notable difference from the paper of the other ancient MSS. excavated by me at this site, as well as at Dandān-Uiliq, is presented by the very peculiar method adopted for making the body of the paper impermeable to liquid ink, and thus its surface more suitable for writing. Professor Wiesner has discovered that this object was attained here not by 'sizing' with a glue of starch, a method of which various fashions can be traced in the early Turkestān papers[8], but by heavily 'loading' the paper with unaltered starch of rice flour, and on that side only which was intended to be written upon. Professor Wiesner has discussed in full detail the interest which this method of 'loading', not previously observed by him in any ancient Asiatic papers, presents for the historical development of early paper manufacture[9]. For our purposes it is also important as further evidence of the importation of this particular MS. Of the paper used for the minor Tibetan text fragments, including the sheet E. i. 11, Professor Wiesner has proved that it agrees in all main characteristics with the paper of the old Brāhmī MSS. and Chinese documents contained in Dr. Hoernle's and my own collections[10]. These texts may, therefore, be supposed to have been actually written within the Khotan region.

Philological importance of Tibetan MS. finds.

The recovered fragments of the Śālistamba-sūtra represent about one half of the text as contained in the Kanjur, and known also from quotations of its Sanskrit original and from Chinese versions[11]. The text differs but little from 'the version of the Kanjur, which indeed

[5] The first identification of the text was made independently by Professors C. Bendall and L. de la Vallée Poussin, the latter recognizing it in the specimens reproduced in Plate XVI of my *Preliminary Report*; see *J.R.A.S.*, 1903, p. 113.

[6] See *J.R.A.S.*, 1903, pp. 109–14.

[7] Compare Professor Wiesner's article *Ein neuer Beitrag zur Geschichte des Papieres*, reprinted from the *Sitzungsberichte* of the Imperial Academy, Vienna, cxlviii, 1904, pp. 14–21.

[8] As, e.g., in the Chinese document D. vii. 3. See Wiesner, loc. cit., pp. 11 sqq.; also Hoernle, *J.R.A.S.*, 1903, p. 678.

[9] Comp. ibid., pp. 18 sqq.

[10] See ibid., p. 21 note.

[11] See *J.R.A.S.*, 1903, pp. 110, 112.

seems to be little more than a later revision of it, with occasional expansions and substitutions of glosses for earlier and more ambiguous terms'. Yet the philological importance of these fragments is great; for together with the smaller Tibetan text fragments and the sgraffiti to be mentioned presently, they are the earliest specimens of Tibetan writing so far known. The archaeological evidence set forth below clearly proves that none of the MS. remains discovered in the Endere shrine can have been deposited there later than the eighth century. In the light of the chronology thus fixed the archaic peculiarities of orthography, first noticed by Dr. Barnett in the Śālistamba-sūtra, and also by the Rev. Mr. Francke in the two religious poems, assume their true significance. But it is, perhaps, even more noteworthy that by the side of these pre-classical spellings the latter pieces, as fully explained in Mr. Francke's *General Note*, furnish also instances of an orthography agreeing with modern dialectal forms. I am not competent to express an opinion as to the very interesting questions which, as forcibly set forth in the remarks of my learned collaborators, these observations are bound to raise as to the real age of the invention of the Tibetan alphabet and of the first introduction of Buddhism into Tibet[12]. But it appears to me that the fact of the Endere texts showing a writing which does not differ from the modern *dbu-can* script may well deserve consideration in this connexion. Seeing that by the evidence of our Endere texts this script is shown to have already assumed in the eighth century that final form in which it continues to the present day, while the classical orthography so intimately connected with it is proved to have been at that time already archaic, it seems difficult to resist a doubt as to the correctness of the tradition which places the invention of both the Tibetan orthography and characters only about a century earlier. Historical questions raised by Tibetan finds.

But these miscellaneous votive deposits of MSS., or pieces of such, in Brāhmī and Tibetan scripts were not the only written remains unearthed in this small temple. A very interesting discovery was that of two tiny fragments of birch-bark, showing each a few characters in Brāhmī, which were found sticking to the plaster of the south wall surface about 1 ft. above the floor and about 6 ft. from the south-east corner. There was nothing to indicate how they had got there, and what the character might have been of the MS. leaf from which they had become detached. The few Akṣaras visible in each fragment seemed Sanskrit, and the writing of an upright Gupta type, which looked older than that of any Brāhmī MS. finds at this site or Dandān-Uiliq. Curiously enough, a few thin scraps of birch-bark, with traces of a character or two in Brāhmī, turned up among the sand and débris of the hole which had been dug into the central base. Is it possible that the latter once contained some deposit, as usually inserted into the base of Stūpas? Fragments of Brāhmī text on birch-bark.

In different places on the floor, but in no case on or near an image pedestal, I discovered three small pieces of paper with Chinese writing (E. i. 8, 36, 44). Though they are mere fragments, M. Chavannes' translation, contained in Part iii. of Appendix *A*, clearly shows that they belonged to secular documents, such as those found in the Dandān-Uiliq dwellings and shrines, and not to religious texts. In E. i. 44 we have manifestly the portion of an official record, for it mentions the petition of a certain functionary, as well as the 'commander-in-chief of the army of *Tso yü lin*, *Wang* (?) *Chih-chiang*'. E. i. 8, 36 are scraps of papers manifestly relating to private affairs, possibly petitions. None of them contains any date, yet even thus they may claim chronological value; for in view of what we have learned above as to the events which finally brought to a close Chinese supremacy over Eastern Tūrkestān as main- Fragments of Chinese documents.

[12] Compare in particular *J.R.A.S.*, 1903, p. 112.

tained by the T'angs[13], these finds of Chinese documents would suffice to show that the Endere temple and the structures surrounding it must have been abandoned not later than the end of the eighth century.

Chinese sgraffito dated 719 A.D.

Fortunately a dated Chinese record inscribed on the walls of the shrine itself supplements this *terminus ad quem* by more precise chronological information. I mean the Chinese sgraffito in three rows of characters which the clearing of the west wall of the cella brought to light close to the left of the north-west corner image, as seen in the photograph reproduced in Plate XI. Owing to the decay of the upper portion of the wall the top of the second and third rows is lost, but luckily the shorter first row on the right, containing the commencement of the record, has escaped mutilation. As shown by M. Chavannes' translation in Part iii. of Appendix *A*, it states the date when the record was inscribed as the seventh year of the K'ai-yüan period, or 719 A.D. Owing to the original surface of the rough and friable wall plaster, into which the characters had been scratched with some blunt-pointed instrument, having peeled off in parts the first character 開 *K'ai* has become indistinct. It was, therefore, at first thought possible that it might read 貞 (*Chêng-yüan*), which would have made the year intended correspond to 791 A.D. But M. Chavannes' reading of it has been confirmed by three Chinese literati as well as by Dr. Bushell, who all, after careful examination of my photographs, arrived independently at the same reading, and it is supported by internal evidence in the text of the sgraffito itself.

Historical bearing of sgraffito record.

The latter, as far as it can be interpreted in its incomplete state, distinctly mentions the 'Four Garrisons' and the 'Great Fan', i.e. the Tibetans, as well as the latter's officers. It further records that 'the high dignitary of the *t'ai-ch'ang* (court of sacrificial worship), *Ch'in Chia-hsing*, returned to the district placed under his orders'. Whether it was this personage or some other Chinese dignitary who, in a preceding passage, is said to have 'heard that his commissioner of troops and of the cavalry was dead' must remain uncertain. Nor has M. Chavannes been able to discover Ch'in Chia-hsing's name in Chinese historical records. We have seen above that in 790 A.D., or at the latest in 791 A.D., Chinese authority in Eastern Turkestān, after having from about 766 A.D. maintained itself only with difficulty and in complete isolation from the rest of the empire, finally succumbed to Tibetan invasion[14]. It seems difficult to believe that in the very year when the effacement of Chinese dominion had been completed and the 'Four Garrisons' had ceased to exist in name as well as in reality, a Chinese dignitary should have recorded on the walls of a Buddhist shrine on the confines of Khotan his return 'to the district placed under his orders'. On the other hand, we know that from 714 A.D. onwards the Tibetans had been annually harassing the Chinese borders, and that in alliance with the Arabs and the rebellious Western Turks they had invaded the 'Four Garrisons' about 717 A.D.[15] From 719 A.D. onwards, however, a series of records in the T'ang Annals attest the gradual reassertion and extension of Chinese power in the Tārīm Basin and the neighbouring regions[16]. These successes of the Emperor Hsüan-tsung's policy seem to have been attained rather by diplomacy than by armed power. Possibly we have some trace of this in the respectful reference our sgraffito makes to the 'Great Fan', the very title by which the Tibetans designate themselves in the Lhasa inscription of 822 A.D. recording a treaty with China.

[13] See above, pp. 63 sqq., 176; also chap. IX. sec. vii., p. 284.

[14] See above pp. 63 sqq., also Appendix *A*., part i.

[15] See above, p. 62; Bushell, *The Early History of Tibet*, p. 26; Chavannes, *Turcs occid.*, p. 284, note 2.

[16] Comp. *Turcs occid.*, p. 292.

However this may be, the prolonged presence of Tibetans at the site is attested not merely by the votive offerings of Tibetan MSS. already described, but also by a series of Tibetan sgraffiti found scratched by various hands, often very cursive, into the north and south walls of the cella. Those on the latter side have been reproduced in Plate XII, while all of them, as far as they are decipherable from the photographs and eye-copies taken by me, will be found translated and learnedly commented upon by the Rev. Mr. Francke in Part iv. of Appendix *B*. They appear to be votive, recording offerings made to certain divinities, and indicating the benefits piously hoped for in return. A journey 'to the land on the other side (Tibet?)', on which the donors of a 'wool-ox' expect to meet with 'wealth, food, and grass', is indicated as the special occasion of the offering recorded in the sgraffito towards the south-west corner (to the right in Plate XII). These Tibetan scrawlings looked on the whole less well-preserved than the Chinese sgraffito, and there was certainly nothing about them to indicate their being later in date. On the contrary, the difference in state of preservation between the Chinese sgraffito on the west wall and an almost completely effaced and illegible Tibetan scrawl on the same wall suggested the reverse conclusion. Tibetan sgraffiti on cella walls.

No direct chronological clue is furnished by the Tibetan sgraffiti; but it can be considered quite certain that the date when they and the Chinese inscription were scratched into the walls could not have preceded by many years the deposition of the various manuscripts and the subsequent abandonment of the shrine. The rough and friable plaster of these cella walls was not a material that could, when exposed, remain intact for a long period without repair, and with its renewal all these casual scratchings as well as the Chinese record would, no doubt, have vanished. Taking into account the plainly attested date of the latter, this consideration fixes approximately the middle of the eighth century as the latest possible time for the abandonment of the shrine, and implicitly also for the production of the manuscripts found in it. This conclusion fully accords with the fact that among the eight Chinese copper coins found either within the Endere fort or in its close vicinity there is not a single specimen of the T'ang issues so common at Dandān-Uiliq[17]. It also accounts for the distinctly older appearance which the sculptural and pictorial remains of Endere seem to bear as compared with those of the Dandān-Uiliq shrines. Date of abandonment of temple.

The possibility of approximately exact dating invests with special interest a series of votive offerings of a humbler kind, which I found scattered in front of the various image bases, and which still remain to be mentioned. The rectangular piece of paper (E. i. 19. a, see Plate LXXIX) found in the sand close to the north-west facet of the central base shows a rapid but spirited sketch in colours of a Bactrian camel suckling her calf. Good observation is displayed in the drawing. On the hindquarter appears a brand. Could it be that of the owner, who, perhaps, deposited the sketch as an ex-voto to recover a lost animal? Two smaller pieces of paper (E. i. 25. a, b, see Plate LXXIX), found rolled up together at the base of the north-west corner image, are mere fragments of larger sketches, one of them interesting on account of the Chinese look of the head represented. Coloured sketches on paper.

But far more numerous were the small strips of textiles, sometimes shaped into little pennons, or else mere shreds, evidently torn from garments, which were found mixed up with the MS. pieces in front of the image bases. They comprised many fabrics, from elaborately-woven silk brocades to simple but strong cotton stuffs resembling the modern 'Khām', and already referred to in Hsüan-tsang's account. Sometimes pieces of the same material turned Strips of textiles deposited as ex-votos.

[17] Among these coins from Endere, most of them fragmentary, three are *wu-chu* pieces, while the rest bear no legend; see App. *D*.

up in different parts of the central base as well as before the corner images, evidently deposited by the same person, anxious to propitiate all the divinities. The variety of this collection vividly reminded me of the wonderful display of rags that graces the approaches to the resting-place of Imām Ja'far Sādiq, and swings from high staffs over the supposed tombs of other saints throughout Turkestān. Islām has indeed little changed the popular type of ex-votos which were in vogue during Buddhist times, and which in this case has provided for us a sample collection of ancient fabrics of no small archaeological interest. Among the specimens carefully described by Mr. Andrews in the list below, and partly reproduced in Plates LXXVI, LXXVII[18], there may be mentioned the little pennons (E. i. 016, 017), made up of silk pieces of different colours and qualities, several of them strongly ribbed. The heavy silk brocades (E. i. 018, 019) show remarkable mastery in weaving technique and in the harmonious mixing of colours. To do full justice to the latter in the reproduction could, owing to the number and delicate blending of the different shades, have been attempted only at a prohibitive cost. Examining in the original even so small a shred as E. i. 020, with a ground of deep gold colour and a pattern in white, red-brown, deep blue, and green, some idea can be gained as to the artistic richness of these garments. Loosely-woven silk fabrics are E. i. 021, 022, and the muslin-like piece E. i. 024. Another silk of this kind (E. i. 027) shows a pattern in satin stitch. Excellent workmanship is displayed by the finely-woven white cotton or linen, E. i. 023, with its lozenge diaper pattern. In E. i. 026 we have a sample of plain cotton cloth dyed dull violet-brown. Finally the peculiar technique of 'knot-dying', still largely practised in North-western India, is illustrated by the piece of blue cotton cloth, E. i. 029, sewn into what seems to have been a small bag, and ornamented by flower-like patterns executed in knot-work, white on blue.

Section III.—THE RUINED FORT AND STŪPA OF THE ENDERE SITE

Excavation of building E. ii.

Immediately after the clearing of the temple had been completed on the morning of February 23, I excavated the row of small rooms marked E. ii, situated to the north of it, at a distance of about 50 ft. The walls here too were built of a timber framework with plaster just as in the cella, but had decayed badly, as the cover of sand was only from 1 to 5 ft. in height. On the north there adjoined a walled-in courtyard which seemed to have been used once in part as a cattle-shed. In the easternmost room there was found a cavity constructed in the floor adjoining the south wall, about 4 ft. long, 3 ft. broad, and 4 ft. deep. Its sides were all carefully plastered. Its purpose could not be definitely established; it might possibly have served as a grain store. That in all probability this dwelling had been tenanted by a small monastic establishment became evident by the finds in the room next to the west.

Fresco in chapel E. ii.

This little apartment, measuring only 8 ft. by $4\frac{1}{2}$ ft. (marked E. ii. in plan, Plate XXXVI), had its narrow south wall decorated with an elaborate fresco painting, still retaining in part its original vivid colours (see Plate X). Of the central figure, which was about life-size and probably represented a Buddha or Bodhisattva, there remained only the feet and the lower portion of the robe, since the wall had broken away above 4 ft. from the floor. The oval vesica painted behind the figure showed small representations of seated Buddhas or saints, all painted

[18] I may note that the Plates show in most cases only small sample cuttings, after the fashion of a draper's catalogue.

alike in red robes on blue ground, each surrounded by a light green halo about 4 in. in diameter. In the triangular space left on each side at the foot of the vesica there appeared nearest to the latter a figure kneeling or seated, more or less effaced, and behind it a figure standing with uplifted sword. The foot of the wall was decorated with a frieze-like painting about 1 ft. 10 in. high, divided into two broad bands by an angular fret. In the upper one there appeared, where the surface had not peeled off, representations of swimming geese and of fleur-de-lys-like leaves, all in blue over a vivid green ground. The lower band, which can be made out in the photograph (Plate X), consisted of triangular imbrications alternately blue and green painted over a rich red ground. A similar style of decoration, but less carefully executed, can be seen in the photograph showing the remains of frescoes on a wall of the Dandān-Uiliq shrine D. vi (Plate IV).

In the east wall of the little apartment, where it adjoined the frescoed south wall, a row of elaborately turned balusters of wood (see Plate X) was found inserted, with their back set into the plaster. The object served by this arrangement is not clear, but the top of the balustrade thus formed may possibly have served as a place of deposit for votive offerings. Within the south-east corner, and lying in loose sand about 4 in. above the floor, I discovered the well-preserved small painted panel (E. ii. 1) reproduced in Plate LXXVIII. It shows a seated Gaṇeśa or Vināyaka, four-armed, all details in the representation of the elephant-headed god being thoroughly Indian. It is well-known that he has remained a favourite in all Buddhist churches of Northern Asia which have been influenced by the Mahāyāna system[1]. A fragment of a planed beam or post, about 3 ft. long with a width of 6 in. and much decayed by exposure, was found lying along the foot of the west wall. It showed faint outlines of part of a Buddha-head drawn in dark brown colour, with a small nimbus behind, and might have formed part of a decorated door jamb. The character of all these painted remains leaves no doubt that this small room served as a chapel. Painted panel of Gaṇeśa.

The remaining rooms westwards, which showed a floor raised 3 ft. above that of E. ii., yielded no finds. Also the detached and in part badly decayed small structures (E. iv) to the north west of the temple were found completely empty but for a well-finished fireplace in plaster.

Apart from the ramparts enclosing the fort, the ruins of the large brick-walled building occupying the eastern portion of the circumvallated area were certainly the most striking feature of the site. The photograph (Fig. 48) taken from the west, shows them as they appeared before any excavation began. The massive walls of sun-dried bricks were in the southern part of the main building almost clear of sand, and still rose in places to heights up to 10 ft. The northern portion, including the half-open hall (E. iii.) and the large apartment forming the north-east corner, was, however, filled with drift-sand, rising from about 6 ft. in the latter to fully 9 ft. in the former. The walls, shown to scale in the plan (Plate XXXVI), had a thickness varying from 4 ft. in the main outside walls to 2 ft. 3 in. in others. The bricks, which contained much straw, with the occasional addition of bits of bone, wood, pottery, and similar refuse, were of two sizes. The usual make measured about 17 in. square with a thickness of 5 in., while another size, about 12 in. square and 3 in. thick, was used in some of the thinner walls. The horizontal rows of bricks were set in layers of mud, about 2 in. thick, and mixed with much straw. In the larger rooms massive posts of wood, about 5 in. square at their foot, but rounded where they projected pilaster-like beyond the wall facing, had been set into the walls at regular intervals, evidently to serve as supports for the roof beams. These posts were more or less intact in the rooms filled by sand (see Fig. 48 on left; also Plate XI), while in the rest cavities left by them Ruins of large building E. iii.

[1] Comp., e. g., Grünwedel-Burgess, *Buddhist art*, p. 183.

in the walls indicated their position, the wood itself having perished through exposure. The remains of massive roof-beams were found lying in the sand above the floor of the north-east corner room. Both there and in the hall (E. iii.) intermediary free-standing posts had helped to carry them[1].

Excavation of northern rooms in E. III.

Owing to the large size of the sand-filled rooms in the northern part of the building their clearing proved a very heavy task, and could, within the limited time available, be accomplished only by keeping every available hand at work until late into the night on the 24th and 25th of February. The offer of additional wages and free rations made the men willing to face the fatigue and the bitter cold of this night labour, while big bonfires kindled with the plentiful dead tamarisk roots gave the light needed. The rooms proved, as I had expected, to have been completely cleared of all movable objects. But in the corner room a sitting platform, 3 ft. broad, as well as a large fireplace, were brought to light, while in the hall (E. iii.), open towards the south, the rough plaster surface of the partly well-preserved east wall proved to be covered with sgraffiti both Tibetan and Chinese. The photograph reproduced in Plate XI shows those to the south of the central pilaster of wood, the scribblings north of it being of a similar type. M. Chavannes' reading and translation, as given in Appendix *A*, Part iii. shows that the Chinese characters nearest to the pilaster contain a mention of 'the imperial envoy *Hsin Li-ch'an*', whose identity, unfortunately, has not been established. The other Chinese scrawls are too rough and effaced to permit of decipherment.

Chinese and Tibetan sgraffiti in hall E. iii.

Contents of Tibetan sgraffiti.

The Tibetan sgraffiti on this wall are on the whole better preserved than the Chinese ones, across which they are partly written, and bear the appearance of being later. They have been deciphered from my photographs and hand-copies by the Rev. Mr. Francke, whose transcripts and annotated translations will be found in Part iv. of Appendix *B*. The most interesting among these scribblings is, undoubtedly, the one which Mr. Francke marks with *C*, and which in the photograph reproduced in Plate XI is seen on the top below some coarse Tibetan and Chinese scrawls. According to Mr. Francke's rendering the main portion of this sgraffito records: 'At Pyagpag (in the) province of Upper *oJom lom* this army was outwitted, and a tiger's meal was obtained (i.e. many were killed).' To this is added, in coarse big letters, evidently by a different hand: '(Now) eat until you are fat.' Have we here a record of some victorious Tibetan engagement, perhaps with a Chinese force? The words, especially those of the postscript, seem to have a strain sufficiently truculent to have been scratched in by some Tibetan brave who took his share in the fighting. But this does not help us to identify either the localities named or the event alluded to. The remaining three sgraffiti translated by Mr. Francke (one of them *A*, scratched in close to a small but well-drawn sketch of a charging tiger) seem to be scribblings of a still more casual character, referring to some property lost, a picture possibly once displayed on the wall, &c.

Purpose of building E. III.

The ruined building, E. III, was undoubtedly the main structure within the area enclosed by the ramparts, and, judging from the size of its halls and rooms, and the general arrangement of its plan, I think it safe to conclude that it had served as the residence of the officer and establishment, for the protection of whom the fort must be supposed to have been primarily intended. Somehow the spaciousness of the whole structure suggested resemblance to a Chinese Ya-mên, but I am unable to judge how far the disposition of the rooms, &c., might be compatible with Chinese architectural convention.

[1] The large beam seen embedded in a slanting position on the right in Fig. 48 probably also served originally as a wall post. If so it would be an indication that the open space along the south wall of the building had once been roofed over like E. iii.

The only other structure of any size of which remains could be traced within the ruined fort was the dwelling (E. v) to the south-east of the temple. The few broken posts protruding from the sand, which first attracted my attention to it (partly visible in the foreground of Fig. 51) proved to have belonged to an upper story which otherwise had almost completely disappeared through erosion. Its floor seemed to have been approximately on a level with that of the temple and of E. iii. There were distinct indications that the walls of this upper story had, apart from the timber framework, consisted of rows of cakes of hard-beaten sheep-dung placed vertically and joined by layers of clay. From my labourers I learned that the use of this queer building material is not altogether unknown at the present day. Below these thin walls the excavation brought to light a lower story or basement, which, as the plan shows, had main walls of stamped clay over 4 feet in thickness. The interior walls dividing some small apartments were of the usual timber and plaster type, and only 6 to 8 in. thick. It is probable that the great thickness of the main walls was intended to meet the pressure of the surrounding soil. There was no entrance leading through any of the outer walls of this basement story, which thus could have been entered only by stairs from above. Probably one or the other of the small apartments separated by timber and plaster walls had contained the ladders that served for stairs. Set into the clay walls were massive posts, which continued into the superstructure. They are seen in the photograph reproduced in Plate X, which shows the south-western corner room after partial clearing. The large and well-preserved fireplace found in this room (also seen in the photograph) showed a projecting top decorated with elaborate mouldings in stucco. In construction it closely resembled some of the fireplaces found in the Dandān-Uiliq dwellings. Its presence proves that the basement rooms, all of which appear to have had a height of 9 ft., must have been partly at least used for living purposes at certain seasons. Such wholly or partially underground apartments were, no doubt, easier to keep warm in the severe winter. They may have thus corresponded in a reverse way to the *Tai-khānas* of Peshāwar and the northern Punjāb, constructed as a cool refuge against the torrid heat of the Indian summer months, which are alluded to, I think, in Philostratus' account of Apollonius of Tyana's Indian journey. The laborious clearing of these basement rooms was not rewarded by any finds of movable objects.

Excavation of dwelling E. v.

Underground rooms of E. v.

No structural remains, besides those already described and the much-decayed walls of a room built against the northernmost part of the ramparts, could be traced within the circumvallation. Nor, having regard to the conditions of the sand and to the available space, did it appear likely that other ruins of any size could lie hidden beneath the drift-sand. In the northern portion of the enclosed area, where a good deal of ground lay almost bare of sand (see plan), compact layers of rubbish, consisting chiefly of horse-dung, stable refuse, and straw, were found in one place to cover the soil to a height of 3 to 4 ft. In order to assure myself of the character of these rubbish accumulations I had a broad trench dug across the whole of them down to the natural ground. But the only finds, besides the refuse described, were broken pieces of coarse pottery and occasional bits of rough cotton fabrics and felt, as well as some cotton seed. It was evident that this portion of the ground had been used for stabling during a prolonged period. Further to the north the bare ground was covered with coarse potsherds, and the same also strewed in profusion the open ground to the south of E. i and E. iii, which erosion, still actually proceeding, had lowered from 3 to 5 ft. below the floor-level of the latter buildings. Just outside the gate of the fort pottery débris was also plentiful.

Refuse accumulations on ground.

Not the least interesting feature of the ruins were the remains of the ramparts which had once defended this small agglomeration of buildings. There was nothing to tell of the attacks

Construction of ramparts.

of enemies which they may have once resisted, but they had certainly helped to ward off that worst danger of ancient remains in the desert region—the erosive action of winds and moving sand. Looking around from the ruined walls it was easy to realize that the original level of the surrounding ground had been lowered at least 10 ft. by erosion. Inside the circumvallation the drift-sand, when once accumulated, was less liable to be shifted by the winds, and thus provided a protecting cover for the ruins. While thus fulfilling for many centuries its defensive function in a fashion the original builders could never have contemplated, the circumvallation itself had suffered badly. The portion facing due south and flanking the gate, which Fig. 51 shows, about 160 ft. long in all, alone still retained distinct indications of its original construction and character. The small segments surviving at other points, as marked in the plan (the background in Fig. 48 and in the photograph of E. v in Plate X shows two of them), had decayed into more or less shapeless mounds, which, however, were useful for determining the original circumference. The circumvallation consisted of a solid rampart of stamped clayey loess, probably built up in rectangular forms, after the fashion which still prevails throughout Central Asia wherever mud walls are used for buildings or fortification [3]. In the north-western segment, however, I noticed that pieces of hard loess held together by calcareous formations had been embedded in horizontal layers into the usual stamped clay.

Dimensions of rampart and parapet.

The width of the rampart appears to have been about 30 ft. at the base, diminishing higher up, as seen also in the section of the ancient walls of Ak-sipil fort (Plate XXXIX). The height in the southern segment, where alone it could be determined with some approach to accuracy, was $17\frac{1}{2}$ ft., measured from the floor-level of E. III. On the top of the ramparts ran a parapet, 5 ft. 6 in. high and about 3 ft. thick, built of sun-dried bricks, and still clearly traceable for a distance of about 40 ft. to the east of the gate. The bricks were of the larger size used in the main walls of E. III. Behind the parapet ran a platform, indicated by horizontal layers of brushwood embedded in the clay, evidently for the sake of giving greater consistency to the top of the rampart. The gate, now marked only by a gap in the rampart, and approximately 18 ft. wide, was flanked on either side by a small square bastion or tower, projecting about 20 ft. beyond the outer foot of the circular rampart. These flanking defences had decayed too far to permit of an accurate survey. To the west of the gate, and close to the inner side of the rampart, rows of massive posts (see Fig. 51) indicated a small room which had been built into the inner escarpment, evidently to serve the purposes of a guard-room. Owing to the erosion proceeding along the inner foot of the circumvallation the floor of the gate now appears in the photograph to be raised above the general ground-level.

Gate of fort.

Circular shape of fort.

The other surviving small segments of the ramparts, varying in height from 5 to 15 ft., showed no indications of any other gate, nor of bastions. Nor were the latter particularly needed, seeing that the circular shape of the fort effectively provided against any 'dead angles'. This circular construction was traced by me again in the wall remnants of the ancient fort of Ak-sipil [4], while modified into an oval it recurs in the probably mediaeval small stronghold found near the site of Uzun-Tati [5]. The manifest advantages of this ground-plan for fortified stations of modest size seem still to be appreciated by the Chinese authorities in Eastern Turkestān, since I found the mud walls of the little military post of Ōpal between Tāshmalik and Kāshgar constructed exactly in the same fashion [5a].

[3] I have noticed it used in many a place of Baluchistān and the Indian N.W. Frontier, from Kalāt to Peshāwar, much after the manner commonly employed throughout Turkestān, both Chinese and Russian.

[4] See below, chap. XIV.

[5] Comp. chap. XIII.

[5a] See *Ruins of Khotan*, p. 117.

50

RUINED STŪPA, ENDERE SITE, SEEN FROM SOUTH-WEST.

51

RUINED RAMPART NEAR GATEWAY, ENDERE FORT, SEEN FROM INTERIOR.

There is nothing in the ruined fort of Endere, the survey of which we have now completed, directly to indicate the purpose for which it was originally placed here. But the approximate date we have ascertained for its abandonment, when compared with what Hsüan-tsang tells us of the condition of this region in the middle of the preceding century, supplies reasonably safe ground for surmises. When the pilgrim had started eastwards from Ni-jang or Niya, he entered 'a great drifting sand desert'[6] the terror and dangers of which for wayfarers he graphically describes, much in the same fashion as Marco Polo, more than six centuries later, described the great desert between Lop Nor and Sha-chou[7]. After having travelled through the moving sands for four hundred li, or four marches, he arrived at 'the old country of the *Tu-huo-lo* 都貨羅'. This country had long been uninhabited. All the towns presented the appearance of an unoccupied waste[8]. Hsüan-tsang's 'old *Tu-huo-lo* country'.

Six hundred li further to the east he reached the old kingdom of *Chê-mo-t'o-na* 折摩駝那, where the city walls still stood high, but there were no inhabitants. This Chê-mo-t'o-na, for which the older name of *Chü-mo* (*Tsiu-mo*) 沮末 is correctly given in Hsüan-tsang's 'Life'[9], can be no other than the oasis of Charchan; the position of the latter is clearly indicated on the one hand by the distance of a thousand li from Ni-jang or Niya, which accurately coincides with the ten marches reckoned at the present day between the two places, and on the other by the bearing and distance to Na-fo-po or Lou-lan, which the pilgrim's narrative places to the north-east and at a thousand li. Na-fo-po or Lou-lan has long ago been identified as the vicinity of Lop-Nor, and a look at the map shows that the oasis of Charchan lies to the south-west of the latter and almost exactly halfway between Lop-Nor and Niya. Hsüan-tsang's *Chê-mo-t'o-na*, or Charchan.

The relative distances which Hsüan-tsang has recorded to Ni-jang or Niya and Chê-mo-t'o-na or Charchan, respectively, oblige us to look for the deserted settlements of what he calls 'the old country of Tu-huo-lo'[10] in the tract surrounding the actual Endere site; for the latter lies within twelve miles or so of the direct route from Niya to Charchan and, as Dr. Hedin's map shows, some sixteen miles nearer to Niya than to Charchan[11]. Seeing that about 645 A.D. this tract was already a waste abandoned to the desert, the question arises how to account for the existence in it of the ruins excavated by me, which were undoubtedly occupied during the early part of the eighth century. Two explanations appear open. Either the tract, perhaps in consequence of the improved conditions following the establishment of Chinese authority throughout Eastern Turkestān, which as we have seen took place within a little over ten years after Hsüan-tsang's passage[12], had again come under cultivation and received a settled population. The Endere fort built in tract deserted at Hsüan-tsang's time.

[6] See *Mémoires*, ii. p. 247; Beal, ii. p. 324 sq.

[7] See Yule, *Marco Polo*, i. p. 196 sq.

[8] See *Mémoires*, ii. p. 247; Beal, ii. 325.

[9] See *Vie de H.-Th.*, p. 290; Chavannes, *Voyage de Song Yun*, p. 13, note 1. The *Hsi-yü-chi* erroneously reads 涅末 Ni-mo. Comp. also *Turcs occid.*, pp. 30, 57, 306, and for the Han Annals' account of the same oasis, Wylie, 'Notes on the Western Regions,' *J. Anthrop. Inst.*, x. p. 28.

[10] Hsüan-tsang's mention of this tract as 'the old country of the *Tu-huo-lo*', i.e. of the tribe of the Tochari which played so important a part in the early history of the Oxus region, and the memory of which long survived in the name of Tokhāristān, has been the subject of much learned speculation: comp., e.g., Marquart, *Ērānšahr*, pp. 206 sqq.; Franke, *Zur Kentniss der Türkvölker*, pp. 28 sqq. A discussion of the various problems connected with the real character and origin of the Tochari (the Tukhāra of Sanskrit texts) does not come within the scope of my task here. But I may mention that, if the conquerors of the Bactrian Greek dominion really came from this region, Dr. Marquart's argument against their having been nomads at the time would find strong support in the physical character of their alleged old home.

That the term Taklamakān, by which all desert ground within the central area of the Tārīm Basin is popularly designated, can neither on linguistic nor on historical grounds be derived from the name *Tu-huo-lo* (see Hedin, *Through Asia*, ii. pp. 784 sq.) scarcely needs to be demonstrated to critical students.

[11] The route followed by me between the Endere Site and Niya, and measured on my map, covers about 110 miles. Between Endere and Charchan Dr. Hedin's map (see *Reisen in Z.-A.*) shows about 110 miles as the road distance without allowance for windings.

[12] See above, pp. 59 sq.

establishment in it of a fortified post, for the accommodation of a small Chinese garrison, and perhaps of some local authority which the former was to support and control, would in that case have probably followed as a matter of course. Or if the tract remained deserted, as Hsüan-tsang saw it in 645 A.D., we should have to recognize in the Endere fort a small stronghold established on the great route from China to Khotan for the express purpose of assuring its safety.

Itinerary of T'ang Annals from Sha-chou to Khotan.

It seems difficult to decide between the two alternatives thus presented. The T'ang Annals furnish us, indeed, with an itinerary for the old route leading from Sha-chou to Khotan as it existed during this period of Chinese supremacy[13]. But though M. Chavannes' notes on it clearly show that the route passed through Chü-mo (Tsiu-mo) or Charchan, where a Chinese garrison is specially mentioned, it is impossible to identify with certainty any of the stages recorded on it further west until we get relatively close to Khotan, as the distances between them are indicated only in one instance[14]. On the whole, I am inclined to favour the first supposition, in view of the analogy offered for it by the case of Chü-mo (Tsiu-mo) or Charchan, which, though wholly deserted in Hsüan-tsang's day, was again inhabited and the seat of a Chinese garrison in T'ang times[15].

Marco Polo on route from *Pein* to *Charchan.*

Marco Polo's description, too, 'of the Province of Charchan' would agree with the assumption that the route west of Charchan was not altogether devoid of settlements even as late as the thirteenth century. After referring to the numerous towns and villages of 'Charchan', he tells us: 'The whole of the Province is sandy, and so is the road all the way from Pein, and much of the water that you find is bitter and bad. However, at some places you do find fresh and sweet water[16].' This account of the route agrees accurately with the conditions now met with between Niya and Charchan. Yet in the passage immediately following the Venetian tells us how 'when an army passes through the land, the people escape with their wives, children, and cattle a distance of two or three days' journey into the sandy waste; and, knowing the spots where water is to be had, they are able to live there, and to keep their cattle alive, while it is impossible to discover them'. It seems to me clear that Marco Polo alludes here to the several river courses which, after flowing north of the Niya-Charchan route, lose themselves in the desert. The jungle belt of their terminal areas, no doubt, offered then, as it would offer now, safe places of refuge to any small settlements established along the route southwards.

Chronological connexion between Endere fort and site

The presumption that the Endere fort had by its side a cultivated area with a settled population would, of course, be definitely established if it were proved that the débris-strewn area to the north-west of it, together with the Stūpa, dated from the same epoch. Unfortunately, conclusive evidence is not at present available on this point. No coins were found

[13] Comp. Chavannes, *Voyage de Song Yun*, p. 12, note 9.

[14] We are told that the route westwards of *Tsiu-mo* 'passed the *Hsi-li-chih* (*Si-li-tche*) wells, the *Yao* wells, the river *Hu-chê* (*Hou-tcho*), and, after 500 li, arrived at the military post of the town of *Lan*, which is east of Yü-t'ien'. If by the latter expression the eastern border of Khotan territory is meant, i.e. Niya, we might be tempted to suggest the location of 'the military post of the town of Lan' at the Endere site. But in the absence of exact evidence there is little profit in such identifications. The mention of wells as stages west of Tsiu-mo suggests that the country traversed was without permanent habitations, yet just as at the present day offered water in wells dug at regular halting stages.

[15] It is noteworthy that Sung Yün, travelling in 519 A.D., describes *Tso-mo*, which M. Chavannes has shown (*Voyage de Song Yun*, p. 13) to be identical with Hsüan-tsang's and the T'ang Annals' *Chü-mo* (*Tsiu-mo*), as 'a town' inhabited by a hundred families only. Charchan, in fact, seems to have passed repeatedly through such successive periods of contraction and expansion, having developed during the last fifty years from a small penal station in the time before Yāqūb Bēg's rebellion into a flourishing settlement of steadily increasing prosperity; comp. *Yarkand Mission Report*, p. 34; Hedin, *Reisen in Z.-A.*, p. 178. Examples of such fluctuations in importance and size of detached oases in this region could probably be greatly multiplied if our records were ampler. They may serve as a warning against any a priori assumption that steady decrease in the size of these oases must always have been the rule during historical times.

[16] See Yule, *Marco Polo*, i. p. 194. For the location of *Pein* in the close vicinity of the present oases of Gulakhma and Domoko west of Keriya, see below, chap. XIII.

except within the ruined fort and its immediate vicinity, and we are still far from having the means which a systematic classification might afford for the approximate dating of coarse pottery fragments. The ground around the Stūpa seemed certainly to have undergone far more general and thorough erosion than that around the fort, but I greatly doubt whether, in view of my previous observations in this respect, it would be safe to draw any chronological conclusion from differences of this kind which a variety of physical factors might account for. *around Stūpa.*

Survey of Endere Stūpa. Nor can, in the present state of our knowledge, the architectural features of the Stūpa itself supply any distinct chronological criteria. The exact survey of the ruin (see plan in Pl. XXXVII) which I effected on February 25, while the excavation of E. III could be safely left to the supervision of the Surveyor, proved a difficult task, owing to the greatly decayed condition of the outer wall-facing and the destruction caused by the burrowing of treasure-seekers. But by carefully plotting the extant masonry and fixing its central line with the help of the square shaft discovered in the interior, I succeeded in ascertaining the approximately accurate dimensions shown in the ground-plan and section. As already stated, wind erosion had considerably lowered the ground generally around the ruin, and had excavated close to its foot hollows which on the south-west and south-east were fully 15 and 10 ft. below the lowest masonry course. This feature is well illustrated by the photograph in Fig. 51, which shows the Stūpa as seen from the south-west. The foot of the ten-foot rod marks the level where the masonry commences. The Stūpa consisted of a square base, approximately orientated with its corners, and of a cylindrical dome rising above it. The base rose in three stories according to the canonical arrangement previously explained, the lowest story measuring 27 ft., with a height of 1½ ft. But these last dimensions, owing to far-advanced decay, can only be approximate. The next story, receding 2 ft. from the first and 6 ft. high, formed the main portion of the base; above it the third, only 1½ ft. high, also receded by 2 ft. The dome had a diameter of 16 ft., rising with its broken top to a height of 14 ft. Its original height could not be ascertained. Along its central line there descended a shaft 1 ft. square, down to a depth of 7 ft. from the extant top.

Cuttings made by treasure-seekers. This shaft, which may have once served for the fixing of a wooden mast carrying the 'Chattras' and other ornaments surmounting the dome, had been rendered accessible from below by a cutting made by treasure-seekers into the south-east face of the dome, evidently a long time ago. Another cutting had been carried from the north-east side of the base towards the centre and upwards to the level of the topmost story. By connecting the two cuttings in the centre I ascertained that no deposit remained in the few feet of masonry intervening. The whole of the Stūpa was constructed of sun-dried bricks, which looked harder and more closely laid than those of the ruin E. III. The prevailing size seemed to be 5 in. thick and about 18 in. square. But owing to the decay and weatherworn condition of the outer faces it was difficult to obtain a sufficient number of measurements. That the surface of the whole structure had once been covered with stucco was proved by a large piece of this coating still adhering on the north side of the dome.

'Tati' remains near Stūpa. The extent of the area over which pottery débris could be traced on the eroded bare loess around the Stūpa suggested a relatively large settlement. That this, like the ruined fort, had received its water supply from the Endere stream still flowing past the site within a little over four miles, is certain. There is no archaeological evidence to indicate the cause or causes which led to its abandonment, and even the chronological relation between it and the fort remains cannot be definitely settled. The specimens enumerated under E. 006 in the list below are representative of the undecorated pottery fragments which were the only ancient remains to be

found at this 'Tati'. More varied were the small pieces of glass and bronze (E. 002, 003, see Plate LII) which were picked up from the sand within the ruined fort and on eroded ground in its immediate vicinity. The fragment of green blue glass (E. 003. e) showing a kind of niello ornamentation deserves special mention. At a point about half a mile to the west Ibrāhīm, my treasure-seeking guide from Niya, alleged that he picked up the two interesting small objects described under E. 001. One of them is a cylindrical piece of hard Chinese ink, drilled for a string at one end (see Plate CV); the other a cubical die of bone (Plate LII) with the incised spots arranged exactly as in classical dice, the sum of any two opposite sides being seven.

Section IV.—LIST OF ANTIQUES FROM THE ENDERE RUINS

MSS. EXCAVATED IN TEMPLE E. i.

MSS. from E. i.

E. i. 1, 2, 3. Brāhmī MS. on paper. No. 39; two small and two large frags.; left sides of (apparently) two leaves, numbered 30 and (29?). See Note xxii below.

E. i. 2. Brāhmī MS. on paper. No. 40. Three frags.; left halves, numbered 11, 12, 13. See Note xxiii.

E. i. 4. Brāhmī MS. on paper. No. 41. Six frags.; two left halves, numbered 44 and 46, and four right halves. See Note xxiii.

E. i. 5. Brāhmī MS. on paper. No. 42. Eight frags.; left halves, numbered 3, 4, 5, 6, 7, 8, 9, 10.

No. 43. Also, two frags., left halves, but numbers not preserved, probably 1 and 2. See Note xxiii.

E. i. 6. Brāhmī MS. on paper; frag. in slanting Central-Asian Brāhmī, measuring about $9'' \times 3''$; found made up into a roll. See Pl. CIX.

E. i. 7. Brāhmī MS. on paper. No. 53. Four leaves, 3 complete, 1 mutilated, found made up into a roll; numbered 19, 145, 149, 152. Comp. Note xxvi. See Pl. CXI.

E. i. 8. Chinese document on paper, small frag. See App. A. iii.

E. i. 9. Brāhmī MS. on paper. No. 44. Two small frags.; one of left half, numbered 114.

No. 45. Also one very small frag. from middle of a leaf. See Note xxv.

E. i. 10. a. Fragment of Śālistamba-sūtra, $8'' \times 2\frac{5}{8}''$. (App. *B*, i. L.)

E. i. 10. b. Fragment of do., $\frac{13}{16}'' \times 1\frac{7}{8}''$. (App. *B*, i. L.)

E. i. 11. Sheet containing 2 poems from the T'eg-mc'og-mdzod, $10\frac{1}{4}'' \times 7\frac{1}{4}''$. (App. *B*, iii.)

E. i. 12. Fragment of Śālistamba-sūtra, $8\frac{7}{8}'' \times 2\frac{5}{8}''$. (App. *B*, i. A.)

E. i. 13. Fragment of do., $8'' \times 2\frac{5}{8}''$. (App. *B*, i. L.)

E. i. 14. Fragment of do., $4\frac{1}{2}'' \times 2\frac{5}{8}''$. (App. *B*, i. J.)

E. i. 15. Fragment of mystic work, $1\frac{1}{2}'' \times 1\frac{1}{2}''$. (App. *B*, ii. D.)

E. i. 16. Fragment of Śālistamba-sūtra, $4\frac{3}{4}'' \times 2\frac{1}{2}''$. (App. *B*, i. O.)

E. i. 17. Fragment of do., $9\frac{1}{8}'' \times 2\frac{1}{2}''$. (App. *B*, i. O.)

E. i. 18. Fragment of do., $4\frac{3}{4}'' \times 2\frac{5}{8}''$. (App. *B*, i. B.)

E. i. 19. Fragment of ritual work, $4\frac{1}{2}'' \times 5\frac{1}{2}''$. (App. *B*, ii. A.)

E. i. 19. a. Fragment of paper, very yellow in tone, on which is a rapid sketch in Indian ink of a Bactrian camel suckling her calf. The drawing shows good general observation. The upper lip, the full eye, and the hairiness of the animal are all well indicated. On near hind quarter a brand (perhaps of the owner). The excitement shown in the drawing of the calf's tail is familiar to those who have watched calves sucking. There is a certain amount of roundness suggested by pink washes round contours; streamers painted red and yellow. Found with E. i. 19. $5\frac{1}{4}'' \times 4\frac{1}{2}''$. See Pl. LXXIX.

E. i. 20. Three fragments of mystical work, $2\frac{1}{2}'' \times 1\frac{7}{8}''$, $1\frac{1}{2}'' \times 1\frac{1}{4}''$, $1\frac{1}{4}'' \times 1\frac{1}{4}''$. (App. *B*, ii. D.)

E. i. 21. Fragment of Śālistamba, $9\frac{3}{8}'' \times 2\frac{5}{8}''$. (App. *B*, i. M.)

E. i. 22. Fragment of do., $6\frac{1}{8}'' \times 2\frac{5}{8}''$. (App. *B*, i. D.)

E. i. 23. Fragment of do., $9\frac{3}{8}'' \times 2\frac{5}{8}''$. (App. *B*, i. D.)

E. i. 24. Fragment of do., $8\frac{7}{8}'' \times 2\frac{5}{8}''$. (App. *B*, i. M.)

E. i. 25. Two fragments of religious work, $4\frac{1}{2}'' \times 1\frac{1}{4}''$ and $3'' \times \frac{3}{4}''$. (App. *B*, ii. C.)

E. i. 25. a. Fragment of paper with very sketchy drawing in ink, done with a brush, showing head, shoulders, and right arm of a figure, and a few other lines the meaning of which is not obvious. Character of drawing rather Chinese. Only colour used a fine vermilion on arm and breast covering. Found rolled up with E. i. 25. b. 5″ × 1⅜″. See Pl. LXXIX.

E. i. 25. b. Fragment of paper with part of rapid sketch in ink, done with a large brush. Looks something like the leg of an animal. Found rolled up with E. i. 25. a. 2¾″ × 3″. See Pl. LXXIX.

E. i. 26. a. Fragment of Śālistamba, 9″ × 2⅝″. (App. *B*, i. J.)

E. i. 26 *. Fragment of do., 1¾″ × 2⅝″. (App. *B*, i. G.)

E. i. 27. a. Fragment of do., 2⅜″ × 1⅜″. (App. *B*, i. G.)

E. i. 27. b. Fragment of do., 2¼″ × 2⅝″. (App. *B*, i. D.)

E. i. 28. a. Fragment of do., 2¾″ × 2⅝″. (App. *B*, i. F.)

E. i. 28. b. Fragment of do., 2 5/16″ × 2⅝″. (App. *B*, i. E.)

E. i. 29. Fragment of note on religious subject, 6⅜″ × 2⅛″. (App. *B*, ii. B.)

E. i. 30. Fragment of Śālistamba, 1⅞″ × 2⅝″. (App. *B*, i. H.)

E. i. 31. a. Fragment of do., 1¼″ × 2⅝″. (App. *B*, i. F.)

E. i. 31. b. Fragment of note on religious subject. (App. *B*, ii. B.)

E. i. 32. a. Fragment of Śālistamba, 9″ × 2⅝″. (App. *B*, i. B.)

E. i. 32. b. Fragment of do., 9½″ × 2⅝″. (App. *B*, i. G.)

E. i. 33. Brāhmī MS. on paper. No. 46. One frag., from middle of leaf. See Note xxiv.

E. i. 34. a. Fragment of Śālistamba, 9¼″ × 2⅝″. (App. *B*, i. c.)

E. i. 34. b. Fragment of do., 2½″ × 2″. (App. *B*, i. E.) **MSS. from E. i.**

E. i. 35. Fragment of do., 3⅛″ × 2⅝″. (App. *B*, i. G.)

E. i. 36. Fragment of religious work, 2¾″ × 1½″. (App. *B*, ii. D.)

E. i. 36. a. Chinese document on paper; frag. See App. *A*. iii.

E. i. 37. Fragment of Śālistamba, 2″ × 2⅝″. (App. *B*, i. F.)

E. i. 38. Fragment of do., 3⅛″ × 2⅝″. (App. *B*, i. N.)

E. i. [no number]. **Fragment of do.,** 1¾″ × ⅝″. (App. *B*, i. K.)

E. i. [no number]. **Three fragments of religious work,** 1¾″ × ⅝″, 1″ × 1⅛″, 1″ × 1⅛″. (App. *B*, ii. D.)

E. i. 39. Brāhmī MS. on paper. No. 47. Three frags.; left halves of leaves, numbered 29, 30, 32.

No. 48. Also, five frags.; left halves, but numbers not preserved. See Note xxiii.

E. i. 40. Brāhmī MS. on paper. No. 50. Twelve frags., right halves.

No. 51. Also, seven small frags.; all of right halves. See Note xxiii.

E. i. 41. Brāhmī MS. on paper. No. 49. One frag., left half, but number not preserved. See Note xxiii.

E. i. 43. Brāhmī MS. on paper. No. 52. One frag.; right half. See Note xxiii.

E. i. 44. Chinese document on paper; frag. See App. *A*. iii.

Notes by Dr. A. F. R. Hoernle on Brāhmī MSS. from E. i.

Note xxii.—No. 39. These pieces belong to left sides of (apparently) two leaves of the same MS., numbered 30 and (29?). The latter number is uncertain. Width of leaf 3¼″, length unknown; no string-hole preserved; 6 lines on page. Language: non-Sanskritic (Proto-Tibetan?). Characters: upright Gupta, of 7th or 8th cent.

Note xxiii.—Nos. 40, 41, 42, 43, 47, 48, 49, 50, 51, 52. All these fragments belong to the same MS. and to 46 leaves (see infra), 16 of which being left-hand portions are numbered (on the obverse sides) 3, 4, 5, 6, 7, 8, 9, 10, 11, 12, 13, 29, 30, 32, 44, 46. The numbers of the remaining leaves are not preserved; but in all probability, as shown by the sequence of the fragments in the discovered bundle, there were the numbers 1, 2, 26, 27, 28, 31, 33, and 45 among them. So that, altogether, there are 24 (actually, or practically) numbered leaves extant, viz. 1–13, 26–33, and 44–46.

The MS. is written in Sanskrit, in upright Gupta characters of the 7th or 8th cent. It is probably a Buddhist canonical work of the *Dhāraṇī* class.

There is one fragment in this set which bears, on its *Obv.*, a remark written in non-Sanskritic language (Proto-Tibetan?). Its *Rev.* commences with *Siddham*; the rest of the text being in Sanskrit. This indicates the beginning of the treatise, and shows that the fragment, which bears no number, belongs to the *first* leaf of the MS. The treatise, accordingly, commenced on the *Rev.* of the initial leaf, while its *Obv.* was inscribed with a non-Sanskritic remark. This remark, probably, recorded the name of the treatise; but, for the present, it is not intelligible. There occurs, however, in it the Sanskrit word *Dhāraṇā*, which may possibly represent the well-known title *Dhāraṇī*. The interlocutor in the treatise is Mañjuśrī.

MSS. from E. i.

The fragments consist of 25 left-hand and 24 right-hand portions of leaves, most of which represent nearly one-half of a leaf. The only six (nearly) left and right halves which (so far) I have been able to piece together with absolute certainty are three left halves of E. i. 39, and two right halves of E. i. 4, and one right half E. i. 42. They make up between them the 29th, 30th, and 31st leaves; whether any more halves can be pieced together appears very doubtful. The probability is that we have three complete leaves, and 22 left-hand and 21 right-hand portions consisting of 43 leaves. Accordingly we have a total of (43 + 3, or) 46 leaves; and this is confirmed by the fact that the left half of the 46th leaf is preserved among the fragments. It would thus appear to be probable that we have the entire MS., which consists of no more than 46 leaves.

As to the three complete leaves, they are not complete in the strict sense. The two 'halves' of which each is made up are not really complete halves; a small portion of the middle of each leaf is wanting. The 'halves' measure about 6" × 3"; the string-hole is at about 4" off the margin; accordingly, the total length of the leaf was about 14", and about 2" are lost. The full size of a leaf must have been about 14" × 3".

Note xxiv.—No. 46. Only a very small fragment, from the middle of a leaf, showing traces of 4 lines of writing. Language uncertain, but probably Sanskrit (*rogyasta* legible).

Note xxv.—Nos. 44, 45.—Two pieces, belonging to a leaf numbered 114, inscribed with large upright Gupta characters of the 7th or 8th cent., in non-Sanskritic language (Proto-Tibetan ?).

Note xxvi.—No. 53. Four leaves of a MS., in non-Sanskritic (Proto-Tibetan ?) language, and in large, upright Gupta characters of the 7th or 8th cent., measuring 18" × 3¼" with 5 lines on a page. Subject, apparently medical or magical, for the term *bhaiṣajya* occurs frequently. The four leaves were found done up in a roll; and as they show the numbers 19, 145, 149, 152, it appears that the MS. was a mere fragment even at the time when it was deposited in the form of a roll. Leaf 152 appears to contain the conclusion of the MS., and probably contained its name; but unfortunately it is much mutilated.

OBJECTS EXCAVATED IN TEMPLE E. i.

Objects from E. i.

E. i. 01. Stucco relief fragment. Red clay; jewelled ornament, elliptical. 2 rows of beads, and within these a fillet, all concentric. Centre plain. 2¼" × 2". See Pl. LXXVIII.

E. i. 02. Stucco relief fragment. Red clay mixed with fibre; jewelled chain ornament; alternately, one large and two small jewels. 3½" × ⅝".

E. i. 03. Stucco relief fragment. Red clay, white-washed, small lotus. 1¼".

E. i. 04. Stucco relief fragments. Red clay, traces of whitewash; 3 fragments of fingers, life-size; nails, long almond-shaped, trimmed to below tip of finger. See Pl. LXXVIII.

E. i. 05. Stucco relief fragment. Red clay mixed with fine vegetable fibre, over grey clay mixed with chopped straw. Portion of ear (life-size) and head; very crude suggestion of hair. 4½" × 3". Friable. See Pl. LXXVIII.

E. i. 06. Stucco relief fragment. Red clay, with coarse rope core. Maṇi-string. 2" long × 1½" greatest diameter. Friable.

E. i. 07. Stucco relief fragment. Red clay; Vajra, tripartite; centre ornamented with diced pattern. 3½" high, 3⅜" across. See Pl. LXXVIII.

E. i. 08. Stucco relief fragment. Red clay, mixed with fine vegetable fibre. 2 Vajras; one affixed to stick by means of similar red clay, but without fibre. The other has impress of stick which has come away from its place. Both coloured pink. Replica E. i. 07. 3⅜" × 3½". Very friable.

E. i. 09. Stucco relief fragment; ornament in 2 pieces. Red clay, mixed with vegetable fibre. Prob. part of personal ornament (necklace); white lotus with smaller turquoise-coloured flowers. The whole evidently was on a cord which has perished. Cf. E. i. 06 and E. i. 010. Length 2½", width 1⅛". Very friable.

E. i. 010. Stucco relief fragment; ornament (7 pieces). Red clay, no fibre; traces of white and turquoise. Portion of necklace (Maṇi-string), moulded with small flowers, with larger ones at intervals; pressed round a core of coarse cord, portions of which remain. Replica E. i. 09. Length (approx.) 4¾", width of large flowers 1¼". Very friable. See Pl. LXXVIII.

E. i. 011. Stucco relief fragment; life-size finger. Red clay, mixed with fibre. Replica E. i. 04. Length 3⅞". Very friable. See Pl. LXXVIII.

E. i. 012. Fresco painting on stucco. Badly damaged fragment of wall surface composed of, at back, coarse plaster mixed with vegetable fibre; over this finer plaster without fibre; and over this again fine stucco which gives the painting surface. About the centre of the fragment is painted a head three-quarters to R. p., surrounded by green nimbus. The shoulders of figure (probably seated) can be made out. Above nimbus seems to be part of green robe of another figure, and to R. p. red robe on shoulder of a third. The ground is Indian red, as are outlines and contours generally. Flesh, yellow pink. There seems to be a vesica behind central figure, the ground of which is pink. Hair, eyebrows, and eyes of figure black. The green field of nimbus has a lighter green line round it, and outside this is a broad Indian red line. Work seems freely and easily executed, and shows complete knowledge of the process. 5¼″ × 4″. See Pl. LXXIX.

E. i. 012. a. Fresco on stucco fragment. Grey clay mixed with coarse vegetable fibre. Surface of clay only covered with thin layer of fine stucco. Conventional floral design painted with great freedom. Colours—green, yellow, orange, deep brown. Appears to be water-colour, and adheres perfectly to the stucco, which was probably wet when paint was applied. Surface much cracked. 5½″ × 5½″.

E. i. 013. Stucco relief fragments.

a. Red clay mixed with fine vegetable fibre, a coarse string passing through length of fragment. Portion of robe from south-west statue showing corner and hanging folds. Robe coloured liver-brown, with border about 1″ wide, bright red. 9″ long, 2¼″ widest part. See Pl. LXXVIII.

b. Clay as above, red wash and traces of white; coarse string forms centre, the two ends appearing at upper side of fragment. Pierced portion of lobe of ear; colossal. From north-west statue. 5½″ length, 1⅞″ at widest part. See Pl. LXXVIII.

E. i. 014. Stucco relief fragment; ornament. Red clay, traces of pink wash. Portion of tiara, or arm ornament; from a band about ¾″ wide, jewelled with square raised jewels at intervals (1 remains), rises a half lotus, from which radiate six ostrich feather-like leaves at regular intervals, forming a semicircular ornament. In centre of outer edge a small round jewel composed of a bead (pearl?) surrounded by a single row of smaller pearls. This jewel is connected with lotus by a row of 4 pearls. At back, a portion of wood support remains. Height 2⅝″, width 3″. Very friable. See Pl. LXXVIII.

E. i. 015. Stucco relief fragment (3 pieces). Red clay, traces of colour. Jewelled amulet, clasped on arm (?), ends of band hanging from jewels. Fragment exhibits on one piece, a square-contoured four-petalled flower, adjoining similar flower, but smaller, within double square moulding; on another piece one flower only; on the third, end of band, hanging against arm (?). Width 2¼″, depth 3½″ (approx.). Very friable. See Pl. LXXVIII.

E. i. 016. Silk fabric. Consists of a kind of pennon composed of four rectangular pieces of silk, joined edge to edge by some kind of paste. To lowest piece are pasted three narrow streamers, the centre one straight and the two side ones at divergent angles. The pennon was originally longer, as is indicated by the paste adhering to the upper edge of the topmost piece. Silk is of usual oriental kind, some pieces being more strongly ribbed than others on account of the warp threads being rather thicker. Length 14″, width about 2¼″. See Pl. LXXVII.

E. i. 017. Silk fabric. Three pieces of silk made into kite-shapes by being doubled over wooden (cane?) stiffeners, and neatly sewn. They are attached to each other, point to tail, by sewing. The silk is of good quality, rather E. i.
strongly ribbed, the upper one having a kind of chevron pattern woven into it. Colour of upper one is drab; that of middle light cream, and of lowest a rich soft pink. Size of each piece about 3¼″ × 1¼″. See Pl. LXXVII.

E. i. 018. Silk fabric. Heavy silk brocade, showing absolute mastery in the art of weaving this class of fabric. It appears to be what is technically known as a double cloth, i.e. it has two warps and two wefts. Texture fine and even; design, a small portion of which is seen, skilfully worked out in diagonal weaving; arrangement of colours most artistic. E.g. the dark pattern is in green, but upon examination the threads used are found to be three or four different shades 'broken' one into another. The ground was originally a rich flesh-pink, the pattern in soft greens and creamy white, and in the ravellings at one corner occurs a cobalt blue. Size 3½″ × 2½″. See Pl. LXXVII.

E. i. 019. Silk fabric. Heavy brocade similar to E. i. 018. Size 3¼ × 2″. See Pl. LXXVII.

E. i. 020. Silk fabric. Piece of silk brocade. Similar in make to E. i. 018 but perhaps not quite so carefully worked. The pattern, of which a very small portion is visible, appears to have been more free and less 'square' than in E. i. 018. The ground appears to have been a deep gold colour, and the pattern in white, red-brown, pale blue, and green. Size 3″ × 1⅛″. See Pl. LXXVII.

E. i. 021. Silk fabric. Loosely woven fine silk fabric, with satin diagonal lines crossing, forming lozenge shapes; in centre of each a flower. Pale steel-blue. 16″ × 1″. See Pl. LXXVII.

E. i. 022. Silk fabric. Thin silk; faded green-blue. 6½″ × 1″. See Pl. LXXVII.

E. i. 023. Woven fabric. Firmly and finely woven white cotton or linen in a lozenge diaper pattern. Still quite strong and beautifully even in workmanship. 5″ × 2⅜″. See Pl. LXXVII.

E. i. 024. Silk fabric. Piece of loosely woven and very flimsy plain muslin dyed pale blue. 9¾″ × 2″. See Pl. LXXVII.

E. i. 025. Silk fabric. Plain silk; has formed part of pennon similar to E. i. 016, and may in fact be part of that. Mucilage visible on two edges. Dull pale brown. 2⅛″ × 2¼″. See Pl. LXXVII.

E. i. 026. Cotton fabric. Coarse, loosely woven plain cotton cloth. Dyed dull violet-brown. 14″ × 1¼″. See Pl. LXXVII.

E. i. 027. Silk fabric. Loosely woven silk, with slight pattern in satin stitch. Colour, pale cream. 5″ × ¾″. See Pl. LXXVII.

E. i. 028. Woven fabric. Cream coloured; not determined. See Pl. LXXVII.

E. i. **E. i. 029. Knot-dyed fabric.** Blue cotton cloth. Seems to have been a small bag, made up of a roughly circular piece hemmed all round for a drawing string, part of which remains. The cloth is ornamented with flower-like patterns executed in knot-work, being white on blue ground. Torn. Diameter of circle about $6\frac{3}{4}''$. See Pl. LXXVI.

PANEL EXCAVATED IN CHAPEL E. ii.

E. ii. **E. ii. 1. Painted panel.** Seated Gaṇeśa, crowned (yellow). Arm, wrist, and ankle ornaments (yellow). Tiger skin Dhōtī (yellow), tight Paijāmas (dark brown), feet bare. Head three-quarters to R. p., yellow. Rest of flesh—pink. Outlines red. Eyebrows, eyelashes, and pupils black. Cushion red and dark-brown vesica; field green; outer edge deep red with fine black line outside, and border of white dots inside, next green. Rosary of white dots. Four arms. R. p. front in lap holding bowl of fruit (?); back, upraised holding spear-head (?). L. p. front, at breast holding long-shaped turnip; back, upraised holding axe. Background of panel white with a few small ornaments. Height $4\frac{5}{8}''$, width $5\frac{1}{4}''$. Wood and painting well-preserved. See Pl. LXXVIII.

MISCELLANEOUS OBJECTS FOUND ON GROUND SURFACE AT ENDERE SITE.

Miscell. objects. **E. 001. a. Piece of Chinese writing ink,** drilled at one end for a string. Length of piece $\frac{7}{8}''$. See Pl. CV.

E. 001. b. Cubical bone die with spots shown by circles with centre dots. The sum of the spots on any two opposite sides is 7. Length of edges $\frac{1}{2}''$. See Pl. LII.

E. 002. Glass and pottery fragments. 7 fragments of pale green-tinted glass with raised pattern. 2 fragments of pale pink-tinted glass with raised pattern. 3 pieces pale green-tinted glass, plain. 1 thick piece pale green-tinted glass, showing cut or ground surface. $1'' \times 1''$. 1 short stick of glass with twist in the grain, $1\frac{1}{4}''$ long. 2 fragments of coarse terra-cotta, glazed green. Found near Endere Fort ruins. See Pl. LII.

[E. 003. **Miscellaneous small objects** found within or near Endere Fort.]

E. 003. a. Rectangular bronze tag, hollow, to take leather or other material, and having 3 rivets remaining in position. One end shaped in a sort of double Ogee; other sides straight. Length $\frac{7}{8}''$, width $\frac{1}{2}''$, thickness (external) $\frac{1}{8}''$. See Pl. LII.

E. 003. b. One side of bronze hinge or clasp, probably from armour. It is slightly concave at its two long edges, convex at the narrow end and straight at the broad end where portions of links of hinge remain. There are three rivet holes, one at narrow end and two at broad end. Whole plate slightly convex on its anterior aspect. Length $1\frac{11}{16}''$, width $\frac{7}{8}''$ to $\frac{1}{2}''$. See Pl. LII.

E. 003. c. Bronze plate, oblong, pierced for a rivet at each end. Probably piece of jazarine. $1\frac{1}{2}'' \times \frac{7}{16}''$.

E. 003. d. Bronze object resembling a spatula. At one end it opens out into a ring, the upper edge of which shows signs of wear, as though from friction with a ring to which it has been suspended. The opposite end broadens and then turns to an obtuse point. Length $1\frac{1}{2}''$, width above point $\frac{3}{8}''$. Thickness at ring end $\frac{1}{16}''$. See Pl. LII.

E. 003. e. Fragment of green-blue glass; portion of edge circular, remainder broken. Pattern incised on one side, the lines having been filled with some metal or pigment (part of which remains) in manner of *niello*. Diameter $\frac{5}{8}''$, thickness $\frac{1}{16}''$. See Pl. LII.

E. 003. f. Piece of yellowish glass; elliptical plano-convex, probably made as a jewel for an ornament. See Pl. LII.

E. 003. g. Fragment of glass vessel, showing cut or ground surfaces.

E. 003. h. Fragment of glass vessel, showing cut or ground surfaces.

E. 003. i. Piece of pink coral or shell, pierced in two places for a cord, and showing a groove worn by the thread between the two piercings. Length $1''$.

E. 005. Miscellaneous beads, bronze rings, &c. (found near Endere Fort). On a woollen string are threaded 1 small coin; 2 blue glass beads; 2 stone or pottery beads; 1 bright yellow glass bead, the glass surface being over some lustrous core; 1 yellow glass or stone bead; 2 fragments of flat bronze ring; 6 fragments of coloured glass and pottery beads; fragment of 1 bronze ring; 1 bronze rivet (length $\frac{3}{16}''$, thickness $\frac{1}{8}''$); 1 bronze, hollow, dome-headed stud (length $\frac{3}{16}''$, diameter $\frac{5}{32}''$); 1 oblong piece of bronze, with lightly incised decorative pattern. $\frac{3}{16}'' \times \frac{5}{16}''$.

E. 006. Terra-cotta fragments, &c., found on eroded ground near Endere Stūpa.

a. Fragment of green glazed terra-cotta vessel; $2\frac{3}{4}'' \times 1\frac{3}{4}''$.

b. Fragment of neck of vessel, of black terra-cotta; $1\frac{3}{4}'' \times 1\frac{3}{4}''$.

c. Fragment of lips of unglazed terra-cotta vessel; $1\frac{1}{4}'' \times 1\frac{3}{8}''$.

d, e. Fragments of coarse terra-cotta.

f. Piece of uncut green jade. $1\frac{7}{8}'' \times 1\frac{7}{8}''$.

CHAPTER XIII

KARA-DONG AND THE SEARCH FOR HSÜAN-TSANG'S P'I-MO

Section I.—EXPEDITION TO THE KARA-DONG RUINS

Departure from Endere.

On February 26 my explorations at the Endere site were completed, thanks to the energy with which the work had been carried on from early morning until after nightfall by the light of bonfires. The information collected during the last weeks showed that there were no ruins known eastwards nearer than Charchan, where, besides the remains of the 'old town' close to the modern oasis which M. Grenard had already examined, some ruins were mentioned to me at a day's journey to the north, probably the same as those referred to by M. Grenard[1]. But the journey alone to and fro would have cost a fortnight, and the time that remained at my disposal seemed none too ample for the expeditions I had yet to make to ancient localities north of Keriya and Khotan. Though the nights were still bitterly cold and the occasional winds light, yet the increasing warmth of the day-time and the prevailing haze intimated that the season of sandstorms and heat was steadily approaching. I knew that its arrival would effectively bar what excavations I had planned in the desert nearer to Khotan. So I reluctantly decided that the time had come to set my face again westwards.

March along Niya-Charchan route.

The rapid marches which, between February 26 and March 2, carried me and my caravan back to Niya, need only to be indicated here in the briefest outlines. The first took me across the Endere river at a point known as Körgach, where I found the main current, about fifteen yards broad, covered with ice still strong enough to bear heavy loads. The deeply-cut bed, some eighty yards wide, demonstrated the extent of the floods which the river carries down in the spring and summer. Beyond, after passing through a belt of closely packed sand-cones covered with tamarisk scrub, we struck an old bed of the river and marched along it through luxuriant Toghrak jungle to a deserted shepherd's hut known as Tokuz-kol. The name means 'Nine Lakes,' but of water there was none. On February 27 we steered due south over a Kumush-covered plain. All traces of the true desert disappeared for a time; and, as if to make the contrast with the scenery seen for the last month still more striking, a temporary lifting of the screen of dust haze allowed us in the morning clearly to sight and fix on the plane-table a series of prominent peaks in the great rampart of the Kun-lun, with the glaciers descending around them. The portion of the range seen lay to the south-east, from sixty to eighty miles distant[2]. The mountain view vanished like a vision when a strong north wind again raised the usual dust haze, just as we had reached the Niya-Charchan route. Along the lonely desert track which has taken the place of what once was the great line of communication to China, we continued over a gravel Dasht, with scarcely any dunes, and bare of vegetation, until after a march of over thirty-two miles Yoka-toghrak was reached. There some brackish water was obtainable in wells about 6 ft. deep, while a small patch of tamarisks and Toghraks offered scanty fuel.

[1] See *Mission D. de Rhins*, i. pp. 183 sq.

[2] Comp. the small-scale map accompanying my *Ruins of Khotan*.

Return to Niya.

On the following day, with the camels tired out by the last forced march, we could only reach the Yārtunguz river. The track led through very low dunes with tamarisk scrub, until close to the left river-bank a broad sandy ridge, about a hundred feet high, had to be crossed. The river flowed here in a bed about twenty yards broad, between steeply-cut banks about 20 ft. high, and carried from 2 ft. to 3 ft. of water. Its ice had evidently melted everywhere up to the foot of the mountains. The two long marches which brought me back to Niya yielded pleasing variety in the little lakes and lagoons we passed. Among them the lakes of Sizütke and Bileklik lay quite close to the route. They are all mainly fed by springs. The water of the latter, which, just released from the grip of frost, was flowing plentifully, comes, no doubt, from the streams that higher up near the mountains are absorbed by the gravel glacis of the Sai. From the Shītala Daryā, a stream then fed by springs but receiving water also from the mountains later in the year, there stretched a continuous expanse of fertile jungle with plentiful Kumush and scrubby undergrowth to within a little over three miles from the Niya river. Ample springs were passed near the grazing grounds of Dong-Ōghil, Kum-Chaklik, and Kalta-qerīn. The eastern bank of the Niya river proved to be flanked just like that of all the other rivers east of Khotan where they enter, or pass through, the desert, by a high 'Dawān' of sand-dunes. This was crossed near the Mazār of Shītala Pādshāhim, a conspicuous collection of poles adorned with fluttering rags. The remainder of the march to Niya, leading past the marshes of the Gilem-kul and through almost continuous stretches of boggy ground, vividly impressed me with the accuracy of the description which Hsüan-tsang has left us of the situation of ancient Ni-jang[3]. Since starting from Niya on January 23, I had covered over 300 miles in a great oval loop. Considering the distance and the deceptive nature of the ground, it was no small satisfaction to me to find, when the positions indicated on the plane-table for the starting and closing points of our route came to be compared, that the difference proved to be only three-fourths of a mile in longitude and a little over a mile in latitude. The astronomical observations for latitude taken at all more important camps have rendered it easy to effect the needful slight adjustments.

Accuracy of survey tested.

Preparations at Keriya.

Leaving my 'goods train' of camels to follow behind, I covered on March 3 and 4 the distance from Niya to Keriya, some eighty miles, in two stages. There I was busy at work with official reports and letters that were to secure for me a minimum of time for preparing a preliminary account of my discoveries on completion of the journey, and with rapid arrangements for my next explorations. Huang-Daloi, the kindly Amban of Keriya or Yü-t'ien, as his district is officially styled, opportunely returned to head quarters from a short tour just after my arrival, and, being endowed like most educated Chinese officials with a keen historical sense, showed very gratifying interest in all my finds and reports. Thanks to his energetic assistance and the unwearied efforts of Ibrāhīm, my excellent Darōgha, a halt of only two days sufficed to make all arrangements for labourers to accompany me for excavation work, for camels to replace part of my worn-out transport, and for the supplies that men and animals needed. I greatly doubt whether all these arrangements could have been carried out more rapidly and effectively at the head quarters of an Indian District—or anywhere else in Asia.

March down Keriya river.

Special care was required about these preparations, in view of the great distance and the total want of local supplies near my next objective. It was the ruined site of *Kara-dong*, situated in the desert some 150 miles north of Keriya, to which Dr. Hedin had paid a short visit in 1896 on his march down the Keriya river[4]. I knew from the accounts of Turdi, whose 'treasure-seeking' expeditions had twice extended to this place, that the structural remains

[3] See above, p. 311.

[4] See *Through Asia*, ii. pp. 816 sq.; *Reisen in Z.-A.*, pp. 43 sq.

as well as the extent of this so-called 'ancient city' (to which he gave the name *Ak-tiken*) were very modest. Yet I felt that, in view of the interest attaching to the site owing to its isolated position so far north, a systematic examination was called for. For the loss of time implied by the distance I endeavoured to make up by hard marching. For an account of the six days' journey, during which I followed the course of the Keriya river downwards, I may refer to my *Personal Narrative*. During the first three marches I retraced the route which had brought me in January from Kochkar-Öghil to Keriya, while a detailed and accurate description of the physical features observed on the next three has already been recorded by Dr. Hedin[5]. The aspect of the river-banks was still as bleak and bare as two months before, but in place of the glassy sheet of ice there now rolled a muddy current, fed by the melting of the ice that had covered the marshes and pools about Keriya. It was the regular spring flood from the *kara-su* or 'black water' feeders of the river, while months would yet pass before the *ak-su* or 'white water' flood would bring down the melting snows of the mountains. At the Mazār of Burhānuddīn-Pādshāhim, where I had a cheerful welcome from the 'Sheikhs', my caravan was joined by Ghāzī Sheikh, the senior of the priestly fraternity. Being himself a large owner of sheep, he knew, of course, every living soul of the little community of nomadic herdsmen who graze the flocks belonging to Keriya 'Bais' in the riverine jungle belt. So it was easy for our Darōgha to strengthen the band of labourers I had brought from Keriya with fresh recruits from among the shepherds. The men joined readily enough; for uncouth and 'jungly' as their appearance was, in rough furs and sandals made of goatskins, they were all quite alive to the chance of earning a little hard cash that might come useful on their periodical visits to Keriya, where many of them have relatives living as cultivators[6]. So my band kept swelling on the way like a small avalanche.

Start from Tonguz-baste.

On the evening of March 12 we had reached the shepherd station of Tonguz-baste, to the north-west of which I knew, from Dr. Hedin's account, the ruins of Kara-dong to be situated. Mullah Shāh, an experienced and intelligent shepherd, who was to guide us—Turdi Khwāja had been sent from Keriya with my mail to Khotan—turned up late at night, and after prolonged protestations of ignorance acknowledged that he had twice visited the ruins. Another shepherd, Muḥammad Shāh, known as 'the hunter' (Merghen), an active young fellow, was to help Mullah Shāh, his 'Ustād', in finding the track. This turned out no easy task. The morning was very hazy, and by the time the water-tanks had been filled and a depôt made of supplies not immediately needed, a stiff north wind sprung up, which by degrees developed into a regular Burān, the first of the season. For about seven miles we steered almost due north through Toghrak jungle invaded by deep sand, following mainly the course of an old river-bed which branches off from the present course of the Keriya Daryā, some six miles above Tonguz-baste, and still receives some water during the summer (see inset of map).

Approach to *Kara-dong*.

After passing a little pool known as Toldāma, with some water left behind by the last flood, our guides struck to the north-west. So far we had marched in a whirl of dust. But now, as the force of the storm increased, the air became so thick that it was difficult to see even for a hundred yards. With the fine sand driving into my face and accumulating under my eyelashes, in spite of goggles, it was difficult to see much of the route. But I noted that after a couple of miles the groups of Toghraks were left behind, and the sand-dunes rose in height, with many tamarisk-covered cones between them. After plodding on among these for another hour our guides, whose local sense was doubly assuring under such trying conditions,

[5] See *Reisen in Z.-A.*, pp. 40 sqq.

[6] See the table of individuals of whom I took anthropometrical measurements on this journey, in *J. Anthrop. Inst.*, xxxiii. p. 322.

declared that we were near the high tamarisk-covered cones that had given the site its name, *Kara-dong* 'the Black Hillocks'. But as in the blinding dust they could not make sure of the exact direction I let them go ahead, while we sought shelter under the lee of a large sand-cone. It was interesting to watch the sand being driven in a thick spray over the crest line of the dunes just as if they were storm-tossed waves. After half an hour Muhammad Shāh returned with the news that the ruins were due west of us, and not far off. In proof he brought a piece of old pottery he had picked up. So the march was resumed just as the force of the storm showed signs of abating, and after another two miles over dunes rising from about 15 to 25 ft. in height we reached easier ground, where pottery fragments appeared in occasional patches between dunes only 5 to 10 ft. high. Going north-west for about half a mile we arrived at a series of remarkably high sand-cones close to the western edge of this relatively open area, and hidden between them sighted the ruins I had come in search of[7].

Survey of ruined quadrangle.

The remains of Kara-dong proved to consist mainly of the ruined quadrangle shown by the plan on Plate XXXVIII. Portions of its southern and eastern faces are seen in Fig. 52, while the photograph (Fig. 53) reproduces the appearance of the ruin from the east near what proved to be the entrance gate. Closer examination showed that the quadrangle was formed by a mud rampart, which was once occupied on its top by rows of rooms built of timber and plaster. Except for lines of posts still rising near the south-east corner (see Fig. 52), splintered beams indicating foundations of cross walls, and similar scanty remains, the walls of the structures once lining the top of the rampart had so badly decayed through erosion that the approximate dimensions and disposition of the quadrangle could be ascertained only after a prolonged examination of all details and repeated surveys. These showed that the quadrangle, as marked by the outer walls of the rooms still traceable on the top of the rampart, had measured approximately 236 ft. square. There were indications both on the northern and western faces that the rows of rooms had been originally double. But as even the foundations of the walls of the second row could be traced only in the form of rotten timber débris strewing the slopes of the rampart where not hidden away by the dunes, no measurements could be obtained. Only in the north-west corner could the dimensions of two small rooms, set back to back, and nearly 11 ft. square each, be exactly ascertained[8]. Here, too, the timber and plaster work of the walls reached nowhere higher than about 1 ft. above the mud flooring.

Height of covering dunes.

But while all extant remains plainly indicated the havoc wrought by excessive erosion, to which the ruin must have been exposed for a prolonged period, its survey was rendered still more difficult by the heavy sand that had subsequently invaded it. Two large dunes stretched diagonally across the quadrangle, rising within it to heights of more than 20 ft. above the original ground-level, and connecting outside with the neighbouring sand-cones, which showed elevations up to 50 ft. (see plan). Where the axis of these dunes lay within the quadrangle it was impossible to see whether they covered any structural remains, and excavation in any case would have been an affair of weeks.

Structures in interior of quadrangle.

But in the receding angle between the two dunes, where the sand was lower, there appeared the timber framework of a building (K. i. in plan), measuring about 48 by 26 ft. It is seen, after excavation, in the centre of Fig. 52. Its clearing proved very difficult, as the plaster of the walls had completely disappeared, and the sand from the slope of the dune behind kept pouring down into the area excavated. The

[7] It may be noted that the route shown on the map is the one followed on my return march to Tonguz-baste, which was more direct. On March 13 survey work with the plane-table was a physical impossibility.

[8] Owing to reduction, the plan in Plate XXXVIII fails to reproduce the distinction between intact and broken walls shown in the original drawing.

INTERIOR OF RUINED QUADRANGLE, KARA-DONG, SEEN FROM NORTH-EAST AFTER PARTIAL EXCAVATION.

WOODEN GATEWAY OF RUINED QUADRANGLE, KARA-DONG, SEEN FROM EAST AFTER EXCAVATION.

heavy timber roofing, which once probably supported a second story, and was still partially preserved, added an element of risk for the labourers, as the rotten posts beneath could not be trusted to support it, and materials for propping it up could not be improvised. In spite of these difficulties my labourers managed, after two days' arduous work, to clear the three rooms on the north down to the original floor. No finds of any kind rewarded this effort. A small square structure close by, to the north-east, was roofless, but had retained parts of its plastered walls; the construction resembled that observed in the Dandān-Uiliq ruins, reed matting and round sticks at regular intervals being used as a backing for the plaster.

Rampart and structures on it.

To the north-east of K. i. the dunes had left a portion of the original ground-level bare, except for the timber débris of some fallen structure (see Fig. 52, on the left). By careful levelling from this point I ascertained that the floor of the rooms occupying the top of the rampart lay 17 ft. above the ground-level in the interior of the quadrangle. The exposed position thus occupied fully explains the complete decay of these rooms. The rampart below them was proved, by a cutting which I had made in an eroded portion of it near the north-east corner, to have had a thickness of about 30 ft. at its foot. On its top a continuous layer of Toghrak and tamarisk branches about 1 ft. thick had served as a general foundation for the floor beams.

Finds of small antiques.

Amidst the sand and timber débris covering the top of the rampart and the outside slopes fragments of coarse pottery, small broken pieces of metal objects, such as rivets and buckles, and of glass; also shreds of felt and coarse cotton fabrics, could be picked up. Among these scanty finds, described under K. 001, 002, two arrow-heads (K. 001. b, K. 002. n), a bronze buckle (K. 002. a), a portion of the lip or foot of a glass vessel (K. 002. j), all shown in Plate LI, may be specially mentioned. An excellently-preserved small comb of sandal-wood, parabolic in shape, and of a pattern exactly the same as still commonly used throughout Northern India (K. 002. o), may have been an importation from that country. It is important to note that of the five Chinese copper coins found among or near the débris of the rampart, all bearing marks of long circulation, two are *Wu-chu* pieces, while the rest show no legend, I may conveniently mention here that, of the other nine coins picked up on eroded ground at some distance to the east and south of the ruin, one belongs to a Wu-chu issue, the rest bearing no legend (see Pl. LXXXIX, 15) or being too defaced for identification (see ibid., 6). In view of the chronological conclusion suggested by these coins, the few tiny pieces of flimsy paper without writing which I found in the débris of the quadrangle enclosure also possess, as we shall see, some antiquarian interest.

Coins found near ruins.

Gateway of quadrangle.

The only relatively well-preserved portion of this ruin was the large gateway (K. ii.), 21 ft. square, which was found to lead through the rampart on the eastern face of the enclosure. Fig. 53 shows it as it appeared, after clearing, seen from the east. The sand was here less high, and had left, even before excavation, some space free below the perfectly intact roof. The latter reached to within a foot of the top level of the rampart, and it was, no doubt, owing to the protection which the latter offered on either side that the timber of the walls and the massive posts supporting the roof had been preserved so well. Besides a central passage, 10 ft. broad, closed on the outside by a massive wooden door of two leaves, there were two side passages, 5 ft. broad. I noticed a somewhat similar arrangement in the gates of all modern Ya-mêns I visited. The folds of the door were 3 in. thick, and strengthened by massive cross-bars.

Construction of walls.

The walls dividing these passages consisted of a timber framework, massively but roughly constructed of Toghrak. The rows of sticks set vertically in it showed that a covering of plaster had once existed. This plaster had, however, completely disappeared both in the inner divisions and in the outer walls. Hence it was most difficult to check the pouring-in of sand and loose earth from all sides. It took my men two days' hard digging—vigorously continued in spite of a fresh sandstorm which visited us on March 17, this time from the south-west—before the floor of the central passage was reached at a depth of 14 ft. from the beams of the ceiling. On examining the latter from below before the clearing had proceeded far down, I ascertained that the thick parallel beams supporting the roof of the gateway were first covered with a thin layer of reeds, evidently intended as a backing for plaster. Above this were laid closely-packed branches of Toghrak, which again were covered with a layer of stamped mud about $1\frac{1}{2}$ ft. in thickness. This had, no doubt, served as the flooring for another story, of which, however, there remained only a few stumps of posts.

Find of ancient cereals. When the men under my direction were proceeding to remove part of this mud flooring, with a view to lightening the weight of the roof and thus reducing the risk of its giving way while the interior was being cleared, we came upon a little store of remarkably well-preserved cereals embedded in the layer of rubbish covering the floor. There were a couple of pounds of 'Tarīgh' or millet (*Setaria italica*), still largely cultivated about Keriya and elsewhere, together with small quantities of rice, barley, lentils, and a capful of large black currants dried perfectly hard. For the determination of the above I am indebted to the Director of the Royal Gardens, Kew. The exact nature of some roots also found here, in which Jasvant Singh, the surveyor's cook, as well as the labourers, thought to recognize some sort of turnip, has not yet been determined. A small ruined structure outside the gate (seen on the left in Fig. 53) was cleared without any finds.

Surroundings of ruined quadrangle. The survey of the surroundings of the ruined quadrangle yielded little of direct antiquarian interest. To the west and north there extended, as far as I could see and the men sent out ascertain, dunes of great height piled up in weird desolation amidst sand-cones quite as big as those encircling the ruin. Such masses of sand would have effectively smothered the remains even of large buildings, if such were likely to have ever existed here. I may note that tracks of wild camels were reported to me on the north. To south and east the ground for about half a mile was relatively open, the dunes stretching across it being only from 5 to 15 ft. in height. But the traces of ancient occupation which the depressions between them revealed were very scanty. Small fragments of coarse pottery appeared indeed over rare patches, but for indications of structural remains such as timber posts or walls, which the low dunes could not have effectively hidden, I looked in vain, except near the southern limit of this débris area, about half a mile to the south-east of the quadrangle, where my guides knew of an 'old house'.

Remains of old dwelling. From the west slope of a dune about 10 ft. high there rose here rough wooden posts marking the position of a compact block of rooms extending about 63 ft. from north to south. On excavation the walls, which seem to have been of mud only, apart from the posts meant to carry the roof, were found to have decayed almost to the ground; the whole structure had evidently been for a prolonged period without the protection of sand. No objects of any kind were found, except three round pots of very coarse pottery set in the mud flooring. Their clay was distinctly inferior to that of the pottery found at the Niya Site, both in point of hardness and uniformity of grain. The largest measured 18 by 18 in., being 7 in. wide at the mouth. The shoulder was decorated with a rough zigzag line, and had three small handles attached to it. Another was 7 in. high, 8 in. broad at its greatest circumference, and was provided with two elongated handles. The mouth, 3 in. wide, had a neck raised 3 in. The third, which was broken at the bottom, measured 14 in. across, with a mouth 5 in. wide. The general character of the remains indicated a relatively large but roughly-built dwelling. The very fact of its survival suggested that what structures may have once existed near by must have been of a still less substantial type.

Use of Toghrak timber.

It seems to me very significant that here, too, just as in all the structures of the ruined quadrangle, only Toghrak was used for the timber-work. This species of wild poplar grows plentifully, as we have seen, in the jungles of all the rivers which lose themselves in the sand; it manages to survive even in desert tracts which surface water never reaches now. But its twisted knotty trunks and branches by no means furnish as good a building material as the Terek or white poplar, the Jigda, and other trees planted in cultivated areas. At Dandān-Uiliq, at the Niya and Endere Sites, only timber of these latter trees appears to have been used for the framework of houses, and this accordingly bore there a far more finished look than at Kara-dong. At those other old sites the dead trunks of Terek and other trees depending on systematic irrigation formed a conspicuous feature. But around Kara-dong I looked for them in vain. Dead trees rising from between the sand-dunes were plentiful on the ground close to the east and south of the main ruin, but they were all old Toghraks or tamarisks, such as are still found growing luxuriantly in a broad belt west of the relatively recent river-bed of Toldāma. I think it justifiable to conclude from this observation that cultivation could not have existed to any considerable extent in the vicinity of the Kara-dong site at the period from which its buildings date.

Purpose of ruined quadrangle.

What then can have been the purpose of the great ruined quadrangle, situated, as it evidently was, in the forest land between the desert and the river, and contrasting so strikingly by its size with the modest character and limited extent of the other remains? Keeping in view the position and the peculiar plan of the structure, I think the suggestion may be hazarded that we have here the ruin of an ancient 'Langar' or roadside Sarai, built primarily to afford accommodation for a large number of people, yet capable of defence, too, if the need should arise. Such a suggestion of course presupposes traffic along the Keriya Daryā at an early period, and for the existence of such, I believe, adequate historical and topographical evidence can be adduced.

Historical indications of Keriya river route.

Mirzā Ḥaidar, when speaking of the rivers which empty their waters into the great lake of the Turkestān desert, i. e. the Lop-Nor, distinctly mentions the Keriya Daryā along with 'the river of Yárkand, the Ak-Kásh (i. e. the Yurung-kāsh), and the Kára-Kásh' and 'the Chárchán'[9]. Seeing how thoroughly well acquainted from personal experience the Moghul historian was with the Khotan region, and how exact his description of it has proved to be, it seems very unlikely that he would have made this statement unless the Keriya river was still in his days believed to reach the Tārīm. Also in the Turkī legendary of Maḥmūd Karam Kābulī, which purports to relate events of the twelfth century A. D., a force of Muhammadan warriors, coming from the conquest of Ak-su, is described as attacking the infidels in the vicinity of Keriya by the route of the Keriya river[10].

Project of re-opening

Whatever the historical value of this tradition may be, M. Grenard was fully justified in quoting it in connexion with the plan which an energetic Amban of Keriya formed in 1893 for

[9] See *Tārīkh-i-Rashīdī*, p. 406. The text calls the lake *Kuk Naur*, i. e. Koko-Nor, but the description given of it shows beyond all doubt that Mirzā Ḥaidar means the Lop-Nor. Subsequently he refers to 'the Kará Murán of Khitái,' i. e. the Huang-ho issuing from one end of the lake,—a reflex, as Mr. Ney Elias duly recognized, of the ancient Chinese legend which supposes a subterraneous connexion between the Lop-Nor and the Huang-ho (see, e. g., Rémusat, *Ville de Khotan*, p. 2). This traditional belief accounts for the confusion of the two names.

[10] See Grenard, *Mission D. de Rhins*, iii. p. 44.

route by Keriya river.

opening up a route to the Tārīm along the Keriya river[11]. Dr. Hedin's journey in 1896 has since proved that the Amban's belief in the possibility of traversing the sands between the present end of the river course and Tārīm without serious difficulty was well founded. On the line followed by the Swedish explorer, which must correspond approximately to the old extension of the river-bed, patches of vegetation and relatively near subsoil water are to be found constantly down to the southernmost branch of the Tārīm[12]. There is good reason to believe that the opening of this route, by the digging of wells and the establishment of small posts, would prove no very formidable task even for the present administration.

Importance of route to Kuchā.

A glance at the map shows that the route along the extant course of the Keriya river and its old extension towards the Tārīm forms the most direct line of communication between the whole Khotan region and the ancient territory of Kuchā, and the other oases further to the north-east[13]. That close relations existed politically between Kuchā and the Khotan kingdom, its immediate neighbour to the south, is attested by the fact that the submission of Khotan to Chinese supremacy in 648 A.D. is described in the T'ang Annals as the direct result of the conquest of Kuchā[14]. It is equally certain that the cultural affinities between the two states, the frontiers of which adjoined in the desert, must have been great[15]. The importance of using the nearest route for communication must have been felt more than ever after the administrative head quarters of the Protectorate of An-hsi, controlling 'the Four Garrisons', and thus also Khotan, had first been established at Kuchā in 648 or 649 A.D.[16]

Kara-dong, a station on Keriya river route.

On the grounds here indicated I think we can safely assume that there lay in early times along the banks of the Keriya Daryā a route regularly followed by traffic, just as that along the Khotan Daryā is at present. Kara-dong lies about halfway between the Tārīm and the main oasis of Khotan, and a small station established here would have conveniently served the double object of affording a safe resting-place to caravans, and of watching the route for customs or police purposes. A small settlement, with a Bāzār, &c., might well have existed by the side of such a post, just as found by the side of many a 'Langar' on modern caravan routes of Eastern Turkestān, and this is all that the traceable remains indicate. The natives may, indeed, call the remains a *kōne-shahr*, using the term which is applied throughout the country to old remains of any kind, even the smallest. But to talk of an 'ancient city' here would imply more imagination than a critical student need care to take credit for.

Period of abandonment.

There is no direct evidence to indicate the exact age of the ruined quadrangle or the date when it was abandoned. But from the fact that among the coins picked up there were no T'ang pieces, it appears probable that the site was deserted earlier than, e.g., Dandān-Uiliq, where coins of the K'ai-yüan period (713–741 A.D.) and of later T'ang issues were common. On the other hand, the pieces, without legend or bearing the characters *wu-chu*, which Kara-dong yielded all show marks of long circulation, like those found at Endere. The finding of

[11] See *Mission D. de Rhins*, i. p. 172.

[12] See Hedin, *Reisen in Z.-A.*, pp. 54 sqq.

[13] The saving in distance, as against the modern route down the Khotan Daryā and thence via Ak-su, becomes still more evident when we remember that the settlements about Dandān-Uiliq allowed travellers to move straight from the Keriya river to the Khotan capital without touching the present oasis of Keriya.

[14] See Chavannes, *Turcs occid.*, p. 126; also above, p. 175.

[15] Comp. Hsüan-tsang's description of Kuchā and its people (*Mémoires*, i. pp. 3 sqq.; Watters, *Yuan Chwang's Travels*, i. p. 59) with that given of Khotan; also the T'ang Annals' accounts of both territories, *Turcs occid.*, pp. 124 sq., 125 sq. For a curious point of contact specially noted by the Chinese annalists, see ibid., p. 115; above, p. 139.

See also the story told by Hsüan-tsang of the Khotan minister at Kuchā and the subsequent miraculous transfer of a Buddha statue from Kuchā; *Mémoires*, ii. p. 230; above, p. 225.

[16] See above, p. 60.

small bits of paper also points to a period not very distant from that ascertained for the ruined fort of the latter site. As to the specific cause of the abandonment it would be hazardous to express any definite opinion. A variety of causes, which we have no means of tracing, might have led to the transfer of the post to another locality, or to its complete desertion.

Lateral shifts of river-course.

It is easy, of course, to suggest a change in the river-course which deprived the station of convenient access to water. That these lateral shifts of the Keriya river must have been in old times quite as frequent and marked as they are now is evident. Yet it is noteworthy that the direct distance between Kara-dong and the Toldāma river-bed, which still receives water during the summer floods, is only about five miles, i.e. considerably less than the distance separating recent beds of the river, e.g. below Kochkar-Öghil or Tonguz-baste. When returning from Kara-dong on March 18 I passed, within only one mile of the main ruin, a depression occupied by a large number of small Toghrak trees, all dead, yet clearly showing by their appearance that the time when they died cannot have been very remote. It would appear that they all sprang up together during a temporary return of a river-branch in this direction, but that the period of revived subsoil moisture was not long enough to permit these hardy pioneers of vegetation to grow up fully. Further on I reached the first groups of living Toghraks within about two and a half miles from the ruin, growing comfortably between the westernmost of the three Dawāns, 50 to 60 feet high, which my map marks. Taking into account all conditions, I am inclined to look upon the position of the Kara-dong ruins as striking evidence how relatively small a change the main direction of the Keriya river course has undergone during the last twelve hundred years or so. A reference to the questions discussed above with reference to the Dandān-Uiliq site will show the antiquarian interest attaching to this observation[17].

OBJECTS FOUND AT KARA-DONG.

K. 001. Small objects in metal, glass, &c., found amidst débris of Kara-dong quadrangle.

a. Small fragment of carved bronze plate. $\frac{5}{8}'' \times \frac{1}{2}''$.

b. Iron point, long square pyramid-shaped, with tang. Probably an arrow-head, but possibly a drill. Length 2″, width of thickest part $\frac{3}{16}''$. See Pl. LI.

c. Wire-like strip of iron. Length 2″.

d. Small irregularly-shaped piece of iron. Length $\frac{1}{2}''$.

e. String of 3 seed pearls, on each side of which a pink (coral) bead, the remainder being greenish glass beads, all strung on a (modern) thread. See Pl. LII.

f. Small fragment of thin yellowish glass; curved.

g. Small copper rivet.

[**K. 002. Objects found on eroded ground east of** Kara-dong quadrangle.]

K. 002. a. Bronze buckle. The portion which was attached to the garment or belt is double, one plate being on each side of material, to which it was secured by four rivets. Loop of buckle rectangular, and lying across it is the tongue, which is broad at base, and rapidly narrows to ordinary thickness, broadening again slightly before reaching angular point. The buckle is flat on inner side, and rounded on outer. In this example there appears to be a raised decorative strengthening piece on the outer plate, very similar in form to the tongue, but broader. Outer plate $1'' \times 1\frac{1}{8}''$. Loop $\frac{3}{4}'' \times 1\frac{1}{8}''$. See Pl. LI.

K. 002. b. Bronze buckle similar to K. 002. a. Outer plate $1\frac{3}{8}'' \times 1\frac{1}{8}''$; loop $\frac{3}{4}'' \times 1\frac{1}{8}''$.

K. 002. c. Double plate of buckle, similar to K. 002. a. Hinge broken and loop, &c. absent. Rivets remain. $\frac{7}{8}'' \times \frac{7}{8}''$.

K. 002. d. Double plate of buckle; inner plate has become detached and is sticking to front of outer plate. Rivets remain. $\frac{7}{8}'' \times \frac{7}{8}''$.

K. 002. e. Outer plate of bronze buckle with 3 rivets remaining. $\frac{7}{8}'' \times \frac{7}{8}''$.

K. 002. f. Thin bronze plate, broken at one corner. One hole in one corner only. $\frac{3}{4}'' \times \frac{11}{16}''$.

K. 002. g. Thin bronze plate; 2 rivets attached. $\frac{3}{4}'' \times \frac{3}{8}''$.

K. 002. h. Thin bronze plate; inner plate of buckle. 4 rivet holes. $\frac{9}{16}'' \times \frac{3}{4}''$.

K. 002. i. Fragment of thin bronze plate. Portion of rivet hole visible. $\frac{5}{16}'' \times \frac{5}{16}''$.

[17] See above, pp. 285 sqq.

K. 002. j. Lip or foot of glass vessel, fragment. Lustreless. Edge, which is intact, rolled over and tubular. $2\frac{1}{4}'' \times \frac{1}{2}''$. See Pl. LI.

K. 002. k. Thin yellowish glass; small fragment.

K. 002. l. Terra-cotta whorl. See Pl. LI.

K. 002. m. Bronze ball, pierced. Perhaps used as a whorl. Diam. $\frac{3}{8}''$ nearly.

K. 002. n. Iron arrow-head. Conical with long thin tang from centre of base. Length of head 1″, diam. $\frac{7}{16}''$. Length of tang $2\frac{1}{8}''$, diam. $\frac{1}{8}''$. See Pl. LI.

K. 002. o. Sandal-wood comb; parabolic-shaped. The teeth, which are rather fine, extend from square end to a length of about $1\frac{3}{8}''$. They are well formed and retain considerable elasticity. The pattern of this comb is the same as that now common in Northern India. Length 3″; width at square end $2\frac{1}{2}''$; thickness at rounded end $\frac{1}{8}''$.

K. 003. Handle of terra-cotta vessel; fragment. Upper surface ornamented, three rows of small impressed rings. Clay very impure. Width $1\frac{1}{4}''$, length $1\frac{3}{16}''$.

K. 004. a. Fragment of coarse terra-cotta vessel, having a rapidly-drawn wavy or nebule pattern upon it.

K. 004. b. Handle of coarse terra-cotta vessel; fragment, ornamented with three rows of small incised rings. $1\frac{3}{8}'' \times 1\frac{1}{4}''$.

SECTION II.—HSÜAN-TSANG'S NOTICE OF P'I-MO AND MARCO POLO'S PEIN

My survey and excavations at Kara-dong were completed on March 17 in the midst of a sandstorm such as had greeted our arrival. Though the force of the wind, this time from the south-west, was somewhat less, the driving sand made the conclusion of the work decidedly trying both to my men and to myself. Next morning I left this desolate spot, just as I had reached it, in an atmosphere thick with dust, and oppressive by its haziness. My eyes were now turned to the south again, where a number of archaeological tasks still awaited me in the vicinity of the inhabited area.

Location of Hsüan-tsang's P'i-mo.

First among them was a search for the ancient town of P'i-mo. Hsüan-tsang had visited it on his way from Khotan to Ni-jang or Niya, and its probable identity with Marco Polo's *Pein*, first suggested by Sir Henry Yule, made me all the more anxious to determine, if possible, its position. The distance and direction which the Chinese pilgrim's narrative records for P'i-mo, and which, together with other archaeological indications we shall presently have occasion to discuss in detail, had long before made me look out for the place somewhere to the north of the small oases extending from Chīra to Keriya. I had been anxious to search for it when leaving Dandān-Uiliq by marching due south through the desert. But the absence of local information and practical considerations connected with the condition of my men and animals prevented the execution of this plan. I was hence much pleased when, on my second visit to Keriya, I heard direct from Huang-Daloi, the kindly Amban, of two 'kōne-shahrs' which had been reported to him to exist in the desert beyond Gulakhma, an oasis on the Khotan road some forty miles to the north-west of Keriya. Rām Singh, too, when marching along this route in December, had heard of old remains in that direction. In order to save time I now decided to reach them, if possible, direct from the Keriya river by striking across the desert south-westwards.

March from Keriya river to Shivul Daryā.

During the four days which saw us returning along the Keriya river as fast as camels and ponies could be got to move, I was surprised to notice that the water in the river had fallen by some 3 ft. as compared with the level of the preceding week. Thus the first spring flood had passed by; yet in the vegetation of the riverine jungle I still looked in vain for any sign of approaching spring. Passing again on March 22 the familiar shrine of Burhānuddīn-Pādshāhim, I picked up *en route* the two men who, under the Amban's orders, had been sent by the Bēg of Gulakhma to guide me to the sites. They looked unusually reticent and stupid, but not till too late did we find out that they knew nothing of such a route as I wished to take. Too timid to admit their ignorance, they thought it safest to guide us further and further south, where at least there was no risk from the dreaded Taklamakān. Thus after leaving on March 23 the left bank of the river at a great bend which the latter makes near the grazing-grounds of Kenk-kiök, we passed into a belt of dunes 20 to 30 ft. high, and further on crossed a big Dawān such as always flanks a riverine area in the desert. After about eight miles we struck the northern edge of a wide area of swampy jungle watered by the stream which flows from the springs and marshes of Shīvul, west of the Keriya oasis.

Terminal swamps of Shivul river.

Passing through thickets of Toghraks and luxuriant undergrowth of scrub we reached a series of large pools, in which this stream seems to find its end during the winter. Half a mile to the south-west of these we came to the edge of a long stretch of boggy marsh, treacherously covered with light sand, through which a safe passage for the ponies could be found only with great difficulty. The ground was quite impracticable for the camels, which accordingly had to be sent northward to turn the obstacle by a détour of many miles. The local knowledge of our guides now quite gave out, and though there were everywhere the tracks of flocks that had grazed here during the winter, we did not succeed in finding a single shepherd to help in guiding us. Fortunately we came at last, after one and a half miles of marsh, upon firm sandy ground near the pasture of Shākün-Öghil. South of this we found the course of the Shīvul Daryā, which flowed here as a limpid and fairly rapid stream, fully 6 to 8 ft. deep, in a winding but well-defined bed about 15 ft. broad. This helped once more to guide our guides. We followed it upwards to a lake-like depression they called Kazān-köl, and ultimately, after a long and tiring day's march, arrived at the solitary little shrine known as Ārish-Mazār ('the Mazār of the Cross-ways'), ensconced in a grove of splendid old Toghraks. A large pond in front of it was said to receive water from another spring-fed stream that traverses the area of sandy jungle between Keriya and Kara-kīr Langar. The camels did not arrive until close on midnight, the big bonfires we had kindled *en route* having helped to guide them.

March to Kara-kīr Daryā.

After the experience gained of the value of our guides, there was no alternative but to resign myself to letting them reach again familiar ground near cultivation before striking off into the desert. Accordingly, on March 24, when the thermometer again registered a minimum temperature close to freezing-point, we made our way south-westwards through a belt of sandy jungle, in which the water of another spring-fed stream, the Kara-kīr Daryā, finally loses itself. Here, at the grazing ground of Kara-chilan, we got hold of some shepherds, who, after prolonged protestations of complete ignorance of any and every route, were prevailed upon to guide us at least to the northernmost portion of the Domoko oasis, the nearest in the direction I was aiming at.

Remains of *Ak-taz*.

The track we now followed led through a maze of tamarisk-covered sand-cones, standing closer together than I had seen them anywhere before on the borders of the true desert. In their midst, at a point about four miles to the north-west of Kara-chilan, where erosion had created a stretch of ground somewhat more open, we unexpectedly came upon the unmistakable remains of some ancient settlement. The shepherds called the spot *Ak-taz*, or simply 'Tatilik'. The latter designation, general as it is, seemed appropriate, seeing that the remains consisted mainly of broken bits of old pottery and similar small débris. The fragments were mostly lying on the top of small loess banks rising 8 to 10 ft. above the rest of the ground, and evidently marking the extent of erosion which the soil had undergone where not protected by remains or otherwise. Close to some tamarisk-covered cones the foundations of a few mud walls could be distinguished, manifestly the remains of small houses. These walls showed bundles of Kumush placed vertically between layers of mud, a mode of construction identical with that which I observed on the day following in the modern ruins of 'Old Domoko', and which still prevails for the ordinary dwellings of villagers throughout the neighbouring oases.

Finds at Ak-taz.

The potsherds, of several varieties, red, black, and a yellowish colour, were hard and undoubtedly old; but neither they (see for specimens D. K. 006 in list below) nor a bronze finger ring with a jewel of opaque glass (D. K. 004, see Plate LI), which was picked up on the slope of a loess bank, furnish any definite indication of date. Several small areas of open ground similarly strewn with pottery débris were passed at short intervals for about one mile, remains of mud walls being traceable, however, only in two places. Other small Tatis are likely enough to be hidden away among the tamarisk-covered cones, but there was no time to search for them. It was clear that, owing to their exposed condition and the vicinity of the inhabited area, these scanty remains of old settlements must have been stripped long ago of any materials of value, and that the chance of finding structural ruins was very slight indeed. Turdi Khwāja, who rejoined me with a Dāk at Ārish-Mazār, had heard the site spoken of by people of Domoko as 'Kōnsamōma's town'[1], and as an object of frequent visits from villagers who hoped to find 'treasure', but were afraid of venturing far into the desert.

Colony of Malakalagan.

Ultimately we struck the well-marked bed of the stream of Domoko, about twenty yards broad, but now completely dry, all its water being absorbed for the time by the irrigation demands of the oasis. Immediately beyond we emerged on land newly brought under cultivation belonging to the village of Malakalagan, where we camped for the night. The colony had been formed about fifteen years before by people from the main oasis of Domoko[2]. The latter, extending on both sides of the Khotan–Keriya road from six to eight miles further south, was said to be separated from Malakalagan by a barren belt of dunes. The reclamation of desert soil going on at the new colony was a sight as cheering as it was instructive. Small irrigation cuts were seen winding along the old tamarisk-covered hillocks of sand that had not yet been levelled down by the combined effect of running water and the cultivators' digging. Between them extended carefully-fenced fields. In order to save all arable ground the colonists had sensibly established their homesteads on the top of the larger sand-cones. Here and there the Toghraks of the desert jungle had been spared, particularly near the huts of the settlers. But it was clear they would soon disappear in a hopeless minority by the side of the avenues of young Tereks, Jigda, mulberry and other fruit trees which were rapidly growing up along all irrigation channels. I regret not to have ascertained, in the midst of many pressing occupations, what specific circumstances had led to the establishment of the new village. But there could be no doubt about the geographically interesting fact of cultivation here successfully invading the desert.

Battle-field noticed by Hsüan-tsang.

Before I proceed to describe my search of the next days for the old sites in the desert northward, it will be convenient to set forth here what Hsüan-tsang tells us of P'i-mo, and what had made me look for possible traces of its position in this vicinity. We learn from the

[1] 'Kōnsamōma' seems to be the name given to a demon of old times whom popular legend of Khotan connects with a number of deserted localities near or within the oases, e. g., above Ujat and at Halālbāgh. I never succeeded in getting hold of a full legend concerning the demon, but gathered that he was credited with feeding on human beings.

[2] This is the form of the local name as I heard it commonly pronounced. M. Grenard, *Mission D. de Rhins*, iii. p. 45, writes it *Doumakou* (i. e. Dumaku), Dr. Hedin, *Reisen in Z.-A.*, p. 201, *Damaku*. The first vowel is certainly *o* or *u*, both sounds being practically interchangeable in Eastern Turkī. The second vowel may be more correctly spelt as *ă*, its sound closely approaching *o*, owing to the preceding labial and the effect of epenthesis from the following *o* or *u*, a phonetic tendency strongly developed in the Khotan dialect.

Hsi-yü-chi that 300 li to the east of the Khotan capital the pilgrim was shown in the middle of a great desert marsh an area of several thousands of acres where the ground was completely bare and of a dark-red colour [3]. Tradition asserted that this was the spot where in old days a large army from the Eastern kingdom, i.e. from China, counting a million of men, was met in battle by the king of Kustana at the head of a hundred thousand horsemen. The troops of Kustana having been defeated, the king was made a prisoner, and all his men slaughtered. Their blood gave to the soil its red colour. Similar local legends, intended, no doubt, to account for pieces of ground of which the striking red colour exercised popular imagination, were heard by Hsüan-tsang elsewhere on his travels [4].

Going thirty li or so to the east of this battle-field, the pilgrim arrived at the town of *P'i-mo* 媲摩. Here there was a miracle-working statue of a standing Buddha carved in sandal-wood, and about twenty feet high [5]. 'Those who have any disease, according to the part affected, cover the corresponding place on the statue with gold-leaf, and forthwith they are healed. People who address prayers to it with a sincere heart mostly obtain their wishes.' The local tradition, which the *Hsi-yü-chi* relates at great length, asserted that this statue had been made by Udayana, king of Kauśāmbī, and that after Buddha's death it came through the air to the town of *Ho-lao-lo-chia* 曷勞落迦. The people of that town were rich and prosperous, but had no regard for Buddhist teaching. Hence, though the image displayed its miraculous power, no worship was paid to it. When subsequently an Arhat came and respectfully saluted the statue, the king, to whom his strange appearance had been reported, ordered him to be covered with sand and earth. The Arhat in this condition was deprived of food, but a pious man who had previously worshipped the statue, secretly supplied him with nourishment. When the Arhat was at the point of departure he predicted to this pious person that in retribution for what he had suffered the town would within seven days be covered by sand and earth and all the people perish.

Sacred image at P'i-mo.

Warned to look to his own safety, the pious man told his relatives and friends of the impending doom, but was treated by all with ridicule. On the second day there arose a great wind, which 'carried before it all the dirty soil, whilst there fell various precious substances' [6]. The pious man, who was thereupon reviled afresh, prudently 'excavated for himself a secret passage leading outside the town. On the night of the seventh day there fell a rain of sand and earth which filled the interior of the town. The pious one escaped through his passage, and going eastwards came to this country, and took up his abode in P'i-mo. At the same time appeared there the statue which he worshipped. Ancient tradition said: "When the law of Śākya is extinct, then this image will enter the dragon-palace." The town of *Ho-lao-lo-chia* is now a great sand mound. The kings of the neighbouring countries and persons in power from distant spots have many times wished to excavate the mound and take away the precious things buried

Legend of destruction of *Ho-lao-lo-chia*.

[3] See *Mémoires*, ii. pp. 242 sq.; Beal, ii. p. 322; Watters, *Yuan Chwang's Travels*, ii. p. 298. There is apparently nothing in the wording of the text to justify Julien's assumption that the battle here related is the one which was supposed to have preceded the foundation of the kingdom of Kustana; see above, p. 157. The very mention of a 'king of Kustana' speaks against it. Rémusat, *Ville de Khotan*, pp. 60 sq., assumes an error in the text and attributes the victory to the Khotan king.

[4] See, e.g., his description of the spot of Buddha's 'body-offering' (*Mémoires*, i. pp. 164 sq.; Beal, i. p. 146), which I have had occasion to discuss in detail in my *Archaeological Report, N.W. Frontier Province*, 1904–05, pp. 41 sqq.

[5] The *Life*, which gives an abbreviated account of P'i-mo, makes the statue thirty feet high, and describes it as 'distinguished at the same time by the beauty of its form and an attitude grave and severe'; see *Vie de H.-Ths.*, p. 289.

[6] This version of Beal's is supported by Watters. Julien, *Mémoires*, ii. p. 245, assumed a corruption of the text and substituted characters meaning 'sand and earth' for 'various precious substances.'

there, but as soon as they have arrived at the borders of the place a furious wind has sprung up, dark clouds have gathered together from the four quarters of heaven, and they have become lost to find their way.'

Topographical data of *Hsi-yü-chi*.

The distance of 330 li east of the Khotan capital indicated by Hsüan-tsang's account clearly points to some locality in the neighbourhood of the closely-adjoining oases of Chira, Gulakhma, and Domoko as the probable position of P'i-mo, seeing that three daily marches from Yōtkan would be counted to any of them. With this location might also be reconciled the *Hsi-yü-chi's* record of the next stage on the pilgrim's onward journey. We are told that, going to the east of 'the valley of P'i-mo' he entered a desert, and after having travelled for about 200 li arrived at the town of Ni-jang or Niya. The distance and character of the ground here indicated are quite correct when referred to the journey from the Keriya river to Niya, a distance still ordinarily reckoned at two marches. But the very distinction made here between the town of P'i-mo and the valley of P'i-mo seemed to me an indication that the town could not have been situated about the present Keriya, quite apart from the fact that the distance between the latter and Yōtkan, about 110 miles by the present road, could not have been treated in Hsüan-tsang's days as a three days' journey, any more than it is at present.

Sung Yün's *Han-mo*.

Sung Yün's narrative, as apparently first recognized by Beal[7], supplies an earlier reference to P'i-mo, but under a different name, and with topographical indications which seem less precise at first sight than those of the *Hsi-yü-chi*[8]. Yet we shall see how important they have proved for the identification of the site. Sung Yün, coming from *Tso-mo*, which, as we have seen above, is identical with Hsüan-tsang's and the T'ang Annals' *Chü-mo* (*Tsiu-mo*), and must be located at the present oasis of Charchan[9], travelled 1275 li westwards, and arrived at the town of *Mo* 未, where flowers and fruits reminded him of those of Lo-yang (the present Ho-nan-fu); but the flat roofs of the mud-built houses formed a contrast.

Legend of sacred image.

'After travelling 22 li to the west of the town of Mo, [Sung Yün] came to the town of *Han-mo* 捍麼 or 捍麽. Fifteen li to the south there is a great temple with more than three hundred monks; in it there is a gilt statue six feet high, of marvellous aspect; on it there are displayed in a manifest fashion the distinctive marks [of Buddha], primary and secondary. Its face is always turned towards the east, and it has refused to turn round to the west. According to the story of the old people it arrived flying through the air; the king of the kingdom of Yü-t'ien came in person to see it, and after having worshipped the statue carried it away on a car, but in the middle of the journey, during a night's halt, it suddenly disappeared; people were sent to search for it, and found that it had returned to its original position; thereupon [the king] erected a temple, and assigned for its maintenance four hundred homesteads; when the people of these families have some disease they apply a gold leaf to the statue in the place where they suffer, and are all at once miraculously healed. Since then thousands of people have by the side of the statue erected statues 16 feet high and all kinds of buildings and shrines.' Sung Yün further records that the banners and canopies of embroidered silk put up there counted by tens of thousands. More than half of them, we are told, were banners presented under the Wei dynasty; many of the Chinese inscriptions on them recorded dates from 495 to 513 A.D., while one of them dated as far back as the period of the Yao Ch'in, 384–417 A.D.[10]

[7] See *Travels of Sung-Yun* in *Si-yu-ki*, i. p. lxxxvi, note 9.

[8] I take all details as to Sung Yün's description from M. Chavannes' *Voyage de Song Yun*, p. 14.

[9] See above, p. 436, note 15.

[10] The banners here referred to were evidently of the same type as the often costly prayer-flags and painted banners still to be found as votive offerings in Buddhist temples of Tibet, China, &c. We have found modest minia-

With the historical interest of the last portion of this record, which plainly indicates close intercourse with China during a period when Chinese political supremacy over the Tārīm Basin was wholly in abeyance, we are not concerned here. What is important for us to note is that the miraculous statue which Sung Yün describes as the object of such extensive worship was undoubtedly the same which Hsüan-tsang saw at P'i-mo, and that accordingly Sung Yün's *Han-mo* must be identified with the latter place. With this conclusion the total distance from Tso-mo, 1275 + 22 li, i.e. approximately thirteen marches, is in remarkably close agreement, seeing that the same number of marches would still be reckoned at the present day from Charchan to Domoko or Gulakhma. In striking contrast with this correct reckoning is the immediately following location of the capital of Yü-t'ien at a distance of 870 li west of Han-mo —an instance of those palpably erroneous measurements which unfortunately are by no means rare in the extant versions of Sung Yün's narrative. *Identity of P'i-mo and Han-mo.*

The clear recognition of the fact that 'Marco Polo's route must have been nearly coincident with that of Hsüan-tsang' led Sir Henry Yule to propose the identification of the former's *Pein* with the Chinese pilgrim's *P'i-mo*[11]. This identification was not based primarily on the similarity of the names—though this is close enough, especially if we consider that, as Sir H. Yule pointed out in a note, '*Pein* may easily have been miscopied for *Pem*, which is indeed the reading of some MSS.', and that 'Ramusio has Peym'—but on the exact correspondence of the topographical facts. For Marco Polo, travelling eastwards, mentions his 'Province of Pein' immediately after Khotan, and his description, already referred to, of the desert route between it and Charchan proves that it must have lain to the west of Niya. Hence Sir Henry Yule, with such geographical information before him as Johnson's journey had furnished, was justified in assuming that Pein or P'i-mo 'cannot have been very far from Kiria'. *Sir H. Yule identifies Pein with P'i-mo.*

The archaeological evidence which I secured, and which greatly strengthens Sir Henry Yule's identification of Pein with P'i-mo, will be detailed hereafter[12]. Here it will suffice to point out that Marco Polo describes Pein as 'a province five days in length, lying between east and north-east. There are a good number of towns and villages, but the most noble is PEIN, the capital of the kingdom'. Its position is not indicated, but the length of five marches given to the province is in close agreement with the distance separating the westernmost outpost of the Chīra oasis from Niya. Of the people who were 'worshippers of Mahommet, and subjects of the great Kaan', he tells us that they 'have plenty of all products, including cotton. They live by manufactures and trade'. The custom he relates of their marital relations has already been quoted in illustration of the laxity of the marriage tie prevailing from old times in the whole Khotan region[13]. *Marco Polo's description of Pein.*

SECTION III.—THE SITES OF UZUN-TATI AND ULŪGH-ZIĀRAT

My previous information had clearly indicated that the ruined sites I was in search of must be situated in the vicinity of the Mazār of Lachin-atā. The people of Malakalagan whom I closely questioned did not deny their knowledge of this popular desert shrine, but none of them would *Start for Uzun-Tati.*

ture representations of them among the deposits in the Endere temple cella; see above, pp. 429 sq.

[11] See Yule, *Marco Polo*, i. pp. 191 sq.

[12] I was glad that a brief abstract of it based upon the remarks in my *Preliminary Report*, pp. 58 sq., with illustrations of some of the antiques from Uzun-Tati, could be embodied by Prof. H. Cordier, Sir Henry Yule's learned editor, in a supplementary note to the new edition of *Marco Polo*, ii. p. 595.

[13] See p. 140.

acknowledge ever having visited the 'kōne-shahr' near it. Our wanderings of the previous two days had left me no illusions as to the value of our two worthy guides. Yet as better were not to be got and time was getting more than ever precious, I decided to set out with them on the morning of March 25. Old Turdi had never visited these sites, and from the 'treasure-seeker's' point of view, did not rate them highly. They were too near, he thought, to the cultivated area to have retained much of either 'treasure' or antiques. Yet he was intelligent enough to realize my interest in them, and I could rely on his instinctive experience of the desert coming to the help of whatever local knowledge the Domoko guides might prove to possess. In order to be prepared for the excavation of any structural remains that might possibly be met, I took along an adequate number of labourers. The heat of the daytime was already becoming troublesome in the desert. But by the precaution of having all the water-tanks filled before the start, I provided against immediate risk resulting from prolonged wanderings. The three days' Odyssey which followed, and which could be traced in detail only on a large scale map, fully bore out my misgivings. Yet there were instructive experiences and interesting antiquarian observations to compensate for the fatigue.

Deserted fields of 'old Ponak'.

The first two miles in a north-westerly direction brought us to the limit of the newly-irrigated land, when to my surprise I came upon unmistakable marks of earlier cultivation beyond. Old fields overgrown with tamarisk and thorny scrub could be clearly distinguished by the little embankments dividing them, as well as by the lines of dry 'Ariks' that once carried water to them. There was but little of drift-sand, and that mainly where rudimentary tamarisk thickets were forming to catch and retain it—the first beginnings, it seemed to me, of the tamarisk-covered sand-cones I have had so often to refer to. The Domoko labourers explained that these were the fields of 'old Ponak' village, which had been abandoned 'in their grandfathers' time', i.e. forty or fifty years before. Passing along the road still frequented by the people visiting the cemeteries of the deserted villages, I arrived, some three miles further north-west, at a point known as Kobza-giram-shāh[1].

Abandoned homesteads of 'old Domoko'.

Here we had reached the southern edge of 'old Domoko', an area covered with the remains of a deserted village group. The crumbling ruins of mud-built dwellings, constructed and arranged exactly as in the now inhabited villages of this tract and forming detached groups, seemed to extend, together with the interspersed orchards, cemeteries and fields, for about three miles from east to west. Going towards the north-west we kept between them for nearly three miles. The mud-walls, strengthened by the insertion of vertical bundles of Kumush, still rose often 4 to 5 ft. above the ground, and the massive fireplaces were intact even to a greater height, as seen in the photograph (Fig. 55) of one of these ruined houses. The deserted homesteads had been stripped of all materials that could be of use, such as beams, wooden doorposts, &c. As scarcely any sand had accumulated about the crumbling ruins, their rapid and complete disappearance seemed inevitable as soon as erosion set in. Of the latter no trace, however, appeared as yet over this area.

Remains on ground of future Tati.

It may be due to this preservation of the surface layer of loess that so little of pottery débris could be seen on the ground. Erosion would probably bring to light plenty of it from different layers. Of other characteristic Tati remains, such as smelting refuse, pieces of charcoal and broken stones from the river-beds, there was a great deal to be seen on the ground. Of the fruit-trees in the orchards only low stumps remained, these trees having evidently died off very soon after irrigation ceased and having then been cut down for timber or fuel. On the

[1] *Kobza-girem-shahr* of the map is a misprint.

other hand, plenty of young Toghraks were growing up among the ruins, indicating the relatively small depth of subsoil water. It was interesting to note at different points a number of big Toghraks, declared by the villagers to be very old, which were still in vigorous life. There could be no doubt that these trees were far older than the ruined dwellings. Their presence in this area, after what I had seen at Malakalagan, distinctly suggested that the ground occupied by the ruin of 'old Domoko' had previously, at no very distant period, been occupied by desert jungle. And with this the fact that most of the dead fruit-trees seemed but of small size and modest age curiously agreed.

The villagers accompanying me, as well as the people I subsequently examined on my return to the oasis, all agreed in asserting that the gradually increasing difficulty of conducting the irrigation water sufficiently far had caused the cultivation of the Domoko oasis to be shifted from this old area to the lands along the Khotan–Keriya road within the memory of living men. Local tradition, in fact, maintained that in the case of Domoko such shifts of the cultivated area, backwards and forwards, had occurred already five times and a sixth was to follow. Whatever the basis of this tradition may be, the explanation given by the villagers as to the cause of these shifts seemed sufficiently matter-of-fact. They attributed the abandonment of 'old Domoko' not to any diminution of cultivation or of the water-supply available for it, but simply to the fact that the particular area could no longer be reached by the water carried in the irrigation canals. Cause of abandonment.

The lands of Domoko, as far as I could ascertain, at all times depend for their water on the Domoko Yār, fed mainly by springs. These may be supposed to bring again to the surface the water which the Sai absorbs higher up from the Nūra river and neighbouring mountain streams. That the level or position in which springs of this kind appear is here subject to considerable changes within short periods I had occasion to learn subsequently in the case of Kara-kir, a small oasis just east of Domoko. There the fresh appearance of abundant springs north of the Khotan–Keriya road some ten years before my visit had promptly been followed by the creation of a new colony, with irrigable land sufficient for 700 to 800 households. With such evidence before me it seemed possible to assume that the position of the Domoko villages might have been similarly affected by earlier changes. But a gradual alteration of ground-levels consequent on irrigation deposits also suggested itself as a possible explanation of the alleged shifts. Nor did the villagers' assertion induce me to overlook the possibility of a diminution in the available quantity of water being the true cause. Irrigation of Domoko oasis.

It was clear that a prolonged and detailed investigation of all local conditions, particularly those connected with the supply of irrigation water, would be needed in order to arrive at any safe conclusion as to the cause or causes at work here. But, however that may be, I could feel no doubt as to the archaeological interest of the ruins I saw here. Modern as they are, they furnish the best illustration of the course of decay through which the 'kōne-shahrs', or Tatis, found along the western route to Khotan and on the outskirts of the oasis, must be supposed to have passed. There, too, village sites were deserted owing to irrigation ceasing from one cause or another, and as they were so much further away from the desert centre than the terminal oases of Dandān-Uiliq or the Niya Site, the heavy drift-sand could not arrive in time and in masses sufficiently large to give effective protection to the ruins. Archaeological interest of modern ruins.

For nearly three miles we traversed the desolate remains of these village homesteads, but it was not until about two miles further to the north-west that the region of true dunes was entered near a little wooden tomb known as Supuji Mazār, and worshipped as the supposed resting-place of a saintly associate of Lachin-atā. The Mazār of the latter was not in view, Desert beyond abandoned settlement.

nor could our guides give any clear idea where we should find the ancient sites previously described as in its vicinity. Amidst the gradually rising dunes we soon left behind the last traces of old cultivation, dead trunks of Jigdas and Tereks, while the tamarisk-covered patches assumed more and more the appearance of isolated cones owing to incipient erosion of the unprotected ground around them.

Legend about Uzun-Tati.

As we plodded on, the villagers I had taken along for eventual excavations became more communicative. They professed never to have seen the sites we were bound for, but they were well aware of the legend concerning them. These same villagers had before shown a clear perception of the direct cause which had led to the abandonment of their old lands. All the greater was my surprise to find that the legend they now proceeded to tell me, of the 'old town' marked by the remains in the desert beyond, was in all substantial points the same as that which more than twelve hundred years ago Hsüan-tsang had heard at P'i-mo of the sand-buried city of Ho-lao-lo-chia. A holy man was treated with contempt by the inhabitants and refused water—as one version of the story asserted, because he had reproved them for unnatural crimes. He thereupon cursed the town, and foretold its approaching destruction. While they still mocked at his prophecy, sand began to rain from the skies, and continued for seven days and nights until the whole of the buildings were buried. Only seven pious people who had shown respect for the *ulūgh ādum* ('the holy man') managed to save their lives through a curious device, which varies from Hsüan-tsang's story. The seven wise men are supposed to have clung to ropes fixed to a high pole after the fashion of a merry-go-round, still a popular diversion throughout Chinese Turkestān. Being whirled round and round by the raging storm they rose higher and higher above the ground while the sand accumulated, and thus escaped.

Continuity of local legend.

Similar stories, no doubt, are current throughout Eastern Turkestān of ruins buried in the Taklamakān, but it is of particular interest to note how the continuity of local tradition had here transferred to the remains of P'i-mo itself the legend which Hsüan-tsang heard at P'i-mo of a still earlier site. For I could safely recognize these remains in the extensive débris-covered area, portions of which we managed to trace in the course of the next two days. On that first evening our luckless guides dragged us aimlessly far out into the desert, until at last the weariness of animals and men and the difficulty of getting the caravan in the darkness over the rising dunes forced us to pitch camp without having come across any indication of an old site. During the night one of the guides deserted. The other, however, a timid young fellow whom Turdi, my desert factotum, kept under his eye, and encouraged by advice drawn from his own lifelong 'treasure-seeking' experience, recovered his bearings, and setting out before daybreak succeeded in discovering one portion of the 'kōne-shahr' about three miles to the south-west.

Site of *Uzun-Tati.*

Uzun-Tati—'the distant Tati'[1], as local tradition appropriately designates this site—proved to consist of several extensive patches of ground, thickly covered with pottery fragments and similar small débris. The northernmost of those which I saw showed an area about one mile from north to south, and over a quarter of a mile across, surrounded by low dunes and tamarisk-covered sand-cones. A few of the latter rose also within the otherwise level débris-strewn area. The effects of long-continued erosion were visible everywhere. Only at relatively few spots little banks of loess showed foundations of mud-walls, nowhere more than a foot or two high. In some of these stumps of round wooden posts remained, with bundles of Kumush placed vertically between them, indicating a mode of construction closely approaching the modern one as observed in the dwellings of 'old Domoko'.

[1] Pronounced with the usual epenthesis, *ta'ti*, the second part of the name sounded here almost like *tĕti*.

Whatever of these scanty structural remains had escaped erosion displayed plain evidence of the destruction dealt here by treasure-seekers' diggings. This was not to be wondered at, seeing how relatively near this ruined site was to the several oases southwards. This easy accessibility and consequent thorough exploitation left no hope of ruins capable of excavation having survived here. In their absence the archaeological indications I needed as to the date when the site was last occupied could be forthcoming only from among the small débris to be picked up on the ground. I accordingly utilized all the available men for a careful search under my own supervision, and this soon yielded interesting chronological evidence. Character of remains.

By the side of small objects in bronze, including a well-shaped clasp, a buckle-like object, some finger rings, and little fragments in the same material (see D. K. 001, 003, 005, shown in Plate LI); small broken pieces of glass and true Chinese porcelain (see for specimens D. K. 002, 003. k, in Plate LII) were common. As porcelain did not make its appearance among the keramic products of China until the beginning of the Sung period (963 A.D.)[1], its discovery here was a clear indication that the site had been inhabited down to a far later time than, e. g. Dandān-Uiliq, Endere, or Kara-dong, where no trace of this material could be found. But even more definite chronological proof was obtained in the shape of the two copper coins which were picked up under my own eyes at the Tati. One of them (see Plate CX, 35) has been recognized as belonging to the Pao-yüan period (1038–39 A.D.) of the Emperor Jên tsung, while the other (see Plate XC, 43) is a piece bearing the name of Muḥammad Arslān Khān, whom we know to have reigned over Khotan and other parts of Eastern Turkestān in the eleventh century. Thus the occupation of the site well into Muhammadan times was established beyond doubt. Chronological indications of débris.

As Uzun-Tati did not offer further scope for systematic work I decided to proceed to the second 'kōne-shahr' of our guides, which the people of Domoko knew of by the name of *Ulūgh-Ziārat* or *Ulūgh-Mazār* 'the holy (or high) shrine.' Though it was known to be situated somewhere between Lachin-atā and Uzun-Tati, and in the end proved only about three miles distant in a direct line to the south-south-east of the latter, it took us nearly two days and very tiring marches and countermarches in the sand before I was able to complete my survey of its remains. Our *soi-disant* guide first led us over fairly high and closely-packed dunes to the south-east. At two points within a mile of Uzun-Tati old pottery débris cropped up again over small patches of bare ground, plainly marking an extension of that site. When we had covered about five miles in a direct line, pottery fragments appeared again in detached spots over large stretches of eroded ground. Our 'guide' now declared that we had got too far towards the Lachin-atā Mazār, and asked to be allowed to turn back once more in quest of the site. Halting the caravan, I accompanied him for some hours with a few of the men in his devious wanderings, but was finally obliged to retrace my steps to the caravan without any clue having been obtained for the location of Ulūgh-Ziārat. As water was getting short I resolved to make for Lachin-atā, where there was said to be a well with brackish water. But after marching for an hour and a half eastwards between tamarisk-covered cones and over dunes 20 to 30 ft. high darkness obliged us to halt. During the night the labourers sent out in search managed to discover Lachin-atā about two miles to the south of our dreary camp. Search for Ulūgh-Ziārat.

Next morning I dispatched our Darōgha with an urgent requisition for efficient guides to the Bēg of Gulakhma, to the north of whose tract we now were. While I was utilizing my enforced halt for anthropometrical measurements on the labourers with me, the luckless 'guide' Remains of Ulūgh-Ziārat.

[1] See Capt. Brinkley's *China, its history, art, &c.*, ix. p. 12.

of the previous day, who had been allowed to keep away during the night for the sake of fresh efforts, turned up after midday with the news that he had found Ulūgh-Ziārat at last. So accompanied by him and old Turdi I started back again to the point where the caravan had been halted on the previous day. Ulūgh-Ziārat proved to be within only two miles of it north-westwards, but effectually hidden behind a series of high tamarisk-covered sand-cones. It was a small eroded plain about half a mile long from north to south, and slightly less broad, resembling in all its features the Uzun-Tati site, and covered like it with plentiful débris of undecorated but well-made old pottery. About its middle rose a loess-bank with what are supposed to be tombs of fifteen or sixteen saints. On its top a small collection of staves, rudely fashioned of tamarisk, clearly marked the spot as still receiving worship. I could not spare time for closer examination of these alleged tombs, as I was anxious to visit before nightfall the 'Sipil' or fortification reported further ahead. After crossing dunes for an approximate direct distance of $1\frac{1}{2}$ miles to the east-north-east, I reached it in the middle of a relatively open sandy plain covered with tamarisk bushes and numerous young Toghrak trees.

Old fort near Ulūgh-Ziārat.

The 'Sipil' proved to consist of the comparatively well-preserved remains of a small fort, built in the form of an oval, having its longer axis from north to south. The greatest length of the interior was about 480 ft., its maximum breadth about 348 ft. The wall, about 11 ft. thick at the base, was constructed of stamped loess with layers of rushes placed horizontally at varying levels to give it consistency. At a height of about 9 ft. from the present ground-level it bore a parapet which seemed to have originally been about 5 ft. 8 in. high, with a thickness of 3 ft. This, however, was broken in numerous places; behind it ran a platform about 5 ft. broad. No loop-holes were visible. The interior of this small fort, approximately approaching in size that of the Endere Fort, showed no remains whatever of buildings, though the sand within it could only be from 3 to 6 ft. high, and in several places left the foot of the walls clear. Except for a gap in the east face, which probably marks the position of the gate, the wall could be followed all round, no portion of it having completely fallen. I noted neither inside nor outside any pottery fragments. In the absence of other remains I was unable to form any definite opinion as to the date of this circumvallation, but its relatively good preservation, its shape different from that of the old forts of Endere and Ak-sipil [4], and the total absence of brickwork seemed all to point to its having been constructed during Muhammadan times. Seeing that the position it occupies lies nearer to the present limits of cultivation, and must even now have subsoil water at a relatively small depth, it appears very improbable that the fort could have been abandoned earlier than Uzun-Tati. I left this ruin at nightfall, and was met on the way while returning to camp by the Bēg of Gulakhma. A close examination of the men he had brought with him showed that the remains I had already examined were all that were known to them.

P'i-mo identified with Uzun-Tati.

That these remains of Uzun-Tati and Ulūgh-Ziārat have a strong claim to be recognized as marking the position of Hsüan-tsang's P'i-mo and Sung Yün's Han-mo can, I believe, be proved by several convincing arguments. In the first place, we find their position in exact agreement with the distance of 330 li, i. e. a little over three marches, and with the eastern bearing from Yōtkan which the *Hsi-yü-chi* indicates for P'i-mo. The coins and porcelain fragments found by me at Uzun-Tati make it quite certain that the site was occupied for centuries after Hsüan-tsang's time, while, on the other hand, there is nothing to preclude us from believing that it existed already long before the pilgrim's visit.

[4] See below, pp. 474 sq.

But the most striking evidence, I think, can be derived from the facts brought to light about the site of Ulūgh-Ziārat. Its position, three miles to the south of Uzun-Tati, accords most accurately with the 15 li to the south at which Sung Yün's narrative places the great temple, with the miracle-working statue of Buddha, relative to the town of Han-mo[5]. After all the evidence which the previous survey of sacred sites in and around the Khotan oasis has furnished of the tenacity of local worship in this region, we can safely recognize the name and sacred character of the 'Ulūgh-Ziārat' as a distinct indication of an important Buddhist shrine having once stood there. The mere fact that the Ziārat, though assumed to be the resting-place of saints, is not called after any one of them is significant. Temple of miraculous statue located at Ulūgh-Ziārat.

But the antiquarian importance of the very name *Ulūgh-Ziārat*, 'the holy Shrine', is still further increased by the fact that we can prove it to be of old date. We have already, in the preceding section, in connexion with the Keriya river route, had occasion to refer to the legendary or Tadhkira of Maḥmūd Karam Kābulī, of which M. Grenard has furnished interesting extracts[6]. This text, which seems to contain certain historical elements going back to the twelfth century of our era, describes in some detail the conquest by the Muhammadan champions of the territory of 'Kenhān', situated between the Keriya river and Khotan. Its ruler, the infidel 'Turk Terkhān', is spoken of as a Jew and as a dependent of Nūdūn Khān, the 'Tersa' or Christian, who held Khotan with his Kirghiz Kalmak or Kara-Khitai. After defeating Turk Terkhān the Muhammadan host is said to have taken and pillaged the rich town of *Ulūgh-Ziārat*, which was close to his capital *Kenhān*[7]. The latter itself vanished through magic, while the Muslim host next occupied Chīra, and victoriously advanced upon Khotan. Whatever interpretation we may care to put upon any historical reminiscences that may possibly have mingled with this legend, it is quite clear on topographical grounds that by 'the province of Kenhān' must be meant the oases stretching from Keriya to Chīra, and by 'the town of Ulūgh-Ziārat' the site of Ulūgh-Ziārat[8]. The 'town of Kenhān', Turk Terkhān's capital, which is said to have vanished, may, at the time not exactly known to us when the legend took the shape recorded in the Tadhkira, have been looked for among the sands of Uzun-Tati. Ulūgh-Ziārat in Muhammadan legend.

Whether the distinction made in the legend between the fate of the two towns indicates that Uzun-Tati was abandoned earlier to the desert than Ulūgh-Ziārat, is a point which cannot be decided, nor one of much consequence for our inquiry. What, however, clearly results is that local tradition assumed both sites to have been inhabited down to the twelfth century, the time of the Kara-Khitai. This brings us still nearer than the archaeological evidence already detailed to the time of Marco Polo. Seeing that Hsüan-tsang's P'i-mo must be located at Uzun-Tati and that old Muhammadan tradition points to the same site as the capital of the tract, I think the conclusion becomes highly probable that Marco's 'Pein, the capital of the kingdom', lay also at this site or in its close vicinity[9]. Uzun-Tati and Marco Polo's 'Pein.'

These days in the desert had convincingly demonstrated the serious difficulties which must always attend a search for scanty remains hidden away among deceptive sand-dunes if made

[5] See above, p. 456.

[6] See *Mission D. de Rhins*, iii. pp. 43 sqq.

[7] See Grenard, loc. cit., iii. pp. 44 sq.

[8] M. Grenard, ibid., p. 45, note 1, expresses his inability to locate 'Ulūgh-Ziārat', though he duly traces, between Keriya and Chīra, the tombs of various saintly warriors of Islām, among them Lachin-atā, whom the legend mentions.

[9] This long-continued importance of the site is a further support for M. Chavannes' view that the town of *K'an* 坎城 mentioned by the T'ang Annals at 300 li east of Yü-t'ien (called town of *Ts'ü* (*T'seu*) 次城 in the T'ang itinerary above quoted, see *Voyage de Song Yun*, p. 13 note) was identical with P'i-mo; see *Voyage de Song Yun*, p. 14, note 4; also above, p. 176.

It may be noted here that Sung Yün's town of *Mo*, which was 22 li east of his *Han-mo* (see above, p. 456), would according to this indication have to be looked for somewhere about 'Old Domoko' or a little to the north of it.

without adequate guidance. The rapidly increasing heat and glare—on March 27 and 28 the air about midday was 88° Fahr. in the shade, though the minimum thermometer had for the corresponding nights still registered 28° and 30° Fahr.—rendered tramps through the sand very trying, and we realized the limitations of the water-supply carried in the tanks when our party was increased by labourers. Hence I felt as glad as my men when the satisfactory conclusion of my task allowed me, on the morning of March 28, to turn back southward to the inhabited area.

Shrine of *Lachin-atā.*

En route I passed the desolate little shrine of Lachin-atā (see Fig. 56), where, besides the usual accumulation of poles bedecked with flags, &c., there was a wretched hut sheltering the brackish well about 20 ft. deep. A small open space closely surrounded by high tamarisk-covered cones showed some pottery débris which looked old (for specimens see L. 002). The saint worshipped here is supposed to have followed the fortunes of Maḥmūd Karam Kābulī[10]. Within about four miles from Lachin-atā our route reached an area of abandoned fields, the cultivation of which a few generations ago was still remembered in village tradition. Some three miles further on we arrived at the scattered hamlet of Ponak, marking the northern edge of the flourishing oasis of Gulakhma. There for the first time I caught sight of the young green of cultivated fields and orchards.

OBJECTS FOUND AT AK-TAZ SITE.

Objects from Ak-taz.

D. K. 004. Bronze finger ring. The ring is made of a flat band of metal, and supports a circular setting in which is a gem apparently made of opaque glass with blue markings. See Pl. LI.

D. K. 006. Two fragments of terra-cotta. 2¼" × 1¼". ¾" × ¾".

OBJECTS FOUND AT UZUN-TATI.

Objects from Uzun-Tati.

D. K. 001. a. Bronze hook, or clasp; cast. The form is like a swan's neck. 1¾" long. See Pl. LI.

D. K. 001. b. Cast bronze buckle-like object; open work. 1¾" × 1⅝". See Pl. LI.

D. K. 002. a. Fragment of Chinese porcelain, finely glazed and decorated in blue outline pattern. ⅞" × ¾". See Pl. LII.

D. K. 002. b. Fragment of Chinese porcelain, glazed, decorated with dirty brown, badly fluxed. 1⅜" × 7/16". See Pl. LII.

D. K. 002. c. Fragment of Chinese porcelain, decorated with edge dark yellow; outside, two lines and pattern outlined in black, filled in with fine green. ½" × ⅜". See Pl. LII.

D. K. 002. d. Fragment of Chinese porcelain, glazed entirely outside with fine turquoise blue. Inside white. ⅜" × 5/16". See Pl. LII.

D. K. 002. e. Fragment of terra-cotta, glazed on inside olive green on which black lines. ⅝" × 3/16".

D. K. 002. f. Two fragments of moulded glass. See Pl. LII.

D. K. 002. g. Denticular setting with gem (dirty yellow glass). Portion of a finger ring.

D. K. 003. Small objects in metal, glass, porcelain, &c.

a. Bronze finger ring, similar to D. K. 004 but with claw (denticular) setting holding piece of flat cloudy-green glass. See Pl. LI.

b. Bronze finger ring, similar to D. K. 004, but larger and gem missing. Metal table for gem broken. On outer surface of flat band a raised design in twisted bronze wire laid on in two opposing zig-zags between parallel borders. See Pl. LI.

c. Table and gem setting of bronze finger ring.

d. Small bronze ring attached to broken bronze fragment, probably of a bell.

e. Several fragments of bronze, and one of iron.

f. Three grape stalks.

g. Small piece of thin bronze wire.

[10] See Grenard, *Mission D. de Rhins*, iii. p. 45, note 5.

h. Two fragments of plain glass. See Pl. LII.

i. Piece of flint. See Pl. LII.

k. Fragment of Chinese porcelain vessel, finely glazed and painted with fret pattern in grey-blue on outside, and two faint lines inside. See Pl. LII.

D. K. 005. Small objects in metal, stone, &c.

a. Large white shell bead. Diameter $\frac{3}{8}$".

b. Broken bronze finger ring, with empty circular gem cavity. See Pl. LI.

c. Small bronze bell. Diameter $\frac{5}{8}$".

Objects from Uzun-Tati.

d. Fragment of small bronze bell.

e. Several small fragments of bronze, some showing engraved patterns.

f. Small piece of bronze with square hole, probably worn-out Chinese coin.

g. Small thin bronze disk; perhaps a coin.

h. Fragment of pale-green stone.

OBJECTS FROM OLD SITE NEAR LACHIN-ATĀ.

L. 001. Small objects in bronze, said to have been found near Lachin-atā Mazār (presented by Bēg of Gulakhma).

a. Piece of cast bronze, to underside of which is attached, by iron rivet, a fragment of leather. It is probably a piece of harness. $1\frac{3}{8}" \times \frac{3}{4}"$.

b. Bronze split ring. Diam. $\frac{5}{8}$".

c. Piece of thin sheet bronze doubled in halves. $\frac{3}{4}" \times \frac{9}{16}"$.

L. 002. Terra-cotta fragments, found on eroded ground near Lachin-atā Mazār.

a. Fragment of rough terra-cotta vessel decorated with lightly scored lines and nebulae. $3" \times 2\frac{3}{4}"$.

b. Fragment of rough terra-cotta vessel decorated with nebular scorings. $2" \times 1\frac{1}{2}"$.

Objects from Lachin-atā.

c. Fragment of neck and shoulder of coarse terra-cotta vessel, with scar showing attachment of handle. Round junction of neck with shoulder two parallel raised mouldings. On neck, as ornament, a few oblique dotted lines. Height $3\frac{3}{8}"$, width $2\frac{1}{8}"$.

L. 003. Chinese seal in bronze, octagonal, said to have been found at old site near Lachin-atā; bought at Gulakhma. *Obv.* shows in cameo one of the hundred forms of the character 壽 = 'longevity' (Dr. Bushell). *Rev.* plain, with shank. Diam. $1\frac{3}{16}"$. See Pl. L.

Section IV.—THE TŪGA-DONG MOUNDS AND THE KERIYA-KHOTAN ROUTE

Halt at Gulakhma.

Gulakhma, which counts over nine hundred homesteads in its several villages, and in its central part near the Keriya-Khotan route bears the look of a thriving small market town, might have tempted me to give to my caravan the rest which it badly needed. But time would not permit of more than a day's halt, which I myself used for the examination of some ancient mounds I had heard of as recently opened by people of Chīra. Etiquette demanded that I should visit them under the guidance of the Bēg of the latter oasis. So on March 29 I rode to Sarīgh, the easternmost village of Chīra, where the Bēg's party joined me, and thence was taken again south-eastwards into the scrub-covered waste which stretches south of the high road connecting the two oases. After about three miles the sandy soil changed to gravel, and three miles beyond, at a point about equidistant from Gulakhma and Chīra, we arrived at the mounds of *Tüga-dong* ('the Camel Hillocks').

Tüga-dong mounds.

There, not far from the edge of the pebble 'Sai,' I found some seventeen roughly circular mounds scattered at irregular intervals, varying from 10 to 60 yards, in the general direction from north to south. The largest measured about 79 ft. in diameter, with a height of 11 ft., while the smallest was about 30 ft. across, rising to 5 ft. above the ground-level. They had all been dug into from the top in the centre, and cuttings had often been made from their sides; but this work of destruction had evidently been carried on at different periods, as indicated by the varying states of decay of the interior materials thus exposed. There were indications of some smaller mounds having been completely levelled in the course of these operations. No recent excavations were acknowledged, except in the case of one of the northernmost mounds, about 48 ft. in diameter and 12 ft. high, which was said to have been opened by Chīra cultivators some two months earlier. The exposed portions of the mounds showed everywhere thin layers of small pieces of Toghrak wood, bark, and thorny scrub, mixed up with the coarse sand and gravel of the surrounding ground. Nowhere could I find in the cuttings remains of walls or traces of any other regular construction. The pieces of wood were far too small to have served for building purposes. Much of it appeared charred, and on digging down to the ground in the second tumulus from the south I came on a layer of wood 6 in. thick which was practically reduced to the condition of charcoal. The ground below this layer consisted of a hard clayey soil which generally seems to underlie here the gravel and pebbles of the 'Sai.'

Human remains in mounds.

The villagers, perhaps by order of the Bēg, who probably was anxious to avoid possible inquiries by the Amban as to finds made, &c., denied all knowledge as to the contents or purpose of the mounds. But when I had picked up two human thigh bones, about 17 in. long, near the recently opened mound on the north, they seemed inclined to admit that human remains had been found also elsewhere. Subsequent inquiries at Gulakhma in fact showed that two skeletons at least had been found in the course of the latest diggings, but the bones had been gathered again and buried somewhere from superstitious fear. From the loose gravel of the same mound where the two above-mentioned bones lay, some ragged fabrics of cotton were picked up by me; of these the pieces illustrated in Pl. LXXVI (see also list below) are specimens. One of them is a coarse cotton stuff resembling in colour mummy cloth, the other of finer make shows buff colour with a small diaper pattern.

Character of tumuli.

I was unable to form any definite opinion as to the original character of these tumuli. But much points to their having been raised as burial mounds, perhaps over partially burned bodies. That they had contained other deposits, too, possibly some of value, appears probable enough, as otherwise the villagers would scarcely have continued so long the laborious task of digging up every mound. But any finds of intrinsic value were stoutly denied by those directly concerned, including the Bēg, no doubt for the reason already indicated, while the gossips of Gulakhma could tell my men only the usual wild stories about ingots of silver, &c., which were supposed to have been found there. As to the age of the mounds, all that can be safely asserted is that the ground near by must at the time of their construction have already borne the character it has now, there being in particular no indication of fertile soil having been removed by deflation. The good preservation of the wood and scrub, which seemed to have been inserted for the sake of giving greater consistency to the mounds, did not suggest very high antiquity.

Farewell visit to Keriya.

On March 30 I sent on the main part of my camp under the Surveyor to reach Khotan by easy stages, while I myself, with the minimum of impedimenta, hurried back to Keriya to bid farewell personally to its kindly Amban. My rapid marches were made pleasant by noticing on all sides exuberant signs of spring, which had come over the land with surprising rapidity. On April 1 I exchanged long visits with Huang-Daloi, and assured him of my due sense of gratitude for his never-failing help, which alone had rendered the extent and success of the last months' explorations possible. I also impressed him with my appreciation of the excellent services which Ibrāhīm, our energetic Darōgha, had rendered, and secured for the latter a public commendation by the Amban and the promise of appropriate promotion.

Antiques purchased at Keriya.

I had previously asked my acquaintances among the Bēgs and Mullahs of Keriya to look out during my absence in the desert for any antiques that might happen to be brought into the Bāzārs by 'treasure-seekers'. But only three small objects of uncertain origin had been secured (see below Ker. 001–003). Among them was the small intaglio, Ker. 003 (Plate XLIX), showing the figures of a man and a running deer, and the bronze finial, Ker. 001 (see Plate LI), which in its decorative motif so closely resembles the styles used in Indian *sūrmadāns* of the present day as to suggest doubts as to its antiquity. The scarcity of these acquisitions seemed to me an indication that the Tatis, which I had heard reported at Kara-khān, to the north-west of Keriya town, and at Tokhtal, near the western edge of the oasis, and to the south of the Khotan road, could not be of great extent or importance. For visits to them I was unable to spare time.

Age of Keriya town.

Keriya town itself is not believed by the people to be a place of great age, and has probably acquired importance only since it was made the head quarters of the separate district of Keriya, newly created after Yāqūb Bēg's rebellion, and now officially styled Yü-t'ien. The oasis itself is, no doubt, of ancient date; but there is nothing to indicate that it was in old times larger or more important than the string of oases which still extend in an almost unbroken line from Chīra to Kara-kīr, and are linked with Keriya by ground that, as the subsequent observations will show, is for a large part capable of cultivation. I am unable to trace any distinct reference to Keriya in the early Chinese texts accessible to me in translations.

Ancient territory of *Yü-mi* (*Han-mi*).

The territory of *Yü-mi* 扜彌, which is mentioned by the Ch'ien Han shu among the small states to the east of Khotan, and subsequently referred to under the slightly different and probably more correct form of *Han-mi* 扞彌 in the Wei lio and T'ang Annals, has indeed been identified by a modern Chinese writer with Keriya[1]. But the topographical indications furnished by the historical texts make it appear far more probable that this territory comprised the whole of the oases between Chīra and Keriya, thus corresponding roughly to Marco Polo's 'Province of Pein', and the tract of which Hsüan-tsang's P'i-mo was the chief place. In the Han Annals' notice *Yü-mi*, then independent of Khotan, is described as being 390 li to the east of the latter[2]. This, assuming that as elsewhere the measurement is taken from capital to capital, points to a location about Gulakhma-Domoko rather than near the town of Keriya. The T'ang Annals, which, as we have seen, enumerate *Han-mi* among the small territories annexed to Yü-t'ien[3], place it to the east of the river of *Chien-tê-li*[4]. The latter itself, being placed 300 li east of Yü-t'ien, can only be the river of Chīra, the distance of which from Yōtkan, about 60 miles, exactly corresponds to the above measurement. We are not told by the notice of the T'ang Annals where in particular was the position of the town of *Ta-tê-li*, its chief place, also known as *Chü-mi*, or earlier as *Ning-mi*. But seeing what we have ascertained above as to the long-continued importance of the town which Sung Yün called *Han-mo* and Hsüan-tsang *P'i-mo*, the location of Ta-tê-li at the same site seems distinctly more probable than at the modern Keriya.

Return journey to Khotan.

On April 2 I started back to Khotan by forced marches. The first, which brought me to Kara-kīr, gave me ample opportunity to observe the abundance of spring water which feeds the numerous marshes skirted by the road between the western edge of the Keriya oasis and the Kara-kīr stream, and which further on runs to waste in the Shīvul Daryā and its terminal swamps. My companions from Keriya, among them the local official whom the attentive Amban had deputed to escort me, declared that the springs were perennial, and believed that with the

[1] See above, p. 167, with note 8; Chavannes, *Turcs occid.*, p. 128, note 1, also p. 311. See now also Chavannes, *Les pays d'occident d'après le Wei lio* in *T'oung-pao*, Ser. ii. vol. vi. p. 538, note 1.

[2] See Wylie, in *J. Anthrop. Inst.*, x. p. 29.

[3] See above, pp. 167, 172.

[4] See Chavannes, *Turcs occid.*, pp. 127 sq.

necessary administrative pressure a good deal of new land could be brought under cultivation besides the small oases of Laisu and Achma recently formed some miles north of the road.

New cultivation at Kara-kir.

At Kara-kīr Langar a new but wholly deserted roadside Bāzār forcibly drew my attention to a curious illustration of the changes affecting cultivation in this tract. About ten years previously, I was told, abundant springs had unexpectedly appeared in the sandy jungle some miles to the south, fed, no doubt, by the Nūra and other hill streams which higher up disappear for the greatest part of the year in the pebble Sai. Before there had been at the spot only a small trickling spring. The water supplied by the new springs was so ample that land sufficient for 700 to 800 households has since been brought under cultivation in what was formerly desert ground two to three miles north of Kara-kīr Langar. The settlement which has sprung up there, and which for the first five years paid no revenue, had now sufficiently developed to have a Bēg of its own, while the wayfarers' custom has been completely transferred to the new main village of Achma. Though the time of *ak-su*, or flood-water, was still distant, the Kara-kīr Daryā was now a lively stream about 15 yards broad, with a depth of about 2 ft.

Oasis of Chira.

My second day's ride brought me over already familiar ground to Chīra, a large and flourishing oasis counting about 3,500 households. The fact that the oases of Gulakhma and Domoko, counting about 2,000 homesteads between them, have now separate Bēgs, whereas before the Muhammadan rising they are known to have been under the Bēg of Chīra, was mentioned to me as an indication of their growth since the material prosperity of the country had been raised by improved administration. In the oasis of Chīra, otherwise wholly dependent on the water brought down by the glacier-fed river which comes from Hāsha, much new land was said to have been taken up northward since the recent supply of spring water. This had been secured by the construction, some fifteen years before my visit, of the canal crossed on the road from Gulakhma, and shown on the map[6]. The dreary desert plain of sand and pebbles over which the march of April 5 lay was a striking contrast to the well-cultivated fields and the blossoming orchards of Chīra. The great depth of the wells dug at the desolate little rest-houses by the route showed that the greatest portion of the ground crossed could never have seen cultivation. All day a strong dust storm was blowing from the west, and the thick haze enveloping us made me thankful for the guidance afforded by the rows of poles marking the road. After thus covering forty odd miles I was glad when, by nightfall, I had regained once more the edge of the Khotan oasis at the large village of Lop.

FABRICS FROM TÜGA-DONG MOUNDS.

T. D. 1. Cotton cloth, harsh, in colour like mummy cloth; plain. 9″×6″. See Pl. LXXVI.

T. D. 2. Cotton cloth, woven in small diaper pattern. One edge oversewn. Dirty-buff colour. 18″×6″. See Pl. LXXVI.

[6] It is somewhere to the north or north-west of Chīra that we should have to look for the 'desert marsh' where Hsüan-tsang, before arriving at Pʻi-mo, had been shown a large stretch of red-coloured soil, in which the legend reproduced above recognized an ancient battle-field. I had no time to search for it in the direction indicated, and it is, of course, not along the present Khotan road but through the sandy jungle 12 to 16 miles north of it that the old route must be supposed to have led. It is curious that on my return journey from Khotan I noticed a large patch of red clayey soil, corresponding exactly to Hsüan-tsang's description just referred to, a quarter of a mile to the west of Kum-rabāt-Pādshāhīm, where the legend heard by the pilgrim, and surviving to the present day, also located a battle-field. There, however, this feature of the ground is not recorded by the pious traveller.

OBJECTS PURCHASED AT KERIYA.

Ker. 001. Bronze finial. On summit of ornamental square shaft, a peacock, with erect crest flattened from before. There is a stump to which probably the spread tail was attached, but the tail is missing. Below the shaft is a round tang, flattened at its lower extremity. The same design of peacock made now by *Koftgārs* of Kōtlī Lōharān, Punjab. It is a motif frequently used for *surmadāns*, and this piece looks very like end of style in surmadān. See Pl. LI.

Ker. 002. Rough terra-cotta vessel, cone-shaped, flat at bottom, broad end uppermost. There seems to have been a permanent cover, but it is much broken. In one side of remaining portion of cover is a small hole about $\frac{3}{16}''$ in diameter and traces of two others in broken cover. The vessel might have been a lamp (*chirāgh*) and the holes made to take the wick. Height $1\frac{3}{4}''$; diam. at top $2\frac{5}{8}''$; diam. at base $1\frac{3}{8}''$.

Ker. 003. Stone seal; wide elliptical, broken at lower side. In intaglio, a running deer to L., with horns like those of the chamois. On further side of deer a man, standing to L., perhaps carrying deer. On his head a cap with heavy brim, or thick hair. $\frac{9}{16}'' \times \frac{1}{2}''$. See Pl. XLIX.

CHAPTER XIV

THE RUINS OF AK-SIPIL AND RAWAK

SECTION I.—HANGUYA TATI AND THE SITE OF TAM-ÖGHIL

FROM Lop, where I had camped, I was able to proceed at once, on April 5, to the easternmost of the ancient sites which still remained to be examined in the vicinity of the Khotan oasis. It was the Tati of Hanguya, from which a number of coins had been brought to me in November, and with which Turdi, from the vicinity of his own village, was, of course, thoroughly familiar. Before setting out for it, and subsequently on my way through the cultivated parts of the Hanguya tract, I had opportunities for interesting glimpses of the flourishing state enjoyed by this canton, and also by the canton of Sampula or Lop adjoining southwards. Both are administratively included in the district of Keriya, of which they form probably the fiscally most valuable portion. The Bēg in charge of it, whose spacious, well-built house, with its pretty garden and orchard, vividly recalled to my mind the ruins of similarly substantial residences explored at the Niya Site, estimated the number of households within both tracts at about 9,000. This number, though nearly double that subsequently communicated to me as the present conventional reckoning[1], did not strike me as extravagant in view of the extent and busy life of Lop Bāzār and the thriving look of the Hanguya villages. Those of Sampula owe their population largely to the flourishing carpet industry which is centred here, and to the early importance of which I have already had occasion to allude[2].

The canton of Sampula.

Immediately to the north of Lop Bāzār stretches a small belt of salty soil, which is liable to be flooded by subsoil water. But immediately beyond richly-cultivated ground is entered, through which the road led for nearly five miles to the central market-place of Hanguya. Water for irrigation was said to be plentiful here, being brought by canals both from the Yurung-kāsh and from springs which appear to the south-west of Lop. Accordingly I was not surprised to find a great deal of ground recently brought under cultivation, i. e. recovered north of Hanguya Bāzār. After riding about three miles north-north-west from the latter place I came to lucerne fields fringed by low dunes, which are slowly but steadily being levelled by irrigation cuts. Half a mile beyond we entered the desert amidst dunes 5 to 10 ft. high, covered at first with thorny scrub, and further on completely bare.

The Hanguya tract.

Following, under Turdi's guidance, a well-marked track which is used by people bringing wood from the jungles along the Yurung-kāsh course, we arrived, after a march of about three miles from the edge of cultivation, at eroded ground covered with red pottery of great hardness. It was the commencement of a typical 'Tati', just as Turdi's report had led me to expect, and which according to his statement extended far away to the north-west right across to the vicinity of Ak-sipil. That the débris-covered area here was large seemed very probable, but the prevailing dust-haze prevented even an approximate estimate.

Hanguya Tati.

[1] See above, p. 132 note.

[2] Comp. above, p. 134.

54

REMAINS OF RUINED STŪPA, ARKA-KUDUK-TIM, HANGUYA TATI, SEEN FROM SOUTH.

55

REMAINS OF DESERTED MODERN DWELLING, AT KOBZAGIRAM-SHAH, DOMOKO.

56

LACHIN-ATA MAZAR, SEEN FROM WEST.

Not far from the southern edge of this area rose the mass of ruined masonry known as *Arka-kuduk-Tim* ('the Mound of the Back Well', from a well in this vicinity sunk by wood-cutters), and shown in Fig. 54, of which Turdi had spoken as the only structural ruin of the site. It proved to be the remains of a Stūpa, but so thoroughly destroyed by erosion, and probably also by diggings for 'treasure', that no approximate idea could be formed of its original shape or dimensions. The extant block of masonry stands 16 ft. high, and measures at its base about 20 ft. from east to west, and 8 ft. from north to south. Great masses of brickwork must have fallen, as the slopes of the conical mound on which the ruin now rises are thickly covered with brick débris. The masonry of the ruin consists mostly of sun-dried bricks measuring 19 by 13 in., and 4 in. thick, but smaller bricks, only $3\frac{1}{2}$ in. thick, and apparently about 13 in. square, could also be distinguished in places. Beneath the lowest course of bricks there was what looked like a foundation or base of hard clayey loess, about $3\frac{1}{2}$ ft. high, resting on the original ground-level. The latter (marked in Fig. 54 by the feet of the upper standing figure), owing to the great erosion that has taken place around the mound, now rises fully 20 ft. above the lowest point immediately to the south of the Stūpa (see the second figure in the photograph). But at a distance of 30 to 40 yards to the south the loess banks rise again in terraces, and the highest of these seemed only 4 to 5 ft. below the ground-level as marked by the foot of the ruin. All the hard loess soil left bare by the sand is covered with pottery fragments. The drift-sand near the Stūpa formed in most places ripple-like dunes, only 2 to 3 ft. high. **Ruin of *Arka-kuduk-Tim.***

Just below the Stūpa the men brought with me from Hanguya picked up a Chinese coin without a legend and a small broken ring in bronze, H. 1 (see Plate LI). From a Hanguya villager, whom Turdi knew and had summoned to the site as a *confrère*, I purchased the small collection of objects in metal, glass, and stone described under H. 001 in the list at the end of this section. Among these, all said to have been found at the site, the massive bronze ring, H. 001. h (see Plate XLIX), with a countersunk device showing a running deer, may be specially mentioned as of evident antiquity. Already in November a set of old coins with a few small antiques had been sold to me at Khotan as coming from the Hanguya Tati. The coins, of which a synopsis will be found in Appendix *D*, comprise over twenty Sino-Kharoṣṭhī pieces, among them one apparently of a unique type. If they were really found here they would prove great antiquity for the site, but the statement as to their origin could, as in most of such cases, not be depended upon with certainty. One of the antiques acquired with them is a small intaglio, H. 002 (see Plate XLIX), similar to those probably coming from Yōtkan. The Tati by its size clearly proves a considerable extension of the ancient cultivated area northward, but there was nothing to indicate, even approximately, when it ceased to be occupied. Seeing how steadily the irrigated area of Hanguya is now again being pushed northward, it is quite possible to suppose that part of the Tati might again be turned into fertile village land. The name *Hanguya* could claim considerable antiquity if its identity with the *Hang-gu-jo* of the Tibetan legend discussed in a previous chapter were to be established[1]. **Antiques from Hanguya Tati.**

I left the site in a mild but sufficiently disagreeable dust storm, and after returning to the northern edge of the Hanguya lands rode south-westwards to rejoin my camp at Yurung-kāsh. For nearly three miles the road led everywhere through young cultivation. The avenues of poplars, willows, and Jigda trees, planted only ten to fifteen years ago, and small enclaves of sandy ground left amidst the fields showed that this area had been recovered from the desert **March to Yurung-kāsh.**

[1] See above, p. 161.

within recent times. West of the Hanguya tract we had to cross a broad belt of sand known as *Arkalik*, which represents an inroad of the desert southwards. There were no regular dunes, and traces of canals and plentiful débris of modern-looking pottery showed that this waste had passed out of cultivation at no very distant period. Isolated plots of new fields in the direction of Yurung-kāsh furnished welcome proof that here, too, efforts at reconquest had started. At the Hanguya-Ūstang we crossed the boundary line dividing the districts of Khotan and Keriya, and riding on towards the town of Yurung-kāsh I soon observed the manifest signs of old and unbroken cultivation in the high level of the fields above the roads and in the large size of the trees lining the latter.

On April 6 I halted in Yurung-kāsh, where fresh supplies and labourers had to be secured and many repairs to be effected in our equipment. Increasing heat by day and recurring dust-storms warned me that the season was close at hand when work in the desert would become impossible. Instead of taking the rest we all by this time felt much in need of, I hastened to set out for the ancient sites which still remained to be explored in the desert north-east of Khotan. So after discharging Ibrāhīm Ākhūn, the worthy Darōgha of the Keriya Ya-mên, with a liberal and amply earned reward in glittering gold roubles, I set the caravan in march again on the morning of April 7.

Excavations at *Tam-Öghil.*

The ruined site known to treasure-seekers as Ak-sipil ('the White Wall'), and situated among high sand-dunes at a distance of about fifteen miles from the right bank of the Yurung-kāsh opposite Khotan, was my first objective. On the march, however, and close to the edge of the cultivated area, I took occasion to examine the interesting small site generally known as *Tam-Öghil*, from an adjoining hamlet, where ancient 'culture-strata', yielding some leaf-gold, besides old coins, terra-cottas, &c., are worked under conditions exactly similar to those described at Yōtkan, but on a far more limited scale. Turning to the east of an isolated and conspicuous sandhill which bears the tomb of Sultān Ḥasan Basrī, some two miles in a direct line to the north-east of Yurung-kāsh, I was taken to the loess banks which form the scene of these excavations. They rise steeply above the level of the fields of lucerne adjoining them on the north, and extend in an irregular line for about 230 yards from west to east.

Ancient culture-strata.

Just as at Yōtkan they were said to have been formed solely by the soil having been gradually cut off for the sake of washing the gold-containing layer below. This proved to be from 3 to 6 ft. in thickness at different points, while the stratum of fertile earth overlying it varied in height from 10 to 18 ft. The riverine loess soil of which this top layer manifestly consists showed here and there distinct traces of stratification. Considering the short distance, less than three miles, which separates this site from the present right bank of the Yurung-kāsh, and the existence in its vicinity of an older bed marked by the jade pits of Kumat[4], it appeared to me possible that these slight layers, 1 to 1½ in. thick, might be due to exceptional floods from the river.

Origin of excavations.

From the local cultivators I ascertained that the excavations, the site of which they called *Mazār-oili-Sai* ('the sai of the Mazār lands'), had been started some twenty years before my visit when the formation of a small 'Yār'—just as at Yōtkan—had revealed the presence of gold in the culture-stratum below the loess soil. The cutting of the banks was said to have begun at a distance of about 150 yards to the north of the present face of the banks, and to have been gradually carried further and further into the rising ground south. The work has been carried on practically only by Tam-öghil villagers, as water for 'washing' the soil is

[4] See for this locality Hedin, *Reisen in Z.-A.*, p. 28.

available for not more than $1\frac{1}{2}$ months in each year and in limited quantity. Apart from the direct profit yielded by the washing of the soil, they also gain additional ground easily capable of irrigation from the existing canals. At the same time the fertile humus of the culture-stratum is used by the villagers as a kind of manure to improve gravelly fields lying further north[4].

It is probable that these subsidiary profits are here a special inducement for the work; for the amount of leaf-gold washed out of the paying seam was declared to be less than at Yōtkan, and the secondary products of antiques and occasional small finds of value are also, judging from general report, proportionately far rarer. However this may be, I succeeded, in spite of diligent inquiries both at the site and previously at Yurung-kāsh, in obtaining only a few specimens of antiques. The single coin obtained from a Tam-ōghil villager was a small Chinese copper piece without legend, while among the other four purchased at Yurung-kāsh two are copper coins of the Yüan-fêng period (1078–85 A.D.). The latter would prove that this site, like Yōtkan, was occupied for at least a century after the Muhammadan conquest. The only other antiques secured are two small terra-cotta grotesques, T. 001. a, b, representing monkeys, and corresponding in type and execution exactly to the similar figurines from Yōtkan. Of the character of the settlement which once occupied the site marked by this layer of ancient rubbish accumulations, and of its extent southwards, nothing definite can be asserted.

Leaf-gold and antiques washed from culture-strata.

OBJECTS FOUND AT, OR SAID TO COME FROM, HANGUYA TATI.

Objects from Hanguya Tati.

H. 1. Small bronze finger ring, broken, found near Stūpa ruin of Hanguya. Diam. about $\frac{5}{8}''$. See Pl. LI.

H 001. Small objects in metal, glass, &c., purchased from treasure-seeker, Hanguya Tati:—

a. Bronze needle or style drilled at thick end, $2\frac{1}{2}'' \times \frac{1}{8}''$ (nearly). See Pl. LI.

b. Square bronze jewel, containing square tablet of yellow agate. On reverse are two small holes probably for attachment of loop of some kind. See Pl. LI.

c. Cast bronze stud. Six-lobed flower, with raised centre. A point is attached to the back, evidently to serve as a rivet or nail. Diam. $\frac{7}{8}''$, length $\frac{3}{4}''$. See Pl. LI.

d. Soapstone ring. Diam. $1\frac{1}{8}''$. See Pl. LII.

e. Rich opaque blue glass bead, Amalaka-shaped, $\frac{1}{2}'' \times \frac{1}{2}''$. See Pl. LI.

f. Yellow glass pipe-like bead. Length $\frac{3}{8}''$, thickness $\frac{1}{4}''$.

g. Fragment of cut octagonal pebble bead.

h. Massive bronze seal ring; countersunk device: a deer running to R. regardant. Trees indicated to R. and at back. Broken. Moulded surface $\frac{7}{8}'' \times \frac{9}{16}''$, elliptical. See Pl. XLIX.

i. Bronze seal ring; broken. Device on elliptical surface, a few crudely incised lines. See Pl. XLIX.

H. 002. Circular seal, in nicolo, bearing, in intaglio, male head to R., wearing moustache and hair rolled. Said to have been found at Hanguya Tati. Diam. $\frac{5}{16}''$. See Pl. XLIX.

H. 003. Fragment of bronze relief. A warrior wearing tunic of mail; arms bare to elbows; in R. hand a spear, in L. what appears to be a small bag, held at the waist (cf. the mail-clad figure of D. II shown in Pl. II). Head and legs missing. $1\frac{3}{8}'' \times 1\frac{1}{16}'' \times \frac{3}{16}''$. Said to have been found at Hanguya Tati. See Pl. LI.

OBJECTS FROM TAM-ÖGHIL.

Objects from Tam-Öghil.

T. 001. a. Grotesque terra-cotta monkey. Owl-like type. Hairless excepting long pigtail. Hands together at breast. Feet together and thrown well back, the whole body having strong forward curve. Umbilicus (?) very pronounced. Height $1\frac{1}{4}''$.

T. 001. b. Terra-cotta grotesque monkey's head, naturalistic type. On top of head a kind of inverted saucer-like cap. $\frac{3}{4}''$.

[4] This use of the culture-stratum finds its exact parallel in the operations by which, at the present day, old mounds and sites, such as Akra, all along the Indian N. W. Frontier are steadily being dug down for the sake of the valuable manuring soil they furnish; see my *Archaeological Survey Report of the N.W. Frontier Province*, 1905, pp. 7 sq., 56.

Section II.—THE REMAINS OF AK-SIPIL AND KIGHILLIK

March to Ak-sipil.

Resuming the march towards Ak-sipil in a north-easterly direction we reached the edge of the cultivated area within less than half a mile from the Tam-öghil excavations. A little beyond the hard loess ground was covered with fragments of ancient pottery showing much effect of erosion, and similar débris was found in patches, at short intervals, for another five miles. As the ground was everywhere greatly eroded, it is impossible to say whether these remains belong to villages existing at the same time or to settlements established at different points during successive periods. Further on we encountered large dunes, which soon grew remarkably steep and high, up to 60 ft. and perhaps more, the heavy coarse sand unmistakably indicating its origin from the silt deposits of the river. The uniform direction of individual dunes, from north-north-west to south-south-east, was here clearly marked, while the general bearing of the sandy range to which they belonged was, as shown in the large-scale inset-map of the Khotan oasis, slightly different. The depression between the large dunes disclosed patches of light sandy soil covered with scrub and Kumush, a plain indication of the nearness of subsoil water, which the shepherds of the Jiya tract who graze their flocks here were said to reach by digging wells only 4 to 5 ft. deep.

Traces of ancient irrigation.

After crossing this succession of dunes for close on four miles, the remains of Ak-sipil came in sight on ground clear of high sand, and showing in places marks of erosion. As we approached the ruined walls which have given the site its name I thought I could distinguish traces of little embankments dividing ancient fields, and of distributing 'Ariks' along them. Close inspection of the patches of open ground in the immediate vicinity of the ruined walls confirmed this impression, and made me inclined to believe that the talk of Turdi and other guides I had brought from Yurung-kāsh as to the existence of an ancient 'Ūstang', visible for some distance along the track usually followed from Hanguya to Ak-sipil, might have foundation in fact.

The remains of Ak-sipil had been visited in 1891 and 1892 by MM. Dutreuil de Rhins and Grenard, and the account published from the notes of the former[1] proved to be as accurate in topographical details and general description as might be expected from so painstaking a traveller. The same can scarcely be said of the data which have been reproduced from the narrative of Mr. Högberg, a Swedish missionary at Kāshgar, who paid a rapid visit to the site in 1897, apparently to look for one of Islām Ākhūn's alleged find-spots of 'ancient books'[2]. An exact archaeological survey of the ruined site still remained to be effected, and to this I devoted one of the two days (April 8–9) during which I camped at Ak-sipil.

Remains of Ak-sipil fort.

The most conspicuous remains are ruined portions of the rampart and parapet of an ancient fort. They rise, among low dunes, 8 to 15 ft. above the original ground-level, and could be recognized at first sight as having belonged to a circumvallation either annular or oval. Fig. 57 shows the best-preserved part of them as seen from the outside, while from Fig. 58 some idea may be formed as to the appearance of almost the whole of them when viewed from some distance inside. Owing to the inadequate cover of sand, by far the greatest part of this circumvallation has completely disappeared through erosion, which has advanced particularly far in the southern portion of the area once covered by the fort. The exact survey made by me,

[1] See *Mission D. de Rhins*, iii. pp. 140 sq.

[2] See Hoernle, *Report of C.-A. ant.*, i. pp. xiv. sq., with some diagrams to be referred to in notes below.

57

PORTION OF NORTH SEGMENT OF RUINED RAMPART, AK-SIPIL, SEEN FROM OUTSIDE.

58

PORTION OF NORTH SEGMENT OF RUINED RAMPART, AK-SIPIL, SEEN FROM INTERIOR.

reproduced in the plan of Plate XXXIX, shows that the extant ruins form the northernmost segment, measuring about 360 ft. at its arc, of a circular rampart which must have originally enclosed an area a little over 800 ft. in diameter[3]. Here, as at Endere, the lower portion of the circumvallation consisted of a rampart of hard stamped loess, such as is still used in Chinese fortifications throughout Eastern Turkestān, and popularly known as *khitailik sighiz*, 'Chinese earth-bank'. This rampart measured approximately 50 ft. at its base, and rose about 11 ft. above the original surface of the ground outside, which was still clearly distinguishable in places free of sand and uneroded.

Parapet of circumvallation. The rampart was surmounted by a parapet built of sun-dried bricks, and resting on a supporting platform of the same construction about 2 ft. high. The thickness of the parapet was 8 ft. at its foot, while the platform below projected beyond it 3 ft. on the outside and 2 ft. on the inside. The parapet, more or less broken on the top, but otherwise in a fair state of preservation, nowhere rose higher than 7 ft. The large size of its sun-dried but relatively hard bricks, about 20 by 15 in., with a thickness of 4 in. on the average, as well as its solid construction, suggest considerable antiquity. The parapet, where intact to a corresponding height, showed loopholes arranged in two uniform levels, one 16 in., the other 5 ft. above the base, but at irregular intervals varying from 5 to 10 ft. The loopholes were about 3 in. square in the lower row, and about 6 in. wide, with a height of 8 in. in the upper one. Only the upper row is likely to have been used for shooting arrows. The top of the rampart on the inside formed a banquette about 4 ft. broad, running along the base of the parapet, and evidently serving as a means of communication for the defenders. At two points of the extant wall-segment the parapet was strengthened by solid brick platforms, projecting about 3 ft. on either side of the base. These probably once bore constructions corresponding to watch-towers. Near the western one (*b* in plan) the remains of a flight of stairs leading to the top from the inside were visible. In the angle formed by this with the adjoining parapet eastwards the red colour of the bricks indicated a sheltered spot where a fire must once have been lit, perhaps by men on guard.

Area within circumvallation. The break visible in the parapet in Fig. 57 does not mark the position of a gate, as Mr. Högberg supposed; for the rampart shows no sign of a cutting. It was easy to see that the parapet had fallen here from some cause. Inside the wall the ground showed considerable erosion down to a depth of 8 ft., as seen in the plan and section. It was this eroded hollow which the same visitor took for the remains of a ditch, with the result that in his description the inside area of the fort figures as the outside[4]. To the north of the extant wall-segment the original surface of the ground was remarkably well-preserved, probably owing to the protection afforded by drift-sand, which the prevailing northerly winds heaped up here against the

[3] Owing to the ruined condition of the extant segment and its relatively small length, the position of the centre, as shown in the plan, could not be determined with absolute exactness. But the position of the isolated wall portion *d* in the reconstructed line of the circumvallation shows that the ascertained radius is sufficiently accurate for practical purposes.

[4] It is this confusion between inside and outside which explains, e. g., the curious observation: 'Near the gate and on the further side of the ditch, the remains of two towers (Stūpas?) are visible, &c.'; see Hoernle, *Report on C.-A. ant.*, i. p. xv. In reality these 'towers' were two of the several little isolated loess banks or 'witnesses' which rise above the eroded ground within the wall. One of them, marked *e* in the plan and visible on the extreme right of the photograph in Fig. 58, was pointed out to me by Turdi (who had accompanied Mr. Högberg and Mr. Bäcklund on their visit), with a twinkle in his eyes, as a spot where 'the Sāhibs' tried some digging for manuscripts, since Islām Ākhūn had indicated Ak-sipil as one of the find-places of his [forged] 'old books.' For a diagram of two similar natural loess banks which had been conjectured to have yielded some of Islām Ākhūn's finds, see ibid.

walls. Over the patches of ground which now lay again clear of sand the small rectangular embankments by which fields are divided at the present day for irrigation could be quite clearly made out. The labourers with me correctly recognized them at once. Just as at the site of 'old Domoko', the ground formerly occupied by fields showed a relatively firm surface[5]. Within the area once enclosed by the circumvallation pottery débris was very scanty, perhaps, as M. Dutreuil de Rhins assumed, owing to the accumulation of drift-sand over the eroded ground. But outside at some distance from the line of walls it cropped up on numerous bare patches.

Marks on bricks.

With the exception of a single Chinese coin showing the characters *wu-chu* (see Plate LXXXIX, 16), which was picked up on the isolated wall portion westwards, no antiques were found during my stay in the immediate vicinity of the ruins. But the latter themselves possibly furnish some indication of their age in the marks which appear on most of the loose-lying bricks, and which M. Dutreuil de Rhins duly noticed. Some of the bricks thus marked, more or less fragmentary, are seen in the photograph reproduced in Plate XVIII[6]. The marks consist of coarse straggling indentures about half an inch broad and a quarter of an inch deep. Several of them seemed to recall Kharoṣṭhī characters, especially *ka* and *ga*, while others showed a resemblance to the general type of that script. As bricklayers' marks of this kind might have remained in use long after current writing in the script had ceased, it would scarcely be possible to draw a definite chronological conclusion from them, even if their Kharoṣṭhī derivation could be established with certainty. But the general impression I gained from the construction of the walls and from the sculptured remains of the neighbouring site of Kighillik, to be discussed presently, favoured the assumption of an early date.

Antiques said to come from Ak-sipil.

'Ak-sipil' serves as a popular designation for the whole of the débris areas stretching north of the Yurung-kāsh and Hanguya. Hence the evidence of the coins and antiques sold to me at Khotan as coming from 'Ak-sipil' would prove little or nothing for the age of this particular ruin, even if such statements could always be accepted with confidence. Among the thirteen old copper coins alleged to have been brought from Ak-sipil Appendix *D* shows one small Sino-Kharoṣṭhī and one *wu-chu* piece, four coins of the Ch'ien-yüan period (758–759 A.D.), two Sung coins of the eleventh century A.D., as well as three early Muhammadan coins, one of these (Plate XC, 42) with the name of Muḥammad Arslān Khān. The range covered by these coins is thus similar to that in the case of Yōtkan, but the local attribution is here quite uncertain. The other antiques comprise some terra-cotta grotesques (A. 001. a, b; A. 006. c), which closely resemble the Yōtkan type, but show marks of erosion. There is among them also a number of seals in stone and bronze, undoubtedly ancient, but difficult at present to fix chronologically, as there are no legends, and the devices are not sufficiently pronounced in character (see A. 001. c; A. 002. a; A. 004. a, b, c; A. 006. a, b, all in Plate L). Among them the stone seal (A. 006. a) is of interest; it is engraved on two sides, showing a humped bull on the one and a fire-altar on the other. Yōtkan, as well as the Niya Site, have furnished seals not unlike this series.

Remains of *Kighillik.*

My surmise that the high sands to the west and south-west of Ak-sipil might possibly hide more structural ruins was verified when, on April 8, Turdi guided me to the remains of what he had previously talked of as 'the Būt-khāna' or temple of Ak-sipil. They were found to be situated at a distance of about $1\frac{1}{2}$ miles to the south-west of the ruined fort, in a small

[5] In the description of the site above quoted these old fields figure as 'thoroughfares and the places where the houses [in the interior of the city] had once been', with more observations of the same value.

[6] For eye-copies of more, see *Mission D. de Rhins*, iii. p. 141.

valley-like depression between two ridges of sand about 60 ft. high, and running approximately from north-west to south-east. The length of the Nullah between them was about 170 yards, and the average breadth of the flat sandy bottom about 80 ft. From a huge heap of consolidated dung and other refuse, referred to hereafter, the spot has received among the treasure-seeking fraternity the designation of *Kighillik* 'the dunghill'. Close to the southern end of the Nullah I found a low mound, about 24 ft. in diameter, of what looked like loess earth mixed up with decomposed wood. It indicated the place of a structure, the original form and dimensions of which could no longer be made out. The mound rose only 2 to 3 ft. above the pottery-strewn ground near by; apart from the thin cover of drift-sand its material proved to consist mainly of decayed plaster such as was used in the walls of the ancient buildings described at Endere, but here quite soft, and often showing a reddish colour. The small pieces of charred wood which also turned up in it suggest fire as the cause of the latter.

Relief fragments in plaster of Paris.

That this débris heap had already been searched through again and again by 'treasure-seekers' was well known to Turdi, and clearly proved by the complete confusion in which I found mixed up drift-sand, decayed plaster, decomposed timber, and small relief fragments in stucco. The latter, of which close on sixty pieces were recovered in the course of a careful search, were found both in the centre of the little mound and within some yards of it. The exceptional hardness of these small fragments, and the peculiar fissured and discoloured appearance of their surface, which in some cases bore clear marks of having been scorched, at once attracted my attention. The assumption that these stucco pieces received their present appearance in a fire that consumed the structure naturally suggested itself. It was confirmed when the closer examination made subsequently by Mr. Andrews showed that the stucco in every one of these fragments, as described under A. 01-057 in the list below, was either vitrified or burnt. But whether this accidental burning would also account for their exceptional hardness was a question that long remained doubtful, as well as the determination of the materials in which the reliefs had been worked.

Effect of fire on relief fragments.

These questions have now been definitely settled by the result of the valuable chemical analysis to which Professor A. H. Church, F.R.S., has subjected specimens of these reliefs. As his Appendix (*F*) shows, the material was essentially 'plaster of Paris', as also in the Dandān-Uiliq stucco reliefs, 'the native gypsum having been prepared in the usual way by moderate heating (called "burning") and then mixed with water, just as is now done in making plaster casts.' But in the case of the plaster of Paris of the 'Kighillik' reliefs three peculiar features, revealed by the chemical analysis, prove 'that it has been subjected, after completion, to a high temperature and to an atmosphere charged with what chemists call "reducing matters"'. With reference to these peculiar features, among which the low percentage of water present deserves special mention, Professor Church has subsequently been kind enough, in a letter dated November 23, 1904, to state his opinion 'that an accidental fire in the presence of vegetable matter has in all probability caused them. It seems a case of chemical reduction hardly contemplated by the stucco-makers'. Thus the fact that the shrine which these stucco-reliefs decorated had been destroyed by fire may now be considered as definitely established.

Remains of temple decoration.

That the structure was indeed a Buddhist shrine, and one adorned in a fashion resembling that of the small Dandān-Uiliq temples, is clearly shown by the reliefs themselves. Fragmentary as they are they could all be recognized as parts of a wall decoration which consisted of a series of small Buddha figures in relief arranged within a large aureole. The edges of the latter were formed by overlapping lotus-leaves in relief with rows of overlapping flame-tongues ranged on the outside. A comparison of the descriptions given of all the pieces, and of the specimens reproduced in Plate LXXX, with the corresponding fragments from the Dandān-Uiliq shrines D. I and D. II (see Plates LIV, LV), will fully bear out this observation. That this aureole in flat relief fixed on the temple wall had served to encircle a large image may be considered as certain. But of this image no recognizable trace has survived; it was in all probability sculptured in soft stucco, just as in the Dandān-Uiliq and Endere shrines.

Style of relief figures. By the side of this general agreement in decorative design and arrangement the Kighillik fragments display unmistakable differences in style and execution, which seem distinctly to indicate an earlier date, and a close connexion also with the relief decoration of the Rawak Stūpa court, to be described hereafter. The small Buddha figures, of which portions were recovered in numerous replicas, undoubtedly represented standing Buddhas, with the right hand raised in the 'Abhayamudrā' attitude, and the left hanging down and holding the robe, as illustrated by the pieces A. 01, A.021 in Plate LXXX. The attitude is the same as in the Dandān-Uiliq aureole reliefs D. II. 34+74, shown in Plate LIV, but the size is far larger, and approaches that of the corresponding reliefs from the Rawak aureoles (see Plate LXXXVI and Figs. 63, 64). Another important point of contact with the latter is the absence among the Kighillik remains of any fragments representing the feet or lower parts of the drapery of these Buddha figures. We shall see that at Rawak the corresponding relief-images were ranged in overlapping rows within the haloes without their lower parts being shown anywhere. The same arrangement may be assumed in the Kighillik wall decoration. In the Dandān-Uiliq aureoles the corresponding figures stood apart, and were hence fully shown.

Period of Kighillik remains. Unfortunately the Kighillik pieces are so fragmentary that we can judge only from small details as to the general characteristics of their style and modelling. But these all indicate unmistakable superiority to the Dandān-Uiliq work, and, as far as care in execution is concerned, also, I think, to the Rawak sculptures. But in comparison with the latter the superiority of the material, hard plaster of Paris as against very friable clayey loess, must not be lost sight of. Hence the chronological relation between the Kighillik and Rawak Stūpa reliefs scarcely permits of definite settlement at present. But that both of them must be assigned to a considerably earlier period than that ascertained for the Dandān-Uiliq shrines is proved by the style of the Kighillik remains quite as clearly as by the more specific evidence which, as we shall see, is available in the case of the Rawak Stūpa sculptures.

Sculptural details of fragments. The great majority of the Kighillik fragments belong to the bodies of small standing Buddhas. One of the largest is the draped torso (A. 01) with the raised right hand (see Plate LXXX). Still better preserved drapery is shown by the replica A. 04 (see ibid.), which retains the original red colour of the robe, now burnt to deep rich brown. Traces of the original colouring remain also on other fragments (A. 05, 06, 09, 010). Portions of the stucco, by which the figures were attached to the wall at the back, are still preserved in A. 012, 017. Numerous fragments prove that they had been broken before they were blackened by fire (see e.g. A. 011, 012). Of the well-modelled left hand grasping the end of drapery, with more of the latter loosely hanging over the forearm (see A. 021, ibid.), we have numerous replicas (A. 016–022). Of the head of the Buddha only two specimens, unfortunately badly injured, survive (see A. 041, ibid.). Of the small nimbus, too, with a raised border, which surrounded the head, also originally coloured, we have but few fragments (A. 050–052, see ibid.). Relatively numerous fragments (A. 035–038, 054–057, see ibid.) belong to the border of a large aureole, formed by overlapping lotus-petals and corresponding in arrangement to the similar relief decorations of the Dandān-Uiliq shrine D. II (see Plate LIV, LV). The latter also show how the overlapping flame-tongues (A. 039, 040, see Plate LXXX) are likely to have been arranged on the outside of the aureole border. The employment of the stucco fragment showing a portion of a jewelled chain (A. 043) remains doubtful. Finally, we may mention some pieces (A. 044) still showing the matrix of a wood core, used here, as in all the larger reliefs of Dandān-Uiliq and Rawak, for strengthening the stucco work.

Ancient pottery; piece of leaf-gold.

Close to the original ground and near the centre of the mound there were found some pieces of perished timber which retained traces of red colouring in spite of their decayed surface. Here, too, was found a tiny piece of leaf-gold which may have peeled off from some gilt statue, supporting the view expressed above as to the origin of the leaf-gold washed from the débris layers of Yōtkan[7]. What purpose a small piece of talc found here may have served remains uncertain. The ground round the destroyed shrine and all over the flat bottom of the Nullah was strewn with ancient pottery débris, much of it being of fine texture and superior in make to the potsherds ordinarily found at the Tatis north of the Khotan oasis. Some of the fragments of terra-cotta vessels described under A. 007 show incised work, others a carefully-smoothed surface or red glaze, while in two pieces (A. 007. g, h) there appear moulded details, probably of grotesque masks such as figure so plentifully among the *appliqué* ornaments of Yōtkan pottery.

Ancient dunghill.

About 70 ft. north of the remains of the shrine there rises from the foot of the dune eastwards a large mound composed mainly of dry dung (*kighik* in Turkī), apparently horse-dung, but containing besides an admixture of bones, charcoal, and chopped bits of fuel. This huge refuse-heap, which measures as far as exposed over 70 ft. from north to south, with a breadth of about 50 ft. and a depth of over 11 ft., has not escaped the attention of 'treasure-seekers'. The regular galleries they have tunnelled into it, two of these reaching down to the bottom and being about 18 to 20 ft. long, enabled me to ascertain its contents with relative ease. This huge accumulation of dung proves that the site, whatever its character, must have been greatly frequented[8]. The dunes close by may cover other remains, but their height precluded any thought of trial excavations on either side of the little depression.

OBJECTS FOUND AT KIGHILLIK.

Objects from Kighillik.

A. 01. Stucco relief fragment, of standing Buddha. Head, legs, and L. p. arm from elbow missing. Plaster blackened and vitrified apparently by fire, probably by accident, as the plaster which attached fragment to wall is equally vitrified. Also the piece was evidently broken before firing. Traces of colour still remain on robe. R. p. hand upraised, palm outwards, wrist bare. Behind L. p. shoulder conventional leaf. Height 5¼", width 4⅞". Very hard. See Pl. LXXX.

A. 02. Stucco relief fragment; robe of Buddha. Plaster vitrified and discoloured. R. p. thigh and portion of torso; part of loose drapery falling from R. p. wrist. Replica A. 01. 3⅞" high, 2⅝" wide. Very hard.

A. 03. Stucco relief fragment; standing Buddha, part of torso. Plaster vitrified and discoloured. Torso from waist downwards. L. p. hand grasping end of robe. Replica A. 01. 2½" high, 2⅞" wide. Very hard.

A. 04. Stucco relief fragment; standing Buddha in 2 pieces. Plaster vitrified and discoloured. Torso L. p. side broken. Robe coloured red, burnt to deep rich brown. Replica A. 01. Length 5¼", width about 3". Very hard. See Pl. LXXX.

A. 05. Stucco relief fragment; drapery. Plaster vitrified and discoloured, traces of colour-wash. Lower R. p. portion, robe of Buddha. Replica A. 01. Very hard.

A. 06. Stucco relief fragment; drapery. Plaster vitrified and discoloured. Traces of colour. Lower L. p. portion, robe of Buddha. Replica A. 01. 2½" x 2 7/16". Very hard.

A. 07. Stucco relief fragment; drapery. Plaster vitrified, much discoloured by fire after being broken. Portion of robe on R. p. thigh of standing Buddha. Replica A. 01. 3¼" x 1½" (app.). Very hard.

[7] See above, p. 194.

[8] Prof. Lóczy, in his instructive description of the parts of Kan-su visited by the Széchenyi expedition, mentions the large accumulations of old manure visible outside the roadside stations, half-buried by sand, on the old desert route leading from An-hsi-fan towards Khāmil; see *Khinai biroda-lom*, p. 494.

Objects from Kighillik.

A. 08. Stucco relief fragment; drapery. Plaster vitrified and discoloured. Portion of robe from R. p. thigh of Buddha. Replica A. 01. $1\frac{3}{8}'' \times 2\frac{3}{8}''$. Very hard.

A. 09. Stucco relief fragment; drapery. Plaster vitrified and discoloured. Traces of pink. Portion of robe from L. p. thigh, standing Buddha. Replica A. 01. $2'' \times 1\frac{3}{4}''$. Very hard.

A. 010. Stucco relief fragment; drapery. Plaster vitrified and discoloured. Traces of colour. Portion of robe, L. p. hip, standing Buddha. Replica A. 01. $1\frac{5}{8}'' \times 1\frac{3}{4}''$. Very hard.

A. 011. Stucco relief fragment; drapery. Plaster vitrified and discoloured by fire after breaking. Portion of robe R. p. thigh, standing Buddha. Replica A. 01. $2\frac{1}{2}'' \times 1\frac{1}{2}''$. Very hard.

A. 012. Stucco relief fragment; drapery. Plaster vitrified and blackened by fire after breaking. On back of fragment, portion of plaster by which it was attached to wall, equally burnt with relief. Portion of robe L. p. thigh, standing Buddha. Replica A. 01. $2\frac{1}{2}'' \times 2\frac{5}{8}''$. Very hard.

A. 013. Stucco relief fragment; drapery. Plaster vitrified and discoloured. Portion of robe and neck, L. p., standing Buddha. Replica A. 01. $1\frac{3}{4}'' \times 1''$. Very hard.

A. 014. Stucco relief fragment; drapery. Plaster vitrified, discoloured, and bent by fire. Portion of undergarment (saṅghāṭi) from R. p. side of standing Buddha. Replica A. 01. $1\frac{3}{4}'' \times 1\frac{1}{8}''$. Very hard.

A. 015. Stucco relief fragment; drapery. Plaster vitrified and discoloured. Traces of colour. L. p. shoulder, draped, of standing Buddha. Replica A. 01. $1\frac{7}{8}'' \times 1\frac{3}{4}''$. Very hard.

A. 016. Stucco relief fragment; drapery. Plaster partially vitrified, discoloured and cracked by fire. L. p. shoulder, hand (upraised palm outwards) and breast of standing Buddha. Replica A. 01. Height $2\frac{1}{2}''$, width $2\frac{7}{8}''$. Hard, and in some parts very hard.

A. 017. Stucco relief fragment; drapery. Plaster vitrified, discoloured, and cracked by fire. Portion of plaster joint remains at back, also burnt. L. p. forearm and hand, loosely hanging, grasping end of drapery. Lower L. p. portion of torso and loose drapery over arm. Little finger of hand curiously crooked (cf. A. 018 and 019). Replica A. 01. Height $2\frac{1}{8}''$, width $2\frac{5}{8}''$. Very hard.

A. 018. Stucco relief fragment; hand of Buddha. Plaster burnt, discoloured and cracked by fire. L. p. hand hanging and grasping drapery of robe; portion of forearm draped. Little finger crooked. Height $1\frac{7}{8}''$, width $1\frac{1}{4}''$. Hard.

A. 019. Stucco relief fragment; hand of Buddha. Plaster vitrified and discoloured by fire. L. p. hand grasping drapery. Little finger crooked (cf. A. 017). Replica A. 020. Height $1\frac{3}{8}''$, width $1\frac{3}{16}''$. Very hard.

A. 020. Stucco relief fragment; hand of Buddha. Plaster vitrified, discoloured, and cracked by fire. L. p. hand (portion) grasping drapery. Drapery over upper part L. p. thigh. Little finger crooked. Replica A. 019. Height $1\frac{1}{2}''$, width $1\frac{3}{4}''$. Very hard.

A. 021. Stucco relief fragment; hand and forearm of Buddha. Plaster vitrified, discoloured, split, and roughened by fire. L. p. hand grasping loose drapery; forearm, over which loose drapery; portion of draped L. p. thigh. Replica A. 022. Height $2\frac{1}{2}''$ (app.), width $2\frac{1}{4}''$. Very hard. See Plate LXXX.

A. 022. Stucco relief fragment: forearm and hand of Buddha. Plaster partially vitrified, slightly discoloured, and cracked by fire. L. p. forearm loosely draped, portion of L. p. side, and L. p. hand (fingers missing), of standing Buddha. Portion of plaster joint, burnt, at back. Replica A. 021. Height $1\frac{7}{8}''$, width $1\frac{5}{8}''$. Very hard.

A. 023. Stucco relief fragment; hand of Buddha. Plaster burnt, slightly discoloured by fire. R. p. hand and wrist upraised, palm outwards. Loose drapery falls on each side of wrist (cf. A. 02). Replica A. 024. Height $2\frac{1}{2}''$. Very hard.

A. 024. Stucco relief fragment; hand of Buddha. Plaster vitrified, cracked at back by heat; colour of beeswax. Replica A. 023. Very hard.

A. 025. Stucco relief fragment; hand of Buddha. Plaster partly vitrified, discoloured almost to black in places, and cracked by heat. Replica A. 023. Very hard.

A. 026. Stucco relief fragment; hand of Buddha. Plaster vitrified; much discoloured and cracked by fire. Fingers missing. Replica A. 023. Very hard.

A. 027. Stucco relief fragment; hand of Buddha. Plaster vitrified, discoloured, and cracked by fire. Hand only. Replica A. 023. Very hard.

A. 028. Stucco relief fragment; wrist of Buddha. Plaster vitrified (partly), discoloured and cracked by fire. Portion of wrist and drapery from R. p. upraised hand. Replica A. 023.

A. 029. Stucco relief fragment; hand of Buddha. Plaster vitrified, discoloured, and cracked by fire. R. p. hand upraised. Thumb and two fingers broken. $1\frac{1}{4}''$ long. Very hard.

A. 030. Stucco relief fragment; portion of Buddha. Plaster burnt, cracked and discoloured by fire. Traces of colour. L. p. shoulder of Buddha, draped. Portion of conventional chrysanthemum leaf at back. Portion of border of nimbus, composed of overlapping lotus-petals, with cincture at intervals. Replica A. 01 (part). Height $2\frac{1}{8}''$, width $2\frac{3}{8}''$. Very hard.

A. 031. Stucco relief fragment; nimbus. Plaster burnt black. L. p. portion of nimbus, with raised border consisting of overlapping lotus-petals, tied at intervals with double cincture. Replica A. 032. $3'' \times 1\frac{1}{2}''$. Very hard.

A. 032. Stucco relief fragment; portion of nimbus. Plaster vitrified and discoloured by fire. Replica A. 031. 2⅜" × ⅞". Very hard.

A. 033. Stucco relief fragment; portion of nimbus. Plaster burnt and discoloured by fire. Replica A. 031. 1⅜" × ⅞". Very hard.

A. 034. Stucco relief fragment; part of nimbus. Plaster vitrified; discoloured and cracked by fire. Portion of L. p. side of border of nimbus (cf. A. 033) with conventional leaf adjoining (cf. A. 01). 1¼" × 2". Very hard.

A. 035. Stucco relief fragment; ornament. Plaster burnt and discoloured. Portion of large aureole border. Overlapping lotus-petals; cf. A. 036. 1⅞" × 2". See Pl. LXXX.

A. 036. Stucco relief fragment; ornament. Plaster vitrified and discoloured by fire. Traces of colour. Portion of large aureole border, of overlapping lotus-petals; cf. A. 035. This piece had a wooden core, which has entirely disappeared, leaving a hard matrix in the stucco. Part of surface is detached. Replica A. 037. 1⅞" × 1½". Very hard.

A. 037. Stucco relief fragment; ornament. Plaster vitrified and discoloured by fire. Portion of large aureole border. Overlapping lotus-petals. Replica A. 036. 1½" × 1⅜". Very hard.

A. 038. Stucco relief fragment; ornament in 2 pieces. Plaster burnt and discoloured by fire. On one traces of green. Portions of large aureole border. Replica A. 036.

A. 039. Stucco relief fragment; ornament. Plaster burnt and much discoloured by fire. Portion of two rows of overlapping flame-tongues (cf. D. II. 89), from large aureole. Replica A. 040. 2⅛" × ⅞". Very hard.

A. 040. Stucco relief fragment; ornament. Plaster burnt and discoloured by fire. Portion of flame-tongues from large aureole. Replica A. 039. 2⅜" × 1". Very hard. See Pl. LXXX.

A. 041. Stucco relief fragment; head of Buddha. Plaster vitrified, discoloured, and cracked by fire. Upper half of head of Buddha; defaced. Above top-knob portion of border of nimbus. 2½" × 1⅜". Very hard. See Pl. LXXX.

A. 042. Stucco relief fragment; head. Plaster vitrified and discoloured by fire. Portion of L. p. side of head showing hair (cf. A. 041). 1$\frac{3}{16}$" × ¾". Very hard.

A. 043. Stucco relief fragment; ornament. Plaster burnt. On convex fragment ¼" thick, portion of jewelled chain. Pattern, a large jewel, two beads each side; this arrangement probably repeated in broken parts. 2½" × 2$\frac{3}{16}$". Hard.

A. 044. Stucco relief fragment; in 3 pieces. Plaster burnt. Three pieces, plain on convex side, and on back showing matrix of wood core. Three pieces together measure 3" × 1¼". Hard.

A. 045. Stucco relief fragment; drapery. Plaster vitrified, discoloured, cracked and twisted by fire. Lower portion of robe of standing Buddha (?). 4⅜" × 2¼". Very hard.

Objects from Kighillik.

A. 046. Stucco relief fragment; hand of Buddha. Plaster burnt, discoloured, and cracked by fire. R. p. hand and forearm of Buddha. Replica A. 026. 2⅝" × 1⅝". Very hard.

A. 047. Stucco relief fragment; torso of Buddha (?). Plaster burnt, discoloured, and cracked by fire. Breast and L. p. shoulder. Well modelled drapery. Width 2⅜" × 1⅜". Very hard.

A. 048. Stucco relief fragment; forearm. Plaster burnt and discoloured by fire. Traces of pink. Forearm of standing Buddha (?), draped. 1⅞" × 1½". Very hard.

A. 049. Stucco relief fragment; drapery. Plaster burnt and discoloured by fire. Traces of pink. Probably L. p. thigh, with loose drapery from L. p. shoulder. 2⅝" × 2". Hard.

A. 050. Stucco relief fragment; nimbus. Plaster vitrified, discoloured, and cracked by heat. Traces of colour. Replica A. 032. 2⅜" × ⅞". Very hard. See Pl. LXXX.

A. 051. Stucco relief fragment; nimbus. Plaster burnt. Traces of colour. Replica A. 032. 1⅝" × ⅞". Very hard.

A. 052. Stucco relief fragment; nimbus. Plaster burnt, split, and discoloured by fire. Replica A. 032. 1½" × 1½". Very hard.

A. 053. Stucco relief fragment; drapery. Plaster burnt, discoloured, and cracked by fire. Traces of colour. Portion of drapery from L. p. thigh of standing Buddha (?). 1⅛" × 1½". Very hard.

A. 054. Stucco relief fragment; ornament. Plaster burnt and discoloured by fire. Replica A. 035. 1⅞" × 1¾". Hard.

A. 055. Stucco relief fragment; ornament. Plaster burnt and discoloured by fire. Portion of border of large aureole. Replica A. 035. 1½" × ⅞". Very hard.

A. 056. Stucco relief fragment; ornament. Plaster vitrified and cracked by fire. Portion of border of large aureole. Replica A. 035. 1" × ⅝". Very hard.

A. 057. Stucco relief fragment; ornament. Plaster burnt. Double row of lotus-petals pointing opposite ways—foot to foot. From border of large aureole (?). Replica A. 035. 1⅝" × 1¼". Very hard.

A. 005. Small piece of leaf-gold found in débris-mound, Kighillik.

A. 007. a. Three terra-cotta fragments of rather flat globular vessel decorated with cut enrichments. Round centre (?), between double incised lines, a row of sunk dots. Above and below this band are flutes. On one side is a hole, and the scar of handle attachment. Width 2¼", height 1¾".

Objects from Kighillik.

A. 007. b. Fragment of terra-cotta vessel, being portion of neck and shoulder. The neck has a rough texture, while the body appears to have been carefully smoothed. Height 1⅞″, width 2¾″.

A. 007. c. d. Terra-cotta fragments of plain vessels, made on wheel and afterwards polished. 1½″ × 1½″ and 2½″ × 2¼″.

A. 007. e. Terra-cotta fragment of coarse vessel, decorated with bold but lightly incised nebulée pattern drawn with a four-pointed tool. 3⅝″ × 2⅜″.

A. 007. f. Fragment of terra-cotta vessel, bearing lightly incised pattern, somewhat resembling that of A. 007. a, but differently treated. 1⅞″ × 2⅜″.

A. 007. g. h. Terra-cotta fragments of moulded details. Probably part of grotesque face and ornament. 1½″ × ⅞″. and 1½″ × ⅞″.

A. 007. i. j. Fragments of terra-cotta vessel, with incised marks. 1½″ × 1¼″ and 2¾″ × ⅜″.

A. 007. k. Fragment of terra-cotta, moulded, which seems to have been painted with a red glaze. 1½″ × 1½″.

A. 007. l. Terra-cotta fragment, moulded, without glaze, and with deep spiral grooves. 1½″ × 1½″.

A. 007. m. Fragment of fine terra-cotta object having incised work upon it. Outer surface quite smooth and polished. 1⅜″ × ⅞″.

A. 007. n. Fragment of terra-cotta of a grey colour, similar to the burnt stuccoes from Kighillik. 1″ × 1¼″.

A. 007. o. Fragment of terra-cotta vessel, overfired, with green glaze which has 'fizzled' in burning. 2¼″ × 1½″.

OBJECTS SAID TO BE FROM AK-SIPIL.

Objects from Ak-sipil.

A. 001. a. Terra-cotta monkey, squatting, considerably eroded, naturalistic. R. hand on breast, L. *veretrum tenens* (?). Back hairy. Cf. Y. 0012. a. iii, Y. 0013. c. Height 1⅝″. See Pl. XLVII.

A. 001. b. Terra-cotta grotesque human head, evidently an ornamental detail from some vessel. Bears some resemblance to Kh. 003. f, but is more human, and more carefully modelled. 1½″ × 1⅜″.

A. 001. c. White marble seal with fret device. ¾″ square, ½″ high. See Pl. L.

A. 002. a. Stone seal, similar in all respects to A. 001. c, but rather smaller. ⅝″ square, ½″ high. See Pl. L.

A. 002. b. Miniature jade battle-axe, pierced for string, probably a necklace ornament. ⅞″ × ⅝″. See Pl. LI.

A. 004. a. Large bronze seal, fragment of; rectangular. Within a simple border an animal (bull ?) couchant to R. All the upper parts of body and head are missing. 1⅝″ × ⅝″. See Pl. L.

A. 004. b. Square seal in black lignite (?), finely engraved with what may be Chinese lapidary characters or fret. At back a semi-cylindrical projection (broken), pierced for cord. 11/16″ square × ½″. See Pl. L.

A. 004. c. Oblong bronze seal, with countersunk device of four conventional half-leaves set in the form of an X. Shank at back in form of a ring. 9/16″ × ½″ × ½″. See Pl. L.

A. 006. a. Black stone seal, pierced for cord. *Obv.* A humped bull with long horns, statant to L. *Rev.* A fire-altar. Rather roughly cut. 13/16″ × ¾″ (nearly) × ⅜″. See Pl. L.

A. 006. b. Bronze irregularly oval seal, with shank at back, pierced with small hole. Device seems to represent an animal (cat ?) seated to R. and in front some small animal (kitten ?). Surface considerably oxidized. ⅝″ nearly × ½″. See Pl. L.

A. 006. c. Terra-cotta grotesque human face. A moulded enrichment for pottery vessel. Much eroded. Face smiling; eyes round, eyebrows meeting in acute downward angle; cheeks prominent; ears high and animal-like. Curled hair surrounds face. 2″ wide, 1⅞″ high.

SECTION III.—THE RAWAK STŪPA

First report of Rawak.

On the morning of April 10 I left Ak-sipil in order to march to the ruin which Turdi and others of his craft had in November reported to me under the name of *Rawak* 'the high mansion'. This designation, and a vague tradition about the images of the *Tokuz-Khākān* or 'Nine Khākāns', sounded promising enough; but no details could then be elicited except that Muḥammad Sharīf, a young 'treasure-seeker', had found near that ruin a short time before a large pot full of Chinese copper coins. Of these I had examined many dozens, with the result that they all proved to be *wu-chu* pieces, most of them in very fair preservation. This pointed to a site of early occupation.

March to Rawak.

The route from Ak-sipil first led, for a direct distance of about three miles, over high dunes of coarse greyish looking sand, which evidently had been deposited from an earlier river-course. The latter itself could be recognized in the broad pebble-covered Sai on which we emerged, and which we proceeded to follow in the general direction towards north-north-east. Further to the south it was said to connect with the Sai near Tam-ōghil and Kumat, referred to above as an ancient river-bed [1]. Though water was alleged never to reach it now, tamarisk growth and other scrub marked its course all along and far away to the north, where I subsequently struck it again near the site of Jumbe-kum. After covering, in great heat and glare, a distance of some six miles, we reached a well with some tamarisks, which Jiya people going for wood use as a resting-place. Thence striking to the north-west, we soon were again among heavy riverine dunes. After about two miles we passed a small open depression, where some old pottery appeared on the bare loess ground. Here I was shown by Muḥammad Sharīf the remains of a large pot of coarse red clay, brought to light from beneath what may have been the mud floor of a completely eroded dwelling. Of timber or other structural débris there was no trace. Muḥammad Sharīf declared the haul of Chinese copper coins made here to have amounted in value to the equivalent of about 300 Khotan Tangas (say somewhat over Rs. 100). From among the 87 coins which I purchased out of this find, and which have all proved to be *wu-chu* pieces, are taken the three specimens reproduced on Plate LXXXIX, 20–22. Most of them show few marks of wear, and thus are not likely to have been long in circulation before the deposit was made.

First view of Rawak Stūpa.

At Rawak, which was reached about a mile beyond, an unexpected and most gratifying discovery awaited me. My honest old guide Turdi had spoken only of 'an old house' to be seen there half buried in the sand; but in reality the first glimpse showed a large Stūpa, with its high base and enclosing quadrangle, by far the most imposing structure I had seen among the extant ruins of the Khotan region. Large dunes of sand rising in places to about 25 ft. covered the quadrangle and parts of the massive square base of the Stūpa on the south-west and north-east faces. But towards the south the drift-sand was lower, and there great portions of the Stūpa base, as well as of the walls marking the quadrangular enclosure of the Stūpa court, could be readily made out. Fig. 60 and Plate XIII.a show the ruins as seen on the morning of April 11, from the east and from near the south corner, respectively. Fragments of coloured stucco, evidently from the heads and upper parts of colossal statues, were lying in the sand on the outside of the enclosing wall near the southern corner, showing with some exposed portions of reliefs on the wall behind that 'treasure-seekers' had been at work here quite recently. I subsequently learned that a party of six men from Jiya had decamped on the morning before my arrival. Fortunately they had cleared away only a few feet of sand, and the damage done by them was not extensive.

Proximity of subsoil water.

I realized at once that there was scope here for extensive excavations, and accordingly lost no time in sending back an urgent requisition to the Bēg of Yurung-kāsh for a reinforcement of labourers. Fortunately the position of the ruin, within a day's march of the oasis, enabled me to secure in quick succession contingents of willing workers from the nearest villages of the Jiya tract. A favourable factor of still greater practical importance was the relative ease with which the question of water-supply for such a number of men was solved. Though the sand-dunes surrounding us looked more formidable and sterile than at any ancient site previously explored, it was possible to dig a well in a depression within two miles of the Stūpa, and there the labourers' camp was conveniently established. A look at the map shows that the distance from the Rawak site

[1] See p. 472.

to the bank of the Yurung-kāsh is only about seven miles. In fact, to this comparative proximity of the present river-bed were due both the forbidding height of the dunes and the slight depth of subsoil water. The early accumulation of the former had, no doubt, been a chief factor in saving the ruins from destruction, whether from erosion or the hand of man. On the other hand, the presence of subsoil moisture had, as we shall see, adversely affected all materials like timber which were likely to attract and absorb it.

Trying conditions of work.

Apart from the question of water, the conditions under which work had to be effected here proved more trying than at any of the other old sites I explored. The season of Burāns had now fully set in; and the gales that were blowing daily, though from different quarters and of varying degrees of violence, carried along with them a spray of light sand that permeated everything. I noticed the frequency with which the wind would shift round to almost opposite directions on successive days, sometimes even between morning and evening, a feature of Burāns well known to all natives living near the Taklamakān [1], and observed also by former travellers. To the discomfort which the constant drifting of sand caused, and which we naturally felt in a most irritating fashion while engaged in excavation, was added the trying sensation of heat and glare all through the daytime. The sun beat down with remarkable intensity through the yellowish dust-haze even while the latter was too thick to permit of satisfactory photographic work, and the reflection of its light from the glittering particles of sand made the heat appear far greater than it really was. The quick radiation that set in as soon as the sun had gone down caused rapid and striking variations in the temperature at different portions of the day. The result of these sudden changes manifested itself in the agues and fevers from which all my followers began to suffer after our start from Yurung-kāsh. It was impossible for me to escape exposure to these adverse atmospheric influences; but I succeeded in keeping their effect in check by liberal doses of quinine until my work at these ruins was done.

Commencement of excavations.

The excavations which I commenced on the morning of April 11 within the south corner of the quadrangle soon revealed evidence that the enclosing wall had been adorned on its inner as well as on its outer face with rows of colossal statues in stucco, thus making the quadrangle correspond to the chapel courts of Buddhist Vihāras in Gandhāra and elsewhere. Those on the inside face of the wall might be expected to be still in a fair state of preservation, owing to the depth of the sand, which even in the most exposed portion of the Stūpa court (between the gate and the south corner) was nowhere less than 5 ft., and greatly increased towards the west and east corners. It was clear that great masses of sand would have to be shifted before these sculptures could be systematically unearthed and examined in safety. For the heavy earthwork implied by this task it was necessary to await the arrival of the reinforcements already summoned. But in the meantime I was able to utilize the dozen labourers already at hand for such clearings as the preliminary survey of the architectural remains demanded.

Plan of Vihāra court.

From the plan on Plate XL which was prepared on the basis of this survey and of subsequent more detailed measurements, it will be seen that the Vihāra court formed a great rectangle, measuring 163 ft. inside on its south-western and north-eastern faces and 141 ft. on its shorter sides towards the north-west and south-east. It was enclosed by a wall about 3 ft. 6 in. thick, solidly built of sun-dried bricks. At the south corner, which of the portions not completely buried under sand was best preserved, this wall rose to a height of 11 ft. but was probably once still higher. The bricks used in it, as well as in the Stūpa, measured on the

[1] Thus, e.g., a strong wind blowing from the north-east on April 13 was followed on the 14th by one from the north-west, which in the afternoon veered round again to the east. On the 15th we had wind from the south-west, later again from the north-west, but the haze raised was not sufficiently thick to prevent the sun shining through.

59

RUINED STŪPA OF RAWAK, SEEN FROM SOUTH PART OF COURT.

60

RUINED STŪPA OF RAWAK WITH SOUTH CORNER OF COURT,
SEEN FROM EAST AT COMMENCEMENT OF EXCAVATION.

average 20 by 14 in., with a thickness of $3\frac{1}{2}$ to 4 in., showing a size practically identical with that of the Ak-sipil bricks. As seen in the plan, the enclosing wall along the whole of the longer sides of the quadrangle, except for a portion near the south corner, was covered by high sand rising to over 20 ft. above the original ground-level, and was thus inaccessible to examination. On the shorter south-east face, however, the top of the wall rose for the greatest part above the sand (as seen in Fig. 60). Here, too, the position of a gate was indicated by a gap in the centre; subsequent excavation proved it to have been $8\frac{1}{2}$ ft. wide. That the opposite side of the enclosure, to the north-west, had no corresponding gate could easily be ascertained without excavation, as in the central part of that face the top of the wall forming a continuous line was just visible above the sand. The latter was here fully 10 ft. high, and no excavation could be attempted. It thus seems probable that the gate in the south-east face was the only entrance to the court.

The centre of the quadrangle is occupied by the imposing Stūpa base which, as shown by the section in Plate XXXIX, rises in three stories to a height of $22\frac{1}{2}$ ft. above the floor of the court. The photograph (Fig. 59) shows the upper portion of this base with the extant part of the dome as seen from the south after some clearing had been effected. Parts of the base are visible also in the background of the photographs reproduced in Plate XVI. a–c. The lowest story, 78 ft. square and $7\frac{1}{2}$ ft. high, rests on a plinth of four steps showing an aggregate elevation of 3 ft. The second story is $45\frac{1}{2}$ ft. square, with a height of 9 ft. It is surmounted by a circular drum, 3 ft. high and receding on the top, which serves as a plinth for another circular drum forming part of the Stūpa dome. The masonry of the latter has remained intact only to a height of about $8\frac{1}{2}$ ft. (see Fig. 59), and it is thus impossible to determine at what elevation the dome proper sprang from this drum. Nor could anything definite be ascertained as to the original shape of the cupola. Base of Stūpa.

Judging from the little débris found on the topmost portions of the base, I doubt whether the dome could have been very high. As the above measurements show, the top of its extant masonry was found to rise to about 31 ft. above the level of the court. The diameter of the circular drum forming part of the dome proved to be a little over 32 ft. It seems to have had an inner chamber about $7\frac{1}{2}$ ft. in diameter; but this could not be exactly determined, as a large cutting had been made, apparently long before my visit, into the dome from the west. Treasure-seeking operations, no doubt, account also to a large extent for the broken state of the top of the dome. The cutting showed clearly that the dome, like the rest of the structure, had been solidly constructed of sun-dried bricks of the same size as those measured in the quadrangle wall. Also the base had been exposed to repeated attacks of treasure-seekers, as was proved by the short galleries which were found to have been tunnelled into both the upper and lower square story on the south-east face (see the plan and Fig. 59, Pl. XVI. a–c). Stūpa dome.

I have referred so far only to those main features in the base in which the typical and orthodox arrangement in three stories found expression. But a look at Plate XL will show that considerable variation and originality was introduced into the ground-plan by a series of bold projections on each face of the base supporting well-proportioned flights of steps. Through these projections the ground-plan of the base has assumed the shape of a symmetrically developed cross, each of the four arms of which extend about 52 ft. on the lowest level, as measured from the centre line of the Stūpa. The broad flights of steps which occupied the centre of each of the four faces of the base, and, carried by the projecting portions, led up without a break from the level of the court to the very foot of the dome (see section on line *AB* in Plate XXXIX; also Fig. 59, Plate XVI. b, c), must have been an imposing architectural feature. The one on the south-east Flights of steps on Stūpa base.

side which faces the entrance gate of the quadrangle could alone be completely cleared. The lowest section of the stairs leading to the top of the first base was fully 14 ft. broad, the one above 9 ft., and the one approaching the dome 6 ft. There can be little doubt that the stairs were in the first place intended to facilitate the circumambulation which was, perhaps, to be performed successively on each of the stories.

Stucco coating of base.

The portions of the lowest base story flanking this flight of steps proved, owing to the protection afforded by sand, to have retained a thick layer of fairly hard stucco bearing a coat of white paint. It may be safely assumed that this stucco coating once covered the whole of the Stūpa. Both the top and the foot of the lowest story were decorated with boldly moulded cornices in stucco about one foot high, as seen in Fig. 59. It was under the bottom cornice, in the receding corner formed to the south of the projection which bears the stairs of the lowest base story on the south-east face, that I discovered, sticking to the plaster at short intervals, four well-preserved *wu-chu* copper coins showing but little wear. Like numerous subsequent finds of such coins, they had manifestly been deposited as votive offerings, and furnished me with the first indication of the probable age of the structure.

Sculptural decoration of Vihāra court.

Interesting as the Stūpa is by its architectural features and imposing dimensions, the great archaeological interest of the ruins does not centre so much in it as in the rich series of relief sculptures decorating the walls of the enclosing Vihāra. These were brought to light by the systematic excavations which I commenced as soon as the bands of additional labourers, quickly collected and dispatched by the Bēg of Yurung-kāsh, began to arrive at my camp on the morning of April 12. In order to avoid the risks of immediate damage to the friable stucco of the sculptures, and to get sufficient room for photographing them, it was necessary to open broad trenches at some little distances from the walls and then to proceed towards the latter, carefully clearing out the sand. The work commenced near the south corner; and it was in the course of this preliminary clearing that the small Stūpa base shown in the plan, and described below (sec. v), together with its interesting deposit of coins, was first reached. From this place the work of excavation was gradually extended along the south-west and south-east walls up to the furthermost points which the high dunes rising over the rest of the south-west side and east corner permitted to be cleared within the time and with the means available[3].

General character of reliefs.

As the work of clearing proceeded I soon realized that the main adornment of the walls, both towards the court and outside, consisted throughout of rows of relief statues in stucco over life-size. All the large reliefs represented Buddhas or Bodhisattvas; but from the varying poses, accessories, &c., still recognizable, a number of groups could be distinguished, arranged apparently with some regard to symmetrical disposition. Between the colossal statues at frequent intervals were smaller representations of attendant divinities or saints. In numerous instances the walls were further decorated with elaborate plaques in stucco, forming haloes above the heads of figures, or, in some cases, where sufficient space had been allowed, even with complete aureoles in relief around them. Here and there remains of small fresco paintings between the statues could also be traced, but generally the latter were too close together to permit of such decoration, at least on the extant lower portion of the walls. The whole of the relief work had been originally coloured, but the layers of paint had in most cases peeled off except where well-protected in drapery folds, &c. Thus the greatest portions of the stucco images presented themselves in the red ground colour of the clay in which they had been modelled.

[3] For a photograph showing a portion of the south-east wall with the trenches in course of excavation, along both its sides, see *Ruins of Khotan*, p. 446. For others helping to convey an impression of the great masses of sand overlying the Vihāra court and surrounding the ruins, see ibid., p. 450, also here Fig. 60 and Pl. XIII.

I found from the first that the excavation of this wealth of statuary was attended with serious difficulty, even in that limited portion of the quadrangle where the conditions of sand permitted operations. Owing probably to the moisture rising from the neighbourhood of subsoil water, the strong wooden framework which once supported internally the masses of stucco and fastened them to beams let into the wall behind, had completely rotted away. The cavities left by these beams, which were apparently 5 in. square, and fixed at a uniform height of about 8 ft. above the floor of the court, could be traced all round the excavated wall portions wherever the latter had retained a corresponding height (see e.g. Figs. 62, 65, 66, 68; Pl. XIII. b; XIV. a, b, &c.). The beams had been set sufficiently deep into the wall to permit of their being covered on the outside with a layer of plaster about 1 to 2 in. thick, flush with the rest of the wall surface. For some distance on the cleared portion of the south-west wall this plaster covering had survived, and it is owing to its presence that in Figs. 61, 62 the cavity left by the perished beams does not appear. Here and at some other points I could still trace small remains of the beams in bits of decomposed wood. The wooden framework of the colossal statues must have been joined to the beam behind usually at the height of the heads or shoulders[4]. The wood of the internal framework had everywhere completely perished, but the position once occupied by portions of it was often still indicated by the matrix this wooden core had left, as visible, e.g. in the arms of colossal images (see Figs. 61, 62, 68, 69; Pl. XV. a, b, c; Pl. XVII. c). **Wooden framework of statues.**

Deprived of this support, the heavy stucco images, especially those still retaining much of their upper portions, threatened to collapse when the protecting sand was being removed. The strong winds blowing day after day greatly added to this risk. They carried away the fine dust of riverine loess which had filled the interstices between the edges of the relief work and the wall behind, and thus increased the danger of the friable masses of clay sliding down through their own weight to immediate destruction. Regrettable experience of this kind was gained in the case of some statues on the inner south-west and south-east walls. It soon showed me that these risks could be obviated only by extreme care in clearing the reliefs—a few hours of exposure to light winds often proved safer than the application of brushes for removing the layers of dust from drapery folds, &c.,—and by covering up again the lowest portions of the statues as soon as they had been examined and photographed. Even so damage could not altogether be prevented. In some instances it was necessary to secure the heads, &c., of statues still intact by means of ropes while they were being photographed. Fig. 69 illustrates this procedure followed in regard to some of the minor statues excavated on the inner side of the south-west wall; it also helps to mark the true size of the colossal image seen to the extreme right by comparison with the labourers. **Risks of excavation.**

The conditions here briefly indicated, which rendered the work so difficult and risky, also explain why many of the colossal statues were found badly damaged in their upper portions and all of them without their heads. Their upper portions, just like the top segments of the great aureoles seen in Figs. 63, 64, had necessarily been left much longer without the protecting cover of sand, and had accordingly fallen away from the wall that formed their backing and support. Numerous fragments of colossal heads which had thus become detached long ago turned up at different depths of sand, especially near the south corner. Their extremely friable condition, which made them break at any attempt to lift or remove them, showed that they had long been exposed to atmospheric vicissitudes. The heads of the smaller images, which the dust accumulations had had time to cover up even while perhaps a roof above gave shelter, were **Damaged state of sculptures.**

[4] See Fig. 62, and in Fig. 61 the series of small holes above, marking the points where the joining pegs or tangs passed through the plaster into the beam behind.

in most cases found intact. I may here note that, among all the sculptural decoration of the Vihāra court as far as excavated, I failed to trace any evidence of wilful destruction by human agency, as distinguished from such casual damage as the spasmodic burrowings of 'treasure-seekers' may have caused at some points of the more exposed portions of the enclosing wall[5]. This observation lends support to the belief, justified, as we shall see, by other abundant evidence, that this great shrine had already been deserted for centuries and the ruins of the quadrangle completely covered up by the time when Islām annexed Khotan.

Original shelter for sculptures.

It appears to me very probable that originally a wooden gallery, or some similar shelter projecting from the roof of the enclosing wall, offered shelter to the sculptures on both of its faces. But this must have been systematically removed, even before the sand had completely invaded the Vihāra court; for only in one place, near the inner south-east face and not far from the gate (see plan), did my excavations bring to light some small pieces of much-perished timber, about 4 in. in thickness, that may have served for such a structure[6]. Considering how comparatively expensive an article building timber is to this day in the immediate vicinity of a large Turkestān town, we can scarcely be surprised at the early removal of this, the most useful material the shrine could offer after its abandonment. It seems impossible to assume that the sculptures along the walls and the fresco decoration of the latter should have been originally exposed to all the influences of the weather. A single heavy downpour of rain, such as is by no means unknown even in this dry region—I experienced one on April 19 soon after I had left Rawak camp—would have sufficed to do great damage to the friable clay of the statuary and the equally soft wall plaster. The existence of some enclosing passage or gallery is indicated also by the foundations of a thin plaster wall, with stucco reliefs facing it, which were brought to light outside the south corner and which will be described at the end of the next section.

Section IV.—THE SCULPTURES OF THE RAWAK VIHĀRA

Number of sculptures excavated.

The excavations which incessant work from early morning till nightfall enabled me to get effected between the 12th and 17th of April, cleared the greatest part of the south-east and a portion of the south-west walls of the Vihāra court, both inside and outside, for an aggregate length of about 155 ft. Though this distance represents only about one-fourth of the perimeter of the Vihāra court, yet the total number of individual reliefs, most of them over life-size, which were brought to light on the cleared wall faces amounted in the end to ninety-one. In addition to these the finds included numerous small reliefs forming parts of sculptured aureoles, &c., or deposited as ex-votos before the main images.

Record of sculptures cleared.

The position of all statues was carefully shown by me in the ground-plan, together with the greatest width of the wall space occupied by them, and a description recorded of every piece of sculpture, as detailed as the available time and the trying conditions prevailing would permit. In addition I used the rare hours, usually in the early morning, when there was less wind and no thick haze or driving dust, for obtaining a complete series of photographs of whatever sculptural work appeared on the excavated wall faces. It was no easy task to collect all these records with the needful accuracy while directing the successive stages of excavation

[5] Such damage would, no doubt, have been more extensive if the stucco reliefs to be obtained at Rawak had not been too heavy or too friable for ready removal to, and disposal in, the antique market of Khotan.

[6] Pl. XV. c shows them as subsequently placed on the top of the enclosing wall.

in atmospheric conditions trying alike to eyes, throat, and lungs. Though Surveyor Rām Singh and Turdi rendered, each in his own way, very intelligent assistance, I had myself to remain in the trenches practically from sunrise until nightfall, feeling all the time permeated, as it were, by the clouds of fine sand which the wind and the digging raised around us.

Survey of sculptured remains.

The photographs taken by me have proved more successful than I could reasonably hope for under such adverse influences. The reproduction from among them of a series showing practically the whole of the cleared wall-faces has appeared to me the best means of illustrating the wealth of interesting materials which these excavations yielded for the study of the ancient sculptural art of Khotan[1]. To the views of the reliefs thus presented I propose here to add such details of archaeological or artistic interest as were recorded by me on the spot, together with any comments which the study of these reliefs as a whole and their comparison with the sculptural remains brought to light at other Khotan sites has suggested to me. We may conveniently commence this survey from the inner south-east wall where my excavations started, and further on follow the numerical order in which the position of the individual sculptures has been shown in the plan of Plate XL.

Colossal statue R. i. on inner south-west wall.

The first statue which the high sand rising towards the west corner permitted to be cleared was that of the colossal standing Buddha, R. i. (see Figs. 61, 69). It was intact up to the shoulders, except for the missing right hand; but as the weight of the extant upper portion caused imminent risk of collapse I did not think it safe to remove entirely the supporting sand layers to the left proper, and thus the left hand does not appear in the photographs. There can be no doubt that it was shown hanging down and resting on the edge of the drapery as seen in the statues, R. v, ix, x (Figs. 61, 62), which are exact replicas of the same image. The right forearm supporting rich folds of drapery is shown in a pose that unmistakably indicates the 'Abhayapāṇimudrā,' with which the Dandān-Uiliq and Kighillik reliefs have already rendered us familiar. The same reappears at Rawak in the numerous small Buddha figures filling the large aureoles of R. xii and R. xiii (see Figs. 63, 64). It is likely to have been the pose also of the numerous over life-size figures of the inner south-east face, all apparently replicas of one type, though only a few of them (R. xxxvi, xxxviii–xli), have retained enough of the right forearm to show its characteristic position. Among the colossal statues of the outer south-east face, most of those preserving the right

[1] The limitations of the photographic instruments at my disposal, and still more the varying levels and distances from which the photographs had to be taken (as determined by the space available within or near the trenches), rendered it quite impossible to attempt anything so ambitious as the production of photographs that would join into a continuous and exactly fitting band or quasi-panoramic view. Hence the angle of view and scale of reduction will be found to vary considerably in the series of photographs reproduced by the plates, while the views presented usually overlap on the sides.

The series comprises: (*a*) four photographs (Figs. 61–64), taken with a half-plate camera showing the sculptures of the inner south-west wall (R. i–v, viii–xii in plan) up to the south corner, also the large relief with vesica (R. xiii) nearest to the latter on the south-east wall; (*b*) nineteen photographs (Pl. XIV. a–d, XV. a–d, XVI. a–d, XVII. a–d, XVIII. a–c, with Pl. XIII. b) taken in continuous succession with the Bridges-Lee photo-theodolite camera and showing the reliefs R. xix–xli on the inner south-east face, R. xlii–lxxi on the outer south-east face, and R. lxxii–lxxviii on the outer south-west face.

The camera serving for the photographs enumerated under (*b*) had the advantage of a superior lens, but on the other hand was not provided with means for raising or lowering its front, being designed only for topographical survey work. As the photographs had to be taken often from the top of the sand cleared from the trenches, this want of a rising front necessitated tilting the camera tripod with consequent foreshortening, appreciable in some instances. I was obliged to use this instrument, since several of the dark slides of my half-plate camera had, probably under the severe climatic changes, developed cracks.

In addition to the above series the photographs in Figs. 65–69 have been reproduced, in order better to illustrate reliefs of special interest or for the sake of giving general views of the sculptural adornment of certain sections of the wall-faces.

arm or indications of it (R. lv, lvii, lix, lxi, lxiii, lxvi; see Pl. XVII. a, b, c) seem likewise to suggest this attitude. Of the size in which the statue R. i was modelled, the measurement of 5 ft. 3 in. from the level of the feet to the bent elbow, which also marks approximately the waist, may afford an indication[1]. The feet rested on a very flat semicircular base of plaster about 3 ft. wide, which on its edges showed traces of having been decorated with lotus-petals in low relief. The robe, judging from surviving traces of colour, appears to have been painted white originally; but almost the whole of the paint had peeled off and thus the surface showed the uniform red colour of the clay.

Drapery after Graeco-Buddhist models.

The most remarkable feature in R. i and its replicas, which we note with minor modifications also in the rest of the Rawak statuary, is the treatment of the drapery. This is almost as Grecian as in the standing Buddha figures of Gandhāra sculpture, and betrays its derivation from the latter in every detail. In order to realize this remarkable agreement, it is sufficient to compare our statues, e. g., with the relief representations of Gautama Buddha from Takht-i-Bahī quoted below, or indeed with any of the finer Gandhāra figures showing Buddha standing in the Abhaya posture[2]. The robe, which covered both shoulders, is laid round the body so as clearly to show its contours. The folds, which are marked with boldly projecting edges, are gracefully disposed and hang in a natural way from the limbs that catch their lines. The shape of the body beneath the robe is in these statues of the inner south-west wall, as well as in the colossal images of the inner south-east face, displayed in a more pronounced fashion than in the Gandhāra sculptures referred to. But the identical arrangement of the drapery is in no way affected thereby. A comparison of the folds gathered over the outstretched left arm (see R. v, ix), or falling from the bent right forearm, both in our statues and in the Gandhāra examples will illustrate this. The 'wonderful tenacity' with which 'the ancient Chinese and Japanese Buddha-figures have preserved the "draping" of the Gandhāra figures in a peculiar way', as duly emphasized by Dr. Grünwedel[4], can no longer surprise us when we see how faithfully old Khotan art in this as in many other respects reproduced its Gandhāra models.

Small statue, R. ii.

Next to R. i. on the right proper we have a small statue, R. ii (see Figs. 61, 69), which was found intact and measured 4 ft. 2 in. to the top of the head. The latter soon became insecure through the force of the winds, and had to be taken off. Though its interior had become hollow owing to the rotting of the wood frame, this head has survived its subsequent transport to London very well, as seen from its reproduction in Plate LXXXII. It measures 10 in. from the chin to the small top-knob, and retains, besides plenty of the original whitewash, traces of black for the eyelashes and of a red *ṭikā*. Much of the whitewash also adhered to the drapery. The latter, by its curious arrangement in several vertical bands of conventional wave lines, strikingly contrasts with the Grecian drapery of the neighbouring colossal statues. The same treatment appears in the lower garment of the Bodhisattva figures R. iv and in R. viii. It strongly recalls the wave lines of the drapery folds seen in a typical Chinese representation of Udayana's Buddha statue referred to in the preceding note, and the explanatory

Wave lines of drapery.

[1] Here, as in all other photographs of Rawak sculptures, the 3 ft. measure placed against the wall indicates the scale.

[2] See Grünwedel-Burgess, *Buddhist art*, Figs. 118, 122; Foucher, *L'Art du Gandhâra*, Figs. 258, 261, 262, 264, &c.

[4] See *Buddhist art*, p. 170. It is interesting to note that a Chinese wooden figure of a standing Buddha, which the above scholar has reproduced for its characteristic drapery, is recognized as 'a replica of a copy which has been preserved in China and is traced back, according to the Chinese tradition, to Udayana's sandal-wood figure of the master'; see ibid., with Fig. 125. The posture of this figure is the same as that of most of our colossal statues (see especially R. xxxvi, xl, xli, where the left hand was slightly more detached from the drapery). Is it possible to suppose that these, too, were intended to reproduce the type of Udayana's image of which, as we have seen (above, p. 455), a miracle-working representative was worshipped at P'i-mo?

61

v iv iii ii i

RELIEF SCULPTURES, R. i—R. v, ON INNER SOUTH-WEST WALL, RAWAK STŪPA COURT.

62

xii xi x ix viii

RELIEF SCULPTURES, R. viii—R. xii, ON INNER SOUTH-WEST WALL, RAWAK STŪPA COURT.

legend 'which a Tibetan historian quotes regarding the Buddha figure of Udayana. He relates how Buddha—in order to lighten the work of the artists, who were blinded by his glory—was mirrored in the water. The artists reproduced this reflection, and thus the waving lines of the robe are accounted for[5]'. By the side of the strange transformation which the classical draping of the Gandhāra models has here undergone, the arrangement, still antique on the whole, of the edges of the drapery falling from the left arm deserves the more attention. Curiously enough, the same arrangement is equally reflected in the Chinese figure of Udayana's Buddha[6].

The robe in R. ii is laid round the neck, as in most of the Rawak Buddha statues; and as no ornaments are displayed, we may assume that the figure was intended to represent a Buddha. The right arm hanging down suggests either the 'Varamudrā' or the 'Bhūmisparśamudrā' attitude, both differing only by the pose of the hand, with the palm turned outwards or inwards[7]. Here the hands are missing. The hair is arranged in rich curls, distinctly after the Gandhāra fashion. The elliptical nimbus formed by a relief plaque behind the head of R. ii was found intact. As it agrees in all essentials with that seen behind the head of R. iv, it may be considered as a typical example of what the completely perished haloes of the larger statues are likely to have been. The elliptical curve on either side was bordered with a cloud scroll. On the top the two scrolls met below a small *appliqué* figure of a Buddha seated within a vesica of lotus-petals (R. ii. 3. c, shown in Plate LXXXVII). Above and by the sides of this *appliqué* figure there rose bundles of flames. Below the top portion of the scroll was a single fleur-de-lis ornament. The field of the nimbus showed on either side two more small seated Buddha figures of a closely similar type, and between them a double fleur-de-lis ornament representing a Vajra, as seen in the portion of the nimbus reproduced on Plate LXXXIII. Small lotuses, as seen in the same piece, appeared above and below this ornament near the scroll of the border. As the whole of the plaque threatened to fall off the wall, I attempted to move it, but owing to the very friable condition of the clay it broke, and only portions of it (R. ii. 2) survived further transport.

Pose and nimbus of R. ii.

R. iii was a colossal statue, which collapsed in the course of excavation. It seemed to have been in all particulars a replica of R. i. On its right stood the relatively well-preserved figure of a Bodhisattva (R. iv), seen in Fig. 69 with its head held in position. This head (R. iv. 1) was found broken at the back and had to be removed, but has in spite of this injury withstood transport to London without further damage. Plate LXXXI shows a full-size reproduction of it. Traces of colouring remain about the eyes and on the lips. The height of the figure was a little over 6 ft., and the modelling of the head seems to show that it was intended to be looked at from below. The nose, mouth, and chin are small, the ears elongated but well-modelled. A triple row of Maṇi strings, as often seen in Gandhāra figures of Prince Siddhārtha, adorns the breast, which appears dressed in a close-fitting garment. Below the very slender waist a broad girdle formed by elliptical jewels supports a skirt-like lower garment, showing elaborate drapery arranged in conventional wave lines, like that described in the case of R. ii. Both arms were found broken below the elbows. On the upper portions appear highly ornamented armlets enriched with an octagonal jewel in the centre and rectangular ones on either side. The nimbus visible in the photographs closely resembles that of R. ii, but shows two Vajras on either side. The band ornaments visible on the lower part probably formed the ends of the jewelled band which, as seen by a fragment above the right ear, encircled the head of the statue.

Bodhisattva statue, R. iv.

[5] See Grünwedel-Burgess, *Buddhist art*, p. 172, with Fig. 125; also above note 4.

[6] See *Buddhist art*, pp. 171 sq.

[7] Foucher, *Iconographie bouddhique*, p. 69.

Statues R. v–viii.

The adjoining colossal statue (R. v) was an exact replica of R. i, but preserved only to below the waist; it collapsed after having been exposed for a day. The next figure (R. vi) was in respect of size, pose, and drapery a replica of the Buddha image R. ii. It was broken by the falling débris of its neighbour. The head, however, had been removed previously and has thus escaped destruction. As seen in Plate LXXXIII, it bears a different expression from that of R. ii; the eyes are almost shut. The hair here, too, is richly curled; traces of colouring about the eyelashes and of a pink *ṭīkā* remain. R. vii, a colossal image and evidently a replica of R. i, was broken before excavation and could not be saved from complete collapse. This damage fortunately did not affect its neighbour R. viii (see Fig. 62), a richly-robed figure, about 5 ft. 6 in. high originally, which was found without head or forearms. The robe leaves the right shoulder bare, and descends in ample folds which are arranged in conventional wave lines after the fashion described in R. ii. There were remains of a nimbus similar to that of R. ii and R. iv; the whole figure retained a good deal of whitewash.

Frescoes between statues R. ix, x.

The statues R. ix and R. x, which appear in the centre of Fig. 62, were replicas of R. i and preserved to above the waist. The wall space between them, about 1½ ft. broad, retained its original coating of plaster with remains of faded fresco decoration. A red-robed figure with a circular nimbus, which can be made out in the reproduction of Fig. 62, about 2½ ft. high, appears in the centre; on its right a standing attendant and on its left a vertical row of four red lotuses could be faintly discerned. Above it there remained the badly effaced outlines of a small seated Buddha within an elliptical vesica, with traces of some still smaller haloed figures on its sides. The wall immediately to the right proper of R. x originally bore small frescoes arranged exactly like those on the left; but besides having badly faded, they had suffered from scratches and cuts apparently before the sand came to cover them up. On the extreme right proper, near the border of the large aureole of R. xii, the original decoration had been painted over, as one of the lotuses is now partly covered by a small fresco representing a seated Buddha, about 6 in. high and faintly visible in Fig. 63 above the left shoulder of the seated Buddha. The fact that the eyes of the painted figure are looking down towards the head of the seated image suggests that the latter itself may also have been a later addition.

Seated Buddha figure R. xi.

The seated Buddha (R. xi) is in several respects of special interest. The statue, which measures 3 ft. across the folded knees and 2 ft. 10 in. from its seat to the top of the head, is well-modelled, and bears a look of individual workmanship for which we search elsewhere in vain among this great array of replicas[8]. This is particularly noticeable in the face, the finely modelled features of which express a smile of resigned contentment. This agrees with the pose of the arms which, though found broken in the foreparts, manifestly rested folded in 'Dhyānamudrā', the posture of 'contemplation'. The head bears a flat crown of hair with a low top-knob. There appears to be no indication of a robe or garment of any kind, which suggests whether the artist's intention may not have been to represent Gautama as practising austerities, while still a Bodhisattva. But the absence of any mark of emaciation, such as is indicated by well-known Gandhāra representations of this stage in the future Buddha's career, does not favour this assumption[9].

Detached reliefs used for adornment.

The whole figure appears to have been whitewashed. It had no nimbus or aureole; but the broken portion of a nimbus plaque (R. xi. 1), similar in type to that of R. ii, yet differing in its cloud scroll (see Plate LXXXVII), was found placed over the left shoulder, no doubt

[8] For a better reproduction of this image see the photogravure serving as frontispiece in my *Ruins of Khotan*.

[9] See Foucher, *L'Art du Gandhâra*, i. p. 379.

xii xi × 63

STATUE OF SEATED BUDDHA, R. xi, BETWEEN COLOSSAL RELIEFS, IN INNER SOUTH CORNER, RAWAK STŪPA COURT.

xiii 64

COLOSSAL STATUE, R. xiii, WITH RELIEF HALO, IN INNER SOUTH CORNER, RAWAK STŪPA COURT.

by the hand of one of the pious who last visited the Vihāra. In the same way a small broken Buddha figure, a replica of the reliefs filling the large aureoles of R. xii and R. xiii, was found placed against the right arm. Sculptured pieces which had fallen off from the haloes of some other images had thus served for pious adornment, an observation made elsewhere in the ruined Vihāra, and clearly showing that the decay of its relief decorations in friable clay must have begun long before the shrine was completely abandoned. More important was the finding of three Chinese coins on the floor just below the base of R. xi. They were all *wu-chu* pieces; and their position, when viewed in the light of subsequent discoveries of the same kind, leaves no doubt that they had been deposited as votive offerings.

Votive coin deposits.

The adjoining south corner of the Vihāra court was occupied by two colossal statues (R. xii and R. xiii), surrounded by very large ornamental vesicas and both exactly alike. The statues themselves had suffered much damage, and the portions from the waist downwards which the excavation brought to light collapsed to about the knees, as seen in Figs. 63, 64. The carefully pleated folds of the robes showed remains of a dark-red paint. The elliptical vesicas, measuring fully 7½ ft. across at their widest and enclosed by a relief border showing a lozenge diaper[10], were decorated with slanting rows, overlapping each other, of small Buddha figures. These were alike in pose, scale, and execution, but, as seen in the photographs, shown to different lengths, according to the exigencies of space. Those displayed down to the thighs measured about 14 in., the rest being shorter. The arrangement of individual figures in the rows on either side of the vesica does not seem to have been everywhere quite symmetrical. The figures actually found adhering to the wall were all exact replicas and evidently from the same mould. The same was the case with most of the small reliefs of this kind, of which many were found in the sand filling the corner, and which manifestly had broken off previously from the higher vesica segments[11].

Reliefs of large vesicas, R. xii, xiii.

The specimen (R. xii. 1) reproduced in Plate LXXXVI, shows the figure of a Buddha standing with the right hand raised in the 'Abhayapāṇimudrā' attitude, as seen from other specimens, and the left on the chest, grasping loose drapery. A round nimbus appears behind the head, the latter showing the hair in simple curls with a top-knob. Traces of whitewash over a red wash appear on all these figures. Besides the latter, there turned up among the débris of the south corner some small stucco heads evidently also representing Buddhas, but differing in expression and modelling from R. xii. 1 and its replicas. Two types could be distinguished, and one of these was represented also among the small detached reliefs which I found placed near R. lxx[12]. It is difficult now to decide whether the stucco figures to which these heads belonged had formed part of no longer extant portions of the vesicas, or had only been placed there after removal from some other group of sculptures. The corner would have formed a likely place for collecting such *disjecta membra* of sacred images from other parts of the Vihāra, and it is, perhaps, significant that a number of such detached pieces were found near the statues of the outside south corner (R. lxx–lxxiv).

Small Buddhas of vesicas.

The sculptural decoration of the inner south-east wall-face, to the south of the gate, showed great uniformity. It consisted mainly of a series of colossal statues (R. xiv, xv, xvi, xvii, xviii, xix, xx, xxi, xxii, xxiii, xxv, xxvi), which appear to have been of identical type, if not exact replicas. They had all suffered much decay, evidently owing to the inadequacy of the

Colossal statues of inner south-east face.

[10] For the description of a specimen piece, see R. 06 in list.

[11] See for such replica fragments, mostly damaged, R. 1, 04, 05, 07, 011, 013, 017, 019.

[12] One type is represented by R. 01, 02, 014, 015, 016 (for a reproduction see *Ruins of Khotan*, p. 464); another by R. 2, 03, which are replicas of R. lxx. 1. b, c (see Pl. LXXXIV).

sand cover which along this wall was only from 5 to 7 ft. in height, and several of the torsos collapsed after excavation. As seen in Plate XIV. a, b, c, the extant portions of these figures closely resemble the colossal Buddha images which were found along the same wall-face north of the gate and are somewhat better preserved (R. xxix, xxx, xxxi, xxxiii, xxxiv, xxxvi, xxxviii, xl, xli; see Plates XIV. d, XV. a, b, c). Judging from the latter, we may safely assume that the posture of all of them was the 'Abhayamudrā'. From the colossal Buddhas on the south-west wall this series seems to have differed only in minor points, such as the bold outward sweep given to the drapery folds below, on the left proper (see e. g. R. xx, xxiii, xxiv, &c.), and the pose of the left hand, which, as shown in R. xxxvi, xl, and xli (see Plate XV. b, c), was probably a little more detached from the drapery of the robe. The height of the statues up to the navel appears to have been about 5 ft. A number of smaller reliefs found by their side probably represent later additions, perhaps donations of individual worshippers. R. xxii is a small male figure, found headless, surmounted by a plaque forming a nimbus after the fashion of that behind R. ii. R. xxiv. a (see Plate XIV. b) is a richly draped figure, also headless, measuring about 3 ft. to the neck, and probably representing a Bodhisattva. The folds of the robe below are arranged in wave lines, while remains of ornaments appear hanging from the neck and around the arms.

Relief plaques as pious deposits.

As seen in Plate XIV. b, small broken Buddha figures, evidently replicas of those in the vesicas of R. xii, xiii, were found placed between or near the feet of several of the colossal images, as noted above in connexion with similar deposits near R. xi. In addition to these, three small plaques representing seated Buddhas were found near R. xxv. Two of them were replicas of the corresponding *appliqué* reliefs in the nimbus of R. ii and R. iv. The third (R. xxv. 3), measuring about 7 by 5 in., is a replica of R. 2 (see Plate LXXXII), a small but well-executed relief, found in the débris of the south corner, showing Buddha seated in the 'Dhyānamudrā' attitude with a small tree, evidently meant for a *Ficus indica*, behind the right shoulder. These two pieces are of particular interest as showing how closely the treatment of the drapery also in seated Buddha figures agrees with that of the corresponding Gandhāra models. The good proportions of the features recall the best type of Gandhāra work, as is easily realized on comparing this small plaque with Gandhāra reliefs representing the same subject[13].

Figures of *dvārapālas* at gate.

The gate was flanked on either side on the inner wall-face by a pair of standing figures, worked almost in the round and about life-size, R. xxvii, xxviii (Plate XIV. c, d). These deserve special attention as the only images on the walls of the Vihāra court which do not present themselves in the orthodox garb of Buddhas or Bodhisattvas. Though the upper portions of their bodies had broken away and were recovered only in fragments, there can be no doubt as to their being intended to represent *Dvārapālas* or 'Guardians of the Gate'. It seems very probable that the four guardians at the Rawak Vihāra gate, like those sculptured at the entrance of many a Buddhist shrine in India, were meant for Yakṣas, and in particular for the four Yakṣa chiefs (Kubera, Virūḍhaka, Virūpākṣa, and Dhṛtarāṣṭra) who as *Lokapālas* or 'World protectors' play so conspicuous a part in the Buddhist mythology of India, as well as of Tibet and the Far East[14]. Buddhist iconography invariably gave to them, as to Yakṣas generally, a wholly human and quasi-secular form, as we noted when identifying with Kubera the armed figure represented by the side of the main image in the small temple cella of D. II at Dandān-Uiliq[15]. It fully accords herewith that the figures at the Rawak gate exhibit

[13] See, e.g., Foucher, *L'Art du Gandhâra*, p. 424, with Fig. 213.

[14] Comp. Grünwedel-Burgess, *Buddhist art*, pp. 45 sq., 136 sq.

[15] See above, pp. 252 sq.

what was manifestly the customary dress of the period and the country, and this gives to them their special archaeological interest.

Costume of Dvārapālas, R. xxvii.

The figure to the left proper in the group south of the gate (R. xxvii; Plate XIV. c) is well preserved from the waist downwards, and is dressed in a double garment descending to the top of high and pointed boots. The latter still retain in parts their dark-red colouring. Of the lower coat only the edges laid in narrow frills are visible. The upper garment is decorated on the edges with a raised border showing wave line ornament, while in the middle between the legs descends a line of broad triangular pleats clearly visible in the reproduction. Hanging from the middle of the waist is shown a knife in a narrow scabbard. The figure nearest to the gate and turned slightly towards the entrance is similarly accoutred in a double garment with broad borders reaching to the top of the red-coloured boots. None of the garments retain any colouring. The folds of the coats are indicated by slightly hollowed lines. In front of this second image were found broken remains of a recumbent figure consisting of a lower portion partly embedded in the floor, a much-decayed torso, and a head. Both are seen in Plate XIV. c as held up by one of the labourers. The head was badly defaced; but features strikingly different from those of the sacred heads, such as broad protruding lips and a flat nose, could still be distinguished on it. The question suggests itself whether this figure, which seems to have been reclining against the feet of the second Dvārapāla, may not represent a demon, Kubera's typical cognizance [16].

Dvārapālas, R. xxviii.

The Dvārapālas to the north of the gate (see Fig. 67, besides Plate XIV. d) differ in dress from the others mainly by showing bulging trousers tucked into boots, which, like the 'Chāruks' worn nowadays, are wide at the top, with an ornamental border on their brim. The boots were originally coloured dark-red. The trousers are for the greater part hidden by two large coats hanging down from the waist, one above the other. The bands of embroidery marked in relief along the hem of the coats show elaborate patterns with small circlets and crochets and are still distinguishable in the original photographs. Below the edges of the upper garment of the figure near the entrance there remained traces of light-brown colouring.

Female bust and head, R. xxvii. 1.

Between the feet of each of the two figures further away from the entrance were found small female busts, visible in Plate XIV. c, d and Fig. 67, and evidently identical. The one (R. xxvii. 1) which could be removed without difficulty broke in transit to London, but its numerous fragments were successfully reunited by Mr. A. P. Ready of the British Museum. Plate LXXXV shows front and side views of the head and bust, both remarkable for graceful outlines and good modelling. The careful and easy treatment of the hair displayed by the side view deserves special notice, as well as the delicate proportions of the breasts. The upward tilt of the head seems an indication that these small sculptures were intended to occupy the position in which they were found, and were not mere deposits from some other part of the shrine. In appearance they curiously recall the female figure which some well-known Gandhāra reliefs, representing the scene of Gautama's final departure from his palace, show rising from the ground between the feet of his horse Kaṇṭhaka [17]. Whether this figure is to be interpreted as the goddess of the earth, according to Prof. Grünwedel's ingenious conjecture, or otherwise to be accounted for, it is clear that our small sculptures can have only a very distant connexion with it. Perhaps they were meant for Yakṣiṇīs, inserted for the purpose of showing that the guardians whose feet they seem to support are Yakṣa kings.

[16] See above, p. 253.

[17] See Grünwedel-Burgess, *Buddhist art*, pp. 100 sqq.; Foucher, *L'Art du Gandhâra*, i. pp. 337 sqq.

Leaf-gold patches on knees of statue R. xxix.

Among the colossal statues which occupied the wall to the north of the gate and which have already been referred to, the one nearest to the group of Dvārapālas (R. xxix), showed on its knees remains of leaf-gold stuck on originally in several small patches. The largest of these, judging from the darkened colour of the plaster surface, seems to have measured about an inch square. I could not have wished for a better illustration of the quaint custom which Hsüan-tsang has recorded of the miracle-working Buddha figure of sandal-wood he saw at P'i-mo. 'Those who have any disease, according to the part affected, cover the corresponding place on the statue with gold-leaf, and forthwith they are healed[18]'. May we conclude from the number of gold-leaf plasters of which the marks remain on this Rawak image that it had enjoyed particular fame for healing power in affections of the knee?

Near the feet of R. xxx was found a small headless figure in flat relief (see Plate XXXVI), representing probably a Buddha which had originally belonged to the decoration of some large vesica like that of R. xii, xiii, and had been transferred here. The pose was similar to, but not quite identical with, that of R. xii. 1 and its numerous replicas. Between the colossal statues the remaining cleared portion of the inner east wall-face also showed four smaller images (R. xxxii, xxxv, xxxvii, xxxix), all of them headless. The last and R. xxxv are replicas of a Bodhisattva, richly adorned with maṇi-strings over the half-uncovered breast. R. xxxvii (Plate XV. b) is also a Bodhisattva, bearing a large jewelled ornament on the breast and a richly decorated girdle knotted with large tassels in the centre. R. xxxv showed but scanty remains of drapery, and had a large piece of stucco of uncertain origin inserted in the place of the missing feet—evidently a rough attempt made by some of the last attendants of the shrine to preserve the mutilated figure from falling.

Reliefs R. xlii–xlvii, on outer south-east wall.

The sculptures facing on the outside the portion of the south-east wall last described (R. xlii–xlvii) had, as Plates XV. d, XVI. a, b show, suffered far more damage than those lining its inner face. Of the two colossal statues (R. xlii, xliii) enough survived to show elaborate and well-arranged drapery. Its folds retained a good deal of whitewash, and behind R. xlii the outlines of part of a painted aureole could still be discerned. The image adjoining (R. xliv) was found broken from above the knees, and the small standing figure by its side was also badly damaged. The large statue (R. xlv) could not be entirely cleared from fear of collapse; but its proportions and the manner in which the drapery is indicated on the part of the body exposed render it probable that it resembled the type represented by the majority of the colossal statues on the other portion of the outer south-east face. This is quite certain of the figure R. xlvii, which on its extant parts displays the pose of the left hand and the stiff vertical lines meant to indicate drapery folds falling over the lower limbs, which are characteristic of that group. R. xliv had collapsed before excavation, and the remainder of the wall space as far as the gate had not retained any of its sculptural decoration. The much decayed surfaces of the masonry within the gate and near it suggested that they may have originally been revetted with wood, which had caused them to be left without proper finish and plaster-covering. Between two small bricks on the right side of the entrance and about $1\frac{1}{2}$ ft. above the ground was found a *wu-chu* coin (see Plate LXXXIX. 17) showing very little wear, evidently a votive deposit.

Statues R. xlviii–liii, south of gate, on outer wall-face.

The reliefs flanking the gate to the left, i.e. towards the outer south corner of the quadrangle, were found completely broken, except for a small figure (R. xlviii), probably a Buddha, standing, with right breast uncovered. Its head, which had to be removed as it threatened to fall (see R. xlviii. 1 in list), is a replica of R. lxxii. R. xlix (see Plate XVI. c, also Fig. 65), broken above the waist, is a colossal draped statue differing by its drapery, gathered in free folds,

[18] See above, p. 455; *Mémoires*, ii. p. 243; Beal, ii. p. 322.

65

lii li l xlix

RELIEF SCULPTURES, R. xlix—lii, ON OUTER SOUTH-EAST WALL, RAWAK STŪPA COURT.

66

lxxiv lxxiii lxxi lxx lxix lxviii lxvii lxvi

COLOSSAL STATUES, R. lxvi—lxxiv, ON OUTER WALLS OF SOUTH CORNER, RAWAK STŪPA COURT, WITH REMAINS OF OUTER PASSAGE IN FOREGROUND.

from the rest of the images on this wall-face. By its side stands the small figure of Bodhisattva, found headless, richly adorned with maṇi-strings across the breast. The horseshoe-shaped nimbus behind the head was a small-scale replica of that of R. ii. The next statue (R. li) representing a colossal Bodhisattva, was broken above the waist, but the greater portion of its head was found lying near its feet, as seen in Fig. 65. This appears to have measured 13 in. from the chin to the knob of the elaborately dressed hair. The right hand hanging down by the side was well-preserved and showed good modelling. The ample robe spreads out skirt-like at the bottom in folds marked by shallow lines, the whole style of the drapery differing in a striking fashion from the quasi-classical arrangement shown by the Buddha statues on the inside walls. Spots of a dark paint remained over the red of the plaster surface. The adjoining figure of a Bodhisattva, R. lii (see Plate XVI. d, also Fig. 65), originally about 5 ft. high, was found without the feet, and had in consequence slipped from its original position, but otherwise was fairly well-preserved. Over the breast it shows a double string of richly-jewelled maṇis and star-shaped ornaments on the arms. Below the narrow waist descends a robe in rich folds marked by wave lines just as in R. iv, which this statue generally resembles. The hair, dressed flat, falls in locks behind the ears. Above the head are the remains of a round or oval nimbus, showing lotus-flowers within a border which seems to consist of a wreath outside a cloud scroll. Spots of some dark paint appear over the red surface of the plaster on breast and arms.

Group of colossal statues, R. liv–lxiii.

Of R. liii only the drapery below the knees remained, showing some resemblance to that of R. xlix. The next image (R. liv) was also broken from above the knees, but enough of the drapery is seen in the photograph (Plate XVI. d) to indicate that it was a replica of the neighbouring colossal statues R. lv, lvii, lix, lxi, lxiii (see Plate XVII. a, b, c, also Fig. 68), which form a large uniform group, and evidently represent Buddhas standing with the right arm raised in the 'Abhayamudrā' attitude. The modelling of these statues, with their narrow waists, projecting hips, badly posed left hands, &c., is decidedly inferior to that of the colossal Buddhas on the inner south-west and south-east wall-faces. The conventional drapery folds hanging like strings across the upper part of the body and descending in stiff parallel lines over the legs show equal degeneracy from the Gandhāra models. The whole effect suggests a distinct step towards the drapery treatment displayed by the Chinese type of Udayana's Buddha statue above discussed[19]. Yet here, too, we see the edges of the drapery folds falling from the left arm still arranged in the antique fashion. In spite of their close agreement in style and execution, the individual statues of this group show certain differences. Thus R. lxi and probably R. lix had the right breast uncovered, while in R. lvii the pose of the left hand is slightly varied. These and other variations may have resulted from restorations effected at different times. There may have been also variations in the colour of the robes, a dark-red paint being partly traceable on the drapery of R. lv, lvii, while elsewhere whitewash seems to have prevailed.

Group of small reliefs, R. lvi–lxii.

The small figures (R. lvi, lviii, lx, lxii) which fill the wall-spaces left between the statues just described[20], show by their uniformity that they were designed for a group. They are raised on lotus-pedestals about one foot from the ground and measure 2 ft. 3 in. in height. They all appear to represent Buddhas. The heads are replicas, but variations are introduced

[19] See above, pp. 490 sq.; comp. Fig. 125 in Grünwedel-Burgess, *Buddhist art*.

[20] Owing to a slight error of the draughtsman, which, I regret, escaped my attention until I came to study the photographs of these sculptures closely, the plan in Pl. XL does not mark the small figure, R. lxvi, in its proper place. The numbers lvi–lxiii (both inclusive) must be read as referring to the sculptures shown nearest on their right, and the mark of a small figure between R. lxiii and lxiv ought to have been omitted. Similarly, the position of the marks for a large and a small figure, R. lxv and R. lxvi, ought to be reversed.

into the dress and pose. R. lx and lxii have the right breast bare, while in R. lvi and lviii the robe is laid round both shoulders. The latter figure is of interest on account of the pose of the right arm which, instead of hanging down as in the *pendants*, is bent towards the breast within the fold of the robe, in a fashion which looks antique. This well-preserved little image also retains a dark-red paint on the robe, with plentiful dark spots distributed over it and the halo, apparently in a kind of rough pattern. The head, hands, neck, and feet are painted white.

Statues R. lxiv–lxvii.

At the foot of R. lxiv, a badly decayed large figure which may have belonged to the preceding group, a well-preserved *wu-chu* coin was found. Next followed the much injured torso of a small image (R. lxv), which, judged by the wave lines of the drapery, may have represented a Bodhisattva. The two adjoining colossal statues, R. lxvi, lxvii (see Plate XVII. c, d; Fig. 66), were preserved up to their necks, about 8 ft. from the ground, but could not be completely excavated, as their heavy masses of stucco threatened to collapse. Both had the right breast uncovered, and the right arm raised from the elbow. The robes retained much of the original paint, white on R. lxvi and light red on R. lxvii, which, together with the relative height of the preserved portions, suggested that these statues may possibly have been better sheltered than the rest on this wall until the drift-sand had sufficiently accumulated to protect the coloured stucco surfaces from atmospheric influences.

Statues of outer south-east wall near south corner.

R. lxviii and R. lxix (see Plate XVII. d) are colossal figures preserved only below the waist. Their drapery and the pose of the left hand show that they must have been modelled after the type represented by the previous group. Between the feet of these images, and leaning against their sides, were found small relief representations of Buddhas, measuring about 1 ft. from the waist, where they were all broken. They had evidently been removed from some other position, and closely resembled three other small figures of this size and type which were grouped near the feet of the adjoining large statue (R. lxx). The latter, though badly injured above the knees, showed remains of an elaborately-draped robe in a rich red colour. The heads of the small Buddha figures at its feet (R. lxx. 1 a, b, c) broke after excavation, but were removed without damage. They are well proportioned, and, as seen in Plate LXXXIV, retain distinct traces of the original colouring. The south corner of the outer wall-face was occupied by the colossal statue R. lxxi (see Plate XVIII. a; Fig. 66), of which, besides the feet and drapery portions from the knees downwards, only a hand was found.

Sculptures of outer south-west face near corner.

On the south-west face outside there adjoined a small Buddha figure (R. lxxii), about 2 ft. 6 in. high (see Plate XVIII. b). Owing probably to the protection afforded by the colossal corner statue, which must have partly hidden it, this little image still showed the red colour of its robe, crossed by white stripes, in good preservation. Its head, which had suffered damage before excavation, is reproduced in Plate LXXXVIII. The nimbus behind resembled that of R. ii, and some of its small plaques, representing seated Buddhas, will be found described in the list. An aureole formed by a cloud-scroll with fleur-de-lis ornaments descended from the nimbus. R. lxxiii, a colossal statue, broken from the thigh, deserves notice on account of its drapery being marked only by shallow outlines, quite different from the deep folds indicated elsewhere. R. lxxiv is a small Buddha figure of about the same size as R. lxxii, to which it seems to have formed a *pendant*. Its head, which broke off on clearing, is, however, of a different type, as seen in Plate LXXXIV. It is well proportioned; the hair is represented in a curious fashion by a series of closely-plaited bands passing from ear to ear. Besides a horseshoe nimbus behind the head the wall showed, by the side of R. lxxxiv, a small plaque representing a seated Buddha like that found near R. xxv.

RELIEF SCULPTURES, R. xxviii, xxix, WITH DVĀRAPĀLAS ON RIGHT, AT GATE OF RAWAK STŪPA COURT.

68

RELIEF SCULPTURES, R. lvi–lxi, ON OUTER SOUTH-EAST WALL, RAWAK STŪPA COURT.

Among the remaining images of the south-west face (see Plate XIII. b), which could be cleared before the slope of the dune rising westwards stopped excavation, R. lxxv proved a replica of R. lxxiii, preserving, like the latter, the original dark-red colour of the robe. R. lxxvi was found broken from the knees upwards; at its feet was placed a small Buddha figure resembling those discovered in a similar position near R. lxx. On R. lxxvii, a large image standing about 6 ft. 3 in. to the shoulders, the surface stucco had peeled off, and with it all indications of the drapery. R. lxxviii, the last of the series, was a small Buddha figure without head, and poorly preserved, but apparently a replica of R. lxxiv.

Remaining images of outer south-west face.

We have now concluded the survey of the sculptural decorations of the enclosing quadrangle as far as my excavations disclosed them. But there still remains to be noticed the relief work discovered on a thin wall of plaster which was brought to light near the south corner. The extant portions run parallel to the south-east and south-west walls for about 20 ft. and 32 ft. respectively, and at a distance of 9 ft. The thickness of this external wall was only about half a foot, and consequently but little survived of the stucco sculptures with which it had been decorated. No continuation of the wall could be traced either to the west or the east, and thus the question whether it once extended like an outer passage or gallery around the whole of the quadrangle cannot be definitely answered. In any case it may be assumed that it was connected with the main wall by a roofing, though no direct indication of the latter survived.

External wall at south corner.

The south-west portion of this wall was originally adorned with reliefs on both faces, but, as seen in Fig. 66, very little remained of the outside statues. On the inside, too, only the lowest portions of the large standing figures (R. lxxix, lxxxi, lxxxiii), from the knees downwards. were preserved. The drapery showed in each case close similarity to that of R. lxxxv, seen on Plate XVIII. c, one of the two large reliefs which still remained on the south-west wall. The skirt-like appearance of what remained of the robes, with the rich folding in wave lines, suggests that the figures represented were Bodhisattvas. Special artistic interest attaches to the remains of the elaborate aureoles which, as seen on Plate XVIII. c, encircled these statues. They varied but little in details. Everywhere there was a border formed by a cloud-scroll, and within it a broad band showing small plaques of seated Buddhas or Bodhisattvas inserted between wreaths of a very graceful bead-ornament. The latter were gathered at intervals into bunches surmounted by a fleur-de-lis. A second band of small seated Buddhas seems to have filled the space left between the highly ornamental border and the edges of the drapery. A number of the small plaques with figures seated within a lotus-petal vesica could be removed and will be found described in the list (see R. lxxix. 1. a–c; lxxxiv. 1, 2; lxxxv. 1, 2); but owing to their very friable material and their thinness few escaped damage in transit. Plate LXXXVII shows two characteristic specimens.

Sculptural decoration of external wall.

Of two small sculptures on the inside (R. lxxx, lxxxii) little more than the feet and bases could be traced; the relief remnants beyond R. lxxxv were also decayed beyond recognition. Of the figures traceable on the outer face of the south-east wall only R. xci retained a small portion of its vesica, showing that its decoration was the same in design as that of the inside reliefs.

SECTION V.—THE DATE OF THE RAWAK REMAINS; THE JUMBE-KUM SITE

Early type of Rawak sculptures.

The remains of the Rawak Vihāra described in the preceding pages enable us better than any other ruins as yet explored in this region to realize what the plan and decorative aspect must have been of large Buddhist shrines in ancient Khotan. The affinity which the Rawak reliefs show in style and most details of execution with the Graeco-Buddhist sculptures of Gandhāra is far closer than that we have had occasion to observe in the plastic remains brought to light elsewhere. These considerations make it all the more important to determine the period, if only within approximate limits, to which the ruined Vihāra belongs. The fact of the Rawak sculptures approaching their Gandhāra models much nearer than those of Dandān-Uiliq or Endere, and the total absence among them of any of those images with multiple limbs, &c., which characterize the later Pantheon of Northern Mahāyāna Buddhism, may at once be accepted as a proof that they are older. But beyond this it would scarcely be safe to draw any further chronological conclusion from the evidence of the artistic remains themselves, seeing how scanty our data are for the chronology of Graeco-Buddhist sculpture in Gandhāra itself, and how little we know as yet of the historical development of its offshoot transplanted to Eastern Turkestān.

Uniform period of stucco sculptures.

It is difficult to say whether the chronological task before us would have been facilitated if the Rawak sculptures had been worked in stone or some other more lasting material instead of friable stucco. In that case we might have been able, perhaps, to trace with more ease successive stages in the local development of this art. But, on the other hand, we could not have assumed as safely as we can in the case of the plaster reliefs that the whole of this wealth of sculptural work must belong to approximately the same period. Large statues in friable clay could not have lasted for many decades, and whatever repairs or restorations became necessary may reasonably be assumed to have been effected in accordance with the style prevailing at the time. Thus a certain uniformity in the sculptural decoration was likely to be maintained even if individual images or groups were put up at different periods.

Numismatic evidence of date.

No epigraphical finds of any kind were made in the course of my excavations, nor has any discovery of manuscript materials ever been reported from this site. The moisture due to the vicinity of subsoil water, which, as we have seen, had caused the wooden framework of the statues, &c., to rot away completely, was not likely to have spared any manuscripts or similar votive deposits, whether of wood, leather, paper, or textile fabric. It was hence particularly fortunate that I was able to secure *in situ* numismatic evidence of distinct chronological interest. I have already mentioned how, in clearing the pedestals of the statues R. xi, xxviii, lxiv, the side-wall of the gate, and the mouldings at the foot of the Stūpa base, we came again and again upon Chinese copper coins bearing the legend *wu-chu*.

Coin finds on small Stūpa base.

But still more conclusive evidence of the same kind was revealed when the base of what I take to have been a small Stūpa, discovered near the inner south corner of the quadrangle (see Plate XL), came to be systematically cleared on April 17. On a platform 8 ft. square and 1 ft. high there rose a mass of brickwork thickly overlaid with plaster, measuring 6 ft. square and about 3 ft. high in its extant condition. The top was completely broken, and a little trench dug through from north to south showed that 'treasure-seekers', probably at an early period, had been at work here. The top of the platform, as well as the lower portions of the sides of the base proper, had been revetted with timber, probably of Terek wood, which was unearthed in a completely rotten condition[1]. Sticking between the plaster of the base on its four sides and the decayed wood of the boarding, at heights varying from $1\frac{1}{2}$ to 2 ft., were found numerous copper coins, which, as far as they have not suffered too much from oxidization, can all be recognized as *wu-chu* pieces. On the south side twelve complete coins were recovered, with numerous fragments of others; on the east side fourteen, with some others which broke into fragments when I attempted to separate pieces which corrosion had stuck together. From the west came twenty-two, with more fragments (for a specimen see Plate LXXXIX, 19), while twelve were discovered on the north face. There could be no doubt that they had been slipped into the places where they were found, through small interstices of the boarding, manifestly as votive offerings, just like the pieces previously discovered in small cavities below images and underneath plaster mouldings.

Type and condition of coins found.

Most of the coins are in good preservation, apart from the oxidization they have undergone, and do not show any marks of long circulation previous to their deposition. Only current coins are likely to have been used for such humble votive gifts; and as no coin finds of a later type have come to my knowledge, we are justified in assuming that the latest known date of these coin-issues marks the lowest chronological limit for the abandonment of the shrine. This conclusion is supported by the evidence of the eighteen coins picked up by my men between the dunes near the ruins, ten of which bear the legend *wu-chu*, while the rest are small square-holed coins without legend. The great find, previously mentioned, of a pot full of *wu-chu* coins at an eroded spot about one mile to the south-east also agrees with it.

Period of *wu-chu* currency.

The *wu-chu* currency belongs properly to the period of the Former and Later Han dynasties. The rule of the second extended from 25 to 220 A.D., but the issue of its coin types appears to have continued in China up to the close of the fourth century, if not to the advent of the Tangs. It is thus difficult to determine the date when the current use of *wu-chu* coins, whether imported from China or locally coined, is likely to have ceased in Khotan. There remain, however, two significant negative observations to guide us. On the one hand, the fact of the numerous coin finds of the Rawak Vihāra not comprising a single later piece makes it probable that the date of these votive deposits could not have been removed by many centuries from the period of the Later Hans, when the *wu-chu* coinage was the recognized currency of the Chinese empire. On the other, the complete absence of Sino-Kharoṣṭhī pieces seems to preclude the assumption that the shrine had existed in the first centuries of our era.

Approximate limits of date of abandonment.

The absence among the votive deposits of the earliest Chinese coin type, that without legend, points also in the same direction. Possibly the complete excavation of the Vihāra may hereafter provide a more definite indication of date. But combining what evidence is at present available in the coins, the style of the sculptures, and such minor antiquarian indications as may be derived from the constructive plan and the materials of the ruins, we can scarcely assert more than that the extant shrine must have been abandoned at some period between the third and seventh centuries of our era.

Removal of larger sculptures impossible.

At the commencement of the excavations I realized with regret that, owing to the extremely friable condition of the stucco and the difficulties of transport, the removal of the larger sculptures was quite impracticable. Those pieces of the colossal images which were found already detached, such as portions of arms, edges of drapery, &c., usually broke when lifted, whatever care was used. An attempt to move the complete statues or torsos from their places without elaborate

[1] A base revetted with timber in exactly the same fashion was excavated in N. xvi, a small structure of the Niya Site which is likely to have been a shrine; see above, pp. 374 sq.

appliances, to improvise which I had neither time nor the technical means, would have meant only Vandal destruction. In any case it would have been a practical impossibility to arrange for the safe transport of such loads over the mountains, whether to Europe or India. All that could be done in the case of these large sculptures was to bury them again safely in the sand after they had been photographed and described, and to trust that they would rest undisturbed under their protecting cover. Of the sculptural pieces already detached and of the smaller relief plaques, I succeeded in bringing away a considerable number as described in the list below[2]. I felt greatly relieved when I found, on my arrival at Kāshgar, and later also in London, that the great trouble and labour which the safe packing of these extremely fragile objects had cost me, was rewarded by their having accomplished the long journey without any serious damage.

Re-burial of sculptures excavated.

By April 18 those portions of the Vihāra had been explored which were not actually buried under sand-dunes. A careful examination of the surrounding area revealed no other structural remains; broken pottery found here and there between the swelling sand-dunes was the only trace left of what probably were modest dwelling-places in the vicinity of the great shrine. The sandstorms which visited us daily, and the increasing heat and glare, had made the work very trying to the men as well as to myself. It was manifestly time to withdraw from the desert. Before, however, leaving the ruins I took care to protect the sculptures of the quadrangle and the foot of the Stūpa base by having the trenches which had exposed them completely filled up again. It was a melancholy duty to perform, strangely reminding me of a true burial.

Visit to *Jumbe-kum*.

While it was still proceeding I took occasion to visit *Jumbe-kum*, the only remaining desert site to the north-east of Khotan from which occasional finds had been reported to me by Turdi and others. I reached its southern edge after going about four miles to the north-east of Rawak. One mile of march had taken us out of the heavy sands surrounding the ruins, and the rest of the way led over low dunes amidst which tamarisks and young Toghrak shoots were growing in plenty. The site itself proved to be a débris-strewn Tati, with only the scantiest of structural remains. Not far from its southern edge I was taken to what Turdi had previously spoken of as a 'Potai'. I found there a small mound of hard stamped loess, about 6 ft. high and some 15 ft. long, surmounted by a foundation of brick masonry, about 2 ft. high and much broken. It was quite impossible to ascertain the original shape and object of the structure that once stood here. The sun-dried bricks appeared to have measured about 19 by 14 in., with a height of 4 in. The ground for about half a mile around was plentifully covered with coarse broken crockery; here three small and much worn copper coins without legend were also picked up.

Ruin of masonry structure.

The pottery débris ceased about half a mile to the north-east of the small mound and did not appear again until, continuing in that direction for about two miles further, we neared Turdi's 'second Potai'. For the greater part of this distance the Kumush-covered Sai which comes from near Tam-öghil, undoubtedly an old river-bed, kept close on our right. About half-way a *wu-chu* coin was picked up by one of the men. The only conspicuous ruin of *Jumbe-kum* or *Jumbe-Kalmak*, as according to Turdi the site is also called[3], proved to consist of a broken mass of masonry about 8 ft. high, which seems to have formed originally a square

[2] Some larger fragments which had become detached, but for the safe transport of which the available means proved inadequate, were deposited at the foot of the small Stūpa base and covered up with sand.

[3] *Kalmak* was said to be another name for the 'Kara-Khitai,' whose memory still vaguely haunts Khotan tradition; for *Jumbe* no explanation was given.

of about 24 ft. Near its approximate centre a shaft, about 3 ft. in diameter, had been sunk to the natural ground, undoubtedly by 'treasure-seekers'. Owing to erosion of the surrounding ground, the ruin seems now to stand on a mound 7 to 8 ft. high. From the large quantity of broken bricks strewing the slopes of the latter and the ground close by, it appears probable that this square mass of masonry may have once served as the base of a small Stūpa. The bricks, formed of fairly hard sun-dried loess, showed the same average dimensions as at the first small ruin. On none of them could I trace any mark.

General character of site.

The general impression I gained was that the eroded site, which was said to cease not far to the north of this structure, was of considerable antiquity. Being on the track of the woodcutters from Yurung-kāsh and the neighbouring tracts, it is often searched for any small objects which the winds may lay bare on the eroded loess surface. The only small antique said to have been obtained here which I was able to acquire at Yurung-kāsh is a terra-cotta handle (J. 001) enriched with an excellently moulded acanthus leaf. In texture and eroded appearance it closely resembles some pieces said to have been brought from 'Ak-sipil', and not much dependence can be placed on its alleged origin.

My visit to Jumbe-kum had sufficed to convince me that there were no remains there capable of excavation. Thus, when on April 19 I started from my Rawak camp back to Khotan in the midst of a violent sandstorm, soon followed by a drenching downpour of rain, I had the satisfaction of knowing that the programme of my explorations in the desert was completed.

SCULPTURES EXCAVATED AT THE RAWAK VIHĀRA.

R. ii–xi.

R. ii. 1. Stucco relief Buddha head, approx. life-size. Red clay, orig. whitewashed over; black for eyelashes, trace of red *ṭikā*; eyes nearly closed. Hair in rich curls; small top-knob; small chin, ears elongated. 10" top-knob to below chin, $5\frac{1}{2}$" ear to ear. Well-preserved, with hollow for wood frame (rotted). See Pl. LXXXII.

R. ii. 2. Stucco relief fragments, 10 pieces. Red clay, traces of white and red washes. Portion of plaque forming halo above head of figure; curve of perfect edge elliptical. Perfect edge bordered with cloud-scroll. Within this upper portion, seated Buddha, cross-legged, hands in lap, L. p. resting in palm of R.; face indistinct; ears elongated and pierced; top-knob; single robe; nimbus at back of head; the whole in vesica formed of lotus, centre of which is treated as rays. Top R. p. of figure a small lotus; below L. p. another. Immediately below figure double fleur-de-lis ornament, representing a Vajra; below and to L. p. of this, small lotus. All ornaments *appliqué*. Portion re-united on restored background. $11\frac{1}{2}$" high, $7\frac{1}{2}$" across. Very friable. See Pl. LXXXIII.

R. ii. 3. Stucco relief fragments.

a. Vesica with seated Buddha. Replica R. ii. 2.

b. Similar to a. See Pl. LXXXVII.

R. iv. 1. Stucco relief; head of Bodhisattva, over life-size. Red clay, orig. colour-washed; white eyeballs; traces of black eyelashes, and dark colour on lips. Single long ringlet of hair in front of each ear. Ears well modelled, elongated, and slit; eyes partly closed; small nose, mouth, and chin. On left of head passing over L. ear, ornamental band, with jewel; R. missing. Top of head broken away. $10\frac{1}{2}$" from below chin upwards, 8" ear to ear. Hollow for wood frame (rotted) and broken at back. Very friable. See Pl. LXXXI.

R. vi. Stucco relief head of Bodhisattva, life-size. Red clay, orig. red-washed; white eyeballs and traces of black eyelashes, and dark colour on lips. Pink *ṭikā*; eyes almost shut; small nose, mouth, and chin; hair rich, curled; ears elongated (broken). Top and back of head broken. Hollow for frame. 9" from top of head to below chin, $6\frac{1}{2}$" between ears. Friable. See Pl. LXXXIII.

R. xi. 1. Stucco relief fragments; 4 principal pieces. Replica of portion of R. ii. 2, but cloud-scrolls much more elaborate; petals forming outer ray of vesica more elaborate; vesica much broken. Vajra and small lotuses (excepting one) absent. Size of vesica $3\frac{5}{8}" \times 2\frac{7}{8}"$. Cloud-scroll $4" \times 2\frac{1}{8}"$. Lotus fragment $1\frac{3}{8}" \times 1\frac{1}{4}"$. Portion of cloud-scroll $1\frac{5}{8}" \times 1\frac{1}{8}"$. Very friable. See Pl. LXXXVII.

R. xii. 1. Stucco relief figure of Buddha, from vesica. Red clay; traces of whitewash over red-wash all over. Standing male figure; portion below middle of thighs missing. Single, flowing robe. R. p. hand raised (broken), L. p. hand raised to chest grasping loose drapery. Head resting against nimbus. Eyes small; nose broad; ears slightly elongated. Hair simple, with top-knob. Height $13\frac{1}{2}$". Width below shoulders 6". Friable. See Pl. LXXXVI.

R. xii. 2. Stucco relief fragment, in 2 pieces. Replica R. xii. 1. Red clay, traces of white and red. Height 9¼", width 6¼". Very friable.

R. xii. 3. Stucco relief fragment, in several pieces. Replica R. xii. 1. Red clay, coloured. Very friable.

R. xxv. 1. Stucco relief fragments. Replica R. ii. 1; outer ray of vesica missing.

R. xxv. 2. Stucco relief fragments. Replica R. ii. 1, broken into many pieces. Very friable.

R. xxv. 3. Stucco relief, seated Buddha, in 2 pieces. Replica R. 2, but without tree. Red clay, red-washed. 7"×5".

R. xxvii. 1. Stucco relief, female head and bust in many fragments (re-united at British Museum). Red clay, with traces of whitewash. Shape of face a full oval, but in re-uniting the fragments the proper and original symmetry of outline has been rather lost. Features small; nose destroyed; depression for eye-pupil; ears slightly elongated; hair long and waved, passing gracefully right and left from beneath a jewelled tiara and falling in full masses over upper parts of ears and round to back of neck. Hair at top of head loosely 'bunched.' Back of head presents a flat surface, where it has been attached to a wall. The neck is rather short (as restored) and thick. The breasts are delicate and small—smaller than is usual even in Northern Indian sculptures—and well-placed, excepting that they are rather high. Shoulders and bust draped. The tilt of the head should be rather more upward than is shown in the reconstruction. Height 8½", width at shoulders 7". See Pl. LXXXV.

R. xxix. 1. Pieces of gold-leaf, two, stuck orig. to L. and R. knee of statue R. xxix.

R. xxx. 1. Stucco relief, standing figure, fragments probably from vesica. Red clay, traces of whitewash on hands and drapery round neck. Trace of red paint on L. p. hand. L. p. hand lowered at side, R. p. hand raised to chest open, palm outwards. Head and legs below knees missing. Single long robe. Hole pierced through for fixing. Height 12½", width below shoulders 6½". Roughly modelled. Friable. See Pl. LXXXVI.

R. xlviii. 1. Stucco relief; head of Buddha; fragment in many pieces. About ⅔ life-size. Replica R. lxxii. 1. Red clay; traces of white in eyes. Nose broken; top-knob missing, much of the hair flaked away. Ears elongated, slit; hair in narrow wavy fillets across head. Very friable.

R. lxx. 1. a. Stucco relief; head of Buddha. Red clay, whitewashed; red contour lines; black eyelashes and pupils. Top-knob to chin 4½", between ears 3½". Colour in good preservation. See Pl. LXXXIV.

R. lxx. 1. b. Stucco relief; head of Buddha; fragment. Replica R. lxx. 1. c. Traces of colour (white, red, black). 5¼" from top-knob to below chin, 3½" between ears. Very friable.

R. lxx. 1. c. Stucco relief; head of Buddha. Red clay, traces of whitewash and red contour lines. Eyes half-closed; features well-proportioned; elongated ears, lobes curved outwards; two deep folds running round neck; hair not modelled, giving appearance of skull-cap with top-knob. 4½" top-knob to below chin, 3½" between ears. See Pl. LXXXIV.

R. lxxii. 1. Stucco relief; head of Buddha with neck; fragment. Red clay, whitewashed. Eyes nearly closed; nose (broken) and mouth small; ears elongated, lobes broken off. Hair simple, much broken; neck, two deep wrinkles. Hollow for wood frame. 6¾" from end of neck to top of head, 4¾" between ears. Friable. See Pl. LXXXVIII.

R. lxxii. 2. a. Stucco relief fragments; 2 pieces. Replica R. lxxxiv. 1. Red clay, traces of whitewash. Seated male figure, legs crossed; hands in lap; turbaned; ornaments in elongated ears. Garment *dhōtī*; nude above middle; neck ornaments. Nimbus at back of head. Remains of vesica under legs. L. p. knee and elbow missing. 4" high, 2½" wide at knees. Very friable.

R. lxxii. 2. b. Stucco relief fragment in many pieces. Seated Buddha with nimbus, vesica broken. Replica R. ii. 2. Red clay, traces of whitewash. 2¾"×1¾". Friable.

R. lxxii. 2. c. Stucco relief fragment. Replica R. lxxxiv. 2. a. Red clay. Much worn, but vesica intact. 3⅜"×3⅜".

R. lxxiv. 1. Stucco relief fragment; head of Buddha. Red clay, traces of whitewash, black pupils to eyes, red contour line visible on R. p. wing of nose. Eyes long, half-closed. Nose, mouth and chin delicate and well-proportioned. Ears broken away. Hair a series of closely-plaited bands commencing forward of ears, each band passing from side to side of head over top. Top-knob consists of similar bands wound horizontally. Hollow for wood frame (rotted). 6½" from top-knob to below chin. 3½" at level of ears. Friable. See Pl. LXXXIV.

R. lxxix. 1. a. Stucco relief fragment. Replica R. ii. 2. Part of outer ray broken. Face broken. Figure seated on bracket. Red clay. 3¾"×3¼".

R. lxxix. 1. b. Stucco relief fragment, Replica R. ii. 2. Portion of outer ray broken; face gone. Red clay. 3½"×3½".

R. lxxix. 1. c. Stucco relief fragment. Replica of R. ii. 2. Hands of figure covered by drapery, also feet. 3"×3". Friable.

R. lxxxiv. 1. Stucco relief fragment (much broken). Replica R. lxxii. a. Red clay. Seated figure, cross-legged, both feet exposed. Hands, palm to palm resting in lap. Garment, *dhōtī*. Upper part of figure nude with the exception of ornament; on each upper arm large armlets of lotus pattern; at wrists, bracelets; in ears, heavy ornaments. Ornamented turban having long streamers crimped end depending over each shoulder, with fringe at end; and folds of drapery resting on front of shoulders. Nimbus at head; portion of vesica. Height from feet to turban 3¾", between knees 2⅛" (approx.). See Pl. LXXXVII.

R. lxxxiv. 2. a. Stucco relief fragment, in 3 pieces. Red clay, coloured rich red-wash. Replica R. ii. 2, but hands clasped holding end of drapery (?) raised towards chest. R. p. foot exposed. R. p. shoulder and breast bare and R. p. arm. Lotus petals very richly modelled. 3½" wide, 3½" high. Very friable. See Pl. LXXXVII.

R. lxxxiv. 2. b. Stucco relief fragment, in several pieces. Replica R. lxxxiv. 2. 3½" × 3½".

R. lxxxv. 1. Stucco relief fragment, in 2 pieces. Replica R. ii. 2. 4" × 4".

R. lxxxv. 2. Stucco relief fragment, in many pieces. Replica R. ii. 2. 4½" high, 4½" across (approx.). Friable.

R. 1. Stucco relief figure, fragment. Replica R. xii. 1, from S. corner. Red clay, traces of white and red washes. Head, legs, and L. p. arm missing. Flowing single robe, open to show R. p. forearm. R. p. hand raised, palm outwards. 6½" height of fragment, 5½" width at shoulders. Very friable.

R. 2. Stucco relief, seated Buddha, in 2 pieces, from S. corner. Red clay. Single flowing robe; hands in lap. L. p. resting in palm of R. p. Ears elongated; eyes drooped in meditation. Features (damaged) well proportioned. Hair (damaged) with top-knob. Nimbus. Small tree grows behind R. p. shoulder. Low pedestal under figure. 7" from top of nimbus to bottom of pedestal. 5" between knees. Friable. See Pl. LXXXII.

R. 3. Stucco relief, head of Buddha; fragment. Replica, R. lxx. 1 b, from S. corner. 4" from top-knob to below chin. 3" between ears.

[**R. 01–019. Stucco reliefs** from inner S. corner.]

R. 01. Stucco relief, head of Buddha; fragment. Replica R. 02. Red clay, red-washed, traces of white in eyes. Eyes small; depression for pupils. Hair simple; top-knob broken away. Fairly large nose, mouth, and chin. Ears slightly elongated, R. ear slit. 3¾" chin to top-knob, 3" between ears. Friable.

R. 02. Stucco relief, head of Buddha; fragment. Replica R. 01. Red clay, traces of whitewash over red-wash. Eyes small, nose and mouth well-proportioned. Ears slightly elongated; L. ear slit. Hair simple; top-knob broken. 3½" top of head to below chin, 3" between ears. Friable.

R. 03. Stucco relief, head of Buddha; fragment. Replica R. lxx. 1. b. Red clay; traces of red-wash; white in eyes; depression for pupil. All features well-proportioned. Ears slightly elongated; R. p. slit, L. broken. R. of face damaged. 4¼" top-knob to below chin, 3½" between ears. Friable.

R. 04. Stucco relief figure, standing Buddha, in several fragments, incomplete. Replica R. xii. 1. Red clay, traces of red-wash over drapery, white on hands. Head and portion below knees missing. Single flowing robe open to expose R. p. forearm. R. p. hand raised palm outwards. L. p. hand raised to chest grasping loose drapery. 8¼" neck to broken knees, 6¼" (approx.) at shoulders. Very friable. R. lxxxiv–013.

R. 05. Stucco relief; figure of standing Buddha, in two fragments; incomplete. Replica R. xii. 1. Red clay, traces of white and red washes. Head, L. p. elbow, and lower portion from above knees missing. Single flowing robe open to expose R. p. forearm. Loose end thrown over L. p. shoulder. R. p. forearm raised, hand missing, L. p. hand raised to chest grasping loose drapery. Height 9"; width below shoulders 5½" (approx.).

R. 06. Stucco relief fragment. Portion of vesica border, curved in section. Red clay, traces of whitewash. Incised ornament consisting of lines crossing diagonally, forming a lozenge diaper; each lozenge having an incised line running from lower angle vertically towards top angle but stopping short; giving an additional effect of light and shade. Lozenge diaper terminated at top by a fillet moulding formed by incising two parallel horizontal lines ⅓" apart. Hollow for wood frame (perished). Length (direction of axis) 5½"; diameter in widest section 3½". Friable.

R. 07. Stucco relief figure. In many fragments, incomplete. Replica R. xii. 1. Red clay, traces of whitewash, and red contour lines. Head and portion from knees downwards missing. Flowing single robe, loose end thrown over L. p. shoulder. Open to show right forearm. R. p. forearm raised (broken). L. p. hand raised on chest grasping loose drapery. Height (approx.) 9"; width 6". Very friable.

R. 08. Stucco relief fragment, in 2 pieces. Replica R. ii. 2, but no trace of whitewash; small lotuses absent. See Pl. LXXXVII.

R. 09. Stucco relief fragments. Replica, portion of cloud R. xi. 1. Principal piece 3" × 2". Very friable.

R. 010. Stucco relief fragment. Red clay. Portion of band ornament; two half-lotuses. 2¾" × 1½". Friable. See Pl. LXXXVII.

R. 011. Stucco relief fragments. Red clay, traces of red and white wash. Two right hands and wrists, raised and turned palm outwards, from replicas R. xii. 1. Drapery falls from wrists. Length of hands 2½".

R. 012. Stucco relief fragments; 2 pieces. Red clay, traces of whitewash. Replica R. ii. 2, much damaged. 3¾" high, 3" wide. Friable.

R. 013. Stucco relief fragments, several pieces. Red clay, traces of white and red wash. Replica R. xii. 1. Head missing and all below middle. Height 6", width below shoulders 6" (approx.). Very friable.

R. 014. Stucco relief, head of Buddha; fragment in 3 pieces. Red clay, traces of white and red wash. Replica R. 01. Portion of nimbus to R. p. of head intact. Both ears represented as pierced. Top-knob to chin 4½", between ears 3". Friable.

R. 015. Stucco relief, head of Buddha, fragments; 4 pieces. Red clay, traces of whitewash and red contours. Replica R. 01. Parts of nimbus remain. Friable.

R. 016. Stucco relief, head of Buddha, fragment in 2 pieces. Red clay, traces of white and red wash. Replica R. 01. L. p. ear missing. Portions of nimbus. Top-knob to chin 4½", ear to side of face 2¾". Friable.

R. 017. Stucco relief figure, in 2 fragments, incomplete. Standing Buddha; replica R. xii. 1. Red clay, traces of white and red wash all over. Portion from middle downwards missing. 12" from top of nimbus downwards, 5¾" below shoulders. Friable.

R. 019. Stucco relief figure, fragment. Standing Buddha; replica R. xii. 1. Red clay, traces of red and white wash. 10½" high, 6½" across shoulders. Friable.

OBJECTS FROM RAWAK.

R. 001. Fragments of terra-cotta vessels, from ground near Vihāra.

a. Fine clay, smoothed outside. 4" × 1½". **b.** Coarse clay. 2" × 1⅞". **c.** Coarse clay, smoothed outside. 2" × 1½". **d.** Very coarse clay. 2½" × 1¼".

R. 002. Piece of sandstone, drilled at one end. Probably a plummet. Found near Rawak Vihāra. 3⅜" × $\frac{9}{16}$" × ½".

R. 003. Gold split ring; probably an earring. Purchased at Khotan. The ring is of round wire, and has two settings for jewels—one opposite the split, and the other at 90° with it, so that if the ring were in the ear, one jewel would be pendant and the other directed forward. Settings very neatly made. See Pl. LI.

POTTERY FROM JUMBE-KUM.

J. 001. Fragment of terra-cotta handle, said to have been found at Jumbe-kum; enriched with an admirably moulded acanthus leaf, quite in Gandhāra style. The clay has the same rich colour, and the same curious eroded appearance as A. 001, A. 006. Height 2", width 1¾".

J. 002. Fragments of terra-cotta vessels, found at Jumbe-kum. Various tints; coarse. Two of them decorated with roughly-incised patterns. **a.** Dark grey. 2¼" × 1½". **b–d.** Dull red inside, Khākī outside. 2¼" × 1⅜", 1⅜" × 1⅛", 1⅜" × ¾", resp.

CHAPTER XV

DEPARTURE FROM KHOTAN

Section I.—ISLĀM ĀKHŪN AND HIS FORGERIES

The eight days' halt that followed my return to Khotan had to be passed by me within doors for most of the time, and partly in bed, owing to an attack of bronchitis, brought on by the exposure of the last weeks in the desert. But the arrangement of my collections, their partial repacking, and the endless little agenda which accumulate after a long season of camp work, kept me so busy that this involuntary confinement was scarcely realized by myself. On the morning after my arrival I still felt well enough to call on Pan Dārin, who received me at his Ya-mên like an old friend, and, as I imagined, somewhat like a fellow-scholar. Much I had to tell him of my excavations and the finds which had rewarded them. When next day the old Amban came to return the visit I had ready a little representative exhibition of my antiques to satisfy his curiosity. I knew the Mandarin to be a man of learning and thoroughly well versed in Chinese history. Nevertheless I was surprised by the historical and critical sense displayed in the questions he put to me regarding the relative age, the import and character of the multifarious ancient documents I had discovered. I felt almost in company of a colleague, and the instinctive comprehension shown by him for my palaeographical and archaeological arguments made me forget for moments the irksome circumlocution and confusion involved in conversation through a not over-intelligent interpreter. Halt at Khotan.

Though the explorations which the learned Amban's unwearying help had done so much to render successful had now been concluded, I still needed his co-operation in view of a curious semi-antiquarian, semi-judicial inquiry which it was important to effect before leaving Khotan. Its success has been greeted with no small satisfaction by fellow-scholars in Europe, besides greatly amusing me at the time. It enabled me to clear up the last doubts as to the strange manuscripts and 'block-prints' in 'unknown characters' which had since 1895 been purchased from Khotan in remarkable numbers, and which not only figured conspicuously in the 'British Collection of antiquities from Central Asia' formed at Calcutta, but had found their way also to public collections in London, Paris, St. Petersburg, and probably elsewhere. Seeing that the first publication of the result of my inquiry sufficed to rid research of this *embarras* of forged riches, I need scarcely regret that limitations of space and time, together with my distance from the collections referred to, now prevent me from attempting either a bibliography of these remarkable acquisitions or a review of the learned endeavours which had been devoted to their analysis and decipherment. But a record of my inquiry which resulted in this *dénouement* may well find a place here, even though it can only by few details supplement the facts published elsewhere [1]. Inquiry into forged MSS. in 'unknown characters'.

I have already had occasion to describe how, on my march to Khotan in the autumn, I had vainly endeavoured to locate the sites where many of the manuscripts and 'block- Previous suspicions of forgeries.

[1] See *Preliminary Report*, pp. 64 sqq.; *Ruins of Khotan*, pp. 471 sqq.

prints' in a variety of 'unknown' characters were alleged to have been found, of which Dr. Hoernle, between 1897 and 1899, published descriptions[2]. In regard to these acquisitions, sold in steadily increasing numbers from Khotan, the suspicion of forgery had before presented itself to some of the Europeans resident at Kāshgar, as well as to competent scholars[3]. But evidence was wanting to substantiate it, and in the meantime these strange texts continued to be edited and to form a subject of learned investigation. On my first arrival at Khotan I had naturally been on the look-out for such 'old books'; but offers in this article were then surprisingly scanty, and curiously enough the very first 'old book' that was then shown to me supplied unmistakable proof of forgery. Hearing of my arrival, a Russian Armenian from Kokand brought to me for inspection a manuscript on birch-bark, consisting of some ten ragged leaves covered with an 'unknown' script, and folded up into 'forms'. He had bought it for forty roubles, undoubtedly as a commercial speculation, and now wished to have his treasure properly appraised.

Scrutiny of forged birch-bark MS.

I saw at once that the birch-bark leaves, consisting each of a single thin layer, had never received the treatment which, as all old Bhūrja codices known to me from Kashmīr show, was needed to make the material suited for permanent record. Nor had the forger attempted to reproduce the special ink required for writing on birch-bark. So when I applied the 'water-test', in accordance with what experience in Kashmīr had taught me, the touch of a wet finger sufficed to take away the queer 'unknown characters' both written and 'block-printed'. It was significant that the 'printed matter' of this manifest forgery, specially abundant on the margins, showed a very close resemblance to the formulas of certain 'block-prints' published by Dr. Hoernle[4]. The inquiries made at the time indicated a close connexion between Ibrāhīm Mullā, the person from whom the Armenian had purchased the leaves, and Islām Ākhūn, the 'treasure-seeker', whose alleged places of discovery I had vainly attempted to locate about Gūma. Local rumour, in fact, credited Islām Ākhūn with having worked a small factory for the production of 'old books'. But at that time he was keeping away from Khotan, and there were reasons for postponing investigations about him.

Forgeries traced to Islām Ākhūn.

The grave suspicions which these local inquiries had led me to entertain about the genuineness of all these 'old books' in 'unknown characters' had been strengthened almost to certainty by my explorations of the winter. Ample as were the manuscript materials which the latter had yielded, and in spite of the great variety of languages and scripts represented among them (Kharoṣṭhī, Indian Brāhmī, Central-Asian Brāhmī, Tibetan, Chinese, and Hebrew), I had failed to trace the smallest scrap of writing in 'unknown characters'. The actual conditions of the sites explored also differed entirely from the conditions under which those queer 'old books' were alleged to have been discovered. There was good reason to believe that Islām Ākhūn, to whom it was possible to trace most of these manuscripts and block-prints that had been purchased on behalf of the Indian Government during the years 1895–98, was directly concerned in the forgeries. He stayed away from Khotan during my first visits. He had been punished some time before on account of other impositions which Captain Deasy and Mr. Macartney had brought to the notice of the Khotan Ya-mên, and he evidently did not think it safe to attempt further deception in my case. I had no reason to regret the wide berth which Islām Ākhūn had given me while I was engaged in my archaeological work about Khotan and at the ancient sites of the desert. But now, when the antiquarian evidence as to the true character of those remarkable literary relics in 'unknown characters' was practically

[2] See above, pp. 100 sqq.

[3] Comp. Hoernle, *Report on C.-A. ant.*, i. pp. 57 sqq.

[4] See *Report on C.-A. ant.*, i. Pls. V–VII, IX–XI, XIII, &c.

69

iv ii i

RELIEF STATUES, R. i—iv, ON INNER SOUTH-EAST WALL, RAWAK STŪPA COURT.

71

ISLĀM ĀKHŪN, FORGER OF ANTIQUES.

70

TURDI KHWĀJA, OF TAWAKKĒL, TREASURE-SEEKER.

complete, and my departure near at hand, I was anxious for a personal examination of that enterprising individual whose productions had engaged so much learned attention in Europe and India.

Pan Dārin, to whom I confidentially communicated my wish to get hold of Islām Ākhūn, readily granted his assistance. As an attempt on the part of Islām Ākhūn to abscond was by no means improbable, and as time was getting short, I took care to impress the Mandarin with the necessity for prompt and discreet action. Nor did he disappoint me in these respects; for on the morning of April 25 Islām Ākhūn (see Fig. 71) was duly produced from Chīra, where he had been practising as a 'Ḥakīm' during the previous winter. He scarcely anticipated being 'wanted' now, as when passing through Chīra some three weeks before I had purposely refrained from making any inquiries about him. The Bēg who escorted him brought also a motley collection of papers, which had been seized partly in Islām Ākhūn's possession and partly in his Khotan house, and which on examination proved rather curious. They were sheets of artificially discoloured paper, covered with impressions of the same elaborate formulas in 'unknown characters' that appeared in some of the last batches of 'ancient block-prints' which had been sold in Kāshgar [5]. A manuscript leaf, also in 'unknown characters', had evidently remained over from the earlier manufacture when the forger was still content to work by mere writing [6]. Arrest of Islām Ākhūn.

The examination of this versatile individual proved a protracted affair, and through two long days I felt as if breathing the atmosphere of an Indian judicial court. When first arraigned in my improvised 'Cutchery', Islām Ākhūn readily and with contrite mien confessed his guilt in having in 1898 obtained money from Badruddīn Khān, the Afghān Ak-sakāl, by a forged note purporting to be in Captain Deasy's handwriting [7]. But in the matter of the 'old books' he for a long time protested complete innocence. He pretended to have acted merely as the Kāshgar sale agent for certain persons at Khotan, since dead or absconded, who, rightly or wrongly, told him that they had picked them up in the desert. When he found how much such 'old books' were appreciated by Europeans he asked those persons to find more. This they did; whereupon he took their finds to Kāshgar, &c. Now, he lamented, he was left alone to bear the onus of the fraud—if such it was. Muḥammad Tārī, one of those who gave the 'books', had previously run away to Yarkand; Muḥammad Siddīq, the Mullā, had absconded towards Ak-su; and a third of the band had escaped from all trouble by dying. First examination of Islām Ākhūn.

It was a cleverly devised line of defence, and Islām Ākhūn clung to it with great consistency and with the wariness of a man who has had unpleasant experience of the ways of the law. I had thought it right to tell him from the first that I was not going to proceed against him at the Amban's Ya-mên in the matter of these happily ended forgeries; for I was aware that such a step, in accordance with Chinese procedure, might at one stage or other lead to the application of some effective means of persuasion, i.e. torture. This, of course, I would not countenance; nor could a confession as its eventual result be to me of any value. Whether it was from Islām Ākhūn's reliance on these scruples of mine, or from his knowledge that direct evidence could not easily be produced within the short time available before my departure, two long cross-examinations, in the interval of which I had Islām Ākhūn's wants hospitably looked to by my own men, failed to bring a solution. However, in the course of his long Prolonged denials of guilt.

[5] For a detailed description of those block-prints which had found their way into the collection formed at Calcutta, see Hoernle, *Report on C.-A. ant.*, i. pp. 45–110, with Pls. V–XVIII.

[6] See for specimens of these written productions, Hoernle, 'Three further collections of ancient manuscripts,' in *J.A.S.B.*, 1897, pp. 250 sqq. with Pls. XI–XX.

[7] See above, p. 102, note 14.

protestations of complete innocence, Islām Ākhūn introduced a denial which seemed to offer some chance of catching my wary defendant. He emphatically denied having seen any of the alleged find-places himself, in fact having ever personally visited any ancient site in the desert.

Islām Ākhūn's early 'narratives'.

I had purposely refrained at the time from showing any special interest in this far-reaching disclaimer. Consequently I had no difficulty in inducing him to repeat it with still more emphasis and in the presence of numerous witnesses when he was brought up 'on remand' for a third time. Whether encouraged by the success he thought to have scored so far, or by the forbearing treatment I had accorded to him, it was evident that the sly, restless-looking fellow was for the time being off his guard. So I promptly confronted him, from the detailed account printed in Dr. Hoernle's *Report*[8], with an exact reproduction of the elaborate stories which he had told, in the course of depositions made before Mr. Macartney in connexion with different sales of 'old books', about his alleged journeys and discoveries in the Taklamakān during the years 1895–98.

Confession of forger.

The effect was most striking. Islām Ākhūn was wholly unprepared for the fact that his lies told years before, with so much seeming accuracy of topographical and other details, had received the honour of permanent record in a scientific report to Government. Hearing them now read out by me in re-translation, he was thoroughly startled and confused. He appeared also greatly impressed by the fact that, with the help of the exact information recorded by Dr. Hoernle from the reports of Mr. Macartney and Captain Godfrey[9], I could enlighten him as to what 'old books' he had sold at Kāshgar on particular occasions, what remarkable statements he had made about the manner of their discovery by himself, &c. He was intelligent enough to realize that he stood self-convicted, and that nothing was to be gained by further protestations of innocence. He now admitted that he had seen manuscripts being written by his above-named employers (*recte* accomplices) at a deserted Mazār near Sampula. Little by little his admissions became more detailed; and ultimately, when assured that no further punishment awaited him, he made a clean breast of it.

Origin of Islām Ākhūn's forging enterprise.

Islām Ākhūn's subsequent confessions proved perfectly correct on many important particulars when checked from the records reproduced by Dr. Hoernle[10], as well as from the evidence of a number of independent witnesses. He showed himself to be possessed of an excellent memory, and readily recognized among the fine photogravure plates accompanying Dr. Hoernle's *Report* those representing specimen pages from the 'block-printed' books in 'unknown characters' which formed his own manufacture. He had, previous to 1894, been engaged at times in collecting coins, seals, and similar antiques from Khotan villages. About that time he learned from Badruddīn Khān and other Afghān traders of the value which the 'Sāhibs' from India attached to ancient manuscripts. Genuine scraps of such had indeed been unearthed by Turdi and some other 'treasure-seekers' at Dandān-Uiliq. But the idea of visiting such dreary desert sites, with the certainty of great hardships and only a limited chance of finds, had no attraction for a person of such parts as Islām Ākhūn. So in preference he conceived the plan of manufacturing the article he was urged to supply the Sāhibs with.

Accomplices in forging.

In this enterprise he had several accomplices, among whom a certain Ibrāhīm Mullā was the leading man. This person appears to have made it his special business to cultivate

[8] See *Report on C.-A. ant.*, i. pp. xvi–xxi.

[9] The accurate 'List of contributions' given by Dr. Hoernle (*Report*, i. pp. iii. sqq.), together with his painstaking description of all peculiar features in the MSS. and 'block-prints,' the queer objects (e. g. a skull) said to have been found with them, &c., was of great utility to me in this connexion.

[10] E. g., Islām Ākhūn distinctly remembered the introduction into certain 'block-printed' volumes of the 'artistic' sketches of heads, discussed in *Report*, i. pp. 62, 85, 105, and seemed rather proud of this proof of cleverness.

the Russian demand for 'old books', while Islām Ākhūn attended chiefly to the requirements of British officers and other collectors. Ibrāhīm Mullā, from whom the Russian Armenian I met on my first arrival at Khotan had purchased his forged birch-bark manuscript, was credited with some knowledge of Russian, a circumstance which explains the curious resemblance previously noticed between the characters used in some of the 'block-prints' and the Greek (*recte* Russian) alphabet[11]. Ibrāhīm Mullā gave proof of his 'slimness', as well as his complicity, by promptly disappearing from Khotan on the first news of Islām Ākhūn's arrest, and could not be confronted with him.

First sales of MSS. in 'unknown' characters.

The first 'old book' produced in this fashion was successfully sold by Islām Ākhūn in 1895 to Munshī Aḥmad Dīn, who was in charge of the Assistant Resident's office at Kāshgar during the temporary absence of Mr. Macartney[12]. These 'books' were written by hand; and an attempt had been made, as also in some others of the earliest products of the factory, to imitate the cursive Brāhmī characters found in fragments of genuine manuscripts which Ibrahīm was said to have secured from Dandān-Uiliq[13]. Though the forgers never succeeded in producing a text showing consecutively the characters of any known script, yet their earliest fabrications were executed with an amount of care and ingenuity which might well deceive for a time even expert scholars in Europe. This may be seen by referring to the facsimiles which are given in part ii of Dr. Hoernle's *Report* of Brāhmī letter-groups from 'codices' belonging to the early output, now deposited, with so many other products of Islām Ākhūn's factory, in the 'forgery' section of the Manuscript Department of the British Museum[14]. The specimen pages reproduced from the 'Codices Högberg' in M. D. Klementz's Report on the Imperial Russian Academy's expedition to Turfān[15], and the facsimile of an 'ancient Khotan manuscript' which appears in the German edition of Dr. Sven Hedin's *Through Asia*, are conveniently accessible illustrations of the factory's output in a somewhat later and less careful phase of its working.

Increased output of factory.

Seeing that remunerative prices could be obtained for such articles at Kāshgar and, through Badruddīn's somewhat careless mediation, also from Ladāk and Kashmīr, the efforts of the forgers were stimulated. As Islām Ākhūn quickly perceived that his 'books' were readily paid for, though none of the Europeans who bought them could read their characters or distinguish them from ancient scripts, it became unnecessary to trouble about imitating the characters of genuine fragments. Thus, apparently each individual factory 'hand' was given free scope for inventing his 'unknown characters'. This explains the striking diversity of these queer scripts, of which the analysis of the texts contained in the 'British collection' at one time revealed at least a dozen—not exactly to the assurance of the Orientalist scholars who were to help in their decipherment.

Production of 'block-printed' books.

The rate of production by the laborious process of handwriting was, however, too slow, and accordingly the factory took to the more convenient method of producing books by means of

[11] See, e.g., Hoernle, *Report on C.-A. ant.*, Pls. V–VII, IX; also a 'Note on some block-prints from Khotan,' in *Proc.A.S.B.*, 1898, pp. 130 sq., where facsimiles of these curious block types are given.

[12] See Hoernle, 'Three further collections,' &c., in *J.A.S.B.*, 1897, pp. 237 sq.

[13] See for such early specimens from the sets of written leaves sold to M. Ahmad Dīn, Hoernle, 'Three further collections,' &c., *J.A.S.B.*, 1897, Pls. XI–XV, XVII; also Pls. XVIII–XIX and *Report*, ii. Pl. I, from sets sold subsequently to Mr. Macartney. For the MS. leaves reproduced in Pls. XI–XIII, ibid., the characters of the genuine Chinese documents from Dandān-Uiliq may be supposed to have served as models.

[14] See *Report on C.-A. ant.*, ii. pp. 4 sq., where the reproduction of the identical Brāhmī letter-group in the formula of one of the 'block-prints', but with the characters reversed owing to the forger's mistake in preparing the block, has also been lucidly explained.

[15] See *Nachrichten über die von der Kais. Akademie der Wissenschaften zu St. Petersburg im Jahre, 1898, ausgerüstete Expedition nach Turfan*, i. Pl. 8.

repeated impressions from a series of wooden blocks. The preparation of such blocks presented no difficulty, as printing from wooden blocks is extensively practised in Chinese Turkestān. This printing of 'old books' commenced in 1896, and its results are partly represented by the forty-five 'block-printed' books which are fully described and illustrated in Dr. Hoernle's *Report*. These, too, showed a remarkable variety of scripts in their ever-recurring formulas, and were often of quite imposing dimensions in size and bulk[16].

Technical details of manufacture of 'old books'.

Islām Ākhūn, when once his defence had collapsed, was not chary about giving technical details about the forgers' methods of work. In fact, he seemed rather to relish the interest I showed in them. Thus he fully described the procedure followed in preparing the paper that was used for the production of manuscripts or 'block-prints', as well as the treatment to which they were subjected in order to give them an ancient look. The fact of Khotan being the main centre of the Turkestān paper industry was a great convenience for the forgers, as they could readily supply themselves with any variety and size of paper needed. The sheets of modern Khotan paper were first dyed yellow or light brown by means of 'Toghrugha', a product of the Toghrak tree, which, when dissolved in water, gives a staining fluid. When the dyed sheets had been written or printed upon they were hung over fireplaces, so as to receive by smoke the proper hue of antiquity. It was, no doubt, in the course of this manipulation that the sheets occasionally sustained the burns and scorchings of which some of the 'old books' transmitted to Calcutta display evident marks[17]. Afterwards they were bound up into volumes. This, however, seems to have been the least efficiently managed department of the concern; for the coarse imitation of European volumes, which is unmistakable in the case of most of the later products, as well as the utter unsuitability of the fastenings employed (usually pegs of copper or mere twists of paper), would *a priori* have justified the gravest suspicions as to their genuineness. Finally, the finished manuscripts or books were treated to a liberal admixture between their pages of the fine sand of the desert, in order to make them tally with the story of their long burial. I well remember how, in the spring of 1898, I had to apply a clothes brush before I could examine one of these forged 'block-prints', forming part of a collection of Khotan 'antiques' that had been purchased for Government by Major Godfrey, then in Kashmīr.

Binding of 'old books'.

Facility of distinguishing forgeries.

All the previously suspected details of this elaborate and, for a time, remarkably successful fraud were thus confirmed by its main operator in the course of a long and cautiously conducted examination. It was a pleasure to me to know, and to be able to tell fellow-scholars in Europe: *habemus confitentem reum*—and that without any resort to Eastern methods of judicial inquiry. Yet I had reason to feel even keener satisfaction at the fact that the positive results of my explorations were sufficient to dispose once for all of these fabrications so far as scholarly interests were concerned, even if Islām Ākhūn had never made his confession. In the light of the discoveries which had rewarded my excavations at Dandān-Uiliq and Endere, and of the general experience gained during my work in the desert, it had become as easy to distinguish Islām Ākhūn's forgeries from genuine old manuscripts as it was to explode his egregious stories about the ancient sites which were supposed to have furnished his 'finds'. Not only in the colour and substance of the paper, but also in arrangement, state of preservation, and a variety of other points, all genuine manuscripts show features never to be found in Islām

[16] See *Report on C.-A. ant.*, i. pp. 45–110; Pls. V–XVIII. Also ibid., ii. pp. 4 sq. and note 14 *supra*.

[17] See Hoernle, *Report*, i. pp. 58 sq. The native information which had reached the late Mr. Bäcklund, of the Swedish Mission at Kāshgar, in regard to this procedure of the forgers proved quite correct. Yet the warning based upon these details did not suffice to stop the continued purchase and study of the productions.

Ākhūn's productions. But apart from this, there was the plain fact that the forgers never managed to produce a text exhibiting consecutively the characters of any known script, while all ancient documents brought to light by my explorations invariably show a writing that is otherwise well known to us. I could thus feel assured that Islām Ākhūn's forgeries would cause no deception hereafter, whatever ingenious arguments might have been used before in defence of their genuineness.

Previous punishments of Islām Ākhūn.

This consideration, as well as the fact of the forgers' work having ceased some three years earlier, decided me not to press for Islām Ākhūn's punishment on the score of this fraud. I knew besides that my kind-hearted friend Pan Dārin was not without reason popularly credited with a pious proneness for pardoning sinners. In fact, I had noticed during our previous interview how relieved the old Amban looked when I told him that I did not consider it a part of my business to demand Islām Ākhūn's punishment for antiquarian forgeries, of which Chinese criminal justice might, perhaps, take a view very different from ours. There was also the manifest difficulty of bringing the other members of the firm to book, not to mention the 'extenuating circumstances' connected with the way in which encouragement had been afforded to the fraud by the undiscriminating competition of purchasers both from the south and the north. Nevertheless, when I remembered the great loss of valuable time and labour which the fabrications of Islām Ākhūn and his associates had caused to scholars of distinction, it was a satisfaction to know that this clever scoundrel had already, on one count or another, received from Chinese justice his well-deserved punishment. For fraudulently obtaining from Badruddīn Khān a sum equivalent to about Rs. 12, on the strength of a scrawl which he pretended to be Captain Deasy's order, he had been made to wear the wooden collar for a good time; for the imposture practised as Mr. Macartney's agent he had suffered corporal punishment as well as a term of imprisonment.

Intelligence of forger.

I had ample opportunity in the course of these prolonged 'interviews' to convince myself that Islām Ākhūn was a man of exceptional intelligence for those parts, and also possessed of a quick wit and humour, equally unusual among the ordinary 'Khotanliks'. He was of slender build, with a face and eyes expressing sharpness as well as sly restlessness. Something in his looks I thought suggested Kashmīrī descent, but this I was not able to trace. He greatly amused me by his witty repartees to honest old Turdi, whom with humorous impudence he adduced as a living demonstration of the fact that 'there was nothing to be got out of the desert'. He was greatly impressed by seeing his own handiwork so perfectly reproduced in the photogravure plates accompanying Dr. Hoernle's *Report* and previous publications in the *Journal* of the Bengal Asiatic Society, and was very anxious to learn how this feat could be accomplished. I have no doubt that he was fully alive to the splendid opportunities for fresh frauds which this 'Wilāyatī' art might provide. I did not care to tell him how much money had been wasted over this superior reproduction of his 'old books', nor of all the painstaking analysis and study which had been bestowed upon them. How much more proud would he have felt if he could but have seen, as I did a few months later, the fine morocco bindings with which a number of his block-printed codices had been honoured in a great European library!

Islām Ākhūn's final attitude.

I represented to Islām Ākhūn that, willing as I was to credit him with a reliable memory concerning the methods and materials employed in his factory, it would still be desirable for me to obtain some tangible memento of them. So he at once volunteered to furnish one or more of the blocks employed in printing those precious 'books'. As all information had by that time been duly recorded, I allowed him to be set free conditionally from the lock-up of the

Ya-mên, and on the following morning he turned up in due course with one of the promised blocks from his own house. The news of his arrest had, of course, long before spread through the town, and hence it was difficult for him to gain access to the homes of his former associates, where more of these materials may have been retained.

Whether it was from a right perception that his rôle was now completely played out, or because he felt that his ignominious collapse in the course of the inquiry had rendered him ridiculous before his old friends, Islām Ākhūn looked far more cowed in the end, though free, than when first brought up as a prisoner. I had told him before in jest that I thought him far too clever a man to be allowed to remain in Khotan among such ignorant people. A curious incident showed that the remark had not passed unappreciated. Shortly before my departure Islām Ākhūn presented himself with a petition, evidently meant to be serious, praying that I would take him along to Europe. It was not clear in what capacity he expected me to utilize his services. But there could be no doubt that the strange request was prompted by the hope of finding in distant 'Wilāyat' a wider sphere for his forging abilities! So I need not regret, perhaps, having shown myself obdurate.

Section II.—LAST DAYS IN THE KHOTAN OASIS

Farewell to Amban Pan Dārin.

On April 28 I left Khotan town, after having on the preceding day paid my farewell visit to the Ya-mên. It meant good-bye to Pan Dārin, who had proved in every way a true friend, and to whose help I owed so much in the course of my explorations. He was unmistakably a man of the old school, not over fond of Western notions and influences. Yet from my first visit I felt assured that he understood my scientific aims and was ready to further them. T'ang-sêng himself, to whom I had so often referred in our interviews as my patron, and who evidently still lives as a glorified Arhat or Bodhisattva in the memory of Chinese Buddhists, could not have favoured my tasks at ancient Kustana more than by placing this learned and kind old administrator in charge for the time being!

Visit to *Kara-kāsh*.

I did not wish to leave the oasis without a farewell visit to the site of the ancient capital, Yōtkan. The march being a short one, I had ample time to collect there samples of soil from the different strata which contain the ancient remains and from the deposit of riverine loess that has buried them. I was also able to acquire that day an additional number of ancient coins, seals, terra-cottas, &c., the owners of which had not come forward on the former occasions. From Yōtkan I proceeded on April 29 to the canton and town of Kara-kāsh, by the route which the inset map of the oasis indicates. I had not found a previous opportunity to visit this important and flourishing centre of the western portion of the oasis, and had now an additional reason to look it up before my departure. Islām Bēg, my faithful Darōgha of the days of Karanghu-tāgh and Dandān-Uiliq, had since been appointed one of the Bēgs of Kara-kāsh. Rightly or wrongly he attributed his good fortune to my recommendation with the Amban. So he was anxious to show me Kara-kāsh, both as his native place and the present sphere of his official functions, while I could use the occasion to secure interesting details about local administration, taxes, &c., from a first-hand authority.

Changes in cultivated area.

While *en route* I enjoyed the rare chance of seeing in perfect clearness the great snowy range to the south which we had surveyed six months before. This distant view, which seemed to extend to the big glacier-crowned main peaks on the watershed towards the Upper Kara-kāsh nearly a hundred miles away, was ascribed to recent snowfall in the mountains and the rain that had cleared away all haze. It was facilitated by the large stretch of open sandy ground between the villages of Tasmache and Khān-arik, which had long been deserted but was beginning to be again brought under cultivation. Also at the point where we struck the right bank of the broad bed of the Kara-kāsh river near the village of Kumuchakar, I had occasion to note a change in the area of cultivation. Here a narrow strip of ground that had evidently been left without irrigation for a prolonged period showed all the characteristics of an incipient Tati. Plentiful pottery, but not of an old look, appeared over loess soil that evidently was undergoing erosion in the immediate vicinity of a fertile village tract.

March through tracts west of Kara-kāsh.

From Kara-kāsh, a lively and comparatively well-built town, I used April 30, my last day within the territory of Khotan, for a long excursion to the Tati known as *Kara-döbe* 'the Black Mound', of which Islām Bēg had obtained information, away to the west on the edge of the desert. In order to reach it we had to traverse the remarkably fertile tracts of Bahrām-su[1], Kayāsh, Mākuya, and Kuya, all stretching in long strips of highly cultivated ground, with many fine orchards and avenues, along their own separate canals fed by the Kara-kāsh. No more pleasing picture could I retain as a souvenir of rural Khotan; but the usual haze had already effaced the vision of its mountain background. Near the Kara-sai stream, which is fed by springs and marshes below Zawa, we left the cultivated area behind and then crossed for about four miles a sandy scrub-covered plain to the Yawa-Üstang. The latter is a permanent stream which receives most of its water from extensive spring-fed marshes north of Zawa-Kurghān, but during flood-times is also reached by water direct from the Kara-kāsh river. Its relatively large volume—where we crossed the stream it was 30 ft. broad, with a depth of about 4 ft., and flowing rapidly—showed that there would probably be no want of water for new colonies in the desert northward.

Remains of *Kara-döbe* site.

At a distance of about two miles to the west I reached, among low dunes, the Tati of *Kara-döbe*, called thus after a neighbouring large tamarisk-covered cone. A stretch of eroded ground, thickly covered with ancient pottery wherever clear of drift-sand, extends here for about a mile from north to south. Owing to high dunes along its east and west edges the full breadth of the débris area could not be ascertained in the short time available. In its midst rises a much-decayed small mound of sun-dried bricks, known as the 'Tim', 9 ft. high, and about 8 ft. square at the base. It has been burrowed into in several places. The intact bricks measured about 19 by 13 in., with an average thickness of $3\frac{1}{2}$ in. Close by fragments of small Chinese copper coins, probably without legend, were picked up by my companions. At a distance of about a hundred yards to the south I could faintly trace the foundations of clay walls, forming apparently a rectangle measuring about 25 ft. from north to south, and 32 ft. from east to west. Within the area thus enclosed there were found several broken pieces of a white and hard stucco, among them some showing traces of relief decoration. Prof. Church's analysis of the latter specimens, in Appendix *F*, has proved that their material was plaster of Paris, just as in the stucco reliefs from Kighillik; their appearance suggests that, like the latter, they had been exposed to accidental fire after having become broken. Two fragments (K. D. 001. d) are of a moulded enrichment, showing a pattern that had evidently formed a border; another (K. D. 001. f) with a surface resembling locks of hair, displays on the back the impression of a fine canvas over which the plaster must have been fixed originally. These pieces might well have belonged to the stucco decoration of a small shrine, of which, as at Kighillik, no structural remains have survived. Fragments of ancient pottery, among them pieces of terracotta with a polished outer surface, were plentiful near the find-place of these stucco fragments.

[1] This canton derives its name from that of its canal, the *Bahrām-su-Üstang*, which tradition ascribes to Bahrām, a son of Afrāsiyāb. The name may be of early date, and in this case a lingering trace of old Īrānian legend.

March to Zawa-Kurghān.

Leaving in the direction of Zawa I passed several small patches of ground covered with old pottery débris up to a distance of about a mile from the mound above mentioned. Beyond, heavy dunes of coarse sand, very trying to our ponies, had to be crossed for some four miles before we struck the western bank of a broad marshy valley in which the stream of Yawa expands between numerous reed-covered lagoons. They vividly recalled to my mind the marshes near Niya in the extreme east of ancient Khotan territory. And when by nightfall I arrived at my camp pitched close to Zawa-Kurghān, I might well feel as if, by these changes of rich village land, sandy jungle, high dunes and marsh, Vaiśravaṇa, the *genius loci* of Buddhist Khotan, had wished to let me once more see, as a parting favour, every type of scenery I had beheld in the land over which he presided.

Farewell to Turdi Khōja.

By daybreak of the 1st of May I set out on my long journey westwards. Cheered as I was by the thought of the road that now lay clear before me to Europe, it was sad to say farewell to a fascinating field of work and to the last of my faithful helpmates. At Zawa itself I had to take leave of Turdi, my faithful old guide, whose experience and local sense had never failed me in the desert. I rewarded his services liberally with more 'treasure', i. e. cash, than he had ever brought back from his wanderings in the Taklamakān. I had also secured for him a promise, through Pan Dārin's favour, that he was to be installed as 'Mīrāb', or steward of irrigation, for his native village near Yurung-kāsh. It was a snug though modest post to which our 'Ak-sakāl of the Taklamakān', as I used to call him, fondly aspired, since he thought that he was getting too old for the desert. When we exchanged our farewell wishes I little thought for how short a time the fateful attraction of the desert would allow the old 'treasure-seeker' to enjoy the comforts of this peaceful retirement. The circumstances of his somewhat tragic end are not yet fully known to me. But it is certain that towards the close of 1903 Muḥammadju, one of my Yarkandī ponymen, when passing through Khotan with Mr. Crosby's caravan, found old Turdi imprisoned, apparently for having been tempted by his love of adventure and 'treasure', to guide some traveller (?) into the desert without the permission of Pan Dārin's successor. A short while later he ended his days, as I am told, while still in prison at Khotan.

Islām Bēg and Badruddīn Khān, who had reason to be satisfied with the rewards their efficient services had earned them, would not leave me until we had reached Tarbugaz, the lonely Langar on the desert edge where I had passed my first night on Khotan soil. It seems that convention, perhaps of old date, makes this the recognized place for the formal welcome and valediction of those who enter or leave the oasis by the western road. When they, too, had bidden me good-bye, and I was riding on alone by the desert track to the 'Pigeons' Shrine', my thoughts were free to turn to a more cheerful theme—the results I was bringing back from Khotan.

POTTERY AND STUCCO FRAGMENTS FROM KARA-DÖBE.

K. D. 001. a. Fragment of moulded terra-cotta vessel; outer surface polished. $2\frac{1}{2}'' \times 2''$.

K. D. 001. b, c. Fragments of plain terra-cotta, polished on outside. $2\frac{1}{2}'' \times 2\frac{1}{2}''$ and $1\frac{1}{4}'' \times 1\frac{1}{2}''$ resp.

K. D. 001. d. Two fragments of plaster of Paris; moulded enrichment. The pattern (a border) consists of 5 or 6 strands of overlapping scales or leaves, bound by crossing bands. $2\frac{1}{4}'' \times 1'' \times \frac{1}{2}''$, $2\frac{3}{4}'' \times 1\frac{1}{8}'' \times \frac{5}{8}''$.

K. D. 001. e. Fragment of plaster of Paris; border, plain, 5-sided.

K. D. 001. f. Fragment of plaster of Paris; enrichment, resembling locks of hair. On the back is the impression of a rather fine canvas. These fragments d, e, f) resemble in material the Ak-sipil burnt stuccos (Cf. A. 024) and, like those, appear to have been burnt by accident after having become broken.

Section III.—FROM KHOTAN TO LONDON

Of the journey which, within two months of my start from Khotan, brought me back to Kāshgar and thence through Russian Turkestān to England, the briefest account will suffice here. Six rapid marches, diversified by Burāns and the rare experience of gathering rain-clouds, carried me to Yarkand, where my caravan had safely preceded me. It was fortunate that, owing to the short stay I was obliged to make at Yarkand for the settlement of my Yarkandī followers' accounts and debts, my collections escaped serious risk of damage from an abnormal burst of rain such as this region had not seen for long years. The downpour of two days and two nights turned all roads into quagmires and caused the mud-built walls of many houses in town and villages to collapse. Return to Kāshgar.

A ride of three days, in advance of my caravan, sufficed to bring me by May 12 to Kāshgar, where, under Mr. Macartney's hospitable roof, the warmest welcome greeted me. It was a source of keen satisfaction to me to show that kind friend, who had from afar followed my explorations with constant interest, what ample results had attended my work, and how much I owed to that local help which his influence and care had mainly assured to me. The kind hospitality I enjoyed at his house made my stay at Kāshgar a period of much needed physical rest, in spite of the multifarious preparations which kept me constantly at work during the next fortnight.

The Government of India in the Foreign Department had obtained for me permission from the authorities in St. Petersburg to travel through Russian Turkestān and to use the Trans-Caspian Railway for my return to Europe. I had also been authorized to take my collections for temporary deposit to England, where alone convenient arrangements could be made for their critical examination. It was hence necessary at Kāshgar to repack all my antiquarian finds with special regard for safe transit on this long journey, while all surveying instruments and other equipment, together with the records of our surveys, were to be sent back to India in charge of the Sub-Surveyor. In addition to the fresh transport arrangements thus necessitated by our different routes, I was kept busy also with the 'demobilization' of my old caravan. Among the items of business connected with the latter, the satisfactory disposal of camels and ponies which had served us so well during the journeys of the previous eight months was an important concern. That after all the hard marching and camping in the desert I was able to sell them with but very small loss to Government from the original prices may serve as a proof of the care we had taken of our animals. 'Demobilization' of caravan.

The arrangements for my onward journey were greatly facilitated by the kind help of M. Petrovsky, late Imperial Consul-General of Russia at Kāshgar, whose acquaintance I was fortunate enough to make on this occasion. During a long official career in Turkestān M. Petrovsky had devoted a great deal of scholarly zeal to the study of the history and antiquities of the country. He now did all in his power to ensure the safe transit of my archaeological finds to England and to procure for me the friendly assistance of the authorities in Russian Turkestān. For the valuable help thus accorded I wish to record here my grateful acknowledgements. Help of Russian Consul-General.

It was a source of genuine satisfaction to me that I had repeated occasions to meet also Huang-kuang-ta, the kindly old Tao-tai of Kāshgar, and to assure him of my gratitude for the most effective co-operation which I had received from the Chinese officials wherever my explorations took me within the districts subordinate to his control. The amiable old administrator did not deny the genuine interest and good-will with which he had followed my work. But he Assistance of Chinese officials.

politely insisted on attributing all the sympathy and support I had enjoyed from him and his Ambans to the benediction of my patron saint, the great 'T'ang-sêng'. While still engaged in the 'demobilization' of my camp and other practical labours, I also succeeded in obtaining, with Mr. Macartney's kind help, a preliminary analysis of the Chinese documents brought to light at the sites of Dandān-Uiliq, Niya, and Endere, from Sun Ssŭ-yeh, the learned Chinese Munshī of the Agency. The close accord between the chronological results derived from these records and my own archaeological conclusions as to the ruined sites from which I had unearthed them, was an assurance greatly appreciated by me at the time.

After a fortnight of busy work my preparations for the rest of the journey were completed, and all my antiques safely packed in twelve large boxes. They were duly presented at the Russian Consulate for customs examination—a most gently conducted one—and then received their seals with the Imperial eagle, which in spite of a succession of Continental customs barriers, I succeeded in keeping intact until I could unpack their contents in the British Museum.

Surveyor Rām Singh.

Owing to Mr. Macartney's unwearying help and hospitality my stay at Kāshgar, full as it was of manifold labours, is still remembered by me with pleasure as the first and practically only rest after my desert wanderings. On the morning of the day when it came to an end I saw Sub-Surveyor (now Surveyor) Rām Singh, the faithful companion of my travels, set out for the return journey to India. He had rendered excellent services in accurately surveying the whole of the ground covered by my journeys, and had in addition to his proper duties been always eager to make himself useful in connexion with my archaeological work. He had cheerfully borne the fatigues inseparable from rapid travelling over difficult ground, and often under very trying climatic conditions, and had also at all times readily helped me in the management of my camp. I had indeed reason to feel grateful to the Survey of India Department for having provided me with so willing and well-trained an assistant. Nor should I omit here to mention Mīān Jasvant Singh, the wiry hill Rājpūt, who had looked after the Surveyor's bodily comforts with exemplary care and devotion, and who now left with him. Cheerful and contented, however long the march or bleak our camping-ground, Jasvant Singh had proved an ideal follower.

Journey to Farghāna.

On May 29, 1901, exactly a year after leaving Srīnagar, I started from Kāshgar for Osh, the nearest Russian town in Farghāna. The feeders of the Kizil-su, which the route repeatedly crosses before reaching Irkesh-tam, the Russian frontier post[1], were, owing to the exceptional rain of the previous weeks and the rapid melting of the snows, all in flood. The passage of my loads of antiques across the swollen streams was thus a daily anxiety, but was effected without loss. The usual route over the Terek Pass was closed by the depth and softness of the snow; so I had to take the circuitous route over the Alai. Though the snow still lay deep on the Taun-murun and Taldik Passes they were crossed without mishap. By keeping in the saddle or on foot from early morning until nightfall I managed to cover the route from Kāshgar to Osh, reckoned at eighteen marches, within ten days.

Assistance of authorities in Russian Turkestān.

At Osh, where I arrived on June 7, I was very kindly received by Colonel Zaytseff, the Chief of the District, and an officer of distinguished attainments. Two days later, at Andijān, I reached the terminus of the Trans-Caspian Railway, which was now to carry me and my antiques in comfort and safety towards Europe. This journey, hurried though it had to be under the circumstances—the year allowed for my deputation had already expired when I left Kāshgar—enabled me to obtain interesting glimpses of a part of Central Asia which, from its historical associations and its ancient culture, has always had a particular fascination for me.

[1] See, for this locality and its historical interest, above, p. 55.

I made short halts at the provincial capitals of Margilān and Samarkand, where I was favoured with much attention by Generals Tchaikovsky and Medinsky, the respective governors, and offered special opportunities for examining the interesting antiquities collected in the local museums.

The few but delightful days which I spent at Samarkand, the Delhi of old Sogdiana, mainly in visits to the great monuments of architecture of Tīmūr's period, were unfortunately too short to permit of more than a glimpse of the important ancient site known as *Afrāsiyāb*, which undoubtedly marks the position of the capital in pre-Muhammadan times. The conditions in which fragments of ancient pottery and similar hard débris are found here under layers of alluvium distinctly recalled my observations at Yōtkan. The incentive of regular finds of gold being absent here, the working of the old culture-strata does not appear to be carried on with the same thoroughness as at Yōtkan, while their extent is probably far greater. Some interesting pieces of ornamented terra-cotta from this site which I was able to acquire will be found reproduced in Plate LXXXVIII, and described in the note below [1]. Among them the large modelled ornament in relief (S. 001) shows a design of remarkably fine conception, and

Antiquities at Samarkand.

[1] It is noteworthy that among the seven pieces of decorated terra-cotta from Afrāsiyāb which I was able to hunt up in the Samarkand Bāzārs, four are fragments of ornamented tiles either modelled or moulded. No remains of this kind seem ever to be found at Yōtkan; their presence among the débris layers of ancient Samarkand is significant, in view of the high artistic development which the production of moulded and enamelled tiles attained here in the period of Moghul grandeur.

S. 001. Modelled terra-cotta 'string' ornament in relief, roughly executed, but of fine design. A bold nebulée moulding meanders horizontally along the face of the tile, crimped with rough 'thumb and finger' work; in the hollow of each downward bend, *appliqué*, a female head of a Madonna-like type, draped and looking upwards. A leaf-like border made by rapid incisions of a blunt modelling tool is indicated near the upper edge, and similar incisions are scattered over the ground below the nebulée. The vertical transverse section is L-shaped, the horizontal member of the L seeming to be constructed for building into a vertical wall or other surface. Broken into three pieces. Length 20″, breadth 7½″, depth (front to back) 1″. See Pl. LXXXVIII.

S. 002. Modelled terra-cotta architectural tile. On a projecting moulding stands a bird in relief with spread tail indicated on background by a few incised strokes of a modelling stick. Ground R. and L. of bird powdered with slightly depressed circular markings; below the moulding, plain. Small fragment, broken at all edges. Height 6″, width 5″, thickness ⅝″. See Pl. LXXXVIII.

S. 003. Terra-cotta fragment in low relief, of moulded tile, representing head and half the body of a pigeon, wearing a ring round the neck. A curved line in relief arches over the pigeon and rests at its visible end upon a system of quarries in raised outline, in the centre of each of which is a four-petalled flower, also raised. From the arched line above spring a few leaves. Two quarries are also visible over the back of the bird. Height 3½″, width 2½″, thickness ½″. See Pl. LXXXVIII.

S. 004. Terra-cotta fragment of moulded border in low arabesque ornament, representing a grotesque bird to L. p., pecking at a bush. Flat mouldings above and below. Well designed. Length 2¾″, width 2¼″, thickness ½″. See Pl. LXXXVIII.

S. 005. Terra-cotta fragment of handle of vessel, impressed with a device in cameo representing within a circular border of depressed bead-like dots a seated lion to L., the upward curved tail dividing into three leaf-like points at tip. Mouth slightly open and tongue visible. Claws rather bird-like. It is quite a good heraldic lion. In front of lion three discs disposed triangularly. Width 2¼″, depth 1½″, thickness ¾″. See Pl. LXXXVIII.

S. 006. Terra-cotta fragment of grotesque animal, probably forming part of handle of a vessel. Legs missing, but there seem to have been four. Round projecting snout; prominent eyes; erect short-cropped mane, curving slightly over at forehead. There seems to have been a kind of saddle, and possibly a rider, as the upper part of saddle presents a broken surface. Width 1¾″, height 2⅛″, thickness 1¼″. See Pl. LXXXVIII.

S. 007. Terra-cotta figurine, roughly modelled, of a *Sitār* player. Head and feet missing. It seems to be clothed in two kurtas, the upper one hanging almost as low as the ankles, and the upper, which has sleeves, coming to about the knees. Round neck a necklace. Sitār held in usual way, and R. hand seems to be plucking the strings. Potter's thumb mark is clearly visible on back, showing how the clay was pressed into the mould. After being turned out of the mould it has been trimmed with a cutting instrument, and is brushed over with a finer clay, giving an egg-shell gloss to the surface. Height 4½″, width 2⅛″, thickness 1⅛″. See Pl. LXXXVIII.

appliqué female heads in a style unmistakably betraying classical influence. The small figurine of a Sitār player (S. 007) and the fragment of a grotesque animal (S. 006) serve to indicate the curious similarity in favourite subjects which seem to connect the Afrāsiyāb terra-cottas with those of Yōtkan. The numerous terra-cotta grotesques seen by me in the Samarkand Museum illustrate this, as well as certain characteristic differences in treatment.

Journey by Trans-Caspian Railway.

A day's stay at Merw allowed me to touch ground full of memories of ancient Īrān. Though the chance of seeing the remains of old Môurwa, the capital of classical Margiana, brought to light from below the alluvium of the oasis, seems scanty indeed, I felt grateful for standing on the soil of a region to which my interest has been attached ever since the commencement of my Orientalist studies. Then past the ruins of Göktepe, an historical site of more recent memories, the railway carried me to Krasnowodsk. From there I crossed the Caspian to Baku, and, finally, after long days in the train, I arrived in London on July 2, 1901.

Arrival in London.

There I was able to deposit the antiques unearthed from the desert sands in the British Museum as a safe temporary resting-place. It was a relief to find that the long and partly difficult transit of close on six thousand miles had caused but slight damage even to those most fragile of objects the reliefs of friable stucco, and that the eight hundred odd negatives on glass plates brought back as the photographic results of my journey were safe. But I soon realized that the successful completion of my exploratory labours, which had been rewarded by results far beyond long-cherished hopes, was also the commencement of a period of toil, the more trying because the physical conditions under which it had to be done were so different from those I had gone through.

Arrangement of collection of antiques.

On the proposal of the Indian Government His Majesty's Secretary of State for India had sanctioned for me a six weeks' period of deputation in London, in order to enable me to make a preliminary arrangement and inventory of my archaeological finds. The authorities of the British Museum, acting on the request submitted for me by Professor E. J. Rapson, agreed to afford accommodation to the collection, pending arrangements for its final distribution between their own museum and the museums of Calcutta and Lahore—a measure decided upon while I was still engaged in my explorations. They also liberally accorded to me all needful assistance towards the cleaning and preservation of the more delicate objects, such as manuscripts, painted tablets, &c. Owing to the great extent of the collections I had succeeded in bringing back the task of arranging and cataloguing proved a very exacting one, and the period of deputation sanctioned for it wholly insufficient. In consideration of these facts the Secretary of State was pleased to extend the latter by another period of six weeks. I had every reason to feel grateful for this concession. Yet it was only at the cost of the greatest exertions and through the devoted help of my friend Mr. Fred. H. Andrews, which I was fortunate enough to secure at this juncture, that I succeeded in accomplishing the most urgent portions of that heavy task and the preparation of my *Preliminary Report* during the allotted period.

Return to Indian duties.

The necessity under which I then was of returning to India for educational duties wholly unconnected with my scientific labours seemed seriously to threaten my hopes of being allowed fully to record and investigate the discoveries which had rewarded my Turkestān explorations. But after long and busy weeks, spent mainly in the basement rooms of the British Museum, where I felt as if immured for the sake of science, I had to welcome even this change as a temporary respite. How much easier would it have been for me, then and since, to face such separation from my true aims of life, could I have felt but assured of the leisure needed for the accomplishment of the work now concluded—and of freedom for fresh explorations!

APPENDIX A

CHINESE DOCUMENTS

FROM THE SITES OF DANDĀN-UILIQ, NIYA AND ENDERE

TRANSLATED AND ANNOTATED

BY

ÉDOUARD CHAVANNES

MEMBRE DE L'INSTITUT, PROFESSEUR AU COLLÈGE DE FRANCE, ETC., ETC.

PART I

LES DOCUMENTS CHINOIS DE DANDĀN-UILIQ

No. 1.

Traduction du document manuscrit chinois No. 1, publié et analysé par M. Hoernle.

(*A Report on the British Collection of Antiquities from Central Asia*, Part ii. p. 24, and Plate iii.)

Lettre officielle. La population de Li-sie (Li-hsieh)[1] ainsi que . . . .

Requête de la population de Li-sie (Li-hsieh) se plaignant de diverses corvées et réquisitions en grains.

Une lettre officielle émanant du général gouverneur de la place[2] s'exprime ainsi: 'J'ai reçu de la population de Li-sie (Li-hsieh) une lettre en écriture barbare; je l'ai fait traduire; elle disait: "La population énumérée ci-dessus[3] . . . .[4] la profonde sollicitude avec laquelle vous subvenez aux besoins de la multitude du peuple. Pendant plusieurs années de suite nous avons souffert des brigands et nos pertes ont été incalculables; dans ces derniers temps, nous avons reçu la faveur qu'on nous fit nous transporter dans

[1] Le caractère [illegible], comme l'a déjà fait remarquer Hoernle (*Report on the British Collection of Antiquities from Central Asia*, Part ii. p. 22), n'existe pas dans les dictionnaires chinois; peut-être faut-il lire 傑 kie. La question ne pourra être élucidée que lorsqu'on aura retrouvé ce nom de lieu dans un texte chinois imprimé. Quant au second caractère figurant dans le nom, c'est le caractère 謝 sie.

[2] La place (鎮, c'est-à-dire la ville avec une garnison militaire) dont il est ici question, est sans doute celle de Li-sie (Li-hsieh); nous voyons en effet par le document No. 2 (traduit plus loin) que Li-sie (Li-hsieh) était considéré comme un 鎮, et qu'il était gouverné par un officier militaire ayant le titre de 'Général commandant la place' 知鎮官將軍. Cet officier était un Chinois (voyez le document No. 2), et c'est pourquoi, comme on va le lire quelques lignes plus bas, lorsqu'il reçoit des gens de Li-sie (Li-hsieh) une requête en écriture barbare 胡, il la fait traduire.

[3] Le mot qui suivait le mot 百 devait sans doute être 姓.

[4] Suppléez: 'Vous est reconnaissante de' — ou: 'fait appel à.'

les Six Cités[5]. L'année dernière, toutes les corvées et les réquisitions en grains que nous devions payer . . .[6], grâce à votre bonté. Cette année, il y a quelques petites corvées et réquisitions en grains pour lesquelles on nous a accordé d'attendre jusqu'à la moisson d'automne, afin qu'à cette date nous nous en acquittions . .[7] et le grain, tout cela est à Li-sie (Li-hsieh) et nous n'osons point aller l'y prendre. Nous espérons humblement que vous discuterez à ce sujet, de manière à décider de nous libérer de tout cela[8].'

— Ce qui est dit dans cette lettre au sujet des hommes (pour les corvées) et du grain (pour les réquisitions) qui sont tous à[9] Li-sie (Li-hsieh), ne donne pas lieu . . . .[10] que la requête écrite en écriture barbare par la

[5] Quelles sont ces Six Cités? On a proposé d'y voir les principales villes du Turkestan oriental, à savoir Koutcha, Aksou, Ouch-Tourfan, Kachgar, Yarkand et Khoten. Mais cette opinion doit être rejetée. En effet, la pièce que nous traduisons en ce moment est adressée à Wei-tch'e, préfet des Six Cités. Ce Wei-tch'e n'est autre, comme nous le montrerons plus loin, que Wei-tch'e Yaò, roi de Khoten, qui fut visité par le pèlerin chinois Wou-k'ong. Or, si Wei-tch'e Yao était roi de Khoten, il n'était pas roi de Kachgar, ni roi de Koutcha, car, au témoignage de Wou-k'ong, c'était P'ei Leng-leng 裴冷冷 qui régnait dans la première de ces deux villes, et Po Houan 白環 dans la seconde. Les Six Cités que gouvernait le roi de Khoten ne comprenaient donc ni Kachgar, ni Koutcha, et il n'y a aucune raison non plus pour faire rentrer dans ce nombre les villes de Aksou, Ouch-Tourfan et Yarkand. — Il faut donc admettre que l'expression 'les Six Cités' désigne la ville capitale du royaume de Khoten et cinq villes voisines. Dans les ouvrages chinois modernes on trouve en effet une énumération de ces six cités qui sont: Iltchi 伊立齊, Kara kach 哈拉哈什, Yurung kach 玉隴哈什, Tchira 齊喇, Thakhaga 塔克 ou 塔喀克 (cette localité se trouve sur la rive gauche du Khoten-daria non loin du confluent du Yurung kach et du Kara kach; elle doit correspondre au Tawakkel de nos cartes) et Keriya 克里雅 (cf. *Si yu t'ou tche*, chap. xix. p. 180; voyez aussi Hoernle, *Report on Central-Asian Antiquities*, Part ii. p. 22, No. 3). Quelles étaient les Six Cités à l'époque des T'ang? Je n'en ai trouvé nulle part une énumération formelle, mais on peut remarquer, en premier lieu, que, dans l'organisation administrative établie par les Chinois après l'année 640, Yu-t'ien (Khoten) eut sous sa dépendance cinq arrondissements et que l'on comptait par conséquent six villes (en y comprenant la capitale) dans la région de Khoten (*T'ang chou*, chap. xliii. b. p. 8 r°); en second lieu, on relèvera, dans les itinéraires qui terminent les chapitres géographiques du *T'ang chou* (chap. xliii. b. p. 15 r°), une liste de cinq villes qui sont mentionnées comme se trouvant dans le voisinage de Khoten et qui toutes sont appelées 鎮; à mes yeux, c'est Yu-t'ien et ces cinq villes qui sont les Six Cités de l'époque des T'ang; voici ce texte: 于闐東三百里有坎城鎮.東六百里有蘭城鎮.南六百里有胡弩鎮.西二百里有固城鎮.西三百九十里有吉良鎮. 'A 300 li à l'Est de Yu-t'ien est la place nommée Ville de K'an (c'est-à-dire "du fossé"); à 600 li à l'Est est la place nommée Ville de Lan (c'est-à-dire "des orchidées"); à 600 li au Sud est la place de Hou-nou; à 200 li à l'Ouest est la place appelée Ville Kou (c'est-à-dire "forte"); à 390 li à l'Ouest est la place de Ki-leang.' — La ville de K'an peut être identifiée avec la ville qui est appelée P'i-mo 媲摩 par Hiuan-tsang, puisqu'elle est, comme elle, à 300 li à l'Est de Yu-t'ien; quant à la ville de Lan, elle pourrait correspondre, avec moins de certitude, à la ville de Ni-jang 泥壤. (Niya) de Hiuan-tsang.

Revenons maintenant au document que nous traduisons: il émane de gens de Li-sie qui, pillés à plusieurs reprises par les brigands, avaient obtenu l'autorisation de se réfugier dans les Six Cités; je suppose que cette expression désigne ici seulement celle des six villes qui commandait aux cinq autres, à savoir la ville même de Yu-t'ien, qui était comme le chef-lieu du territoire appelé les Six Cités. D'autre part, la ville de Li-sie n'était point cependant entièrement abandonnée puisqu'il y était resté l'officier chinois chargé de la commander et un certain nombre de gens de basse condition qu'on employait pour les corvées. La pièce que nous traduisons est une lettre que le commandant chinois de la place de Li-sie adresse au roi de Khoten pour le prier de prendre certaines mesures administratives afin que les gens aisés de Li-sie, réfugiés dans la ville de Yu-t'ien, puissent venir chercher le grain qu'ils ont laissé à Li-sie et payer les réquisitions auxquelles ils sont astreints.

[6] Suppléez: 'ont été diminuées' ou 'ont été supprimées'.

[7] La suite du texte prouve que les deux mots manquants doivent être les mots 其人; la phrase signifie donc: 'les hommes (pour les corvées) et le grain (pour les réquisitions), tout cela est à Li-sie'.

[8] En d'autres termes, les gens de Li-sie qui avaient quitté cette ville pour se réfugier dans les Six Cités, déclarent qu'ils ne peuvent pas s'acquitter des corvées et des taxes qu'on réclame d'eux, car leurs hommes et leurs grains sont restés à Li-sie et ils ne peuvent pas aller les prendre. Ils demandent donc qu'on les libère purement et simplement des corvées et des taxes. Ici prend fin la requête des gens de Li-sie; c'est maintenant l'officier chinois qui va parler.

[9] Le mot 在 a dû être oublié ici par le scribe.

[10] Suppléez: 'à ce qu'on trouve juste ou admissible.'

population se plaigne de diverses corvées et réquisitions, afin qu'on prenne la décision de lui délivrer une lettre officielle par laquelle elle soit libérée [11]. Les hommes (pour les corvées) et le grain (pour les réquisitions), tout cela étant à Li-sie (Li-hsieh), je désire qu'on y envoie des hommes pour aller prendre le grain ; mais que nul ne se permette de le faire de son autorité privée ; pour que, quand on aura demandé une autorisation, dans chaque cas une lettre officielle soit le moyen par lequel (on sera autorisé à le faire), et pour que je vous invite à décider en outre que ces gens auront eux-mêmes la charge d'opérer ce transport, je fais cette lettre officielle.'

Lettre officielle de l'officier [12] Tch'eng Sien (Ch'êng Hsien), datée du 23ᵉ jour du 3ᵉ mois de la 3ᵉ année ta-li (768).

Adressée à Wei-tch'e (Wei-ch'ih), tche-lo (chih-lo) préfet des Six Cités et a-mo-tche (a-mo-chih) [13].

1. 牒傑謝百姓幷
2. 傑謝百姓狀訴雜差科等
3. 被鎭守軍牒稱。得傑謝百姓胡書。翻稱上件百
4. □□□□深憂獲蒼生。頻年被賊損莫知
5. 計。近日蒙差使移到六城。去載所着差科並納
6. □□□慈流。今年有小小差科放至秋熟依限輸
7. 納。□□粮並在傑謝。未敢就取。伏望商量者
8. 使判一切並放者。其人粮狀稱並傑謝未有處
9. □□□□百姓胡書狀訴雜差科准使判牒所
10. 由放。其人粮並在傑謝欲往使人就取粮。未敢
11. □擅執□諮取處分。說各牒所由者使又
12. 判任自般運者故牒
13. 大曆三年三月廿三日典成銑牒
14. 六城質邏刺史阿摩支尉遲

[11] Ainsi l'officier chinois refuse d'admettre que l'excuse invoquée par les ex-habitants de Li-sie pour ne pas payer les taxes soit valable. Il veut que ces gens viennent chercher leur grain à Li-sie ; mais, pour prévenir toute fraude, il entend que ces personnes, d'une part soient tenues d'obtenir du roi de Khoten une autorisation écrite, et d'autre part soient chargées de faire elles-mêmes le transport des grains. C'est pour ces deux raisons qu'il écrit au roi de Khoten.

[12] Le mot 典 paraît avoir le sens d' 'officier'. Cf. E. i. 44.

[13] Wei-tch'e 尉遲 est le nom de famille des rois de Khoten. Le roi qui régnait à Khoten en 768 était Wei-tch'e Yao 尉遲曜. En effet, dans la biographie de Wei-tch'e Cheng 尉遲勝 (*Kieou T'ang chou*, chap. cxliv. p. 6 vᵒ, *T'ang chou*, chap. cx. p. 7 rᵒ), nous lisons que Wei-tch'e Cheng, roi de Khoten, vint au début de la période tche-tö (756–757) en Chine et qu'il y resta ; en quittant Khoten, il avait confié le gouvernement à son frère cadet, Wei-tch'e Yao. Pendant la période kouang-tö (763–764), l'Empereur voulut renvoyer Wei-tch'e Cheng dans son pays ; mais celui-ci refusa et obtint qu'un édit impérial conférât le titre de roi de Khoten à Wei-tch'e Yao. Au début de la période tcheng-yuan (785–804), Wei-tch'e Yao demanda à la cour de Chine de désigner pour son successeur éventuel le fils de Wei-tch'e Cheng ; mais ce jeune homme préféra rester à la cour impériale. Lorsque le pèlerin Wou-k'ong revint en Chine, il séjourna six mois à Khoten ; c'était alors, semble-t-il, l'année 788 ou l'année 789, et le roi était encore Wei-tch'e Yao.

Le titre d'a-mo-tche 阿摩支 est cité par le *T'ang chou* (chap. ccxxi. a. p. 9 vᵒ) comme le titre du roi de Sou-le (Kachgar) et il semble bien que ce soit le même mot qu'on retrouve sous la transcription a-mi-kiue 阿彌厥 dans la notice du *Souei chou* (chap. lxxxiii. p. 5 vᵒ). En outre, dans le texte des deux brevets d'investiture par lesquels le gouvernement chinois conféra, en l'an 728, le titre de roi au roi de Khoten et au roi de Kachgar (*Ts'ö fou yuan kouei*, chap. dcdlxiv. p. 16 rᵒ), le roi de Khoten a le titre d'a-mo-tche de Yu-t'ien, tandis que le roi de Kachgar est appelé a-mo-tche de Sou-le (cf. mes *Documents sur les Tou-kiue occidentaux*, pp. 207–208).

Enfin le titre de ts'eu che 刺史 est celui que nous traduisons généralement par 'préfet' et nous voyons par l'histoire de Chine que ce titre était fréquemment donné à des princes barbares; mais je n'ai rencontré nulle part l'expression tche-lo 質羅 qui précède ici le titre de 'préfet' et qui paraît être une transcription d'un nom étranger.

A la fin de ce document (voyez Hoernle, *Report . . .*, II. pl. 3), on remarque le caractère 信 'digne de foi'; ce caractère, qui est écrit d'une main étrangère, est sans doute la marque autographe que l'officier Tch'eng Sien a apposée sur la copie de son scribe afin de l'authentiquer.

No. 2.

Traduction du document manuscrit chinois No. 2, analysé par M. Hoernle.

(*A Report on the British Collection of Antiquities from Central Asia*, Part ii. p. 24.)

Garnison de Li-sie (Li-hsieh). — Notification au magistrat civil.

Une peau de bœuf pour tapis de selles et pour tambours, et des plumes de caille[1] sont gâtées et ne valent plus rien; elles sont pourries et il serait difficile de les conserver. Cette peau, ainsi que les sabots et les os, je les mets ci-joints à votre disposition. Les objets précités sont des fournitures militaires à l'usage de la garnison. Ils sont endommagés, et, en ce qui concerne les flèches, on n'a pas de plumes pour en fabriquer. Quand cette notification vous sera parvenue, je compte que vous vous procurerez (ces objets) dans les quantités requises et que, dans un délai de cinq jours, vous nous les enverrez. Quand cette notification vous sera parvenue veuillez me répondre. Le 23ᵉ jour du 12ᵉ mois, notification faite.

Le général Yang Tsin-k'ing (Yang Chin-ch'ing), gouverneur de la place.

1. 傑謝鎮 帖知事
2. 韉鼓牛皮一張雜鳥鴿破碎不堪。焦爛難蓄。皮并蹄骨等
3. 右奉處分。上件等物爲鎮器械。破折損箭
4. 無鴿修造。帖至仰准數採覓。限五日內送納。
5. 帖至准覆。十二月二十三日帖
6. 知鎮官將軍楊晉卿

[1] Ces plumes devaient servir à empenner des flèches.

No. 3.

Analyse du document manuscrit chinois No. 3, publié et décrit par M. Hoernle.

(*A Report on the British Collection of Antiquities from Central Asia*, Part ii. pp. 24–5, and Plate iv.)

Cette pièce, qui est une convention relative à un prêt d'argent est très difficile à lire. Elle est datée du 7e mois de la 7e année kien-tchong (786) 建中七年七月; le jour était aussi indiqué mais n'est plus visible; on peut cependant déchiffrer encore le mot 'vingt' 廿, qui prouve que ce jour était compris entre le vingtième et le vingt-neuvième. — Un homme appelé Sou Men-ti (Su Mên-ti) 蘇門悌 (?), ayant un besoin pressant d'argent 爲切要錢用, fait maintenant un emprunt de quinze mille pièces de monnaie 錢壹拾伍仟文. Le propriétaire de cet argent 其錢主 stipule que, dans un délai de huit mois 限八月內, on lui rendra une somme dont le premier terme (dernier caractère de la ligne 3) est malheureusement fort indistinct; les mots qui suivent sont 拾陸仟文 'Seize mille pièces de monnaie'; mais, si le mot précédent est 'deux', par exemple, il faudrait lire 'vingt-six mille pièces de monnaie'. Si l'emprunteur outrepasse ce délai sans faire la restitution de cette somme, celle-ci produira un intérêt mensuel 如違限不付每月頭分生利. S'il ne paie pas, il sera alors permis au prêteur d'emporter son mobilier de manière à compenser le capital et les intérêts 卽任牽掣家資將充本利. S'il est insolvable, on pourra exiger que le garant restitue le prêt à sa place 保人代還. La pièce se termine par la formule habituelle: 'les deux parties ont ensemble trouvé cela équitable et clair et ont apposé l'empreinte de leurs doigts pour servir de marque' 兩共平章畫指爲記. — Ensuite sont mentionnés 1° le propriétaire de l'argent 錢主; 2° l'emprunteur, Sou Men-ti (?), âgé de trente-neuf ans 舉錢人蘇門悌年卅九; 3° le garant, Ngan □, âgé de trente ans 保人安·年卅. — Une note ajoutée plus tardivement indique que le cinquième jour du dixième mois de la septième année kien-tchong 建中七年十月五日, une personne originaire de Li-sie 桀謝, mais dont le nom est indéchiffrable, a payé sur la somme prévue dans le contrat dix mille pièces de monnaie 契內行錢壹拾阡文.

No. 4.

D. v. 6.

(Planche CXV.)

TRADUCTION.

(Contre) le scribe[1] en écriture barbare A-che-nai (A-shih-nai)[2] et le scribe en écriture barbare . . ., (qui dépendent)[3] de l'a-mo-tche Che-tseu[4] (a-mo-chih Shih-tzŭ) . . . Requête. Moi, Sseu-lio (Ssŭ-lüeh), dans le cinquième mois de l'année dernière, j'ai donné aux deux hommes précités un âne; nous avons convenu du prix de six mille pièces de monnaie.

Moi, Sseu-lio (Ssŭ-lüeh), j'avais accordé un délai[5]; il s'est écoulé maintenant un délai de dix mois et je n'ai pas reçu (l'argent), et l'âne n'a pas été rendu; j'espère humblement que vous les poursuivrez et les punirez et que vous réglerez cette affaire.

[1] On a vu plus haut (p. 523, n. 12), que le mot 典 paraît avoir le sens 'd'officier'. Un 胡書典, c'est donc un scribe officiel qui écrit en écriture barbare.

[2] La lecture de ce nom est douteuse.

[3] Après le mot 子 et avant le mot 胡, il y a un caractère illisible qui doit avoir le sens de 'subordonné à' ou 'dépendant de'.

[4] Si ma lecture est correcte, le mot 師子 pourrait être la traduction du nom turc Arslan. On a vu plus haut, à propos du document No. 1, qu'a-mo-tche est le titre du roi de Khoten qui était le chef suprême de la circonscription des Six Cités.

[5] Le mot 丁, d'après la suite des idées, paraît avoir le sens de 'délai'; mais c'est une valeur qu'il n'a pas ordinairement en chinois.

Requête respectueuse.

Le copiste qui a copié la déposition verbale.

Le deuxième mois de la seizième année ta-li[6] (781), l'homme du peuple Sseu[-lio], (Ssŭ-lüeh), originaire de Li-sie (Li-hsieh) dans la circonscription des Six Cités.

1. 阿磨支師子□胡書典阿施捺胡書典□□□□□
2. 牒思略去年五月內與上件二人驢准作錢六阡
3. 思略放丁經今十箇月丁不得驢不還伏望
4. □追徵處分謹牒　　抄□抄人
5. 大曆十六年二月六城桀謝百姓思□

[6] La période ta-li s'étend de 766 à 779. L'année 781 devrait être désignée sous le nom de 'seconde année kien-tchong'.

No. 5.

D. VII. 2.

(Planche CXV.)

TRADUCTION.

La troisième année kien-tchong (chien-chung) (782), le douzième jour du septième mois, le soldat Ma Ling-tche, (Ma Ling-chih), ayant un besoin pressant d'argent, et ne trouvant aucun moyen de s'en procurer, s'est alors adressé à K'ien-ying (Ch'ien-ying), religieux du temple Hou-Kouo (Hu-kuo)[1] et lui a emprunté mille pièces de monnaie. Cette somme produira mensuellement un intérêt[2] de cent pièces de monnaie. Si K'ien-ying (Ch'ien-ying) a lui-même un besoin pressant d'argent, il demandera à Ma Ling-tche (Ma Ling-chih) de lui rendre intégralement le capital et l'intérêt. Si (Ma Ling-tche) est incapable de le faire, il sera entièrement permis à K'ien-ying (Ch'ien-ying) d'emporter le mobilier de (Ma) Ling-tche, ses bœufs et son bétail pour compenser la valeur de la somme d'argent. Ce qui dépassera cette valeur, il ne le réclamera pas[3]. Dans

[1] Hou-kouo signifie 'qui protège le royaume'.

[2] Les mots 分生利 sont ajoutés en surcharge à droite d'un mot qui est malheureusement déchiré. Si ce mot était un nom de nombre, l'intérêt était peut-être supérieur à cent pièces de monnaie par mois, quoique cet intérêt de 10 % par mois soit déjà formidable.

[3] C'est-à-dire que le prêteur ne pourra faire une saisie des biens de son débiteur que jusqu'à concurrence de la valeur de sa créance.

la crainte que quelqu'un ne soit pas (de bonne foi)[4], on a donc dressé ce contrat privé. Les deux parties ont toutes deux trouvé cela juste et clair et y ont apposé l'empreinte de leurs doigts pour servir de marque.

Le propriétaire de l'argent.

L'emprunteur Ma Ling-tche (Ma Ling-chih), âgé de vingt ans.

La personne qu'il a prise avec lui, sa mère, la seconde fille Tang, âgée de cinquante ans.

La personne qu'il a prise avec lui, sa sœur cadette, la seconde fille Ma, âgée de douze ans[5].

1. 建中三年七月十二日。健兒馬令痣爲急
2. 要錢用。交無得處。遂於護國寺僧虔
3. 英邊舉錢壹阡文。其錢每月頭口分生利
4. 佰文。如虔英自要錢用。即仰馬令痣本
5. 利並還。如不得。一任虔英牽掣令痣家
6. 資牛畜。將充錢直。有剩不追。恐人無。故
7. 立私契。兩共平章畫指爲記。
8. 錢主
9. 舉錢人馬令痣年廿
10. 同取人母党二娘年五十
11. 同取人妹馬二娘年十二

[4] Le mot 信 paraît avoir été omis après le mot 無.

[5] Ce contrat offre un intérêt tout particulier parce que, de tous ceux que nous avons étudiés, il est le seul qui soit entier. On y voit que les religieux boudhiques du temple Hou-kouo faisaient valoir leur argent en le prêtant à un intérêt mensuel qui était de 10 %, peut-être même davantage si le dernier mot de la ligne 3 est un nom de nombre autre que 1. Ces religieux étaient coutumiers du fait, car, dans le contrat D. VII. 4. a (Document No. 10), c'est le même K'ien-ying, religieux du temple Hou-kouo, qui apparaît comme le prêteur d'une certaine quantité de grain. Enfin, les deux dernières lignes de la pièce traduite ci-dessus nous montrent que les témoins pouvaient être des femmes et même de toutes jeunes filles.

No. 6.

D. VII. 3. a.

(Planche CXVI.)

TRADUCTION.

1ère ligne: (En tel mois de telle année de la période) ta-li (766–779) . . , Hiu Che-sseu (Hsü Shih-ssŭ) ayant un besoin pressant d'argent et . [1].

2e ligne: ne trouvant aucun moyen de s'en procurer, alors . . . . . . un peigne[2], en tout formant un gage de cinq cents pièces de monnaie.

3e ligne: mensuellement . . . . . . pièces de monnaie. Hiu Che-sseu (Hsü Shih-ssŭ) elle-même.

4e ligne: . prendre l'argent du capital et de l'intérêt pour racheter (le gage). Si elle outrepasse le terme sans racheter (le gage), ses . . . . . .

[1] Le mot manquant ici doit être le mot 交.

[2] Ce terme paraît être le dernier d'une énumération d'objets qui sont livrés par la dame Hiu Che-sseu en gage de la somme d'argent qu'elle a empruntée; peut-être faut-il supposer une lacune plus considérable que celle que j'ai admise et placer avant le mot 梳 une ou deux lignes en blanc qui auraient contenu cette sorte d'inventaire.

5e ligne: seront entièrement confisqués et il sera absolument permis de les acheter[3]. Dans la crainte que quelqu'un ne soit pas de bonne foi, on a donc dressé ce contrat privé. Les deux parties ont toutes deux trouvé cela équitable

6e ligne: et clair . . . .

7e ligne: Le propriétaire de l'argent.

8e ligne: L'emprunteur, dame Hiu Che-sseu (Hsü Shih-ssŭ)[4], âgée de vingt-six ans.

9e ligne: La personne qu'elle a prise avec elle, son fils, Tsin-kin (Chin-chin), âgé de huit ans.

10e ligne: Le témoin.

1. 大曆□□□□□□許十四爲急要錢用□
2. 無得處遂□□□□□□梳一共典錢伍佰□
3. 每月頭□□□□□□錢許十四自□□□□
4. □將本利錢贖如違限不贖其□□□□□□
5. 竝沒一任將買恐人無信故立私契兩共平
6. 章□□□□
7. 錢主
8. 舉人女婦許十四年卄六
9. 同取人男進金年八歲
10. 見人

[3] On attendrait plutôt: 'et il sera entièrement permis de les faire vendre'. C'est une vente judiciaire en effet qui permettrait au créancier de recouvrer son argent. Je proposerais donc de substituer 賣 à 買.

[4] Ce nom indique que cette femme était le quatorzième enfant de sa famille.

No. 7.

D. VII. 3. b.

En haut, un fragment de contrat relatif à un prêt d'argent; on y lit les mots: 伯文如 'Cent pièces de monnaie; comme . . .'; puis les mots 休梅如先 et enfin le commencement de la formule 恐人無信 'de crainte que quelqu'un ne soit pas de bonne foi'.

No. 8.

D. VII. 3. c.

Dans le bas, les seuls mots qui offrent un sens sont, à gauche la phrase 雇驢一頭 'louer un âne', puis la date qui commence par la mention de la période 貞元 tcheng-yuan (chêng-yüan) (785–804).

No. 9.

D. vii. 3. d.

(Planche CXVI.)

Traduction.

1re ligne: . . . . . résident habituel[1]. Quinze mille pièces de monnaie . . . . .

2e ligne: . . . . . terme. Si on outrepasse ce (terme) sans rendre (l'argent prêté), on aura alors le droit . . . . .

3e ligne: . . . . . la huitième année kien-tchong (chien-chung) (787)[2], le quatrième mois, le vingtième jour.

4e ligne: L'emprunteur, la femme A-souen (A-sun)[3].

5e ligne: Le fils de l'emprunteur, Sou Kia-yi (Su Chia-i), âgé de vingt ans.

6e ligne: Le fonctionnaire, le tchong-lang (chung-lang) Lien K'i (Lien Ch'i).

1. □常住錢壹拾伍仟
2. □會如違不還即任
3. 建中八年四月廿日
4. 負錢人妻阿孫
5. 負錢人男蘇嘉依年廿
6. 行官中郎廉奇

[1] Dans D. viii. 1, nous verrons de même la formule 'résident habituel de ce temple' 本寺常住.

[2] La période kien-tchong ne comprend en réalité que les années 780 à 783.

[3] Le mot 孫 est douteux.

No. 10.

D. vii. 4. a.

(Planche CXV.)

Traduction.

La dix-septième année ta-li (782)[1]. . . . . , ayant besoin de grain et ne sachant (où en trouver)[2], s'est adressé à K'ien-ying (Ch'ien-ying), religieux du temple Hou-kouo (Hu-kuo), et lui a emprunté dix-sept . [3] de grain. Ce grain, Ho Hin-yue (Ho Hsin-yüeh) a lui-même fixé un terme, (stipulant que), dans un délai de neuf mois, il le rendrait. S'il outrepasse ce terme (et ne rend pas le grain)[4], par le religieux K'ien-ying (Ch'ien-ying) seront emportés le mobilier de Ho Hin-yue (Ho Hsin-yüeh), ses bœufs et ses bestiaux pour compenser la valeur du grain. Ce qui dépassera cette valeur, il ne le réclamera pas. Dans la crainte que quelqu'un ne soit pas de bonne foi, on a donc dressé ce contrat privé. Les deux parties, étant en présence l'une de l'autre, l'ont trouvé équitable et clair et ont apposé l'empreinte de leurs doigts pour servir de marque.

Le propriétaire du grain. — L'emprunteur du grain, le fonctionnaire Ho Hin-yue (Ho Hsin-yüeh), âgé de trente-sept ans. — Avec lui, la femme de l'emprunteur, la troisième fille Ma, âgée de trente-cinq ans. — La personne qu'ils ont prise avec eux, la fille aînée Ho, âgée de quinze ans.

[1] La période ta-li ne comprend en réalité que les années 766 à 779.

[2] Les trois mots manquants sont vraisemblablement: 得處遂.

[3] Ici un nom de mesure que je ne puis pas lire.

[4] Suppléez ici les deux mots 不還.

1. 大曆十七年
2. □粮用交無□□□於護國寺僧虔英
3. 邊便粟壹拾柒□其粟霍昕悅自立限至
4. 九月內還如違限□□由僧虔英牽掣霍
5. 昕悅家資牛畜將充粟直有剩不追恐人
6. 無信故立私契兩共對兩平章畫指爲記
7. 粟主
8. 便粟人行官霍昕悅年卅七
9. 同便人妻馬三娘年卅五
10. 同取人女霍大娘年十五

No. 11.

D. VII. 4. b.

A droite, en haut, la fin d'un contrat relatif à un prêt d'argent. On y lit deux fois les mots 同負 qui devaient commencer la formule: 'En même temps, de l'emprunteur . . .' Cf. le document n° 9 où se trouve l'expression 負錢人 'l'emprunteur'. Le fragment qui suit est collé à l'envers; on y distingue les deux mots 錢 'monnaie' et 到 'arrivé'. Puis vient un fragment avec les mots 見人 'le témoin'; cf. D. VII. 3.

No. 12.

D. VII. 4. c.

A gauche, en haut, les fragments en gros caractères présentent les mots suivants:

在城
將
追并
由必

會

在
兩
决

限當日
大譬速
如所

Ces mots n'offrent aucun sens suivi. Je me bornerai à faire remarquer que le nom du religieux Ta-p'i se retrouve dans D. VIII. I (document n° 17), où on voit que ce personnage avait le titre de wei-na ou karmadâna.

No. 13.

D. VII. 4. d.

Ces fragments paraissent appartenir à un seul et même document, mais ils sont disposés dans le plus grand désordre. Tout en bas, à gauche, on lit le nom de la période tcheng-yuan 貞元; à peu près au milieu de la planche, on voit d'autre part les mots 五年; je crois qu'il faut réunir ces deux fragments et lire 貞元五年 'la cinquième année tcheng-yuan (789)'. Les autres fragments contiennent pour la plupart des noms de religieux tels que: 應清 Ying-ts'ing (Ying-ch'ing), 玄應 Hiuan-ying (Hsüan-ying), 義法 Yi-fa (I-fa), 善意 Chan-yi (Shan-i), 法幽 Fa-yeou (Fa-yu).— Enfin, en bas, à gauche, on déchiffre les mots: 望商量請處 'espère qu'on discutera, et demande qu'on règle l'affaire'.

No. 14.

D. VII. 4. e.

. . . . . pièces de monnaie. Cet argent, depuis . . . . Emporter . . . . .

文其錢自

牽掣

No. 15.

D. VII. 4. f.

Compte . . [1].— Pétition.
Les religieux: P'ou (P'u).— . . — Pao-ming.—. . —
. . .— Fa-tsin (Fa-chin).— Tao-tch'ao (Tao-ch'ao) [2].

計□□ 狀上

僧普□ 寶明 □□ □□ 法進 道超

[1] Les deux caractères qui suivent le mot 計 ne se trouvent pas dans les dictionnaires.

[2] Ce document paraît avoir contenu une liste de noms de religieux.

No. 16.

D. vii. 7.

(Planche CXVI.)

Traduction.

1re ligne: Du temple Hou-kouo (Hu-kuo). . . . . le religieux Ta-yen qui est chargé de la surveillance de l'extérieur

2e ligne: . . . . . combien.

3e ligne: Quand l'ordre précité arrivera à Yang-ling, les domestiques susnommés couperont l'herbe pendant trois

4e ligne: jours; on laissera un seul homme pour arroser les champs; les autres hommes s'en iront tous et ne pourront pas

5e ligne: s'occuper à leur fantaisie; si on viole cet ordre il faudra certainement qu'on soit condamné. Le huitième mois, le vingt-septième

6e ligne: jour, notification faite. — Le wei-na (karmadâna) général, le religieux Houei-ta (Hui-ta).

7e ligne: Le chang-tso (shang-tso) (sthavira), le religieux Houei (Hui). . — Le sseu-tchou (ssŭ-chu) (vihârasvâmin), le religieux Houei-yun (Hui-yün)[1].

1. 護國寺□□□□外巡僧大言
2. 先果□□□多少等
3. 右帖至仰嶺前件家人刈草參
4. 日留一人澆田餘人盡將去不得
5. 妄作事故違必宜科決八月廿七
6. 日帖　　　　都維那僧惠達
7. 上座僧惠□　　　　寺主僧惠雲

[1] Il est intéressant de constater que cet ordre donné à des gens qui dépendaient du temple Hou-kouo est signé par les trois principaux dignitaires de ce temple, à savoir: le wei-na ou karmadâna (cf. I-tsing, *Record of the Buddhist Religion*, trad. Takakusu, p. 148, n. 1), le chang-tso ou sthavira et le sseu-tchou ou vihârasvâmin.

No. 17.

D. viii. 1.

(Planche CXVI.)

Traduction.

4e ligne: Résident habituel de ce temple; un écrit qu'il a laissé voir . . .

5e ligne: ne pas, le wei-na (karmadâna)[1], le religieux Ta-p'i[2], prenant le contrat relatif à la servante donnée en gage . . .

6e ligne: s'est perdu en route, il faut craindre que ce . . .

[1] Sur le wei-na ou karmadâna, cf. D. vii. 7 (document no 16), note 1.

[2] Ce Ta-p'i est mentionné dans D. vii. 4. c (document no 12).

1. 來去
2. 身在自
3. 仙
4. 本寺常住其遺書見
5. 不維那僧大辟將本典婢契於
6. 中路遺失當恐此

No. 18.

D. IX. i.

(Planche CXVI.)

'La sixième année tcheng-yuan (chêng-yüan) (790)'. . . . .

貞元六年

Note additionnelle.

Les dates qu'on peut relever dans ces divers documents sont les suivantes:

768, troisième année ta-li, dans Ms. A. (document n° 1).

781, inexactement indiquée par D. v. 6 comme étant la seizième année ta-li (766–779) (document n° 4).

782, inexactement indiquée par D. VII. 4 comme étant la dix-septième année ta-li (766–779), mais exactement mentionnée par D. VII. 2 comme étant la troisième année kien-tchong (documents n°s 10 et 5).

786, inexactement indiquée par Ms. C. comme étant la septième année kien-tchong (780–783) (document n° 3).

787, inexactement indiquée par le quatrième fragment de D. VII. 3 comme étant la huitième année kien-tchong (780–783) (document n° 9).

789, cinquième année tcheng-yuan (785–804), dans D. VII. 4 (document n° 13).

790, sixième année tcheng-yuan (785–804), dans D. IX. 1 (document n° 18).

Par leur âge, ces documents représentent les dernières traces de l'influence chinoise dans le Turkestan oriental sous la dynastie T'ang. Le commissaire de la Chine dans le Turkestan oriental était alors Kouo Hin 郭昕, qui résidait à Koutcha, où le vit en effet le pèlerin Wou-k'ong peu avant l'année 789 (*Journ. asiatique*, Sept.–Oct. 1895, p. 367). Ce Kouo Hin, qui était apparenté au célèbre Kouo Tseu-yi 郭子儀, est l'objet d'une courte notice dans l'histoire des T'ang (*Kieou T'ang chou*, chap. cxx. p. 11 v°; *T'ang chou*, chap. cxxxvii. p. 7 r°); voici en effet ce qu'on lit à son sujet dans le *Kieou T'ang chou*:

'Kouo Hin, à la fin du règne de Sou-tsong (756–762), fut nommé Résident des Quatre Garnisons

四鎮留後[1]. A partir du moment où la région de Kouan et de Long 關隴[2] fut tombée au pouvoir des Tibétains, ses communications avec la Chine furent interceptées par les barbares. Ceux qui étaient en titre commissaires des Quatre Garnisons et de Pei-t'ing, à savoir Li Sseu-ye 李嗣業 et Li-fei Yuan-li 茘非元禮, se contentaient tous de les commander de loin. Après que Kouo Hin eut été ainsi tenu dans l'isolement pendant quinze années, la deuxième année kien-tchong (781), lui et Li Yuan-tchong 李元忠, qui avait le titre de tsie-tou-che de Yi 伊 (Hami), Si 西 (Tourfan) et Pei-t'ing 北庭 (près de Goutchen), envoyèrent tous deux des messagers à la cour. (L'empereur) Tö-tsong les félicita et rendit un décret en ces termes: "Les Quatre Garnisons et les Deux Cours royales[3] ont la charge générale des cinquante-sept peuples barbares du Hia occidental 西夏[4] et des Dix Tribus 十姓部落[5]. Depuis que notre dynastie a été fondée, ces peuples sont venus constamment (rendre hommage à la cour) et se sont tous acquittés de leurs devoirs. Mais, à partir du moment où la région de Kouan et de Long a cessé d'être bien gardée, l'Est et l'Ouest se sont trouvés séparés; les hommes loyaux et fidèles[6], en versant des larmes de sang, se sont maintenus à leur poste; avec vigilance ils ont défendu le territoire qui leur était confié; ils ont conservé et honoré la règle qui leur avait été imposée par l'empereur. Ce résultat a été obtenu parce que tous, chefs suprêmes, gouverneurs et généraux, ont uni leurs efforts et ont été d'accord. Le tsie-tou-che de Yi, de Si et de Pei-t'ing, Li Yuan-tchong, sera nommé grand Protecteur du Pei-t'ing; le résident gouverneur des Quatre Garnisons, Kouo Hin, sera nommé grand Protecteur du Ngan-si, et tsie-tou-che des Quatre Garnisons. Quant aux autres fonctionnaires, généraux, officiers et gens de grades inférieurs, on les promouvra en comptant leur temps de service comme valant sept fois (le même temps de service dans des conditions normales)." — Li Yuan-tchong avait à l'origine le nom de famille Ts'ao 曹, et le nom personnel Ling-tchong 令忠; c'est à cause des services qu'il avait rendus qu'on le gratifia d'un autre nom de famille[7] et d'un autre nom personnel. — En ce temps, les envoyés de Kouo Hin passèrent par (le territoire des) Houei-ho 廻紇 (Ouïgours)[8] et traversèrent les diverses tribus barbares et c'est ainsi qu'ils parvinrent à la cour[9].'

Voici maintenant ce que le *Tseu tche t'ong kien* de *Sseu-ma Kouang* nous apprend sur ces mêmes évènements:

Année 781: 'Depuis que les T'ou-po (Tibétains) s'étaient emparés de la région de Ho 河 et de Long 隴, le Pei-t'ing (près de Goutchen) et le Ngan-si (Koutcha) s'étaient vu intercepter toute communication (avec la Chine). Le tsie-tou-che de Yi (Hami), Si (Tourfan) et Pei-t'ing (près de Goutchen), Li Yuan-tchong 伊西北庭節度使李元忠, et le résident des Quatre Garnisons, Kouo Hin 四鎮留後郭昕,

[1] Les quatre garnisons sont Karachar, Koutcha, Kachgar et Khoten. L'ensemble de ces territoires formait le Protectorat de Ngan-si 安西 dont le centre administratif était à Koutcha.

[2] La région de Kouan et de Long, qu'on appelle aussi la région de Ho et de Long 河隴, comprenait le Kan-sou et l'extrême ouest du Chàn-si. C'est pendant la révolte de Ngan Lou-chan (756–758) que les Tibétains, profitant des embarras causés à la dynastie T'ang par cette rébellion qui la mit à deux doigts de sa perte, commencèrent à occuper ces territoires. Quand ils s'y furent définitivement établis, c'est-à-dire vers 766, ils interceptèrent toute communication entre la Chine et les deux Protectorats de Ngan-si (Koutcha) et de Pei-t'ing (près de Goutchen).

[3] La Cour antérieure était à Kao-tch'ang (Yar-khoto, près de Tourfan); la Cour postérieure était à Pei-t'ing (Dsimsa, près de Goutchen).

[4] Le Hia occidental est l'ancien pays de Ta-hia 大夏; cette expression désigne ici les contrées d'occident où le royaume qui exerçait une sorte d'hégémonie sur tous les autres était le Tokharestan. Le pèlerin Wou-k'ong parle en effet, vers 780 p.C., des 'cinquante-sept peuples du Tokharestan'; le *Tseu tche t'ong kien* (chap. 431. p. 6 r°) mentionne, en 784, les 'cinquante-sept royaumes des contrées d'occident' 西域五十七國; le *T'ang chou* (chap. lxxvii. p. 7 r°) dit que, en 767, on rétablit le nom de Ngan-si 安西 (qui avait été changé en celui de Tchen-si 鎮西 à la date de 757) et que, par la suite, on rattacha à la juridiction du Protecteur du Ngan-si les cinquante-sept commissaires des pays barbares 其後増領五十七蕃使.

[5] Les cinq tribus Tou-lou (Dzoungarie) et les cinq tribus Nou-che-pi (région de Talas et de Tokmak) qui formaient la nation des *T'ou-kiue* (Turcs) occidentaux.

[6] Éloge de Kouo Hin, de Li Yuan-tchong et de leurs subordonnés qui, quoique séparés pendant de longues années de la mère-patrie, restèrent courageusement au poste qui leur avait été confié.

[7] Le nom de famille Li 李 était celui de la famille impériale des T'ang.

[8] Les Ouïgours étaient alors alliés aux Chinois. En 789, le pèlerin Wou-k'ong passa lui aussi par le territoire des Ouïgours lorsqu'il voulut revenir de Pei-t'ing (près de Goutchen) à Tch'ang-ngan (Si-ngan-fou).

[9] Cf. Bushell, *The early history of Tibet*, Journ. R. As. Soc., N.S., vol. xii. p. 504, lignes 4–6.

s'étaient mis à la tête de leurs officiers et de leurs soldats, avaient fermé leur territoire et s'étaient tenus sur la défensive; à plusieurs reprises ils avaient envoyé des messagers apporter des requêtes (à l'Empereur), mais aucun d'eux n'était parvenu à destination et (ces deux hauts fonctionnaires) étaient restés sans nouvelles depuis plus de dix ans. En cette année, ils envoyèrent des messagers qui, en prenant des chemins détournés et en traversant les divers peuples *Hou* 胡, arrivèrent en venant du pays des Houei-ho 回紇 (Ouïgours); l'Empereur les félicita; le septième mois, le jour wou-wou, qui était le premier du mois, il promut (Li) Yuan-tchong à la dignité de Grand protecteur général de Pei-t'ing 北庭大都護 et lui conféra le titre nobiliaire de Roi de la commanderie de Ning-sai (près de Si-ning) 寧塞郡王; il nomma (Kouo) Hin Grand protecteur général du Ngan-si 安西大都護, tsie-tou-che des Quatre Garnisons 四鎭節度使, et lui conféra le titre nobiliaire de Roi de la Commanderie de Wou-wei (Leang tcheou) 武威郡王. Les officiers et les soldats furent promus en comptant au septuple leur temps de service.'

'Le jour ping-tseu, on conféra à Yuan Kouang-t'ing 袁光庭, préfet défunt de l'arrondissement de Yi 伊 (Hami), le titre posthume de président du ministère des travaux publics. A la fin de la période t'ien-pao (742–755), (Yuan) Kouang-t'ing était préfet de l'arrondissement de Yi (Hami) lorsque les Tibétains s'emparèrent de la région de Ho 河 et de Long 隴. (Yuan) Kouang-t'ing se défendit avec énergie pendant plusieurs années; les T'ou-po (Tibétains) eurent recours à cent moyens pour l'attirer, mais ne purent le soumettre. Lorsque les vivres furent épuisés, que les armes de guerre eurent été toutes employées et que la ville fut sur le point de tomber au pouvoir de l'ennemi, Yuan Kouang-t'ing commença par tuer sa femme et ses enfants, puis il se brûla lui-même. C'est quand les envoyés de Kouo Hin arrivèrent à la cour qu'on fut pour la première fois informé de ces faits et c'est alors qu'on conféra un titre posthume à Yuan Kouang-t'ing.'

En 783, révolte de Tchou Ts'eu 朱泚 qui s'empare de la capitale *Tch'ang-ngan* (Si-ngan fou) et qui se proclame Empereur de la grande dynastie Ts'in 大秦皇帝, puis, l'année suivante, Premier empereur céleste de la dynastie Han 漢元天皇. Il est vaincu et tué du cinquième au sixième mois de l'année 784.

A la suite de ces évènements, le *Tseu tche t'ong kien* ajoute (automne 784):

'Auparavant, quand l'empereur avait lancé les T'ou-po (Tibétains) à l'attaque de Tchou Ts'eu 朱泚, il leur avait promis que, s'ils réussissaient dans leur entreprise, il leur donnerait les territoires de Yi 伊 (Hami), de Si 西 (Tourfan) et de Pei-t'ing 北庭 (près de Goutchen). Quand Tchou Ts'eu eut été mis à mort, les T'ou-po (Tibétains) vinrent réclamer ces territoires; l'Empereur était disposé à envoyer aux commandants des deux garnisons (de Ngan-si et de Pei-t'ing), Kouo Hin 郭昕 et Li Yuan-tchong 李元忠, l'ordre de revenir à la cour, et il était prêt à livrer ces territoires (aux Tibétains). Mais Li Pi[1] 李泌 tint ce discours: "Les gens de Pei-t'ing et de Ngan-si ont un caractère fier et brave; ils tiennent en respect les cinquante-sept royaumes des contrées d'occident[2] ainsi que les T'ou-kiue des Dix Tribus. En outre ils séparent en deux la puissance des T'ou-po (Tibétains) et les empêchent de masser tous leurs soldats pour faire des empiètements du côté de l'est. Comment donc les livrerions-nous en restant les mains jointes[3]? En outre, les habitants de ces deux garnisons, quoique se trouvant isolés dans une région lointaine, ont agi avec le plus entier loyalisme et ont déployé tous leurs efforts pour défendre avec énergie depuis près de vingt ans les intérêts du gouvernement chinois; ils sont vraiment dignes de compassion. Si un beau matin on les abandonne pour les livrer aux barbares, ils en éprouveront certainement un profond ressentiment dans leur cœur, et, quelque autre jour, ils viendront à la suite des T'ou-po (Tibétains) faire des ravages comme s'ils avaient à venger un grief personnel. D'ailleurs, récemment, les T'ou-po (Tibétains) ont gardé l'expectative et n'ont pas avancé, hésitant secrètement entre deux partis[4]; ils ont fait un grand pillage à Wou-kong 武功 et ne se sont retirés qu'après avoir reçu une rançon[5]. Quel service nous ont-ils rendus?"— Les membres de l'assemblée délibérante approuvèrent les paroles de Li Pi et l'Empereur ne livra donc pas ces territoires.'

[1] Les paroles de Li Pi se retrouvent dans la biographie de cet homme d'état (*T'ang chou*, chap. cxxxix. p. 5 r°).

[2] Cf. p. 534, n. 4.

[3] C'est-à-dire sans agir, sans faire aucun effort pour conserver ces territoires.

[4] Leur dévouement à la cause de l'Empereur n'était pas sincère.

[5] Le quatrième mois de l'année 784, les Tibétains avaient rançonné la ville de Wou-kong, qui était à 40 li à l'est de la sous-préfecture actuelle de Mei 郿 (préf. de Fong-siang, prov. de Chàn-si).

Année 789, douzième mois: 'Auparavant, (les gouverneurs de) Ngan-si (Koutcha) et de Pei-t'ing (près de Goutchen) avaient tous deux emprunté le chemin des Houei-hou (Ouïgours)[1] 回鶻 pour présenter leurs rapports à l'empereur, et c'est pourquoi ils entretenaient des relations d'amitié avec ce peuple; cependant, comme le Pei-t'ing (près de Goutchen) était très voisin des Houei-hou (Ouïgours), ceux-ci lui adressaient sans se lasser des réquisitions extraordinaires. Il y avait aussi plus de six mille tentes des Cha-t'o 沙陀 qui s'appuyaient sur le Pei-t'ing; puis les trois (tribus) Ko-lou 葛祿 (Karlouks) et les T'ou-kiue (Turcs) à vêtements blancs 白服突厥 qui tous se rattachaient aux Houei-hou (Ouïgours). Les Houei-hou (Ouïgours) ayant à diverses reprises envahi et pillé (les Karlouks et les Turcs), les T'ou-po (Tibétains) employèrent les troupes des Ko-lou (Karlouks) et des (Turcs à) vêtements blancs pour attaquer le Pei-t'ing; le grand conseiller des Houei-hou (Ouïgours), Hie-kan-k'ia-sseu 頡干迦斯, vint à la tête d'une armée secourir (le Pei-t'ing).'

Année 790, cinquième mois: 'Hie-kan-k'ia-sseu, (général des) Houei-hou (Ouïgours), livra bataille aux T'ou-po (Tibétains), mais il ne fut pas vainqueur; les T'ou-po (Tibétains) attaquèrent avec un redoublement d'énergie le Pei-t'ing; les gens du Pei-t'ing, qui d'ailleurs étaient excédés des réquisitions extraordinaires que leur adressaient les Houei-hou (Ouïgours), vinrent avec le chef des Cha-t'o, Tchou-ye Tsin-tchong 朱邪盡忠, faire leur soumission aux T'ou-po (Tibétains). Le tsie-tou-che Yang Si-kou[2] 楊襲古, à la tête des deux mille soldats qui composaient son armée, s'enfuit dans l'arrondissement de Si 西州 (près de Tourfan).'

'En automne (790), Hie-kan-k'ia-sseu mit en campagne tous les soldats de son pays, au nombre de plusieurs myriades, et alla tenter de reprendre le Pei-t'ing; mais il fut encore une fois battu par les T'ou-po (Tibétains); plus de la moitié de ses soldats périrent; Yang Si-kou rassembla les quelques centaines d'hommes qui lui restaient et se disposa à regagner l'arrondissement de Si (près de Tourfan); cependant Hie-kan-k'ia-sseu, usant de fourberie, lui dit: "Venez pour un moment avec moi à l'ordo (du kagan Ouïgour)." Mais alors on le retint et on ne le renvoya pas; en définitive on le mit à mort. A partir de ces évènements, le Ngan-si (Koutcha) fut entièrement isolé et nul ne sut quel avait été son sort[3]; cependant l'arrondissement de Si (près de Tourfan) continua à se bien défendre pour rester fidèle aux T'ang.'

De tous ces textes historiques il résulte que, dès l'année 766 environ, les Tibétains réussirent à isoler presque entièrement de la Chine le Turkestan oriental et la région de Tourfan et de Goutchen; les fonctionnaires chinois établis dans ces possessions lointaines de l'empire réussirent cependant à s'y maintenir, et, en 781, ils purent faire parvenir de leurs nouvelles à la cour. En 784, l'empereur, pour reconnaître les services que lui avaient rendus les Tibétains en l'aidant à triompher du rebelle Tchou Ts'eu, était disposé à leur céder ces débris d'une grandeur déchue, mais il en fut empêché par les remontrances de Li Pi. Lorsque le pèlerin Wou-k'ong revint des Pamirs en Chine en passant par Kachgar, Khoten, Aksou, Saïram (?), Koutcha, Karachar et Dsimsa (près de Goutchen), pour partir de cette dernière ville dans le neuvième mois de l'année 789 et arriver à Si-ngan fou dans le deuxième mois de l'année 790, il trouva encore, dans toutes les localités que nous venons d'énumérer, des résidents chinois dont il nous a conservé les noms. Cependant, dès le douzième mois de l'année 789, les Tibétains étaient venus assiéger Pei-t'ing (près de Goutchen); les Ouïgours tentèrent vainement de secourir cette place; ils furent vaincus le cinquième mois de l'année 790 et le résident chinois Yang Si-kou fut obligé de s'enfuir à Tourfan; cinq mois plus tard les Ouïgours alliés à Yang Si-kou faisaient un nouvel effort pour reprendre le Pei-t'ing, mais ils échouaient cette fois encore et Yang Si-kou allait mourir chez les Ouïgours qui trouvaient plus expédient de se débarrasser de lui que de continuer à le soutenir. A partir de ce moment, toute relation cessa entre la Chine et le Turkestan oriental. Les documents de Dandān Uiliq dont les dates s'échelonnent de 768 à 790 se rapportent à cette période où l'influence chinoise subsistait encore dans tout le Turkestan oriental, bien qu'elle n'eût déjà presque plus de communications avec le gouvernement central.

[1] Ceci est en parfait accord avec ce que nous apprend la notice sur Wou-k'ong; nous savons en effet que ce pèlerin, pour se rendre en 789 de Pei-t'ing (près de Goutchen) à Tch'ang-ngan (Si-ngan fou), dut prendre le chemin qui traversait le pays des Ouïgours (*Journ. asiatique*, Sept.–Oct. 1895, p. 370).

[2] Cet officier chinois est mentionné dans la notice sur Wou-k'ong (*Journ. asiatique*, Sept.–Oct. 1895, p. 369).

[3] Le Pei-t'ing (près de Goutchen) étant tombé au pouvoir des Tibétains, le Ngan-si (Koutcha) fut dès lors entièrement isolé et on n'eut plus aucune nouvelle de la petite garnison chinoise qui y était restée.

PART II

LES DOCUMENTS SUR BOIS DE NIYA

N. xv. 326. (Planche CXII.) 泰始五年十月戊午朔廿日丁丑敦煌太守...

'La cinquième année t'ai-che, le vingtième jour ting-tch'eou du dixième mois dont le premier jour est le jour wou-wou, le préfet de Touen-houang, . . .'

Note.—Comme l'a reconnu S. W. Bushell, la cinquième année t'ai-che est l'année 269 p. C. L'exactitude de cette date est confirmée par les textes historiques; nous lisons en effet dans le *Tsin-chou* (chap. iii), que, la cinquième année t'ai-che, le premier jour du cinquième mois fut le jour ting-hai, 24ᵉ du cycle. Il résulte de là que, en la cinquième année t'ai-che, le premier jour du dixième mois dut être le jour wou-wou, 55ᵉ du cycle; or cette indication est précisément celle que nous trouvons sur cette fiche de bois. — Touen-houang se confond pratiquement avec Cha-tcheou 沙州, à l'Ouest de la province de Kan-sou. — Après le mot 守, on voit la partie de gauche d'un caractère qui est vraisemblablement le mot 都; c'est peut-être le commencement d'un titre qui pourrait être analogue à celui que nous trouvons à la date de 270 dans le *Tsin-chou* (chap. iii): 都督雍涼二州諸軍事 'surintendant général des affaires militaires des deux arrondissements de *Yong* et de *Leang*.'

N. xv. 93 a. b. (Planche CXIII.) 晉守侍中大都尉奉晉大侯親晉鄯善焉耆龜茲疏勒...

'(Un tel, ayant par délégation de la dynastie) Tsin (les titres de) ta-tou-wei exerçant les fonctions de che-tchong, grand marquis investi par les Tsin, (protecteur? de) Chan-chan (au sud du Lop-nor), Yen-k'i (Karachar), K'ieou-tseu (Koutcha), Sou-le (Kachgar) . . . ., allié aux Tsin . . .'

Note.—Dans la formule 晉守侍中大都尉, le mot 守 signifie que le fonctionnaire ayant le titre réel de ta-tou-wei *exerce* les fonctions de che-tchong. Le mot 守 a donc la même valeur que le mot 行; mais le mot 守 paraît être employé quand un homme exerce des fonctions *supérieures* à celles de son titre réel, tandis que le mot 行 donnerait à entendre qu'il est chargé de fonctions *inférieures* à celles qu'implique son titre réel.—Le *Wei lio* 魏略, publié entre 239 et 265 p. C. par Yu Houan, nous fournit un autre exemple de la formule que nous avons ici; parlant en effet du roi du Kiu-che postérieur 車師後部, il dit: 魏賜其王壹多雜守魏侍中號大都尉受魏王印 'La dynastie Wei conféra au roi de ce pays, Yi-to-tsa, le titre de ta-tou-wei exerçant les fonctions de che-tchong nommé par les Wei; il reçut le sceau de roi nommé par les Wei.'—D'autre part, la formule 親晉 'allié aux Tsin' que nous avons ici rappelle celle que nous trouvons à la date de 229 p. C. dans le *San-kouo-tche* (chap. ii, p. 3 r°), lorsqu'il est question de l'investiture accordée à Po-t'iao 波調 (Vāsudeva?, ap. Parker, *As. Quart. Review*, July 1902), roi des Ta Yue-tche 大月氏 (Indoscythes); Po-t'iao fut en effet nommé 親魏大月氏王 'roi des Ta Yue-tche allié aux Wei.' De même, en 238 p. C., une lettre de l'empereur de Chine à la reine du Japon (l'impératrice Jingô) commence par les mots: 'Édit adressé à Pi-mi-hou, reine de Wo, alliée aux Wei . . .' 制詔親魏倭王卑彌呼 (*San-kouo-tche*, chap. xxx. p. 11 v°). Enfin en 126 p. C., un personnage de la tribu postérieure de Kiu-che 車師 fut nommé 'marquis, allié aux Han, de la tribu postérieure' 後部親漢侯. — On voit dans les Annales principales du *Tsin-chou* (chap. iii) que l'empereur Wou (265–289) de la dynastie Tsin sut faire reconnaître sa suprématie dans les pays d'Occident: en 270, les pays de Ta-yuan 大宛 (Ferghânah) et de Yen-k'i 焉耆 (Karachar) vinrent lui apporter tribut; en 283, l'état de Chan-chan 鄯善 (au sud du Lop-nor), en 285, ceux de K'ieou-tseu 龜茲 et de Yen-k'i 焉耆 (Karachar), envoyèrent des jeunes gens de familles princières pour qu'ils prissent du service à la cour de l'Empereur; en 287, le K'ang-kiu 康居 (Samarkand) chargea un ambassadeur d'offrir des présents. Après le règne de l'empereur Wou, ces relations avec les pays d'Occident cessent entièrement; quoique non datée, la fiche que nous étudions en ce moment doit donc selon toute vraisemblance remonter à l'époque de l'empereur Wou (265–289). Dès lors, elle ne peut guère désigner que le roi de Yen-k'i (Karachar), Long Houei 龍會, qui, vers cette époque, semble-t-il, prit l'hégémonie sur tous les Hou d'Occident (voir plus loin la note additionnelle, p. 543); seul en effet Long Houei eut un pouvoir assez étendu pour être désigné comme dominant à la fois sur Chan-chan, Yen-k'i (Karachar), K'ieou-tseu (Koutcha) et Sou-le (Kachgar).

N. xv. 188. (Planche CXII.) 武威西平西郡張掖酒泉敦(煌)...

'Wou-wei, Si-p'ing, commanderie de Si, Tchang-ye, Tsieou-ts'iuan, Touen-houang . . .'

Note.—Nous avons ici une énumération partielle des huit commanderies qui dépendaient de Leang-tcheou 涼州. Le *Tsin-chou* (chap. xiv. p. 14 v°) nous fournit la liste complète qui est la suivante: 'Commanderies: 1° de Kin-tch'eng 金城 (auj. préfecture de Lan-tcheou 蘭州, prov. de Kan-sou), 2° de Si-p'ing 西平 (auj. préfecture de Si-ning 西寧), 3° de Wou-wei 武威 (auj. préfecture de Leang-tcheou 涼州), 4° de Tchang-ye 張掖 (auj. préfecture de Kan-tcheou 甘州), 5° de Si 西 (auj. sous-préfecture de Chan-tan 山丹, dans la préfecture de Kan-tcheou), 6° de Tsieou-ts'iuan 酒泉 (auj. préfecture sec. de Sou 肅), 7° de Touen-houang 敦皇 (auj. sous-préfecture de Touen-houang), 8° de Si-hai 西海 (en un point mal déterminé du territoire de la préfecture de Kan-tcheou 甘州).'

N. xv. 116. (Planche CXIII.) 張掖酒泉會

Tchang-ye et Tsieou-ts'iuan, à (telle date). Cf. N. xv. 188.

N. xv. 85. (Planche CXIV.) 長史苞下

'Le tchang-che Pao a rendu (cet édit).'

Note.—Il est vraisemblable que, avant les mots 長史, se trouvaient les deux mots 西域: nous retrouverions alors le titre de tchang-che des pays d'Occident qui apparaît dans le document N. xv. 328.

N. xv. 328 et N. xv. 75. (Planches CXII, CXIII.) (Ces deux fragments se font suite.) 西域長史督寫鴻臚書到如書羅捕言會十一月廿日如詔書律令

'Le tchang-che des pays d'Occident a écrit dans son camp (?) que, lorsque la lettre du Hong-lou serait arrivée, on devrait se conformer à cette lettre. Lo-pou-yen, le 20e jour du 11e mois, s'est conformé à l'ordre de l'édit.'

Note.—Le titre de tchang-che des pays d'Occident apparaît dans le *Tsin-chou* (chap. lxxxvi. p. 7 r°), où on voit mentionné, vers l'an 324 p. C., un certain Li Po, tchang-che des pays d'Occident 西域長史李柏.—Le Hong-lou sseu est l'administration chinoise qui était chargée des relations avec les peuples barbares.—Lo-pou-yen paraît être un nom d'homme; cf. N. xv. 348.

N. xv. 348. (Planche CXII.) 寫下詔書到羅捕言會三月卅日如詔書

'. . . . a écrit et rendu (un édit); lorsque cet édit est arrivé, Lo-pou-yen, le 30e jour du 3me mois, s'est conformé à l'édit.'

N. xv. 345. (Planche CXIV.) 鄯善王

'Le roi de Chan-chan.'

Note.—Chan-chan était au sud du Lop-nor.—A droite de ces mots, il semble qu'on puisse lire le mot 詔 'édit'.

N. xv. 73. (Planche CXII.) 于寘王寫下詔書到奉

'Le roi de Yu-t'ien a écrit et rendu (un édit); cet édit est arrivé et a été reçu . . . .'

N. xv. 109. (Planche CXIII.) 去三月一日騎馬詣元城收責期行當

'Le premier jour du troisième mois dernier, il est monté à cheval et s'est rendu à Yuan-tch'eng pour y recueillir (le montant de cette) dette; au terme fixé, il est parti; il faudra . . .'

Note.—On trouve mentionné dans le *Tsin-chou* (chap. xiv. p. 6 v°) une sous-préfecture de Yuan-tch'eng qui existe encore aujourd'hui sous ce nom dans la province de Tche-li; mais il est vraisemblable que cette ville n'a rien de commun avec celle dont il est question ici.

N. xv. 353. (Planche CXII.) 連會不還或安別牧私行糴買無通所啟信前各私從吏周

'. . . en plusieurs occasions successives n'a pas rendu; peut-être Ngan-pie-mou en secret a-t-il pratiqué des achats de grain sans en donner aucun avis. Avant que la lettre eût été rédigée, chacun secrètement, le fonctionnaire Tcheou . . .'

Note.—Traduction très incertaine. Au lieu de Ngan-pie-mou, on pourrait aussi lire: 'Le préfet de Ngan-pie'.

N. xv. 314. (Planche CXIII.) 上言府普告絶逐捕不得使經家而不禽獲已牧𢷬皆先問前所經

'Il dit à ses supérieurs que dans toute l'étendue de la préfecture on avait déclaré que les poursuites pour l'arrêter avaient été vaines; on lui permit ainsi de passer dans sa famille sans être pris. Le préfet et son adjoint ont tous deux demandé d'abord par où il avait passé auparavant.'

Note.—Traduction très incertaine. — Le mot 府 désigne une circonscription administrative que nous traduisons par le mot 'préfecture'.

N. xv. 189. (Planche CXIII.) 獲得牧𢷬經付

'. . . a été arrêté; le préfet et son adjoint ont déjà livré . . .'

N. xv. 362. (Planche CXII.) 右八人篤啟辭前已言府逐捕今重下普下

'Les huit hommes susnommés, Tou, avant qu'il eût rédigé par écrit son accusation, avait dit au préfet de les poursuivre et de les arrêter; maintenant, de nouveau cet ordre est rendu et il est promulgué partout.'

N. xv. 176. (Planche CXIII.) 捕未獲

'. . . . . l'arrêter; on ne l'a pas encore pris.'

N. xv. 315. (Planche CXIII.) 言被都官從軍苻此牒胡與繫者辭連胡

'. . . . dit que, par Fou, officier militaire dépendant du gouverneur, a été déposée cette accusation contre Hou (le barbare?) et les prisonniers; l'accusation implique Hou . . .'

Note.—Traduction très incertaine.

N. xv. 37. (Planche CXIII.) 生矣與繫者辭連苻問友荅辭吶儉

'Cheng-yi (nom d'homme?) fut impliqué avec les prisonniers dans l'accusation; Fou interrogea son ami; la réponse fut lente et brève.'

Note.—La mention du personnage appelé Fou semble indiquer que ce document se rapporte à la même affaire que N. xv. 315.

N. xv. 203. (Planche CXII.) 右一人屬典內□□□織錢佛居中自齎
敦煌太守往還通

'L'homme susnommé appartient à . . . . . . habits et pièces de monnaie, dans le stûpa (?), va le porter lui-même au préfet de Touen-houang; à l'aller et au retour . . .'

Note.—Traduction très incertaine.

N. xv. 53. (Planche CXIII.) 月支國胡支柱年四十九中人黑色

'Hou-tche-tchou (c.-à.-d. le barbare Indoscythe Tchou), (originaire) du royaume des Yue-tche (Indoscythes), âgé de 49 ans, taille moyenne, teint foncé.'

N. xv. 191. (Planche CXII.) 月支國胡支

'. . . (originaire) du royaume des Yue-tche, Hou-tche (c.-à-d. le barbare Indoscythe . . .).'

N. xv. 337. (Planche CXII.) 丑年十四短小同着布袴褶挾

'Tch'eou (ce mot paraît faire partie d'un nom d'homme), âgé de quatorze ans; de petite taille; tous sont vêtus de pantalons et de tuniques de toile; ils tiennent sous le bras (?) . . . .'

N. xv. 152. (Planche CXIII.) 卅中人黑色大目有髭鬚

. . (âgé de (trente) ans); taille moyenne, teint foncé; yeux grands; il porte la moustache et la barbe.'

N. xv. 08. (Planche CXII.) 異年五十六一名奴中人髮鬚倉白着

'Yi (ce mot doit faire partie d'un nom d'homme), âgé de cinquante-six ans; on l'appelle aussi Nou; de taille moyenne; ses cheveux et sa barbe sont grisonnants; il est vêtu . . .'

N. xv. 192. (Planche CXIV.) 異年五十六一名奴中人髮鬚倉白色

'Yi, âgé de cinquante-six ans; on l'appelle aussi Nou; de taille moyenne; ses cheveux et sa barbe sont grisonnants; son teint . . .'

Note.—Voyez la fiche précédente, N. xv. 08.

N. xv. 09. (Planche CXIV.) 有髭鬚着白布

'. . . il porte la moustache et la barbe; et il est vêtu de . . . . en toile blanche.'

N. xv. 02. (Planche CXII.) 着布袴褶緼履

'. . . il est vêtu d'un pantalon et d'une tunique de toile et porte des chaussures de chanvre.'

N. xv. 339. (Planche CXIII.) 有髭鬚

'. . . il porte la moustache et la barbe.'

N. xv. 169. (Planche CXIII.) 髭鬚

'. . . la moustache et la barbe.'

N. xv. 139. (Planche CXIII.) 昌縣

'sous-préfecture de □ — tch'ang.'

N. xv. 145. (Planche CXIV.) 州下郡推辟

'le gouverneur a ordonné au préfet d'appeler à cette charge . . .'

N. xv. 010. (Planche CXIV.) 推辟掾錄

'. . . appeler à cette charge; l'adjoint l'a noté par écrit.'

N. xv. 59. b. (Planche CXIII.) 日從事到上

'le . . . jour, la lettre du ts'ong-che (nom de fonctions) est parvenue à son supérieur.'

N. xv. 117. (Planche CXIII.) 從事峻書召詡

'le ts'ong-che Siun manda par écrit Hiu.'

N. xv. 324. (Planche CXII.) 白駁馬故素鞍勒

'. . . un cheval blanc tacheté; une selle et une bride vieilles et ordinaires.'

N. xv. 101. a. (Planche CXIV.) 別下所在郡縣牧送若辟

'. . . par un acte séparé, il ordonna à tous les préfets et sous-préfets que cela concernait de fournir des escortes à celui qu'on avait nommé à cette charge.'

N. xv. 100. (Planche CXII.) 右二人從去正月廿三日

'. . . les deux hommes susnommés, depuis le vingt-troisième jour du premier mois dernier . . .'

N. xv. 207. (Planche CXIII.) 十二月卌日違

'. . . le trentième jour du douzième mois a enfreint . . .'

N. xv. 123. (Planche CXII.) 還不克期日私行無通

'. . . rendre; il ne l'a pas pu au jour de l'échéance; il est parti secrètement, sans en donner avis . . .'

N. xv. 175. (Planche CXIV.) 活從奴一人

'Houo (nom d'homme), accompagné d'un serviteur . . .'

N. xv. 125 & 127. (Planche CXIV.) 側不在□□所者所在郡徑言府逐

'Ts'ö n'est pas au nombre de ceux qui . . .; tous les préfets que cela concerne ont déjà dit au gouverneur de le poursuivre . . .'

N. xv. 351. (Planche CXIV.) 八月卌日

'Le trentième jour du huitième mois . . .'

N. xv. 61 & 62. (Planche CXIII.) 男生年卄五車牛二乘黄犅牛二頭

'Nan-cheng (nom d'homme?), âgé de 25 ans; deux chars à bœufs; deux bœufs jaune-rouge (?)'.

N. xv. 82. (Planche CXII.) 所行治生

'là où aller, gagner sa vie . . .'

N. xv. 78. (Planche CXII.) 畠薑南椰貨物

'le gingembre de?, la noix d'arec du sud, ces marchandises . . .'

N. xv. 69. (Planche CXIII.) 莞劇前已表言

'Houan Ki déjà auparavant a exposé dans ses discours que . . .'

N. xv. 72. (Planche CXIV.) 印是兄

'Yin est son frère aîné (?).'

N. xv. 34. (Planche CXII.) 官一印餘皆錄送舊

'. . . de fonctionnaire un sceau; tout le reste a été inscrit et envoyé à l'ancien . . .'

N. xv. 349. (Planche CXII.)
N. xv. 82. a. (Planche CXIV.) } illisibles.

Note additionnelle.

Les fiches qu'on vient de lire se rapportent au début de la dynastie Tsin, qui commença de régner en 265 p.C. Je crois donc utile de donner ici la traduction des notices fort succinctes que le *Tsin-chou* (chap. xcvii. p. 5 v°—6 v°) a consacrées aux royaumes de Yen-k'i (Karachar), K'ieou-tseu (Koutcha), Ta-yuan (Oura-tépé) et K'ang-kiu (Soghd)[1].

ROYAUME DE YEN-K'I 焉耆 (KARACHAR).

Le royaume de Yen-k'i (Karachar) est à 8200 li à l'ouest de Lo-yang; son territoire s'étend dans la direction du sud jusqu'à Wei-li[2] 尉犁; dans la direction du nord, il est limitrophe des Wou-souen[3] 烏孫; il a une superficie de quatre cents li de côté. Sur ses quatre faces il y a de hautes montagnes; les chemins y sont difficiles d'accès et resserrés; si cent hommes les défendent, mille hommes ne pourraient passer. Pour ce qui est de leurs mœurs, les hommes se coupent les cheveux; les femmes portent des vestes et mettent de grands pantalons. Les coutumes du mariage y sont les mêmes qu'en Chine. Ces gens sont âpres au gain et pratiquent la tromperie. Le roi a pour gardes du corps quelques dizaines d'hommes qui sont tous arrogants et n'ont aucun égard pour les personnes, de quelque rang qu'elles soient.

[1] Si on joint à ces notices la notice sur le pays de Ta Ts'in 大秦 qui n'ajoute rien à ce que renfermait déjà le *Heou Han-chou*, et la longue notice sur les T'ou-yu-houen 吐谷渾 du Koukou nor, qui sont en dehors du cadre de notre étude, on aura la totalité des pages que le *Tsin-chou* consacre aux barbares d'Occident.

[2] Près de Kourla (cf. *T'oung-pao*, 1905, p. 552, n. 5).

[3] Dans les vallées de la Tékès et de l'Ili.

Sous le règne de l'empereur Wou, pendant la période t'ai-k'ang (280–289 p. C.), Long Ngan 龍安, roi de ce pays, envoya son fils (à la cour des *Tsin*) pour qu'il fût enrôlé dans les gardes du corps [1].

La femme de (Long) Ngan était une fille de (la peuplade des) Kouai Hou [2] 獪胡; elle fut enceinte pendant douze mois, puis son côté se fendit et elle mit au monde un fils qui fut nommé Houei 會 et qui devint héritier présomptif. Dès sa jeunesse, Houei se distinguait par sa bravoure; (Long) Ngan, se sentant gravement malade, dit à Houei: 'J'ai été autrefois insulté par le roi de K'ieou-tseu (Koutcha), Po Chan 白山, et mon cœur ne saurait l'oublier; si vous êtes capable de me venger, vous serez véritablement mon fils.' Quand Houei fut monté sur le trône, il attaqua à l'improviste et vainquit Po Chan; il s'établit alors dans le royaume de ce dernier et envoya son propre fils, Hi 熙, régner dans son ancien pays.

Houei était audacieux et avisé; il prit donc l'hégémonie sur tous les Hou 胡 d'occident [3]; à l'est des Ts'ong-ling 葱嶺 (Pamirs), il n'y eut personne qui ne lui fût soumis. Cependant, confiant dans sa bravoure, il se conduisait sans précautions; une fois qu'il était sorti et passait la nuit dehors, il fut assassiné par un homme du pays de K'ieou-tseu (Koutcha) nommé Lo-yun 羅雲.

Après ces évènements, Tchang Kiun [4] 張駿 envoya (345 p. C.) le préfet de Cha-tcheou 沙州, Yang Siuan 楊宣, à la tête d'une multitude d'hommes, pour administrer les pays d'occident; (Yang) Siuan donna le titre de commandant d'avant-garde à son lieutenant Tchang Tche 張植; partout où allait (Tchang Tche), les habitants disparaissaient comme le vent. Cette armée étant arrivée dans le royaume (de Yen-k'i), Hi 熙 lui tint tête et livra bataille auprès de la ville de Pen-louen 賁崙城, mais il fut battu par (Yang) Tche. (Yang) Tche continua sa marche en avant pour camper au (défilé des) Portes de fer [5] 鐵門; il en était encore à plus

[1] Dans les annales principales de l'empereur Wou (*Tsin-chou*, chap. iii. p. 10 r°), nous lisons que le 10e mois de la sixième année t'ai-k'ang (285) les rois de K'ieou-tseu (Koutcha) et de Yen-k'i (Karachar) envoyèrent chacun un de leurs fils à la cour des Tsin pour y être au service de l'empereur.

[2] Nous retrouvons le nom des Kouai Hou mentionné cent ans plus tard; en effet, lorsque, en l'année 384, Lu Kouang 呂光, général de Fou Kien 苻堅, fit sa célèbre expédition dans le Turkestan oriental, le roi de Koutcha, Po Chouen 帛純, implora contre lui le secours du roi des Kouai Hou; le *Tseu tche t'ong kien* (7e mois de l'année 384) se borne à dire à ce propos que les Kouai Hou étaient plus loin dans l'ouest que Koutcha.

[3] Le passage nous fait croire que Long Houei est précisément le personnage qui est mentionné dans la fiche N xv. 93. a, b avec une titulature qui ne peut convenir qu'à un prince ayant l'hégémonie sur tout le Turkestan oriental.

[4] Tchang Kiun 張駿 était un membre de cette famille Tchang qui, pendant plusieurs siècles, devait continuer à fournir des chefs locaux à la région de Leang tcheou 凉州 (voyez l'inscription de 894 consacrée au gendre de Tchang Yi-tch'ao 張義潮 dans '*Dix inscriptions chinoises de l'Asie Centrale* d'après les estampages de M. Ch. E. Bonin', p. 77 et suiv.). En l'année 334, Tchang Kiun avait reçu des Tsin les titres de 'général en chef, surintendant général des affaires militaires dans la région à l'ouest de Chăn et dans les arrondissements de Yong, Ts'in et Leang' 大將軍都督陝西雍秦凉州諸軍事 (*Tseu tche t'ong kien*, 2e mois de l'année 334). En fait Tchang Kiun était un petit souverain et se conduisait comme tel. C'est en 345 que sa puissance atteignit son apogée, après qu'il eut conquis Yen-k'i (Karachar); voici en effet ce que nous lisons dans le *Tseu tche t'ong kien*, à la date de 345: 'Le douzième mois, Tchang Kiun attaqua Yen-k'i (Karachar) et le soumit. Cette même année, Tchang Kiun détacha les onze commanderies qui sont Wou-wei 武威 (Leang tcheou), etc., pour en faire l'arrondissement de Leang 凉州, et il y nomma comme préfet son fils héritier, (Tchang) Tch'ong-houa 重華; il détacha les huit commanderies qui sont Tsin 晉, etc., pour en faire l'arrondissement de Ho 河州; il y nomma comme préfet Tchang Kouan 張瓘, commandant militaire, Pacificateur des Jong; il détacha les trois commanderies qui sont Touen-houang 燉煌, etc., ainsi que les trois camps du Protecteur général des régions d'occident 西域都護三營, et en fit l'arrondissement de Cha 沙州; il y nomma comme préfet Yang Siuan 楊宣, commandant militaire des Hou occidentaux 西胡校尉. Tchang Kiun lui-même se proclama grand surintendant général et général en chef et s'arrogea le titre de roi de Leang 凉王; il eut la haute direction des trois arrondissements.'

[5] Le défilé des Portes de fer était à 50 li à l'ouest de Yen-k'i (Karachar); il est indiqué sur une des deux cartes chinoises gravées en 1137 et conservées dans le Pei lin de Si-ngan fou (*Documents sur les Tou-kiue Occidentaux*, p. 7, lignes 38–39; Bulletin de l'École française d'Extrême Orient, tom. iii, carte A, en regard de la p. 214).

de dix li lorsque Houei, s'étant mis de nouveau à la tête de ses troupes, le devança pour l'attendre dans la vallée de Tchö-lieou 遮留谷; au moment où (Yang) Tche allait y parvenir, quelqu'un lui dit: '(L'empereur Kao)-tsou, de la dynastie Han, s'effraya (du nom) de Po-jen[1] 柏人 et Ts'en P'ong 岑彭 mourut à P'ong-wang[2] 彭亡; maintenant le nom de cette vallée est Tchö-lieou[3] 遮留; peut-être va-t-il y avoir là une embuscade.' (Yang) Tche s'avança seul à cheval pour s'assurer de ce qui en était; il y avait en effet des gens embusqués qui se levèrent; (Yang) Tche les attaqua au galop et les vainquit; puis il s'avança et s'empara de Wei-li 尉犁[4]. Hi 熙, à la tête des quarante mille hommes de son peuple, vint, le buste dénudé, se soumettre à (Yang) Siuan 楊宣. (Plus tard,) quand Lu Kouang 呂光 fit la conquête des pays d'occident (383), Hi 熙 se soumit derechef à lui[5]; puis, lorsque (Lu) Kouang se fut arrogé la dignité impériale, Hi 熙 envoya un de ses fils pour entrer au nombre de ses gardes du corps.

ROYAUME DE K'IEOU-TSEU 龜茲 (KOUTCHA).

Le royaume de K'ieou-tseu (Koutcha) est à 8280 li à l'ouest de Lo-yang. Les mœurs des habitants sont les suivantes: ils ont une ville murée avec des faubourgs et les remparts y sont triples; à l'intérieur se trouvent mille stūpas et temples du Buddha. Les gens s'occupent d'agriculture et d'élevage. Les hommes et les femmes coupent leurs cheveux et les font pendre sur le cou[6]. Le palais du roi est magnifique et resplendit comme une demeure céleste.

Sous le règne de l'empereur Wou, pendant la période t'ai-k'ang (280–289 p.C.), le roi de ce pays envoya son fils (à la cour des Tsin) pour qu'il entrât dans les gardes du corps. A la fin des règnes des empereurs Houei 惠 (290–306) et Houai 懷 (302–312), à cause des troubles qui désolaient l'Empire du Milieu, il envoya des ambassadeurs offrir en tribut des produits de son pays à Tchang Tch'ong-houa[7] 張重華. A l'époque de Fou Kien 苻堅, celui-ci envoya (383 p.C.) son général Lu Kouang 呂光, à la tête de soixante-dix mille hommes, pour attaquer ce pays[8]: le roi, Po Chouen 白純, défendit son territoire et ne se rendit pas[9]; (Lu) Kouang, faisant alors avancer ses troupes, vainquit et conquit (cette principauté).

[1] En 199 av. J.-C., l'empereur Kao-tsou arriva dans la localité de Po-jen et devait y passer la nuit; on avait fait des préparatifs pour l'y assassiner; mais l'empereur eut un pressentiment et ne séjourna pas en cet endroit (cf. *Sseu-ma Ts'ien*, trad. fr., t. ii. p. 391–392). D'après le texte que nous traduisons, c'est le nom même de Po-jen qui aurait été la cause du pressentiment qui sauva Kao-tsou; Po-jen en effet, si on l'écrit 迫人 signifie 'qui met les gens en péril'.

[2] Ts'en P'ong est un général de l'empereur Kouang-wou, fondateur de la dynastie des Heou Han; il périt en l'année 35 p.C., dans la localité de P'ong-wang, dont le nom signifie précisément 'la perte de P'ong' (*Heou Han chou*, chap. xlvii. p. 8 v°).

[3] C.-à-d. 'qui arrête'.

[4] Cf. p. 542, n. 2.

[5] Il est vraisemblable que Hi, qui était roi de Yen-k'i en 345, devait avoir cessé de régner en 383; en effet, d'après d'autres textes, le roi de Yen-k'i qui fit sa soumission à Lu Kouang se nommait Ni-lieou 泥流 (*Tsin-chou*, chap. cxxii. p. 1 v°).

[6] Cette phrase me paraît signifier qu'hommes et femmes coupent leurs cheveux à ras du cou.

[7] Nous avons vu plus haut (p. 543, lignes 21–22 de la n. 4) que Tchang Tch'ong-houa était le fils de Tchang Kiun. Il succéda à son père comme roi de Leang et mourut le jour ting-mao du onzième mois de l'année 353 (*Tseu tche t'ong kien*.

[8] Lu Kouang était guidé dans son expédition par le roi du Kiu-che 車師 antérieur (Tourfan), Mi-t'ien 彌寘, et par le roi de Chan-chan 鄯善, Hieou-mi-t'o 休密馱, qui tous deux étaient dévoués à la dynastie Ts'in 秦.

[9] En 384, le roi de Koutcha, Po Chouen 帛純 (c'est ainsi que le nom est écrit dans le *Tseu tche t'ong kien*), se voyant menacé par Lu Kouang, demanda du secours à ses voisins de l'ouest, les Kouai Hou 獪胡 (cf. p. 543, n. 2); le roi des Kouai Hou envoya à sa rescousse son frère cadet, le marquis de Nou-long 吶龍, et le général K'ouei 馗, à la tête de plus de 200,000 cavaliers; ces chefs entraînèrent en outre avec eux plus de 200,000 soldats des pays de Wen-sou 温宿 (Ouch Tourfān), Wei-t'eou 尉頭 (Safyr bay, au S.-O. d'Ouch Tourfān), &c., mais cette immense armée fut vaincue par Lu Kouang, qui pénétra en vainqueur à Yen-k'i (Karachar) et soumit ainsi à son autorité tout le Turkestan oriental. On sait que, parmi les prisonniers que Lu Kouang fit à Karachar, se trouvait le célèbre religieux Kumārajīva (cf. Bunyiu Nanjio, *Catalogue*, App. II. n° 59).

ROYAUME DE TA-YUAN 大宛 (OURA-TÊPÊ).

Le Ta-yuan est à 13350 li à l'ouest de Lo-yang; vers le sud, il arrive jusqu'aux Ta Yue-tche 大月氏; vers le nord, il touche au K'ang-kiu 康居 (Soghd). Il possède soixante-dix villes grandes ou petites: le sol est favorable à la culture du riz et du blé; on y fait du vin de raisin; il s'y trouve beaucoup d'excellents chevaux; ces chevaux suent du sang. Les habitants ont les yeux profondément enfoncés et ont beaucoup de barbe. Voici ce qui concerne leurs mœurs: quand un homme veut se marier, il commence par donner en présent à sa fiancée une bague d'accordailles en or; puis on le met à l'épreuve en lui livrant trois servantes; si aucune d'elles n'a de fils, on rompt le mariage. Quand des enfants sont issus d'un commerce illégitime, on méprise leurs mères. Quand quelqu'un donne un cheval à monter à un homme, si le cheval est indocile et si l'homme tombe et se tue, c'est le propriétaire du cheval qui doit faire les frais de la mise en bière. Ils sont tous commerçants et se disputent pour un profit de la moitié d'un vingt-quatrième d'once. Quand ils ont de l'or et de l'argent de l'Empire du Milieu, ils en font aussitôt des ustensiles et ne s'en servent pas comme de monnaie.

La sixième année t'ai-k'ang (285 p. C.), l'empereur Wou envoya l'ambassadeur Yang Hao 楊顥 conférer au roi de ce pays nommé Lan-seou 藍庾 le titre de roi de Ta-yuan. A la mort de Lan-seou, son fils Mo-tche 摩支 prit le pouvoir; il envoya un ambassadeur offrir en tribut des chevaux qui suaient du sang.

ROYAUME DE K'ANG-KIU 康居 (SOGHD).

Le royaume de K'ang-kiu est à environ 2000 li au nord-ouest du Ta-yuan. Il est limitrophe (des pays) de Sou-yi[1] 粟弋 et de Yi-lie 伊列. Le roi réside dans la ville de Sou-hiai[2] 蘇薤城. Pour les mœurs, pour l'aspect extérieur des hommes et pour les vêtements, ce royaume est à peu près identique au Ta-yuan. Ce pays est tempéré; il produit en abondance le peuplier, les saules et la vigne; il s'y trouve beaucoup de bœufs et de moutons; d'excellents chevaux en proviennent.

Pendant la période t'ai-che (265–274 p. C.), le roi de ce pays, nommé No-pi 那鼻, envoya un ambassadeur présenter à l'empereur une requête scellée et en même temps lui offrir d'excellents chevaux.

[1] Le *Heou Han chou* (chap. cxviii. p. 6 r°) écrit Li-yi 栗弋.

[2] Comme l'a établi Marquart (*Die Chronologie der alttürkischen Inschriften*, p. 52, et *Ērānšahr*, pp. 303–304), Sou-hiai doit être la transcription du nom de Soghd. C'était alors la ville de Kesch qui était la capitale du Soghd (cf. *Documents sur les Tou-kiue occidentaux*, p. 146).

PART III

LES SGRAFFITI ET DOCUMENTS CHINOIS DU FORT D'ENDERE

Sgraffito du temple d'Endere (Planche XI).

開元七年記
至建。聞其兵馬使死。及四鎮大蕃
和大蕃官。太常卿秦嘉興歸本道。

Note écrite la septième année k'ai-yuan (719 p. C.).

. . . tche kien. Il apprit que son commissaire des troupes et de la cavalerie était mort; puis les Quatre Garnisons [1] et les grands Tibétains [2] . . .

. . . avec les officiers des grands Tibétains. Le haut dignitaire du t'ai-tch'ang [3], Ts'in Kia-hing [4], revint dans le district placé sous ses ordres.

[1] En 719, les Quatre Garnisons étaient Koutcha, Khoten, Kachgar, Karachar. Cf. *Documents sur les Tou-kiue occidentaux*, p. 113, n. 2.

[2] Ta Fan, 'les grands Fan,' est le nom sous lequel les Tibétains se désignent eux-mêmes dans le traité entre la Chine et le Tibet gravé sur pierre à l'époque des T'ang. Cf. Bushell, 'The early history of Tibet,' *Journ. Roy. As. Soc.*, N.S., vol. xii. p. 536.

[3] Le t'ai-tch'ang-sseu est l'administration qui préside aux sacrifices de l'État.

[4] Je n'ai pas retrouvé le nom de ce personnage dans les histoires chinoises.

Sgraffito de la salle E. iii d'Endere (Planche XI).

國 (écrit 囯).
使
辛 (écrit 辛, ce qui est une orthographe admise).
利
川

'L'envoyé impérial Sin Li-tch'ouan.'

Fragments de documents trouvés dans le temple d'Endere.

E. i. 36 a.

妄相羅織人前妻李十三
右希俟先婚前
同君

1re ligne: D'une manière injuste, il a travaillé à impliquer dans l'affaire l'ancienne femme de cet homme, Li Che-san . . .

2e ligne: par ce qui précède espère attendre jusqu'avant le mariage . . .

E. i. 44.

日典周玄福牒

候守左羽林軍大將軍王直將

1re ligne : . . . jour. Requête du fonctionnaire Tcheou Hiuan-fou[1].
2e ligne : heou-cheou[2], général en chef du corps d'armée des tso yu lin, *Wang* (?) *Tche-tsiang*.

[1] Nom d'homme. Le mot 典 signifie 'fonctionnaire'.

[2] Ces deux mots doivent faire partie d'un titre chinois.

E. i. 8.

今從城 來請

Maintenant, venant de la ville . . . venir demander.

APPENDIX B

TIBETAN MANUSCRIPTS AND SGRAFFITI

DISCOVERED BY DR. M. A. STEIN AT ENDERE

EDITED BY

L. D. BARNETT, M.A., D.LITT.,

ASSISTANT IN THE DEPARTMENT OF ORIENTAL PRINTED BOOKS AND MSS.
OF THE BRITISH MUSEUM

AND

A. H. FRANCKE

MORAVIAN MISSION, LEH

PREFATORY NOTE

In the following pages are published the fragments of Tibetan literature and the inscriptions discovered by Dr. Stein at Endere. A brief estimate of their historical significance has been given by me in the *Journal of the Royal Asiatic Society* for January, 1903, and I hope that the publication of this material will enable scholars to supplement and, where necessary, to correct the views which I have there expressed.

In the division of the work between the Rev. Mr. Francke and myself, the task of editing the whole of sections i and ii fell to my lot, while most of the philological material forming sections iii and iv has been contributed by Mr. Francke. When I began to study the MSS. and photographs placed in my hands by Dr. Stein, it became plain to me that the dialectal peculiarities of sections iii and iv presented problems soluble only by a scholar in close touch with the living dialects of Tibet. I therefore sent to Mr. Francke at Leh a transcription of the poems forming section iii and my readings of the inscriptions photographed by Dr. Stein (section iv), and in response to my request Mr. Francke generously contributed the valuable notes which appear under his name in those sections.

Finally I have to record my deep gratitude to two scholars who have unselfishly aided me. The weary labour of piecing together the multitudinous fragments of the Śālistamba-sūtra was lightened by the generosity of Professor de la Vallée Poussin, who placed at my disposal the Sanskrit version of that text prepared by him; and on many points, notably in the study of section iii, Mr. F. W. Thomas has given me the benefit of his scholarly judgement.

L. D. BARNETT.

London, 1905.

I

FRAGMENTS OF THE ŚĀLISTAMBA-SŪTRA

THE following fragments originally formed part of a large and well-written paper *pothī* about 18¼ inches in width and 2⅝ inches in height. The text is written on one side of the page only, and has been revised, apparently by the scribe himself. The pages are numbered in the usual way, seemingly by a second hand, on the left-hand margin.

As will be seen from the appended notes, the text preserved in these fragments substantially agrees with that of the Kanjur, *Mdo* xvi. It would seem that the text of the Kanjur represents a later revision of that given in our fragments, a few passages having been slightly expanded and ambiguous words changed to simpler language.

A number of roots which in modern Tibetan end in *-r*, *-l*, or *-n* have in our MS. a final *-d* appended. This *-d* is the so-called *drag* of the native grammarians, and is found in other ancient MSS.[1] In isolated instances it is omitted in our MS. from roots that elsewhere have it, a fact indicating that it was already beginning to be dropped in actual speech. It can only be traced with certainty once in the other MSS. of Endere, which are apparently of a less canonical literary character; and the popular sgraffiti on the same site also have it but once.

We also find a *y* between *m* and the high vowels *i* and *e*, for example *myi*, *mye*, *myed*. An exception is *me t'og*. After vowels, where the modern literary language writes *aṅ*, we find here always *yaṅ*. Before a short pause a final *-a* sometimes appears to be lengthened to *-ā*, the letter འ being added on the line[2], whereas at the end of sentences the syllable *o* is added, as in the modern language. Other peculiarities need not be specially mentioned.

As regards form of writing, there is no difference from the modern *dbu-can* script. The letter འ sometimes has a small tick attached to it on the top, at the right side; but this form is found in other old MSS. and books.

The fragments are numerous, and often minute; but with the aid of a block-print of *Mdo* xvi of the Kanjur, kindly lent by the India Office Library, it has been possible to assign to each its place. In the text given below a thick vertical bar marks the point where the reading of a line in a fragment ends and is continued in another fragment. The original owner of the MS. tore each page into two or more pieces, and distributed these as offerings before the statues of divers deities [see above, p. 425]. Hence it is seldom that all the component parts of a page have survived. As a rule, however, the morsels are sufficiently numerous to make up the larger part of each of the pages represented by them; and in such cases the gaps in the pages have been filled up by inserting the missing words from the Kanjur, enclosing them in square brackets and altering their spelling to suit that of our text. Before each page or fragment of a page is added a reference to the numbers of the pages in the India Office copy of the Kanjur, *Mdo* xvi, upon which the corresponding passage is to be found. In collation I have disregarded variations of no importance, e.g. where the Kanjur gives *pa* or *ba* as compared with *ba* or *pa* of our MS., or when the

[1] See the Rev. Mr. Francke's *Kleine Beiträge zur Phonetik und Grammatik des Tibetischen*, in the *Zeitschrift der Deutschen Morgenländischen Gesellschaft*, Bd. lvii. p. 295.

[2] I have indicated this added sound in transcription by a circle at the end of the word, corresponding to the transcription of the འ at the beginning of words.

latter uses contracted forms, as *tiṅe*, for the full spelling of the Kanjur (*tiṅ ṅe*). My aim has been to show the more significant points of difference between our fragments and the Kanjur, and their bearing upon the original Sanskrit text.

On the subject of this Sūtra the essays of Professor de la Vallée Poussin should be consulted (*Journal asiatique*, 1902, p. 237 ff., 1904, p. 357 ff.).

(A. Fol. 2.)

(*Mdo* xvi, ff. 191 b–192 a.)

Line 1. . . . rky[e]nd kyis reg pa | reg pai rkyend kyis ts'or bao[1] | ts'or bai rkyend kyis sred pa | sred pai rkyend kyis lend pa | [lend pai rkyend kyis srid pa | srid pai rkyend kyis skye ba | skye bai rkyend kyis rga ši daṅ | mya ṅan daṅ | smre sṅags ○dond pa daṅ |]

2. sdug bsṅal ba daṅ | yi mug pa[2] daṅ | ok'rug pa rnams ○byuṅ ste | de ltar sdug bsṅal gyi p'uṅ po c'en po[3] ○di ○b[yuṅ bar ○gyur ro || de la ma rig pa ○gags pas ○du byed ○gag | ○du byed ○gags pas rnam par šes pa]

3. ○gag | rnam par šes pa ○gags pas myiṅ daṅ gzugs ○gag | myiṅ dan gzugs ○gags pas [skye mc'ed drug ○gag | skye mc'ed drug ○gags pas reg pa ○gag | reg pa ○gags pas ts'or ba ○gag |]

4. ts'or ba ○gags pas sred pa ○gag | sred pa ○gags pas lend pa ○gag | lend pa ○gags pas srid pa ○gag [| srid pa ○gags pas skye ba ○gag | skye ba ○gags pas rga ši daṅ | mya ṅan daṅ | smre sṅags ○dond pa daṅ | sdug]

5. bsṅal ba daṅ | yi mug pa daṅ ok'rug pa rnams ○gag par ○gyur te | de ltar sdug bsṅal gyi p'uṅ po c'en [p]o [○di ○gag par ○gyur ro | ○di ni bcom ldan ○das kyis rtend ciṅ ○breld par ○byuṅ ba]

A single fragment (E. i. 12), numbered on the margin *gñis*, and forming part of page 2 in the original MS.; width, about $8\frac{7}{8}$ in.; height, $2\frac{5}{8}$ in.

[1] This lengthening seems due to the short pause following; compare *nao* (B 1 and 4, E 1), *pao* (C 2, M 2), *myio* (D 2).

[2] The Kanjur here, and again in l. 5, reads *yid mi bde ba*.

[3] After *po* here and in l. 5 the Kanjur adds *○ba žig po*, i.e. केवलस्य.

(B. Fol. 3.)

(*Mdo* xvi, ff. 192 a–b.)

Line 1. žes gsuṅs so || c'os gaṅ že nao | op'ags pai lam yan lag brgyad pa ste | ○dio[1] lta ste yaṅ da|g pai lta ba daṅ | yaṅ dag pai rtog pa daṅ | yaṅ dag pai ṅag [daṅ | yaṅ dag pai las kyi mt'a daṅ | yaṅ dag pai]

2. ○ts'o ba daṅ | yaṅ dag pai rtsol ba daṅ | yaṅ dag pai dran ba daṅ | yaṅ dag pai tiṅe ○dzind to | ○di ni op'ag|s pai lam yan lag brgyad pa žes bya ste ○bras bu rñed[2] pa [daṅ mya ṅan las ○das pa gcig tu bsdus]

3. te bcom ldan ○das kyis c'os so žes bka stsald to || | de la saṅs rgyas bcom ldan ○das | gaṅ že na | sus c'os t'ams cad t'ugsu c'ud pai p'yir saṅs rgyas že[s bya ste | des op'ags pai šes rab kyi spyan daṅ]

4. c'os kyi skus[3] byaṅ cub[4] byed pa daṅ | slob pa daṅ myi slob pai c'os rnams gzigs so[5] || de la ji ltar na rtend ciṅ ○breld par ○byuṅ ba mt'oṅ že nao[6] | ○di la bcom ldan ○[das kyis sus rtend ciṅ ○breld par ○byuṅ ba rtag pa]

5. daṅ srog myed pa daṅ srog daṅ bral ba daṅ ji lta bu ñid daṅ ma nord pa daṅ ma skyes pa daṅ ma byuṅ ba daṅ ma byas pa daṅ[7] t'og|s pa myed pa daṅ dmyigs pa myed pa daṅ ži ba daṅ myi ○jigs pa[8] daṅ [myi op'rogs pa daṅ rnam par ži ba ma yin pai raṅ bžin du mt'oṅ ba ste |]

Two fragments (E. i. 32 a, 18), forming the greater part of the third page of the original MS.; width, 9 in. and $4\frac{3}{4}$ in. respectively; height, $2\frac{5}{8}$ in.

[1] The spelling འདིའ་ is probably a mere blunder.

[2] In the Kanjur *t'ob*.

[3] The Kanjur reads *sku daṅ ldan pa daṅ slob pa*, &c. Our text seems to point to an original reading something like आर्यप्रज्ञाचक्षुर्धर्मकायेन बोधितः, whereas the words in the Kanjur represent *कायसम्पन्नः.

[4] The unaspirated *cub* may be a mere error; compare *nam ka*, M 4.

[5] The Kanjur reads *c'os de dag* (i.e. ताण् . . . धर्माण्) *gjigs pao*.

[6] This clause in the Kanjur reads thus: *de la rten ciṅ obrel pa ɔbyuṅ ba mt'oṅ ba ji lta bu že na*. The Sanskrit has कथं . . . पश्यति.

[7] Here the Kanjur adds *odus ma byas pa daṅ*, i.e. असंस्कृतम्. The omission here is probably due to error.

[8] The Kanjur reads *ojigs pa med pa*.

(C. Fol. 4.)

(*Mdo* xvi, ff. 192 b–193 a.)

Line 1. [gaṅ gis ts'ul odi odra bar c'os la yaṅ rtag pa | srog myed pa daṅ srog daṅ bral ba daṅ ji lta bu ñid] daṅ ma nord pa daṅ ma skyes pa daṅ ma byuṅ ba daṅ ma byas pa daṅ odus ma byas pa daṅ t'ogs pa myed pa[1] daṅ dmyigs pa myed pa daṅ ži [ba]

2. [daṅ myi ojigs pa daṅ myi op'rogs pa daṅ rnam par ži ba ma yin pai raṅ bžin du mt'oṅ ba de op'ags pai c'os mṅon bar rt]ogs šiṅ[2] yaṅ dag pai ye šes t'ob pas[3] bla na myed pai c'os kyi skur saṅs rgyas mt'oṅ ṅo žes gsuṅs so || smras pao

3. [cii p'yir rtend ciṅ obreld par obyuṅ ba žes bya | smras pao | rgyu daṅ bcas rkyend daṅ bcas pa la byai | rgyu myed rkyen]d myed pa la[4] ma yin te | dei p'yir rtend ciṅ obreld par obyuṅ ba žes byao || de la bcom ldan odas kyis rtend ciṅ obrel par

4. [obyuṅ bai mts'an ñid mdor bsdus te bka stsald pa rkyend odi ñid kyi obras bu ste | de bžin gšegs pa rnams byuṅ ya]ṅ ruṅ ma byuṅ yaṅ ruṅ c'os rnams kyi c'os ñid odi ni odug pao žes bya ba nas | gaṅ odi c'os ñid daṅ c'os gnas pa ñid daṅ

5. [c'os myi ogyur ba ñid daṅ rtend ciṅ obreld par obyuṅ ba mt'un ba daṅ de bžin ñid daṅ ma nord pa de bžin ñid da]ṅ | gžan ma yin ba de bžin ñid daṅ | yaṅ dag pa ñid daṅ bden ba k'o na daṅ | de k'o na ñid[5] daṅ ma nord pa ñid daṅ p'yin ci ma

A single fragment (E. i. 34 a), forming about half of the fourth page in the original MS.; width, about 9¼ in.; height, 2⅜ in.

[1] The Kanjur, which immediately before agreed with our text, has here the words *t'og ma med pa*, i.e. अनादि.

[2] *Te* in the Kanjur.

[3] In the Kanjur *ye šes daṅ ldan pas*, as on the preceding page. Our text suggests a reading स आर्यधर्माभिसमये सम्यग्ज्ञानलब्धितः (ज्ञानोपनयनेन ?) &c.

[4] The Kanjur reads *la ni*.

[5] The words *de k'o na ñid daṅ*, probably a translation of तत्त्वता or तत्त्वम्, are wanting in the Kanjur.

(D. Fol. 7.)

(*Mdo* xvi, f. 194 a–b.)

Line 1. ba byao sñam du myi sems so | de bžin du c'ui k'ams kyaṅ odi ltar[1] bdag [gi]|s sa bon rlan[2] par [byao sñam du myi] | sems so | myei k'ams kyaṅ odi ltar bdag gis sa bon ts'os par byao[3] sñam du myi sems so | rluṅ gi k'ams kyaṅ odi

2. ltar bdag gis sa bon obu bar byao sñam du myi sems so | nam mk'ai | k'ams kyaṅ odi ltar b[dag gis sa bon] | la myi sgrib pai bya ba byao sñam du myi sems so | dus kyaṅ odi ltar bdag gis sa bon bsgyur bai bya ba byao sñam du myio |

3. sems so | sa bon yaṅ odi ltar bdag gis myi gu[4] mṅon bar bsgrub | bo sñam du myi se[ms so | myi gu ya]|ṅ odi ltar bdag ni rkyen odi dag gis mṅon bar bsgrubs so sñam du myi sems mod kyi | oond kyaṅ rkyend odi dag yod

4. la sa bon ogag pa na myi gu mṅon bar ogrub par ogyur ro | d[e] bžin du | me t'og[5] yod na obras bui b[ar du yaṅ m]|ṅon bar ogrub par ogyur te | myi gu de yaṅ bdag gis ma byas | p'a rold kyis[6] ma byas gñis kas ma byas dbaṅ pos ma spruld[7] |

5. dus kyis ma bsgyurd raṅ bžin las ma [byuṅ] rgyund[8] myed pa las kyaṅ ma s[ky]e[s te | o]|ond kyaṅ sa daṅ c'u daṅ mye [daṅ rluṅ daṅ nam] | mk'a[9] daṅ dus kyi k'ams rnams odus pa las[10] myi gu skye žiṅ obyuṅ bar[11] ogyur te | de ltar p'yi rold kyi rtend ciṅ obrel par obyuṅ bai[12]

Two large fragments, with the gap between them partly filled up by a smaller piece (E. i. 22, 27 b, 23). As arranged, they measure respectively 6$\frac{1}{8}$, 1$\frac{3}{4}$–2$\frac{1}{4}$, and 9$\frac{3}{8}$ in. in width; in height, 2$\frac{5}{8}$ in.

[1] Our MS. has here and in all similar cases *odi ltar* corresponding to *odi sñam du* in the Kanjur.

[2] The Kanjur reads *brlan*.

[3] Instead of *ts'os par byao* (पाचयति) the Kanjur reads *yoṅs su dro bar byao*, i.e. परिपाचयति, as in the Sanskrit.

[4] Our MS. in all cases reads *myi gu*, where the Kanjur has the modern form *myu gu*.

[5] The aspirate of *me t'og* is noteworthy.

[6] The Kanjur reads *gžan gyis*, with much the same meaning (परहतो).

[7] This represents नेश्वरनिर्मितो. The Kanjur has *dbaṅ p'yug gis ma byas*, which is perhaps due to a reviser who wished to avoid confusion between the two meanings (*indriya* and *īśvara*) of *dbaṅ po*.

[8] Apparently a mistake for the Kanjur reading *rgyu*.

[9] Written [illegible].

[10] In the Kanjur *odus nas*, after which are added the words *sa bon ogag pa na*, in the Sanskrit text बीजे निरुध्यमाने.

[11] In the Kanjur *myu gu mṅon par ogrub par*, supported by the Sanskrit चक्षुरस्याभिनिर्वृत्तिर्भवति. Our reading points to a variant something like जातिसमुत्पत्तिर्भवति.

[12] The Kanjur reads *p'yi rol gyis rten ciṅ obrel par ogyur ba rkyen daṅ obrel bar bltao* ||

(E. Fol. 10.)

(*Mdo* xvi, ff. 195 b–196 a.)

Line 1. . . . daṅ[1] gaṅ odus pai p'yir že nao | . . .

2. [. . . gaṅ odi odus nas lus kyi sra bai dṅos po mṅon bar ogrub par] byed pa odi ni sai k'am[s] šes byao . . .

3. [. . . gaṅ lus kyi dbugs] p'yi naṅ du rgyu bai bya ba byed pa odi ni [rluṅ gi k'ams šes byao | . . .]

4. [. . . gaṅ mdud k'yim gyi ts'ul du lus kyi myiṅ daṅ gzugs kyi myi gu mṅon bar ogrub par byed pa rnam par šes pai ts'ogs] lṅa daṅ ldan ba[2] daṅ | zag pa daṅ bcas [pai yid kyi rnam par šes pa . . .]

5. [. . . nam mk'a daṅ rnam par šes] pai k'ams rnams kyaṅ ts'aṅ bar gyur[3] . . .

One fragment (E. i. 28 b); width, c. 2$\frac{5}{16}$ in.; height, 2$\frac{5}{8}$ in.

[1] The Kanjur has *k'ams drug pa gaṅ dag odus*, &c.

[2] The Kanjur reads *lṅa odus pa*, with practically the same meaning.

[3] Apparently *gyurd*.

(F. Fol. 11.)

(*Mdo* xvi, ff. 196 b–197 a.)

Line 1. . . . s]ñam du myi sems so | myei k'ams kya[ṅ odi ltar bdag gis lus kyi zos pa daṅ ot'uṅs] | pa daṅ oc'os [pa daṅ] | myaṅs pa rnams odzu [bar byao] | sñam du myi sems so | rluṅ gi k'am[s . . .]

2. . . . nam mk'ai k'ams kyaṅ odi ltar bdag [gis lus kyi naṅ sbubs yod par byao sñam du myi se]|ms so | rnam | par šes pai k'ams kyaṅ o[di ltar] | bdag[1] gis lus mṅon du ogrub pa[r ogyur ro sñam du myi sems so | . . .]

3. . . . bskyed do sñam du myi sems mod kyi [|oond kyaṅ rkyend odi dag yod na lus obyu]|ṅ[2] bar ogyur ro || de la sai k'ams[3] bdag ma [yi]|n | sems can ma yin [sr]o[g ma yin | . . .]

4. . . . yin | skyes pa ma yin | ma niṅ ma [yin | ṅa ma yin | bdag gi ma yin te | gžan sui yaṅ] | ma yin no | de b|žin du c'ui k'ams daṅ myei [k'ams] | daṅ rluṅ gi [k'ams . . .]

5. . . . ma yin | srog ma yin | skye ba[4] ma yin | [šed las skyes pa ma yin | šed bu ma yin | bud mye]|d ma yin | skye|s pa ma yin | ma niṅ ma yin | . . .

Four fragments (E. i. 28 a, 31 a, 37, 34 b), of which the last contains nothing of the fifth line; width 2$\frac{1}{4}$, 1–1$\frac{1}{4}$, 1$\frac{1}{2}$–2, and 1–2$\frac{1}{2}$ in. respectively; height, 2$\frac{5}{8}$, 2$\frac{5}{8}$, 2$\frac{5}{8}$, 2 in.

[1] This clause appears in the Kanjur as *bdag gis lus kyi miṅ daṅ gzugs mṅon par bsgrub bo*; both represent चक्षुं कायस्य नामरूपमभिनिर्वर्तयामि. The clause is absent in the Śikṣāsamuccaya.

[2] *Skye* in the Kanjur.

[3] *K'ams ni* in the Kanjur.

[4] *Skye pa po* in the Kanjur.

(G. Fol. 12.)

(*Mdo* xvi, f. 197 a–b.)

Line 1. ༄ ‖ yaṅ ma yi[n no ‖ de la ma rig pa gaṅ že na | k'ams drug po odi dag ñid la ga]|ṅ gcig pur odu šes pa[1] | ril por odu šes [pa da]|ṅ rtag par odu šes pa[2] | brtan bar odu šes pa | t'er zug du[3] odu šes pa | bde bar odu šes pa[4] | sems can daṅ srog daṅ skye ba daṅ[5]

2. gsō[6] ba daṅ skyes [bu daṅ . . .] | du odu šes pa[7] daṅ | ṅa žes bya ba daṅ bdagi | žes bya bar odu šes pa odi lta bu las stsogs[8] pa myi šes pa rnam pa sna ts'ogs pa odi lta bu[9] ni ma rig pa žes byao | de ltar ma rig

3. pa yod pas yul | [r]na[ms la o]dod c'ags [daṅ že sdaṅ daṅ gti mug pa] | ojug go[10] | de la yul rnams la odod | c'ags daṅ že sdaṅ daṅ gti mug pa[11] gaṅ yin ba odi ni[12] odu byed rnams so | dṅos po so sor rnam par rig pa rnam par

4. šes pao | rna]m par šes pa daṅ lhan c'ig [sky]e[13] [ba ñe bar lend pai] | p'uṅ po gzugs myed pa bži po de dag ni[14] | myiṅ daṅ gzugs so | myiṅ daṅ gzugs la rten[15] pai dbaṅ po rnams ni skye mc'ed drug go | c'os gsum odus pa ni reg pao |

5. reg pa myoṅ ba ni | ts'or bao | ts'or ba las[16] žen [ba ni sred pao | sred pa] | op'eld pa ni lend pao | len p'a las | skyes pa yaṅ srid[17] pa skyed pai las ni srid pao | rgyu de las p'uṅ po byuṅ ba ni skye bao | skyes nas p'uṅ po smyind pa ni rgas pao[18]

Four fragments (E. i. 26*, 27 a, 35, 32 b), of which the smallest (27 a) forms only the second division of lines 3–5; width, c. 1¼, 2⅜, 3⅛, and 9½ in.; height 2⅝, 1⅜, 2⅝, 2⅝ in.

[1] *Pa daṅ* in the Kanjur.

[2] In the Kanjur *pa daṅ*, and so in the following three categories.

[3] In the Kanjur *tu*.

[4] After this follow in the Kanjur the words *bdag tu odu šes pa daṅ*, corresponding to आत्मसंज्ञा in the Sanskrit text.

[5] In the Kanjur *skye pa po*.

[6] The lengthening of the vowel is noteworthy, the word being written གསཱོ. See note on A 1.

[7] These words do not agree with the Kanjur, which reads *skyes bu daṅ | gaṅ zug tu odu šes pa daṅ | šed las skyes pa daṅ | šed bur odu šes pa daṅ | ṅa žes bya ba daṅ | bdag gi žes bya bar odu šes pa ste.*

[8] In the Kanjur *la sogs*.

[9] The Kanjur has merely *sna ts'ogs odi ni*; *odi ni* agrees with the Sanskrit इयम्.

[10] In the Kanjur *gti mug ojug ste*.

[11] *Pa* is not in the Kanjur.

[12] After *ni* follow in the Kanjur the words *ma rig pai rkyen gyis*, which are not represented in the Sanskrit. Instead of *so* the Kanjur has *žes byao*, agreeing with उच्यन्ते of the Sanskrit.

[13] *Byuṅ* in the Kanjur. Note the spelling *c'ig*.

[14] This clause agrees with the words of the Sanskrit विज्ञानसहभुवश्चत्वारोऽरूपिण उपादानस्कन्धाः; the Kanjur, however, reads *p'uṅ po bži po odi dag ni*. The clause following this in the Sanskrit text (चत्वारि च महाभूतानि चोपादाय उपादाय रूपमेकध्यमभिसंक्षिप्य तन्नामरूपम् in Śikṣāsamuccaya) is omitted here and in the Kanjur.

[15] In the Kanjur *brten*, agreeing with संनिश्रितानि in the excerpt in the Śikṣāsamuccaya.

[16] In the Kanjur *la*.

[17] For this and the next three words the Kanjur gives merely *bskyed pai*.

[18] In the Kanjur *rga bao*.

(H. Fol. 13.)

(*Mdo* xvi, ff. 197 b–198 a.)

Line 1. rgas nas p'uṅ po . . .

2. myi bde ba myoṅ ba ni sdu[g] . . .

3. mṅon bar odu byed . . .

4. skom bai p'yir sred . . .

5. smre sṅags odond . . .

One fragment (E. i. 30), 1⅞ × 2⅜ in., orming the left-hand side of the page immediately following the preceding (G). There are no divergences from the Kanjur.

(J. Fol. 14.)

(*Mdo* xvi, f. 198 a–b.)

Line 1. ∽ || de ltar ma rig pa yod na ᴑdu byed rnam gsum mṅon bar ᴑgru[b ste] | bsod namsu ñe bar ᴑgro ba daṅ | bsod [nams ma yin bar] | ñe bar ᴑgro ba daṅ | myi gyo bar ñe bar ᴑgro bao[1] | de la bsod na[ms su ñe bar ᴑgro bai ᴑdu byed rnams las bsod]

2. nams su ñe bar ᴑgro bai rnam par šes pa ñid du ᴑgyur ba daṅ | b[sod] nams ma yin bar ñe bar ᴑgro [bai ᴑdu byed rnam]|s las bsod nams ma yin bar ñe bar ᴑgro bai rnam [par šes pa ñid du ᴑgyur ba daṅ | myi gyo bar ñe]

3. bar ᴑgro bai ᴑdu byed rnams las myi gyo bar ñe bar ᴑgro bai [rnam] par šes pa ñid du ᴑgyur ba ᴑ[di ni ᴑdu byed kyi] | rky[e]nd kyis rnam par šes pa žes byao | de bžin du[2] rnam pa[r šes pai rkyend kyis myiṅ daṅ gzugs šes byao[3] |]

4. myiṅ daṅ gzugs rnam par op'eld pas skye mc'ed drug gi sgo nas bya ba byed pa rnams ᴑbyuṅ ste ᴑ[di[4]] n[i] myi[ṅ daṅ gzugs ky]|i [rky]e[nd ky]i[s] skye mc'ed drug ces byao | skye mc'ed drug [po dag la reg pai ts'ogs drug ᴑbyuṅ ste | ᴑdi ni]

5. skye mc'ed drug gi rkyend kyis reg pa žes byao | ji lta bur reg pa ᴑbyuṅ ba de lta[5] bur ts'or myoṅ ste[6] de ni reg pai rkyend ky|is ts'or bai[7] [by]e brag de dag myoṅ ba daṅ . . .

Two fragments (E. i. 26, 14), forming the greater part of the folio which directly followed the preceding (H); width 9 and 4½ in.; height, 2⅜ in.

[1] In the Kanjur *ᴑgroo*.

[2] These three words are not in the Kanjur or Sanskrit.

[3] Such appears to have been the original reading of our text, and it nearly agrees with the variant quoted in Prof. Bendall's note to the Śikṣāsamuccaya, viz. तदेव विज्ञानप्रत्ययं नामरूपम्, the other variant being simply एवं ना°. The Kanjur gives *rnam par šes pa daṅ lhan cig skyes pa p'uṅ po gzugs can ma yin pa bži daṅ | gzugs gaṅ yin pa de ni rnam par šes pai rkyen gyis miṅ daṅ gzugs žes byao*, representing विज्ञानसहजाश्चत्वारोऽरूपिणः स्कन्धा यच्च रूपं तदेव विज्ञानप्रत्ययं नामरूपमिति, as in the Śikṣāsamuccaya. Hence it would seem that the original reading was simply एवं (or तथा) विज्ञानप्रत्ययं नामरूपमिति.

[4] In the Kanjur *de*; and so in the next sentence.

[5] The Kanjur gives wrongly *ltar bur*.

[6] For अनुभूयते? The Kanjur has *ts'or ba ᴑbyuṅ ste* (प्रवर्तते), agreeing with the Sanskrit.

[7] In the Kanjur *ts'or ba žes byao || ts'or bai*, &c. This points to या तस्याः स्पर्शप्रत्ययाया वेदनायास्तान् विशेषानास्वादयति or the like; but the actual Sanskrit has यस्तां वेदयति विशेषेणास्वादयति.

(K. Fol. 15.)

(*Mdo* xvi, ff. 198 b–199 a.)

Line 3. . . . ba de ni . . .
4. . . . yin . . .
5. . . . su zug[s] . . .

One small fragment, 1¾ in. high × ⅝ in. wide.

(L. Fol. 16. See Plate CXVII.)

(*Mdo* xvi, f. 199 a–b.)

Line 1. || g rkyend gcig[1] | rtag pa ma yin myi rtag pa ma yin | ᴑdus byas ma yin ᴑdus ma byas ma yin[2] | myoṅ [ba yod pa ma yin | rgyu myed pa las byuṅ ba |] | rky[e]nd myed pa las byuṅ ba ma yin | zad pai c'os ma yin | ᴑjig pai c'os ma yin | ᴑgog pai c'os ma yin

2. te t'og[3] ma myed pai dus nas kluṅ gi rgyun bžin du rgyund ma c'ad par žugs pai rjesu žugs mo[d kyi | ᴑond kyaṅ yan lag bži] | po ᴑdi dag ni[4] de bsdu bar bya bai rgyur ᴑgroo[5] | bži gaṅ že na ᴑdi lta ste | ma rig pa daṅ | sred pa daṅ | las daṅ

3. rnam par šes pao | de la rnam par šes pa ni sa bon gyi raṅ bžin gyis rgyu byed do | la[s ni žiṅ

gi raṅ] | bžin gyi[s rgyu byed do | ma ri]|g pa daṅ sred pa ni ñon moṅs pai raṅ bžin gyis rgyu byed do | de la las daṅ ñon moṅs pa dag ni sa bon

4. rnam par šes pa skyed do | de la las ni sa bon rnam par šes pai žiṅ gi bya ba byed do | [sred pa ni sa bo]|n rnam pa[r šes pa rlan[6] bar bye]|d do | ma rig pa ni sa bon rnam par šes pa odebs te | rkyend odi dag myed na sa bon rnam par šes pa mṅon bar

5. ogrub par myi ogyur ro | de la las kyaṅ odi ltar bdag gis sa bon rnam par šes pai žiṅ gi [bya ba byao sñam] | du myi [s]e[ms so | sred pa yaṅ o]|di ltar bdag gis sa bon rnam par šes pa rlan bar byao sñam du myi semso | ma rig pa yaṅ odi ltar

Two considerable fragments (E. i. 10, 13), with a small piece partially filling up the gap between them in lines 3–5. As arranged, they measure respectively $7\frac{1}{8}$–8, $\frac{1}{2}$–$1\frac{3}{8}$, and 8 in. in width, and in height $2\frac{3}{8}$, $1\frac{7}{8}$, and $2\frac{5}{8}$ in.

[1] The Kanjur for these words reads *rgyu gžan daṅ gžan las byuṅ ba | rkyen gžan daṅ gžan las byuṅ ba*, i.e. अन्योन्यहेतुकान्यन्योन्यप्रत्ययानि.

[2] After this word the Kanjur adds *rgyu med pa ma yin | rkyen med pa ma yin*, i.e. नाहेतुकानि नाप्रत्ययानि. This is probably a variant for the words *rgyu myed pa las byuṅ ba rkyend myed pa las byuṅ ba ma yin* in our text, which are not found in the Kanjur.

[3] This and the following words as far as *rjesu žugs* appear in the Kanjur as *t'og ma med pai dus nas žugs pa | rgyun ma c'ad par kluṅ gi rgyun bžin du rjes su žugs*, i.e. अनादिकालसमवृत्तान्यनुच्छिन्नान्यनुप्रवर्तन्ते (more exactly अनुच्छिन्नप्रवृत्तानुप्र° or °प्रवृत्तिमनुप्र°) नदीस्रोतवत्.

[4] After *ni* follow in the Kanjurt he words *rten ciṅ obrel par obyuṅ bai yan lag bcu gñis po*, in Sanskrit द्वादशाङ्गस्य प्रतीत्यसमुत्पादस्य ; and for *de* the Kanjur has *de dag*.

[5] In the Kanjur *ogyur ro*; the Sanskrit has हेतुत्वेन प्रवर्तन्ते.

[6] Here and in the next line the Kanjur spells this word *brlan*.

(M. Fol. 17. See Plate CXVII.)

(*Mdo* xvi, ff. 199 b–200 a)

Line 1. bdag gis sa bon rnam par šes pa gdab bo sñam du myi sems so | sa bon rnam par šes pa yaṅ odi ltar bdag ni rkyen odi da|g gis bskyed do sñam du myi sems te | oond kyaṅ sa bon rnam par šes pa las kyi žiṅ la rtend[1] pa sred pai rland kyis brlan

2. pao[2] | ma rig pai ltad kyis btab pa[3] | skye ba na[4] mai mṅal de daṅ der myiṅ daṅ gzugs kyi myi gu mṅon bar ogrub ste | myi|ṅ daṅ gzugs kyi myi gu de yaṅ bdag gis ma byas | p'a rold kyis[5] ma byas | gñis kas ma byas | dbaṅ pos ma spruld[6] | dus

3. kyis ma bsgyurd | raṅ bžin las ma byuṅ | byed pa la rag las pa ma yin | rgyu myed pa las kyaṅ ma skyes | te[7] | p'a daṅ ma p'rad pa daṅ dus[8] daṅ ldan ba daṅ rkyend gžan yaṅ ts'ogs pa na[9] | bdag po myed pai c'os bdagi myed pa odzind

4. pa myed pa | nam ka[10] daṅ mts'uṅs pa sgyu mai mts'an ñid daṅ[11] raṅ bžin dag la rgyu daṅ rkyend ma ts'aṅ ba myed pai p'yi|r | skye bai gnas ñid mts'ams sbyor ba | mai mṅal de daṅ der myoṅ ba daṅ ldan bai sa bon rnam par šes pa myiṅ daṅ gzugs

5. kyi myi gu mṅon bar sgrub bo[12] ‖ odi lta ste myig gi rnam par šes pa ni lṅai p'yir skyeo[13] | lṅa gaṅ že na myig la brtend | pa daṅ | gzugs daṅ | snaṅ ba daṅ | nam mk'a daṅ | de skyed pa[14] yid la byed pa la yaṅ brtend nas myig gi rnam par šes pa skyeo[15]

Two fragments (E. i. 24, 21), forming a complete page of $18\frac{1}{4} \times 2\frac{5}{8}$ in.; their respective widths are $8\frac{7}{8}$ and $9\frac{3}{8}$ in.

[1] Spelt in the Kanjur *brten*.

[2] See note on A 1.

[3] This clause apparently means अविद्याद्ध्यावकीर्णं. The Kanjur reads *ma rig pai lus kyis bran pa*, in which *lus* seems to be a corruption of *ltas*. The Mādhyamika-vṛtti has अविद्यया स्ववकीर्णं, supported by its Tibetan version, and the Śikṣāsamuccaya reads अविद्यावकीर्णं.

[4] After this the Kanjur inserts *skye bai gnas ñid mts'ams sbyor ba*, supported by the Sanskrit तत्र तत्रोत्पत्त्यायतनसन्धौ (or °प्रतिसन्धौ).

[5] In the Kanjur *gžan gyis*, as before.

[6] In the Kanjur *dbaṅ p'yug gis ma byas*, as before.

[7] Followed in the Kanjur by *oon kyaṅ*.

[8] For this word the Kanjur substitutes *zla mts'an*, which looks like a gloss. The Sanskrit has ऋतु.

[9] In the Kanjur *ots'ogs na*.

[10] Sic!

[11] In the Kanjur *kyi*.

[12] In the Kanjur *ogrub bo*.

[13] These three words appear in the Kanjur as *rgyu lṅas obyuṅ ste*; the Sanskrit has पञ्चभिः कारणैः.

[14] In the Kanjur *pai*.

[15] In the Kanjur *obyuṅ ṅo*.

(N. Fol. 19.)

(*Mdo* xvi, f. 201 a-b.)

Line 1. [. . . de la c'os gaṅ yaṅ ojig rten odi nas oji]g rten p'a roldu myi op'o mod kyi . . .

2. [. . . me loṅ gi dkyil ok'ord tu ma op'os] mod kyi rgyu daṅ rkyend ma ts'aṅ ba myed [pai p'yir bžin du mṅon ba yaṅ yod do | . . .]

3. [. . . odi lta ste |] dper na zla bai dkyil ok'ord ni [dpag ts'ad bži k'ri ñis stoṅ nas ogro ste | . . .]

4. [. . . snod c'uṅ ṅu c'us] gaṅ bai naṅ du soṅ ba yaṅ myed mod [kyi . . .]

5. [. . . rgyu daṅ rkyend rnams ma ts'aṅ ba myed pai p'yir las] kyi obras bur[1] mṅon ba yaṅ yo[d] do | o[di lta ste | . . .]

One fragment (E. i. 38), 2⅛-3⅛ in. wide; 2⅝ in. high.

[1] The Kanjur reads *bu*.

(O. Fol. 20.)

(*Mdo* xvi, ff. 201 b-202 a.)

Line 1. [. . . nam mk']a daṅ mts'uṅs pa sgyu mai mts'an ñid daṅ[1] raṅ bžin dag la | rgyu daṅ [rkyend[2]] | ma ts'aṅ ba myed pai p'yir skye bai gnas ñid mts'ams sbyor ba | mai mṅal de daṅ der sa bon rnam par šes pa las daṅ ñon

2. [moṅs pa rnams kyis bskyed pa myiṅ daṅ gzugs] kyi myi gu mṅon bar sgrub[3] ste | de ltar naṅ gi rtend ciṅ o[brel]]d par obyuṅ bai[4] rkyend obreld par bltao || de la naṅ gi rtend ciṅ obreld par obyuṅ ba lṅai p'yir[5] blta ste | lṅa gaṅ že na |

3. [rtag par ma yin ba daṅ | c'ad par ma yin ba daṅ | op'o bar] ma yin ba daṅ | rgyu c'uṅu las obras bu c'en po obyuṅ ba da[ṅ | de] | daṅ edra bai rgyud duo | ji ltar rtag par ma yin že na | gaṅ gi p'yir t'a mai oc'i bai p'uṅ po rnams kyaṅ gžan la skye bai c'ar

4. [gtogs pa rnams kyaṅ gžan te | t'a ma oc'i ba]i p'uṅ po gaṅ yin ba de ñid skye bai c'ar gtogs pa rnam]s ma yin gyi t'a ma oc'i bai p'uṅ po rnams kyaṅ ogag la | skye bai c'ar gtogs pai p'uṅ po rnams kyaṅ obyuṅ

5. [bas dei p'yir rtag par ma yin no || ji ltar c'ad par ma] yin že na | t'a ma[6] oc'i bai p'uṅ po rnams ogags[7] pa las | | skye bai c'ar gtogs pai p'uṅ po rnams obyuṅ ba ma yin | ma ogags pa las kyaṅ ma yin gyi | t'a ma oc'i bai p'uṅ

Two large fragments (E. i. 16, 17); width, 4¼-4¾ and 9⅛ in. wide; height, 2½ in.

[1] In the Kanjur *kyi*.

[2] The Kanjur adds here *rnams*.

[3] In the Kanjur *ogrub*.

[4] *Ba* in the Kanjur, which also inserts *daṅ* after *rkyend*.

[5] Instead of *lṅai p'yir* the Kanjur gives *rnam pa lṅar*. The Sanskrit has पञ्चभिः कारणैः here (but directly afterwards पञ्चभिः only), corresponding to the former; *rnam pa* stands for पञ्चभिराकारैः.

[6] In the Kanjur *mai*, and the same below; it also reads *ogags*.

[7] Before *ogag* the Kanjur adds *sṅon*, which finds some support in the Sanskrit versions (पूर्वं).

II

MISCELLANEOUS FRAGMENTS OF RELIGIOUS WORKS

A FRAGMENT of paper, imperfect on both the right and the left sides (E. i. 19), width, 4–4½ in.; height, 5½ in. See Plate CXVIII. The first passage is written in a good square hand; the second is in a coarse cursive script. Both relate to the ritual of sacrifice.

A. 1. *Recto.*

1. . . . བཟང་ཁྲི་ཀུ་ལོད། ལྷ་སྲིན་གྱི་མཆིད་གསོལ . . .
2. . . . . པ་སྨོན་པ་བཞིན་ནས་གློབ་དགའ། བ . . .
3. . . . . . . ན་པར་ཁྱི་ཁལ་བྱེད་བྱིན་བད། བདགི་ར་ཞི . . .
4. . . . . . . སླངས་སྟེ་དཀྲུང་ལོ་གསུམ་ལོན་ནས་བདས . . .
5. . . . . ལཁ་བསྟན་བ། ཁྱོད་བུ་ཡང་ཕྱུགས་རྫིར་སྟུ[ལ ?] . . .
6. . . . . སནས། ཁྱི་ཡང་ཚང་བར་མ་ཐོབ། སྟེ་ཁྱོད . . .
7. . . . . . . རིན་སླ་ར་ཕྲོགས་ཟེར་ན་བདེན་རྫུ་ན་ཅིར་གྱུར་བ . . .
8. . . . . ས། ཆེད་པོ་སློན་སྟོང་སྲུམལ། ཁྱི་ཁལ . . .
9. . . . . པདེ་ནམཁལ་བྱེད་དང་གསུམ་དགེ་ལེགས[ས] . . .
10. . . . . ནམཁལ་བྱེད་དང་གསུམ་འབྲུ་བྱད་བ . . .

3. The ར at the end of the line is very uncertain.—5. *bstan?* only *s* and *n* are clear.—10. Above ཁལ is written ནས.

The dog has a place in the σφάγια of Asia. We read of white dogs being sacrificed to ratify covenants, for example, in the treaty between the Chinese and the Tibetans mentioned in the *Journal of the Royal Asiatic Society*, New Series, vol. xii. p. 128 f.

2. *Verso.*

1. །འཕན་ངོ . . . . . .
2. . . . ར་གསྲོ་ཁྲོད་ལ་བཙོང་དུ་བཙོལ་པ། ལོ་གཅིག[ག] . . .
3. . . . པ ༎ ལྷ་སྲུམ་དང་ངེད་ཅད་པར་བྱས་ཟིན་གྱིས . . .

4. . . . ཁྲི་ཁམ་གཅིག་དང་། ཇོ་བཙན་ལེགས་ . .

5. . . བགྱིད་ཀྱིས་མཁན་ཐུས་ནས་། ང་ལ་ཐྲི . . . .

6. . . . དང་གཉིས་། བ་ལེག་དུ་ཕྱོད་བུའི་ . . . .

7. . . . ཁྲིབ་ཅི . . ང་ཅེད་པར་། ཁ་ལྷ་ཐྲིན་ . . . .

8. . . . ང་དང་ཞིར་སུམ་དུ་མ་ཐུག་གྱི . . . .

3. ཅད has been added afterwards.—5. Perhaps ཐྲིན.—6. After ཕྱོད་ the writer has inserted the letters དཀྱི.

B.

Two small fragments of a slip of coarse brownish paper daubed with red paint, about two inches in height, which seems to have originally contained two lines of writing in a very coarse half-cursive hand (E. i. 31 b, 29).

(*a*) ༄ ། མཁན་པན་ . . .

སང་རྫོས་ཆིས་ . . .

3½ in. wide, 1 in. high. The letter after པན་ is either ན or མ.

(*b*) . . . རྟོལ་ལ་གར་ཀྱིར་སོག་ལ་ཁྲབ་ཞ . . .

. . ཁྱེ་སྐྱེབས་སྐྲ་སང་ཀྱིར་བསུ་འཐི་ནུགས་ཤ . . .

6¾ in. wide, 2½ high. In line 1 རྟོལ is preceded by an illegible letter. At the end of the line are traces of a letter not unlike ཤ.

In line 2 the vowel of ཁྱེ་ is uncertain; it may be *i*.

C.

Two fragments of a page of white paper, written in a small and somewhat affected cursive hand (E. i. 25).

(*a*) བབས་

ཤུ་དུ་མཛལ་དེ། ད་དཀའ་ཐྲ་ . .

འདྲའན་ ༎

སྐྱབབ་ལ་བབ་ཅེས་བགྱི་བ་ . . .

. [བ]ཀའ་སྟུ་བ་འདིབས་གཏང . . . .

4½ in. wide, 1½ in. high. It is difficult to distinguish between *b* and *c*.

Over སྐྱ is a horizontal line, perhaps intended for the vowel *e*. After the following ལ is an ས, which the writer has struck out. The ང in the last line is uncertain, and after it are traces of another letter.

On the reverse are the following words:—

ཉེ་ལམ། །སྒུགས་པ ...

... འཛྲའན།

... གསོལ༎ བ

... བདག་ངན ...

The last letter is doubtful.

(b) ... . དང་ ...

... འང་འདི་ལགས་ཏེ༎ ...

. . . . . ཛོ་ལ་དཔལ ...

3 in. wide, ¾ in. high. Only the vowel and the upper part of the *akṣara* ཛོ is visible. After དཔལ are traces of བི or པི, perhaps བྱི.

On the reverse:—

... བདག ..

D.

Fragments of a work on mystical rites, written in a good hand similar to that in which the Śālistamba-sūtra is written, but with slightly larger letters (E. i. 15, 20, 36). See Plate CXVIII.

(a) ... ར་ཀུང་ནས ...

... དར་སྐྱབས་ཟ་བ་ད[ང་] ...

... དང༎ སེམ ...

E. i. 15. Nearly 1½ in. wide, 1⅓ in. high. In the last line only the tops of the letters survive.

(b) ཀ་ར་ཏུ་ཙུམ་འདྲུམ། ...

བྱིད་དུ་སམ་བྷུ་ཏྲན་ཀ ...

E. i. 20 a. 2½ in. wide, 1⅞ in. high. Incomplete on the right side.

(c) ... [ད]ང་དཀར་མོ ...

... ང་བ་མཛོད་ད ...

E. i. 20 b. Nearly 1½ in. wide, 1¼ in. high.

(d) ... པད་མོ ...

... སོལ༎ ཕྱོམ ...

E. i. 20 c. Nearly 1¼ in. wide, 1¼ in. high. The scribe has written པཎ, and then struck out the ཎ.

(*e*) . . . འགྲོག་པ་ད[ང་] . . .

. . . བབ་པ་འབྱུང་བ་དང་ . . .

. . . [ད]ང་ ‖ ཟས་སུ་ཚོམ . . .

. . . ས་འགྲོག་པ་ད[ང་] . . .

E. i. 36. 2¾ in. high, 1¾ in. wide. At the beginning of the first line there survives the shaft of a letter that may be *k* or *g*, with a fragment of a *mātrā* over it. In the last line the first letter seems to be *p* or *b*, but only the shaft is visible.

(*f*) . . . ཟ་བ་དང་ | ཟས་སུ . . .

1¾ in. high, ¾–⅞ in. wide. Only the left-hand corner of the syllable *su* remains.

(*g*) . . . བདཔག . . .

1 in. high, 1⅛ in. wide. A fragment of the last line of a page. The letters are imperfect on top.

(*h*) . . . དང ‖ བས . . .

A scrap similar to the preceding, and of the same size.

III

TWO RELIGIOUS POEMS

Two poems from the T'eg-mc'og-mdzod (परमयानकोश ?)[1]. Written on a nearly rectangular sheet of native paper about 10¼ inches by 7¼ inches, in a careless, semi-cursive script (E. i. 11). See Plate CXVIII.

1.

༄ ‖ ཀྱེ་མ་འོ། ¹ ཁམས་གསུམ་ལྷ་མི་འགྲོ་བ་རྣམས ‖ ² དྲི་ཟའི་གྲོང་ཁྱེར་ཇི་བཞིན་དུ ‖ ³ དེ་བཞིན་གནས་
ཤིང་དེ་བཞིན་འཇིག ‖ ⁴ དེ་བཞིན་དུ་ནི་འཇིག་འགྱུར་བས ‖ ⁵ རྨི་ལམ་གྱི་བདེན་སྨུག་ཀྱི་འདྲ ‖ ⁶ སྲིད་དང་སྣ་ཀུན་སྟོང་
པ་ཡིན ‖ ⁷ ནོར་དང་ཡུལ་ས་སྡུག་བསྔལ་སྐྱོན ‖ ⁸ ཁྱིམ་དང་བུ་སྨད་ཀྱེ་བུར་འདྲ ‖ ⁹ མཁས་པ་གང་ཞིག་དེ་ལ་
གནས ‖ ¹⁰ ཇི་ལྟར་ལྕགས་ཀྱི་བྱ་ཁྲ་བཞིན ‖ ¹¹ སྒང་བུ་དབེན་པའི་ཅེ་ཁྲིམས་ནས ‖ ¹² སྤྲིན་ཀླུང་བཞིན་དུ་དགའ་དགུར་
འདོད ‖ ¹³ བུ་སྨད་གཉེན་ཀྱིས་བསླུས་སྣ་རེ ‖ ¹⁴ འཁོར་བ་ཞགས་པས་ཟིན་ཏ་རེ ‖ ¹⁵ སྨྱོས་པའི་གླང་གིས་བརྫིས་སྣ་རེ ‖ ¹⁶
ལོག་པ་ང་ཤེས་མ་བཏང་ན ‖ ¹⁷ སྡིག་ལས་གསུམ་ལས་ཐར་པ་སྐྱེད ‖ ¹⁸ སྡིག་མོས་གཡེལ་དུ་གྲུ་ཡང་སྐྱེད ‖ ¹⁹ དེ་བས་
དམ་ཆོས་སྤྱོད་པའི་རིགས ‖ ²⁰

5. After ནི the scribe has written འག by mistake, and then put marks of erasure.—9. ཀྱེ་ apparently by error for ཀྱི་.—15. The ཏ་ is very uncertain; perhaps པ་ should be read.—18. At the beginning of this and the next verse the scribe has written སྡུ, and then scored out the letters.

Mr. Francke has kindly furnished the following translation and notes. By a remarkable coincidence, these poems were a part of the studies once pursued by the evangelist of the Moravian Mission at Leh, T'ar-ñed C'os-op'el, formerly a Lama at the Tashilhunpo monastery, who learnt them and other works by heart[2]; and Mr. Francke has likewise communicated this version, which, as will be seen, differs very little from that of the Endere MS.

The version of T'ar-ñed C'os-op'el, apart from details of spelling, presents the following divergences:— v. 8, *nor dan p'a yul*; v. 9, *mi bur*; v. 11, *ci ltar*; v. 12, *sgon bu*; v. 13, *dga mgu odod*; v. 14, *bu smad gñen gyis slu bar red*; v. 15, *ok'or ba žags pas odzin par red*; v. 16, *smyo bai glan gis* [*o*]*rdzis par red*; v. 17, *log pa na rjes mi gton na*; v. 19, *grva.*

[1] The first poem is quoted by Sarat Chandra Das, *Tibetan Dictionary*, p. 653, as Mr. F. W. Thomas first perceived.

[2] This gentleman confirms the statement of Sarat Chandra Das, as he informs us that *both* poems are from the T'eg-mc'og-mdzod.

Mr. Francke contributes the following remarks on the text of the Endere MS.:—

'V. 1, *kye ma* is not an exclamation of misery, as stated in Das's dictionary, but expresses something like "now listen!" In the same sense it occurs in the Kesar-saga.

'V. 6, *rmyi lam*; it is interesting that in the Lower Ladākhi pronunciation *rnyi-lam* or *snyi-lam*, the ancient *y* is preserved. *Smyug rgyu*, or perhaps *dmyug rgyu*, is stated by T'ar-rñed C'os-op'el to be a well-known word meaning "phantom"[3].

'V. 8, *yul sa*, he says, is used in the same sense as *p'a yul*.

'V. 8, *mi bur* is also known to him as meaning "unreal".

'V. 11, *lcags kyi bya k'ra*, a name of the eagle, on account of his beak, which is like iron.

'V. 12, *sgoṅ bu* is the Tibetan rendering of the Sanskrit अण्डज; but *sgoṅ skyes* would have been better. The vowel has probably been forgotten. *Brims pa* is given by T'ar-rñed C'os-op'el as "go", here "soar".

'V. 13, *bžin du* seems to stand here for the ordinary *la*; *ᴐdoṅ* is a mistake for *ᴐdod*.

'Vv. 14–16. The terminations *ā re* or *ta re* (the *t* in the latter may be the last remnant of a *drag*[4]) are ordinary Central Tibetan formations of the present tense with auxiliary, parallel to Ladākhi formations like *taṅṅa mi dug*.

'V. 15, *zin*, for *ᴐdzin*, is due to assimilation to *bzuṅ*, *zum*.

'V. 17, *žes* is a development of *rjes*. Compare *Ladakhi Grammar*, Laws of Sound, No. 1, which in full is thus:—*s* or $r+c=sh$; *s* or $r+j=zh$; *s* or $r+ts=s$; *s* or $r+dz=z$.

'V. 19, *gru* is a way of writing *grva* "wisdom", "knowledge". In Khalatse the word *rva* "horn", is pronounced *ru* as well as *ra*. *Sdig mos* T'ar-rñed C'os-op'el explains as meaning "many sins"; but it may be simply "delight in sin".'

Mr. Francke supplies the following translation:—

1. Now listen! 2. The three realms—gods, men, and animals— 3. are like the town of the Scent-eaters[5]; 4. like that they exist and perish. 5. As they are perishing in this way, 6. they are like an unreal dream, or like a phantom. 7. All power and fame are empty [conceptions]. 8. Suppress [love for] wealth and home, like a misfortune! 9. House and family are as unrealities. 10. Those who are wise live according to these [considerations]. 11. They are like the eagle [with the] iron [beak], 12. the egg-born one, who soars about solitary peaks, 13. and experiences ninefold joys from clouds and air. 14. Family and friends deceive us, so 15. that we are caught by the noose of rebirth, 16. so that we are trampled down as by mad elephants. 17. If you will not turn away from them and follow me (Buddha), 18. there is no deliverance from the three ways of sin[6]; 19. there is no knowledge to ford [the river of] sin. 20. Therefore it is right to perform the holy religion[7].

2.

༄ ། གསེར་གྱི་རི་ཡོད་དངུལ་གྱི་གླིན་བྱུང་ཡང ། 1 ནོར་གྱི་དབྱིག་དུ་ཇི་ཡང་ཚོག་ཤེས་འཛིན ། 2 དཔལ་གྱི་རི་མཐོ་
སྲིད་ཀྱི་ཚ་བྲུག་ཀྱང ། 3 དགོས་པའི་དོན་དུ་བླ་མེད་སངས་རྒྱས་སྒྲུབ ། 4 འཇིག་རྟེན་དཀ་ན་ཆེ་བཙན་སྲུ་བྱམས་ཀྱང ། 5

[3] [More exactly, it is the कदली of Sanskrit literature.—L. D. B.]

[4] Above, p. 549.

[5] [The *Gandharva-nagara*, Fata Morgana or mirage, familiar in Sanskrit writings. On the term *Gandharva* see especially Prof. de la Vallée Poussin's note in the *Journal asiatique*, Sept.–Oct. 1902, pp. 294 f.—L. D. B.]

[6] Explained by T'ar-rñed C'os-op'el as referring to the sins done by gods, men, or animals.—A. H. F.

[7] [There is a striking parallelism in thought and expression between this poem and the Maitri Upanishad, IV. 2, as has been pointed out to me by Mr. F. W. Thomas.—L. D. B.]

གང་ཞིག་ལེགས་པར་འདོམས་པའི་བཤེས་མཆོག་འབའ་ཞིག་བརྟེན་པར་བྱ ॥6 མཛོད་ཕྱུག་གཅིག་དང་གཙུགས་པའི་གྲོགས
མང་ཡང་ ॥7 གང་ཞིག་དེ་ལ་དགེ་སླུལ་བ་གྲོགས་མཆོག་ཡིན ॥8 སེམས་ཀྱི་དམུ་གོད་འདོད་ཁམས་ཉམས་ལ་གཡེངས ॥9
ཡུལ་དྲུག་ཅི་དོག་བཟང་ལ་སྐྱེགས་པའི་ཕྱིར ॥10 ངམ་གྲོག་གཡང་ས་ཉོན་མོངས་འབྱུང་ཆུ་ལ ॥11 བག་མེད་འཕོ་ཞིང་འགྲོ
བར་འདོད་པ་ལ ॥12 ཕན་དང་བདེ་མཛད་སྟབ་ཀྱི་འཁྲུག་སྟབ་འདི ॥13 སྟོན་པ་ཆོས་ཀྱི་རྗེ་སྟེ། དུས་གསུམ་ཀུན་བས
གསུངས ॥14 ཚུལ་ཁྲིམས་བཙན་པོའི་ཅང་གིས་མཐའ་བསྐོངས་ཤིང ॥15 ཤེས་རབ་མཐུར་གིས་ལམ་གི་བཀྲི་དགུ་བྱེས ॥16
ཉོན་མོངས་ཀྱི་བདེ་འཁོར་བའི་ལམ་འདི་ལ ॥17 ཐབས་དང་ཤེས་རབ་ལྡན་བའི་ས་མཁན་གྱིས ॥18 གཟེབ་ཤུལ་ཏེ་ལོ་དྲང་པོ
བསྟན་པའི ॥19 སྤྲོ་ཞིང་དད་བས་འཁོར་བའི་སྡུག་བསྔལ་སྐྱེད ॥20 མཆོ་དང་གཙང་ཆེན་མཐར་ཡས་འབྱུས་ལ ॥21 བྱང་
ཆུབ་སེམས་ཀྱི་གྲུ་དང་ཙིང་བཅས་ཤིང ॥22 བརྩོན་གྲུས་གཏོས་དང་སྐྱ་བས་རྒྱུར་དུ་སྐྱོད ॥23

གནམ་ཆེད་པོས་བགྱིས་པ་ ལགས་སོ། མཆོག་གི་དཔལ་ལགས་པ་སྙོ་ ལགས་རེ་མོད ॥

In v. 7 གཅིག and in v. 8 དགེ have been added as later corrections, in the same hand.—11. བྱང has the vowel *i* written over it.

The divergences in T'ar-ñed C'os-op'el's version, communicated by Mr. Francke, are as follows:—v. 1, *dnul*; v. 2, *tu* and *mc'og*; v. 4, *bla mai*; v. 5, *mts'an*; v. 6, *šes*, *žig rten*; v. 7, *p'yugs poi* for *gc'ugs pai*; v. 10, *t'og* for *dog*; v. 12, *op'o žin*; v. 15, *gtsan* and *skons*; v. 16, *gyis* and *gyi*; v. 18, *ldan pai*; v. 19, *bstan pas*; v. 20, *spro žin*; v. 21, *gtsan* and *ebyes*; v. 22, *kyi*, *bcas*; v. 23, *brtson grus no rtog skye bas*. According to this version also the postscript reads *gnam p'yed pos bris* [*bgyis?*] *pa lags so* | *mc'og gi dpal du lags par sñam lags red mod* ||

Mr. Francke has also communicated the following notes:—

'V. 2, *mt'o srid* means "paradise" or "realm of the gods".

'V. 5, *bts'an* is the modern *btsan*.

'V. 13, *bde mdzad* "producer of welfare".

'V. 14, *rje ste* "being a high one". The "three times" are past, present, and future.

'V. 15, *rtsan*, instead of *gtsan*, corresponds to the modern Ladākhi pronunciation.

'V. 18, *sa mk'an* is for *sa yi mk'an po*.

'V. 19, *gzeb šul*, according to T'ar-ñed C'os-op'el, means "a steep and difficult place". *Bstan p'ai* I take for *bstan pas*, at any rate for a gerund; in Lower Ladākhi letter-writing the gerund in *pai* or *bai* is very common.

'V. 21, *gts'an* for *gtsan*.

'V. 22, *rtsin* is a ritual bell, as T'ar-ñed C'os-op'el says. He takes the word to stand for all ritual. I cannot express an opinion, and leave the word untranslated.'

The following translation is furnished by Mr. Francke:—

1. There is a hill of gold, and although there also arises a spring of silver, 2. as his highest wealth he (Buddha) seizes superior wisdom, whatever it be. 3. Although water is flowing from the glorious hill of heaven, 4. the all-surpassing Buddha fulfils the needed end. 5. Although certain men spread their power over all the worlds, 6. he will only keep to the well-admonishing wisdom, whatever it be.

7. Although there are many pleasant friends, relations, and friendly companions, 8. with whom is admonition to virtue, he is his highest friend, whoever he be. 9. Connected with hatred and all lust is inattention, 10. on account of coveting the goodly fruits of the six lands[8]. 11. [But] at the steep shores, abysses, and sea of misfortune 12. be there no fear! For him who wishes to migrate, 13. this bridle [of the passions] for advantage and welfare 14. was pronounced by the Teacher, the greatest in religion, the conqueror of the three times. 15. 'Filled to the brim with the purity of the powerful moral law, 16. behave in your passage [through life] according to the descended wisdom. 17. On this road of misery, unhappiness, and rebirths, 18. the wisest of the earth, who has means and wisdom, 19. shows a short and straight way [wherever there is] a steep place. 20. [Then] there is joy, and through faith the misery of rebirth does not exist. 21. On the journey over the sea [of rebirth] and the endless river, 22. being furnished with holiness and wisdom of soul, 23. take heed, and you will be quickly freed from birth!'

Written [or made] by Gnam-c'ed-p'o. It is [written] for the glory of the High One, thus I think.

Mr. Francke adds the following remarks:—

'V. 3 may refer to Buddhacarita, XV. 73, and Lalitavistara, XXIV. Although this scene does not appear to be of great importance, it was represented in ancient art. It may not yet have been discovered in the Gandhāra art; but the Sanskrit original of the song may go back to earlier times. In vv. 2, 6, 8 "whatever" and "whoever" may refer to the fact that Buddha tried at first all sorts of teachers in order to be taught the truth.'

The following General Note on the language of the two poems is added by Mr. Francke:—

'*Instances of archaic (pre-classical) orthography* are (1) the occurrence of the *drag*[9]; (2) the frequent occurrence of *y* joined to *m* before *i* and *e*[10]; (3) the frequent use of *p'a* instead of *pa* for the article. I can imagine that the modern *p* was preceded by *p'*. In this case the change of *p'* to *b* between two vowels and in similar cases would simply become an instance of No. 6 of the Laws of Sound in *Ladakhi Grammar*. A remnant of the tenuis aspirata in the article we probably have still in the occasional occurrence of *p'o*, for instance *bya p'o*, *rta p'o*, &c.; (4) in the words *bts'an* (for *btsan*), *gc'ugs* (for *gcugs*), *gts'aṅ* (for *gtsaṅ po*), *bc'as* (for *bcas*), we notice that the tenuis aspirata suffers a *g* or *b p'ul* to be written before it. At the present time the tenuis aspirata can only be furnished with *m* or འ *p'ul*. It is not improbable that in these cases the tenuis aspirata is older than the modern tenuis. But one thing would then have to be conceded. The *b* and *g p'ul* must then be allowed to form a short syllable with a perhaps indistinct vowel. Later, when the pronunciation of the *p'ul* became interchanged with the prefixed *s* or *r* (compare Song ii. line 15, *rtsaṅ* for *gtsaṅ* or *gts'aṅ*), the ancient tenuis aspirata had to become tenuis, because the pronunciation *rthsa* or *rcha* is impossible according to Tibetan phonetics.

Instances of modern dialectal orthography.—(1) In Song i. 14–16 we have specimens of present tenses as modern as possible. The *p* and *r* of the syllable *par* have entirely disappeared; so has the *d* of *red*. (2) In No. i. 17 we find the modern *źes* instead of classical *rjes*. (3) In the postscript we find the modern pronunciation *c'ed* in a place where *p'yed* must almost necessarily be expected. *Gnam p'yed po* is a house or family name. It is given to houses and families which would receive only the midday sun in winter. (4) The words *bslus*, *brdzis*, *btaṅ*, *brten*, *bsgoṅs*, *bstan* are perfect tenses used as presents; the modern dialects of Central Tibet as well as of Western Tibet make use of perfect tenses to express the present.

'In conclusion, what shall we say? There can be no doubt that in the eighth century A.D. the modern pronunciation of Central Tibet was already developed, at least to a very far degree. But what shall we do then with T'onmi-sambhota, who is the supposed inventor of the Tibetan alphabet as well as of the Tibetan orthography?

[8] [The six *viṣayas*; see Jaeschke, p. 513.—L. D. B.]

[9] [See above, p. 549.]

[10] As observed in the *Preliminary Notice*, in the *Journal of the Royal Asiatic Society*, 1903, pp. 110 f.

'Or should we say that the Tibetan orthography was constructed in places of learning in Turkestān, for instance at Endere, and that the Tibetan dialect of Endere was very archaic, even more archaic than the present dialects of Purig and Baltistān? No. Think of the contrast of dialects that we should then have to accept for Central Tibet and Endere—in Central Tibet a dialect not very different from modern Central Tibetan, and at Endere a dialect more archaic than even the classical language.

'I think a way out of this difficulty is suggested by Mr. Barnett in his article (*J.R.A.S.*, 1903, p. 112). We are simply compelled to consider the story of Sroṅ btsan sgam po as legendary, and to ascribe a much more ancient date to the introduction of Buddhism into Tibet, as well as to the invention of the Tibetan orthography and characters.

'There is also the possibility that Tibet was in possession of an archaic sacred language from time immemorial, that it was this language which was first reduced to writing, and that this already sacred language was accepted as the language of Buddhism.

'*Palaeographic Note.*—It is possible that the *dbu can* alphabet of Tibet was preceded by the *dbu med* alphabet (*Indian Antiquary*, vol. xxxii. 1903, pp. 361 ff.), and also that the two songs were copied from a *dbu med* MS., in which *d* and *ṅ* look entirely alike. Thus in No. i. 13 *odoṅ* was wrongly written where *odod* was meant. Of special interest is the form of ཞ, which is differentiated from ཤ only by a little stroke. The same form is found in the Ladākhi rock-carvings.

'*Sandhi Laws.*—Although the genitives and instrumentals are not confused in the two songs, the Sandhi laws are in many cases violated. This need not necessarily lead us to the conclusion that Sandhi laws did not exist in those times. That they exist in the vernaculars of to-day is beyond doubt. But there are many persons at the present day who, although they observe them very well in their speech, do not trouble about them when writing. Then I may repeat another observation of my own. Just as the Sandhi laws of Leh and Khalatse are not entirely the same, so also in ancient times the Sandhi laws of certain districts may have been different, and monasteries, where people from various districts were gathered, may have been just the places to destroy them.'

IV

SGRAFFITI OF ENDERE

THE following sgraffiti were found scratched on the walls of the temple cella (E. i) and the open hall east of the latter (E. iii) at Endere. Photographs and eye-copies were made by Dr. Stein, from which it has been possible to decipher a certain amount of the text. Many of the characters, however, are uncertain, and not a few are wholly illegible. Doubtful characters are bracketed.

I. *Inscriptions in the Temple Cella.*

A. On the northern wall, near the north-eastern corner. The letters are from two to four inches high, of a fairly good majuscule type.

1. . . . . ཟ[ར] . . . . . [ཝུ]ག་ལྒ་ཕྱུགི

2. . . ཡོན་ . . . [མཆི]སོ། དར་སྣ་དང་ནོར་བུ་བཟང།

3. . [ཕྲུལ]སྣ་གསོལྡུ་ལོ་ཕྱག་ཀྱང་འཚོལ

1. In ཝུ [*sic!*] only the vowel is distinctly visible.

2. The third and fourth letters were read by Dr. Stein as མཆ; the photographs do not render them very clearly, but on the whole support the reading.

Mr. Francke renders the inscription into modern Tibetan as follows:—

1. . . . [ཝུ]ག་ལྒལ་(or རྒལ་)ཕྱག་གི་

2. . . . ཡོན་ . . . མཆིས་སོ། དར་སྣ་དང་ནོར་བུ་བཟང།

3. . . . [ཕྲུལ]སྣ་གསོལ་དུ་ལོ་ཕྱག་ཀྱང་འཚལ་ལོ།

Mr. Francke translates thus:—

1. . . . 'of the hand [of] *Wug lgal*' (or *ₒBrug rgal?*)
2. . . . 'an offering . . . it is; divers silks and good jewels';
3. . . . 'different offerings; as a prayer (offering) the annual rent also [will be] offered.'

Upon this Mr. Francke remarks:—'In the name *Wug lgal* (or *ₒbrug rgal*) the interchange of *r* and *l* (although with regard to this name there is no security at all) would not be impossible according to West-Tibetan phonetics. With regard to གསོལྡུ, I believe that it stands for གསོལ་དུ, and that གསོལྡེ in Inscription I. B. is for གསོལ་དེ. The final consonant of the syllable is written below the first consonant. An exactly parallel case we find in my *Dritte Sammlung von Felszeichnungen von Unter-Ladakh* (Harrassowitz, Leipzig, 1902), Tafel I. 2, where in the name *Rinchen* the second consonants of the two syllables are written below the first consonants, placed slightly to the right. In the case of གསོལྡུ the *l* is also placed slightly to the right, but in གསོལྡེ it is placed exactly below the *s*. This may be due to careless writing; the two words are apparently the same. The frequent occurrence of the word *gsol ba* in the votive records of erection of ancient Ladākhi stūpas is also remarkable.'

Below this inscription appear what seem to be traces of the word ཇོས་; and still further down, somewhat to the left, is a rude drawing of an ox or cow, beneath which the letters བ་མ་ are faintly visible.

B. On the southern wall, towards the south-eastern corner. See Plate XII. The letters vary from 1½ to 3 inches, and are some 2 feet above the level of the floor.

1. . . . གཞི་བ[སར་]ཀུ[ཧླ]ཡོན་གསྡོད་ཕྱག་འཚལད་པའི།
2. . . ཞི་བས་མ་བ་བལྟ་ན་གྲུབ་[བ་]སྡི་ཕྱུགས་ལ་ནོངས་ཡོ་མ་མཆིས་པ་སོ་ནཏོ

1. The first ས in the bracket is not at all clear, but is not improbable. The ར following is a mere conjecture. The reading ཀུ་ is likewise very uncertain, only the lower part of the character being plainly visible. In ཧླ་ the *l* only is distinct.

Above the first line may be deciphered the letters . . . ས་ལེགས་ . . . scrawled in large characters, apparently by another hand.

Mr. Francke gives the following version in modern Tibetan:—

1. . . . ཞི་བ[ས་རོ་]ཀུ[ལྷ་?]ལ་ཡོན་གསོལ་ཏེ། ཕྱག་འཚལ་བའི་[or བས་]
2. . . ཞི་བས་དམའ་བར་བལྟ་ན། གྲྭ་པ་མི་ཕྱུགས་ལ་ནོངས་གཡོ་མ་མཆིས་པར་[གསོན་ཏོ་?]

Mr. Francke translates thus:—

1. . . . 'an offering being offered by *Žiba* to *Roku* [*lha*?][1] and salutations made'

2. . . . 'if . . . *Žiba* looks lowly' (i.e. if he is humble), 'the monks, men, and cattle will be without fault or deceit' (i.e. without harm), 'and [live?]'

C. To the right of this inscription, towards the south-western corner, may be read the following lines, written in a bold flowing hand, in letters from three to five inches in height.

1. རོ་ཀུ་འདོ་བྲོ་ལ་སྩོགས་པས་ཡོན་དུ
2. བལ་ཀྱི་གླང་པོ་ནི་ཕྱལ་ཏེ་གླ་ཡུལ་དུ
3. ཀྱི་རྟ་ལ་ནོ་ཡོ་སྐྱེད་ནོར་ཟས
4. ཙ་ཕྲད་ཡུལ་དུ་ཕྱིར་པར་སོ་ན་

1. The པ has been added below the line.

3. After ནོར་ is the upper part of a letter that has been begun and then struck out. Perhaps the writer intended to write ཇོས, and then, changing his mind, proceeded to write ཟས་ instead.

There is no distinct trace of any dots dividing the syllables.

Of these lines Mr. Francke supplies the following modern version:—

1. རོ་ཀུ་འདོ་བྲོ་ལ་ཚོགས་པས་ཡོན་དུ
2. བལ་གྱི་གླང་པོ་ནི་ཕྱལ་ཏེ། གླ་ཡུལ་དུ
3. མི་རྟ་ནོངས་གཡོ་མེད། ནོར་ཟས
4. རྩ་འཕྲད་དེ། ཡུལ་དུ་ཕྱིར་པར་[སོན་?]

[1] Mr. Francke reports, on the authority of T'ar-rñed C'os-op'el, that *Ro-ku* is still a well-known *devatā*. Compare the next inscription.

Mr. Francke's translation is as follows:—'The pioneers' (perhaps of merchants), 'having offered a wool-ox (Yak? or 'a full load of wool') to *Ro ku odam bro*, their men and horses will be without fault or deceit (harm) [on their journey] to *Sla yul*. After having met with wealth, food, and grass [for their horses, they will come] again to the land on the other side.' He adds:—'Although the names of the gods addressed cannot be identified with recorded deities, this need not be a stumbling-block. These *nāgas* may be of an entirely local character. With regard to *so na* (perhaps instead of *so na to*), I dare not express an opinion. ཉོ་ཡོ་ is probably ཉོངས་གཡོ་. The dropping of final *ṅ* and *ṅs* is an often-recurring fact. Thus the name of the monastery *Lamayuru* is spelt even nowadays གླ་མ་གཡུང་དྲུང་; but in our case the spelling may be due to careless writing.

'སླ་ཡུལ་ is possibly ལྷས། With regard to ལྷ་ it must be said that the spelling *lha* is incorrect. To be in accordance with the actual pronunciation, the spelling ought to be *hla*. *Hla* is a god of the pre-Buddhistic *gliṅ c'os* of Tibet. At the present time all letters preceding an *l*, at least in Western Tibet, are pronounced as *h*; thus we have *glad* = *hlad*, *glu* = *hlu*, *klu* = *hlu*, &c. We may suppose that the letter *h* in the word *hla*, as properly pronounced, is a substitute for a different letter which stood originally before the *l*. I am inclined to believe that it stands for an original *z* or *s*. The word *la*, which also indicates the dative case, means originally not only "pass", but simply "elevation". It is used in this sense in Western Tibet. By prefixing an *s* or *z* before the word *la*, we make causative-denominative forms (according to Prof. A. Conrady, *Eine indochinesische Causativ-denominativ-Bildung*). These forms *zla* or *sla* would have to be translated "the elevated ones". *Zla* is the present word for "planet" and "moon"; *sla* I take to be the original form of *hla* "god". As I have already mentioned in my *A Lower Ladakhi Version of the Kesar-saga* (*Bibl. Ind.*), we meet with an *s* or *z* (it is difficult to decide which) before the word *lha* "god", in such dialectal forms as *p'a sla* (*zla*?) "paternal deity", *ma sla* "maternal deity". The words *yul* and *sa* are synonyms used interchangeably in West-Tibetan dialects in the sense of "place". From West-Tibetan also cases of change in the local names can be adduced, e.g. *T'iṅmogaṅ* instead of the ancient *T'iṅbraṅ*; see my article on *Balu mk'ar* (*Indian Antiquary*, 1905).

'Dr. Barnett makes the suggestion that the vowel *o* was forgotten, and that *sla yul* stands for *lho yul*. This is quite possible, and the change of *sl* to *lh* would have to be explained in a similar way.'

II. *Inscriptions on the Eastern Wall of the Hall of Endere.*

A. To the right of a figure of a charging tiger, in letters of 1½ to 2 inches in height:—

རྐྱང་པོ་དབྱི་མིས། རྨགས་པོ་བའ།

Mr. Francke gives the modern version as:—

རྐྱང་པོ་དབྱི་མའི་རྨགས་པོ་འབའ་[ཞིག་]

This he translates as: 'This is only the sign (picture) of the *rkyaṅ*, the lynx, and the peacock.' He remarks: 'There may have been a Buddhist picture painted on silk hung up in the hall, on which the above creatures were represented (among others). The *rkyaṅ* can often be seen on Lamaist pictures; most of the horses have its shape (perhaps through the influence of the *gliṅ c'os*). *Rmas* is instead of *rmai*; did the confusion between genitive and instrumental begin already at these early times?'

B. Underneath the preceding, in rude letters 3 to 5 inches high:—

ཞལྡྲེ། ཟིགས་ཀན། གྲིག

དུ། མཆིསསྟེ

Mr. Francke regards the words ཟིགས་ . . . སྟེ་ as a later addition to the first three words, and renders the whole as: ཞུ་[?]ཟླ་མེད། གཟིགས་མཁན། གྲིག་དུ་མཆིས་དེ།, which he translates: 'five caps (?) are lost' (literally, 'are not')—'were seen, it being all right'. He remarks: 'At first the announcement was made that five caps (or other articles) were lost; when they were found, the second part of the inscription was added. ཤྭ་ instead of ཞྭ་ reminds us of the West-Tibetan form *shnga*. *Mye* is perhaps defective writing for *myed*. In *zigskan* the inverted vowel-sign is interesting; many Ladākhi rock-carvings also show it. *Kan* instead of *mk'an* may be due to the influence of the preceding *s*; if the word is pronounced quickly, a full *k'* cannot easily be pronounced after *s*. ཀྲིག་དྲ་ may stand for ཁྲིག་དྲ་ (spelling of dictionaries), or ཤྲིག་དྲ་ (modern West-Tibetan), which is always used for "all right" or "in good order".'

Above A are faint traces of some letters, of which ལྷ་སྨ may be deciphered. To the right of A and B is a Chinese inscription, and still further to the right are the following sgraffiti, C and D (see Plate XI):—

C. ༄༎ འཇོམ། ལོམ། སྟོད། ཀྱི། སྡེ། ཕྱག།
ཕག། ཅ། མགའདྲི། ལས། རྐོལ། རྐོལ། སྟེ། སྟག། ཀྱི། ཟའ། ཅན། ཕྲོབ། འོ། ཀྱག
པ། ལ། ཟོ།

The last four syllables are added by a different hand in letters 3 to 4 inches high. The preceding words are in letters of 1 to 1½ inches in height.

Mr. Francke's version is as follows:—

འཇོམ་ལོམ་སྟོད་ཀྱི་སྡེ། ཕྱག་ཕག་ཅར། དམག་འདྲི་ལས་རྐོལ་རྐོལ་སྟེ། སྟག་གི
ཟ་ཅན་ཐོབ་བོ།
ཀྱགས་པ་ལ་ཟོ།

This Mr. Francke renders as follows:—

'At Pyagpag [in the] province of Upper o*Jom lom* this army was outwitted, and a tiger's meal was obtained' (i. e. many were killed). [Addition, probably by a different person:—] '[Now] eat until you are fat!'

Mr. Francke points out that this is apparently a record of a battle, and adds: '*rk'ol* is very probably the same word as *bkol* in "Ladakhi Proverbs", *Journal of the Asiatic Society of Bengal*, vol. lxix, pt. i. p. 146, no. 44. The orthography of *bkol* in the proverb is that of the writer of the proverb, who wished to express the deep aspirated guttural sound of *ch* in the Scottish *loch*, and wrote *bk*. The writer of the Endere inscription probably wished to express the same sound when he wrote *rk'*. The ablative before *rk'ol ste* is parallel to the ablative with *rgyal ba* "to conquer". ཟ་ཅན་ is West-Tibetan for "food". It is remarkable that a number of West-Tibetan words occur in these inscriptions.'

D. Below the beginning of the second line of C, in letters 2 to 3 inches high.

སེ། ཡུ། འདུས། དང
ལེན། བ། ཕྱི། མ།

Mr. Francke gives as the modern equivalent for སེ་ཡུ་ the word སེའུ་, and tentatively translates: 'Collect and take pomegranates, [O] *Bap'yima*!' He adds: 'This is very uncertain; perhaps it was left incomplete. སེའུ་ may just as well be translated as 'cherries', 'figs', or even 'roses'. It is used for several kinds of fruit. From a Tibetan point of view a name like *Bap'yima* or *Bak'yima* is hardly possible. It may be incomplete, or not of Tibetan origin. But the *ba* (= *va*) may stand for *wa* "Oh!" and the name may be *P'yima*.'

Above C may be deciphered the words མཆོད་པར་, written in large letters; and beneath D, a little to the left, appear the words:—

དཀྲའ
བཟོ

The འ seems to have been left unfinished.

APPENDIX C

THE JUDAEO-PERSIAN DOCUMENT

FROM DANDĀN-UILIQ

EDITED BY

D. S. MARGOLIOUTH, D.Litt.

PROFESSOR OF ARABIC IN THE UNIVERSITY OF OXFORD, ETC.

The document which is printed, transliterated, and provisionally translated below, was discovered by Dr. Stein in circumstances which have been recorded in his own words. The evidence obtained by different lines appears to point to this document being not later than the eighth century A.D., which would make it more than 200 years earlier than the oldest document in Judaeo-Persian as yet known, viz. the Law Report of the year 1020, preserved in the Bodleian Library, and published by me in the *Jewish Quarterly Review* for 1899. But it would also be the earliest document in modern Persian of any sort, since the earliest manuscript of a prose work in ordinary Persian appears to be the Vienna copy (dated 1055 A.D.) of the treatise by Muwaffaḳ Ibn 'Alī of Herat, composed between 961 and 976 A.D., and the earliest specimen of a Persian poem ostensibly bears the date 808 A.D., but has been shown by convincing grounds to be centuries later[1].

It seems, however, that this document does not belong to the end of the eighth century, but to the beginning. For in line 23 there occurs a sentence 'since Yazid sent a . . . . to the Ispahbad,' and something more is told us of the Ispahbad, of which, however, owing to the loss of words, the sense is not quite clear. Now the Ispahbad was the prince of Tabaristan, and in spite of Moslem conquests the title was retained in that region for many centuries: a prince who bore this title is mentioned by Ya'kubi (ob. circ. 912; *Bibl. Geogr. Arab.*, vii. 276), and also by an author of the seventh century (Schefer, *Chrest. Persane*, ii. 99)[2]. When therefore the letter mentions communications between Yazid and the Ispahbad, it is natural to suppose the reference to be to those of *Yazid son of Al-Muhallab* son of Abu Ṣufrah, who conquered parts of Tabaristan in the days of Sulayman son of 'Abd al-Malik, about 99 A.H. or 717 A.D. The history of his exploits is told by Baladhuri, pp. 335-9; Ṭabari, ii. 1320, &c.; Yāḳūt, s.v. *Ṭabaristān*; Ibn Khallikan, s.v. *Yazīd*: and correspondence between Yazid and the Ispahbad is mentioned by these authorities (e.g. Ṭabari, p. 1324). Now this Yazid was imprisoned by Omar son of Abd al-'Aziz, who reigned from 99 to 101 A.H., and died about 102 A.H. His message to the Ispahbad cannot have been later than 99 A.H., or 717 A.D. And the context of the Judaeo-Persian document (so far as it is intelligible) implies that the writer is describing an event, not of the distant past, but of the immediate past. Hence (if this identification be correct) the date of our letter will be about 100 A.H. or 718 A.D.

That Persian (in the ordinary sense of the word) was at this time a commonly used vernacular is quite certain, though there may have been no books written in it. But that would not prevent it from being a vehicle for correspondence. Arabic writers frequently put Persian sentences into the mouths of persons of

[1] Pizzi, *Storia della poesia Persiana* (1899), i. 66.

[2] See also E. G. Browne, *Ibn Isfandiyār's History of Tabaristan*, 1905.

the first century. Makḥul (ob. 112 A.H.) is thus quoted by Tirmidhi (ob. A.H. 279; Jāmi', Lucknow ed., p. 12) as saying *nadānam* for 'I do not know'. Jāḥiẓ, or a not much later writer confused with him, quotes from Anushirwan (ob. 577 A.D.) a Persian sentence (*Maḥāsin*, ed. Vloten, 169) occurring in a work bearing the Arabic name *Tawḳī'āt*. Some early verses in which Persian is mixed with Arabic are given by Jāḥiẓ in his *Bayān* (i. 61). There is therefore nothing surprising about one Jew corresponding with another in Persian about the year 718 A.D.; we should indeed expect the Persian of that period to be practically free from Arabic words, and this expectation is justified, since in the fragments of thirty-two lines which we can read, besides the proper name Yazid there are only two words which are Arabic, viz. רקיבא, which in the context where it occurs certainly means *stirrups*, and is therefore the Arabic رِكاب, which in Persian is written رِكيب; and the double misspelling is probably evidence of early borrowing—perhaps through Aramaic; and مسما (line 23) 'named', if that word be correctly read. Since these thirty-two lines are all fragmentary, we cannot indeed be sure that no other Arabic words occurred in the document when it was intact; but the chances are greatly against their occurrence, since in documents dating from the time when Persian was commonly written, when the authors do not purposely avoid them, Arabic words occur too frequently to admit the possibility of their exclusion to the extent which this document displays. The treatise of about 990 A.D. which Schefer puts at the commencement of his *Chrestomathie* gives evidence of this statement; and the Jewish-Persian of all periods is no less full of Arabic than the Mohammedan. Hence their absence from this document seems both to agree with the above identification and to confirm it. Bacher has with justice called attention to the further absence of *Hebrew* words, a sprinkling of which we should expect in a communication from one Jew to another; but there appears to be no such sprinkling: and most surprising is the designation of the Deity by the Persian names *Ized Khudā*, instead of by one of the familiar abbreviations or periphrases to be found in ordinary books. Perhaps the writer of the letter was a sectarian, or a non-Israelite, who for some reason employed the Hebrew script.

Too much is lost for the editor to endeavour to make out a continuous sense. In lines 25 and 26 some one seems to be describing a prescription (magical or otherwise) which a handmaiden is to be taught: perhaps the writer had been employed as physician or magician by the Ispahbad, and explains how easily the charm can be wrought. The greater part of the letter is occupied with some details about the sale of sheep, in which the writer appears to be complaining of unfair treatment. He is evidently writing to some one who is superior to himself, and who in the writer's opinion is able to do him some commercial or pecuniary service. His correspondent was probably purveyor to the Ispahbad, or at any rate administrator of some branch of business at his court. The writer of the document was a merchant of sheep, and complains that worthless animals had been bought, in consequence of which a number of sheep had been left on his hands: these he requests the government purveyor to buy.

Apparently he had also been compelled to sell some other property, but the nature of his complaint can scarcely be guessed. The name of a city was mentioned, but the part of the leaf bearing it has been lost.

Further evidence for the antiquity of the document is furnished by Dr. Stein's account of its discovery and by Professor Wiesner's report on the paper, which both give the *end of the eighth century* as a *terminus ad quem* for the document.

Mr. A. E. Cowley, Sub-librarian of the Bodleian Library and Fellow of Magdalen College, has made the following observations on the palaeography, which give the same kind of date:—

'Some of the characters are ordinary and give no indication of age.

The most interesting are: א, ג, ה, ז, ל, נ, פ, צ, ק.

Of these א, ה, and to some extent ל, פ, ק bear a slight resemblance to characters used (according to Lidzbarski) in Babylonia and Persia in the twelfth century.

The date of the document cannot be so late as that. The writing is throughout more archaic than that of the Persian deed of 1021. Some letters approach more nearly to forms in pre-Christian papyri and inscriptions, as א, ג, ה (inscription form), ק, and less closely פ, צ.

The א seems to be half-way between the Egyptian-Aramaic (papyrus) form (𐡀 𐡀) and the later N. It is sometimes less developed than the א on the Nash papyrus.

The ג is almost identical with the Eg.-Ar. ʌ, and identical with the form on a papyrus (sixth century) in the Bodleian.

The closed ה is the form found in inscriptions of the first to third centuries, and hardly distinguishable from ח, which does not occur. It is later than the form in the Nash papyrus.

The מ is also later than that of the Nash papyrus.

The פ, with its tail sometimes nearly straight, is very archaic, and approaches the forms used before final and initial letters were differentiated.

The צ is unusually large and clumsy. It is nearer to the form used in inscriptions (fourth to fifth century B.C.) ץ than to the later צ.

The ק has a small head and is like the Eg.-Ar. form P or ק, but its long tail is later.

Taking the character of the writing generally, there seems no reason to doubt, and much reason to believe, that it dates from about the eighth century. If it were on anything but paper it might be even a little earlier. It stands midway between the eleventh-century Persian deed and the remains of the third and fourth centuries.'

TEXT AND TRANSLITERATION [1].

(1) יזיד כודה אי יאר באשר זוד רחק
ایزذ خذائی یار باشذ زوذ روزگ

(2) דה וביסת נאמה ביש קרדום ביש
ذه وبیست نامه بیش کرذم بیش[مار]

(3) ר קו צאמק אי מן פא צי רסד ופא דת
کی جامگی من با چه رسذ وپاذست

(4) רסד יש בי פרמי דארון אני מרא ב
رسذش به فرمای ذارن انی مرا

(5) וברין כרי תא מן אבר כסתומי פרוד אמדומי
وبرین خری تا من ابر خستمی فروذ امذمی

(6) ן אי כוש תורא יזיד כודה פריש מחד אי
زیا]ن خویش تورا ایزذ خذا بذش مزذی [ذهاذ

(7) דור בי ופתאר תא מה מר מה ותא דה מר
ذور بیفتاذ تا مه مر مه وتا ذه مر

(8) ספנד בי בוד וסוסת כרנד יזיד כודה סכת
گو]سپند بی بوذ وسست خرنذ ایزذ خذا سخت

(9) מבאד צי קס אז ישאן . . . מא בוד צון ישאן
مباذ چه کس ازیشان . . . ما بوذ چون ایشان

(10) צמה אי פרוכתה בודאם אין פא רוי ימא אב
جامهٔ فروخته بوذم این با روی ما اب[زرذن

(11) רוכתה בוד קס נא בוד סטרדומאן אי שהר פר
فر]وخته بوذ کس نبوذ . . ستردمانِ شهر

(12) סיה פתקו אי כראם וציז בד נא פידה הסת
سه پتکوئی خرم وچیز بذ نه پیذا هست

(13) צון אני מן פא תו ומיד דארום יקי קא
چون انی من با تو امیذ ذارم یکی کا[ر]

(14) סוד זיאן אי מן בי שנאכתן וש שבילי
سوذ وزیان من بشناختن و[ش] شپیلی

(15) ש גוספנד אז סויאי מנרא כרידן תא
شـ]ش گوسپند از سوئ منرا خریذن تا

(16) אידון גופתי קו אבי סיה
ایذون گفتی که ابی سه

(17) וסכת זיאן ומנד הס
وسخت زیانمنذ هس[ت]

(18) ורא אז סוי אי מן יז תר
ذ]ورا از سوئ من یاز

[1] A considerable amount is lost at *both ends* of the lines.

(19) קו כוד כרידי וכוד פרוכתי וכוד רכ
كه خون خريدى وخون فروختى وخون

(20) אגר מן רא סוד בכת באיסתי בודן מן
اگر من را سود بخت بايستى بودن من

(21) בי תו אז שמר כויש ציז אנדוה מא פר
بى تو از شمر خويش چيز اندوه ما پد[ير

(22) יסתידי יני אידר בוד וספ⁖ אזי גוספנד ידון דרוס[ת
[فر] ستيدى اين ايدر بود وسف[ند] از گوسپند
ايدون درس[ت

(23) תו רסד צון יזיד פרסתד ותן מסמא נזדיק סבאבד
تو رسد چون يزيد فرستاد وتن تسما نزديك سپابد

(24) מרא סבאבד ידון קופתר קו מרא צמכוי יקי
مرا سپابد ايدون گفت كه مرا چمخوئ يكى

(25) צמכוי ארי מן קניזק רא אמוזם וצנד ציסת
چمخوي آرى من كنيزك را آموزم وچند چيست

(26) בה בינרום נא בינדארם בי אב גורבק יקי
به بينرم نه بگذارم بياب گربك يكى

(27) דהום תא בגירי רא בא אמוזד אנדריק אי סי
دهم تا بگيرى را به آموزد اندريكى س[ه

(28) קו נאמה ישמא יפתום בי יקי ביה אז אן גופתיד
كه نامهٔ شما يافتم به يكى به از ان گفتيد

(29) קאר אי פרמודי סכת קונום תא קרדה בוד
كارى فرمودى سخت كنم تا كرده بود

(30) ושום ציז אנדוה מא כור צי הושום רא כוסתה בוד
ه]وشم چيزاندوه مخورچى هوشم را خسته بود

(31) אז סו⁖ושת רופתה פורסידום ידון קופת
از سوى [د]شت رفته پرسيدم ايدون گفت

(32) צא פרואן יכוד אז אן סוי כואסתאדן פרמי
جا پروان خود از ان سوى خواستادن فرماى

(33) אגר נאמה ישמא פריסתידי קרדא
اگر نامهٔ شما فرستادى كرده

(34) פיציהאי אן פשיז אין גוספנד
پيچهٔ ان پيش از اين گوسپند

(35) בירון נאמדה אז סוי
بيرون نآمده از سوى

(36) נמוד אז זין ורקיבא ודואל
نمود از زين وركيبا ودوال

(37) אז הר צי היז אי בדין אן
از هر چه هيزى [چيزى] بدين

TRANSLATION.

(1) *If* the Lord God help, quickly the salary
(2) ten and twenty letters I wrote to no purpose without number
(3) Saying 'What has become of my stipend, and into the hand *of whom*
(4) *if* convenient, command him to give it to me.
(5) and buy with that' till I stopped, desisted.
(6) to your own *hurt*, may the Lord God *give* you a reward therefor.
(7) it got delayed, till month *was added* to month and from decade to decade
(8) that they should buy worthless and lean sheep. The Lord God *give them* very
(9) God forefend that one of them should be our . . . since they
(10) *when* I had sold a garment, they *cast* it in my face
(11) had sold: there was no-one of the men of the city PD
(12) that I should buy three *petkū*: and no mischief came of it.
(13) since I expect this of you: first the affair
(14) to be acquainted with my profit and loss, and my office
(15) to buy the six sheep of me, in order that
(16) you said thus: . . . three
(17) and it is very detrimental
(18) try and it from me
(19) you yourself bought, and you yourself sold, and you yourself
(20) if I am to be successful

(21) to you. Be in no way distressed on your own account
(22) nor (?) was it here. And the bond for the sheep was right
(23) came. When Yazid sent the person mentioned to the Ispahbad
(24) the Ispahbad said: *bring* me a . . .
(25) *if* you bring . . . I will teach the girl. And as many things as
(26) I have seen, I will not transgress. Take a *gurbak*
(27) I will give, that she may learn the sense. To each one three
(28) that I received your letter, one better than that you said
(29) had you ordered anything, I would work hard that it might be done
(30) my feelings, do not trouble at all if my feelings be hurt.
(31) going behind, I asked: he said thus
(32) command that your minister (?) be interrogated concerning that
(33) if you had sent your letter *it would have been* done
(34) its young before these sheep
(35) it did not come out from the direction of
(36) he showed, such as saddle, stirrups, and a hide
(37) of everything that

Notes.

1. יזיד כודה אי. This formula occurs again in line 6 and line 8; יזיד by itself in line 23. In the last place چون يزيد فرستاد . . . نزديك سپابد 'when Yazid sent . . . to the Governor', it certainly is the Arabic proper name Yazid. But in the first three places it is followed by the word כודה, which can only be the Persian خدا 'God'. And, indeed, the context in the first two lines is in favour of the meaning 'God'. '[If] the Lord God befriend' is like a line ap. Vullers, i. 194 بخت پاينده خدا يار وسعادت باقى; and 'May the Lord God [give] the reward' is too common to need illustration. The third passage is too obscure to serve as an argument.

2. ביש קרדום. Compare بيشكار, ap. Vullers.

3. צאמק אי. In line 10 we read of a צמה being sold. The Jewish dialect confuses ج and چ, e.g. in the Law Report ציםלא stands for جمله.

ופא דות 'perhaps into the hand of'. Bacher suggests that this is the compound پادست 'goods taken on credit'. Perhaps what is meant is پاداش 'remuneration'.

5. The writer is evidently clear that the perfect should have a *u* in the first person.

6. כוש seems to be خويش, ordinarily pronounced *khīsh*.

7. דור בי ופתאר. It is noticeable that the writer makes no difficulty about beginning words with vowels: cf. line 13, ומיר.

תא מה מור מה. מור stands for רא as often. The sense of דה is not quite clear: if it means 'decade', it must have been used for the jingle.

11. The letter before מרדומאן is like an incomplete ם with the head of a ל. Perhaps it is an abbreviation, and with the previous words should be read نه باور كرد 'did not believe'.

12. פחקי. This seems to mean some sort of animal.

14. בי שנאכתן. In certain Persian dialects ب is regularly prefixed to the infinitive.

שבילי. The letter before שבילי appears to be a ש with the middle stroke omitted: perhaps a letter wrongly written twice, and intended to be erased.

15. The combination סויאי appears to be the Persian سو followed by the *yā* of *izāfet*.

16. The last word but one commences with a letter consisting of an upright stroke crossed by a wedge, like an ancient form of א: I am unable to say what it was meant for.

23. ותן. Since the writer begins words with vowels, this should stand for אותן. Another suggestion would be آتن, according to Vullers 'a governess', which would suit the context.

24. קופתור, doubtless an error (cp. בינרום for ديدم in l. 26). צמכרי perhaps for جام خار 'tray'.

26. נורבק perhaps for گربه 'a cat'; yet a herb of some sort seems more likely. Vullers says, *res felis capiti similis e salice proveniens*.

The chief archaism in the document appears to be the separation of the *yā* of *izāfet* in the form of אי, the use of which, however, appears to be rather irregular; it is identical in form with the indefinite article, and apparently יקי (lines 13, 26, 28) and יק אי (line 27) can be written indifferently. The double spelling of the word گفتن with נ and ק seems to point to the writer being either a foreigner or unpractised in writing: it even suggests suspicion of transliteration from Arabic writing, with which the confusion between ب, پ, and ج, چ would agree.

27. בנירי. Perhaps for پيكار 'purpose, intent'.

APPENDIX D

INVENTORY LIST OF COINS FOUND OR PURCHASED

PREPARED FROM NOTES OF

S. W. BUSHELL, M.D., C.M.G.,

AND

E. J. RAPSON, M.A.

PROFESSOR OF SANSKRIT IN THE UNIVERSITY OF CAMBRIDGE

I. COINS FROM TOGUJAI, MOJI.

(See above, pp. 110 sqq.)

a. 78 Muhammadan copper coins of the type described by Dr. Hoernle, *Report on C.-A. ant.*, i. pp. 32 sqq., and tentatively attributed by him to 'Sulaymān Khāqān'. They seem to bear the name of the Caliph Al-Musta'ṣim, 1242–1258 A.D. See Pl. XC, Nos. 45, 46 (both belonging to Dr. Hoernle's second variety, comp. his Pl. I, fig. 31).

9 Muhammadan copper coins, probably of Muḥammad Arslān Khān (see Hoernle, loc. cit., i. p. 30).

3 small copper coins not identified.

b. 28 Muhammadan copper coins, apparently mostly of 'Sulaymān Khāqān'; see Pl. XC, No. 47 (cf. Hoernle, Pl. I, fig. 31).

c. 19 Chinese copper coins, small, with the legend *Ch'ien-yüan chung pao*, of Ch'ien-yüan period (758–759 A.D.).

4 Chinese copper coins, with same legend, large.

12 non-Chinese copper coins, not identified.

II. COINS PURCHASED AT YŌTKAN.

(See above, pp. 203 sqq.)

a. **Copper coins purchased at site, Oct. 16, 1900.**

1 coin of Kujula-kara-kadphises (comp. Cunningham, *Numism. Chronicle*, 1892, Pl. IV, 9); see Pl. LXXXIX, No. 1.

2 coins of Kaniṣka; probably *Brit. Mus. Cat.*, No. 52.

1 Chinese coin unknown to Chinese numismatists; first character *yü* 于, second perhaps *t'un*; see Pl. LXXXIX, No. 5, and above, p. 205.

1 Chinese coin with the legend *Hou-ch'üan* (comp. Lockhart, i. p. 14, Nos. 95–102); see Pl. LXXXIX, No. 11.

2 Chinese coins with legend *Wu-chu*.

26 Chinese coins of Ch'ien-yüan period (758–759 A.D.); see Pl. XC, Nos. 28, 29.

5 Chinese coins of Ta-li period (766–779 A.D.); see Pl. XC, No. 32.

26 coins undetermined, owing to very poor preservation; several among them may be small Sino-Kharoṣṭhī pieces.

b. **Copper coins purchased at Yōtkan, Nov. 27, 1900.**

2 Sino-Kharoṣṭhī coins, small.

1 Chinese coin with legend *Wu-chu*.

9 Chinese coins of Ch'ien-yüan period (758–759 A.D.), large.

4 Chinese coins of Ch'ien-yüan period, small.

1 Chinese coin of Ta-li period (766–779 A.D.), small.

2 Chinese coins of T'ang dynasty, undeciphered.

4 Chinese coins of Sung dynasty, periods Yüan-fêng (1078–85 A.D.) and Yüan-yu (1086–93 A.D.); see Pl. XC, Nos. 36, 37.

1 Muhammadan coin, probably of Muḥammad Arslān.

7 Muhammadan coins, small, uncertain.

c. **Copper coins purchased at Yōtkan, April 28, 1901.**

1 Chinese coin with legend *K'ai-yüan t'ung pao*, of type first issued by Kao tsu—and by his successors for nearly a hundred years. See Pl. LXXXIX, No. 24.

17 Chinese coins of Ch'ien-yüan period (758–759 A.D.), several fragmentary.

3 Chinese coins of Ta-li period (766–779 A.D.).

2 Chinese coins, illegible.

1 unperforated coin, illegible.

d. Copper coins purchased at Yōtkan, April 29, 1901.

6 Sino-Kharoṣṭhī coins, with traces of Kharoṣṭhī legend.
1 coin of Kaniṣka (MAO); see *Brit. Mus. Cat.*, No. 45.
3 Chinese coins, ancient, uninscribed.
6 Chinese coins of Ch'ien-yüan period (758–759 A.D.). See Pl. XC, No. 30.
2 small coins, probably Muhammadan, but quite indeterminable.
3 coins illegible.

III. COINS PURCHASED AT KHOTAN.

Most of them probably coming from Yōtkan (see above, pp. 203 sqq.)

a. Copper coins, batch Y. 001.

2 Sino-Kharoṣṭhī coins, large. The Chinese legend reads *Chung nien ssŭ chu lü ch'ien* 'Engraved money weighing twenty-four *chu*.' For the Kharoṣṭhī legend, illegible on one coin, comp. Hoernle, *Report*, i. pp. 5 sqq. See Pl. LXXXIX, No. 3.

1 Sino-Kharoṣṭhī coin, small. For the Chinese legend, see Hoernle, *Report*, i. p. 10. Khar. legend obliterated.

b. Copper coins, batch Y. 0025.

2 Sino-Kharoṣṭhī coins, large. One (see Pl. LXXXIX, No. 2) shows the Khar. legend [. . . . *sa*][1] *ra*[*ja*]*tirajasa* [/// /// ///] *sa*[2] *Gug'amo*[*ya*] *sa*.

c. 2 Sino-Kharoṣṭhī copper coins, large, batch Y. 0026. Traces only of legends on both.

d. Copper coins, batch Kh. 004.

2 Sino-Kharoṣṭhī coins, large;
4 Sino-Kharoṣṭhī coins, small; fragments of Khar. legend.
2 coins undeciphered.

e. Copper coins, batch Kh. 002.

1 large and 14 small Sino-Kharoṣṭhī coins, with fragments of Khar. letters.

f. 6 Sino-Kharoṣṭhī coins, Æ, bought Nov. 11, 1900. Traces of Khar. legend.

g. Copper coins, presented by Wang-Daloi, Nov. 23, 1900, as coming from Yōtkan.

3 Sino-Kharoṣṭhī coins, small. See Pl. LXXXIX, No. 4.
1 coin of Kaniṣka (MAO); comp. *Brit. Mus. Cat.*, No. 45.

h. Copper coins, batch Y. 005.

1 coin of Kaniṣka, reverse undetermined.
1 Muhammadan coin, probably of Muḥammad Arslān.

i. Copper coins, batch Y. 007.

2 Chinese coins of Ch'ien-yüan period (758–759 A.D.).
1 Chinese coin of Shao-shêng period (1094–97 A.D.); see Pl. XC, No. 38.
1 Chinese coin of Sung Hui tsung Shêng-sung (1101 A.D.); see Pl. XC, No. 39.

k. Copper coins, batch B.D. 001.

1 large Sino-Kharoṣṭhī coin.
1 small Sino-Kharoṣṭhī coin.
1 Chinese coin of Hsi-ning period (1068–77 A.D.).
1 Chinese coin of Shao-shêng period (1094–97 A.D.).
1 Chinese coin of Tsung-ning period (1102–06 A.D.).
2 Muhammadan coins of 'Sulaymān Khāqān'; see Pl. XC, No. 44.
1 coin of Muḥammad Arslān.
1 Muhammadan coin, of Yarkand (see Hoernle, *Report*, i. p. 35).
7 Muhammadan coins, uncertain.

l. Copper coins, batch Y. 006.

5 Muhammadan coins, uncertain ('Sulaymān Khāqān'?).

m. 13 Copper coins said to come from Halāl-bāgh. All in very poor preservation. *Perhaps* 4 large Sino-Kharoṣṭhī, 7 small Sino-Kharoṣṭhī, 2 Chinese.

IV. COINS SAID TO HAVE BEEN FOUND AT CHALMA-KAZĀN.

(See above, p. 234.)

a. Copper coins presented by Wang-Daloi, Nov. 23, 1900.

2 Chinese coins of K'ai-yüan period (713–741 A.D.); see Pl. LXXXIX, No. 26.
3 Chinese coins of Ch'ien-yüan period (758–759 A.D.).
2 Chinese coins of Ta-li period (766–779 A.D.); see Pl. XC, No. 33.
1 Chinese coin of Chien-chung period (780–783 A.D.); see Pl. XC, No. 34.
1 Chinese coin of Tsung-ning period (1102–06 A.D.); see Pl. XC, No. 40.

b. Copper coins sold as coming from Chalma-Kazān, batch C. 001.

4 large Sino-Kharoṣṭhī coins.
3 small Sino-Kharoṣṭhī coins.
2 Muhammadan coins, apparently of Muḥammad Arslān.
1 Muhammadan coin, uncertain.

[1] From Sir D. Forsyth's coin the first two words of the inscription *maha*° and *raja*° are certain.

[2] Traces of three or four Akṣaras. *tratarasa* seems to be a very probable restoration.

On the other coin only traces of legend.

V. COINS FROM DANDĀN-UILIQ.

(See above, p. 283.)

a. **Copper coins brought by Turdi, Nov. 17, 1900.**

5 Chinese coins with legend *Wu-chu.*
3 Chinese coins, ancient, uninscribed.
1 Chinese coin of Ch'ien-yüan period.

b. **Copper coins found at site.**

Dec. 18, 1900: 1 Chinese coin, uninscribed.
1 Chinese coin of K'ai-yüan period (713–741 A. D.).
Dec. 25: 1 Chinese coin of K'ai-yüan period.
Dec. 27–28: 6 Chinese coins, without inscription; see Pl. LXXXIX, No. 23.
1 fragmentary coin, with *Wu-chu* legend.
Jan. 3, 1901: 4 Chinese coins of K'ai-yüan period (713–741 A. D.); see Pl. LXXXIX, No. 25.
3 Chinese coins of Ch'ien-yüan period (758–759 A. D.); see Pl. LXXXIX, No. 27.

VI. COPPER COINS FOUND AT RAWAK BEYOND DANDĀN-UILIQ.

(See above, p. 306.)

5 Chinese coins, small, uninscribed; see Pl. LXXXIX, No. 9.
1 Chinese coin of K'ai-yüan period (713–741 A. D.).

VII. COPPER COINS FOUND AT NIYA SITE.

(See above, p. 369.)

a. 1 Chinese coin, undetermined, found near Stūpa, Jan. 28, 1901.

b. 1 Chinese coin, with traces of legend *Wu-chu,* found west of Stūpa, Jan. 30, 1901.

c. Coins found near ruin N. v., Feb. 7–8, 1901.
5 Chinese coins, ancient, uninscribed; see Pl. LXXXIX, No. 7.
3 Chinese coins, with legend *Wu-chu*; see Pl. LXXXIX, No. 12.

d. 1 Chinese coin of Ch'ien-yüan period, brought by Turdi to camp near N. v., Feb. 8, 1901; comp. above, p. 369, note 29.

e. Coins found near ruin N. VIII, Feb. 11, 1901.
2 Chinese coins, small, uninscribed; see Pl. LXXXIX, Nos. 8, 10.
2 Chinese coins, with legend *Wu-chu*; see Pl. LXXXIX, No. 14.

f. 2 Chinese coins, with legend *Wu-chu,* found by Hassan Ākhūn, Feb. 1, 1901, circ. three miles north of Stūpa; see Pl. LXXXIX, No. 13.

VIII. COPPER COINS FOUND AT ENDERE SITE.

(See above, p. 429.)

a. **Coins found inside Endere Fort, Feb. 22–25, 1901.**

2 Chinese coins, uninscribed.
2 Chinese coins, with legend *Wu-chu.*

b. **Coins found near Endere Fort, Feb. 21–23, 1901.**

3 Chinese coins (two fragmentary), uninscribed.
1 Chinese coin, fragmentary, with legend *Wu-chu.*

IX. COPPER COINS FOUND AT KARA-DONG SITE.

(See above, p. 447.)

a. **Coins found within ruined quadrangle, March 13–15, 1901.**

3 Chinese coins, uninscribed.
2 Chinese coins, with legend *Wu-chu.*

b. **Coins found near Kara-dong ruin, March 15–17, 1901.**

8 Chinese coins, uninscribed, too defaced for reading; see Pl. LXXXIX, Nos. 6, 15.
1 Chinese coin, with remains of legend *Wu-chu.*

X. COPPER COINS FOUND AT UZUN-TATI SITE, MARCH 26, 1901.

(See above, p. 461.)

1 Chinese coin of Pao-yüan period (1038–39 A. D.), broken; see Pl. XC, No. 35.
1 Muhammadan coin, of Muḥammad Arslān (comp. Hoernle, *Report,* i. p. 30; Pl. I, 22); see Pl. XC, No. 43.

XI. COPPER COINS FROM HANGUYA SITE.

(See above, p. 471.)

a. 1 Chinese coin found near Hanguya Stūpa ruin, April 5, 1901, uninscribed.

b. Coins said to have been brought from Hanguya Tati; purchased at Khotan, Nov. 18, 1900.

1 Sino-Kharoṣṭhī coin, large.

20 (circ.) Sino-Kharoṣṭhī coins, small.

16 coins indeterminable. Among them there is one coin of a curious and apparently unknown type. *Obv.* Large figure of horse walking to L.; above some symbol (?). *Rev.* Uncertain, ? Æ or Lead; ·9; weight 213 (much corroded).

1 Chinese coin of Ch'ien-yüan period (758–759 A.D.).

XII. COPPER COINS FROM TAM-ÖGHIL.

(See above, p. 473.)

a. 1 Chinese coin, uninscribed, small, bought on April 7, 1901, at site of excavations.

b. 4 Chinese coins, of Yüan-fêng period (1078–85 A.D.), both in seal character and running-hand script, purchased at Yurung-kāsh, April 4, 1901, and said to have been obtained at Tam-Öghil.

XIII. COPPER COINS FROM AK-SIPIL SITE.

(See above, p. 476.)

a. 1 Chinese coin, with legend *Wu-chu*, picked up on remains of Ak-sipil rampart, April 8, 1901; see Pl. LXXXIX, No. 16.

b. Coins said to have been found near Ak-sipil; purchased at Khotan, Nov. 17, 1900.

1 Chinese coin, with legend *Wu-chu*.

2 Chinese coins of Ch'ien-yüan period (758–759 A.D.), small.

1 Chinese coin of Ch'ien-yüan period (?), large.

c. Coins said to have been brought from Ak-sipil Tati; purchased at Khotan, Nov. 19, 1900.

1 Sino-Kharoṣṭhī coin, small.

1 Chinese coin of Ch'ien-yüan period (758–759 A.D.), large.

1 Chinese coin of Ch'ing-li period (1041–48 A.D.).

1 Chinese coin of Yüan-yu period (1086–93 A.D.).

1 coin of Muḥammad Arslān Qākhān (*sic*); see Pl. XC, No. 42 (comp. Hoernle, *Report*, i. p. 30; Pl. I, 22–24).

3 small Muhammadan coins, uncertain ('Sulaymān Khāqān'?).

XIV. COPPER COINS FROM RAWAK STŪPA AND VICINITY.

(See above, pp. 500, 501.)

a. 4 Chinese copper coins, with legend *Wu-chu* readable on three, found at SE. foot of Stūpa base, April 13, 1901; see Pl. LXXXIX, No. 18.

b. 2 Chinese coins, with legend *Wu-chu* readable on one, found below relief R. xi, April 12, 1901.

c. 1 Chinese coin, with legend *Wu-chu*, found behind relief R. xxviii, April 15, 1901.

d. 1 Chinese coin, with legend *Wu-chu*, found in wall on north side of gateway, Rawak Stūpa court, April 16, 1901; see Pl. LXXXIX, No. 17.

e. 1 Chinese coin, probably inscribed *Wu-chu*, found at foot of statue R. lxiv, April 16, 1901.

f. 24 Chinese coins (circ. 12 fragm.) much corroded, apparently most with legend *Wu-chu*; found on south side of small Stūpa base, April 11, 1901.

g. 20 Chinese coins (six fragm.), much corroded, probably all with legend *Wu-chu*; found on east side of small Stūpa base, April 17, 1901.

h. 12 Chinese coins (one fragm.), much corroded, probably all of *Wu-chu* type; found on north side of small Stūpa base, April 17, 1901.

i. 28 Chinese coins (six fragm.), much corroded, probably all reading *Wu-chu*; found on west side of small Stūpa base, April 17, 1901. See Pl. LXXXIX, No. 19.

j. 8 small Chinese coins, without legend, found between dunes near Rawak Stūpa, April 11, 1901.

k. 10 Chinese coins, corroded and sticking together, probably all with legend *Wu-chu*; found by labourers near Rawak camp, April 15, 1901.

l. 23 Chinese coins, probably all inscribed *Wu-chu*, bought at Khotan, Nov. 30, 1900, from large find made near Rawak Stūpa. See Pl. LXXXIX, No. 21.

m. 6 Chinese coins, with legend *Wu-chu*, selected Dec. 6, 1900, from large find made near Rawak Stūpa. See Pl. LXXXIX, No. 20.

n. 59 Chinese coins, some uninscribed, others with legend *Wu-chu*, said to have been brought from Ak-sipil, Rawak, and Jumbe-kum. Purchased at Yurung-kāsh, April 6, 1901; see Pl. LXXXIX, No. 22.

XV. COPPER COINS FOUND AT JUMBE-KUM SITE, APRIL 14, 1901.

(See above, p. 502.)

4 Chinese coins, uninscribed.

1 Chinese coin, with legend *Wu-chu*.

XVI. COPPER COINS FOUND NEAR KARA-DÖBE, APRIL 30, 1901.

(See above, p. 515.)

2 Chinese coins, fragments, undetermined.

XVII. COPPER COINS SAID TO COME FROM MAZĀR-TĀGH SITE,

near lower Yurung-kāsh; purchased at Khotan, Nov. 18, 1900.

1 Sino-Kharoṣṭhī coin, small.

3 Chinese coins of Ch'ien-yüan period (758–759 A. D.), large.

8 Chinese coins of Ch'ien-yüan period, small; see Pl. XC, No. 31.

2 Chinese coins of Ta-li period (766–779 A. D.).

1 Chinese coin of Hsi-ning period (1068–77 A. D.).

6 Chinese coins of Tsung-ning period (1102–06 A. D.); see Pl. XC, No. 41.

5 Chinese coins, undetermined.

TABLE OF COIN SPECIMENS REPRODUCED IN PLATES LXXXIX AND XC.

PLATE LXXXIX.

No.	*Description.*	*Reign, period, or legend.*	*Weight.*	*Place of find or purchase.*
1	Indo-Scythian	Kujula-kara-kadphises	46·5	Yōtkan
2	Sino-Kharoṣṭhī, large	. . . .	263·	Khotan
3	do. do.	. . . .	240·3	do.
4	do. small	. . . .	49·8	do.
5	Chinese, archaic	. . . .	51·	Yōtkan
6	Chinese	Uninscribed (?)		Kara-dong
7	do.	Uninscribed	16·	Niya Site
8	do.	do.	12·	do.
9	do.	do.	3·6	Rawak, Dandān-Uiliq
10	do.	do.		Niya Site
11	do.	*Hou-ch'üan*	84·	Yōtkan
12	do.	*Wu-chu*	20·	Niya Site
13	do.	do.	36·	do.
14	do.	do.	49·5	do.
15	do.	*Wu-chu* (? defaced)	49·	Kara-dong
16	do.	*Wu-chu*	22·6	Ak-sipil
17	do.	do.	50·3	Rawak
18	do.	do.	27·5	do.
19	do.	do.	62·7	do.
20	do.	do.	43·5	do. (?)
21	do.	do.	42·4	do (?)
22	do.	do.	13·4	do. (?)
23	do.	Uninscribed	13·5	Dandān-Uiliq
24	do.	K'ai-yüan (713–741 A. D.).	broken	Yōtkan
25	do.	do.	68·6	Dandān-Uiliq
26	do.	do.	50·6	Chalma-kazān
27	do.	Ch'ien-yüan (758–759 A.D.).	124·2	Dandān-Uiliq

PLATE XC.

No.	*Description.*	*Reign, period, or legend.*	*Weight.*	*Place of find or purchase.*
28	Chinese	Ch'ien-yüan (758–759 A.D.)	103·5	Yōtkan
29	do.	do.	43·3	do.
30	do.	do.	116·3	do.
31	do.	do.	33·8	Mazār-tāgh (?)
32	do.	Ta-li (766–779 A.D.)	43·1	Yōtkan
33	do.	do.	54·3	Chalma-kazān (?)
34	do.	Chien-chung (780–783 A.D.)	60·	do.
35	do.	Pao-yüan (1038–39 A.D.)	broken	Uzun-Tati
36	do.	Yüan-yu (1086–93 A.D.)	101·4	Yōtkan
37	do.	do.	87·	do.
38	do.	Shao-shêng (1094–97 A.D.)	118·2	Khotan
39	do.	Shêng-sung (1101 A.D.)	180·3	do.
40	do.	Tsung-ning (1102-06 A.D.)	144·7	Chalma-kazān (?)
41	do.	do.	199·7	Mazār-tāgh (?)
42	Muhammadan	Muḥammad Arslān	90·8	Ak-sipil (?)
43	do.	do.	98·7	Uzun-Tati
44	do.	Sulaymān Khāqān (?)	65·5	Khotan
45	do.	do.	68·2	Togujai
46	do.	do.	50·	do.
47	do.	do.	67·	do.

APPENDIX E

EXTRACTS FROM TIBETAN ACCOUNTS OF KHOTAN

COMMUNICATED AND ANNOTATED

BY

F. W. THOMAS, M.A.

LIBRARIAN OF THE INDIA OFFICE

PART I

NOTES ON ROCKHILL'S 'LIFE OF THE BUDDHA', PP. 230 SQQ., AND THE WORKS THERE QUOTED.

(See also 'Buddhist and Other Legends about Khotan', by BABU SARAT CHANDRA DAS, C.I.E., in the *Journal of the Asiatic Society of Bengal*, 1886, Part I, pp. 193–203.)

P. 236, *sub fin.*—The letters resemble closely those of 'India (*Rgya*)'. But *Rgya* by itself generally means China: cf. pp. 234, l. 16 sqq., p. 236, l. 20, &c. &c. *Rgya-gar* is the usual term for India, *Rgya* or *Rgya-nag* for China (f. 429 b).

P. 237.—From Kustana (Sa-nu) onwards there were fifty-six kings and one regent. The capital of Khotan was built by Ye-u-la, son of Kustana (f. 430 a).

P. 237, *sub fin.*—In the reign of Vijayasambhava many other Vihāras (besides *Tsar-ma*) were built (f. 430 a–b).

P. 238, ll. 21 sqq.—With Buddhaduta (*sic*=°dūta) are mentioned *Kha-ga-ta*, *Kha-ga-drod*, *Bžin-byuṅ*, and others (f. 431 b). The more exact name of the Vihāra is *Ḫgeḫu-to-śan* (apparently also named *Mo-rgu-bde-śi*). The name of the princess is *Puṇyeśvara* (Pu-ñe-śva-ra *sic*) (f. 432 b).

P. 239, l. 8.—Saṅghagoṣa (*sic*) (f. 433 a).

Ibid., ll. 14 sqq.—The three sons of Vijayajaya were, in order of age, *Ḫdon-dros* (who went to India), *Parama* (who became a Bande with the name *Dharmānanda*, and also went to India), and *Vijayadharma*. 'The venerable *Mantasidhi*' should be '*Samantasiddhi*'. *Ḫdon-dros*, on returning to Khotan (Li-yul) dwelt in *Saṅ-tir*, because he was disliked by the king, and even there was persecuted by him. Samantasiddhi came from India and effected a reconciliation: but *Ḫdon-dros* does not appear ever to have been king (ff. 433 b–436 a).

P. 240, ll. 1–2.—This king, who became a Buddhist, received the name *Ānandasena*. He was sent to *Śu-lig* (Kāshgar), and Vijayasiṃha built for him the Vihāra *Sum-ña* (f. 436 b).

Ibid.—*Vijayakīrti* I was succeeded by his son *Vijayasaṅgrāma*, who acquired the name *seṅ-ge* 'lion', and by whom (or at whose instance by a certain Dharmakīrti) a Vihāra was built. After his time, during fourteen generations of kings, Li-yul was sometimes invaded by enemies, and sometimes vice versa. The fourteen generations begin with Vijayadharma (son of Vijayasaṅgrāma ?), and end with Vijayakīrti. In *this* period Li-yul was invaded by the *Drug-gu* king *'A-no-śos*, who destroyed the Vihāras as far as *Ḥgeḥu-to-śan* (*sic*) (f. 437 a–b).

P. 240, ll. 1–2.—*Vijayakīrti* II was the fourteenth from *Vijayadharma*, who was probably the successor of Vijayasaṅgrāma I, and not the original Vijayadharma. It is related in the *Saṅghavardhana Vyākaraṇa* how Vijayasaṅgrāma (II, son of Vijayakīrti II), who became king when seven years old, hearing of the former invasions by the Drug-gu *'A-no-mo-śoṅ* (*sic*) and others, invaded their kingdoms, and caused great slaughter, to atone for which he built the Vihāra *Ḥgu-gžan*, which was then called *Ḥgu-žan-ta* (ff. 437 b–438 b).

Ibid.—The successors of *Vijayakīrti* II are as follows:—

1. *Vijayasaṅgrāma*, a minor, who carried war into the land of the *Drug-gu*, and built the *Ḥgu-gžan* Vihāra (see above).
2. *Vijayasiṃha*, who built the *Bžaḥ-ser-ma* Vihāra: a contemporary of an Arhat *Dharmapāla* (f. 438 b).
3. *Vijaya*, during whose reign, and in the five following reigns, no Vihāras were built (f. 438 b).
4. *Vijayapāla*, and his son *Vijayasaṅgrāma* succeeding him. The latter built the *Bžaḥ-saṅ-gre-re-ma* Vihāra (ff. 438 b–439 b).
5. *Vijayaśātra*, during whose and the three following reigns no Vihāras were built (f. 439 b).
6. *Vijayakīrti*, in *Ḥgum-tir*, built the Vihāra *Bha-va-ña* (f. 439 b).
7. Then *Vijayasaṅgrāma*, and his son of the like name (perhaps, however, the son's name was *Vijayavikrama*): in the course of a visit to China the father was killed by *Drug-gus*, and while the son was a minor, the minister *Ama-la-khe-meg* governed for twelve years, and built the Vihāra and Stūpa of *Ma-na-ḥdi* (ff. 439 b–440 a).
8. *Vijayavikrama*, on becoming king, built the Vihāra *Byi-ža-gra-ma* for the Arhat *Devendra*, his Kalyāṇamitra (f. 440 a).
9. Then the Chinese minister (?=Ambassador or Amban) *Ser-the-śi* and king *Vijayadharma* built the Vihāra of *Maitra* (f. 440 a).
10. Then the Chinese minister *Ka-the-śi* and king *Vijayasambhava* built the Vihāra *Khe-gan-tsa* for two Arhats, his Kalyāṇamitras *'U-dren-dra-rod-ci* and *Dharmānanta* (°*da*?). They also built a Stūpa called *Su-stoṅ-ña* (f. 440 a–b).
11. Then *Vijaya-bohan-chen-po* rebuilt the *Su-stoṅ-ña* Stūpa (f. 440 b).
12. Then *Vijayasiṃha*, a youth, was oppressed by his ministers and the queen mother, whom on growing up he punished. He then built the Vihāra *Ro-bya*. Afterwards, his queen having been helped by certain Arhats (*Manya* (?), *Satrajñā*, *Sthyaśuta*, *Siṃhanāda*, and others) from India, he built the Vihāra *Nu-bo-ña* (ff. 440 b–441 a).
13. Then a bhikṣuṇī named *Śoḥi-dzaya*, the elder sister of *Śo-rgya*, queen of *Ḥdon-ḥdros*, being arrived from China to act as Kalyāṇamitra of her younger sister, the queen built in her honour the Vihāra *Yo-zo-ḥjo* (f. 441 a).
14. *Vijayaśata*, contemporary of an Arhat *Buddhānanta* (*sic*), built the *Zer-ro-ḥjo* Vihāra (f. 441 a–b).
15. Then *Vijayānanta* (*sic*) built for a female Arhat *Dharmapāla* (*sic*), who came from *Śu-lig*, the *Po-len-to* Vihāra (f. 441 b).
16. Then *Vijayasaṅgrāma*, also named *seṅ-ge* (see p. 240 *supra*), with his mother *Dharmā*, for whom, as a female Arhat, he built the *Dro-mo-ḥdza* Vihāra (ff. 441 b–442 a).
17. Then *Vijayasaṅgrāma's* younger brother *Bre-sa-ya-stu-lag* married the *Drug-gu* queen *A-žu-ka-su-ma*, and built the *Ta-ke-ḥjo* Vihāra (f. 442 a).
18. Then the wife of king Vijayasaṅgrama, who built the *Ḥgu-gžan* Vihāra [? see above No. (1)], daughter of king *Phrom-ge-sar*, and another wife, the daughter of (?) *Ḥu-rod-ga*, being Arhats, by name *Śilamata* and *Go-ḥu-śa-ra* (*sic*), came from Kashmīr and became Kalyāṇamitras of the queen mother *Ḥu-rod*, who built for them the Vihāra *Ho-roṅ-ḥjo* (f. 442 a).

19. *Vijayakīrti*, whose daughter *Vi-śa-dza-ya* built for her elder sister, the female Arhat *Vi-te-śe*(?)-*ci*, the Vihāra *Po-blo-na-jo* (f. 442 a).

20. *Vijaya-ḥzaḥ-la*, whose daughter *Vi-śa-phra-ba*, wife of the king of the *Gu-zin* [for *Gu-zan*?], built for the bhikṣuṇī Arhat Vijaya the *'Er-mo-no* Vihāra (f. 442 a–b).

21. *Vijayapāla* built for his wife, who had become an Arhat, the Vihāra *Kho-mo-no-no* (f. 442 b).

22. The minister *'A-ma-cag*(*ca-ga*? *ca-kha*?, see p. 582, l. 22)-*meg* built the Vihāra *Gus-sde-re-ma* (f. 442 b).

23. *Vijayajaya* having married the daughter of the king of *'O-sku*, the queen built for her Kalyāṇamitra, the female Arhat *Gže-ma*, the Vihāra *'O-ka-no* (f. 442 b).

24. *Vijayasaṅgrāma* built the Vihāra *Kus-gyi-'or-myoṅ-nag* (f. 442 b).

25. *Vijaya*, married to *Śu-to-ka*, daughter of the king of *Beu-gun-pan* (cf. p. 584, l. 5 from end), built for her elder sister, the female Arhat *'A-śo-ko-śi-la*, the Vihāra *Geo-lu*(?)-*na* (f. 442 b).

26. *Vijayasiṃha*, married to *'A-lyo-ḥjaḥ*, the *Ga-ḥjag* queen, built for his queen's Kalyāṇamitra, the female Arhat *Bdun-gśegs* from *Śu-lig*, the Vihāra *Khye-śo-na* (ff. 442 b–443 a).

27. *Vijayasaṅgrāma* built at the advice of two bhikṣuṇīs, *Śi-la-ha*(*ta*?) and *Ḥgeḥu-śu-su-ga*(*ya*?), from Kashmīr, the Vihāra *Na-mo-ḥbu-gdoṅ* (f. 443 a).

28. A minister *Ḥdaḥ-no-ya* set up in the lower market a great image, afterwards repaired by the minister *A-ma-ca-vi-dad*, who also built for the bhikṣuṇī *Koṅ-śed* [written Koṅ-śeṅ] (*or* bhikṣuṇīs of *Koṅ-śed*?) the Vihāra *Ḥdaḥ-no-yo-no* (f. 443 a).

A. *Sum of the Vihāras, &c. in 'U-then*[1] (*Khotan*), *in the Sku-mkhar* (*=the fortress or capital*?) *and outside*:—

1. 68 great Vihāras, 95 middle ones, 148 small ones, 3,688 other temples (*sgo-sgoḥi-mchod-paḥi-lha-khaṅ*?) and minor foundations, *chortens*, &c.
2. In a certain year (*byi-ba-lo* 'mouse year'[2]) there were 10,000 persons of both sexes in the Saṅghas (*dge-ḥdun-spos-ḥtsho*).

B. *Sum of the Vihāras, &c. in Mdo-lo and Me-skar*:—

1. 4 great Vihāras, over 100 other temples, &c.
2. 20 Saṅghas.

C. *Sum of the Vihāras, &c. in Kam-śed, Pha-ña, Be-rka-ḥb*(*d*?)*ra and 'O-sku to Ji-la, in the Sku-mkhar* (*the fortress, capital*?) *and outside*:—

1. 23 large Vihāras, 21 middle ones, 23 small, 839 minor foundations.
2. 438 Saṅghas (persons in the Saṅghas).

D. *Sum of the Vihāras, &c. from Gyil-kyaṅ to Koṅ-śed and Du-rya*, in the *Sku-mkhar* (*fortress, capital*?) and outside:—

1. 15 Vihāras, and a number (not stated) of minor foundations.
2. 963 persons in the Saṅghas (f. 443 a–b).

The *Li-yul Vyākaraṇa* was composed in accordance with the request of Ārya *Zla-ba i-sñiṅ-po* (Candragarbha) and Devī *Dri-ma-med-paḥi-ḥod* (*Amalaprabhā*?) and with the *Vyākaraṇa* of Saṅghavardhana in the last intercalary month of the autumn of the Khyi (dog) year, 1256 years from the introduction of the Dharma by Vijayasambhava. Its subject is how the Āryas were invited by the kings and Buddhism was taught, how Vihāras and temples were built, and how the Saṅghas spread (fol. 444 a).

[1] For an etymology of the name *'U-then* see S. C. Das, *op. cit.*, p. 195, n.

[2] The mouse year is merely the first of the 12 years cycle.

PART II

NOTES FROM THE GOŚṚṄGA VYĀKARAṆA.

(Cf. 'Notes Chinoises sur l'Inde', V, by M. SYLVAIN LÉVI, in the *Bulletin de l' École Française d' Extrême-Orient*, vol. v. pp. 256 sqq.)

THIS work is an account of the *Gośṛṅga* mountain and country, put into the mouth of Buddha as prophecy. It is found in the *Bkaḥ-ḥgyur* (Mdo xxx, ff. 336 b–354 b; see Rockhill, p. 231).

Foll. 337–340 a.—Buddha visits the site of the future kingdom of *Dge-ba*, or *Li-yul*, and sinks in dhyāna on the peak of Mt. *Gośṛṅga*, on which occasion we find mentioned in the north a great lake, a little distance from which Buddha meditated, in the west the Caitya *Go-ma-sa-la-gan-da* (338 b) and the Vihāras *Spoṅ-byed* (*Vṛji*), *Ḥjigs-tshogs-spoṅ-byed*, *Ḥod-can*, *Vi-si-mo-ña*, *Ye-śes-ri*, *Va-*(?)*-no-co*, and *Dge-ḥdun-skyoṅ* founded by the blessing of Mañjuśrī, Ākāsagarbha [1] (?), Avalokiteśvara, Maitreya, Bhūmigarbha [2] (?), Vaidyarāja [3] (?), and Samantabhadra respectively, as well as other Vihāras by other divinities (340 a).

Foll. 340 a–b.—Here is told the incident of the 353 (*sic*) lotuses (Rockhill) and it is prophesied that in a like number of Vihāras the Ghaṇṭā would be sounded in this country. The *Go-ma-sa-la-gan-da* country receives the blessing of Buddha.

Foll. 341 a–342 b.—One hundred years after the Nirvāṇa comes the story of *Kustana* and the Chinese Minister *Ḥjan-śo* (sic), whose sovereign was named *Cha-yaṅ* (acc. to S. C. Das, *op. cit.*, pp. 196–7 = the great emperor *Che-he-Wang*, or *Che-Hwang-te*). In the time of *Kustana* arose many Vihāras, Caityas, &c. The name *Go-ma-sa-la-gan-da* (according to the Sūtrapiṭaka derived from being on the bank of the river *Go-ma*, cf. Lévi, *op. cit.*, p. 258) and that of *Dge-ba* are explained (342 a). When the *Go-ma-sa-la-gan-da* and the *Ḥod-sruṅ* (Kaśyapa) caityas sink, the country is to sink.

Certain divinities are assigned as protectors of the country. The Sūtra (i.e. the Gośṛṅga Vyākaraṇa) is to be read in times of trouble and danger (342 b).

Foll. 343 a–345 b.—In future times the *Sum-pa* and *Bod-pa*, and also the Chinese, would invade the country, but would be diverted by the Sūtra (343 a) from harming it. In order that the Sūtra should be firmly established in Li-yul, it was taught by Buddha, on *Gośṛṅga*, to the Aparānta king and his ministers from Li-yul.

Foll. 345 b–347 a.—In future invasions by the *Sum-pa*, *Drug-gu*, *Hor*, and others, the country would be defended by images of Buddha brought from various other countries (345 b).

From the fortress *Dge-ba-can* (capital of Dge-ba) would come the image of Buddha *Bde-baḥi-ḥbyuṅ-gnas* (Puṇyasambhava ?), which would protect the fortress and district of the West, *Ku-Śed*.

The Northern district *Śen-ža* would be protected by an image *Śen-ža*, which would arise from the nether world.

In the East the fortress and district of *Phye-ma* would be protected by the image *Kiḥu-laṅ* (v. l. *Kaḥi-laṅ*).

In the North (*sic* for South ?), before the Kaśyapa Caitya on Mt. *Gośṛṅga* the image *Phye* (v. l. *Phyi*)*-se* would protect the religion and country (346 a).

In the fortress *Dge-ba-can*, in the market-place *Kla-*(? *Ka'a*)*-sta-ḥdi-ze*, in the Royal Palace, an image of Buddha would protect the fortress and district.

In the region *Ño-mo-ña*, where king Yol (= *Ye-u-la* of p. 581 ?) proposed ('will propose' *gros-byed-par-ḥgyur*) to build a fortress, an image called *Cu-gon-pan* (cf. p. 583, l. 10) would protect the fortress *Dge-ba-can* and district.

In the Vihāra *Di-na-*(*Ñi-na* ?)*-dzya* would be an image of Dīpaṃkara Buddha, surpassing all others, made in *Thor-koṅ* (*khoṅ* ? *khuṅ* ?) of Jambudvīpa. The people of Li-yul would call it *Ḥkhor-baḥi-mar-me-mdzad* (Cakradīpaṃkara ?), and it would be a model for all the Li-yul images made in *Thor-koṅ*. It would guard the religion and district (346 b).

[1] Nam-mkhaḥi-sñiṅ-po. [2] Saḥi-sñiṅ-po. [3] Sman-gyi-rgyal-po.

There would be many other great images in Dge-ba-can unrivalled in the world, and so long as they remained the country would never be deserted. The merchants coming from all parts would pass unharmed (346 b–347 a).

Buddha speaks of these great images as *Śiṅ-rtas-ḥdren-pa-byas* (*śakaṭena nīta kṛta* 'brought in a chariot'? or contains the name of the maker?) (347 a).

Other images of Buddha, 208 in number, in connexion with the Kaśyapa Caitya, would all be miracle-working (347 a).

Foll. 347b–348 a.—To protect the country kings and ministers must acquire puṇya in the *Tsar-ma* Caitya, because there were the first believers. In the time of evil kings and ministers the people must go to that part of *Gośṛṅga* where Buddha had been. When the people do what is wrong, atonement must be made in the Caitya *Go-ma-sa-la-gan-da*. Kings and ministers taking vows of penance must resort to that part of *Gośṛṅga* where is the image *Phye-se* and the Saṅgha of the Kaśyapa Caitya. People of Li-yul who are sentenced to punishment must resort to the Saṅgha of *Go-ma-sa-la-gan-da* on *Gośṛṅga*.

Foll. 348 a–351 a.—The virtues of *Go-ma-sa-la-gan-da* and the circumstances under which the country would be protected by Buddha from external and internal enemies. The religion would be Mahāyānist, for which reason the kings, when weak, must rely upon the kings of Bde, China and other Mahāyānist countries (350 b).

Fol. 351 b.—Foreigners coming into the country would be made mild and peaceable.

Foll. 354 a-b.—The blessing and naming by Buddha of the *Go-ma-sa-la-gan-da* Caitya, the *Gośṛṅga* Mountain, the district *Dge-ba*, the fortress *Dge-ba-can* and the *Dgon-pa* (monastery).

At the command of Buddha Śāriputra and Vaiśravaṇa divide the mountain *Śa* (*śa-ri*—have we here some popular etymologizing in relation to Śāriputra?), and thus bring together into being the Caitya *Go-ma-sa-la-gan-da*, Mt. *Gośṛṅga* and Li-yul (see Rockhill, p. 233).

APPENDIX F

NOTES ON SPECIMENS OF ANCIENT STUCCO FROM KHOTAN SITES

BY

A. H. CHURCH, F.R.S.

PROFESSOR OF CHEMISTRY, ROYAL ACADEMY OF ARTS

ELEVEN specimens of ancient stucco, from ruined sites in the Khotan territory, Chinese Turkestān, have been examined by me chemically. They fall into two groups, one, soft and friable, being essentially silicious, and the other, hard and crystalline, being what is now known as plaster of Paris.

The following specimens belong to Group I :—

1. Stucco from statues in Rawak Stūpa Court.
2. Stucco from image, Rawak Stūpa Court, showing red surface and grey backing.
5. Stucco from detached clay-seal mixed with vegetable fibre from ancient dwelling N. xv, Niya Site.
6. Stucco from red clay-seal (N. xv. 133 a) from ancient dwelling N. xv, Niya Site.
7. Stucco from relief decoration of Buddhist temple, Endere ruins.
8. Backing earth mixed with vegetable fibre from stucco relief (E. i. 015), Endere ruins.
10. Scraping from painted slab of wall in Buddhist shrine (D. II), Dandān-Uiliq Site.

The material of which these specimens consist resembles adobe or loess, but, as will be seen from the quantitative analyses of Nos. 1 and 7 given further on, it contains a rather notable amount of combined water. This feature points to the presence of a considerable proportion of true clay (hydrous aluminium silicate) in these stuccoes. The variation in the colour of the specimens is not important; it depends upon the state of the iron present. In the yellow material the iron exists as ferric hydroxide, in the red as ferric oxide, and in the grey as a lower oxide or as disulphide.

The two analyses here given of the specimens Nos. 1 and 7 belonging to Group I are not exhaustive, as several of the minor constituents, such as phosphoric acid and alkalies, were not separately determined, while the residue, included under 'silica and silicates insoluble in hydrochloric acid' comprises felspathic and micacious minerals. It should be added that small amounts of the lime and magnesia found were actually in combination with silica, but it has been thought expedient to exhibit the lime as carbonate and the magnesia as free. Here are the percentage results of the two analyses :—

	No. 1	No. 7
Water given off at 100° C.	3·06	2·00
Water given off on ignition	8·58	6·20
Silica, and silicates insoluble in hydrochloric acid	57·40	67·30
Ferric oxide ($Fe_2 O_3$)	5·59	5·20
Alumina ($Al_2 O_3$)	6·81	7·50
Lime, as Calcium carbonate ($Ca CO_3$)	14·72	8·66
Magnesia ($Mg O$)	3·22	3·12
Sulphuric anhydride	*none*	*trace*

The remaining specimens, four in number, belong to our second Group. They are these:—

3. Fragment of hard stucco from ancient relief, Ak-sipil Site.
4. Fragment of hard stucco from Kara-döbe Site.
9. Stucco from relief decoration of Buddhist shrines, Dandān-Uiliq Site.
11. Hard stucco, probably fragment of large image, found at Rawak, beyond Dandān-Uiliq.

These four specimens are essentially plaster of Paris. The native gypsum from which they have been formed has been prepared in the usual way by moderate heating (called 'burning') and then been mixed with water just as is now done in making plaster casts. These processes were known to classical antiquity in Europe. There are two observations that should be made in connexion with specimens 3 and 11. The latter is unusually crystalline; the former shows that it has been subjected, after completion, to a high temperature and to an atmosphere charged with what chemists call 'reducing' matters. The evidence for this fact is furnished by three peculiar features shown by specimen 3. These are:—the low percentage of water present; the existence of sulphur in the form of a sulphide, probably an oxy-sulphide; and the grey discoloration which has penetrated deeply from the surface inwards. Here are the percentages of the two analyses of specimens 3 and 9:—

	No. 3	No. 9
Water given off on ignition	13·85	20·46
[of this was given off by long heating at 100° C.	10·64	17·87]
Silica	2·72	2·16
Ferric oxide and Alumina	0·95	0·55
Calcium Sulphate ($Ca SO_4$)	79·96	75·34
Calcium Sulphide (Ca S)	2·13	*none*
Calcium Carbonate ($Ca CO_3$)	0·36	0·85

A cast from pure plaster of Paris would when quite air-dry contain about 80 per cent. calcium sulphate and 20 per cent. water.

APPENDIX G

NOTES ON SAND AND LOESS SPECIMENS

BROUGHT BY DR. M. A. STEIN FROM THE REGION OF KHOTAN

BY

L. DE LÓCZY

PROFESSOR OF GEOGRAPHY IN BUDAPEST UNIVERSITY, ETC.

I.—YŌTKAN.

THE specimens of soil from Yōtkan were obtained from two strata of the banks excavated in the course of gold-washing operations.

(i) The upper stratum, identical with the fertile soil of light yellowish colour at present under cultivation, shows a thickness varying from 9–20 ft. near the hamlet of Khalche. The three specimens from this stratum consist of a very uniform loess-like substance. One was taken west of Khalche from a depth of 6 ft.; the second east of it from a depth of 16 ft.; the third from the same place at a depth of 10 ft.

The first specimen (marked No. 3) contains grains of gypsum, and remains of charcoal, straw, and grass. Under the microscope the material proves to consist of angular quartz and mica. The diameter of the angular quartz grains varies from 0·009–0·09 mm. The second specimen (No. 10) is also angular quartz sand with plentiful mica; the grains, finer even than in the first specimen, show a diameter of 0·004–0·060 mm. Among the fine sand are found here and there mica scales, 0·19–0·28 mm. in diameter. Under hydrochloric acid the two specimens undergo scarcely any effervescence. On the whole the upper stratum at Yōtkan is formed by very fine quartz sand containing a very slight quantity of clayey dust. It is this fine dust, almost unmeasurable in diameter, which effervesces under hydrochloric acid; it is hence chalky. The fine clayey and chalky dust binds the quartz grains and mica into small clumps showing slight cohesion.

The third specimen (Nos. 6 and 7), found to the east of Khalche, in the immediate vicinity of a sandy layer (No. 5), agrees completely with the two previous ones. The angular quartz and mica grains of No. 6 vary from 0·0095 to 0·09 mm. in diameter. Its coarser grains attain 0·19 mm. in diameter. In specimen No. 5, taken from the sandy layer (10–12 inches thick) referred to below, the fine dust grains measure 0·0019–0·0097 mm., the coarser grains 0·028–0·381 mm. Otherwise this sandy layer, in regard to its substance and character, is identical with the three previous specimens marked as 'silt.' Specimen No. 7 consists of quartz and mica, 0·0066–0·0476 mm. in diameter, with an admixture of coarser grains (0·0095–0·095 mm.) and of a somewhat larger quantity of chalk dust.

(ii) Of the 'culture-stratum' of Yōtkan, which lies under the light-yellowish upper stratum in a thickness of 5–14 ft., and shows a darker (brownish) colour, there are two specimens in the collection. One (No. 4) was taken from a depth of 17 ft., the other (No. 9) from a depth of 18 ft. Specimen No. 4 is a fine sandy substance full of angular chalk concretions. These concretions have formed around plant stems or roots, and resemble those small, irregularly angular concretions which are common in the mud of the ancient alluvium along the Tisza and Maros rivers. The interior of the concretions is formed by a grey, clayey and sandy fresh-water chalk, full of biotite mica scales. Outside, the concretions are covered with a fine dust. Under the microscope this dust proves to consist of angular quartz and mica. A magnification of 600 is needed to distinguish its particles, the larger grains measuring 0·0007–0·007 mm., and the smaller ones 0·0004–0·0007 mm.

Specimen No. 9 is an angular quartz dust with plentiful mica. I found in it also some very tiny pieces of gypsum. It is considerably more chalky than the soil of the light-coloured upper layer. The finest grains have a diameter of 0·007–0·070 mm. Very rarely there appear in it also grains of 0·014 mm. diameter.

All the specimens briefly described so far are fine mud which is formed mainly by angular quartz and besides by mica, both black and white (biotite and muscovite). Of clay or chalk dust it contains very little.

This quartz dust is finer than the fine clayey and chalky dust of the Chinese loess (Lan-chou 0·004–0·007 mm., Hui-ning 0·0014–0·0028 mm.). It cannot be doubted that this fine mud has formed by the same process in the darker culture-stratum as in the cultivated soil of the surface. It belongs to that category of desert deposits which I have described in Count Béla Széchenyi's work on the scientific results of his East-Asiatic expedition, i. pp. 478–89, i.e. to the group of riverine loess.

It seems scarcely doubtful that the strata of Yōtkan owe their origin more to the dust which has been deposited among the vegetation of old cultivated ground and there retained, than to mud carried by the river. The origin of the mud has to be sought for in riverine deposit; this, however, could not have brought to the area watered by the Kara-kāsh and Yurung-kāsh such fine mud as we have recognized in the mud strata of Yōtkan, exceeding 20 ft. in thickness. The two rivers debouch from the hills on to the alluvial fan formed by the bare pebble *Dasht*, at a distance of about 10–12 miles south of Khotan. Their water must have a rapid flow, not only in their beds but also in the irrigation canals distributed from the foot of the hills. It is probable that Khotan is situated on an alluvial fan similar to those occupied by Liang-chou, Kan-chou, &c., near the Gobi of Kan-su. I think I may identify the Yōtkan soil with the thin, horizontally stratified riverine loess which I observed on the alluvial fans of all hill streams at the northern foot of the Nan-shan and also near the Su-la-ho river.

My view is confirmed by the 'sandy' layer, 10–12 inches in thickness, found between the upper layer and the culture-stratum. This, in spite of its fineness, may be more readily taken for mud deposit from river or canal than the very fine soil of the layers above and below. Even this sandy material is still very fine; chiefly 0·00952–0·095 mm., with the coarsest grains 0·19 mm. in diameter. It is thus far finer than, e.g., the mud of the Tisza near Szeged (0·04–0·10 mm.).

I am inclined to attribute to the Yōtkan deposits an origin similar to that of the stratified riverine loess near the banks of the Kāshgar Tārīm of which Dr. Futterer has given a good account (*Durch Asien*, i. p. 89; Pl. v). Such banks of stratified loess are to be found also below Khotan, and near the hamlet of Tam-ōghil, towards Ak-sipil.

The Yōtkan sand represented by specimen No. 11 consists mainly of angular quartz and mica scales. Its finest grains measure 0·128 mm. in diameter, its coarsest (chiefly biotite and muscovite) 0·76–0·95 mm. Its greatest portion is made up of angular quartz flakes and fresh biotite, measuring 0·38–0·57 mm. in diameter. This sand is typical river sand.

II.—AK-SIPIL.

The sand specimen, taken from the crest of a dune, is composed of quartz, mainly angular, with a few rounded grains; there is a plentiful admixture of large elliptical biotite scales. Magnetite and here and there worn amphibole show also in it.

The finest portion, about 0·095 mm. in diam., consists of quartz, mainly angular, with little mica and still less magnetite and pyrite; the medium-sized portion, which preponderates, 0·14–0·38 mm., has the same composition; the large-grained portion, 0·6–0·76 mm., is made up chiefly of worn biotite and muscovite scales.

The sand of Ak-sipil closely resembles that of Yōtkan, being only slightly finer. It can scarcely be doubted that this sand is alluvium from the Yurung-kāsh river; nor is it difficult to trace its origin. South of Khotan, within the drainage area of the river, rises the Tikelik range, in which Bogdanowich found gneiss and granite. It is evident that the sand of Yōtkan and Ak-sipil is derived from these mountains.

The sand of Khotan differs strikingly from the sand specimens collected by me in the vicinity of the Kia-yü-kuan Gate, and around An-si-fan and Sha-chou which show rounded, smoothed and polished grains. Not only the quartz but also the shales and the carboniferous limestone grains are of similar size; mica is almost completely absent; of dust there is none here. It is clear that in the drift-sand of the Kan-su Gobi the river-mud does not play as large a part as in that of the Khotan region, but that, particularly in the vicinity of Sha-chou (Tung-huan), the decomposition of the desert subsoil has supplied the sand of polygenous substance and uniform grain. Near An-si-fan the sand is more angular, and rare mica scales are found in it; composed of otherwise similar minerals the riverine angular sand shows there more cohesion, i.e. more clayey admixture (comp. *Gróf Széchenyi utazásának tudományos eredményei*, i. p. 483; German edition, i. p. 522).

III.—DANDĀN-UILIQ.

Specimen No. 1.—Sand from Buddhist shrine D. II, sticking to stucco reliefs found 3–4 ft. above floor.
Specimen No. 12.—Sand sticking to MS. from ruin D. VII.
Specimen No. 13.—Sand with decomposed paper from the floor of shrine D. X.

The specimen No. 1 is relatively large-grained sand containing a good deal of gypsum particles and reed fragments. The sand grains are mostly angular quartz flakes; half-rounded grains as well as magnetite and pyrite are rare among them. The grains vary about the average diameters of 0·047–0·07–0·14 mm. The specimen contains fine dust with a diameter of 0·0035–0·0095 mm. The coarser grains consist chiefly of mica scales.

The fine sand from the Dandān-Uiliq ruins, in view of its angularity, and still more on account of its large mica contents, must be considered river deposit, which has scarcely as yet undergone wind transport, and which has been carried into the ruined structures from immediately adjoining ground. It can be distinguished at a glance from true drift-sand. This sand is homogeneous with the material composing the strata of loess mud at Yōtkan, though its grains are far larger.

The sand of specimens Nos. 12, 13, which stuck to the MSS. lying on the floor of ruined structures, does not differ materially from the upper sand. It contains angular quartz grains, 0·0685–0·14 mm. in diam.; plentiful mica, 0·135–0·20 mm. in diam.; also fine dust, with diameters 0·004–0·014–0·058 mm. Worn grains of quartz are very scarce in it. The fibres of the paper rags bind the sand into a thin crust resembling felt. The sand sticking to the paper in specimen No. 13 contains fragments of tree leaves, turned brown, and plant fibres. Small pieces of straw or grass are also found in it. It appears that moisture has reached this vegetable material.

There is no difference between the finer portions of the upper and lower sand specimens from Dandān-Uiliq. Even in the finest dust there is so much mica as to place its character as river deposit beyond all doubt. The wind could have transported it only for a very short time, and from a very short distance.

IV.—NIYA SITE.

Specimen No. 8. Sand found in wooden tablet, from N. XV.
Specimen No. 14. Sand sticking to ancient guitar, from N. XII.
Specimen No. 20. Sand sticking to wooden tablet, from N. IV.

All three specimens contain identical sand, plentifully mixed with vegetable fibres, straw, grain-husks, &c. The sand is composed of angular quartz grains and mica scales. In No. 8 fine quartz grains prevail, varying from 0·009 to 0·047 mm. in diam. The larger quartz and biotite grains measure 0·09–0·14 mm. in diam. In specimen No. 14 the sand adheres in small brownish clumps and scales. No. 20 shows coarser grains with much mica. The diameter of the larger quartz and mica grains varies between 0·095–0·38 mm., that of the finer grains between 0·047–0·095 mm.

The sand from the Niya Site closely resembles that of Dandān-Uiliq, and is also river deposit which the wind has scarcely yet carried and abraded.

INDEX OF OBJECTS FOUND, PURCHASED, ETC.

* *See* Corrigenda.

* *See* Corrigenda.

* *See* Corrigenda.

* *See* Corrigenda.

† *See also* N. xv. 328+75.

* *See* Corrigenda.

* *See* Corrigenda.

PLATES, DESCRIBED, OR REFERRED TO, IN THE TEXT

* *See* Corrigenda.

* *See* Corrigenda. † *For* CX *read* XC.

GENERAL INDEX

THE spelling of Chinese names and terms conforms to Wade's system of transliteration, except in Appendix A, where M. Chavannes has followed the system adopted by the École Française d' Extrême-Orient; the first occasion on which a Chinese word is given in the original characters is indicated in the Index by printing the number of the page in italics. In the transcription of Sanskrit, Prākrit, Persian, and Arabic words the system followed is that adopted by the International Congress of Orientalists in 1894. In the case of Turkestān local names, no attempt has been made to restore the original form of any Arabic elements contained in them, but the actual forms heard have been reproduced in phonetic transcription. Where discrepancies have occurred in transcription, the Index is to be taken as a final criterion.

JUDAEO-PERSIAN DOCUMENT. (7/10)

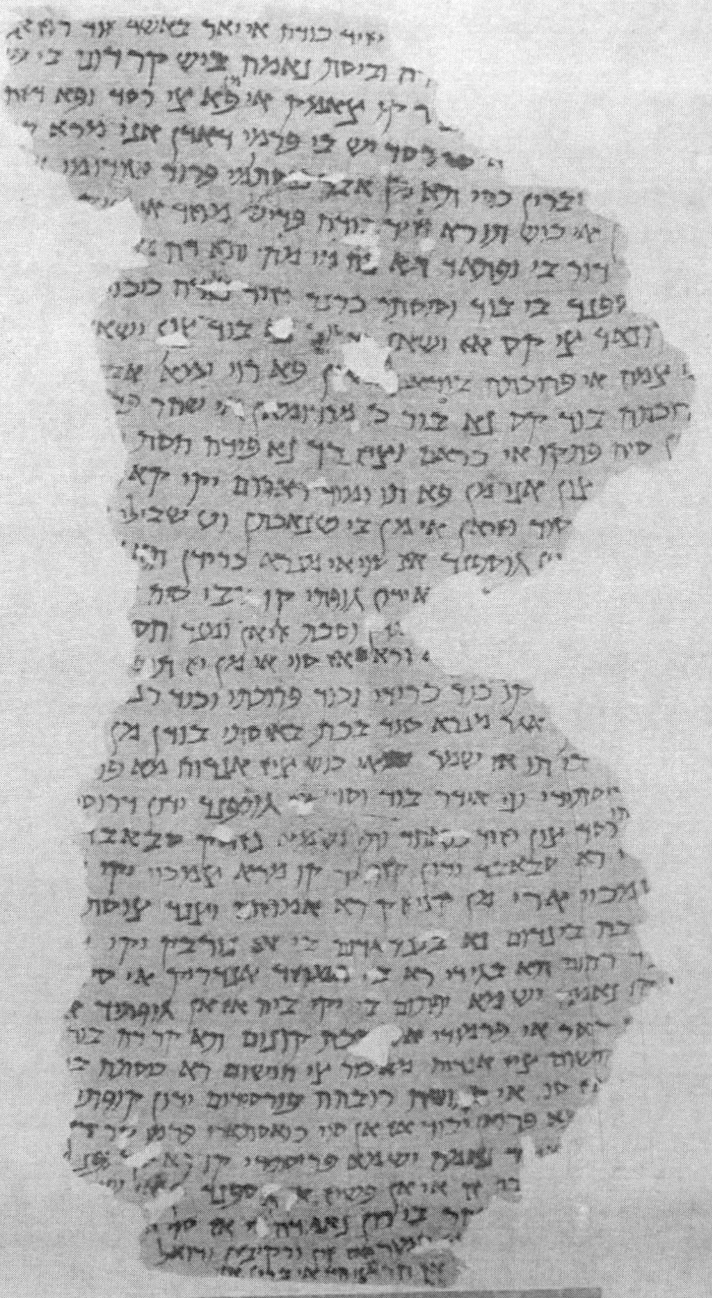

JUDAEO-PERSIAN DOCUMENT BROUGHT FROM DANDĀN-UILIQ

See Chap. X. sec. iii; Appendix C.

E. i. 11. (3/4)

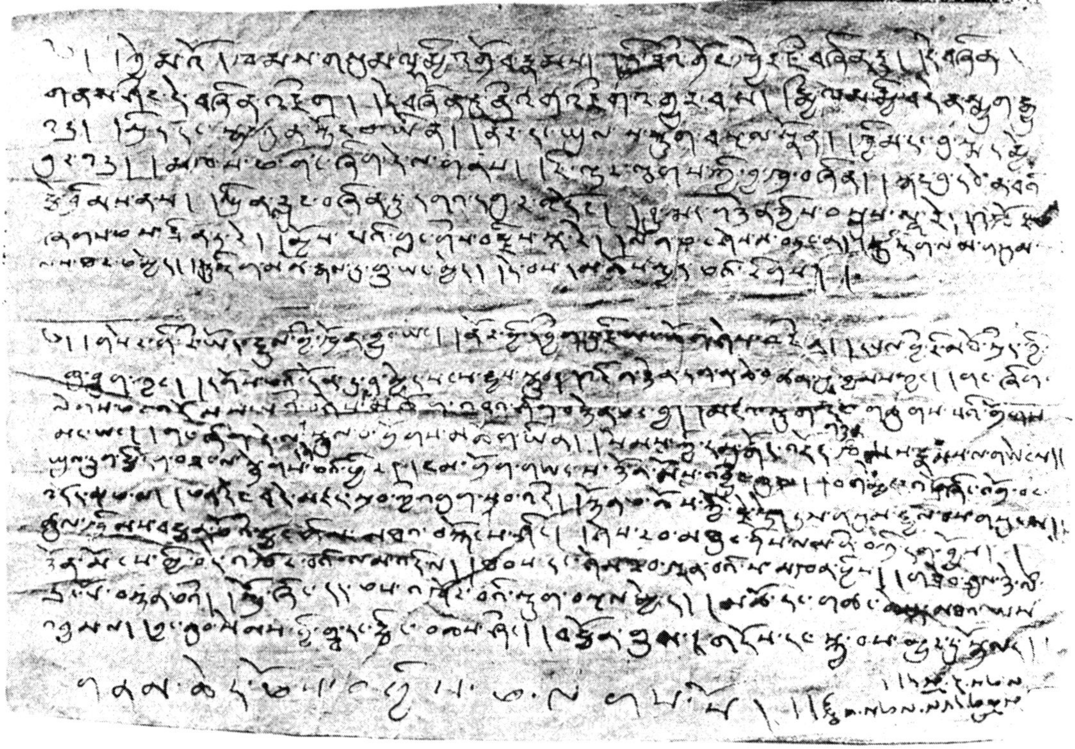

TWO BUDDHIST POEMS, IN TIBETAN, FOUND IN ENDERE TEMPLE.

(See Chap. XII. sec. ii, iv; Appendix B, part iii.)

E. i. 19. (3/4)

E. i. 15. (3/4)

E. i. 20. (3/4)

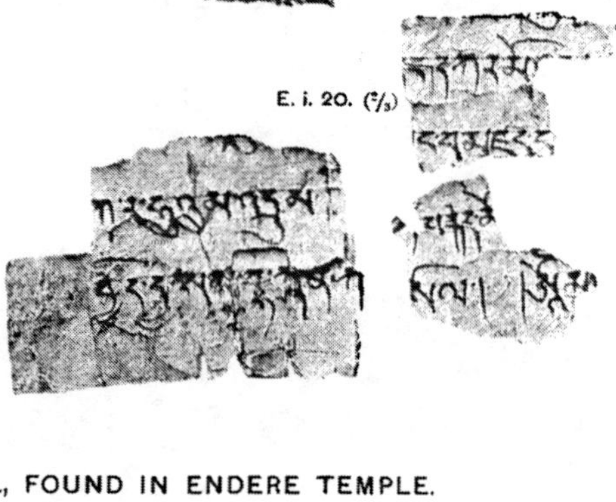

FRAGMENTS OF TIBETAN TEXTS ON RITUAL, FOUND IN ENDERE TEMPLE.

(See Chap. XII. sec. ii, iv; Appendix B, part ii.)

E. i. 24. FOL. 17, LEFT HALF.

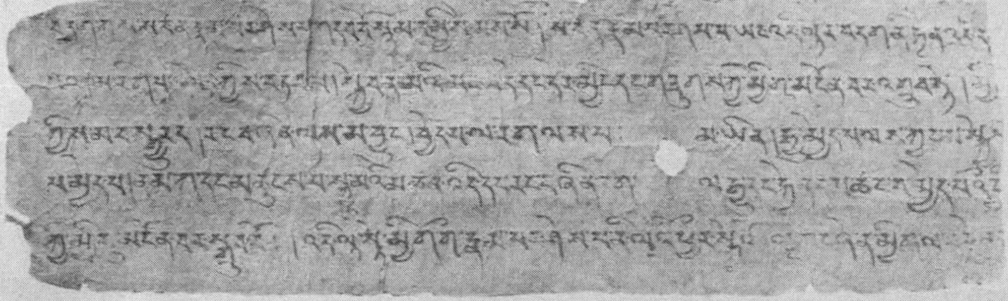

E. i. 21. FOL. 17, RIGHT HALF.

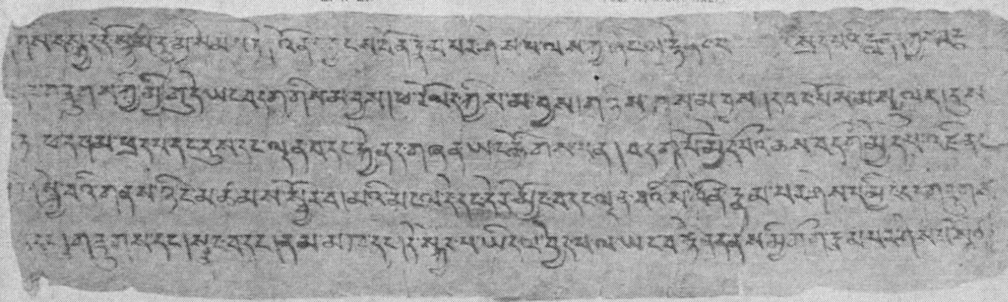

E. i. 10. FOL. 18, LEFT HALF.

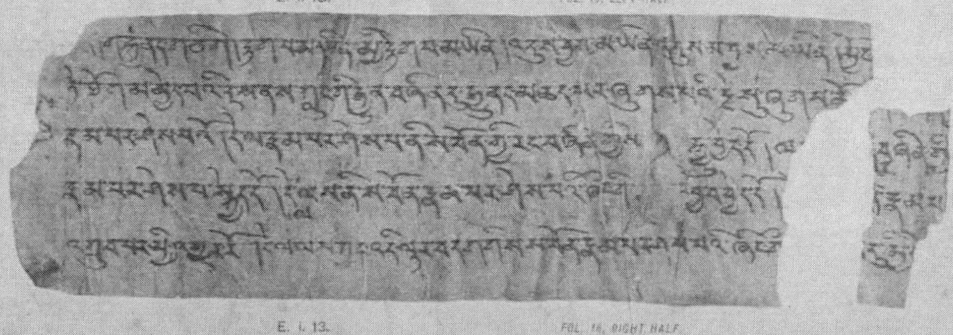

E. i. 13. FOL. 18, RIGHT HALF.

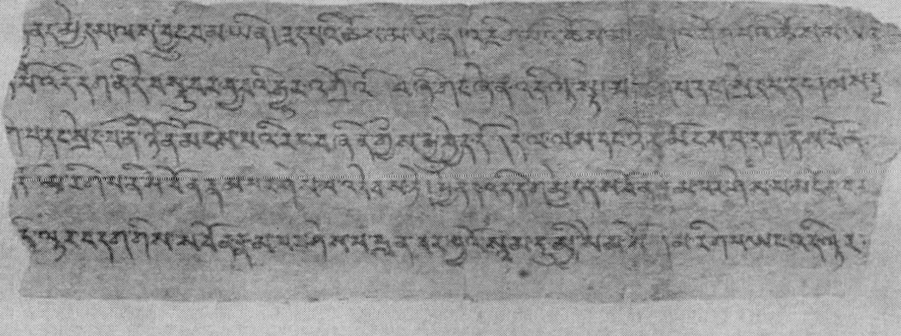

PORTIONS OF LEAVES OF ŚĀLISTAMBA-SŪTRA, IN TIBETAN, FOUND IN ENDERE TEMPLE.

(See Chap. XII. sec. ii, iv; Appendix B, part i.)

SCALE 10/11

D. VII. 3. a

D. VIII. 1.

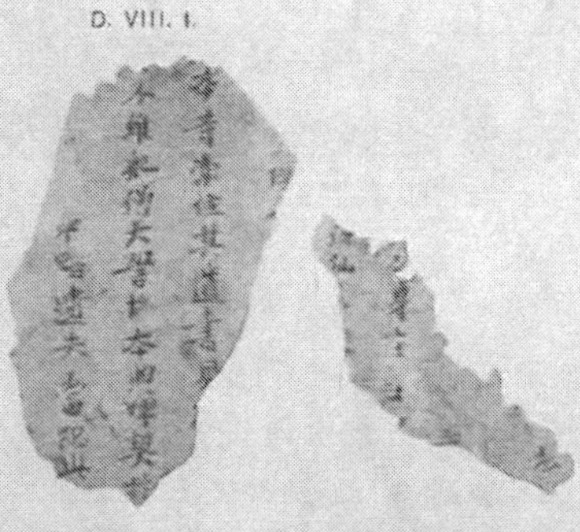

D. IX. 1.
790 A.D.

D. VII. 7.

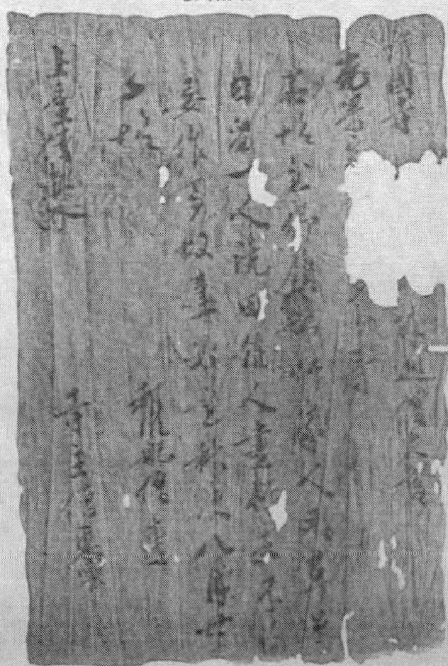

D. VII. 3. d. 787 A.D.

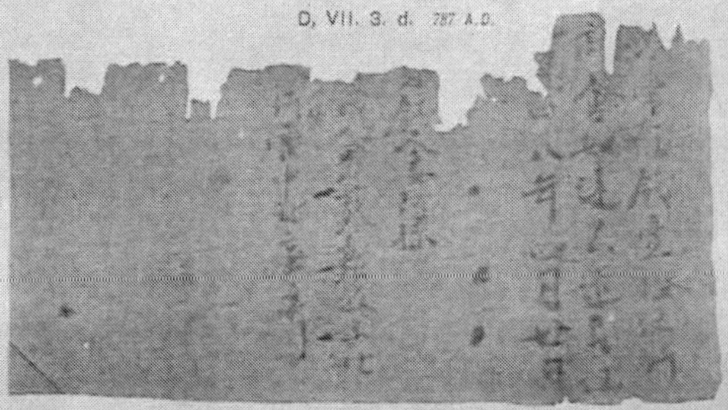

CHINESE RECORDS ON PAPER, TWO DATED 787, 790 A.D., FROM RUINS D. VII, D. VIII, D. IX, DANDĀN-UILIQ.

(See Chap. IX. sec. vi, vii, viii; Appendix A, part i.)

SCALE ¹/₄

D. VII. 4. a.
IN ORIGINAL FOLDED STATE.

D. VII. 4. a. UNFOLDED.

D. V. 6.

D. VII. 2.

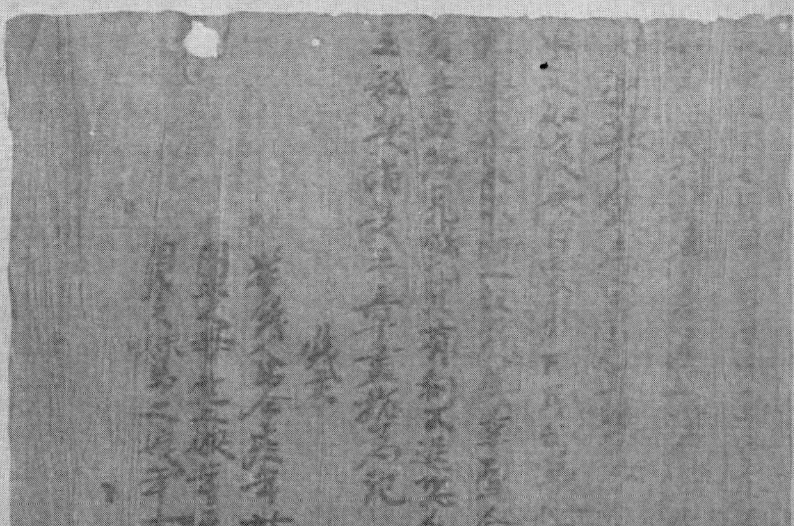

CHINESE RECORDS ON PAPER, DATED 781-782 A.D., FROM RUINS D. V. VII. DANDĀN-UILIQ.

(See Chap. IX. sec. v, vi, viii; Appendix A, part i.)

SCALE ½

CHINESE RECORDS ON WOODEN SLIPS AND A TABLET FOUND IN ANCIENT RUBBISH-HEAP N. xv, NIYA SITE.

(See Chap. XI. sec. v, viii; Appendix A, part ii.)

SCALE 1/1

CHINESE RECORDS ON WOODEN SLIPS, FOUND IN ANCIENT RUBBISH-HEAP N. xv, NIYA SITE.

(See Chap. XI. sec. v. viii; Appendix A, part ii.)

SCALE 1/1

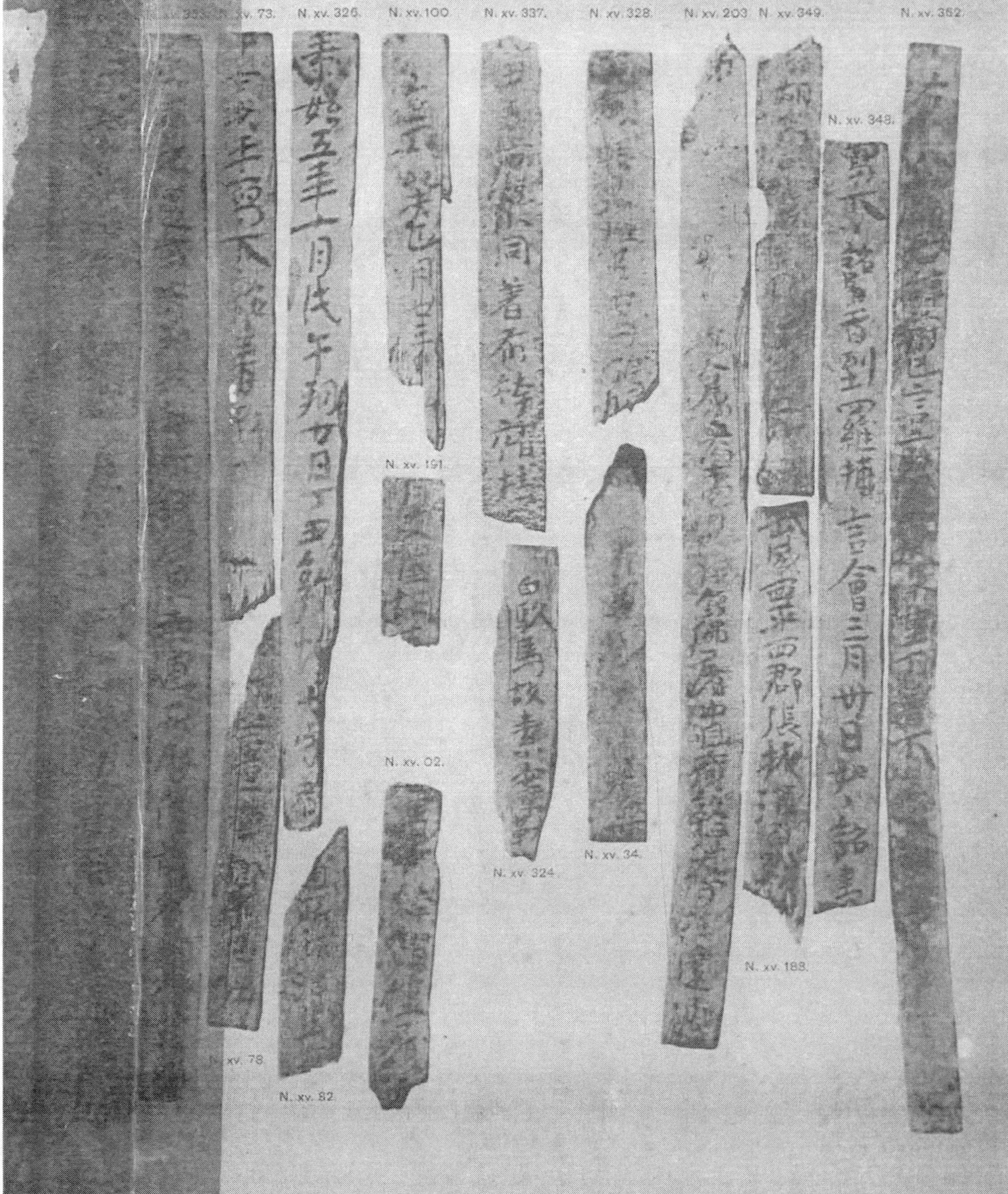

CHINESE RECORDS ON WOODEN SLIPS, FOUND IN ANCIENT RUBBISH-HEAP
N. xv, NIYA SITE.

(See Chap. XI, sec. v, viii; Appendix A, part ii.)

SCALE 1/1

E. i. 7. (2/3)

OBVERSE OF FOL. 19.

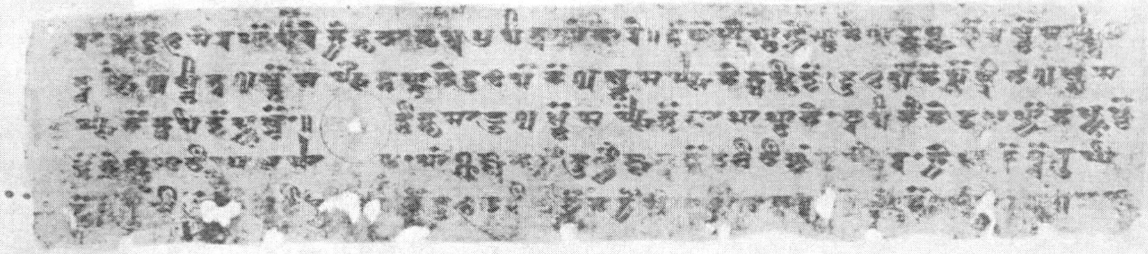

REVERSE OF FOL. 19.

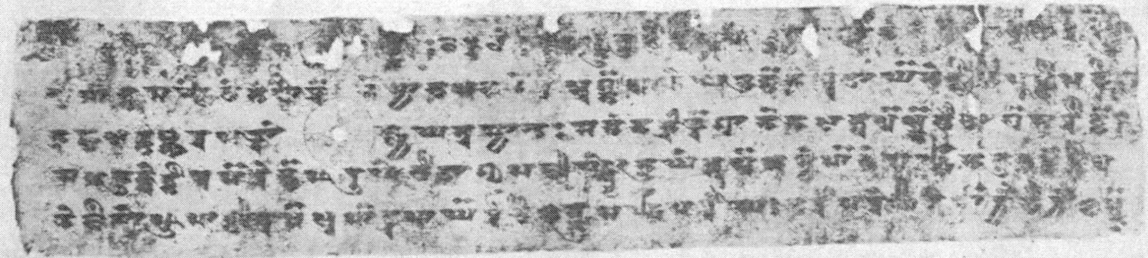

OBVERSE OF FOL. 149.

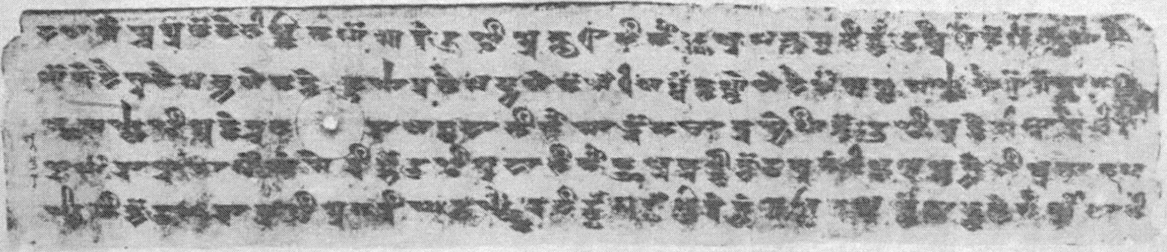

REVERSE OF FOL. 149.

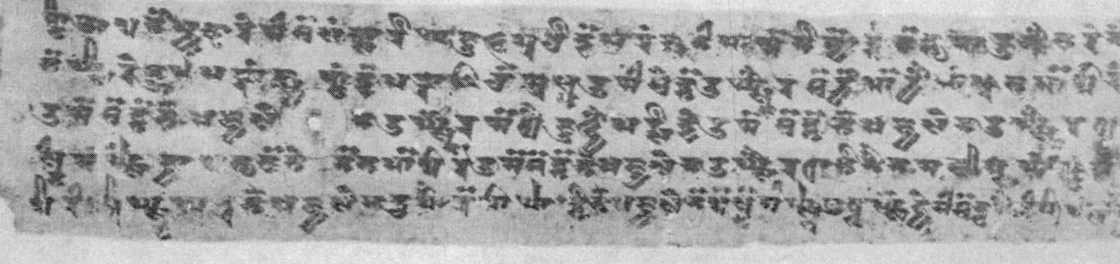

TWO LEAVES FROM A ROLL CONTAINING PORTIONS OF A BUDDHIST CANONICAL WORK, IN NON-SANSKRITIC LANGUAGE, FROM ENDERE TEMPLE.

(See Chap. XII. sec. ii, iv.)

D. III. 1. *OBVERSE OF FOL. 3.*

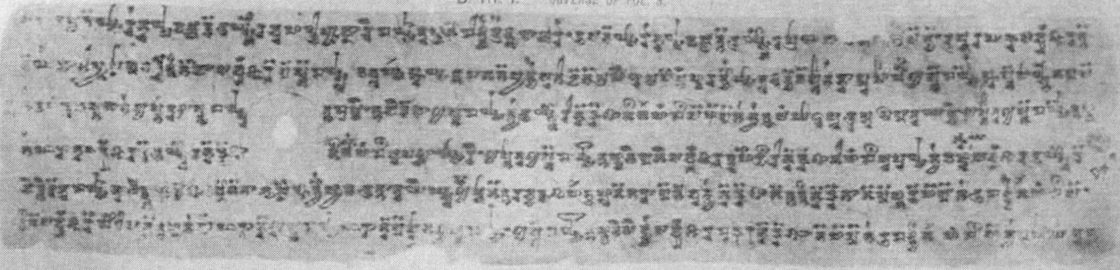

D. III. 1. *REVERSE OF FOL. 3.*

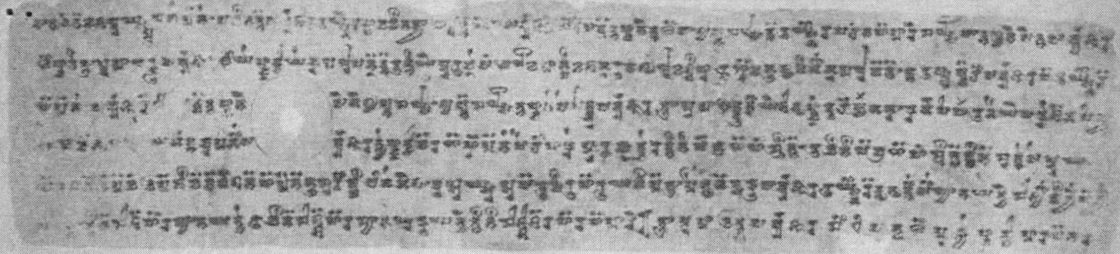

LEAF OF A BUDDHIST CANONICAL WORK, IN NON-SANSKRITIC LANGUAGE, FOUND IN RUIN D. III, DANDĀN-UILIQ.

(See Chap. IX. sec. iv, viii.)

D. III. 12
OBVERSE.

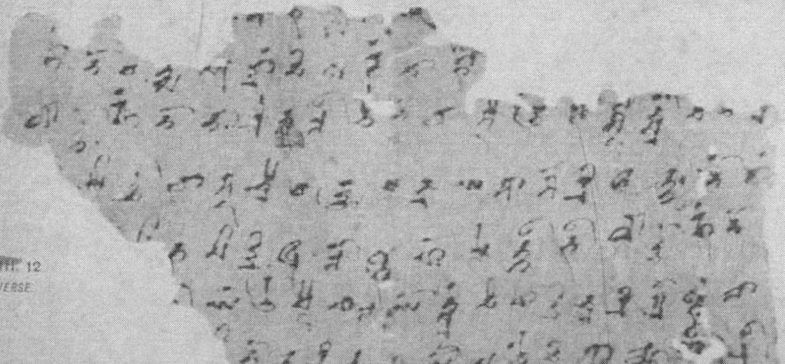

FRAGMENT OF DOCUMENT, IN CURSIVE BRĀHMĪ AND NON-SANSKRITIC LANGUAGE, FOUND IN RUIN D. III, DANDĀN-UILIQ.

(See Chap. IX. sec. iv, viii.)

SCALE [illegible]

E. i. 39. OBVERSE OF FOL. 30.

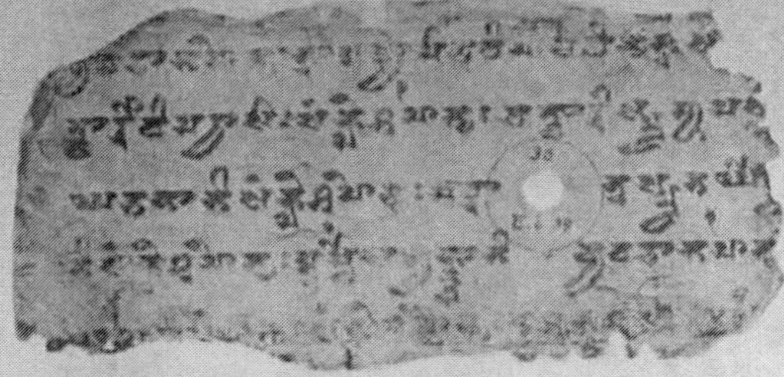

E. i. 4.

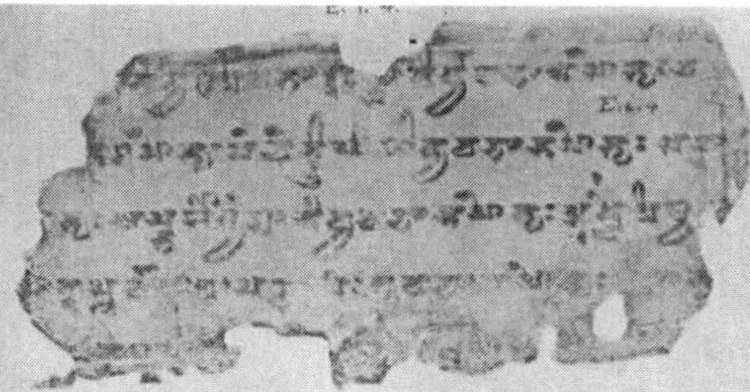

E. i. 39. REVERSE OF FOL. 30.

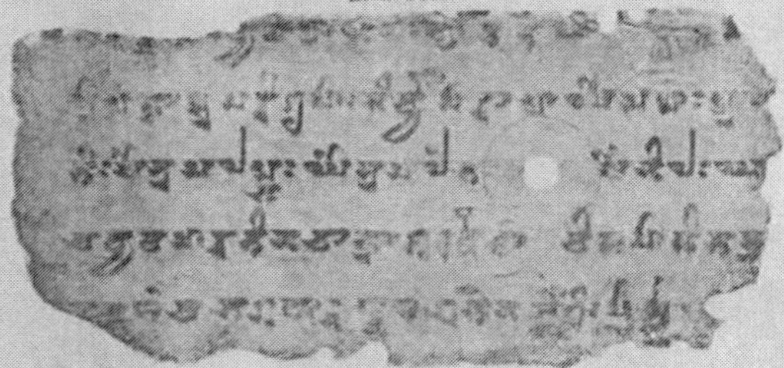

E. i. 4.

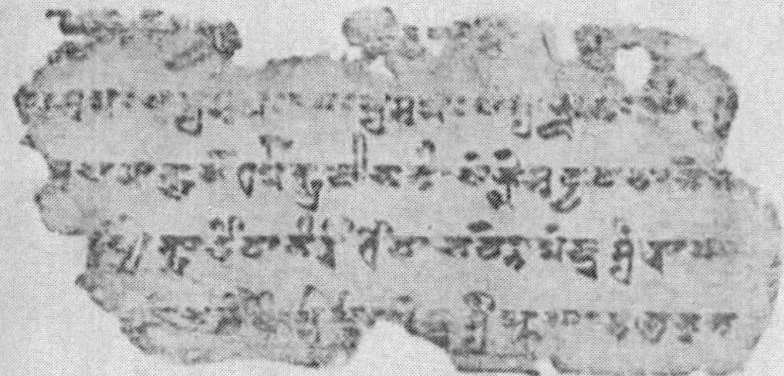

LEAF OF A BUDDHIST DHĀRAṆĪ, IN SANSKRIT, BROKEN IN TWO HALVES, FOUND IN ENDERE TEMPLE.

(See Chap. XII. sec. ii, iv.)

D. III. 9.

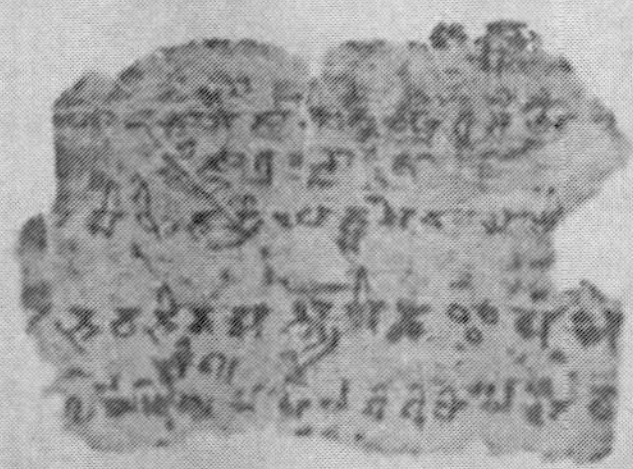

FRAGMENT OF LEAF, IN SANSKRIT, DANDĀN-UILIQ.

(See Chap. IX. sec. iv, viii.)

E. i. 6. OBVERSE

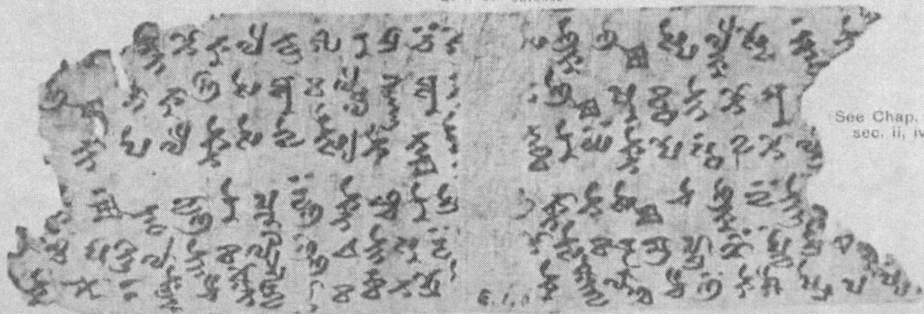

(See Chap. XII. sec. ii, iv.)

FRAGMENT OF LEAF, IN SLANTING CENTRAL-ASIAN BRĀHMĪ AND NON-SANSKRITIC LANGUAGE, FROM ENDERE.

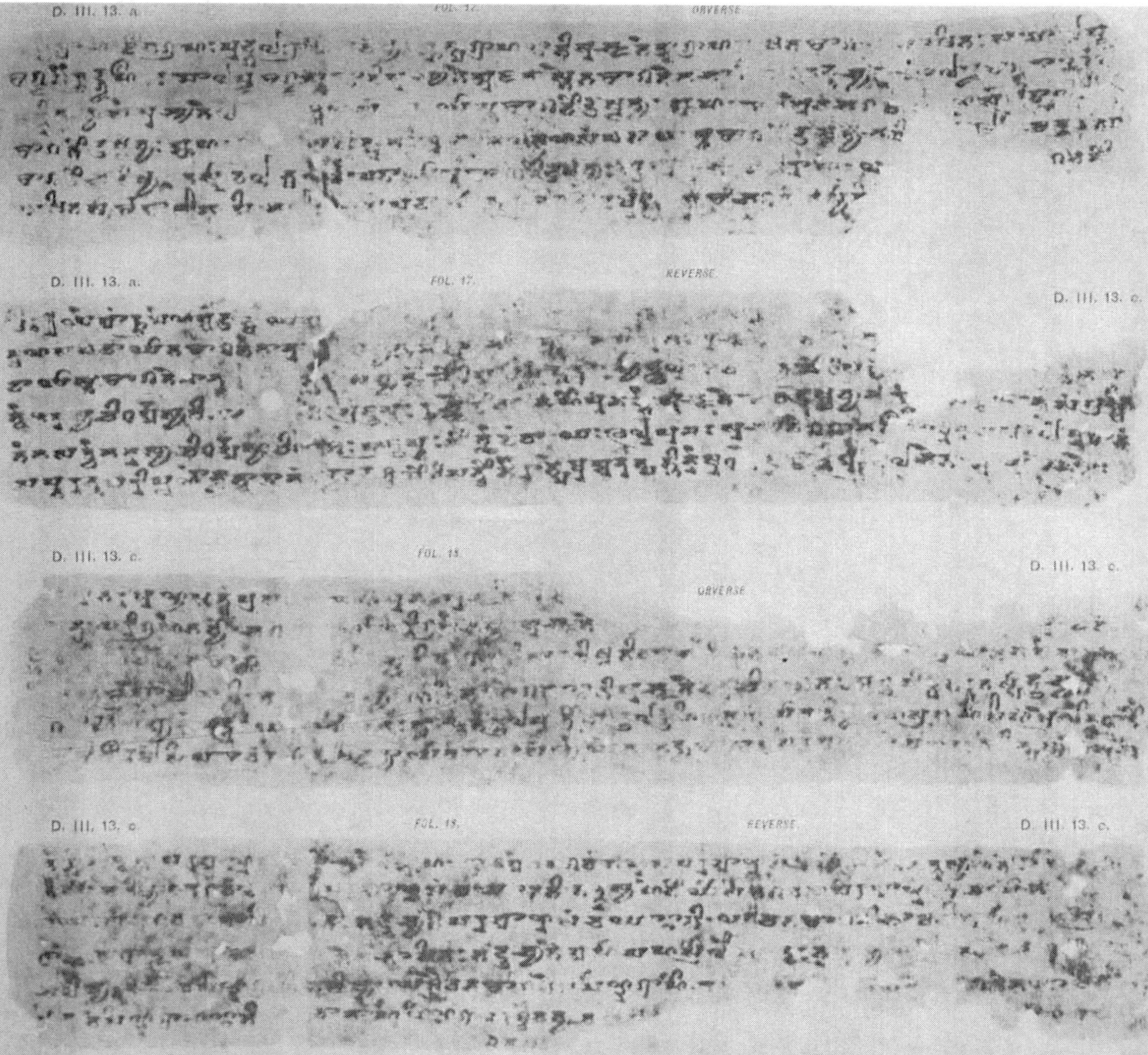

TWO LEAVES OF VAJRACCHEDIKĀ MS., IN SANSKRIT, FOUND IN RUIN D. III, DANDĀN-UILIQ.

See Chap. IX. sec. iv, viii; comp. text in 'Anecdota Oxoniensia,' p. 42, l. 12—p. 45, l. 7.

SCALE ½.

D. III. 7. Fol. 132. OBVERSE. D. III. 8.

LEAF OF SOME MAHĀYĀNA WORK IN SANSKRIT, FOUND IN RUIN D. III, DANDĀN-UILIQ.

(See Chap. IX. sec. iv. viii.)

D. III. 7. Fol. 132. REVERSE. D. III. 8.

SCALE 1/2

D. V. 5 (1/1). REVERSE OF STICK WITH CHINESE INSCRIPTION.

D. R. 01 + 02 (5/8). BRĀHMĪ DOCUMENT ON WOOD.

OBVERSE OF UNDER-TABLET.

OBVERSE OF COVERING-TABLET (FRAGM.)

D. XII. 7 (1/1). WOODEN BOARD OF POTHĪ WITH REMAINS OF PAPER.

D. V. 3 (3/10). OBVERSE OF BRĀHMĪ TABLET.

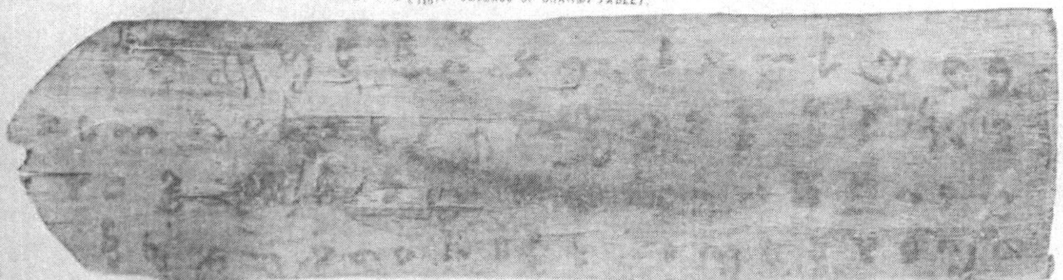

BRĀHMĪ AND CHINESE RECORDS ON WOOD, FROM DANDĀN-UILIQ AND RAWAK.

(See Chap. IX. sec. iv, v, viii; Chap. X. sec. i.)

CV
[DOC.]

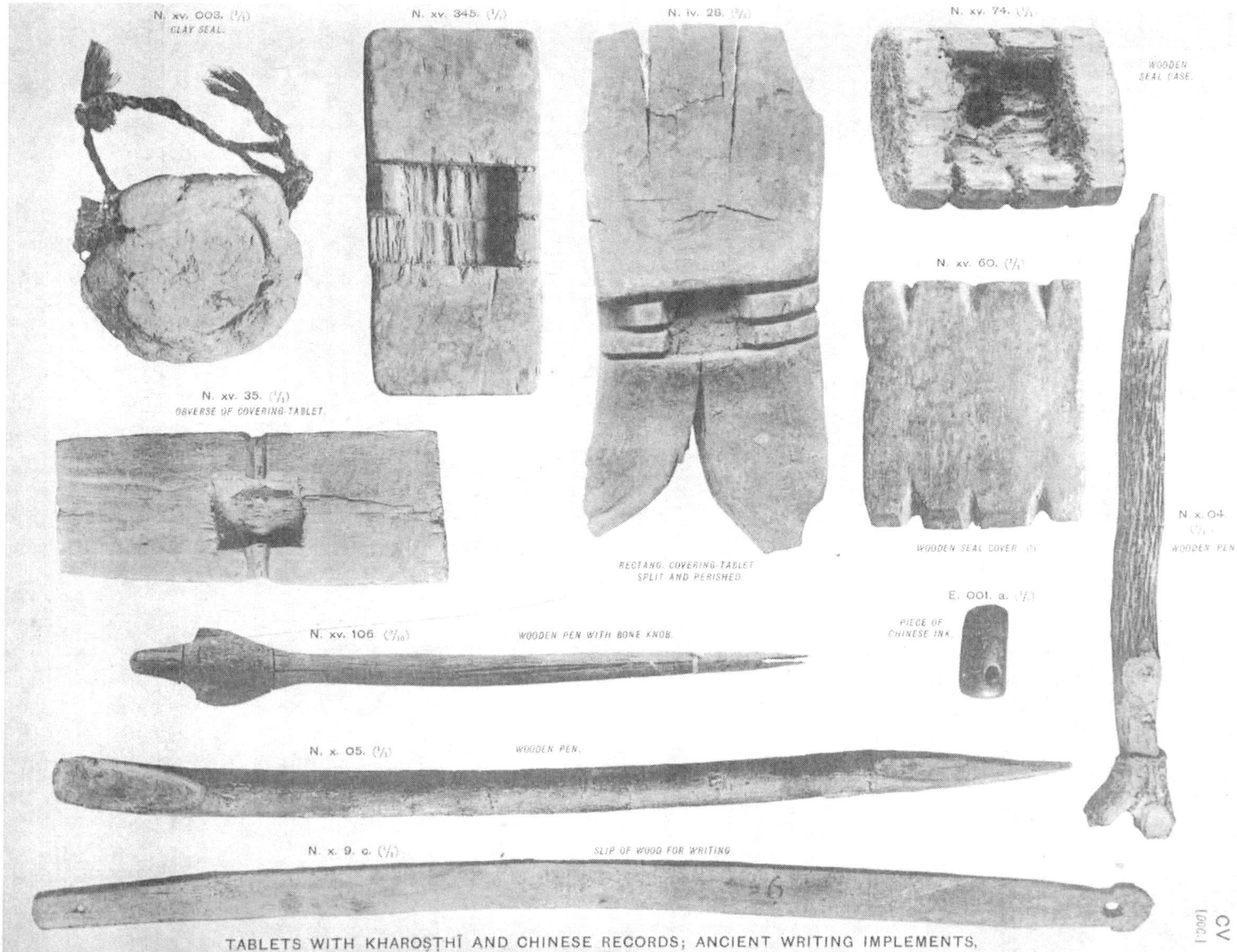

TABLETS WITH KHAROṢṬHĪ AND CHINESE RECORDS; ANCIENT WRITING IMPLEMENTS, MAINLY FROM NIYA SITE.

(See Chap. XI. sec. ii, iv-vi, viii.)

KHAROṢṬHĪ DOCUMENTS ON TABLETS, PARTLY DEFACED; SPECIMENS OF WOOD FROM ANCIENT FRUIT TREES, NIYA SITE.

(See Chap. XI. sec. i, ii, vii, viii.)

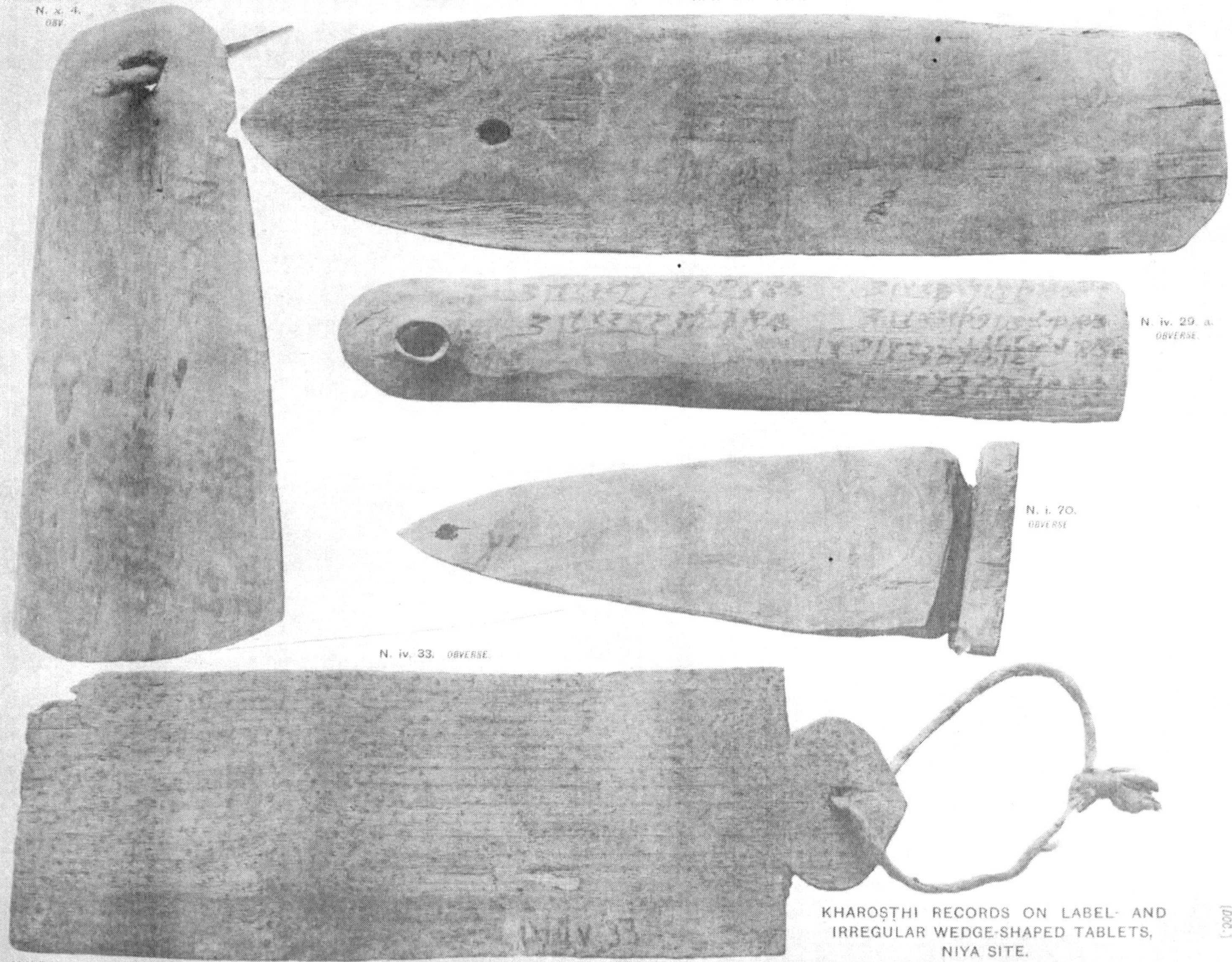

KHAROṢṬHI RECORDS ON LABEL- AND IRREGULAR WEDGE-SHAPED TABLETS, NIYA SITE.

(See Chap. XI. sec. i, ii, iv, viii.)

CII
[DOC.]

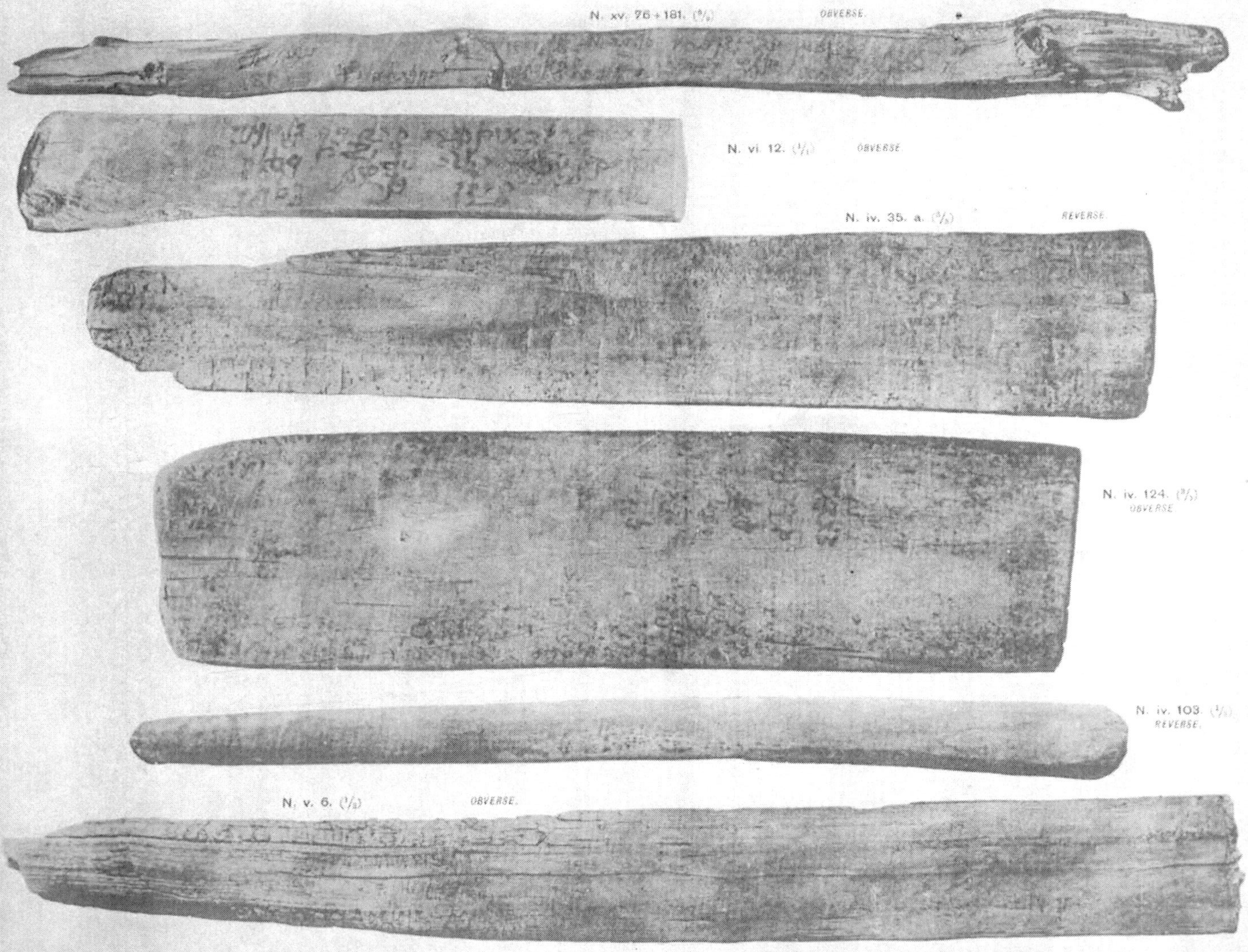

KHAROṢṬHĪ RECORDS ON OBLONG AND STICK-LIKE TABLETS, FROM NIYA SITE.

(See Chap. XI, sec. i, ii, iv, viii.)

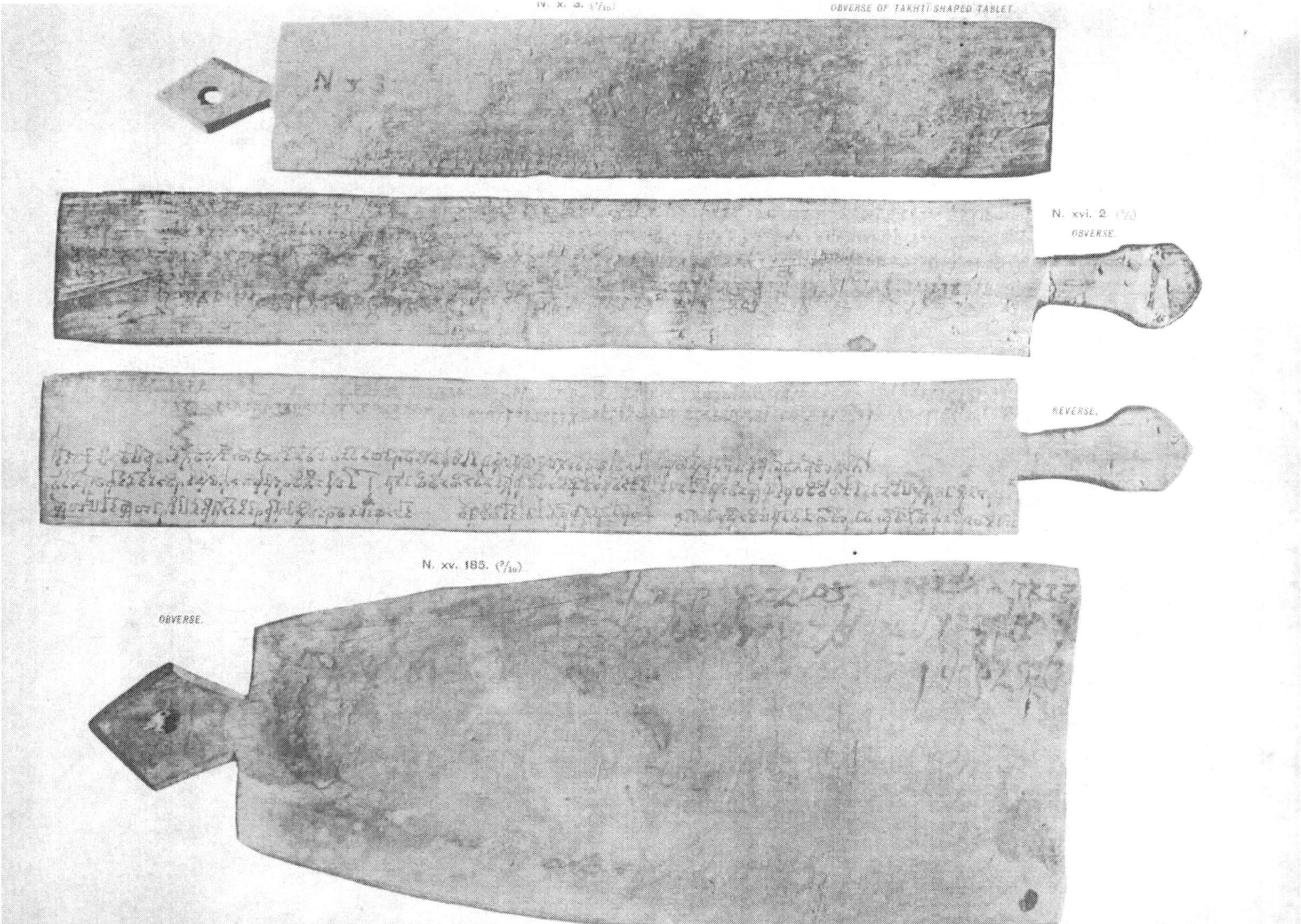

KHAROṢṬHI DOCUMENTS ON TAKHTĪ-SHAPED TABLETS, FROM NIYA SITE.

(See Chap. XI. sec. ii, iv, vi, viii.)

C
[DOC.]

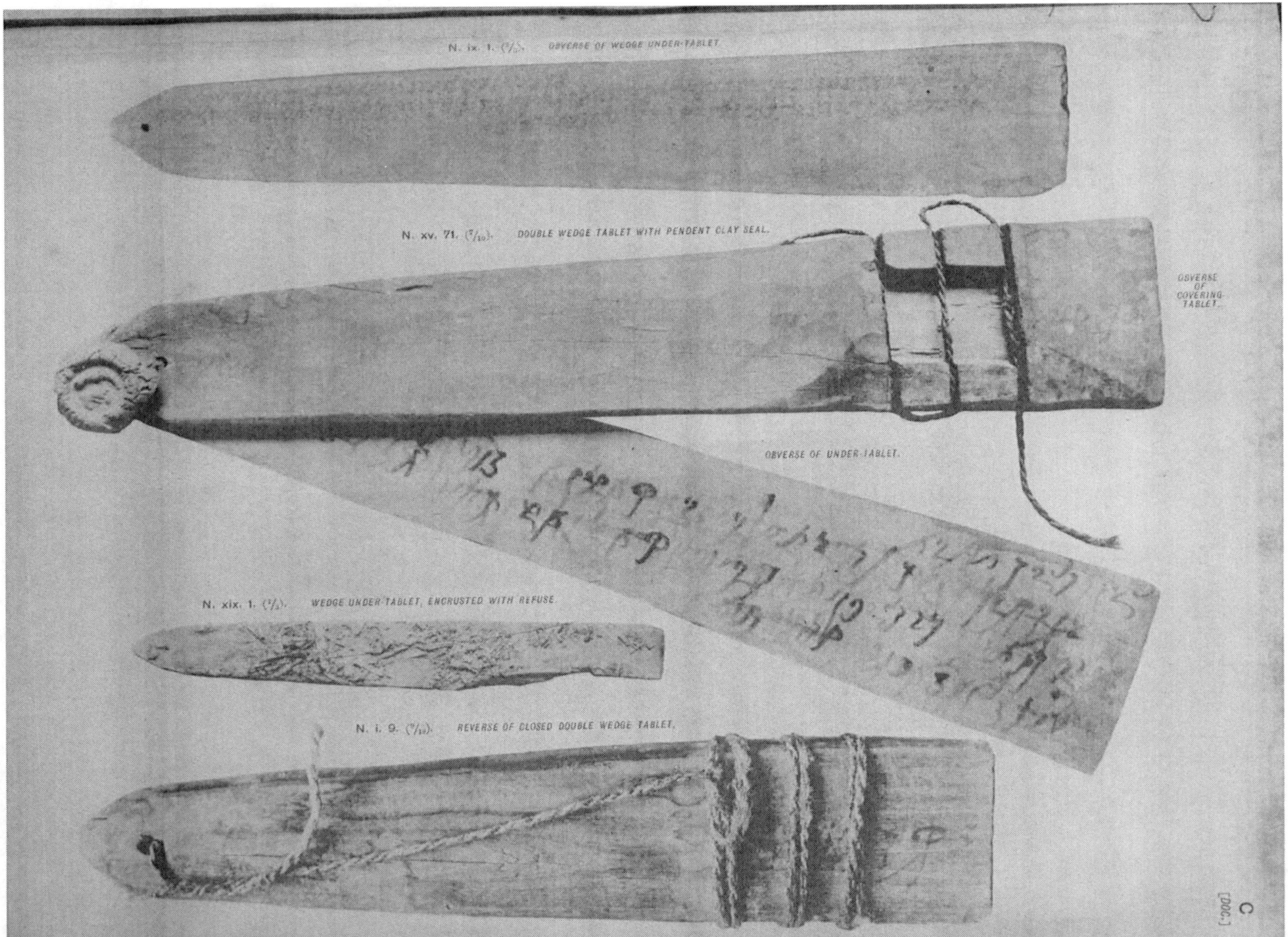

N. ix. 1. (1/2). OBVERSE OF WEDGE UNDER-TABLET.

N. xv. 71. (7/10). DOUBLE WEDGE TABLET WITH PENDENT CLAY SEAL.

OBVERSE OF COVERING-TABLET.

OBVERSE OF UNDER-TABLET.

N. xix. 1. (1/2). WEDGE UNDER-TABLET, ENCRUSTED WITH REFUSE.

N. i. 9. (9/10). REVERSE OF CLOSED DOUBLE WEDGE TABLET.

KHAROṢṬHĪ DOCUMENTS ON WEDGE TABLETS, FROM NIYA SITE.

(See Chap. XI. sec. i, ii, iv, vii, viii.)

KHAROṢṬHĪ DOCUMENTS ON DOUBLE WEDGE TABLETS, FROM NIYA SITE.

(See Chap. XI. sec. i, iv, viii.)

SCALE ½

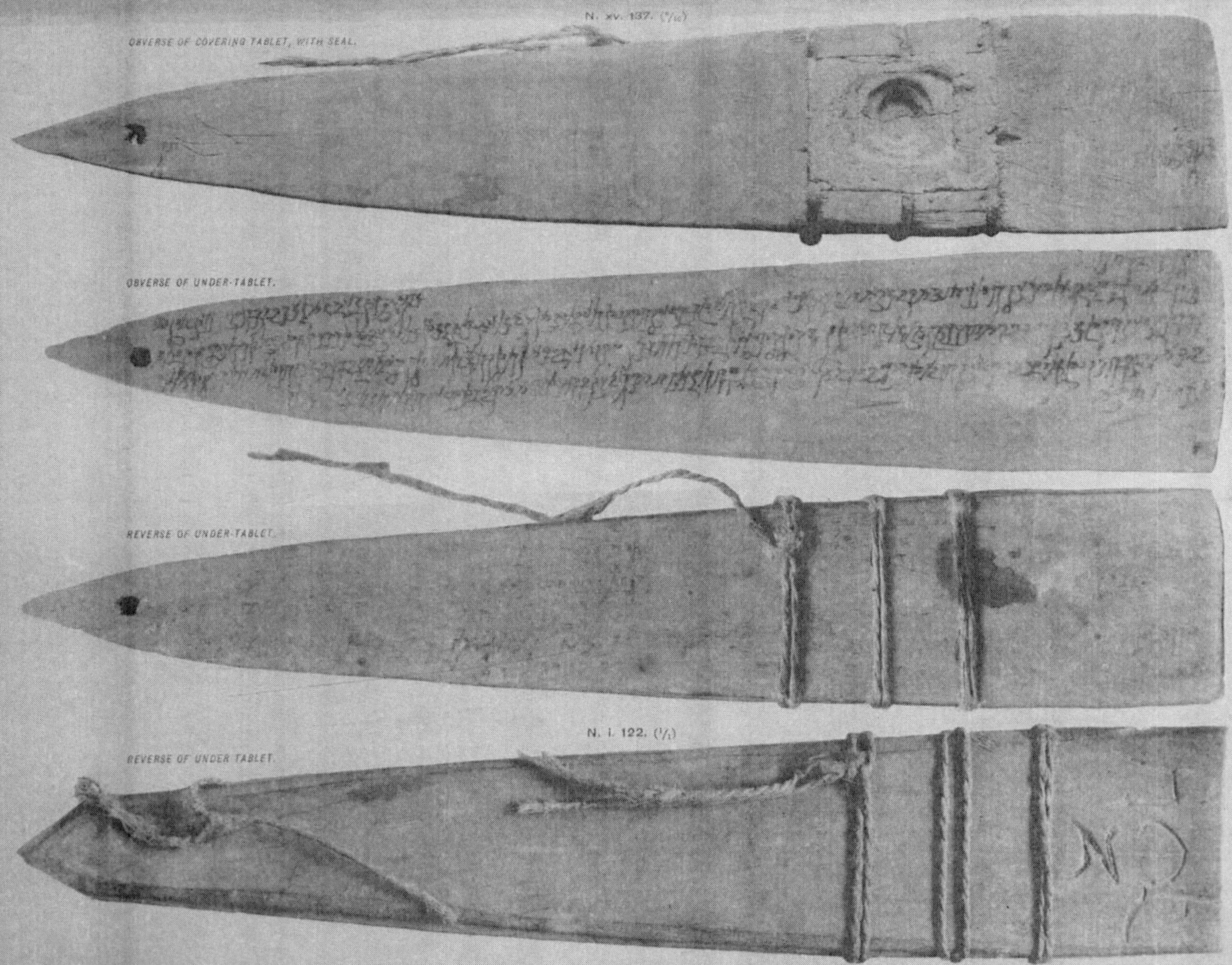

KHAROṢṬHI DOCUMENTS ON DOUBLE WEDGE TABLETS, FROM NIYA SITE.

(See Chap. XI. sec. i, iv, viii.)

N. xv. 122. REVERSE.

N. xv. 154. OBVERSE OF COVERING TABLET.

REVERSE OF COVERING TABLET.

KHAROṢṬHĪ DOCUMENTS ON OBLONG AND RECTANGULAR TABLETS, FROM NIYA SITE.

(See Chap. XI. sec. ii, iv, viii.)

SCALE 1/1

N. xv. 160. *REVERSE OF COVERING TABLET.*

N. iv. 139. *OBVERSE OF UNDER-TABLET.*

KHAROṢṬHĪ DOCUMENTS ON RECTANGULAR TABLETS, NIYA SITE.

(See Chap. XI. sec. i, iv, viii.)

SCALE ⅓.

N. xv. 166.

DOUBLE TABLET UNOPENED.

REVERSE OF COVERING-TABLET.

OBVERSE OF UNDER-TABLET.

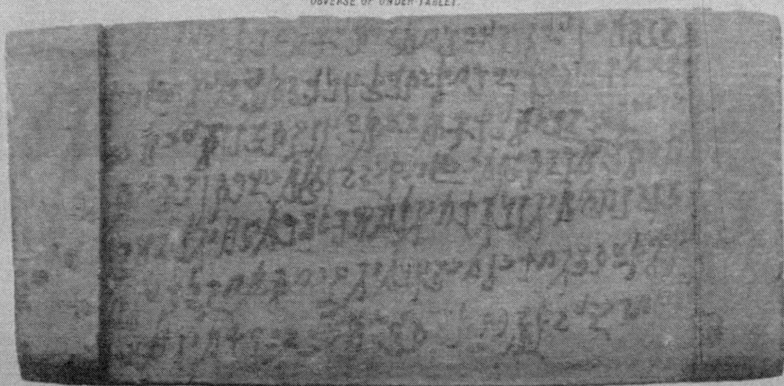

KHAROṢṬHĪ DOCUMENT ON RECTANGULAR DOUBLE TABLET, FROM N. xv, NIYA SITE.

(See Chap. XI. sec. iv, viii.)

N. xv. 334.

OBVERSE.

REVERSE.

RECTANGULAR COVERING TABLET OF KHAROṢṬHĪ DOCUMENT, NIYA SITE.

SCALE ¹/₁

XCIV

[DOC.]

N. xv. 155.

DOUBLE TABLET UNOPENED.

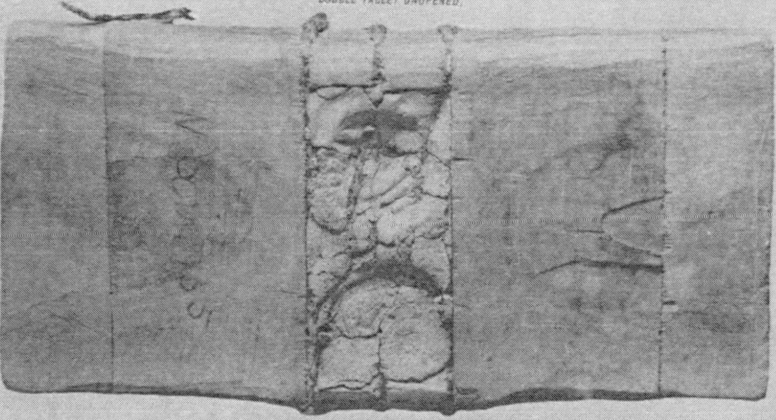

REVERSE OF COVERING-TABLET.

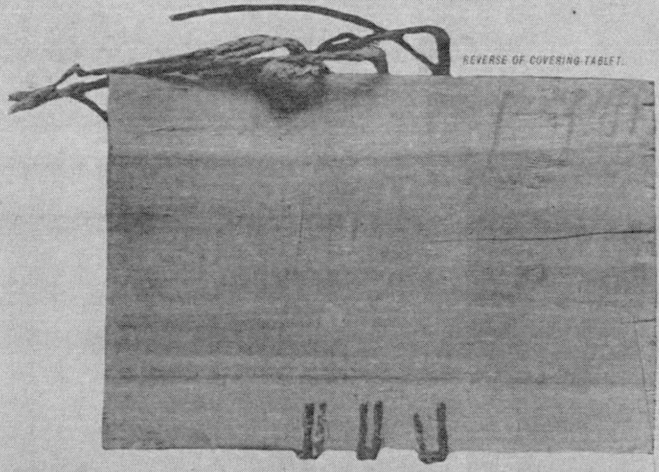

OBVERSE OF UNDER-TABLET.

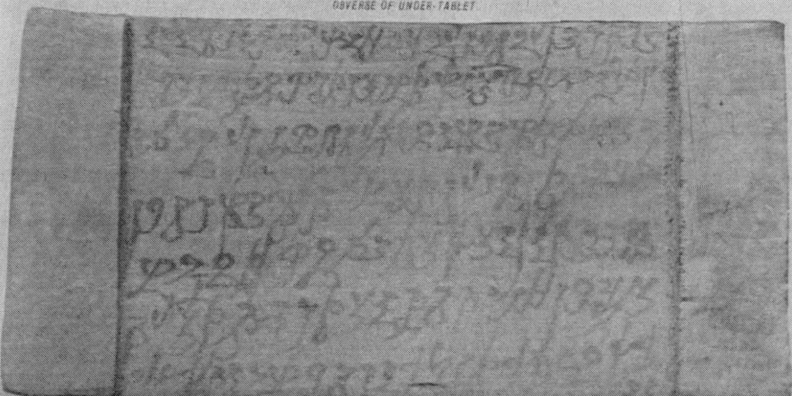

N. xv. 196.

REVERSE OF RECTANG. DOUBLE TABLET, UNOPENED, CONTAINING KHAROṢṬHĪ DOCUMENT.

(See Chap. XI. sec. iv, viii.)

KHAROṢṬHĪ DOCUMENT ON RECTANGULAR DOUBLE TABLET, FROM N. xv, NIYA SITE.

(See Chap. XI. sec. iv, viii.)

SCALE 1/2

N. iv. 133.

KHAROṢṬHĪ DOCUMENT ON WOODEN TABLET, NIYA SITE.
(See Chap. XI. sec. ii, viii.)

N. xv. 305.

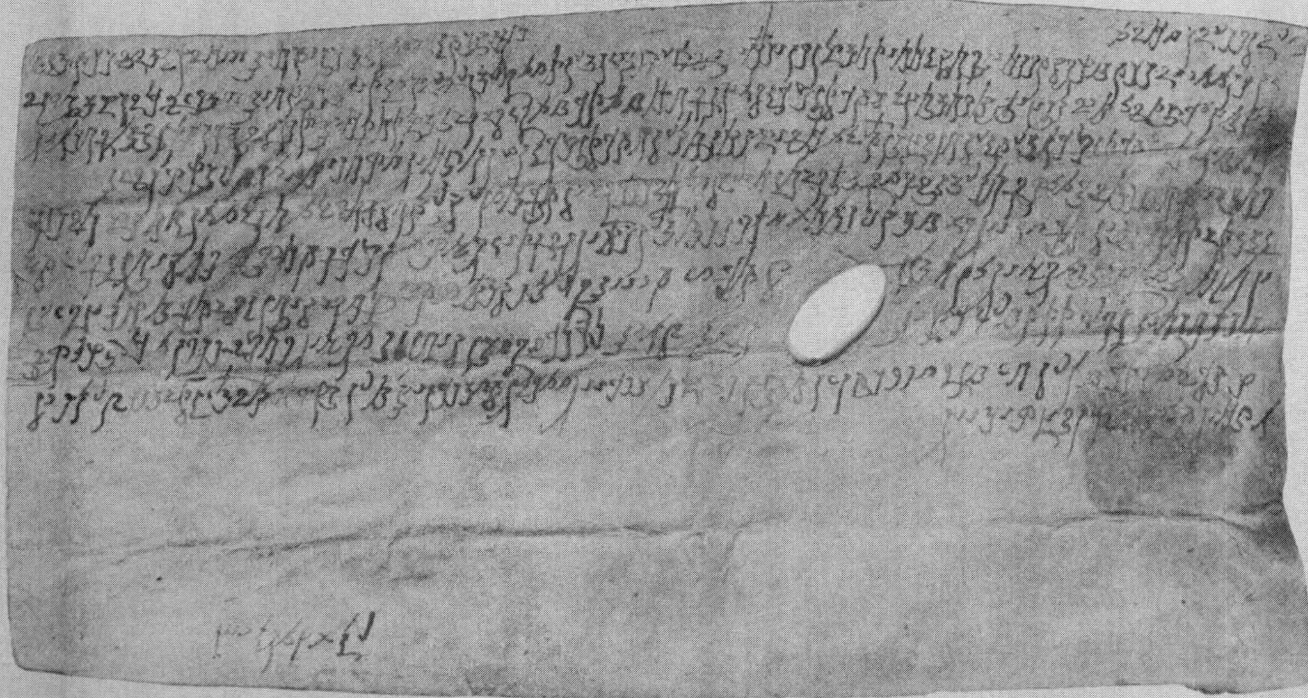

KHAROṢṬHĪ DOCUMENT ON LEATHER, FROM RUBBISH-HEAP N. xv, NIYA SITE.
(See Chap. XI. sec. iv, viii.)

SCALE 1/1

N. xv. 168.

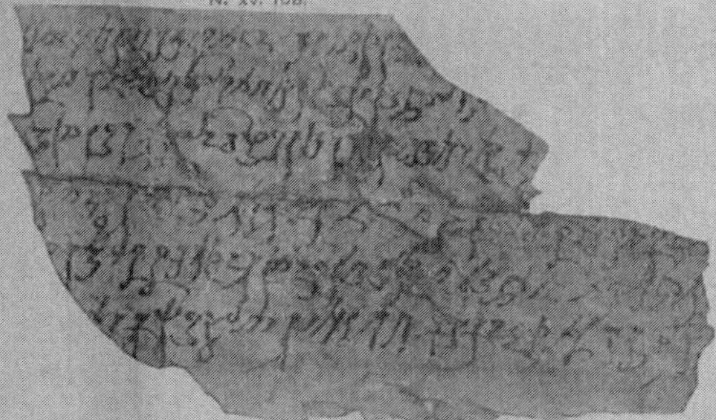

N. xv. 333.

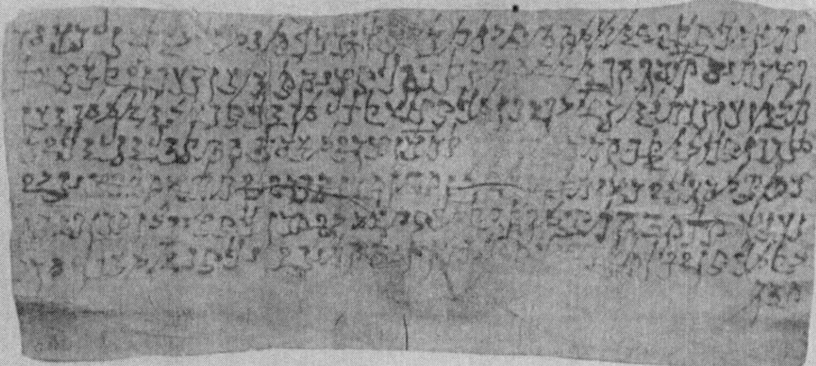

N. xv. 88.

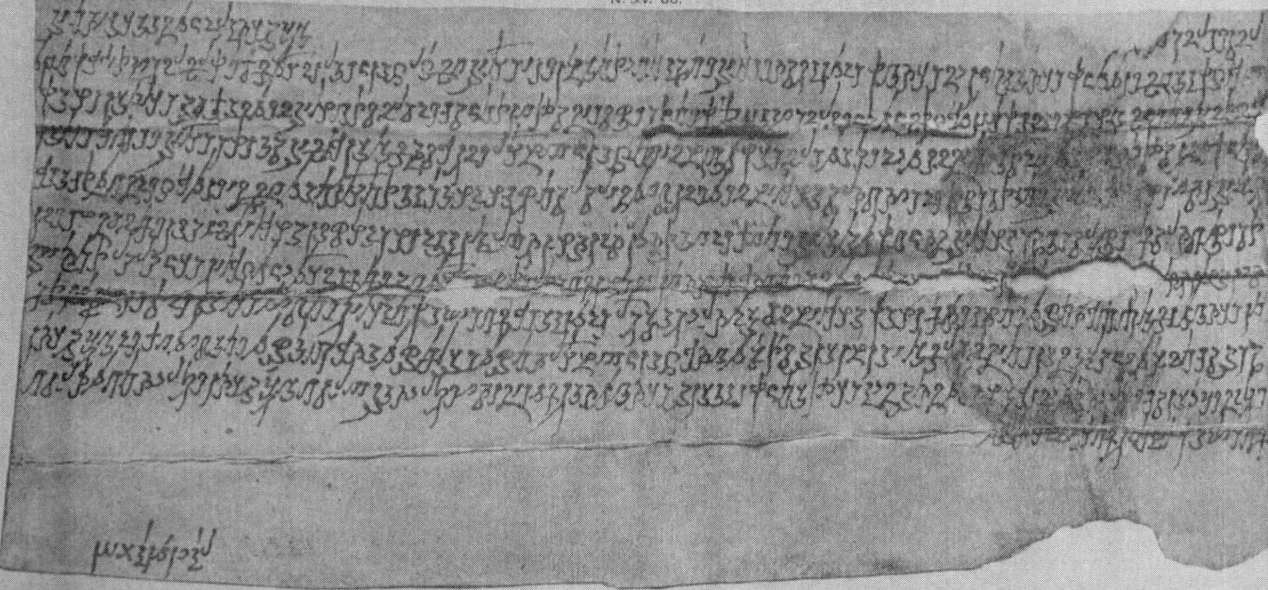

KHAROṢṬHĪ DOCUMENTS ON LEATHER, FROM RUBBISH-HEAP N. xv, NIYA SITE.

(See Chap. XI. sec. iv, viii.)

SCALE ¹/₁

XCII
[ANT.]

N. xv. 310. UNFOLDED.

N. xv. 101

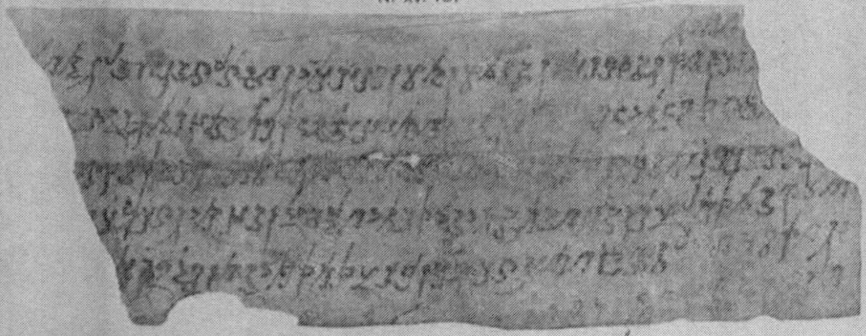

N. xv. 310.
IN ORIGINAL FOLDED STATE.

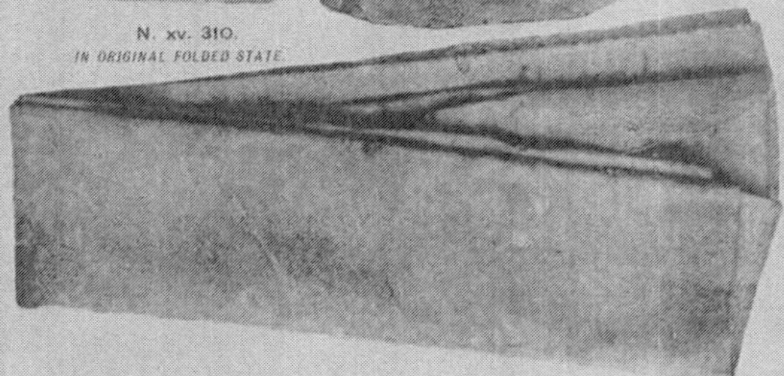

KHAROṢṬHĪ DOCUMENTS ON LEATHER, FROM RUBBISH-HEAP N. xv, NIYA SITE.

(See Chap. XI. sec. iv, viii.)

SCALE 1/1

MEDIAEVAL CHINESE AND MUHAMMADAN COINS FROM DANDAN-UILIQ, YŌTKAN, AND OTHER SITES.

(See Appendix D, with table.)

LXXXIX
[COINS]

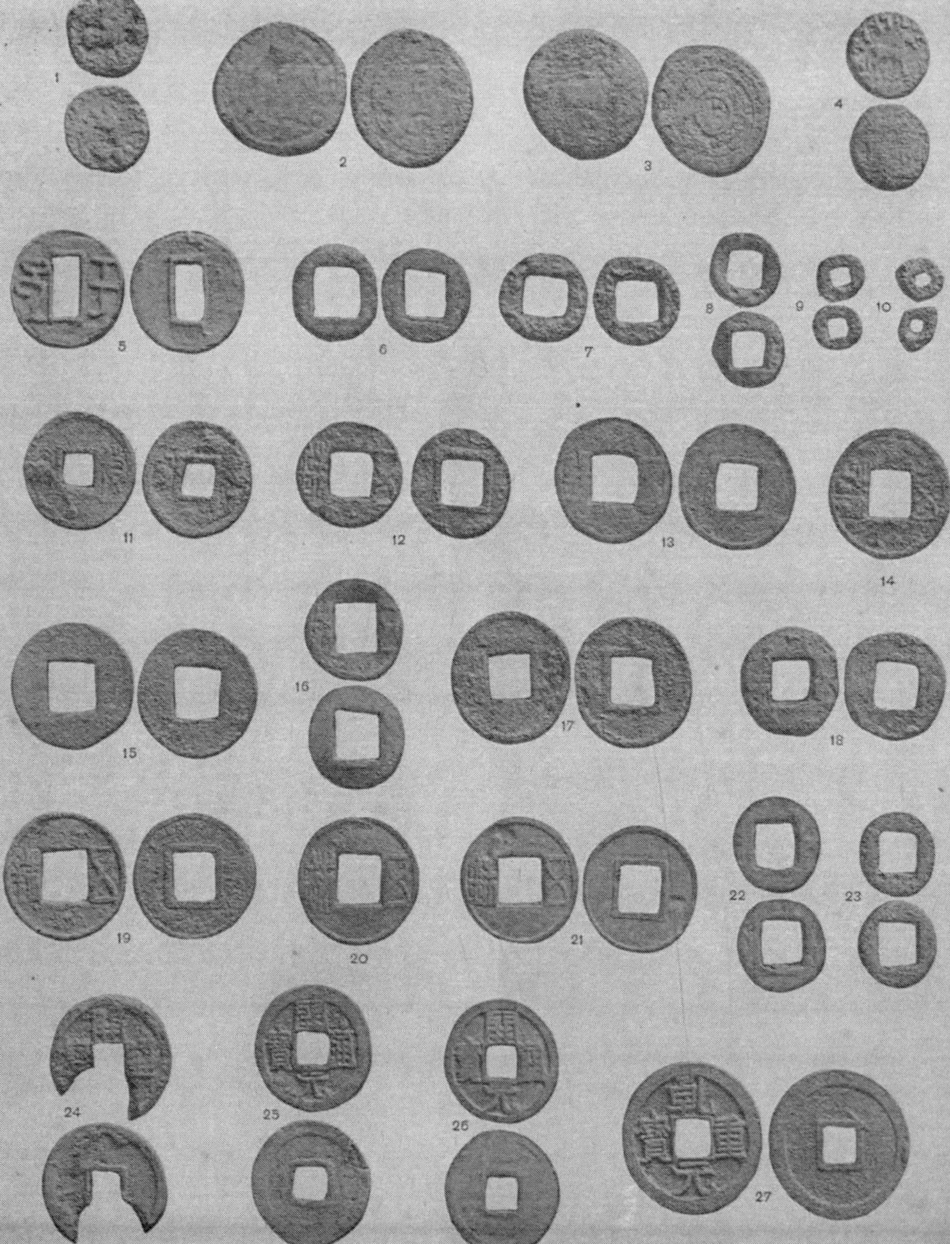

SINO-KHAROṢṬHĪ, INDO-SCYTHIAN, AND EARLY CHINESE COINS FROM YŌTKAN, NIYA, RAWAK, AND OTHER SITES.

(See Appendix D, with table.)

LXXXVIII

[ANT.]

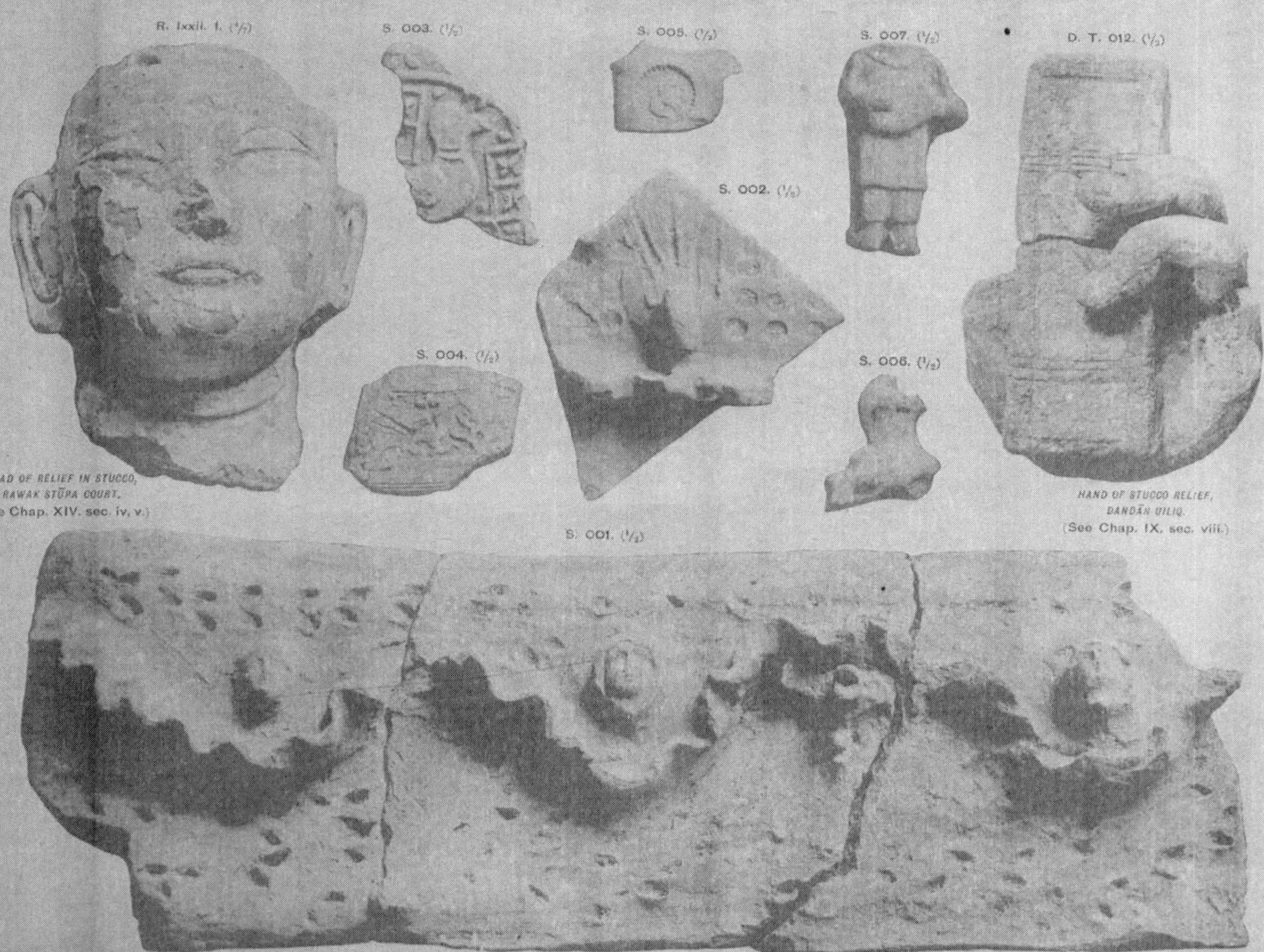

R. lxxii. 1. (1/2)

HEAD OF RELIEF IN STUCCO, RAWAK STŪPA COURT.

(See Chap. XIV. sec. iv, v.)

S. 003. (1/2)

S. 004. (1/2)

S. 005. (1/2)

S. 002. (1/2)

S. 007. (1/2)

S. 006. (1/2)

D. T. 012. (1/2)

HAND OF STUCCO RELIEF, DANDĀN UILIQ.

(See Chap. IX. sec. viii.)

S. 001. (1/2)

FRAGMENTS OF DECORATIVE POTTERY, FROM AFRĀSIYĀB SITE, SAMARKAND.

(See Chap. XV. sec. iii.)

PLAQUES IN STUCCO AND OTHER RELIEF FRAGMENTS FROM DECORATIVE HALOS OF STATUES, RAWAK STŪPA COURT.

(See Chap. XIV. sec. iv, v.)

R. xii. 1.

R. xxx. i.

RELIEF FIGURES OF BUDDHA, IN STUCCO, FROM DECORATIVE HALOS OF COLOSSAL STATUES, RAWAK STŪPA COURT.

(See Chap. XIV. sec. iv, v.)

SCALE 1/[illegible]

R. xxvii. 1. (4/7)

PROFILE

FRONT.

RELIEF HEAD AND BUST IN STUCCO (RESTORED), FROM RAWAK STŪPA COURT.

(See Chap. XIV. sec. iv, v.)

LXXXV

[ANT.]

LXXXIV
[ANT.]

R. lxx. 1. a. (4/5)

R. lxxiv. 1. (5/8)

R. lxx. 1. c. (4/5)

HEADS OF COLOURED RELIEFS IN STUCCO,
FROM RAWAK STŪPA COURT.

(See Chap. XIV. sec. iv, v.)

R. vi. 1. (¼)

HEAD OF LIFE-SIZE
STATUE AND DECORATIVE
RELIEF FRAGMENT IN STUCCO.
RAWAK STŪPA COURT.
(See Chap. XIV. sec. iv, v.

LXXXIII
[ANT.]

R. ii. 1. (3/4)

HEAD OF LIFE-SIZE RELIEF IN STUCCO FROM RAWAK STŪPA COURT.

(See Chap. XIV. sec. iv, v.)

R. 2. (2/7)

STUCCO RELIEF FROM WALL DECORATION OF RAWAK STŪPA COURT.

(See Chap. XIV. sec. iv, v.)

LXXXII

[ANT.]

R. iv. 1. (1/1)

HEAD OF COLOSSAL RELIEF IN STUCCO, FROM RAWAK STŪPA COURT

(See Chap. XIV. sec. iv, v.)

LXXX
[ANT.]

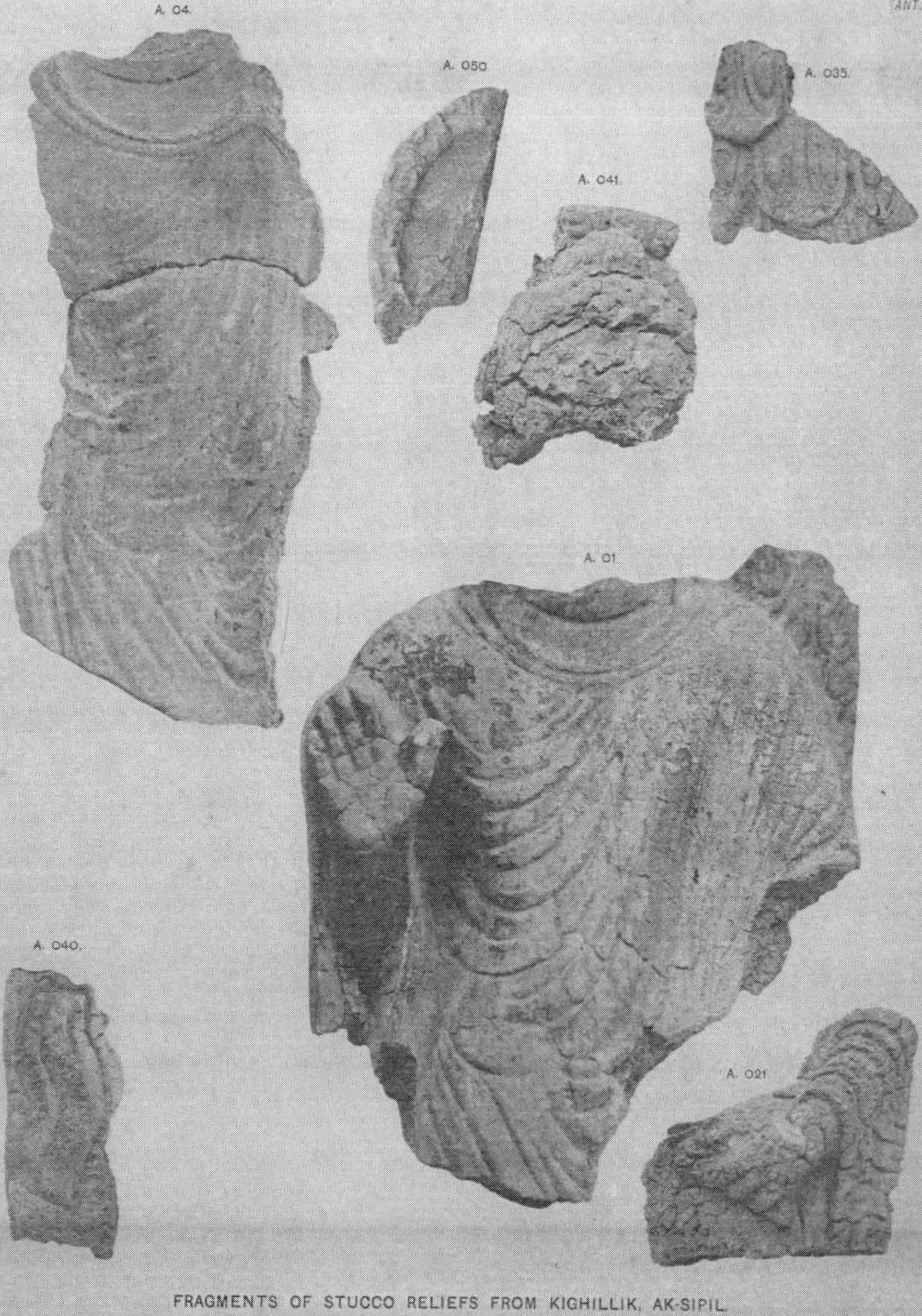

FRAGMENTS OF STUCCO RELIEFS FROM KIGHILLIK, AK-SIPIL.

(See Chap. XIV. sec. ii.)

SCALE 1/3

E. i. 25. a.

E. i. 25. b.

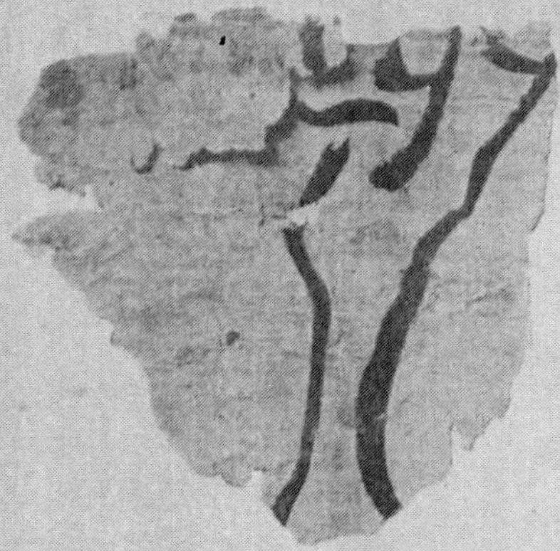

E. i. O12.

E. i. 19. a.

FRESCO FRAGMENT ON STUCCO AND INK SKETCHES ON PAPER, FROM ENDERE TEMPLE.

(See Chap. XII. sec. ii, iv.)

SCALE 1/1

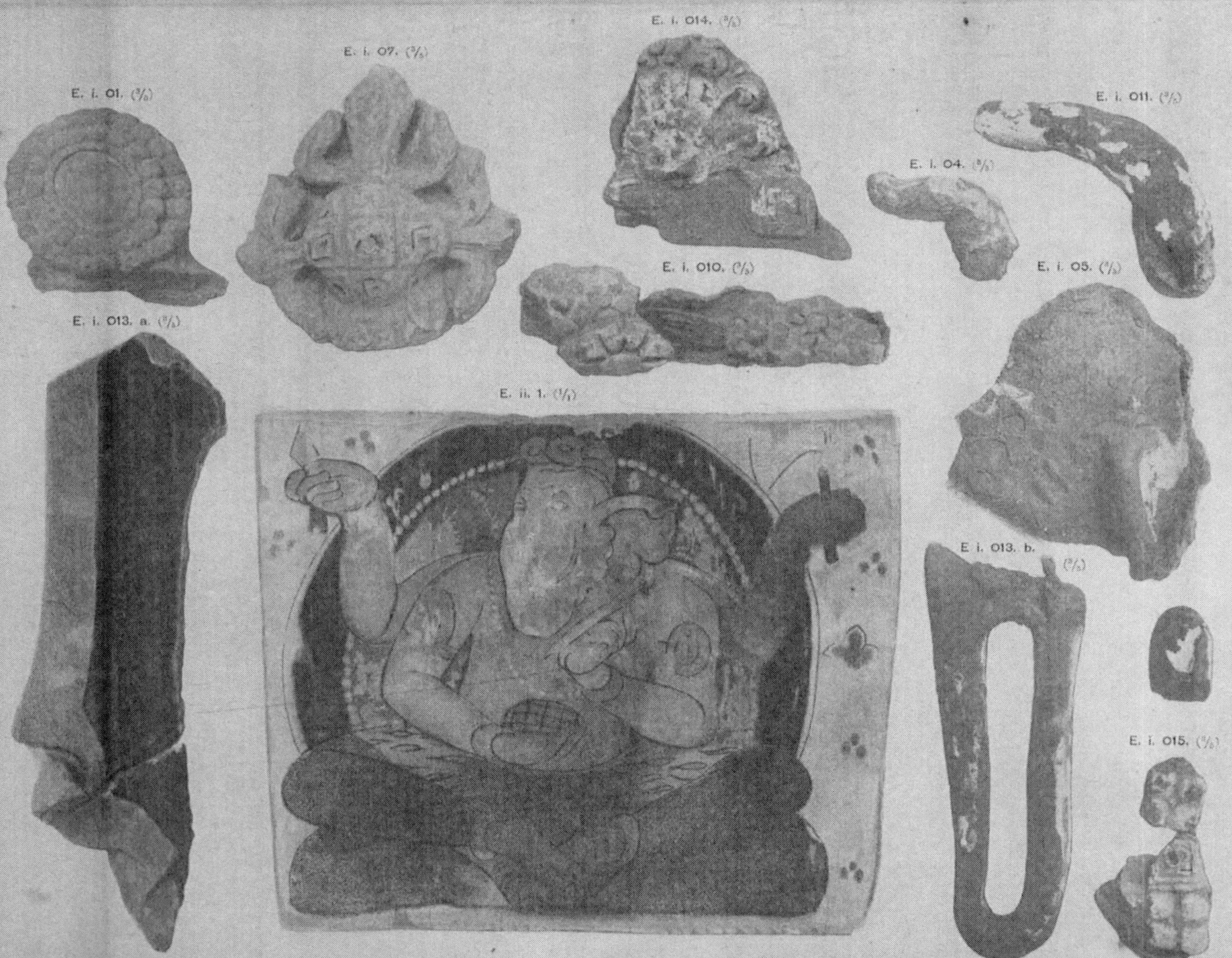

LXXVIII
[ANT.]

PAINTED TABLET AND FRAGMENTS OF STUCCO RELIEFS FROM ENDERE RUINS.

(See Chap. XII. sec. ii–iv.)

SAMPLES OF ANCIENT FABRICS IN SILK AND COTTON, FROM ENDERE TEMPLE.

(See Chap. XII. sec. ii, iv.)

SCALE 1/1

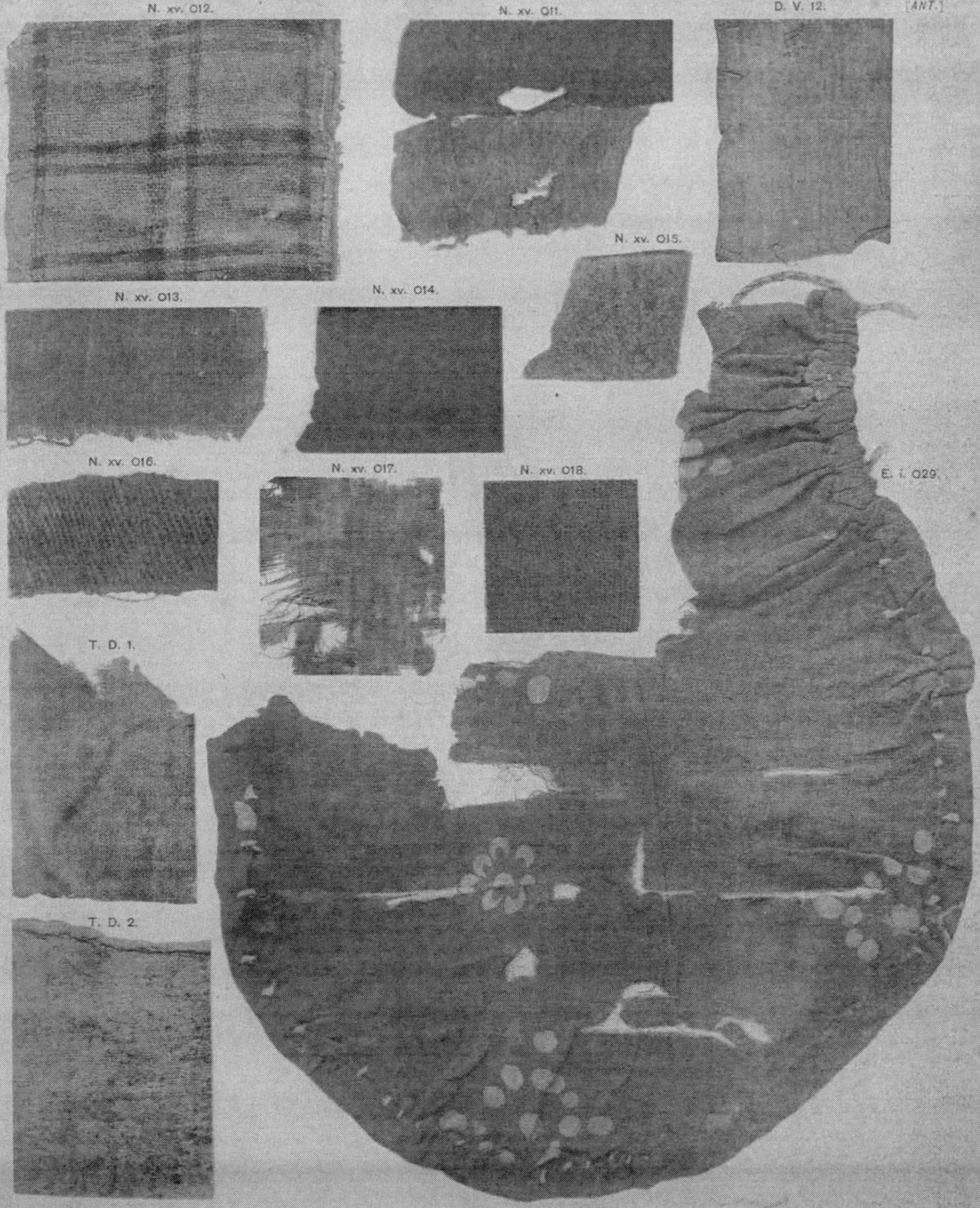

SAMPLES OF ANCIENT FABRICS IN SILK, COTTON, FELT, FROM NIYA SITE, ENDERE, DANDĀN-UILIQ, TŪGA-DONG.

(See Chap. IX. sec. v, viii; Chap. XI. sec. vi, viii; Chap. XII. sec. ii, iv; Chap. XIII. sec. iv.)

SCALE 1/1

LXXV
[ANT.]

N. vii. 3. (1/2)

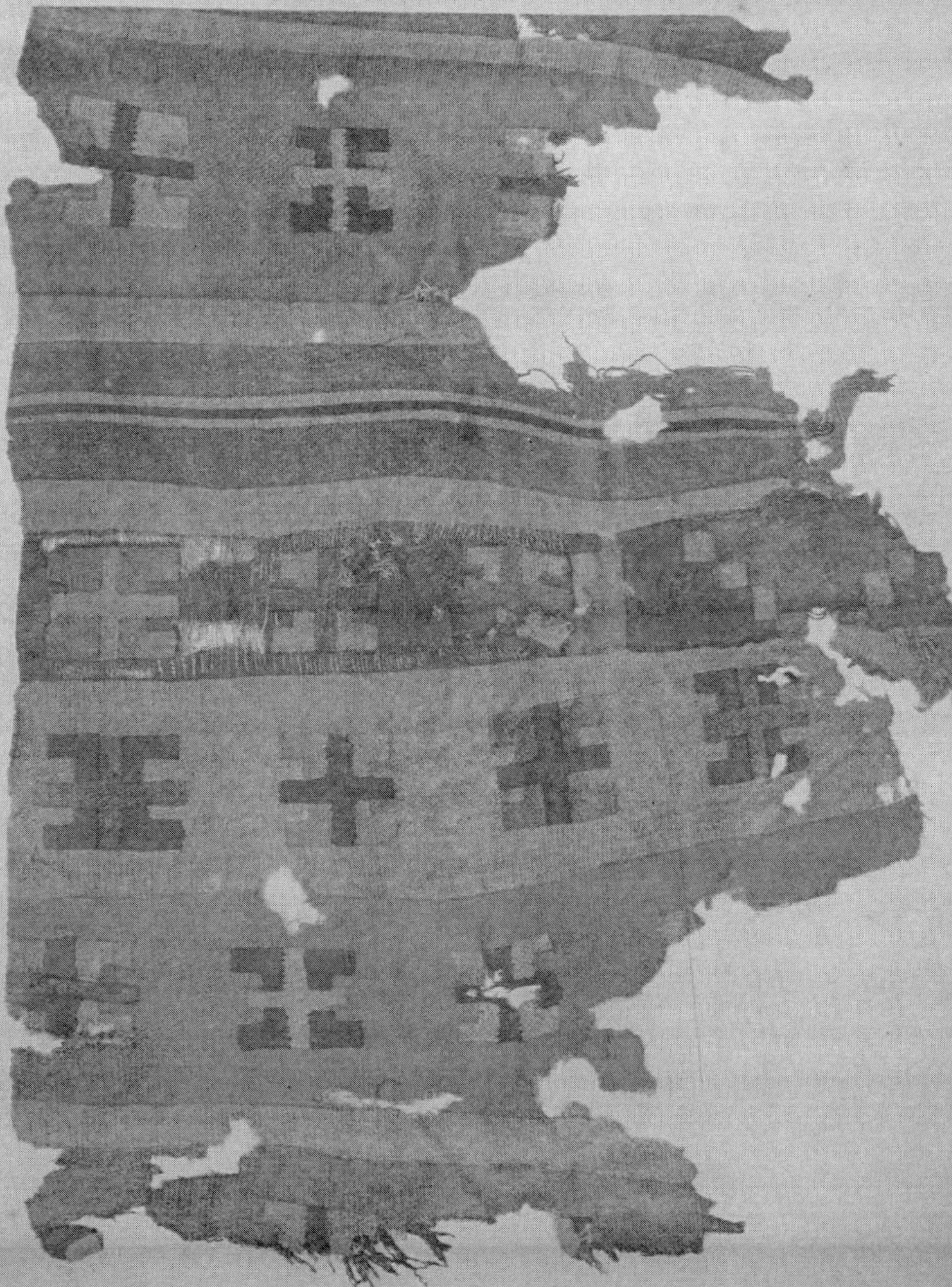

PORTION OF COLOURED RUG FROM DWELLING N. III. NIYA SITE.

(See Chap. XI. sec. ii, viii.)

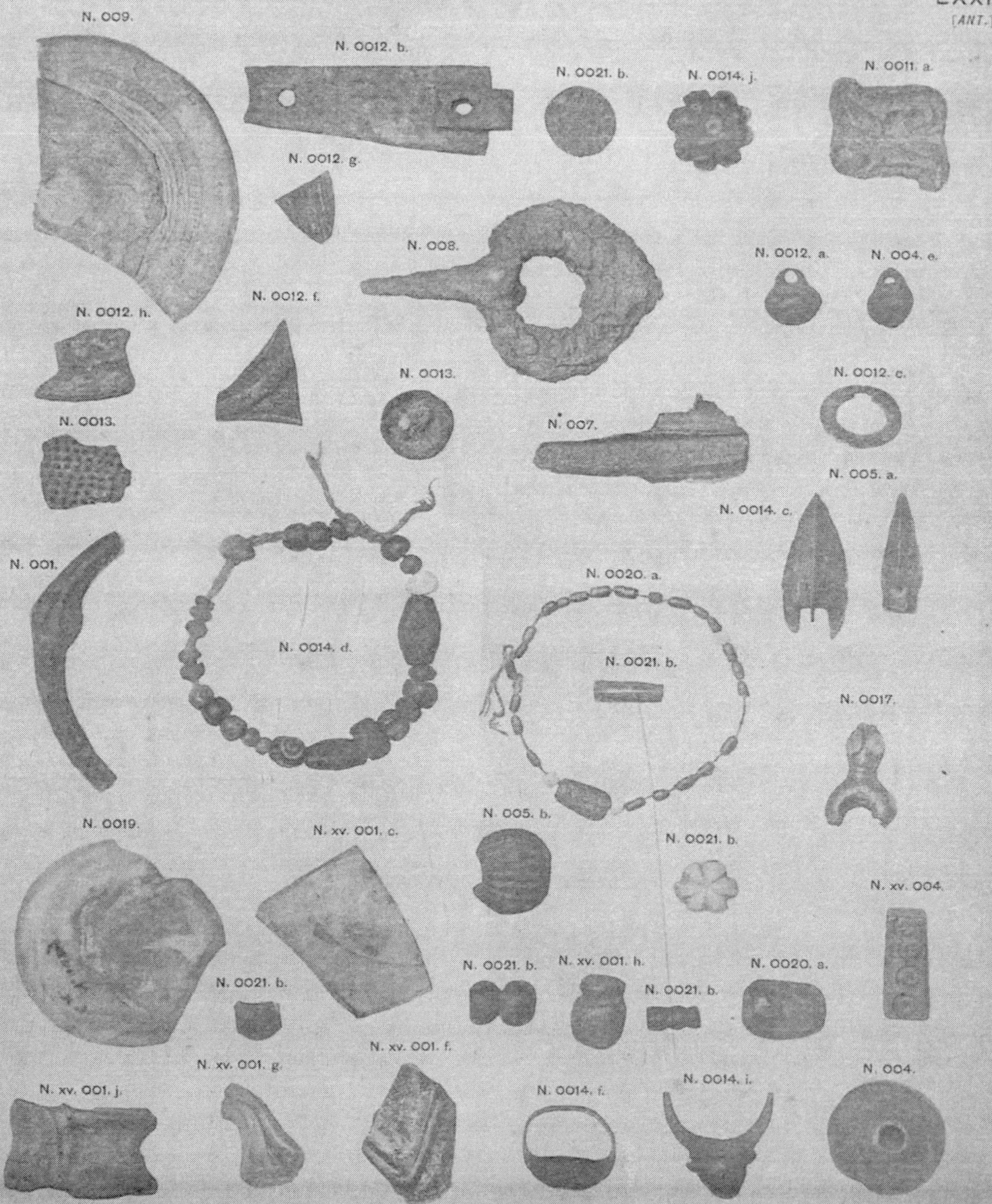

MISCELLANEOUS SMALL OBJECTS AND FRAGMENTS IN GLASS, METAL, BONE, &c., FROM NIYA SITE.

(See Chap. XI. sec. vi, vii, viii.)

SCALE 1/1

LXXIII
[ANT.]

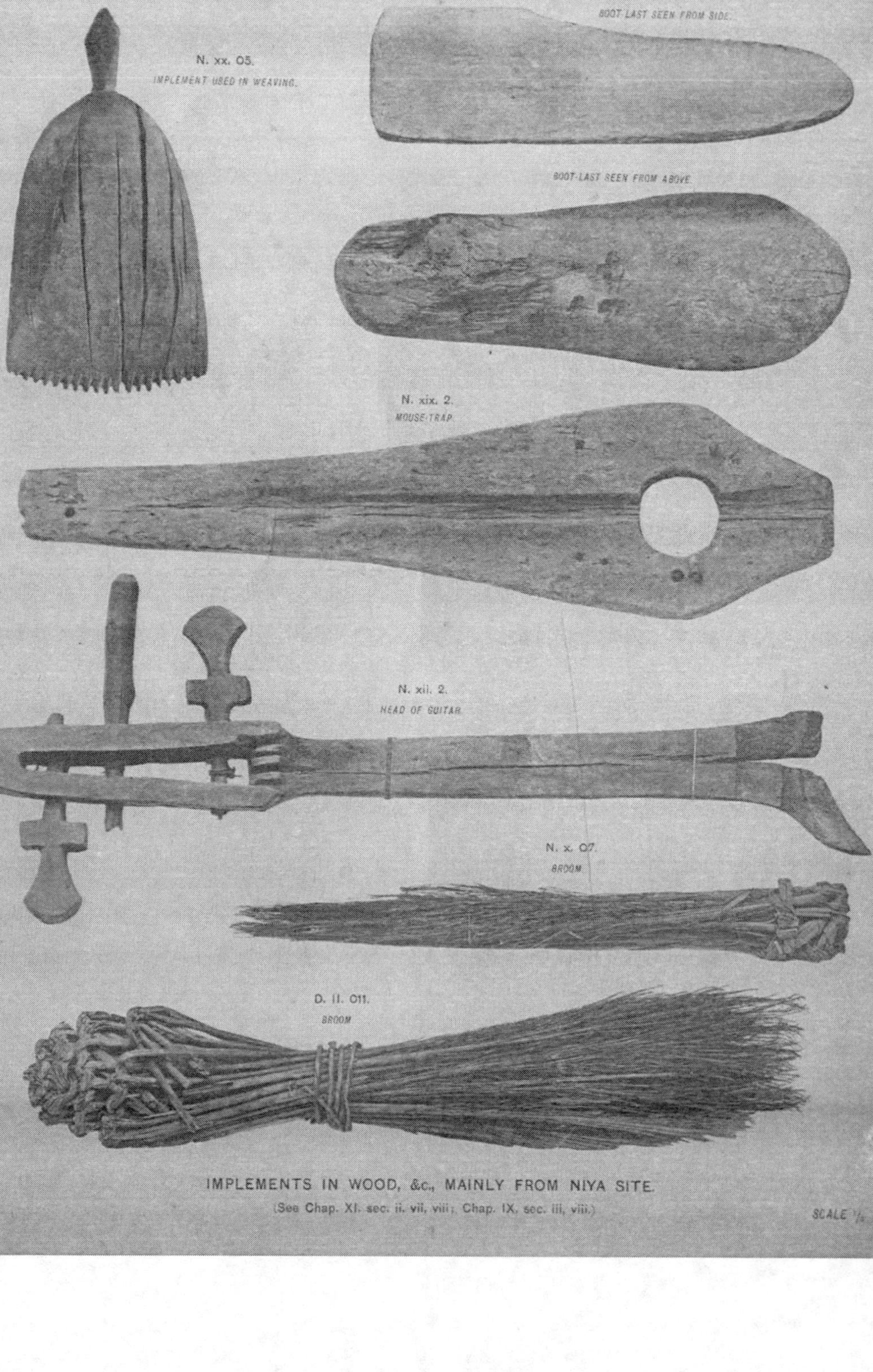

IMPLEMENTS IN WOOD, &c., MAINLY FROM NIYA SITE.

(See Chap. XI. sec. ii. vii, viii; Chap. IX. sec. iii, viii.)

SCALE 1/3

N. xv. 155 (enlarged).

N. xv. 163.

N. xv. 38.

N. xv. 133. a.

N. xv. 71.

N. xv. 196.

N. xv. 334.

N. xv. 133. b

N. xxiii. 1.

N. xv. 167 (enlarged).

CLAY SEALS OF KHAROṢṬHI DOCUMENTS ON WOOD, NIYA SITE.

(See Chap. XI. sec. iv, vii, viii.)

SCALE ¹/₁

LXXI
[ANT.]

CLAY SEALS OF KHAROṢṬHĪ DOCUMENTS ON WOOD, NIYA SITE.

Mostly enlarged 40 60 per cent. (See Chap. XI, sec. iv, viii.)

LXX
[ANT.]

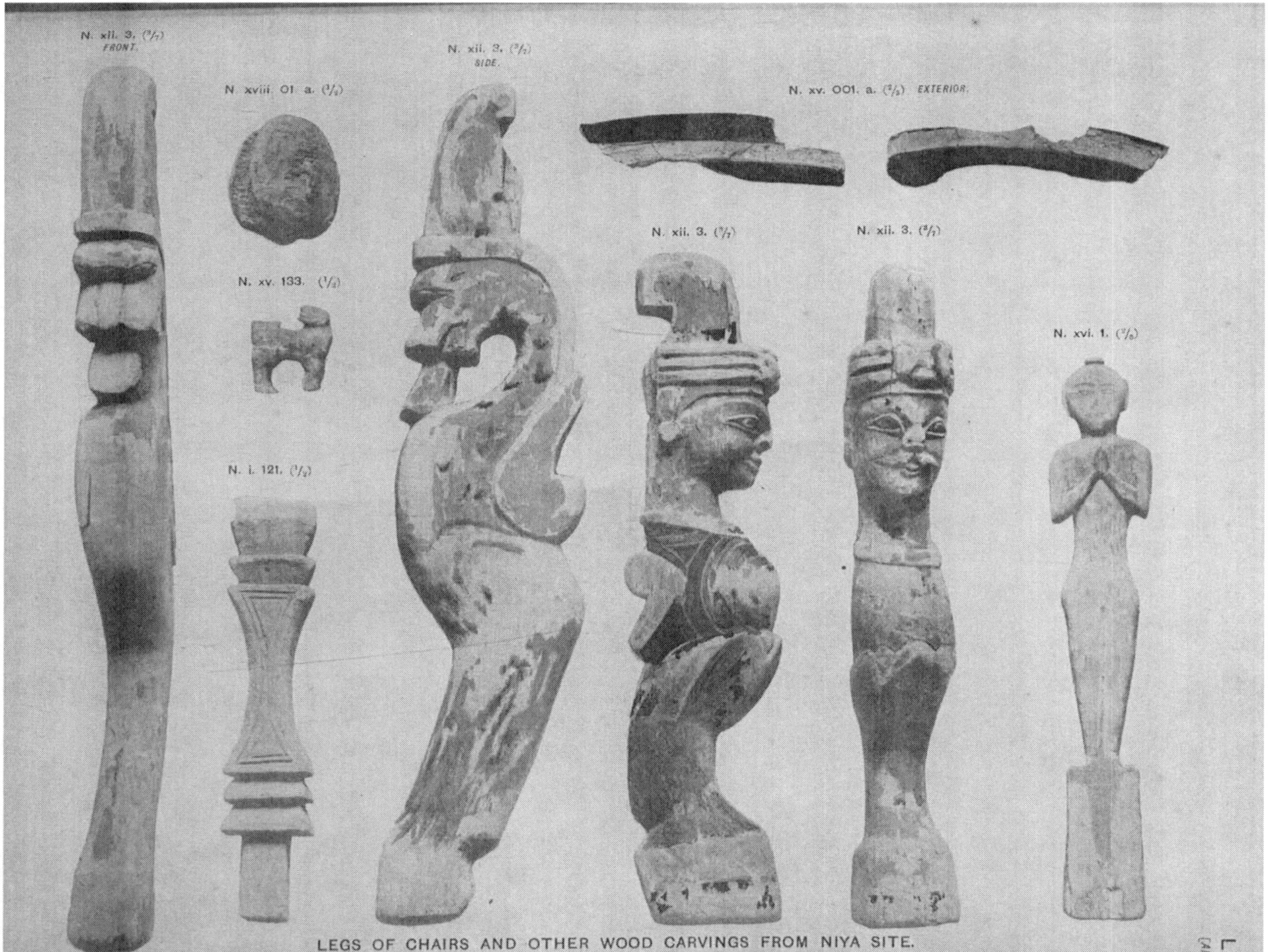

LEGS OF CHAIRS AND OTHER WOOD CARVINGS FROM NIYA SITE.

(See Chap. XI. sec. i, ii, vi, vii, viii.)

CARVED DOUBLE BRACKETS IN WOOD, FROM DWELLING N. VIII, NIYA SITE.

(See Chap. XI. sec. vii, viii.)

SCALE [illegible]

N. xx. 01. (⅔)

CARVED WOOD PANEL FROM DWELLING N. VIII, NIYA SITE.

(See Chap. XI. sec. vii, viii.)

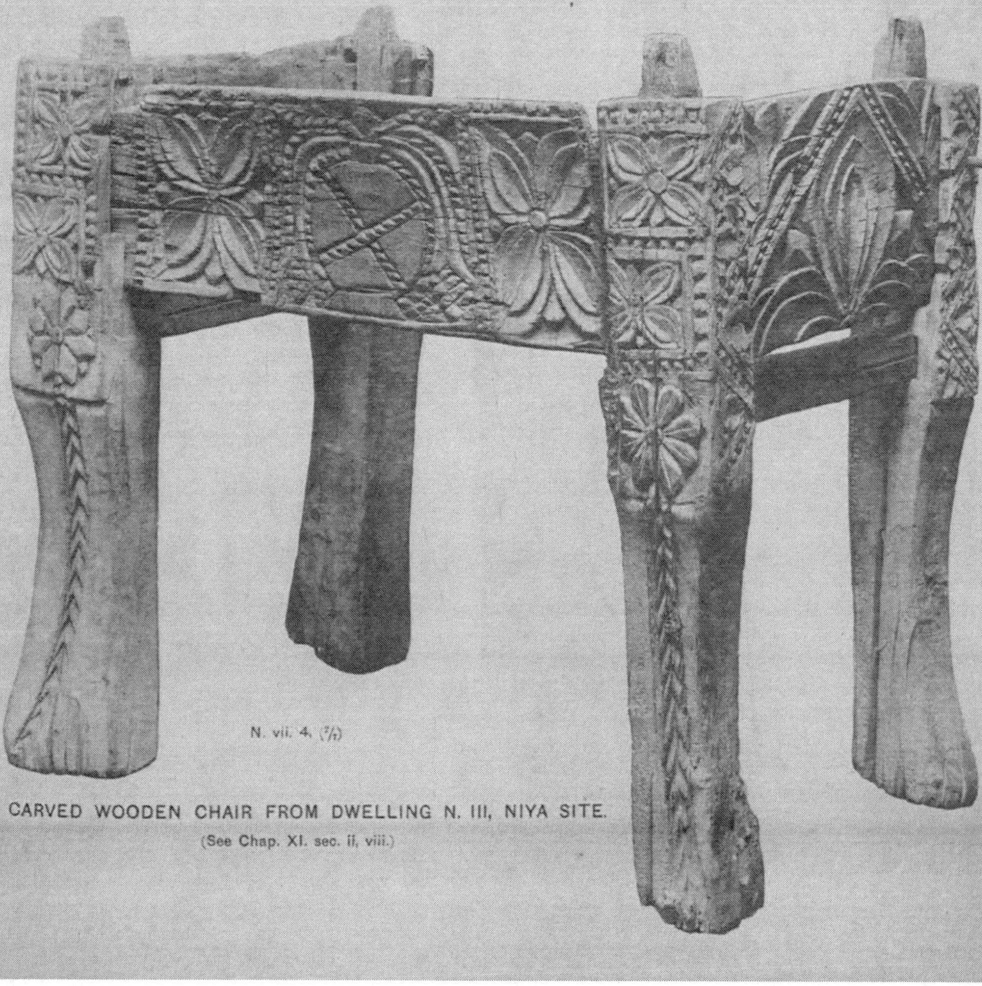

N. vii. 4. (⅔)

CARVED WOODEN CHAIR FROM DWELLING N. III, NIYA SITE.

(See Chap. XI. sec. ii, viii.)

D. VI. 3

PORTION OF PAINTED PANEL FROM SHRINE D. VI, DANDĀN-UILIQ.
(See Chap. IX. sec. vi, viii.)

D. II. 010.

PAINTED PANEL FROM SHRINE D. II, DANDĀN-UILIQ.
(See Chap. IX. sec. iii, viii.)

SCALE ½

LXVI
[ANT.]

D. II. 79.

SCALE 2/3

OBVERSE OF PAINTED PANEL FROM SHRINE D. II, DANDĀN-UILIQ.

(See Chap. IX. sec. iii, viii.)

D. VII. 1.

PAINTED PANEL FROM DWELLING D. VII, DANDĀN-UILIQ.

(See Chap. IX. sec. vi, viii.)

OBV. D. IV. 4. (1/3) REV.

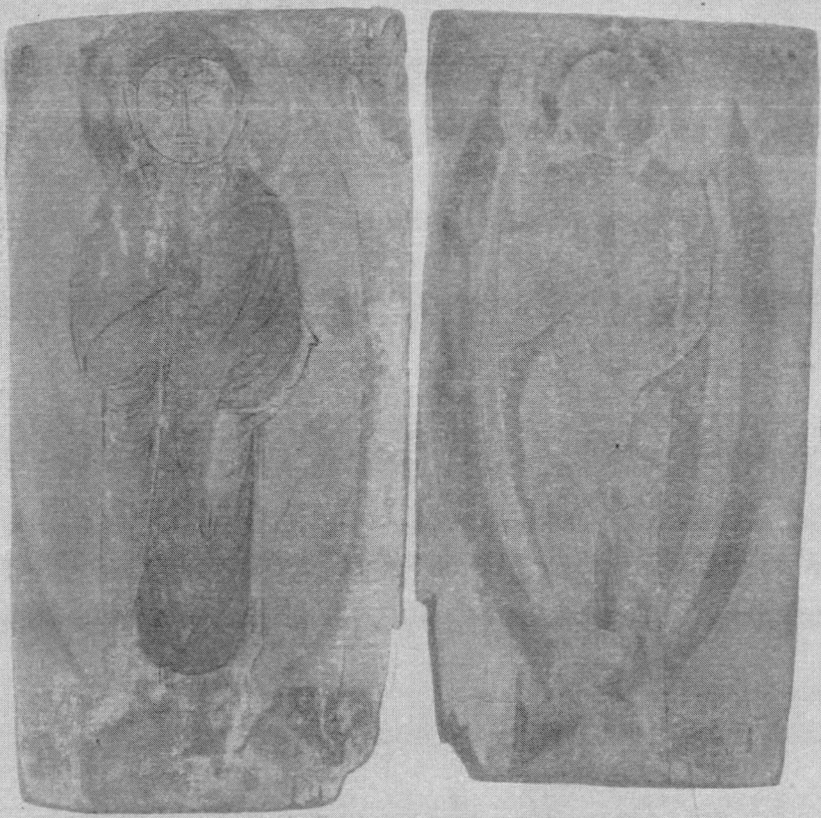

PAINTED PANEL FROM SHRINE D. IV, DANDĀN-UILIQ.
(See Chap. IX. sec. v, viii.)

D. V. 9. (1/2)

FRAGMENT OF PAINTED BOWL (INSIDE), DANDĀN-UILIQ.
(See Chap. IX. sec. v, viii.)

N. xv. 001. a. (1/2)

FRAGMENT OF LACQUERED BOWL (INSIDE), FROM RUBBISH-HEAP N. xv, NIYA SITE.
(See Chap. XI. sec. vi, viii.)

D. I. 04. (2/3)

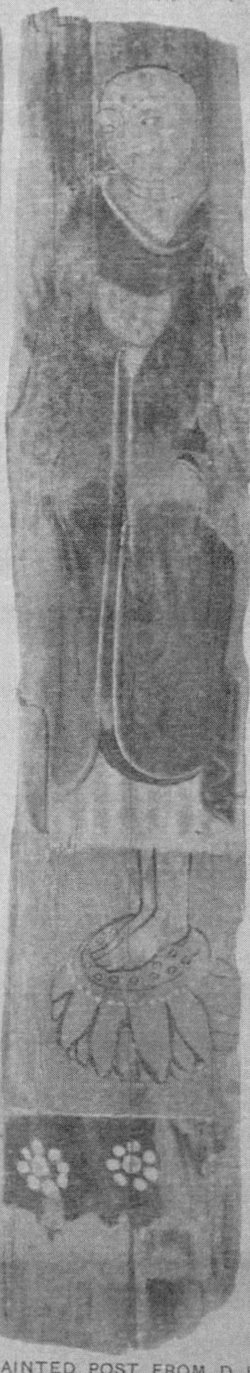

PAINTED POST FROM D. I. DANDĀN-UILIQ.
(See Chap. IX. sec. i, viii.)

D. X. 3. (1/1)

PAINTED PANEL FROM SHRINE D. X, DANDĀN-UILIQ.

(See Chap. IX. sec. iv, viii.)

LXIV

[ANT.]

LXIII
[ANT.]

D. X. 4.

PAINTED PANEL FROM SHRINE D. X, DANDĀN-UILIQ.
(See Chap. IX. sec. iv, viii.)

D. IV. 5.

OBVERSE OF PAINTED PANEL FROM SHRINE D. IV, DANDĀN-UILIQ.
(See Chap. IX. sec. v, viii.)

SCALE 2/3

D. X. 8.

D. X. 5.
REV.

PAINTED PANELS FROM SHRINE D. X, DANDĀN-UILIQ.

(See Chap. IX. sec. iv, viii.)

SCALE $^3/_4$

REVERSE OF PAINTED PANEL D. VII. 6 (⅓) FROM RUINED DWELLING, DANDĀN-UILIQ.

(See Chap. IX. sec. vi, viii.)

OBVERSE OF PAINTED PANEL D. VII. 6 (1/2) FROM RUINED DWELLING, DANDĀN-UILIQ.

(See Chap. IX. sec. vi, viii.)

D. II. 13.

PORTION OF FRESCO PAINTING ON STUCCO, WITH INSCRIPTION, FROM WALL OF SHRINE D. II, DANDĀN-UILIQ.

(See Chap. IX. sec. ii, viii.)

D. VII. 5.

PAINTED PANEL, FROM RUINED DWELLING, DANDĀN-UILIQ.

(See Chap. IX. sec. vi, viii.)

LVIII
[ANT.]

D. T. 017. (1/3)

D. II. 08. (2/3)

D. T. 016. (1/3)

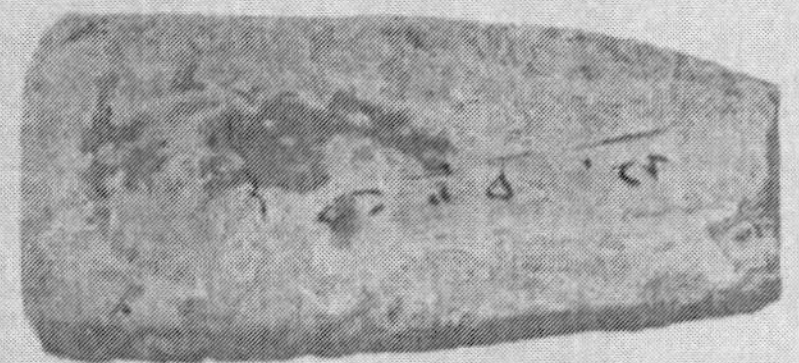

FRAGMENTS OF FRESCO PAINTINGS, WITH INSCRIPTIONS, BROUGHT FROM DANDĀN-UILIQ.

(See Chap. IX. sec. ii, vii, viii.)

PIECE OF FRESCO PAINTING WITH INSCRIPTION, FROM WALL OF SHRINE D. II, DANDĀN-UILIQ.

(See Chap. IX. sec. ii, viii.)

LVII
[ANT.]

FRAGMENTS OF WALL DECORATION IN STUCCO RELIEF
FROM SHRINE D. I, DANDĀN-UILIQ.

(See Chap. IX. sec. ii, viii.)

SCALE 1/1

D. VI. 5.

D. XII. 1.

D. T. O2.

D. XII. 5.

STUCCO RELIEFS FROM WALL DECORATION OF SHRINES D. I, D. VI, D. XII, DANDĀN-UILIQ.

(See Chap. IX. sec. ii, iv, viii.)

SCALE $^1/_4$

LV
[ANT.]

STUCCO RELIEFS FROM WALL DECORATION OF SHRINES D. I, D. II, DANDĀN-UILIQ.

(See Chap. IX. sec. ii, viii.)

SCALE 1/2

STUCCO RELIEFS FROM WALL DECORATION OF
SHRINES D. I, D. II, DANDĀN-UILIQ.

(See Chap. IX. sec. ii, viii.)

SCALE 1/10

D. II. 09. (3/8)

STUCCO RELIEF OF BUDDHA OR BODHISATTVA,
FROM SHRINE D. II, DANDĀN-UILIQ.

(See Chap. IX. sec. iii, viii.)

FRAGMENTS OF GLASS WARE AND OF MISCELLANEOUS CERAMIC AND METAL OBJECTS FROM MOJI, ENDERE, &c.

SCALE 1/1

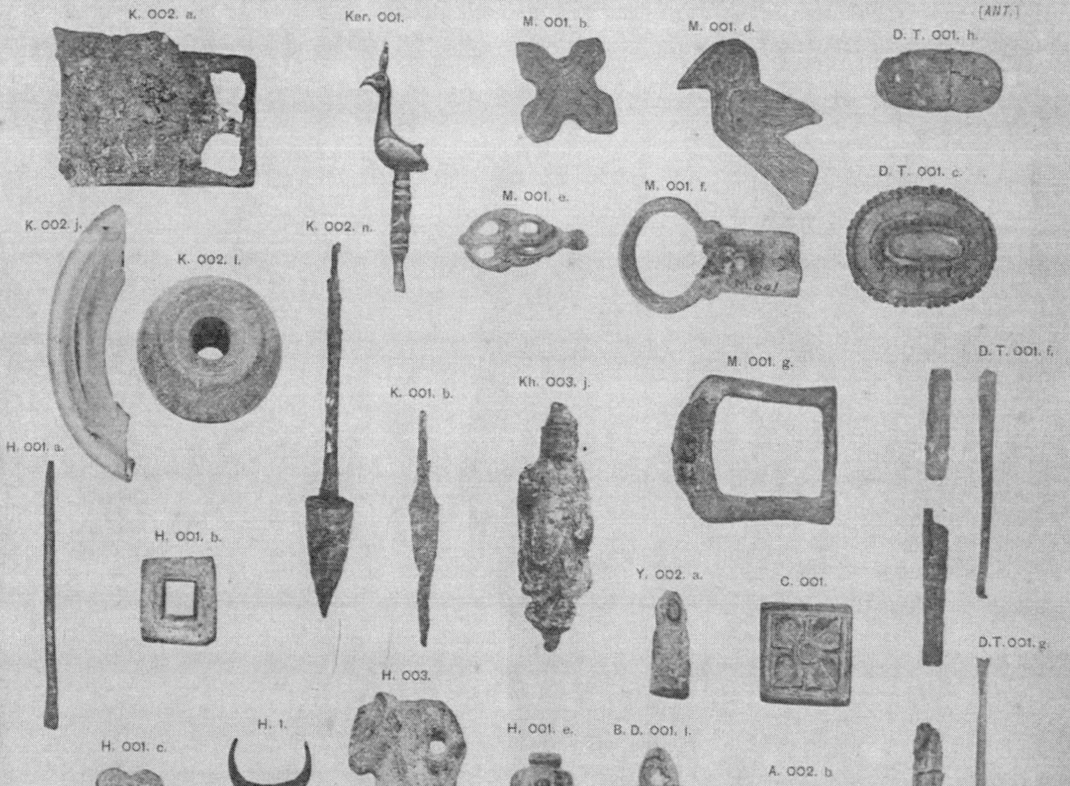

MISCELLANEOUS OBJECTS, CHIEFLY IN METAL, FROM DANDĀN-UILIQ, KARA-DONG, HANGUYA, AND OTHER SITES.

81408

SCALE 1/1

L
[ANT.]

SEALS IN STONE, METAL, &c., FROM YÖTKAN AND OTHER SITES.

(See Chap. V. sec. iii, iv; Chap. VIII. sec. iv; Chap. XI. sec. viii, &c.

SCALE 1/1

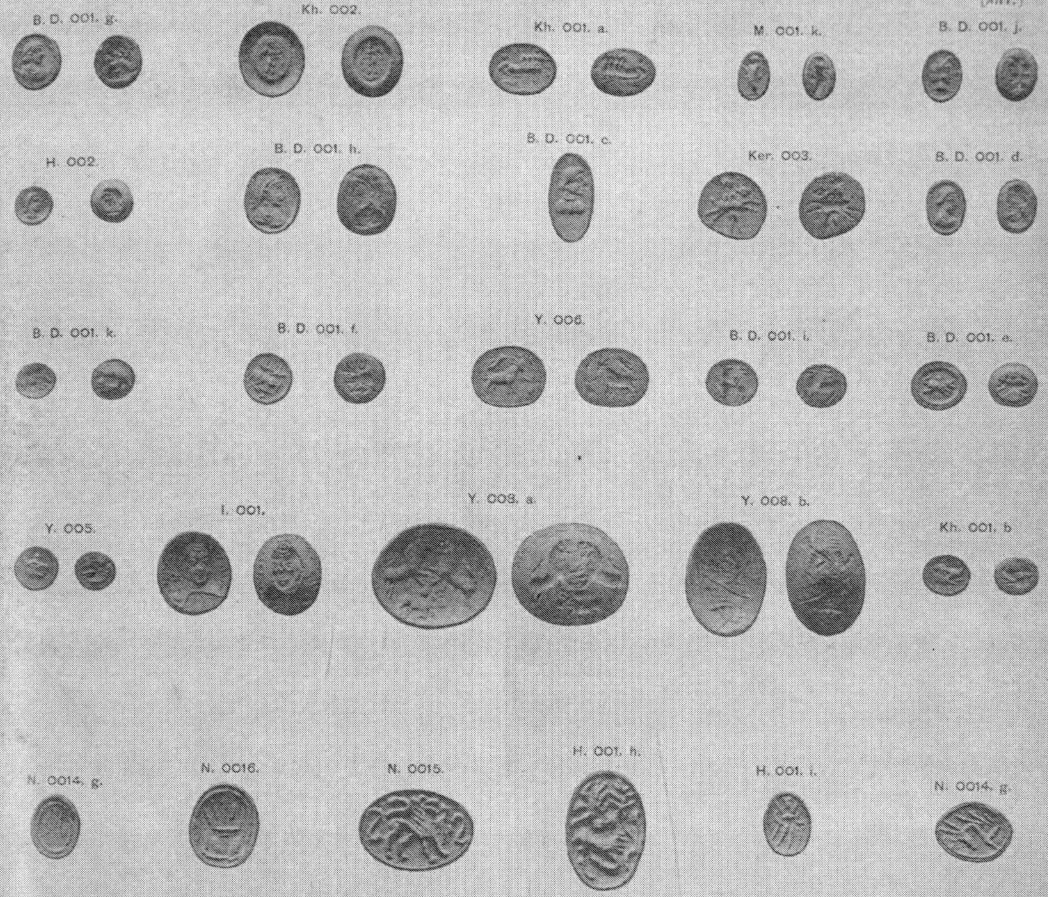

SEALS IN STONE AND METAL FROM YŌTKAN, NIYA SITE, &c.

(See Chap. VIII. sec. iv; Chap. XI. sec. viii.)

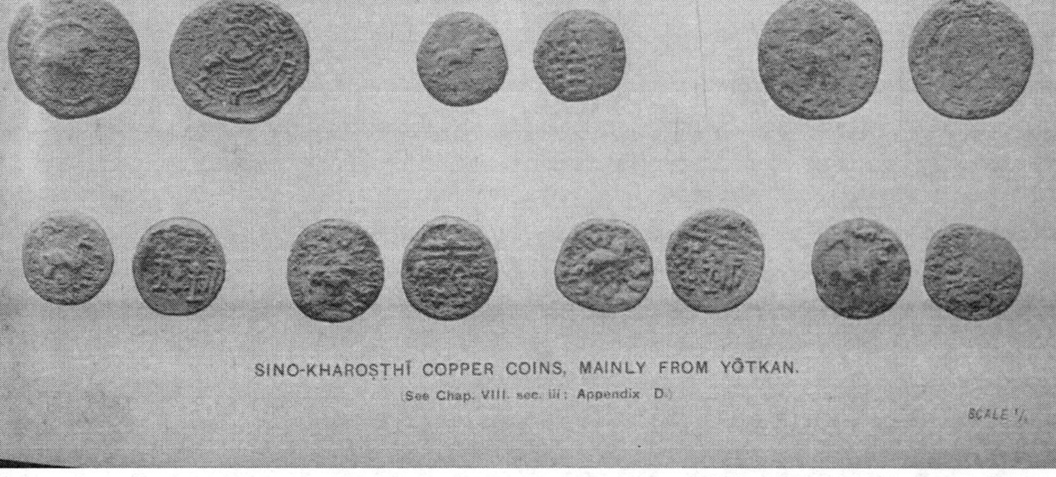

SINO-KHAROṢṬHĪ COPPER COINS, MAINLY FROM YŌTKAN.

(See Chap. VIII. sec. iii; Appendix D.)

SCALE 1/1

XLVIII

[ANT.]

RELIEF CARVINGS IN STONE AND IVORY, MAINLY FROM YŌTKAN.

(See Chap. VIII. sec. iv.)

SCALE 1/1

XLVII
[ANT.]

TERRA-COTTA FIGURINES FROM YŌTKAN.

(See Chap. VIII. sec. iv.)

SCALE $^1/_1$

TERRACOTTA FIGURINES FROM YŌTKAN.

(See Chap. VIII. sec. iv.)

SCALE 1/1

XLV
[ANT.]

TERRACOTTAS FROM YŌTKAN

(See Chap. VIII. sec. iv.)

SCALE 1/1

XLIV
[ANT.]

TERRACOTTAS FROM YŌTKAN.

(See Chap. VIII. sec. iv, v.)

SCALE 4/5

TERRACOTTAS FROM YŌTKAN.

(See Chap. VIII. sec. iv.)

XLII
[ANT.]

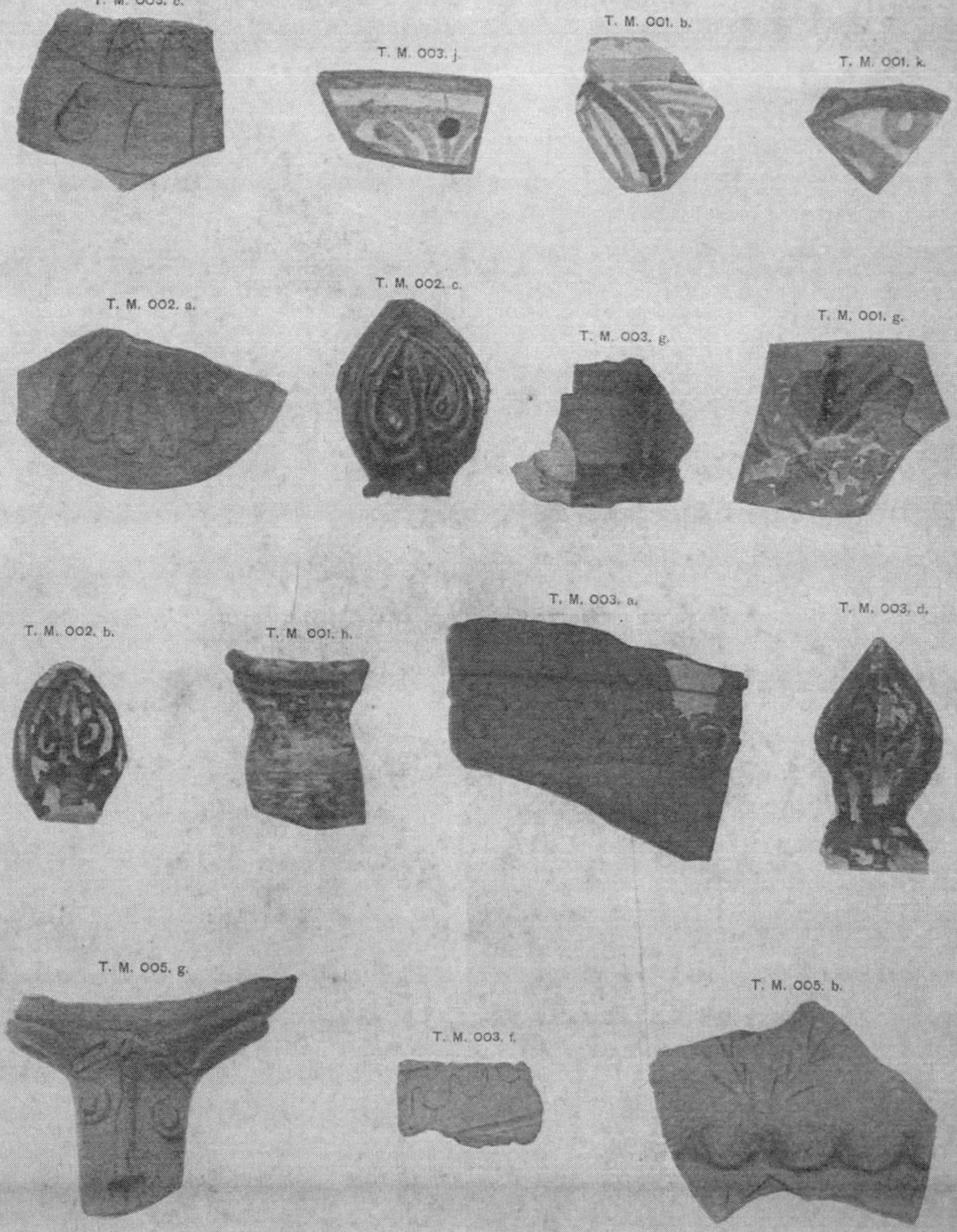

FRAGMENTS OF GLAZED POTTERY FROM MOJI.

(See Chap. V. sec. iv.)

SCALE 1/

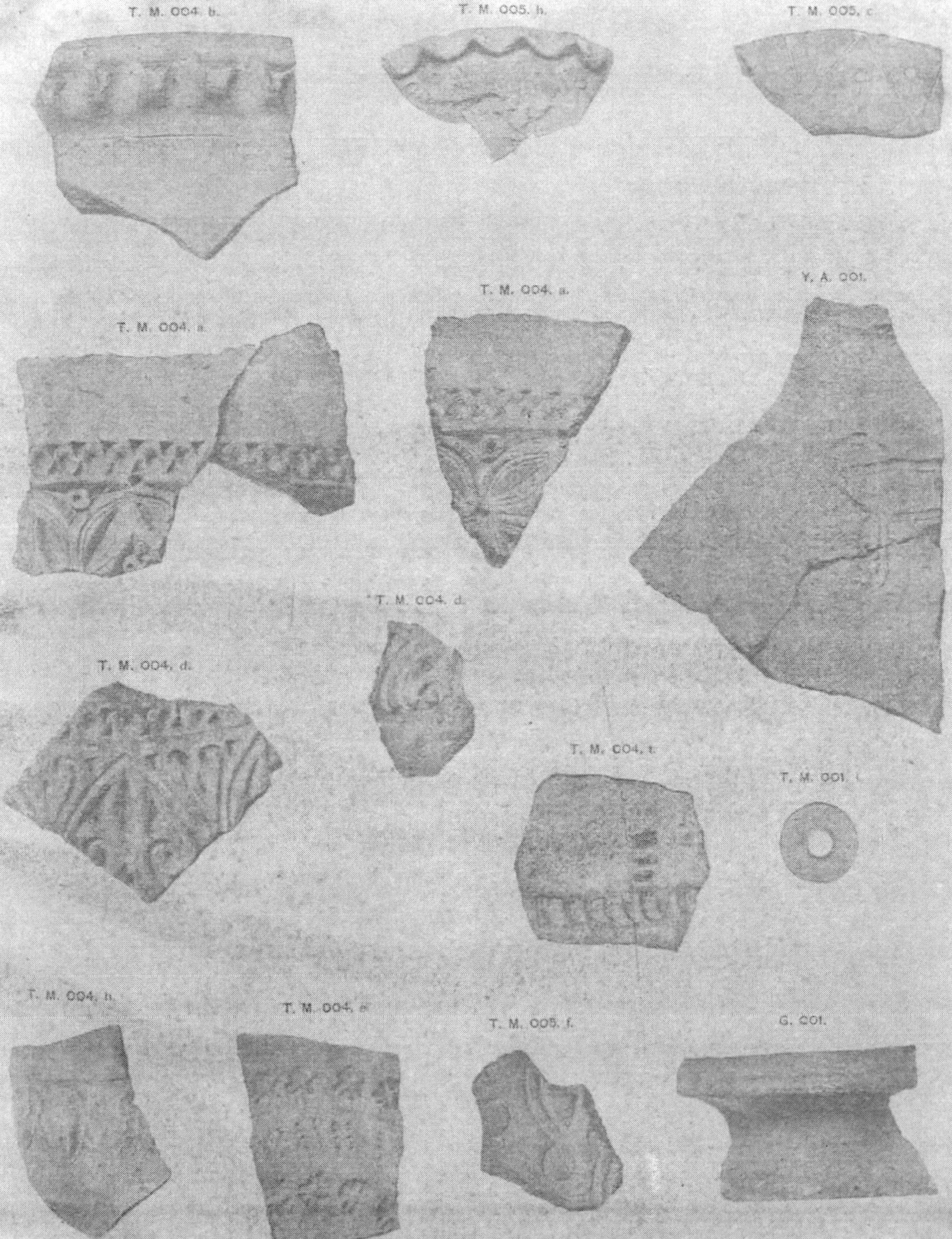

POTTERY FRAGMENTS FROM MOJI AND KAKSHAL TATI.

(See Chap. V. sec. iii, iv.)

SCALE $\frac{1}{3}$

XL
[PLAN.]

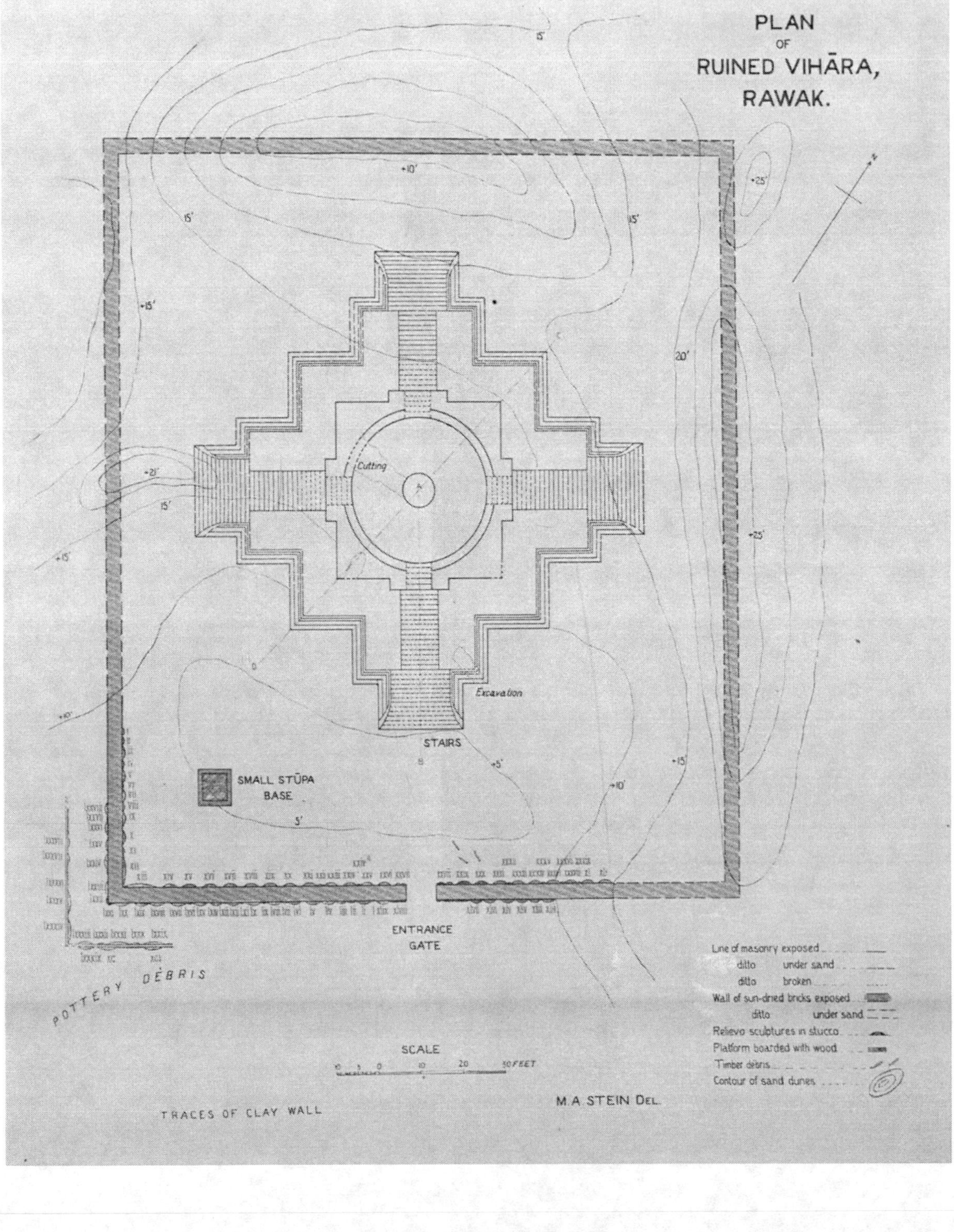
PLAN
OF
RUINED VIHĀRA,
RAWAK.
Cutting
Excavation
STAIRS
SMALL STŪPA
BASE
ENTRANCE
GATE
POTTERY DÈBRIS
Line of masonry exposed
ditto under sand
ditto broken
Wall of sun-dried bricks exposed
ditto under sand
Relievo sculptures in stucco
Platform boarded with wood
Timber débris
Contour of sand dunes
SCALE
10 5 0 10 20 30 FEET
M.A. STEIN DEL.
TRACES OF CLAY WALL

PLAN OF EXTANT WALL SEGMENT OF RUINED FORTIFICATION, AK-SIPIL.

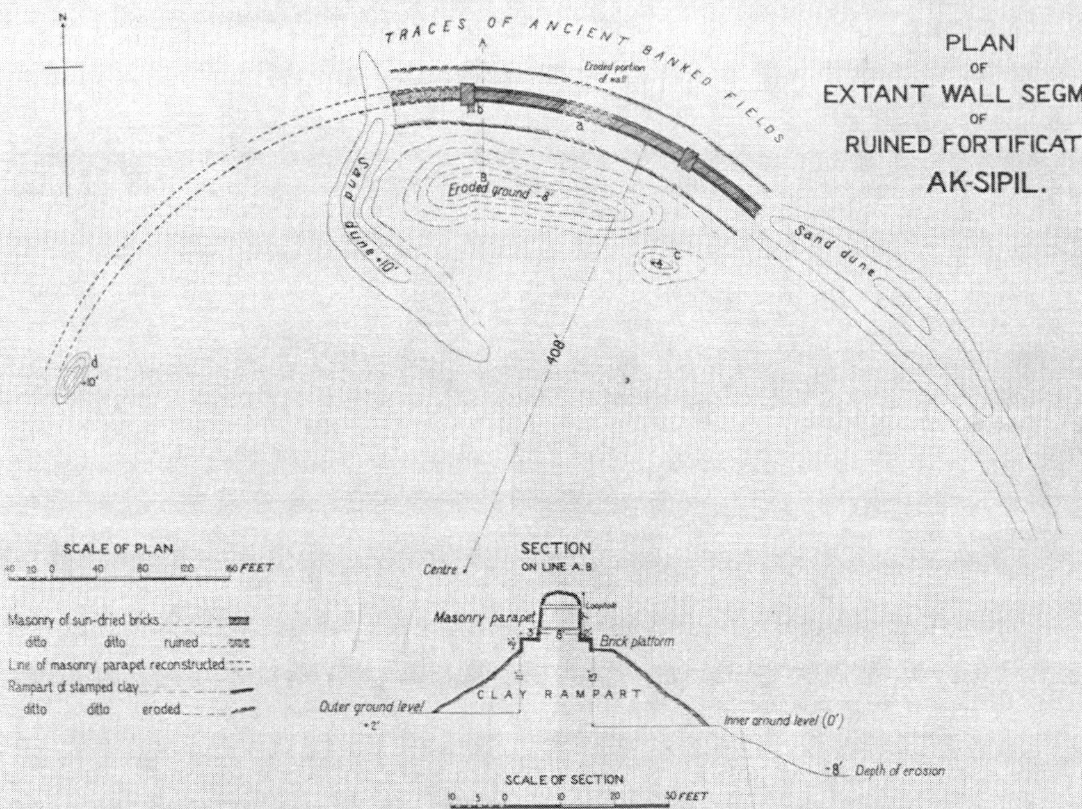

SECTIONS OF RUINED STŪPA, RAWAK.

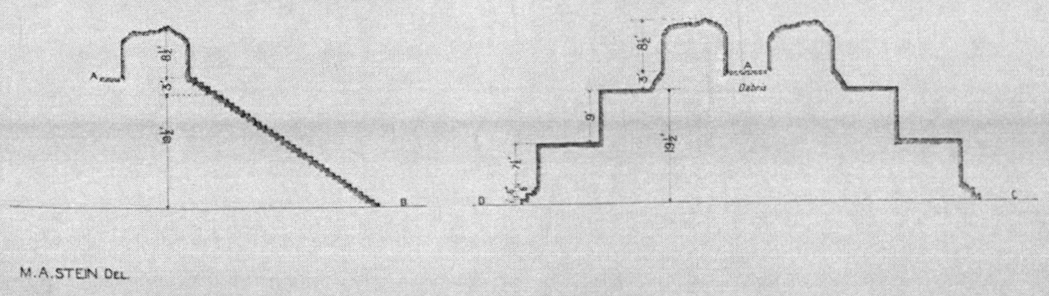

M.A.STEIN DEL.

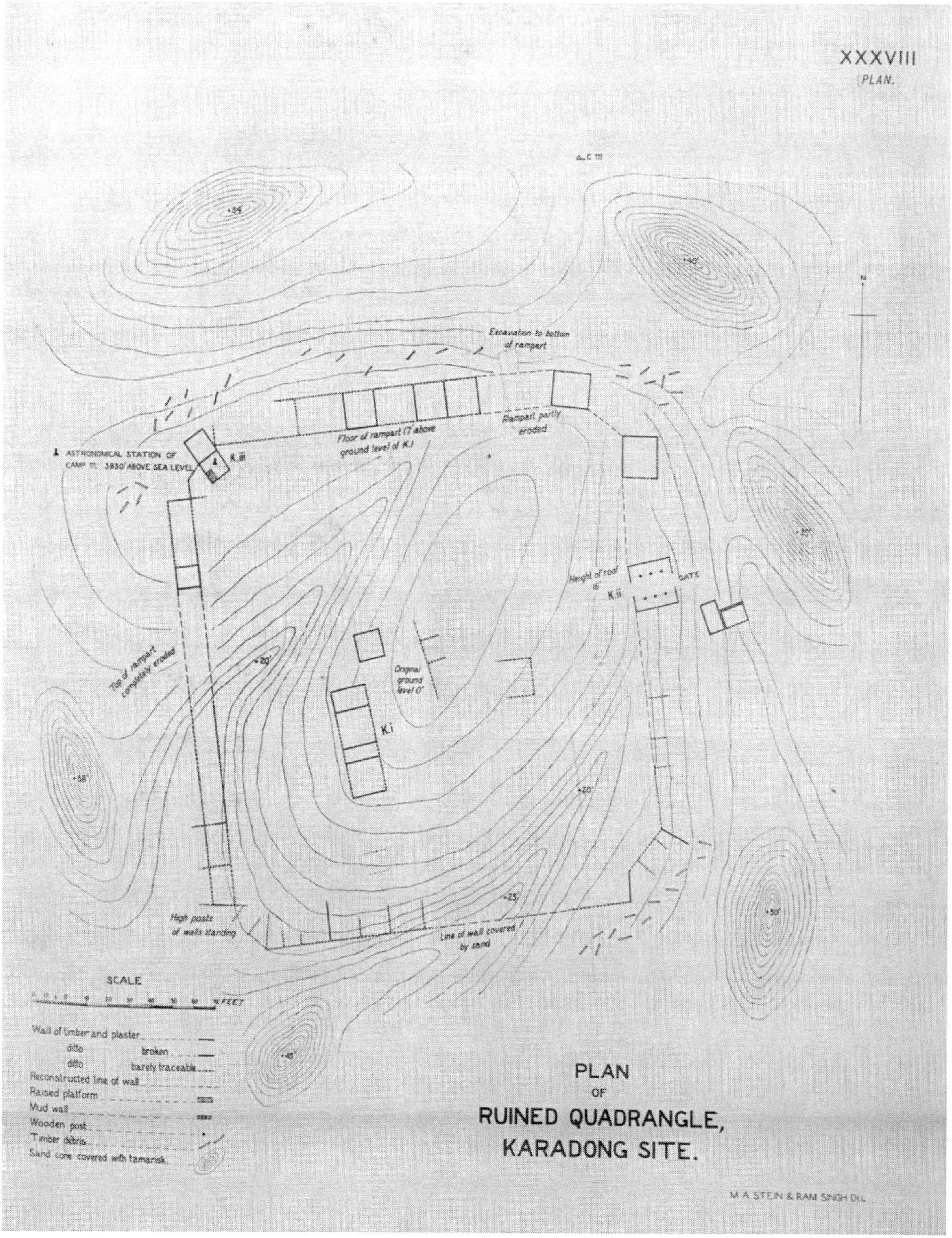

PLAN
OF
RUINED QUADRANGLE,
KARADONG SITE.

RUINED STŪPA, ENDERE SITE.

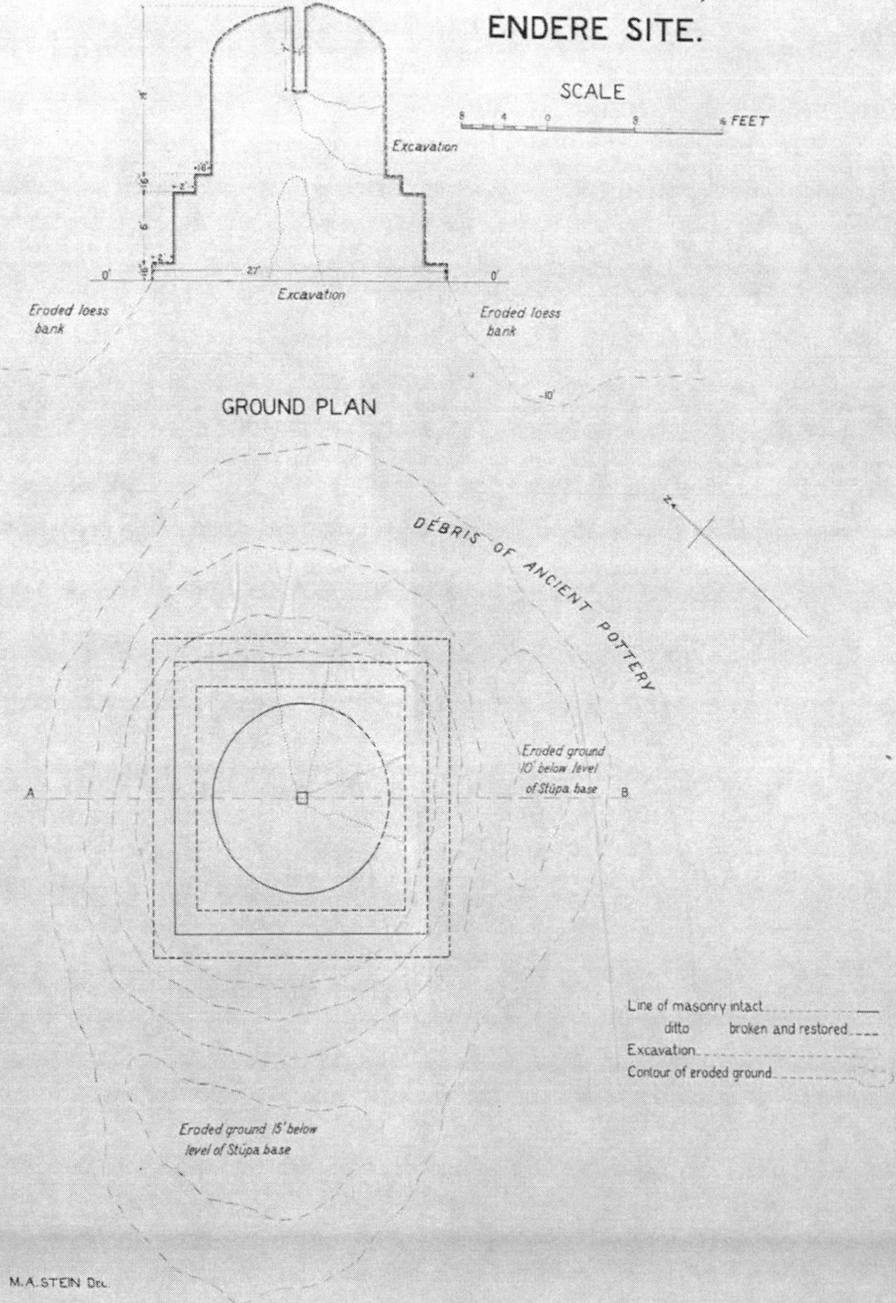

M.A. STEIN DEL.

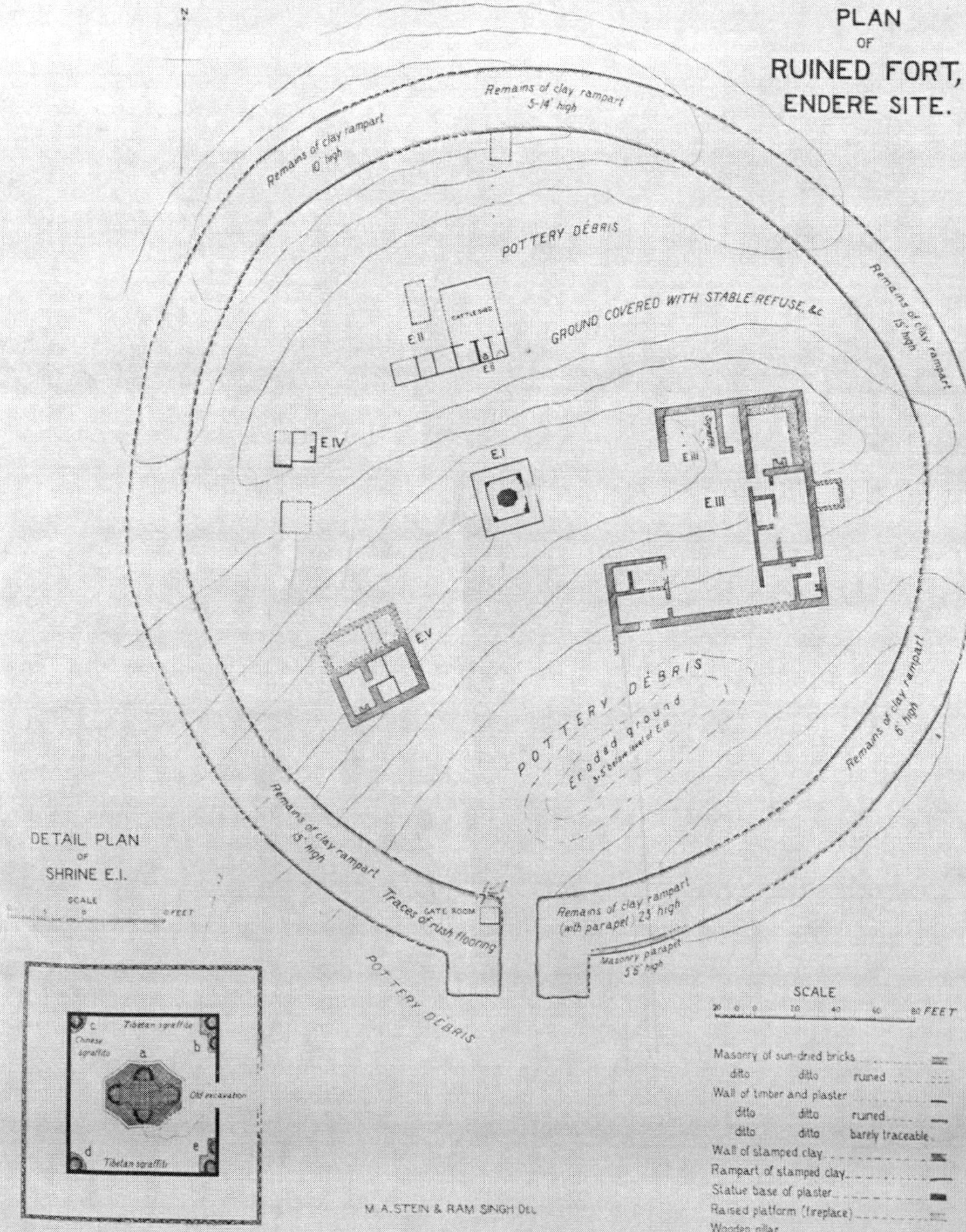
PLAN
OF
RUINED FORT,
ENDERE SITE.
Remains of clay rampart
5-14' high
Remains of clay rampart
10' high
POTTERY DÉBRIS
CATTLE SHED
E.II
GROUND COVERED WITH STABLE REFUSE, &c
Remains of clay rampart
15' high
E.IV
E.I
E.III
E.III
E.V
POTTERY DÉBRIS
Eroded ground
Remains of clay rampart
6' high
Remains of clay rampart
15' high
Traces of rush flooring
GATE ROOM
Remains of clay rampart
(with parapet) 25' high
Masonry parapet
5'6" high
POTTERY DÉBRIS
DETAIL PLAN
OF
SHRINE E.I.
SCALE
FEET
Tibetan sgraffito
Chinese sgraffito
Old excavation
Tibetan sgraffito
SCALE
FEET
Masonry of sun-dried bricks
ditto ditto ruined
Wall of timber and plaster
ditto ditto ruined
ditto ditto barely traceable
Wall of stamped clay
Rampart of stamped clay
Statue base of plaster
Raised platform (fireplace)
Wooden pillar
M. A. STEIN & RAM SINGH DEL.

XXXV
[PLAN.]

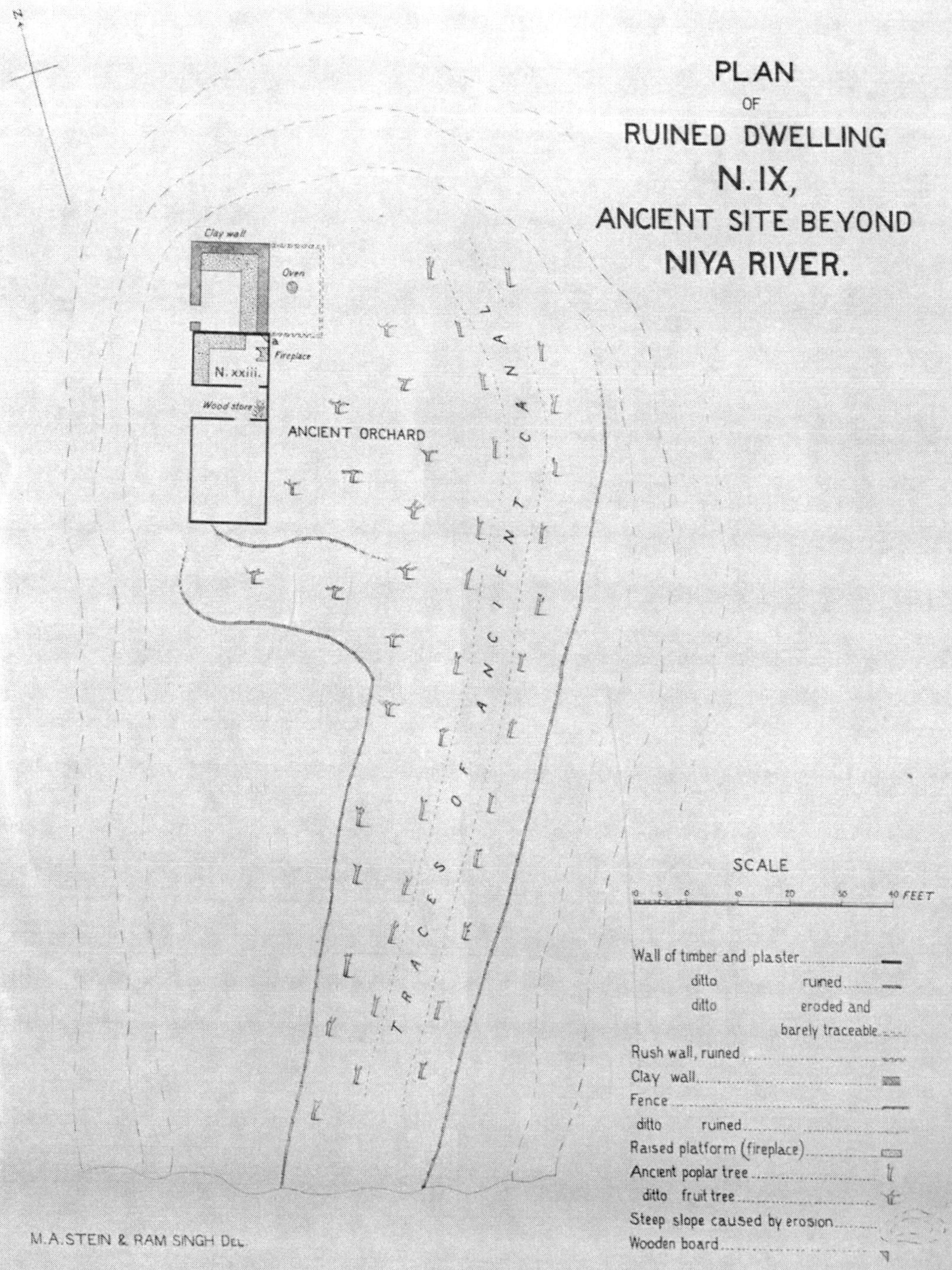

PLAN
OF
RUINED DWELLING
N. IX,
ANCIENT SITE BEYOND
NIYA RIVER.
N
Clay wall
Oven
a
Fireplace
N. xxiii.
Wood store
ANCIENT ORCHARD
TRACES OF ANCIENT CANAL
SCALE
10 5 0 10 20 30 40 FEET
Wall of timber and plaster
ditto ruined
ditto eroded and barely traceable
Rush wall, ruined
Clay wall
Fence
ditto ruined
Raised platform (fireplace)
Ancient poplar tree
ditto fruit tree
Steep slope caused by erosion
Wooden board
M. A. STEIN & RAM SINGH DEL.

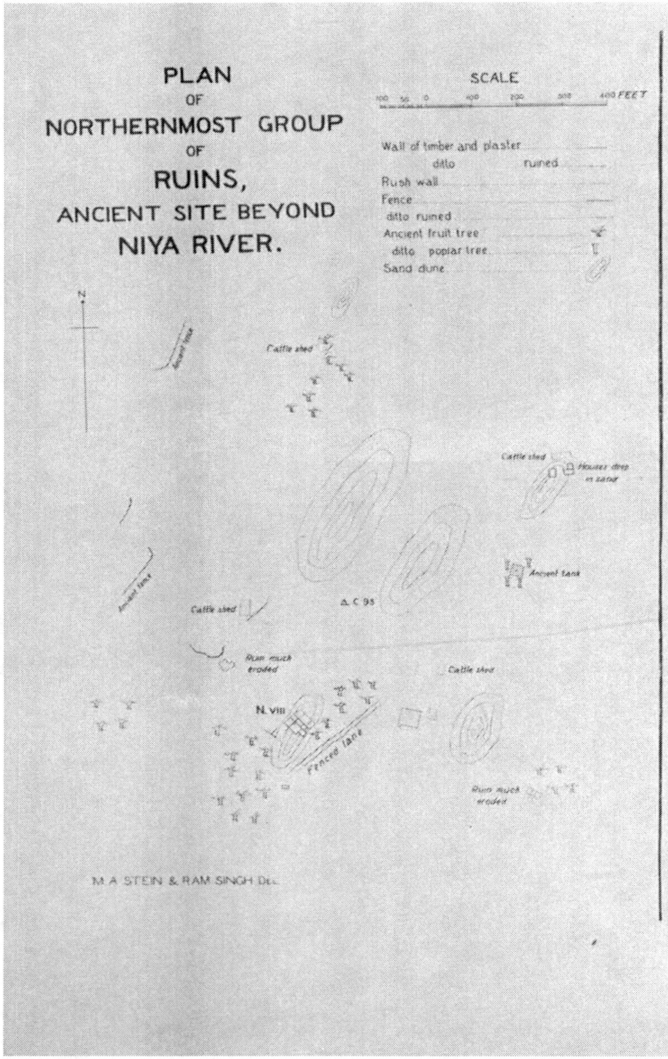
PLAN
OF
NORTHERNMOST GROUP
OF
RUINS,
ANCIENT SITE BEYOND
NIYA RIVER.
SCALE
100 50 0 100 200 300 400 FEET
Wall of timber and plaster
ditto ruined
Rush wall
Fence
ditto ruined
Ancient fruit tree
ditto poplar tree
Sand dune
N
Cattle shed
Houses deep in sand
Ancient tank
Δ C 95
Ruin much eroded
Cattle shed
N. VIII
Fenced lane
M. A. STEIN & RAM SINGH Del.

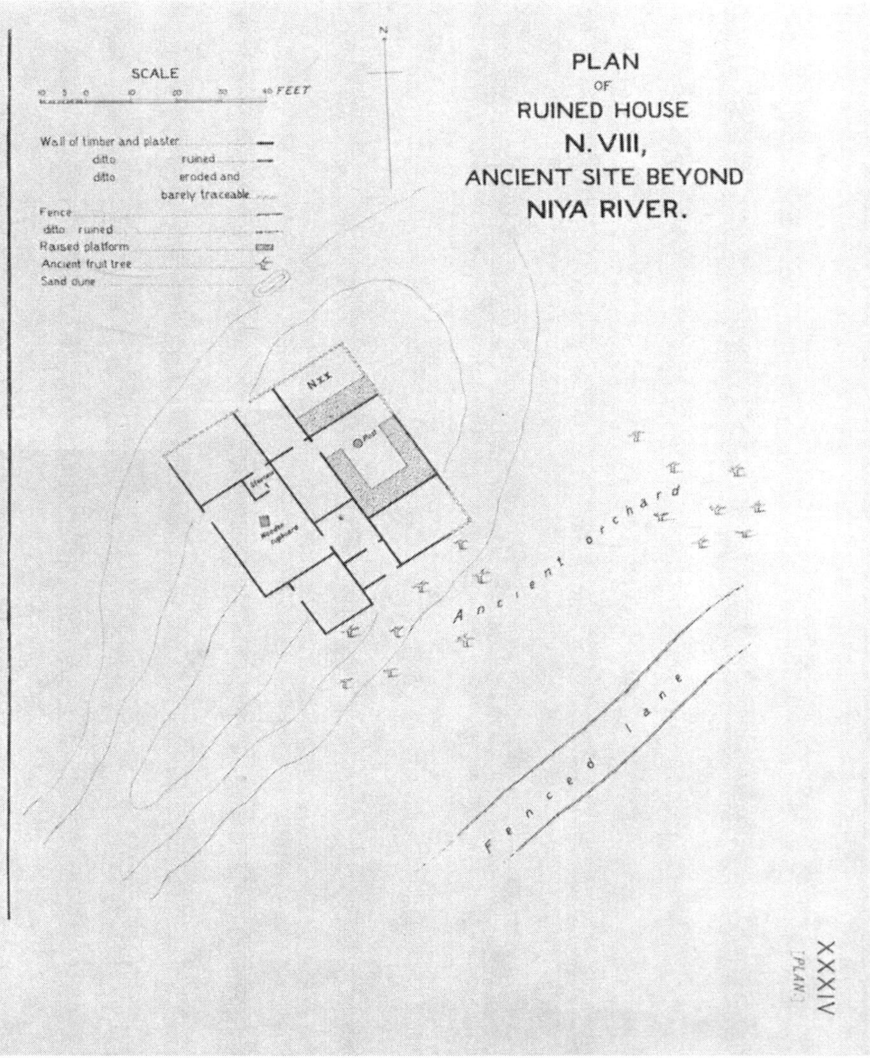
PLAN
OF
RUINED HOUSE
N. VIII,
ANCIENT SITE BEYOND
NIYA RIVER.
SCALE
10 5 0 10 20 30 40 FEET
Wall of timber and plaster
ditto ruined
ditto eroded and barely traceable
Fence
ditto ruined
Raised platform
Ancient fruit tree
Sand dune
N
N.XX
Post
Ancient orchard
Fenced lane

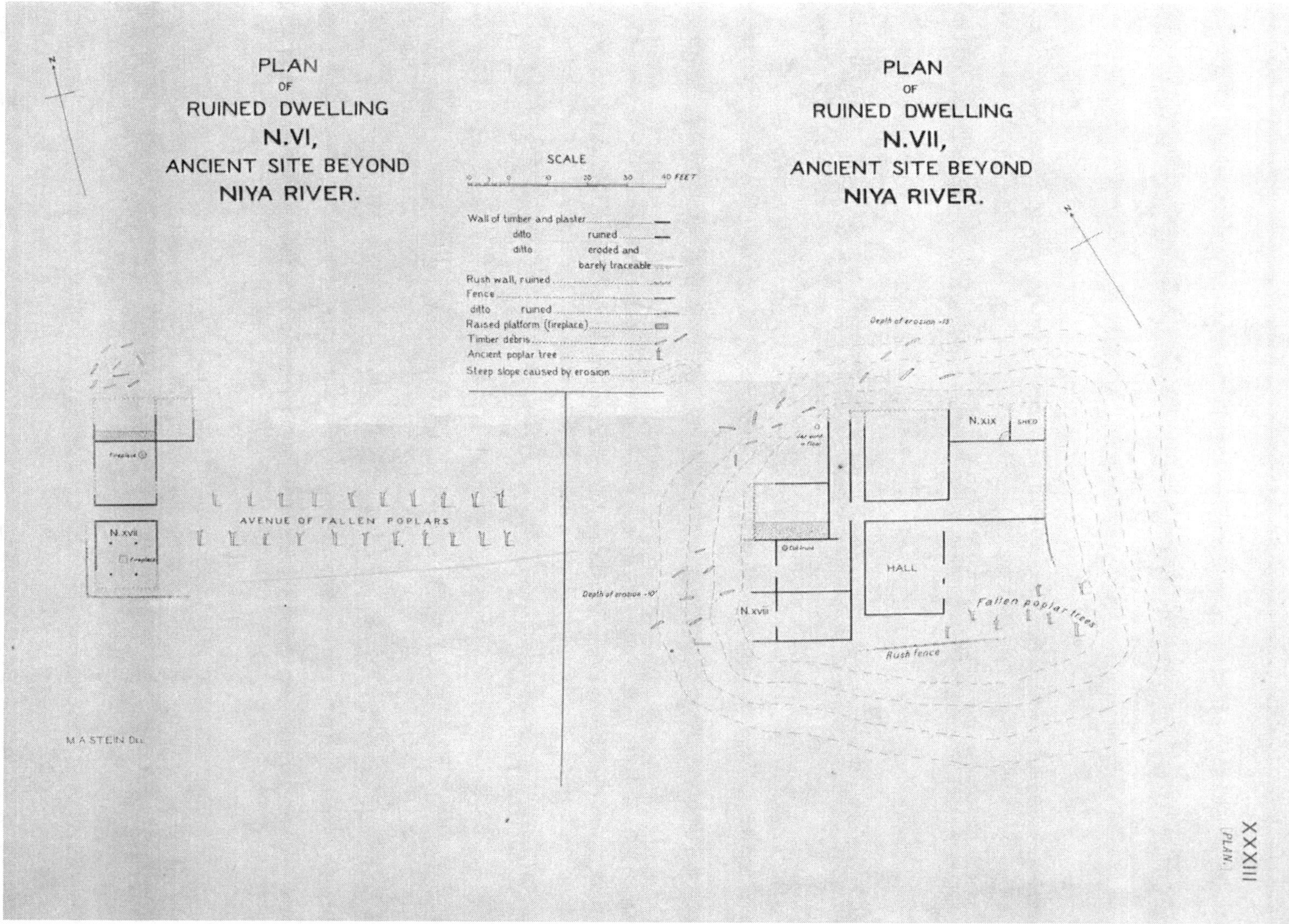
PLAN
OF
RUINED DWELLING
N.VI,
ANCIENT SITE BEYOND
NIYA RIVER.
N
SCALE
10 5 0 10 20 30 40 FEET
Wall of timber and plaster
ditto ruined
ditto eroded and barely traceable
Rush wall, ruined
Fence
ditto ruined
Raised platform (fireplace)
Timber debris
Ancient poplar tree
Steep slope caused by erosion
Fireplace
N.xvii
Fireplace
AVENUE OF FALLEN POPLARS
M.A.STEIN DEL.
PLAN
OF
RUINED DWELLING
N.VII,
ANCIENT SITE BEYOND
NIYA RIVER.
N
Depth of erosion -15'
N.xix
SHED
HALL
N.xviii
Depth of erosion -10'
Fallen poplar trees
Rush fence
XXXIII
[PLAN.]

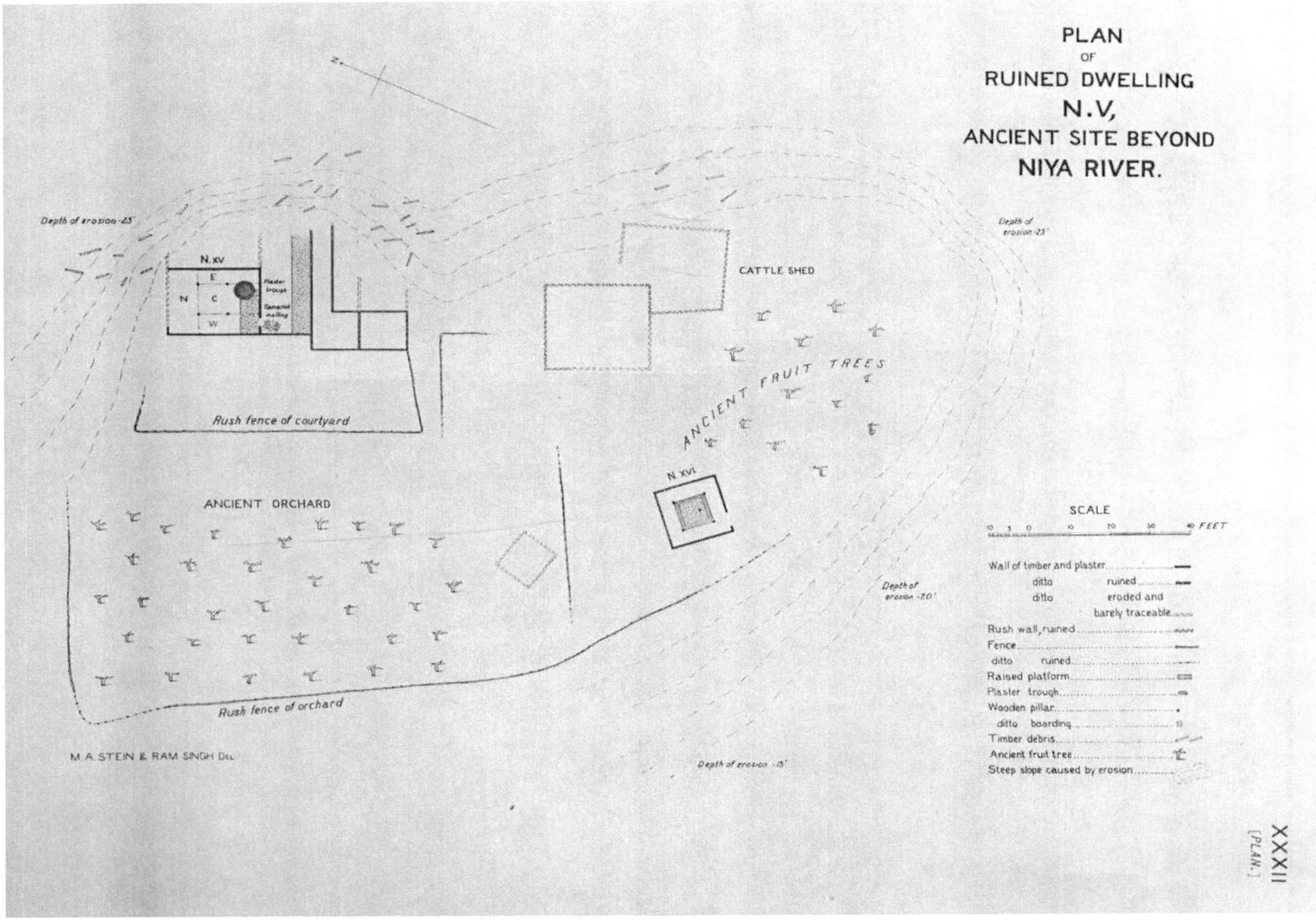
PLAN
OF
RUINED DWELLING
N.V,
ANCIENT SITE BEYOND
NIYA RIVER.
N.
Depth of erosion ·25'
N.XV
E
N
C
W
Plaster trough
Tamarisk matting
Rush fence of courtyard
CATTLE SHED
ANCIENT FRUIT TREES
N.XVI
ANCIENT ORCHARD
Rush fence of orchard
Depth of erosion ·25'
Depth of erosion ·20'
Depth of erosion ·15'
M.A. STEIN & RAM SINGH DEL.
SCALE
10 5 0 10 20 30 40 FEET
Wall of timber and plaster
ditto ruined
ditto eroded and barely traceable
Rush wall, ruined
Fence
ditto ruined
Raised platform
Plaster trough
Wooden pillar
ditto boarding
Timber debris
Ancient fruit tree
Steep slope caused by erosion

XXXI
[PLAN.]

PLAN OF RUINED DWELLING N.IV, ANCIENT SITE BEYOND NIYA RIVER.

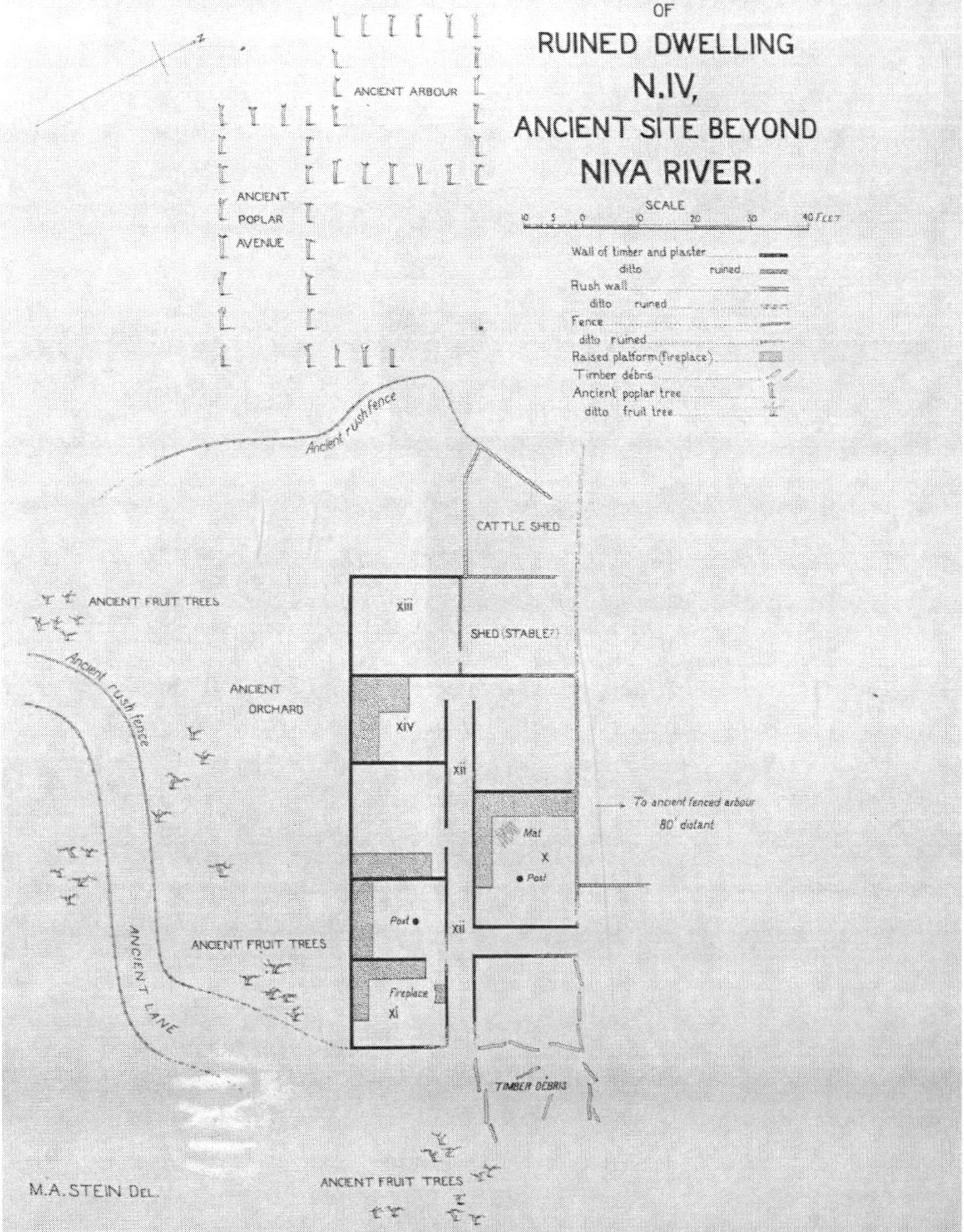

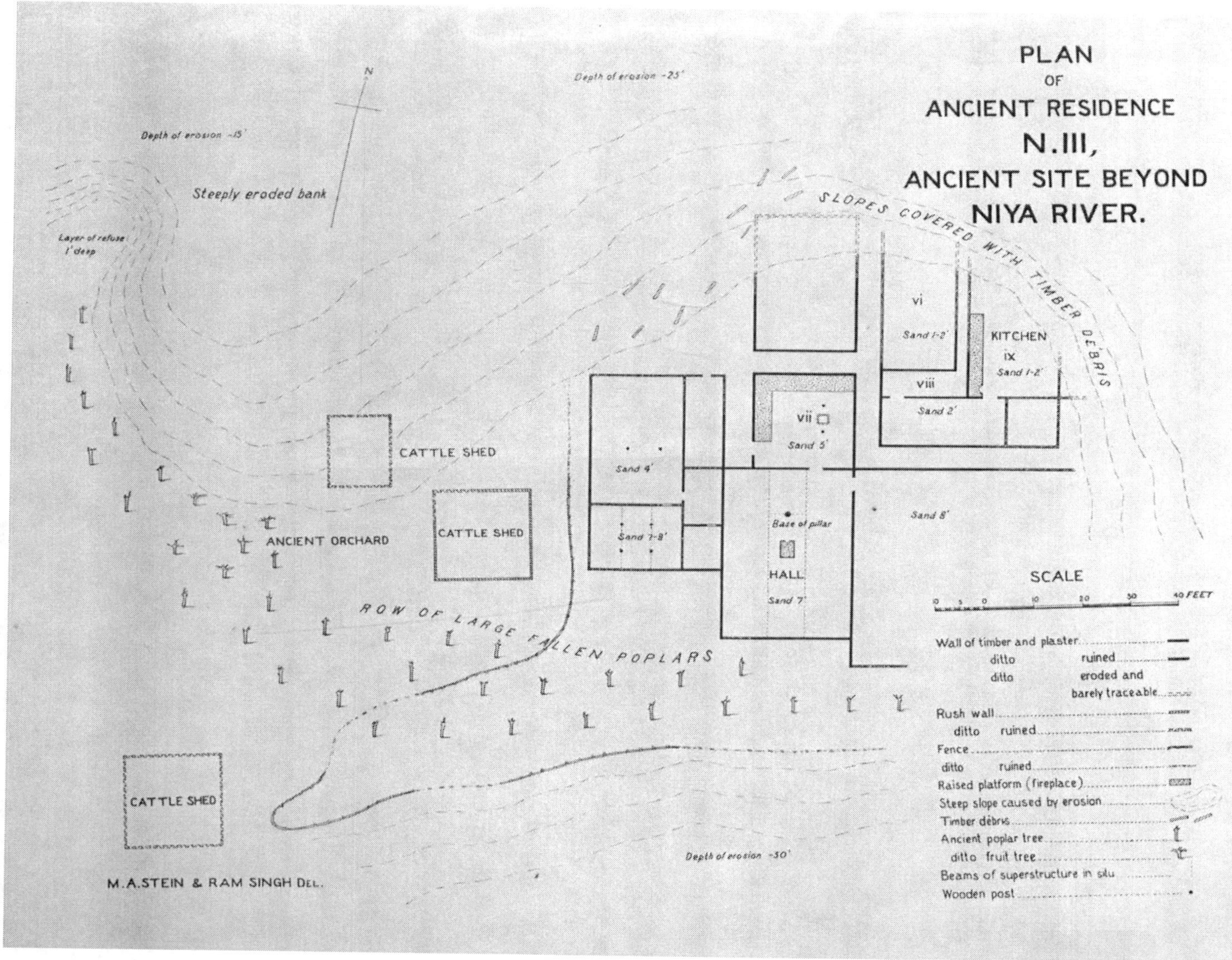
PLAN
OF
ANCIENT RESIDENCE
N.III,
ANCIENT SITE BEYOND
NIYA RIVER.
N
Depth of erosion -15'
Depth of erosion -25'
Steeply eroded bank
Layer of refuse 1' deep
SLOPES COVERED WITH TIMBER DÉBRIS
vi
Sand 1-2'
KITCHEN
ix
Sand 1-2'
viii
Sand 2'
vii
Sand 5'
Sand 4'
Sand 7-8'
Base of pillar
HALL
Sand 7'
Sand 8'
CATTLE SHED
CATTLE SHED
CATTLE SHED
ANCIENT ORCHARD
ROW OF LARGE FALLEN POPLARS
Depth of erosion -30'
SCALE
0 5 0 10 20 30 40 FEET
Wall of timber and plaster
ditto ruined
ditto eroded and barely traceable
Rush wall
ditto ruined
Fence
ditto ruined
Raised platform (fireplace)
Steep slope caused by erosion
Timber débris
Ancient poplar tree
ditto fruit tree
Beams of superstructure in situ
Wooden post
M.A.STEIN & RAM SINGH DEL.

RUINED STŪPA, ANCIENT SITE BEYOND NIYA RIVER.

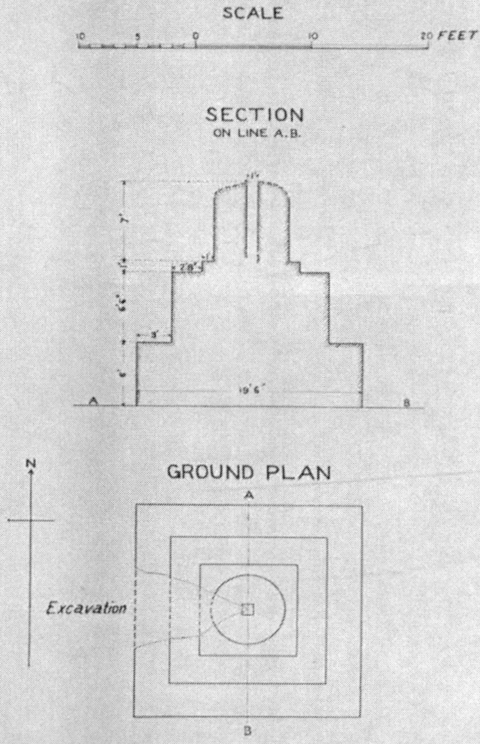

M. A. STEIN DEL.

SKETCH PLAN OF GROUP OF RUINED DWELLINGS N. II, ANCIENT SITE BEYOND NIYA RIVER.

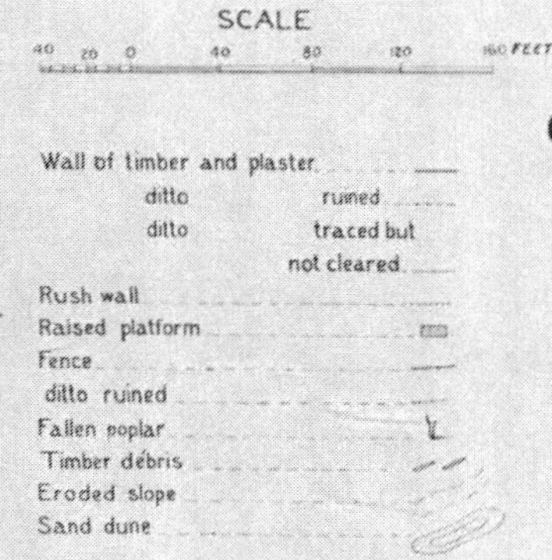

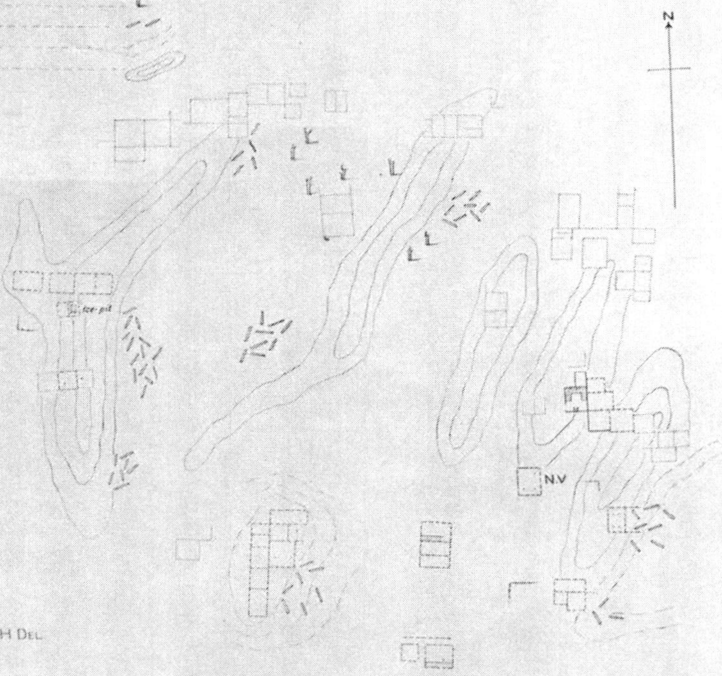

RAM SINGH DEL.

XXVIII
[PLAN.]

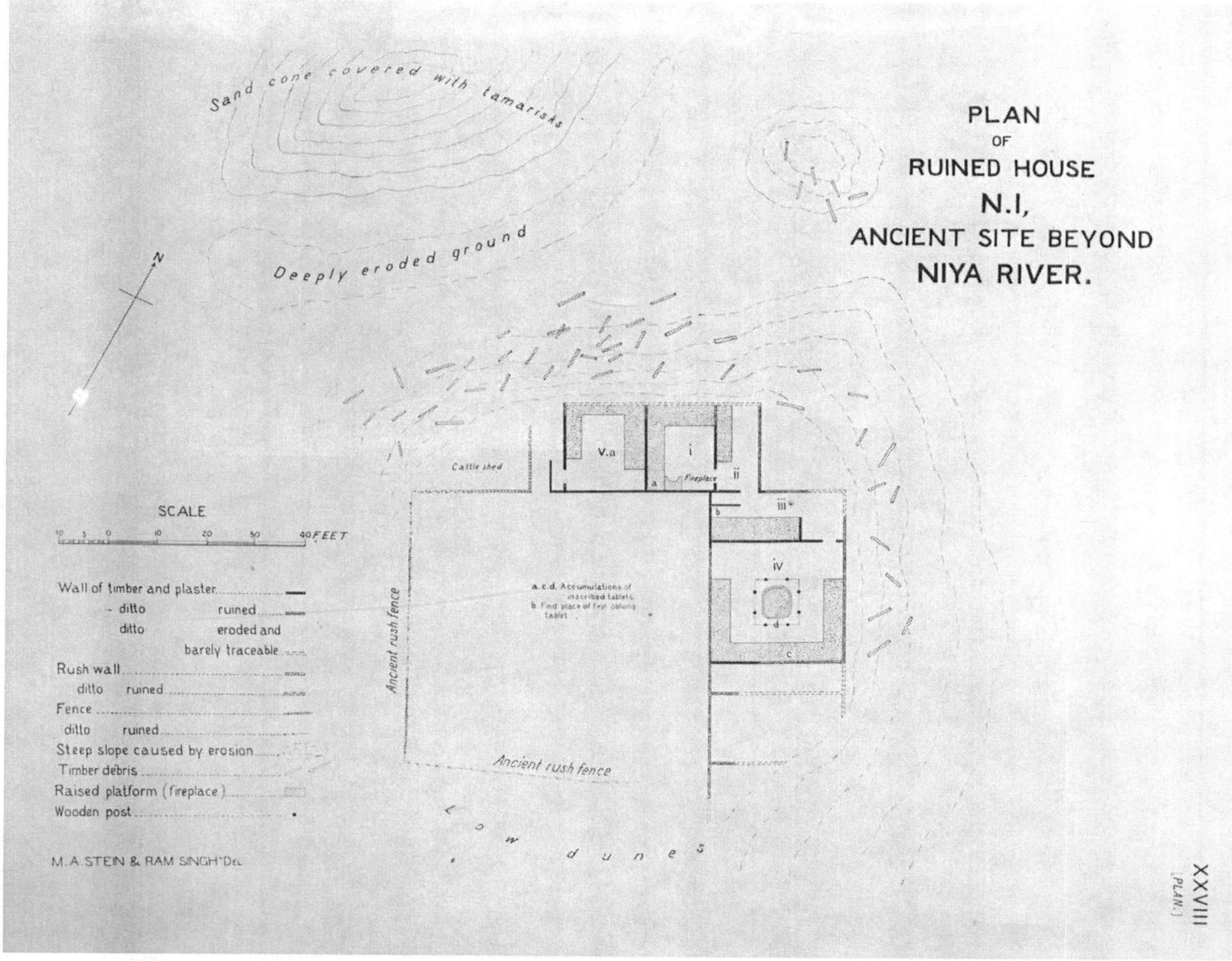
PLAN
OF
RUINED HOUSE
N.I,
ANCIENT SITE BEYOND
NIYA RIVER.
Sand cone covered with tamarisks
Deeply eroded ground
N
Cattle shed
V.a
i
ii
iii*
iv
a
b
c
d
Fireplace
a. c. d. Accumulations of inscribed tablets.
b. Find place of first oblong tablet
Ancient rush fence
Ancient rush fence
Low dunes
SCALE
10 5 0 10 20 30 40 FEET
Wall of timber and plaster
ditto ruined
ditto eroded and barely traceable
Rush wall
ditto ruined
Fence
ditto ruined
Steep slope caused by erosion
Timber debris
Raised platform (fireplace)
Wooden post
M. A. STEIN & RAM SINGH DEL.

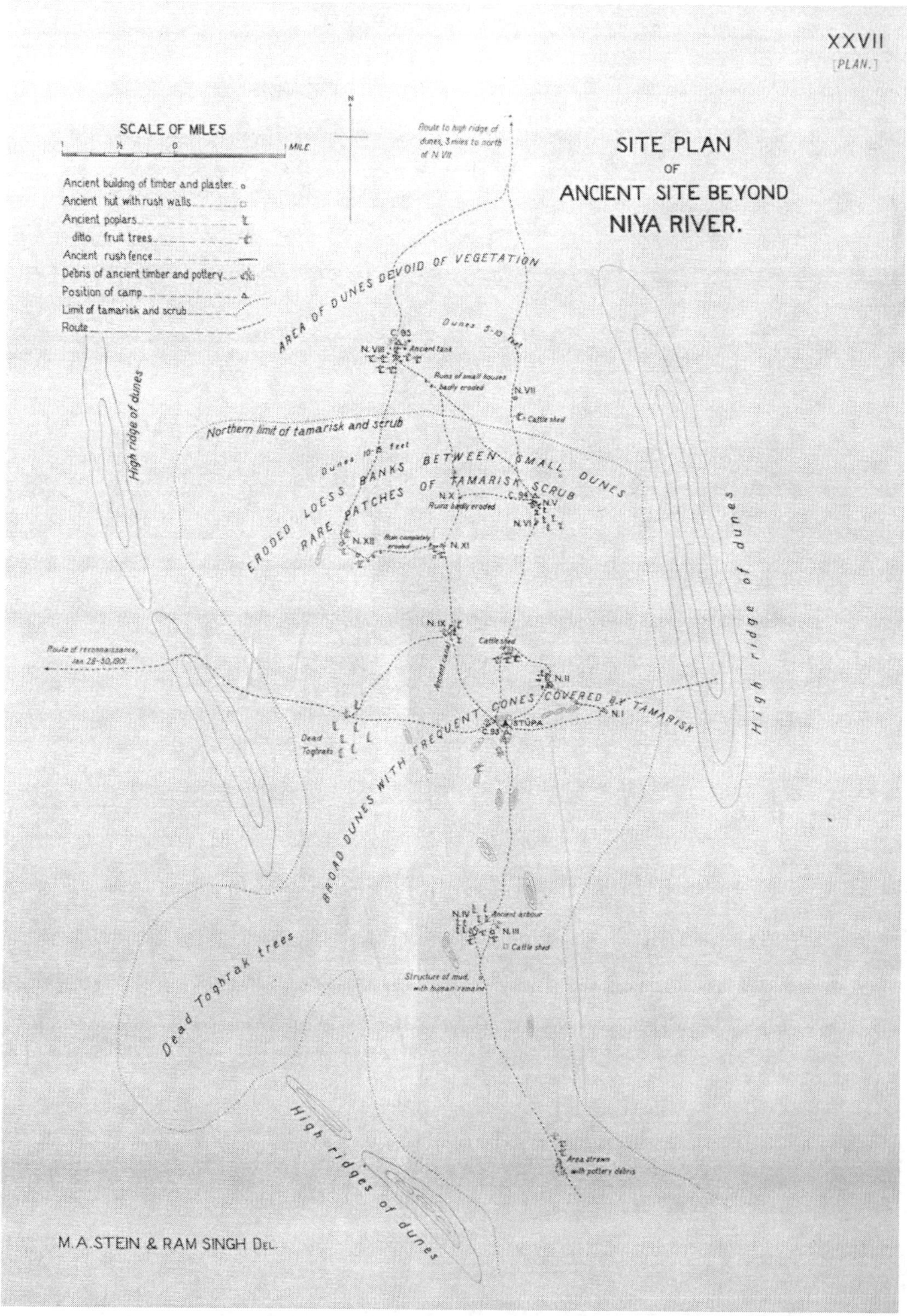
SITE PLAN
OF
ANCIENT SITE BEYOND
NIYA RIVER.
SCALE OF MILES
1 ½ 0 1 MILE
Ancient building of timber and plaster
Ancient hut with rush walls
Ancient poplars
ditto fruit trees
Ancient rush fence
Debris of ancient timber and pottery
Position of camp
Limit of tamarisk and scrub
Route
N
Route to high ridge of dunes, 3 miles to north of N. VII
AREA OF DUNES DEVOID OF VEGETATION
Dunes 5-10 feet
C. 93
N. VIII
Ancient tank
Ruins of small houses badly eroded
N. VII
Cattle shed
High ridge of dunes
Northern limit of tamarisk and scrub
Dunes 10-15 feet
ERODED LOESS BANKS BETWEEN SMALL DUNES
RARE PATCHES OF TAMARISK SCRUB
N. X
Ruins badly eroded
C. 95
N. V
N. VI
N. XII
Ruin completely eroded
N. XI
High ridge of dunes
N. IX
Cattle shed
Route of reconnaissance, Jan. 28-30, 1901
Ancient canal
N. II
CONES COVERED BY TAMARISK
N. I
STŪPA
C. 93
Dead Toghraks
BROAD DUNES WITH FREQUENT CONES
N. IV
Ancient arbour
N. III
Cattle shed
Structure of mud, with human remains
Dead Toghrak trees
High ridges of dunes
Area strewn with pottery debris
M. A. STEIN & RAM SINGH DEL.

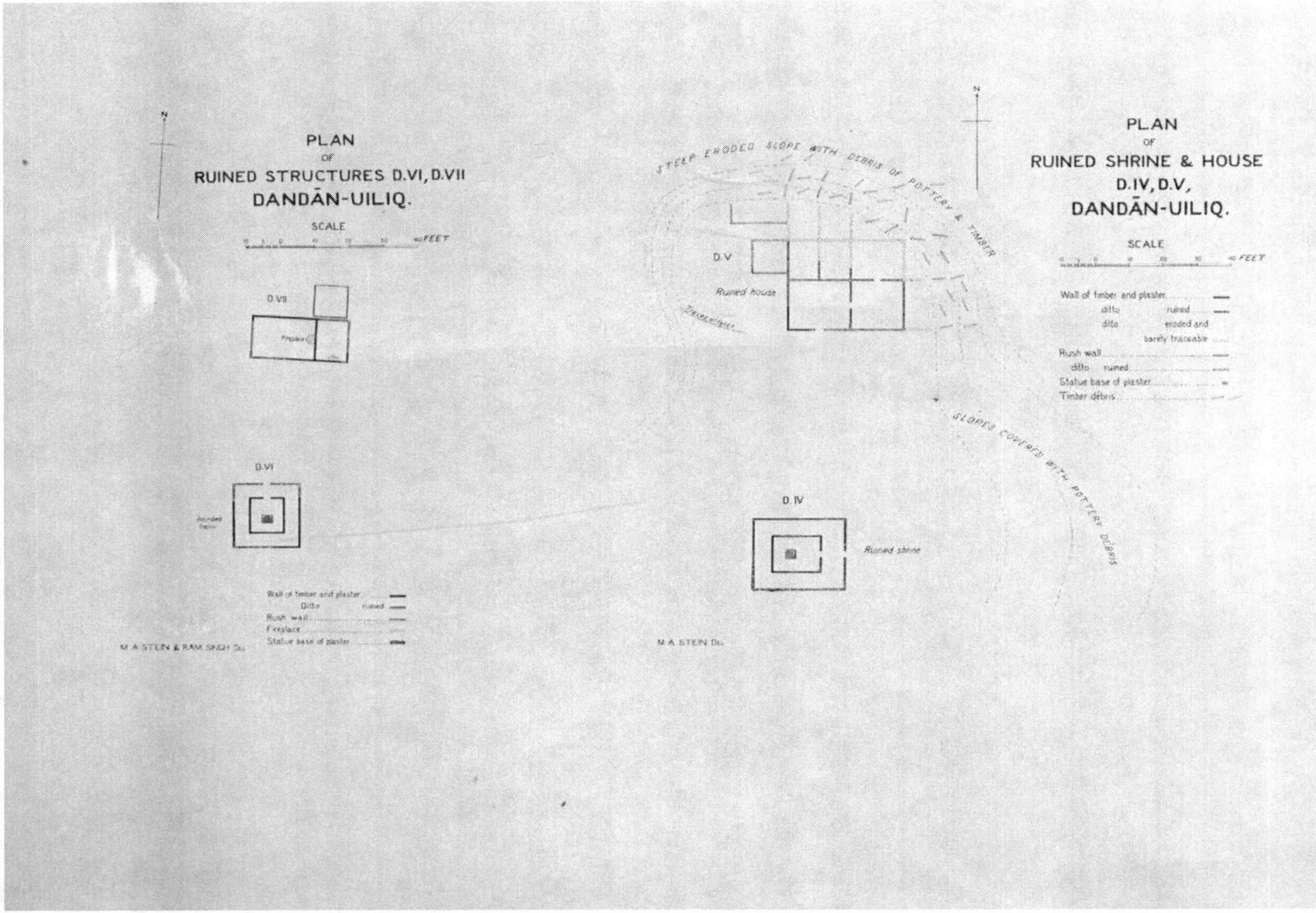
PLAN
OF
RUINED STRUCTURES D.VI, D.VII
DANDĀN-UILIQ.
SCALE
FEET
D.VII
Fireplace
D.VI
Wall of timber and plaster
Ditto ruined
Rush wall
Fireplace
Statue base of plaster
M. A. STEIN & RAM SINGH Dr.
N
PLAN
OF
RUINED SHRINE & HOUSE
D.IV, D.V,
DANDĀN-UILIQ.
SCALE
FEET
Wall of timber and plaster
ditto ruined
ditto eroded and barely traceable
Rush wall
ditto ruined
Statue base of plaster
Timber débris
STEEP ERODED SLOPE WITH DEBRIS OF POTTERY & TIMBER
D.V
Ruined house
SLOPES COVERED WITH POTTERY DÉBRIS
D.IV
Ruined shrine
M. A. STEIN Dr.

XXVI
[PLAN.]

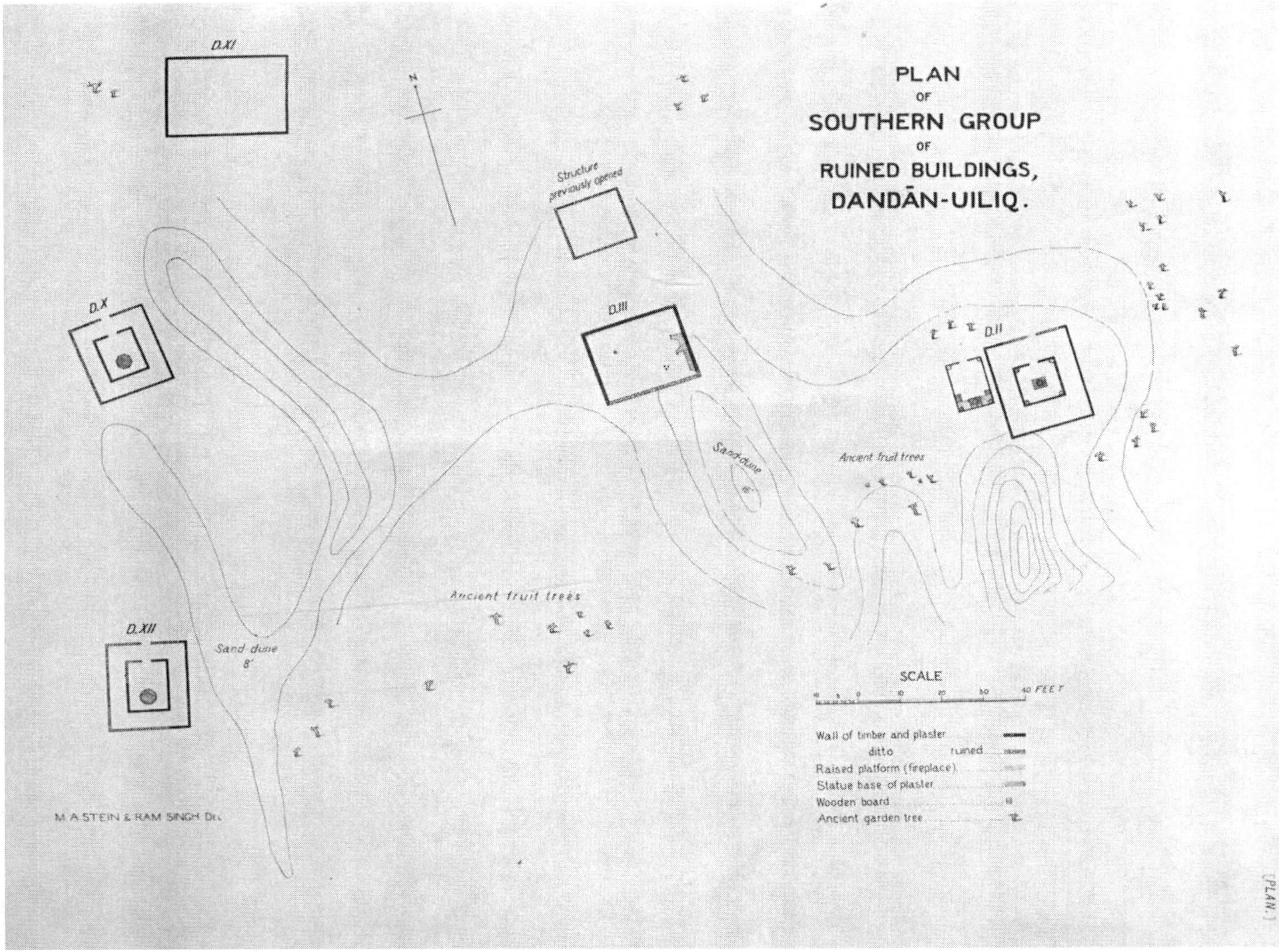
PLAN
OF
SOUTHERN GROUP
OF
RUINED BUILDINGS,
DANDĀN-UILIQ.
N
D.XI
D.X
D.XII
D.III
D.II
Structure previously opened
Sand-dune
Sand-dune 8'
Ancient fruit trees
Ancient fruit trees
SCALE
10 5 0 10 20 30 40 FEET
Wall of timber and plaster
ditto ruined
Raised platform (fireplace)
Statue base of plaster
Wooden board
Ancient garden tree
M. A. STEIN & RAM SINGH Del.

XXIV
[PLAN.]

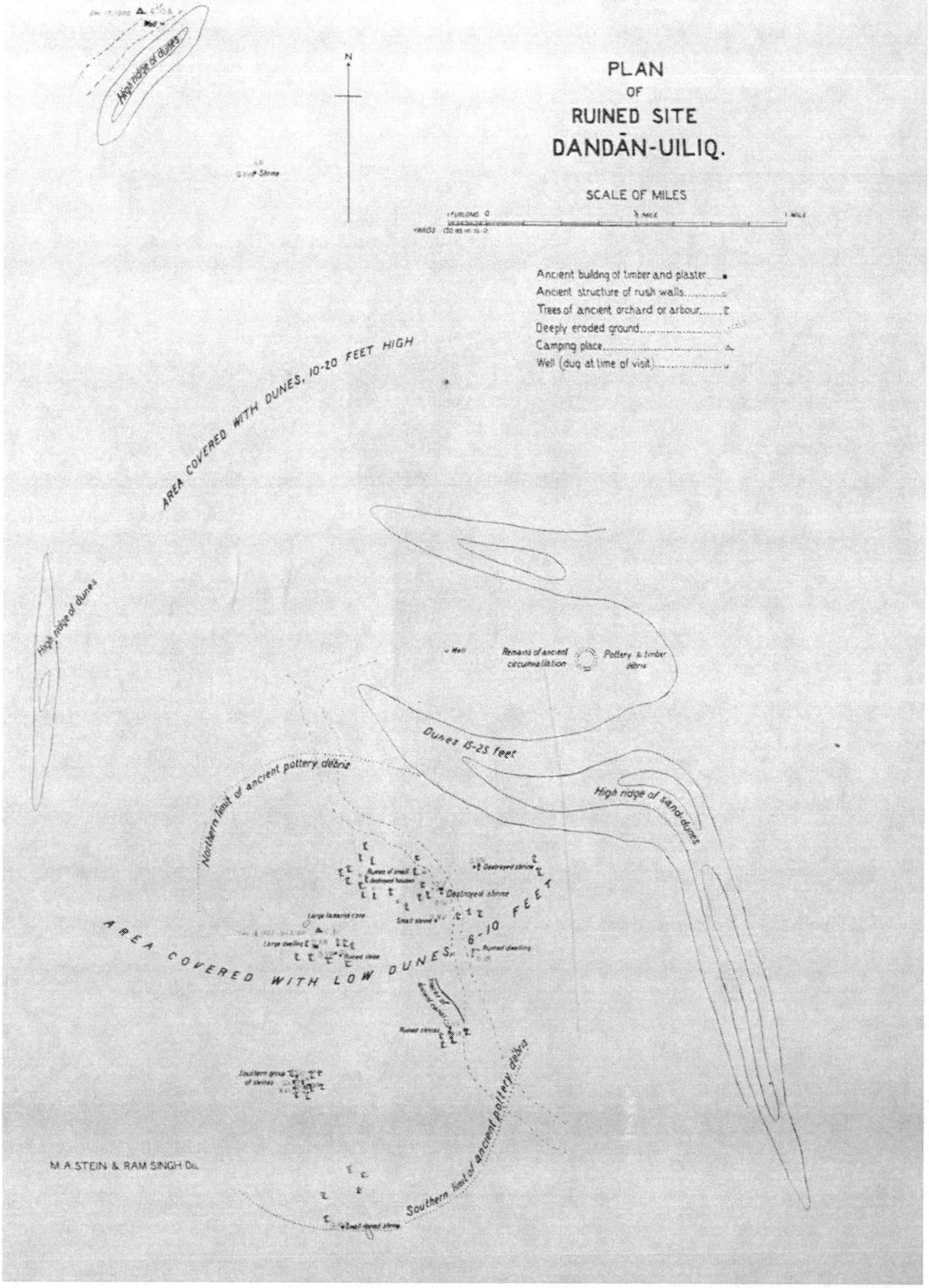

PLAN
OF
RUINED SITE
DANDĀN-UILIQ.
SCALE OF MILES
N
Ancient building of timber and plaster
Ancient structure of rush walls
Trees of ancient orchard or arbour
Deeply eroded ground
Camping place
Well (dug at time of visit)
High ridge of dunes
AREA COVERED WITH DUNES, 10-20 FEET HIGH
High ridge of dunes
Well
Remains of ancient circumvallation
Pottery & timber débris
Dunes 15-25 feet
High ridge of sand-dunes
Northern limit of ancient pottery débris
Destroyed shrine
Destroyed shrine
Large tamarisk cone
AREA COVERED WITH LOW DUNES, 6-10 FEET
Ruined dwelling
Ruined shrine
Southern group of shrines
Southern limit of ancient pottery débris
M.A. STEIN & RAM SINGH Del.

XXIII
[PLAN.]

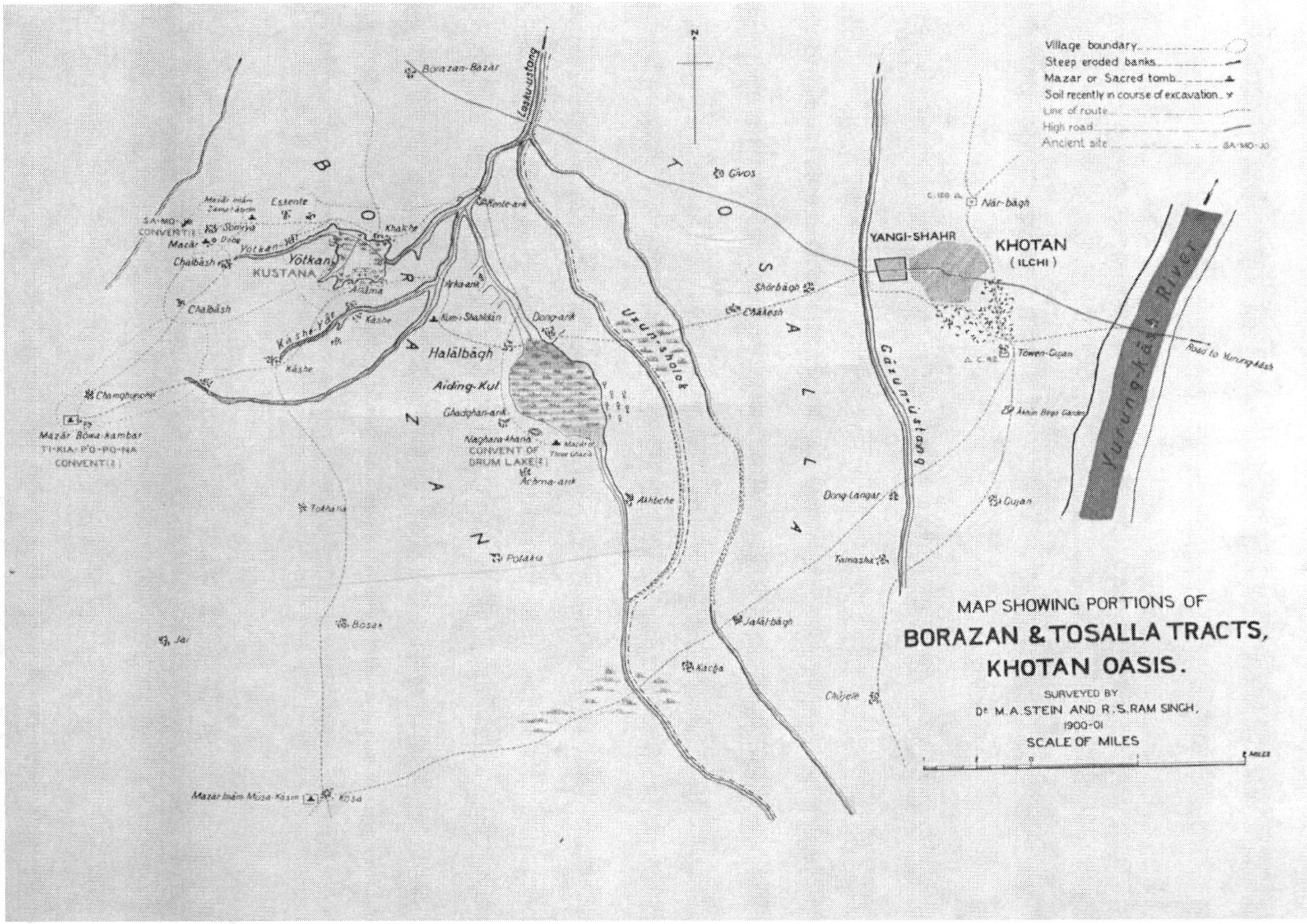
MAP SHOWING PORTIONS OF
BORAZAN & TOSALLA TRACTS,
KHOTAN OASIS.
SURVEYED BY
Dr M.A. STEIN AND R.S. RAM SINGH,
1900-01
SCALE OF MILES
Village boundary
Steep eroded banks
Mazar or Sacred tomb
Soil recently in course of excavation
Line of route
High road
Ancient site
KHOTAN
(ILCHI)
YANGI-SHAHR
Nār-bāgh
Yurung-kāsh River
Road to Yurung-kāsh
Tōwen-Gujan
Gujan
Gāzun-ūstang
Dong-Langar
Tamasha
Chījele
Jalālbāgh
Kachā
Shōrbāgh
Chākesh
Givos
Uzun-sholok
Akhtche
Lāshu-ūstang
Borazan-Bazār
Konte-arik
Dong-arik
Halālbāgh
Aiding-Kul
Ghadghan-arik
Naghara-khāna
CONVENT OF
DRUM LAKE(?)
Āchma-arik
Polākia
Bosan
Tokhaila
Jai
Mazār Imām Mūsa-Kāsim
Kōsa
Chamghuchoq
Mazār Bōwa-kambar
TI-KIA-P'O-PO-NA
CONVENT(?)
SA-MO-
CONVENT(?)
Mazār
Chalbāsh
Yōtkan
KUSTANA
Yōtkan-Yār
Khakhe
Essente
Somiya
Kāshe-Yār
Kāshe
Arka-arik
B O R A Z A N
T O S A L L A

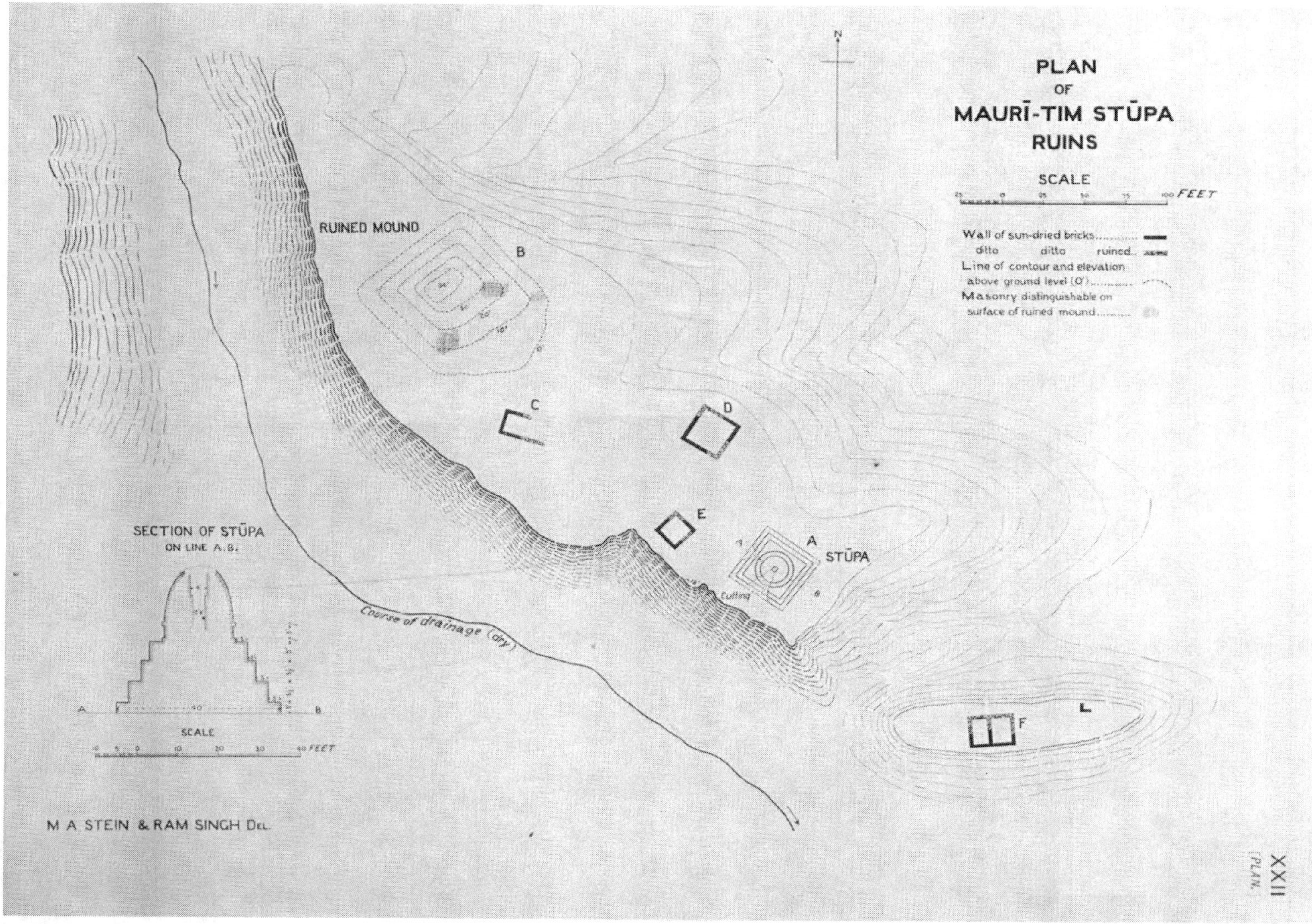

PLAN
OF
MAURĪ-TIM STŪPA
RUINS
SCALE
FEET
Wall of sun-dried bricks
ditto ditto ruined
Line of contour and elevation above ground level (0')
Masonry distinguishable on surface of ruined mound
N
RUINED MOUND
B
C
D
E
A
STŪPA
Cutting
F
Course of drainage (dry)
SECTION OF STŪPA
ON LINE A.B.
SCALE
FEET
M A STEIN & RAM SINGH Del.

XXII
[PLAN.]

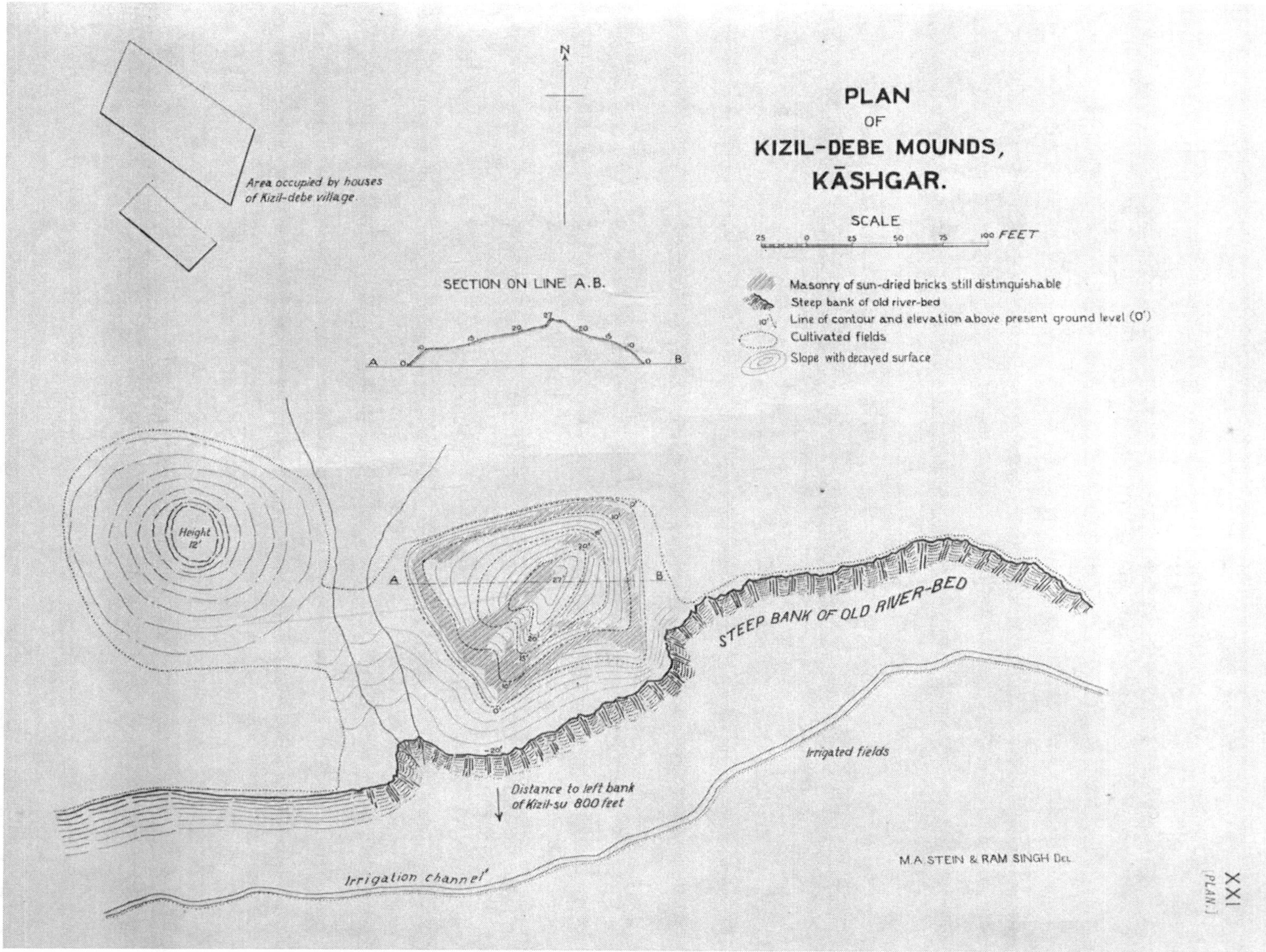
PLAN
OF
KIZIL-DEBE MOUNDS,
KĀSHGAR.
SCALE
25 0 25 50 75 100 FEET
N
Area occupied by houses of Kizil-debe village
SECTION ON LINE A.B.
A
B
Masonry of sun-dried bricks still distinguishable
Steep bank of old river-bed
Line of contour and elevation above present ground level (0')
Cultivated fields
Slope with decayed surface
Height 12'
STEEP BANK OF OLD RIVER-BED
Distance to left bank of Kizil-su 800 feet
Irrigated fields
Irrigation channel
M.A. STEIN & RAM SINGH Del.

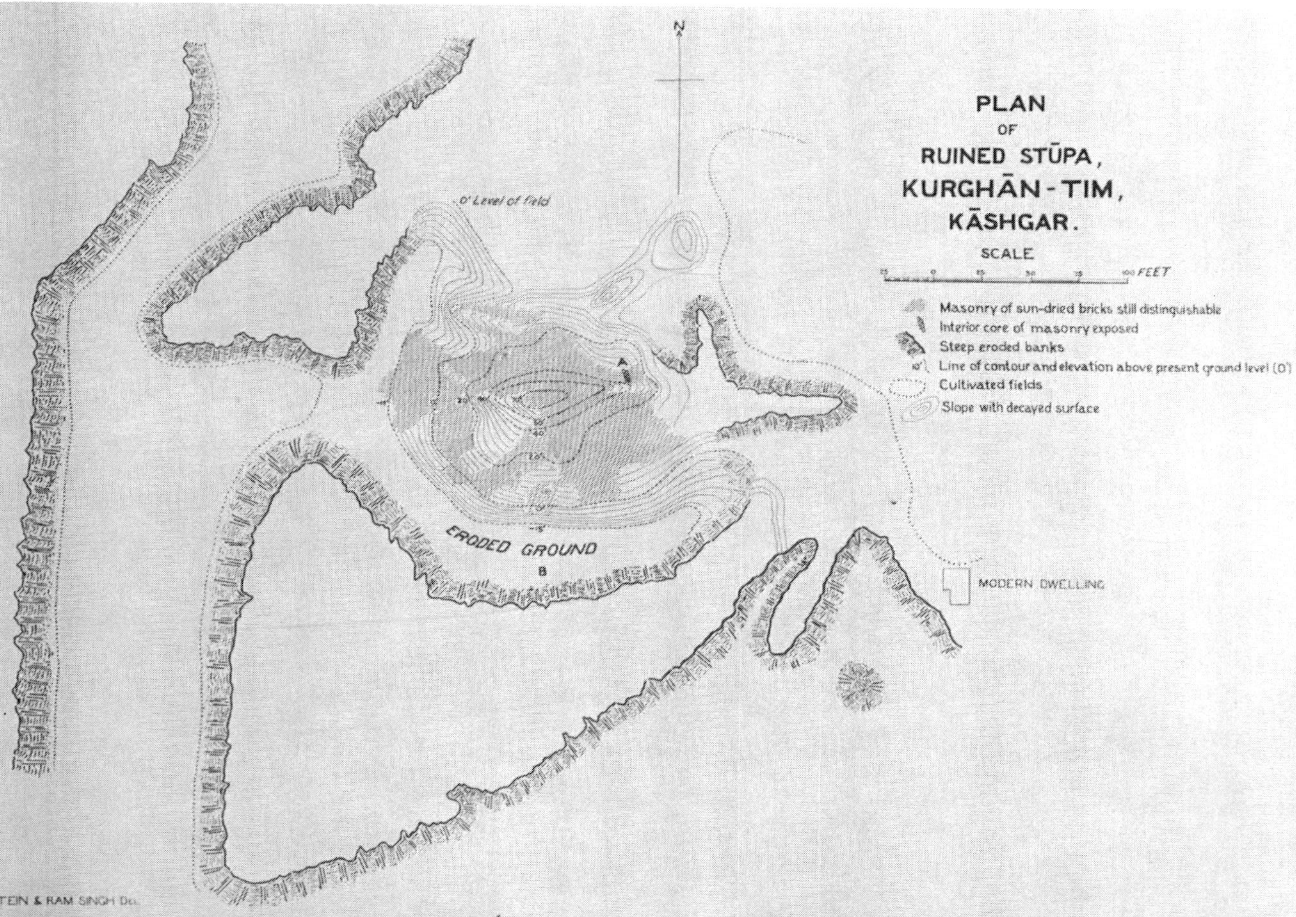
PLAN
OF
RUINED STŪPA,
KURGHĀN - TIM,
KĀSHGAR.
SCALE
25 0 25 50 75 100 FEET
Masonry of sun-dried bricks still distinguishable
Interior core of masonry exposed
Steep eroded banks
10' Line of contour and elevation above present ground level (0')
Cultivated fields
Slope with decayed surface
N
0' Level of field
A
ERODED GROUND
B
MODERN DWELLING
M.A. STEIN & RAM SINGH Del.

XIX
[PLAN.]

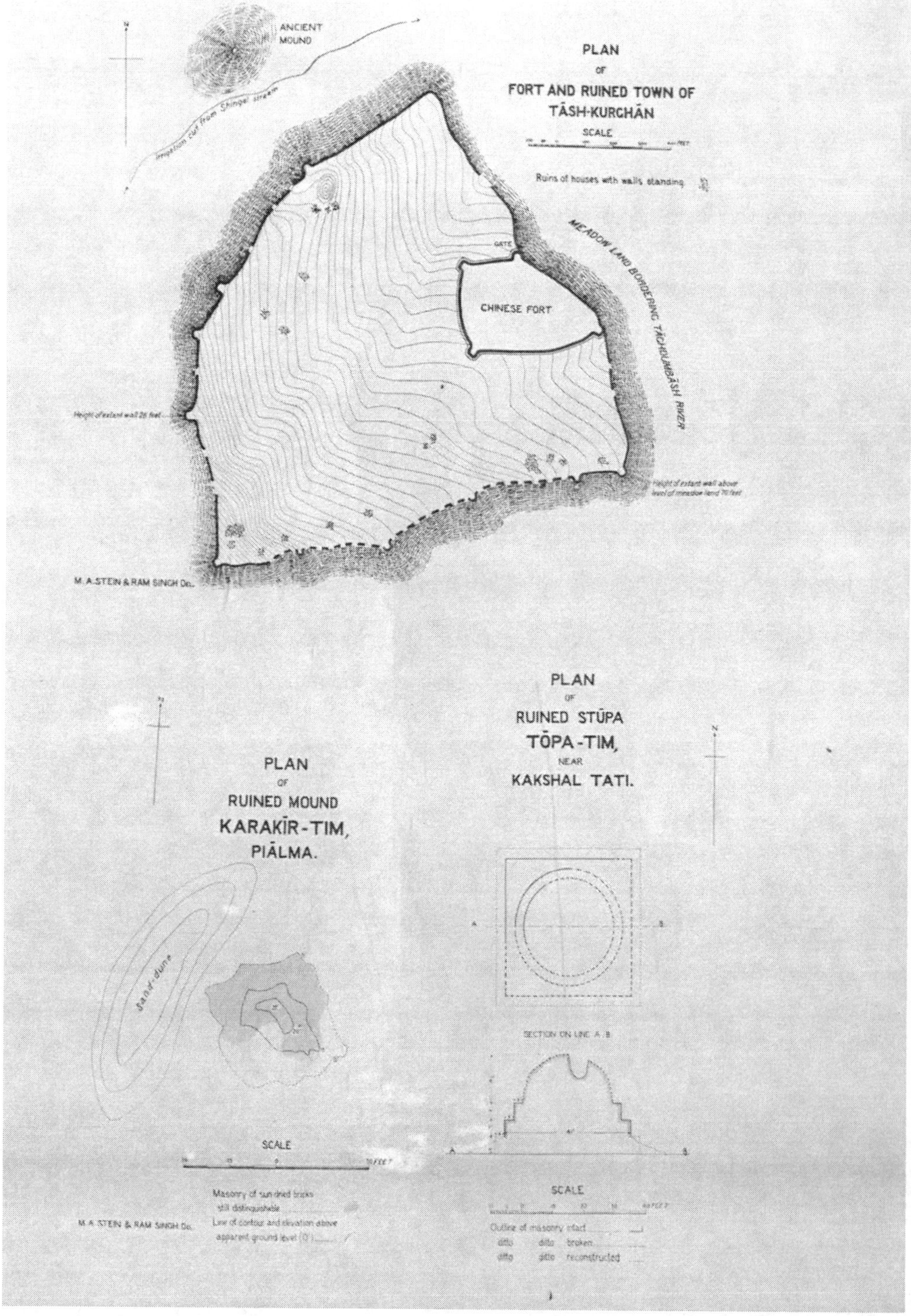

PLAN
OF
FORT AND RUINED TOWN OF
TĀSH-KURGHĀN
SCALE
Ruins of houses with walls standing
ANCIENT
MOUND
Irrigation cut from Shingal stream
MEADOW LAND BORDERING TĀGHDUMBĀSH RIVER
GATE
CHINESE FORT
Height of extant wall 28 feet
M. A. STEIN & RAM SINGH Del.
PLAN
OF
RUINED MOUND
KARAKĪR-TIM,
PIĀLMA.
Sand-dune
SCALE
Masonry of sun-dried bricks still distinguishable
Line of contour and elevation above apparent ground level (0')
M. A. STEIN & RAM SINGH Del.
PLAN
OF
RUINED STŪPA
TŌPA-TIM,
NEAR
KAKSHAL TATI.
SECTION ON LINE A.B
A
B
SCALE
Outline of masonry intact
ditto ditto broken
ditto ditto reconstructed

XVIII
[PHOT.]

lxxvii lxxvi lxxv lxxiv lxxiii lxxii b

lxxi a

RELIEF SCULPTURES, R. lxxi–R. lxxvii, AT OUTER SOUTH CORNER, RAWAK STUPA COURT.

(See Chap. XIV. sec. iv.)

d

BRICKS WITH MARKS FROM RAMPART, AK-SIPIL.

(See Chap. XIV. sec. ii.)

lxxxv c

REMAINS OF STUCCO RELIEF, R. lxxxv, WITH DECORATIVE HALO, ON OUTER PASSAGE WALL, RAWAK STŪPA COURT.

(See Chap. XIV. sec. iv.)

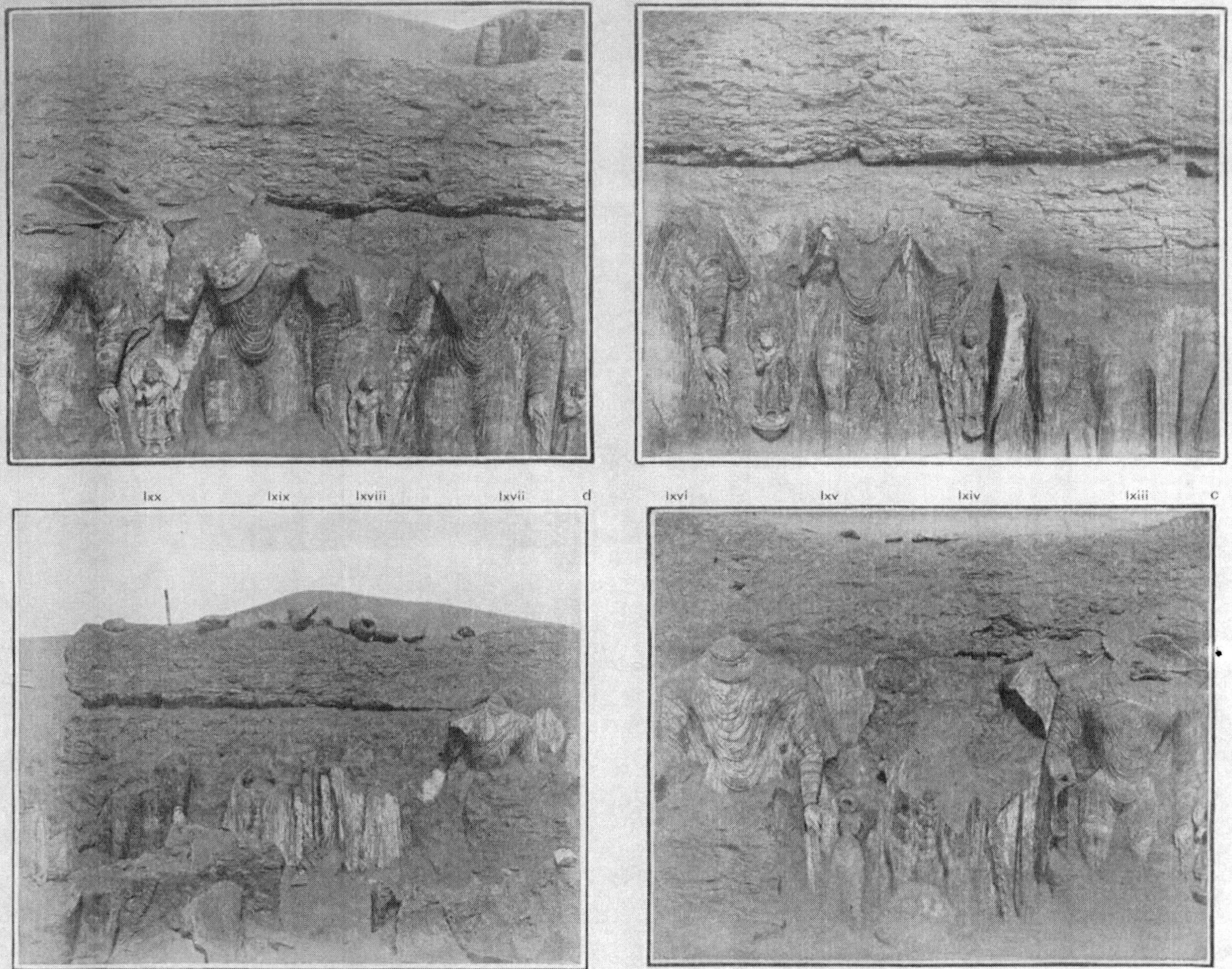

RELIEF SCULPTURES, R. lv–R. lxx, ON OUTER SOUTH-EAST WALL, RAWAK-STUPA COURT.

(See Chap. XIV. sec. iv.)

ENTRANCE GATE b

xlvii xlvi xlv a

liv liii lii d

li l xlix xlviii c

RELIEF SCULPTURES, R. xlv–R. liv, ON OUTER SOUTH-EAST WALL, RAWAK STŪPA COURT. (See Chap. XIV. sec. iv.)

b — xxxvii xxxvi xxxv xxxiv

a — xxxiv xxxiii xxxii xxxi

d — xliv xliii xlii

c — xli xl xxxix xxxviii

RELIEF SCULPTURES, R. xxxi–R. xliv, ON INNER AND OUTER SOUTH-EAST WALL, RAWAK STŪPA COURT. (See Chap. XIV, sec. iv.)

xxv xxiv a xxiv xxiii b

xxii xxi xx xix a

xxx xxix xxviii d

xxvii xxvii xxvi c

RELIEF SCULPTURES, R. xix–R. xxx, ON INNER SOUTH-EAST WALL, RAWAK STŪPA COURT.

(See Chap. XIV. sec. iv.)

XIII

a [PHOT.]

RAWAK STŪPA AND ENCLOSING QUADRANGLE, SEEN FROM SOUTH.

(See Chap. XIV. sec. iii.)

b

lxxvii lxxvi lxxv lxxiv lxxiii lxxii

RELIEF SCULPTURES, R. lxxii–R. lxxvii, ON OUTER SOUTH-EAST WALL, RAWAK STŪPA COURT.

(See Chap. XIV. sec. iv.)

TIBETAN SGRAFFITI ON SOUTH WALL OF TEMPLE CELLA E. i, ENDERE FORT.

(See Chap. XII, sec. ii; Appendix B, part iv.)

CHINESE SGRAFFITO OF A.D. 719, ON WALL OF TEMPLE CELLA E. i, ENDERE FORT.

(See Chap. XII. ii; Appendix A. part iii.)

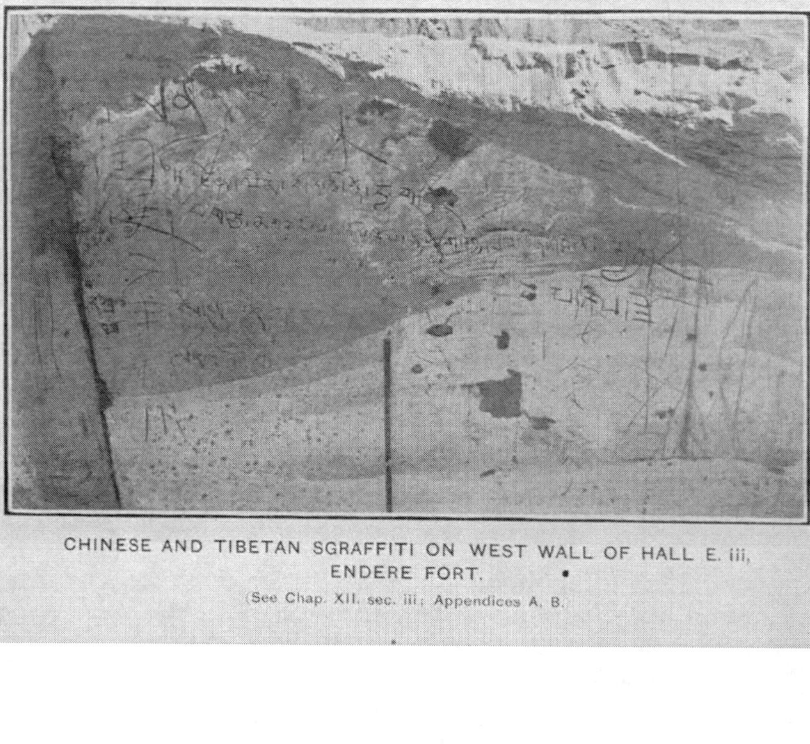

CHINESE AND TIBETAN SGRAFFITI ON WEST WALL OF HALL E. iii, ENDERE FORT.

(See Chap. XII. sec. iii; Appendices A. B.)

FRESCO IN CHAPEL E. ii, ENDERE FORT.

(See Chap. XII. sec. iii.)

UNDERGROUND ROOM, WITH FIRE-PLACE, IN DWELLING E. V, ENDERE FORT.

(See Chap. XII. sec. iii.)

IMPLEMENTS IN WOOD AND POTTERY, FOUND IN DWELLINGS N. III, V, VII, VIII, NIYA SITE.

(See Chap. XI. sec. ii, vi-viii.)

ANCIENT JAR BROUGHT FROM NIYA SITE.

(See Chap. X. sec. iii.)

WOODEN CUPBOARD FOUND IN RUINED DWELLING N. XI, NIYA SITE.

(See Chap. XI. sec. vii.)

ARCHITECTURAL WOOD-CARVING FROM RUINED HOUSE, N. VIII, NIYA SITE.

(See Chap. XI. sec. vii.)

ANCIENT CHAIR AND OTHER WOOD-CARVINGS FOUND IN HOUSE N. III, NIYA SITE.

(See Chap. XI. sec. ii.)

NORTH WALL OF HALL IN RUINED HOUSE N. III, NIYA SITE, AFTER EXCAVATION.

(See Chap. XI. sec. ii.)

SOUTH WALL IN HALL OF RUINED HOUSE, N. III, NIYA SITE, AFTER EXCAVATION.

(See Chap. XI. sec. ii.)

ROOM N. i, FIRST FIND-PLACE OF INSCRIBED TABLETS, NIYA SITE,
SEEN FROM NORTH AFTER EXCAVATION.

(See Chap. XI. sec. i.)

ROOM N. i, IN RUINED HOUSE N. I, NIYA SITE.
SEEN FROM SOUTH BEFORE EXCAVATION.

(See Chap XI. sec. i.)

FRESCO ON THE SOUTH WALL OF CELLA D. XII, DANDĀN-UILIQ.

(See Chap. IX. sec. iv.

RUINED CELLA D. XII, DANDĀN-UILIQ, AFTER EXCAVATION.

(See Chap. IX. sec. iv.)

FRESCOES ON SOUTH WALL OF TEMPLE CELLA D. VI, DANDĀN-UILIQ.

(See Chap. IX, sec. vi.)

FRESCOES ON INNER SOUTH WALL OF PASSAGE, SHRINE D. VI, DANDĀN-UILIQ.

(See Chap. IX, sec. vi.)

FRESCO FROM OUTER SOUTH WALL OF PASSAGE, SHRINE D. II, DANDĀN-UILIQ.

(See Chap. IX. sec. ii.)

(SCALE ONE-EIGHTH OF ORIGINAL)

FRESCOES ON INNER WALLS OF PASSAGE, SHRINE D. II, DANDĀN-UILIQ, SEEN FROM SOUTH-EAST.

(See Chap. IX. sec. ii.)

FRESCO AND STUCCO RELIEF IN SMALL CELLA OF SHRINE D. II, DANDĀN-UILIQ.

(See Chap. IX. sec. iii.)

II [PHOT.]

I
[PHOT.]

MAURĪ-TIM STŪPA, SEEN FROM SOUTH.

(See Chap. IV. sec. ii.)

RUINED STŪPA AND MOUND OF MAURĪ TIM, SEEN FROM SOUTH-EAST.

(See Chap. IV. sec. ii)

LIST OF PLATES

ANCIENT KHOTAN

DETAILED REPORT OF ARCHAEOLOGICAL EXPLORATIONS IN CHINESE TURKESTAN

CARRIED OUT AND DESCRIBED UNDER THE
ORDERS OF H.M. INDIAN GOVERNMENT BY

M. AUREL STEIN
INDIAN EDUCATIONAL SERVICE

VOL. II PLATES

OF

PHOTOGRAPHS, PLANS, ANTIQUES AND MSS.
WITH A MAP OF THE TERRITORY OF KHOTAN
FROM ORIGINAL SURVEYS

OXFORD
AT THE CLARENDON PRESS
1907

Published by

Gyan Publishing House
5, Ansari Road
Daryaganj, New Delhi-110002
Phone: 011-47034999, 9811692060
E-mail: books@gyanbooks.com

Distribution Network
gyanbooks.com
India, USA, Canada, UK, Australia, France

ISBN : 978-81-212-2424-6 (Set)
ISBN : 978-81-212-2673-8 (HB)
First Published, 1907

2nd Impression 2021

Printed at: Gyan Press, Delhi.

Ancient Khotan Vol. II
Author: M. Aurel Stein

ANCIENT KHOTAN

DETAILED REPORT OF ARCHAEOLOGICAL
EXPLORATIONS IN CHINESE TURKESTAN

CARRIED OUT AND DESCRIBED UNDER THE
ORDERS OF H.M. INDIAN GOVERNMENT BY

M. AUREL STEIN

INDIAN EDUCATIONAL SERVICE

VOL. II PLATES

OF

PHOTOGRAPHS, PLANS, ANTIQUES AND MSS.
WITH A MAP OF THE TERRITORY OF KHOTAN
FROM ORIGINAL SURVEYS

GYAN